Probabilistic Tsunami Hazard and Risk Analysis
Towards Disaster Risk Reduction and Resilience

Probabilistic Tsunami Hazard and Risk Analysis
Towards Disaster Risk Reduction and Resilience

Edited by

Katsuichiro Goda
Department of Earth Sciences, Western University, London, ON, Canada

Raffaele De Risi
School of Civil, Aerospace and Design Engineering, University of Bristol, Bristol, United Kingdom

Aditya Gusman
GNS Science, Lower Hutt, Wellington, New Zealand

Ioan Nistor
Department of Civil Engineering, University of Ottawa, Ottawa, ON, Canada

ELSEVIER

Elsevier
Radarweg 29, PO Box 211, 1000 AE Amsterdam, Netherlands
125 London Wall, London EC2Y 5AS, United Kingdom
50 Hampshire Street, 5th Floor, Cambridge, MA 02139, United States

Copyright © 2025 Elsevier Inc. All rights are reserved, including those for text and data mining, AI training, and similar technologies.

Publisher's note: Elsevier takes a neutral position with respect to territorial disputes or jurisdictional claims in its published content, including in maps and institutional affiliations.

No part of this publication may be reproduced or transmitted in any form or by any means, electronic or mechanical, including photocopying, recording, or any information storage and retrieval system, without permission in writing from the publisher. Details on how to seek permission, further information about the Publisher's permissions policies and our arrangements with organizations such as the Copyright Clearance Center and the Copyright Licensing Agency, can be found at our website: www.elsevier.com/permissions.

This book and the individual contributions contained in it are protected under copyright by the Publisher (other than as may be noted herein).

MATLAB® is a trademark of The MathWorks, Inc. and is used with permission. The MathWorks does not warrant the accuracy of the text or exercises in this book. This book's use or discussion of MATLAB® software or related products does not constitute endorsement or sponsorship by The MathWorks of a particular pedagogical approach or particular use of the MATLAB® software.

Notices

Knowledge and best practice in this field are constantly changing. As new research and experience broaden our understanding, changes in research methods, professional practices, or medical treatment may become necessary.

Practitioners and researchers must always rely on their own experience and knowledge in evaluating and using any information, methods, compounds, or experiments described herein. In using such information or methods they should be mindful of their own safety and the safety of others, including parties for whom they have a professional responsibility.

To the fullest extent of the law, neither the Publisher nor the authors, contributors, or editors, assume any liability for any injury and/or damage to persons or property as a matter of products liability, negligence or otherwise, or from any use or operation of any methods, products, instructions, or ideas contained in the material herein.

ISBN: 978-0-443-18987-6

For Information on all Elsevier publications
visit our website at https://www.elsevier.com/books-and-journals

Publisher: Candice Janco
Acquisitions Editor: Jennette McClain
Editorial Project Manager: Aleksandra Packowska
Production Project Manager: Paul Prasad Chandramohan
Cover Designer: Miles Hitchen

Typeset by MPS Limited, Chennai, India

Contents

List of contributors   xiii
Preface   xv

1. **Introduction of probabilistic tsunami hazard and risk analysis—toward disaster risk reduction and resilience**   1

    Katsuichiro Goda, Raffaele De Risi, Aditya Gusman and Ioan Nistor

    1.1 **Introduction**   1
    1.2 **Historical events and lessons learned**   4
        1.2.1 2004 Indian Ocean Tsunami   6
        1.2.2 2010 Maule Chile Tsunami   7
        1.2.3 2011 Tohoku Japan Tsunami   8
        1.2.4 2018 Indonesian tsunamis in Sulawesi and Sunda Strait   9
        1.2.5 2024 Noto Peninsula Tsunami   12
    1.3 **Tsunami hazard and risk assessments**   12
        1.3.1 Tsunami hazard assessment   12
        1.3.2 Tsunami risk assessment   15
    1.4 **Tsunami disaster risk reduction and management**   16
        1.4.1 Mitigation   16
        1.4.2 Preparedness   17
        1.4.3 Response   17
        1.4.4 Recovery   18
    1.5 **Scope and aims of the book**   18
    References   19

Section 1
Fundamentals of probabilistic tsunami hazard and risk analysis

2. **Tsunami generation**   27

    Katsuichiro Goda, Aditya Gusman, Raffaele De Risi and Ioan Nistor

    2.1 **Introduction**   27
    2.2 **Seismotectonic characteristics of active seismic regions**   28
    2.3 **Earthquake occurrence**   30
        2.3.1 Statistical analysis of earthquake catalog and time-independent occurrence model   31
        2.3.2 Time-dependent occurrence model using renewal process   33
        2.3.3 Characteristic earthquake model for large events   37
    2.4 **Earthquake rupture model**   39
        2.4.1 Finite-fault models   39
        2.4.2 Analysis of finite-fault models   42
        2.4.3 Prediction of earthquake source parameters   47
    2.5 **Source modeling for tsunamigenic earthquakes**   51
        2.5.1 Characterization of seismic tsunami sources   51
        2.5.2 Stochastic tsunami source modeling   53
        2.5.3 Kinematic stochastic tsunami source modeling   54
        2.5.4 Seafloor displacement due to earthquake rupture   57
    2.6 **Nonseismic sources of tsunamis**   63
        2.6.1 Sources from landslides   64
        2.6.2 Sources from volcanic activities   67
    2.7 **Research needs**   72
    References   72

3. **Tsunami propagation and runup modeling**   79

    Aditya Gusman, Katsuichiro Goda, Raffaele De Risi and Ioan Nistor

    3.1 **Introduction**   79
    3.2 **Tsunami characteristics**   79
        3.2.1 Harmonic wave   83
        3.2.2 Tsunami ray tracing   84
        3.2.3 Tsunami travel time   87
    3.3 **Tsunami propagation and inundation modeling**   88
        3.3.1 Equation of mass conservation   88

3.3.2　Equation of motion: Euler's
　　　　　　equation　　　　　　　　　　89
　　　3.3.3　Shallow-water wave equations　89
　3.4　**Numerical method**　　　　　　　　95
　　　3.4.1　Finite difference scheme　　　96
　　　3.4.2　Spatial grid refinements　　　99
　　　3.4.3　Outermost grid boundary
　　　　　　condition　　　　　　　　　100
　　　3.4.4　Moving boundary scheme　　102
　3.5　**Tsunami simulations**　　　　　　　102
　　　3.5.1　Tsunami source　　　　　　102
　　　3.5.2　Nearshore and offshore tsunami
　　　　　　modeling　　　　　　　　　108
　　　3.5.3　Tsunami inundation modeling　110
　3.6　**Research needs**　　　　　　　　　118
　References　　　　　　　　　　　　　　118

4. Tsunami effects on built environment　　　　　　　　　121

Ioan Nistor, Seyed Abbas Jazaeri, Joseph Kim, Raffaele De Risi, Katsuichiro Goda and Aditya Gusman

　4.1　**Introduction**　　　　　　　　　　121
　4.2　**Existing design guidelines, standards, and codes**　　　　　　　　　122
　　　4.2.1　Differences among guidelines,
　　　　　　standards, and codes　　　　122
　　　4.2.2　Early development of tsunami
　　　　　　guidelines, standards, and codes　122
　　　4.2.3　Japan　　　　　　　　　　123
　　　4.2.4　United States of America　　123
　　　4.2.5　Other countries　　　　　　124
　4.3　**Tsunami-induced loads and effects**　124
　　　4.3.1　Hydrostatic loads　　　　　126
　　　4.3.2　Hydrodynamic loads　　　　127
　　　4.3.3　Debris impact loads　　　　134
　4.4　**Differences in tsunami-induced loads between American and Japanese design codes**　　　　　　　　141
　　　4.4.1　Loading conditions　　　　　141
　　　4.4.2　Hydrostatic forces　　　　　141
　　　4.4.3　Hydrodynamic forces　　　　142
　　　4.4.4　Debris impact loads　　　　142
　4.5　**Tsunami structural vulnerability assessment**　　　　　　　　　142
　　　4.5.1　Tsunami damage states　　　143
　　　4.5.2　Forms and typologies of
　　　　　　vulnerability functions　　　144
　　　4.5.3　Empirical fragility functions　145
　　　4.5.4　Analytical fragility functions　150
　4.6　**Research needs**　　　　　　　　　151
　References　　　　　　　　　　　　　　152

5. Probabilistic tsunami hazard and risk assessments　　　　　　157

Raffaele De Risi, Katsuichiro Goda, Ioan Nistor and Aditya Gusman

　5.1　**Introduction**　　　　　　　　　　157
　5.2　**Overview of tsunami hazard and risk assessment**　　　　　　　　158
　　　5.2.1　Sensitivity analysis　　　　　158
　　　5.2.2　Worst case scenario approach　158
　　　5.2.3　Probabilistic approach　　　159
　　　5.2.4　General probabilistic formulation　160
　　　5.2.5　Logic tree approach　　　　162
　5.3　**Probabilistic seismic tsunami hazard analysis**　　　　　　　　163
　　　5.3.1　Source characterization and
　　　　　　magnitude-frequency distribution　164
　　　5.3.2　Scaling relationships of earthquake
　　　　　　source parameters and stochastic
　　　　　　source models　　　　　　　168
　　　5.3.3　Tsunami modeling　　　　　169
　　　5.3.4　Empirical tsunami hazard
　　　　　　curve based on tsunami simulation
　　　　　　results　　　　　　　　　　170
　　　5.3.5　Hazard maps　　　　　　　170
　　　5.3.6　Optimal number of simulations　172
　　　5.3.7　Disaggregation　　　　　　172
　5.4　**Probabilistic seismic tsunami risk analysis**　173
　　　5.4.1　Exposure characterization　　174
　　　5.4.2　Vulnerability characterization　176
　　　5.4.3　Numerical evaluation of tsunami
　　　　　　risk equation　　　　　　　177
　　　5.4.4　Risk metrics　　　　　　　178
　5.5　**Extension to multihazard risks**　　　179
　　　5.5.1　Compound earthquake–tsunami
　　　　　　hazard and risk assessment　　179
　　　5.5.2　Earthquake simulation　　　179
　　　5.5.3　Development of joint
　　　　　　earthquake–tsunami hazard
　　　　　　curves　　　　　　　　　　180
　　　5.5.4　Earthquake–tsunami uniform
　　　　　　hazard maps　　　　　　　182
　　　5.5.5　Earthquake–tsunami disaggregation　182
　　　5.5.6　Earthquake–tsunami risk　　183
　5.6　**Research needs**　　　　　　　　　186
　References　　　　　　　　　　　　　　186

6. Tsunami disaster risk reduction and management　　　　　　　191

Katsuichiro Goda, Raffaele De Risi, Ioan Nistor and Aditya Gusman

　6.1　**Introduction**　　　　　　　　　　191

6.2	Tsunami hazard–risk mapping for disaster preparedness planning	192
	6.2.1 Risk assessment procedure and relative risk scoring	192
	6.2.2 Identification of critical tsunami scenarios and hazard–risk mapping	196
	6.2.3 Tsunami fatality risk and societal risk tolerability	202
6.3	Emergency response	205
	6.3.1 Tsunami early warning system	205
	6.3.2 Tsunami evacuation	212
6.4	Tsunami risk financing	217
	6.4.1 Financial risk metrics for natural catastrophe risk management	218
	6.4.2 Insurance-reinsurance system for tsunami risk	220
	6.4.3 Risk transfer for tsunami	224
6.5	Research needs	226
References		226

Section 2
Advanced topics and applications related to probabilistic tsunami hazard and risk analysis

7. Historical tsunami records and paleotsunamis 233

Kenji Satake

7.1	Introduction	233
7.2	Tsunami waveforms recorded on instruments	235
	7.2.1 Coastal sea-level measurements	235
	7.2.2 Offshore tsunami measurements	236
	7.2.3 Cabled bottom pressure gauges	236
	7.2.4 Deep ocean measurements: offline and deep-ocean assessment and reporting of tsunamis	238
	7.2.5 Satellite observations	240
	7.2.6 Oceanographic radar measurements	240
7.3	Source models of the 2011 Tohoku Earthquake based on the inversion of tsunami data	241
7.4	Field surveys to measure tsunami heights	242
7.5	Historical data	244
	7.5.1 The 1854 Ansei Earthquakes and Tsunamis along Nankai Trough	245
	7.5.2 The 1896 Sanriku Tsunami	245
	7.5.3 The 1611 Keicho Tohoku Tsunami	247
	7.5.4 The 869 Jogan Tsunami	247

7.6	Geological data	248
	7.6.1 Tsunami deposits in Sendai Plain	249
	7.6.2 Sediment transport modeling	250
7.7	Conclusions	252
References		252

8. Informing megathrust tsunami source models with knowledge of tectonics and fault mechanics 257

Kelin Wang, Matías Carvajal, Yijie Zhu, Tianhaozhe Sun, Jiangheng He and Matthew Sypus

8.1	Introduction	257
8.2	Subduction megathrust and tsunamigenic earthquakes	258
	8.2.1 Fault zone structure affected by seafloor morphology and sediment subduction	258
	8.2.2 Fault zone rheology and fault friction	259
	8.2.3 Updip and downdip limits of megathrust rupture	260
	8.2.4 Intriguing strike dimension	262
8.3	Slip distribution in predictive source scenarios	262
	8.3.1 Static kinematic models informed by fault mechanics	263
	8.3.2 Slip distribution in the dip direction	265
	8.3.3 Off-megathrust permanent deformation	269
	8.3.4 Along-strike variability of rupture behavior	271
8.4	Concluding remarks	271
References		272

9. Tsunamis triggered by splay faulting 277

Mohammad Mokhtari and Katsuichiro Goda

9.1	Introduction	277
9.2	Tectonic background	277
9.3	Seismic expression of offshore splay faults	278
	9.3.1 Makran subduction zone	278
	9.3.2 Nankai Trough zone	279
9.4	Splay fault and tsunami generation	281
9.5	Numerical example	282
9.6	Discussions	284
9.7	Conclusions	286
References		286

10. Tsunami hazard from subaerial landslides — 289

Finn Løvholt, Sylfest Glimsdal and Carl Bonnevie Harbitz

- 10.1 Introduction — 289
- 10.2 Method for subaerial landslide probabilistic hazard analysis — 291
  - 10.2.1 Numerical modeling — 291
  - 10.2.2 Probabilistic treatment — 291
  - 10.2.3 Shortcomings — 292
  - 10.2.4 Suggested improved methodology using hindcasting to calibrate parameter probabilities — 293
- 10.3 Results—model calibration for constraining uncertainty analysis — 294
- 10.4 Discussion—implications for future methods — 298
- 10.5 Conclusions — 299
- References — 299

11. Dense tsunami monitoring system — 303

Yuichiro Tanioka

- 11.1 Introduction — 303
- 11.2 Accurate earthquake source estimation — 304
- 11.3 Data assimilation — 306
  - 11.3.1 Tsunami data assimilation — 306
  - 11.3.2 Green's function-based tsunami data assimilation — 308
  - 11.3.3 Near-field tsunami data assimilation — 310
- 11.4 Tsunami early warning — 312
  - 11.4.1 Tsunami forecasting method based on inversion for initial sea-surface height — 312
  - 11.4.2 Tsunami forecast using multiindex method for a scenario database — 315
- 11.5 Observed tsunamis from other sources — 316
  - 11.5.1 Meteotsunamis — 316
  - 11.5.2 Analysis of the 2022 Tonga Tsunami — 317
- 11.6 Summary — 321
- References — 321

12. Machine learning approaches for tsunami early warning — 325

Iyan E. Mulia

- 12.1 Introduction — 325
- 12.2 Theoretical framework — 326
- 12.3 Existing studies — 327
- 12.4 Advantages, limitations, and future direction — 330
- References — 333

13. Global tsunami hazards and risks — 339

Yong Wei

- 13.1 Introduction — 339
- 13.2 Short-term hazard assessment — 340
  - 13.2.1 Methods — 340
  - 13.2.2 Short-term hazard assessment: the March 11, 2011 Japan Tsunami — 343
  - 13.2.3 Enhancements to short-term hazard assessments — 343
  - 13.2.4 Short-term tsunami hazard assessment systems in other countries and regions — 350
- 13.3 Long-term hazard assessment — 352
  - 13.3.1 Challenges for long-term assessment — 352
  - 13.3.2 Methods — 353
  - 13.3.3 Deterministic tsunami hazard assessment — 356
  - 13.3.4 Probabilistic tsunami hazard assessment—the American Society of Civil Engineers framework of tsunami design zone — 360
- 13.4 Conclusions — 367
- References — 368

14. Probabilistic tsunami hazard assessment for New Zealand — 373

William Power, Aditya Gusman, David Burbidge and Xiaoming Wang

- 14.1 Introduction — 373
- 14.2 Overview of statistical modeling approach — 374
- 14.3 Earthquake source models — 378
  - 14.3.1 Plate interface earthquake models — 378
  - 14.3.2 Crustal fault earthquake models — 381
- 14.4 Tsunami numerical simulation — 381
- 14.5 Tsunami height estimation — 383
  - 14.5.1 Regional and distant sources — 383
  - 14.5.2 Local subduction zones — 384
  - 14.5.3 Local crustal faults — 384
- 14.6 Tsunami hazard model results — 384
  - 14.6.1 Tsunami hazard curves and disaggregation results — 385

	14.6.2 Tsunami inundation hazard modeling	385
14.7	Conclusions	392
Acknowledgments		392
References		392

15. Tsunami hazard and risk in the Mediterranean Sea — 397

Anita Grezio, Marco Anzidei, Alberto Armigliato, Enrico Baglione, Alessandra Maramai, Jacopo Selva, Matteo Taroni, Antonio Vecchio and Filippo Zaniboni

15.1	Introduction	397
15.2	Regional tsunami hazard	398
	15.2.1 Seismic tsunami hazard	399
	15.2.2 Submarine and subaerial landslides tsunami hazard	399
	15.2.3 Volcanic tsunami hazard	401
	15.2.4 Meteotsunami hazard	402
	15.2.5 Effects of climate change and vertical land movements on tsunamis hazard	404
15.3	Multisource probabilistic tsunami hazard analysis in the Gulf of Naples	406
15.4	Regional tsunami risk index	406
15.5	Discussion and final remarks	407
References		410

16. Tsunami hazard assessment in Chile — 417

Patricio Andrés Catalán and Natalia Zamora

16.1	Introduction	417
16.2	Tsunami occurrence in Chile	419
16.3	Tsunami hazard assessments in Chile	420
	16.3.1 Boundary conditions and local hydrodynamics	420
	16.3.2 Initial conditions: seismotectonic segmentation along Chile	421
	16.3.3 Scenario-based tsunami hazard assessment	423
	16.3.4 Seismic probabilistic tsunami hazard assessment	423
	16.3.5 Other susceptibility estimation studies	425
	16.3.6 Short-term assessment: applications to early warning	426
16.4	Discussion	428
16.5	Conclusions and perspectives	428
References		429

17. Uncertainty in empirical tsunami fragility curves — 437

Fatemeh Jalayer and Hossein Ebrahimian

17.1	Introduction	437
17.2	Definitions and limitations of empirical fragility curves	438
	17.2.1 Class of asset at risk	439
	17.2.2 Tsunami intensity measure	439
	17.2.3 Tsunami damage scale	439
	17.2.4 Representation of empirical fragility curves for a damage scale	440
17.3	Methods for empirical tsunami fragility assessment	441
	17.3.1 Sources of uncertainties in tsunami intensity characterization	442
	17.3.2 Sources of uncertainties related to tsunami damage evaluation	443
	17.3.3 Sources of uncertainties related to the choice of fragility models	444
	17.3.4 Sources of uncertainties related to fragility model parameters	444
17.4	Conclusions	445
References		446

18. Analytical tsunami fragility curves — 449

Tiziana Rossetto, Marta Del Zoppo, Marco Baiguera and Jonas Cels

18.1	Introduction	449
18.2	Tsunami actions on buildings	450
	18.2.1 Characterization of tsunami inundation flows	450
	18.2.2 Characterization of tsunami actions on buildings	450
	18.2.3 Tsunami intensity measures for fragility analysis	454
18.3	Tsunami structural assessment approaches	454
	18.3.1 Tsunami nonlinear time-history dynamic analysis	455
	18.3.2 Tsunami nonlinear static analyses	456
18.4	Structural modeling for tsunami analysis	458
18.5	Tsunami damage assessment	459
18.6	Development of analytical fragility curves	462
18.7	Conclusions	463
References		463

19. Modeling and uncertainty in probabilistic tsunami hazard and risk assessment   465

Nobuhito Mori and Takuya Miyashita

- 19.1 Introduction   465
- 19.2 Uncertainty classification   467
  - 19.2.1 Uncertainties related to tsunami source   467
  - 19.2.2 Uncertainty related to tsunami propagation   468
  - 19.2.3 Uncertainty related to tsunami inundation   470
  - 19.2.4 Uncertainty related to exposure and damage   470
- 19.3 Uncertainty propagation   470
  - 19.3.1 Earthquake source modeling   471
  - 19.3.2 Tsunami modeling   472
  - 19.3.3 Fragility modeling   472
  - 19.3.4 Compound and cascading hazard modeling   473
  - 19.3.5 Incorporating different uncertainty models   474
- 19.4 Summary and conclusions   474
- References   475

20. Multihazard risk assessments   479

Hyoungsu Park

- 20.1 Introduction   479
- 20.2 Multihazards, damage, and risk assessments at Seaside, Oregon   480
  - 20.2.1 Study site   480
  - 20.2.2 Probabilistic seismic and tsunami hazard analysis   481
  - 20.2.3 Probabilistic seismic and tsunami damage assessment   488
  - 20.2.4 Loss and risk assessment   495
- 20.3 Conclusions   496
- References   497

21. Dynamic agent-based evacuation   501

Tomoyuki Takabatake and Miguel Esteban

- 21.1 Introduction   501
- 21.2 Literature review on agent-based tsunami evacuation simulations   501
- 21.3 Development of an agent-based tsunami evacuation simulation model   503
- 21.4 Application of an agent-based tsunami evacuation simulation model   505
- 21.5 Conclusions   508
- References   508

22. Sea-level rise and tsunami risk   513

Miguel Esteban, Tomoyuki Takabatake, Ryutaro Nagai, Kentaro Koyano and Tomoya Shibayama

- 22.1 Introduction   513
- 22.2 Methodology   514
  - 22.2.1 Sea-level rise scenarios   514
  - 22.2.2 Simulation methodology   515
- 22.3 Simulation results   516
  - 22.3.1 Increase in tsunami risk around Tokyo Bay due to sea-level rise   516
  - 22.3.2 Increase in tsunami risk around the eastern coast of the Kanto region due to sea-level rise   519
- 22.4 Discussion   520
- 22.5 Conclusions   522
- References   522

23. Long-term tsunami risk considering time-dependent earthquake hazard and nonstationary sea-level rise   525

Katsuichiro Goda and Raffaele De Risi

- 23.1 Introduction   525
- 23.2 Tofino and physical environment   526
  - 23.2.1 Cascadia subduction zone   526
  - 23.2.2 District of Tofino   526
  - 23.2.3 Tides and relative sea-level rise in Tofino   527
- 23.3 Long-term probabilistic tsunami risk model for Tofino   528
  - 23.3.1 Occurrence and magnitude models of megathrust Cascadia earthquakes   529
  - 23.3.2 Conditional tsunami risk curves   529
  - 23.3.3 Long-term tsunami risk analysis with tidal variations and relative sea-level rises   533
- 23.4 Long-term tsunami risk assessment for Tofino   533
  - 23.4.1 Effects of baseline sea levels   533
  - 23.4.2 Tsunami risk curves for different elapsed times and relative sea-level rise scenarios   537
- 23.5 Conclusions   539
- References   540

24. Digital twin paradigm for coastal disaster risk reduction and resilience — 543

Shunichi Koshimura, Nobuhito Mori, Naotaka Chikasada, Keiko Udo, Junichi Ninomiya, Yoshihiro Okumura and Erick Mas

- 24.1 Introduction — 543
- 24.2 Concept of coastal digital twin — 543
- 24.3 Tsunami disaster digital twin in Japan — 544
  - 24.3.1 Real-time tsunami inundation and damage forecast — 546
  - 24.3.2 Tsunami exposure analysis using mobile spatial statistics — 548
  - 24.3.3 Multiagent modeling of disaster response activities — 550
- 24.4 Towards establishing coastal digital twin paradigm — 552
- 24.5 Summary and future challenges — 554
- References — 555

Appendix — 561
Index — 573

List of contributors

Marco Anzidei Osservatorio Nazionale Terremoti, Istituto Nazionale di Geofisica e Vulcanologia, Rome, Italy

Alberto Armigliato Dipartimento di Fisica e Astronomia "Augusto Righi", Alma Mater Studiorum—Università di Bologna, Bologna, Italy

Enrico Baglione Sezione di Bologna, Istituto Nazionale di Geofisica e Vulcanologia, Bologna, Italy

Marco Baiguera School of Engineering, University of Southampton, Southampton, United Kingdom

David Burbidge GNS Science, Lower Hutt, Wellington, New Zealand

Matías Carvajal Instituto de Geografía, Pontificia Universidad Católica de Valparaíso, Valparaíso, Valparaiso, Chile

Patricio Andrés Catalán Centro Nacional de Investigación para la Gestión Integrada de Desastres (CIGIDEN), Santiago, Chile; Departamento de Obras Civiles, Universidad Tecnica Federico Santa Maria, Valparaiso, Chile

Jonas Cels UCL EPICentre, University College London, London, United Kingdom

Naotaka Chikasada National Research Institute for Earth Science and Disaster Resilience, Tsukuba, Ibaraki, Japan

Raffaele De Risi School of Civil, Aerospace and Design Engineering, University of Bristol, Bristol, United Kingdom

Marta Del Zoppo Department of Structures for Engineering and Architecture, University of Naples Federico II, Naples, Italy

Hossein Ebrahimian Department of Structures for Engineering and Architecture, University of Naples Federico II, Naples, Italy

Miguel Esteban Faculty of Science and Engineering, Waseda University, Tokyo, Japan

Sylfest Glimsdal Natural Hazards Department, Norwegian Geotechnical Institute, Oslo, Norway

Katsuichiro Goda Department of Earth Sciences, Western University, London, ON, Canada

Anita Grezio Sezione di Bologna, Istituto Nazionale di Geofisica e Vulcanologia, Bologna, Italy

Aditya Gusman GNS Science, Lower Hutt, Wellington, New Zealand

Carl Bonnevie Harbitz Natural Hazards Department, Norwegian Geotechnical Institute, Oslo, Norway

Jiangheng He Geological Survey of Canada, Sidney, BC, Canada

Fatemeh Jalayer Department of Risk and Disaster Reduction, University College London, London, United Kingdom

Seyed Abbas Jazaeri Department of Civil Engineering, University of Ottawa, Ottawa, ON, Canada

Joseph Kim Department of Civil Engineering, University of Ottawa, Ottawa, ON, Canada

Shunichi Koshimura International Research Institute of Disaster Science, Tohoku University, Sendai, Miyagi, Japan; RTi-cast, Inc., Sendai, Miyagi, Japan; Graduate School of Engineering, Tohoku University, Sendai, Miyagi, Japan

Kentaro Koyano Faculty of Science and Engineering, Waseda University, Tokyo, Japan

Finn Løvholt Offshore Energy Department, Norwegian Geotechnical Institute, Oslo, Norway

Alessandra Maramai Sezione Roma1, Istituto Nazionale di Geofisica e Vulcanologia, Rome, Italy

Erick Mas International Research Institute of Disaster Science, Tohoku University, Sendai, Miyagi, Japan; RTi-cast, Inc., Sendai, Miyagi, Japan; Graduate School of Engineering, Tohoku University, Sendai, Miyagi, Japan

Takuya Miyashita Disaster Prevention Research Institute, Kyoto University, Uji, Kyoto, Japan

Mohammad Mokhtari International Institute of Earthquake Engineering and Seismology, Tehran, Iran

Nobuhito Mori Disaster Prevention Research Institute, Kyoto University, Uji, Kyoto, Japan

Iyan E. Mulia Hydrography Research Group, Faculty of Earth Sciences and Technology, Bandung Institute of Technology, Bandung, Indonesia; Research Center for Disaster Mitigation, Bandung Institute of Technology, Bandung, Indonesia

Ryutaro Nagai Faculty of Science and Engineering, Waseda University, Tokyo, Japan

Junichi Ninomiya Institute of Science and Engineering, Kanazawa University, Kanazawa, Ishikawa, Japan

Ioan Nistor Department of Civil Engineering, University of Ottawa, Ottawa, ON, Canada

Yoshihiro Okumura Faculty of Societal Safety Sciences, Kansai University, Osaka, Osaka, Japan

Hyoungsu Park Department of Civil, Environmental, and Construction Engineering, University of Hawaii at Manoa, Honolulu, HI, United States

William Power GNS Science, Lower Hutt, Wellington, New Zealand

Tiziana Rossetto UCL EPICentre, University College London, London, United Kingdom

Kenji Satake Earthquake Research Institute, The University of Tokyo, Tokyo, Japan

Jacopo Selva Dipartimento di Scienze della Terra, dell'Ambiente e delle Risorse, Università di Napoli—Federico II, Napoli, Italy

Tomoya Shibayama Faculty of Science and Engineering, Waseda University, Tokyo, Japan

Tianhaozhe Sun Geological Survey of Canada, Sidney, BC, Canada; School of Earth and Ocean Sciences, University of Victoria, Victoria, BC, Canada

Matthew Sypus School of Earth and Ocean Sciences, University of Victoria, Victoria, BC, Canada

Tomoyuki Takabatake Department of Civil and Environmental Engineering, Kindai University, Higashiosaka, Japan

Yuichiro Tanioka Institute of Seismology and Volcanology, Faculty of Science, Hokkaido University, Sapporo, Hokkaido, Japan

Matteo Taroni Sezione Roma1, Istituto Nazionale di Geofisica e Vulcanologia, Rome, Italy

Keiko Udo Graduate School of Engineering, Tohoku University, Sendai, Miyagi, Japan

Antonio Vecchio Department of Astrophysics/IMAPP, Radboud University, Nijmegen, The Netherlands

Kelin Wang Geological Survey of Canada, Sidney, BC, Canada; School of Earth and Ocean Sciences, University of Victoria, Victoria, BC, Canada

Xiaoming Wang GNS Science, Lower Hutt, Wellington, New Zealand

Yong Wei Cooperative Institute for Climate, Ocean, & Ecosystem Studies, University of Washington, Seattle, WA, United States; National Oceanic & Atmospheric Administration Pacific Marine Environmental Laboratory, Seattle, WA, United States

Natalia Zamora Wave Phenomena Group, Barcelona Supercomputing Center, Barcelona, Spain

Filippo Zaniboni Dipartimento di Fisica e Astronomia "Augusto Righi", Alma Mater Studiorum—Università di Bologna, Bologna, Italy

Yijie Zhu School of Earth and Ocean Sciences, University of Victoria, Victoria, BC, Canada

Preface

Powerful tsunamis, triggered by earthquakes, landslides, and volcanic eruptions, can devastate coastal areas and bring misery to numerous people globally. At present, it is difficult to predict when and where destructive tsunamis will occur. Under such circumstances, our best course of action is creating resilient coastal communities that are aware and informed by the latest tsunami science and engineering and equipped with effective disaster risk reduction measures. We need to build tsunami-resistant structures using sound coastal engineering techniques to prepare for future high-impact tsunami events. If tsunami disasters occur, people must be vigilant in responding and adapting to them as soon as possible.

Probabilistic tsunami hazard analysis (PSHA) and probabilistic tsunami risk analysis (PTRA) are indispensable tools for forecasting and quantifying the impacts of tsunamis to coastal communities and built environments and have advanced significantly over the last three decades. Their applications facilitate improved disaster risk mitigation and management. Uncertainties associated with forecasted tsunami impacts can be substantial, and practitioners and policymakers need guidance on implementing disaster risk reduction actions at all levels (local, regional, national, and international). In communicating broad ranges of possible consequences with stakeholders, disaster scenarios need to be selected carefully and their risks need to be communicated clearly.

PTHA and PTRA involve a wide range of knowledge related to tsunami generation, propagation, and inundation and its effects on people and structures. To appreciate what PTHA and PTRA can offer, broad knowledge of seismology, geophysics, oceanography, coastal engineering, and disaster risk management, as well as mastery of mathematics, probability and statistics, and geomatics is necessary. Moreover, the expertise and skills of computer modeling and data analytics are essential for understanding the multidisciplinary fields of tsunami science and engineering deeply.

This book aims at providing the state-of-the-art review of PTHA and PTRA and discussing their applications to tsunami disaster risk reduction. This book is written and edited by keeping two groups of audience in mind. The first group includes upper-division undergraduate and graduate students focusing their studies on tsunamis as well as professionals who are new to the topics of PTHA and PTRA. The second group includes experienced researchers and professionals who are interested in acquiring information about the latest developments related to PTHA and PTRA.

This book is composed of two sections. Section 1 covers the introduction to tsunami hazard and risk assessments (Chapter 1), the fundamental theories and physics of tsunami generation (Chapter 2), tsunami propagation and inundation (Chapter 3), and its impact on the built environment (Chapter 4). Building upon these basic subjects, practical approaches to probabilistic tsunami hazard and risk analyses are explained in Chapter 5, whereas examples of tsunami disaster risk management are discussed in Chapter 6. To grasp the contents covered in the introductory chapters, readers should be familiar with calculus, differential equations, and statistics, which are typically taught in university courses for basic physics, geophysics, hydrodynamics, and engineering mechanics. The authors strongly believe that deeper understanding of the concepts, theories, and case studies can be gained by analyzing data and visualizing the results. For this purpose, MATLAB® codes/tools and datasets, which can be used to generate some of the numerical examples shown in Chapters 2, 3, and 5, are provided in the **Appendix**.

Section 2 of this book is devoted to applications and case studies for tsunami hazard and risk analyses, which are contributed by the tsunami experts worldwide. Chapter 7 presents an overview of tsunami data (e.g., tidal gauges, ocean bottom sensors, and field surveys) and their use in tsunami source inversion analysis to estimate the tsunami generation processes of the historical and prehistorical events. Chapter 8 explains the roles of tectonics and fault mechanics (e.g., fault roughness and temperature) on the tsunami generation process. On the other hand, Chapter 9 sheds light on the splay faulting that generates tsunamis efficiently due to steep dip angles and low rock rigidity. Chapter 10 discusses the approaches of incorporating landslide-triggered tsunamis into PTHA and highlights key uncertainties associated with the triggering mechanisms. Chapter 11 summarizes the recent developments of dense tsunami monitoring systems (e.g., S-net in Japan) and discusses the new insights that can be gained from such systems (e.g., improved tsunami source modeling and early warning). Moreover, Chapter 12 provides insights on the advancement of tsunami early warning systems using new machine learning techniques.

Chapters 13 through 16 in Section 2 cover the state-of-the-art summary of tsunami hazard and risk assessments in different parts of the world. Chapter 13 is focused on tsunami hazard and risk assessments for the Pacific Ocean, led by the National Oceanic and Atmospheric Administration (NOAA). On the other hand, Chapter 14 presents the tsunami hazard and risk assessments for New Zealand, conducted by the GNS Science. Chapter 15 provides an overview of tsunami hazards in the Mediterranean Sea, including nonseismic tsunami sources due to volcanic activity and landslides as well as meteotsunamis. Chapter 16 summarizes the recent developments of tsunami hazard assessments in Chile and discusses the future extensions to achieve improved tsunami risk mapping and design of coastal structures.

Chapters 17 through 20 in Section 2 elaborate on the topics related to tsunami risk assessments. Chapter 17 provides an overview of developing tsunami fragility functions based on empirical tsunami damage data, whereas Chapter 18 discusses the emerging approaches of analytical tsunami fragility modeling using numerical models of coastal structures subjected to tsunami loadings. Furthermore, Chapter 19 offers an integrated perspective of the uncertainties inherent in PTHA and PTRA, starting from tsunami source to exposure-vulnerability characterizations. Chapter 20 presents an innovative new multihazard risk assessment methodology for coastal cities and towns subjected to ground shaking and tsunami inundation.

Chapters 21 through 24 in Section 2 discuss the advanced topics related to PTHA and PTRA. Chapter 21 introduces a dynamic tsunami evacuation framework by accounting for complex evacuation behavior and interactions among evacuees, which will inform the development of more effective tsunami evacuation planning in coastal cities and towns. Chapter 22 presents the tsunami inundation simulations by considering the future sea-level rise scenarios and discusses the importance of accounting for the climate change effects in coastal environments. Moreover, Chapter 23 performs the long-term tsunami risk assessments for a coastal community by incorporating variable tidal level and future projections of relative sea-level rise. Finally, Chapter 24 provides an overview of emerging coastal digital twin models that create a cyber-physical system of coastal environments and generate the new insights for optimal solutions in the physical world.

Finally, the editors of the book would like to thank all chapter contributors for their passions and commitments to making this book possible. Experiences and lessons learned in this process have been inspiring and rewarding and the editors hope that the readers of this book will also experience such stimulating and enlightening learning opportunities.

Katsuichiro Goda, Raffaele De Risi, Aditya Gusman, and Ioan Nistor
July 2024

Chapter 1

Introduction of probabilistic tsunami hazard and risk analysis—toward disaster risk reduction and resilience

Katsuichiro Goda[1], Raffaele De Risi[2], Aditya Gusman[3] and Ioan Nistor[4]

[1]Department of Earth Sciences, Western University, London, ON, Canada, [2]School of Civil, Aerospace and Design Engineering, University of Bristol, Bristol, United Kingdom, [3]GNS Science, Lower Hutt, Wellington, New Zealand, [4]Department of Civil Engineering, University of Ottawa, Ottawa, ON, Canada

1.1 Introduction

A tsunami, a combination of two Japanese words translated into English as "wave in harbor," is a series of water waves generated by several potential sources, such as seismic activities, aerial and submarine landslides, volcanic eruptions, and rare meteoric impacts. Earthquakes are the principal source of over 90% of tsunamis, whereas landslides and volcanic eruptions have triggered devastating tsunamis in history. During an earthquake rupture, tsunamis are generated by transforming large-scale seabed deformation into the potential energy of displaced water above the deformation (Levin & Nosov, 2016). The initial dislocation of a large volume of water at the surface then propagates across oceans due to gravity and, depending on its magnitude, nearshore bathymetry, and coastal topography, can cause substantial onshore inundation along coastlines, making tsunamis one of the most destructive and deathly natural hazards (Bernard & Titov, 2015). Coastal flooding is initiated by a shoaling process during which, as the tsunami approaches the coast, the wave propagation speed decreases and, while the tsunami height increases, the wave period remains unchanged. The increase in the tsunami height occurs because the wave amplitude is a function of the propagation velocity, which depends on the ocean depth along its propagation path. On the other hand, the tsunami wave loses some of its energy due to bottom friction and turbulence. The inundation spatial extent depends on the topography of the nearshore coastline and on the wave height, breaking type (bore or surge), and its period at the shore.

Globally, numerous cities and towns are located along the world's coastline around active seismic regions, and, increasingly, more people migrate to coastal areas for socio-economic reasons. As a result, the population in tsunami-prone regions has been steadily increasing over the last three decades (Løvholt et al., 2014). Fig. 1.1 highlights the significant exposure of the world's population and countries to earthquakes and tsunamis. As more population migrate to coastal regions, the demand for buildings and urban infrastructures increases rapidly. In coastal areas, there are conventional and critical infrastructures (Fig. 1.2). Conventional infrastructure comprises roads, railways, power, and habitable accommodation (houses). On the other hand, critical infrastructures in coastal areas typically include hospitals, police and fire stations, harbors, gas and oil storage, nuclear power plants, and facilities for tsunami early warning systems. Both conventional and critical infrastructures underpin socio-economic activities in urban areas.

Historical records show that tsunamis caused significant socio-economic impacts on coastal communities, as evidenced by the major destruction of coastal communities. Recent major tsunami events have stressed the importance of tsunami hazard assessment, preparedness, and risk mitigation. For example, the 2004 Indian Ocean Earthquake, which had a moment magnitude (M_w) of 9.3, triggered a massive tsunami that reached a maximum runup of 30 m or more (Titov et al., 2005; Wang & Liu, 2006). This tsunami left hundreds of thousands of deaths and billions of dollars in damages across 19 countries. The lack of an early warning system in the Indian Ocean and of risk management strategies made the 2004 Indian Ocean Tsunami the second most devastating natural disaster of the past century (Ghobarah et al., 2006). The devastating event led to an astounding death toll of 227,000 spread across 14 countries surrounding the Indian Ocean. On the other hand, the 2011 Tohoku Tsunami left more than 19,000 fatalities and caused hundreds of billions of dollars in

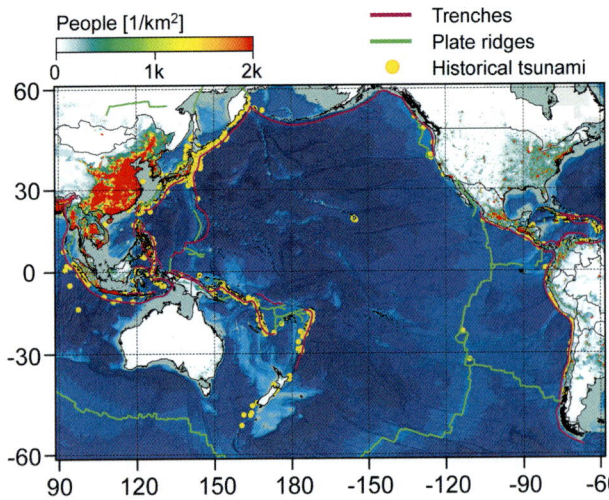

FIGURE 1.1 Population distribution around the Pacific Ocean Rim. Subduction trenches and plate boundaries are shown with magenta and green lines, respectively, whereas historical seismically triggered tsunamis are indicated by yellow dots. Data are obtained from the National Geophysical Data Center (https://www.ngdc.noaa.gov/hazard/tsu_db.shtml). Population data are based on NASA's Earth Observing System Data and Information System (http://sedac.ciesin.columbia.edu/). Map lines delineate study areas and do not necessarily depict accepted national boundaries.

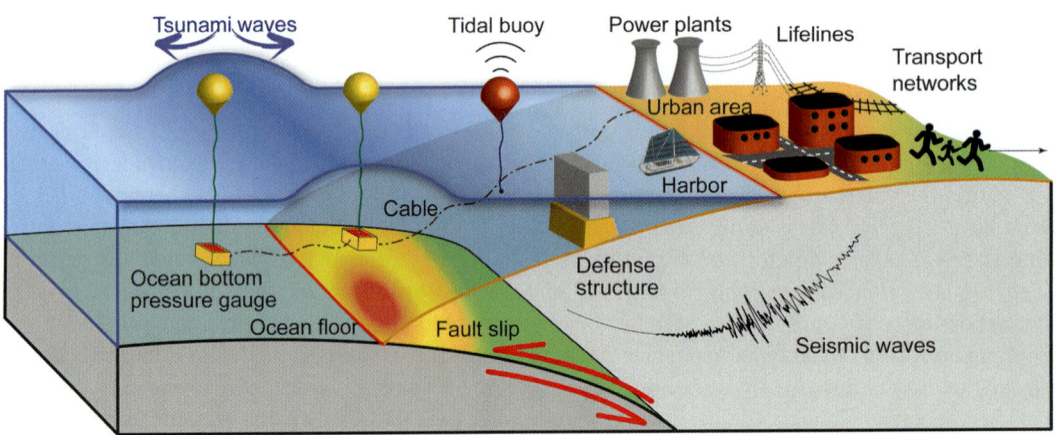

FIGURE 1.2 Coastal urban infrastructure under tsunami threat in a subduction zone.

damages despite extensive protective structures along the Japanese coastlines (Fraser et al., 2013; Latcharote et al., 2018). This event highlighted another issue in tsunami hazard and risk assessments, that is, the difficulty of predicting the characteristics of future events based on historical records alone. The Tohoku event was exceptional because the actual event was significantly greater than what scientists and engineers had previously thought that this subduction zone could generate. The uncertainty of future events affects risk management decisions, which could ultimately fail to prevent major devastation and human casualties. These catastrophes prompted scientists and engineers to develop improved tsunami design codes, early warning systems, and new tsunami analysis techniques (Chock, 2016; Mori et al., 2022).

Tsunami hazard analysis is essential for designing critical buildings and infrastructure and safeguarding people and assets in coastal areas. It determines the relevant tsunami intensity metrics (e.g., offshore wave amplitudes, inundation flow depth, flow velocity, and momentum flux) that must be considered for calculating tsunami loads and effects impacting the infrastructure. Defining tsunami hazard scenarios for engineering design purposes is critical because, regardless of how structural systems located in tsunami-prone areas are designed, if the tsunami hazard is underestimated, catastrophic damage and loss may occur, such as the ones experienced after the 2004 Indian Ocean and the 2011 Tohoku Tsunamis. Existing tsunami design codes and guidelines for buildings in the United States (American Society of Civil Engineers, 2022) and Japan (Japan Society of Civil Engineers, 2016) have sections that outline prescriptive technical procedures for determining tsunami design parameters as well as prescriptive methods, written in mandatory language, for the design of critical infrastructure. A standard approach to defining tsunami design scenarios for buildings and coastal infrastructure is to identify critical historical events that had occurred in the region and to include their characteristics in the definition of the tsunami design procedures. As such, various types of paleo-seismic/tsunami records and geological/geophysical data should be consulted to ensure that possible tsunami sources and scenarios are not missed. These tsunami scenarios can be used deterministically to calculate the critical tsunami amplitudes, tsunami inundation depths, and tsunami flow velocities for coastal tsunami design purposes. A shortcoming of this approach is its high reliance on the completeness and correctness of the available data records, which are often limited in quality, duration, accuracy, and scope.

Over the past two decades, probabilistic methods have been introduced in tsunami structural design codes and guidelines as alternatives to conventional deterministic methods. The necessity of probabilistic approaches was motivated by the catastrophic tsunami events in the Indian and Pacific Oceans and by the similarity to the earthquake engineering standards and practices that have long considered such an approach. Historical data alone were proven insufficient to define possible extreme scenarios (Kagan & Jackson, 2013). Instead, probabilistic tsunami hazard analysis (PTHA) offers a systematic way to include uncertainties associated with tsunami sources, occurrence probability, tsunami generation, wave propagation, and inundation of land areas (Behrens et al., 2021; Mori et al., 2018). The PTHA methods provide probabilistic estimates of offshore tsunami wave amplitudes along coasts, tsunami inundation flow depth and velocity within the inundation zone, and the upper limit of runup. Early PTHA investigations started in the 1980s by using the available information on historical events (Rikitake & Aida, 1988) with close similarity to probabilistic seismic hazard analysis (Baker et al., 2021; Cornell, 1968). After the 2004 Indian Ocean Tsunami, the second generation of PTHA studies has expanded globally (Annaka et al., 2007; Geist & Parsons, 2006; Power et al., 2007; Thio et al., 2012). Moreover, after the 2011 Tohoku Tsunami, the third generation of PTHA studies has accelerated by considering details of fault characteristics (e.g., fault geometry and slip distribution) and their uncertainties more comprehensively through a logic tree and stochastic source modeling (De Risi & Goda, 2017; Fukutani et al., 2015; Miyashita et al., 2020; Mueller et al., 2015; Park & Cox, 2016). It is essential to recognize that PTHA is not a solution; rather, it is a framework to be explicit about what needs to be adopted in line with the current tsunami engineering design practices. PTHA also transparently incorporates uncertainties related to tsunami hazard assessments and facilitates sensitivity analysis to examine the influence of adopted assumptions.

Quantitative risk assessments of catastrophic tsunami events are a prerequisite for achieving effective disaster risk reduction (DRR) and disaster risk management (DRM). To evaluate the tsunami impacts on society and the built environment, PTHA can be extended to probabilistic tsunami risk analysis (PTRA) (Goda & De Risi, 2017). Key elements of quantitative risk assessments are hazard, exposure, vulnerability, and loss (Fig. 1.3), and uncertainties associated with these components are integrated into the final risk assessments (Beven et al., 2018). This risk framework has been adopted widely across different natural hazards, such as earthquakes (Goulet et al., 2007) and landslides (Lee & Jones, 2014), and has been referred to as natural catastrophe modeling (or CAT model) in the insurance industry (Mitchell-Wallace et al., 2017; Woo, 2011). It has become a vital tool not only for the insurance and reinsurance industry but also for governmental agencies that are responsible for implementing DRM policies at local, regional, and national levels. Standard outputs from quantitative risk models are often obtained in the form of an exceedance probability (EP) curve and average annual loss (AAL). The framework is particularly useful for defining the long-term objectives in reducing the consequences of future disasters and for promoting risk-based management decisions (Liel & Deierlein, 2013; Yoshikawa & Goda, 2014).

Resilience and sustainability are fundamental requirements for coastal communities and modern infrastructures, and major progress has been made in DRM (Ayyub, 2014; Bruneau & Reinhorn, 2007; United Nations Office for Disaster Risk Reduction, 2022). A key to achieving effective disaster resilience and sustainability is the implementation of a holistic risk management strategy that integrates all phases of a disaster cycle (i.e., mitigation, preparedness, response, and recovery; Fig. 1.4) across different administrative levels. Resilience is the capability of people and infrastructures

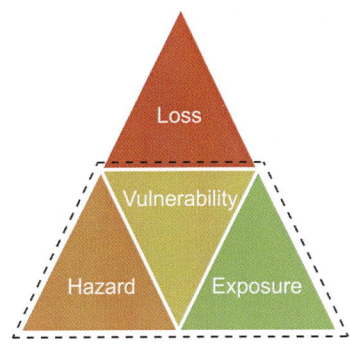

FIGURE 1.3 Elements of disaster risk assessment. The dotted line enclosure indicates that loss assessment is based on and facilitated by integrating hazard, exposure, and vulnerability.

FIGURE 1.4 Disaster cycle framework.

to survive, adapt, and grow against unforeseen extreme events, such as earthquakes and tsunamis. There are three main aspects of infrastructure resilience. The first is system resistance and robustness, representing the capacity to withstand and survive external disturbances. The second is system response and recovery, which is the capability of taking actions initially to reduce the adverse impact due to disturbances and eventually to return the community to a normal condition (although it may not be identical to the predisaster situation) or even better condition (i.e., the *Build Back Better* principle). The third is system preparedness, which is critical for the subsequent disaster response—recovery—mitigation cycle. At the same time, modern infrastructure should be sustainable by maintaining human equity, diversity, and the functionality of natural systems and, therefore, not jeopardizing future generations' well-being. Building infrastructure sustainably increases resilience since it improves the overall system performance in terms of satisfaction of people's need for basic services. It also strengthens various intangible values that are the main drivers for long-term social development, such as social vitality, well-being, social and environmental health, and safety. Moreover, protecting environmental assets reduces the risk and provides resources to facilitate recovery.

This chapter presents an overview of PTHA and PTRA, particularly emphasizing disaster risk mitigation and resilience. Section 1.2 mentions several tsunami disasters to introduce the historical context of current PTHA and PTRA approaches and summarizes the lessons learned from these events. In Section 1.3, tsunami hazard and risk assessments and their key elements (e.g., hazard, exposure, and vulnerability) are introduced, while in Section 1.4, a summary of tsunami DRR and DRM is given. Sections 1.3 and 1.4 serve as a primer for the later chapters (i.e., fundamental sections and advanced research topics). Section 1.5 formally introduces the scope of the book by relating the contents/topics covered in the "fundamental" chapters and the "advanced" chapters.

1.2 Historical events and lessons learned

Large tsunamis typically occur in major subduction zones and claim numerous fatalities when they hit populated coastal areas. Major destructive tsunami events have been documented in historical records since ancient times. Table 1.1 lists historical tsunami events that caused more than 2000 fatalities. Many deadly tsunamis occurred in Japan, Indonesia, Chile, and Peru. The two most recent large tsunamis, namely, the 2004 Indian Ocean Tsunami and the 2011 Tohoku Tsunami, were particularly destructive.

Most tsunami sources coincide with the fast-moving convergent subduction boundaries in the Pacific and Indian Oceans (Fig. 1.1). Tsunamis also occurred in the Mediterranean and Caribbean Seas, whereas not many tsunamis were recorded in the Atlantic Ocean due to the divergent nature of tectonics. Since the beginning of instrumental recordings of earthquakes in the early 1900s, many tsunamis have been observed to identify tsunamigenic zones from an empirical perspective. Nevertheless, the period of instrumental records is short compared to the geological time of major subduction earthquakes, and not all tsunami sources have been firmly confirmed. For instance, several zones have been recently recognized as potential sources capable of causing large tsunamigenic earthquakes, such as the Makran region off the coasts of Pakistan and Iran (Heidarzadeh et al., 2008; Mokhtari et al., 2008) and the Cascadia region off the western United States and Canada (Goldfinger et al., 2012; Satake et al., 2003). Future scientific discoveries will update the current understanding of tsunami hazards and risks.

In the following, six recent tsunami events are discussed to highlight the key features of each destructive tsunami. These field observations are particularly insightful in understanding the complex patterns of earthquake—tsunami damage and their effects on coastal communities and infrastructure. This is useful for improving the current approaches to tsunami hazard and risk assessments.

TABLE 1.1 Regional and local tsunamis that caused 2000 or more deaths.

Year	Month	Day	Source location	Estimated casualties and missing
365	7	21	Crete, Greece	5000
887	8	2	Niigata, Japan	2000
1341	10	31	Aomori Prefecture, Japan	2600
1498	9	20	Enshunada Sea, Japan	5000
1570	2	8	Central Chile	2000
1605	2	3	Nankaido, Japan	5000
1611	12	2	Sanriku, Japan	5000
1674	2	17	Banda Sea, Indonesia	2244
1687	10	20	Southern Peru	5000[a]
1692	6	7	Port Royal, Jamaica	2000
1703	12	30	Boso Peninsula, Japan	5233[a]
1707	10	28	Enshunada Sea, Japan	2000
1707	10	28	Nankaido, Japan	5000[a]
1741	8	29	Hokkaido, Japan	2000
1746	10	29	Central Peru	4800
1751	5	20	Northwest Honshu, Japan	2100
1755	11	1	Lisbon, Portugal	50,000[a]
1771	4	24	Ryukyu Islands, Japan	13,486
1792	5	21	Kyushu Island, Japan[b]	14,524
1854	12	24	Nankaido, Japan	3000[a]
1868	8	13	Northern Chile[a]	25,000
1877	5	10	Northern Chile	2282
1883	8	27	Krakatau, Indonesia[b]	34,417
1896	6	15	Sanriku, Japan	27,122[a]
1899	9	29	Banda Sea, Indonesia	2460[a]
1908	12	28	Messina Strait, Italy	2000
1923	9	1	Sagami Bay, Japan	2144
1933	3	2	Sanriku, Japan	3022
1945	11	27	Makran Coast, Pakistan	4000[a]
1952	11	4	Kamchatka, Russia	10,000
1960	5	22	Southern Chile	2223
1976	8	16	Moro Gulf, Philippines	6800
1998	7	17	Papua New Guinea	2205
2004	12	26	Banda Aceh, Indonesia	227,899[a,c]
2011	3	11	Tohoku, Japan	18,453[a,c]
Total				508,014

[a]May include earthquake deaths.
[b]Tsunami generated by volcanic eruption.
[c]Includes dead/missing near and outside source region.
Source: International Tsunami Information Center.

1.2.1 2004 Indian Ocean Tsunami

On December 26, 2004, the M_w 9.1 Indian Ocean Tsunami caused tremendous devastation and deaths throughout the Indian Ocean region. Its origin was where the Indo-Australian Plate subducts beneath the Eurasian Plate at its eastern margin. Fig. 1.5 shows a map of the Indian Ocean region; the rupture occurred along the fault between the Burma Plate and the Indian Plate with a length exceeding 1000 km. More than 227,000 people were either killed or missing in fourteen countries around the Indian Ocean, and 1127,000 were displaced by the earthquake and subsequent tsunami. The worst hit country was Indonesia, with 167,540 listed as dead or missing, followed by Sri Lanka (35,322), India (16,269), and Thailand (8212). The total estimated material losses in the Indian Ocean region were US$ 10 billion, and insured losses were US$ 2 billion. The 2004 event provided unique and valuable opportunities to evaluate the performance of various structures to the ground motion and investigate the impact of the accompanying large tsunami waves on structures. These observations inspired research into the development of tsunami design and mitigation procedures (Macabuag et al., 2018).

- The photo taken in Nang Thong, Thailand (Fig. 1.5A), shows damage to columns and structural members due to the tsunami-involved impact by objects, such as boats and cars, as well as static and dynamic effects of debris in increasing the area subjected to water pressure and impact on structures. Ensuring structural redundancy is important so that the building will not collapse as a consequence of the failure of one or two of its columns.
- Anchoring roofs, precast concrete floor slabs, precast dock concrete slabs, bridge decks, and storage tanks to the beams, piers, abutments, and foundations are essential to maintain the integrity of structures. The design tension forces on the anchoring systems caused by uplift and buoyancy forces of water required re-evaluation.

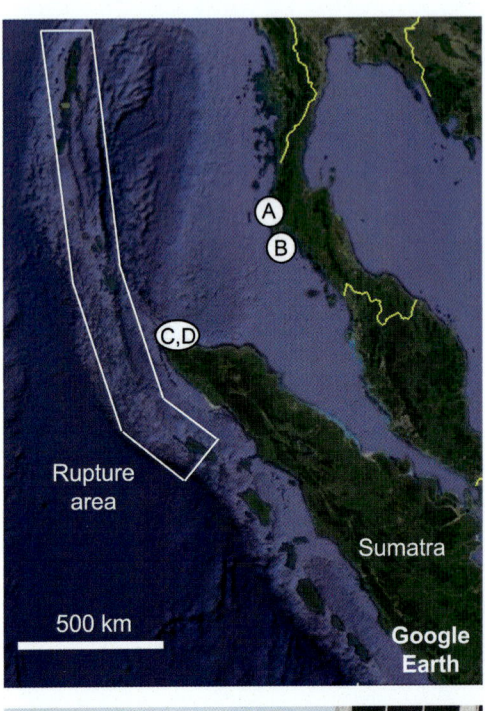

FIGURE 1.5 Tsunami damage in Thailand and Indonesia after the 2004 Indian Ocean Tsunami.

- The photo taken in Phuket, Thailand (Fig. 1.5B), shows the loss of foundation soil under footings observed on the beaches that occurred due to water scour around buildings. Scour was limited to shallow foundations on granular and sandy soil. This type of failure can be avoided by designing deep foundations for buildings near the shore in tsunami-vulnerable coastal areas or by ensuring adequate protection around the perimeter of the building.
- Many engineered structures suffered severe damage and collapse due to the earthquake ground motion in Banda Aceh, northern Sumatra, Indonesia (Fig. 1.5C and D). Most structural failures and collapses can be attributed to inadequate design, nonductile detailing, and construction deficiencies.
- Well-designed and constructed buildings survived up to 12-m tsunami wave runup with minimum structural damage, although damage to windows, infill walls, balcony rails, fixtures, and building content was extensive. Such buildings can be used as shelter; repairing the damage to nonstructural elements is simple. Therefore the building can return to service in a short time.

1.2.2 2010 Maule Chile Tsunami

On February 27, 2010, the M_w 8.8 Maule Earthquake occurred offshore Maule and Biobio regions of central Chile and generated a tsunami with significant runup heights (e.g., 29 m in Constitucíon and 10.5 m in Dichato). The earthquake was generated along the subduction zone at the boundary between the Nazca Plate and the South America Plate (Fig. 1.6). The fault rupture had a width of over 100 km and a length of approximately 500 km and was parallel to the central

FIGURE 1.6 Tsunami damage along the coast of Chile after the 2010 Maule Tsunami.

Chilean coast. The tsunami waves arrived within 30 minutes at many locations along the coast of Chile. This earthquake and tsunami caused US$ 30 billion loss and 521 deaths, of which the tsunami was directly responsible for 156 deaths. For this scale of the earthquake, the number of fatalities was small, despite the unfavorable occurrence time of the earthquake at 3:34 a.m. local time. This is largely attributed to preevent preparedness, awareness, and education. Elders who lived through the 1960 Chile Tsunami passed on their experience and wise advice to younger generations, and, additionally, the 2004 Indian Ocean Tsunami raised tsunami awareness. Factors that helped reduce the consequences include (1) limited earthquake damage due to well-engineered structures, (2) tsunami signage, (3) tsunami-prepared police and fire responders, and (4) education and training of inhabitants in tsunami-prone areas (Palermo et al., 2013).

- Robust-engineered structures performed well during the tsunami flooding with only nonstructural damage, mostly to glazing units.
- A number of possible sources contributed to the observed failures, including hydrodynamic (drag) forces, impulse loading from the leading edge of the tsunami, hydrostatic forces, buoyant forces, and debris impact loading.
- Residential housing consisting of light timber frame and concrete frame construction with brick masonry infill walls experienced widespread damage throughout the surveyed coastal region in Chile. Damage to residential dwellings included the punching of brick masonry infill walls, the partial and complete failure of reinforced concrete columns, and the sliding and unseating of second-story and roof levels. These can be seen in the photos taken in Constitucíon (Fig. 1.6A) and Concepcion (Fig. 1.6C).
- The photo taken in Talcahuano (Fig. 1.6C) shows that small debris resulting from collapsed timber housing as well as large debris, such as automobiles, fishing vessels, and shipping containers, were present throughout the coastal regions.
- Critical infrastructure, including roads and reinforced concrete utility poles, were extensively damaged during the tsunami. The photo taken in Dichato (Fig. 1.6B) shows the erosion of the paved road surface. The collapse of hydro poles led to the disruption of hydro services and affected the recovery of the communities from the tsunami.

1.2.3 2011 Tohoku Japan Tsunami

On March 11, 2011, the M_w 9.0 Tohoku Earthquake generated a massive tsunami due to coseismic ruptures on the plate interface along the Japan Trench. The direct loss from the tsunami exceeded US$ 210 billion, while only a small portion of this economic loss (US$ 37 billion) was covered by insurance (Kajitani et al., 2013). In Fig. 1.7, a map of the Tohoku region is shown with an estimated fault rupture area of 500 km long and 200 km wide. The estimated slip near the Japan Trench was about 30–40 m. In this figure, the runup height is also included (Mori et al., 2011). The maximum runup height of 40 m was observed in Miyako; tsunami inundation height near the shore exceeded 10–15 m at several locations, causing tremendous damage to buildings and infrastructure. An important feature in understanding tsunami runup along the Tohoku coast was the difference between rias and coastal plains. The rias are submerged river valleys and have a jagged shape (see the coastal line between (A) and (B) in Fig. 1.7). During a tsunami event, these valleys acted as amplifiers of the tsunami height, primarily due to their narrowing shape near the shoreline. This funneling effect led to a substantial increase in the height of the tsunami inundation and runup. On the other hand, in the coastal plains, as there were not many obstructions near the shore and the land profile was flat, tsunami height was not significantly amplified, but the tsunami inundated far inland horizontally. Such differences in the nearshore coastal bathymetry and topography can be observed in the runup data shown in Fig. 1.7 and result in very different tsunami damage patterns.

- The photo taken in Taro (Fig. 1.7A) shows a catastrophic failure of the famous seawall, "The Great Wall"; it consisted of two layers of concrete walls 10–11 m high above mean sea level, and the total length of the two wall layers was 2.4 km. The nearshore (outer) layer was almost completely destroyed by the tsunami; only part of the wall around the gates remained. The tsunami wave overtopped the inner seawall and destroyed the town. The interviews with survivors indicated that the existence of massive tsunami defense infrastructure provided them with a false sense of security and might have resulted in a delay in evacuation.
- The photo taken in Minamisanriku (Fig. 1.7B), located in a ria coastal region, shows a devastated three-story steel building that was used as the Crisis Management Department headquarters. This structure was located about 500 m from the shoreline and was inundated by a 10-m tsunami. In this building, 30 staff were trapped when the tsunami arrived, and tragically, 20 of those lost their lives. Tremendous tsunami damage occurred elsewhere in Minamisanriku, and several concrete buildings were destroyed (in addition to numerous wooden houses). One of the vertical evacuation buildings survived; however, it suffered from significant foundation damage due to scouring (Fraser et al., 2013).

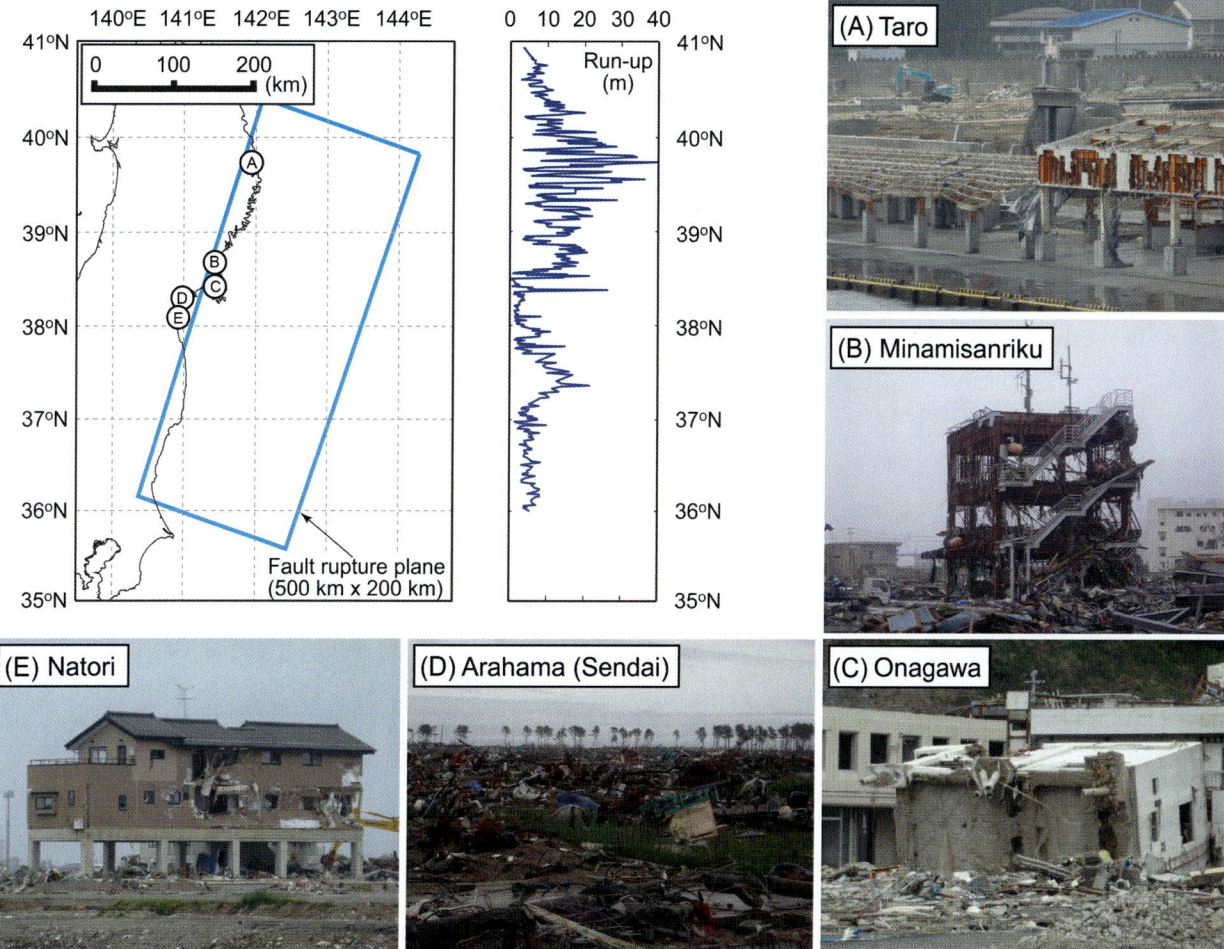

FIGURE 1.7 Tsunami damage along the Tohoku coast of Japan.

- The photo taken in Onagawa (Fig. 1.7C) shows an overturned two-story concrete shear wall structure with a pile foundation. One of the most significant features of the tsunami wave in Onagawa was the very high tsunami inundation (up to about 20 m) with fast inundation flow. This building was subjected to a sequence of external loads, including strong shaking, liquefaction and shear forces (which had caused damage to piles), and tsunami (hydrostatic and hydrodynamic forces, buoyant forces, etc.). The effects of scouring were evident at this building's location.
- The photo taken in Arahama (Fig. 1.7D) demonstrates the devastation of local residential buildings in the coastal plains. The tsunami height ranged there from 5 to 10 + m. An inundation depth exceeding about 2 m is critical for typical residential wooden houses, as their chance of being washed away under such conditions increases significantly. The photo in Natori (Fig. 1.7E) shows an elevated residential house, which survived the tsunami inundation. Another noticeable feature of the same photo is the impact of debris, such as cars, ships, and lumber. Other typical damage features in the coastal plains include deep scouring of foundations.

1.2.4 2018 Indonesian tsunamis in Sulawesi and Sunda Strait

On September 28, 2018, a M_w 7.5 earthquake struck Sulawesi Island of Indonesia. The earthquake occurred along the Palu−Koro fault. The city of Palu was devastated by a sequence of multiple hazards—strong shaking due to the mainshock, triggered tsunamis, and large-scale mudflows. The number of deaths and injuries exceeded 2100 and 4400, respectively, with more than 1300 people missing. The earthquake rupture was mainly of strike−slip, but with some dipping components on a steeply inclined fault plane. An analysis of synthetic aperture radar from Sentinel 1 satellite images suggested that the North−South offsets along the main strand of the fault of 3−4 m with subsidence of up to 1 m observed within the Palu basin and at the coast (Gusman et al., 2019). The earthquake and multiple

FIGURE 1.8 Earthquake and tsunami damage around Palu Bay.

subaerial—submarine landslides generated tsunamis with up to 3.6 m flow depth and 350 m horizontal inundation within the city of Palu (Paulik et al., 2019). The tsunamis washed away numerous houses along the coastal line of Palu Bay and destroyed port facilities. Fig. 1.8 shows a map of Palu Bay and photos taken at several locations affected by this earthquake and tsunami (Goda, Mori, Yasuda, Prasetyo, et al., 2019).

- At Donggala Port, near the mouth of Palu Bay (Fig. 1.8A), major damage to port facilities and buildings was observed due to lateral spreading of the ground, where evidence of liquefaction was noted. In the south of Donggala Port, there was a section of coastal areas that was protected by a mangrove forest, which mitigated some of the tsunami damage to houses in these areas.
- At several locations along the western coast of Palu Bay, local communities were destroyed due to tsunami waves (Fig. 1.8B).
- Along the bay area of Palu (Fig. 1.8C and D), major destruction to buildings was observed, which was caused by intense ground shaking, coastal landslides, and subsequent tsunamis. The foundation part of some of the buildings suffered scouring damage, whereas some notable tilt was caused by the lateral push due to inundation. Due to the subsidence of the bay area, normal tidal variations resulted in the inundation of this area. This kind of physical condition made it challenging to recover and reconstruct this important area.
- In inland areas of Palu, several buildings (e.g., a large shopping complex, hospital, and hotel) had collapsed due to violent ground shaking, whereas massive destruction of residential areas occurred due to large-scale mudflows in Petobo (Fig. 1.8E), Balaroa, and Jono Oge.
- At places on the eastern side of Palu Bay (Fig. 1.8F and G), severe destruction of local communities was observed due to tsunami inundation. The tsunami inundation depth that was experienced at these damaged buildings appeared

to be at the roof level of the first floor or higher (3–5 m range; Fig. 1.8F). At Pantoloan Port (south of the location shown in Fig. 1.8G), a tidal measurement was recorded during the tsunami, where the maximum wave height of about 2 m was recorded. In the north of Pantoloan Port, many vessels, including a large one shown in Fig. 1.8G, were swept ashore, and local communities were severely damaged by the tsunami.

Approximately 3 months after the Palu Earthquake and Tsunami, a deadly silent tsunami occurred on December 22, 2018, claimed more than 440 lives and 32,000 injuries, and displaced 16,000 people across the Sunda Strait region. The tsunami was associated with the flank collapse of the Anak Krakatau Volcano (Fig. 1.9). In the past, the Krakatau Volcano caused the disastrous volcanogenic tsunami in 1883, and it has been active since then. In 1927 the Anak Krakatau, meaning "the child of Krakatau" in Bahasa Indonesia, emerged above the sea from the remnants of the

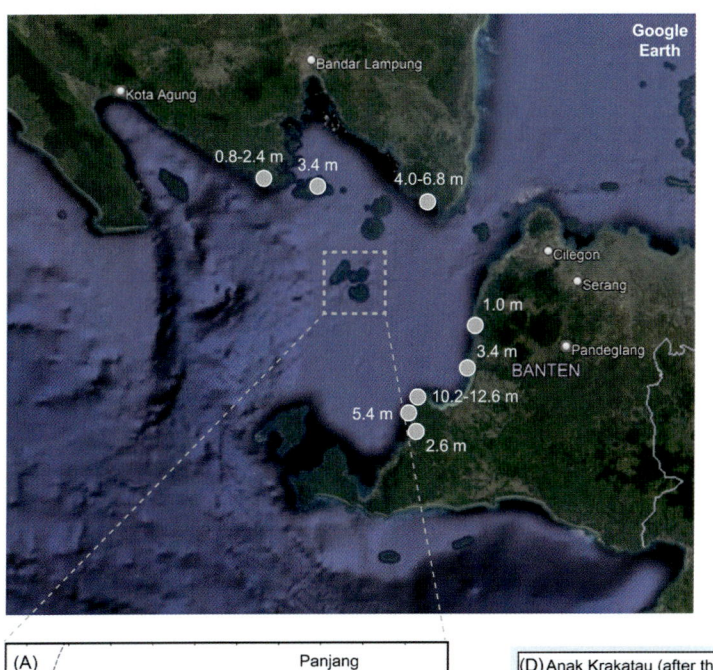

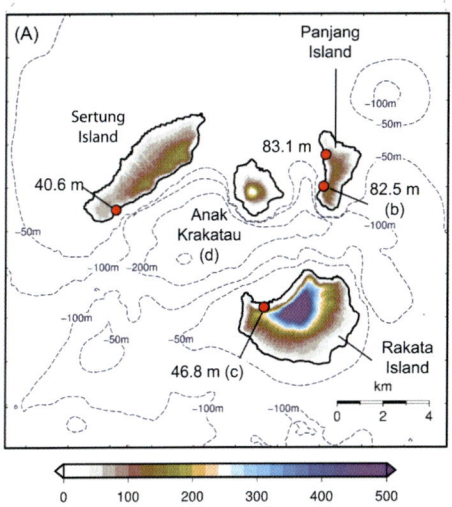

FIGURE 1.9 Maps of the Sunda Strait region and the Anak Krakatau. On the regional map of Sunda Strait, tsunami runup heights measured by Takabatake et al. (2019) are shown. (A) Map of the Anak Krakatau, including tsunami runup heights measured by Esteban et al. (2021) on three surrounding islands. (B) Photo of the runup survey point on Panjang Island, (C) photo of the runup survey point on Rakata Island, and (D) aerial photo of the Anak Krakatau. Map lines delineate study areas and do not necessarily depict accepted national boundaries. *Adapted from (B)–(D) Esteban, M., Takabatake, T., Achiari, H., Mikami, T., Nakamura, R., Gelfi, M., Panalaran, S., Nishida, Y., Inagaki, N., Chadwick, C., Oizumi, K., Shibayama, T., (2021). Field survey of flank collapse and run-up heights due to Anak Krakatau tsunami.* Journal of Coastal and Hydraulic Structures*. 1. https://doi.org/10.48438/jchs.2021.0001 with permission.*

original volcano. During the 2018 event, the edifice formed by the accumulation of pyroclastic material partly collapsed and slid down into the water, triggering impulsive tsunami waves.

- Takabatake et al. (2019) conducted tsunami reconnaissance surveys along the Sunda Strait coasts. The runup values measured are indicated in the regional map of Fig. 1.9. At Cipenyu on Java Island, the tsunami runup exceeded 10 m. Tsunami runups were affected by the directivity of tsunamis, which were influenced by the local bathymetry near the Anak Krakatau and the sheltering effects of the islands around the Anak Krakatau Volcano (Fig. 1.9A).
- Esteban et al. (2021) estimated the volume of the collapsed flank to be 0.29–0.57 km^3 based on the digital elevation model before the collapse (Fig. 1.9A) and the drone survey after the collapse (Fig. 1.9D). To accurately simulate the effects of landslides on tsunamis, a reliable estimate of the volume of the displaced mass is essential.
- At the neighboring islands of the Anak Krakatau Volcano, very high runups were measured (Esteban et al., 2021). On Panjang Island (west of the volcano), runups of up to 80 + m were observed (Fig. 1.9B), whereas on Rakata Island (south of the volcano), runups of approximately 50 m were observed (Fig. 1.9C).
- Evacuation of the coastal communities during the tsunami was difficult, as most tsunami early warning systems were primarily designed to provide warning for tsunamis triggered by earthquake sources, and there was no proper warning released before the first wave hit the area. Modification and enhancement of tsunami early warnings are required to anticipate a volcanogenic tsunami in the future, as Anak Krakatau will remain active and pose threats to the surrounding area.

1.2.5 2024 Noto Peninsula Tsunami

On January 1, 2024, the M_w 7.5 Noto Peninsula Earthquake occurred on the Sea of Japan side of Honshu Island of Japan, causing 245 deaths. The mechanism was reverse faulting and ruptured very near the ground surface or seabed. The inverted fault rupture model, developed by the Geospatial Information Authority of Japan (https://www.gsi.go.jp/cais/topic20240101Noto.html), is shown in Fig. 1.10. The strike of the earthquake rupture ran parallel with the northern coast of Noto Peninsula, and the major slip concentrations were estimated at the northwestern corner of the Noto Peninsula. Due to the proximity of the rupture to coastal towns and villages on the Noto Peninsula, very intense ground shaking was observed at Suzu, Wajima, Noto, and Anamizu (peak ground accelerations exceeding 1.0 g). Tsunamis were also observed with inundation and runup heights up to 6 m and inundated coastal areas of the towns along the northern shoreline of the Noto Peninsula (Yuhi et al., 2024).

- Many buildings in the Noto Peninsula region were affected by a sequence of loadings caused by the earthquake (i.e., shaking, tsunami, and fire damage). The cascading and compounding aspects of multiple types of hazards originating from a single event were significant.
- One of the remarkable aspects of the physical effects of this event was significant coseismic displacement. In Kaiso village at the northwestern corner of Noto Peninsula (Fig. 1.10A), a 3–4 m uplift was experienced. Although the village was not affected by the major tsunami waves, the elevated port area is no longer suitable for fishery, and port facilities need to be redeveloped completely.
- In Wajima (Fig. 1.10B), a traditional city famous for its oldest morning market in Japan, a massive fire broke out and burned the entire old district of Wajima. Recovery from the complete devastation of the city is very difficult and will have a long-term impact on residents and communities.
- The arrival of tsunami waves along the towns and villages of the Sea of Japan was immediately after the earthquake due to proximity to the rupture area. In such situations, tsunami evacuations can be challenging, and the key aspects for taking action swiftly depend on the levels of preparedness by individuals.

1.3 Tsunami hazard and risk assessments

1.3.1 Tsunami hazard assessment

A numerical model is required to evaluate tsunami hazards. Such a model for tsunami hazard analysis involves various components (e.g., earthquake occurrence, earthquake rupture, tsunami generation, wave propagation, and inundation) with varying degrees of uncertainty. Available data quality to develop these components varies from region to region. Many model components can be improved by gathering more data and by adopting more advanced physical and statistical methods. Conventionally, deterministic tsunami source scenarios have been used for tsunami hazard assessments. Challenges in defining critical scenarios for tsunami design purposes are quantifying uncertainties associated with these

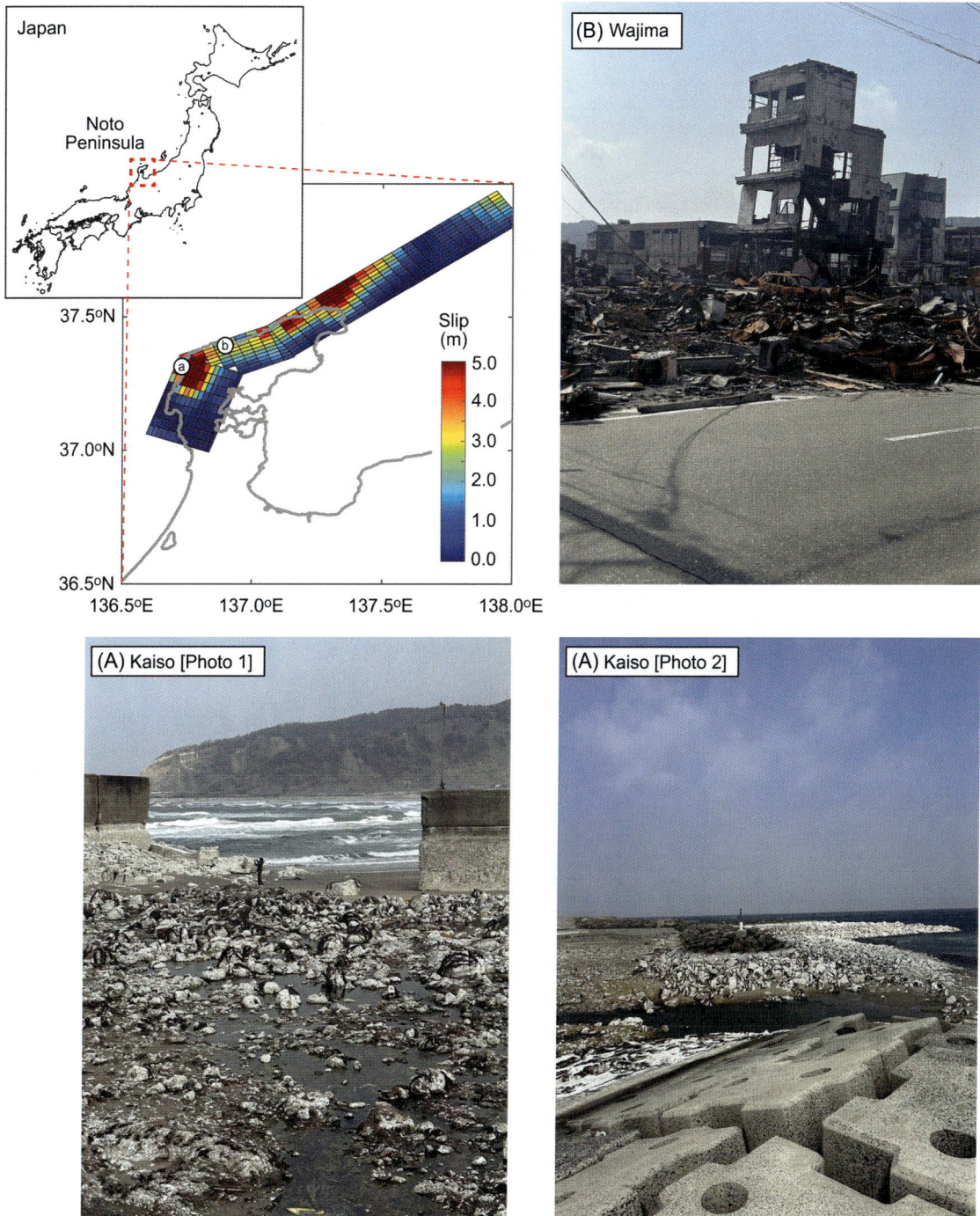

FIGURE 1.10 Map of the Noto Peninsula region of Japan. The earthquake source model shown on the map is generated based on the Geospatial Information Authority of Japan. Map lines delineate study areas and do not necessarily depict accepted national boundaries.

historical events (magnitudes, locations, and earthquake rupture characteristics) and accounting for possibilities that extreme events may be missed in the history of major tsunami events in the region. Another difficulty is to assign their probability of occurrence without a formal assessment of tsunami hazards from various tsunami sources. The two recent catastrophic tsunami disasters in the Indian Ocean (2004) and Japan (2011) have persuaded tsunami scientists and engineers to adopt different approaches.

PTHA offers a computational platform to incorporate new scientific models and data and to improve the capability of predicting future tsunami impacts on coastal communities and the built environment. Focusing on seismic tsunami

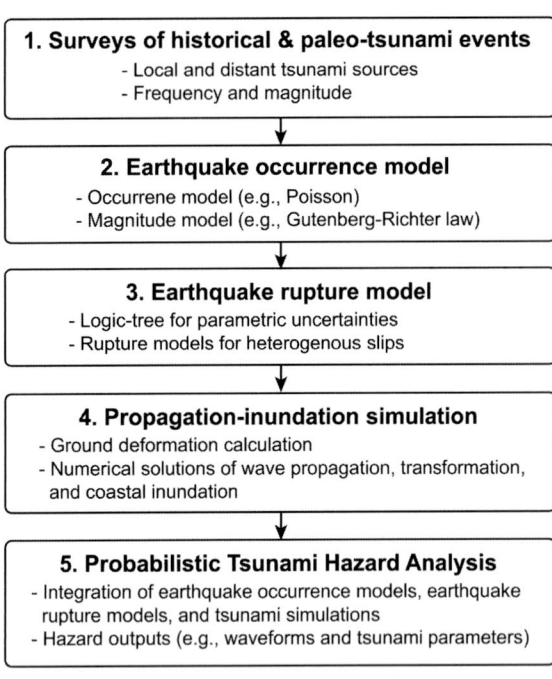

FIGURE 1.11 Flowchart of probabilistic tsunami hazard analysis.

genesis, the process of PTHA comprises (1) the literature review of historical and geological/geophysical information, (2) frequency—magnitude modeling, (3) earthquake rupture modeling, (4) simulation of tsunami waves, and (5) integration of different components to obtain tsunami hazard estimates at different return period levels. Fig. 1.11 presents the computational steps of the modern PTHA methodology. It is noteworthy that the method can be extended to nonseismic tsunami sources as well. In the following, brief descriptions of the PTHA method are provided; mathematical details and computational procedures will be explained in later chapters.

The first step of PTHA is to identify all seismic sources capable of producing damaging tsunamis. The identified tsunamigenic sources can be classified into near-field (local) and far-field (distant). Tsunamis triggered by near-field seismic sources can be regarded as the main contributors to the tsunami impact, and they should be studied in detail. Compared to local tsunamis, a simpler parameterization is usually sufficient for far-field tsunamis because earthquake magnitude, focal mechanism, and radiation pattern are more influential than earthquake slip distribution on a rupture plane (Geist & Parsons, 2006). Gathering information on historical events and their geological/geophysical evidence is critical. When reliable historical records are available, the occurrence times of these events can be determined, while the locations and sizes of these events are only loosely constrained. A standard approach to improve the estimates of the earthquake characteristics of historical tsunami events is to develop a numerical tsunami model and compare their predicted tsunami wave characteristics with recorded ones. It is essential to recognize that many model components in this "validation" exercise involve uncertainties.

The second step of PTHA is to determine the frequency—magnitude relationship based on historical tsunami events. It is important to recognize that the number of tsunami events is usually small and needs to be supplemented by seismic events with smaller magnitudes (and the frequency distribution needs to be extrapolated to larger magnitudes). For modeling the temporal occurrence of historical events, a common approach is to adopt a stationary Poisson process with the mean occurrence rate. The mean occurrence rate can be estimated by fitting the relationship proposed by Gutenberg and Richter (1956) to the data. For many subduction zones, the maximum magnitude allowed in PTHA is an important parameter but is challenging to constrain with the historical data alone. This parameter is often incorporated in a logic tree of a PTHA model (Annaka et al., 2007). There are several alternatives to the standard approach. A renewal process can be adapted to perform time-dependent PTHA (Goda, 2019). The main differences between the Poisson and renewal processes are that the occurrence rate of major tsunami events is not constant and evolves over time, and that the elapsed time since the last event can be formally incorporated into tsunami hazard analysis. For the magnitude distribution, the characteristic magnitude model of Youngs and Coppersmith (1985) can be used to consider the nonexponential magnitude distribution in PTHA.

The third step of PTHA is to characterize different configurations of tsunami generation due to earthquake rupture. There are three common approaches for reflecting a variety of earthquake ruptures in PTHA, that is, empirical, logic tree, and stochastic. The empirical approach uses a combination of historical or known source scenarios based on expert opinions (González et al., 2009). This is a straightforward approach, but the number of events and variations of sources are limited. The logic tree approach specifies a combination of earthquake slip scenarios and geometric slip parameters, and each branch represents a possible tsunami rupture model (Annaka et al., 2007; Fukutani et al., 2015). The stochastic rupture approach can also capture the uncertainty of the earthquake characteristics in PTHA (Li et al., 2016). It aims to characterize the uncertainty of the fault plane using earthquake scaling relationships and heterogeneous earthquake slip distributions, reflecting the key features of past source inversion models (Goda et al., 2016). By generating many synthetic rupture models and by running tsunami simulations for each of the generated rupture models, uncertainties of the tsunami generation from various seismic sources can be captured, and this uncertainty can be further passed on to tsunami propagation and inundation in determining the tsunami intensity metrics.

The fourth step of PTHA is to perform tsunami propagation and runup simulations. Once earthquake rupture scenarios are selected, initial tsunami conditions can be estimated using analytical formulae (Okada, 1985) or numerical models (Grilli et al., 2013). Subsequently, to calculate the tsunami time histories and tsunami intensity metrics, tsunami simulations can be performed by numerically solving the governing equations, such as linear/nonlinear shallow water equations (Goto et al., 1997) and dispersive Boussinesq equations (Baba et al., 2017; Kirby et al., 2013). These steps require boundary conditions, which are bathymetry and onshore topographic data of the coastal region, to be known. Adequate spatial resolutions must be chosen to achieve the necessary resolution in the tsunami simulations. Another important aspect is the realistic representation of surface friction, vegetation, and built environment in the governing equations (Kaiser et al., 2011).

The fifth step of PTHA is to integrate all model components and produce outputs. Important tsunami hazard parameters are the maximum tsunami amplitude at an offshore/nearshore location or the tsunami parameters (inundation depth and flow velocity) at an inland location. Multiple tsunami hazard curves can be obtained for each model combination when the logic tree model is adopted. The tsunami simulation results corresponding to a branch in the logic tree are combined with the lognormal distribution that captures the discrepancy between model predictions and the observation data. To construct fractile tsunami hazard curves (e.g., mean and median), the weights of the logic tree branches are considered. The final tsunami hazard estimates for a given site can be expressed as an EP curve or a tsunami hazard curve. The maximum considered tsunami can be identified by specifying the target return periods for tsunami design.

1.3.2 Tsunami risk assessment

A quantitative risk assessment involves not only tsunami hazard characterization but also exposure model and vulnerability assessment (i.e., loss = hazard × exposure × vulnerability; Fig. 1.3). A standard approach to conducting such quantitative risk assessments for tsunamis is to generate a stochastic event set that contains various possible tsunamigenic seismic events in a region of interest and then to carry out tsunami risk assessments for the target communities and structures (Mitchell-Wallace et al., 2017).

The exposure model characterizes the population, properties, infrastructure, and assets that are potentially at risk. It involves quantifying the geographical distribution of affected people and estimates of the costs associated with the interruption of normal socio-economic activities. The exposure data are usually a critical input to CAT models and define the scope of CAT modeling and natural catastrophe assessments. Exposure modeling evaluates the number, replacement cost, and type of assets at some granular geographical resolution and involves geomatics, geographical information systems, remote sensing, economics, as well as social and environmental sciences. The datasets that go into exposure modeling include cartographic, land-use, survey, or socio-economic sources that are collected for various purposes. For tsunami risk assessments, the spatial resolution of the exposure data and their elevation information are critical.

The vulnerability component is the interface among hazard, exposure, and loss and involves quantifying the relationship between the severity of a hazard at a given location (e.g., inundation depth) and the resulting loss. The vulnerability assessment is sometimes divided into two stages, tsunami fragility functions and damage−loss functions. The intermediate outputs of this two-step approach are tsunami damage states that are often associated with tsunami damage observation data. When tsunami damage surveys are conducted, inspectors often assign damage states (e.g., minor, moderate, extensive, wash-away/collapse) to individual buildings. The tsunami fragility functions describe the relationship between the tsunami hazard parameters and the occurrence (or exceedance) of damage states. Then, the damage−loss functions are used to relate the damage states to the incurred tsunami loss in terms of total replacement cost (e.g., loss ratio).

Generally, there are three approaches to developing vulnerability and fragility models. The expert opinion approach uses elicitation methods by relying on the knowledge of thematic experts, which can be subjective. An empirical method uses postevent damage data together with statistical analysis (Koshimura et al., 2009). The collection and availability of extensive tsunami damage data from the 2011 Tohoku Tsunami have created new research opportunities to include various attributes, such as coastal topography, material type, and the number of building floors (Suppasri et al., 2013). Moreover, large empirical data for tsunami fatalities have also become available for developing the predictive equations of tsunami fatality rates (Latcharote et al., 2018). An analytical approach uses computational/physical models to simulate the expected hazard intensity and measure the expected structural response to physical forces (Attary et al., 2017; Petrone et al., 2017). In this approach, a challenge still exists in generating realistic synthetic sequences of earthquake ground motions and tsunami waves and accounting for the cumulative damage of such sequential loading conditions to structures.

Once the tsunami losses for individual properties and/or the building portfolio are evaluated for all tsunami scenarios considered for the tsunami hazard component, the final tsunami risk curve or the EP curve can be derived. The tsunami risk curve should account for the probability of occurrence for each tsunami scenario. Using the tsunami risk curve, various risk metrics can be derived. The AAL is the area under the tsunami risk curve and can be used as a fundamental metric of the financial loss distribution. The value at risk (VaR) is the loss value at a specific quantile of the relevant loss distribution. In the insurance industry, VaR is sometimes referred to as probable maximum loss and is often used for managing financial risk exposure to extreme events. Alternatively, a coherent risk measure, such as tail VaR or tail conditional expectation, which is the conditional expected value of loss above a specific quantile of the relevant loss distribution, can be derived from the tsunami risk curve. It is also important to recognize that when the tsunami fatality is taken as the tsunami risk measure, the tsunami loss curve represents a so-called $F-N$ curve, that is, the annual frequency (F) of experiencing N fatalities or more (Evans & Verlander, 1997; Lee & Jones, 2014). The tsunami $F-N$ curve can be used to investigate the societally acceptable level of tsunami risks in comparison with other natural hazards and technological risks.

1.4 Tsunami disaster risk reduction and management

Coastal communities in active subduction regions are exposed to significant risks from megathrust earthquakes and tsunamis. Such risks are low-probability, high-consequence events and often involve joint occurrences of multiple perils that are triggered by a single initiating event. Cascading chains of hazard events can consist of numerous combinations of ground surface rupture, strong ground motion, tsunami inundation, landslide, liquefaction, and aftershocks, and their combined effects can significantly worsen consequences compared to individual impacts of such hazard events. To mitigate consequences due to multiple concurrent hazards, integrated hazard and risk assessment and management policies are necessary (Tilloy et al., 2019). Improved risk quantification tools for earthquakes and tsunamis will promote mapping and forecasting of cascading and compounding earthquake-tsunami risks in developing community-focused solutions, such as early warning systems, evacuation planning, and land-use planning. They can also be used for deriving disaster risk financing tools, such as insurance rate-making and alternative risk transfer instruments, and for comparing the benefits and costs associated with different alternatives (Scolobig et al., 2017). Moreover, the new approaches can incorporate maintenance and inspection costs, and other environmental impacts from "cradle to grave" to further improve both the resilience and sustainability of society and built environments for coastal communities and infrastructures (Akiyama et al., 2020).

This section discusses tsunami DRR and DRM by adopting a disaster management cycle, that is, mitigation, preparedness, response, and recovery, as a guiding framework (Fig. 1.4). Enhanced disaster management aims to reduce/avoid potential losses from cascading hazards, assure prompt and appropriate response and support/assistance for victims, and achieve rapid and effective recovery (Bruneau & Reinhorn, 2007). By integrating these key elements of earthquake—tsunami risk mitigation measures from a holistic risk management perspective, future resilience-based approaches for earthquakes and tsunamis will emerge. Ultimately, resilient capacity building in coastal communities will empower residents, responders, organizations, communities, governments, and society to share the responsibility to prevent disasters and adapt to future perils (Ayyub, 2014; United Nations Office for Disaster Risk Reduction, 2022).

1.4.1 Mitigation

The mitigation (prevention) phase aims to minimize disaster effects (e.g., retrofitting and zoning, vulnerability analyses, and public education). Protection via physical measures is the primary approach for controlling disaster risks to built environments, and building codes have played a key role (Chock, 2016). In promoting more resilient building design and construction practices, life-cycle benefit—cost assessments based on suitable risk models are essential (Akiyama et al., 2020; Liel & Deierlein, 2013). Additionally, quantitative comparisons of the benefits and costs facilitate the more efficient use of available resources and budgets for DRR.

Improving the accuracy of the hazard and risk assessment methods is important, which will eventually lead to enhanced actions for risk mitigation. Currently, a critical research gap is the multihazard fragility that accounts for the damage accumulation effects from multiple concurrent perils (Attary et al., 2021). Resolving this research bottleneck will require interdisciplinary collaboration between earthquake−tsunami scientists and seismic-coastal engineers. The inputs to numerical structural models must come from advanced multihazard simulators of strong motion and tsunami waves (Goda et al., 2017; Maeda et al., 2013), accounting for uncertainties of the multihazard processes and scenarios. These hazard inputs can be used to conduct the multihazard damage assessments of the structural models via analytical approaches, such as sequential nonlinear static analyses (Petrone et al., 2017) and more computationally demanding nonlinear dynamic analyses (Attary et al., 2021).

1.4.2 Preparedness

The preparedness phase is focused on planning how to respond to a disaster and includes preparedness planning, emergency exercises/training, and warning systems. The multihazard risk models for earthquakes and tsunamis can contribute to this phase of the disaster management cycle in various ways. For instance, joint multihazard mapping in relation to an overall portfolio risk promotes the direct association of possible disaster scenarios and consequences by considering uncertainties and helps visualize these multihazard risk results. This is particularly useful for risk communications with residents and community responders.

The developed multihazard risk models can be used as multihazard scenario simulators considering different hazard interactions (e.g., triggering, exacerbating, catalyzing, and impeding relationships) (Dunant et al., 2021; Gill & Malamud, 2016). When the cumulative impacts from multiple damaging perils are fully developed in the models, deteriorating vulnerability, which represents the diminishing capacity of society and infrastructure subjected to the cascade of multiple compounding risks, can be incorporated into the risk assessments. The changing exposure can also be integrated into the risk assessments (Mesta et al., 2022). Hazard interactions, which are likely to be affected by natural, anthropogenic, and technological processes, are complex. In the above context, natural hazards are not limited to geological risks (strong motion, tsunami, landslide, liquefaction, and aftershock) but include climate and meteorological risks.

1.4.3 Response

The response phase of the disaster management cycle strives to minimize the hazard impacts created by a disaster and can involve search and rescue, emergency relief, early warning announcements and evacuations, and rapid risk assessment. In implementing DRR actions for this phase, it is important to consider the impact on human fatalities (Latcharote et al., 2018), physical damage, and economic losses. In conducting the research related to the response, key competing requirements are accuracy and time (e.g., issuing warning messages), and one must consider uncertainties and errors associated with such assessments and decisions.

Enhanced tsunami preparedness, tsunami early warning systems, and response plans are effective tools to reduce victims (Harig et al., 2020). In creating a comprehensive database for a tsunami early warning system and developing a viable algorithm for issuing accurate warning messages, it is important that the system is tested to work well for a variety of possible earthquake ruptures. Because it is uncertain how future tsunami events may unfold, it is sensible to use synthetic tsunami wave data to calibrate the tsunami early warning algorithm based on the conventional earthquake information (e.g., magnitude estimate and epicenter location) as well as offshore wave data (e.g., S-net in Japan). When the multihazard risk outputs are included, not only conventional hazard-based warning systems but also risk-based warning systems can be developed (Li & Goda, 2022). Furthermore, the use of synthetically simulated hazard and risk data allows more advanced statistical and machine-learning methods to be applied to the development of early warning systems (Mulia et al., 2022).

The use of multihazard simulators also presents new research avenues for carrying out earthquake−tsunami evacuation. Agent-based evacuation models can capture the dynamics of tsunami inundation, interaction with the built environment (e.g., road network capacity), and complex human interactions (Lämmel et al., 2009; Takabatake et al., 2020; Wood et al., 2016). When agent-based models are implemented within a probabilistic tsunami framework, the effectiveness of current and proposed evacuation systems (different routes and destinations, including vertical evacuation shelters) can be evaluated by fully accounting for uncertainties of different multihazard scenarios (Muhammad et al., 2021). The results from such integrated hazard-evacuation simulations are insightful as they produce the community-level performance metrics of different evacuation systems and strategies.

Rapid tsunami impact assessment methods are currently lacking. This contrasts sharply with rapid earthquake impact assessment tools implemented globally in the USGS PAGER (Prompt Assessment of Global Earthquakes for Response) system (Wald et al., 2010). Conventionally, the earthquake and tsunami impact can be assessed quickly based on earthquake magnitude and location. With the expansion of recording networks of earthquakes (e.g., K-NET and KiK-net in Japan) and tsunamis (e.g., S-net in Japan), real-time recorded shaking and wave information could be employed (Li & Goda, 2023). Moreover, the recent availability of satellite imageries and semiautomated image processing, combined with machine-learning techniques, can be exploited to develop multihazard rapid impact assessment tools (Goda, Mori, Yasuda 2019; Koshimura et al., 2020; Moya et al., 2018; Mulia et al., 2022; Naito et al., 2020). The fusion of remote sensing technology and advanced data analytics is a promising research field for postdisaster hazard monitoring and risk management.

1.4.4 Recovery

The recovery phase concerns returning to normality for the affected communities as soon as possible by providing victims with temporary housing, financial aid, and medical/mental care. Insurance can be viewed as a financial means for affected households to recover from the disaster impacts and to expedite the recovery process.

Although earthquake insurance is usually available in many earthquake-prone countries and regions (Organisation for Economic Co-operation and Development, 2018), tsunami insurance coverage is not widely offered for coastal areas that are exposed to potential tsunami risks. The development of tsunami insurance products requires the fair quantification of tsunami risks for their pricing. In this regard, capable tsunami risk models are essential (Song & Goda, 2019). Furthermore, when multihazard insurance products for shaking and tsunami risks are to be offered, accurate multihazard risk models will be needed to differentiate insurance rates for multihazard shaking-tsunami loss coverage by focusing on strong shaking intensity and effects of land elevation and topography.

To mitigate the economic impact of catastrophic shaking-tsunami hazards for insurers and local/central governments, financial risk transfer instruments offer alternative ways to diversify the financial risk exposures due to natural catastrophes. For instance, the insurance/reinsurance industry and governments can use parametric catastrophe bonds to transfer catastrophic risks to the financial markets (Goda, Franco, et al., 2019). The advantages of parametric catastrophe bonds are low moral hazard and swiftness/transparency of the payment (unlike conventional insurance). The disadvantage is the basis risk (i.e., the discrepancy between the payment and actual loss). Accurate catastrophe models are necessary for designing effective bond triggers and reducing the basis risk. In this context, rapid risk assessment methods for earthquakes and tsunamis will facilitate the new development of disaster risk financing tools.

1.5 Scope and aims of the book

Tsunamis can cause catastrophic loss to coastal cities and communities globally, and developing effective DRR and DRM strategies is urgently needed. Recent advances in quantitative tsunami hazard and risk analyses call for synthesizing a broad knowledge basis and a solid understanding of interdisciplinary fields, spanning seismology, tsunami science, coastal engineering, and DRM. Such new approaches are essential for enhanced disaster resilience of society under multiple hazards and changing climates. They are required in various related sectors, including governments, local communities, and the insurance and reinsurance industry.

This book is the first comprehensive textbook on probabilistic tsunami hazard and risk analyses with applications to disaster risk resilience and sustainability. The book consists of two parts: "fundamentals of PTHA and PTRA" (Chapters 2–6) and "advanced topics and applications related to PTHA and PTRA" (Chapters 7–24). Moreover, numerical examples (MATLAB® codes and datasets) that are discussed in Chapters 2, 3, and 5 are included in the Appendix. The target audience of the fundamental chapters is graduate students and professionals who are relatively new to this topic, whereas the advanced chapters are particularly for more experienced researchers and professionals who are interested in specific topics. This book is not intended to provide design guidance for tsunami-resistant infrastructure or be an encyclopedia of tsunami research. Rather, it is an attempt to present state-of-the-art in the field of quantitative tsunami hazard and risk analyses, as well as disaster risk mitigation and management.

The main aim of the "fundamentals of PTHA and PTRA" is to provide a broad coverage of various technical contents related to PTHA and PTRA, including:

- Chapter 2 explains earthquake occurrence, earthquake rupture, tsunami source characterization, and nonseismic triggers of tsunamis.
- Chapter 3 presents wave characteristics, wave propagation in deep waters, wave propagation in shallow waters, tsunami inundation, and computational methods for tsunami propagation and inundation.

- Chapter 4 summarizes tsunami design codes, tsunami effects on structures, structural design against tsunami loading, and tsunami fragility and vulnerability.
- Chapter 5 focuses on PTHA and PTRA, and their extensions to combined earthquake–tsunami methods.
- Chapter 6 presents hazard identification and planning, physical risk mitigation, emergency response, risk management, and risk financing.

The advanced chapters cover cutting-edge research topics related to specific elements of PTHA and PTRA. Chapters 7–12 provide a wide range of topics related to tsunami records, tsunami generation, tsunami propagation, and tsunami runup. Chapters 13–16 present global and regional examples of tsunami hazard and risk assessments. Chapters 17 and 18 focus on topics related to tsunami impacts on the built environment. Chapters 19 and 20 focus on the methodologies related to probabilistic tsunami hazard and risk assessments. Chapters 21–24 focus on advanced topics related to tsunami risk management.

- Chapter 7 focuses on historical tsunami records that are the fundamental inputs to tsunami hazard characterizations.
- Chapter 8 probes into the details of tectonic and subduction zone characteristics and their implications on tsunami generation.
- Chapter 9 presents tsunami generation mechanisms by splay faults and their effects on generated tsunamis.
- Chapter 10 explains tsunami generation mechanisms by landslides and presents approaches to account for uncertainties associated with landslide-triggered tsunamis.
- Chapter 11 presents an overview of modern dense tsunami monitoring systems and algorithms that can be used for tsunami early warning purposes.
- Chapter 12 introduces applications of advanced machine learning techniques for tsunami early warning.
- Chapter 13 summarizes tsunami hazard assessments and tsunami risk mitigation tools for the Pacific Ocean.
- Chapter 14 presents tsunami hazard assessments in New Zealand by considering near and far tsunami sources.
- Chapter 15 presents tsunami hazard characterization for regions in the Mediterranean Sea.
- Chapter 16 presents tsunami hazard characterization and tsunami risk mitigation approaches in Chile.
- Chapter 17 focuses on statistical methods for developing tsunami fragility functions using tsunami damage data.
- Chapter 18 summarizes analytical approaches that can be used to quantify the vulnerability of buildings subjected to tsunamis.
- Chapter 19 presents the characterization of epistemic uncertainty associated with tsunami hazard and risk assessments.
- Chapter 20 presents a multihazard risk assessment for the Cascadia subduction zone in the Pacific Northwest.
- Chapter 21 presents a computational approach for conducting dynamic agent-based simulations for tsunami evacuation purposes.
- Chapter 22 introduces a scenario-based methodology to consider sea-level rise effects on tsunami risk.
- Chapter 23 presents a probabilistic methodology for long-term tsunami risk assessment by considering time-dependent earthquake hazards, tidal level variability, and relative sea-level rise.
- Chapter 24 presents a digital twin framework for improved coastal DRM.

References

Akiyama, M., Frangopol, D. M., & Ishibashi, H. (2020). Toward life-cycle reliability-, risk- and resilience-based design and assessment of bridges and bridge networks under independent and interacting hazards: Emphasis on earthquake, tsunami and corrosion. *Structure and Infrastructure Engineering*, *16*(1), 26–50. Available from https://doi.org/10.1080/15732479.2019.1604770, http://www.tandf.co.uk/journals/titles/15732479.asp.

American Society of Civil Engineers. (2022). *Minimum design loads and associated criteria for buildings and other structures*. American Society of Civil Engineers.

Annaka, T., Satake, K., Sakakiyama, T., et al. (2007). Logic-tree Approach for Probabilistic Tsunami Hazard Analysis and its Applications to the Japanese Coasts. *Pure and Applied Geophysics*, *164*, 577–592. Available from https://doi.org/10.1007/s00024-006-0174-3.

Attary, N., Unnikrishnan, V. U., van de Lindt, J. W., Cox, D. T., & Barbosa, A. R. (2017). Performance-based tsunami engineering methodology for risk assessment of structures. *Engineering Structures*, *141*, 676–686. Available from https://doi.org/10.1016/j.engstruct.2017.03.071, http://www.journals.elsevier.com/engineering-structures/.

Attary, N., van de Lindt, J. W., Barbosa, A. R., Cox, D. T., & Unnikrishnan, V. U. (2021). Performance-Based tsunami engineering for risk assessment of structures subjected to multi-hazards: Tsunami following earthquake. *Journal of Earthquake Engineering*, *25*(10), 2065–2084. Available from https://doi.org/10.1080/13632469.2019.1616335, http://www.tandf.co.uk/journals/titles/13632469.asp.

Ayyub, B. M. (2014). Systems resilience for multihazard environments: Definition, metrics, and valuation for decision making. *Risk Analysis*, *34*(2), 340–355. Available from https://doi.org/10.1111/risa.12093.

Baba, T., Allgeyer, S., Hossen, J., Cummins, P. R., Tsushima, H., Imai, K., Yamashita, K., & Kato, T. (2017). Accurate numerical simulation of the far-field tsunami caused by the 2011 Tohoku earthquake, including the effects of Boussinesq dispersion, seawater density stratification, elastic loading, and gravitational potential change. *Ocean Modelling, 111*, 46−54. Available from https://doi.org/10.1016/j.ocemod.2017.01.002, http://www.elsevier.com/inca/publications/store/6/0/1/3/7/6/index.htt.

Baker, J., Bradley, B., & Stafford, P. (2021). *Seismic hazard and risk analysis*. Cambridge University Press. Available from https://doi.org/10.1017/9781108425056.

Behrens, J., Løvholt, F., Jalayer, F., Lorito, S., Salgado-Gálvez, M. A., Sørensen, M., Abadie, S., Aguirre-Ayerbe, I., Aniel-Quiroga, I., Babeyko, A., Baiguera, M., Basili, R., Belliazzi, S., Grezio, A., Johnson, K., Murphy, S., Paris, R., Rafliana, I., De Risi, R., … Vyhmeister, E. (2021). Probabilistic tsunami hazard and risk analysis: A review of research gaps. *Frontiers in Earth Science, 9*. Available from https://doi.org/10.3389/feart.2021.628772, https://www.frontiersin.org/journals/earth-science.

Bernard, E., & Titov, V. (2015). Evolution of tsunami warning systems and products. *Philosophical Transactions of the Royal Society A: Mathematical, Physical and Engineering Sciences, 373*(2053), 20140371. Available from https://doi.org/10.1098/rsta.2014.0371.

Beven, K. J., Almeida, S., Aspinall, W. P., Bates, P. D., Blazkova, S., Borgomeo, E., Freer, J., Goda, K., Hall, J. W., Phillips, J. C., Simpson, M., Smith, P. J., Stephenson, D. B., Wagener, T., Watson, M., & Wilkins, K. L. (2018). Epistemic uncertainties and natural hazard risk assessment—Part 1: A review of different natural hazard areas. *Natural Hazards and Earth System Sciences, 18*(10), 2741−2768. Available from https://doi.org/10.5194/nhess-18-2741-2018, http://www.nat-hazards-earth-syst-sci.net/volumes_and_issues.html.

Bruneau, M., & Reinhorn, A. (2007). Exploring the concept of seismic resilience for acute care facilities. *Earthquake Spectra, 23*(1), 41−62. Available from https://doi.org/10.1193/1.2431396, https://journals.sagepub.com/home/eqs.

Chock, G. Y. K. (2016). Design for tsunami loads and effects in the ASCE 7-16 standard. *Journal of Structural Engineering, 142*(11). Available from https://doi.org/10.1061/(asce)st.1943-541x.0001565.

Cornell, C. A. (1968). Engineering seismic risk analysis. *Bulletin of the Seismological Society of America, 58*(5), 1583−1606. Available from https://doi.org/10.1785/bssa0580051583.

De Risi, R., & Goda, K. (2017). Simulation-based probabilistic tsunami hazard analysis: Empirical and robust hazard predictions. *Pure and Applied Geophysics, 174*(8), 3083−3106. Available from https://doi.org/10.1007/s00024-017-1588-9, http://www.springer.com/birkhauser/geo + science/journal/24.

Dunant, A., Bebbington, M., & Davies, T. (2021). Probabilistic cascading multi-hazard risk assessment methodology using graph theory, a New Zealand trial. *International Journal of Disaster Risk Reduction, 54*, 102018. Available from https://doi.org/10.1016/j.ijdrr.2020.102018.

Esteban, M., Takabatake, T., Achiari, H., Mikami, T., Nakamura, R., Gelfi, M., Panalaran, S., Nishida, Y., Inagaki, N., Chadwick, C., Oizumi, K., & Shibayama, T. (2021). Field survey of flank collapse and run-up heights due to Anak Krakatau tsunami. *Journal of Coastal and Hydraulic Structures, 1*. Available from https://doi.org/10.48438/jchs.2021.0001, https://journals.open.tudelft.nl/jchs/article/view/4822.

Evans, A. W., & Verlander, N. Q. (1997). What is wrong with criterion FN-lines for judging the tolerability of risk? *Risk Analysis, 17*(2), 157−168. Available from https://doi.org/10.1111/j.1539-6924.1997.tb00855.x.

Fraser, S., Raby, A., Pomonis, A., Goda, K., Chian, S. C., Macabuag, J., Offord, M., Saito, K., & Sammonds, P. (2013). Tsunami damage to coastal defences and buildings in the March 11th 2011 Mw 9.0 Great East Japan earthquake and tsunami. *Bulletin of Earthquake Engineering, 11*(1), 205−239. Available from https://doi.org/10.1007/s10518-012-9348-9, https://rd.springer.com/journal/10518.

Fukutani, Y., Suppasri, A., & Imamura, F. (2015). Stochastic analysis and uncertainty assessment of tsunami wave height using a random source parameter model that targets a Tohoku-type earthquake fault. *Japan Stochastic Environmental Research and Risk Assessment, 29*(7), 1763−1779. Available from https://doi.org/10.1007/s00477-014-0966-4, http://link.springer-ny.com/link/service/journals/00477/index.htm.

Geist, E. L., & Parsons, T. (2006). Probabilistic analysis of tsunami hazards. *Natural Hazards, 37*(3), 277−314. Available from https://doi.org/10.1007/s11069-005-4646-z.

Ghobarah, A., Saatcioglu, M., & Nistor, I. (2006). The impact of the 26 December 2004 earthquake and tsunami on structures and infrastructure. *Engineering Structures, 28*(2), 312−326. Available from https://doi.org/10.1016/j.engstruct.2005.09.028.

Gill, J. C., & Malamud, B. D. (2016). Hazard interactions and interaction networks (cascades) within multi-hazard methodologies. *Earth System Dynamics, 7*(3), 659−679. Available from https://doi.org/10.5194/esd-7-659-2016, http://www.earth-syst-dynam.net/volumes_and_issues.html.

Goda, K. (2019). Time-dependent probabilistic tsunami hazard analysis using stochastic rupture sources. *Canada Stochastic Environmental Research and Risk Assessment, 33*(2), 341−358. Available from https://doi.org/10.1007/s00477-018-1634-x, http://link.springer-ny.com/link/service/journals/00477/index.htm.

Goda, K., & De Risi, R. (2017). Probabilistic tsunami loss estimation methodology: Stochastic earthquake scenario approach. *Earthquake Engineering Research Institute, United Kingdom Earthquake Spectra, 33*(4), 1301−1323. Available from https://doi.org/10.1193/012617EQS019M, http://earthquakespectra.org/doi/pdf/10.1193/012617EQS019M.

Goda, K., Yasuda, T., Mori, N., & Maruyama, T. (2016). New scaling relationships of earthquake source parameters for stochastic tsunami simulation. *Coastal Engineering Journal, 58*(3). Available from https://doi.org/10.1142/S0578563416500108, https://www.tandfonline.com/loi/tcej20.

Goda, K., Petrone, C., De Risi, R., & Rossetto, T. (2017). Stochastic coupled simulation of strong motion and tsunami for the 2011 Tohoku, Japan earthquake. *Stochastic Environmental Research and Risk Assessment, 31*(9), 2337−2355. Available from https://doi.org/10.1007/s00477-016-1352-1, http://link.springer-ny.com/link/service/journals/00477/index.htm.

Goda, K., Mori, N., & Yasuda, T. (2019). Rapid tsunami loss estimation using regional inundation hazard metrics derived from stochastic tsunami simulation. *International Journal of Disaster Risk Reduction, 40*, 101152. Available from https://doi.org/10.1016/j.ijdrr.2019.101152.

Goda, K., Mori, N., Yasuda, T., Prasetyo, A., Muhammad, A., & Tsujio, D. (2019). Cascading geological hazards and risks of the 2018 Sulawesi Indonesia earthquake and sensitivity analysis of tsunami inundation simulations. *Frontiers in Earth Science, 7*. Available from https://doi.org/10.3389/feart.2019.00261, https://www.frontiersin.org/journals/earth-science.

Goda, K., Franco, G., Song, J., & Radu, A. (2019). Parametric catastrophe bonds for tsunamis: Cat-in-a-box trigger and intensity-based index trigger methods. *Earthquake Spectra*, *55*(1), 113−136. Available from https://doi.org/10.1193/030918EQS052M, https://earthquakespectra.org/doi/pdf/10.1193/030918EQS052M.

Goldfinger, C., Nelson, C. H., Morey, A. E., Johnson, J. E., Patton, J., Karabanov, E., Gutierrez-Pastor, J., Eriksson, A. T., Gracia, E., Dunhill, G., Enkin, R. J., Dallimore, A., & Vallier, T. (2012). *Turbidite event history: Methods and implications for Holocene paleoseismicity of the Cascadia subduction zone. Professional Paper 1661*. Geological Survey. Available from https://pubs.er.usgs.gov/publication/pp1661F, https://doi.org/10.3133/pp1661F.

González, F. I., Geist, E. L., Jaffe, B., Kânoğlu, U., Mofjeld, H., Synolakis, C. E., Titov, V. V., Arcas, D., Bellomo, D., Carlton, D., Horning, T., Johnson, J., Newman, J., Parsons, T., Peters, R., Peterson, C., Priest, G., Venturato, A., Weber, J., ... Yalciner, A. (2009). Probabilistic tsunami hazard assessment at Seaside, Oregon, for near- and far-field seismic sources. *Journal of Geophysical Research*, *114*(C11). Available from https://doi.org/10.1029/2008jc005132.

Goto,C., Ogawa,Y., Shuto,N., & Imamura,F. (1997). *Numerical method of tsunami simulation with the leap-frog scheme*. IOC Manual.

Goulet, C. A., Haselton, C. B., Mitrani-Reiser, J., Beck, J. L., Deierlein, G. G., Porter, K. A., & Stewart, J. P. (2007). Evaluation of the seismic performance of a code-conforming reinforced-concrete frame building—From seismic hazard to collapse safety and economic losses. *Earthquake Engineering and Structural Dynamics*, *36*(13), 1973−1997. Available from https://doi.org/10.1002/eqe.694, http://onlinelibrary.wiley.com/journal/10.1002/(ISSN)1096-9845.

Grilli, S. T., Harris, J. C., Tajalli Bakhsh, T. S., Masterlark, T. L., Kyriakopoulos, C., Kirby, J. T., & Shi, F. (2013). Numerical simulation of the 2011 Tohoku tsunami based on a new transient FEM co-seismic source: Comparison to far- and near-field observations. *Pure and Applied Geophysics*, *170*(6−8), 1333−1359. Available from https://doi.org/10.1007/s00024-012-0528-y, http://www.springer.com/birkhauser/geo + science/journal/24.

Gusman, A. R., Supendi, P., Nugraha, A. D., Power, W., Latief, H., Sunendar, H., Widiyantoro, S., Daryono., Wiyono, S. H., Hakim, A., Muhari, A., Wang, X., Burbidge, D., Palgunadi, K., Hamling, I., & Daryono, M. R. (2019). Source model for the tsunami inside Palu bay following the 2018 Palu earthquake, Indonesia. *Geophysical Research Letters*, *46*(15), 8721−8730. Available from https://doi.org/10.1029/2019GL082717, http://agupubs.onlinelibrary.wiley.com/hub/journal/10.1002/(ISSN)1944-8007/.

Gutenberg, B., & Richter, C. F. (1956). Magnitude and energy of earthquakes. *Annals of Geophysics*, *9*, 1−15.

Harig, S., Immerz, A., Weniza., Griffin, J., Weber, B., Babeyko, A., Rakowsky, N., Hartanto, D., Nurokhim, A., Handayani, T., & Weber, R. (2020). The tsunami scenario database of the Indonesia tsunami early warning system (InaTEWS): Evolution of the coverage and the involved modeling approaches. *Pure and Applied Geophysics*, *177*(3), 1379−1401. Available from https://doi.org/10.1007/s00024-019-02305-1, http://www.springer.com/birkhauser/geo + science/journal/24.

Heidarzadeh, M., Pirooz, M. D., Zaker, N. H., Yalciner, A. C., Mokhtari, M., & Esmaeily, A. (2008). Historical tsunami in the Makran Subduction zone off the southern coasts of Iran and Pakistan and results of numerical modeling. *Ocean Engineering*, *35*(8−9), 774−786. Available from https://doi.org/10.1016/j.oceaneng.2008.01.017.

Japan Society of Civil Engineers, *Tsunami assessment method for nuclear power plants in Japan*. Japan Society of Civil Engineers, (2016), Available from: https://committees.jsce.or.jp/ceofnp/node/84.

Kagan, Y. Y., & Jackson, D. D. (2013). Tohoku earthquake: A surprise? *Bulletin of the Seismological Society of America*, *103*(2B), 1181−1194. Available from https://doi.org/10.1785/0120120110.

Kaiser, G., Scheele, L., Kortenhaus, A., Løvholt, F., Römer, H., & Leschka, S. (2011). The influence of land cover roughness on the results of high resolution tsunami inundation modeling. *Natural Hazards and Earth System Science*, *11*(9), 2521−2540. Available from https://doi.org/10.5194/nhess-11-2521-2011.

Kajitani, Y., Chang, S. E., & Tatano, H. (2013). Economic impacts of the 2011 Tohoku-oki earthquake and tsunami. *Earthquake Spectra*, *29*(1), 457−478. Available from https://doi.org/10.1193/1.4000108.

Kirby, J. T., Shi, F., Tehranirad, B., Harris, J. C., & Grilli, S. T. (2013). Dispersive tsunami waves in the ocean: Model equations and sensitivity to dispersion and Coriolis effects. *Ocean Modelling*, *62*, 39−55. Available from https://doi.org/10.1016/j.ocemod.2012.11.009.

Koshimura, S., Oie, T., Yanagisawa, H., & Imamura, F. (2009). Developing fragility functions for tsunami damage estimation using numerical model and post-tsunami data from Banda Aceh, Indonesia. *Coastal Engineering Journal*, *51*(3), 243−273. Available from https://doi.org/10.1142/S0578563409002004, https://www.tandfonline.com/loi/tcej20.

Koshimura, S., Moya, L., Mas, E., & Bai, Y. (2020). Tsunami damage detection with remote sensing: A review. *Geosciences*, *10*(5), 177. Available from https://doi.org/10.3390/geosciences10050177.

Lämmel, G., Rieser, M., Nagel, K., Taubenböck, H., Strunz, G., Goseberg, N., Schlurmann, T., Klüpfel, H., Setiadi, N., & Birkmann, J. (2009). *Emergency preparedness in the case of a Tsunami—Evacuation analysis and traffic optimization for the Indonesian city of Padang*. Springer Science and Business Media LLC. Available from https://doi.org/10.1007/978-3-642-04504-2_13.

Latcharote, P., Leelawat, N., Suppasri, A., Thamarux, P., & Imamura, F. (2018). Estimation of fatality ratios and investigation of influential factors in the 2011 Great East Japan tsunami. *International Journal of Disaster Risk Reduction*, *29*, 37−54. Available from https://doi.org/10.1016/j.ijdrr.2017.06.024, http://www.journals.elsevier.com/international-journal-of-disaster-risk-reduction/.

Lee, E. M., & Jones, D. K. C. (2014). Landslide risk assessment (509). ICE Publishing.

Levin, B. W., & Nosov, M. (2016). *Physics of tsunamis* (388). Springer.

Li, L., Switzer, A. D., Chan, C. H., Wang, Y., Weiss, R., & Qiu, Q. (2016). How heterogeneous coseismic slip affects regional probabilistic tsunami hazard assessment: A case study in the South China Sea. *Journal of Geophysical Research: Solid Earth*, *121*(8), 6250−6272. Available from https://doi.org/10.1002/2016JB013111, http://onlinelibrary.wiley.com/journal/10.1002/(ISSN)2169-9356.

Li, Y., & Goda, K. (2022). Hazard and risk-based tsunami early warning algorithms for ocean bottom sensor S-net system in Tohoku, Japan, using sequential multiple linear regression. *Geosciences*, *12*(9), 350. Available from https://doi.org/10.3390/geosciences12090350.

Li, Y., & Goda, K. (2023). Random forest-based multi-hazard loss estimation using hypothetical data at seismic and tsunami monitoring networks. *Geomatics, Natural Hazards and Risk*, *14*(1). Available from https://doi.org/10.1080/19475705.2023.2275538, http://www.tandfonline.com/toc/tgnh20/current.

Liel, A. B., & Deierlein, G. G. (2013). Cost-benefit evaluation of seismic risk mitigation alternatives for older concrete frame buildings. *Earthquake Spectra*, *29*(4), 1391–1411. Available from https://doi.org/10.1193/030911EQS040M, http://earthquakespectra.org/doi/pdf/10.1193/030911EQS040M.

Løvholt, F., Glimsdal, S., Harbitz, C. B., Horspool, N., Smebye, H., de Bono, A., & Nadim, F. (2014). Global tsunami hazard and exposure due to large co-seismic slip. *International Journal of Disaster Risk Reduction*, *10*, 406–418. Available from https://doi.org/10.1016/j.ijdrr.2014.04.003, http://www.journals.elsevier.com/international-journal-of-disaster-risk-reduction/.

Macabuag, J., Raby, A., Pomonis, A., Nistor, I., Wilkinson, S., & Rossetto, T. (2018). Tsunami design procedures for engineered buildings: A critical review. *Proceedings of the Institution of Civil Engineers: Civil Engineering*, *171*(4), 166–178. Available from https://doi.org/10.1680/jcien.17.00043, http://www.icevirtuallibrary.com/content/serial/cien.

Maeda, T., Furumura, T., Noguchi, S., Takemura, S., Sakai, S., Shinohara, M., Iwai, K., & Lee, S. J. (2013). Seismic- and tsunami-wave propagation of the 2011 off the Pacific Coast of Tohoku earthquake as inferred from the tsunami-coupled finite-difference simulation. *Bulletin of the Seismological Society of America*, *103*((2B), 1456–1472. Available from https://doi.org/10.1785/0120120118Japan, http://www.bssaonline.org/content/103/2B/1456.full.pdf+html.

Mesta, C., Cremen, G., & Galasso, C. (2022). Urban growth modelling and social vulnerability assessment for a hazardous Kathmandu Valley. *Scientific Reports*, *12*(1). Available from https://doi.org/10.1038/s41598-022-09347-x, http://www.nature.com/srep/index.html.

Mitchell-Wallace, K., Jones, M., Hillier, J., & Foote, M. (2017). *Natural catastrophe risk management and modelling: A practitioner's guide* (536). Wiley-Blackwell.

Miyashita, T., Mori, N., & Goda, K. (2020). Uncertainty of probabilistic tsunami hazard assessment of Zihuatanejo (Mexico) due to the representation of tsunami variability. *Coastal Engineering Journal*, 413–428. Available from https://doi.org/10.1080/21664250.2020.1780676, https://www.tandfonline.com/loi/tcej20.

Mokhtari, M., Abdollahie Fard, I., & Hessami, K. (2008). Structural elements of the Makran region, Oman sea and their potential relevance to tsunamigenisis. *Natural Hazards*, *47*(2), 185–199. Available from https://doi.org/10.1007/s11069-007-9208-0.

Mori, N., Takahashi, T., Yasuda, T., & Yanagisawa, H. (2011). Survey of 2011 Tohoku earthquake tsunami inundation and run-up. *Geophysical Research Letters*, *38*(18). Available from https://doi.org/10.1029/2011GL049210, http://onlinelibrary.wiley.com/journal/10.1002/(ISSN)1944-8007/issues?year=2012.

Mori, N., Goda, K., & Cox, D. (2018). Recent process in probabilistic tsunami hazard analysis (PTHA) for mega thrust subduction earthquakes. *Advances in Natural and Technological Hazards Research*, *47*. Available from https://doi.org/10.1007/978-3-319-58691-5_27, http://www.springer.com/series/6362?detailsPage=titles.

Mori, N., Satake, K., Cox, D., Goda, K., Catalan, P. A., Ho, T. C., Imamura, F., Tomiczek, T., Lynett, P., Miyashita, T., Muhari, A., Titov, V., & Wilson, R. (2022). Giant tsunami monitoring, early warning and hazard assessment. *Nature Reviews Earth and Environment*, *3*(9), 557–572. Available from https://doi.org/10.1038/s43017-022-00327-3, http://nature.com/natrevearthenviron/.

Moya, L., Mas, E., Adriano, B., Koshimura, S., Yamazaki, F., & Liu, W. (2018). An integrated method to extract collapsed buildings from satellite imagery, hazard distribution and fragility curves. *International Journal of Disaster Risk Reduction*, *31*, 1374–1384. Available from https://doi.org/10.1016/j.ijdrr.2018.03.034, http://www.journals.elsevier.com/international-journal-of-disaster-risk-reduction/.

Mueller, C., Power, W., Fraser, S., & Wang, X. (2015). Effects of rupture complexity on local tsunami inundation: Implications for probabilistic tsunami hazard assessment by example. *Journal of Geophysical Research: Solid Earth*, *120*(1), 488–502. Available from https://doi.org/10.1002/2014JB011301, http://onlinelibrary.wiley.com/journal/10.1002/(ISSN)2169-9356.

Muhammad, A., De Risi, R., De Luca, F., Mori, N., Yasuda, T., & Goda, K. (2021). Are current tsunami evacuation approaches safe enough? *Stochastic Environmental Research and Risk Assessment*, *35*(4), 759–779. Available from https://doi.org/10.1007/s00477-021-02000-5, http://link.springer-ny.com/link/service/journals/00477/index.htm.

Mulia, I. E., Ueda, N., Miyoshi, T., Gusman, A. R., & Satake, K. (2022). Machine learning-based tsunami inundation prediction derived from offshore observations. *Nature Communications*, *13*(1). Available from https://doi.org/10.1038/s41467-022-33253-5, http://www.nature.com/ncomms/index.html.

Naito, S., Tomozawa, H., Mori, Y., Nagata, T., Monma, N., Nakamura, H., Fujiwara, H., & Shoji, G. (2020). Building-damage detection method based on machine learning utilizing aerial photographs of the Kumamoto earthquake. *Earthquake Spectra*, *36*(3), 1166–1187. Available from https://doi.org/10.1177/8755293019901309, https://journals.sagepub.com/home/eqs.

Okada, Y. (1985). Surface deformation due to shear and tensile faults in a half-space. *Bulletin of the Seismological Society of America*, *75*(4), 1135–1154. Available from https://doi.org/10.1785/bssa0750041135.

Organisation for Economic Co-operation and Development. (2018). *Financial management of earthquake risk*. Organisation for Economic Co-operation and Development. Available from: https://www.oecd.org/finance/insurance/Financial-management-of-earthquake-risk.pdf.

Palermo, D., Nistor, I., Saatcioglu, M., & Ghobarah, A. (2013). Impact and damage to structures during the 27 February 2010 Chile tsunami. *Canadian Journal of Civil Engineering*, *40*(8), 750–758. Available from https://doi.org/10.1139/cjce-2012-0553.

Park, H., & Cox, D. T. (2016). Probabilistic assessment of near-field tsunami hazards: Inundation depth, velocity, momentum flux, arrival time, and duration applied to Seaside, Oregon. *Coastal Engineering*, *117*, 79–96. Available from https://doi.org/10.1016/j.coastaleng.2016.07.011, http://www.elsevier.com/inca/publications/store/5/0/3/3/2/5/.

Paulik, R., Gusman, A., Williams, J. H., Pratama, G. M., Lin, Sl, Prawirabhakti, A., Sulendra, K., Zachari, M. Y., Fortuna, Z. E. D., Layuk, N. B. P., & Suwarni, N. W. I. (2019). Tsunami hazard and built environment damage observations from Palu city after the September 28 2018 Sulawesi earthquake and tsunami. *Pure and Applied Geophysics*, *176*(8), 3305−3321. Available from https://doi.org/10.1007/s00024-019-02254-9, http://www.springer.com/birkhauser/geo + science/journal/24.

Petrone, C., Rossetto, T., & Goda, K. (2017). Fragility assessment of a RC structure under tsunami actions via nonlinear static and dynamic analyses. *Engineering Structures*, *136*, 36−53. Available from https://doi.org/10.1016/j.engstruct.2017.01.013, http://www.journals.elsevier.com/engineering-structures/.

Power, W., Downes, G., & Stirling, M. (2007). Estimation of Tsunami Hazard in New Zealand due to South American Earthquakes. *Pure and Applied Geophysics*, *164*, 547−564. Available from https://doi.org/10.1007/s00024-006-0166-3.

Rikitake, T., & Aida, I. (1988). Tsunami hazard probability in Japan. *Bulletin - Seismological Society of America*, *78*(3), 1268−1278.

Satake, K., Wang, K., & Atwater, B. F. (2003). Fault slip and seismic moment of the 1700 Cascadia earthquake inferred from Japanese tsunami descriptions. *Journal of Geophysical Research: Solid Earth*, *108*(11). Available from http://onlinelibrary.wiley.com/journal/10.1002/(ISSN)2169-9356.

Scolobig, A., Komendantova, N., & Mignan, A. (2017). Mainstreaming multi-risk approaches into policy. *Geosciences*, *7*(4), 129. Available from https://doi.org/10.3390/geosciences7040129.

Song, J., & Goda, K. (2019). Influence of elevation data resolution on tsunami loss estimation and insurance rate-making. *Frontiers in Earth Science*, *7*. Available from https://doi.org/10.3389/feart.2019.00246, https://www.frontiersin.org/journals/earth-science.

Suppasri, A., Mas, E., Charvet, I., Gunasekera, R., Imai, K., Fukutani, Y., Abe, Y., & Imamura, F. (2013). Building damage characteristics based on surveyed data and fragility curves of the 2011 Great East Japan tsunami. *Natural Hazards*, *66*(2), 319−341. Available from https://doi.org/10.1007/s11069-012-0487-8, http://www.wkap.nl/journalhome.htm/0921-030X.

Takabatake, T., Shibayama, T., Esteban, M., Achiari, H., Nurisman, N., Gelfi, M., Tarigan, T. A., Kencana, E. R., Fauzi, M. A. R., Panalaran, S., Harnantyari, A. S., & Kyaw, T. O. (2019). Field survey and evacuation behaviour during the 2018 Sunda Strait tsunami. *Coastal Engineering Journal*, *61*(4), 423−443. Available from https://doi.org/10.1080/21664250.2019.1647963, https://www.tandfonline.com/loi/tcej20.

Takabatake, T., Nistor, I., & St-Germain, P. (2020). Tsunami evacuation simulation for the district of Tofino, Vancouver Island, Canada. *International Journal of Disaster Risk Reduction*, *48*, 101573. Available from https://doi.org/10.1016/j.ijdrr.2020.101573.

Thio, H. K., Somerville, P., & Ichinose, G. (2012). Probabilistic analysis of strong ground motion and tsunami hazards in Southeast Asia. *Journal of Earthquake and Tsunami*, *1*(2), 119−137. Available from https://doi.org/10.1142/s1793431107000080.

Tilloy, A., Malamud, M. D., Winter, H., & Joly-Laugel, A. (2019). A review of quantification methodologies for multi-hazard interrelationships. *Earth-Science Reviews*, *196*(102881). Available from https://doi.org/10.1016/j.earscirev.2019.102881, https://www.sciencedirect.com/science/article/pii/S001282521930025X.

Titov, V., Rabinovich, A. B., Mofjeld, H. O., Thomson, R. E., & González, F. I. (2005). The global reach of the 26 December 2004 Sumatra tsunami. *Science*, *309*(5743), 2045−2048. Available from https://doi.org/10.1126/science.1114576.

United Nations Office for Disaster Risk Reduction. (2022). *Technical guidance on comprehensive risk assessment and planning in the context of climate change*. United Nations Office for Disaster Risk Reduction, United Nations Office for Disaster Risk Reduction. Available from: https://www.undrr.org/publication/technical-guidance-comprehensive-risk-assessment-and-planning-context-climate-change.

Wald, D. J., Jaiswal, K. S., Marano, K. D., Bausch, D. B., & Hearne, M. (2010). PAGER—Rapid assessment of an earthquake's impact. *U.S. Geological Survey Fact Sheet*. Available from https://doi.org/10.3133/fs20103036, https://pubs.usgs.gov/fs/2010/3036/pdf/FS10-3036.pdf.

Wang, X., & Liu, P. L. F. (2006). An analysis of 2004 Sumatra earthquake fault plane mechanisms and Indian Ocean tsunami. *Journal of Hydraulic Research*, *44*(2), 147−154. Available from https://doi.org/10.1080/00221686.2006.9521671, http://www.tandfonline.com/toc/tjhr20/current.

Woo, G. (2011). Calculating catastrophe. Imperial College Press, Undefined Imperial College Press, 1−356, Available from http://www.worldscientific.com/worldscibooks/10.1142/P786t = toc, https://doi.org/10.1142/P786.

Wood, N., Jones, J., Schmidtlein, M., Schelling, J., & Frazier, T. (2016). Pedestrian flow-path modeling to support tsunami evacuation and disaster relief planning in the U.S. Pacific Northwest. *International Journal of Disaster Risk Reduction*, *18*, 41−55. Available from https://doi.org/10.1016/j.ijdrr.2016.05.010, http://www.journals.elsevier.com/international-journal-of-disaster-risk-reduction/.

Yoshikawa, H., & Goda, K. (2014). Financial seismic risk analysis of building portfolios. *Natural Hazards Review*, *15*(2), 112−120. Available from https://doi.org/10.1061/(ASCE)NH.1527-6996.0000129.

Youngs, R. R., & Coppersmith, K. J. (1985). Implications of fault slip rates and earthquake recurrence models to probabilistic seismic hazard estimates. *Bulletin of the Seismological Society of America*, *75*(4), 939−964. Available from https://doi.org/10.1785/BSSA0750040939, https://pubs.geoscienceworld.org/ssa/bssa/article-abstract/75/4/939/118730/Implications-of-fault-slip-rates-and-earthquake?redirectedFrom = fulltext.

Yuhi, M., Umeda, S., Arita, M., Ninomiya, J., Gokon, H., Arikawa, T., Baba, T., Imamura, F., Kawai, A., Kumagai, K., Kure, S., Miyashita, T., Suppasri, A., Nobuoka, H., Shibayama, T., Koshimura, S., & Mori, N. (2024). Post-event survey of the 2024 Noto Peninsula earthquake tsunami in Japan. *Coastal Engineering Journal*, 1−14. Available from https://doi.org/10.1080/21664250.2024.2368955.

Section 1

Fundamentals of probabilistic tsunami hazard and risk analysis

Chapter 2

Tsunami generation

Katsuichiro Goda[1], Aditya Gusman[2], Raffaele De Risi[3] and Ioan Nistor[4]

[1]Department of Earth Sciences, Western University, London, ON, Canada, [2]GNS Science, Lower Hutt, Wellington, New Zealand, [3]School of Civil, Aerospace and Design Engineering, University of Bristol, Bristol, United Kingdom, [4]Department of Civil Engineering, University of Ottawa, Ottawa, ON, Canada

2.1 Introduction

Tsunamis are generated by the abrupt changes in the seabed that displace the water column above it. Large earthquakes are the main tsunami sources (Levin & Nosov, 2016), whereas landslides and volcanic eruptions can also generate a large volume of displaced water column due to the sliding or submerged land mass, causing destructive tsunamis (Gusman et al., 2020; Løvholt et al., 2019). Since the main triggers of large tsunamis are subduction earthquakes, seismic tsunami sources are primarily the focus of this chapter. In conducting probabilistic tsunami hazard analysis (PTHA) and probabilistic tsunami risk analysis (PTRA), key characteristics and uncertainties of tsunami sources need to be captured in terms of their occurrence, location, spatial extent/magnitude, and earthquake rupture process (e.g., kinematic earthquake slip distribution). This chapter presents the fundamentals and standard approaches to characterizing the tsunami generation processes.

Section 2.2 provides a summary of seismotectonic characteristics of tsunamigenic earthquakes by focusing on subduction zones. The basic terminology related to the earthquake source characteristics is mentioned (e.g., earthquake magnitude and faulting mechanism).

Section 2.3 focuses on earthquake occurrence by analyzing earthquake catalog data for deriving standard seismological models, such as the Poisson occurrence model and the Gutenberg−Richter (G−R) relationship (Section 2.3.1). This conventional approach of time-independent earthquake occurrence modeling can be extended by introducing renewal earthquake occurrence models, enabling time-dependent earthquake occurrence characterization (Section 2.3.2). In Section 2.3.3, the characteristic earthquake model is introduced to allow the nonexponential magnitude−frequency distribution for large events.

Section 2.4 provides a summary of rupture features of tsunamigenic earthquakes. In Section 2.4.1, fault geometry and key source parameters are introduced by analyzing finite-fault models of past earthquakes obtained by conducting earthquake source inversion analysis. Existing databases of finite-fault models, such as SRCMOD (http://equake-rc.info/srcmod/; Mai & Thingbaijam, 2014) and the U.S. Geological Survey (Hayes, 2017), can be utilized. Subsequently, empirical scaling relationships of the earthquake source parameters (e.g., fault length and fault width) are explained, and their use for tsunami source characterization is discussed in Sections 2.4.2 and 2.4.3.

Section 2.5 summarizes approaches to tsunami source modeling (Section 2.5.1) and presents a synthetic method of generating earthquake slip distributions by adopting wavenumber representations of heterogeneous earthquake slips. The simulation-based methods can be used for characterizing both instantaneous and kinematic rupture processes of future earthquakes (Sections 2.5.2 and 2.5.3). In Section 2.5.4, standard approaches to converting earthquake ruptures to displacements at the ocean bottom are explained (Okada, 1985; Tanioka & Satake, 1996). The ocean-bottom displacements are often taken as input (i.e., initial boundary condition) in numerical tsunami simulations (Chapter 3).

Section 2.6 provides a summary of tsunami source modeling resulting from nonseismic sources. Section 2.6.1 focuses on modeling techniques for tsunamis caused by submarine landslides. Volcanic eruptions can initiate various tsunami-inducing mechanisms. In Section 2.6.2, modeling approaches for tsunamis arising from pyroclastic flows, volcanic earthquakes, caldera collapses, and shock waves are mentioned.

Section 2.7 provides a summary of research gaps and needs related to tsunami source modeling, which affects tsunami source generation significantly.

2.2 Seismotectonic characteristics of active seismic regions

Seismic activity is high at plate boundaries. Plate boundaries can be classified into convergent (compression), divergent (tension), and transform (shear). Fig. 2.1A shows the major tectonic plates defined by Bird (2003). For example, the oceanic Pacific Plate creates convergent tectonic environments at plate boundaries with the Eurasian Plate and the North American Plate and subducts underneath these continental plates. Numerous earthquakes occur near the locations of the major plate boundaries. This can be observed by comparing the locations of the major tectonic plate boundaries (Fig. 2.1A) with the locations of major earthquakes (Fig. 2.1B). The earthquake locations shown in Fig. 2.1B are based on the Harvard Global Centroid Moment Tensor catalog (https://www.globalcmt.org/CMTsearch.html) with earthquake magnitudes of 6.0 or greater. The moment magnitude M_w is the measure of the earthquake size and can be defined by (Hanks & Kanamori, 1979)

$$M_w = \frac{2}{3}\left(\log_{10}M_0 - 9.1\right) \tag{2.1}$$

where the seismic moment M_0 (Nm) can be expressed as

$$M_0 = \mu_s A_f S_a \tag{2.2}$$

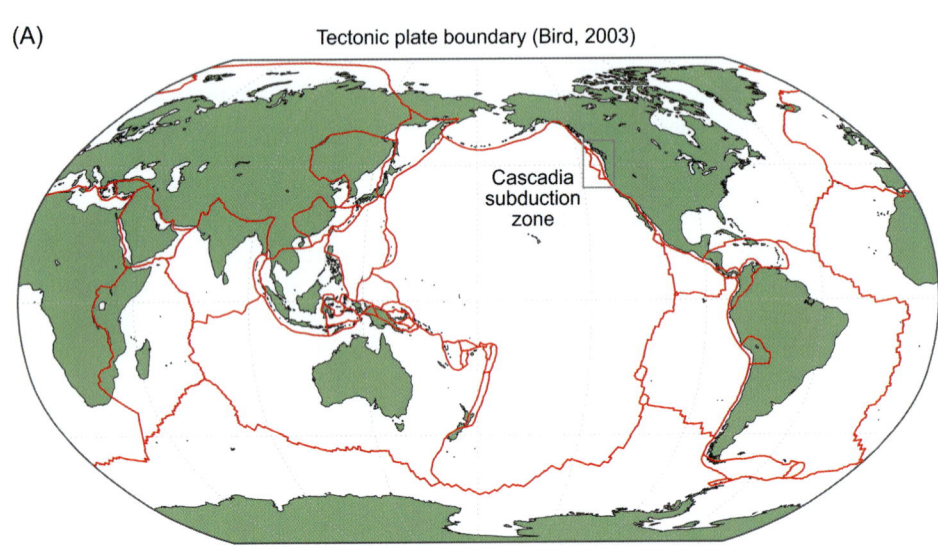

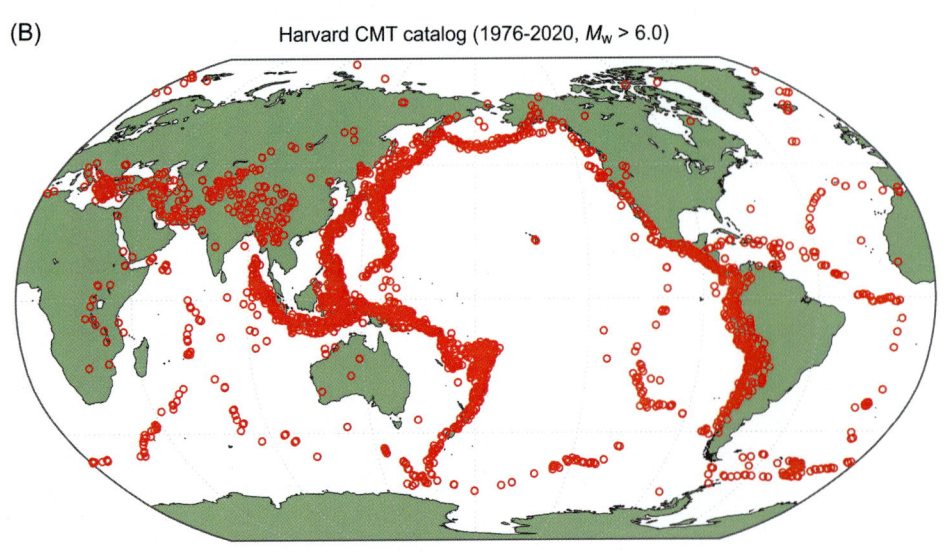

FIGURE 2.1 (A) Major tectonic plate boundaries by Bird (2003) and (B) locations of earthquakes with moment magnitudes greater than 6.0 based on the Harvard Global CMT catalog. *CMT*, Centroid moment tensor. Map lines delineate study areas and do not necessarily depict accepted national boundaries.

in which μ_S is the shear modulus (Pa), A_f is the fault area (which can be expressed by the fault length L times the fault width W for a rectangular fault plane; m^2), and S_a is the fault slip (m). Importantly, the moment magnitude can be directly related to the earthquake source parameters, such as L, W, and S_a (see Section 2.4).

In the subduction region, colliding motions of two or more tectonic plates create different tectonic regimes and faulting mechanisms (Fig. 2.2). A schematic of different fault movements is shown in Fig. 2.3. Tectonic regimes largely control the distributions of fault rupture depth and fault dip. For continental earthquakes, the slip centroids are well confined within depths less than 20–30 km (i.e., the seismogenic thickness of the upper crust). On the other hand, earthquakes in subduction zones can occur at significant depths. Subduction interface events occur within depths less than 40–50 km, while in-slab (or intraslab) events can be observed at depths over 100 km. Reverse faulting involves the upward movement of a hanging wall block with respect to a footwall block; strike–slip faulting involves the horizontal sliding of two fault blocks with nearly vertical fault plane; and normal faulting involves the downward movement of a hanging wall block with respect to a footwall block. Furthermore, tectonic regimes are correlated with the average fault dip angle δ, transitioning from steeper to shallower from strike–slip ($\delta \sim 70$–90 degrees), to normal faulting ($\delta \sim 50$–60 degrees), to shallow crustal reverse faulting ($\delta \sim 40$–50 degrees), and finally to subduction interface ($\delta \sim 10$–30 degrees) events.

In the compressional tectonic regime, reverse faulting typically occurs. At the shallow part of the subduction interface, relative plate motions are locked, and strain is accumulated over the fault interface. When accumulated strain is released suddenly, low-angle thrust earthquakes tend to be triggered. Such a megathrust interface subduction earthquake can generate very large earthquakes and tsunamis ($M_w > 8.0$). On the other hand, in the tensional tectonic regime, normal faulting becomes more frequent. At the deep part of the subducting oceanic lithosphere, in-slab earthquakes with normal faulting tend to occur inside the subducting slab at depths greater than 40 km, where the temperature of the subducting plate becomes hot and ductile ruptures, rather than brittle ruptures, tend to occur more frequently. The normal faulting is also dominant in the outer-rise region as a result of the bending of the subducting plate on the seaward side of the trench. Generally, large outer-rise earthquakes are tsunamigenic, but their ground shaking is not always felt by people on the shore due to relatively large distances from the source. For this reason, large outer-rise events are described as "silent tsunamigenic earthquakes." An example of such a silent but devastating tsunami earthquake was the M_w 8.4 1933 Showa Sanriku Earthquake. Another faulting mechanism that occurs within the continental plate is a

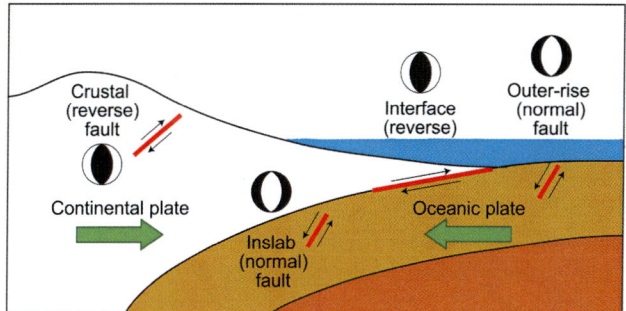

FIGURE 2.2 A schematic of a subduction seismotectonic environment.

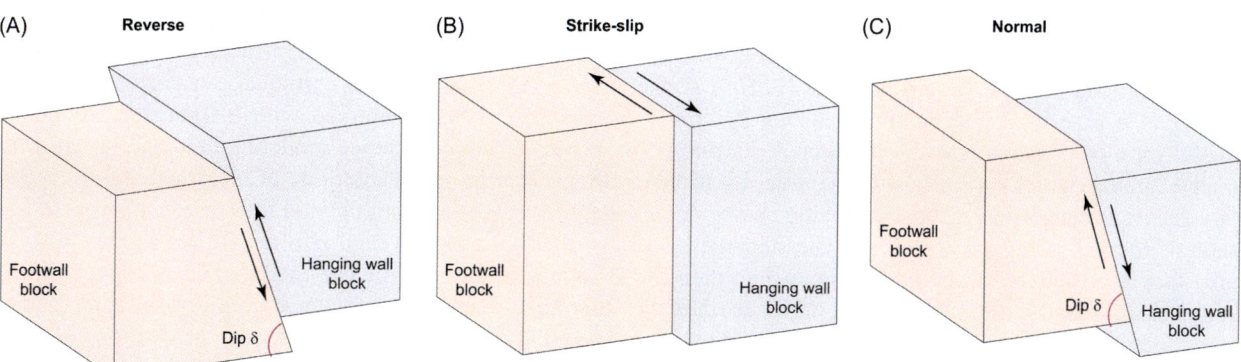

FIGURE 2.3 Fault motions: (A) reverse, (B) strike–slip, and (C) normal faulting.

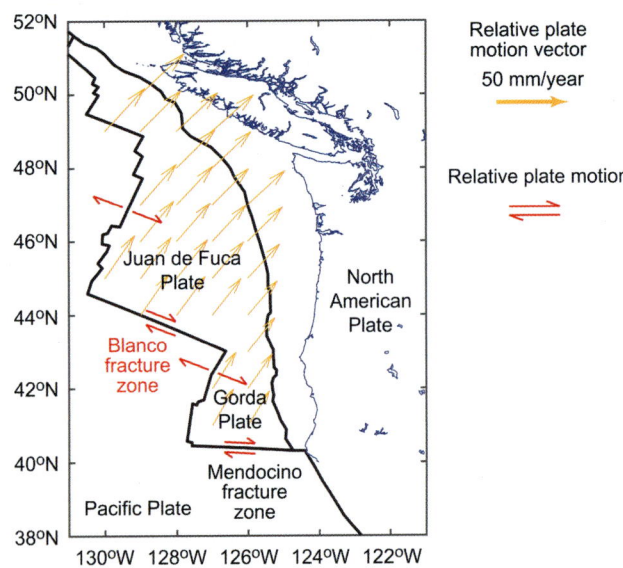

FIGURE 2.4 Relative plate motion velocity vectors of the Juan de Fuca Plate and the Gorda Plate with respect to the North American Plate in the Cascadia subduction zone.

strike−slip type. Generally, strike−slip earthquakes are not considered to be tsunamigenic. However, when the fault cuts across a submerged steep valley with some oblique faulting components, large strike−slip earthquakes can cause devastating tsunamis due to the fault displacement and due to submarine landslides.

The current movements of tectonic plates can be seen by plate motion velocities (https://www.unavco.org/software/geodetic-utilities/plate-motion-calculator/plate-motion-calculator.html). Plate motion velocities provide important information on relative plate motions at the plate interfaces. The directions of plate motion velocities are also useful for anticipating the dominant types of earthquakes. The information on regional plate motion velocities can be used to estimate the earthquake recurrence periods and to constrain the earthquake source characteristics of the subduction zone based on slip deficits (Kimura et al., 2019) and serves as a critical input to PTHA (Small & Melgar, 2021).

Plate motion velocities are not always perpendicular to the plate boundaries, and the fault parallel components of the relative plate motions need to be accommodated by different tectonic movements. An example of the oblique relative fault motions of the oceanic plate with respect to the continental plate at the subduction interface margin is the Cascadia subduction zone in the Pacific Northwest (Fig. 2.1). In the Cascadia subduction zone, the Juan de Fuca Plate and the Gorda Plate are subducting under the North American Plate. The relative plate velocities of the Juan de Fuca Plate and the Gorda Plate with respect to the North American Plate are shown in Fig. 2.4 based on DeMets et al. (2010). The relative velocities are between 40 and 50 mm/year, and the directions of the vectors in the southern part of the subduction zone are oblique to the plate boundary. The fault parallel component of the plate motion is accommodated by the Mendocino fracture zone and the Blanco fracture zone.

2.3 Earthquake occurrence

Earthquake occurrence is one of the most influential elements in PTHA and involves significant uncertainty. Similar to probabilistic seismic hazard analysis (PSHA) (Baker et al., 2021), a standard earthquake occurrence model in PTHA is a time-independent homogeneous Poisson process and is often combined with a truncated exponential model for a $G-R$ magnitude distribution. A starting point of earthquake occurrence modeling is to analyze an earthquake catalog, which typically lists the time, location, depth, and earthquake magnitude of historical and instrumental seismic events. The earthquake catalog can be downloaded from various national and international organizations, such as the U.S. Geological Survey. The statistical analysis of an earthquake catalog often adopts a seismic source area that is considered to be homogenous, and then the $G-R$ model can be fitted to the data. The output from this standard approach is the occurrence rate for earthquakes that have magnitudes greater than a certain value and the so-called b-value, which is a slope parameter of the $G-R$ model. The b-value controls relative proportions of small-to-moderate versus moderate-to-large earthquakes in the magnitude distribution. This approach is explained in Section 2.3.1.

For mature fault systems and subduction zones, the occurrence rates of earthquakes are non-Poissonian and quasiperiodic (Griffin et al., 2020; Ogata, 1999; Sykes & Menke, 2006; Williams et al., 2019). Both physics-inspired occurrence models (Console et al., 2008; Shimazaki & Nakata, 1980) and statistical renewal models (Abaimov et al., 2008; Cornell & Winterstein, 1988; Matthews et al., 2002) have been adopted in seismic and tsunami hazard assessments. A renewal process can characterize the evolution of occurrence probability with time by specifying an interarrival time (IAT) distribution of earthquakes. Typically, the IAT distribution can be characterized by three parameters: mean recurrence time, coefficient of variation of occurrence time (also referred to as aperiodicity), and elapsed time since the previous event. The challenging aspect of the renewal earthquake occurrence modeling is the scarcity of the IAT data for a given fault system or subduction zone. This method is explained in Section 2.3.2.

When the renewal process is adopted for earthquake occurrence, the corresponding magnitude distribution can be specified by the conventional G–R exponential model or by the characteristic magnitude model (Youngs & Coppersmith, 1985). The latter is typically used for modeling the earthquake size distribution of a specific fault system or subduction zone. It consists of the exponential decay part for small-to-moderate magnitudes and the uniform, constant part for large magnitudes. An alternative to the uniform magnitude distribution part of the characteristic earthquake model is to use the normal distribution. For a given earthquake source, different magnitude models can be calibrated by satisfying regional seismic moment release constraints. In other words, the distribution of the seismic moment release is different (e.g., some models release more seismic moments in the larger magnitude range), while the total seismic moment release is kept constant. The characteristic earthquake model is explained in Section 2.3.3.

2.3.1 Statistical analysis of earthquake catalog and time-independent occurrence model

Gutenberg and Richter (1956) proposed that the magnitude–frequency distribution of earthquakes can be represented by a log–linear relationship between the annual number of seismic events with the magnitude M greater than a particular value (note: M denotes a general earthquake magnitude scale and can be changed to M_w), $N_A(M \geq m)$, and the threshold magnitude m. The equation is given by

$$\log_{10} N_A(M \geq m) = a - bm \tag{2.3}$$

The total number of seismic events above $m = 0$, or seismic activity rate, is obtained as 10^a, while the decay rate of seismic events as a function of earthquake magnitude is represented by the slope parameter b. Various regional and global studies have been conducted to characterize the seismicity using the G–R model. Although the value of a (i.e., activity rate) varies from region to region, the b-value is relatively stable and typically takes a value close to 1. This means that the earthquake activity rate drops by a factor of 10 when a threshold magnitude is increased by one unit.

The functional form of Eq. (2.3) follows the exponential distribution. This becomes obvious when the logarithmic base is changed from 10 to natural log e. The probability density function is given by

$$f_M(m) = \beta \exp(-\beta m) \tag{2.4}$$

where $\beta = b\ln(10)$. When the minimum magnitude $M_{\min}$ and the maximum magnitude $M_{\max}$ are introduced (i.e., the earthquake magnitude is exponentially distributed between $M_{\min}$ and $M_{\max}$), the probability density function of earthquake magnitude can be expressed as follows

$$f_M(m) = \frac{\beta \exp(-\beta(m - M_{\min}))}{1 - \exp(-\beta(M_{\max} - M_{\min}))} \tag{2.5}$$

The truncated exponential model of Eq. (2.5) can be given in the form of a cumulative distribution function

$$F_M(m) = \frac{1 - \exp(-\beta(m - M_{\min}))}{1 - \exp(-\beta(M_{\max} - M_{\min}))} \tag{2.6}$$

whereas the complementary cumulative distribution function is defined as

$$G_M(m) = 1 - F_M(m) \tag{2.7}$$

A basic procedure for analyzing the seismicity data in the form of the G–R model consists of the following steps:

Step 1: Define a region of interest and a time window. The minimum magnitude and the depth range may be specified. In a seismic catalog, search for seismic events that occurred within the spatial and temporal windows.

Step 2: Define a magnitude range and bin. For instance, the magnitude range can be specified from M 4.0 to M 9.5 by setting a bin width of 0.1 units. The lowest value of the magnitude range depends on the completeness of the earthquake catalog.

Step 3: For each value of the lower bound of the magnitude bin m, count the number of events whose magnitudes are equal to or greater than m, $N_T(M \geq m)$. Divide the earthquake event counts by the duration of the earthquake catalog T to obtain $N_A(M \geq m) = N_T(M \geq m)/T$.

Step 4: Plot the data points of m and $N_A(M \geq m)$ by taking the logarithmic scale for $N_A(M \geq m)$.

An illustration of the $G-R$ model for the Tohoku region of Japan is shown in Figs. 2.5 and 2.6. Fig. 2.5 shows the NEIC catalog data (https://earthquake.usgs.gov/earthquakes/search/) by considering the minimum magnitude of four and the catalog duration between 1976 and 2012. The spatial window of the earthquake event selection is shown in Fig. 2.5; 6500 events that are within the spatial window are shown with variable sizes and colors in terms of earthquake magnitude, whereas events that fall outside of the spatial window are shown with black dots. Fig. 2.6 shows an empirical $G-R$ plot of the Tohoku seismicity data (i.e., blue squares). The magnitude bin size is 0.1 units, spanning from M 4.0 to M 9.0. Note that the data point shown at M 9.0 corresponds to the mainshock. For large earthquakes whose recurrence periods are longer than the catalog duration, the assigned annual frequency should be interpreted as an upper bound or conservative estimate. The empirical $G-R$ data of the Tohoku region can be fitted with Eq. (2.3). A standard approach is to use the maximum likelihood method (Aki, 1965). A computer code for this analysis is provided in Example 2-1 (Appendix). Although the simple formula of Aki (1965) is used to fit the $G-R$ model by assuming that the catalog completeness coincides with the catalog duration, unequal observation periods can be taken into account using the method by Weichert (1980). More details of the statistical analysis of the seismicity data can be found in Baker et al. (2021).

Monte Carlo simulations of seismic events that follow the Poisson process with the $G-R$ magnitude–frequency distribution are straightforward. For simulating a T-year period of seismic events, earthquake occurrence times can be generated by following one of the two methods:

- *Method 1* simulates the number of seismic events for the T-year period with the annual activity rate λ_A $(=N_A(M \geq M_{\min}))$. Then earthquake event times are uniformly distributed within the T-year period.
- *Method 2* simulates the IATs of seismic events over the T-year period. The mean recurrence period between the two successive seismic events is $1/\lambda_A$, and the IAT distribution is exponentially distributed. The occurrence times of earthquakes can be obtained by adding the simulated IATs cumulatively.

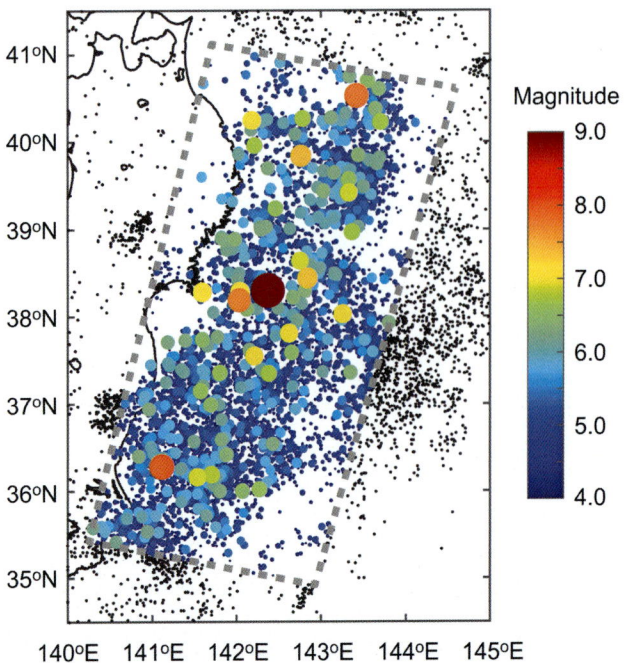

FIGURE 2.5 Earthquake data for the Tohoku region of Japan. The earthquake magnitudes of the events are represented by the marker size and the color. The spatial window is shown with the dotted gray line.

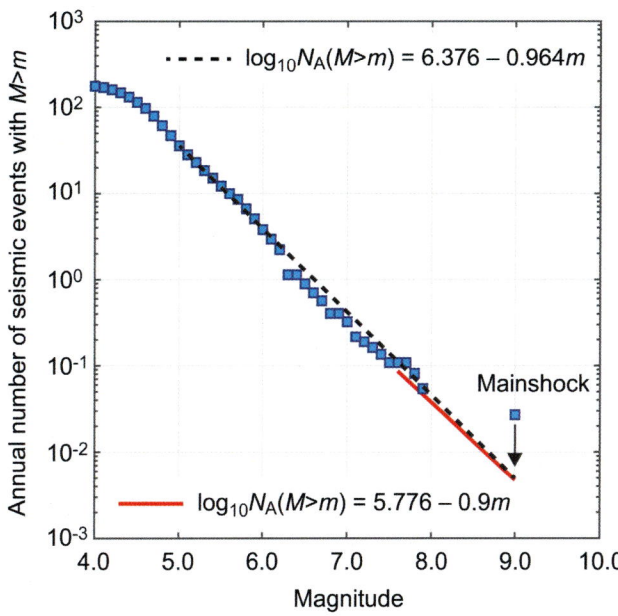

FIGURE 2.6 Gutenberg–Richter plot of the earthquake catalog data for the Tohoku region of Japan. The red log–linear relationship is the G–R relationship for the offshore interface subduction zone, developed by the Headquarters for Earthquake Research Promotion of Japan.

On the other hand, the earthquake magnitude can be simulated from the complementary cumulative distribution function shown in Eq. (2.7). Applying the inverse transformation method, a random sample of earthquake magnitude m_{simu} is generated by

$$m_{simu} = -\ln(\exp(-\beta M_{max}) + u(\exp(-\beta M_{min}) - \exp(-\beta M_{max})))/\beta \tag{2.8}$$

where u is a random sample from the standard uniform distribution that ranges between 0 and 1.

The overall simulation procedure is also straightforward:

Step 1: Determine the duration of seismic event simulation and the number of Monte Carlo simulations (i.e., the number of stochastic event catalogs). For the earthquake source of interest, the activity rate λ_A $(= N_A(M \geq M_{min}))$ and the slope parameter β $(= b\ln(10))$ are estimated.
Step 2: Simulate the earthquake occurrence times of the seismic events over T years. *Method 1* or *Method 2* can be used for this purpose.
Step 3: Simulate the earthquake magnitude from Eq. (2.8).
Step 4: Repeat *Steps 2* and *3* for the required number of stochastic event catalogs.

Fig. 2.7 shows five realizations of stochastic event catalogs for $T = 10$ years. Note that $\lambda_A = 10^{6.376 - 0.964 \times 7} = 0.4218$ and $\beta = 0.964 \times \ln(10) = 2.221$. A computer code for this analysis is provided in Example 2-1 (Appendix).

2.3.2 Time-dependent occurrence model using renewal process

A stochastic renewal process can be adopted for characterizing earthquake occurrence, where the IAT between successive earthquakes is modeled by some suitable probabilistic models (Ogata, 1999). The probability density function and cumulative distribution function of the IAT are denoted by $f_{IAT}(t)$ and $F_{IAT}(t)$, respectively. To characterize the IAT distribution, three model parameters are used: the mean recurrence time μ, the coefficient of variation ν, and the elapsed time since the last event T_E.

Four IAT distributions are introduced in this section. The exponential distribution is most popular and corresponds to a memory-less Poisson process (Section 2.3.1). The lognormal distribution is often adopted for practical reasons (e.g., mathematical simplicity). The Brownian Passage Time (BPT) distribution can be related to the physical phenomena of loading and unloading processes of stress along fault rupture planes (Matthews et al., 2002). The Weibull distribution is often used for modeling failure times of engineering products and is suitable for representing a process having an increasing hazard rate function since the last failure (Abaimov et al., 2008). The probability density functions and their parameters of the four models are summarized in Table 2.1.

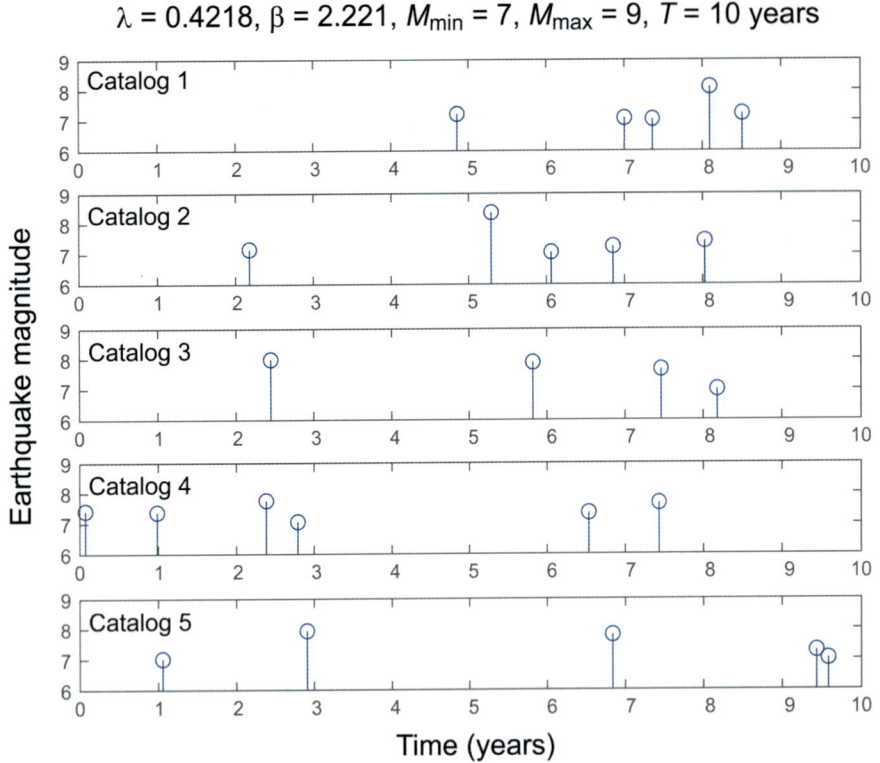

FIGURE 2.7 Five stochastic event catalogs. The parameters of the stochastic event simulations are based on the fitted Gutenberg–Richter model.

TABLE 2.1 Probabilistic models and parameters of the four interarrival time distributions.

Distribution type	Density function	Parameters
Exponential	$f_{IAT}(t) = (1/\mu)\exp(-t/\mu)$	μ
Lognormal	$f_{IAT}(t) = \frac{1}{\sqrt{2\pi}\sigma_{\ln}t}\exp\left(-\frac{(\ln(t)-\mu_{\ln})^2}{2\sigma_{\ln}^2}\right)$	$\mu_{\ln} = \ln(\mu/\sqrt{1+v^2})$ $\sigma_{\ln} = \sqrt{\ln(1+v^2)}$
Brownian Passage Time	$f_{IAT}(t) = \sqrt{\frac{\mu}{2\pi v^2 t^3}}\exp\left(-\frac{(t-\mu)^2}{2\mu v^2 t}\right)$	μ, v
Weibull	$f_{IAT}(t) = (k/u)(t/u)^{k-1}\exp(-(t/u)^k)$	$\mu = u/\Gamma(1+1/k)$ $v^2 = \frac{\Gamma(1+2/k)}{[\Gamma(1+1/k)]^2} - 1$

The mean recurrence time μ depends on how major events are defined in earthquake and tsunami hazard assessments. It needs to be estimated based on various empirical and physical evidence from historical earthquake records and geological records (Ogata, 1999). The v parameter determines the periodicity of earthquake occurrence. When v is small (e.g., less than 0.2), the process becomes more periodic, whereas when v is large and close to 1.0, the process becomes more random, and the clustering of the events tends to occur more frequently. Suitable v values for major fault systems and subduction zones have been studied in the literature. Sykes and Menke (2006) investigated the global statistics of the v parameter for large subduction events and reported it as 0.5 ± 0.2. Working Group on California Earthquake Probabilities (2008) used v values between 0.2 and 0.8 for the assessment. Relatively, small values of v ($\approx 0.2-0.3$) were used for probabilistic seismic hazard assessments in Japan (Headquarters for Earthquake Research Promotion, 2019; Ogata, 1999). The parameter T_E is relevant for evaluating the occurrence time of the first event because the start time of the hazard analysis and the last event time of the stochastic process do not usually coincide. In such cases, the IAT distribution needs to be modified as

$$f'_{IAT}(t|T_E) = \frac{f_{IAT}(t + T_E)}{1 - F_{IAT}(T_E)} \tag{2.9}$$

The modified distribution $f'_{IAT}(t|T_E)$ is left-truncated at T_E and is referred to as a hazard rate function.

To illustrate the different behavior of the IAT distributions, probability density and cumulative distribution functions of the four IAT models are compared in Fig. 2.8A and B, respectively, by considering $\mu = 100$ years, $\nu = 0.25$, and $T_E = 0$ years. Two representations of the same IAT distribution can capture the key features of the models differently. The exponential model is very different from the three other models, which have distinct peaks near the mean recurrence period. Moreover, Fig. 2.8C and D show the BPT models with different values of ν and T_E (with $\mu = 100$ years); as a benchmark, the exponential model with the same μ value is included in the figures. The increase of the ν parameter results in a flatter distribution, whereas the larger T_E value shifts the distribution toward the left and rescales it upwards. Importantly, the combination of distribution type and parameter values can represent a wide range of temporal behavior of earthquake occurrence. Another way to show the differences in IAT distributions is to compare the hazard rate

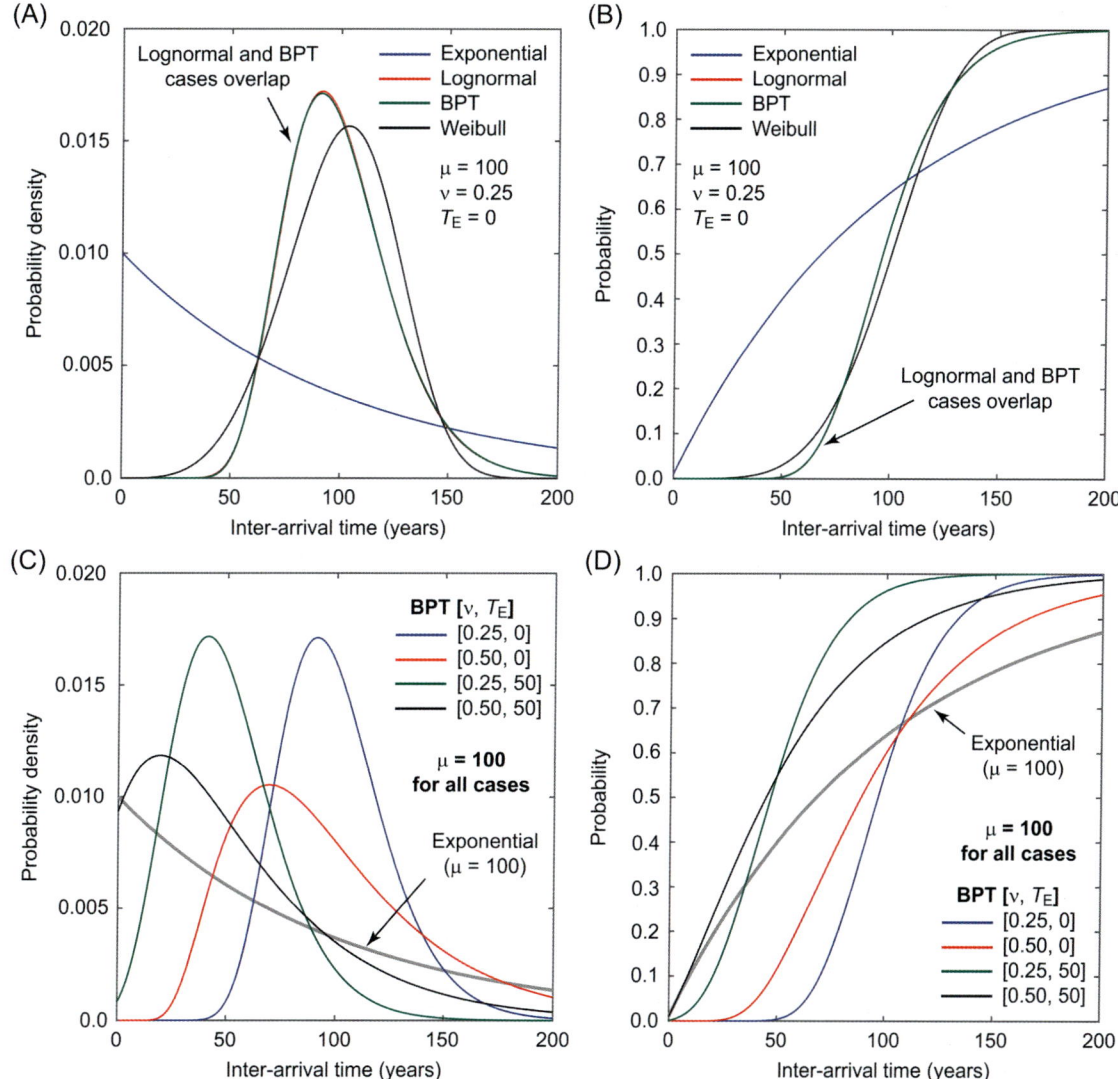

FIGURE 2.8 Renewal models for earthquake occurrence. (A) Probability density functions and (B) cumulative distribution functions of the four interarrival time distributions ($\mu = 100$ years, $\nu = 0.25$, and $T_E = 0$ years). (C) Probability density functions and (D) cumulative distribution functions of the Weibull model with different values of ν and T_E (with $\mu = 100$ years).

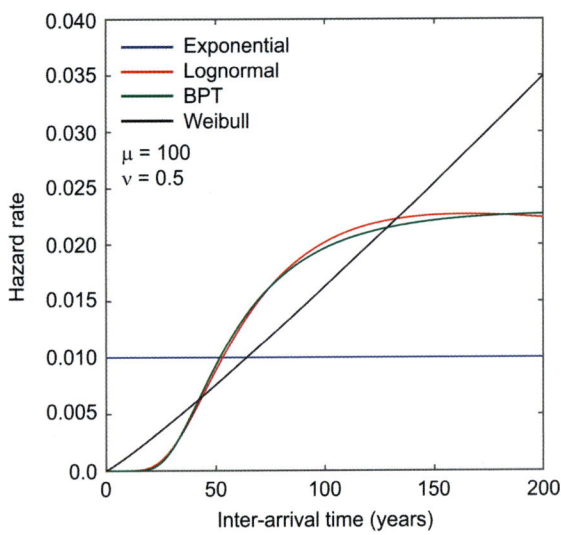

FIGURE 2.9 Hazard rate functions of the four interarrival time distributions (μ = 100 years and ν = 0.5).

functions (Eq. 2.9). Such a comparison of the four IAT distributions is shown in Fig. 2.9. The hazard rate function for the exponential distribution is constant, the hazard rate functions for the lognormal and BPT distributions show similar behavior (i.e., increase sharply and decrease gradually after the peak value is reached), and the hazard rate function for the Weibull distribution increases continuously. A computer code for this analysis is provided in Example 2-2 (Appendix).

For given IAT data, the parameters of the IAT distribution can be estimated using the method of moments and the maximum likelihood method. The method of moments is straightforward; once the mean and standard deviation of the IAT data are calculated, these statistics can be used to estimate the model parameters (see the *Parameters* column of Table 2.1). Many probability distributions have two parameters, and hence, two statistical moments need to be matched, whereas a one-parameter distribution, such as the exponential distribution, only requires the matching of the first statistical moment (i.e., mean). The maximum likelihood method is more statistically rigorous and is often preferred to the method of moments. The method determines the model parameters by maximizing the likelihood function (or more practically log-likelihood function). The likelihood function can be defined as the product of the probability density function for the observed data.

In PTHA, an evaluation of the renewal process can be facilitated through Monte Carlo simulations. In the simulation-based approach, random numbers from a specified IAT distribution are generated using an inverse transformation method or a rejection method. For the renewal process, special attention is necessary to distinguish the first event and subsequent events (i.e., $T_E \neq 0$ vs $T_E = 0$). This is illustrated in Fig. 2.10. The simulation procedure is similar to *Method 2*, mentioned in Section 2.3.1:

Step 1: Determine the IAT distribution type and its parameters (μ, ν, and T_E). In addition, set the duration for the hazard assessment T_D.

Step 2: For simulating the occurrence time of the first event (Fig. 2.10A), the modified distribution $f'_{IAT}(t|T_E)$ should be used by taking into account T_E. When the simulated time t_{IAT} is less than T_D, the simulated event should be registered as $t_1 = t_{IAT}$ in a stochastic event catalog and proceed to the second event; otherwise, the simulation process for this catalog realization is stopped (i.e., no event occurs over a period of T_D).

Step 3: For the second event (Fig. 2.10B), the elapsed time is reset to 0, and an IAT t_{IAT} is sampled from $f_{IAT}(t)$. When $t_2 = t_1 + t_{IAT}$ is less than T_D, the second event is registered in the stochastic event catalog; otherwise, the simulation for this catalog realization is stopped.

Step 4: Continue *Step 3* until the updated time of the most recent event exceeds T_D.

Step 5: Repeat *Steps 2–4* N times to generate a set of N stochastic event catalogs, each with the duration T_D. When T_D is short relative to μ, many catalogs contain no event.

It is important to emphasize that for large earthquakes, the IAT data are often based on historical or paleo-seismic/tsunami data, whose event times are uncertain and ambiguous. For instance, Headquarters for Earthquake Research Promotion (2019) of the Cabinet Office in Japan evaluated the long-term probability of earthquake occurrence along

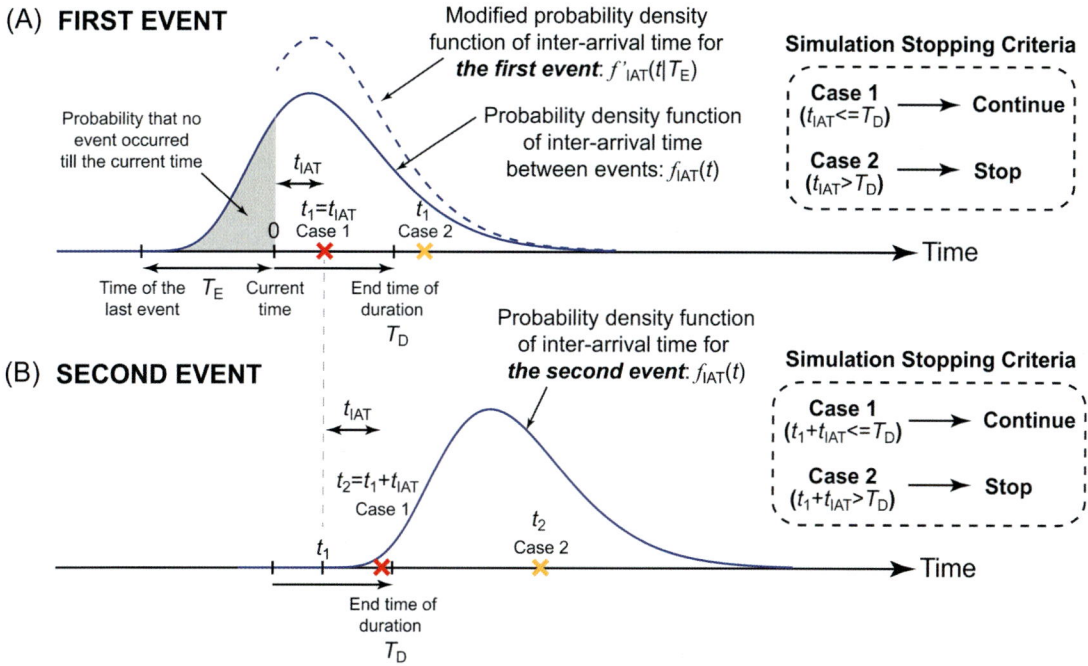

FIGURE 2.10 Renewal process for major earthquakes: (A) first event and (B) second (subsequent) event.

the Japan Trench, where the 2011 Tohoku Earthquake and Tsunami occurred. In the evaluation, five megathrust subduction events were considered to develop the BPT-based renewal earthquake occurrence model. The five events were the 2011 Tohoku event (known precisely), the 1611 Keicho/1454 Kyotoku event, the 869 Jogan event (known with high confidence), an event that occurred between CE 300 and 400, and an event that occurred between 400 and 300 BCE. The tsunami sediment records and offshore tsunami turbidite data indicated one major event occurred between the 2011 and 869 events, but which of the Keicho and Kyotoku events caused massive tsunamis in the Tohoku region was indecisive and thus these two events were considered to be equally likely. On the other hand, two prehistoric events were uncertain and were best identified at 100-year intervals. For the Keicho/Kyotoku and the prehistoric events, Monte Carlo simulations were conducted to generate resampled earthquake catalogs, and then the mean and standard deviation of the IAT distributions were obtained as 590 years and 120 years, respectively ($\nu = 0.20$). For this dataset, the non-Poissonian behavior can be justified on the basis of the hypothesis testing of the Poisson process (Kulkarni et al., 2013), in which the critical value of the coefficient of variation of the IAT data can be established as $\nu = 0.37$ at 5% significance level. The observed coefficient of variation ($\nu = 0.20$) was less than this critical value, and thus the null hypothesis that the IAT data follow the Poisson process can be rejected at the considered significance level. It is also noteworthy that in the long-term probability evaluation carried out by Headquarters for Earthquake Research Promotion (2019), the IAT distribution (i.e., BPT) is "assumed"—four IAT data are insufficient to determine which IAT distributions are suitable (Rhoades et al., 1994). In this regard, epistemic uncertainty related to the IAT distribution type should be captured in the logic tree (Goda, 2019).

2.3.3 Characteristic earthquake model for large events

For large earthquakes from a mature fault system or subduction zone, the characteristic model by Youngs and Coppersmith (1985) is often considered to be applicable. The starting point in characterizing the magnitude−recurrence relationship is to specify the slip rate within the fault zone. The slip rate serves to control how active the fault system is in terms of seismic moment release rate

$$\dot{M}_0 = \mu_S A_f S_r \tag{2.10}$$

where μ_S is the shear modulus, A_f is the area of the fault zone, and S_r is the slip rate of the fault zone. On the other hand, different earthquake magnitude models characterize how the seismic moment release is distributed over the earthquake magnitude range.

The model formulations introduced by Youngs and Coppersmith (1985) facilitate the consistent seismic moment release from the two alternative magnitude models. The probability density function for the characteristic magnitude model, which consists of the exponential part and the characteristic part, is expressed as (Convertito et al., 2006)

$$f_M(m) = \begin{cases} 0 & \text{for } m < M_{\min} \\ \dfrac{\beta \exp(-\beta(m - M_{\min}))}{(1+C)[1 - \exp(-\beta(M_{\max} - M_{\min} - \Delta m_2))]} & \text{for } M_{\min} \leq m < M_{\max} - \Delta m_2 \\ \dfrac{\beta \exp(-\beta(M_{\max} - M_{\min} - \Delta m_1 - \Delta m_2))}{(1+C)[1 - \exp(-\beta(M_{\max} - M_{\min} - \Delta m_2))]} & \text{for } M_{\max} - \Delta m_2 \leq m \leq M_{\max} \\ 0 & \text{for } m > M_{\max} \end{cases} \quad (2.11)$$

where the constant C represents the total probability mass for the characteristic part and is given by

$$C = \frac{\beta \exp(-\beta(M_{\max} - M_{\min} - \Delta m_1 - \Delta m_2))}{1 - \exp(-\beta(M_{\max} - M_{\min} - \Delta m_2))} \Delta m_2 \quad (2.12)$$

In the above equations, β, $M_{\min}$, and $M_{\max}$ are defined in the same way as the G–R relationship (Eq. 2.5); Δm_1 is the magnitude interval that is used to specify the probability density value for the characteristic part; and Δm_2 is the magnitude interval for the characteristic part. Youngs and Coppersmith (1985) suggested Δm_1 and Δm_2 equal to 1.0 and 0.5, respectively. The interpretation of Eq. (2.11) is that in the magnitude range between $M_{\min}$ and $M_{\max} - \Delta m_2$, the magnitude distribution follows the exponential distribution, whereas in the magnitude range between $M_{\max} - \Delta m_2$ and $M_{\max}$, the magnitude distribution follows the uniform distribution (i.e., constant probability density). It is noted that Eq. (2.11) accommodates the truncated exponential magnitude model (i.e., G–R model) by setting $\Delta m_2 = 0$ (i.e., $C = 0$). The seismic activity rate for the exponential part of $f_M(m)$ is given by (Convertito et al., 2006)

$$\alpha_{\exp} = \frac{\mu_S A_f S_r [1 - \exp(-\beta(M_{\max} - M_{\min} - \Delta m_2))]}{K M_0^{\max} \exp(-\beta(M_{\max} - M_{\min} - \Delta m_2))} \quad (2.13)$$

where the seismic activity rate for the characteristic part of $f_M(m)$ is given by

$$\alpha_{\text{char}} = \alpha_{\exp} \frac{\beta \Delta m_2 \exp(-\beta(M_{\max} - M_{\min} - \Delta m_1 - \Delta m_2))}{1 - \exp(-\beta(M_{\max} - M_{\min} - \Delta m_2))} \quad (2.14)$$

In Eqs. (2.13) and (2.14), $M_0^{\max}$ is the seismic moment that corresponds to the maximum magnitude $M_{\max}$, and the constant K is given by

$$K = \frac{b 10^{-1.5 \Delta m_2}}{1.5 - b} + \frac{b \exp(\beta \Delta m_1)(1 - 10^{-1.5 \Delta m_2})}{1.5} \quad (2.15)$$

The characteristic magnitude model requires the following model parameters: μ_S, A_f, S_r, β (or b), Δm_1, Δm_2, $M_{\min}$, and $M_{\max}$. The shear modulus can be selected from a range between 30 and 40 GPa (depending on a seismic region of interest), whereas the fault zone areas can be determined from the fault length and width (see Section 2.4). The slip rate for the fault zone should be determined from geodetic measurements around the fault, and multiple values may need to be considered in the logic tree as the slip rate is highly variable. The maximum magnitude is determined by first evaluating the magnitude value that corresponds to the fault length and width of the rupture zone using a scaling relationship (Murotani et al., 2013; Strasser et al., 2010; Thingbaijam et al., 2017) and by adding half of the characteristic magnitude range (e.g., $\Delta m_2/2 = 0.25$).

To illustrate how the characteristic earthquake model compares with the G–R model, Fig. 2.11 shows two magnitude–frequency relationships by considering $\mu_S = 35$ GPa, $S_r = 40$ mm/year (typical value for an active subduction region), $A_f = 500 \times 200$ km (M 9 subduction fault), $M_{\min} = 6.5$, $M_{\max} = 9.0$, $\Delta m_1 = 1.0$, and $\Delta m_2 = 0.5$. The probability density function of the characteristic magnitude model has a uniform magnitude distribution for the range between M 8.5 and M 9.0 (note: M $8.5 = $ M $9.0 - \Delta m_2$), and its amplitude is equal to the exponential portion at M 7.5 ($=$ M $8.5 - \Delta m_1$). The magnitude–frequency relationship of the characteristic magnitude model has lower occurrence rates for $<$M 8.5 but has higher occurrence rates for $>$M 8.5. It is noted that the relative likelihood of the truncated exponential (G–R) model versus the characteristic model is difficult to determine explicitly, and both models may be considered to be applicable in conducting seismic hazard analysis (Page & Felzer, 2015; Stirling & Gerstenberger, 2018). A computer code for this analysis is provided in Example 2-3 (Appendix).

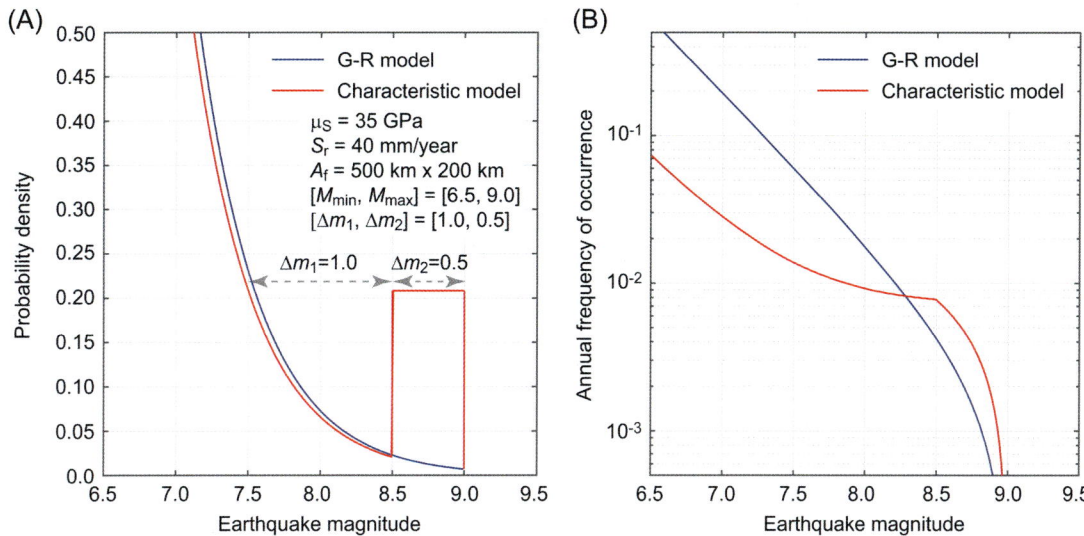

FIGURE 2.11 Comparison of Gutenberg–Richter and characteristic magnitude–frequency relationships: (A) probability density function and (B) annual frequency of occurrence.

2.4 Earthquake rupture model

Earthquake rupture is a complex process that is influenced by various factors, such as seismotectonic settings, heterogeneous frictional properties, and prerupture stress conditions of a fault. The governing equations of physical phenomena (e.g., elastic dislocation theory and elastodynamic theory; Aki and Richards (2002)) can be used to generate synthetic data for prediction, which is called a *forward* problem. On the other hand, by recording and analyzing seismic waves, ground deformations, and tsunami waves from an actual fault rupture, the earthquake source process can be inferred by matching synthetic data with key features of observations (Ji et al., 2002; Satake, 1989; Wald et al., 1996), which is called an *inverse* problem (Chapter 7). Therefore, inverted source models that capture important source characteristics of past earthquakes are useful for predicting the behavior of future earthquakes. It is important to recognize that the solution for an inverse problem is not unique and multiple models/parameters can achieve an equally good fit to the observations. This section introduces a finite-fault representation of a seismic source for tsunamis in terms of geometry and earthquake slip distribution. In the context of PTHA, the earthquake source characterization and its uncertainty are critical elements in simulating tsunami waves and tsunami intensity parameters due to major subduction earthquakes (Chapter 3).

Section 2.4.1 introduces finite-fault models that have been developed for past major earthquakes through source inversion analysis. The parameters of finite-fault models can be broadly classified into geometric parameters and slip parameters. The slip parameters can include the mean, maximum, and earthquake slip distribution on the fault plane. When the earthquake rupture is characterized kinematically, additional parameters, such as rise time and rupture propagation speed, become relevant.

Section 2.4.2 presents a procedure for analyzing a finite-ault model and determining the earthquake source parameters by focusing on a finite-fault model for the 2005 Sumatra Earthquake, developed by Shao and Ji (2005). This section aims to demonstrate the spectral analysis of an earthquake slip distribution through an example.

Section 2.4.3 presents a summary of empirical scaling relationships of earthquake source parameters that are obtained from the spectral analysis of finite-fault models. Earthquake source scaling relationships can be used to predict the earthquake characteristics for a specified earthquake magnitude value. Therefore, in the context of PTHA, given a seismic event (i.e., earthquake occurrence time, earthquake location, and earthquake magnitude), the earthquake rupture can be characterized using the scaling relationships.

2.4.1 Finite-fault models

Finite-fault models of major historical earthquakes are published in the literature and by organizations, such as the U.S. Geological Survey and the Geospatial Information Authority of Japan. SRCMOD (http://equake-rc.info/srcmod/) is a growing online database of published finite-fault models (Mai & Thingbaijam, 2014) and offers access to a collection of rupture models with visualization tools and supplementary information. The finite-fault models that are included in the SRCMOD database are based on source inversion analyses of geophysical data observed during major seismic events in the past. The data type ranges from teleseismic, strong motion, geodetic, tsunami, and joint use of different types of data. The database is useful for analyzing the key features of the source parameters statistically and for developing scaling relationships of the source parameters (Mai & Beroza, 2000, 2002; Thingbaijam et al., 2017) (Section 2.4.3).

A typical finite-fault model can be defined by its fault geometry and earthquake slip information. Fig. 2.12 shows a schematic of finite-fault geometry. For a rectangular fault plane, the fault area can be defined by the fault length L and the fault width W. Typically, the fault plane is buried, and the depth to the top of the fault plane is specified by H_{top}. On the other hand, the surface rupture can be represented by allowing the top of the fault plane to reach the ground surface. The strike angle ϕ is the clockwise orientation of the fault trace measured with respect to the North, whereas the dip angle δ is the inclination of the fault plane with respect to the horizontal plane. The strike angle is defined between 0 and 360 degrees, and the dip angle is defined between 0 and 90 degrees. Following the convention by Aki and Richards (2002), the fault dips to the right side of the trace. In a simple form, the earthquake slip information can be characterized by the average slip S_a and the rake angle λ. The rake angle defines the direction that the hanging wall moves during rupture, measured relative to the footwall, and takes a value between -180 and 180 degrees. The rake angles close to 0 and ± 180 degrees correspond to strike−slip fault motions (left lateral motion or right lateral motion); the rake angles close to 90 degrees correspond to the reverse fault motions; and the rake angles close to -90 degrees correspond to the normal fault motions (Fig. 2.3). As defined in Eq. (2.2), the fault area ($= L \times W$) and the average slip are the primary parameters in determining the seismic moment of the earthquake rupture.

For many of the past major seismic events, more detailed information on the earthquake rupture process is available. Fig. 2.13A shows a finite-fault model for the 2005 Sumatra Earthquake (Shao & Ji, 2005). This model consists of a single fault plane, and the fault plane is divided into gridded cells, each having its own slip value and rake angle. The earthquake slip distribution provides additional information on the earthquake rupture, such as the concentration and spatial features of earthquake slip within the fault plane. The asperity is an area on a fault plane where there is increased friction and thus locked. Generally, the asperity location coincides with a local area where earthquake rupture begins and is associated with a large slip during rupture (Mai et al., 2005). For some earthquakes, the finite-fault models can be characterized on the basis of multiple fault segments/subplanes. An example of such a complex finite-fault model is shown in Fig. 2.13B for the 2010 Darfield Earthquake (Atzori et al., 2012). For this finite-fault model, there are eight segments with different geometrical parameters.

For well-recorded earthquakes, multiple finite-fault models are developed in different studies individually. Because their purposes, data selection strategies, and modeling approaches are different, the final finite-fault models are not identical. Therefore, it is important to recognize this nonuniqueness of the inverted finite-fault models in developing predictive equations of the earthquake source parameters (Section 2.4.3). To illustrate this, Fig. 2.14 shows six

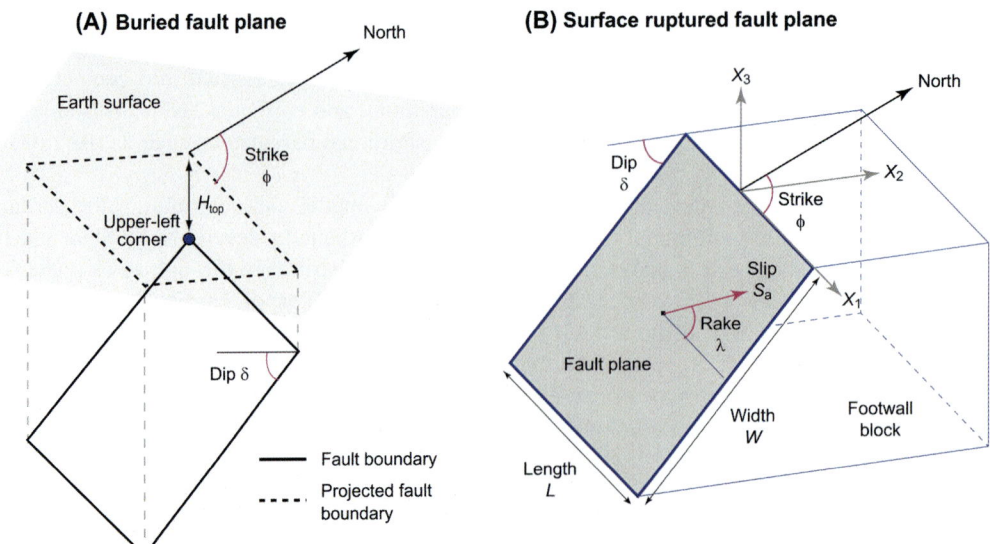

FIGURE 2.12 Parameters for fault geometry: (A) buried fault plane and (B) surface rupture fault plane.

Tsunami generation Chapter | 2 | 41

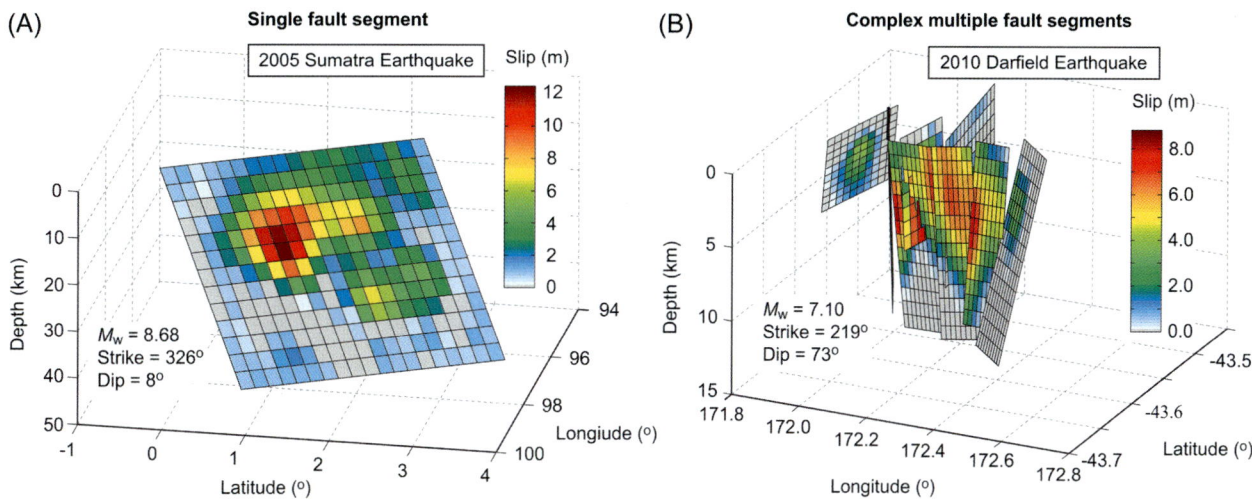

FIGURE 2.13 Examples of finite-fault models: (A) the 2005 Sumatra Earthquake (Shao & Ji, 2005) and (B) the 2010 Darfield Earthquake (Atzori et al., 2012).

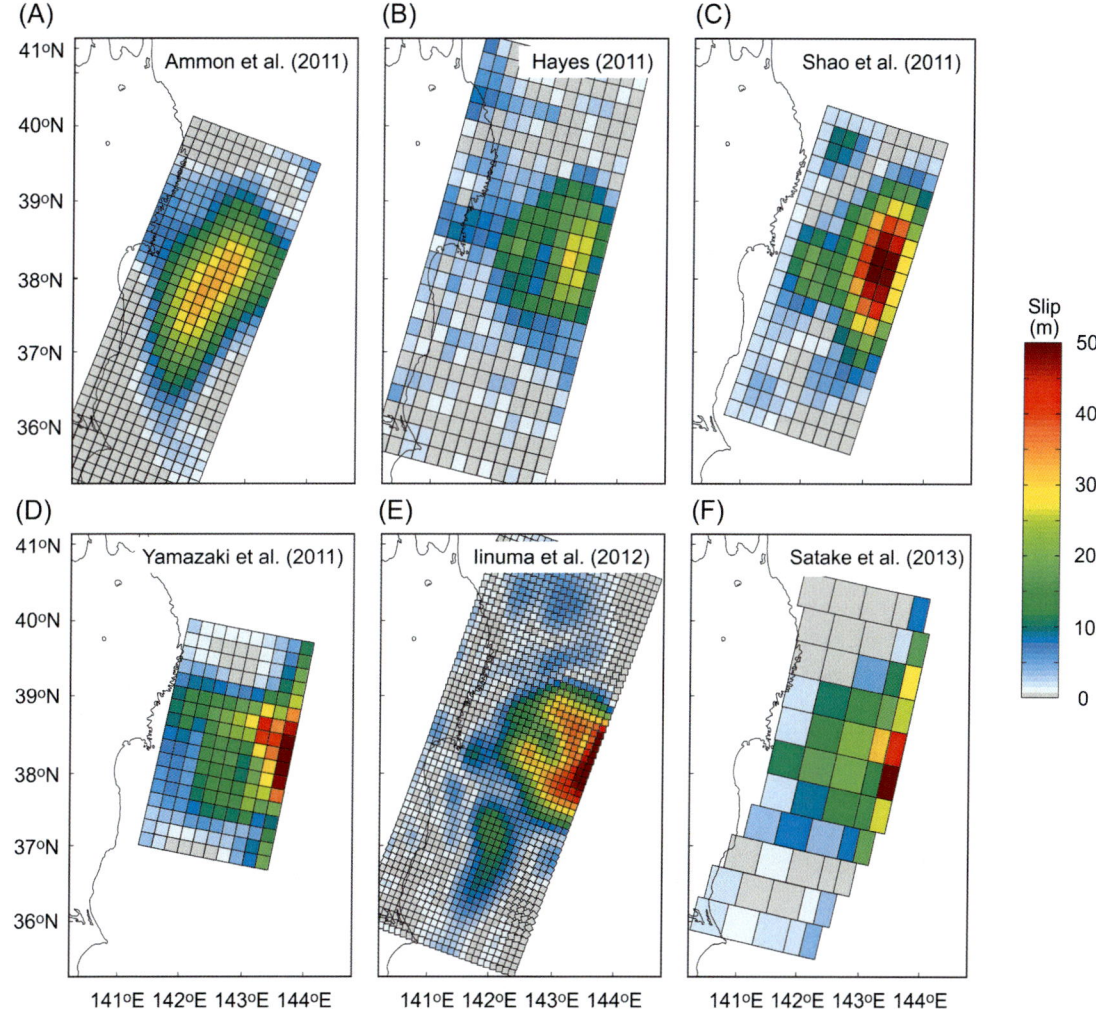

FIGURE 2.14 Finite-fault models (earthquake slip distributions) for the 2011 Tohoku Earthquake and Tsunami developed by six different studies: (A) Ammon et al. (2011), (B) Hayes (2011), (C) Shao et al. (2011), (D) Yamazaki et al. (2011), (E) Iinuma et al. (2012), and (F) Satake et al. (2013).

finite-fault models for the 2011 Tohoku Earthquake and Tsunami. Shao et al. (2011) and Hayes (2011) used teleseismic data; Iinuma et al. (2012) used geodetic data; Satake et al. (2013) used tsunami data; Ammon et al. (2011) used teleseismic and geodetic data jointly in their inversion; and Yamazaki et al. (2011) used teleseismic and tsunami data. The maximum slip values for the six models vary significantly (from 35 to 75 m), and the areas/shapes of the main slip asperity differ among these models. These individual models reflect some aspects of the complex rupture process of the 2011 Tohoku Earthquake and Tsunami, but not all features completely. For specific observations (e.g., tsunami waveform at a particular tidal gauge), some models may perform better than others. In predicting future tsunami scenarios, postevent model evaluations are not possible and a priori performance tests of individual models are not always feasible. The resulting variability in tsunami predictions can be interpreted as epistemic uncertainty due to modeling errors when using different data and methods.

For the finite-fault models shown in Figs. 2.13 and 2.14, the final earthquake slip distributions are presented. Some studies carry out source inversion analysis by considering the temporal and spatial evolution of earthquake slip during the rupture. The developed models are called kinematic finite-fault models. The additional earthquake source parameters include the rupture nucleation point on the fault plane, rupture propagation speed, and rise time (local duration of slip) (Mai et al., 2005; Melgar & Hayes, 2017; Ye et al., 2016). For illustration, the finite-fault model by Satake et al. (2013) for the 2011 Tohoku Earthquake and Tsunami is shown in Fig. 2.15. The nucleation point is close to the latitude of 38°N and the longitude of 138°E. The kinematic rupture profiles are obtained every 30 seconds. The rupture front expands in all directions until the rupture front reaches the boundaries of the finite-fault model. Large slips along the Japan Trench occur in time intervals between 150 and 180 seconds, which is the main trigger of massive tsunamis and extensive coastal inundations along the Tohoku coast of Japan.

2.4.2 Analysis of finite-fault models

Spectral representations of earthquake slip models, each derived using different data, analysis methods, and parameterizations, have been investigated in the literature (Herrero & Bernard, 1994; Lavallée et al., 2006; Mai & Beroza, 2002; Somerville et al., 1999). These studies showed that the spatial distribution of earthquake slip can be characterized as a power spectral density in the wavenumber domain, and the autocorrelation function can be used to parameterize the complexity of earthquake slip. For instance, Mai and Beroza (2002) investigated the applicability of Gaussian, exponential, von Kármán, and fractal autocorrelation functions by considering 44 slip distributions for worldwide crustal earthquakes. In their study, parameters of an autocorrelation function model (i.e., fractal dimension and correlation length) were related to macroscopic earthquake source parameters, such as moment magnitude and fault size. They suggested that the anisotropic von Kármán model is most suitable among the tested models for characterizing the heterogeneity of earthquake slip. There are alternative approaches to modeling the heterogeneous earthquake slip distribution. For instance, LeVeque et al. (2016) and Melgar et al. (2016) proposed the Karhunen−Loeve expansion technique, whereas Park and Cox (2016) adopted the two-dimensional Gaussian distribution.

Adopting the spectral method by Mai and Beroza (2002), the source parameters (i.e., geometry, slip statistics, and spatial slip distribution) can be evaluated through (1) effective dimension analysis (Mai & Beroza, 2000), (2) marginal earthquake slip statistics and Box-Cox analysis (Goda et al., 2014), and (3) Fourier spectral analysis (Mai & Beroza, 2002). For illustration of the analysis procedure, the finite-fault model for the 2005 Sumatra Earthquake developed by Shao and Ji (2005) is considered. The earthquake slip distribution of this finite-fault model is shown in Fig. 2.13A. A computer code for this analysis is provided in Example 2-4 (Appendix).

First, subfaults having zero slips along the edges of the rupture plane are removed from the original source models as they are considered unimportant features of the rupture model. To focus on the major slip features of the original source models, effective source dimensions are evaluated from the autocorrelation dimensions W_{AC} and L_{AC} along dip or strike. W_{AC} and L_{AC} are calculated as the area under the autocorrelation function of the one-dimensional slip function normalized by the zero-lag value (Mai & Beroza, 2000)

$$W_{AC} = \frac{\int_{-\infty}^{\infty} (f_W * f_W) dx}{f_W * f_W |x=0} \tag{2.16}$$

and

$$L_{AC} = \frac{\int_{-\infty}^{\infty} (f_L * f_L) dx}{f_L * f_L |x=0} \tag{2.17}$$

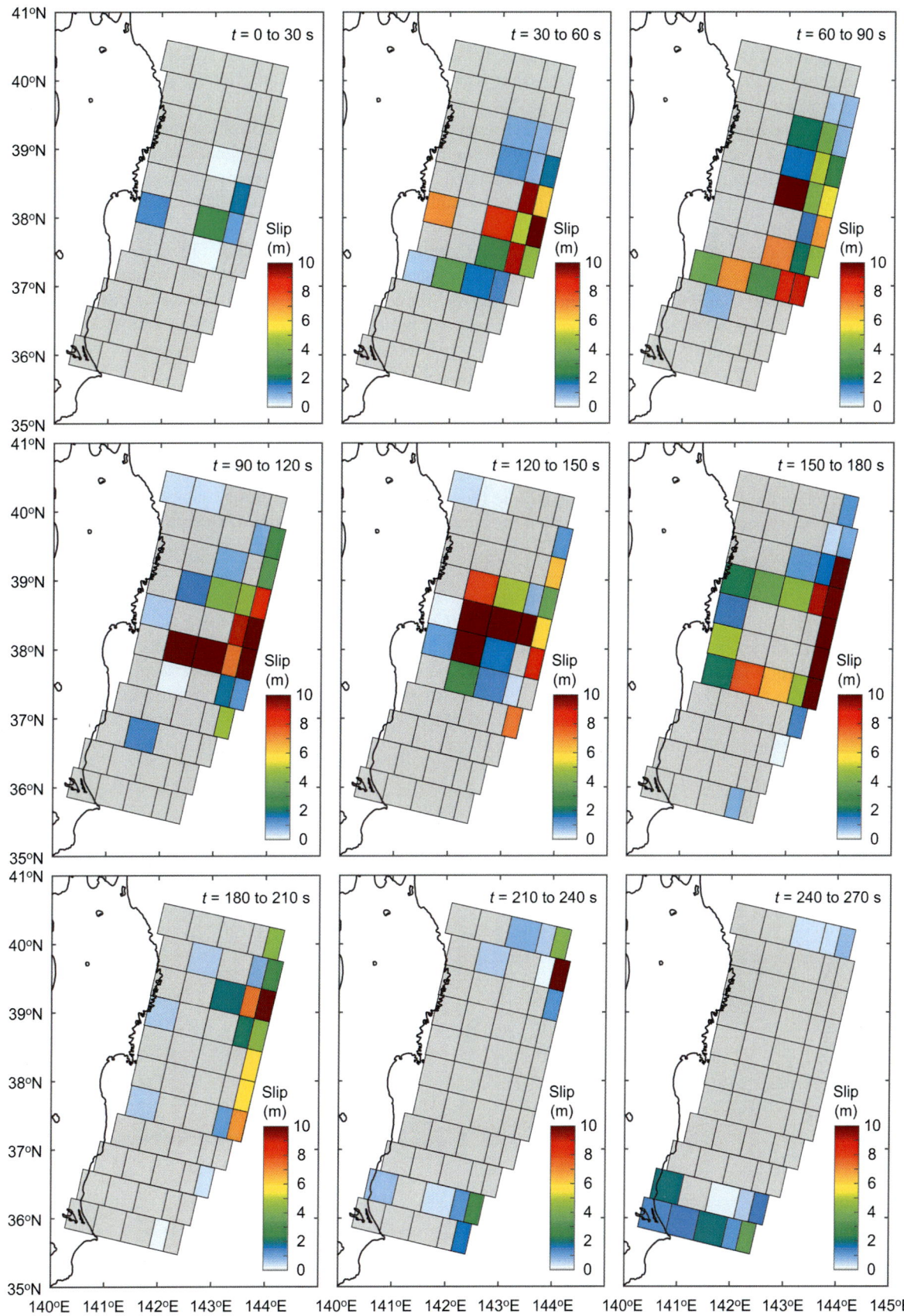

FIGURE 2.15 Temporal evolution of earthquake slip distribution for the 2011 Tohoku Earthquake and Tsunami (Satake et al., 2013).

where f_W and f_L are the one-dimensional slip function along dip and strike, respectively (which can be computed from the original slip distribution by summing up the slip values along strike or dip), and f_W*f_W and f_L*f_L represent the autocorrelation function of the one-dimensional slip along dip and strike, respectively. The estimation of the final effective source dimensions is iterative. The largest dimension that fits the autocorrelation width/length (as in Eqs. 2.16 and 2.17) is determined such that the difference between the two is less than or equal to the subfault size. The trimming process is done by removing any row/column at the fault edge from the slip distribution if the maximum slip in this row/column is less than or equal to a threshold value of 0.01 km. Note that when the top edge of the rupture plane is located on the Earth surface and a large-slip asperity is encountered within 5 km depth, the top edge of the fault plane is not trimmed. Depending on the slip distribution of the finite-fault model, estimated effective dimensions can be smaller than the original source dimensions.

An illustration of effective dimension analysis of a finite-fault model is shown in Fig. 2.16. Using the original finite-fault model, the one-dimensional slip functions along dip and strike directions can be evaluated and are shown in the side panels to the left-hand side and bottom of the slip distribution; gray lines shown in the side panels are the effective dimensions that are estimated through the iterative row/column removing procedures (Fig. 2.16A). For the slip distribution of the 2005 Sumatra Earthquake (Fig. 2.13A), two rows are removed from the bottom (i.e., downdip edge), whereas one column each is removed from both right- and left-hand sides of the fault plane (i.e., strike edges). The dotted rectangle represents the effective source dimension for the slip distribution. The effects of considering effective dimensions are that the fault dimensions become smaller, whereas the mean slip values become greater, compared with the original slip distribution. Subsequently, slip values of the original finite-fault model, specified at individual subfaults (i.e., cell-based slip distribution), are transformed into a corner-based slip distribution (Fig. 2.16B). In this step, a quarter of the slip value for each subfault is assigned to the four corners, and the sum of the assigned slip values at the corner grids of the slip distribution is calculated. Then the corner-based slip distribution is interpolated with a smaller grid size. The grid size for interpolation is not smaller than one-fifth of the original grid resolution. This is illustrated in Fig. 2.16B.

Second, using the effective source model, mean slip, S_a, and maximum slip, S_m, are evaluated. A histogram of the effective source model from Fig. 2.16A is shown in Fig. 2.17A. Box-Cox analysis of the slip distribution is then carried out to characterize the probability distribution of slip values within the fault plane. It identifies the best power parameter such that a nonnormal random variable (original data) can be transformed into a normal random variable. Generally, many inverted finite-fault models have slip values that are distributed with the right tail heavier than the normal distribution (i.e., positive skewness). The motivation of this approach is to generate random fields using the Fourier integral method (Mai & Beroza, 2002) (Section 2.5.2), whose output slip distribution is normally distributed. To achieve the

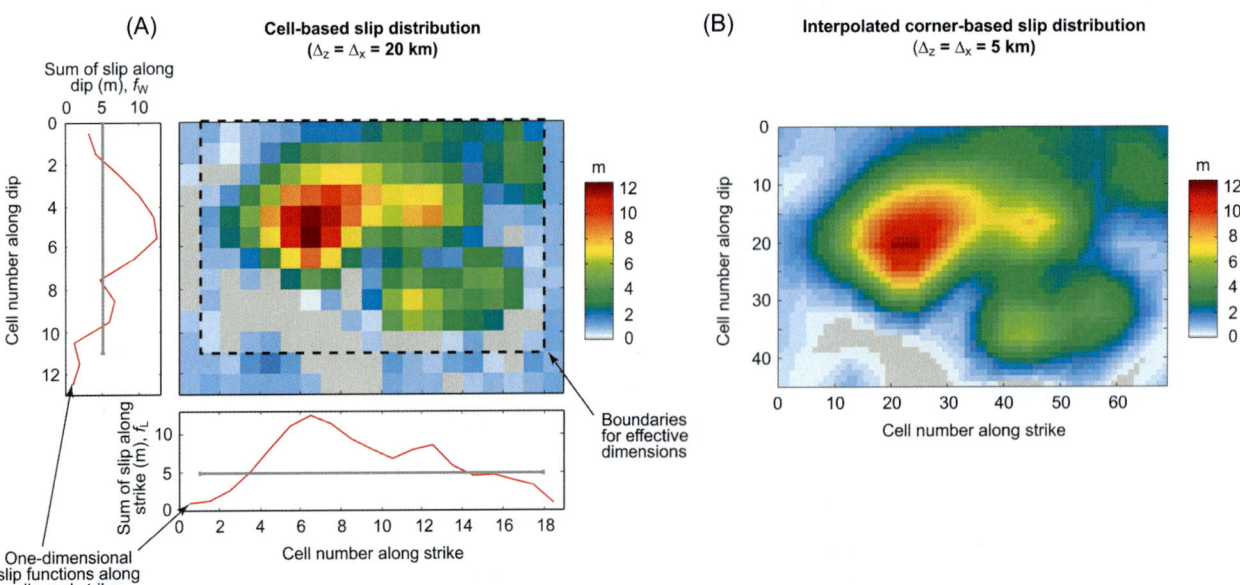

FIGURE 2.16 Effective dimension analysis of the earthquake slip distribution: (A) cell-based slip distribution and (B) interpolated corner-based distribution. The slip distribution is based on Shao and Ji (2005).

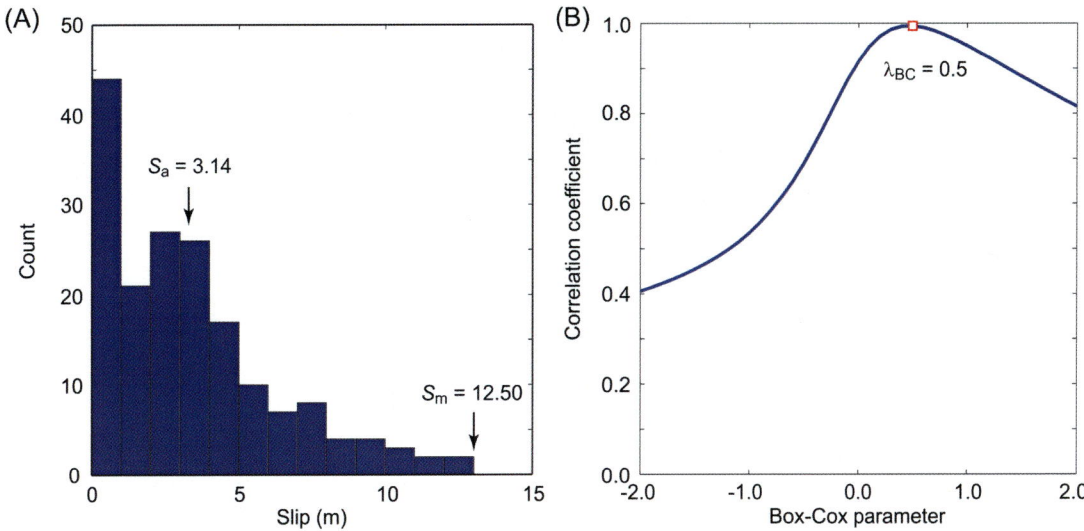

FIGURE 2.17 (A) Histogram of the earthquake slip data and (B) Box-Cox analysis of the earthquake slip data. The slip distribution is based on Shao and Ji (2005).

nonnormal distribution, inverse Box-Cox transformation can be applied (Goda et al., 2014). The Box-Cox transformation is defined as follows

$$Y = \frac{X^{\lambda_{BC}} - 1}{\lambda_{BC}} \quad (2.18)$$

where X is the original variable (i.e., nonnormal), and Y is the transformed variable (i.e., normal). λ_{BC} is the power parameter and is not equal to 0 in the above equation. When $\lambda_{BC} = 0$, the Box-Cox transformation corresponds to the lognormal transformation. The best power parameter can be identified by calculating the linear correlation coefficient of the standard normal variable and the transformed variable of the slip values (after standardization) for a range of λ_{BC} values (typically −2 to 2). The value of λ_{BC} that achieves the maximum linear correlation coefficient is adopted. For the slip data, $\lambda_{BC} = 0.5$ is obtained as the best power parameter (Fig. 2.17B).

Third, the spectral analysis of the earthquake slip distribution is carried out based on the procedures given by Mai and Beroza (2002) and Goda et al. (2014). The von Kármán wavenumber spectrum $P(k)$ is considered for characterizing the spatial slip distribution (Mai & Beroza, 2002)

$$P(k) \propto \frac{CL_z CL_x}{(1+k^2)^{H+1}} \quad (2.19)$$

where k is the wavenumber, $k = (CL_z^2 k_z^2 + CL_x^2 k_x^2)^{0.5}$ (note: wavenumber is proportional to the reciprocal of wavelength), CL_z and CL_x are the correlation lengths along dip and strike directions, respectively, and H is the Hurst number. The correlation length determines the absolute level of the power spectrum in the low wavenumber range and captures the anisotropic spectral features of the slip distribution when different correlation lengths are specified for dip and strike directions. The Hurst number controls the slope of the power spectral decay in the high wavenumber range and is theoretically constrained to range between 0 and 1. The spectral analysis method consists of several steps as follows:

Step 1. Starting with the interpolated corner-based distribution (Fig. 2.16B), the slip distribution is tapered using a window function by adding two rows/columns to all sides of the source model. For this purpose, a Hanning window function can be adopted. Tapering ensures that the earthquake slip decays gradually along the edges of the fault plane.

Step 2. The two-dimensional Fast Fourier Transform (FFT) of the interpolated and tapered slip distribution is computed, and the amplitude spectrum is normalized with respect to the maximum value (Fig. 2.18A). The grid number of the two-dimensional FFT is the nearest higher power of 2 based on the larger source dimension of the fault plane.

Step 3. The circular average of the normalized wavenumber spectrum is calculated (Anguiano et al., 1993), and the fractal dimension D is determined on basis of the least squares fitting. The obtained fractal dimension is then converted to the Hurst number: $H = 3 - D$. The estimated value of D is constrained to lie between 2 and 3

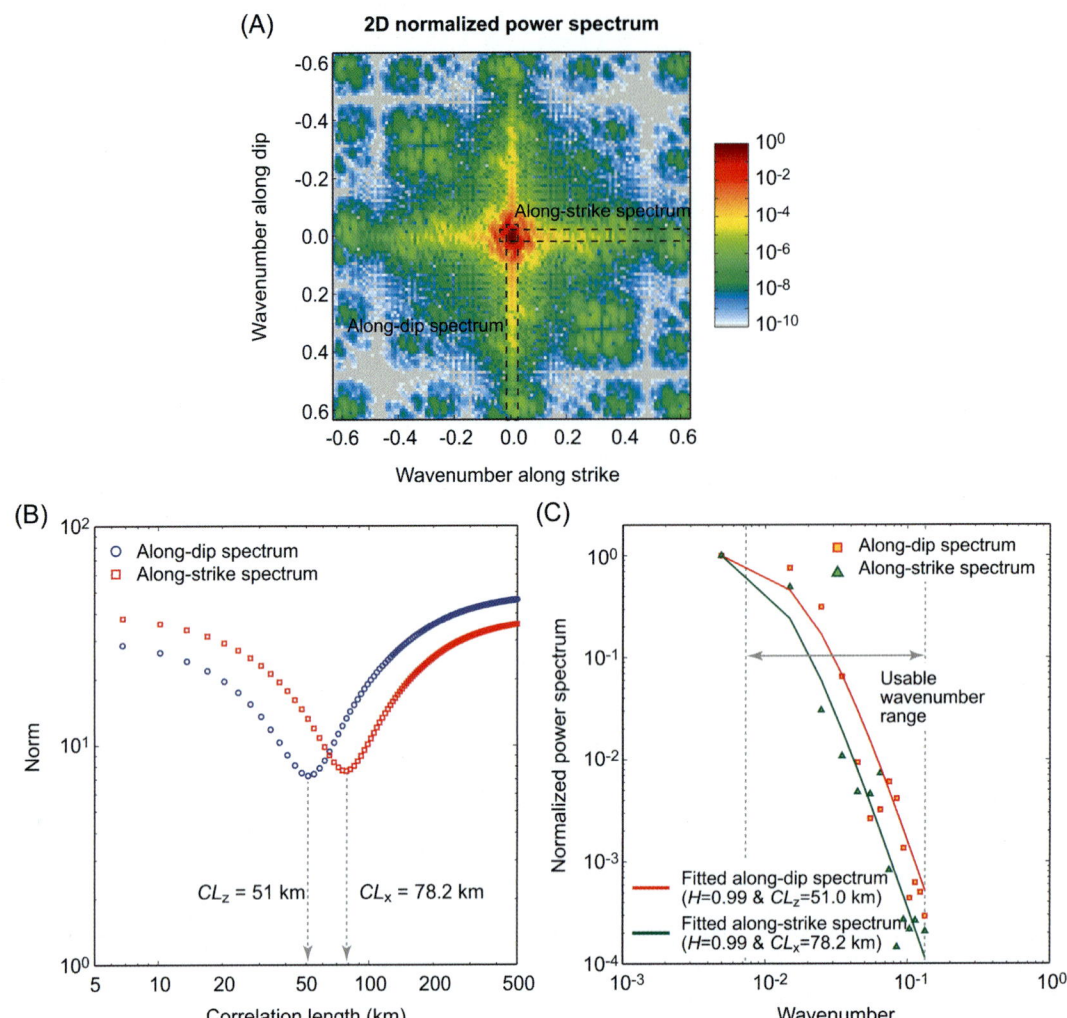

FIGURE 2.18 (A) Two-dimensional wavenumber power spectrum of the earthquake slip distribution, (B) misfit functions for estimating the correlation lengths in the downdip and along-strike directions, and (C) observed and modeled one-dimensional power wavenumber spectra in the downdip and along-strike directions. The slip distribution is based on Shao and Ji (2005).

(as H is theoretically constrained to range between 0 and 1), noting that $H = 1$ corresponds to "k-squared" model (Herrero & Bernard, 1994; Mai & Beroza, 2002). Accordingly, the minimum and maximum values of H are set to 0.01 and 0.99, respectively, as practical limiting values. For the identified value of H, a one-dimensional search of suitable correlation lengths for dip and strike directions is carried out by minimizing the norm between the along-dip/along-strike wavenumber spectrum and the analytical von Kármán model (Fig. 2.18B). The spectral fitting for the dip and strike directions is illustrated in Fig. 2.18C. For this example, the Hurst number is estimated to be 0.99, whereas the correlation lengths along dip and strike directions are 51 and 78.2 km, respectively.

Several remarks regarding the abovementioned spectral analysis are necessary. In *Step 1*, interpolation and tapering of the slip distribution before the two-dimensional FFT affect the estimation of the Hurst number and the correlation lengths. The interpolation essentially introduces additional spectral components in the large wavenumber range. On the other hand, tapering may alter the spectral components of the slip distribution when it has large slip concentrations near the edges of the fault plane. In such cases, tapering forces the slip distribution to decay to zero over grid-size distances as specified by the window function. This may lead to bias in the spectral decay feature due to wavenumber spectral content that is artificially introduced by tapering, more significantly in the large wavenumber range. Although large slip concentrations along the edges of the fault plane and abrupt termination of the slip distribution (when tapering is not considered) may also be regarded as unrealistic, tapering of the slip along the top edge may adversely introduce biases in the estimated spatial slip distribution parameters when large asperities along the top-edge are real features of the earthquake rupture (e.g., Fig. 2.15). In such cases, more careful investigations are required. The sensitivity of these different approaches was investigated by Goda et al. (2014, 2016).

2.4.3 Prediction of earthquake source parameters

Earthquake rupture modeling aims at predicting key characteristics of a fault rupture based on past major earthquakes and scaling relationships for earthquake source parameters can be used for this purpose (Blaser et al., 2010; Goda et al., 2016; Leonard, 2010; Mai & Beroza, 2000, 2002; Murotani et al., 2013; Somerville et al., 1999; Strasser et al., 2010; Thingbaijam et al., 2017; Wells & Coppersmith, 1994). They are useful for defining a range of uncertain earthquake source features in PTHA. For such applications, various parameters need to be specified, including geometry (in accordance with regional seismotectonic setting), slip statistics (mean, maximum, and distribution types), spatial slip distribution, and temporal rupture evolution (Sections 2.4.1 and 2.4.2).

This section explains the development process of earthquake source scaling relationships for tsunamigenic earthquakes by following Goda et al. (2016). It is noted that there are different approaches that can be adopted to achieve the same objectives. The underlying database for developing the scaling relationships is the SRCMOD. The parameterization of the finite-fault model is consistent with the method shown in Section 2.4.2, and the eight earthquake source parameters, that is, L, W, S_a, S_m, λ_{BC}, CL_x, CL_z, and H, are considered. The consideration of effective fault plane dimensions, that is, L_{AC} and W_{AC}, for L and W essentially reduces the geometrical dimensions of some of the original finite-fault models and affects the mean slip, because they are estimated on the basis of "effective" finite-fault models. It is arguable as to which of the dimensions, that is, effective versus original, is suitable for earthquake source modeling. The fundamental difficulties of this problem are that the boundaries of the rupture models are defined only loosely in source inversion analysis and that these small slip values may not be robust features of the physical rupture process, involving relatively large errors (Satake et al., 2013).

The first step in developing the scaling relationships of earthquake source parameters is to select a set of suitable finite-fault models based on criteria, such as earthquake characteristics, model resolution, and model complexity. In Goda et al. (2016), the starting dataset was 317 finite-fault models that were available in the SRCMOD database at the time of the model development. The spectral analysis of a finite-fault model requires that an earthquake slip distribution is relatively simple, typically mapped onto a single fault plane (Fig. 2.13A), or multiple fault planes are aligned well with one another. On the other hand, multisegment finite-fault models with complex configurations (Fig. 2.13B) are not suitable and thus are excluded from the spectral analysis. By applying other criteria, such as minimum magnitude, exclusion of deep in-slab earthquakes (which do not usually cause large tsunamis), spatial slip distribution resolution, and limiting to one model per event from the same developers, 226 finite-fault models were identified as usable. Keeping the objective of applying the scaling relationships to tsunami hazard studies, the selected finite-fault models are divided into 100 tsunamigenic events and 126 nontsunamigenic events. This distinction differs from the conventional classification of earthquakes based on faulting mechanisms (Thingbaijam et al., 2017; Wells & Coppersmith, 1994). Tsunamigenic events typically occur at the subduction interface, and the dominant mechanism for the tsunamigenic events is reverse faulting. Different model classifications can be considered. Fig. 2.19A shows the spatial distribution of the selected 226 finite-fault models, distinguishing tsunamigenic, and nontsunamigenic models. The tsunamigenic models are located in ocean areas, whereas the majority of the nontsunamigenic models are located in inland areas. Fig. 2.19B shows a histogram of the 226 source models in terms of moment magnitude. The earthquake magnitudes for the tsunamigenic models are greater than those for the nontsunamigenic models (a range from 7.0 to 9.2 vs a range from 5.8 to 8.0). Fig. 2.19C shows the magnitude–dip angle relationship. The tsunamigenic models have gentler dip angles (typically less than 40 degrees, except for the outer-rise normal faulting events).

Subsequently, the selected 226 finite-fault models were uniformly and systematically analyzed by following the procedure outlined in Section 2.4.2, and the estimated earthquake source parameters were used to develop scaling relationships. For each of the eight source parameters, the dependency of the parameters on M_w was examined. When a clear dependency on M_w was observed, regression analysis was carried out by considering the following functional form

$$\log_{10}\theta = I_T(a_T + b_T M_w) + I_{NT}(a_{NT} + b_{NT} M_w) + \sigma\varepsilon \tag{2.20}$$

where θ is the source parameter of interest (e.g., L and W); a_T, b_T, a_{NT}, and b_{NT} are regression parameters for tsunamigenic (T) and nontsunamigenic (NT) models; σ is the standard deviation of regression residuals; ε is the standard normal variable (i.e., zero mean and unit standard deviation) and represents the randomness associated with the developed equation; and I_T and I_{NT} are the indicator variables and take a value of 1 when the model is classified as tsunamigenic and nontsunamigenic, respectively (otherwise zero).

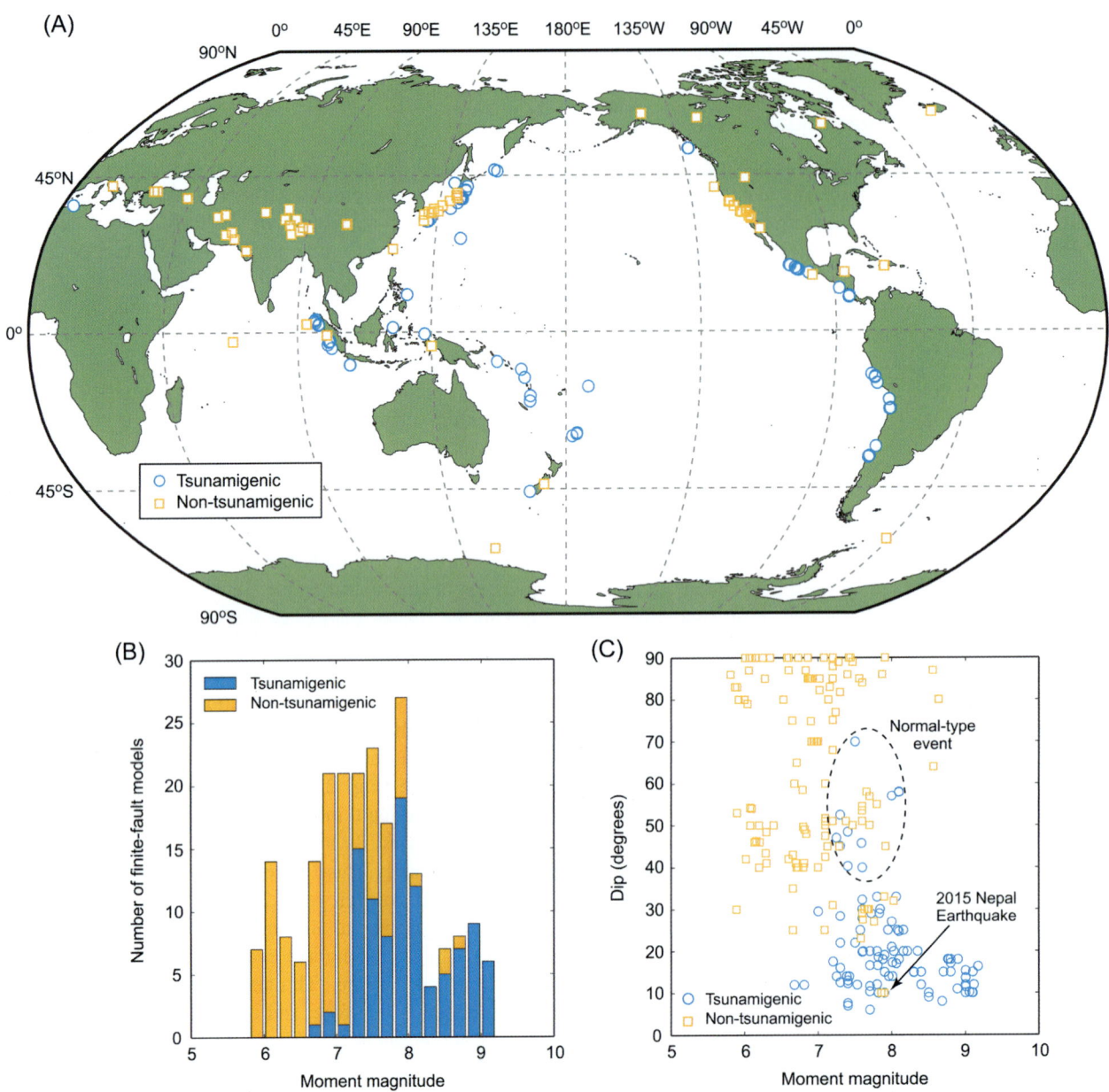

FIGURE 2.19 Characteristics of the finite-fault models used by Goda et al. (2016) for developing statistical scaling relationships of the earthquake source parameters: (A) locations, (B) histogram in terms of moment magnitude, and (C) moment magnitude–dip angle relationship. Map lines delineate study areas and do not necessarily depict accepted national boundaries.

For the earthquake source parameters, L, W, S_a, S_m, CL_x, and CL_z, the magnitude dependency was considered in the scaling relationships, and the equations were obtained as follows

$$\log_{10}L = I_T(-1.5021 + 0.4669M_w) + I_{NT}(-2.1621 + 0.5493M_w) + 0.1717\varepsilon_L \qquad (2.21)$$

$$\log_{10}W = I_T(-0.4877 + 0.3125M_w) + I_{NT}(-0.6892 + 0.2893M_w) + 0.1464\varepsilon_W \qquad (2.22)$$

$$\log_{10}S_a = I_T(-5.7933 + 0.7420M_w) + I_{NT}(-4.3611 + 0.6238M_w) + 0.2502\varepsilon_{Sa} \qquad (2.23)$$

$$\log_{10}S_m = I_T(-4.5761 + 0.6681M_w) + I_{NT}(-3.7393 + 0.6151M_w) + 0.2249\varepsilon_{Sm} \qquad (2.24)$$

$$\log_{10}CL_x = I_T(-1.9844 + 0.4520M_w) + I_{NT}(-2.4664 + 0.5113M_w) + 0.2204\varepsilon_{CLx} \qquad (2.25)$$

$$\log_{10}CL_z = I_T(-1.0644 + 0.3093M_w) + I_{NT}(-1.3350 + 0.3033M_w) + 0.1592\varepsilon_{CLz} \qquad (2.26)$$

The developed scaling relationships are statistical prediction models because the epsilon terms in Eqs. (2.21)–(2.26) capture the randomness of the predictive relationships. Since epsilon terms for different source parameters (e.g., ε_W and ε_{CLz}) are correlated, the linear correlation coefficients of the epsilon terms were obtained as shown in Table 2.2. When the correlation of the regression residuals is considered, the scaling relationships can be implemented as multivariate prediction models of the source parameters. In contrast to the preceding six earthquake source parameters, no clear dependency on M_w was observed for λ_{BC} and H, and these parameters are characterized as independent random variables from other parameters. More specifically, λ_{BC} is a normal variable with a mean equal to 0.312 and a standard deviation equal to 0.278, whereas H takes a value of 0.99 with a probability of 0.43 and a normal variable with a mean equal to 0.714 and a standard deviation equal to 0.172 with a probability of 0.57. More details of the developed scaling relationships can be found in Goda et al. (2016).

To demonstrate the developed scaling relationships for the geometrical parameters, Fig. 2.20 shows the results for the fault width and the fault length as a function of M_w by distinguishing tsunamigenic and nontsunamigenic models. The M_w–L scaling behavior for the tsunamigenic models is similar to that for the nontsunamigenic models, whereas the M_w–W scaling behavior for the tsunamigenic models differs from that for the nontsunamigenic models. For the same M_w values, W for the tsunamigenic models is greater than W for the nontsunamigenic models. This is because the fault planes of the nontsunamigenic models are dipping more steeply than those of the tsunamigenic models (Fig. 2.19C). Therefore, the fault plane can be extended in the downdip direction. The downdip limit of the seismogenic zone is mainly controlled by the thermal condition of the subduction zone (Hyndman & Wang, 1995), noting that the temperature increases with the depth, thereby brittle rupture is less likely to occur. Fig. 2.20B clearly shows that the distinction

TABLE 2.2 Linear correlation coefficients of regression residuals of the source parameters L, W, S_a, S_m, CL_x, and CL_z (Goda et al., 2016).

Variable	ε_L	ε_W	ε_{Sa}	ε_{Sm}	ε_{CLx}	ε_{CLz}
ε_L	1.0	0.139	−0.595	−0.516	0.734	0.249
ε_W	0.139	1.0	−0.680	−0.545	0.035	0.826
ε_{Sa}	−0.595	−0.680	1.0	0.835	−0.374	−0.620
ε_{Sm}	−0.516	−0.545	0.835	1.0	−0.337	−0.564
ε_{CLx}	0.734	0.035	−0.374	−0.337	1.0	0.288
ε_{CLz}	0.249	0.826	−0.620	−0.564	0.288	1.0

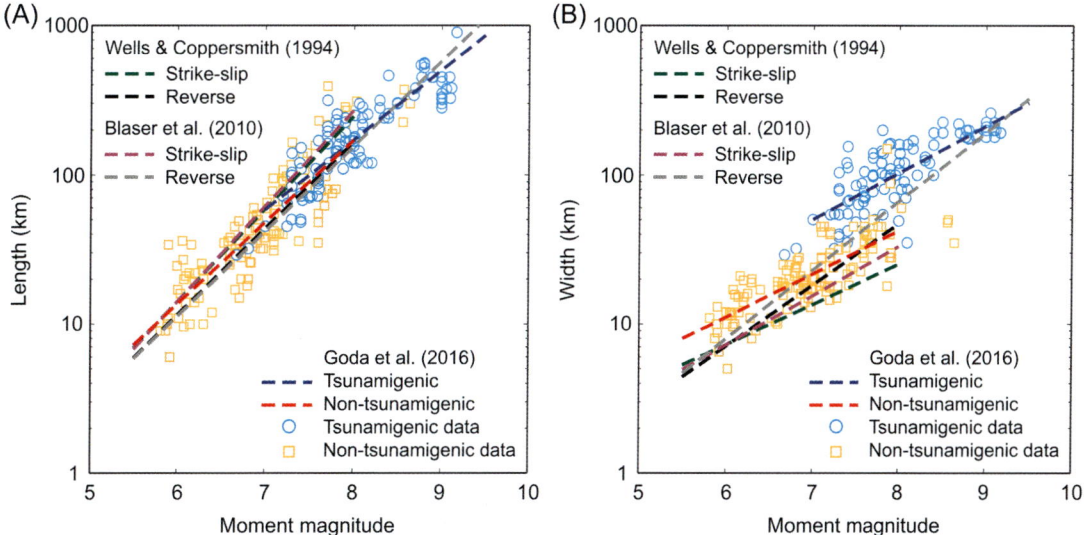

FIGURE 2.20 (A) Moment magnitude–fault length relationship and (B) moment magnitude–fault width relationship by Goda et al. (2016) in comparison with existing relationships by Wells and Coppersmith (1994) and Blaser et al. (2010).

of model types significantly improves the fitting performance of the developed scaling relationships for W. Note also that ε_L and ε_W are only weakly correlated (Table 2.2). To compare the scaling relationships for L and W by Goda et al. (2016) with other existing relationships, the equations by Wells and Coppersmith (1994) [WC94] and Blaser et al. (2010) [B10] are included in Fig. 2.20. The comparison of the scaling relationships for W indicates that the B10 reverse relationship is similar to the prediction model for all earthquake types, whereas the WC94 reverse relationship is consistent with the nontsunamigenic data and is similar to the nontsunamigenic model. On the other hand, existing relationships for L are consistent with the data and the scaling models by Goda et al. (2016).

Next, Fig. 2.21 shows the mean slip and the maximum slip as a function of M_w. The results for the M_w–S_a relationship clearly show that the data for the tsunamigenic and nontsunamigenic models are distributed differently; for the same M_w values, S_a for the tsunamigenic events is smaller than S_a for the nontsunamigenic models. The differences in the M_w–S_a scaling behavior for the two datasets can be attributed to differences in stress drop for these data (note: for a circular-crack model, mean slip is proportional to stress drop). The statistical analysis carried out by Allmann and Shearer (2009) indicates that the stress drop for strike–slip events is three to five times greater than other types of earthquakes and that the stress drop for intraplate events is two times greater than interplate events. The nontsunamigenic models include all strike–slip events and are of intraplate type, whereas the tsunamigenic models correspond to interplate events. Consequently, the stress drop for the nontsunamigenic models is significantly greater than the stress drop for the tsunamigenic models. Another contributing factor to the differences in S_a values is the fault area (= $L \times W$; Fig. 2.20). The M_w–W relationship for the tsunamigenic models is larger than that for the nontsunamigenic models. Thus the M_w–S_a relationship for the tsunamigenic models is smaller. The results for the M_w–S_m relationship also show different scaling behaviors for the tsunamigenic and nontsunamigenic models. To compare the scaling relationship for S_a developed by Goda et al. (2016) with other equations, the scaling relationship by Wells and Coppersmith (1994) for all event types and the scaling relationship by Murotani et al. (2013) [M13] are included in Fig. 2.21. The comparison between the WC94 relationship and the model for the nontsunamigenic events by Goda et al. (2016) is consistent, whereas the M13 relationship is in good agreement with the fitted model without model-type distinction.

As additional remarks on the scaling relationships of the earthquake source parameters, Goda et al. (2016) investigated the effects of the manipulation of the finite-fault models, noting that their final equations are based on the earthquake source parameters estimated by considering the trimming, interpolation, and tapering (Section 2.4.2). The sensitivity analysis carried out by Goda et al. (2016) suggests that the effects of the manipulations (i.e., tapering and interpolation) on the estimated spatial slip distribution parameters are complex and some models are affected more significantly than others. Their results indicate that overall, the scaling relationships of the correlation lengths for the adopted manipulation set-up are representative of several other analysis settings of the manipulation of the slip distribution. More detailed discussions can be found in Goda et al. (2016).

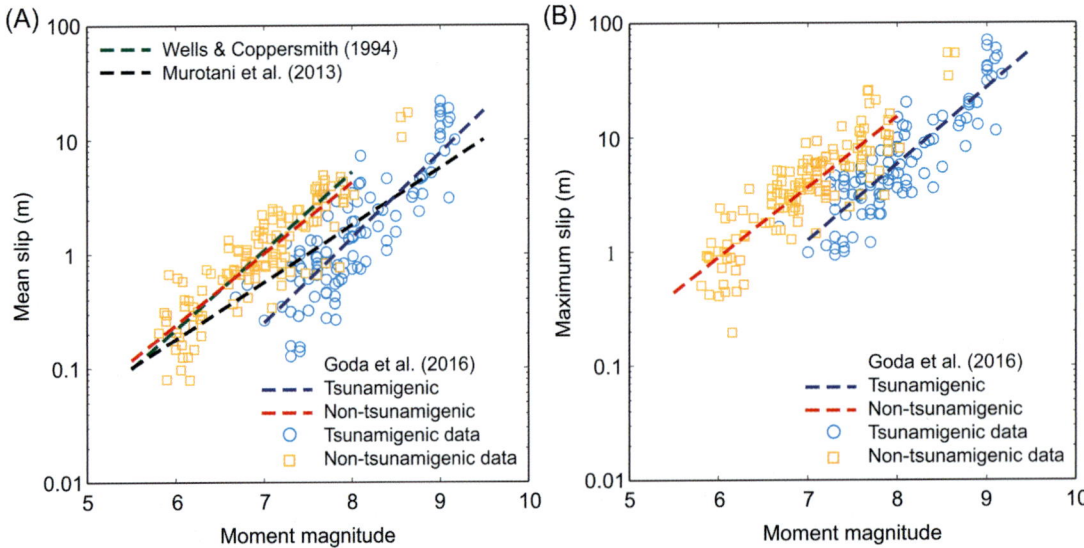

FIGURE 2.21 (A) Moment magnitude–mean slip relationship and (B) moment magnitude–maximum slip relationship by Goda et al. (2016) in comparison with existing relationships by Wells and Coppersmith (1994) and Murotani et al. (2013).

2.5 Source modeling for tsunamigenic earthquakes

Source modeling for tsunamigenic earthquakes requires various inputs from geological and seismological data and models. First, the tsunamigenic source region needs to be specified and characterized as an area where such earthquakes can occur (Section 2.2). Depending on available information in the target region, different approaches will need to be taken in specifying tsunami sources for PTHA and PTRA. With the delineated tsunami source region, one must decide how the uncertainty of the earthquake rupture characteristics is represented in PTHA and PTRA (Chapter 5). For instance, the variable fault plane geometry and earthquake slip distribution can be modeled using a logic tree approach (Fukutani et al., 2015), which consists of discrete sets of geometrical parameters and characteristic slip models. Alternatively, the same uncertainty can be modeled using scaling relationships of earthquake source parameters (Section 2.4.3). When the latter approach is taken, it is important to account for the variability of earthquake slip distribution for future events. As demonstrated in Section 2.4.1, even for well-recorded historical tsunami events, the earthquake rupture modeling involves significant variability (Fig. 2.14). To capture this uncertainty, earthquake slip random fields can be synthesized (Geist, 2002; Løvholt et al., 2012; Mueller et al., 2015). Subsequently, for each specified/synthesized slip distribution, the initial tsunami conditions need to be evaluated using analytical formulae or numerical models. Typically, coseismic ground deformation due to an earthquake rupture is calculated using methods proposed by Okada (1985) and Tanioka and Satake (1996), and the computed deformation is used as input initial boundary conditions for tsunami propagation and inundation simulation (Chapter 3), assuming the incompressibility of water. To account for the hydrodynamic response of water, appropriate filters, such as Kajiura (1963), or nonhydrostatic equations for tsunami propagation and inundation simulation need to be implemented (Løvholt et al., 2012; Yamazaki et al., 2023).

Section 2.5.1 summarizes different approaches that can be adopted for tsunami source modeling. The selection of the approach depends on the amount of information that is available for the subduction region of interest. Knowledge of seismotectonic features and historical events helps choose the approach. The tsunami source may be modeled by a set of finite-fault models that capture the geometrical and rupture characteristics using a logic tree. Adopting a more numerically extensive approach, a stochastic source modeling approach can be implemented.

Section 2.5.2 presents a synthesis method for generating stochastic source models for a given seismotectonic region. For this purpose, earthquake source parameters are generated from applicable scaling relationships. The stochastic methods are implemented to generate a second-order stationary random field whose covariance structure is defined by the von Kármán model (Mai & Beroza, 2002; Pardo-Iguzquiza & Chica-Olmo, 1993). Moreover, Section 2.5.3 introduces a method to simulate the kinematic earthquake rupture models. These methods are demonstrated for the finite-fault model for the 2005 Sumatra Earthquake, developed by Shao and Ji (2005).

Section 2.5.4 presents the analytical method of computing horizontal and vertical deformations of seafloor due to earthquake fault rupture based on Okada (1985). To account for the effects of horizontal deformation of steeply inclined slopes underwater, the method by Tanioka and Satake (1996) can be considered. Moreover, to account for the hydrodynamic response of seawater, smoothing spatial filters, such as Kajiura (1963), can be considered. This section aims to demonstrate the essential calculation steps to generate the input water surface conditions for the numerical simulation of tsunamis (Chapter 3).

2.5.1 Characterization of seismic tsunami sources

There are broadly four approaches to specifying seismic tsunami sources for PTHA and PTRA, with increasing complexity and necessary information. Key features of these approaches are summarized in this section.

The first approach is to define a (usually rectangular) source region within a target subduction zone at convergent plate boundaries. The fault plane size, strike, dip, rake, and (uniform) slip (Fig. 2.12) are determined on the basis of a general seismotectonic constraint. The geometry and average slip should be consistent with the specified moment magnitude (Eq. 2.2). Most simply, the variation of the moment magnitude for a specified fault rupture area can be considered by varying the average slip. By adopting an earthquake scaling law based on a circular-crack model (Eshelby, 1957), the mean slip S_a can be related to the static stress drop $\Delta\sigma$

$$\Delta\sigma = \frac{7}{16}\frac{M_o}{(A_f/\pi)^{1.5}} = \frac{7\pi^{1.5}}{16}\frac{\mu_S A_f S_a}{A_f^{1.5}} = \frac{7\pi^{1.5}}{16}\frac{\mu_S S_a}{A_f^{0.5}} \qquad (2.27)$$

Empirical constraints of the static stress drop (Allmann & Shearer, 2009) can be used to define a possible variation of the mean slip. For instance, Central Disaster Management Council (2012) reported $\Delta\sigma = 1.2$ MPa as a mean value, and $\Delta\sigma = 0.7$ and 2.2 MPa as mean minus/plus one standard deviation by analyzing six global megathrust subduction events (in total, 12 finite-fault models). Alternatively, the scaling relationship for the mean slip (Section 2.4.3) can be used. Thereby, for different magnitude values, different fault rupture planes and corresponding mean slips (with some variations) can be used. These individual models can be integrated into a logic tree to account for the uncertainty in tsunami source modeling.

The second approach is to utilize inverted finite-fault models from historical tsunami events (Section 2.4.1). In this approach, a heterogeneous (deterministic) earthquake slip distribution can be specified. Some variations of the earthquake slip can be considered by changing the static stress drop (as in the first approach above) and the earthquake slip distribution can be multiplied by the determined factors. Caution needs to be exercised in representing the uncertainty associated with the earthquake slip distribution following this approach, because a small set of finite-fault models from historical events may not capture the uncertainty of the future earthquake rupture fully.

The third approach is to develop a set of characteristic tsunami source models by reflecting various types of data and models from historical events. In this development process, existing earthquake scaling relationships and other region-specific characteristics are incorporated, involving a group of experts. The developed tsunami source models typically have heterogeneous earthquake slip distributions and accommodate several earthquake rupture patterns. These rupture patterns can be specified as single-asperity, multiple-asperity, trench-rupture, and splay-rupture models (Gao et al., 2018).

To illustrate such characteristic tsunami source scenarios, tsunami rupture models for the megathrust rupture scenarios in the Nankai–Tonankai Trough region can be examined. The Nankai–Tonankai Trough accommodates tectonic movements between the Philippines Sea and the Eurasian Plates. The accumulated strain is released periodically during major subduction earthquakes. Since 1700, five M_w 8+ earthquakes happened (Ando, 1975). The most recent ones were the 1944 M_w 8.1 Tonankai Earthquake and the 1946 M_w 8.4 Nankai Earthquake. In 1854 large dual earthquakes (Ansei Earthquakes, M_w 8.4 and M_w 8.5) occurred in sequence, separated by 32 hours. In 1707 the M_w 8.7+ Hoei Earthquake ruptured the entire Nankai–Tonankai region. The recurrence interval of large subduction events in the Nankai–Tonankai Trough varies from 90 years to 264 years (Ando, 1975). In light of uncertainty, Central Disaster Management Council (2012) developed multiple characteristic tsunami source models for the M_w 9.1 megathrust subduction scenarios in the Nankai–Tonankai Trough. Nine tsunami source models are shown in Fig. 2.22. The models consider single-asperity rupture patterns (Cases 1–5) and two-asperity rupture patterns (Cases 6–9). For the single-asperity models, the large slip areas (i.e., asperity) shift from East to West along the trench line. The locations of the asperities correspond to the five major seismotectonic segments, that is, Tokai, Tonankai, Kii Peninsula, Shikoku, and Kyushu, identified for the Nankai–Tonankai Trough. On the other hand, for the two-asperity models, two large slip concentrations exist in different segments of the Nankai–Tonankai Trough. In specifying the average slip over the fault plane, the scaling law in Eq. (2.27) was used with an upper bound estimate of the static stress drop of 3.0 MPa. In the source models, large-slip areas take up about 20% of the entire fault plane with an average slip value twice as large as the average slip over the fault plane and are positioned at depths shallower than 20 km. On the other hand, very large-slip areas take up about 5% of the entire fault plane with an average slip-value four times as large as the average slip over the fault plane. It is noteworthy that the kinematic earthquake slip distributions are implemented for the tsunami source models by Central Disaster Management Council (2012) by considering the rupture propagation velocity of 2.5 km/second and the subfault rise time of 60 seconds.

The fourth approach is the stochastic source method (Geist, 2002; Goda et al., 2016; Løvholt et al., 2012; Mueller et al., 2015), which captures the uncertainty in geometry as well as earthquake slip distribution by implementing statistical scaling relationships of earthquake source parameters and synthesizing random-field earthquake distributions. The method allows the incorporation of the seismotectonic characteristics of the subduction region by specifying the desirable features of the synthesized stochastic source models, such as asperity regions for the overall earthquake slip concentrations. This approach is illustrated in detail in Section 2.5.2.

The last remark of this section is that the uniform earthquake slip model, which has been adopted in past tsunami hazard studies, tends to underestimate the regional tsunami hazard significantly. Therefore, when PTHA and PTRA are conducted using the first tsunami source modeling approach, the final tsunami hazard values tend to be significantly lower than those evaluated using the heterogeneous slip distributions for the same scenario magnitude. For example, Melgar et al. (2019) reported that the differences between the uniform slip model and the heterogenous slip model are a factor of 2 or greater. Another important aspect that is not fully discussed in this section is the existence of other earthquake rupture mechanisms that generate tsunamis more effectively, such as splay faulting that has steeper dip angles, thereby more effective in causing larger vertical deformation for the same amount of earthquake slip (Chapter 9). A recent study for the Makran subduction zone off Iran and Pakistan by Momeni et al. (2023) demonstrated that the effects of splay fault rupture could result in increased tsunami wave heights by 3–9 times, compared to ruptures exclusively on the plate boundary.

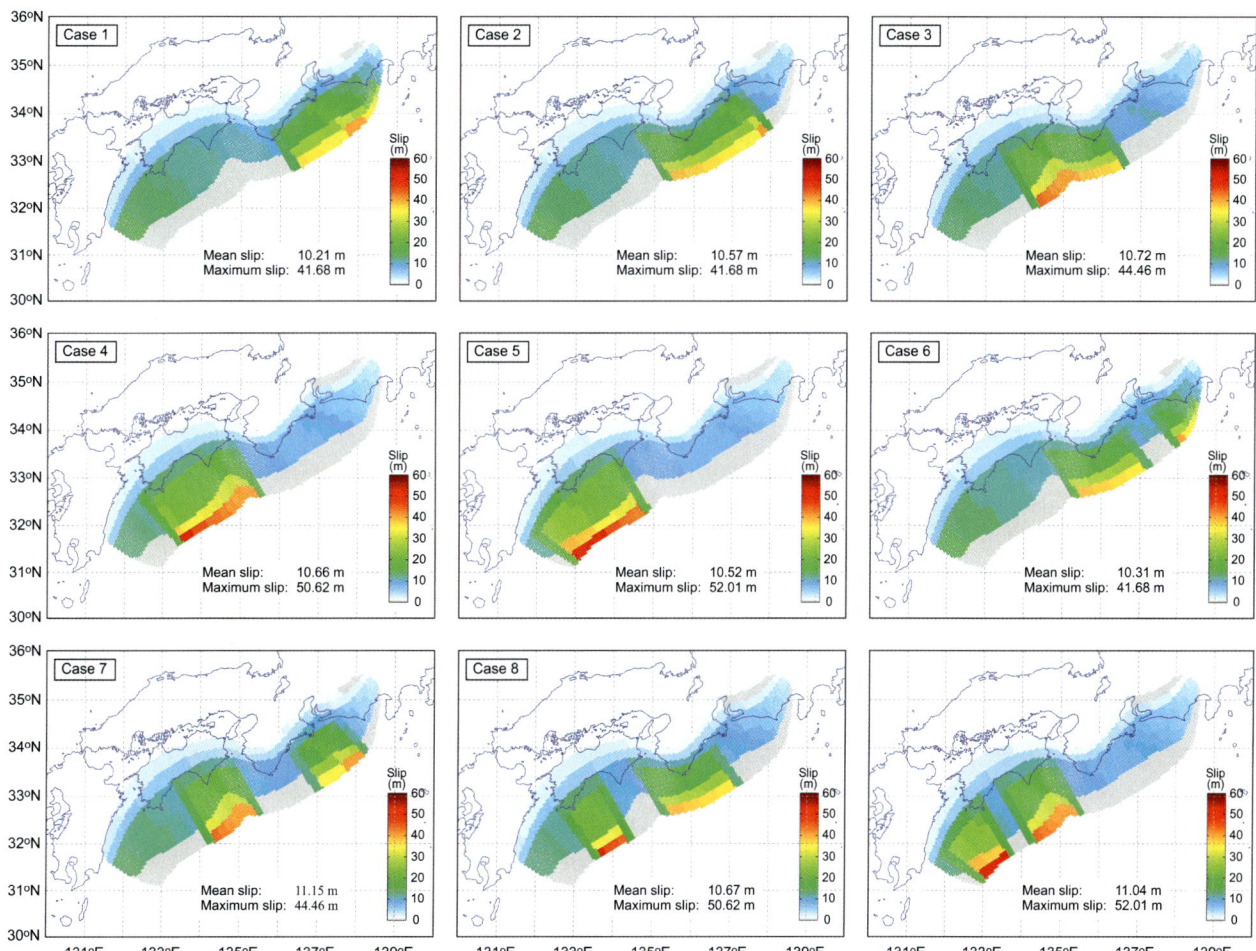

FIGURE 2.22 Characteristic tsunami source models for the Nankai−Tonankai Trough megathrust scenarios developed by Central Disaster Management Council (2012).

2.5.2 Stochastic tsunami source modeling

In this section, a numerical procedure of stochastic tsunami source modeling (i.e., the fourth tsunami source modeling approach in Section 2.5.1) is illustrated. The explanations follow the procedure outlined in Goda et al. (2016). The main procedure consists of five steps.

Step 1: Define an earthquake scenario of interest by specifying the moment magnitude, source region/fault model, and asperity zone. The asperity zone can be used to constrain the desirable patterns of the earthquake slip distribution, such as more concentrated slip distribution along the trench. The specified scenario parameters essentially reflect the seismological knowledge of earthquake rupture in the target region.

Step 2: Generate random samples of earthquake source parameters, that is, L, W, S_a, S_m, λ_{BC}, CL_x, CL_z, and H, from suitable statistical scaling relationships. In the case of Goda et al. (2016) (Section 2.4.3), Eqs. (2.21)−(2.26) can be used. The epsilon terms in Eqs. (2.21)−(2.26) should be simulated as correlated normal variables (in the logarithmic space) by using the correlation coefficient matrix shown in Table 2.2.

Step 3: Using the simulated spatial slip distribution parameters (i.e., CL_x, CL_z, and H), a random slip field is generated using a Fourier integral method (Mai & Beroza, 2002; Pardo-Iguzquiza & Chica-Olmo, 1993). The amplitude spectrum is specified by the von Kármán model (Eq. 2.19), whereas the phase spectrum is randomly generated by sampling values between 0 and 2π. The constructed matrix of complex Fourier coefficients is transformed into the spatial domain via a two-dimensional inverse FFT. To achieve slip distribution with realistic skewness, the synthesized slip distribution is converted via Box-Cox transformation (Goda et al., 2014). The transformed slip distribution is tapered to avoid large slip values at the edges of the earthquake slip distribution and is adjusted to achieve the target mean slip S_a. Then the position of the synthesized fault plane is determined randomly within the overall fault plane (note: when the scenario magnitude is not very large with respect to the overall fault plane area, the synthesized earthquake slip distribution will not fill the entire fault plane area).

Step 4: The candidate earthquake slip distribution generated in *Step 3* is examined for its suitability as one of the final accepted stochastic source models. Typically, several criteria are implemented, including the earthquake magnitude check (i.e., consistency between the simulated values of L, W, and S_a and the target seismic moment), the maximum slip check (i.e., rejecting a slip distribution with very large slip values exceeding the specified slip limit), the asperity area ratio check (i.e., the proportion of the subfault areas with 1.5 times the mean slip with respect to the simulated fault area falls between 0.2 and 0.3 (Murotani et al., 2013)), and the slip concentration check (i.e., a certain proportion of the slips is concentrated in the specified area of the fault plane).

Step 5: If the simulated earthquake slip distribution does not meet all the criteria set in *Step 4*, the candidate earthquake slip distribution is discarded and a new candidate stochastic source model is generated by returning to *Step 2*. This process is repeated until the required number of acceptable stochastic source models is synthesized.

To illustrate the abovementioned five-step procedure, a numerical example is discussed below by focusing on the 2005 Sumatra Earthquake. A computer code for this analysis is provided in Example 2-5 (Appendix).

For *Step 1* of the stochastic source modeling procedure, a suitable fault plane is specified. For the 2005 Sumatra Earthquake, three finite-fault models are available (Ji, 2005; Konca et al., 2007; Shao & Ji, 2005) and are shown in Fig. 2.23. The finite-fault model by Shao and Ji (2005) is adopted to define the overall fault plane in this illustration. The fault plane model by Shao and Ji (2005) consists of 19 by 13 subfaults, whose size is 20 km by 20 km (i.e., $L = 380$ km and $W = 260$ km). The strike, dip, and rake angles are 326, 8, and 117 degrees, respectively. The scenario magnitudes that are considered for this illustration range between M_w 8.5 and M_w 8.7, which are consistent with the existing finite-fault models. To place a broad constraint on the synthesized earthquake slip distributions, the asperity region is specified as shaded subfaults in Fig. 2.23D. The specified slip concentration proportions for the asperity area are between 60% and 90%, which is similar to the slip concentrations of the existing finite-fault models. Note that these settings are adopted for the demonstration of the stochastic source modeling only.

For *Steps 2–4* of the stochastic source modeling procedure, the constrained earthquake slip random fields are generated iteratively, and their suitability is evaluated by considering the earthquake magnitude check (falling between M_w 8.5 and M_w 8.7), the maximum slip check ($S_m < 25$ m), the asperity area ratio check (falling between 0.2 and 0.3), and the slip concentration check (falling between 60% and 90%). One of the accepted stochastic source models is shown in Fig. 2.24. The moment magnitude is 8.60, and the length and width are 340 and 220 km, respectively. The mean and maximum slips are obtained as 4.12 and 18.32 m, respectively.

By repeating the stochastic synthesis procedure for 100 acceptable models, earthquake source characteristics of the synthesized source models can be examined, in comparison with the earthquake scaling relationships. The results are shown in Fig. 2.25. The 100 data points from the synthesized stochastic source models are consistent with the scaling relationships by Goda et al. (2016). This overall agreement of the stochastic source models and the underlying statistical relationships that are derived from the past finite-fault models (Section 2.4.3) is an important feature of stochastic source modeling.

It is insightful to examine the individual synthesized earthquake slip distributions. For this purpose, three realizations of the accepted stochastic source models are shown in Fig. 2.26A–C, whereas the average slip distribution of the 100 accepted stochastic source models is shown in Fig. 2.26D. The three realizations show that individual stochastic source models have different earthquake slip patterns with asperities at different locations within the overall fault plane and their intensities are also different. Some realizations have multiple asperities. When all stochastic source models are averaged, the consistent tendency with the predefined asperity zone (Fig. 2.23D) becomes obvious and its intensity agrees with the underlying scaling relationship for the mean slip (Fig. 2.25C). In short, stochastic source modeling provides an effective approach to generating numerous slip random fields, whose spatial distributions exhibit different features but whose ensemble characteristics are consistent with the underlying seismological data and models.

2.5.3 Kinematic stochastic tsunami source modeling

The preceding section focused on the stochastic slip synthesis of instantaneous earthquake slip distributions. This procedure can be extended to kinematic earthquake rupture models by generating additional source parameters. More specifically, *Step 2* of the abovementioned procedure can include the random sampling of the subfault rise time and the

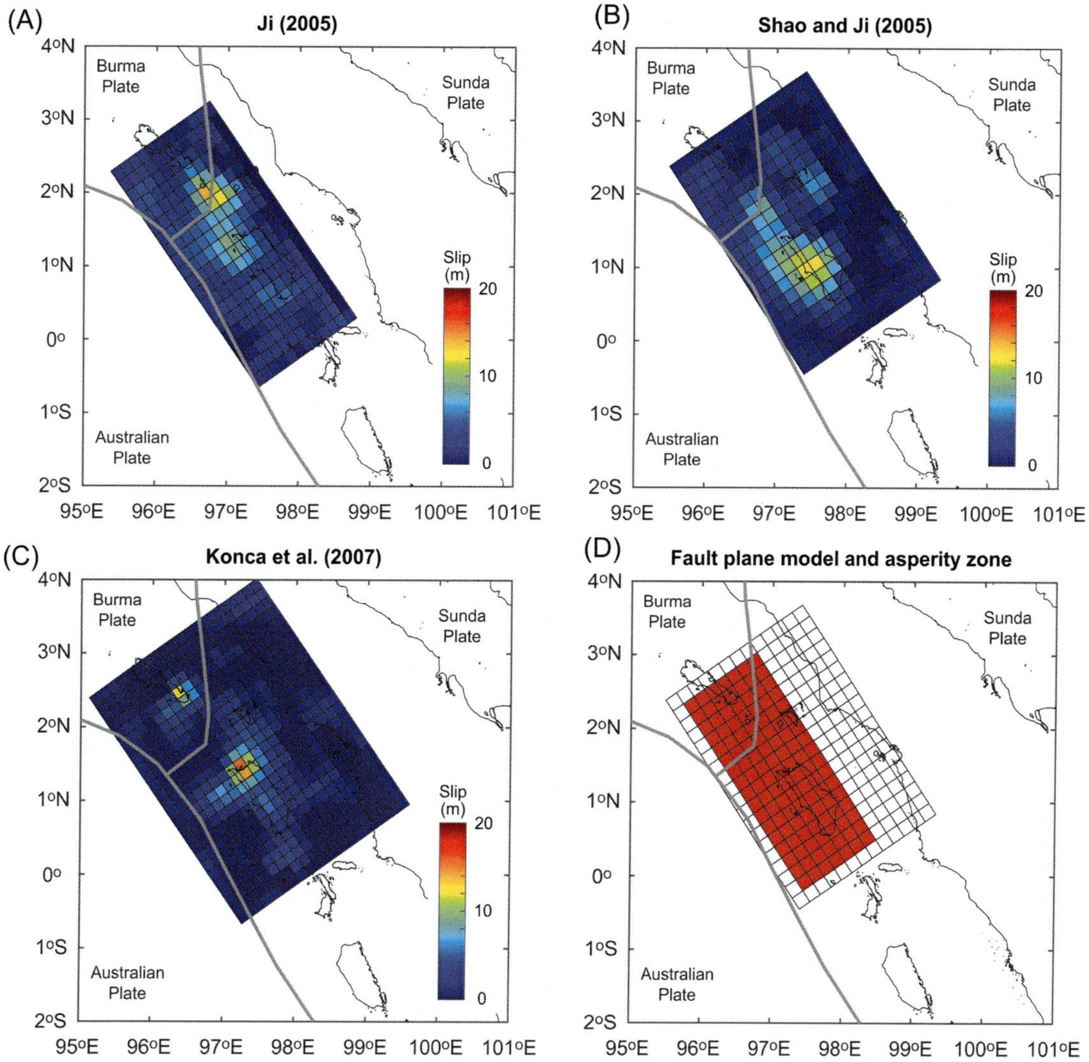

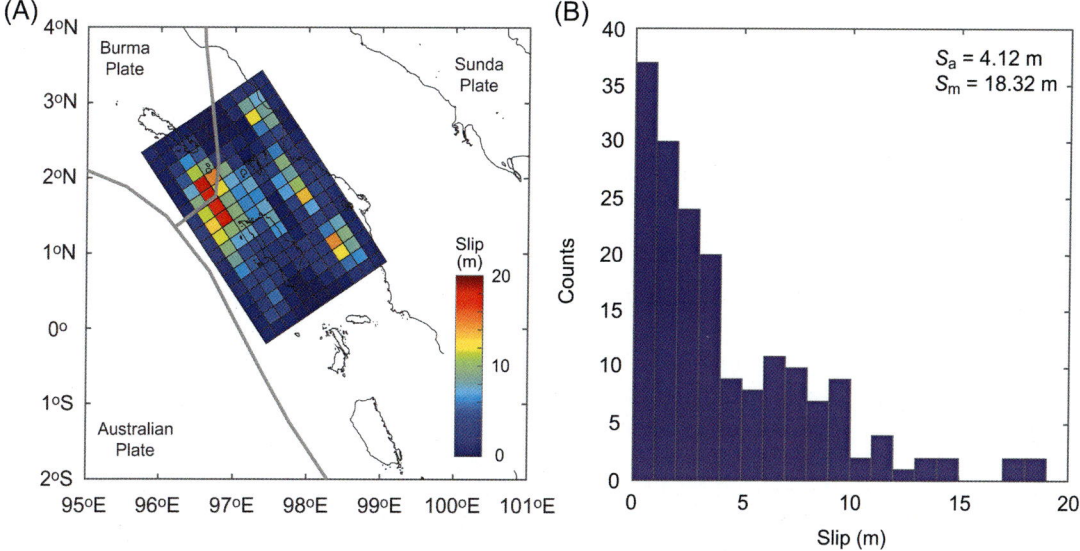

FIGURE 2.23 Finite-fault models for the 2005 Sumatra Earthquake by (A) Ji (2005), (B) Shao and Ji (2005), and (C) Konca et al. (2007). (D) Fault plane model and asperity zone (shaded subfaults) for the stochastic source modeling demonstration.

FIGURE 2.24 (A) Acceptable earthquake source model and (B) histogram of the simulated earthquake slips.

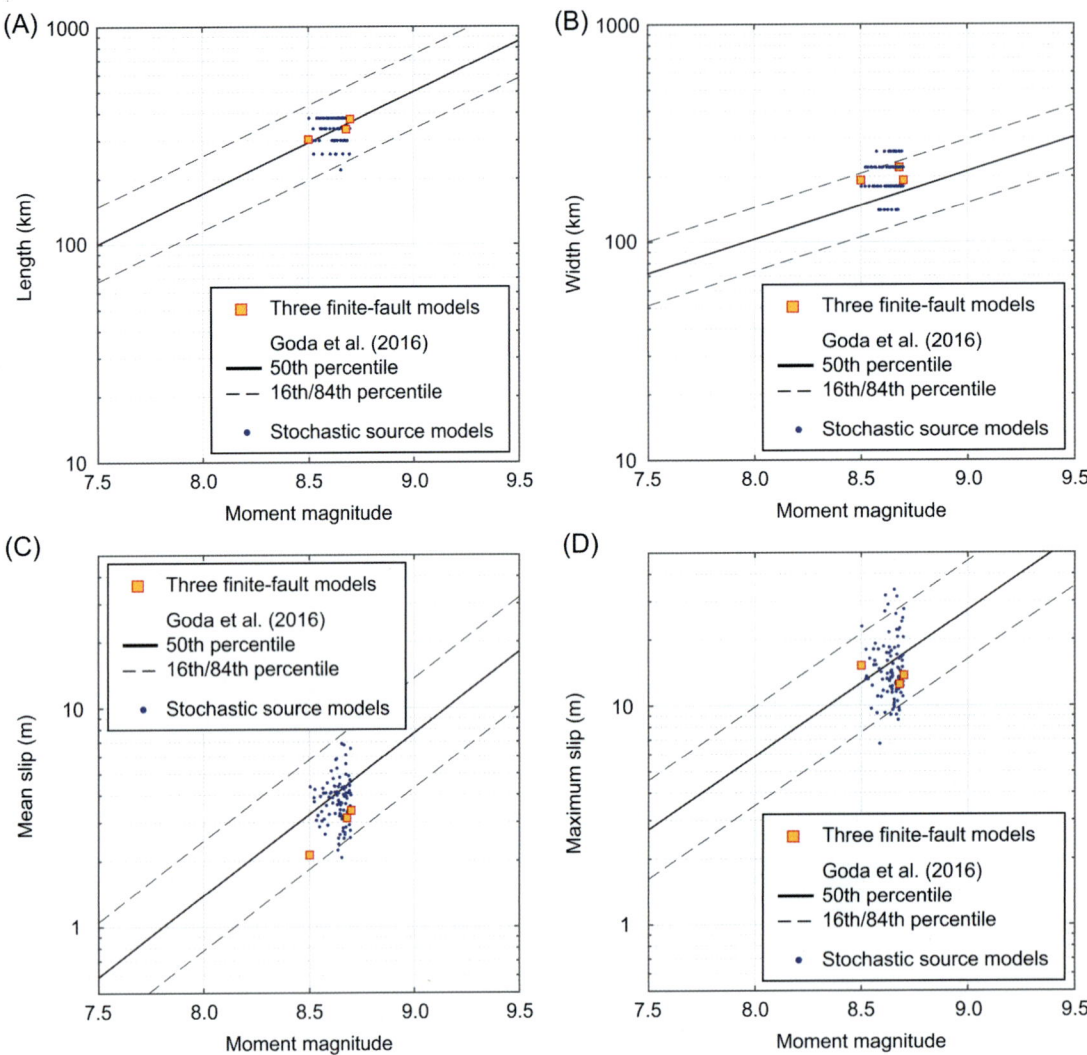

FIGURE 2.25 Comparison of the earthquake source parameters of the 100 synthesized earthquake source models with the corresponding scaling relationships by Goda et al. (2016): (A) fault length, (B) fault width, (C) mean slip, and (D) maximum slip.

rupture propagation speed. For this purpose, the statistical relationship for the rise time proposed by Melgar and Hayes (2017) can be adopted

$$\log_{10}(rise\ time) = -5.323 + 0.293\log_{10}M_0 \qquad (2.28)$$

The result for the rupture propagation speed shown by Melgar and Hayes (2017) indicates that this parameter is not dependent on seismic moment and typically distributed between 2.0 and 3.0 km/second. The rupture propagation speed can be approximated by a truncated normal variable with mean = 2.5 km/second, standard deviation = 0.5 km/second, and lower and upper limits of 1.5 and 3.5 km/second.

To obtain the kinematic earthquake rupture profiles (e.g., Fig. 2.15), the rupture initiation/nucleation point needs to be determined. Mai et al. (2005) investigated the location of the hypocenter (which can be considered a rupture initiation point) with respect to the fault geometry and the slip distribution and developed statistical models. The method by Mai et al. (2005) for generating a random location of hypocenter based on an earthquake source model is illustrated in Fig. 2.27. This example is based on realization 1 of the stochastic source model for the 2005 Sumatra Earthquake (Fig. 2.26A). This slip distribution is also shown in Fig. 2.27A and will be used as part of the simulation process of a hypocenter. The preliminary weighting functions for hypocenter locations are specified on the basis of the fault dimensions and the mean/maximum slip ratios (Fig. 2.27B). This weighting function reflects the higher chance of earthquake nucleation at the center of the fault plane. Subsequently, further constraints are placed to exclude unlikely locations for

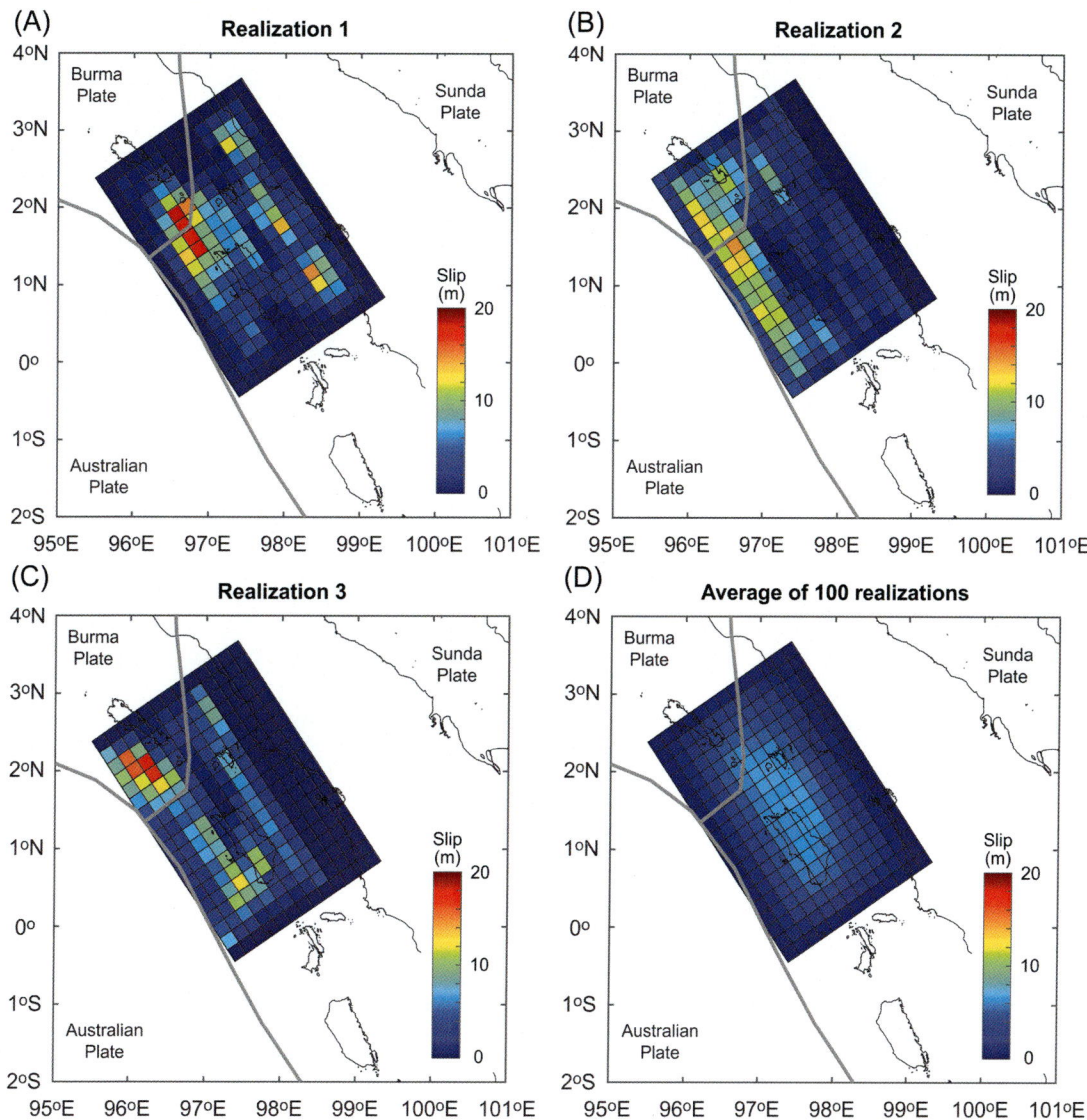

FIGURE 2.26 (A–C) Three realizations of the accepted stochastic source models and (D) average earthquake slip distribution based on the 100 accepted realizations.

a given slip distribution, using empirical findings and constraints (Fig. 2.27C; the spatial pattern of Fig. 2.27C resembles that of Fig. 2.27A). By multiplying the preliminary weighting functions by the empirical weighting constraints, the final weighting function for hypocenter locations is obtained (Fig. 2.27D). From this two-dimensional weighting function, the location of a hypocenter can be determined on the fault plane (see a gray circle in Fig. 2.27D). Subsequently, using the simulated rupture propagation speed and the rise time, the kinematic rupture models can be obtained.

2.5.4 Seafloor displacement due to earthquake rupture

The last step before performing tsunami propagation and inundation simulations (Chapter 3) is to compute the initial water surface condition due to an earthquake rupture. When the earthquake slip distribution is specified (instantaneous and kinematic ruptures for Sections 2.5.2 and 2.5.3, respectively), the most popular approach is to use the analytical formulae provided by Okada (1985). The theoretical basis of the Okada formulae is the elastic theory of dislocation for a semiinfinite elastic medium (Steketee, 1958). The main reasons for this popularity are twofold: the theoretical soundness of the method and the computational efficiency. As shown in the paragraphs below, the analytical formulae have

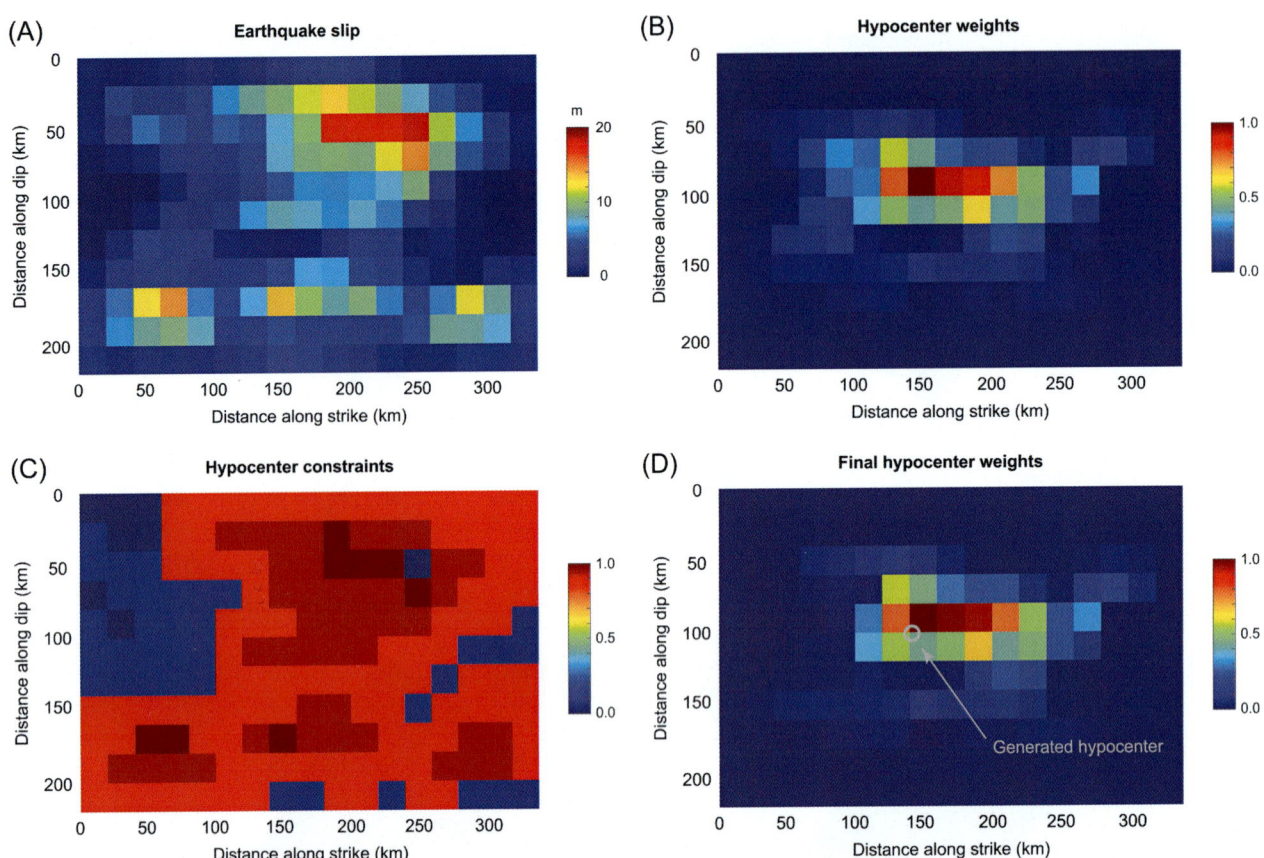

FIGURE 2.27 Generation of hypocenter locations (the color range is between 0 and 1): (A) synthesize slip distribution, (B) preliminary weighting function for hypocenter locations; (C) constraints based on asperity areas; and (D) final weighting function for hypocenter locations.

been used to calculate the seafloor deformation due to an earthquake rupture by considering an elastic homogeneous half-space. Because of the assumed elasticity of the medium, the displacements at surface grid locations can be calculated by summing the contributions from all subfault ruptures. Compared with more rigorous but computationally expensive finite element modeling (Grilli et al., 2013), the use of Okada formulae substantially reduces the computational efforts and required information on subsurface properties. These advantages of the reduced computation are critical when numerous earthquake rupture cases need to be performed over a very large subduction region. In this section, the method based on Okada (1985) is focused upon.

For a rectangular finite-fault source (with the fault length L, the fault width W, the fault depth d, and the dip angle δ), the following equations can be used to compute the displacement field (u_x, u_y, and u_z) at an arbitrary point on the surface or inside the medium due to the dislocations U_1, U_2, and U_3. The three dislocation quantities correspond to the strike–slip, dip–slip, and tensile components of the dislocation (i.e., the movement of the hanging wall relative to the footwall). For the calculations of the displacements at the point of interest, the directions x, y, and z are defined as the fault strike orientation, the orientation perpendicular to the fault strike, and the vertical orientation, respectively. For strike–slip, the displacements can be calculated as follows

$$\begin{cases} u_x = -\frac{U_1}{2\pi}\left[\frac{\xi q}{R(R+\eta)} + \tan^{-1}\left(\frac{\xi \eta}{qR}\right) + I_1 \sin\delta\right]\Big|_C \\ u_y = -\frac{U_1}{2\pi}\left[\frac{\tilde{y} q}{R(R+\eta)} + \left(\frac{q\cos\delta}{R+\eta}\right) + I_2 \sin\delta\right]\Big|_C \\ u_z = -\frac{U_1}{2\pi}\left[\frac{\tilde{d} q}{R(R+\eta)} + \left(\frac{q\sin\delta}{R+\eta}\right) + I_4 \sin\delta\right]\Big|_C \end{cases} \quad (2.29)$$

For dip−slip, the displacements can be calculated as follows

$$\begin{cases} u_x = -\dfrac{U_2}{2\pi}\left[\dfrac{q}{R} - I_3\sin\delta\cos\delta\right]\Big|_C \\ u_y = -\dfrac{U_2}{2\pi}\left[\dfrac{\tilde{y}q}{R(R+\xi)} + \cos\delta\tan^{-1}\left(\dfrac{\xi\eta}{qR}\right) - I_1\sin\delta\cos\delta\right]\Big|_C \\ u_z = -\dfrac{U_2}{2\pi}\left[\dfrac{\tilde{d}q}{R(R+\xi)} + \sin\delta\tan^{-1}\left(\dfrac{\xi\eta}{qR}\right) - I_5\sin\delta\cos\delta\right]\Big|_C \end{cases} \quad (2.30)$$

For tensile−slip, the displacements can be calculated as follows

$$\begin{cases} u_x = \dfrac{U_3}{2\pi}\left[\dfrac{q^2}{R(R+\eta)} - I_3\sin^2\delta\right]\Big|_C \\ u_y = \dfrac{U_3}{2\pi}\left[\dfrac{-\tilde{d}q}{R(R+\xi)} - \sin\delta\left\{\dfrac{\xi q}{R(R+\eta)} - \tan^{-1}\left(\dfrac{\xi\eta}{qR}\right)\right\} - I_1\sin^2\delta\right]\Big|_C \\ u_z = \dfrac{U_3}{2\pi}\left[\dfrac{\tilde{y}q}{R(R+\xi)} + \cos\delta\left\{\dfrac{\xi q}{R(R+\eta)} - \tan^{-1}\left(\dfrac{\xi\eta}{qR}\right)\right\} - I_5\sin^2\delta\right]\Big|_C \end{cases} \quad (2.31)$$

In Eqs. (2.29)−(2.31), ξ and η are the variables that are used for integration over the fault dimensions, and Chinnery's notation $|_C$ is used to represent the substitution

$$f(\xi, \eta)|_C = f(x, p) - f(x, p - W) - f(x - L, p) + f(x - L, p - W) \quad (2.32)$$

The terms I_1 to I_5 used in Eqs. (2.29)−(2.31) are expressed as follows

$$\begin{cases} I_1 = \dfrac{\mu_L}{\lambda_L + \mu_L}\left[\dfrac{-1}{\cos\delta}\dfrac{\xi}{R+\tilde{d}}\right] - \dfrac{\sin\delta}{\cos\delta}I_5 \\ I_2 = \dfrac{\mu_L}{\lambda_L + \mu_L}[-\ln(R+\eta)] - I_3 \\ I_3 = \dfrac{\mu_L}{\lambda_L + \mu_L}\left[\dfrac{1}{\cos\delta}\dfrac{\tilde{y}}{R+\tilde{d}} - \ln(R+\eta)\right] + \dfrac{\sin\delta}{\cos\delta}I_4 \\ I_4 = \dfrac{\mu_L}{\lambda_L + \mu_L}\dfrac{1}{\cos\delta}[\ln(R+\tilde{d}) - \sin\delta\ln(R+\eta)] \\ I_5 = \dfrac{\mu_L}{\lambda_L + \mu_L}\dfrac{2}{\cos\delta}\tan^{-1}\left(\dfrac{\eta(X+q\cos\delta) + X(R+X)\sin\delta}{\xi(R+X)\cos\delta}\right) \end{cases} \quad (2.33)$$

in which μ_L and λ_L are Lame's constants. Note that when $\delta = 90$ degrees, Eq. (2.33) is modified to

$$\begin{cases} I_1 = -\dfrac{\mu_L}{2(\lambda_L + \mu_L)}\dfrac{\xi q}{(R+\tilde{d})^2} \\ I_3 = \dfrac{\mu_L}{2(\lambda_L + \mu_L)}\left[\dfrac{\eta}{R+\tilde{d}} + \dfrac{\tilde{y}q}{(R+\tilde{d})^2} - \ln(R+\eta)\right] \\ I_4 = -\dfrac{\mu_L}{\lambda_L + \mu_L}\dfrac{q}{R+\tilde{d}} \\ I_5 = -\dfrac{\mu_L}{\lambda_L + \mu_L}\dfrac{\xi\sin\delta}{R+\tilde{d}} \end{cases} \quad (2.34)$$

Other variables and parameters used in Eqs. (2.29)–(2.31) are defined as follows

$$\begin{cases} p = y\cos\delta + d\sin\delta \\ q = y\sin\delta - d\cos\delta \\ \tilde{y} = \eta\cos\delta + q\sin\delta \\ \tilde{d} = \eta\sin\delta - q\cos\delta \\ R^2 = \xi^2 + \eta^2 + q^2 = \xi^2 + \tilde{y}^2 + \tilde{d}^2 \\ X^2 = \xi^2 + q^2 \end{cases} \quad (2.35)$$

Computer codes for calculating the ground surface displacements due to an earthquake rupture are available. One of the useful resources is https://www.bosai.go.jp/e/dc3d.html.

Using the calculated deformation fields from the Okada formulae as initial boundary conditions of water surface in tsunami simulations is popular. However, this approach ignores the hydrodynamic response of seawater. The modification to the deformation fields, before the tsunami simulations, can be made by filtering out nonphysical sharp peaks of seabed deformation (Glimsdal et al., 2013; Kajiura, 1963; Løvholt et al., 2012). The displacement field of the water surface $\eta(x,y)$ at constant depth d can be given by

$$\eta(x,y) = d^{-2} \int_{-\infty}^{\infty} \int_{-\infty}^{\infty} D(x', y') G(|r - r'|/d) dx' dy' \quad (2.36)$$

where $D(x,y)$ is the uplift displacement field, and r is the position vector. The normalized Green function is given by

$$G(r) = \frac{1}{\pi} \sum_{n=0}^{\infty} \frac{(-1)^n (2n+1)}{\{(2n+1)^2 + r^2\}^{1.5}} \quad (2.37)$$

The $G(r)$ function decays rapidly with r. Eqs. (2.36) and (2.37) are often referred to as Kajiura filter. An alternative to the Kajiura filter is to apply a spatial smoothing filter with a selected filter length (e.g., mean and median filters). For instance, the mean filter was adopted by Central Disaster Management Council (2012) in the tsunami simulations for the Nankai–Tonankai Trough.

There is one more important remark on the use of Okada formulae for the calculations of initial boundary conditions of water surface in tsunami simulations. When the seafloor is flat, the horizontal deformation of the ocean bottom does not affect the vertical profile of the water surface, as illustrated in Fig. 2.28A. However, when the seafloor has steep gradients and the horizontal displacement due to earthquake rupture is relatively large, causing the slope to shift, this movement creates the vertical displacement of water due to the horizontal movement of the slope (Tanioka & Satake, 1996). Given the bathymetry of the region, the effects of the horizontal movement of a translated submerged slope on the vertical displacement of water can be calculated as

$$u_h = u_x \frac{\partial H}{\partial x} + u_y \frac{\partial H}{\partial y} \quad (2.38)$$

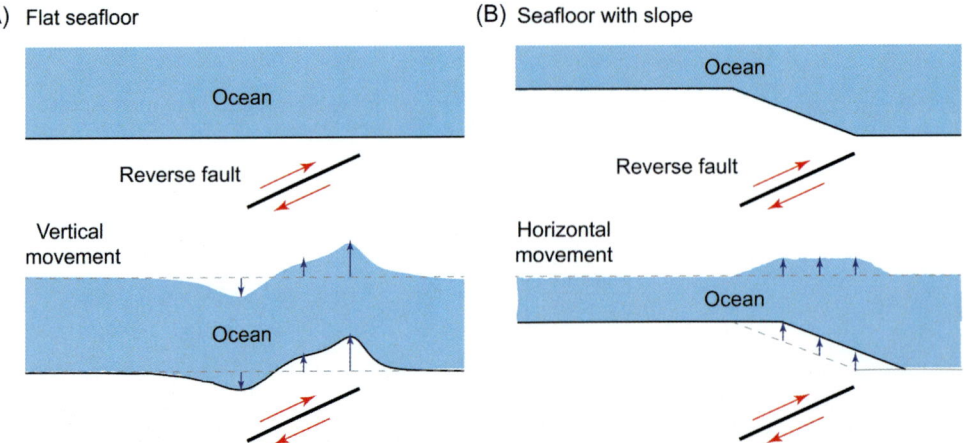

FIGURE 2.28 Tsunami initial conditions for reverse faulting earthquakes: (A) vertical displacement due to faulting with flat seafloor and (B) horizontal displacement with sloping seafloor. The top illustrates the ocean bottom before faulting and the bottom illustrates the ocean bottom after faulting.

This effect of horizontal displacement depends on the slopes of the bathymetry $\partial H/\partial x$ and $\partial H/\partial y$ along the two horizontal directions. This is illustrated in Fig. 2.28B.

To understand how the Okada formulae and the Kajiura filter work in calculating the ground displacement, a parametric example is useful. For this purpose, consider an earthquake rupture, having the following fault geometry and slip: fault length = 100 km, fault width = 50 km, strike = 180 degrees (i.e., north–south direction; origin is at the northern end), dip = 10 degrees, and a unit slip vector with a rake of 90 degrees. This can be regarded as a typical situation for low-angle reverse-faulting earthquakes in a subduction zone. The vertical deformation at numerous locations around the fault plane is evaluated using Okada equations. Fig. 2.29 shows two cases: one with rupture to the seabed (H_{top} = 0 km) and the other with a buried fault plane (H_{top} = 5 km). The contour plot shows the spatial distribution of vertical deformation, and the figure panel below the contour shows three cross-sectional profiles of vertical deformation perpendicular to the fault strike. The vertical deformation is greatest at the center of the top edge of the fault plane (cross-section 1), and it decreases sharply at both ends of the fault length (cross-sections 2 and 3). A remarkable difference between the cases with H_{top} = 0 km and H_{top} = 5 km is that the vertical deformation profile for H_{top} = 0 km is discontinuous at the strike and does not increase as the distance to the fault strike approaches zero (from the left-hand side), while that for H_{top} = 5 km is continuous and has a peak at the fault strike. The comparison of the vertical deformation profiles for H_{top} = 0 km and H_{top} = 5 km indicates that noticeable increases in vertical deformation profiles are caused near the peak along the fault strike and outside of the projected fault plane, where no vertical deformation exists for H_{top} = 0 km. Consequently, when deformation profiles with a buried fault plane are used in tsunami modeling, the uplifted parts (i.e., peak along the strike and outer fault plane region) generate additional tsunami waves, in comparison with the seabed surface rupture case (Goda, 2015).

To examine the effects of dip angle and top depth, Figs. 2.30A and B show the vertical deformation profiles (along cross-section 1) for three dip angles, that is, 5, 10, and 20 degrees, by considering H_{top} = 0 km and H_{top} = 5 km (for the same fault geometry and slip in Fig. 2.29). A notable change due to dip variations is the increased deformation at the top edge. The relative effects due to steeper dip angles on vertical deformation are different for H_{top} = 0 km and H_{top} = 5 km. When the rupture to the seabed is considered, the maximum vertical deformation along the fault strike

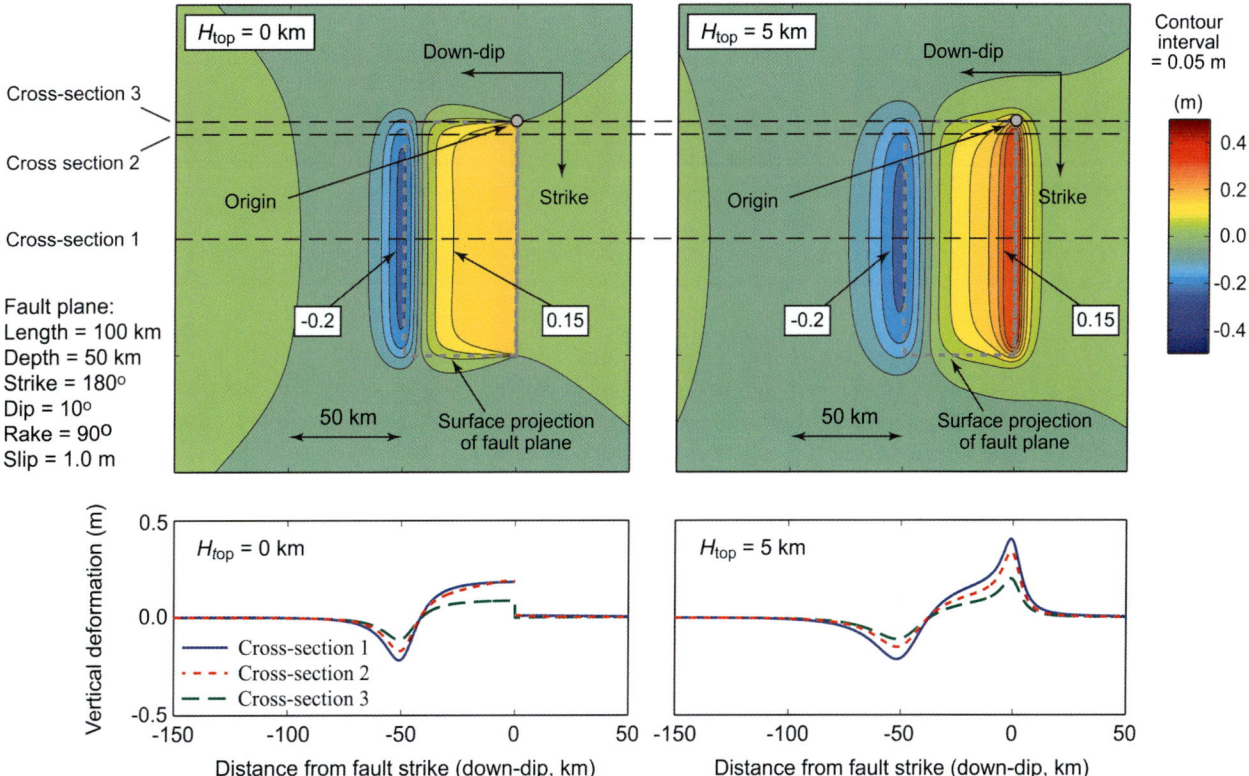

FIGURE 2.29 Vertical deformation profiles (top: contour plot and bottom: cross-sectional profiles) for two cases with H_{top} = 0 km (left) and H_{top} = 5 km (right).

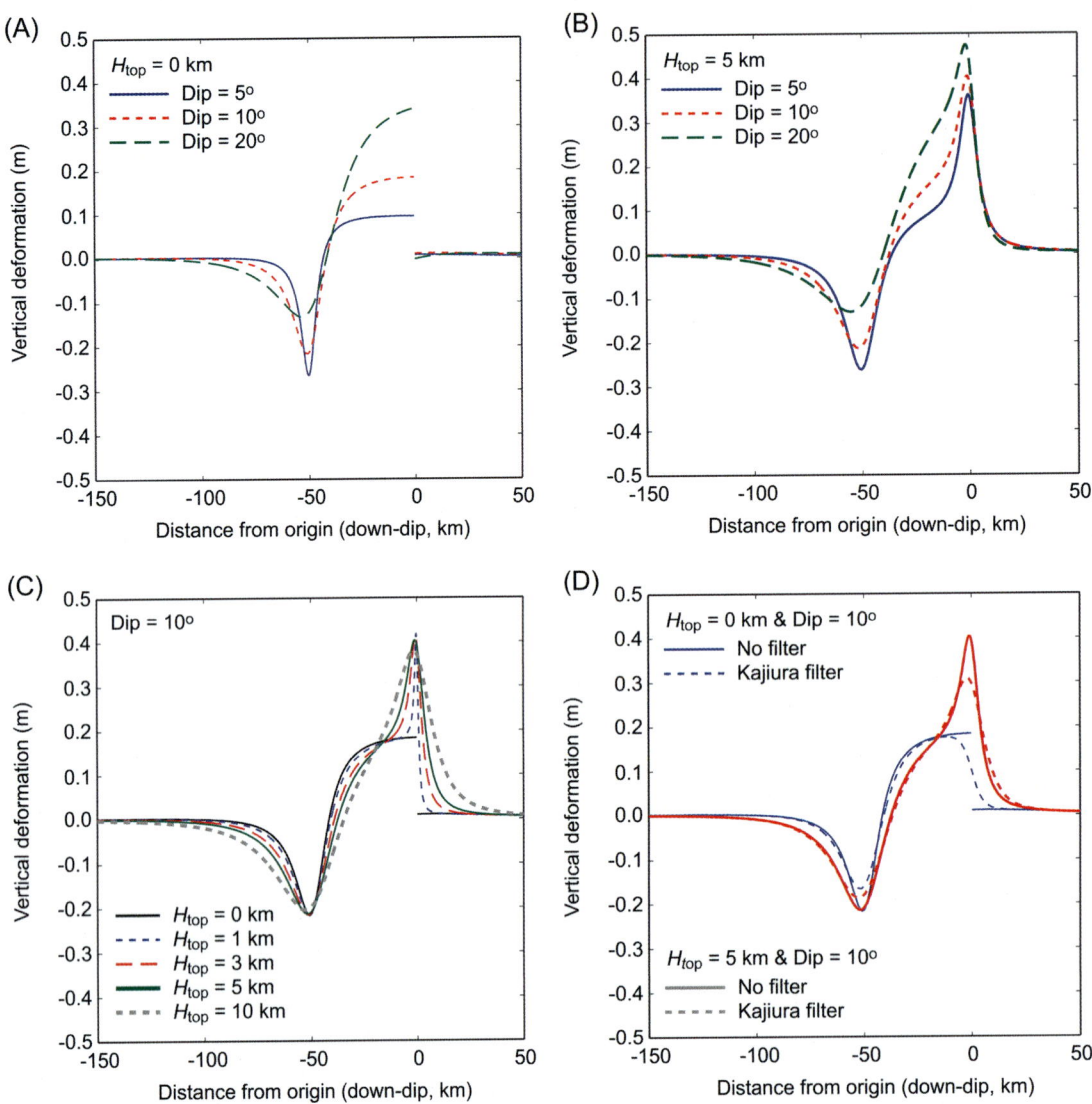

FIGURE 2.30 Comparison of vertical deformation profiles: (A) varying dip angles with $H_{top} = 0$ km, (B) varying dip angles with $H_{top} = 5$ km, (C) varying top-edge depths of the fault plane with dip = 10 degrees, and (D) effects of the hydrodynamic filter by Kajiura (1963).

increases significantly with similar discontinuous cross-sectional profiles. This indicates that the total displaced volume of water above the fault plane has changed significantly. This is not the case for the buried rupture case. The increase of the maximum vertical deformation profile at the top edge is less dramatic, having a similar cross-sectional profile. Consequently, the impact due to steeper dip angles is less pronounced in this case. Moreover, Fig. 2.30C shows the vertical deformation profiles (along cross-section 1) for different values of H_{top}, ranging from 0 to 10 km. A large spike is generated along the top edge. As the depth becomes shallower, the spike becomes sharper (note: the maximum height of the spike is not significantly changed), indicating that short-wavelength components tend to be generated due to the top-edge depth effects. Finally, Fig. 2.30D compares the vertical deformation profiles along cross-section 1 for dip = 10 degrees and $H_{top} = 0$ or 5 km by considering the Kajiura filter. The hydrodynamic response filter makes the vertical deformation profile smoother (and continuous even for $H_{top} = 0$ km). Notably, for the buried rupture case, the height of the peak along the top edge is decreased, whereas the uplift of water outside of the fault rupture plane is similar.

Finally, the use of the Okada formulae is illustrated for more complex earthquake slip distributions. The computer code of Example 2-5 includes the calculation of coseismic deformation distributions for the simulated stochastic source models (Appendix). By taking the earthquake slip distributions for the 2005 Sumatra Earthquake, shown in Fig. 2.26, as an example, the vertical displacement fields are calculated and shown in Fig. 2.31. It is noteworthy that the Okada formulae calculate the elastic response of the half-space due to an earthquake rupture. Therefore, the vertical displacement

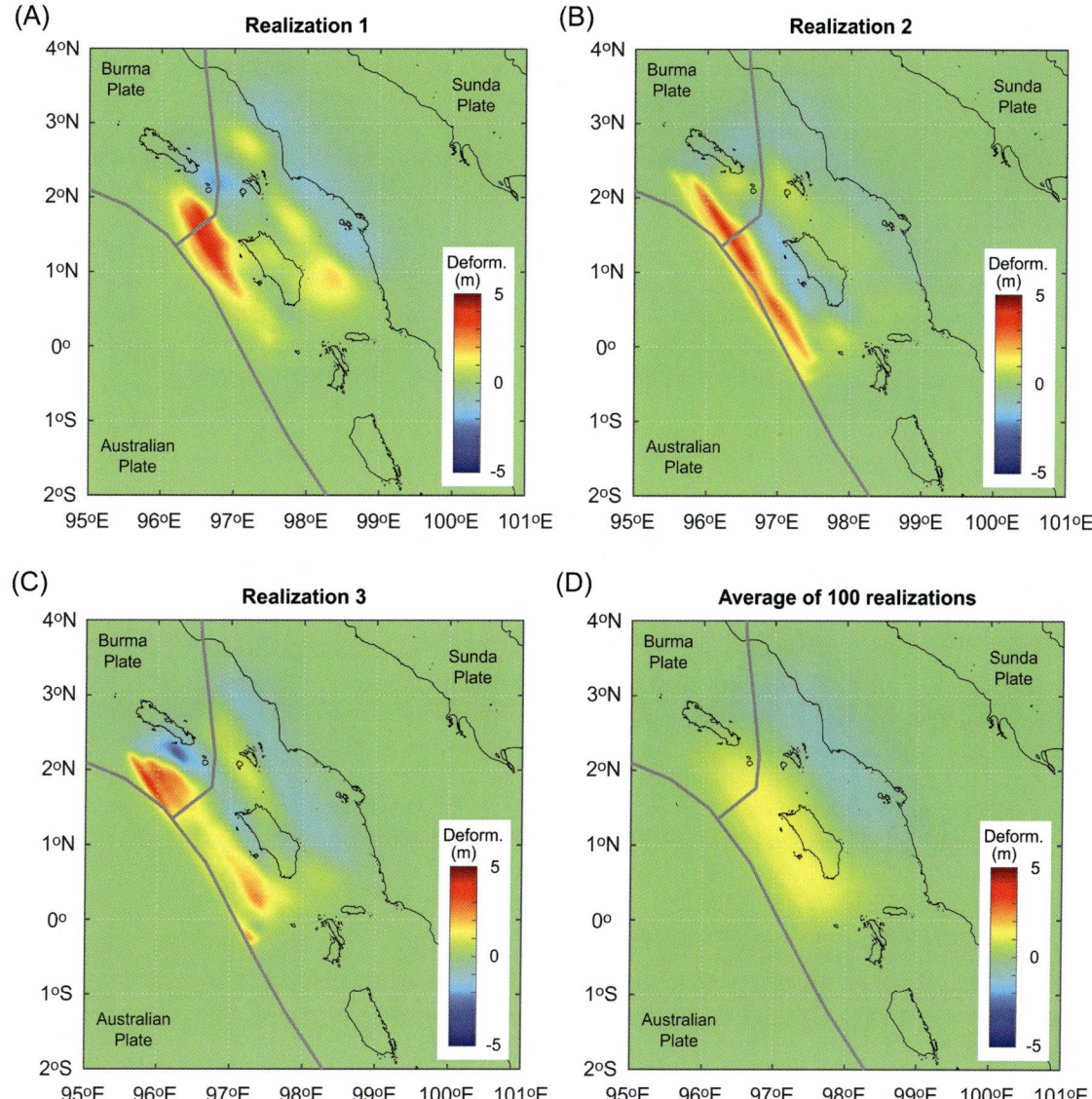

FIGURE 2.31 The vertical displacement fields for the 2005 Sumatra Earthquake rupture scenarios. (A–C) Three realizations of the accepted stochastic source models and (D) average earthquake slip distribution based on the 100 accepted realizations.

fields due to individual subfault ruptures (in this case, 20 km by 20 km, with a specific slip value) can be evaluated for the common grid locations, and their displacement values at the same locations can be summed to obtain the total vertical displacement field. The vertical displacement fields for different earthquake slip distributions (i.e., three stochastic realizations and the average of the 100 stochastic realizations) show different deformation patterns but are similar to the underlying earthquake slip distributions. These vertical deformation profiles, after taking into account the hydrodynamic responses of seawater, can be used as initial boundary conditions of water surfaces in tsunami simulations (Chapter 3).

2.6 Nonseismic sources of tsunamis

This section introduces the modeling approaches used to simulate tsunami sources associated with landslides and volcanic eruptions. Landslides that can generate tsunamis are typically associated with earthquakes and volcanic eruptions as main triggers. These two events have the potential to induce rapid displacement of large sliding volumes of debris into water. On the other hand, volcanic tsunamis can be caused by various mechanisms, including volcanic earthquakes, slope instabilities (landslides), pyroclastic flows, underwater explosions, air pressure waves, and caldera collapse. By exploring these nonseismic source mechanisms, this section aims to provide insights into the scientific methods employed to understand the dynamics of tsunami generation originating from these sources.

Section 2.6.1 revolves around understanding tsunamis generated by underwater landslides and their formations and behavior. The section also summarizes different approaches that can be adopted for tsunami source modeling. The sliding solid block model can be used to estimate the submarine landslide source for tsunami simulations. The most important factors in the model include the landslide material dimension, incline angle, water depth, initial acceleration, and sliding path. A comprehensive discussion on a model that couples the landslide and tsunami layers is presented in Chapter 3.

Section 2.6.2 summarizes various tsunami-generating mechanisms triggered by volcanic eruptions. It provides formulae to calculate sea surface displacements caused by underwater explosions, pyroclastic flows, atmospheric pressure waves (airwaves), caldera collapses, and volcano earthquakes. While most of these formulae yield static sea surface displacements for the tsunami source, models for pyroclastic flows and airwaves require a coupled approach between the source and water layers. In-depth discussions on these coupled models are included in Chapter 3.

2.6.1 Sources from landslides

According to the National Geophysical Data Center at the National Oceanic and Atmospheric Administration (NGDC–NOAA), globally, there have been 262 recorded landslide tsunamis, including landslide (121), volcano–landslide (14), earthquake–landslide (125), and volcano–earthquake–landslide (2) events. A tsunami caused by a landslide can be particularly dangerous because it can strike shorelines with little or no warning, making it difficult for people in coastal areas to evacuate in time. Some notable examples include

- Lituya Bay, Alaska (1958): On July 10, 1958, a massive landslide triggered by an earthquake with a magnitude of 7.8–8.3 caused a huge wave in Lituya Bay. The landslide impacted the water at high speed generating a giant tsunami with a maximum runup of 524 m, causing total forest destruction (Fritz et al., 2009).
- Vajont Dam, Italy (1963): A massive landslide, entering into the reservoir of the Vajont Dam, generated a tsunami with a maximum runup of 200 m on the mountainside opposite the landslide. The tsunami also overtopped the dam, causing devastating floods and the loss of 1917 lives (Manenti et al., 2016).
- Papua New Guinea (1998): A submarine landslide off the coast of Papua New Guinea generated a tsunami that devastated three coastal villages, causing the loss of 2200 lives (Tappin et al., 2008). The submarine landslide was most likely to be generated by an underwater earthquake with a magnitude of 7.1.

The process of generating tsunamis due to underwater landslides is complicated. Understanding the behavior of underwater landslides requires a careful analysis of their acceleration, deformation, and fluid flow. Watts et al. (2003) focused on examining the shape, motion, and deformation of underwater landslides, as these factors directly contribute to tsunami generation. This section introduces a solid block model for underwater landslides, which was initially proposed by Watts (2000) and later refined by Watts et al. (2003). The theoretical initial acceleration of a solid block can be approximated by

$$a_0 = \frac{g(\gamma - 1)(\sin\theta - C_n\cos\theta)}{\gamma + C_m} \quad (2.39)$$

where g is the acceleration of gravity; $\gamma = \rho_b/\rho_w$ in which ρ_b is the bulk density of the landslide material and ρ_w is the water density; C_m is the added mass coefficient; C_n is the Coulomb friction coefficient; and θ is the angle of the assumed planar incline. The terminal velocity of a solid block is approximated by

$$u_t = \sqrt{\frac{4bg(\gamma - 1)(\sin\theta - C_n\cos\theta)}{3C_d}} \quad (2.40)$$

where b is the initial landslide length parallel to the slope. The drag coefficient C_d depends on the solid block shape through the incline angle, and the Coulomb friction coefficient C_n is determined by the solid block and incline material. The solution for that provides the position of the center of mass of the sliding block along the incline as a function of time

$$S(t) = S_0 \ln\left[\cosh\left(\frac{t}{t_0}\right)\right] \quad (2.41)$$

in which S_0 ($\equiv u_t^2/a_0$) denotes a characteristic distance of motion, and t_0 ($\equiv u_t/a_0$) denotes a characteristic duration of motion (Watts, 2000). Watts et al. (2003) suggested the added mass coefficient $C_m \approx 1.76$ (in Eq. 2.39) and the drag coefficient $C_d \approx 1.53$ (in Eq. 2.40) obtained from the experimental work of Grilli et al. (2002) performed with an ellipsoidal slide shape. When an underwater mass slides with no significant basal friction, the Coulomb coefficient C_n is zero, which is a value often assumed (Watts et al., 2005). The representative Coulomb friction coefficient can be calculated with the following equation

$$C_n = \frac{3S_u}{2(\rho_b - \rho_w)Tg\cos\theta} \qquad (2.42)$$

where T is the landslide thickness, and $S_u = 0.19(\rho_b - \rho_w)Tg$ is the effective residual shear strength.

The COMCOT tsunami simulation program (Wang & Power, 2011) has incorporated the sliding block model as one of the mechanisms for generating tsunamis. To understand how the Watts formulae work in approximating the water displacement, a parametric example is useful. For this purpose, consider a landslide source with a half ellipsoidal shape with a length of 500 m, width of 500 m, and thickness of 10 m. The volume and mass of the solid material can be calculated from the length, width, thickness, and density parameters. The formulae alone cannot determine where the solid block will stop; therefore, a stopping point should be assumed. The start and stopping points for the sliding solid material can be determined on the basis of the bathymetry of the study area. To make the movement more realistic, the location at which the solid block starts to decelerate can be assumed. Both the deceleration point and the endpoint of the landslide can be determined from bathymetry data. The stopping point may be located at relatively flat bathymetry, while the deceleration point can be at the breakpoint between the steep slope and the flat bathymetry. For the following example, the deceleration is assumed to be located at a 70% distance between the start and stopping points. In this example, the initial center-of-mass position of the landslide is at a longitude and latitude of (0 degree, 0 degree) and the end position is at (0.01 degrees, 0 degree), which gives the sliding trajectory with a path distance of about 1 km. The water depth at the start point is 500 m. To examine the effects of incline angle, parts (A) and (B) of Fig. 2.32 show the tsunami elevation profiles for two incline angles of 10 degrees and 20 degrees. For these models, the added mass coefficient is 1.0, the friction coefficient is 1.0, the specific gravity ($\gamma = \rho_b/\rho_w$) is 2.0, and the location at which deceleration happens is at 70% of the total distance. Considering the model with the incline angle of 20 degrees, the landslide moves faster, and the wave is more energetic than the one with the incline angle of 10 degrees. This is because the solid material slides down a slope under gravity. Moreover, the first peak wave amplitude from the 20 degrees model of 50 cm is higher than the one from the 10 degrees model of 15 cm. The location of the landslide source has a notable effect on the resulting tsunami amplitude. Specifically, when the source is situated at greater depths, the tsunami amplitude is reduced, as depicted in Fig. 2.32A and C. The dimension of the landslide source also plays a significant role in determining the resulting tsunami amplitude. An increase in the thickness of the landslide source leads to an amplitude increase in the resulting tsunami, as demonstrated in Fig. 2.32B and D.

The above model considers the landslide as a block moving with an assumed path, shape, and slope. This kind of model has been implemented in previous studies (Enet & Grilli, 2007; Tinti et al., 1999, 2000; Watts et al., 2003). However, the behavior of the landslide and the interaction with the topography are not captured in this modeling approach.

In certain cases, there is a need for a static water surface displacement model to investigate tsunami sources originating from landslides. Watts et al. (2005) suggested a model of static water displacement due to an underwater landslide with a characteristic wave amplitude expressed as follows

$$\eta_c = 0.0286T(1 - 0.75\sin\theta)\left(\frac{b\sin\theta}{d}\right)^{1.25} \qquad (2.43)$$

where T is the thickness, b is the length of the sliding mass, d is the submergence depth, and θ is the angle of the slope. These model parameters are illustrated in Fig. 2.33. As described in Heidarzadeh et al. (2014), the characteristic amplitude equation was also utilized by Synolakis (2003) to model the water displacement pattern for underwater landslides using the following equation

$$\eta(x,y) = \frac{w}{w+\lambda}\left(\text{sech}\left(\frac{3y}{w+\lambda}\right)\right)^2 \left(-1.2Z_{\min}exp\left(-\left(1.2Z_{\min}\frac{x-X_{\min}}{\lambda Z_{\max}}\right)^2\right)\right) + Z_{\max}exp\left(-\left(\frac{x-X_{\min}-\Delta x}{\lambda}\right)^2\right) \qquad (2.44)$$

where $Z_{\min}$ is the maximum depression of the water surface calculated using $Z_{\min} = 2.1\eta_c$, $Z_{\max}$ is the maximum elevation obtained from $Z_{\max} = 0.64\eta_c(0.8 + 0.2d/(b\sin\theta))$, w is the width of the slide, g is the gravitational acceleration, u_t

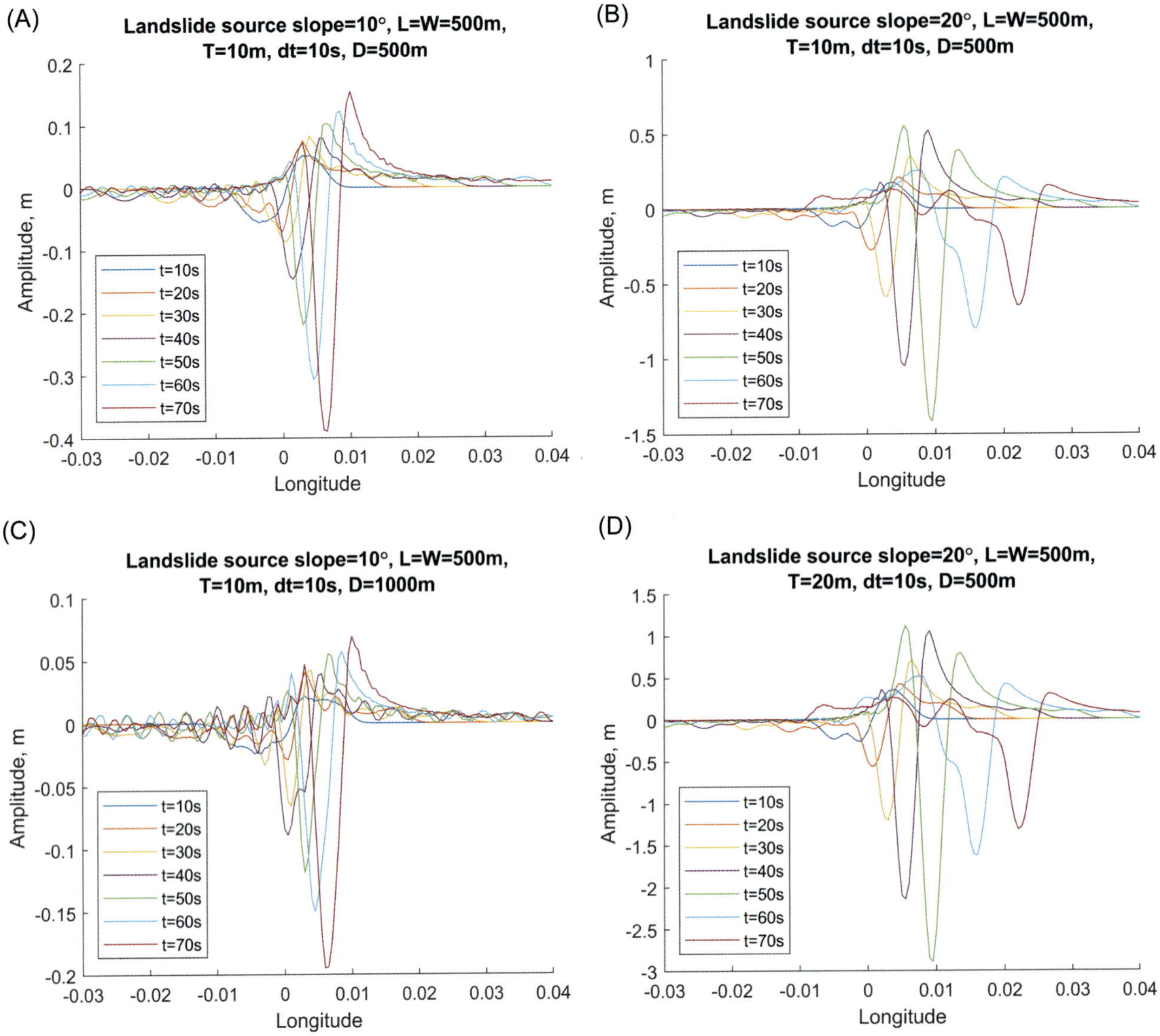

FIGURE 2.32 Comparison of tsunami amplitude at 10, 20, 30, 40, 50, 60, and 70 seconds from landslide models with different incline angles, water depths, and material thickness. The sliding block is assumed to be a solid material in the half ellipsoidal shape with a length of 500 m and a width of 500 m. (A) Landslide model with a thickness of 10 m, incline angle of 10 degrees, and water depth of 500 m. (B) Landslide model with a thickness of 10 m, incline angle of 20 degrees, and water depth of 500 m. (C) Landslide model with a thickness of 10 m, incline angle of 10 degrees, and water depth of 1000 m. (D) Landslide model with a thickness of 20 m, incline angle of 20 degrees, and water depth of 500 m.

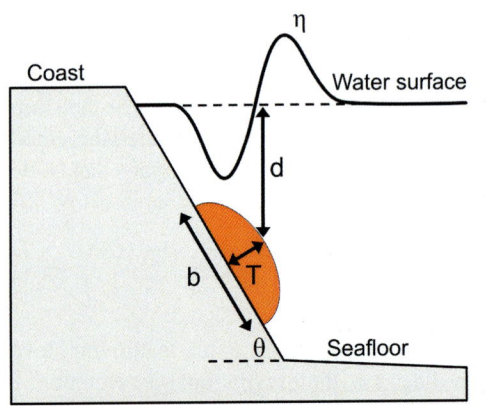

FIGURE 2.33 Illustration for parameters used in the model by Watts et al. (2005) for static water displacement due to an underwater landslide.

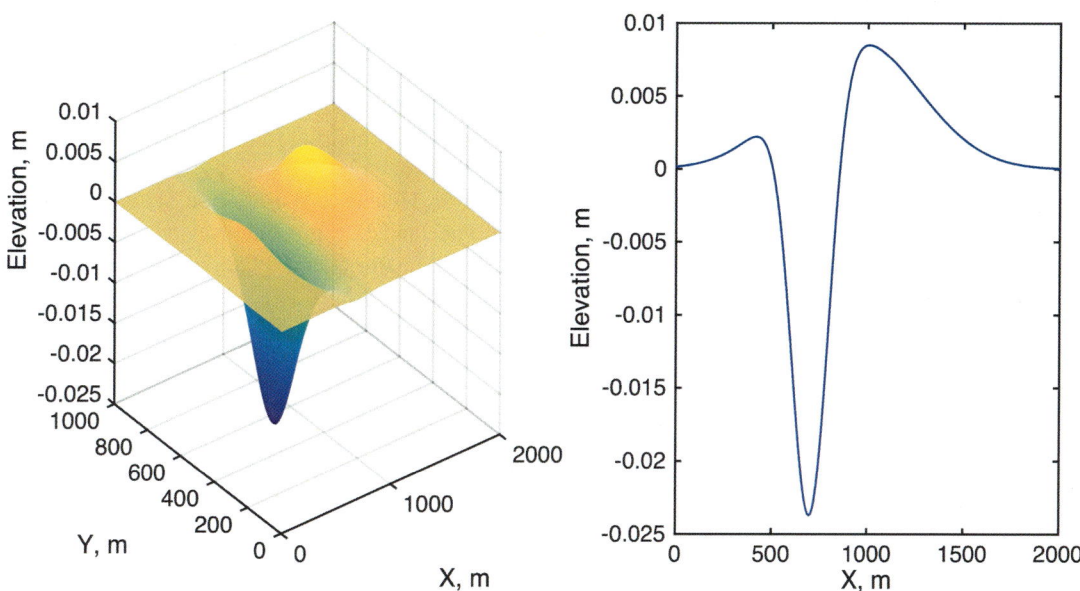

FIGURE 2.34 Static water surface displacement from an underwater landslide model with parameters of thickness of 10 m, slope angle of 10 degrees, landslide length of 100 m, landslide width of 50 m, and submergence depth of 30 m.

is the terminal velocity of the slide, and a_0 is the initial acceleration of the slide. The characteristic wavelength is given by $\lambda = t_0\sqrt{gd}$, where $t_0 \ (\equiv u_t/a_0)$ is the characteristic duration of motion (Watts et al., 2005). The distance between the crest of the elevation wave and the trough of the depression wave is given by $\Delta x = 0.5\lambda$, and $X_{\min}$ is the distance between the trough of the depression wave and the shoreline. The static water surface displacement from a specific landslide model is shown in Fig. 2.34.

2.6.2 Sources from volcanic activities

Volcanic eruptions have the potential to generate tsunamis through various mechanisms. It is important to keep in mind that not all volcanic eruptions result in tsunamis. The occurrence of a tsunami relies on several variables, including the size and type of eruption, the proximity of the volcano to water bodies, and the specific characteristics of the volcanic event. Here are some of the ways in which volcanic activity can lead to tsunami generation:

- Underwater explosions can be caused by the eruption of an underwater volcano.
- Pyroclastic flows are hot, fast-moving flows of ash, gas, and rock that can be generated by volcanic eruptions.
- Air pressure waves are high-pressure waves that can be generated by explosive volcanic eruptions.
- Caldera collapse is the collapse of the summit of a volcano, which can cause a large volume of water to be displaced and generate a tsunami.
- Volcano−earthquakes are earthquakes that are caused by the movement of magma or other fluids beneath a volcano.
- Landslides are failures of volcanic slopes that can be caused by earthquakes, eruptions, or other phenomena. This mechanism has been described in Section 2.6.1.

The number of tsunamis generated by volcanic activities is generally low, compared to those caused by earthquakes. However, some volcanic tsunami events in the past have been exceptionally catastrophic. Some examples of volcanic tsunamis include

- Mount Unzen, Japan (1792): A major eruption of Mount Unzen caused a catastrophic landslide and subsequent tsunami in the Ariake Sea, Kumamoto Prefecture. The tsunami reached up to 25 m in height and struck the surrounding coastal areas of Nagasaki and Shimabara. The event resulted in the deaths of approximately 15,000 people.
- Krakatau, Indonesia (1883): The eruption of Krakatau Volcano generated a series of devastating tsunamis. The waves reached up to 40 m in height and caused widespread destruction along the coasts of Java and Sumatra. The event resulted in the deaths of approximately 36,000 people. The air pressure waves from the massive explosion also generated unusually fast tsunamis that were recorded in far places, such as San Francisco.

- Ritter Island, Papua New Guinea (1888): The collapse of the volcanic island of Ritter Island triggered a powerful tsunami. The waves reached up to 15 m in height and impacted nearby coastal areas, including the island of New Britain. The tsunami caused significant loss of life, with an estimated death toll of over 3000 people.
- Stromboli, Italy (1930): The eruption of Stromboli Volcano triggered a large-scale landslide and subsequent tsunami. The waves reached 2–3 m in height and impacted the coasts of Calabria, Italy. The tsunami caused the deaths of two individuals.
- Mount St. Helens, United States (1980): The eruption of Mount St. Helens in Washington State resulted in a massive landslide and the collapse of the volcano's northern flank. The landslide generated a localized tsunami in Spirit Lake, causing the water to surge over the surrounding areas. The tsunami claimed the lives of 57 people.
- Anak Krakatau, Indonesia (2018): An eruption of Anak Krakatau, the child of Krakatau Volcano, triggered a tsunami in the Sunda Strait. The tsunami struck the coasts of Java and Sumatra Islands, resulting in significant damage and the loss of 400 lives (Muhari et al., 2019).

According to the NGDC–NOAA, there have been 164 documented tsunami events that are attributed solely to volcanic activities or a combination of volcanic eruptions, earthquakes, and landslides. These volcanic-related events account for approximately 5.8% of the total number of recorded tsunami events, specifically out of a total of 2824 recorded events.

2.6.2.1 Underwater explosion

The water crater produced by a near-surface explosion may expose the seafloor of the caldera. After reaching its maximum size, the water crater collapses and the water rushes inward under the influence of gravity onto the crater. Several tsunami studies (Maeno & Imamura, 2011; Torsvik et al., 2010; Ulvrová et al., 2014) used a formula to estimate the initial water displacement model for underwater explosions proposed by Le Méhauté (1971). Le Méhauté and Wang (1996) described that the initial conditions of water displacements were determined numerically by inverse transformation from experimental wave records. The theoretical solutions for simplified cases lead to the initial water disturbance that approximates the explosion source being a parabolic crater with a vertical steep water rim. The mathematical model for the initial water displacement that fits most of the data suggested by Le Méhauté (1971) is given as follows

$$\eta(r) = \begin{cases} \eta_0 \left[2\left(\frac{r}{R}\right)^2 - 1 \right] & \text{if } r \leq R \\ 0 & \text{if } r > R \end{cases} \quad (2.45)$$

where R is the initial water crater radius, r is the distance from the center of the explosion, and η_0 is the maximum height located along the water crater rim. The study also recommended setting the initial condition for current velocity to zero ($u(r) = 0$). The formula utilized by Le Méhauté (1971) is not appropriate as an initial condition for long wave models due to the discontinuity it introduces in the water level at $r = R$. To address this issue, Torsvik et al. (2010) made modifications to the formula for the initial water displacement as follows

$$\eta(r) = \begin{cases} \eta_0 \left[2\left(\frac{r}{R}\right)^2 - 1 \right] & \text{if } r \leq R \\ \eta_0 \left[2\left(\frac{r}{R}\right)^2 - 1 \right] e^{P_r(1 - r/R)} & \text{if } R < r \leq 2R \\ 0 & \text{if } r > 2R \end{cases} \quad (2.46)$$

where the factor P_r controls the rate of decline in surface elevation for $r > R$. In the simulations for small amplitude initial disturbance, Torsvik et al. (2010) used $P_r = 200$.

In this model, an initial water elevation is assumed to have a crater shape. The maximum water elevation along the rim (η_0) can be estimated from the explosion energy as follows

$$\eta_0 = cE^{0.24} \quad (2.47)$$

where c is constant and has a value of 0.024. Explosion energy (E in Joule) is generally proportional to the third power of the crater radius. Sato and Taniguchi (1997) suggested an empirical relationship

$$E = 3.56 \times 10^7 R^3 \quad (2.48)$$

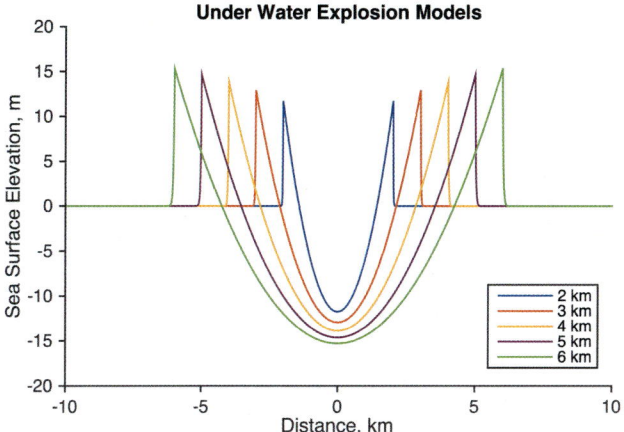

FIGURE 2.35 Initial sea surface elevation models for underwater explosions with 2, 3, 4, 5, and 6 km source radii.

where data from chemical, nuclear, and volcanic explosions varying over 14 orders of magnitude up to $E = 10^{17}$ J (corresponding to a crater radius up to $R = 1.5$ km) were used for fitting. The volcanic explosion energy typically ranges between 10^{12} and 10^{17} J, corresponding approximately to a volcanic crater radius between 100 and 1500 m. Using Eq. (2.47), Fig. 2.35 provides a comprehensive view of the sea surface elevation models resulting from underwater explosions with 2, 3, 4, 5, and 6 km source radii.

Eqs. (2.46)–(2.48) imply that the water depth of the explosion does not affect the initial water displacement and thus oversimplifies the phenomenon. Shen et al. (2021) proposed a more detailed exploration of waves generated by underwater gas eruptions through physical experiments. They discussed the classification of underwater eruptions based on the scaled water depth and the relationship between water depth and wave generation. Three types of eruptions are identified: deep, intermediate, and shallow water. The threshold between deep and intermediate water eruptions is determined by the ability of an eruption to produce noticeable waves at the water surface. The presence or absence of waves depends on the plume's gas and water mixture as well as the cooling, condensation, and dissolution of gases as the plume rises. The research focuses specifically on waves generated by underwater eruptions in shallow and intermediate water depths. In intermediate water depths, decreasing water depth increases wave height due to reduced energy loss. However, in shallow water eruptions, decreasing water depth leads to decreased wave heights as more energy dissipates into the air. The critical effective scaled water depth serves as the threshold between shallow and intermediate water explosions, representing the point at which maximum wave heights are generated.

To account for the influence of water depth, Ulvrová et al. (2014) suggested incorporating Eq. (2.47), which involves adjusting the values of the coefficient c. This approach allows for the estimation of the maximum water elevation by considering the water depth of the explosion as a factor in the calculation. When dealing with shallow water explosions, characterized by the condition $0 < dc/W^{1/3} < 0.076$, where dc represents the water depth of explosion in m and W is the explosion yield in pounds of TNT, the suggested value of c is 0.029. For deeper explosions, where the range $0.076 < dc/W^{1/3} < 2.286$ holds, the suggested value for c is 0.014.

Eq. (2.46) provides the maximum water height for the shallow water explosion condition. However, this is not completely consistent with recent laboratory experiments that the maximum water height typically occurs between shallow and intermediate water explosions (Shen et al., 2021). During volcanic eruptions occurring in deep water, most energy is dispersed within the water column before the eruption plume reaches the surface. As a result, the generation of waves is minimal. However, in intermediate water eruptions, the reduction in water depth limits the dissipation of energy within the water column, resulting in a greater amount of energy available for wave generation. Consequently, wave heights increase as the water becomes shallower, up to a certain point. In cases where the water is sufficiently shallow, the depth becomes small such that much of the energy from the eruptive jet or plume passes through the water and dissipates into the air. Therefore, while Eq. (2.47) with the coefficient adjustment offers an estimation method, it may not fully capture the actual water height patterns. Further research and refinement are necessary to better model the maximum water elevation for underwater volcanic explosions.

2.6.2.2 Pyroclastic flow

Highly mobile mixtures of gases and volcanic ash from volcanic eruptions, known as pyroclastic flow, can generate tsunamis. The eruption of the Krakatau Volcano in Indonesia in 1883 stands as one of the most explosive volcanic events and one of the most devastating tsunamis in recorded history. One of the primary hypotheses regarding the mechanisms behind the generation of the accompanying tsunami is the pyroclastic flow. Maeno and Imamura (2011) developed a numerical model for the pyroclastic flow hypothesis, and its validity was tested against the data from the volcanic tsunami event. They developed two types of two-layer shallow water models, known as dense and light pyroclastic flow models. As this method involves coupling between the pyroclastic flow layer and the water layer, it will be discussed in Chapter 3.

2.6.2.3 Air pressure wave

Sharp changes in atmospheric pressure can induce ocean waves with characteristics similar to tsunamis. These tsunamis of atmospheric origin, known as meteotsunamis, have been observed at various locations worldwide (Defant, 1960; Monserrat et al., 2006; Nomitsu, 1935; Pattiaratchi & Wijeratne, 2015; Rabinovich & Monserrat, 1998; Vilibić et al., 2016). Among these, the first documented tsunami, caused by atmospheric Lamb waves or air pressure waves from a volcanic eruption, dates back to the 1883 Krakatau eruption. Meteorological stations globally captured the air pressure anomalies stemming from this eruption. Generally, tsunamis generated by such air pressure waves travel faster than those caused by localized volcanic mechanisms. These airwaves can move almost at the speed of sound, continuously exciting the tsunami waves as they propagate across the ocean. The most significant eruption of Krakatau in 1883 was noted at 3:02 UTC. Meanwhile, a tide gage in San Francisco registered the tsunami at 17:40 UTC on August 27, 1883. Given the approximately 14,000 km distance between Krakatau and San Francisco, the inferred average speed of this tsunami is about 265 m/s. This speed surpasses that of a typical tsunami traversing solely through an oceanic path from its volcanic origin. Therefore, it is probable that the tsunami observed in San Francisco resulted from the air pressure wave emanating from the Krakatau eruption.

A more recent volcanic eruption event that produced a tsunami generated by the air pressure wave from the explosion was the 2022 Hunga Tonga-Hunga Ha'apai Volcano eruption, in the Tongan Archipelago. For this case, the air wave speed was estimated from air pressure gauges and satellite image data to be between 317 and 322 m/s (Gusman et al., 2022; Omira et al., 2022). The air wave amplitude decays because of the geometric spreading of the wave proportionally to $1/r^{0.5}$ for the distance within the region ± 90 degrees from the source (Dogan et al., 2023; Gusman et al., 2022). The formula for the air wave peak and trough amplitudes derived from the 2022 Hunga Tonga-Hunga Ha'apai eruption data is given as

$$P(r) = \begin{cases} \alpha & 0 < r \leq 1 \\ \alpha \dfrac{1}{\sqrt{r}} & 1 < r \leq \dfrac{\pi}{2} R_E \\ \alpha \dfrac{1}{\sqrt{\pi R_E - r}} & \dfrac{\pi}{2} R_E < r \leq \pi R_E \end{cases} \tag{2.49}$$

where R_E is the radius of the Earth (6371 km), and α is an empirical constant determined by air pressure gage data. More details on how to consider these kinds of sources in tsunami simulation are discussed in Chapter 3.

2.6.2.4 Caldera collapse

Tsunamis can be generated by the collapse or reduction in depth of an underwater caldera following a large eruption. However, the specific duration of this process is still largely unknown, which contributes to the uncertainty involved in simulating tsunamis resulting from this mechanism. After a significant eruption, the underwater caldera can experience a reduction in depth or collapse. However, the timeframe in which this occurs can vary significantly. The collapse process may occur within hours. The variability in the collapse duration adds complexity to accurately model and predict the resulting tsunami (Maeno et al., 2006). Wilson and Hildreth (1997) estimated that the collapse of the Long Valley caldera (30 km $\times$ 15 km) took approximately 98 hours. The caldera-forming eruption of Pinatubo (2.5 km in diameter) in 1991 lasted approximately 9 hours (Wolfe & Hoblitt, 1996), while Lavallée et al. (2004) used a mean value of 17 hours based on the scaling of their analog experiments.

The size of the caldera collapse of an eruption can be obtained by measuring the bathymetry before and after the eruption. For example, the Hunga Tonga-Hunga Ha'apai caldera has undergone significant changes following the 2022 catastrophic eruption. The caldera is now 4 km wide and descends to a depth of 850 m below sea level, compared to its previous depth of about 150 m. Multibeam mapping is used to collect data from the RV Tangaroa research vessel and from an unmanned remotely operated research vessel.

Geometrical changes can be modeled using the following equations of free fall and uniform velocity, respectively (Maeno et al., 2006)

$$h(t) = h_{\text{before}} - \frac{1}{2}gt^2 \tag{2.50}$$

$$h(t) = h_{\text{before}} + \left\{ (h_{\text{after}} - h_{\text{before}}) \frac{t}{t_s} \right\} \tag{2.51}$$

where $h(t)$ is the time dependent still water depth, h_{before} is the constant still water depth before the caldera collapse, and h_{after} is the constant still water depth after the collapse. The variable t represents time, while t_s corresponds to the duration of the collapse.

Typically, the propagation of a tsunami within the vicinity of a caldera follows a process characterized by the following sequence of events (Maeno et al., 2006; Nomikou et al., 2016). Initially, as the collapse begins, the sea level experiences a rapid drawdown due to the influx of seawater into the void created by the collapse. Then the potential energy drop decreases as the caldera fills. Subsequently, the sea level gradually recovers and starts to rise as water waves enter the collapsed area, interacting and causing the overall wave height to increase. This leads to the formation of a large water wave.

2.6.2.5 Volcanic earthquake

Most shallow earthquakes have seismic radiation patterns that are consistent with the double-couple model for shear failure on planar faults (Dziewonski & Woodhouse, 1983; Frohlich, 1995; Isacks et al., 1968; Sykes, 1967). However, in volcanic and geothermal areas, other processes, such as the migration of magmatic and/or hydrothermal fluids or rupture on nonplanar faults, can produce earthquakes with significant nondouble-couple components. Shuler, Nettles, et al. (2013) and Shuler, Ekström, et al. (2013) investigated the link between volcanic unrest and the occurrence of nondouble-couple earthquakes with dominant vertical tension or pressure axes, known as vertical compensated-linear-vector-dipole (vertical-CLVD) earthquakes. The Smith submarine caldera near the Izu−Bonin Trench in Japan and the Curtis submarine volcano in the Kermadec subduction zone in New Zealand have been observed to produce repeating tsunami earthquakes originating from vertical CLVD earthquakes (Gusman et al., 2020). These earthquakes have demonstrated an unusual characteristic of generating disproportionately large tsunamis relative to their moderate magnitudes (Shuler, Ekström, et al., 2013). Typically, the magnitude of tsunamis, known as tsunami magnitude (M_t), corresponds to the M_w estimates of the source earthquake. However, in the case of the Smith Caldera Earthquake (M_w 5.6), it produced an M_t 7.3 tsunami (Abe, 1995; Satake & Kanamori, 1991). Two different earthquakes that generate almost identical tsunami and seismic waves are extremely rare. This rare occurrence also took place in the Kermadec Ridge, New Zealand, on February 17, 2009 (M_w 6.0) and December 8, 2017 (M_w 5.9), with both tsunamis surpassing expectations based on their moment tensor solutions. In the Smith caldera along the Izu−Bonin arc, tsunamigenic CLVD earthquakes occurred multiple times in the past, approximately every 10 years, on June 13, 1984, September 4, 1996, January 1, 2006, and May 2, 2015.

The physical processes causing vertical CLVD earthquakes are not well understood. Previous studies proposed four possible types of physical mechanisms. The first type includes mechanisms caused by fluid flow but with insignificant net volumetric change. An example of this type is a mass exchange between two magma chambers with an implosive component for the deeper chamber and an explosive component for the shallower chamber (Tkalčić et al., 2009). The second type includes mechanisms that have a volumetric change, where the isotropic component might not be clearly observed because of the trade-off of long-period seismic waveforms between the isotropic and vertical CLVD components (Kawakatsu, 1996). Examples of this type of mechanism include a tensile crack opening and hydrofracturing caused by injection of supercritical water heated by magma (Kanamori et al., 1993). The third type is a cone-faulting mechanism in which vertical CLVD earthquakes are produced by failure on curved or cone-shaped faults triggered by inflation or deflation of shallow magma chambers (Nettles & Ekström, 1998). Global observations of vertical CLVD earthquakes suggest that vertical-T (dominant tension axes) CLVD earthquakes usually occur before volcanic unrest, whereas vertical-P (dominant pressure axes) CLVD earthquakes follow episodes of volcanic unrest or eruptions (Shuler, Ekström, et al., 2013; Shuler, Nettles, et al., 2013). The fourth type was suggested by Sandanbata et al. (2023) in which the inside of a curved fault system beneath a submerged volcano with a caldera structure suddenly moved upward, together with a large intra-caldera fault slip and a volume increase of a shallow magma reservoir. The overpressure created by magma accumulation beneath the submarine caldera can induce meter-scale tsunamis without significant ground motions.

For a CLVD source with no volumetric change, Fukao et al. (2018) suggested a rectangular prim model and formulae for the source parameters. The relationship between horizontal compression (D_H) and vertical expansion (D_V) is expressed by

$$D_V L^2 = 2 D_H L T \qquad (2.52)$$

while the seismic moment can be calculated by

$$M_0 = (3/2)\mu D_V L^2 \qquad (2.53)$$

The seafloor displacement can be calculated from a distribution of z-, x-, and y-normal tensile fault planes using the Okada (1985) model (Section 2.5.4). The vertical expansion for each fault plane can be obtained by dividing the total vertical expansion amount (D_V) by the number of z-normal planes. Similarly, the horizontal compression for each x- or y-normal fault plane can be obtained by dividing the total horizontal compression (D_H) amount by the number of x- or y-normal planes.

Ring faults are dip–slip faults that exhibit a curved or cone-shaped geometry. These faults are formed due to the inflation and deflation of shallow magma chambers and are closely associated with caldera collapse and resurgence processes (Gottsmann & Martí, 2011). While observing ring faults can be challenging as they are often concealed by lava, pyroclastic flow deposits, or crater lakes, both inward- and outward-dipping ring faults have been identified in eroded volcanic areas (Cole et al., 2005; Geyer & Martí, 2008; Lipman, 1997). Typically, ring faults are characterized by steep dip angles, approximating subvertical orientations (Gudmundsson & Nilsen, 2007). Slips along curved normal faults can generate vertical-P earthquakes, while slips along curved reverse ring faults can produce vertical-T earthquakes (Ekström, 1994). Meade (2007) introduced analytical algorithms to compute displacements, strains, and stresses resulting from slip on a triangular dislocation element within a homogeneous elastic half-space. These algorithms can be utilized to simulate displacements caused by dip–slip motions on ring faults in earthquake scenarios.

2.7 Research needs

The characterization of seismic tsunami sources has progressed significantly. Various refined models for earthquake occurrence and rupture have been developed in the literature. Importantly, the uncertainty of these models has been modeled explicitly and has been incorporated into PTHA and PTRA (Chapter 5). Nevertheless, there are several important and influential model components to be further advanced in the future. Incorporating available information from the historical and paleo tsunamis is essential (Chapter 7). The time dependency of the earthquake occurrence has a major influence on the PTHA and PTRA results (Chapter 23). The source characterization is still challenging, and improvements are necessary. For instance, more complex fault geometry and variable source parameters (e.g., depth-dependent rigidity), especially in the accretionary prism of a subduction zone, could be adopted (Chapter 8). Moreover, different earthquake rupture types, such as splay faulting (Chapter 9) need to be characterized. Eventually, these additional tsunami sources should be incorporated into PTHA.

For the nonseismic tsunami sources, the basic understanding of the trigger mechanisms of tsunamis due to landslides and volcanic activities needs to be improved. The uncertainty in specifying the initial conditions of the landslide-triggered and volcanic eruption-triggered tsunamis is very large. For some trigger mechanisms, the coupling of the trigger mechanisms and the tsunami generation must be taken into account and needs to be implemented properly in the numerical schemes (Chapter 3). For landslide-triggered tsunamis, numerous parameters define the initial conditions of tsunami generation, and the comprehensive consideration of these different cases is not an easy task (Chapter 10). The same applies to volcanic eruptions. Accounting for different trigger mechanisms and their initial conditions in PTHA is an open area for future research.

References

Abaimov, S. G., Turcotte, D. L., Shcherbakov, R., Rundle, J. B., Yakovlev, G., Goltz, C., & Newman, W. I. (2008). Earthquakes: Recurrence and interoccurrence times. *Pure and Applied Geophysics, 165*(3–4), 777–795. Available from https://doi.org/10.1007/s00024-008-0331-y.

Abe, K. (1995). *Tsunami: Progress in prediction. Disaster prevention and warning estimate of tsunami run-up heights from earthquake magnitudes* (pp. 21–35). Science and Business Media LLC. Available from 10.1007/978-94-015-8565-1_2.

Aki, K. (1965). Maximum likelihood estimate of b in the formula log$N = a - bm$ and its confidence limits. *Bulletin of the Earth Research Institute, 43*, 237–239.

Aki, K., & Richards, P. G. (2002). *Quantitative Seismology, 2*. Available from https://www.ldeo.columbia.edu/~richards/Aki_Richards.html.

Allmann, B. P., & Shearer, P. M. (2009). Global variations of stress drop for moderate to large earthquakes. *Journal of Geophysical Research: Solid Earth*, *114*(1). Available from https://doi.org/10.1029/2008JB005821, http://onlinelibrary.wiley.com/journal/10.1002/(ISSN)2169-9356.

Ammon, C. J., Lay, T., Kanamori, H., & Cleveland, M. (2011). A rupture model of the 2011 off the Pacific coast of Tohoku earthquake. *Earth, Planets and Space*, *63*(7), 693–696. Available from https://doi.org/10.5047/eps.2011.05.015.

Ando, M. (1975). Source mechanisms and tectonic significance of historical earthquakes along the Nankai Trough, Japan. *Tectonophysics*, *27*(2), 119–140. Available from https://doi.org/10.1016/0040-1951(75)90102-X.

Anguiano, E., Pincorbo, M., & Aguilar, M. (1993). Fractal characterization by frequency analysis I: Surfaces. *Journal of Microscopy*, *172*(3), 223–232. Available from https://doi.org/10.1111/j.1365-2818.1993.tb03416.x, https://onlinelibrary.wiley.com/doi/abs/10.1111/j.1365-2818.1993.tb03416.x.

Atzori, S., Tolomei, C., Antonioli, A., Merryman Boncori, J. P., Bannister, S., Trasatti, E., Pasquali, P., & Salvi, S. (2012). The 2010–2011 Canterbury, New Zealand, seismic sequence: Multiple source analysis from InSAR data and modeling. *Journal of Geophysical Research: Solid Earth*, *117*(8). Available from https://doi.org/10.1029/2012JB009178, http://onlinelibrary.wiley.com/journal/10.1002/(ISSN)2169-9356.

Baker, J., Bradley, B., & Stafford, P. (2021). *Seismic hazard and risk analysis*. Cambridge University Press. Available from 10.1017/9781108425056.

Bird, P. (2003). An updated digital model of plate boundaries. *Geochemistry, Geophysics, Geosystems*, *4*(3). Available from https://doi.org/10.1029/2001GC000252.

Blaser, L., Kruger, F., Ohrnberger, M., & Scherbaum, F. (2010). Scaling relations of earthquake source parameter estimates with special focus on subduction environment. *Bulletin of the Seismological Society of America*, *100*(6), 2914–2926. Available from https://doi.org/10.1785/0120100111.

Central Disaster Management Council 2012 *Unpublished content Working group report on mega-thrust earthquake models for the Nankai Trough, Japan*. Cabinet Office of the Japanese Government. https://www.bousai.go.jp/jishin/nankai/nankaitrough_info.html

Cole, J. W., Milner, D. M., & Spinks, K. D. (2005). Calderas and caldera structures: A review. *Earth-Science Reviews*, *69*(1-2), 1–26. Available from https://doi.org/10.1016/j.earscirev.2004.06.004, http://www.sciencedirect.com/science/journal/00128252.

Console, R., Murru, M., Falcone, G., & Catalli, F. (2008). Stress interaction effect on the occurrence probability of characteristic earthquakes in Central Apennines. *Journal of Geophysical Research: Solid Earth*, *113*(8). Available from https://doi.org/10.1029/2007JB005418, http://onlinelibrary.wiley.com/journal/10.1002/(ISSN)2169-9356.

Convertito, V., Emolo, A., & Zollo, A. (2006). Seismic-hazard assessment for a characteristic earthquake scenario: An integrated probabilistic-deterministic method. *Bulletin of the Seismological Society of America*, *96*(2), 377–391. Available from https://doi.org/10.1785/0120050024.

Cornell, C. A., & Winterstein, S. R. (1988). Temporal and magnitude dependence in earthquake recurrence models. *Bulletin of the Seismological Society of America*, *78*(4), 1522–1537.

Defant, A. (1960). *The harmonic analysis of tidal observations. Physical oceanography* (pp. 299–319). Pergamon Press.

DeMets, C., Gordon, R. G., & Argus, D. F. (2010). Geologically current plate motions. *Geophysical Journal International*, *181*(1), 1–80. Available from https://doi.org/10.1111/j.1365-246X.2009.04491.x.

Dogan, G. G., Yalciner, A. C., Annunziato, A., Yalciner, B., & Necmioglu, O. (2023). Global propagation of air pressure waves and consequent ocean waves due to the January 2022 Hunga Tonga–Hunga Ha'apai eruption. *Ocean Engineering*, *267*113174. Available from https://doi.org/10.1016/j.oceaneng.2022.113174.

Dziewonski, A. M., & Woodhouse, J. H. (1983). An experiment in systematic study of global seismicity: Centroid-moment tensor solutions for 201 moderate and large earthquakes in 1981. *Journal of Geophysical Research*, *88*(4), 3247–3271. Available from https://doi.org/10.1029/JB088iB04p03247.

Ekström, G. (1994). Anomalous earthquakes on volcano ring-fault structures. *Earth and Planetary Science Letters*, *128*(3-4), 707–712. Available from https://doi.org/10.1016/0012-821X(94)90184-8.

Enet, F., & Grilli, T. (2007). Experimental study of tsunami generation by three-dimensional rigid underwater landslides. *Journal of Waterway, Port, Coastal and Ocean Engineering*, *133*(6), 442–454. Available from https://doi.org/10.1061/(ASCE)0733-950X(2007)133:6(442).

Eshelby, J. D. (1957). The determination of the elastic field of an ellipsoidal inclusion, and related problems. *Proceedings of the Royal Society of London, Series A, Mathematical and Physical Sciences*, *241*(1226), 376–396. Available from https://doi.org/10.1098/rspa.1957.0133, https://royalsocietypublishing.org/doi/10.1098/rspa.1957.0133.

Fritz, H. M., Mohammed, F., & Yoo, J. (2009). Lituya bay landslide impact generated mega-tsunami 50th anniversary. *Pure and Applied Geophysics*, *166*(1-2), 153–175. Available from https://doi.org/10.1007/s00024-008-0435-4.

Frohlich, C. (1995). Characteristics of well-determined non-double-couple earthquakes in the Harvard CMT catalog. *Physics of the Earth and Planetary Interiors*, *91*(4), 213–228. Available from https://doi.org/10.1016/0031-9201(95)03031-Q.

Fukao, Y., Sandanbata, O., Sugioka, H., Ito, A., Shiobara, H., Watada, S., & Satake, K. (2018). Mechanism of the 2015 volcanic tsunami earthquake near Torishima, Japan. *American Association for the Advancement of Science, Japan Science Advances*, *4*(4). Available from https://doi.org/10.1126/sciadv.aao0219, http://advances.sciencemag.org/content/4/4/eaao0219/tab-pdf.

Fukutani, Y., Suppasri, A., & Imamura, F. (2015). Stochastic analysis and uncertainty assessment of tsunami wave height using a random source parameter model that targets a Tohoku-type earthquake fault. *Stochastic Environmental Research and Risk Assessment*, *29*(7), 1763–1779. Available from https://doi.org/10.1007/s00477-014-0966-4, http://link.springer-ny.com/link/service/journals/00477/index.htm.

Gao, D., Wang, K., Insua, T. L., Sypus, M., Riedel, M., & Sun, T. (2018). Defining megathrust tsunami source scenarios for northernmost Cascadia. *Natural Hazards*, *94*(1), 445–469. Available from https://doi.org/10.1007/s11069-018-3397-6, http://www.wkap.nl/journalhome.htm/0921-030X.

Geist, E. L. (2002). Complex earthquake rupture and local tsunamis. *Journal of Geophysical Research: Solid Earth*, *107*(5). Available from http://onlinelibrary.wiley.com/journal/10.1002/(ISSN)2169-9356.

Geyer, A., & Martí, J. (2008). The new worldwide collapse caldera database (CCDB): A tool for studying and understanding caldera processes. *Journal of Volcanology and Geothermal Research, 175*(3), 334–354. Available from https://doi.org/10.1016/j.jvolgeores.2008.03.017.

Glimsdal, S., Pedersen, G. K., Harbitz, C. B., & Løvholt, F. (2013). Dispersion of tsunamis: Does it really matter? *Natural Hazards and Earth System Sciences, 13*, 1507–1526. Available from https://doi.org/10.5194/nhess-13-1507-2013, http://www.nat-hazards-earth-syst-sci.net/volumes_and_issues.html.

Goda, K. (2015). Effects of seabed surface rupture versus buried rupture on tsunami wave modeling: A case study for the 2011 Tohoku, Japan, earthquake. *Bulletin of the Seismological Society of America, 105*(5), 2563–2571. Available from https://doi.org/10.1785/0120150091, http://www.bssaonline.org/content/105/5/2563.full.pdf.

Goda, K. (2019). Time-dependent probabilistic tsunami hazard analysis using stochastic rupture sources. *Stochastic Environmental Research and Risk Assessment, 33*(2), 341–358. Available from https://doi.org/10.1007/s00477-018-1634-x, http://link.springer-ny.com/link/service/journals/00477/index.htm.

Goda, K., Mai, P. M., Yasuda, T., & Mori, N. (2014). Sensitivity of tsunami wave profiles and inundation simulations to earthquake slip and fault geometry for the 2011 Tohoku earthquake. *Earth, Planets and Space, 66*(1). Available from https://doi.org/10.1186/1880-5981-66-105, http://rd.springer.com/journal/40623.

Goda, K., Yasuda, T., Mori, N., & Maruyama, T. (2016). New scaling relationships of earthquake source parameters for stochastic tsunami simulation. *Coastal Engineering Journal, 58*(3). Available from https://doi.org/10.1142/S0578563416500108, https://www.tandfonline.com/loi/tcej20.

Gottsmann, J., & Martí, J. (2011). *Caldera volcanism: Analysis, modelling and response*. Elsevier.

Griffin, J. D., Stirling, M. W., & Wang, T. (2020). Periodicity and clustering in the long-term earthquake record. *Geophysical Research Letters, 47*(22). Available from https://doi.org/10.1029/2020GL089272, http://agupubs.onlinelibrary.wiley.com/hub/journal/10.1002/(ISSN)1944-8007/.

Grilli, S. T., Harris, J. C., Tajalli Bakhsh, T. S., Masterlark, T. L., Kyriakopoulos, C., Kirby, J. T., & Shi, F. (2013). Numerical simulation of the 2011 Tohoku tsunami based on a new transient FEM co-seismic source: Comparison to far- and near-field observations. *Pure and Applied Geophysics, 170*(6-8), 1333–1359. Available from https://doi.org/10.1007/s00024-012-0528-y, http://www.springer.com/birkhauser/geo+science/journal/24.

Grilli, S. T., Vogelmann, S., & Watts, P. (2002). Development of a 3D numerical wave tank for modeling tsunami generation by underwater landslides. *Engineering Analysis with Boundary Elements, 26*(4), 301–313. Available from https://doi.org/10.1016/S0955-7997(01)00113-8.

Gudmundsson, A., & Nilsen, K. (2007). Ring-faults in composite volcanoes: Structures, models and stress fields associated with their formation. *Geological Society, London, Special Publications, 269*(1), 83–108. Available from https://doi.org/10.1144/gsl.sp.2006.269.01.06.

Gusman, A. R., Kaneko, Y., Power, W., & Burbidge, D. (2020). Source process for two enigmatic repeating vertical-T CLVD tsunami earthquakes in the Kermadec ridge. *Geophysical Research Letters, 47*(16). Available from https://doi.org/10.1029/2020GL087805, http://agupubs.onlinelibrary.wiley.com/hub/journal/10.1002/(ISSN)1944-8007/.

Gusman, A. R., Roger, J., Noble, C., Wang, X., Power, W., & Burbidge, D. (2022). The 2022 Hunga Tonga–Hunga Ha'apai volcano air-wave generated tsunami, New Zealand. *Pure and Applied Geophysics, 179*(10), 3511–3525. Available from https://doi.org/10.1007/s00024-022-03154-1, https://www.springer.com/journal/24.

Gutenberg, B., & Richter, C. F. (1956). Magnitude and energy of earthquakes. *Annals of Geophysics, 9*, 1–15.

Hanks, T. C., & Kanamori, H. (1979). A moment magnitude scale. *Journal of Geophysical Research: Solid Earth, 84*(B5), 2348–2350. Available from https://doi.org/10.1029/JB084iB05p02348.

Hayes, G. P. (2011). Rapid source characterization of the 2011 M_w 9.0 off the pacific coast of tohoku earthquake. *Earth, Planets and Space, 63*(7), 529–534. Available from https://doi.org/10.5047/eps.2011.05.012, http://rd.springer.com/journal/40623.

Hayes, G. P. (2017). The finite, kinematic rupture properties of great-sized earthquakes since 1990. *Earth and Planetary Science Letters, 468*, 94–100. Available from https://doi.org/10.1016/j.epsl.2017.04.003, http://www.sciencedirect.com/science/journal/0012821X/321-322.

Headquarters for Earthquake Research Promotion. (2019). *Evaluation of long-term probability of earthquake occurrence along the Japan Trench*.

Heidarzadeh, M., Krastel, S., & Yalciner, A. C. (2014). *Submarine mass movements and their consequences. The state-of-the-art numerical tools for modeling landslide tsunamis: A short review* (pp. 483–495). Springer. Available from http://link.springer.com/book/10.1007%2F978-3-319-00972-8, 10.1007/978-3-319-00972-8_43.

Herrero, A., & Bernard, P. (1994). A kinematic self-similar rupture process for earthquakes. *Bulletin of the Seismological Society of America, 84*(4), 1216–1228. Available from https://doi.org/10.1785/bssa0840041216.

Hyndman, R. D., & Wang, K. (1995). The rupture zone of Cascadia great earthquakes from current deformation and the thermal regime. *Journal of Geophysical Research: Solid Earth, 100*(B11), 22133–22154. Available from https://doi.org/10.1029/95jb01970.

Iinuma, T., Hino, R., Kido, M., Inazu, D., Osada, Y., Ito, Y., Ohzono, M., Tsushima, H., Suzuki, S., Fujimoto, H., & Miura, S. (2012). Coseismic slip distribution of the 2011 off the Pacific coast of Tohoku earthquake (*M* 9.0) refined by means of seafloor geodetic data. *Journal of Geophysical Research: Solid Earth, 117*(7). Available from https://doi.org/10.1029/2012JB009186, http://onlinelibrary.wiley.com/journal/10.1002/(ISSN)2169-9356.

Isacks, B., Oliver, J., & Sykes, L. R. (1968). Seismology and the new global tectonics. *Journal of Geophysical Research, 73*(18), 5855–5899. Available from https://doi.org/10.1029/jb073i018p05855.

Ji, C. (2005). Preliminary result 05/03/28 (Mw8.7) Sumatra earthquake. http://www.tectonics.caltech.edu/slip_history/2005_sumatra/sumatra.html

Ji, C., Wald, D. J., & Helmberger, D. V. (2002). Source description of the 1999 Hector Mine, California, earthquake, part I: Wavelet domain inversion theory and resolution analysis. *Bulletin of the Seismological Society of America, 92*(4), 1192–1207. Available from https://doi.org/10.1785/0120000916.

Kajiura, K. (1963). *Bulletin of the Earthquake Research Institute. The leading wave of a tsunami* (41, pp. 535–571). Tokyo: University of Tokyo.

Kanamori, H., Ekström, G., Dziewonski, A., Barker, J. S., & Sipkin, S. A. (1993). Seismic radiation by magma injection: An anomalous seismic event near Tori Shima, Japan. *Journal of Geophysical Research: Solid Earth*, *98*(B4), 6511–6522. Available from https://doi.org/10.1029/92jb02867.

Kawakatsu, H. (1996). Observability of the isotropic component of a moment tensor. *Geophysical Journal International*, *126*(2), 525–544. Available from https://doi.org/10.1111/j.1365-246X.1996.tb05308.x, http://gji.oxfordjournals.org/.

Kimura, H., Tadokoro, K., & Ito, T. (2019). Interplate coupling distribution along the Nankai Trough in Southwest Japan estimated from the block motion model based on onshore GNSS and seafloor GNSS/A observations. *Journal of Geophysical Research: Solid Earth*, *124*(6), 6140–6164. Available from https://doi.org/10.1029/2018JB016159, http://agupubs.onlinelibrary.wiley.com/hub/jgr/journal/10.1002/(ISSN)2169-9356/.

Konca, A. O., Hjorleifsdottir, V., Song, T. A., Avouac, J., Helmberger, D. V., Ji, C., Sieh, K., Briggs, R., & Meltzner, A. (2007). Rupture kinematics of the 2005, Mw 8.6, Nias-Simeulue earthquake from the joint inversion of seismic and geodetic data. *Bulletin of the Seismological Society of America*, *97*.

Kulkarni, R., Wong, I., Zachariasen, J., Goldfinger, C., & Lawrence, M. (2013). Statistical analyses of great earthquake recurrence along the Cascadia subduction zone. *Bulletin of the Seismological Society of America*, *103*(6), 3205–3221. Available from https://doi.org/10.1785/0120120105, http://www.bssaonline.org/content/103/6/3205.full.pdf + html.

Lavallée, D., Liu, P., & Archuleta, R. J. (2006). Stochastic model of heterogeneity in earthquake slip spatial distributions. *Geophysical Journal International*, *165*(2), 622–640. Available from https://doi.org/10.1111/j.1365-246X.2006.02943.x.

Lavallée, Y., Stix, J., Kennedy, B., Richer, M., & Longpré, M. A. (2004). Caldera subsidence in areas of variable topographic relief: Results from analogue modeling. *Journal of Volcanology and Geothermal Research*, *129*(1-3), 219–236. Available from https://doi.org/10.1016/S0377-0273(03)00241-5, http://www.sciencedirect.com/science/journal/03770273.

Le Méhauté, B. (1971). *Theory of explosion-generated water waves* (pp. 1–79). Elsevier BV. Available from 10.1016/b978-0-12-021807-3.50006-0.

Le Méhauté, B., & Wang, S. (1996). Water waves generated by underwater explosion. *Advanced Series on Ocean Engineering*. *10*. Available from https://doi.org/10.1142/2587, https://apps.dtic.mil/sti/pdfs/ADA304244.pdf.

Leonard, M. (2010). Earthquake fault scaling: Self-consistent relating of rupture length, width, average displacement, and moment release. *Bulletin of the Seismological Society of America*, *100*(5 A), 1971–1988. Available from https://doi.org/10.1785/0120090189, http://www.bssaonline.org/cgi/reprint/100/5A/1971.

LeVeque, R. J., Waagan, K., González, F. I., Rim, D., & Lin, G. (2016). Generating random earthquake events for probabilistic tsunami hazard assessment. *Pure and Applied Geophysics*, *173*(12), 3671–3692. Available from https://doi.org/10.1007/s00024-016-1357-1, http://www.springer.com/birkhauser/geo + science/journal/24.

Levin, B. W., & Nosov, M. (2016). *Physics of tsunamis*. Springer.

Lipman, P. W. (1997). Subsidence of ash-flow calderas: Relation to caldera size and magma-chamber geometry. *Bulletin of Volcanology*, *59*(3), 198–218. Available from https://doi.org/10.1007/s004450050186, http://link.springer.de/link/service/journals/00445/index.htm.

Løvholt, F., Pedersen, G., Bazin, S., Kühn, D., Bredesen, R. E., & Harbitz, C. (2012). Stochastic analysis of tsunami runup due to heterogeneous coseismic slip and dispersion. *Journal of Geophysical Research: Oceans*, *117*(3). Available from https://doi.org/10.1029/2011JC007616, http://onlinelibrary.wiley.com/journal/10.1002/(ISSN)2169-9291.

Løvholt, F., Schulten, I., Mosher, D., Harbitz, C., & Krastel, S. (2019). Modelling the 1929 Grand Banks slump and landslide tsunami. *Geological Society, London, Special Publications*, *477*(1), 315–331. Available from https://doi.org/10.1144/sp477.28.

Maeno, F., & Imamura, F. (2011). Tsunami generation by a rapid entrance of pyroclastic flow into the sea during the 1883 Krakatau eruption, Indonesia. *Journal of Geophysical Research: Solid Earth*, *116*(9). Available from https://doi.org/10.1029/2011JB008253, http://onlinelibrary.wiley.com/journal/10.1002/(ISSN)2169-9356.

Maeno, F., Imamura, F., & Taniguchi, H. (2006). Numerical simulation of tsunamis generated by caldera collapse during the 7.3 ka Kikai eruption, Kyushu, Japan. *Earth, Planets and Space*, *58*(8), 1013–1024. Available from https://doi.org/10.1186/BF03352606, http://rd.springer.com/journal/40623.

Mai, P. M., & Beroza, G. C. (2000). Source scaling properties from finite-fault-rupture models. *Bulletin of the Seismological Society of America*, *90*(3), 604–615. Available from https://doi.org/10.1785/0119990126, http://www.bssaonline.org/.

Mai, P. M., & Beroza, G. C. (2002). A spatial random field model to characterize complexity in earthquake slip. *Journal of Geophysical Research: Solid Earth*, *107*(11). Available from http://onlinelibrary.wiley.com/journal/10.1002/(ISSN)2169-9356.

Mai, P. M., Spudich, P., & Boatwright, J. (2005). Hypocenter locations in finite-source rupture models. *Bulletin of the Seismological Society of America*, *95*(3), 965–980. Available from https://doi.org/10.1785/0120040111.

Mai, P. M., & Thingbaijam, K. K. S. (2014). SRCMOD: An online database of finite-fault rupture models. *Seismological Research Letters*, *85*(6), 1348–1357. Available from https://doi.org/10.1785/0220140077, http://srl.geoscienceworld.org/content/85/6/1348.full.pdf + html.

Manenti, S., Pierobon, E., Gallati, M., Sibilla, S., D'Alpaos, L., Macchi, E., & Todeschini, S. (2016). Vajont disaster: Smoothed particle hydrodynamics modeling of the post-event 2D experiments. *Journal of Hydraulic Engineering*, *142*(4). Available from https://doi.org/10.1061/(ASCE)HY.1943-7900.0001111, http://ascelibrary.org/journal/jhend8.

Matthews, M. V., Ellsworth, W. L., & Reasenberg, P. A. (2002). A Brownian model for recurrent earthquakes. *Bulletin of the Seismological Society of America*, *92*(6), 2233–2250. Available from https://doi.org/10.1785/0120010267.

Meade, B. J. (2007). Algorithms for the calculation of exact displacements, strains, and stresses for triangular dislocation elements in a uniform elastic half space. *Computers & Geosciences*, *33*(8), 1064–1075. Available from https://doi.org/10.1016/j.cageo.2006.12.003.

Melgar, D., & Hayes, G. P. (2017). Systematic observations of the slip pulse properties of large earthquake ruptures. *Geophysical Research Letters*, *44*(19), 9691–9698. Available from https://doi.org/10.1002/2017GL074916, http://onlinelibrary.wiley.com/journal/10.1002/(ISSN)1944-8007/issues?year = 2012.

Melgar, D., LeVeque, R. J., Dreger, D. S., & Allen, R. M. (2016). Kinematic rupture scenarios and synthetic displacement data: An example application to the Cascadia subduction zone. *Journal of Geophysical Research: Solid Earth*, *121*(9), 6658−6674. Available from https://doi.org/10.1002/2016JB013314, http://onlinelibrary.wiley.com/journal/10.1002/(ISSN)2169-9356.

Melgar, D., Williamson, A. L., & Salazar-Monroy, E. F. (2019). Differences between heterogenous and homogenous slip in regional tsunami hazards modelling. *Geophysical Journal International*, *219*(1), 553−562. Available from https://doi.org/10.1093/gji/ggz299, http://gji.oxfordjournals.org/.

Momeni, P., Goda, K., Mokhtari, M., & Heidarzadeh, M. (2023). A new tsunami hazard assessment for eastern Makran subduction zone by considering splay faults and applying stochastic modeling. *Coastal Engineering Journal*, *65*(1), 67−96. Available from https://doi.org/10.1080/21664250.2022.2117585, https://www.tandfonline.com/loi/tcej20.

Monserrat, S., Vilibić, I., & Rabinovich, A. B. (2006). Meteotsunamis: Atmospherically induced destructive ocean waves in the tsunami frequency band. *Natural Hazards and Earth System Science*, *6*(6), 1035−1051. Available from https://doi.org/10.5194/nhess-6-1035-2006, http://www.nat-hazards-earth-syst-sci.net/volumes_and_issues.html.

Mueller, C., Power, W., Fraser, S., & Wang, X. (2015). Effects of rupture complexity on local tsunami inundation: Implications for probabilistic tsunami hazard assessment by example. *Journal of Geophysical Research: Solid Earth*, *120*(1), 488−502. Available from https://doi.org/10.1002/2014JB011301, http://onlinelibrary.wiley.com/journal/10.1002/(ISSN)2169-9356.

Muhari, A., Heidarzadeh, M., Susmoro, H., Nugroho, H. D., Kriswati, E., Supartoyo., Wijanarto, A. B., Imamura, F., & Arikawa, T. (2019). The December 2018 Anak Krakatau volcano tsunami as inferred from post-tsunami field surveys and spectral analysis. *Pure and Applied Geophysics*, *176*(12), 5219−5233. Available from https://doi.org/10.1007/s00024-019-02358-2, http://www.springer.com/birkhauser/geo + science/journal/24.

Murotani, S., Satake, K., & Fujii, Y. (2013). Scaling relations of seismic moment, rupture area, average slip, and asperity size for $M \sim 9$ subduction-zone earthquakes. *Geophysical Research Letters*, *40*(19), 5070−5074. Available from https://doi.org/10.1002/grl.50976.

Nettles, M., & Ekström, G. (1998). Faulting mechanism of anomalous earthquakes near Bárdarbunga Volcano, Iceland. *Journal of Geophysical Research: Solid Earth*, *103*(8), 17973−17983. Available from https://doi.org/10.1029/98jb01392, http://agupubs.onlinelibrary.wiley.com/hub/jgr/journal/10.1002/(ISSN)2169-9356/.

Nomikou, P., Druitt, T. H., Hübscher, C., Mather, T. A., Paulatto, M., Kalnins, L. M., Kelfoun, K., Papanikolaou, D., Bejelou, K., Lampridou, D., Pyle, D. M., Carey, S., Watts, A. B., Weiß, B., & Parks, M. M. (2016). Post-eruptive flooding of *Santorini caldera* and implications for tsunami generation. *Nature Communications*, *7*(1). Available from https://doi.org/10.1038/ncomms13332.

Nomitsu, T. (1935). A theory of tunamis and seiches produced by wind and barometric gradient. *Memoirs of the College of Science. Kyoto Imperial University Series A*, *18*(4), 201−214.

Ogata, Y. (1999). Estimating the hazard of rupture using uncertain occurrence times of paleoearthquakes. *Journal of Geophysical Research: Solid Earth*, *104*(8), 17995−18014. Available from https://doi.org/10.1029/1999jb900115, http://agupubs.onlinelibrary.wiley.com/hub/jgr/journal/10.1002/(ISSN)2169-9356/.

Okada, Y. (1985). Surface deformation due to shear and tensile faults in a half-space. *Bulletin of the Seismological Society of America*, *75*(4), 1135−1154. Available from https://doi.org/10.1785/bssa0750041135.

Omira, R., Ramalho, R. S., Kim, J., González, P. J., Kadri, U., Miranda, J. M., Carrilho, F., & Baptista, M. A. (2022). Global Tonga tsunami explained by a fast-moving atmospheric source. *Nature*, *609*(7928), 734−740. Available from https://doi.org/10.1038/s41586-022-04926-4, https://www.nature.com/nature/.

Page, M., & Felzer, K. (2015). Southern San Andreas fault seismicity is consistent with the Gutenberg−Richter magnitude−frequency distribution. *Bulletin of the Seismological Society of America*, *105*(4), 2070−2080. Available from https://doi.org/10.1785/0120140340, http://www.bssaonline.org/content/105/4/2070.full.pdf.

Pardo-Iguzquiza, E., & Chica-Olmo, M. (1993). The Fourier integral method: An efficient spectral method for simulation of random fields. *Mathematical Geology*, *25*(2), 177−217. Available from https://doi.org/10.1007/bf00893272.

Park, H., & Cox, D. T. (2016). Probabilistic assessment of near-field tsunami hazards: Inundation depth, velocity, momentum flux, arrival time, and duration applied to Seaside, Oregon. *Coastal Engineering*, *117*, 79−96. Available from https://doi.org/10.1016/j.coastaleng.2016.07.011, http://www.elsevier.com/inca/publications/store/5/0/3/3/2/5/.

Pattiaratchi, C. B., & Wijeratne, E. M. S. (2015). Are meteotsunamis an underrated hazard? *Philosophical Transactions of the Royal Society A: Mathematical, Physical and Engineering Sciences*, *373*(2053)20140377. Available from https://doi.org/10.1098/rsta.2014.0377.

Rabinovich, A. B., & Monserrat, S. (1998). Generation of meteorological tsunamis (large amplitude seiches) near the Balearic and Kuril Islands. *Natural Hazards*, *18*(1), 27−55. Available from https://doi.org/10.1023/A:1008096627047.

Rhoades, D. A., Van Dissen, R. J., & Dowrick, D. J. (1994). On the handling of uncertainties in estimating the hazard of rupture on a fault segment. *Journal of Geophysical Research: Solid Earth*, *99*(7), 13701−13712. Available from https://doi.org/10.1029/94jb00803, http://agupubs.onlinelibrary.wiley.com/hub/jgr/journal/10.1002/(ISSN)2169-9356/.

Sandanbata, O., Watada, S., Satake, K., Kanamori, H., & Rivera, L. (2023). Two volcanic tsunami events caused by trapdoor faulting at a submerged caldera near Curtis and Cheeseman Islands in the Kermadec Arc. *Geophysical Research Letters*, *50*(7). Available from https://doi.org/10.1029/2022GL101086, http://agupubs.onlinelibrary.wiley.com/hub/journal/10.1002/(ISSN)1944-8007/.

Satake, K. (1989). Inversion of tsunami waveforms for the estimation of heterogeneous fault motion of large submarine earthquakes: The 1968 Tokachi-oki and 1983 Japan Sea earthquakes. *Journal of Geophysical Research*, *94*(5), 5627−5636. Available from https://doi.org/10.1029/JB094iB05p05627.

Satake, K., Fujii, Y., Harada, T., & Namegaya, Y. (2013). Time and space distribution of coseismic slip of the 2011 Tohoku earthquake as inferred from tsunami waveform data. *Bulletin of the Seismological Society of America*, *103*(2B), 1473−1492. Available from https://doi.org/10.1785/0120120122Japan, http://www.bssaonline.org/content/103/2B/1473.full.pdf + html.

Satake, K., & Kanamori, H. (1991). Abnormal tsunamis caused by the June 13, 1984, Torishima, Japan, earthquake. *Journal of Geophysical Research: Solid Earth*, *96*(B12), 19933−19939. Available from https://doi.org/10.1029/91jb01903.

Sato, H., & Taniguchi, H. (1997). Relationship between crater size and ejecta volume of recent magmatic and phreato-magmatic eruptions: Implications for energy partitioning. *Geophysical Research Letters*, *24*(3), 205−208. Available from https://doi.org/10.1029/96GL04004, http://onlinelibrary.wiley.com/journal/10.1002/(ISSN)1944-8007/issues?year = 2012.

Shao, G., & Ji, C. (2005). Preliminary result of the Mar 28, 2005 *Mw* 8.68 Nias Earthquake. http://www.geol.ucsb.edu/faculty/ji/big_earthquakes/2005/03/smooth/nias.html.

Shao, G., Li, X., Ji, C., & Maeda, T. (2011). Focal mechanism and slip history of the 2011 *Mw* 9.1 off the Pacific coast of Tohoku earthquake, constrained with teleseismic body and surface waves. *Earth, Planets and Space*, *63*(7), 559−564. Available from https://doi.org/10.5047/eps.2011.06.028, http://rd.springer.com/journal/40623.

Shen, Y., Whittaker, C. N., Lane, E. M., White, J. D. L., Power, W., & Nomikou, P. (2021). Laboratory experiments on tsunamigenic discrete subaqueous volcanic eruptions. Part 2: Properties of generated waves. *Journal of Geophysical Research: Oceans*, *126*(5). Available from https://doi.org/10.1029/2020JC016587, http://agupubs.onlinelibrary.wiley.com/agu/jgr/journal/10.1002/(ISSN)2169-9291/.

Shimazaki, K., & Nakata, T. (1980). Time-predictable recurrence model for large earthquakes. *Geophysical Research Letters*, *7*(4), 279−282. Available from https://doi.org/10.1029/GL007i004p00279.

Shuler, A., Ekström, G., & Nettles, M. (2013). Physical mechanisms for vertical-CLVD earthquakes at active volcanoes. *Journal of Geophysical Research: Solid Earth*, *118*(4), 1569−1586. Available from https://doi.org/10.1002/jgrb.50131, http://onlinelibrary.wiley.com/journal/10.1002/(ISSN)2169-9356.

Shuler, A., Nettles, M., & Ekström, G. (2013). Global observation of vertical-CLVD earthquakes at active volcanoes. *Journal of Geophysical Research: Solid Earth*, *118*(1), 138−164. Available from https://doi.org/10.1029/2012JB009721, http://onlinelibrary.wiley.com/journal/10.1002/(ISSN)2169-9356.

Small, D. T., & Melgar, D. (2021). Geodetic coupling models as constraints on stochastic earthquake ruptures: An example application to PTHA in Cascadia. *Journal of Geophysical Research: Solid Earth*, *126*(7). Available from https://doi.org/10.1029/2020JB021149, http://agupubs.onlinelibrary.wiley.com/hub/jgr/journal/10.1002/(ISSN)2169-9356/.

Somerville, P., Irikura, K., Graves, R., Sawada, S., Wald, D., Abrahamson, N., Iwasaki, Y., Kagawa, T., Smith, N., & Kowada, A. (1999). Characterizing crustal earthquake slip models for the prediction of strong ground motion. *Seismological Research Letters*, *70*(1), 59−80. Available from https://doi.org/10.1785/gssrl.70.1.59.

Steketee, J. A. (1958). On Volterra's dislocations in a semi-infinite elastic medium. *Canadian Journal of Physics*, *36*(2), 192−205. Available from https://doi.org/10.1139/p58-024.

Stirling, M., & Gerstenberger, M. (2018). Applicability of the Gutenberg-Richter relation for major active faults in New Zealand. *Bulletin of the Seismological Society of America*, *108*(2), 718−728. Available from https://doi.org/10.1785/0120160257, https://pubs.geoscienceworld.org/ssa/bssa/article-pdf/108/2/718/4109524/bssa-2016257.1.pdf.

Strasser, F. O., Arango, M. C., & Bommer, J. J. (2010). Scaling of the source dimensions of interface and intraslab subduction-zone earthquakes with moment magnitude. *Seismological Research Letters*, *81*(6), 941−950. Available from https://doi.org/10.1785/gssrl.81.6.941, http://srl.geoscienceworld.org/cgi/reprint/81/6/941.

Sykes, L. R. (1967). Mechanism of earthquakes and nature of faulting on the mid-oceanic ridges. *Journal of Geophysical Research*, *72*(8), 2131−2153. Available from https://doi.org/10.1029/jz072i008p02131.

Sykes, L. R., & Menke, W. (2006). Repeat times of large earthquakes: Implications for earthquake mechanics and long-term prediction. *Bulletin of the Seismological Society of America*, *96*(5), 1569−1596. Available from https://doi.org/10.1785/0120050083.

Synolakis, C.E. (2003). Chapter 9 - Tsunami and seiche. Earthquake Engineering Handbook. https://www.taylorfrancis.com/chapters/mono/10.1201/9781420042443-13/tsunami-seiche-charles-scawthorn-wai-fah-chen?context = ubx&refId = eb7aaf87-3182-4e56-98f5-0480150df554.

Tanioka, Y., & Satake, K. (1996). Tsunami generation by horizontal displacement of ocean bottom. *Geophysical Research Letters*, *23*(8), 861−864. Available from https://doi.org/10.1029/96GL00736, http://onlinelibrary.wiley.com/journal/10.1002/(ISSN)1944-8007/issues?year = 2012.

Tappin, D. R., Watts, P., & Grilli, S. T. (2008). The Papua New Guinea tsunami of 17 July 1998: Anatomy of a catastrophic event. *Natural Hazards and Earth System Sciences*, *8*(2), 243−266. Available from https://doi.org/10.5194/nhess-8-243-2008.

Thingbaijam, K. K. S., Mai, P. M., & Goda, K. (2017). New empirical earthquake source-scaling laws. *Bulletin of the Seismological Society of America*, *107*(5), 2225−2246. Available from https://doi.org/10.1785/0120170017, https://pubs.geoscienceworld.org/bssa/article-pdf/107/5/2225/3712063/BSSA-2017017.1.pdf.

Tinti, S., Bortolucci, E., & Armigliato, A. (1999). Numerical simulation of the landslide-induced tsunami of 1988 on Vulcano Island, Italy. *Bulletin of Volcanology*, *61*(1-2), 121−137. Available from https://doi.org/10.1007/s004450050267.

Tinti, S., Bortolucci, E., & Romagnoli, C. (2000). Computer simulations of tsunamis due to sector collapse at Stromboli, Italy. *Journal of Volcanology and Geothermal Research*, *96*(1-2), 103−128. Available from https://doi.org/10.1016/S0377-0273(99)00138-9.

Tkalčić, H., Dreger, D. S., Foulger, G. R., & Julian, B. R. (2009). The puzzle of the 1996 Bárdarbunga, Iceland, earthquake: No volumetric component in the source mechanism. *Bulletin of the Seismological Society of America*, *99*(5), 3077−3085. Available from https://doi.org/10.1785/0120080361, http://www.bssaonline.org/cgi/reprint/99/5/3077.

Torsvik, T., Paris, R., Didenkulova, I., Pelinovsky, E., Belousov, A., & Belousova, M. (2010). Numerical simulation of a tsunami event during the 1996 volcanic eruption in Karymskoye lake, Kamchatka, Russia. *Natural Hazards and Earth System Science*, *10*(11), 2359−2369. Available from https://doi.org/10.5194/nhess-10-2359-2010.

Ulvrová, M., Paris, R., Kelfoun, K., & Nomikou, P. (2014). Numerical simulations of tsunamis generated by underwater volcanic explosions at Karymskoye lake (Kamchatka, Russia) and Kolumbo volcano (Aegean Sea, Greece). *Natural Hazards and Earth System Sciences*, *14*(2), 401–412. Available from https://doi.org/10.5194/nhess-14-401-2014.

Vilibić, I., Šepić, J., Rabinovich, A. B., & Monserrat, S. (2016). Modern approaches in meteotsunami research and early warning. *Frontiers in Marine Science*, *3*(May). Available from https://doi.org/10.3389/fmars.2016.00057.

Wald, D. J., Heaton, T. H., & Hudnut, K. W. (1996). A dislocation model of the 1994 Northridge, California, earthquake determined from strong-motion, GPS, and leveling-line data. *Bulletin of the Seismological Society of America*, *86*(1B), S49–S70.

Wang, X., & Power, W.L. (2011). COMCOT: A tsunami generation propagation and run-up model. *GNS Science Report 2011/43*, 121 p.

Watts, P. (2000). Tsunami features of solid block underwater landslides. *Journal of Waterway, Port, Coastal and Ocean Engineering*, *126*(3), 144–152. Available from https://doi.org/10.1061/(ASCE)0733-950X(2000)126:3(144).

Watts, P., Grilli, S. T., Kirby, J. T., Fryer, G. J., & Tappin, D. R. (2003). Landslide tsunami case studies using a Boussinesq model and a fully nonlinear tsunami generation model. *Natural Hazards and Earth System Science*, *3*(5), 391–402. Available from https://doi.org/10.5194/nhess-3-391-2003, http://www.nat-hazards-earth-syst-sci.net/volumes_and_issues.html.

Watts, P., Grilli, S. T., Tappin, D. R., & Fryer, G. J. (2005). Tsunami generation by submarine mass failure. II: Predictive equations and case studies. *Journal of Waterway, Port, Coastal and Ocean Engineering*, *131*(6), 298–310. Available from https://doi.org/10.1061/(ASCE)0733-950X(2005)131:6(298).

Weichert, D. H. (1980). Estimation of the earthquake recurrence parameters for unequal observation periods for different magnitudes. *Bulletin of the Seismological Society of America*, *70*(4), 1337–1346. Available from https://doi.org/10.1785/bssa0700041337.

Wells, D. L., & Coppersmith, K. J. (1994). New empirical relationships among magnitude, rupture length, rupture width, rupture area, and surface displacement. *Bulletin of the Seismological Society of America*, *84*(4), 974–1002.

Williams, R. T., Davis, J. R., & Goodwin, L. B. (2019). Do large earthquakes occur at regular intervals through time? A perspective from the geologic record. *Geophysical Research Letters*, *46*(14), 8074–8081. Available from https://doi.org/10.1029/2019GL083291, http://agupubs.onlinelibrary.wiley.com/hub/journal/10.1002/(ISSN)1944-8007/.

Wilson, C. J. N., & Hildreth, W. (1997). The Bishop Tuff: New insights from eruptive stratigraphy. *Journal of Geology*, *105*(4), 407–440. Available from https://doi.org/10.1086/515937.

Wolfe, E.W., & Hoblitt, R.P. (1996). Overview of the eruptions. In Fire and mud: Eruptions and lahars of Mount Pinatubo, Philippines. https://pubs.usgs.gov/pinatubo/wolfe/.

Working Group on California Earthquake Probabilities. (2008). *The Uniform California earthquake rupture forecast, version 2 (UCERF2)*. Available from: https://pubs.usgs.gov/of/2007/1437/.

Yamazaki, Y., Bai, Y., Goo, L. L., Cheung, K. F., & Lay, T. (2023). Nonhydrostatic modeling of tsunamis from earthquake rupture to coastal impact. *Journal of Hydraulic Engineering*, *149*(9). Available from https://doi.org/10.1061/JHEND8.HYENG-133, https://ascelibrary.org/doi/10.1061/JHEND8.HYENG-13388.

Yamazaki, Y., Lay, T., Cheung, K. F., Yue, H., & Kanamori, H. (2011). Modeling near-field tsunami observations to improve finite-fault slip models for the 11 March 2011 Tohoku earthquake. *Geophysical Research Letters*, *38*(20). Available from https://doi.org/10.1029/2011GL049130, http://onlinelibrary.wiley.com/journal/10.1002/(ISSN)1944-8007/issues?year=2012.

Ye, L., Lay, T., Kanamori, H., & Rivera, L. (2016). Rupture characteristics of major and great (Mw ≥ 7.0) megathrust earthquakes from 1990 to 2015: 1. Source parameter scaling relationships. *Journal of Geophysical Research: Solid Earth*, *121*(2), 826–844. Available from https://doi.org/10.1002/2015JB012426, http://onlinelibrary.wiley.com/journal/10.1002/(ISSN)2169-9356.

Youngs, R. R., & Coppersmith, K. J. (1985). Implications of fault slip rates and earthquake recurrence models to probabilistic seismic hazard estimates. *Bulletin of the Seismological Society of America*, *75*(4), 939–964. Available from https://doi.org/10.1785/BSSA0750040939, https://pubs.geoscienceworld.org/ssa/bssa/article/75/4/939/118730/Implications-of-fault-slip-rates-and-earthquake.

Chapter 3

Tsunami propagation and runup modeling

Aditya Gusman[1], Katsuichiro Goda[2], Raffaele De Risi[3] and Ioan Nistor[4]

[1]*GNS Science, Lower Hutt, Wellington, New Zealand*, [2]*Department of Earth Sciences, Western University, London, ON, Canada*, [3]*School of Civil, Aerospace and Design Engineering, University of Bristol, Bristol, United Kingdom*, [4]*Department of Civil Engineering, University of Ottawa, Ottawa, ON, Canada*

3.1 Introduction

Tsunami numerical models are tools that can be used to simulate the propagation and behavior of tsunamis. These models are typically based on the shallow-water wave equations. Numerical methods, such as the finite difference method, can be employed to solve the equations. With input data, such as the tsunami source, bathymetry, and topography data, a tsunami numerical model can be run. Tsunami numerical models are essential in various types of studies, including (1) tsunami source studies, (2) tsunami inundation mapping, (3) tsunami early warning, (4) tsunami hazard assessments, (5) coastal engineering and structural design, and (6) tsunami impact and damage assessments. The objective of this chapter is to describe numerical methods for modeling tsunamis.

Section 3.2 discusses the features and properties of tsunamis, such as wavelength, period, height, and speed. This section presents how various simple equations and models can be used to estimate tsunami parameters, such as runup height, inundation distance, and inland flow depth. The section also briefly discusses harmonic wave theory, tsunami ray tracing method, and tsunami travel time calculation, which can provide useful insights into the behavior of tsunamis.

Section 3.3 introduces the basic principles and shallow-water equations used in tsunami numerical modeling, including those that describe wave propagation and inundation. The shallow-water equations are a set of partial differential equations that can describe the behavior of tsunamis. Several formulations (e.g., linear versus nonlinear and nondispersive vs dispersive) are introduced in this section.

Section 3.4 presents the finite difference scheme that can be used for solving shallow-water equations, which are an essential tool for building tsunami models. By applying the finite difference method, the equations can be discretized and then solved numerically. This section also provides a technique that allows simulating the tsunami at a fine resolution for some coastal regions where a detailed inundation map is necessary, while other regions of less interest may have less refinement.

Section 3.5 provides examples of how tsunami numerical models have been used in research and practical applications, such as for understanding tsunami events, tsunami warnings, and risk mitigation. Examples in this chapter illustrate some considerations involved in running a tsunami numerical model, such as selecting appropriate input data and parameter values and validating the model results.

3.2 Tsunami characteristics

The most common type of ocean waves is wind-generated waves. These waves are created by the transfer of energy from the wind to the ocean surface. The stronger the wind, the more powerful the waves can become. There are other types of waves in the ocean that are generated by different forces, such as tsunamis. Tsunamis are usually generated by sudden and large displacements of water in the ocean caused by earthquakes, landslides, volcanic eruptions, underwater explosions, and pressure waves propagating in the atmosphere (Chapter 2).

Ocean waves can be characterized by their length, height, velocity, and period. A classification can be made based on the wave period, as shown in Fig. 3.1 (Munk, 1950). The wave period is the time interval between the passage of

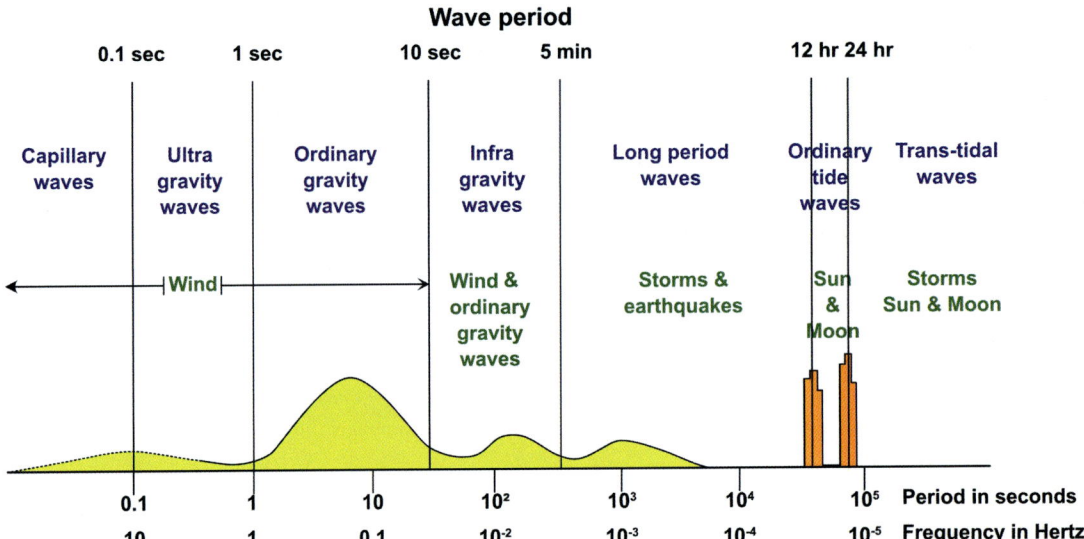

FIGURE 3.1 Classification of ocean waves according to wave period. The forces responsible for various portions of the spectrum are shown. The relative amplitude is indicated by the curve. *Modified from Munk, W. (1950). Origin and generation of waves.* Proceedings of the International Conference on Coastal Engineering, *1(1), 1. https://doi.org/10.9753/icce.v1.1.*

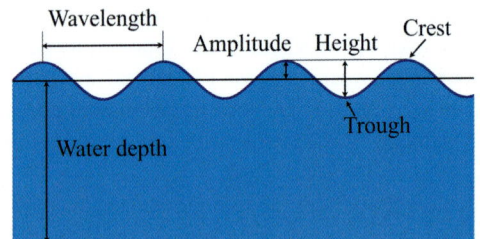

FIGURE 3.2 Parameters of water waves.

successive crests at a fixed point. Tsunamis have periods ranging from several minutes to over an hour, whereas wind waves have significantly shorter periods ranging from a few seconds to several minutes. In general, tsunamis have longer periods than wind waves because they have longer wavelengths.

Wavelength is another important parameter, which is the distance between two crests or troughs (Fig. 3.2). The wavelength of a tsunami typically ranges from tens to hundreds of kilometers. When the tsunami propagated across the Pacific Ocean, which is deeper than the water in the source area, its wavelength increased to around 500 km, while its wave height decreased. As an example, the wavelength of the 2011 Tohoku Tsunami in Japan was around 300 km in the earthquake source area (Gusman et al., 2012; Saito et al., 2014).

Tsunamis are typically categorized as shallow-water waves because they have long wavelengths relative to the water depth in which they are traveling. In general, a wave is considered a shallow-water wave when the ratio of the depth of the water to the wave length is typically less than 1/20. Most tsunamis have wavelengths in the range of 100–500 km, whereas the average depth of the world's oceans is only 4 km.

The period of a single tsunami event is relatively consistent, with a characteristic period that reflects the dominant frequency of the source event. However, there may still be some variability in the period of the tsunami due to factors, such as the bathymetry and coastal geography. It is possible to estimate the wavelength of a tsunami at the source using observations far from the source, especially using data measured by instruments in the deep ocean, as the bathymetric effects become minimal (Heidarzadeh & Satake, 2015). Tsunami source periods are usually dictated by the water depth at the location of the source and by the dimensions of the tsunami source and can be calculated using the following formula

$$C = \sqrt{gd} = \frac{\lambda}{T} \tag{3.1}$$

where λ is the tsunami wavelength (m), g is the gravitational acceleration of 9.81 m/second^2, d is the ocean water depth at the location of the tsunami source (m), and T is the tsunami period (second). From this equation, the estimated speed

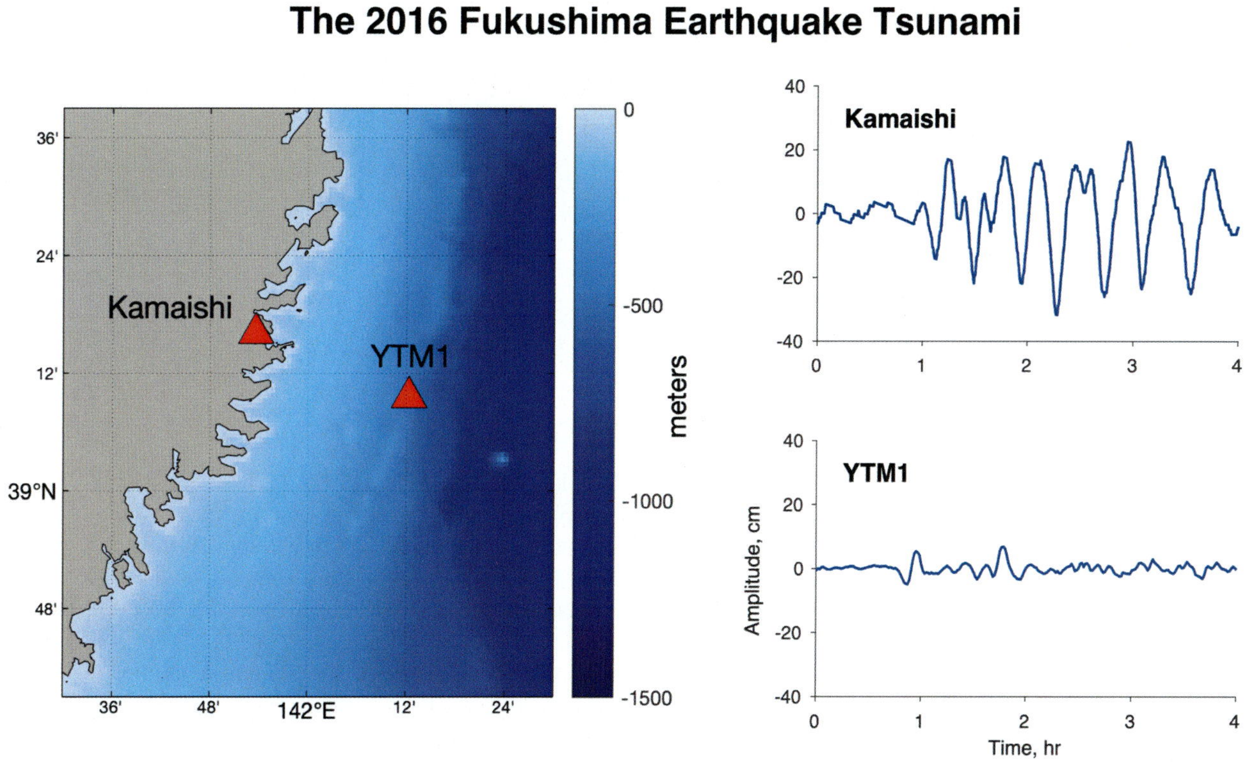

The 2016 Fukushima Earthquake Tsunami

FIGURE 3.3 Tsunami observations from the 2016 Fukushima Earthquake in Japan, recorded at offshore pressure gauge and coastal gauge. Map lines delineate study areas and do not necessarily depict accepted national boundaries.

of a tsunami propagating at a water depth of 4 km is 200 m/second. Alternatively, Eq. (3.1) can be used to estimate the tsunami period at a depth for a known wavelength. For instance, when considering a hypothetical tsunami source wavelength of 200 km at two varying water depths of 100 and 2000 m, the tsunami periods are estimated to be 106 and 24 minutes, respectively.

In the open ocean, tsunami waves may not be visible, and their height may be only a few centimeters or less. However, as they approach the shore and the ocean floor becomes shallower, the waves slow down, and their heights increase. For example, the tsunami of the 2016 Fukushima Earthquake (M_w 7.4) recorded at YTM1 station off the coastal town of Kamaishi was 6 cm, while the recorded tsunami at Kamaishi station near the coast was 20 cm (Fig. 3.3) (Gusman et al., 2017).

To gain an understanding of how water depth may influence the height of a tsunami, one can utilize Green's law. Green's law, which is a conservation law in fluid dynamics based on linearized shallow-water equations, can be used to approximate the tsunami amplitude near the coast from the tsunami amplitude at a location several kilometers away from the coast and at a deeper water depth. It is important to note, however, that this approximate approach is not suitable for estimating tsunami inundation. Tsunami amplitudes at two different depths can be calculated using the following formula (Satake, 2015)

$$A_2 = \sqrt{\frac{b_1}{b_2}} \sqrt[4]{\frac{d_1}{d_2}} A_1 \qquad (3.2)$$

where A_1 and A_2 are the tsunami amplitudes at location 1 and location 2, respectively, d_1 and d_2 are the water depths at the two locations, and b_1 and b_2 are the distances between two neighboring wave rays. The distance between two rays is constant if the wave propagates toward a straight coast with depth contours parallel to the coast. The wave refraction effects can be incorporated by including b into the formula.

Consider that a tsunami is observed at a water depth (d_1) of 500 m by an offshore station with an amplitude (A_1) of 0.3 m. If the wave ray paths of a plain wave from the station area to the coast are known, the distances of two rays at the station (b_1) and those at the beach (b_2) can be calculated. Assume that the measured b_1 and b_2 are 100 and 150 m, respectively. By applying Eq. (3.2), the tsunami amplitude (A_2) near the coast at a water depth (d_2) of 10 m is estimated

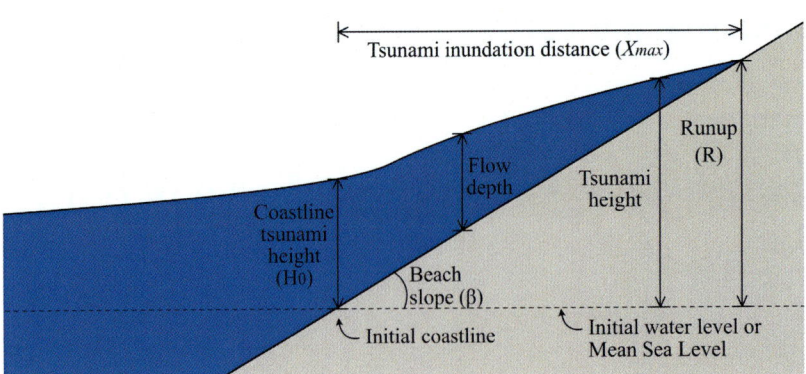

FIGURE 3.4 Definitions of runup, tsunami height, tsunami flow depth, and tsunami inundation distance.

to be about 0.65 m. The evolution described by Green's law is valid over a wide range of slopes and for finite-amplitude waves at least in the region of gradual shoaling (Synolakis & Skjelbreia, 1993).

Green's law should not be used to estimate the tsunami amplitude at the coast from the maximum sea surface displacement of the source or vice versa. This is because at the source, the tsunami energy is radiated from the source in different directions and not just to the direction of the evaluated location. However, Eq. (3.2) may be useful to estimate the tsunami amplitude in the vicinity of the source and in areas outside of it based on observations along the coast (Abe, 1973; Satake, 1988).

Green's law is also not applicable to estimating the inland tsunami inundation or the tsunami amplitude at locations very close to complex coastal geomorphology as the nonlinear effects become important. The tsunami runup (R) is one of the most important parameters in the study of tsunamis (Fig. 3.4). A tsunami runup is defined as a difference between the elevation of the maximum tsunami penetration (inundation line) and the sea level at the time of the tsunami. In practical terms, runup is only measured where there is clear evidence of the inundation limit on the shore. Where the elevation is not measured at the maximum of horizontal inundation, this is often referred to as the inundation height.

The height of a tsunami can vary significantly along the coast depending on several factors. One of the main factors is the shape of the coastline. In areas where the coastline has deep bays or inlets, the tsunami can be amplified, and its height can be much higher than that in areas where the coastline is straight and uniform. The height of the 2011 Tohoku Tsunami that struck Japan was particularly devastating. For example, the tsunami that hit Kamaishi reached heights of up to 20 m in some very steep areas, while the tsunami around a gentler slope was approximately 10 m high (Gusman et al., 2012; Mori et al., 2012). The impact of topography on tsunami runup heights was found to be significant, indicating a strong influence of local geomorphology on the severity of the tsunami's impact. Therefore, it is necessary to use high-quality and high-resolution bathymetry and topography data when studying tsunami inundation (MacInnes et al., 2013).

Tsunami runup can be estimated from the tsunami height at the coastline (H_0) using Eq. (3.3), which is based on the linear solution for the wave height over the sloping beach (Synolakis, 1987, 1991).

$$R = 2.831 \sqrt{\cos\beta}\, H_0^{5/4} \tag{3.3}$$

where β is the slope of the beach (Fig. 3.4). This formula is valid when $(H_0)^{0.5} \gg 2.88\tan\beta$ and for waves that do not break during runup.

The following equation can be used to estimate the tsunami inundation distance (X_{max}) on flat-lying coast from the wave height at the coast H_0 (Bretschneider & Wybro, 1977)

$$X_{max} = 0.06 H_0^{4/3}/n^2 \tag{3.4}$$

where n is Manning's roughness coefficient. Eq. (3.4) is further modified to include a slope factor and estimate the tsunami flow depth (McSaveney & Rattenbury, 2000) as given in the following equation

$$H_{loss} = \left(167 n^2 / H_0^{1/3}\right) + 5\sin S \tag{3.5}$$

where H_{loss} is the loss in wave height per meter of inundation distance, and S is the ground slope. The equation can be implemented using the cost-distance function available in GIS software (Berryman, 2005). The evaluation for the function requires a grid of cells that represents the sea, while the cost surface is another grid of cells that shows the loss in wave height (H_{loss}). It is worth noting that the cost associated with roughness is direction-independent, while the cost linked with slope depends on the direction of travel. However, a potential issue arises when determining the slope as the absolute value of the maximum slope using the digital elevation model (DEM), which may or may not correspond to the direction of the tsunami. This mismatch can result in the underestimation of the inundation distance, leading to potential inaccuracies in tsunami inundation distance.

Bernoulli's principle for fluid dynamics can be used to develop an equation for predicting the maximum tsunami depth profiles and inundation distances (Smart et al., 2016). The following equation is suggested to estimate the flow depth

$$D = (H_0 + aS_0)e^{-\frac{2x}{3a}} - aS_0 \qquad (3.6)$$

where D is the inland flow depth, a is the roughness aperture, and S_0 is a uniform ground slope. The roughness can be calculated using $a = 2\Delta/f$, where f is the Darcy friction factor and Δ is the protrusion spacing. A typical f value of 0.05 (for fully turbulent flow in rough conduits) with a typical protrusion spacing of 2 m (e.g., the distance between coconut palms or buildings) gives an indicative roughness aperture a value of 80 m (Smart et al., 2016).

3.2.1 Harmonic wave

Tsunamis are more complex than simple harmonic waves. Nevertheless, harmonic wave theory can provide useful insights into the behavior of tsunamis. A progressive (propagating) wave can be described as a harmonic wave, and the water displacement $h(x,t)$ can be described as a function of both position and time

$$h(x,t) = A \sin(kx \pm \omega t) \qquad (3.7)$$

A harmonic wave is characterized by its amplitude A, the angular frequency ω, and the wavenumber k. The harmonic wave solution can give a periodic function of time at a position x_0. Because the function returns to the same value when ωt changes by 2π, the oscillation is characterized by the period, $T = 2\pi/\omega$, the time over which it repeats. The periodicity can also be described by the frequency, $f = 1/T = \omega/2\pi$, which is the number of oscillations within a unit time or by the angular frequency. The period has the dimensions of time (T), so the frequency and angular frequency have dimensions of T^{-1}.

Alternatively, $h(x,t)$ at a fixed time, t_0, can be examined by plotting $h(x,t_0) = A \sin(kx - \omega t_0)$ as a function of position. The displacement is periodic in space over a distance equal to the wavelength, $\lambda = 2\pi/k$, the distance between two corresponding points in a cycle. How the oscillation repeats in space can also be described by k, the wavenumber or spatial frequency, which is 2π times the number of cycles occurring in a unit distance. The wavelength has units of distance, so the wavenumber has dimensions of M^{-1}. A list of relationships between wave variables is shown in Table 3.1.

The harmonic wave function is useful for comprehending wave characteristics. To illustrate this, let us consider a transverse wave with an amplitude of 0.1 m, wavelength of 2 m, and period of 0.1 seconds. The wavenumber of this wave is $k = 2\pi/2 = \pi$, and its angular frequency is $\omega = 2\pi/0.1 = 20\pi$. The speed of the wave can be found using the wavenumber and the angular frequency or using the wavelength and the wave period. In this case, the wave speed is $C = 2/0.1 = 20$ m/second. To visualize the wave, its displacement can be plotted on the y-axis and its position on the x-axis at different times (Fig. 3.5). This graphical representation provides a clear understanding of the wave's behavior, including its shape, amplitude, and velocity.

TABLE 3.1 Relationship between wave variables.

Variable	Unit	Relationships
Speed	Distance/time	$C = \frac{\lambda}{T} = f\lambda = \frac{\omega}{k}$
Period	Time	$T = \frac{1}{f} = \frac{\lambda}{C} = \frac{2\pi}{\omega}$
Angular frequency	Time^{-1}	$\omega = 2\pi f = kC = \frac{2\pi}{T}$
Frequency	Time^{-1}	$f = \frac{1}{T} = \frac{C}{\lambda} = \frac{\omega}{2\pi}$
Wavelength	Distance	$\lambda = \frac{C}{f} = CT = \frac{2\pi}{k}$
Wavenumber	Distance^{-1}	$k = \frac{\omega}{C} = \frac{2\pi f}{C} = \frac{2\pi}{\lambda}$

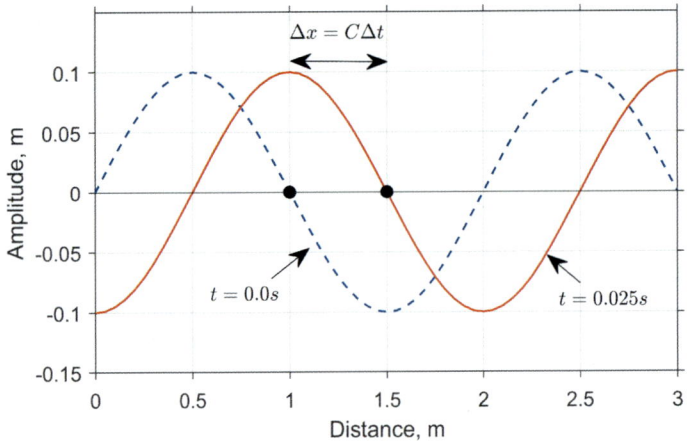

FIGURE 3.5 Two waves plotted as a function of distance from the origin for two different times using Eq. (3.7). The dashed blue line indicates the wave at $t = 0.0$ second, and the solid red line indicates the wave at $t = 0.025$ seconds. Black dots indicate the distance (0.5 m) traveled by the wave with a speed of 20 m/second in 0.025 seconds.

3.2.2 Tsunami ray tracing

Tsunami (long wave) speed depends on water depth, propagating slower in shallower water. It is well known that spatial variations of wave speed allow waves to change directions. The direction of propagation of tsunami waves can be determined from the wave rays. A wave ray is the closest path traveled by a wave. By tracking the paths of the wave rays, the areas where the tsunami waves will concentrate their energy can be determined. Tsunami ray tracing can also be used to evaluate the bathymetry effect along the tsunami propagation path (Gusman et al., 2017; Satake, 1988). The differential equations for tsunami ray paths are (Satake, 1988)

$$\frac{d\theta}{dt} = \frac{1}{sR}\cos\zeta \tag{3.8}$$

$$\frac{d\varphi}{dt} = \frac{1}{sR\sin\theta}\sin\zeta \tag{3.9}$$

$$\frac{d\zeta}{dt} = -\frac{\sin\zeta}{s^2R}\frac{\partial s}{\partial\theta} + \frac{\cos\zeta}{s^2R\sin\theta}\frac{\partial s}{\partial\varphi} - \frac{1}{sR}\sin\zeta\cot\theta \tag{3.10}$$

where θ and φ are the colatitude (90 degrees − latitude) and longitude (in radians) of the ray at time t, s is the slowness calculated from the phase velocity ($s = 1/C$), R is the radius of the Earth (6371 km), and ζ is the ray direction measured counterclockwise from the south (in radians). A tsunami can be assumed to be a shallow-water wave with a phase velocity of $c = \sqrt{gd}$. The differential equations can be solved by the fourth-order Runge–Kutta method. Ray paths are typically computed using initial directions ranging from 0 degree to 360 degrees. Utilizing a 0.1 degrees interval results in a total of 3600 distinct ray paths.

The tsunami energy transmission for the initial wave from the source location can be represented by the tsunami ray tracing result, as shown in Fig. 3.6 for the case of the M_w 7.4, 2016 Fukushima Earthquake (Gusman et al., 2017). A plot of tsunami rays can show how the wave is refracted and how it keeps changing direction until it reaches the coastline. The ray tracing technique was used to show that much of the wave from the tsunami source was refracted back to the coast due to the bathymetric gradient off the eastern coast of Japan and around the epicenter. In this case, the bathymetry around the earthquake source allows a large portion of the tsunami energy to be refracted back to the island of Honshu, Japan, as indicated by the tsunami rays.

The long-wave theory is commonly used in tsunami ray tracing, where the speed of the waves is affected by the depth of the water. It is important to note that a tsunami is a dispersive wave whose propagation depends not only on the depth of the water but also on the frequency of the waves (Sandanbata et al., 2018). The speed of a long-period wave with a period longer than about 1000 seconds at a given water depth is faster than waves with shorter periods (Fig. 3.7). The phase velocity (C) of a surface gravity wave can be well approximated by

$$C = \frac{\omega}{k} = \sqrt{\frac{g}{k}\tanh kd} = \sqrt{\frac{g\lambda}{2\pi}\tanh\frac{2\pi d}{\lambda}} \tag{3.11}$$

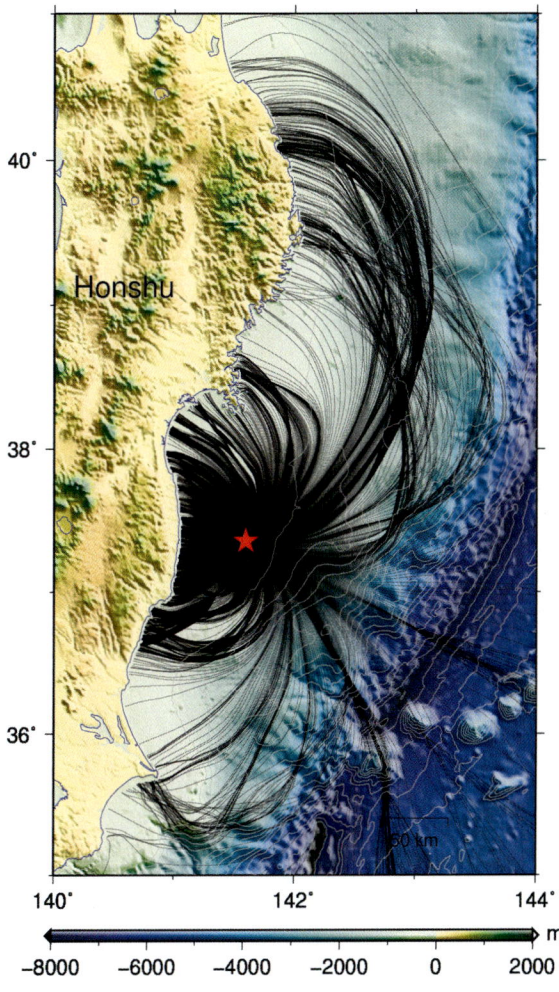

FIGURE 3.6 Tsunami rays (*black* lines) from the epicenter (*red* star) of the 2016 Fukushima (M_w 7.4) Earthquake in Japan. Map lines delineate study areas and do not necessarily depict accepted national boundaries. *From Gusman, A. R., Satake, K., Shinohara, M., Sakai, S., & Tanioka, Y. (2017). Fault slip distribution of the 2016 Fukushima earthquake estimated from tsunami waveforms.* Pure and Applied Geophysics, 174, *2925–2943. https://doi.org/10.1007/s00024-017-1590-2.*

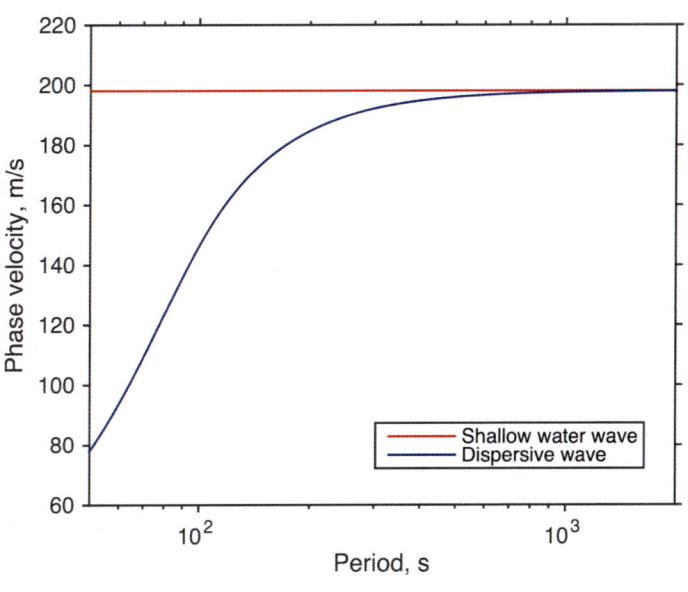

FIGURE 3.7 Phase velocities of waves with different periods traveling at a water depth of 4000 m.

Another form for the phase velocity can be written as follows

$$C = \frac{\omega}{k} = \frac{2\pi f}{k} \tag{3.12}$$

To find ray paths of dispersive tsunamis, the phase velocity $C(f,d)$ for a given pair of frequency f and depth d is needed. The unknown parameter k can be first calculated by combining Eqs. (3.11) and (3.12). Once k is known, the phase velocity can be solved using the following formula

$$\frac{2\pi f}{k} = \sqrt{\frac{g}{k} \tanh kd} \tag{3.13}$$

Eq. (3.13) can be solved iteratively using a recursive algorithm with an initial phase velocity of

$$C_0 = \sqrt{gd} \tag{3.14}$$

The calculation of

$$k_n = \frac{2\pi f}{C_{n-1}}, n = 1, 2, 3, \ldots \tag{3.15}$$

and

$$C_n = \sqrt{\frac{g}{k_n} \tanh(k_n D)}, n = 1, 2, 3, \ldots \tag{3.16}$$

can be repeated until the dimensionless condition for convergence of

$$\frac{|k_n - k_{n-1}|}{k_n} < \Delta C \frac{k_n}{2\pi f} = \frac{\Delta C}{C_{n-1}} \tag{3.17}$$

is satisfied with a very small value of ΔC (e.g., 0.00001 m/second). A solution can also be obtained graphically.

Let us consider an example of the phase velocity and wavenumber for a wave with a frequency of 20 mHz traveling at a water depth of 4000 m. The phase velocities for waves with a wave frequency of 20 mHz and different wavenumbers can be calculated using Eq. (3.12), and the result is represented by the blue line in Fig. 3.8. On the other hand, phase velocities for waves with different wavenumbers that travel at the water depth of 4000 m can be calculated using Eq. (3.11), and the result is represented by the red line in Fig. 3.8. The intersection of these two lines marked by the plus sign is the solution to the problem which means that the wave traveling at a frequency of 20 mHz and at a water depth of 4000 m has $C = 78.06$ m/second and $k = 0.0016$. While the speed of shallow-water waves traveling at a water depth of 4000 m is 198 m/second.

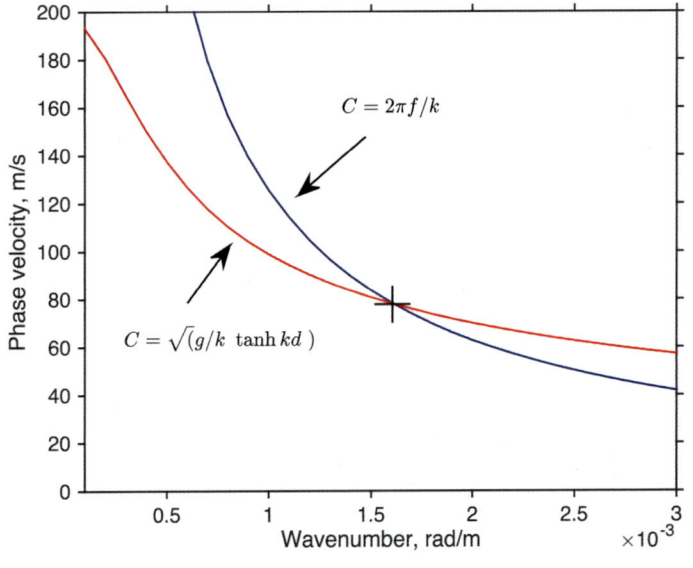

FIGURE 3.8 Phase velocities for different wavenumbers. The blue line represents phase velocities of waves with a frequency of 20 mHz at different wavenumbers. The red line represents phase velocities for waves at 4000 m depth and at different wavenumbers.

3.2.3 Tsunami travel time

Tsunamis triggered by earthquakes have been recorded by tide gauges, particularly at locations around the Pacific Ocean since the late 19th century. The National Oceanic and Atmospheric Administration (NOAA) coastal tide gauge data collection includes tsunami marigrams from 1854 to 1994 (https://www.ngdc.noaa.gov/hazard/tide.shtml). The timing of tsunami arrival can be determined using records of coastal water-level gauges. Estimating the arrival time of a tsunami is one of the most important pieces of information that authorities can provide to coastal communities following a submarine earthquake. For instance, the Pacific Tsunami Early Warning Center typically uses the estimated time of arrival to issue timely warnings and alerts to coastal communities, allowing people to evacuate to safer areas before the tsunami arrives.

Tsunami travel time estimates can be obtained by using the shallow-water wave approximation, in which the tsunami propagates at shallow-water wave velocity. A velocity field can be calculated from a bathymetry grid. Huygens's principle can be used to calculate the travel time at every cell of the grid. The principle states that "Every point on a wavefront is the source of spherical wavelets which spread out in the forward direction at the speed of light (shallow water wave speed in case of a tsunami). The sum of these spherical wavelets forms the wavefront." Geoware's tsunami travel time program (Wessel, 2009) computes the velocities by utilizing an input bathymetry grid and employs Huygens's constructions to propagate the wavefront from the epicenter to all cells on the grid. Wessel (2009) showed the accuracy of tsunami travel time predictions using historical data for tsunamigenic earthquakes in the Pacific and Indian Oceans. More than 1500 tsunami travel-time records for 127 earthquakes were analyzed, and the observed travel times were compared with predictions using Huygens' method and the long-wave approximation. A high correspondence between predicted and reported travel times was found. However, significant discrepancies still existed, including outliers with slower propagation speeds than predicted travel times.

Fig. 3.9 shows the predicted tsunami travel time originating from the epicenter of the 2021 Raoul Island Earthquake (M_w 8.1) in the Kermadec subduction zone. In most cases, the agreement between the predicted tsunami travel times

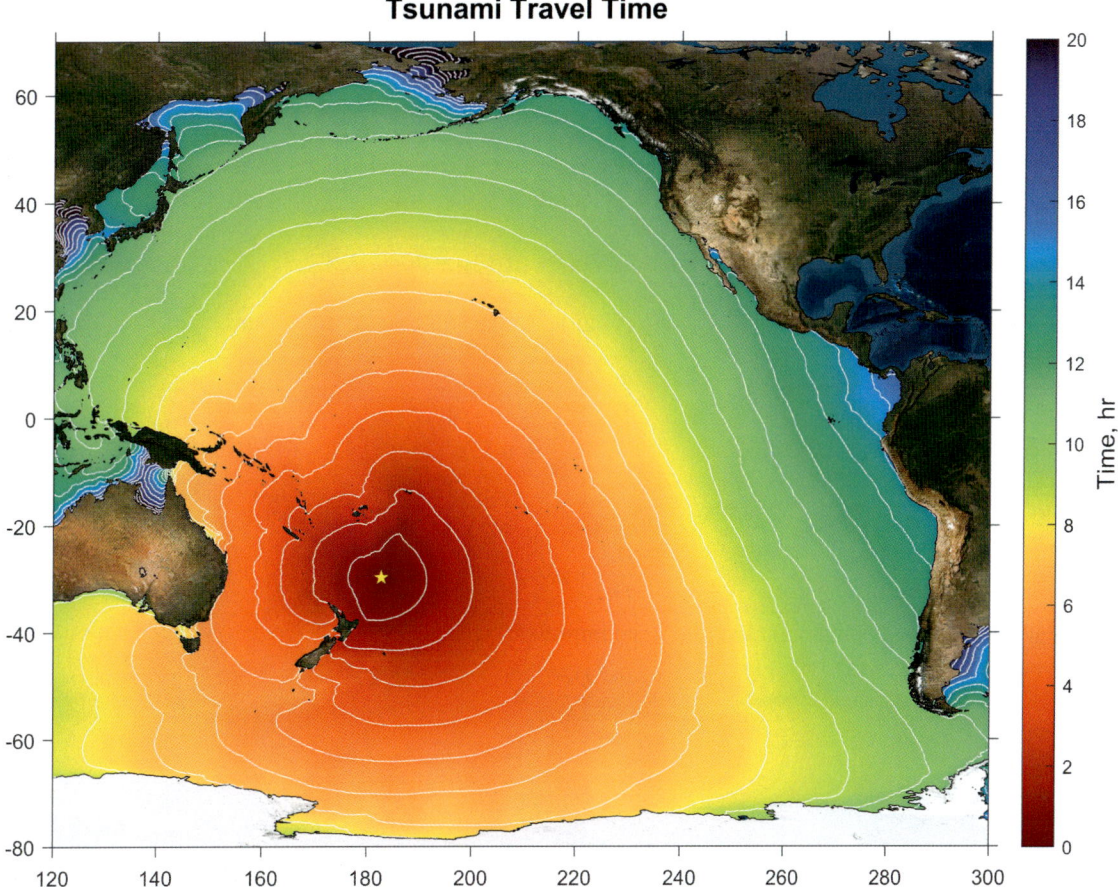

FIGURE 3.9 Predicted tsunami travel time using Huygens' method. The contour interval for the tsunami travel time map is 1 h. The yellow star represents the epicenter of the 2021 Raoul Island Earthquake (M_w 8.1). Map lines delineate study areas and do not necessarily depict accepted national boundaries.

and the actual observations is high, especially at offshore stations. The predicted estimated times of arrival of a tsunami using Huygens' method may not always align with the observed arrival times due to various factors, including but not limited to the following (Wessel, 2009):

- Inaccurate bathymetry grid.
- Poorly located epicenter or uncertain origin time.
- Selection of a pseudo-epicenter off the coast when the epicenter is on land.
- Insufficient representation of the rupture zone due to a point approximation of the epicenter.
- A significant impact of nonlinear propagation effects in shallow waters.

The tsunami ray tracing method can be used to estimate the tsunami travel time from the source. The method of tracing tsunami rays from the tsunami source can be seen in Section 3.2.2. The tsunami travel time along the path is available as a time step and is used to calculate the progression of the path for each iteration. With this method, the timing along the ray paths can be derived through calculations, and subsequently, the wavefront at specific time instances can be plotted on the basis of these computed points.

3.3 Tsunami propagation and inundation modeling

Shallow-water (long) wave theory facilitates the analysis of tsunami characteristics. In the shallow-water wave theory, the vertical acceleration of water particles is considered negligible compared to the gravitational acceleration. The theory applies to wavelengths significantly longer than the water depth. The amplitude of tsunamis offshore is relatively small compared to the ocean depth and wavelength, an assumption that is also used in the theory.

In the following, the equations of mass conservation and motion in the absence of thermal changes, which are irrelevant to the propagation of water waves, are summarized. Any energy equation is a consequence of only the motion through Newton's second law without any contribution from the thermodynamics of the fluid.

3.3.1 Equation of mass conservation

Under the fundamental assumption that matter (mass) is neither created nor destroyed anywhere in the fluid, the change in mass within a control volume is solely due to the rate of mass flowing into the volume across its boundary. A common form of the equation of mass conservation or the continuity equation is

$$\frac{\partial \rho}{\partial t} + \rho(\nabla \cdot \mathbf{u}) + (\mathbf{u} \cdot \nabla)\rho = 0 \tag{3.18}$$

The velocity of the fluid is $u(\mathbf{x},t)$ where $\mathbf{x}$ is the position vector and t is the time coordinate. The density (mass/unit volume) of the fluid is $\rho(\mathbf{x},t)$.

The convective derivative operator

$$\frac{D}{Dt} \equiv \frac{\partial}{\partial t} + \mathbf{u} \cdot \nabla \tag{3.19}$$

can be used to rewrite Eq. (3.18) as follows

$$\frac{D\rho}{Dt} + \rho \nabla \cdot \mathbf{u} = 0 \tag{3.20}$$

Noting that the following equation holds for an incompressible flow

$$\frac{D\rho}{Dt} = 0 \tag{3.21}$$

Eq. (3.20) becomes

$$\nabla \cdot \mathbf{u} = 0 \tag{3.22}$$

In Cartesian coordinates, $\mathbf{x} \equiv (x,y,z)$, with $\mathbf{u} \equiv (u,v,w)$, Eq. (3.22) becomes

$$\frac{\partial u}{\partial x} + \frac{\partial v}{\partial y} + \frac{\partial w}{\partial z} = 0 \tag{3.23}$$

Eq. (3.22), which shows the divergence of **u** is zero, describes that the velocity in an incompressible fluid has a solenoidal distribution. The constancy of ρ on individual fluid shown in Eq. (3.21) can be interpreted as ρ is constant everywhere for an incompressible fluid, which is an adequate assumption for fluids like water.

3.3.2 Equation of motion: Euler's equation

Newton's second law can be applied to a fluid that is assumed to be inviscid, that is, it has zero viscosity. Newton's second law dictates that the rate of change of momentum of the fluid is balanced against the resultant force acting on the fluid. Two types of force are relevant in fluid mechanics: a body force, which has its source exterior to the fluid, and a local force, which is the force exerted on a fluid element by other elements nearby. The body force that needs to be considered in studying water waves is gravity. The general body force is defined as $\mathbf{F}(\mathbf{x},t)$ per unit mass. If **F** is due solely to the acceleration of gravity (g), $\mathbf{F} \equiv (0,0,-g)$ in Cartesian coordinates with z measured positive upwards. The local force comprises a pressure contribution that is present and is represented by the stress tensor in the fluid. Here, only the pressure (P) is retained in the formulation, which produces a normal force acting on any element of the fluid. The Euler's equation for an inviscid fluid can be written as follows

$$\frac{D\mathbf{u}}{Dt} = -\frac{1}{\rho}\nabla P + \mathbf{F} \tag{3.24}$$

In Cartesian coordinates (i.e., $\mathbf{x} \equiv (x,y,z)$, with $\mathbf{u} \equiv (u,v,w)$ and $\mathbf{F} \equiv (0,0,-g)$), Eq. (3.24) becomes

$$\frac{Du}{Dt} = -\frac{1}{\rho}\frac{\partial P}{\partial x} \tag{3.25}$$

$$\frac{Dv}{Dt} = -\frac{1}{\rho}\frac{\partial P}{\partial y} \tag{3.26}$$

$$\frac{Dw}{Dt} = -\frac{1}{\rho}\frac{\partial P}{\partial z} - g \tag{3.27}$$

where

$$\frac{D}{Dt} \equiv \frac{\partial}{\partial t} + u\frac{\partial}{\partial x} + v\frac{\partial}{\partial y} + w\frac{\partial}{\partial z} \tag{3.28}$$

Eqs. (3.25), (3.26), and (3.27) can be rewritten as follows

$$\frac{\partial u}{\partial t} + u\frac{\partial u}{\partial x} + v\frac{\partial u}{\partial y} + w\frac{\partial u}{\partial z} = -\frac{1}{\rho}\frac{\partial P}{\partial x} \tag{3.29}$$

$$\frac{\partial v}{\partial t} + u\frac{\partial v}{\partial x} + v\frac{\partial v}{\partial y} + w\frac{\partial v}{\partial z} = -\frac{1}{\rho}\frac{\partial P}{\partial y} \tag{3.30}$$

$$\frac{\partial w}{\partial t} + u\frac{\partial w}{\partial x} + v\frac{\partial w}{\partial y} + w\frac{\partial w}{\partial z} = -\frac{1}{\rho}\frac{\partial P}{\partial z} - g \tag{3.31}$$

3.3.3 Shallow-water wave equations

Tsunami wavelengths are usually much longer than the water depth ($\lambda \gg d$). For example, consider the 2011 Tohoku Tsunami, where the recorded period at the deep-ocean assessment and reporting of Tsunamis (DART) 21418 station was approximately 35 minutes (Gusman et al., 2012). The DART station is situated at a water depth of 5.7 km. According to Eq. (3.1), the estimated wavelength for this particular tsunami event at the station is about 500 km, significantly surpassing the water depth at which this wave was recorded. In this situation, the vertical acceleration of water is much smaller than the acceleration of gravity. This means that the horizontal motion of water particles is almost uniform from the ocean bottom to the surface. As the vertical acceleration and velocity are neglected (where $Dw/Dt = 0$ and $w = 0$), the vertical component in Eq. (3.31) becomes

$$\frac{1}{\rho}\frac{\partial P}{\partial z} = -g \tag{3.32}$$

By assuming the density (ρ) is uniform, and the pressure is hydrostatic, the vertical pressure can be expressed as follows

$$P(z) = \rho g(\eta + d) + p \tag{3.33}$$

where η is the water level, d is the water depth, and p is the atmospheric pressure.

The pressure derivatives in the horizontal directions, as stated in Eqs. (3.29) and (3.30), are now dependent on the water level and atmospheric pressure

$$\frac{\partial P}{\partial x} = \rho g \frac{\partial \eta}{\partial x} + \frac{\partial p}{\partial x} \tag{3.34}$$

$$\frac{\partial P}{\partial y} = \rho g \frac{\partial \eta}{\partial y} + \frac{\partial p}{\partial y} \tag{3.35}$$

All terms containing w in Eqs. (3.29) and (3.30) can be omitted as the velocity of water particles in the vertical direction is neglected. As shown in Mader (2004), the integration of Eqs. (3.29) and (3.30) over the range from $z = d(x,y)$ to $z = \eta(x,y)$, and the introduction of vertically averaged velocity, Eqs. (3.34) and (3.35) become

$$\frac{\partial u}{\partial t} + u\frac{\partial u}{\partial x} + v\frac{\partial u}{\partial y} + g\frac{\partial \eta}{\partial x} + \frac{1}{\rho}\frac{\partial p}{\partial x} = 0 \tag{3.36}$$

$$\frac{\partial v}{\partial t} + u\frac{\partial v}{\partial x} + v\frac{\partial v}{\partial y} + g\frac{\partial \eta}{\partial y} + \frac{1}{\rho}\frac{\partial p}{\partial y} = 0 \tag{3.37}$$

The effects of atmospheric pressure, temporal changes, and spatial gradient on tsunami propagation under normal weather conditions are very small ($\partial p/\partial x \approx 0$, $\partial p/\partial y \approx 0$) and can be ignored. However, a large atmospheric pressure disturbance, such as a fast-moving air pressure wave with a mesoscale atmospheric pressure spatial gradient, can cause a displacement of the water body, thus generating a tsunami. Tsunamis of atmospheric origin, known as meteotsunamis, have been observed in many places around the world (Gusman et al., 2022; Monserrat et al., 2006; Pattiaratchi & Wijeratne, 2015; Rabinovich & Monserrat, 1998; Vilibić et al., 2016). For instance, atmospheric pressure waves generated by a volcanic eruption can trigger a sudden pressure variation in the atmosphere and can be the source of a tsunami.

Next, the incompressibility of water is assumed with the finite depth of the ocean. The equation of continuity can be integrated over the vertical direction with boundary conditions at the free surface of

$$w(\eta) = \frac{\partial \eta}{\partial t} + u\frac{\partial \eta}{\partial x} + v\frac{\partial \eta}{\partial y} \tag{3.38}$$

and at the bottom of

$$w(-d) = -u\frac{\partial d}{\partial x} - v\frac{\partial d}{\partial y} \tag{3.39}$$

The vertical integration of the continuity equation (Eq. (3.23)) can be performed with these free surface and bottom boundaries (Mader, 2004). Then, the linear shallow-water equations can be written as follows

$$\frac{\partial \eta}{\partial t} + \frac{\partial (u(\eta + d))}{\partial x} + \frac{\partial (v(\eta + d))}{\partial y} = 0 \tag{3.40}$$

$$\frac{\partial u}{\partial t} + g\frac{\partial \eta}{\partial x} = 0 \tag{3.41}$$

$$\frac{\partial v}{\partial t} + g\frac{\partial \eta}{\partial y} = 0 \tag{3.42}$$

3.3.3.1 Linear tsunami model

The linear shallow-water equations can be solved to simulate tsunami propagation. Typically, a linear tsunami model is applicable to simulating tsunamis in locations that are sufficiently deep and far enough from the coast. A location is considered sufficiently deep if the tsunami elevation is much smaller than the water depth. For simulating nearshore tsunamis, it is more appropriate to utilize the nonlinear shallow-water equations, which are explained in the next section.

The above equations utilize the velocity fields u and v. These equations can also be written in terms of discharge fluxes (M and N) in the x and y directions. An alternative formulation of the shallow-water equations can adopt the total water depth $D = \eta + d$ by introducing the discharge fluxes

$$M = \int_{-d}^{\eta} u\,dz = u(\eta + d) \tag{3.43}$$

$$N = \int_{-d}^{\eta} v\,dz = v(\eta + d) \tag{3.44}$$

The shallow-water equations can be written in terms of discharge fluxes as follows

$$\frac{\partial \eta}{\partial t} + \frac{\partial M}{\partial x} + \frac{\partial N}{\partial y} = 0 \tag{3.45}$$

$$\frac{\partial M}{\partial t} + gd\frac{\partial \eta}{\partial x} = 0 \tag{3.46}$$

$$\frac{\partial N}{\partial t} + gd\frac{\partial \eta}{\partial y} = 0 \tag{3.47}$$

A linear tsunami propagation model in which the water depth d is considered instead of the total water depth D. This is still valid for simulating tsunamis in the open ocean because the one-dimensional water elevation is much smaller than the water depth ($\eta \ll d$).

In tsunami simulations, the seafloor or sea surface displacement from an earthquake source is typically assumed to happen instantaneously. In this case, the initial condition for the simulation is the sea surface displacement caused by the earthquake. Fig. 3.10 shows how a tsunami from a sudden sea surface displacement propagates across the open ocean. The tsunami amplitude decreases, and its wavelength increases as it propagates into deeper water. On the other hand, the tsunami amplitude increases, and its wavelength decreases as it propagates into shallower water.

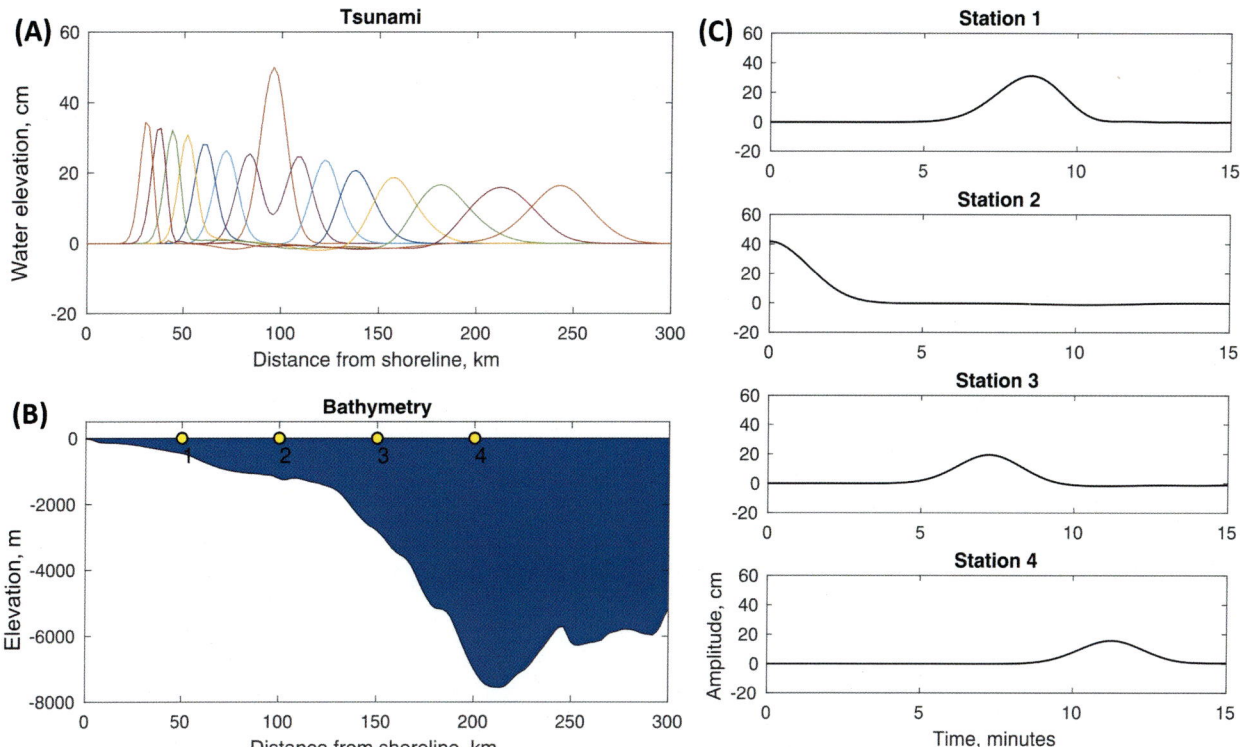

FIGURE 3.10 One-dimensional tsunami simulation. (A) Simulated tsunami profiles at different times. (B) Bathymetry data used in the tsunami simulation. Yellow circles indicate the observation points. (C) Simulated tsunami time series at the four observation points.

To model the transient seafloor motions, such as displacement from submarine earthquake rupturing process or landslides, a bathymetric variation term ($\partial \xi / \partial t$) needs to be introduced. The continuity equation, when incorporating the seafloor motion, can be expressed as follows

$$\frac{\partial \eta}{\partial t} + \frac{\partial M}{\partial x} + \frac{\partial N}{\partial y} = \frac{\partial \xi}{\partial t} \tag{3.48}$$

where ξ is the displacement at the seafloor or the sea surface.

3.3.3.2 Nonlinear tsunami model

To simulate a tsunami along the coast and tsunami inundation more realistically, the nonlinear shallow-water equations that include the convection and bottom friction terms should be considered. In the coastal area, both terms become significant. The bottom stress term (τ_x/ρ), which is proportional to squared velocity, can be introduced to the momentum equations (i.e., Eqs. (3.46) and (3.47)). Manning's roughness coefficient can be used to establish the relationship between the squared velocity and the bottom stress. For this case, the shallow-water equation formulation should adopt the total water depth/inland flow depth $D = \eta + d$. Then the bottom friction terms can be expressed as follows

$$\frac{\tau_x}{\rho} = \frac{gn^2}{D^{7/3}} M \sqrt{M^2 + N^2} \tag{3.49}$$

$$\frac{\tau_y}{\rho} = \frac{gn^2}{D^{7/3}} N \sqrt{M^2 + N^2} \tag{3.50}$$

where n is Manning's roughness (second/m$^{1/3}$) that should be selected depending on the condition of the bottom surface. These coefficients represent the resistance of the underlying surface to the flow of water and are dependent on factors, such as surface roughness and vegetation cover. A common value of Manning's roughness used for bare land in tsunami studies is 0.025, while other values and surface classes can be considered as listed in Table 3.2. By introducing the bottom friction terms, the tsunami governing equations become

$$\frac{\partial \eta}{\partial t} + \frac{\partial M}{\partial x} + \frac{\partial N}{\partial y} = 0 \tag{3.51}$$

$$\frac{\partial M}{\partial t} + \frac{\partial}{\partial x}\left(\frac{M^2}{D}\right) + \frac{\partial}{\partial y}\left(\frac{MN}{D}\right) + gD\frac{\partial \eta}{\partial x} + \frac{gn^2}{D^{7/3}} M \sqrt{M^2 + N^2} = 0 \tag{3.52}$$

$$\frac{\partial N}{\partial t} + \frac{\partial}{\partial x}\left(\frac{MN}{D}\right) + \frac{\partial}{\partial y}\left(\frac{N^2}{D}\right) + gD\frac{\partial \eta}{\partial y} + \frac{gn^2}{D^{7/3}} N \sqrt{M^2 + N^2} = 0 \tag{3.53}$$

TABLE 3.2 Roughness values for different land cover groups used in tsunami inundation studies.

Land cover group	Roughness (second/m$^{1/3}$)
Bare land	0.025
Built-up area	0.06
Cropland	0.03
Low vegetation	0.03
Scrub	0.04
Sea	0.011
Tall vegetation	0.04
Urban open area	0.025
Water	0.011

The above shallow-water wave equations in the Cartesian coordinate system can be used to model tsunami propagation and inundation at the regional scale. To simulate the tsunami propagation over a wide area across the ocean and around the globe, spherical coordinates, $\mathbf{x} \equiv (r,\theta,\varphi)$, which define the radial distance, polar angle, and azimuthal angle, respectively, must be used. If the Earth is represented as a sphere, r is constant and equal to the Earth's radius of 6371 km.

Because the Earth spins, Earth-bound observers need to account for the Coriolis force to correctly analyze the motion of water particles. The Coriolis effects generally become noticeable only for tsunami propagation occurring over large distances across the ocean and long periods of time. The Coriolis force coefficient is given by

$$f = 2\Omega \sin\theta \tag{3.54}$$

where Ω is the angular rotation rate of the Earth (7.27×10^{-5} rad/second), and θ is the latitude.

As the Earth spins eastward, the x-axis is positive toward the east. The incorporation of the Coriolis force coefficient (f) into the two-dimensional system on the spherical coordinate results in

$$\frac{\partial \eta}{\partial t} + \frac{1}{R\cos\theta}\left[\frac{\partial M}{\partial \varphi} + \frac{\partial(N\cos\theta)}{\partial \theta}\right] = 0 \tag{3.55}$$

$$\frac{\partial M}{\partial t} + \frac{1}{R\cos\theta}\left[\frac{\partial}{\partial \varphi}\left(\frac{M^2}{D}\right) + \frac{\partial}{\partial \theta}\left(\frac{MN}{D}\right)\cos\theta + gD\frac{\partial h}{\partial \varphi}\right] = fN - \frac{gn^2}{D^{7/3}}M\sqrt{M^2+N^2} \tag{3.56}$$

$$\frac{\partial N}{\partial t} + \frac{1}{R}\left[\frac{1}{\cos\theta}\frac{\partial}{\partial \varphi}\left(\frac{MN}{D}\right) + \frac{\partial}{\partial \theta}\left(\frac{N^2}{D}\right) + gD\frac{\partial h}{\partial \theta}\right] = -fM - \frac{gn^2}{D^{7/3}}N\sqrt{M^2+N^2} \tag{3.57}$$

where φ is the longitude, θ is the latitude, M is the flux along the lines of longitude with east being in the positive direction, and N is the flux along the lines of latitude with north being in the positive direction.

3.3.3.3 Dispersive tsunami model

Water waves with different wavelengths (corresponding to different frequencies) travel at different velocities, resulting in the separation of wave components in time and space. This phenomenon is known as frequency dispersion. Far-field tsunami observations in the open ocean, such as those from the 2010 Chile and 2011 Tohoku Earthquakes, revealed that tsunami is a dispersive wave (Kirby et al., 2013; Saito et al., 2014; Watada et al., 2014). Meanwhile, in linear long-wave theory (shallow-water wave), waves are considered nondispersive, and the velocity of the wave is independent of the wave's frequency. A Boussinesq-type approach that adds a dispersion term to the shallow-water equations can be used to simulate dispersive tsunamis (Baba et al., 2015). The momentum equations (without advection, bottom friction, pressure, and Coriolis terms) with Boussinesq (dispersion) terms in Cartesian coordinates are expressed as follows

$$\frac{\partial M}{\partial t} + gd\frac{\partial \eta}{\partial x} = \frac{d^2}{3}\frac{\partial}{\partial x}\left(\frac{\partial^2 M}{\partial x \partial t} + \frac{\partial^2 N}{\partial y \partial t}\right) \tag{3.58}$$

$$\frac{\partial N}{\partial t} + gd\frac{\partial \eta}{\partial y} = \frac{d^2}{3}\frac{\partial}{\partial y}\left(\frac{\partial^2 M}{\partial x \partial t} + \frac{\partial^2 N}{\partial y \partial t}\right) \tag{3.59}$$

The governing equations for dispersive tsunami modeling on the spherical coordinates can be expressed as follows

$$\frac{\partial M}{\partial t} + \frac{gd}{R\cos\theta}\frac{\partial \eta}{\partial \varphi} = \frac{d^2}{3\cos\theta}\frac{\partial}{\partial \varphi}\left[\frac{1}{R\cos\theta}\left(\frac{\partial^2 M}{\partial \varphi \partial t} + \frac{\partial^2 (N\cos\theta)}{\partial \theta \partial t}\right)\right] \tag{3.60}$$

$$\frac{\partial N}{\partial t} + \frac{gd}{R}\frac{\partial \eta}{\partial \theta} = \frac{d^2}{3R}\frac{\partial}{\partial \theta}\left[\frac{1}{R\cos\theta}\left(\frac{\partial^2 M}{\partial \varphi \partial t} + \frac{\partial^2 (N\cos\theta)}{\partial \theta \partial t}\right)\right] \tag{3.61}$$

To illustrate the difference between simulations using a linear tsunami model and a dispersive tsunami model, simulation results from a hypothetical M_w 8.9 earthquake in the Kermadec subduction zone are shown in Fig. 3.11 and Fig. 3.12. The maximum tsunami patterns of both models are similar, although the pattern from the dispersive tsunami models shows more concentrated amplitudes in the far field. The simulated tsunami waveforms (Fig. 3.12) show that the linear model gives slightly larger amplitudes at far-field stations. For a typical large plate interface earthquake, the frequency dispersion effects are not so strong.

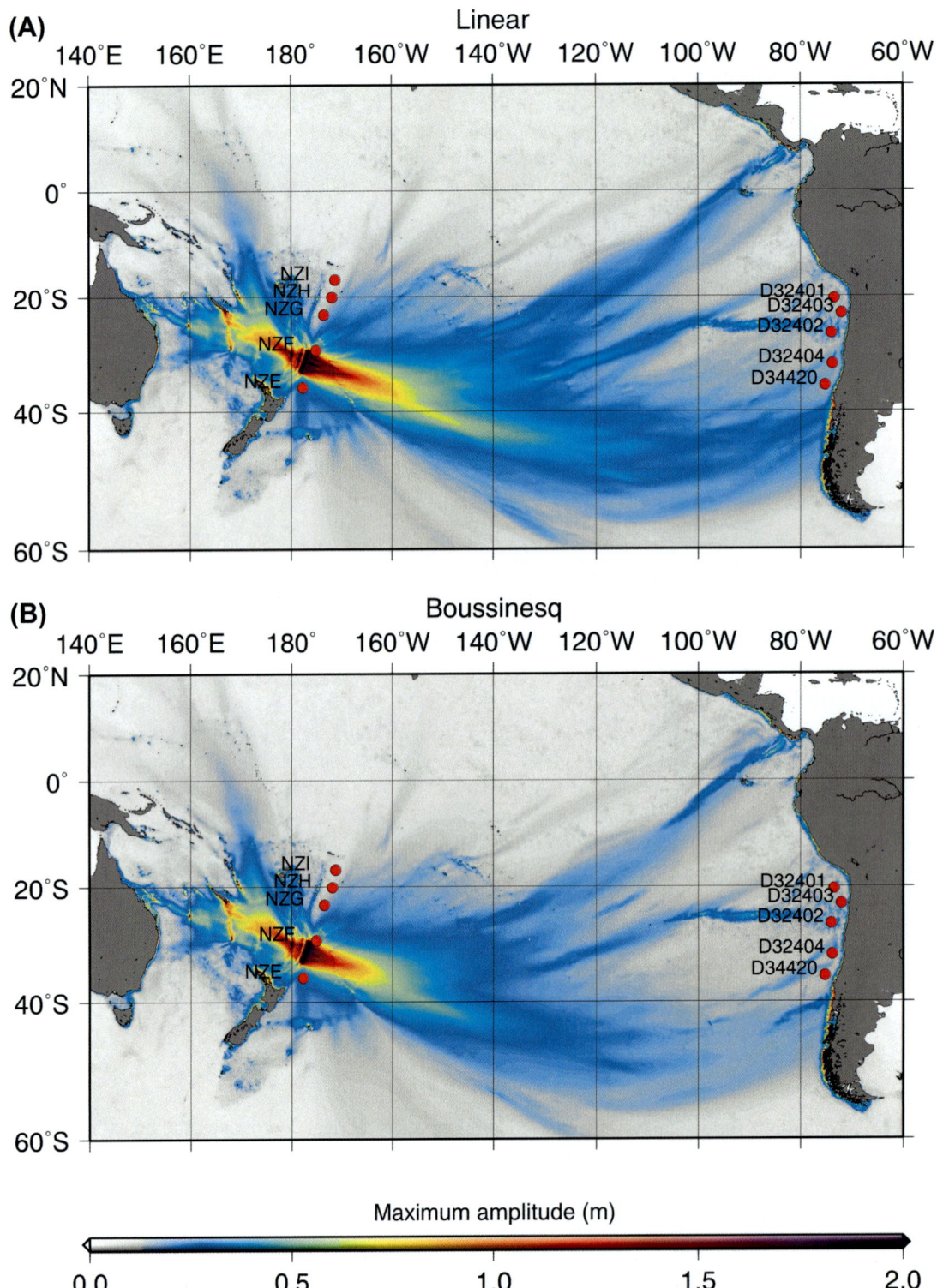

FIGURE 3.11 Comparison of simulation results from a hypothetical M_w 8.9 earthquake in the Kermadec subduction zone using a linear tsunami model and a dispersive tsunami model. (A) Maximum tsunami amplitude simulated by solving the linear shallow-water equations. (B) Maximum tsunami amplitude simulated by solving the linear shallow-water and Boussinesq equations. Map lines delineate study areas and do not necessarily depict accepted national boundaries.

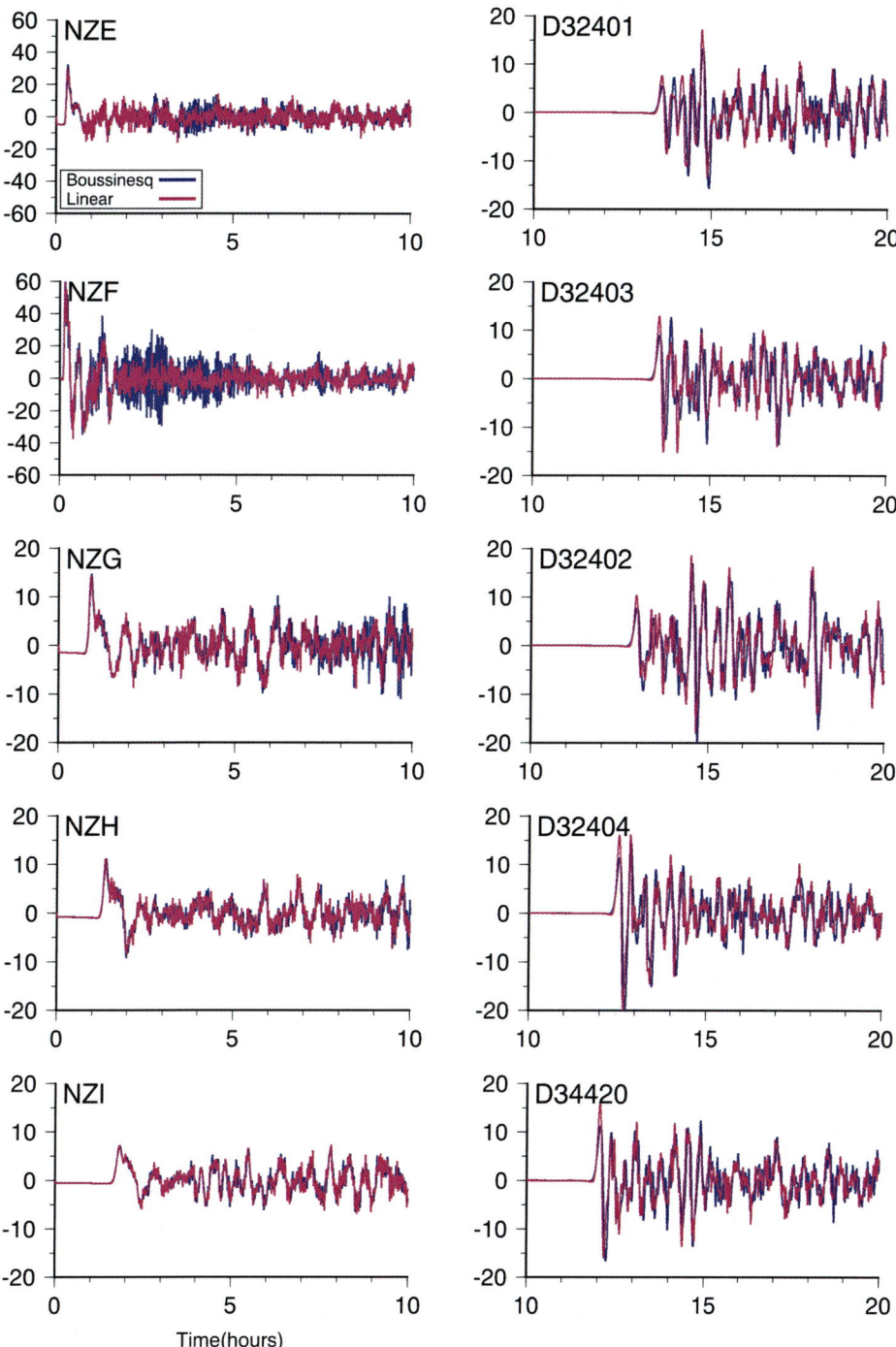

FIGURE 3.12 Comparison of simulated tsunami waveforms from a hypothetical M_w 8.9 earthquake in the Kermadec subduction zone using a linear tsunami model and a dispersive tsunami model.

3.4 Numerical method

Existing tsunami numerical models, such as COMCOT (Wang & Liu, 2006), TUNAMI (Imamura, 2009; Imamura et al., 2006), MOST (Titov et al., 2016), and JAGURS (Baba et al., 2017), solve the tsunami governing equations using the finite difference method. This section provides an overview of the finite difference method, modeling grid refinement, and boundary conditions. Detailed descriptions of model development methodologies can be found in the publications for each numerical model.

3.4.1 Finite difference scheme

Here is an example of the discretization applied in COMCOT as explained in Wang and Power (2011). An explicit staggered leap-frog finite difference scheme is used in COMCOT to solve shallow-water equations both in Cartesian and spherical coordinate systems. The evaluations of water level (η) and volume fluxes (M,N) are staggered in both time and space. In a staggered grid configuration (staggered in space), water surface fluctuations are presented at the center of a grid cell, and the volume flux components are evaluated at the centers of cell edges (Fig. 3.13). The staggered time method updates the unknown variables of water level (η) and volume fluxes (M,N) at staggered time steps, rather than at the same time step, to avoid computing the values of all the unknown variables at the same time step (Fig. 3.13). For the discretization, the time is denoted as $n\Delta t$ and the center of the cell is ($i\Delta x, j\Delta y$). The discretized form of the linear part of the shallow-water wave equations can be expressed as follows

$$\eta_{i,j}^{n+1/2} = \xi_{i,j}^{n+1/2} - \xi_{i,j}^{n-1/2} + \eta_{i,j}^{n-1/2} - \frac{\Delta t \left(M_{i+1/2,j}^{n} - M_{i-1/2,j}^{n} \right)}{\Delta x} - \frac{\Delta t \left(N_{i,j+1/2}^{n} - N_{i,j-1/2}^{n} \right)}{\Delta y} \quad (3.62)$$

$$M_{i+1/2,j}^{n+1} = M_{i+1/2,j}^{n} - g d_{i+1/2,j} \Delta t \frac{\eta_{i+1,j}^{n+1/2} - \eta_{i,j}^{n+1/2}}{\Delta x} \quad (3.63)$$

$$N_{i,j+1/2}^{n+1} = N_{i,j+1/2}^{n} - g d_{i,j+1/2} \Delta t \frac{\eta_{i,j+1}^{n+1/2} - \eta_{i,j}^{n+1/2}}{\Delta y} \quad (3.64)$$

The leap-frog scheme with a staggered grid is used to calculate the water level at the center of every modeling grid cell ($\eta_{i,j}^{n+1/2}$). In this computation, the volume flux components ($M_{i+1/2,j}^{n}, M_{i-1/2,j}^{n}, N_{i,j+1/2}^{n}, N_{i,j-1/2}^{n}$) are staggered with respect to the water level. Another information required for the calculation is the water level at the evaluated grid cell from the previous time step ($\eta_{i,j}^{n-1/2}$). The volume flux components are located at the center of the four edges of the grid cell. The calculations for the water-level and the volume flux components are also staggered in time (Fig. 3.14). The water level is evaluated at time steps $t = (n - ½)\Delta t$ and $t = (n + ½)\Delta t$, while the fluxes are calculated at time steps $t = n\Delta t$ and $t = (n + 1)\Delta t$.

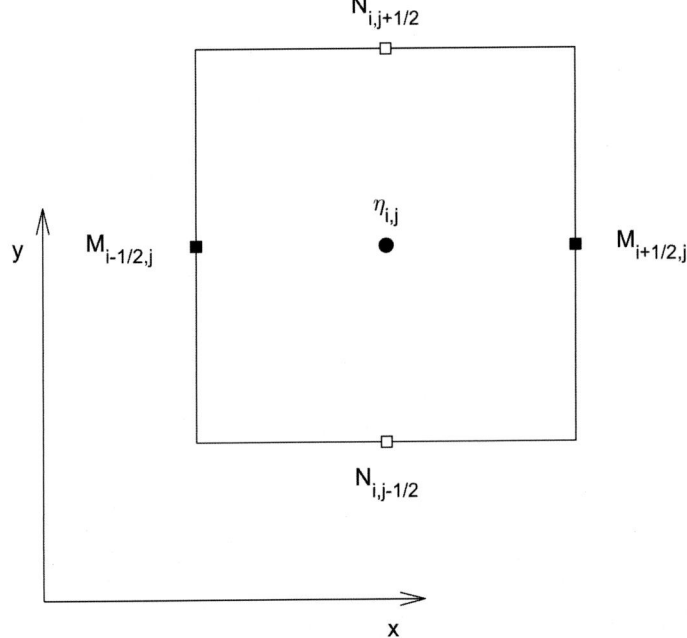

FIGURE 3.13 A staggered grid setup in which water elevation is presented at the center of a grid cell, indicated by circles, and the volume flux components, represented by squares, are evaluated at the centers of cell edges.

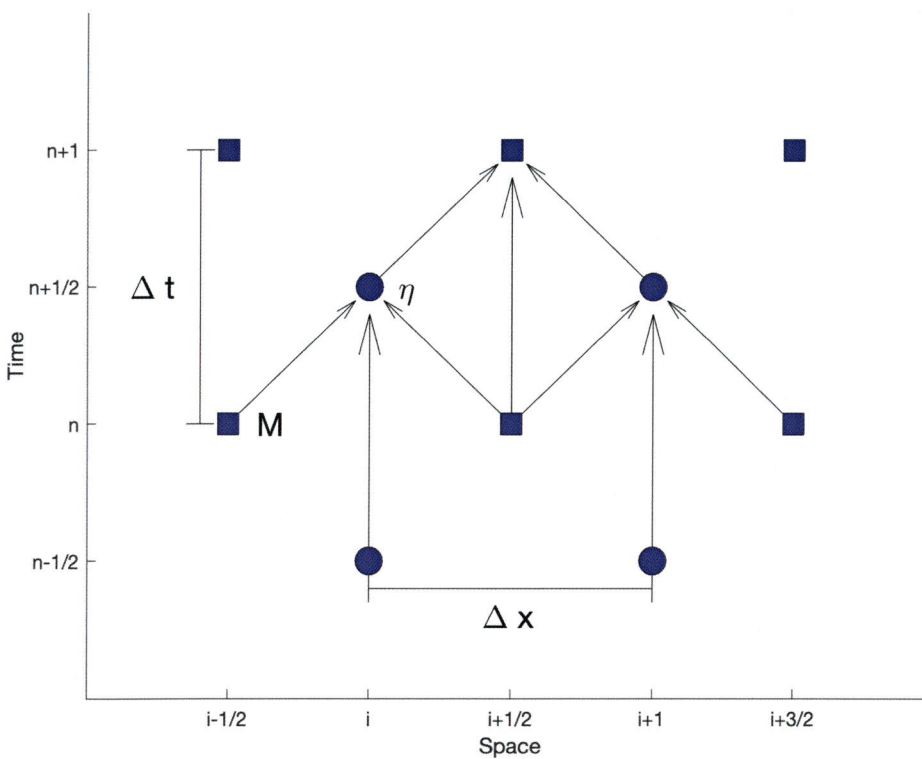

FIGURE 3.14 A staggered time setup. Circles represent the water elevation and squares represent the volume flux.

The computation will be stable if the Courant–Friedrichs–Lewy (CFL) condition is satisfied. Because a time-marching solver is used in this numerical scheme, the dimensionless Courant number (C_r) should be smaller than 1. The CFL condition for the computation can be written in the following form

$$C_r = \frac{\Delta t}{\Delta x}\sqrt{2gd} < 1 \tag{3.65}$$

For the nonlinear shallow-water equations, the nonlinear convection terms shown in Eqs. (3.56) and (3.57) are discretized with an upwind finite difference scheme in COMCOT (Wang & Power, 2011). To simulate nearshore and onshore tsunamis, the total water depth/inland flow depth ($D = \eta + d$) that changes at each iteration should be considered instead of a constant but nonuniform water depth (d). The upwind scheme is preferable as it requires a small computational effort when the CFL condition is satisfied, although it may introduce some numerical dissipation. The nonlinear convective terms in the momentum equations that are discretized by using an upwind scheme can be written as follows

$$\frac{\partial}{\partial x}\left\{\frac{M^2}{D}\right\} = \frac{1}{\Delta x}\left\{\lambda_{11}\frac{\left(M^n_{i+3/2,j}\right)^2}{D^n_{i+3/2,j}} + \lambda_{12}\frac{\left(M^n_{i+1/2,j}\right)^2}{D^n_{i+1/2,j}} + \lambda_{13}\frac{\left(M^n_{i-1/2,j}\right)^2}{D^n_{i-1/2,j}}\right\} \tag{3.66}$$

$$\frac{\partial}{\partial y}\left\{\frac{MN}{D}\right\} = \frac{1}{\Delta y}\left\{\lambda_{21}\frac{(MN)^n_{i+1/2,j+1}}{D^n_{i+1/2,j+1}} + \lambda_{22}\frac{(MN)^n_{i+1/2,j}}{D^n_{i+1/2,j}} + \lambda_{23}\frac{(MN)^n_{i+1/2,j-1}}{D^n_{i+1/2,j-1}}\right\} \tag{3.67}$$

$$\frac{\partial}{\partial x}\left\{\frac{MN}{D}\right\} = \frac{1}{\Delta x}\left\{\lambda_{31}\frac{(MN)^n_{i+1,j+1/2}}{D^n_{i+1,j+1/2}} + \lambda_{32}\frac{(MN)^n_{i,j+1/2}}{D^n_{i,j+1/2}} + \lambda_{33}\frac{(MN)^n_{i-1,j+1/2}}{D^n_{i-1,j+1/2}}\right\} \tag{3.68}$$

$$\frac{\partial}{\partial y}\left\{\frac{N^2}{D}\right\} = \frac{1}{\Delta y}\left\{\lambda_{41}\frac{\left(N^n_{i,j+3/2}\right)^2}{D^n_{i,j+3/2}} + \lambda_{42}\frac{\left(N^n_{i,j+1/2}\right)^2}{D^n_{i,j+1/2}} + \lambda_{43}\frac{\left(N^n_{i,j-1/2}\right)^2}{D^n_{i,j-1/2}}\right\} \tag{3.69}$$

where the coefficients λ are determined from

$$\begin{cases} \lambda_{11} = 0, \ \lambda_{12} = 1, \ \lambda_{13} = -1, & \text{if } M^n_{i+1/2, j} \geq 0 \\ \lambda_{11} = 1, \ \lambda_{12} = -1, \ \lambda_{13} = 0, & \text{if } M^n_{i+1/2, j} < 0 \end{cases}$$

$$\begin{cases} \lambda_{21} = 0, \ \lambda_{22} = 1, \ \lambda_{23} = -1, & \text{if } N^n_{i+1/2, j} \geq 0 \\ \lambda_{21} = 1, \ \lambda_{22} = -1, \ \lambda_{23} = 0, & \text{if } N^n_{i+1/2, j} < 0 \end{cases}$$

$$\begin{cases} \lambda_{31} = 0, \ \lambda_{32} = 1, \ \lambda_{33} = -1, & \text{if } M^n_{i, j+1/2} \geq 0 \\ \lambda_{31} = 1, \ \lambda_{32} = -1, \ \lambda_{33} = 0, & \text{if } M^n_{i, j+1/2} < 0 \end{cases}$$

$$\begin{cases} \lambda_{41} = 0, \ \lambda_{42} = 1, \ \lambda_{43} = -1, & \text{if } N^n_{i, j+1/2} \geq 0 \\ \lambda_{41} = 1, \ \lambda_{42} = -1, \ \lambda_{43} = 0, & \text{if } N^n_{i, j+1/2} < 0 \end{cases} \quad (3.70)$$

The bottom frictional terms can be discretized as follows

$$F_x = \nu_x \left(M^{n+1}_{i+1/2, j} + M^n_{i+1/2, j} \right) \quad (3.71)$$

$$F_y = \nu_y \left(N^{n+1}_{i, j+1/2} + N^n_{i, j+1/2} \right) \quad (3.72)$$

where ν_x and ν_y are given by

$$\nu_x = \frac{1}{2} \frac{gn^2}{\left(D^n_{i+1/2, j} \right)^{7/3}} \left[\left(M^n_{i+1/2, j} \right)^2 + \left(N^n_{i+1/2, j} \right)^2 \right]^{\frac{1}{2}} \quad (3.73)$$

$$\nu_y = \frac{1}{2} \frac{gn^2}{\left(D^n_{i, j+1/2} \right)^{7/3}} \left[\left(M^n_{i, j+1/2} \right)^2 + \left(N^n_{i, j+1/2} \right)^2 \right]^{\frac{1}{2}} \quad (3.74)$$

The discretized form of the governing equations for the linear and dispersive tsunami models with Boussinesq terms shown in Section 3.3.3 (Eqs. (3.58) and (3.59)) using the staggered leap-frog finite difference scheme can be written as

$$\eta^{n+1/2}_{i, j} = \eta^{n-1/2}_{i, j} - \frac{\Delta t \left(M^n_{i+1/2, j} - M^n_{i-1/2, j} \right)}{\Delta x} - \frac{\Delta t \left(N^n_{i, j+1/2} - N^n_{i, j-1/2} \right)}{\Delta y} \quad (3.75)$$

$$-\frac{d^2}{3\Delta x^2} M^{n+1}_{i+1/2, j} + \left(1 + \frac{2d^2}{3\Delta x^2}\right) M^{n+1}_{i-1/2, j} - \frac{d^2}{3\Delta x^2} M^{n+1}_{i-3/2, j} - \frac{d^2}{3\Delta x \Delta y} N^{n+1}_{i, j+1/2} + \frac{d^2}{3\Delta x \Delta y} N^{n+1}_{i-1, j+1/2}$$

$$+ \frac{d^2}{3\Delta x \Delta y} N^{n+1}_{i, j-1/2} - \frac{d^2}{3\Delta x \Delta y} N^{n+1}_{i-1, j-1/2}$$

$$= M^{n-1}_{i-1/2, j} - gd \frac{\Delta t}{\Delta x} \left(\eta^{n+1/2}_{i, j} - \eta^{n+1/2}_{i-1, j} \right) - \frac{d^2}{3\Delta x^2} \left(M^{n-1}_{i+1/2, j} - 2M^{n-1}_{i-1/2, j} + M^{n-1}_{i-3/2, j} \right)$$

$$- \frac{d^2}{3\Delta x \Delta y} \left(N^{n-1}_{i, j+1/2} - N^{n-1}_{i-1, j+1/2} - N^{n-1}_{i, j-1/2} + N^{n-1}_{i-1, j-1/2} \right) \quad (3.76)$$

$$-\frac{d^2}{3\Delta y^2} N^{n+1}_{i, j+1/2} + \left(1 + \frac{2d^2}{3\Delta y^2}\right) N^{n+1}_{i, j-1/2} - \frac{d^2}{3\Delta y^2} N^{n+1}_{i, j-3/2} - \frac{d^2}{3\Delta x \Delta y} M^{n+1}_{i+1/2, j} + \frac{d^2}{3\Delta x \Delta y} M^{n+1}_{i-1/2, j}$$

$$+ \frac{d^2}{3\Delta x \Delta y} M^{n+1}_{i+1/2, j-1} - \frac{d^2}{3\Delta x \Delta y} M^{n+1}_{i-1/2, j-1}$$

$$= N^{n-1}_{i, j-1/2} - gd \frac{\Delta t}{\Delta x} \left(\eta^{n+1/2}_{i, j} - \eta^{n+1/2}_{i, j-1} \right) - \frac{d^2}{3\Delta y^2} \left(N^{n-1}_{i, j+1/2} - 2N^{n-1}_{i, j-1/2} + N^{n-1}_{i, j-3/2} \right)$$

$$- \frac{d^2}{3\Delta x \Delta y} \left(M^{n-1}_{i+1/2, j} - M^{n-1}_{i-1/2, j} - M^{n-1}_{i+1/2, j-1} + M^{n-1}_{i-1/2, j-1} \right) \quad (3.77)$$

At time $t = n\Delta t$, first, Eq. (3.75) is solved to obtain $\eta^{n+1/2}_{i, j}$. Then $\eta^{n+1/2}_{i, j}$ is used as an input to solve Eqs. (3.76) and (3.77). These two equations can be solved to obtain M^{n+1} and N^{n+1} iteratively, using the Gauss–Seidel method.

The Gauss−Seidel method works by starting with an initial guess for the solution and then iteratively refining that guess until a desired level of accuracy is achieved. In Eq. (3.76), there are seven unknown variables ($M^{n+1}_{i+1/2,j}, M^{n+1}_{i-1/2,j}, M^{n+1}_{i-3/2,j}, N^{n+1}_{i,j+1/2}, N^{n+1}_{i-1,j+1/2}, N^{n+1}_{i,j-1/2}, N^{n+1}_{i-1,j-1/2}$). To begin the iterative process, the first variable (e.g., $M^{n+1}_{i+1/2,j}$) can be computed using the initial values of the other six unknowns. These six initial values can be obtained from a linear model computation without the Boussinesq terms. Then, the remaining six variables can be computed similarly. Eq. (3.77) should be solved using the same approach. In the next iteration, all variables in Eqs. (3.76) and (3.77) are computed again. This process is repeated until the solution converges to the desired level of accuracy.

JAGURS is a tsunami numerical model that employs the Gauss−Seidel method to solve for the Boussinesq terms and simulate dispersive tsunamis. In JAGURS, the iteration is terminated when the maximum number of iterations or a threshold is reached. For example, a threshold of 1×10^{-8} m/second indicates that the iteration will be stopped when the difference in velocities between iterative computations falls below this value. The resulting M^{n+1} and N^{n+1} are then used in Eq. (3.75) to solve $\eta^{n+1/2}_{i,j}$ for the next time step $t = (n+1)\Delta t$.

3.4.2 Spatial grid refinements

To assess the hazards posed by tsunamis, it is essential to implement simulation models with more detailed spatial and temporal resolutions in certain areas. A nested grid configuration can be implemented in the model to balance computational efficiency and numerical accuracy. It uses a relatively large grid spacing to efficiently simulate transoceanic tsunami propagation in the deep ocean and switches to refined grid spacing in nearshore and coastal regions to account for the shortening of tsunami wavelength, due to the shallowness of water depth. This technique allows a fine resolution of some coastal regions where a detailed inundation map is desired, while other regions of less interest may have less refinement.

In a two-way interactive nested grid system, an inner (child) grid layer with finer grids is nested within the outer (parent) grid layer with coarser grids (Fig. 3.15). The communication from the coarse grid to the nested grid is typically

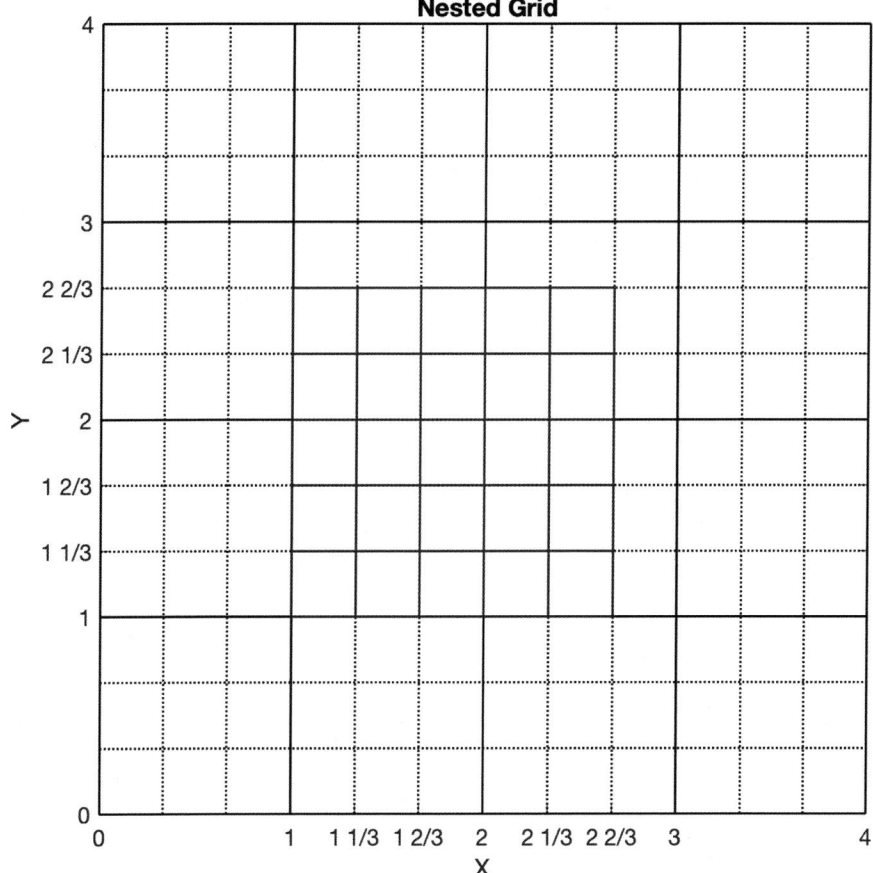

FIGURE 3.15 A nested grid system with a refinement ratio of 3. The outer (parent) grid (solid *black* grid) size is 1, and the inner (child) grid (*blue* grid) size is 1/3. The boundary of the inner grid must match the boundary of the fraction of the outer grid (dashed *black* grid). The division of the outer grid depends on the refinement ratio.

through the specification of the boundary condition of the nested grid. The solution on the coarse grids is continually updated by that on the nested grids wherever the two grids coincide. In contrast, in a one-way nested grid system, the solution on the coarse grid layer is independent of that on the finer grid layer.

A grid ratio can be used to determine the grid sizes for the two layers. The refinement factor ranges from 3 to 5 to have a smooth transition between the coarse and fine grids. Good computational efficiency and stability can be achieved by using a ratio within this range. The nested grid layer can also enclose a grid layer with a smaller grid size. A nested grid system can be set to have a hierarchy of grid layers where the outermost grid layer covers the whole ocean with a grid size of several kilometers and nested grid layers with different grid sizes in which the finest grid layer covers a coastal community with a grid size of several meters.

In a tsunami simulation using a nested grid system, the simulation should start with the computation of the water level in the outer grid layer by solving the continuity equation. To compute the water level in the inner grid layer, the flux along its boundary needs to be evaluated. The flux on the inner grid cells along the boundary can be obtained by interpolating the fluxes on the connecting cells of the outer grid layer. The computed water levels in the inner grid layer are then used to update the water levels in the outer grid layer. Next, the fluxes in the outer and then in the inner grid layers are computed by solving the momentum equations. Finally, the fluxes in the outer grid layer are updated by averaging the values in the inner grid layer.

Since the inner grid layer has a grid size smaller than the outer grid layer, the time step for the embedded inner layer has to be sufficiently small for stability consideration. The time step for the inner that should be used to satisfy the CFL condition depends on the grid size and the maximum water depth of the grid layer. As a result, when the calculation of the outer grid layer advances one step forward, the inner grid layer needs to be computed more than once to match the time in the outer grid layer. Because the calculation is staggered in time too, depending on the ratio of the time refinement, the computed values in the inner layer may need to be interpolated in the time domain to match the time of the outer grid layer.

In the following example, a tsunami simulation is performed using an earthquake source model offshore Peru (Fig. 3.16). To study the tsunami impact from such an earthquake in coastal towns and communities along the coast of New Zealand, a grid resolution of 15 arcseconds for the study area is desired. It is computationally expensive if a 15-arcsecond grid size is used for the whole modeling domain that covers the Pacific Ocean. Fig. 3.16 shows the maximum simulated tsunami amplitude plots for Layer 1, Layer 2 (child grid of Layer 1), Layer 3 (child grid of Layer 2), and Layer 4 (child grid of Layer 2). In this case, the grid resolutions of Layer 1 and Layer 2 are 240 and 60 arcseconds, respectively. While the grid resolution for Layer 3 and Layer 4 is 15 arcseconds. The time interval of the simulation for Layers 1, 2, 3, and 4 is 2.53, 2.53, 0.63, and 0.84 seconds, respectively. One-step forward computation of Layer 1 is followed by one-step forward computation of Layer 2 as they have the same time interval. Each one-step forward computation of Layer 2 must be followed by the four-step forward computation of Layer 3 and the three-step forward computation of Layer 4. This approach of using multiple time intervals improves the computation efficiency.

3.4.3 Outermost grid boundary condition

An open boundary condition in tsunami modeling is a computational boundary that allows waves generated in the interior domain to pass through the boundary of the outermost grid without reflection. Here are approaches that can be used for an open ocean boundary condition.

3.4.3.1 Radiation boundary condition

The radiation boundary condition for open ocean is a standard boundary condition used in tsunami models. This boundary condition determines boundary values using theoretical analysis, allowing water waves to travel outward through the computational domain's boundaries as if no barrier exists. The water surface fluctuation is calculated at a boundary grid using the principles of linear long-wave theory (Dean & Dalrymple, 1991) as follows

$$\eta = \frac{\sqrt{M^2 + N^2}}{\sqrt{gd}} \qquad (3.78)$$

The variables M and N are the volume flux components through the inner edges of the boundary grid cell in the x and y directions, respectively. The gravitational acceleration is represented by g, and d represents the still water depth at the boundary grid cell. The water surface fluctuation's sign is determined by the volume flux component through the inner edge of the boundary grid cell aligned with the boundary. If the flux component points inward from the boundary, η is given a negative sign. Conversely, if the flux component points outward toward the boundary, a positive sign is

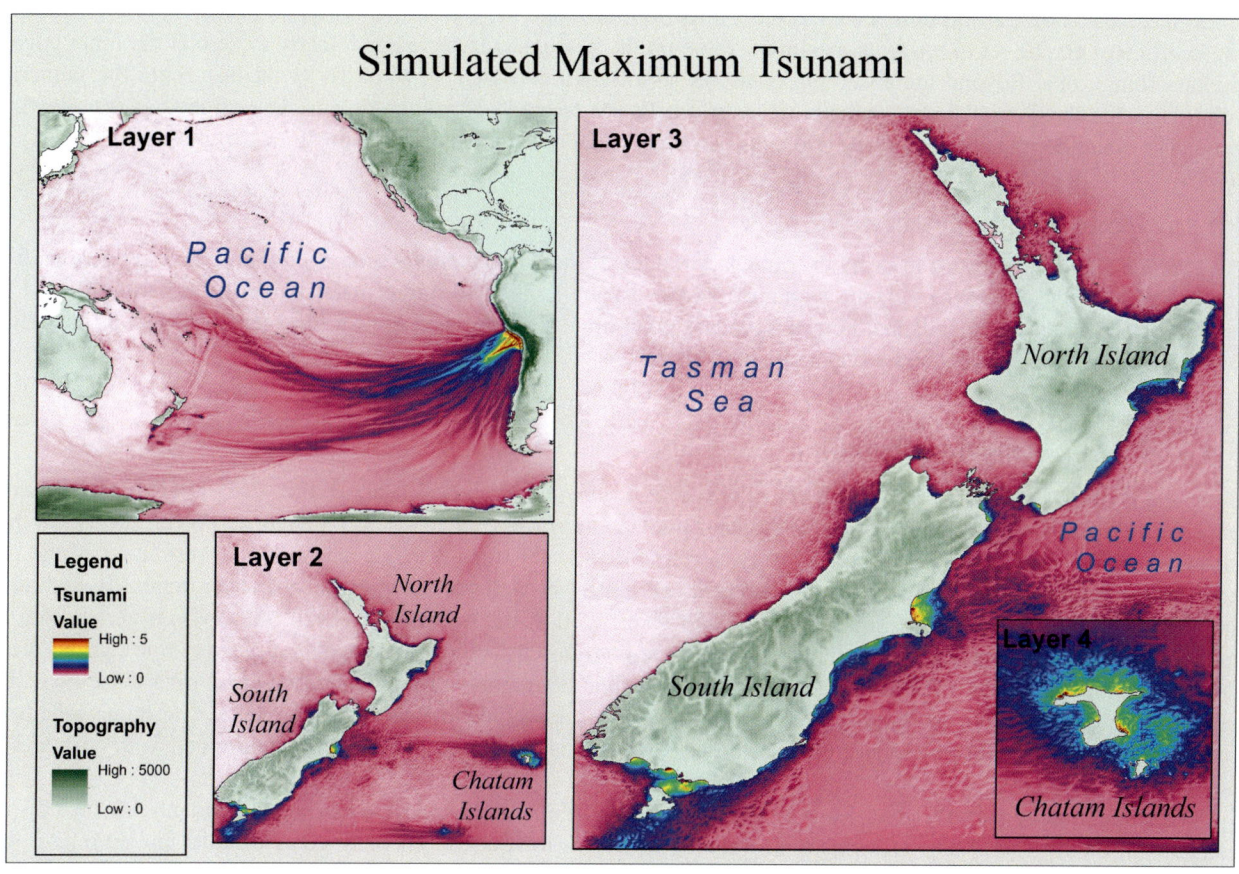

FIGURE 3.16 An example of tsunami simulation using multiple layers with different grid sizes. The earthquake source is located offshore Peru, and the higher resolution simulation is for the area around New Zealand. Map lines delineate study areas and do not necessarily depict accepted national boundaries.

used. This treatment only allows for water waves to travel outward through the boundaries, and ideally, no waves should be reflected from the boundaries. However, this approach is based on linear long-wave theory and assumes that the boundaries are in open water with flat bathymetry, which may not be realistic. Reflections can occur if waves are short compared to the water depth, if nonlinearity cannot be ignored, or if the boundary intersects complicated bathymetry, such as regions with steep slopes or those close to the shore. These reflections can accumulate and ultimately degrade the numerical solution inside the domain.

3.4.3.2 Absorbing boundary condition

The absorbing boundary condition is formed by combining an absorption scheme with an external solution that gives boundary values, including the radiation boundary condition mentioned earlier. The absorption scheme alone cannot generate a boundary solution. Instead, it is devised to dampen the solution inside the computational domain toward the external solution within a predetermined distance called the sponge or absorption zone, utilizing an absorption function. In COMCOT, the governing equations in Cartesian coordinates within the sponge zone can be expressed as a function of η, M, and N, with corresponding external solutions denoted by η_e, M_e, and N_e, respectively, and absorbing coefficients σ_x and σ_y that vary only with x and y, respectively

$$\frac{\partial \eta}{\partial t} + \frac{\partial M}{\partial x} + \frac{\partial N}{\partial y} = -(\sigma_x + \sigma_y)(\eta + \eta_e) \tag{3.79}$$

$$\frac{\partial M}{\partial t} + gd\frac{\partial \eta}{\partial x} = -\sigma_x(M + M_e) \tag{3.80}$$

$$\frac{\partial N}{\partial t} + gd\frac{\partial \eta}{\partial y} = -\sigma_y(N + N_e) \tag{3.81}$$

In these equations, absorption occurs in the x direction for η and M, and in the y direction for η and N. The absorbing coefficient can be set to reach its maximum value on the boundary. It decreases linearly to zero at the inner edge of the absorbing zone. Beyond the absorbing zone, the absorbing coefficient is zero. To avoid disturbing the numerical solution inside the computational domain, ghost grid cells are created outside the domain and utilized as the absorbing zone. The water depth at the boundary grid cells can be copied outward as the water depth of the ghost grid.

3.4.3.3 Forced boundary condition

An external solution is forced to be as the boundary values of the computational domain. To minimize reflections from the boundaries, the absorbing scheme explained above can be applied to dampen the numerical solution toward the external solution on the boundaries. Forced boundary conditions can be specified as a time series in tsunami modeling, which allows the wave to penetrate the modeling domain.

3.4.4 Moving boundary scheme

Tsunami numerical models like COMCOT implement a moving boundary scheme that can describe the dynamic changes in water level and flow during the tsunami inundation process. In COMCOT (Wang & Power, 2011), the precise positioning of the interface between wet (water) and dry (land) areas depends on the spatial resolution of the grid cells into which the model area is divided. Each grid cell has a water elevation (η) and a still water depth (d), both measured at the cell's center relative to a reference water level, such as the mean water level. Initially, η is zero across the model, and d is measured from the reference level to the bottom. Grid cells are categorized as wet (water grids) if d is positive, or dry (land grids) if d is negative, with the coastline defined along the boundaries separating these two types of cells. During a tsunami, η changes, affecting the total water depth/inland flow depth D, which determines whether a cell is wet or dry, and thereby where the coastline lies. A cell is wet when D is positive, and a cell is dry when D is negative.

The simulation updates the model using the continuity equation and boundary conditions to compute surface displacements for the next time step, affecting both wet and dry cells. However, momentum equations are applied only in wet cells to update volume fluxes. Dry cells have no volume fluxes, so their free surface displacement remains zero. A moving boundary scheme is employed to adjust the shoreline dynamically, deciding when and how to flood dry cells or drain wet cells based on changes in total water depth (D) at each time step, thus simulating the advance or retreat of the coastline.

3.5 Tsunami simulations

3.5.1 Tsunami source

A tsunami can be generated by submarine or coastal geological processes, such as earthquakes, landslides, and volcanic eruptions (Chapter 2). Most tsunamis are generated by the seafloor displacement due to large submarine earthquakes. The most basic assumption regarding a tsunami source is that it involves an instantaneous displacement of the seafloor. Surface displacement from earthquake source parameters can be calculated using the formulae explained in Okada (1985). The model assumed an isotropic homogeneous half-space to calculate the surface displacement earthquake parameters of fault length, width, depth, strike angle, rake angle, dip angle, tensile amount, and slip amount. For most tsunamigenic earthquakes, the seafloor displacement calculated by the Okada formulae can be assumed to be the same as the sea surface displacement. This assumption holds as long as the wavelength of the seafloor displacement is much longer than the water depth. In cases where this assumption no longer holds, a filter proposed by Kajiura (1963) can be used to calculate the sea surface displacement from the seafloor displacement (Chapter 2). When an earthquake occurs in an area characterized by abrupt changes in bathymetry (steep bathymetry), it becomes important to consider the horizontal displacement and the effects of bathymetric slope on the sea surface displacement. Fig. 3.17 illustrates the coseismic displacement calculated from earthquake fault parameters and incorporating topographic and bathymetric slopes for an earthquake fault rupture in New Zealand.

Most tsunami source types can be included in a simulation by adding the term ($\partial \xi / \partial t$), which represents the transient motion of the seafloor or the sea surface, to the governing equations. The seafloor displacement model for an earthquake, landslide, underwater volcanic eruption, or caldera collapse can be assumed to occur instantaneously or over a period of time, and the tsunami can also be assumed to have no feedback on how the seafloor displacement process

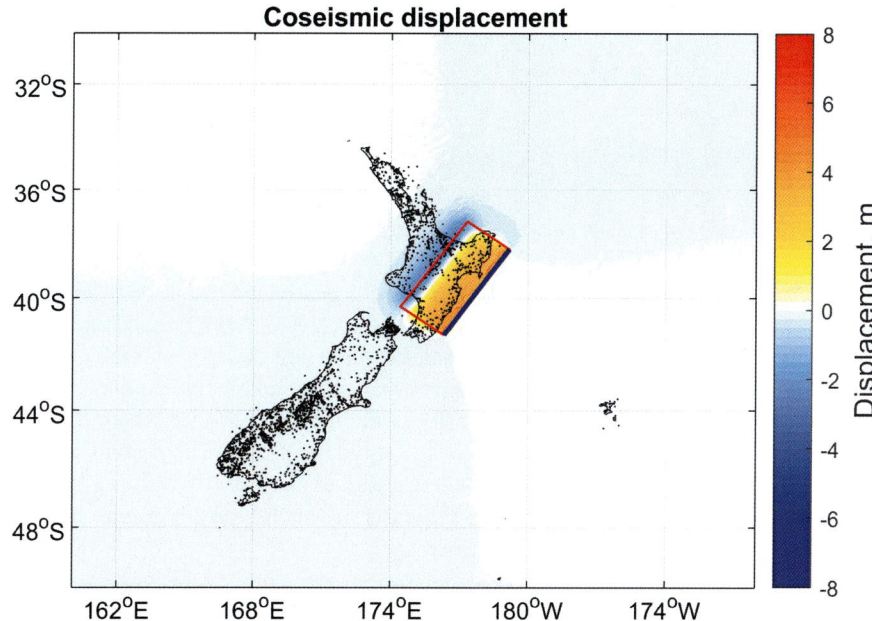

FIGURE 3.17 Vertical surface displacement is calculated from earthquake fault parameters, taking into account the topographic and bathymetric slopes. Red lines indicate the fault plane, while the blue line indicates the top edge of the fault.

occurs. However, a coupled model between tsunami and landslide layers or tsunami and pyroclastic flow layers may be needed to simulate the interaction of the two fluids.

3.5.1.1 Two-layered landslide-tsunami model

A coupled model provides a more realistic simulation compared to methods that impose a static initial water surface or assume the sliding motion of a rigid body (Enet et al., 2003; Watts et al., 2003; Watts, 2000). Both landslide debris motion and sea surface displacement can be simulated using the shallow-water equations of mass conservation and momentum balance. The VolcFlow code (https://lmv.uca.fr/volcflow/) has the capability of simulating the interaction of landslide and tsunami propagations at each time step (Kelfoun et al., 2010). The water affects the landslide, and the landslide influences the water. The code is based on the 2D depth-averaged method, but it is improved to incorporate 3D interactions between the sea and the landslide to achieve greater accuracy. The model assumes that there is no mixing between the landslide material and the water, and the densities of both landslide and water are constant over time. However, where the landslide is underwater, a reduced density of the landslide under the water is considered (i.e., the density of the landslide mass less the density of the water). This assumption rules out the mixing of the landslide with the seawater, which could result in turbidity currents and affect the water wave dynamics (Giachetti et al., 2012). The model also assumes that the water depth does not influence the underlying landslide dynamics. The landslide thickness and morphology affect the water by changing the bottom of the ocean and, thus, the slope.

In the VolcFlow code, the landslide is simulated by the following momentum balance and mass conservation

$$\frac{\partial}{\partial t}(h_a u_x) + \frac{\partial}{\partial x}(h_a u_x^2) + \frac{\partial}{\partial y}(h_a u_x u_y) = g h_a \sin\alpha_x - \frac{1}{2}k_{\text{act/pass}}\frac{\partial}{\partial x}(g h_a^2 \cos\alpha) + \frac{T_x}{\rho} \qquad (3.82)$$

$$\frac{\partial}{\partial t}(h_a u_y) + \frac{\partial}{\partial y}(h_a u_y^2) + \frac{\partial}{\partial x}(h_a u_x u_y) = g h_a \sin\alpha_y - \frac{1}{2}k_{\text{act/pass}}\frac{\partial}{\partial y}(g h_a^2 \cos\alpha) + \frac{T_y}{\rho} \qquad (3.83)$$

$$\frac{\partial h_a}{\partial t} + \frac{\partial}{\partial x}(h_a u_x) + \frac{\partial}{\partial y}(h_a u_y) = 0 \qquad (3.84)$$

The variable h_a is the landslide thickness, α is the ground slope, and ρ is the relative density equal to the landslide density ρ_a (= 2000 kg/m^3) where the landslide is subaerial and $\rho_a - \rho_w$ where it is submarine, and ρ_w is the water density (= 1000 kg/m^3). Variables u_x and u_y are the landslide velocities along the x-axis and y-axis, respectively, $k_{\text{act/pass}}$ is the Earth pressure coefficient (ratio of ground-parallel stress to ground-normal stress) used with basal and internal friction angles (Iverson & Denlinger, 2001), and g is the gravity. The terms on the right-hand side of the equations for momentum balance indicate, from left to right, the effect of the weight, the pressure gradient, and the total retarding stress T.

The water wave is simulated using the following momentum and mass conservation

$$\frac{\partial}{\partial t}(\eta v_x) + \frac{\partial}{\partial x}(\eta v_x^2) + \frac{\partial}{\partial y}(h_a v_x v_y) = g\eta \sin \beta_x - \frac{1}{2}\frac{\partial}{\partial x}(g\eta^2 \cos \beta) + \frac{R_x}{\rho_w} - 3\frac{\mu_w}{\rho_w \eta}v_x \quad (3.85)$$

$$\frac{\partial}{\partial t}(\eta v_y) + \frac{\partial}{\partial y}(\eta v_y^2) + \frac{\partial}{\partial x}(\eta v_x v_y) = g\eta \sin \beta_y - \frac{1}{2}\frac{\partial}{\partial y}(g\eta^2 \cos \beta) + \frac{R_y}{\rho_w} - 3\frac{\mu_w}{\rho_w \eta}v_y \quad (3.86)$$

$$\frac{\partial \eta}{\partial t} + \frac{\partial}{\partial x}(\eta v_x) + \frac{\partial}{\partial y}(\eta v_y) = 0 \quad (3.87)$$

where η is the water height, μ_w is the water viscosity (1.14×10^{-3} Pa/second), β is the ocean bottom slope calculated from the initial topography and the simulated landslide thickness, and v_x and v_y are the tsunami velocities along the x- and y-axis, respectively. The terms on the right-hand side of the momentum equations indicate, from left to right, the effect of the weight, the pressure gradient, the drag between water and landslide, and the drag between water and the ocean bottom.

The landslide is affected by the water in two ways. First, the reduced density of the landslide is used where the landslide is underwater. This reduces the driving forces and, thus, the velocity of the submarine flow. The second effect is related to the drag exerted by the water on the landslide T_{aw}. T can be first expressed as the sum of the drag between the water and the landslide T_{aw} and the stress between the landslide and the ground T_{ag}, which typically ranges between 20 and 100 kPa (Kelfoun et al., 2010). The frictional-retarding stress is defined by

$$T_{ag} = -\rho h \left(g \cos\alpha + \frac{u^2}{r} \right) \tan\varphi_{bed} \times \frac{u}{\|u\|} \quad (3.88)$$

where φ_{bed} is the basal frictional angle. The drag between the water and the landslide is defined by

$$T_{aw} = -\frac{1}{2}\rho \left(\tan\beta_m C_f + \frac{1}{\cos\beta_n} C_s \right) \|u - v\|(u - v) \quad (3.89)$$

The coefficients C_f and C_s determine the drag on the surface of the landslide normal and parallel to the displacement, respectively.

The water is displaced by the landslide in two ways. It can be accelerated by the displacement of the landslide. The condition $\mathbf{R} = -\mathbf{T}_{aw}$ allows the landslide to push the water near the shoreline. The transfer of momentum has a small effect on the velocity of the water at depth, where the mass of the landslide is small relative to the mass of the surrounding ocean. The second effect is due to the elevation of the base of the water by the landslide, which is expressed by a change in the basal slope β.

The flank collapse model of the 2018 Anak Krakatau Volcano eruption (Mulia et al., 2020) is shown here as an example. The Anak Krakatau flank collapse simulation results for the landslide thickness and the tsunami height at 1 minute after the origin time of the collapse are plotted in Fig. 3.18. The submarine part of the volcano initially deforms, while the water is affected by momentum transfer from the subaerial part, resulting in a maximum water elevation exceeding 50 m. The landslide moves southwest corresponding to the unstable edifice of the volcano. Within 1 minute, the center of mass of the landslide reaches the bottom of the caldera, which has a water depth of 200 m. The estimated landslide volume of 0.24 km³ has a duration of 3–5 minutes, which is relatively short. The evolution of the simulated landslide propagation and tsunami propagation profiles is shown in Fig. 3.18.

3.5.1.2 Two-layered pyroclastic flow-tsunami model

To calculate pyroclastic flows and tsunamis simultaneously, two types of two-layer shallow-water models, a dense-type (DPF) and a light-type (LPF) (Fig. 3.19), have been developed (Maeno & Imamura, 2011). The dense-type two-layer shallow-water model is used for a pyroclastic flow that is denser than seawater, while the light-type two-layer shallow-water model is used for a pyroclastic flow that is lighter than seawater.

3.5.1.2.1 High-density pyroclastic flow (dense-type)

The governing equations include full nonlinearity under the assumption of a long-wave approximation. Subscripts 1 and 2 indicate the upper water layer and the lower pyroclastic flow layer, respectively. Equations for the upper layer of the water in Cartesian coordinates of the dense-type (DPF) model are

$$\frac{\partial(\eta_1 - \eta_2)}{\partial t} + \frac{\partial M_1}{\partial x} + \frac{\partial N_1}{\partial y} = 0 \quad (3.90)$$

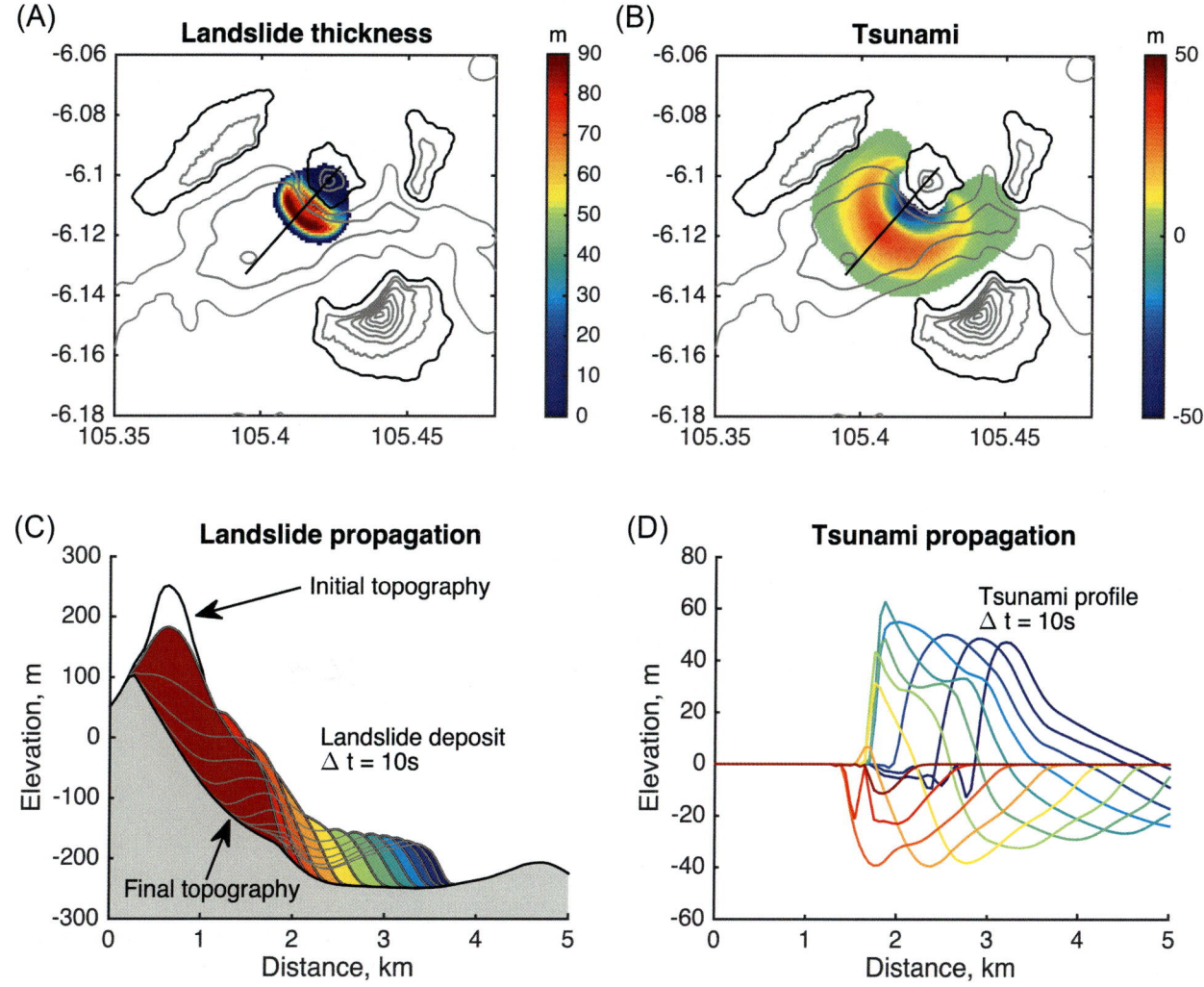

FIGURE 3.18 Simulation results for the 2018 Anak Krakatau Volcano flank collapse using the VolcFlow code. (A) Simulated landslide thickness at 1 min after the initial collapse. (B) Simulated tsunami as a result of the subaerial and submarine flank collapse. (C) Simulated landslide propagation shown as topography profiles at times between $t = 0$ and $t = 2$ minutes with a time interval of 10 seconds. (D) Simulated propagation of the tsunami generated by the landslide at times between $t = 0$ and $t = 2$ minutes with a time interval of 10 seconds.

$$\frac{\partial M_1}{\partial t} + \frac{\partial}{\partial x}\left(\frac{M_1^2}{D_1}\right) + \frac{\partial}{\partial y}\left(\frac{M_1 N_1}{D_1}\right) + gD_1\frac{\partial \eta_1}{\partial x} + \frac{gn_m^2}{D_1^{7/3}}M_1\sqrt{M_1^2 + N_1^2} - INTF_x = 0 \qquad (3.91)$$

$$\frac{\partial N_1}{\partial t} + \frac{\partial}{\partial x}\left(\frac{M_1 N_1}{D_1}\right) + \frac{\partial}{\partial y}\left(\frac{N_1^2}{D_1}\right) + gD_1\frac{\partial \eta_1}{\partial y} + \frac{gn_m^2}{D_1^{7/3}}N_1\sqrt{M_1^2 + N_1^2} - INTF_y = 0 \qquad (3.92)$$

Those for the lower layer (pyroclastic flow) of the DPF model are

$$\frac{\partial \eta_2}{\partial t} + \frac{\partial M_2}{\partial x} + \frac{\partial N_2}{\partial y} = 0 \qquad (3.93)$$

$$\frac{\partial M_2}{\partial t} + \frac{\partial}{\partial x}\left(\frac{M_2^2}{D_2}\right) + \frac{\partial}{\partial y}\left(\frac{M_2 N_2}{D_2}\right) + gD_2\left(\alpha\frac{\partial D_1}{\partial x} + \frac{\partial \eta_2}{\partial x} - \frac{\partial h_1}{\partial x}\right) + \frac{gn_p^2}{D_2^{7/3}}M_2\sqrt{M_2^2 + N_2^2} + \alpha INTF_x = DIFF_x \qquad (3.94)$$

$$\frac{\partial N_2}{\partial t} + \frac{\partial}{\partial x}\left(\frac{M_2 N_2}{D_2}\right) + \frac{\partial}{\partial y}\left(\frac{N_2^2}{D_2}\right) + gD_2\left(\alpha\frac{\partial D_1}{\partial x} + \frac{\partial \eta_2}{\partial x} - \frac{\partial h_1}{\partial x}\right) + \frac{gn_p^2}{D_2^{7/3}}N_2\sqrt{M_2^2 + N_2^2} + \alpha INTF_y = DIFF_y \qquad (3.95)$$

where η_1 is the water surface elevation, η_2 is the thickness of a pyroclastic flow, h_1 is the still water depth, and $D_1 = h_1 + \eta_1$ is the total depth, and for a dense pyroclastic flow $D_2 = \eta_2$, M and N are the discharge in the x and y

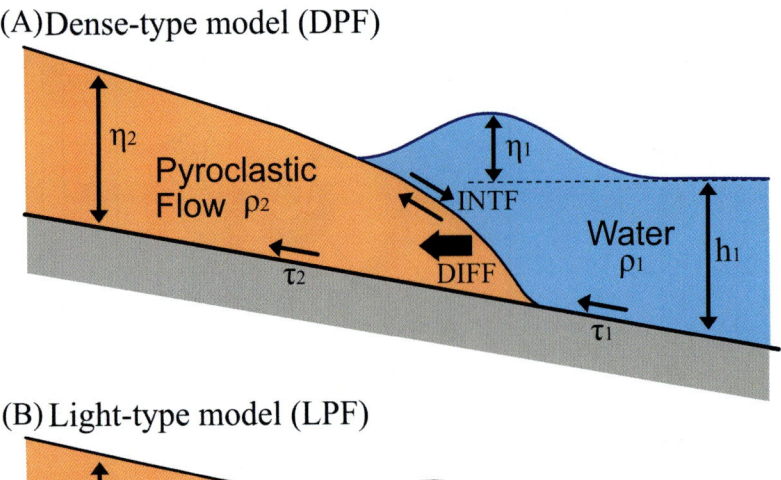

FIGURE 3.19 Schematic of two types of two-layer shallow-water models describing pyroclastic flows entering the sea. (A) Dense-type model for a pyroclastic flow that is heavier than water. (B) Light-type model for a pyroclastic flow that is lighter than water. *From Maeno, F., & Imamura, F. (2011). Tsunami generation by a rapid entrance of pyroclastic flow into the sea during the 1883 Krakatau eruption, Indonesia. Journal of Geophysical Research: Solid Earth, 116(B9). https://doi.org/10.1029/2011JB008253.*

directions, respectively, α is the density ratio ($= \rho_1/\rho_2$), n_m is the bottom friction coefficient (Manning's roughness coefficient) for the water layer and set to 0.025, and n_p is the bottom friction coefficient for the pyroclastic flow. The bottom friction coefficient for the pyroclastic flow on land can be set to be between 0.01 and 0.06. When the pyroclastic flow is propagating underwater, the bottom friction coefficient can be set to be between 0.06 and 0.08. *INTF* is the term for interfacial shear stress, and *DIFF* is the term for turbulent diffusion force. Interfacial shear stress and turbulent diffusion stress for the x and y directions are given as follows

$$INTF_x = f\bar{u}\sqrt{\bar{u}^2 + \bar{v}^2} \tag{3.96}$$

$$INTF_y = f\bar{v}\sqrt{\bar{u}^2 + \bar{v}^2} \tag{3.97}$$

$$DIFF_x = k\sqrt{\frac{\partial^2 M}{\partial x^2} + \frac{\partial^2 M}{\partial y^2}} \tag{3.98}$$

$$DIFF_y = k\sqrt{\frac{\partial^2 N}{\partial x^2} + \frac{\partial^2 N}{\partial y^2}} \tag{3.99}$$

where f is the interfacial drag coefficient between the density current and water, $\bar{u}$ and $\bar{v}$ are the relative velocities between flow and water in the x and y directions, respectively, and k is the turbulent diffusion coefficient. The interfacial drag coefficient between the pyroclastic flow and the water can be set to be between 0.06 and 0.2.

3.5.1.2.2 Low-density pyroclastic flow (light-type)

The low-density pyroclastic flow (LDF) (light-type) model that also adopts the two-layer representation is developed to characterize the situation where the pyroclastic flow has a lower density than the water (LPF). In this case, the pyroclastic flow propagates on top of the water. Its fundamental physics and governing equations are similar to the dense-type model, and the order of layers is switched. Subscripts 1 and 2 indicate the upper light pyroclastic flow layer and the lower water layer, respectively. Equations for the upper pyroclastic flow layer for the LDF model are given as follows

$$\frac{\partial \eta_1}{\partial t} + \frac{\partial M_1}{\partial x} + \frac{\partial N_1}{\partial y} = 0 \tag{3.100}$$

$$\frac{\partial M_1}{\partial t} + \frac{\partial}{\partial x}\left(\frac{M_1^2}{D_1}\right) + \frac{\partial}{\partial y}\left(\frac{M_1 N_1}{D_1}\right) + gD_1\left(\frac{\partial \eta_1}{\partial x} + \frac{\partial \eta_2}{\partial x}\right) + \frac{gn_p^2}{D_1^{7/3}} M_1 \sqrt{M_1^2 + N_1^2} + INTF_x = DIFF_x \tag{3.101}$$

$$\frac{\partial N_1}{\partial t} + \frac{\partial}{\partial x}\left(\frac{M_1 N_1}{D_1}\right) + \frac{\partial}{\partial y}\left(\frac{N_1^2}{D_1}\right) + gD_1\left(\frac{\partial \eta_1}{\partial y} + \frac{\partial \eta_2}{\partial y}\right) + \frac{gn_p^2}{D_1^{7/3}} N_1 \sqrt{M_1^2 + N_1^2} + INTF_y = DIFF_y \tag{3.102}$$

Those for the lower layer of a light-type mode are given as follows

$$\frac{\partial \eta_2}{\partial t} + \frac{\partial M_2}{\partial x} + \frac{\partial N_2}{\partial y} = 0 \tag{3.103}$$

$$\frac{\partial M_2}{\partial t} + \frac{\partial}{\partial x}\left(\frac{M_2^2}{D_2}\right) + \frac{\partial}{\partial y}\left(\frac{M_2 N_2}{D_2}\right) + gD_2\left(\alpha\frac{\partial \eta_1}{\partial x} + \frac{\partial \eta_2}{\partial x}\right) + \frac{gn_m^2}{D_2^{7/3}} M_2 \sqrt{M_2^2 + N_2^2} - \alpha INTF_x = 0 \tag{3.104}$$

$$\frac{\partial N_2}{\partial t} + \frac{\partial}{\partial x}\left(\frac{M_2 N_2}{D_2}\right) + \frac{\partial}{\partial y}\left(\frac{N_2^2}{D_2}\right) + gD_2\left(\alpha\frac{\partial \eta_1}{\partial x} + \frac{\partial \eta_2}{\partial x}\right) + \frac{gn_m^2}{D_2^{7/3}} N_2 \sqrt{M_2^2 + N_2^2} - \alpha INTF_y = 0 \tag{3.105}$$

where η_1 is the thickness of a light pyroclastic flow, η_2 is the water surface elevation, d is the still water depth, $D_2 = \eta_2 + d$ is the total water depth, for a dense pyroclastic flow $D_1 = \eta_1$, and the definitions of the other parameters and the terms are the same as for the dense-type model above.

The bottom friction for the dense pyroclastic flow adopts Manning's roughness coefficient, which is commonly used for characterizing the bottom friction in water flow. Ioki et al. (2019) enhanced the model by introducing the concept of Coulomb friction, which accounts for the frictional interaction between two solid surfaces. The modified bottom friction terms (τ/ρ) become

$$\frac{\tau_{x2}}{\rho_2} = \frac{gn_p^2}{D_2^{7/3}} M_2 \sqrt{M_2^2 + N_2^2} + \frac{M_2 gD_2 \tan\phi}{\sqrt{M_2^2 + N_2^2}} \tag{3.106}$$

$$\frac{\tau_{y2}}{\rho_2} = \frac{gn_p^2}{D_2^{7/3}} N_2 \sqrt{M_2^2 + N_2^2} + \frac{N_2 gD_2 \tan\phi}{\sqrt{M_2^2 + N_2^2}} \tag{3.107}$$

where ϕ is the apparent friction angle, and n_p is Manning's roughness coefficient in the lower layer.

3.5.1.3 Air pressure wave source

Air pressure waves propagating in the atmosphere can continuously excite tsunami waves. In simulating tsunamis caused by tsunamigenic earthquakes, the pressure terms in the shallow-water equations are often neglected. This omission is due to atmospheric pressure changes occurring at a much slower speed compared to tsunamis under normal weather conditions. However, for tsunamis generated by airwaves, it is crucial to include the pressure terms in the momentum equations of the tsunami model

$$\frac{\partial M}{\partial t} + gd\frac{\partial \eta}{\partial x} + \frac{d}{\rho}\frac{\partial p}{\partial x} = 0 \tag{3.108}$$

$$\frac{\partial N}{\partial t} + gd\frac{\partial \eta}{\partial y} + \frac{d}{\rho}\frac{\partial p}{\partial y} = 0 \tag{3.109}$$

where p is the atmospheric pressure, which can be substituted with the airwave pressure in Pascal units, and ρ is the water density. The tsunami can be simulated by forcing the pressure from the airwave continuously.

The eruption of Hunga Tonga - Hunga Ha'apai Volcano in 2022 generated atmospheric Lamb waves (airwaves), which are nondispersive atmospheric waves, whose energy is optimally transmitted far away from the source with minor losses (Amores et al., 2022). Lamb waves have purely horizontal motions, occupying the full depth of the troposphere and with a maximum pressure signal at the surface. These waves are only slightly affected by the Earth's rotation and travel at the speed of sound in the media (Gossard & Hooke, 1975). Assuming an isothermal troposphere, the phase velocity of the Lamb waves, C_T, is only affected by the air temperature and is defined as follows

$$C_T = \sqrt{\frac{\gamma RT}{M_a}} \qquad (3.110)$$

where γ ($=1.4$) is the ratio of specific heat of air corresponding to the range of atmospheric temperatures, R ($=8314.36$ J/kmol/K) is the universal gas constant, M_a ($=28.966$ kg/kmol) is the molecular mass for dry air, and T is the absolute temperature.

The propagation of Lamb waves through the atmosphere with spatially varying temperatures is analogous to the behavior of oceanic long-waves propagating over an ocean with variable depth. Long waves in the ocean are also nondispersive barotropic waves traveling with a phase velocity, given by

$$C = \sqrt{gd} \qquad (3.111)$$

Given these similarities between atmospheric Lamb waves and oceanic shallow-water waves, the atmospheric Lamb wave generated after the 2022 Hunga Tonga–Hunga Ha'apai Volcano explosion can be simulated using the same approach as the tsunami model or a vertically integrated hydrodynamic ocean model (Amores et al., 2022). To do so, the Lamb wave velocity is substituted with the long-wave velocity to obtain the representative depth for Lamb wave simulation

$$D_T = \frac{\gamma RT}{M_a g} \qquad (3.112)$$

The average surface temperature T of 288K can be assumed for the simulation. The shallow-water equations for the Lamb wave propagation can be written in terms of discharge fluxes as follows

$$\frac{\partial p}{\partial t} + \rho_a g \left(\frac{\partial M_p}{\partial x} + \frac{\partial N_p}{\partial y} \right) = 0 \qquad (3.113)$$

$$\frac{\partial M_p}{\partial t} + \frac{D_T}{\rho_a} \frac{\partial p}{\partial x} = 0 \qquad (3.114)$$

$$\frac{\partial N_p}{\partial t} + \frac{D_T}{\rho_a} \frac{\partial p}{\partial y} = 0 \qquad (3.115)$$

where p is the atmospheric pressure, ρ_a is the density of the air ($=1.225$ kg/m^3 in 15°C), and M_p and N_p are the discharge fluxes for the atmospheric wave in the x and y directions, respectively.

To illustrate how Lamb waves generate tsunamis, a simulation is conducted by solving the above continuity and momentum equations for meteotsunamis. For the simulation, a simple atmospheric pressure anomaly source model is used. The Lamb wave speed in this simulation is fixed to 320 m/second. The bathymetric data used for the simulation are typical for areas around subduction zones with continental shelf, continental slope, trench, and deep ocean. Since there is no initial sea surface displacement, the only generating source is the Lamb wave. The simulated Lamb wave and tsunami profiles at 2-minute intervals are plotted in Fig. 3.20A, while the time series for the Lamb wave and tsunami waveforms at the four positions is shown in Fig. 3.20B.

The simulation results show that the airwave can effectively excite a tsunami in deeper waters, as demonstrated by comparing the waveforms at stations 1 and 3 (Fig. 3.20). Initially, the excited tsunami propagates at the same speed as the Lamb wave, acting as the leading tsunami wave. However, due to the complexity of the bathymetry, the typical tsunami traveling speed causes the tsunami to gradually separate from the Lamb wave. This decoupled tsunami then travels at a slower pace, lagging behind the leading tsunami wave, which can be seen in the Lamb wave and tsunami profile plots as well as the tsunami waveforms observed at station 4 (Fig. 3.20).

3.5.2 Nearshore and offshore tsunami modeling

Tsunami models using the linear shallow-water equations are well suited for accurately simulating tsunamis in the open ocean, particularly in areas with water depths more than 50 m. Linear tsunami models are also desirable when a quick

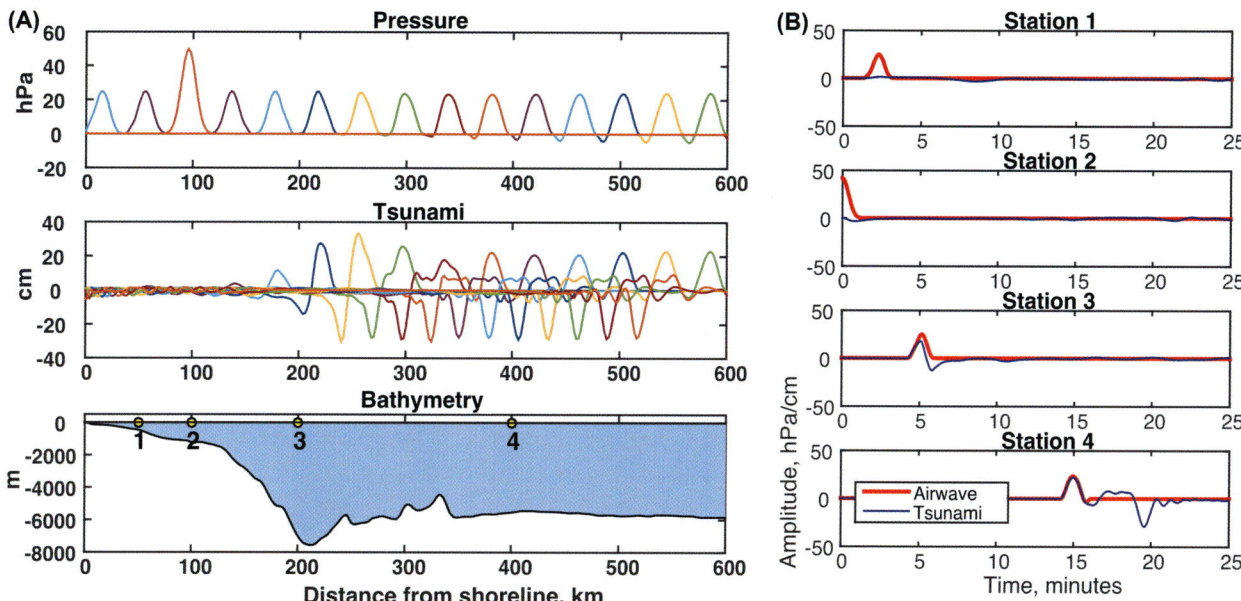

FIGURE 3.20 Simulated atmospheric Lamb wave and the generated tsunami. (A) Snapshots of Lamb wave and tsunami profiles at different times. The bathymetry used for the simulation is shown in the lower panel. (B) Time series of Lamb wave and tsunami waveforms at the four positions indicated in the bathymetry plot.

tsunami prediction is needed in a tsunami warning situation. In tsunami hazard assessments for a coastal area, a large set of scenarios is often required. A nonlinear tsunami model would be a better choice in terms of accuracy compared to the linear one for running scenarios of tsunami hazard assessments. For coastal areas with complex bathymetric features, a linear tsunami model may not be sufficient to accurately simulate the tsunami near the coast. In these cases, a nonlinear tsunami model may be needed.

When the goal is to simulate the general tsunami propagation pattern across the ocean, a grid size between 1 and 5 arcminutes can be employed. In many cases, such a linear model can replicate the first few wave cycles at tide gauge locations within harbors where water depths typically range from 5 to 10 m. For simulating tsunamis along the coastline, grid sizes finer than 15 arcseconds are typically adequate. The General Bathymetric Chart of the Oceans (GEBCO) provides a global terrain model for ocean and land on 15 arcsecond interval grids (GEBCO). GEBCO bathymetric data have been widely used to simulate tsunamis, particularly in the open ocean.

To simulate the tsunami at the coast with precision, it is necessary to use high-quality and high-resolution grids. In this case, ensuring that detailed bathymetric features are comprehensively represented in the DEM is crucial. The DEM creation should involve the incorporation of gridded bathymetry data and nautical charts from local authorities, hydrographic agencies, or other relevant organizations. When the tsunami source is situated on the opposite side of the ocean or located further offshore the area of interest, it is suggested to use a nested grid system for simulation efficiency. This setup involves utilizing a larger grid size for the entire modeling domain and implementing finer grids within the area of interest around the coastal gauges.

Tsunami observations from coastal gauges and offshore gauges serve as validations for a tsunami scenario. This is achieved by comparing the observational data with the tsunami simulation results. Water-level data from some selected coastal gauges in countries around the world are available from the Sea-Level Station Monitoring Facility website (https://www.ioc-sealevelmonitoring.org/) provided by the Intergovernmental Oceanographic Commission of the United Nations Educational, Scientific and Cultural Organization. Every country usually has its domestic agency and website that provides coastal gauge data. For offshore observations, an example is the DART buoy observation system, which was designed to detect and monitor tsunamis in the deep ocean. The system is maintained, and the data are provided by the NOAA (https://nctr.pmel.noaa.gov/Dart/).

Observed tsunami waveforms can be incorporated into a source inversion technique (Satake, 1987), which is discussed in Chapter 7. The tsunami waveform inversion method aims to determine the tsunami source mechanism by evaluating waveforms recorded at coastal and offshore water-level gauges. Essential to this technique are synthetic tsunami waveforms, produced by the numerical model. Many studies of significant tsunami events used tsunami waveforms at observation stations. For example, the source of the 2011 Tohoku Earthquake (M_w 9.0) was estimated using tsunami waveforms recorded at offshore and coastal gauges (Fujii et al., 2011; Gusman et al., 2012). The source model for the 2014 Chile Earthquake (M_w 8.2) estimated from teleseismic data was validated by tsunami data recorded at both coastal gauges and DART stations (Heidarzadeh et al., 2016). The tsunami source of the 2021 Raoul Island Earthquake (M_w 8.1) in the Kermadec subduction zone was estimated by inverting the tsunami waveforms recorded at DART stations and coastal gauges (Romano et al., 2021).

3.5.2.1 Example: linear tsunami simulation

An example of COMCOT setup for simulating the tsunami generated by the 2021 Raoul Island Earthquake (M_w 8.1) is provided in Example 3-1 (Appendix). The tsunami simulation involves solving the linear shallow-water equations. Two nested grid layers are utilized. Layer 1 covers the entire Pacific Ocean with a grid resolution of 10 arcminutes, while Layer 2 covers New Zealand with a grid resolution of 5 arcminutes. For the initial condition, the sea surface displacement is utilized (Romano et al., 2021). The simulated maximum tsunami amplitude distribution on Layer 1 and the simulated travel time contours are shown in Fig. 3.21. Fig. 3.22 shows snapshots of the tsunami propagation at different times. The simulated tsunami waveforms at DART stations are compared with observations in Fig. 3.23. The simulation results demonstrate that the source model used in the simulation can replicate the observations.

3.5.3 Tsunami inundation modeling

To simulate tsunami inundation, the nonlinear shallow-water equations must be solved. In coastal areas, the friction and convection terms in the shallow-water equations become significant. Therefore, these terms must be included in the computation. For the case of simulating the tsunami propagation in the ocean, no major concerns are necessary. On the other hand, for the case of simulating the tsunami inundation process on land, the computation on both land and offshore cells is necessary. Tsunami numerical models, such as COMCOT, implement a moving boundary scheme that describes the dynamic changes in water level and flow as the tsunami propagates inland. In simulating tsunami inundation, coseismic coastal deformation, land cover for the study area, tidal level, and sea-level rise (SLR) factors are often considered. These factors are explained in the following.

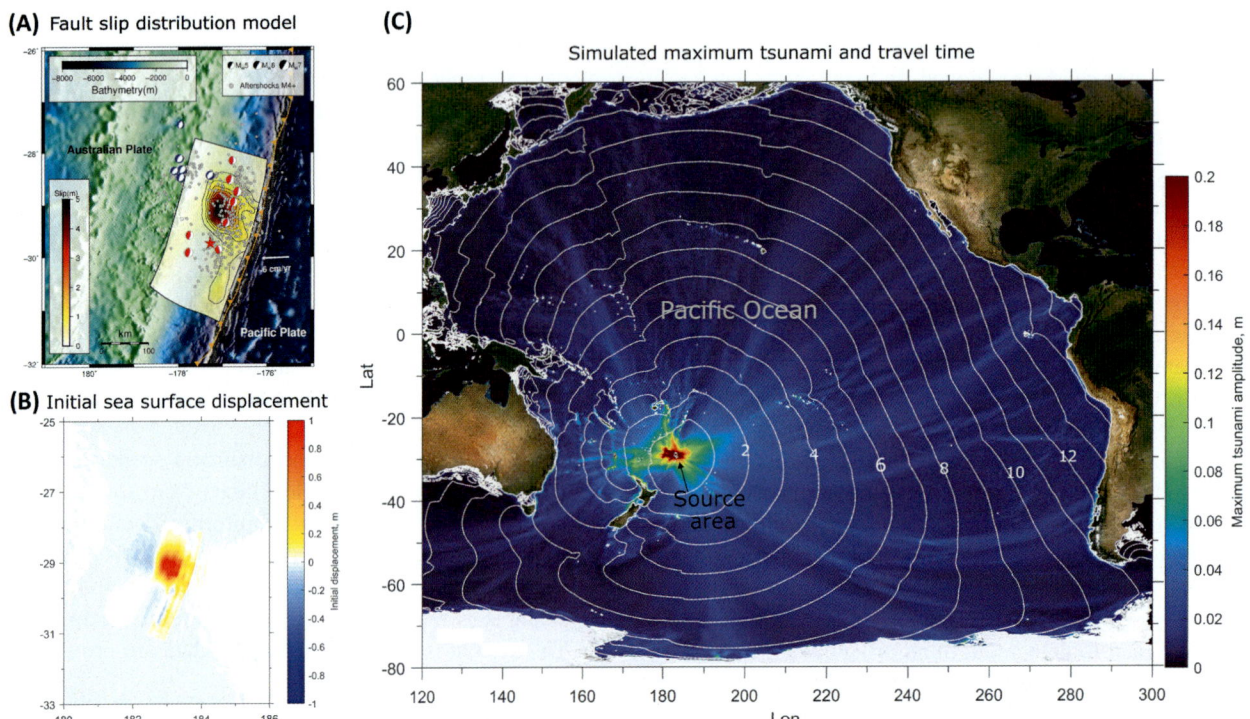

FIGURE 3.21 (A) Fault slip distribution of the 2021 Raoul Island Earthquake (M_w 8.1) estimated from tsunami waveforms by Romano et al. (2021). (B) The sea surface displacement from the earthquake source model calculated using Okada formulae and considering the horizontal displacement effects. (C) The simulated maximum tsunami amplitude distribution. White contours indicate the simulated tsunami travel time, with a time interval of 1 hour. Map lines delineate study areas and do not necessarily depict accepted national boundaries.

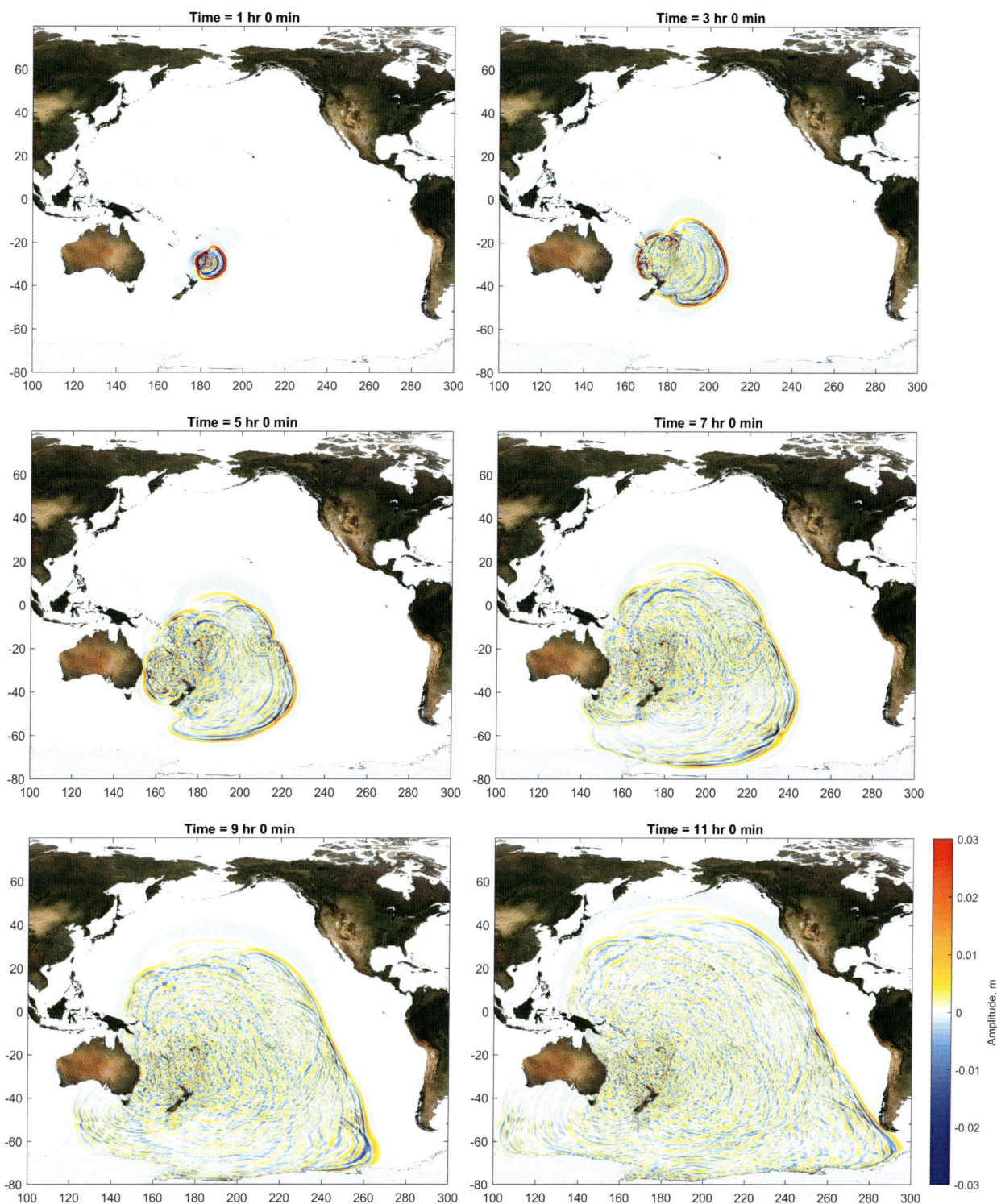

FIGURE 3.22 Snapshots of tsunami propagation from the 2021 Raoul Island Earthquake (M_w 8.1) at 1, 3, 5, 7, 9, and 11 hours after the earthquake's origin time. Map lines delineate study areas and do not necessarily depict accepted national boundaries.

3.5.3.1 Coseismic deformation in coastal areas

Large earthquakes can make the nearby coastal area subside or uplift. The lowered ground level due to coseismic deformation can increase the tsunami impact. This was evident in the case of the 2004 Sumatra−Andaman Earthquake that

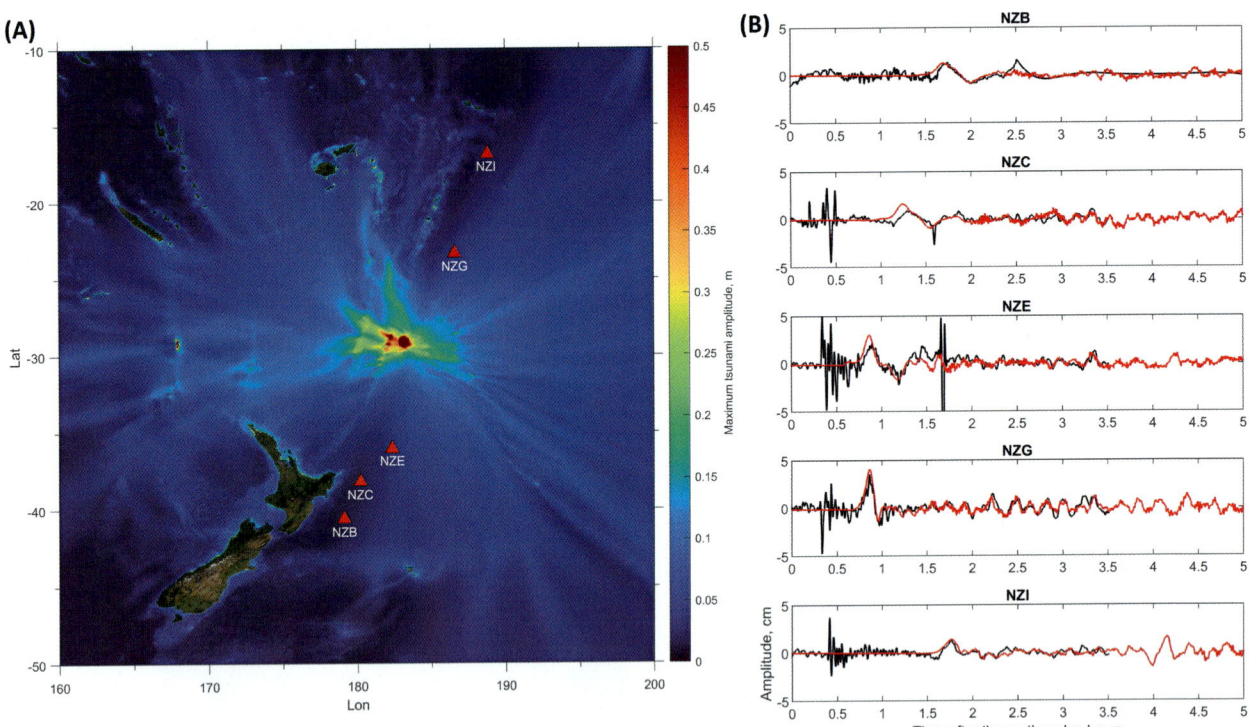

FIGURE 3.23 (A) The maximum tsunami resulting from the 2021 Raoul Island Earthquake (M_w 8.1), simulated using the source model estimated by Romano et al. (2021). Red triangles indicate the DART stations. (B) Simulated (*red* lines) and observed (*black* lines) tsunami waveforms recorded at the DART stations. The tsunami was simulated by solving the linear shallow-water equations using COMCOT. Map lines delineate study areas and do not necessarily depict accepted national boundaries.

produced significant subsidence along the Northern Sumatran mainland. The subsidence and scour permanently moved inland the shoreline of Banda Aceh as much as 1.5 km, and the generated tsunami penetrated up to 4 km inland (Borrero, 2005), causing massive flooding and destruction. The 2016 Kaikōura Earthquake ruptured multiple faults across and subparallel to the coastline, reaching into the sea, and generated a tsunami. Along ~110 km of coastline, field surveying, Lidar data differencing, and satellite geodesy revealed highly variable vertical displacements, ranging from −2.5 to 6.5 m (Clark et al., 2017). These studies show that deformation in coastal areas can be particularly severe and can have significant impacts on the behavior of tsunami waves. Therefore, it is important to incorporate this deformation into tsunami simulations to accurately model the potential impacts of a tsunami. This can be accomplished by adjusting the DEM data used in the simulation to account for earthquake-triggered deformation.

3.5.3.2 Bottom roughness and built area

The influence of bottom roughness, caused by vegetation and built environments, on the behavior of inundation can be significant. In nonlinear tsunami models, wave attenuation caused by bottom roughness is typically represented by Manning's roughness coefficient (n). To obtain the appropriate Manning's n value, it is necessary to incorporate land cover information (Kaiser et al., 2011). Past research highlighted the effect of bottom roughness on the extent of tsunami inundation, with higher roughness levels leading to reduced inundation (Dao & Tkalich, 2007; Gibbons et al., 2022). Considering that building damage depends on flow depth and speed, the selection of Manning's n significantly affects the formulation of building codes for essential structures and lifelines, as well as the planning of no-build zones (Bricker et al., 2015). Additionally, a higher Manning's n causes a delay in the arrival time of the tsunami front in the lee of vegetated or urban areas, a critical consideration for evacuation planning.

To create a realistic tsunami inundation model, significant steps can be taken to generate a high-resolution topographic dataset that incorporates building height and land cover information. However, this process can be costly and time-consuming due to the need for fine computational grids, which result in longer computational times. In general, there are three types of ground surface models for tsunami inundation modeling (Muhari et al., 2011). The first model, the simplest among the three, is called the Constant Roughness Model (CRM). It applies a uniform bottom roughness coefficient throughout the model area. The second is the Topographic Model (TM), which integrates the building height as well as other important coastal and river structures into the DEM. The third model is the Equivalent Roughness Model (ERM), which uses different bottom roughness coefficients that are a function of land use and the percentage of building occupancy on each grid cell (Imamura, 2009; Leschka et al., 2009). In the ERM for a built-up area, the friction terms in the momentum equations are given as follows

$$\frac{g}{D^{7/3}}\left(n^2 + \frac{C_D}{2gk}\frac{\theta}{100-\theta}D^{4/3}\right)M\sqrt{M^2+N^2} \tag{3.116}$$

$$\frac{g}{D^{7/3}}\left(n^2 + \frac{C_D}{2gk}\frac{\theta}{100-\theta}D^{4/3}\right)N\sqrt{M^2+N^2} \tag{3.117}$$

where k is the average building's width, θ is the percentage of the bottom area occupied by the building in a grid cell, and C_D is the drag coefficient. Depending on the ratio between the width of the building and the flow depth, the drag coefficient usually ranges from 1 to 2.

If building height information is not available, an equivalent bottom roughness can be used depending on the percentage of building occupancy. The results obtained from the CRM should be considered representing the maximum potential extent of tsunami inundation but with high uncertainty. Generally, it can be concluded that without high-resolution data on land cover or buildings, inundation maps can be expected to be less accurate with respect to inundation depths (Kaiser et al., 2011). On the other hand, results from the TM and ERM can be interpreted as the maximum potential extent of tsunami inundation if no buildings were destroyed (Muhari et al., 2011). In the TM and ERM, the dynamic roughness, which changes over time during the inundation process, is ignored. Further research on dynamic roughness may improve our understanding of its influence on the inundation process.

The ERM model can be made by using land-cover data to build a roughness grid. This approach can still be classified as an ERM method, which adopts a simplified approach by assigning a single roughness value to the entire built-up area instead of using Eqs. (3.116) and (3.117), and assigning different values to other land covers. Common values of Manning's roughness and land cover groups used in tsunami studies are listed in Table 3.2. These values were derived by comparing roughness values found in the literature, grouping and averaging roughness values for similar land-cover types (Bricker et al., 2015; Fraser et al., 2014; Gayer et al., 2010; Imamura et al., 2006; Kaiser et al., 2011; Matsutomi et al., 2001; Wang & Liu, 2007; Arcement, and Schneider, 1989) but leaning slightly toward the lower end of the value ranges. For example, Fig. 3.24 shows the land cover in Wellington, New Zealand, and provides a visual representation of the various types of land cover in the area, including urban areas, forests, grasslands, and water bodies. The land cover data were used to create a roughness grid for tsunami modeling input (Fig. 3.25). The simulated tsunami inundation from a hypothetical plate interface earthquake (M_w 8.7) scenario in the Hikurangi subduction zone is shown in Fig. 3.26.

3.5.3.3 Tidal state and sea-level rise

The tidal state at the time of the tsunami's arrival can have a significant effect on the maximum nearshore water level and consequent inundation. Tsunami simulations can be conducted by assuming that the tsunami arrives at mean high water springs (MHWS) levels to provide conservative inundation estimates. Mean sea level (MSL) is commonly used as the reference level to interpret input data for elevation information in the DEM and to create output data, such as tsunami elevations.

In the modeling of tsunami inundation in Wellington (Fig. 3.26), a static level of 0.69 m above local MSL is used to represent MHWS. This level is assumed to remain constant and not vary with tide fluctuations during the modeling process. The simulation results show that coastal areas around Lyall Bay, Evans Bay, and Lambton Harbor are widely inundated with flow depths of up to 2 m. Another simulation is conducted by assuming that the tsunami occurs at the current MHWS plus 1 m of SLR, as shown in the right panel of Fig. 3.26. As expected, the result of adding 1 m of SLR increases the tsunami impact. The inundation area is wider than that from the simulation at the current MHWS, and significantly larger areas are inundated with flow depths of up to 3 m.

According to the definition, MSL is the average level of the sea surface over a long period or the average level that would exist in the absence of tides. MHWS and mean low water springs (MLWS) are defined as the average

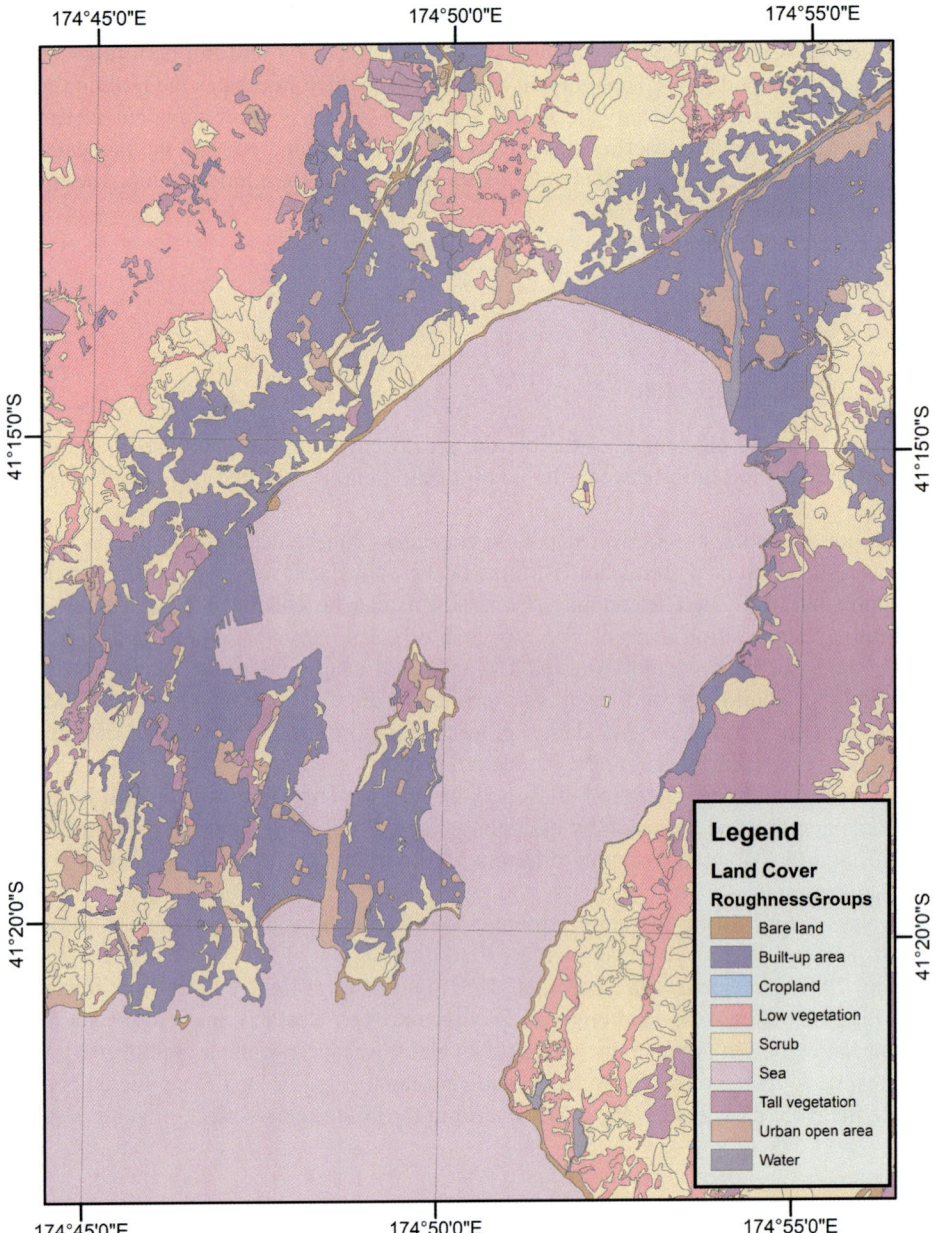

FIGURE 3.24 Land cover in Wellington, New Zealand. Land cover data can be used to create a roughness grid for tsunami modeling input. Map lines delineate study areas and do not necessarily depict accepted national boundaries.

levels of each pair of successive high waters and low waters, respectively, during approximately 24 hours in each semilunation (approximately every 14 days) when the range of the tide is greatest (spring range). Mean high water neaps (MHWN) and mean low water neaps (MLWN) are the average levels of each pair of successive high and low waters, respectively, during approximately 24 hours in each semilunation when the range of the tide is least (neap range). Chart datum (CD) is a water level that is low enough that the tide will seldom fall below it. When meteorological conditions lower the sea level, the tide may fall below the predicted low water heights. In places where the CD is relatively high, the actual depths at or near low water may be less than the charted water level. The highest astronomical tide (HAT) and lowest astronomical tide (LAT) are the highest and lowest tidal levels that can be predicted to occur under average meteorological conditions over 18 years. Fig. 3.27 illustrates the relationships among MSL, MHWS, MLWS, MHWN, MLWN, HAT, LAT, and CD. Modern CDs are set at the approximate level of LAT, and tide tables list the predicted height of the tide above the CD. However, in abnormal meteorological conditions, the water level may fall below the LAT.

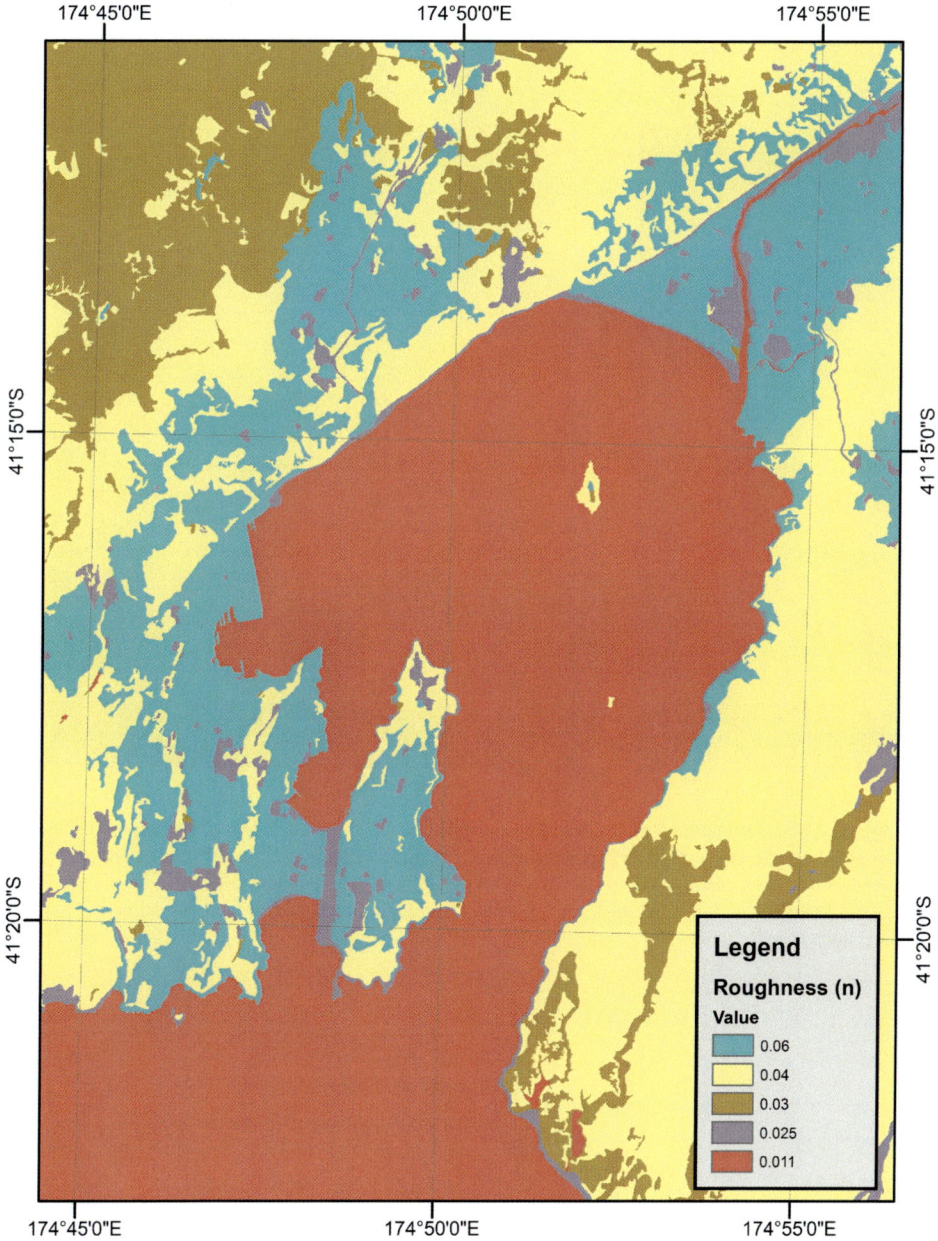

FIGURE 3.25 Manning's roughness (*n*) values in Wellington, New Zealand, created on the basis of the land cover data of the region. Map lines delineate study areas and do not necessarily depict accepted national boundaries.

3.5.3.4 Example: tsunami inundation modeling

An example of COMCOT setup for simulating the inundation from a series of tsunamis generated during the 2018 Palu Earthquake (M_w 7.5) in Indonesia is provided in Example 3-2 (Appendix). The simulation involves solving the nonlinear shallow-water equations and incorporating the moving boundary scheme to simulate the tsunami inundation. A series of tsunamis was generated by the earthquake and by the subsequent subaerial–submarine landslides that occurred around the bay. The earthquake source and landslide source models estimated in a previous study by Gusman et al. (2019) are utilized in the simulation. The tsunami occurred during high tide at approximately 1 m above MSL that also increases the tsunami potential impact (Fig. 3.28B). Therefore, a tidal offset of +1 m is included in the simulation. The estimated coseismic seafloor vertical displacement pattern is complex with up to 0.8-m uplift in the middle of the bay and two subsidence areas on the northern and southern parts of the bay (Fig. 3.28A). The tsunami source model reproduced the tsunami waveform at the Pantoloan tide gauge station with a tsunami wave amplitude of 2 m (Fig. 3.28C). The observed tsunami runup of up to 4 m above MSL and ∼400-m limit of inundation were reproduced by the combined coseismic vertical displacement and multiple landslide models (Fig. 3.29).

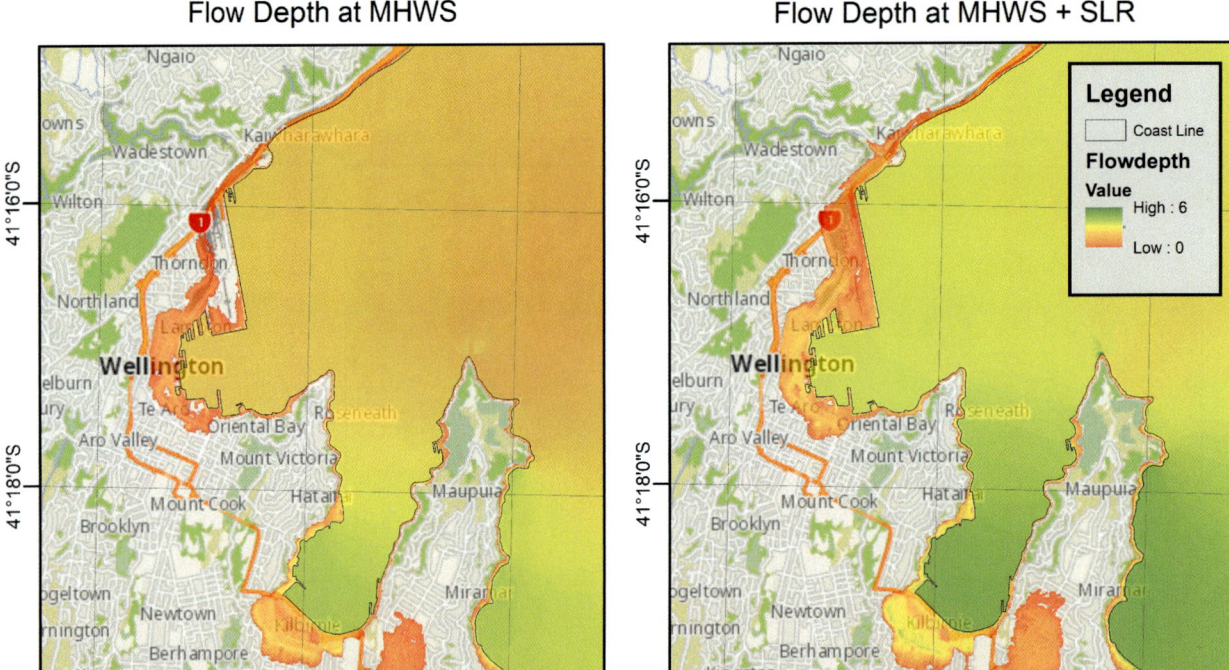

FIGURE 3.26 Simulated tsunami inundation height in Wellington, New Zealand for a hypothetical plate interface earthquake (M_w 8.7) scenario in the Hikurangi subduction zone. Offshore values refer to maximum tsunami heights, while onshore values refer to flow depths. The left panel shows simulated tsunami inundation with an assumption that the tsunami occurs at MHWS. The right panel shows simulated tsunami inundation with an assumption that the tsunami occurs at the current MHWS plus 1 m of sea-level rise. *MHWS*, Mean high water spring. Map lines delineate study areas and do not necessarily depict accepted national boundaries.

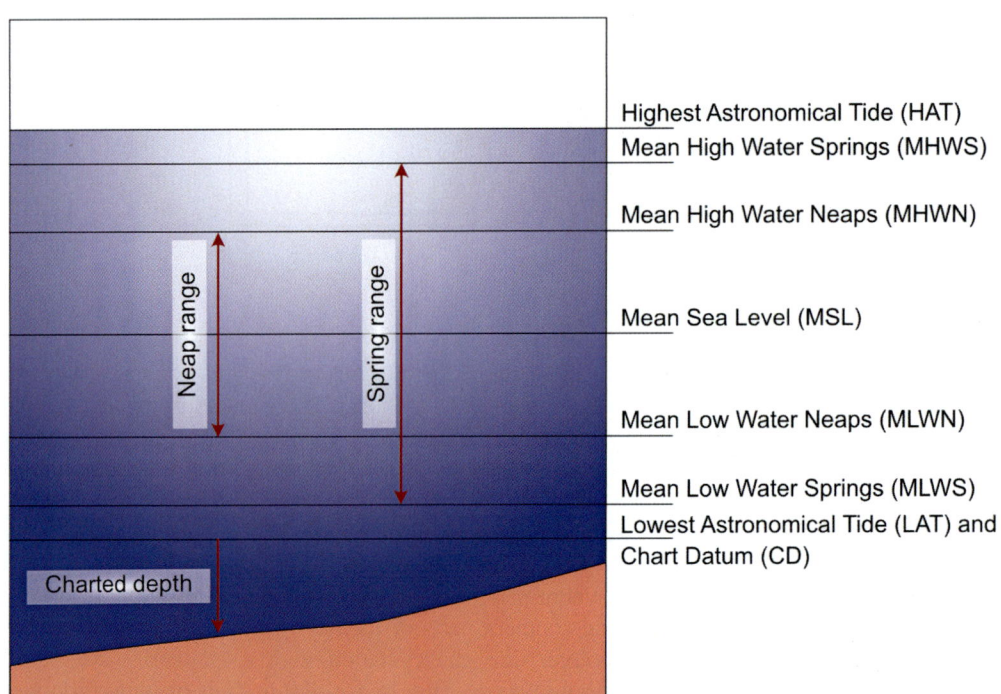

FIGURE 3.27 Illustration of the relationships among MSL, MHWS, MLWS, MHWN, MLWN, HAT, LAT, and CD. *CD*, Chart datum; *HAT*, highest astronomical tide; *LAT*, lowest astronomical tide; *MHWN*, mean high water neaps; *MHWS*, mean high water springs; *MLWN*, mean low water neaps; *MLWS*, mean low water springs; *MSL*, mean sea level.

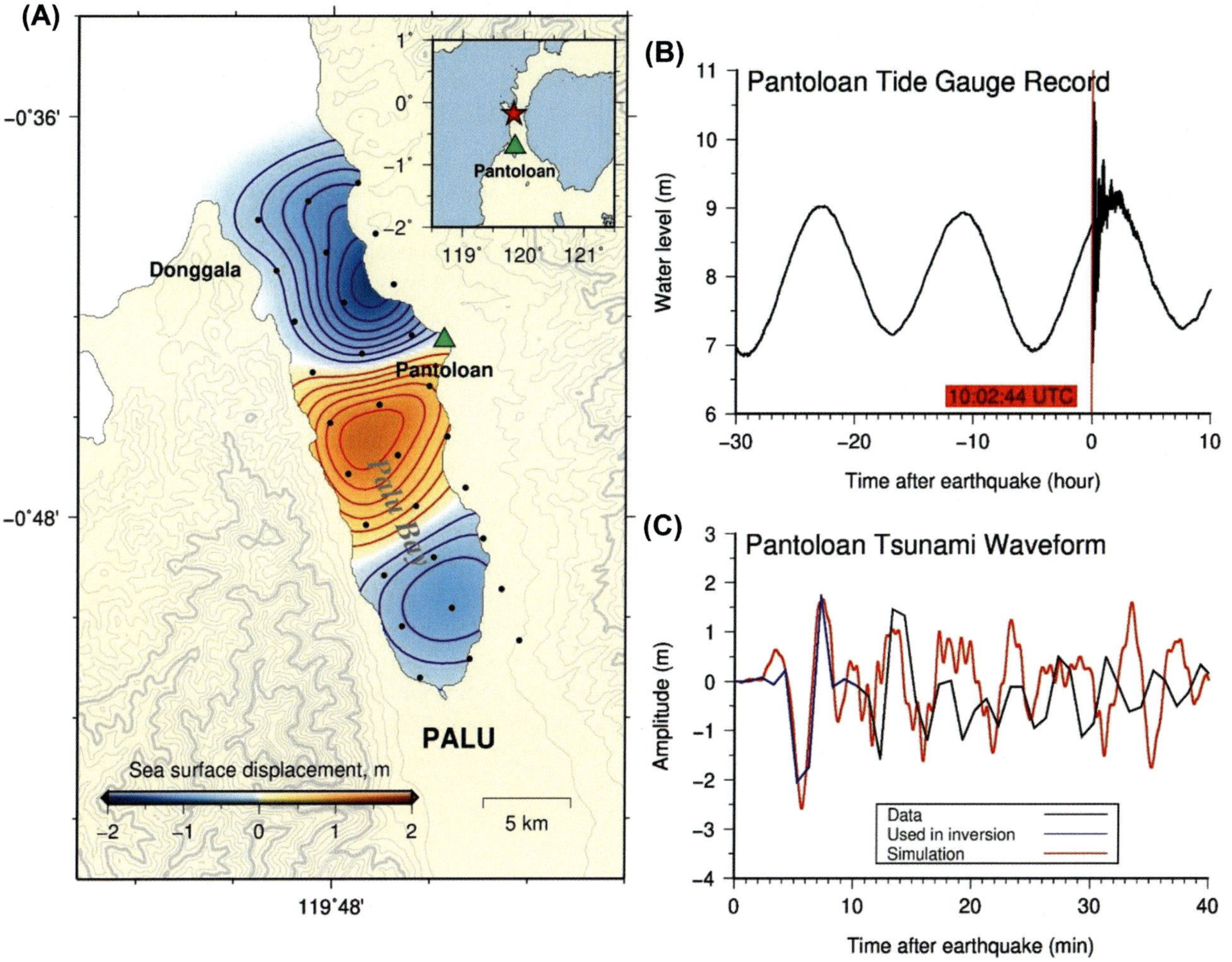

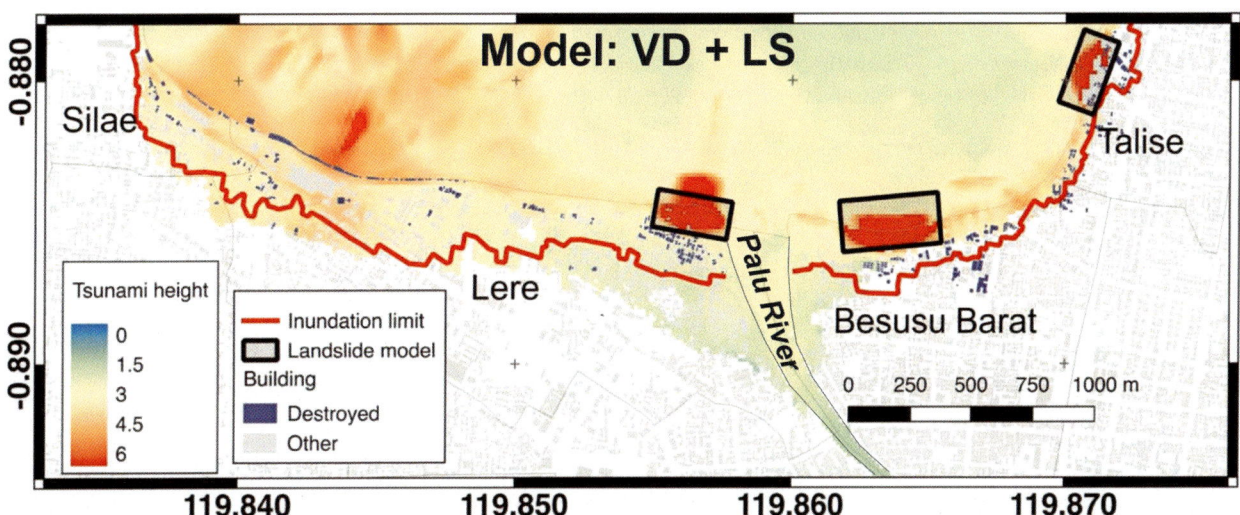

FIGURE 3.28 (A) Sea surface vertical displacement estimated by a tsunami waveform inversion using data at the Pantoloan tide gauge station (Gusman et al., 2019). (B) Tide gauge record at Pantoloan station. (C) Time series for the observed (*black* line) and simulated (*red* line) tsunami waveforms. The data used in the inversion is shown as a blue line. Map lines delineate study areas and do not necessarily depict accepted national boundaries. *From Gusman, A. R., Supendi, P., Nugraha, A. D., Power, W., Latief, H., Sunendar, H., Widiyantoro, S., Daryono, Wiyono, S. H., Hakim, A., Muhari, A., Wang, X., Burbidge, D., Palgunadi, K., Hamling, I., & Daryono, M. R. (2019). Source model for the tsunami inside Palu Bay following the 2018 Palu earthquake, Indonesia.* Geophysical Research Letters, 46(15), 8721–8730. https://doi.org/10.1029/2019GL082717.

FIGURE 3.29 Comparison of observed and simulated tsunami inundation from the combined coseismic vertical displacement and multiple landslides models of the 2018 Palu Earthquake. Red lines represent the limit of tsunami inundation as interpreted from tsunami debris visible on satellite images. Black rectangles represent the landslide block models. Map lines delineate study areas and do not necessarily depict accepted national boundaries. *From Gusman, A. R., Supendi, P., Nugraha, A. D., Power, W., Latief, H., Sunendar, H., Widiyantoro, S., Daryono, Wiyono, S. H., Hakim, A., Muhari, A., Wang, X., Burbidge, D., Palgunadi, K., Hamling, I., & Daryono, M. R. (2019). Source model for the tsunami inside Palu Bay following the 2018 Palu earthquake, Indonesia.* Geophysical Research Letters, 46(15), 8721–8730. https://doi.org/10.1029/2019GL082717.

3.6 Research needs

Over the past several decades, scientists and engineers have made efforts to refine and enhance methods for simulating tsunamis. At present, such sophisticated models are capable of replicating the dynamics of tsunami propagation across the ocean as well as the tsunami inundation processes. These models presented in this chapter are largely based on shallow-water theory. The linear shallow-water equations can be used to simulate tsunamis in the open ocean, while the nonlinear shallow-water equations can simulate tsunami inundation. With today's computational capabilities, simulating tsunami propagation in the open ocean over a vast area can be done in a matter of a few minutes. It still takes a significant amount of time to simulate tsunami inundation on high-resolution grids. With further advancements in computer technology, the computational time for high-resolution tsunami simulations will be decreased.

Numerical tsunami models allow the assessment of the potential tsunami hazards and impacts in coastal regions. These models enable in-depth analyses of various tsunami sources, including earthquake scenarios and other potential mechanisms, such as underwater landslides. When conducting tsunami inundation modeling, the availability of high-resolution topographical data is essential, ensuring reliable results. Moreover, comprehensive studies of tsunami inundation often require data concerning land cover, coastal and river structures, such as seawalls, breakwaters, and stop banks, and building footprints within the study area, all of which contribute to the model's accuracy.

Tsunami impacts, and their origins, have been extensively investigated through the utilization of tide gauge records, providing insights into events spanning over the past 150 years. Within the past two decades, the number of offshore tsunami gauges has increased rapidly, expanding our capacity to investigate tsunami sources and dynamics. These offshore tsunami observations serve a dual purpose, not only enhancing our ability to study the origins of tsunamis but also proving instrumental in the operation of tsunami early warning systems.

References

Abe, K. (1973). Tsunami and mechanism of great earthquakes. *Physics of the Earth and Planetary Interiors*, 7(2), 143–153. Available from https://doi.org/10.1016/0031-9201(73)90004-6.

Amores, A., Monserrat, S., Marcos, M., Argüeso, D., Villalonga, J., Jordà, G., & Gomis, D. (2022). Numerical simulation of atmospheric Lamb waves generated by the 2022 Hunga-Tonga volcanic eruption. *Geophysical Research Letters*, 49(6), e2022GL098240.

Arcement, G. J., & Schneider, V. R. (1989). *Guide for selecting Manning's roughness coefficients for natural channels and flood plains*. U.S. Geological Survey. Available from https://doi.org/10.3133/wsp2339.

Baba, T., Allgeyer, S., Hossen, J., Cummins, P. R., Tsushima, H., Imai, K., Yamashita, K., & Kato, T. (2017). Accurate numerical simulation of the far-field tsunami caused by the 2011 Tohoku earthquake, including the effects of Boussinesq dispersion, seawater density stratification, elastic loading, and gravitational potential change. *Ocean Modelling*, 111, 46–54. Available from https://doi.org/10.1016/j.ocemod.2017.01.002.

Baba, T., Takahashi, N., Kaneda, Y., Ando, K., Matsuoka, D., & Kato, T. (2015). Parallel implementation of dispersive tsunami wave modeling with a nesting algorithm for the 2011 Tohoku tsunami. *Pure and Applied Geophysics*, 172, 3455–3472. Available from https://doi.org/10.1007/s00024-015-1049-2.

Berryman, K. (2005). *Review of tsunami hazard and risk in New Zealand*. Institute of Geological & Nuclear Sciences. Available from https://www.gdc.govt.nz/__data/assets/pdf_file/0008/13031/review-of-tsunami-hazard-and-risks-in-nz-sept-05.pdf.

Borrero, J. C. (2005). Field data and satellite imagery of tsunami effects in Banda Aceh. *Science*, 308(5728), 1596. Available from https://doi.org/10.1126/science.1110957.

Bretschneider, C. L., & Wybro, P. G. (1977). Tsunami inundation prediction. *Coastal Engineering Proceedings*, 1006–1024. Available from https://doi.org/10.9753/icce.v15.59.

Bricker, J. D., Gibson, S., Takagi, H., & Imamura, F. (2015). On the need for larger Manning's roughness coefficients in depth-integrated tsunami inundation models. *Coastal Engineering Journal*, 57(2). Available from https://doi.org/10.1142/S0578563415500059, https://www.tandfonline.com/loi/tcej20.

Clark, K. J., Nissen, E. K., Howarth, J. D., Hamling, I. J., Mountjoy, J. J., Ries, W. F., Jones, K., Goldstien, S., Cochran, U. A., & Villamor, P. (2017). Highly variable coastal deformation in the 2016 Mw7.8 Kaikōura earthquake reflects rupture complexity along a transpressional plate boundary. *Earth and Planetary Science Letters*, 474, 334–344. Available from https://doi.org/10.1016/j.epsl.2017.06.048.

Dao, M. H., & Tkalich, P. (2007). Tsunami propagation modelling: A sensitivity study. *Natural Hazards and Earth System Sciences*, 7(6), 741–754. Available from https://doi.org/10.5194/nhess-7-741-2007.

Dean, R. G., & Dalrymple, R. A. (1991). *Water wave mechanics for engineers and scientists* (2). World Scientific Publishing Company. Available from http://doi.org/10.1142/1232.

Enet, F., Grilli, S. T., & Watts, P. (2003). *Laboratory experiments for tsunamis generated by underwater landslides: Comparison with numerical modeling*. International Ocean and Polar Engineering Conference, ISOPE-I-03-255.

Fraser, S. A., Power, W. L., Wang, X., Wallace, L. M., Mueller, C., & Johnston, D. M. (2014). Tsunami inundation in Napier, New Zealand, due to local earthquake sources. *Natural Hazards*, 70, 415–445. Available from https://doi.org/10.1007/s11069-013-0820-x.

Fujii, Y., Satake, K., Sakai, S., Shinohara, M., & Kanazawa, T. (2011). Tsunami source of the 2011 off the Pacific coast of Tohoku Earthquake. *Earth, Planets and Space*, 63(7), 815–820. Available from https://doi.org/10.5047/eps.2011.06.010, http://rd.springer.com/journal/40623.

Gayer, G., Leschka, S., Nöhren, I., Larsen, O., & Günther, H. (2010). Tsunami inundation modelling based on detailed roughness maps of densely populated areas. *Natural Hazards and Earth System Sciences*, *10*(8), 1679–1687. Available from https://doi.org/10.5194/nhess-10-1679-2010.

Giachetti, T., Paris, R., Kelfoun, K., & Ontowirjo, B. (2012). 1*Tsunami hazard related to a flank collapse of Anak Krakatau volcano, Sunda Strait, Indonesia* (361, pp. 79–90). Geological Society. Available from http://doi.org/10.1144/SP361.7.

Gibbons, S. J., Lorito, S., de la Asunción, M., Volpe, M., Selva, J., Macías, J., Sánchez-Linares, C., Brizuela, B., Vöge, M., Tonini, R., Lanucara, P., Glimsdal, S. T., Romano, F., Meyer, J. C., & Løvholt, F. (2022). The sensitivity of tsunami impact to earthquake source parameters and Manning friction in high-resolution inundation simulations. *Frontiers in Earth Science*, *9*. Available from https://doi.org/10.3389/feart.2021.757618.

Gossard, E. E., & Hooke, W. H. (1975). *Waves in the atmosphere: Atmospheric infrasound and gravity waves—Their generation and propagation* (2). Elsevier.

Gusman, A. R., Roger, J., Noble, C., Wang, X., Power, W., & Burbidge, D. (2022). The 2022 Hunga Tonga-Hunga Ha'apai volcano air-wave generated tsunami. *Pure and Applied Geophysics*, *179*(10), 3511–3525. Available from https://doi.org/10.1007/s00024-022-03154-1.

Gusman, A. R., Satake, K., Shinohara, M., Sakai, S., & Tanioka, Y. (2017). Fault slip distribution of the 2016 Fukushima earthquake estimated from tsunami waveforms. *Pure and Applied Geophysics*, *174*, 2925–2943. Available from https://doi.org/10.1007/s00024-017-1590-2.

Gusman, A. R., Supendi, P., Nugraha, A. D., Power, W., Latief, H., Sunendar, H., Widiyantoro, S., Daryono Wiyono, S. H., Hakim, A., Muhari, A., Wang, X., Burbidge, D., Palgunadi, K., Hamling, I., & Daryono, M. R. (2019). Source model for the tsunami inside Palu Bay following the 2018 Palu earthquake, Indonesia. *Geophysical Research Letters*, *46*(15), 8721–8730. Available from https://doi.org/10.1029/2019GL082717, http://agupubs.onlinelibrary.wiley.com/hub/journal/10.1002/(ISSN)1944-8007/.

Gusman, A. R., Tanioka, Y., Sakai, S., & Tsushima, H. (2012). Source model of the great 2011 Tohoku earthquake estimated from tsunami waveforms and crustal deformation data. *Earth and Planetary Science Letters*, *341*, 234–242. Available from j.epsl.2012.06.006.

Heidarzadeh, M., Murotani, S., Satake, K., Ishibe, T., & Gusman, A. R. (2016). Source model of the 16 September 2015 Illapel, Chile, M_w 8.4 earthquake based on teleseismic and tsunami data. *Geophysical Research Letters*, *43*(2), 643–650. Available from https://doi.org/10.1002/2015GL067297, https://doi.org/10.1002/2015GL067297.

Heidarzadeh, M., & Satake, K. (2015). New insights into the source of the Makran tsunami of 27 November 1945 from tsunami waveforms and coastal deformation data. *Pure and Applied Geophysics*, *172*, 621–640. Available from https://doi.org/10.1007/s00024-014-0948-y.

Imamura, F. (2009). *The Sea tsunami modeling: Calculating inundation and hazard maps*. Harvard University Press.

Imamura, F., Yalciner, A. C., & Ozyurt, G. (2006). *Tsunami modelling manual*. UNESCO IOC International Training Course on Tsunami Numerical Modelling. https://www.tsunami.irides.tohoku.ac.jp/media/files/_u/project/manual-ver-3_1.pdf.

Ioki, K., Tanioka, Y., Yanagisawa, H., & Kawakami, G. (2019). Numerical simulation of the landslide and tsunami due to the 1741 Oshima–Oshima eruption in Hokkaido, Japan. *Journal of Geophysical Research: Solid Earth*, *124*(2), 1991–2002. Available from https://doi.org/10.1029/2018JB016166.

Iverson, R. M., & Denlinger, R. P. (2001). Flow of variably fluidized granular masses across three-dimensional terrain: 1. Coulomb mixture theory. *Journal of Geophysical Research: Solid Earth*, *106*(B1), 537–552. Available from https://doi.org/10.1029/2000JB900329.

Kaiser, G., Scheele, L., Kortenhaus, A., Løvholt, F., Römer, H., & Leschka, S. (2011). He influence of land cover roughness on the results of high resolution tsunami inundation modeling. *Natural Hazards and Earth System Sciences*, *11*(9), 2521–2540. Available from https://doi.org/10.5194/nhess-11-2521-2011.

Kajiura, K. (1963). Leading wave of a tsunami. *Bulletin of the Earthquake Research Institute*, *41*, 535–571.

Kelfoun, K., Giachetti, T., & Labazuy, P. (2010). Landslide-generated tsunamis at Réunion Island. *Journal of Geophysical Research: Earth Surface*, *115*(F4). Available from https://doi.org/10.1029/2009JF001381.

Kirby, J. T., Shi, F., Tehranirad, B., Harris, J. C., & Grilli, S. T. (2013). Dispersive tsunami waves in the ocean: Model equations and sensitivity to dispersion and Coriolis effects. *Ocean Modelling*, *62*, 39–55. Available from https://doi.org/10.1016/j.ocemod.2012.11.009.

Leschka, S., Pedersen, C., & Larsen, O. (2009). *On the requirements for data and methods in tsunami inundation modelling: Roughness maps and uncertainties*. Proceedings of the South China Sea Tsunami Workshop (pp. 3–5).

MacInnes, B. T., Gusman, A. R., LeVeque, R. J., & Tanioka, Y. (2013). Comparison of earthquake source models for the 2011 Tohoku event using tsunami simulations and near-field observations. *Bulletin of the Seismological Society of America*, *103*(2B), 1256–1274. Available from https://doi.org/10.1785/0120120121.

Mader, C. L. (2004). *Numerical modeling of water waves*. CRC Press.

Maeno, F., & Imamura, F. (2011). Tsunami generation by a rapid entrance of pyroclastic flow into the sea during the 1883 Krakatau eruption, Indonesia. *Journal of Geophysical Research: Solid Earth*, *116*(B9). Available from https://doi.org/10.1029/2011JB008253.

Matsutomi, H., Kawata, Y., Shuto, N., Tsuji, Y., Fujima, K., Imamura, F., Matsuyama, M., Takahashi, T., Maki, N., & Han, S. S. (2001). *Flow strength on land and damage of the 1998 Papua New Guinea tsunami. Tsunami research at the end of a critical decade* (pp. 179–195). Springer. Available from https://doi.org/10.1007/978-94-017-3618-3_13.

McSaveney, M., & Rattenbury, M. (2000). *Tsunami impact in Hawke's Bay. Institute of Geological and Nuclear Sciences*, *77*. Available from https://cms.hbrc.govt.nz/assets/Uploads/Tsunami-Impacts-in-Hawkes-Bay.pdf.

Monserrat, S., Vilibić, I., & Rabinovich, A. B. (2006). Meteotsunamis: Atmospherically induced destructive ocean waves in the tsunami frequency band. *Natural Hazards and Earth System Sciences*, *6*(6), 1035–1051. Available from https://doi.org/10.5194/nhess-6-1035-2006.

Mori, N., & Takahashi, T. (2012). 2011 Tohoku earthquake Tsunami Joint Survey Group, Nationwide post event survey and analysis of the 2011 Tohoku earthquake tsunami. *Coastal Engineering Journal*, *54*(01), 1250001. Available from https://doi.org/10.1142/S0578563412500015.

Muhari, A., Imamura, F., Koshimura, S., & Post, J. (2011). Examination of three practical run-up models for assessing tsunami impact on highly populated areas. *Natural Hazards and Earth System Sciences*, *11*(12), 3107–3123. Available from https://doi.org/10.5194/nhess-11-3107-2011.

Mulia, I. E., Watada, S., Ho, T. C., Satake, K., Wang, Y., & Aditiya, A. (2020). Simulation of the 2018 tsunami due to the flank failure of Anak Krakatau volcano and implication for future observing systems. *Geophysical Research Letters*, *47*(14). Available from https://doi.org/10.1029/2020GL087334, e2020GL087334.

Munk, W. (1950). Origin and generation of waves. *Proceedings of the International Conference on Coastal Engineering*, *1*(1), 1. Available from https://doi.org/10.9753/icce.v1.1.

Okada, Y. (1985). Surface deformation due to shear and tensile faults in a half-space. *Bulletin of the Seismological Society of America*, *75*(4), 1135−1154. Available from https://doi.org/10.1785/BSSA0750041135.

Pattiaratchi, C. B., & Wijeratne, E. M. S. (2015). Are meteotsunamis an underrated hazard? *Philosophical Transactions of the Royal Society A*, *373* (2053), 20140377. Available from https://doi.org/10.1098/rsta.2014.0377.

Rabinovich, A. B., & Monserrat, S. (1998). Generation of meteorological tsunamis (large amplitude seiches) near the Balearic and Kuril Islands. *Natural Hazards*, *18*, 27−55. Available from https://doi.org/10.1023/A:1008096627047.

Romano, F., Gusman, A. R., Power, W., Piatanesi, A., Volpe, M., Scala, A., & Lorito, S. (2021). Tsunami source of the 2021 MW 8.1 Raoul Island earthquake from DART and tide-gauge fata inversion. *Geophysical Research Letters*, *48*(17). Available from https://doi.org/10.1029/2021GL094449, http://agupubs.onlinelibrary.wiley.com/hub/journal/10.1002/(ISSN)1944-8007/.

Saito, T., Inazu, D., Miyoshi, T., & Hino, R. (2014). Dispersion and nonlinear effects in the 2011 Tohoku-Oki earthquake tsunami. *Journal of Geophysical Research: Oceans*, *119*(8), 5160−5180. Available from https://doi.org/10.1002/2014JC009971.

Sandanbata, O., Watada, S., Satake, K., Fukao, Y., Sugioka, H., Ito, A., & Shiobara, H. (2018). Ray tracing for dispersive tsunamis and source amplitude estimation based on Green's Law: Application to the 2015 volcanic tsunami earthquake near Torishima, South of Japan. *Pure and Applied Geophysics*, *175*, 1371−1385. Available from https://doi.org/10.1007/s00024-017-1746-0.

Satake, K. (1987). Inversion of tsunami waveforms for the estimation of a fault heterogeneity: Method and numerical experiments. *Journal of Physics of the Earth*, *35*(3), 241−254. Available from https://doi.org/10.4294/jpe1952.35.241.

Satake, K. (1988). Effects of bathymetry on tsunami propagation: Application of ray tracing tracing to tsunamis. *Pure and Applied Geophysics*, *126*, 27−36. Available from https://doi.org/10.1007/BF00876912.

Satake, K. (2015). Tsunamis. Elsevier, Treatise on Geophysics. *Earthquake Seismology*, *4*. Available from https://doi.org/10.1016/B978-0-444-53802-4.00086-5.

Smart, G. M., Crowley, K. H. M., & Lane, E. M. (2016). Estimating tsunami run-up. *Natural Hazards*, *80*, 1933−1947. Available from https://doi.org/10.1007/s11069-015-2052-8.

Synolakis, C. E. (1987). The runup of solitary waves. *Journal of Fluid Mechanics*, *185*, 523−545. Available from https://doi.org/10.1017/S002211208700329X.

Synolakis, C. E. (1991). Tsunami runup on steep slopes: how good linear theory really is. *Tsunami Hazard: A Practical Guide for Tsunami Hazard Reduction* (pp. 221−234). Kluwer Academic Publishers.

Synolakis, C. E., & Skjelbreia, J. E. (1993). Evolution of maximum amplitude of solitary waves on plane beaches. *Journal of Waterway, Port, Coastal, and Ocean Engineering*, *119*(3), 323−342. Available from https://doi.org/10.1061/(ASCE)0733-950X(1993)119:3(323).

Titov, V., Kânoğlu, U., & Synolakis, C. (2016). Development of MOST for real-time tsunami forecasting. *Journal of Waterway, Port, Coastal, and Ocean Engineering*, *142*(6), 03116004. Available from https://doi.org/10.1061/(ASCE)WW.1943-5460.000035.

Vilibić, I., Šepić, J., Rabinovich, A. B., & Monserrat, S. (2016). Modern approaches in meteotsunami research and early warning. *Frontiers in Marine Science*, *3*, 57. Available from https://doi.org/10.3389/fmars.2016.00057.

Wang, X., & Liu, P. L. F. (2006). An analysis of 2004 Sumatra earthquake fault plane mechanisms and Indian Ocean tsunami. *Journal of Hydraulic Research*, *44*(2), 147−154. Available from https://doi.org/10.1080/00221686.2006.9521671.

Wang, X., & Liu, P. L. F. (2007). Numerical simulations of the 2004 Indian Ocean tsunamis: Coastal effects. *Journal of Earthquake and Tsunami*, *1*(03), 273−297. Available from https://doi.org/10.1142/S179343110700016X.

Wang, X., & Power, W. L. (2011). COMCOT: A tsunami generation propagation and run-up model. *GNS Science*. Available from https://www.gns.cri.nz/data-and-resources/comcot-a-tsunami-generation-propagation-and-run-up-model/.

Watada, S., Kusumoto, S., & Satake, K. (2014). Traveltime delay and initial phase reversal of distant tsunamis coupled with the self-gravitating elastic Earth. *Journal of Geophysical Research: Solid Earth*, *119*(5), 4287−4310. Available from https://doi.org/10.1002/2013JB010841.

Watts, P. (2000). Tsunami features of solid block underwater landslides. *Journal of Waterway, Port, Coastal, and Ocean Engineering*, *126*(3), 144−152. Available from https://doi.org/10.1061/(ASCE)0733-950X(2000)126:3(144).

Watts, P., Grilli, S. T., Kirby, J. T., Fryer, G. J., & Tappin, D. R. (2003). Landslide tsunami case studies using a Boussinesq model and a fully nonlinear tsunami generation model. *Natural Hazards and Earth System Sciences*, *3*(5), 391−402. Available from https://doi.org/10.5194/nhess-3-391-2003.

Wessel, P. (2009). Analysis of observed and predicted tsunami travel times for the Pacific and Indian Oceans. *Pure and Applied Geophysics*, *166*, 301−324. Available from https://doi.org/10.1007/s00024-008-0437-2.

Chapter 4

Tsunami effects on built environment

Ioan Nistor[1], Seyed Abbas Jazaeri[1], Joseph Kim[1], Raffaele De Risi[2], Katsuichiro Goda[3] and Aditya Gusman[4]

[1]Department of Civil Engineering, University of Ottawa, Ottawa, ON, Canada, [2]School of Civil, Aerospace and Design Engineering, University of Bristol, Bristol, United Kingdom, [3]Department of Earth Sciences, Western University, London, ON, Canada, [4]GNS Science, Lower Hutt, Wellington, New Zealand

4.1 Introduction

Over the past decades, numerous coastal communities around the world have suffered massive fatalities and economic losses due to extreme tsunami events (Chapter 1). Comprehensive forensic engineering field surveys conducted after these catastrophes provided empirical evidence of the damage that tsunami waves can cause to coastal structures (Chock, Carden, et al., 2013; Chock, Robertson, et al., 2013; Saatcioglu et al., 2006; Sassa & Takagawa, 2019). After the 2011 Tohoku Earthquake and Tsunami in Japan, Mikami et al. (2012) investigated the tsunami's reach along the coast and the resultant damage to structures and infrastructure. Their findings highlighted the survival of two evacuation buildings in Minami Sanriku and Shizugawa, built in response to the 1960 Chile Tsunami, adhering to the Structural Design Method of Buildings for Tsunami Resistance as outlined by Okada et al. (2015) under the auspices of the Building Center of Japan. These cases underscored the critical need for coastal construction to follow tsunami-specific safety standards to mitigate damage and ensure resilience against natural disasters. Tsunamis may have catastrophic effects on the built environment, and a proper engineering approach is needed to quantify them rigorously.

Section 4.2 summarizes the developments of existing guidelines, standards, and codes to design tsunami-resistant structures and infrastructure. They are essential to safeguard people who live in coastal areas by minimizing fatalities and economic losses during extreme events. Specifically, historical developments of the guidelines, standards, and codes in Japan and the United States of America are reviewed.

Section 4.3 introduces tsunami loads and their effects on the built environment by adopting the most recent tsunami design provisions of the American Society of Civil Engineers (ASCE) standard (American Society of Civil Engineers, 2022). Throughout this chapter, when the ASCE standard is mentioned, Chapter 6 should be referred to. Hydrostatic, hydrodynamic, and debris loads are focused upon to quantify the tsunami effects on structures. It should be noted that this chapter does not address tsunami-induced scour and its effects on the foundations of structures. The methods and equations for calculating the tsunami loadings are based on the accumulated knowledge to date from empirical observations, experimental investigations, and numerical analyses. Their implementations in designing coastal structures are crucial for the development of robust design strategies and mitigation measures aimed at safeguarding structures against the devastating forces unleashed by tsunamis.

Section 4.4 presents a comparison of two existing approaches for designing tsunami-resistant buildings and other structures in the United States (American Society of Civil Engineers, 2022) and Japan (Ministry of Land, Infrastructure, Transport and Tourism, 2011a). Such a comparison is useful for contrasting the differences in the calculations of tsunami loadings on structures, mentioned in Section 4.3.

Section 4.5 introduces the concept of damage states (*DS*s) of structures and presents the methodology for developing tsunami vulnerability/fragility models. Tsunami vulnerability and fragility functions are useful for assessing the extent of physical damage to the built environment during tsunamis and facilitate the quantitative tsunami risk analysis by considering the uncertainty associated with tsunami damage (Chapter 5).

4.2 Existing design guidelines, standards, and codes

4.2.1 Differences among guidelines, standards, and codes

Guidelines, standards, and codes outline the methods required to design tsunami-resistant infrastructure. While often used interchangeably in structural engineering, these documents are distinct and should not be confused with one another. They have different levels of scope, specificity, and legal obligations, each representing a unique phase in the evolutionary process of standardizing the design of tsunami-resistant infrastructure.

Guidelines provide a foundation of recommendations and best practices. They serve an advisory role when designing tsunami-resistant infrastructure by synthesizing the current state of research, highlighting industry practices, providing expert insights, summarizing historical data, and offering a comprehensive resource for designers and planners.

Standards establish specific technical criteria, methodologies, and procedures. Generally, national or international certification bodies develop standards through a consensus process involving various stakeholders and experts. Once adopted by authorities and national or local governments, depending on the legislation, standards often become legally mandatory, aiming to provide consistent and reliable practices to ensure a uniform level of safety and performance throughout the industry.

Lastly, codes explicitly define the minimum acceptable standards for designing and constructing tsunami-resistant infrastructure. Codes are adopted by governments, making them legal requirements. This makes compliance with codes mandatory, and failing to comply can lead to penalties, fines, and criminal charges. Often, codes refer to standards and guidelines as the minimum requirements for design and construction, thus highlighting the symbiotic relationship among guidelines, standards, and codes.

4.2.2 Early development of tsunami guidelines, standards, and codes

On June 15th, 1896, a magnitude 8.5 earthquake originated 150 km off the Sanriku Coast of Japan. As the earthquake generated a weak ground tremor on land and caused minimal damage to buildings, the event caused little concern to residents. The earthquake generated a tsunami (known as the Meiji Sanriku Tsunami), which arrived about 35 minutes after the earthquake. The maximum wave height was recorded to be 38.2 m. The tsunami caused 22,066 casualties and 8891 destroyed houses. While an investigation group on the earthquake was created at the Council on Earthquake Disaster Prevention of the Japanese Ministry of Education, tsunami science was in its infancy; thus no tsunami disaster prevention measures were implemented (Nakao, 2005).

It was not until March 3, 1933 when the Showa Sanriku Tsunami caused significant damage and 3000 fatalities on the Sanriku Coast again, that tsunami countermeasures were proposed. The Council on Earthquake Disaster Prevention recommended 10 countermeasures as the first set of formal guidelines on tsunami countermeasures (Council on Earthquake Disaster Prevention, 1933; Shuto & Fujima, 2009):

1. *Relocation of dwelling houses to high ground: This is the best measure against tsunamis.*
2. *Coastal dikes: Dikes against tsunamis may become too large and financially impractical.*
3. *Tsunami control forests: Vegetations may dampen the power of tsunamis.*
4. *Seawalls: These could be effective for smaller tsunamis.*
5. *Tsunami-resistant areas: If the tsunami height is not so high in a busy quarter, solid concrete buildings are to be built in the front line of the area.*
6. *Buffer zone: Dammed by structures, a tsunami inevitably increases its height. To receive the flooding, rivers and lowlands are to be designated as a buffer zone to be sacrificed.*
7. *Evacuation routes: Roads to safe high ground are required for every village.*
8. *Tsunami watch: Because it takes 20 minutes for a tsunami to arrive at the Sanriku Coast, we may detect an approaching tsunami and prepare for it.*
9. *Tsunami evacuation: The elderly, children, and the weak should be evacuated to safe, higher ground where they could wait for about 1 hour. Ships more than a few hundred meters offshore should move farther offshore.*
10. *Memorial events: Holding memorial services, erecting monuments, etc. may help keep events alive in people's minds.*

Of these 10 recommendations, the first formal commentary on the structural design of tsunami-resistant infrastructure was included in the 5th recommendation. Since then, there has been enormous progress in outlining the design methodology of tsunami-resistant buildings and other structures through the work of the international tsunami engineering and science community. In the following, major milestones and work in the field of guidelines, standards, and codes in tsunami structural engineering are reviewed for Japan, the United States of America, and other parts of the world.

4.2.3 Japan

Japan's history of repeated tsunami events has led to a robust collection of guidelines, standards, and codes in addition to an extensive catalog of academic and industry publications, such as *"Random Seas and Design of Maritime Structures"* (Goda, 2011). Table 4.1 summarizes major guidelines, standards, and codes after 1950, which were created to enhance the resilience and safety of tsunami-resistant buildings and other structures.

While design requirements provided through the Structural Design Method of Buildings for Tsunami Resistance (Okada et al., 2015) and the Tsunami Evacuation Building Guidelines (Cabinet Office of Japanese Government, 2005) existed, few buildings were designed and constructed on the basis of these requirements before 2011 (Nakano, 2017). After the 2011 Tohoku Earthquake and Tsunami, significant efforts were made to update the methodology, drawing on recent observational data and newfound experience. In November 2011, the Ministry of Land, Infrastructure, Transport, and Tourism (MLIT) officially adopted the recommendation and issued the Guidelines on the Structural Design of Tsunami Evacuation Buildings (Ministry of Land, Infrastructure, Transport and Tourism, 2011a) as MLIT 2570. MLIT 2570 is the primary design resource in Japan and sets the legal minimum requirements for tsunami-resistant buildings and other structures. In 2014, the Japan Electric Association published its code on tsunami design specifically for sea defense structures protecting nuclear power plants (Japan Electric Association, 2015, Japan Electric Association, 2021).

4.2.4 United States of America

Coastal communities located along the Pacific coastlines and several islands of the United States are exposed to the threat of tsunamis, originating from near and distant sources. The Pacific coast is particularly vulnerable due to the

TABLE 4.1 Guidelines, standards, and codes related to the design of tsunami-resistant buildings and other structures in Japan.

Name	First published	Last revised
Concerning the Prevention of Disasters to Buildings from Storm and Flood Damage (Housing Bureau Notification No. 42) (Pomonis et al., 2013)	1959	1959
Building Standard Law of Japan	1960	2015
Technical Standards for Port and Harbor Facilities (Overseas Coastal Area Development Institute of Japan, 2009)	1978	2009
Committee on Coastal Engineering, Japan Society of Civil Engineers Design Manual for Coastal Facilities (Japan Society of Civil Engineers, 2000)	2000	2000
Structural Design Method of Buildings for Tsunami Resistance (SMBTR) (Okada et al., 2015)	2005	2005
Japan Cabinet Office Tsunami Evacuation Building Guidelines (Cabinet Office of Japanese Government, 2005)	2005	2005
Technical Guidelines for Seismic Design of Nuclear Power Plants (JEAG-4601) (Japan Electric Association, 2015)	2008	2021
Guidelines on the Structural Design of Tsunami Evacuation Buildings (MLIT Technical Advice No. 2570) (Ministry of Land, Infrastructure, Transport and Tourism, 2011a)	2011	2011
Concerning Setting the Safe Structure Method for Inundation (MLIT Notification No. 1318) (Ministry of Land, Infrastructure, Transport and Tourism, 2011b)	2011	2011
Tsunami-Resistant Design Guideline for Breakwaters (Ministry of Land, Infrastructure, Transport and Tourism, 2013)	2013	2013
Technical Code for Tsunami Design of Nuclear Power Plants (JEAC-4629) (Japan Electric Association, 2021)	2014	2021

TABLE 4.2 Guidelines, standards, and codes related to the design of tsunami-resistant buildings and other structures in the United States of America.

Name	First published	Last revised
The City and County of Honolulu Building Code (Tsunami Loading) (City and County of Honolulu, 2013)	1980	2021
Coastal Construction Manual (Federal Emergency Management Agency, 2011)	2005	2011
Guidelines for Design of Structures for Vertical Evacuation from Tsunamis (Federal Emergency Management Agency, 2019)	2008	2019
ASCE 7–22 Chapter 6: Tsunami Loads and Effects (American Society of Civil Engineers, 2022)	2016	2022

ASCE, American Society of Civil Engineers.

seismic activity associated with the Pacific Ring of Fire, highlighting pronounced risk of tsunamis in these areas. Table 4.2 lists major guidelines, standards, and codes related to the design of tsunami-resistant buildings and other structures.

In the United States, preliminary details on tsunami-induced loading were introduced through Hawaii's Building Code in the 1980s (City and County of Honolulu, 2013). Major tsunami events, such as the 2004 Indian Ocean Tsunami and the 2011 Tohoku Tsunami, highlighted the urgency to develop detailed design procedures for tsunami-resistant buildings and other structures. This culminated in the publication of ASCE 7–16 Chapter 6: *"Tsunami Loads and Effects"* (American Society of Civil Engineers, 2017), the first standard in the United States for designing tsunami-resistant buildings and other structures. ASCE 7–16 was updated in a subsequent revision cycle as ASCE 7–22 (American Society of Civil Engineers, 2022) with additional provisions for horizontal pipelines, debris damming and impact, and enhanced methods for assessing sediment transport and scour (Robertson, 2023).

4.2.5 Other countries

Despite the tsunami risk in Europe (Dawson et al., 2004), no standardized provisions for tsunami loading are currently provided in the European building code (EuroCode). New Zealand launched its first guideline on *Tsunami Loads and Effects on Vertical Evacuation Structures* in 2020 (New Zealand Government, 2020). This guideline contextualizes the concepts in ASCE 7–16 in relation to New Zealand's building code and standards.

4.3 Tsunami-induced loads and effects

Evaluating strength and stability of a structure is imperative to ascertain its resistance against tsunami forces. The tsunami loading on a structure can be calculated using tsunami wave hydrodynamic conditions at the structure's location. For this purpose, the tsunami inundation analysis is often performed to determine the tsunami wave profile affecting a structure of interest (Chapter 3). For tsunami design purposes, the tsunami wave profile is characterized by several parameters, such as the maximum wave amplitude and tsunami wave period. In Chapter 6 of ASCE 7–22 (American Society of Civil Engineers, 2022), three cases are defined to capture several critical moments regarding tsunami loading. Long-term tsunami hazard assessments that have been carried out for the ASCE 7–22 tsunami design provisions can be found in Chapter 13.

In Load Case 1, the evaluation encompasses a scenario where the external inundation depth does not surpass the maximum inundation depth or the lesser of the elevation of a single story or the elevation of the upper boundary of the first-floor windows. Under this case, it is essential to evaluate the minimum combined impact of hydrodynamic and buoyant forces, considering the water depth inside the structure. This particular load case is not mandatory for open structures or those whose soil characteristics, along with foundation and structural design, mitigate the risk of adverse hydrostatic pressure beneath the foundation and the lowest structural level. The objective of this load case is to ascertain the maximum buoyant force acting upon the structure in conjunction with the lateral hydrodynamic force, with the primary aim of verifying the structural and foundational stability against net uplift forces.

Load Case 2 delves into a scenario where the depth reaches two-thirds of the maximum inundation depth, positing that the peak velocity and specific momentum flux are likely to manifest in either the incoming or receding directions. The essence of Load Case 2 lies in computing the maximal hydrodynamic forces exerted on the structure.

Lastly, Load Case 3 considers the scenario of the maximum inundation depth, assuming that the velocity will be a third of its maximum, regardless of the direction of water movement. The aim of Load Case 3 is to evaluate the hydrodynamic forces that come into play at the maximum inundation depth.

The parameters for inundation depths and velocities pertinent to Load Cases 2 and 3 are derived from Fig. 4.1. Similar to approaches for tsunami inundation simulations, outlined in Chapter 3, the inundation characteristics prescribed by American Society of Civil Engineers (2022) are obtained through a procedure described in that document. This figure presents the normalized inundation depth and depth-averaged velocity time−history (TH) curves, formulated on the basis of tsunami footage analyses. These curves are consistent with numerical modeling outcomes, particularly in delineating the critical phases of structural loading for design considerations (Ngo & Robertson, 2012).

The primary tsunami forces and effects must be integrated with other specified loads through the application of the load combination formulas

$$0.9D + F_{TSU} + H_{TSU}$$

$$1.2D + F_{TSU} + 0.5L + 0.2S + H_{TSU} \tag{4.1}$$

In these formulas, D and L are the dead load and the live load, respectively. F_{TSU} represents the effect of tsunami loads during the ingress and egress flow directions, while H_{TSU} denotes the load resulting from lateral foundation

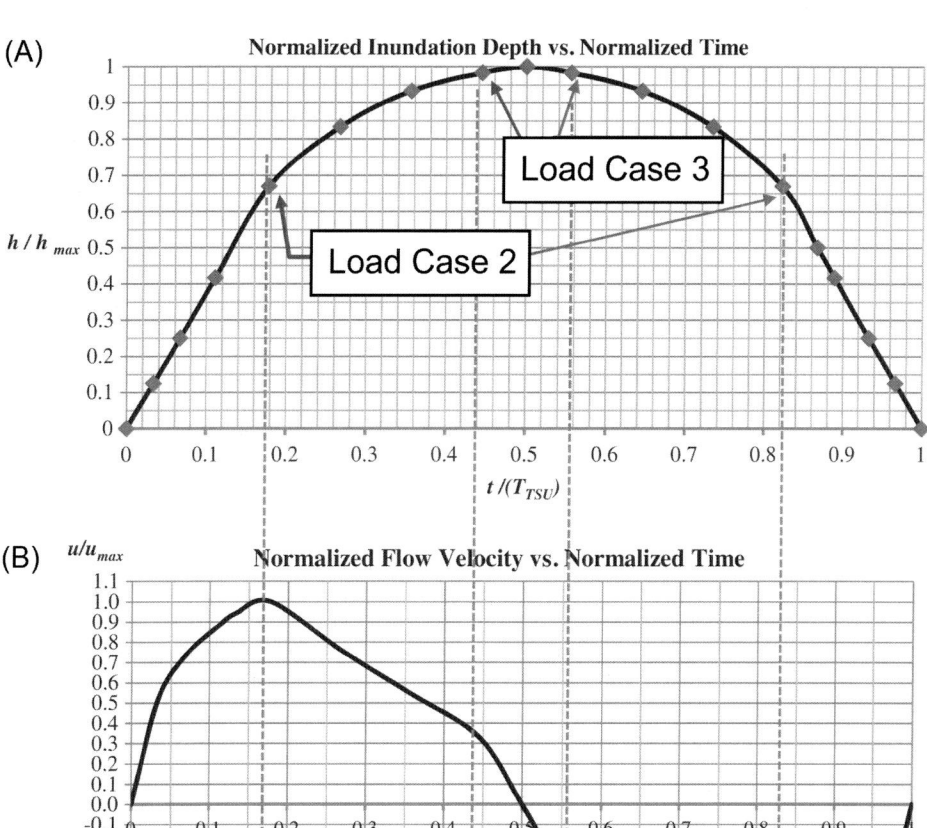

FIGURE 4.1 (A) Normalized inundation depth versus normalized time plot and (B) normalized flow velocity versus normalized time plot for Load Cases 2 and 3. *From American Society of Civil Engineers. (2022). Minimum design loads and associated criteria for buildings and other structures (Chapter 6) Tsunami loads and effects ASCE/SEI7-22. https://doi.org/10.1061/9780784415788.*

pressures induced by a tsunami under submergence conditions. When the net impact of H_{TSU} serves to mitigate the primary load effect, a load factor of 0.9 shall be applied to H_{TSU}. This approach ensures that the structural design adequately accounts for the dynamic nature of tsunami forces in conjunction with other relevant loads, providing a comprehensive basis for ensuring structural resistance.

This section delves into three predominant categories of tsunami-induced loads based on the latest version of ASCE 7–22 (American Society of Civil Engineers, 2022) and covers hydrostatic, hydrodynamic, and debris loads, each of which presents unique challenges to the integrity and resistance of structures in the face of such extreme events. Hydrostatic loads pertain to the pressures exerted by the static water column, hydrodynamic loads arise from water movement and its interaction with structures, and debris loads are associated with the impact forces from objects carried by tsunami waves. The elucidation of these load types is pivotal for developing robust design strategies and mitigation measures aimed at safeguarding structures against the devastating forces unleashed by tsunamis.

4.3.1 Hydrostatic loads

4.3.1.1 Buoyancy

The phenomenon of buoyancy-induced uplift has precipitated numerous instances of structural failure during historical tsunamis, notably affecting concrete and steel buildings that were not designed to withstand such conditions (Chock, Carden, et al., 2013; Chock, Robertson, et al., 2013). The principle of buoyancy operates on structures akin to its effect on ships or submerged bodies, engendering an upward force equivalent to the weight of the water displaced by submerged sections of the structure. The swift inundation of soil and elevated water table levels during tsunamis can exacerbate the soil pore pressure beneath a structure, augmenting the uplift force. The extent of the buoyancy force is contingent upon the volume of water displaced, encompassing all submerged structural components and any enclosed spaces. In the computation of the uplift force, it is imperative to account for every component of the building capable of displacing water, including structural components, spaces beneath the flood level, floor soffits, and any areas susceptible to air entrapment. These elements collectively contribute to the buoyancy effect.

Nonstructural elements, such as walls designed to detach under tsunami forces and conventional windows, are generally expected to succumb to tsunami forces, facilitating water ingress into the structure. This water influx balances the internal and external pressures, thereby diminishing the buoyancy effect. Conversely, windows engineered for resistance against high-impact forces, such as those found in hurricane-prone regions or designed for blast resistance, are presumed to withstand the event, thereby sustaining the buoyancy effect unless a detailed analysis suggests otherwise. Enclosures with openings or designed-to-fail elements that constitute a minimum of 25% of the submerged envelope or that are engineered to give way under tsunami forces are expected to become inundated, effectively neutralizing the buoyancy effect for these areas.

The reduction in net weight due to buoyancy (F_V) is determined by the equation involving the displaced water volume (V_W) and the minimum specific weight density of the fluid (γ_s), which is derived from the specific weight density of seawater adjusted by a fluid density factor (k_s) of 1.1, following (American Society of Civil Engineers, 2022)

$$F_V = V_W \gamma_s \tag{4.2}$$

The buoyancy effects resulting from the submersion of elevated floors reduce the axial load exerted on columns and walls at lower levels. This decrement in axial load must be factored into the load combinations employed for assessing the design strength of such components. This consideration is important because of the potential detrimental impact on the bending and shear capacities of the components, especially in the case of reinforced concrete members (Del Zoppo et al., 2023).

4.3.1.2 Unbalanced lateral hydrostatic force

Unbalanced lateral forces represent another category of hydrostatic pressures exerted on structures during tsunamis, originating from disparities in water levels across a wall. Such disparities can emerge during tsunami inundation, where swift alterations in water elevation can impose significant hydrostatic pressures on edifices. The genesis of this force is independent of the wall's alignment with the flow direction of the tsunami, underscoring that the differential in water levels is the pivotal factor rather than the trajectory of the flow. For walls that are either slender or feature openings that account for more than 10% of the wall's surface area, it is posited that water levels will stabilize on both sides, thus diminishing the unbalanced force. This reduction in pressure differential is attributed to the openings facilitating water transit, thereby equalizing the pressures. Conversely, in instances involving broad walls or configurations where perpendicular barriers impede water movement, the wall will be subjected to unbalanced hydrostatic loads. The impediment to water level equalization, due to physical obstructions, results in a persistent pressure differential.

Two distinct equations are employed to compute the lateral hydrostatic force exerted on a wall, contingent upon whether the flow surmounts the wall. For scenarios where the flow does not exceed the wall's height, the lateral hydrostatic force (F_h) is determined by

$$F_h = \frac{1}{2}\gamma_s bh^2 \qquad (4.3)$$

where b is the width of the wall subjected to force, and h is the tsunami inundation depth above the grade plane at the structure. In contrast, when the flow overtops the wall, the calculation of the lateral hydrostatic force incorporates an additional factor of 1.2 to address the absence of aeration beneath the overtopping flow, which results in negative pressure. The equation for this case is given by

$$F_h = 1.2\frac{1}{2}\gamma_s(2h - h_w)bh_w \qquad (4.4)$$

where h_w is the wall height. This modification accounts for the enhanced pressure exerted by the unventilated, overtopping flow (Patil et al., 2018).

4.3.1.3 Residual water surcharge load

In the context of tsunami inundation and subsequent drawdown phases, it is plausible for disparate water levels to manifest on either side of a wall, building, or analogous structure. This discrepancy engenders a differential in hydrostatic surcharge pressure impacting the foundations, necessitating consideration within the foundation design procedure. The surcharge pressure (p_s) attributable to this differential can be quantified using

$$p_s = \gamma_s h_{\max} \qquad (4.5)$$

where $h_{\max}$ denotes the peak inundation depth above the grade plane at the structure in question.

During the drawdown phase of a tsunami, elevated floor slabs, particularly those encircled by perimeter structural components, such as an upturn beam, masonry, or a concrete wall or parapet, may impede water drainage. This obstruction can lead to surcharge loads on the floor slab, potentially surpassing its load-bearing capacity. The conceivable depth of water retained on the slab is contingent upon the maximum inundation depth of the tsunami. Still, it is constrained by the height of any continuous perimeter structural components capable of withstanding tsunami forces and thereby retaining water on an inundated floor. The pressure due to this residual water surcharge (p_r) is computed as follows

$$p_r = \gamma_s h_r \qquad (4.6)$$

where $h_r = h_{\max} - h_s$, with h_s representing the elevation of the top of the floor slab.

4.3.2 Hydrodynamic loads

Hydrodynamic loads arise from the interaction of fluid flow with objects obstructing its path. The phenomenon of tsunami inundation manifests either as a swiftly escalating tidal surge or a fragmented bore. This discussion encompasses both scenarios. However, since tsunami waves typically break offshore, the loads associated with wave breaking are not addressed in this context. The latter are usually dealt with in other design guidelines specific to coastal structures affected by a combination of storm surges and wind waves. The hydrodynamic forces exerted on structures can manifest as lateral pressures impacting walls and vertical forces acting upon slabs and horizontal elements. In the context of lateral hydrodynamic forces, American Society of Civil Engineers (2022) delineates two methodological approaches: a simplified approach and a detailed approach for the assessment.

4.3.2.1 Simplified approach

Structures are typically engineered to withstand lateral forces, such as those arising from wind and seismic loading. In scenarios involving large or tall buildings, particularly within regions of significant seismic risk, these alternate forms of lateral stress may impose more substantial demands on the lateral-force-resisting system than those attributed to tsunami-induced loads. Consequently, in such instances, a simplified yet conservative methodology is advantageous for determining the potential impact of tsunami-generated forces on structural integrity. As delineated in American Society of Civil Engineers (2022), the peak hydrodynamic loads are postulated to coincide with a water height (h) that is two-thirds of the maximum inundation depth (h_{max}), a conservative Froude number of $\sqrt{2}$, and a drag coefficient (C_d) of 2.0. Under these presumptions, the lateral load per unit width (f_w) imposed on a structure is given by

$$f_w = 1.5 \times \frac{1}{2} k_s \rho_{sw} I_{tsu} C_d C_{cx} \left(h(1.25u)^2 \right)_{max} \tag{4.7}$$

where u is the flow velocity, C_{cx} is the proportion of closure coefficient equal to 0.7 in a conservative assumption, ρ_{sw} is the effective mass density of seawater (typically, 1,025 kg/m³), and I_{tsu} is the importance factor for tsunami forces to account for additional uncertainty in estimated parameters. The application of a 1.25 amplification factor to the flow velocity is intended to encapsulate the most severe outcomes of flow concentration, as delineated in American Society of Civil Engineers (2022). Substituting the predetermined values as per the aforementioned assumptions, the equation simplifies to express the lateral pressure per unit width as follows

$$f_w = 1.6 k_s \rho_{sw} g I_{tsu} h_{max}^2 \tag{4.8}$$

where g is the gravitational acceleration. This derived pressure assumes a rectangular distribution over a height of $1.3 h_{max}$, leading to the cumulative pressure exerted across the full width of the structure being calculated as follows

$$P_{us} = 1.25 I_{tsu} \gamma_s h_{max} \tag{4.9}$$

All structural elements situated below a height of $1.3 h_{max}$ should be assessed by evaluating the impact of this pressure relative to the tributary width of the projected area.

4.3.2.2 Detailed approach for evaluating overall drag force

In the detailed approach to assessing hydrodynamic forces during the design phase, it is essential to account for multiple factors, the foremost being the total drag force exerted on the structure. When fluid flow envelops a building or structure, the resultant unbalanced lateral load due to hydrodynamic effects (F_{dx}) can be quantified by the following equation, rooted in fluid dynamics principles

$$F_{dx} = \frac{1}{2} \rho_s I_{tsu} C_d C_{cx} B \left(h_{sx} u^2 \right) \tag{4.10}$$

Here, B denotes the overall width of the structure perpendicular to the flow direction, while h_{sx} represents the height of the specific story under consideration. The closure coefficient, C_{cx}, quantifies the ratio of the vertical projected area of structural components to the vertical projected area of the submerged portion of the building, with a minimum threshold of 0.7 to accommodate potential debris accumulation.

The drag coefficient values are specified in Table 4.3 and vary depending on the ratio of the building's width perpendicular to the flow to the depth of inundation. A broader structure will experience a larger water level accumulation at its forefront. When the building exhibits horizontal setbacks, the drag force coefficient will differ across each inundated segment with uniform width. For such scenarios, the B/h ratio is calculated for each segment independently, with h representing the total height of the inundated stories, including any partially submerged ones, for each specific segment with a consistent building width. For instance, as depicted in Fig. 4.2, the drag force coefficient applied to two flooded stories of a narrow tower will be lower compared to that on two flooded stories of a broader podium beneath it due to a smaller B/h ratio in the submerged tower section relative to the podium.

TABLE 4.3 Drag coefficients for rectilinear structures (American Society of Civil Engineers, 2022).

Width to inundation depth ratio (B/h)	Drag coefficient (C_d)
≤ 12	1.25
60	1.75
≥ 120	2.0

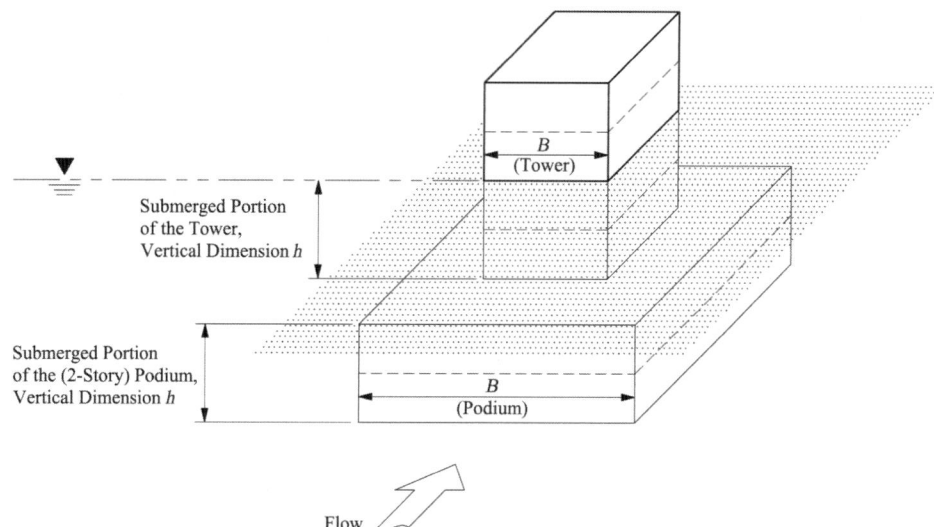

FIGURE 4.2 Illustration of dimensional parameters B and h applied to a building with a setback. *From American Society of Civil Engineers. (2022). Minimum design loads and associated criteria for buildings and other structures (Chapter 6) tsunami loads and effects ASCE/SEI7-22. https://doi.org/10.1061/9780784415788.*

4.3.2.3 Detailed approach for evaluating drag force on individual components

The second hydrodynamic load of significance in structural design pertains to the drag forces exerted on individual structural components. Each structural component or external wall assembly beneath the inundation depth endures hydrodynamic drag forces, calculated as follows

$$F_d = \frac{1}{2}\rho_s I_{\text{tsu}} C_d b \left(h_e u^2\right) \tag{4.11}$$

where h_e is the inundated height of the individual element. This formula incorporates an empirically derived drag coefficient, C_d, which varies according to the shape of the individual element in question. Table 4.4 provides typical C_d values for various common cross-sectional shapes of members (Blevins, 1984; Newman & Landweber, 1978; Overseas Coastal Area Development Institute of Japan, 2009; Sarpkaya, 2012). It is crucial to recognize that these drag coefficients are not applicable to the design of vertical load-bearing elements within storage warehouses and truck/bus garages due to the high risk of debris accumulation. For these specific elements, a conservative drag coefficient of 2.0 is recommended, with defined widths of 9 ft (2.74 m) and 40 ft (12.19 m) for storage warehouses and truck/bus garages, respectively. Importantly, the drag force calculated for individual elements should not be cumulatively added to the overall drag force assessed for the whole building.

The force F_w acting on vertical structural components is determined by hydrodynamic drag forces, utilizing the steady-state drag expression

$$F_w = \frac{1}{2}\rho_s I_{\text{tsu}} C_d b \left(h_e u^2\right) \tag{4.12}$$

Nonetheless, empirical evidence from laboratory experiments indicates that the initial impact of a tsunami surge, often manifesting as a broken bore, against a wide wall element, generates a transient impulsive load surpassing the hydrodynamic drag force derived from Eq. (4.12) (Arnason et al., 2009; Ramsden, 1996; Robertson et al., 2013; Paczkowski, 2011). This impulsive load typically exceeds the steady-state drag force by approximately 50%, necessitating the application of an amplification factor of 1.5 to the steady-state drag expression

$$F_w = \frac{3}{4}\rho_s I_{\text{tsu}} C_d b \left(h_e u^2\right)_{\text{bore}} \tag{4.13}$$

Given the variability in tsunami inflow, characterized by multiple bores of different magnitudes, assessing the bore loading effects under the most critical conditions is imperative. These include (1) scenarios where the inflow velocity is at maximum and affects all wall elements wider than three times the inundation depth and (2) instances where the inundation depth attains a level one-third of the structural width of each wall element, accompanied by corresponding flow velocities. This criterion is grounded in the bore height to specimen width ratio from the experiments of Arnason et al. (2009), which demonstrated impulsive forces surpassing the subsequent steady-state drag force. The transient nature of this impulsive load presupposes the integrity of windows and doors until the culmination of the peak load.

TABLE 4.4 Drag coefficients for structural components (American Society of Civil Engineers, 2022).

Structural element section	Drag coefficient (C_d)
Round column or equilateral polygon with six sides or more	1.2
Rectangular column of at least 2:1 aspect ratio with longer face oriented parallel to flow	1.6
Triangular column pointing into the flow	1.6
Freestanding wall submerged in flow	1.6
Square or rectangular column with longer face oriented perpendicular to flow	2.0
Triangular column pointing away from flow	2.0
Wall or flat plate, normal to flow	2.0
Diamond-shape column pointed into the flow (based on face width, not projected width)	2.5
Rectangular beam, normal to flow	2.0
I, L, and channel shapes	2.0

4.3.2.4 Detailed approach for evaluating drag force on vertical structural components

An advanced methodology for estimating the maximal impulsive force upon a bore's collision with a wall was formulated by Robertson et al. (2013), predicated on extensive experiments at Oregon State University. This approach posits that the lateral load per unit width of a wall can be approximated by

$$F_w = k_s \rho_{sw} \left(\frac{1}{2} g h_b^2 + h_j v_j^2 + g^{\frac{1}{3}} (h_j v_j)^{\frac{4}{3}} \right) \tag{4.14}$$

Here, h_b denotes the bore height (sum of the still water depth, d_s, and jump height, h_j), while the bore velocity, v_j, is deduced through hydraulic jump theory as follows

$$v_j = \sqrt{\frac{1}{2} g h_b \left[\frac{h_b}{d_s} + 1 \right]} \tag{4.15}$$

This force manifests as a triangular pressure distribution over a height h_p with a base pressure of $2F_w/h_p$, where the instantaneous ponding height at peak load, h_p, is calculated as follows

$$h_p = \left(0.25 \frac{h_j}{d_s} + 1 \right)(h_b + h_r) \leq 1.75(h_b + h_r) \tag{4.16}$$

with $h_r = (v_j h_j/g^{0.5})^{2/3}$. Employing this formula on a significantly impacted structural wall during the 2011 Tohoku Tsunami revealed a remarkable correlation with the observed damage patterns (Chock, Carden, et al., 2013; Chock, Robertson, et al., 2013). These equations become particularly relevant when detailed data on bore strike scenarios are available from field observations or can be inferred from site-specific inundation numerical analyses. For design objectives, juxtaposing the detailed approach with the measured data for all bore-on-wall experimental cases conducted at Oregon State University consistently aligned the predicted outcomes and experimental findings (Fig. 4.3).

4.3.2.5 Detailed approach for evaluating drag force on perforated walls

The impulsive force exerted on a solid wall can be mitigated if the wall features openings that allow the flow to pass through. This reduction can be quantified by incorporating the closure coefficient C_{cx}, leading to the following formula for computing the impulsive force on a perforated wall (Santo & Robertson, 2010)

$$F_{pw} = (0.4 C_{cx} + 0.6) F_w \tag{4.17}$$

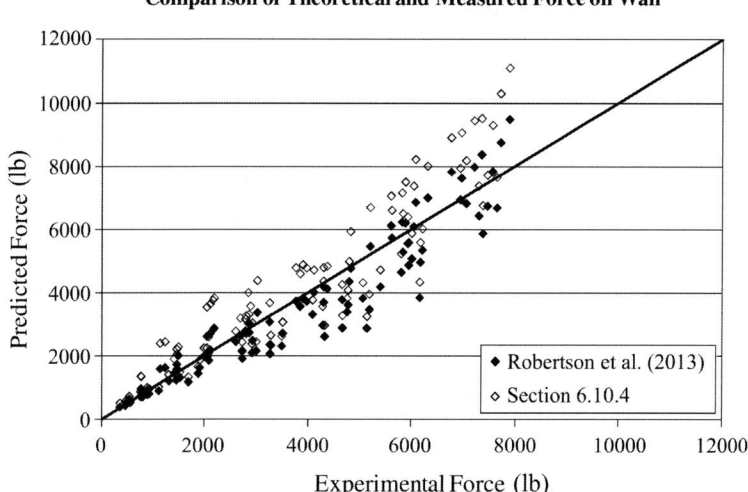

FIGURE 4.3 Predicted versus experimental bore loading on a wall. *From American Society of Civil Engineers. (2022). Minimum design loads and associated criteria for buildings and other structures (Chapter 6) tsunami loads and effects ASCE/SEI7-22. https://doi.org/10.1061/9780784415788.*

For walls that are positioned at an angle less than 90 degrees relative to the direction of the flow, the transient lateral load per unit width, $F_{w\theta}$, should be calculated using

$$F_{w\theta} = F_w \sin^2\theta \tag{4.18}$$

where θ represents the angle between the wall and the flow direction. This formulation implies a reduction in hydrodynamic loads for walls that are oblique to the flow, highlighting the importance of considering wall orientation in assessing tsunami-induced forces.

4.3.2.6 Detailed approach for evaluating drag force on elevated horizontal slabs

In the design of structures, it is imperative to consider the hydrodynamic pressure exerted on slabs. The observations from the 2011 Tohoku Tsunami revealed that structurally enclosed spaces, bounded by structural walls on three sides and a structural slab, can experience pressurization due to the ingress of flow into wall-enclosed spaces lacking openings in the side or leeward walls (Chock, Carden, et al., 2013; Chock, Robertson, et al., 2013). Analyses of such conditions, particularly within reinforced concrete structures, confirmed that the internal pressure can escalate to the theoretical flow stagnation pressure

$$P_p = \frac{1}{2}\rho_s I_{tsu} u^2 \tag{4.19}$$

where u represents the maximum free flow velocity at the given location. Consequently, it is crucial for the walls and slabs of spaces within buildings that may be subjected to flow stagnation pressurization, to be engineered to withstand pressures determined by this equation.

Further experiments conducted at Oregon State University on horizontal slabs, devoid of any flow obstructions above or below, indicated the potential for uplift pressures of 20 lb/ft^2 (0.958 kPa) to develop on the slab's soffit (Ge & Robertson, 2010). In scenarios involving a horizontal elevated slab positioned over the sloping ground without flow obstructions, the uplift hydrodynamic pressure necessitates an adjustment to incorporate the impact of the grade slope. In such cases, the flow acquires a vertical velocity component $u_v = u_{max}\tan\varphi$, where φ denotes the average slope of the ground beneath the slab. The following equation estimates the uplift pressure induced by this vertical velocity component

$$P_u = 1.5 I_{tsu} \rho_s u_v^2 \tag{4.20}$$

On the other hand, historical data from previous tsunamis indicate that substantial uplift forces may accumulate beneath horizontal structural elements (e.g., floors) when barriers like walls obstruct the flow beneath these elements (Chock, Robertson, et al., 2013; Saatcioglu et al., 2006). The mechanism involves redirecting the incoming water flow upwards upon encountering the wall. Yet, this upward flow is subsequently impeded by the overlying slab, culminating in significant pressures exerted on the wall and the underside of the slab in proximity to the wall's surface. A series of scaled experiments conducted at Oregon State University and the University of Hawaii revealed that this phenomenon becomes pronounced when the depth of water inundation surpasses two-thirds of the slab's height, delineated by the ratio $h_s/h \leq 1.5$ (Ge & Robertson, 2010). The findings, depicted in Fig. 4.4, encompass results from small-scale (utilizing the tsunami wave basin at Oregon State University and a dam break setup at the University of Hawaii at 1:10 scale) and large-scale experiments (conducted in the large wave flume at Oregon State University at 1:5 scale), with the large-scale bore heights being double those of the small-scale tests. An analytical envelope encapsulating 90% of the experimental data was constructed, as illustrated in Fig. 4.4. Despite the inherent variability in uplift pressures for identical wave conditions, a degree of consistency was observed across small- and large-scale experiments, suggesting that the laboratory findings apply to real-world scenarios. The estimation of uplift pressures necessary for the failure of pier access slabs during the 2011 Tohoku Tsunami yielded values ranging between 180 and 250 lb/ft^2 (8.62–12.0 kPa), as delineated by the shaded region in Fig. 4.4 (Chock, Robertson, et al., 2013). While the precise inundation depth at the moment of slab failure remains uncertain, the observed uplift pressures at full scale corroborate the validity of the proposed analytical envelope depicted in Fig. 4.4.

Observations of pressure distributions at the zenith of uplift revealed that the lateral pressure exerted on the wall and the uplift pressure beneath the slab in close proximity to the wall were approximately one-third higher than the mean pressure. Conversely, the lateral pressure at locations distanced from the wall was reduced to roughly half this magnitude. In scenarios where the wall possesses a finite width (l_w) and allows for the circumvention of water around its extremities, it is posited that the uplift pressure on the slab beyond a distance of $h_s + l_w$ diminishes to a baseline uplift pressure of 30 lb/ft^2 (1.44 kPa). Consequently, the entire wall and the adjacent slab area extending from the wall to h_s should be engineered to withstand an uplift pressure of 350 lb/ft^2 (16.76 kPa). For the slab section extending from h_s to $h_s + l_w$, an uplift pressure resistance of 175 lb/ft^2 (8.38 kPa) is required. Beyond the $h_s + l_w$ demarcation, the slab's design specification should accommodate an uplift pressure of 30 lb/ft^2 (1.44 kPa).

Under certain conditions, the hydrodynamic forces acting upon wall-slab recesses can be mitigated as per the guidelines outlined in American Society of Civil Engineers (2022). Illustratively, Fig. 4.4 demonstrates that when the depth of water inundation falls below two-thirds of the slab's elevation, or the ratio h_s/h exceeds 1.5, there is a linear reduction in the uplift pressure envelope. In such instances, the previously delineated uplift pressures are eligible for a reduction in alignment with the subsequent formula, which delineates the proportional linear decrement for the 350 lb/ft^2 (16.76 kPa) pressure scenario

$$P_u = I_{\text{tsu}}\left(28.25 - 7.66\frac{h_s}{h}\right) \tag{4.21}$$

Moreover, the empirical data suggest a minimal average uplift pressure threshold of 20 lb/ft^2 (0.96 kPa) for slabs that are up to five times taller than the inundation depth. Beyond this proportion, wherein the slab's elevation quintuples the depth of inundation, the vertically oriented flow fails to impact the slab, nullifying the need to account for uplift pressure.

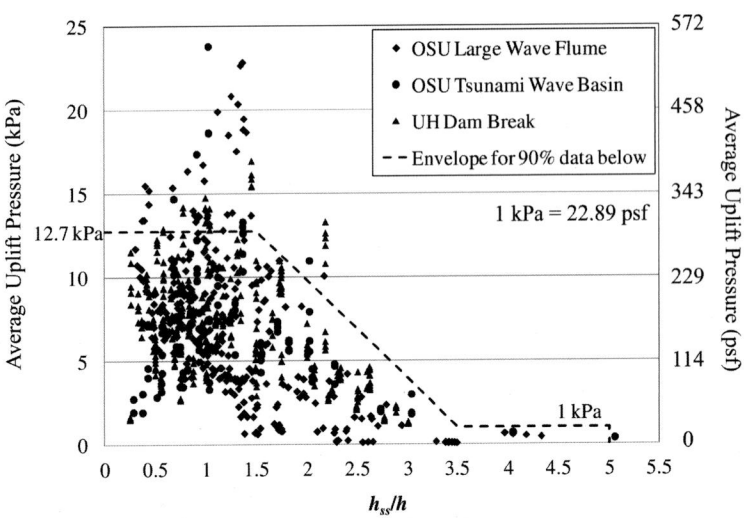

FIGURE 4.4 Average uplift pressures on slab soffit when the flow is blocked by a solid wall. *From American Society of Civil Engineers. (2022). Minimum design loads and associated criteria for buildings and other structures (Chapter 6) tsunami loads and effects ASCE/SEI7-22. https://doi.org/10.1061/9780784415788.*

Investigations conducted within the tsunami wave basin at Oregon State University, employing an identical experimental framework as previously delineated but substituting the solid barrier with a perforated one behind the slab, shed light on the influence of wall perforations on slab uplift pressures (Ge & Robertson, 2010). These experiments revealed a linear decrement in slab uplift pressures correlating with the degree of obstruction offered by the perforated wall, as depicted in Fig. 4.5. Notably, a basal uplift pressure of 20 lb/ft² (0.96 kPa) was recorded in scenarios where the wall was entirely absent. Consequently, the diminished pressures exerted on the wall, and the slab under these modified conditions can be quantitatively expressed by

$$P_{ur} = C_{cx} P_u \qquad (4.22)$$

where C_{cx} represents the proportion of the wall's impermeable surface relative to the entire submerged vertical cross-section of the wall at the corresponding elevation.

In an alternative strategy to mitigate uplift pressures on structural slabs, the introduction of clearance between the slab and an adjacent solid barrier has been examined. Research conducted at the University of Hawaii has elucidated that such a configuration can substantially diminish the uplift forces exerted on the slab. This effect is graphically represented in Fig. 4.6, derived from Takakura and Robertson (2010), which contrasts the variations in uplift pressure, P_u, for slabs equipped with an opening gap against those without any gap, in relation to the gap width, w_g, proportionate to

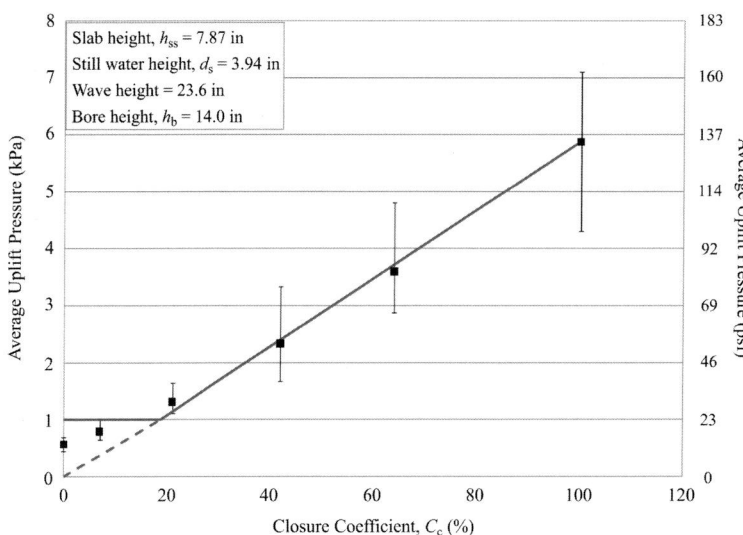

FIGURE 4.5 Mean slab uplift pressure reduction caused by a perforated wall. *From American Society of Civil Engineers. (2022). Minimum design loads and associated criteria for buildings and other structures (Chapter 6) tsunami loads and effects ASCE/SEI7-22. https://doi.org/10.1061/9780784415788.*

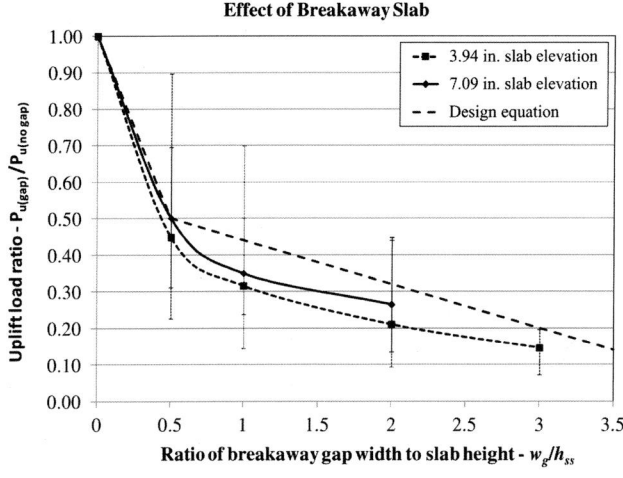

FIGURE 4.6 Reduction in slab uplift pressure caused by the presence of an opening gap or breakaway slab. *From American Society of Civil Engineers. (2022). Minimum design loads and associated criteria for buildings and other structures (Chapter 6) tsunami loads and effects ASCE/SEI7-22. https://doi.org/10.1061/9780784415788.*

the slab soffit height, h_{ss}. The stipulated guideline, as per American Society of Civil Engineers (2022), is depicted by the dashed line in Fig. 4.6 and is mathematically articulated as follows

$$P_{ur} = C_{bs}P_u \qquad (4.23)$$

wherein, for $w_g < 0.5h_{ss}$, $C_{bs} = 1.0 - w_g/h_s$ and for $w_g \geq 0.5h_{ss}$, $C_{bs} = 0.56 - 0.12w_g/h_s$. This phenomenon is presumed to be replicable by deploying a panel engineered to detach under uplift pressures lower than 175 lb/ft² (8.38 kPa). Observations from various structures that experienced inundation during the 2011 Tohoku Tsunami revealed that mechanisms, such as pressure relief grates and loosely fitted access panels, served effectively as detachable elements to alleviate uplift pressures on the residual slab structures (Chock, Robertson, et al., 2013). Furthermore, in instances where the flow-restricting barrier is conceptualized as a tsunami-dissipative wall, the calculation of slab uplift may be conducted in adherence to Eq. (4.19).

4.3.3 Debris impact loads

Tsunamis can mobilize an extensive array of debris, encompassing virtually any object within their inundation path that is buoyant enough at a given depth and unable to resist the force of the water flow. Typical debris includes trees, wooden utility poles, automobiles, and parts of wood-framed buildings. Additionally, denser materials, such as boulders and concrete fragments, may be transported if the tsunami's flow possesses sufficient force. Calculating the impact forces of debris on structures within a tsunami's reach is imperative, considering the potential debris from the surrounding area that could be propelled toward the site. Special attention is warranted for structural elements situated along the perimeter and aligned perpendicular to the flow, as these are particularly susceptible to debris impacts. The compromise or failure of these elements could critically undermine the structural integrity needed for bearing gravitational loads. Generally, internal structural supports, such as columns and walls, are not considered at risk from debris impact forces. Nevertheless, under specific circumstances, it becomes imperative to design internal load-bearing components to withstand debris impacts:

- Vehicles within inundated parking garages may become buoyant and exert forces on interior structural components. This applies to all flooded levels, including subterranean ones, where vehicles may be displaced due to water flow entering via ramps.
- At the ground level, failure of slabs-on-grade may precipitate the generation of tumbling debris, posing a risk to internal structural components (Chock, Robertson, et al., 2013; Robertson et al., 2009; Stolle et al., 2020). Notably, such phenomena are unlikely in basement slabs due to lower flow velocities. To mitigate uplift forces resulting from buoyancy, ground-level slabs-on-grade can be intentionally designed with isolation joints to promote failure under uplift conditions, necessitating the design of adjacent vertical structural elements to accommodate the consequent impact loads.
- Concrete slab panels designed to break away and alleviate uplift pressures on elevated slabs will become dislodged under significant uplift forces, transforming into hazardous debris. Columns and structural walls near these panels must be engineered to absorb the impact of such debris.

The magnitude of impact forces is contingent upon the velocity at which debris strikes, which, in the case of floating debris, is postulated to be synonymous with the velocity of the flow itself. The specific locations at which these impact forces, conceptualized as concentrated loads, are applied should be strategically selected to represent the most severe conditions of shear and bending moments for each structural component under consideration. This assessment should consider the depth of inundation and the associated flow velocity to ensure a comprehensive evaluation of the structural integrity within the tsunami-affected zone.

The prevalent presence of materials, such as logs and/or poles, passenger vehicles, boulders, and concrete debris, necessitates the presumption of their potential impact on structures, contingent upon whether the depth and velocity of inundation render such interactions feasible. Moreover, the propensity of closed shipping containers to remain buoyant, even when laden, implies that, in proximity to container yards, the possibility of impacts from such containers must be accounted for in structural assessments. Additionally, larger maritime vessels, including ships, ferries, and barges, represent substantial debris risks, particularly to structures situated near ports and harbors where such vessels are moored or operate.

Given the substantial structural demands and potential economic impracticality of designing for such extraordinary debris impacts, only facilities categorized as *critical facilities* (Tsunami Risk Category III and Tsunami Risk Category IV in American Society of Civil Engineers (2022)) must consider these impacts in their design. A comprehensive site hazard assessment is imperative to ascertain the necessity of factoring in impacts from shipping containers at a given site, as delineated in the following procedural outline.

4.3.3.1 Detailed approach
4.3.3.1.1 Site hazard assessment

The methodology for site hazard assessment is based on the hypothesis that debris disperses from a linear origin, with the subsequent hazard zone associated with this source being delineated through geometric principles. Empirical studies, such as those examining debris distribution following dam breaches, have indicated that this approach tends to be conservative (Stolle et al., 2018). When a structure is situated within the defined debris impact hazard zone, potential impacts from shipping containers, ships, and barges should be considered.

The procedural steps for the site hazard assessment are as follows:

1. The geographic nucleus of the debris source must be pinpointed in conjunction with the predominant direction of the flow.
2. A line is to be drawn traversing the center of the source, oriented perpendicularly to the main flow direction, the length of which should mirror the anticipated width of the debris source.
3. Commencing from the extremities of this width line, additional lines are to be extended at angles of ± 22.5 degrees relative to the tsunami's approach direction, extending these lines until they intersect with the grounding limit, as depicted in Fig. 4.7. The choice of the 22.5-degree angle for debris dispersion is informed by field observations documented by Naito et al. (2014) and further corroborated by experimental studies conducted by Nistor et al. (2017).

The dissemination of debris during tsunami events can be mitigated by tsunami-resistant infrastructures, natural geographical barriers, and limitations posed by the depth of inundation. For instance, shipping containers within a yard encircled by robust structural steel and concrete barriers are unlikely to be dispersed beyond these confines, provided the containers cannot surmount these barriers. Similarly, vessels with a specific draft, such as 4 ft (1.2 m), are unlikely

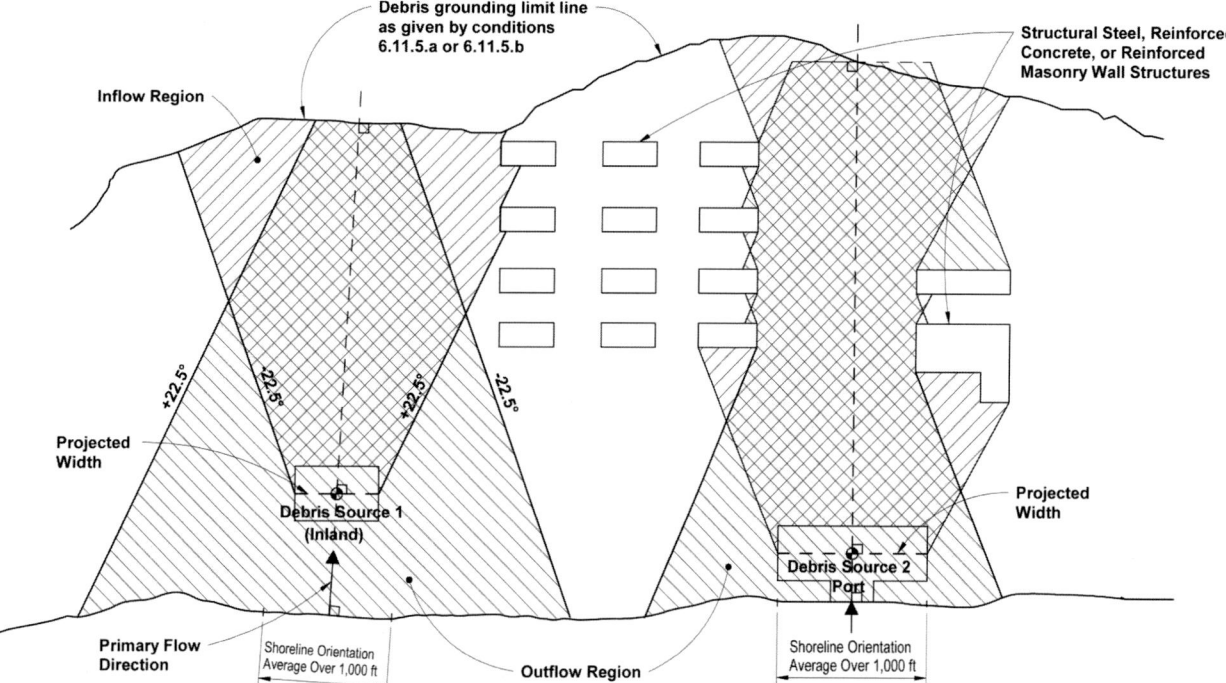

FIGURE 4.7 Illustration of determination of float debris impact hazard region. *From American Society of Civil Engineers. (2022). Minimum design loads and associated criteria for buildings and other structures (Chapter 6) tsunami loads and effects ASCE/SEI7-22. https://doi.org/10.1061/9780784415788.*

to be transported extensively or at considerable speeds in water depths equal to their draft. Based on these considerations, the delineation of the inland boundary of the debris impact hazard area should be adjusted as follows:

- Where the peak inundation depth does not exceed 3 ft (0.91 m), or for ships where the inundation depth is less than the vessel's ballasted draft plus an additional 2 ft (0.61 m).
- In areas where barriers comprise structural steel, reinforced concrete, or solidly grouted masonry, walls impede debris transport. For a barrier to be deemed effective, its height must at least equal the inundation depth minus 2 ft (0.61 m) for containers and barges, or for ships, the inundation depth minus the combined height of the ballasted draft and an additional 2 ft (0.61 m).
- The utilization of numerical debris transport modeling is endorsed to ascertain the inland boundary of the debris impact hazard area. This modeling should employ massless tracers, reflecting debris movement within the fluid, as derived from site-specific inundation analyses. The modeling approach should adhere to established methodologies, such as using massless tracers (van Sebille et al., 2018), to estimate the inland reach of debris conservatively. The model's parameters should align with best practices in existing literature (Lebreton et al., 2012), including implementing release and grounding thresholds contingent upon flow depth and velocity. Ensuring numerical convergence for the inland limit of debris transport is crucial, as well as ensuring that this limit is not influenced by the numerical time step or the initial spacing of tracers. Recommended numerical settings include a time step of one second or less for tracer trajectory updates and an initial tracer distribution of at least one tracer per every 110 ft^2 (10.2 m^2) of the debris source area.

In instances where topographical features, such as hills, serve to confine the water flow within the designated sector, adjustments to the sector's orientation are necessary to align with the contours of these topographical barriers, or alternatively, the sector's breadth may be reduced in areas where it is encircled by topographical constraints on multiple fronts. When a site-specific inundation analysis provides a detailed mapping of how the topography influences the flow direction of inundation across the tsunami design zone, it is acceptable to redefine the primary flow direction to align with the inundation flow field vectors identified in the analysis, extending from the site to the grounding limit. This redefined flow direction should incorporate a lateral dispersion of 22.5 degrees from the central flow field vectors to adequately capture the potential spread of water and debris due to the tsunami.

To ascertain the debris impact hazard zone attributable to outflow, a procedure involves establishing a line perpendicular to the inward flow's central axis at the point where the flow ceases, known as the grounding limit. This line's length should mirror the anticipated width of the debris source. Commencing from each terminus of this width, lines should be extended at an angle of 22.5 degrees away from the direction of outflow, consistent with the illustrative guidelines provided in Fig. 4.7.

Regarding structural design imperatives, structures situated solely within the inflow hazard zone must be engineered to withstand impacts from containers or vessels transported by the inflow. Conversely, structures located exclusively within the outflow hazard zone are to be designed to anticipate impacts from containers or vessels associated with the outflow. Structures that find themselves within the inflow and outflow zones must be robustly designed to endure impacts from containers or vessels moving in either direction, ensuring comprehensive protection against the dynamic nature of tsunami-induced debris transport.

4.3.3.1.2 Impact force by general objects

Historically, guidelines for calculating debris impact forces, such as those outlined in Section C5.4.5 of American Concrete Institute (2013), were predominantly grounded in the impulse-momentum theory, necessitating an assumption regarding the impact's duration. In contrast, the newly suggested approach for determining the nominal maximum instantaneous debris impact force (F_{ni}) transitions to focusing on stress wave propagation within the debris and the structural member. This shift acknowledges the potential deformability of both the debris and the structural element upon impact. The formula for F_{ni} is expressed as follows

$$F_{ni} = u_{max} \sqrt{km_d} \tag{4.24}$$

In this equation, u_{max} represents the peak flow velocity at the site that is capable of floating the debris. The term m_d denotes the mass of the debris, calculated as W_d/g, where W_d is the weight of the debris, and g is the gravitational acceleration. A minimum debris weight of 1000 lb (454 kg) is stipulated, a benchmark derived from flooding considerations in Section C5.4.5 of American Society of Civil Engineers (2013). This baseline weight approximately equates to a log measuring 30 ft (9.14 m) in length with a diameter of 1 ft (30.5 cm), although it is acknowledged that larger trees may be encountered depending on the specific geographical context, necessitating adjustments based on regional and local conditions.

The variable k in Eq. (4.24) stands for the effective stiffness, which could either pertain to the stiffness of the debris or the lateral stiffness of the structural element impacted, adopting the lesser of the two values. For logs or poles, the stiffness calculation is $k = EA/L$, where E denotes the longitudinal modulus of elasticity of the log, A is the cross-sectional area, and L is the length. A minimum stiffness value for the log is set at 350 kips/in (61,300 kN/m), a value that correlates with the minimum weight of 1000 lb (454 kg) and assumes a modulus of elasticity of approximately 1100 ksi (7580 MPa).

The foundational premises of Eq. (4.24) are based on the occurrence of an elastic impact and the orientation of the impact being longitudinal. Specifically, this implies that in the context of a pole or log, the impact with a structural entity occurs at the log's end (butt end), as opposed to a sideway (transverse) collision. Similarly, for a shipping container, the underlying assumption is that the impact involves the terminal point of one of the container's bottom longitudinal rails contacting the structural component.

This theoretical framework and the accompanying equation have been empirically validated through comprehensive testing conducted on a full-scale basis at Lehigh University. These experiments included assessments of a utility pole and a 20 ft (6.1 m) shipping container, ensuring that the conditions mirrored the assumptions of a longitudinal strike. The findings from these investigations, as documented in Piran Aghl et al. (2014) and Riggs et al. (2014), lend credence to the applicability and reliability of Eq. (4.24) under the specified impact scenarios.

The design instantaneous debris impact force (F_i) is calculated using

$$F_i = I_{tsu} C_o F_{ni} \tag{4.25}$$

In this equation, C_o represents the orientation coefficient, set to 0.65 for logs and poles. This coefficient is derived from the empirical data presented by Haehnel and Daly (2004), encapsulating the mean value augmented by one standard deviation of the log debris impact force from experiments that encompassed glancing and direct impacts by freely floating logs.

Furthermore, the duration of the impulse for an elastic impact (t_d) is determined by

$$t_d = \frac{2 m_d u_{max}}{F_{ni}} \tag{4.26}$$

This equation is also based on the principle of elastic impact and presupposes a constant impact force throughout the duration, thereby implying a rectangular shape for the force–time curve. While this approach might lead to a slight underestimation of the actual impulse duration, it is designed to err on the side of caution, ensuring that the calculated total impulse remains conservatively high, thereby contributing to the overall safety and resistance of the structure against debris impact forces.

While debris impact is inherently dynamic, necessitating a response from structural elements that is equally dynamic, applying an equivalent static analysis method is deemed acceptable within the proposed framework. This approach amplifies the static displacement by a dynamic response factor (R_{max}), as delineated in Table 4.5. This adjustment factor, which serves as a scaling mechanism, depends upon the ratio of the impact duration (t_d) to the natural period of the structural element (T), effectively encapsulating the shock spectrum characteristics. The shock spectrum is influenced by this ratio and the specific form of the force–time curve.

The shock spectrum outlined in Table 4.5, which is derived from Section C5.4.5 of American Society of Civil Engineers (2013), represents a modified version of the shock spectrum, tailored to a half-sine wave profile, with a notable distinction being its constant factor for scenarios where $t_d/T > 1.4$, in contrast to the declining trend observed in the conventional half-sine wave shock spectrum. This modification is substantiated by the experimental findings from Lehigh University, which indicate a closer alignment with the tabulated factors due to the force–time history not conforming precisely to a half-sine wave shape (Piran Aghl et al., 2014).

For analytical purposes, particularly in the context of walls, the impact force is presumed to act at the midpoint along the horizontal axis of the wall. Furthermore, determining the natural period may be simplified by equating it to the fundamental period of a hypothetical column, the width of which is set to half the wall's vertical dimension. This approximation aids in the pragmatic assessment of the wall's dynamic response to debris impact.

4.3.3.1.3 Impact by vehicles

Passenger vehicles, prevalent and capable of buoyancy, are susceptible to transport by floodwaters. The threshold for such buoyancy is identified in American Society of Civil Engineers (2022) as an inundation depth of 3 ft (0.91 m), beyond which vehicles are considered at risk of impact. The studies by National Crash Analysis Center (2011) and

TABLE 4.5 Dynamic response ratio for impulsive loads (American Society of Civil Engineers, 2022).

Ratio of impact duration to the natural period of the impacted structural element	R_{max} (response ratio)
0.0	0.0
0.1	0.4
0.2	0.8
0.3	1.1
0.4	1.4
0.5	1.5
0.6	1.7
0.7	1.8
0.9	1.8
1.0	1.7
1.1	1.7
1.2	1.6
1.3	1.6
≥ 1.4	1.5

National Crash Analysis Center (2012) offer an insightful examination of the dynamic response of a 2400 lb (1090 kg) subcompact passenger vehicle during a frontal collision with a barrier at a velocity of 35 mph (15.6 m/s), from which the vehicle's initial stiffness was deduced to be 5700 lb/in (1 kN/mm). Employing an assumed velocity of 9 mph (4 m/s), Eq. (4.24) yields an impact force in the vicinity of 30 kips (133 kN) (Naito et al., 2014). Considering a more probable oblique impact with a reduced contact area, an impact force of 30 kips (133 kN) is considered a conservatively prescriptive load encompassing a spectrum of potential vehicle impact scenarios. The impact may manifest at any point ranging from 3 ft (0.91 m) to the maximum inundation depth.

The incipient motion, transforming the vehicles into floating debris, might commence at inundation depths lower than 3 ft (0.91 m), influenced by the specific vehicle characteristics and the flow velocity (Fig. 4.8). Martínez-Gomariz et al. (2018) examined existing literature on small-scale laboratory experiments that quantified the motion threshold for vehicles in steady-state flood conditions. The results depicted in Fig. 4.8 vary based on vehicle dimensions and ground clearance. The findings suggest that for inundation depths exceeding 3 ft (0.91 m) or a flow velocity surpassing the baseline 10 ft/second (3.0 m/second), all categories of passenger vehicles are likely to transition into floating debris.

4.3.3.1.4 Impact by shipping containers

Calculating impact forces exerted by shipping containers should follow the methodologies outlined in Eqs. (4.24) and (4.25). Piran Aghl et al. (2014) demonstrated that the mass of the contents within a container does not markedly influence the impact force, provided these contents are not rigidly affixed to the container's structural framework. Consequently, for shipping containers, the calculation in Eqs. (4.24) and (4.25) will utilize the container's empty mass. An orientation coefficient, C_o, with a value of 0.65, is recommended, based on the assumption of a comparable level of randomness in the alignment of the shipping containers' lower corner steel chords with the impacted structural element.

The presence of contents may extend the impact duration. In this case, the pulse duration for loaded shipping containers is calculated using

$$t_d = \frac{(m_d + m_{contents})u_{max}}{F_{ni}} \quad (4.27)$$

In this equation, $m_{contents}$ is recommended to be considered 50% of the shipping container's maximum rated capacity for contents. The minimum mass values for $m_d + m_{contents}$ for loaded shipping containers are provided in Table 4.6. The design process must account for empty and loaded states of shipping containers. The force applied in this equation

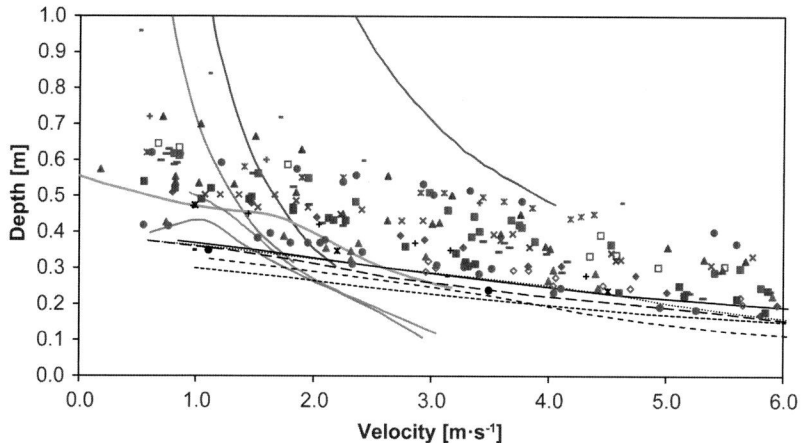

FIGURE 4.8 Threshold-of-motion criteria for vehicles as a function of flow depth and velocity. *From American Society of Civil Engineers. (2022). Minimum design loads and associated criteria for buildings and other structures (Chapter 6) tsunami loads and effects ASCE/SEI7-22. https://doi.org/10.1061/9780784415788.*

TABLE 4.6 Weight and stiffness of shipping container waterborne floating debris (American Society of Civil Engineers, 2022).

Type of debris	Weight	Debris stiffness (k)
20 ft (6.1 m) standard shipping container oriented longitudinally	Empty: 5000 lb (2270 kg); Loaded: 29,000 lb (13,150 kg)	245 kips/in (42,900 kN/m)
40 ft (12.2 m) standard shipping container oriented longitudinally	Empty: 8400 lb (3810 kg); Loaded: 38,000 lb (17,240 kg)	170 kips/in (29,800 kN/m)

remains consistent with that for an empty container. Eq. (4.27) presupposes a primarily plastic impact resulting from the plastic deformation of the container, implying that the container is presumed to adhere to the structural element without rebounding.

The stiffness of the container can be calculated as $k = EA/L$, where E represents the modulus of elasticity of the container's bottom rail, A is the cross-sectional area of this bottom rail, and L is the length. Although shipping containers are standardized in dimensions, variations in weight and structural specifications may arise among different manufacturers. The estimates for weight and stiffness presented in Table 4.6 are deemed accurate for most standard International Organization for Standardization (ISO) shipping containers. The assumption for loaded weights is that the containers are filled to approximately 50% capacity. Therefore, when converted to mass, these values can be directly applied in Eq. (4.27) for $m_d + m_{contents}$.

The nominal design impact force, F_{ni}, derived from Eq. (4.24) for shipping containers, should not exceed 220 kips (980 kN). This value does not represent the ultimate force a container might exert but is based on empirical data from the experiments conducted at a speed of 8.5 mph (3.8 m/second). Recent simulation results, such as those of

Madurapperuma and Wijeyewickrema (2013), suggest that the maximal force could be greater, contingent upon the specific impact scenario. Nonetheless, the force of 220 kips (980 kN) has been selected as a pragmatically reasonable value for design purposes.

Eq. (4.24) lacks a component to account for potential increases in force resulting from the fluid dynamics altered by the abrupt halt of the debris, a factor included in some alternative models. For the case of a log impacting longitudinally, such an escalation in force is not anticipated to be substantial. Experimental investigations using scale-model shipping containers within the tsunami wave flume at Oregon State University revealed that fluid dynamics minimally influenced longitudinal impacts (Riggs et al., 2014). Consequently, the force determined by Eq. (4.24) is deemed adequately conservative, permitting the omission of the transient fluid "added mass" effect without compromising safety or accuracy in the assessment.

4.3.3.2 Alternative detailed approach

Dynamic analysis is sanctioned to evaluate the structural response to debris impact forces. In instances where the impact is substantial enough to induce inelastic behavior within the structure, the utilization of an equivalent single-degree-of-freedom mass-spring system featuring nonlinear stiffness to account for the ductility of the impacted structure is approved for the dynamic analysis. For linear elastic analysis of a single-degree-of-freedom system, the peak response exceeds that of an equivalent nonlinear analysis, attributed to the disparity between the rectangular shock and the half-sine wave spectra. When inelastic behavior is factored in, the structural ductility serves to mitigate the force demands, enabling the application of a work-energy approach for analysis.

The work-energy method for assessing the impact of large, essentially rigid debris on structures while accounting for the structure's ductility is encapsulated by

$$F_{\text{cap}} = u_{\max}\left[\frac{1+e}{\left(1+\frac{M_m}{m_d}\right)\sqrt{2\mu-1}}\right]\sqrt{k_e M_e} \tag{4.28}$$

where F_{cap} represents the structural capacity, e is the coefficient of restitution between the debris and the structure, which is assumed to be 1 unless proven otherwise, M_e is the effective mass of the structure at the impact point, k_e is the initial elastic stiffness of the structure at the impact point, and μ is the permissible ductility ratio, as might be specified in standards, such as Appendix C and Section C.3.7 of American Concrete Institute (2013).

This formulation is predicated on an elastic-perfectly plastic force—displacement relationship, where k_e is the elastic stiffness, F_{cap} is the yield force, and μ represents the ratio of the maximum displacement to F_{cap}/k_e. The mass terms are computed as $M_m = \Sigma m_i \Delta_i$ and $M_e = \Sigma m_i \Delta^2{}_i$ with Δ_i symbolizing the static displacements of the structure due to a force at the impact point, normalized to 1 at the impact point. This method is a nuanced adaptation of the approach presented by Kuilanoff and Drake (1991). Selection of the structural segment for the determination of k_e, M_m, and M_e is critical, as is the consideration of the impact's duration. For instance, impacts by logs or shipping containers are likely to occur within a brief timeframe, making it improbable for the entirety of a large structure to react within such a time span. Consequently, it may be more appropriate to consider a specific section of the structure, potentially down to an individual structural member. Conversely, impacts by ships, which may unfold over a longer period, might necessitate considering the response of the entire structure.

For other work-energy methods, where forces are absorbed through inelastic behavior up to a certain ductility limit, the structure's initial stiffness is adjusted to an effective stiffness that accounts for the deformation. Instead of conducting a nonlinear time-history analysis, employing secant or effective stiffness to linearize the modeled response analysis

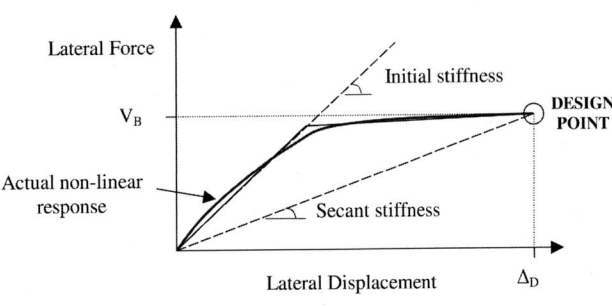

FIGURE 4.9 Stiffness definitions for initial elastic and effective secant stiffness. *From American Society of Civil Engineers. (2022). Minimum design loads and associated criteria for buildings and other structures (Chapter 6) tsunami loads and effects ASCE/SEI7-22. https://doi.org/10.1061/9780784415788.*

based on a performance point within the inelastic range is a recognized method, as depicted in Fig. 4.9. Preliminary analyses suggest that for achieving consistent results across a wide spectrum of inelastic behavior, the velocity used in the work-energy analysis should be $u_{\max}$ augmented by the product of the importance factor, I_{tsu}, and the orientation factor, C_o.

4.3.3.3 Simplified approach

In addition to the detailed guidance on debris impact load on structures, two alternative methodologies are also available for consideration. The first approach adopts a conservative, prescriptive load to avoid detailed analysis of impacts by diverse debris types, such as logs, poles, vehicles, boulders, concrete debris, and shipping containers. In this context, the debris impact can be quantified as follows to represent a maximal static load, substituting the specific loads delineated for each category of debris

$$F_i = 330 C_o I_{tsu} \text{ [kips]} = 1470 C_o I_{tsu} \text{ [kN]} \tag{4.29}$$

This force will be applied at critical points for flexure and shear within all members in the inundation zone. The selected value of 330 kips (1470 kN) is derived from the previously mentioned cap of 220 kips (980 kN) for shipping containers, augmented by a dynamic amplification factor of 1.5, deliberately chosen to be less than the maximum from Table 4.5 to account for the dampening of peak forces by the inelastic behavior of the impacted element.

Should it be established on the basis of a site hazard assessment that the location is not susceptible to impacts from containers or ships, the impact force is presumed to stem from a direct hit by a wooden log. Piran Aghl et al. (2014) examined a standard wood log weighing 450 lb (204 kg). In such instances, a basic direct strike force of 165 kips (734 kN) may be utilized, halving the force of 330 kips (1470 kN) associated with shipping containers, thereby reducing the simplified debris impact force to 50% of that proposed by Eq. (4.29). The nominal design impact force for wooden logs or poles is capped by the material's crushing strength. This stipulated load of 165 kips (734 kN), inclusive of a structural dynamic response factor of 1.5, pertains to poles and logs possessing a parallel-to-grain crushing strength of 5000 psi (34.5 MPa) for Southern Pine or Douglas Fir as per (ASTM International, 2011), with an effective contact area of 22 in^2 (142 cm^2), approximating 20% of the frontal area of a 1 ft (30.5 cm) diameter pole, in alignment with Piran Aghl et al. (2014). The orientation factor, C_o, is applied in all scenarios. This simplified force is deemed conservative in relation to laboratory findings, which in turn are considered conservative relative to typical field conditions.

4.4 Differences in tsunami-induced loads between American and Japanese design codes

As ASCE 7−22 (American Society of Civil Engineers, 2022) and MLIT2570 (Ministry of Land, Infrastructure, Transport and Tourism, 2011a) are the only two *prescriptive* codes for tsunami-resistant buildings and other structures, this section summarizes the approaches used by the two standards to characterize tsunami-induced loads and effects for structural design. The codes should be reviewed in their entirety for the full set of design provisions and equations.

4.4.1 Loading conditions

ASCE 7−22 prescribes three loading conditions that must be evaluated (Fig. 4.1): Load Case 1 considers a depth not exceeding the maximum inundation depth, one story, or the height of the top of the first-story window; Load Case 2 considers a depth at two-thirds of the maximum inundation depth, where maximum velocity and specific momentum flux are expected in both inflow and outflow cycles; and Load Case 3 considers a depth at maximum inundation depth, with velocities assumed to be one-third of the maximum in both inflow and outflow cycles. On the other hand, MLIT2570 does not include velocity in its procedures explicitly. Instead, it only has one loading condition required where the maximum inundation depth is attained.

4.4.2 Hydrostatic forces

Three hydrostatic forces are considered in ASCE 7−22: buoyancy, unbalanced lateral hydrostatic force, residual water surcharge load on floors and walls, and hydrostatic surcharge pressure on the foundation. Design equations are available for each of these forces. MLIT2570 includes two buoyancy scenarios: superstructure design, which considers the water displaced by air trapped below the floor framing system, and foundation design, where no trapped air is considered.

4.4.3 Hydrodynamic forces

Design equations for drag forces on the overall structure and individual components, impulsive forces from tsunami bores on walls, slabs, and aboveground pipelines, flow stagnation pressure, and shock pressure in structural wall-slab recesses due to entrapped tsunami bore flow are provided in ASCE 7−22. In MLIT2570, hydrodynamic forces are simplified by approximating the flow condition only in terms of inundation depth. This is due to the complexity of accurately determining tsunami velocity fields, as the local conditions may present existing structures or other local effects (Nakano, 2017). MLIT2570 takes the approach described in Asakura et al. (2000) where the hydrodynamic forces are approximated and included in the evaluation of the hydrostatic force. Asakura et al. (2000) proposed a "water depth coefficient" where the hydrostatic pressure profile is increased by a factor in the form of a coefficient. The water depth coefficient was determined through experimental investigations where the hydrodynamic forces are captured by using a factor of 1.5−3, depending on the distance that the energy dissipation structure is located from the shoreline or riverbank.

4.4.4 Debris impact loads

ASCE 7−22 requires that gravity-load-carrying structural components located on the principal structural axes perpendicular to the range of inflow and outflow directions be designed to withstand impact from floating wood poles, logs, vehicles, tumbling boulders, and concrete debris. Sites close to ports or container yards include strikes from shipping containers, ships, and barges. Explicit design guidance is provided for each of the debris types. ASCE 7−22 also offers progressive collapse avoidance (i.e., ensuring that the collapse of one column does not lead to further collapse of the entire structural system) as an alternative method to reach the acceptance criteria of tsunami-resistant structural components. In MLIT2570, several formulae are provided to calculate debris impact forces, but there is significant variability in the calculated debris impact load, depending on the selected formula (Nakano, 2017). As such, progressive collapse avoidance (i.e., ensuring that the collapse of one column does not lead to further collapse of the entire structural system) has been widely applied in favor of the debris impact formulae.

4.5 Tsunami structural vulnerability assessment

The tsunami design approaches presented in Sections 4.3 and 4.4 demonstrate the source of the vulnerability of buildings to tsunami loadings and consequential impacts. Such an interaction between tsunamis and structures can be described systematically and studied in engineering terms by defining structural and geotechnical models that can be examined under forcing models. The availability of such models facilitates the reliability assessment of a structural system to withstand tsunami impact. A reliability study is traditionally focused on computing the probability of failure of a structural system or its components.

A comprehensive reliability analysis should account for external or internal disturbances, such as loads and settlements, which can compromise the integrity of the structural system. This leads to computing engineering demand parameters (*EDP*s), denoted by *Demand* (*D*), describing how the structural system responds to the perturbances (e.g., interstory drift and shear force in the columns). These *EDP*s can then be compared against *Capacity* (*C*) to determine Damage Measures under certain conditions. *Demand* and *Capacity* must be defined consistently. For instance, for reinforced concrete structures, this could mean comparing the shear force in a section with its corresponding shear strength. In the context of reliability analysis, *Failure* is recognized as occurring when *Demand* exceeds *Capacity*. Moreover, in assessing the structural vulnerability of a system (e.g., a bridge or a building), uncertainties must be accounted for. This is even more relevant considering the abovementioned definition of reliability as linked to the probability of failure.

When a structural system or a structural component fails, it can result in consequences typically categorized into consequence classes. To conduct a reliability analysis, it is essential to establish a commonly accepted set of limit states (*LS*s). These *LS*s describe conditions of a system or component beyond which it no longer meets the relevant design and use criteria (e.g., lack of bearing capacity), which has been referred to as failure. Exceeding the limit state indicates damage occurrence, which is characterized by the *DS*. This distinction offers a straightforward means of categorizing the extent of damage, effectively separating the desired state from the adverse state of the system or component.

4.5.1 Tsunami damage states

*DS*s are discrete variables classifying the damage that a structure can be subjected to. It is crucial to observe that structures, before being subjected to an earthquake-triggered tsunami, may be affected by the earthquake; therefore, in principle, the *DS* is a combination of the two hazards. According to Hill and Rossetto (2008), the definition of the *DS*s must respect six main rules:

1. *Mutual exclusivity*: Each *DS* should represent damage conditions that are mutually exclusive and, therefore, cannot be contained in full or in part in any other classification.
2. *Damage progression*: *DS*s should be sorted according to the increasing severity of the damage.
3. *Ease of measurement*: *DS*s should be clearly distinguishable and should be general enough to be applied to a population of buildings.
4. *Coverage*: Descriptions of the *DS*s should capture the entire possible range of damage that a building typology could experience.
5. *Global and local damage*: *DS*s should capture global (i.e., systemic) and local (i.e., single component) damage conditions.
6. *Structural and nonstructural*: *DS*s should capture damage to the structural components and to the nonstructural elements.

Existing definitions of *DS*s are mainly based on posttsunami surveys; therefore, they can be country-specific. For example, the MLIT of the Japanese Government assessed *DS*s of more than 200,000 buildings after the 2011 Tohoku Tsunami (Fig. 4.10). Seven *DS*s were adopted: no damage (*DS1*), minor damage (*DS2*), moderate damage (*DS3*), major damage (*DS4*), complete damage (*DS5*), collapse (*DS6*), and washed away (*DS7*). In engineering terms, *DS1*, *DS2*, and *DS3* identify nonstructural damage (i.e., minor flooding or slight damage to nonstructural components), whereas *DS4* to *DS7* refer to structural damage, such as damage to frames/walls and overturning/translation of buildings. When the damage classifications do not conform to the rules mentioned above, multiple *DS*s can be combined to make sure that they are compliant. For example, many studies that have used the MLIT database combined *DS6* and *DS7* as they were two different descriptions of complete loss of the structure.

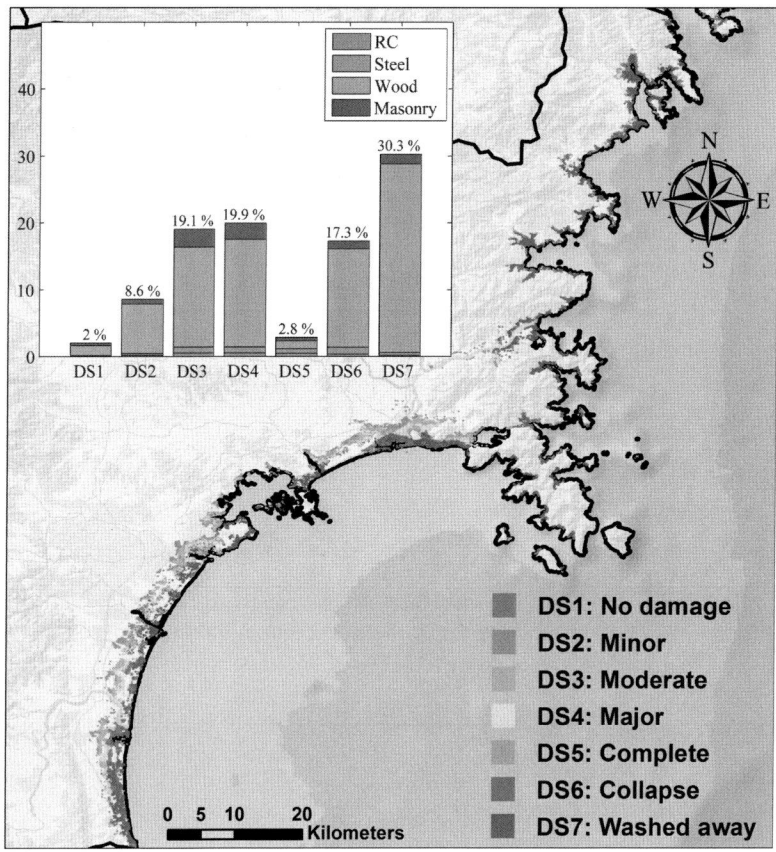

FIGURE 4.10 Spatial distribution of the MLIT damage data for the 2011 Tohoku Tsunami and distribution of damage states for different structural typologies. *MLIT*, Ministry of Land, Infrastructure, Transport, and Tourism. Map lines delineate study areas and do not necessarily depict accepted national boundaries.

TABLE 4.7 Damage state classification, descriptions, and conditions adopted by the Ministry of Land, Infrastructure, Transport, and Tourism.

DS	Classification	Description	Condition
1	No damage	There is no structural or nonstructural damage	Usable as usual
2	Minor damage	There is no significant structural or nonstructural damage, possibly only minor flooding. Specifically, only inundation below the ground floor.	Possible to be used immediately after minor floor and wall clean-up
3	Moderate damage	Slight damage to nonstructural components. The building is inundated less than 1 m above the floor.	Possible to use after moderate repair
4	Major damage	Heavy damage to some walls but no damage to columns. The building is inundated more than 1 m above the floor but below the ceiling.	Possible to use after major repairs
5	Complete damage	Heavy damage to several walls and some columns	Major work (retrofit) is required for the reuse of the building
6	Collapsed	Destructive damage to walls (more than half of the wall density) and several columns	Complete loss of functionality. Not repairable or retrofitting nonfinancially convenient.
7	Washed away	Only the foundation remained	Total reconstruction is needed

Fig. 4.10 also shows that the *DS* is independent of the structural typology (i.e., the definition of the *DS* remains the same for all investigated structural typologies). Therefore, definitions must be general enough to be applied to the entire portfolio of structures. For example, the definitions and conditions associated with each *DS* of the MLIT database are reported in Table 4.7, containing elements from the review done by Suppasri et al. (2015, 2014). Other studies (Dias et al., 2009; Tarbotton et al., 2015) considered more simplified *DS*s consisting mainly of three possible outcomes: partial damage but usable (*DS1*), partial damage but unusable (*DS2*), and complete damage (*DS3*).

In conclusion, when defining *DS*s, it is critical to ensure that the definitions meet the six criteria mentioned above to avoid inconsistent and site-specific *DS* classifications. Homogenizing *DS* classifications presented in multiple databases and used by multiple countries is still an unresolved issue.

4.5.2 Forms and typologies of vulnerability functions

The vulnerability analysis must be probabilistic. In performance-based tsunami engineering, vulnerability is often defined by functions that return some conditional probability of exceeding or attaining a *DS* for given values of the tsunami intensity measure (*IM*). Such functions are usually provided for specific building categories (i.e., timber, concrete, masonry, and steel buildings). Two possible functions can be used to describe the tsunami damage of buildings, that is, vulnerability and fragility.

The vulnerability functions relate the probability (*P*) that an expected loss (*L*) exceeds a specific level (*l*) to a specific *IM*

$$P(L \geq l | IM) \tag{4.30}$$

Eq. (4.30) is a typical analytical form of a vulnerability function. Fig. 4.11A shows the shape of the typical vulnerability function. The values of the vulnerability function should be interpreted as the mean damage ratio, that is, the percentage of the total value of the asset at risk that can be lost for a specific level of *IM*. Insurance and reinsurance companies often use vulnerability functions as they are directly derived from postevent claim data.

A fragility curve evaluates the probability (*P*) of reaching or exceeding specific *DS*s (*ds*) for a given hazard intensity (Porter et al., 2007):

$$P(DS \geq ds | IM) \tag{4.31}$$

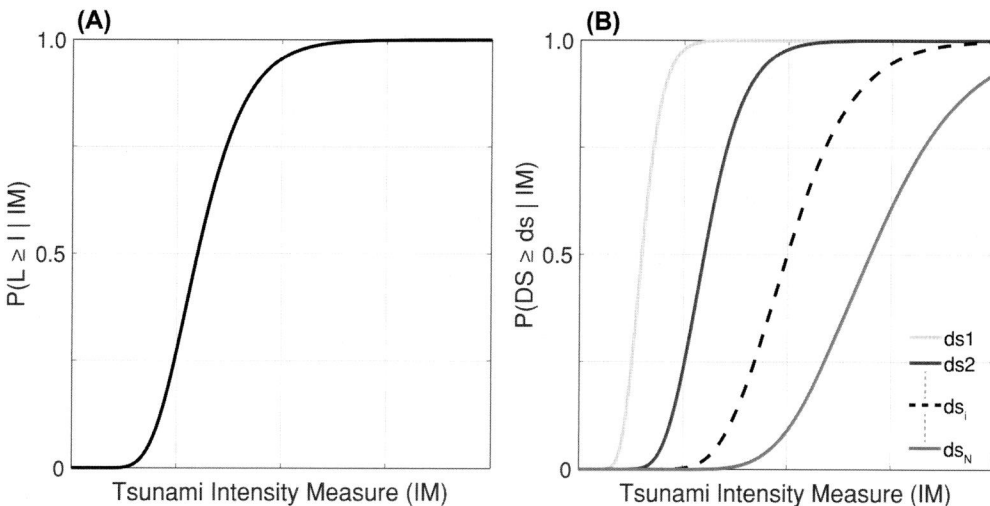

FIGURE 4.11 (A) Vulnerability function and (B) fragility functions.

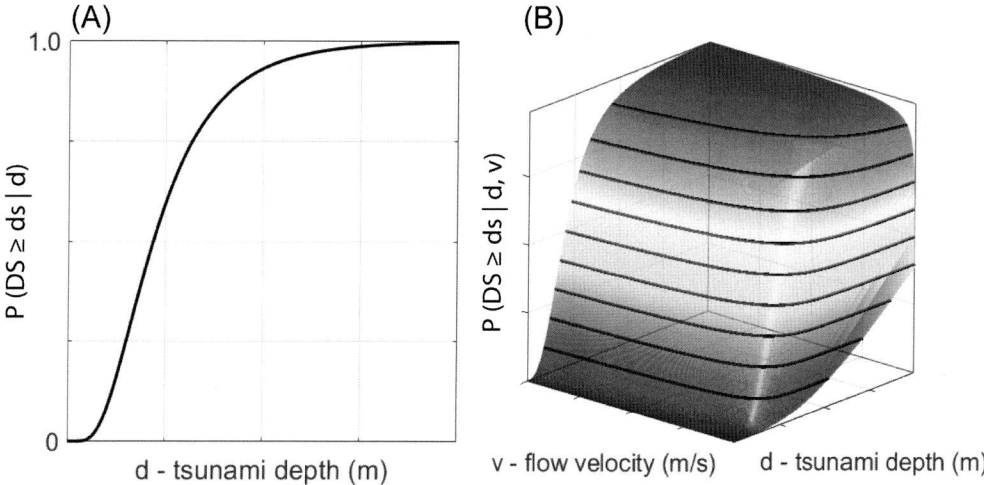

FIGURE 4.12 (A) Univariate and (B) multivariate fragility functions.

Eq. (4.31) is a typical form of a fragility function. Fig. 4.11B shows the shape of typical fragility functions associated with multiple *DS*s. The difference between two consecutive fragilities is the probability of being in a specific *DS*. Combining fragility curves and consequence (damage–loss) functions makes it possible to derive vulnerability curves (Rossetto & Ioannou, 2018).

The remaining part of this section will focus on the tsunami fragility curves and their derivation. Tsunami fragility functions can be univariate (i.e., a function of a single scalar *IM*) or multivariate (i.e., a function of multiple *IM*s or vectorial *IM*s). Fig. 4.12 shows generic examples of univariate and multivariate fragility curves, respectively. The tsunami inundation depth is the most adopted *IM* for the univariate fragility function for two reasons: first, because it is easy to measure this quantity in the aftermath of a tsunami, and second, because the tsunami depth relates to the simplest quantification of the tsunami force and therefore damage. For multivariate models, the flow velocity or momentum flux is usually also added. The typology of fragility is defined by the way the function is derived. In the following, two families of fragility functions, that is, empirical (Chapter 17) and analytical (Chapter 18), are focused on.

4.5.3 Empirical fragility functions

An empirical model consists of the relationship between some input variables describing the problem (known as explanatory variables or input variables) and some descriptive variables representative of the response of a structure subject to the perturbance (known as response variables or output variables). In the case of the fragility function, the output corresponds to the probability of reaching or exceeding a specific *DS* (Eq. (4.31)).

The input variables represent the hazard and the structure or asset at risk. In the case of tsunamis, the hazard-related input variables are the *IM*s, such as tsunami depth, velocity, and momentum flux. The input-related variables depend on the asset; for the case of buildings, these variables are the construction material (e.g., timber, concrete, steel, and masonry), the number of stories, etc. Among the input variables, it is also possible to introduce other factors that may influence the hazard, such as the coastal type (i.e., ria and plain coasts).

To build fragility models, databases detailing damage to structures following tsunami occurrences are necessary. These databases must encompass input variables (i.e., *IM*s and structural typology) and response variables (i.e., observed *DS*s). The collected data and the corresponding developed models are typically tailored to specific tsunami events and specific structural typologies available in particular countries.

If empirical estimation is impractical, simulation techniques are usually employed. For instance, assessing tsunami flow velocity from video recordings or buoy sensors is difficult, especially on urban or regional scales; thus tsunami inundation simulations are used together with empirical values. Characterizing affected assets involves a variety of governmental databases on built structures and volunteered geographic information. These data and information become crucial when structures are destroyed, making on-site characterization impossible during damage surveys. Similarly, geomorphological features are examined through topographical observations.

A detailed literature review of empirical vulnerability models for buildings is provided by Charvet et al. (2017). Several statistical models exist to develop fragility curves. The most adopted are the ordinary least squares method (OLSM) (Dias et al., 2009; Gokon et al., 2009; Hayashi et al., 2013; Koshimura & Gokon, 2012; Koshimura & Kayaba, 2010; Koshimura et al., 2009; Maruyama et al., 2014; Mas et al., 2012; Murao & Nakazato, 2010; Narita & Koshimura, 2015; Nihei et al., 2012; Peiris & Pomonis, 2004; Suppasri et al., 2009; Suppasri et al., 2011; Suppasri et al., 2014; Valencia et al., 2011) and the generalized linear model (GLM) (Charvet, Ioannou, et al., 2014; Charvet, Suppasri, et al., 2014; De Risi, Goda, Mori, et al., 2017; De Risi, Goda, Yasuda, et al., 2017; Lahcene et al., 2021; Macabuag et al., 2018; Macabuag, Rossetto, & Ioannou, 2016; Macabuag, Rossetto, Ioannou, et al., 2016; Reese et al., 2011; Song et al., 2017). Machine learning algorithms have recently been used (Mas et al., 2020; Rasheed et al., 2022).

OLSMs, rooted in the probability-paper approach, are the most conventional and frequently employed approach. This proves practical when dealing with limited data availability and when deriving univariate models (Fig. 4.12A). This approach involves binning data for each *DS* according to specific *IM* values, which are fitted with a normal or lognormal functional shape. GLMs allow univariate (Fig. 4.12A) and multivariate (Fig. 4.12B) model implementations, enabling the simultaneous utilization of multiple input variables. GLMs establish a link between damage probability and linear predictors (e.g., probit and logit). The predictors represent a straightforward linear combination of the explanatory variables, as the regression coefficients dictate. However, it is worth noting that GLMs require large databases to perform effectively. Two key differences exist between OLSMs and GLMs. OLSMs need a preliminary database sorting to identify specific fragilities for predefined *DS*s and other input variables, such as structural typology and number of stories. GLMs, on the other hand, allow all the input variables at once, employing dummy variables that work as categorical inputs. Finally, among machine-learning algorithms, fragility models are developed by using an artificial neural network, where the input explanatory variables are organized into the input layer. In contrast, the output layer delivers the damage classification.

The availability of a plethora of methods and models requires model diagnosis and selection. These approaches can involve conducting statistical tests, such as the likelihood ratio test, and assessing indicators, such as the Bayesian information criterion (BIC) (Schwarz, 1978), the Akaike information criterion (AIC) (Akaike, 1974), and residual deviance. Typically, models with the smallest diagnostic indicators are preferred. In the following, three of the most conventional fitting approaches are explained.

4.5.3.1 Lognormal method

In the lognormal method, exceedance probabilities for *DS*s are calculated, and median values are plotted against a range of equally spaced bins of tsunami hazard parameters (e.g., the inundation depth interval of 0.5 m). The probability of occurrence of damage is

$$P(DS \geq ds | IM) = \Phi\left(\frac{\log IM - \log \eta}{\beta}\right) \tag{4.32}$$

where $\Phi(\bullet)$ is the standard normal distribution function, η is the median, and β is the logarithmic standard deviation (or dispersion parameter). The two parameters η and β are obtained by plotting the logarithm of inundation depth versus the inverse cumulative distribution function of the exceedance probability and by performing a linear regression analysis according to the following relation (Fig. 4.13A)

$$\ln IM = \ln \eta + \beta \cdot \Phi^{-1}[P(DS \geq ds|IM)] + \varepsilon_R \qquad (4.33)$$

where ε_R is the term representing the regression error, which is distributed according to the normal distribution with zero mean and standard deviation σ_R. In Eq. (4.33), $P(DS \geq ds|IM)$ can be obtained by analyzing damage outcomes for each inundation depth bin. It follows that ln IM is normally distributed with a mean function of parameters η and β, and standard deviation equal to σ_R. The linear regression is generally carried out through a least squares fitting procedure by minimizing the squares of residuals between empirical data and calculated values. Fig. 4.13B shows an example of a logarithmic fragility curve for which the confidence interval is also presented on the basis of the error of the linear regression as explained above.

4.5.3.2 Binomial logistic method

Logistic regression is a special case of a GLM and can be used for developing empirical fragility functions using binomial data. For each DS, individual building damage survey results provide a binary indication of whether the considered DS is exceeded or not and the maximum water depth at the building site. Unlike the lognormal model, the data cannot be organized in bins necessarily. However, the regression works also for binned data. Fig. 4.14 shows the two cases of binned and unbinned fitting.

In the *binned* approach, let π_i denote the probability that for the ith bin of observed M_i, structures are diagnosed as attaining a specific damage state ds. The probability that all observed bins are classified with ds is

$$\prod_{i=1}^{n} \binom{M_i}{y_i} \cdot \pi_i^{y_i} \cdot (1-\pi_i)^{M_i - y_i} \qquad (4.34)$$

where M_i is the total number of structures populating a specific bin, and y_i is the number of structures of the bin that can be categorized with the ith DS. Therefore, Eq. (4.34) is the likelihood function representing the probability of observed data. The term π may assume different forms, such as probit, logit, and loglog (Hosmer et al., 2013). For example, the logit function is

$$\pi_i = \frac{\exp(b_1 + b_2 \cdot \log IM_i)}{1 + \exp(b_1 + b_2 \cdot \log IM_i)} \qquad (4.35)$$

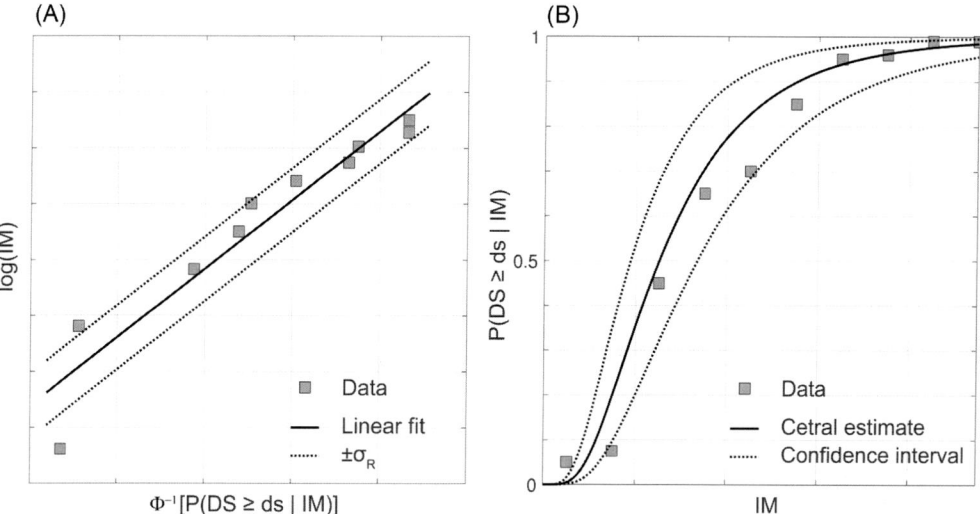

FIGURE 4.13 (A) Probability paper and (B) final fragility with a confidence interval.

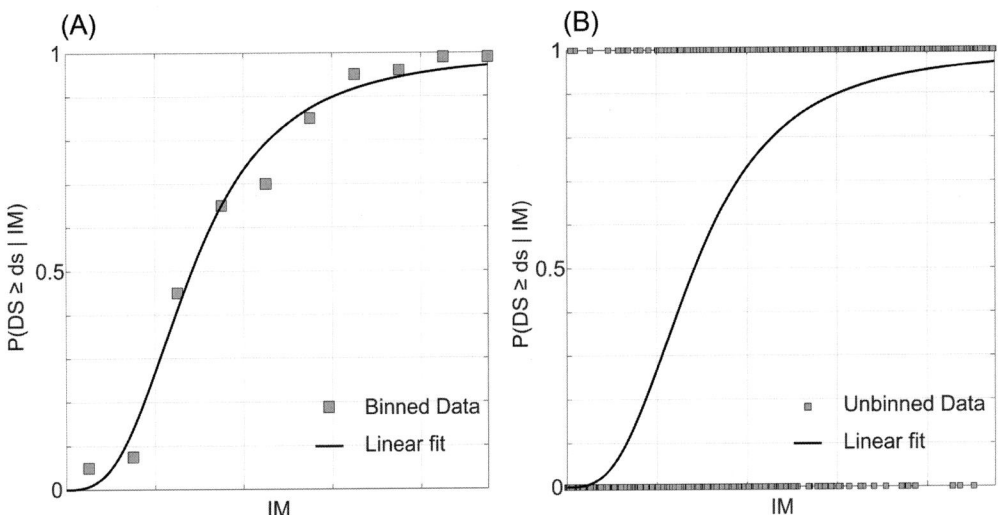

FIGURE 4.14 (A) Binned regression and (B) unbinned regression.

where b_1 and b_2 are the model parameters. The point estimation is carried out through the maximum likelihood estimation (MLE).

In the *unbinned* approach, let π_i denote the probability that the ith observed building is diagnosed as attaining a specific damage state ds. The probability that all observed buildings are classified with ds is

$$\prod_{i=1}^{n} \binom{1}{y_i} \cdot \pi_i^{y_i} \cdot (1-\pi_i)^{1-y_i} \tag{4.36}$$

where y_i is equal to 1 if the ith observation falls in the examined DS, and it is zero otherwise. Therefore, Eq. (4.36) is the likelihood function representing the probability of observed data. Also, in this case, the term π may assume different forms, such as probit, logit, and loglog (Hosmer et al., 2013).

4.5.3.3 Multinomial logistic method

This method is an extension of binomial logistic regression presented above to the case of multiclass problems (i.e., *polytomous* response), belonging to the family of GLMs. This regression can handle more than two DSs simultaneously and does not require data binning. It considers the ordered and hierarchical nature of DSs, avoiding inconsistent results, such as the intersection of fragility functions. Previous studies typically aggregate the damage data in bins having similar tsunami intensity values, and then they fit a simple or generalized linear statistical model to the binned data. This approach may lead to biased estimates of the regression parameters due to binning. For instance, grouping in bins can affect fragility curves, especially in the tails, and bins that present extreme probabilities (i.e., 0 and 1, which correspond to no damage and complete damage, respectively) are systematically dismissed during the classical statistical linear fitting (Charvet, Ioannou, et al., 2014; Charvet, Suppasri, et al., 2014). Furthermore, the classical linear least squares fitting is not recommended in the case of discrete and ordinal variables, such as tsunami DSs.

In the *binned* approach, denoting the probability that structures corresponding to the ith observation data bin fall in the jth DS by π_{ij}, the probability that all buildings of the ith bin fall in the respective DS class is given by the multinomial probability distribution (Charvet, Suppasri, et al., 2014; De Risi, Goda, Yasuda, et al., 2017)

$$\frac{m_i!}{\prod_{j=1}^{k} y_{ij}!} \prod_{j=1}^{k} \pi_{ij}^{y_{ij}} \tag{4.37}$$

where m_i is the total number of structures composing the ith observation bin, k is the number of DSs, and y_{ij} is the number of structures for the ith observation bin attaining ds_j. In Eq. (4.37), π_{ij} may take different functions (e.g., probit, logit, and loglog) and may assume different features (e.g., nominal, ordinal, and hierarchical), as described in McCullagh and Nelder (1983). For example, the logit function can be considered, and the hierarchical partially ordered approach can be adopted for regression. In the hierarchical method, π_{ij} assumes the following form

$$\pi_{ij} = \frac{\exp(b_{1,j} + b_{2,j} \cdot \ln IM_i)}{1 + \exp(b_{1,j} + b_{2,j} \cdot \ln IM_i)} \cdot \left(1 - \sum_{l=1}^{j-1} \pi_{il}\right) \quad (4.38)$$

where $b_{1,j}$ and $b_{2,j}$ are the model parameters (i.e., the intercept and the slope of the GLM, respectively); the approach is partially ordered because the model parameters are different for all considered DSs. Considering k damage states, it is possible to write $k-1$ sets of Eq. (4.38). The first equation presents only the fraction (i.e., no term in parentheses), and Eq. (4.38) for the kth DS is equal to one. Point estimates of the model parameters are calculated, in accordance with the MLE approach, by computing the first and second derivatives of the likelihood (or log-likelihood) function that is expressed as follows

$$\prod_{i=1}^{n} \prod_{j=1}^{k} \pi_{ij}^{y_{ij}} \quad (4.39)$$

where n is the number of bins.

In the *unbinned* approach, consider that the DS that takes, for example, one of six discrete values, that is, $ds1, ds2, \ldots,$ and $ds6$. Let

$$\pi_{ij} = P(DS_i = ds_j) \quad (4.40)$$

denote the probability that the ith observation falls in the jth category. As the DSs are mutually exclusive and completely exhaustive, the sum of the DS probabilities equals one for each observation. The probability that the ith building falls in the respective DS is given by the multinomial probability distribution

$$\frac{1}{\prod_{j=1}^{6} y_{ij}!} \cdot \prod_{j=1}^{6} \pi_{ij}^{y_{ij}} \quad (4.41)$$

where y_{ij} is 1 or 0 for DS attainment or not, respectively. The distribution shown in Eq. (4.41) represents the random component of the model, which describes the distribution of the response around the central value.

The systematic component of the model relates the probability π_{ij} and a vector of explanatory variables x and is represented by a link function. Usually, it is a linear function of explanatory variables

$$f(\pi_{ij}) = \theta_{j,0} + \sum_{k=i}^{p} \theta_{j,k} \cdot x_k \quad (4.42)$$

where θ is the vector of the model regression parameters $\theta_{j,k}$, and p is the number of explanatory variables. When the model regression parameters are the same for all DSs (except for the intercept $\theta_{j,0}$) and the resultant fragility curves have the same slope, the model is referred to as ordered. Relaxing the previous constraint, the model is called partially ordered. Several link functions are suggested in the literature (Hosmer et al., 2013), including probit and logit models

$$\text{Probit: } f(\pi_{ij}) = \Phi^{-1}(\pi_{ij})$$

$$\text{Logit: } f(\pi_{ij}) = \log\left(\frac{\pi_{ij}}{1 - \sum_{k=1}^{j} \pi_{ik}}\right) \quad (4.43)$$

where $\Phi(\bullet)$ represents the standard normal cumulative distribution function. Depending on the link function, the regression procedure is commonly known as multinomial logit or multinomial probit regression.

The point estimates for the model parameters are obtained on the basis of the MLE approach by computing the first and second derivatives of the likelihood function that is expressed as follows

$$L(\theta|x,y) = \prod_{i=1}^{n} \prod_{j=1}^{6} \pi_{ij}^{y_{ij}} \quad (4.44)$$

where n is the number of data points.

4.5.3.4 Functional forms of the link function

The functional form presented in Eq. (4.42) allows the creation of probabilistic models with explanatory variables that are not necessarily continuous. In fact, they can be used to implement categorical aspects (e.g., structural typology and coastal typology) using dummy variables. Also, cross-terms can be implemented by representing the interaction between the different explanatory variables. As an example, a functional form used in De Risi, Goda, Yasuda, et al. (2017) is reported as follows

$$f(\pi_{ij}) = \theta_{j,0} + \theta_{j,1}\log(IM1_i) + \theta_{j,2} \cdot d_W + \theta_{j,3} \cdot d_M + \theta_{j,4} \cdot d_S + \cdots$$
$$+ \theta_{j,5}\log(IM1_i) \cdot d_W + \theta_{j,6}\log(IM1_i) \cdot d_M + \theta_{j,7}\log(IM1_i) \cdot d_S + \cdots$$
$$+ \theta_{j,8}\log(IM2) + \theta_{j,9}\log(IM2_i) \cdot d_W + \theta_{j,10}\log(IM2_i) \cdot d_M + \theta_{j,11}\log(IM2_i) \cdot d_S \quad (4.45)$$

where $IM1$ is the tsunami depth, $IM2$ is the tsunami velocity, d_W, d_M, and d_S are the dummy variables for wood, masonry, and steel structures, and $\theta_{j,k}$ are the regression parameters. Note that this model included also reinforced concrete structures simply imposing d_W, d_M, and d_S to zero.

4.5.3.5 Model selection

Different explanatory variables (and their combinations) can be taken into account for the systematic component, leading to several statistical models. In particular, it is necessary to identify the model that provides the best fit in comparison with the available alternatives. A diagnostic analysis or model selection can be performed by assessing the relative goodness-of-fit of the candidate models. The model selection can be conducted at two levels, identifying the best link function for the same model (i.e., two models with the same linear predictors but different link functions) and the best model among several candidate nested models for the same link function (i.e., the more complex model includes at least all the parameters of its simpler counterpart).

Two criteria, that is, the BIC and the AIC, can be used

$$\text{BIC} = -2\ln[L(x,y|\theta)] + k\ln n \quad (4.46)$$

$$\text{AIC} = -2\ln[L(x,y|\theta)] + 2k \quad (4.47)$$

where $L(x,y|\theta)$ denotes the data likelihood under the MLE of a candidate model, k is the number of regression parameters θ, and n is the number of observations. The model that presents the smallest value of BIC or AIC is considered to provide a better fit to the available data. For selecting a suitable nested model, several statistical tests can be used to assess the relative goodness-of-fit (e.g., the likelihood-ratio test, the F test, and the analysis of variance). These tests are unreliable for unbinned binary data (Hosmer et al., 2013; McCullagh & Nelder, 1983). For this reason, the preceding diagnostic criteria are used consistently. In the context of model selection, generally, BIC tends to choose models that are more parsimonious than those favored by AIC.

In addition, the residual deviance G^2 can be adopted to measure the overall model performance with respect to the data. This measure compares the proposed model (i.e., a model with a small number of parameters) to a saturated one (i.e., a model with parameters equal to the number of observations) and is expressed as follows

$$G^2 = 2\sum_{i=1}^{n}\sum_{j=1}^{6} y_{ij}\ln\frac{y_{ij}}{\pi_{ij}} \quad (4.48)$$

A model with smaller deviance is preferred (i.e., the model is close to the saturated one, presenting zero residual deviance). It is worth noting that for unbinned data, there are no overdispersion issues (i.e., model dispersion is greater than data dispersion). In fact, when overdispersion occurs, the overall goodness-of-fit is distorted. This problem has never been treated rigorously in the previous studies related to tsunami fragility model selection.

4.5.4 Analytical fragility functions

When referring to analytical fragility functions (or, for brevity, analytical fragilities), it is essential to mention that those are mainly analytical "physical" functions. They are "analytical" because they are provided as analytical functional forms that are based on the results obtained from a physical model through numerical simulations.

The first aspect that distinguishes the analytical fragilities from the empirical ones is that these fragilities are generally computed for single structures for which specific design details and blueprints are available. Therefore, they can be produced for evacuation shelters, for example, or critical infrastructure. If the analyzed model structure represents an entire portfolio, extending the curves to the population of elements under scrutiny is possible.

The first step is to create a robust physical model of the structure under assessment. Typical physical models of buildings are realized with finite element modeling (FEM) software. Typical structural software allowing an FEM implementation includes OpenSees (McKenna et al., 2010; McKenna, 2011), SAP2000 (Computers & Structures, Inc., 1997), MIDAS (2020), and SeismoStruct (2022). The models should be first validated to make sure they are technically sound and capture realistic structural behavior under different load conditions. Then, they should be capable of modeling the tsunami loading as described in the previous part of this chapter.

The structural analysis can be implemented as static or dynamic and linear or nonlinear. In principle, it should be possible to study the problem in the most sophisticated way possible. During the analysis, multiphysics should be implemented by considering at the same time the dynamic nature of the input forcing coming from the tsunami wave and the dynamic nonlinear behavior of the structure under scrutiny. In such a sophisticated approach, the dynamic hydrodynamic behavior of the tsunami is affected by the dynamic response of the structure that can change its geometry during the analysis due to possible local, partial, or global failures. However, such an approach is complicated and takes a long time to perform; therefore, simplified approaches must be adopted. The forcing systems presented in Section 4.3 are the most important aspects to consider when defining the simplification presented below.

Four main structural analysis typologies are available in the literature: (1) the constant depth pushover (CDPO), (2) the variable depth pushover (VDPO), (3) the modified VDPO2, and (4) the time-history analysis. It is critical to observe that all the methods for fitting the fragility functions were originally developed for earthquake engineering applications (i.e., cloud analysis and incremental dynamic analysis) (Jalayer et al., 2015). Regarding seismic actions, it is also important to realize that a tsunami may have previously struck structures subjected to an earthquake. Only a few studies have considered the sequence of earthquakes and tsunamis (Petrone et al., 2020). The CDPO (Alam et al., 2018; Attary et al., 2017) adopts a specific maximum inundation depth and transforms this to a known loading shape (i.e., hydrostatic or hydrodynamic). Then a traditional displacement-based pushover analysis is performed. In this analysis, the forces are adapted in magnitude to better capture the nonlinear behavior of the structure due to the progressive distribution of the plasticity among the structural elements. The VDPO, proposed by Petrone et al. (2017), consists of increasing the inundation level progressively (with a consequent adaptation of the corresponding flow velocity by keeping the Froude number constant) and analyzing the structure under force-based control, that is, only the peak strength can be captured and not the possible softening behavior. This was initially deemed acceptable, considering the impulsive action of the first wave loading. The VDPO2, proposed by Baiguera et al. (2022) and suggested by American Society of Civil Engineers (2022), consists of a two-step nonlinear analysis. First, the conventional VDPO is executed, and then the analysis switches from force-based to displacement-based so that it is possible to capture the softening behavior. Finally, the time-history analysis can be performed when tsunami simulations are available in terms of wave depth and velocity time histories (Goda et al., 2017). These time histories can be converted into dynamic loading on the structural system in a dynamic regime; therefore, the dynamic behavior of the structural system is also accounted for.

4.6 Research needs

This review of the state-of-the-art tsunami effects on infrastructure and of the current guidelines used to design critical infrastructure in tsunami-prone areas provided several potential avenues for future research that would hopefully improve the current design recommendations and prescriptions. These potential directions of research are outlined as follows:

1. The present formulae used in the calculations of the various tsunami-induced forces exerted on structures only consider rigid structures. Future investigations using flexible, nonrigid structures would provide a more realistic understanding and quantification of these forces.
2. It may be necessary to incorporate a variable-depth nonlinear pushover tsunami analysis technique in the provisions for performance-based design options. In addition, the effect of the preceding seismic-induced nonlinear behavior of structures should be considered in future research on tsunami response.
3. At present, the effect of debris impact assumes a deterministic approach; all structures in the potential impact area are considered for the maximum impact probability. However, the randomness of debris propagation and impact

occurrence hints at the need for research toward establishing a probabilistic framework for debris generation, propagation, and impact.
4. Due to limited field data and inadequacies of the current experimental facilities, present design standards and guidelines require significant improvements in adequately assessing the spatiotemporal extent and magnitude of scour as well as its effects on structures subjected to tsunami inundation. In addition, the influence of soil liquefaction on scour and structural stability needs to be investigated to a much greater extent.
5. There is an acute need to investigate scaling laws for experimental work on tsunami effects in structures: from correctly reproducing the duration of tsunami inundation to the use of properly scaled-down structures and surrounding foundation soil characteristics.

References

Akaike, H. (1974). A new look at the statistical model identification. *IEEE Transactions on Automatic Control, 19*(6), 716–723. Available from https://doi.org/10.1109/TAC.1974.1100705.

Alam, M. S., Barbosa, A. R., Scott, M. H., Cox, D. T., & van de Lindt, J. W. (2018). Development of physics-based tsunami fragility functions considering structural member failures. *Journal of Structural Engineering, 144*(3). Available from https://doi.org/10.1061/(ASCE)ST.1943-541X.000195.

American Concrete Institute. (2013). *Code requirements for nuclear safety-related concrete structures and commentary ACI CODE-349-13*.

American Society of Civil Engineers. (2013). *Minimum design loads and associated criteria for buildings and other structures ASCE/SEI7-10*. https://doi.org/10.1061/9780784412916.

American Society of Civil Engineers. (2022). *Minimum design loads and associated criteria for buildings and other structures (Chapter 6) tsunami loads and effects ASCE/SEI7-22*. https://doi.org/10.1061/9780784415788.

American Society of Civil Engineers. (2017). *Minimum design loads for buildings and other structures Tsunami loads and effects ASCE/SEI 7-16*. https://doi.org/10.1061/9780784414248.

Arnason, H., Petroff, C., & Yeh, H. (2009). Tsunami bore impingement onto a vertical column. *Fuji Technology Press, Iceland Journal of Disaster Research, 4*(6), 391–403. Available from https://doi.org/10.20965/jdr.2009.p0391.

Asakura, R., Iwase, K., Ikeya, T., Takao, M., Kaneto, T., Fujii, N., & Omori, M. (2000). An experimental study on wave force acting on on-shore structures due to overflowing tsunamis. *Proceedings of Coastal Engineering*, 911–915.

ASTM International. (2011). *Practice for establishing clear wood strength values ASTMD2555-17*. https://doi.org/10.1520/D2555-06.

Attary, N., Unnikrishnan, V. U., van de Lindt, J. W., Cox, D. T., & Barbosa, A. R. (2017). Performance-based tsunami engineering methodology for risk assessment of structures. *Engineering Structures, 141*, 676–686. Available from https://doi.org/10.1016/j.engstruct.2017.03.071, http://www.journals.elsevier.com/engineering-structures/.

Baiguera, M., Rossetto, T., Robertson, I. N., & Petrone, C. (2022). A procedure for performing nonlinear pushover analysis for tsunami loading to ASCE 7. *Journal of Structural Engineering, 148*(2). Available from https://doi.org/10.1061/(ASCE)ST.1943-541X.0003256.

Blevins, R. D. (1984). *Applied fluid dynamics handbook*. Van Nostrand Reinhold Co.

Cabinet Office of Japanese Government. (2005). Guidelines concerning the tsunami evacuation building.

Charvet, I., Ioannou, I., Rossetto, T., Suppasri, A., & Imamura, F. (2014). Empirical fragility assessment of buildings affected by the 2011 Great East Japan tsunami using improved statistical models. *Natural Hazards, 73*(2), 951–973. Available from https://doi.org/10.1007/s11069-014-1118-3, http://www.wkap.nl/journalhome.htm/0921-030X.

Charvet, I., Macabuag, J., & Rossetto, T. (2017). Estimating tsunami-induced building damage through fragility functions: Critical review and research needs. *Frontiers in Built Environment, 3*. Available from https://doi.org/10.3389/fbuil.2017.00036, https://www.frontiersin.org/articles/10.3389/fbuil.2017.00036/pdf.

Charvet, I., Suppasri, A., & Imamura, F. (2014). Empirical fragility analysis of building damage caused by the 2011 Great East Japan tsunami in Ishinomaki city using ordinal regression, and influence of key geographical features. *Stochastic Environmental Research and Risk Assessment, 28*(7), 1853–1867. Available from https://doi.org/10.1007/s00477-014-0850-2, http://link.springer-ny.com/link/service/journals/00477/index.htm.

Chock, G., Carden, L., Robertson, I., Olsen, M., & Yu, G. (2013). Tohoku tsunami-induced building failure analysis with implications for U.S. tsunami and seismic design codes. *Earthquake Spectra, 29*, 99–126. Available from https://doi.org/10.1193/1.4000113.

Chock, G., Robertson, I. N., Kriebel, D. L., Francis, M., & Nistor, I. (2013). *Tohoku, Japan, Earthquake and Tsunami of 2011: Performance of structures under tsunami loads*. American Society of Civil Engineers.

City and County of Honolulu. (2013). *Regulations within flood hazard districts and development adjacent to drainage facilities (Article 11) city and county of Honolulu building code*. (pp. 115–120).

Computers & Structures, Inc. (1997). *SAP2000 Structural analysis and design*. https://www.csiamerica.com/products/sap2000.

Council on Earthquake Disaster Prevention. (1933). *Note on prevention against tsunamis*.

Dawson, A. G., Lockett, P., & Shi, S. (2004). Tsunami hazards in Europe. *Environment International, 30*(4), 577–585. Available from https://doi.org/10.1016/j.envint.2003.10.005, https://www.sciencedirect.com/science/article/pii/S0160412003002101.

De Risi, R., Goda, K., Mori, N., & Yasuda, T. (2017). Bayesian tsunami fragility modeling considering input data uncertainty. *Stochastic Environmental Research and Risk Assessment, 31*(5), 1253–1269. Available from https://doi.org/10.1007/s00477-016-1230-x, http://link.springer-ny.com/link/service/journals/00477/index.htm.

De Risi, R., Goda, K., Yasuda, T., & Mori, N. (2017). Is flow velocity important in tsunami empirical fragility modeling? *Earth-Science Reviews*, *166*, 64–82. Available from https://doi.org/10.1016/j.earscirev.2016.12.015, http://www.sciencedirect.com/science/journal/00128252.

Del Zoppo, M., Rossetto, T., Di Ludovico, M., & Prota, A. (2023). Effect of buoyancy loads on the tsunami fragility of existing reinforced concrete frames including consideration of blow-out slabs. *Scientific Reports*, *13*(1). Available from https://doi.org/10.1038/s41598-023-36237-7, https://www.nature.com/srep/.

Dias, W. P. S., Yapa, H. D., & Peiris, L. M. N. (2009). Tsunami vulnerability functions from field surveys and Monte Carlo simulation. *Civil Engineering and Environmental Systems*, *26*(2), 181–194. Available from https://doi.org/10.1080/10286600802435918, https://www.tandfonline.com/doi/full/10.1080/10286600802435918, Sri Lanka.

Federal Emergency Management Agency. (2011). *Coastal construction manual*. https://www.fema.gov/sites/default/files/2020-08/fema55_vol1_combined.pdf.

Federal Emergency Management Agency. (2019). *Guidelines for design of structures for vertical evacuation from tsunamis*. https://www.fema.gov/sites/default/files/2020-08/fema_earthquakes_guidelines-for-design-of-structures-for-vertical-evacuation-from-tsunamis-fema-p-646.pdf.

Ge, M., & Robertson, I. N. (2010). *Uplift loading on elevated floor slab due to a tsunami bore UHM/CEE/10-03*. University of Hawaii. http://www.cee.hawaii.edu/wp-content/uploads/UHM-CEE-10-03.pdf.

Goda, K., Petrone, C., De Risi, R., & Rossetto, T. (2017). Stochastic coupled simulation of strong motion and tsunami for the 2011 Tohoku, Japan earthquake. *Stochastic Environmental Research and Risk Assessment*, *31*(9), 2337–2355. Available from https://doi.org/10.1007/s00477-016-1352-1, http://link.springer-ny.com/link/service/journals/00477/index.htm.

Goda, Y. (2011). *Random seas and design of maritime structures* (15). World Scientific. Available from https://doi.org/10.1142/3587.

Gokon, H., Koshimura, S., Matsuoka, M., & Namegaya, Y. (2009). Developing tsunami fragility curves due to the 2009 tsunami disaster in American Samoa. *Journal of Japan Society of Civil Engineers (Coastal Engineering)*, *67*(2), 1321–1325. Available from https://doi.org/10.2208/kaigan.67.I_1321.

Haehnel, R. B., & Daly, S. F. (2004). Maximum impact force of woody debris on floodplain structures. *Journal of Hydraulic Engineering*, *130*(2), 112–120. Available from https://doi.org/10.1061/(ASCE)0733-9429(2004)130:2(112).

Hayashi, S., Narita, Y., & Koshimura, S. (2013). Developing tsunami fragility curves from the surveyed data and numerical modeling of the 2011 Tohoku earthquake tsunami. *Journal of Japan Society of Civil Engineers (Coastal Engineering)*, *69*(2), 386–390. Available from https://doi.org/10.2208/kaigan.69.i_386.

Hill, M., & Rossetto, T. (2008). Comparison of building damage scales and damage descriptions for use in earthquake loss modelling in Europe. *Bulletin of Earthquake Engineering*, *6*(2), 335–365. Available from https://doi.org/10.1007/s10518-007-9057-y.

Hosmer, D. W., Lemeshow, S., & Sturdivant, R. X. (2013). *Applied logistic regression*. Wiley. Available from https://doi.org/10.1002/9781118548387.

Jalayer, F., De Risi, R., & Manfredi, G. (2015). Bayesian cloud analysis: Efficient structural fragility assessment using linear regression. *Bulletin of Earthquake Engineering*, *13*(4), 1183–1203. Available from https://doi.org/10.1007/s10518-014-9692-z, https://rd.springer.com/journal/10518.

Japan Electric Association. (2015). *Technical guidelines for seismic design of nuclear power plants JEAC 4601-2015*.

Japan Electric Association. (2021). *Technical code for tsunami design of nuclear power plants JEAC 4629-2021*.

Japan Society of Civil Engineers. (2000). *Design manual for coastal facilities*.

Koshimura, S., & Gokon, H. (2012). Structural vulnerability and tsunami fragility curves from the 2011 Tohoku earthquake tsunami disaster. *Journal of Japan Society of Civil Engineers (Coastal Engineering)*, *68*(2), 336–340. Available from https://doi.org/10.2208/kaigan.68.i_336.

Koshimura, S., & Kayaba, S. (2010). Tsunami fragility inferred from the 1993 Hokkaido Nansei-Oki earthquake tsunami disaster. *Journal of Japan Association for Earthquake Engineering*, *10*(3), 87–101. Available from https://doi.org/10.5610/jaee.10.3_87.

Koshimura, S., Oie, T., Yanagisawa, H., & Imamura, F. (2009). Developing fragility functions for tsunami damage estimation using numerical model and post-tsunami data from Banda Aceh, Indonesia. *Coastal Engineering Journal*, *51*(3), 243–273. Available from https://doi.org/10.1142/S0578563409002004.

Kuilanoff, G., & Drake, R.M. (1991). *Design of DOE facilities for wind-generated missiles*. In Third DOE Natural Phenomena Hazards Mitigation Conference. https://inis.iaea.org/collection/NCLCollectionStore/_Public/26/001/26001517.pdf.

Lahcene, E., Ioannou, I., Suppasri, A., Pakoksung, K., Paulik, R., Syamsidik, S., Bouchette, F., & Imamura, F. (2021). Characteristics of building fragility curves for seismic and non-seismic tsunamis: case studies of the 2018 Sunda Strait, 2018 Sulawesi—Palu, and 2004 Indian Ocean tsunamis. *Natural Hazards and Earth System Sciences*, *21*(8), 2313–2344. Available from https://doi.org/10.5194/nhess-21-2313-2021.

Lebreton, L. C. M., Greer, S. D., & Borrero, J. C. (2012). Numerical modelling of floating debris in the world's oceans. *Marine Pollution Bulletin*, *64*(3), 653–661. Available from https://doi.org/10.1016/j.marpolbul.2011.10.027.

Macabuag, J., Rossetto, T., & Ioannou, I. (2016). *Investigation of the effect of debris-induced damage for constructing tsunami fragility curves for buildings*. In Proceedings of the 1st International Conference on Natural Hazards & Infrastructure.

Macabuag, J., Rossetto, T., Ioannou, I., Suppasri, A., Sugawara, D., Adriano, B., Imamura, F., Eames, I., & Koshimura, S. (2016). A proposed methodology for deriving tsunami fragility functions for buildings using optimum intensity measures. *Natural Hazards*, *84*(2), 1257–1285. Available from https://doi.org/10.1007/s11069-016-2485-8, http://www.wkap.nl/journalhome.htm/0921-030X.

Macabuag, J., Rossetto, T., Ioannou, I., & Eames, I. (2018). Investigation of the effect of debris-induced damage for constructing tsunami fragility curves for buildings. *Geosciences*, *8*(4), 117. Available from https://doi.org/10.3390/geosciences8040117.

Madurapperuma, M. A. K. M., & Wijeyewickrema, A. C. (2013). Response of reinforced concrete columns impacted by tsunami dispersed 20′ and 40′ shipping containers. *Engineering Structures*, *56*, 1631–1644. Available from https://doi.org/10.1016/j.engstruct.2013.07.034.

Martínez-Gomariz, E., Gómez, M., Russo, B., & Djordjević, S. (2018). Stability criteria for flooded vehicles: A state-of-the-art review. *Journal of Flood Risk Management*, *11*, S817–S826. Available from https://doi.org/10.1111/jfr3.12262.

Maruyama, Y., Kitamura, K., & Yamazaki, F. (2014). *Tsunami damage assessment of buildings in Chiba Prefecture, Japan using fragility function developed after the 2011 Tohoku-Oki Earthquake*. In Proceedings of Safety, Reliability, Risk and Life-Cycle Performance of Structures & Infrastructures (pp. 4237–4244). 978-1-138-00086-5. https://doi.org/10.1201/B16387-613.

Mas, E., Koshimura, S., Suppasri, A., Matsuoka, M., Matsuyama, M., Yoshii, T., Jimenez, C., Yamazaki, F., & Imamura, F. (2012). Developing tsunami fragility curves using remote sensing and survey data of the 2010 Chilean Tsunami in Dichato. *Natural Hazards and Earth System Science*, *12*(8), 2689–2697. Available from https://doi.org/10.5194/nhess-12-2689-2012.

Mas, E., Paulik, R., Pakoksung, K., Adriano, B., Moya, L., Suppasri, A., Muhari, A., Khomarudin, R., Yokoya, N., Matsuoka, M., & Koshimura, S. (2020). Characteristics of tsunami fragility functions developed using different sources of damage data from the 2018 Sulawesi earthquake and tsunami. *Pure and Applied Geophysics*, *177*(6), 2437–2455. Available from https://doi.org/10.1007/s00024-020-02501-4.

McCullagh, P., & Nelder, J. A. (1983). *Generalized linear models*. Chapman and Hall.

McKenna, F. (2011). OpenSees: A framework for earthquake engineering simulation. *Computing in Science and Engineering*, *13*(4), 58–66. Available from https://doi.org/10.1109/MCSE.2011.66.

McKenna, F., Scott, M. H., & Fenves, G. L. (2010). Nonlinear finite-element analysis software architecture using object composition. *Journal of Computing in Civil Engineering*, *24*(1), 95–107. Available from https://doi.org/10.1061/(ASCE)CP.1943-5487.0000002.

MIDAS. (2020). *Gen & civil: Online manuals* https://globalsupport.midasuser.com/helpdesk/KB/View/32610193-midas-gen-manuals-and-tutorials.

Mikami, T., Shibayama, T., Esteban, M., & Matsumaru, R. (2012). Field survey of the 2011 Tohoku earthquake and tsunami in Miyagi and Fukushima prefectures. *Coastal Engineering Journal*, *54*(1). Available from https://doi.org/10.1142/S0578563412500118.

Ministry of Land, Infrastructure, Transport and Tourism. (2011a). *Concerning setting the safe structure method for tsunamis which are presumed when tsunami inundation occurs*. MLIT Notification No. 1318.

Ministry of Land, Infrastructure, Transport and Tourism. (2011b). *Guidelines on the structural design of tsunami evacuation buildings*. MLIT Technical Advice No. 2570.

Ministry of Land, Infrastructure, Transport and Tourism. (2013). *Tsunami resistant design guideline for breakwaters*.

Murao, O., & Nakazato, H. (2010). *Vulnerability functions for buildings based on damage survey data in Sri Lanka after the 2004 Indian Ocean tsunami*. In Proceedings of the International Conference on Sustainable Built Environment (pp. 371–378).

Naito, C., Cercone, C., Riggs, H. R., & Cox, D. (2014). Procedure for site assessment of the potential for tsunami debris impact. *Journal of Waterway, Port, Coastal and Ocean Engineering*, *140*(2), 223–232. Available from https://doi.org/10.1061/(ASCE)WW.1943-5460.0000222.

Nakano, Y. (2017) *Structural design requirements for tsunami evacuation buildings in Japan*. In Proceedings of the First ACI & JCI Joint Seminar: Design of Concrete Structures Against Earthquake and Tsunami Damage (pp. 1–12). https://doi.org/10.14359/51689683.

Nakao, M. (2005). The Great Meiji Sanriku Tsunami. Failure knowledge database. https://www.shippai.org/fkd/en/hfen/HA1000616.pdf.

Narita, Y., & Koshimura, S. (2015). Classification of tsunami fragility curves based on regional characteristics of tsunami damage. *Journal of Japan Society of Civil Engineers (Coastal Engineering)*, *71*(2), 331–336. Available from https://doi.org/10.2208/kaigan.71.I_331.

National Crash Analysis Center. (2011). *Development and validation of a finite element model for the 2010*. Toyota Yaris Passenger Sedan NCAC 2011-T-001.

National Crash Analysis Center. (2012). *Extended validation of the finite element model for the 2010*. Toyota Yaris Passenger Sedan NCAC 2012-W-005.

New Zealand Government. (2020). *Tsunami loads and effects on vertical evacuation structures*. https://www.building.govt.nz/assets/Uploads/building-code-compliance/geotechnical-education/tsunami-vertical-evacuation-structures.pdf.

Newman, J. N., & Landweber, L. (1978). Marine hydrodynamics. *Journal of Applied Mechanics*, *45*(2), 457–458. Available from https://doi.org/10.1115/1.3424341.

Ngo, N., & Robertson, I. N. (2012). *Video analysis of the March 2011 tsunami in Japan's coastal cities UHM-CEE-12-11*. University of Hawaii. http://www.cee.hawaii.edu/wp-content/uploads/UHM-CEE-12-11.pdf.

Nihei, Y., Maekawa, T., Ohshima, R., & Yanagisawa, M. (2012). Evaluation of fragility functions for tsunami damage in coastal district in Natori City, Miyagi Prefecture and mitigation effects of coastal dune. *Journal of Japan Society of Civil Engineers (Coastal Engineering)*, *68*(2), 276–280. Available from https://doi.org/10.2208/kaigan.68.i_276.

Nistor, I., Goseberg, N., Stolle, J., Mikami, T., Shibayama, T., Nakamura, R., & Matsuba, S. (2017). Experimental investigations of debris dynamics over a horizontal plane. *Journal of Waterway, Port, Coastal and Ocean Engineering*, *143*(3). Available from https://doi.org/10.1061/(ASCE)WW.1943-5460.0000371.

Okada, T., Sugano, T., Ishikawa, T., Ohgi, T., Takai, S., & Hamabe, C. (2015). *Structural design methods of buildings for tsunami resistance*. The Building Center of Japan.

Overseas Coastal Area Development Institute of Japan. (2009). *Technical standards and commentaries for port and harbour facilities in Japan*.

Paczkowski, K. (2011). *Bore impact upon vertical wall and water-driven, highmass, low-velocity debris impact*. Available from https://scholarspace.manoa.hawaii.edu/items/50af013e-bb95-46d6-9df4-896e41f7f96a.

Patil, A., Mudiyanselage, S. D., Bricker, J. D., Uijttewaal, W., & Keetels, G. (2018). Effects of overflow nappe non-aeration on tsunami breakwater failure. *Coastal Engineering Proceedings*, *36*, 18. Available from https://doi.org/10.9753/icce.v36.papers.18.

Peiris, N., & Pomonis, A. (2004). *Indian Ocean tsunami: Vulnerability functions for loss estimation in Sri Lanka*. In Proceedings of the International Conference of the Geotechnical Engineering for Disaster Mitigation and Rehabilitation (pp. 411–416). https://doi.org/10.1142/9789812701602_0045.

Petrone, C., Rossetto, T., Baiguera, M., De la Barra Bustamante, C., & Ioannou, I. (2020). Fragility functions for a reinforced concrete structure subjected to earthquake and tsunami in sequence. *Engineering Structures*, *205*, 110120. Available from https://doi.org/10.1016/j.engstruct.2019.110120.

Petrone, C., Rossetto, T., & Goda, K. (2017). Fragility assessment of a RC structure under tsunami actions via nonlinear static and dynamic analyses. *Engineering Structures*, *136*, 36−53. Available from https://doi.org/10.1016/j.engstruct.2017.01.013.

Piran Aghl, P., Naito, C. J., & Riggs, H. R. (2014). Full-scale experimental study of impact demands resulting from high mass, low velocity debris. *Journal of Structural Engineering*, *140*(5). Available from https://doi.org/10.1061/(ASCE)ST.1943-541X.0000948.

Pomonis, M. A., Macabuag, M. J., Hutt, M. C. M., Alexander, D., Andonov, M. A., Crawford, D. C., Platt, D. S., & Raby D. A. (2013). *Recovery two years after the 2011 Tohoku Earthquake and Tsunami: A return mission report*. Earthquake Engineering Field Investigation Team.

Porter, K., Kennedy, R., & Bachman, R. (2007). Creating fragility functions for performance-based earthquake engineering. *Earthquake Spectra*, *23*(2), 471−489. Available from https://doi.org/10.1193/1.2720892.

Ramsden, J. D. (1996). Forces on a vertical wall due to long waves, bores, and dry-bed surges. *Journal of Waterway, Port, Coastal, and Ocean Engineering*, *122*(3). Available from https://doi.org/10.1061/(ASCE)0733-950X(1996)122:3(134).

Rasheed, A., Usman, M., Zain, M., Iqbal, N., & Yaseen, Z. M. (2022). Machine learning-based fragility assessment of reinforced concrete buildings. *Computational Intelligence and Neuroscience*, *2022*, 1−12. Available from https://doi.org/10.1155/2022/5504283.

Reese, S., Bradley, B. A., Bind, J., Smart, G., Power, W., & Sturman, J. (2011). Empirical building fragilities from observed damage in the 2009 South Pacific tsunami. *Earth-Science Reviews*, *107*(1−2), 156−173. Available from https://doi.org/10.1016/j.earscirev.2011.01.009.

Riggs, H. R., Cox, D. T., Naito, C. J., Kobayashi, M. H., Piran Aghl, P., Ko, H. T.-S., & Khowitar, E. (2014). Experimental and analytical study of water-driven debris impact forces on structures. *Journal of Offshore Mechanics and Arctic Engineering*, *136*(4). Available from https://doi.org/10.1115/1.4028338.

Robertson, I. N. (2023). Recent advances in tsunami design of coastal structures. *Coastal Engineering Proceedings*, *37*(83). Available from https://doi.org/10.9753/icce.v37.structures.83.

Robertson, I. N., Carden, L., Riggs, H. R., Yim, S., Young, Y. L., & Paczkowski K. (2009). *Reconnaissance following the September 29, 2009 Tsunami in Samoa UHM/CEE/10-01*. University of Hawaii.

Robertson, I. N., Paczkowski, K., Riggs, H. R., & Mohamed, A. (2013). Experimental investigation of tsunami bore forces on vertical walls. *Journal of Offshore Mechanics and Arctic Engineering*, *135*(2). Available from https://doi.org/10.1115/1.4023149.

Rossetto, T., & Ioannou, I. (2018). *Risk modeling for hazards and disasters. Empirical fragility and vulnerability assessment: Not just a regression* (pp. 79−103). Elsevier. Available from https://www.sciencedirect.com/science/article/pii/B9780128040713000045, https://doi.org/10.1016/B978-0-12-804071-3.00004-5.

Saatcioglu, M., Ghobarah, A., & Nistor, I. (2006). Performance of structures in Indonesia during the December 2004 Great Sumatra earthquake and Indian Ocean tsunami. *Earthquake Spectra*, *22*(3), 295−319. Available from https://doi.org/10.1193/1.2209171.

Santo, J., & Robertson, I.N. (2010). *Lateral loading on vertical structural elements due to a Tsunami Bore UHM/CEE/10-02*. University of Hawaii. http://www.cee.hawaii.edu/wp-content/uploads/UHM-CEE-10-02.pdf.

Sarpkaya, T. S. (2012). *Wave forces on offshore structures* (pp. 1−322). Cambridge University Press. Available from http://doi.org/10.1017/CBO9781139195898, https://doi.org/10.1017/CBO9781139195898.

Sassa, S., & Takagawa, T. (2019). Liquefied gravity flow-induced tsunami: First evidence and comparison from the 2018 Indonesia Sulawesi earthquake and tsunami disasters. *Landslides*, *16*(1), 195−200. Available from https://doi.org/10.1007/s10346-018-1114-x.

Schwarz, G. (1978). Estimating the dimension of a model. *The Annals of Statistics*, *6*(2), 461−464. Available from https://doi.org/10.1214/aos/1176344136, https://www.jstor.org/stable/2958889.

SeismoStruct. (2022). *A computer program for static and dynamic nonlinear analysis of framed structures*. https://seismosoft.com/product/seismostruct/.

Shuto, N., & Fujima, K. (2009). A short history of tsunami research and countermeasures in Japan. *Proceedings of the Japan Academy*, *85*(8), 267−275. Available from https://doi.org/10.2183/pjab.85.267.

Song, J., De Risi, R., & Goda, K. (2017). Influence of flow velocity on tsunami loss estimation. *Geosciences*, *7*(4), 114. Available from https://doi.org/10.3390/geosciences7040114.

Stolle, J., Goseberg, N., Nistor, I., & Petriu, E. (2018). Probabilistic investigation and risk assessment of debris transport in extreme hydrodynamic conditions. *Journal of Waterway, Port, Coastal and Ocean Engineering*, *144*(1). Available from https://doi.org/10.1061/(ASCE)WW.1943-5460.0000428.

Stolle, J., Krautwald, C., Robertson, I., Achiari, H., Mikami, T., Nakamura, R., Takabatake, T., Nishida, Y., Shibayama, T., Esteban, M., Nistor, I., & Goseberg, N. (2020). Engineering lessons from the 28 September 2018 Indonesian tsunami: Debris loading. *Canadian Journal of Civil Engineering*, *47*(1), 1−12. Available from https://doi.org/10.1139/cjce-2019-0049.

Suppasri, A., Charvet, I., Imai, K., & Imamura, F. (2015). Fragility curves based on data from the 2011 Tohoku-oki Tsunami in Ishinomaki City, with discussion of parameters influencing building damage. *Earthquake Spectra*, *31*(2), 841−868. Available from https://doi.org/10.1193/053013EQS138M, http://earthquakespectra.org/doi/pdf/10.1193/053013EQS138M.

Suppasri, A., Koshimura, S., & Imamura, F. (2011). Developing tsunami fragility curves based on the satellite remote sensing and the numerical modeling of the 2004 Indian Ocean tsunami in Thailand. *Natural Hazards and Earth System Science*, *11*(1), 173−189. Available from https://doi.org/10.5194/nhess-11-173-2011.

Suppasri, A., Koshimura, S., Imamura, F., Ruangrassamee, A., & Foytong, P. (2014). A review of tsunami damage assessment methods and building performance in Thailand. *Journal of Earthquake and Tsunami*, *07*(05), 1350036. Available from https://doi.org/10.1142/s179343111350036x.

Suppasri, A., Koshimura, S., & Imamura, F. (2009). *Tsunami fragility curves and structural performance of building along the Thailand coast*. In Proceedings of the 8th International Workshop on Remote Sensing for Disaster Management (pp. 3−8).

Takakura, R., & Robertson, I.N. (2010). *Reducing Tsunami Bore uplift forces by providing a breakaway panel UHM/CEE/10-06*. University of Hawaii.

Tarbotton, C., Dall'Osso, F., Dominey-Howes, D., & Goff, J. (2015). The use of empirical vulnerability functions to assess the response of buildings to tsunami impact: Comparative review and summary of best practice. *Earth-Science Reviews*, *142*, 120–134. Available from https://doi.org/10.1016/j.earscirev.2015.01.002, http://www.sciencedirect.com/science/journal/00128252.

Valencia, N., Gardi, A., Gauraz, A., Leone, F., & Guillande, R. (2011). New tsunami damage functions developed in the framework of SCHEMA project: Application to European-Mediterranean coasts. *Natural Hazards and Earth System Sciences*, *11*(10), 2835–2846. Available from https://doi.org/10.5194/nhess-11-2835-2011.

van Sebille, E., Griffies, S. M., Abernathey, R., Adams, T. P., Berloff, P., Biastoch, A., Blanke, B., Chassignet, E. P., Cheng, Y., Cotter, C. J., Deleersnijder, E., Döös, K., Drake, H. F., Drijfhout, S., Gary, S. F., Heemink, A. W., Kjellsson, J., Koszalka, I. M., Lange, M., ... Zika, J. D. (2018). Lagrangian ocean analysis: Fundamentals and practices. *Ocean Modelling*, *121*, 49–75. Available from https://doi.org/10.1016/j.ocemod.2017.11.008.

Chapter 5

Probabilistic tsunami hazard and risk assessments

Raffaele De Risi[1], Katsuichiro Goda[2], Ioan Nistor[3] and Aditya Gusman[4]

[1]School of Civil, Aerospace and Design Engineering, University of Bristol, Bristol, United Kingdom, [2]Department of Earth Sciences, Western University, London, ON, Canada, [3]Department of Civil Engineering, University of Ottawa, Ottawa, ON, Canada, [4]GNS Science, Lower Hutt, Wellington, New Zealand

5.1 Introduction

Probabilistic tsunami hazard assessment (PTHA) and probabilistic tsunami risk assessment (PTRA) are rigorous and structured methodologies enabling robust quantitative assessments of tsunami hazards and risks. These methodologies are (1) rigorous because they are based on a precise and rational mathematical formulation; (2) structured because they can be solved by implementing an algorithm and therefore following explicit computational steps; and (3) robust because they account for and propagate important uncertainties involved in the problem, such as earthquake nucleation, magnitude, recurrence, and built environment vulnerability. The Total Probability Theorem (Jaynes, 2003) is commonly adopted as a basis for PTHA and PTRA and is used to propagate uncertainties.

The input components for PTHA and some elements of PTRA have been described in detail in the previous chapters of this book. Building upon these, this chapter focuses on their integration or, more formally, convolution (i.e., *risk = hazard × exposure × vulnerability*). For PTHA, the input consists of various data (see Chapter 2 and 3), such as the identification of the potential tsunamigenic sources, their characterization (e.g., seismotectonic context and seismicity for the subduction areas), the bathymetry, the digital elevation model or the digital terrain model, and the surface roughness of coastal areas. PTRA requires additional inputs (see Chapter 4 and the following sections of this chapter), such as a clear definition of the exposed assets (e.g., buildings and infrastructures), their vulnerability to tsunamis, and their value, considering any potential costs incurred due to tsunami damage. As elaborated in the subsequent sections, vulnerability can be represented using fragility or vulnerability curves, while the cost model can be described using consequence functions.

A key advantage of PTHA and PTRA is their ability to provide information that aids long-term decision-making for coastal communities. The outputs of PTHA include tsunami hazard curves, hazard maps, and hazard disaggregation, which will be explained in detail later in this chapter. Similarly, the results of PTRA encompass tsunami risk curves, risk maps, and risk metrics, such as the average annual loss (AAL). These tools can also be employed to characterize potential future natural catastrophes probabilistically (Mitchell-Wallace et al., 2017), enabling mitigation strategies to improve societal resilience, as discussed in Chapter 6.

PTHA and PTRA build upon methods that were initially developed for the seismic ground-motion case, that is, probabilistic seismic hazard analysis (PSHA; Cornell, 1968). PSHA provides a probabilistic definition of the expected seismic hazard at a specific site in a seismic region. The original formulation of PSHA builds upon the Total Probability Theorem (McGuire, 2004). Such a formulation has been further expanded to transform from the hazard into the risk. Specifically, the seismic risk assessment formulation relies on the performance-based framework (Fig. 5.1) advocated by Cornell and Krawinkler (2000) at the Pacific Earthquake Engineering Research (PEER) Center. Such a framework is versatile and has been used for many hazard typologies (e.g., earthquakes, floods, and hurricanes); its

FIGURE 5.1 Quantitative risk assessment framework by the Pacific Earthquake Engineering Research Center (Cornell & Krawinkler, 2000).

flexibility is because the risk can be decomposed in a modular manner considering single-nested physical phenomena that can be ordered hierarchically. Hence, a highly complex risk assessment problem can be decomposed into modules interacting with each other via interfacing intermediate variables. This point will become clearer in the following sections of this chapter.

Although PTHA and PTRA are rigorous and can be expressed by precise analytical formulations, it is not practically possible to solve them analytically in a closed form. A broadly accepted approach for conducting such quantitative hazard and risk assessments in the field of natural hazards and within the performance-based framework is to generate a set of stochastic events (or chains of events) for the sites of interest where the target buildings and infrastructures are located. For tsunami hazards, after modeling and simulating the triggering phenomena at the source level (Chapter 2), a numerical solution of the governing equations of tsunami wave propagation and runup for a specific initial water dislocation is computationally tractable and common (Park et al., 2019) (Chapter 3). Therefore, temporal and spatial variations of the tsunami hazard must be simulated based on the physical governing equations of wave propagation and runup processes. This chain of simulations can be computationally costly; thus the adoption of appropriate simulation methods depends on the availability of (1) existing models, (2) validation data, and (3) computational resources.

In the following, the state-of-the-art tsunami hazard and risk assessments are presented. First, the basics of tsunami hazard analysis are provided in Section 5.2, focusing on its historical evolution. Section 5.3 provides more details about the seismic PTHA and its evaluation with a simulation-based approach. Subsequently, Section 5.4 extends the probabilistic framework to PTRA; this section will also present how to characterize the exposure and vulnerability components. Section 5.5 presents how to extend the PTHA and PTRA methodologies to the earthquake—tsunami joint hazard and risk assessment. Finally, Section 5.6 identifies several research gaps for future research. It is important to note that although the exposure component is common to multiple hazards (e.g., storm surges, floods, earthquakes, tsunamis, and hurricanes), the vulnerability needs to be characterized differently as the response of the structures (e.g., buildings and bridges) to these types of external loading is different.

5.2 Overview of tsunami hazard and risk assessment

Several tsunami hazard-risk assessment methodologies exist in the literature and have historically evolved following the most recent tsunami events. Finding a commonly accepted generalization for such methods is difficult. However, it is important to agree on a distinction among methods for the sake of communication and to adopt a representative technical term. To this aim, the authors refer to González et al. (2009) (see Chapter 13). They identified three main categories of tsunami hazard assessment methodologies:

- Sensitivity analysis
- Worst case scenario approach
- Probabilistic approach

5.2.1 Sensitivity analysis

Sensitivity analysis is often used to understand what factors govern the hazard and the risk most. More specifically, the sensitivity analysis, also known as response analysis, provides an improved understanding of tsunami hazard and risk by exploring, for example, the relative importance of different tsunamigenic sources for different coastal communities. A typical approach used in the past consists of modeling known historical tsunami events and identifying which event poses the greatest threat to a specific region; an example can be found in Tang et al. (2006). There are a few limitations to sensitivity analyses. First, they may be computationally heavy depending on the type of algorithms used to assess uncertainties (Beven, Almeida, et al., 2018; Beven, Aspinall, et al., 2018; Pianosi et al., 2016). Second, the sensitivity does not encompass the likelihood of each considered scenario. On the positive side, a preliminary sensitivity analysis can help optimize more rigorous and computationally expensive worst case scenarios and probabilistic approaches. As explained later in this chapter, tsunami hazard disaggregation, a byproduct of PTHA, can be considered a sensitivity analysis tool for optimizing hazard and risk assessments.

5.2.2 Worst case scenario approach

The worst case scenario is computationally efficient as it considers only the most adverse hypothetical conditions, often leading to overestimating the consequences. Such an approach generally refers to plausible, credible, and scientifically defensible worst case scenarios involving significant tsunamigenic sources. To be scientifically sound, such an approach often refers to and is based on paleotsunami events. Since this approach refers to the worst conditions that a coastal community could experience, worst case scenarios are often used to develop emergency management planning tools, such as determining evacuation routes and assembly points. Unlike the sensitivity analysis, a worst case scenario can be characterized by a probability of occurrence. Indeed, it can be selected as the most intense scenario among all those considered in the probabilistic assessment procedure. The worst case scenario can also be identified via a preliminary sensitivity analysis.

An example of a worst case scenario was provided by Schlurmann et al. (2010), where the tsunami risk and the evacuation features are assessed for Padang, Western Sumatra, Indonesia. The tsunami hazard was based on the observations of coral microatolls corresponding to two events of moment magnitude (M_w) 8.5 and 8.7 (Natawidjaja et al., 2006).

5.2.3 Probabilistic approach

A probabilistic approach accounts for all possible scenarios, which is based on the principle that it is unknown *where* the next event will be, *when* it will happen, and *how intense* it will be exactly. Therefore, these components of "when," "where," and "how intense" are convoluted together to produce transparent results accounting for the sensitivity of the hazard and risk to the uncertainties. An advantage of such a probabilistic approach is that the results are not affected by subjective components, such as selecting a single worst case scenario. In the following, different probabilistic approaches are discussed in the context of tsunamis also looking at their historical evolution.

Existing PTHA can be classified into three categories. In the first category (Fig. 5.2), PTHA is conducted by employing tsunami catalogs (Burroughs & Tebbens, 2005; Kulikov et al., 2005; Orfanogiannaki & Papadopoulos, 2007; Tinti et al., 2005). Tsunami catalogs, similarly, to earthquake catalogs, are lists of events that can be used as basic data for statistical modeling and interpretation. In the second category, scenario-based PTHA methods are adopted (Burbidge et al., 2008; Downes & Stirling, 2001; Farreras et al., 2007; Geist & Dmowska, 1999; González et al., 2009; Liu et al., 2007; Løvholt et al., 2012; Power et al., 2007; Yanagisawa et al., 2007). In the third category, a combination of the preceding two is suggested (Annaka et al., 2007; Burbidge et al., 2008; Fukutani et al., 2015; Geist & Parsons, 2006; Geist, 2005; Grezio et al., 2010, 2012; Horspool et al., 2014; Parsons & Geist, 2008; Thio et al., 2012). The level of sophistication of PTHA has progressively increased over the last four decades. Specifically, three main generations of PTHA can be recognized:

- The first generation started in the 1980s (Rikitake & Aida, 1988).
- The second generation was developed after the 2004 Indian Ocean Tsunami (Annaka et al., 2007; Geist & Parsons, 2006; Power et al., 2007; Thio et al., 2012).
- The third generation started after the 2011 Tohoku Tsunami (De Risi & Goda, 2017; Fukutani et al., 2015; Grezio et al., 2017; Miyashita et al., 2020; Mueller et al., 2015; Park & Cox, 2016).

The main objective of PTHA is to estimate the probability of exceeding a specific threshold of a tsunami metric in a particular time interval at a given location (or for an entire geographical region). Similar to other natural hazards, tsunami metrics are called intensity measures (*IM*) and represent a physical feature of the hazard that can be quantified (e.g., flow depth, flow velocity, wave amplitude, and momentum flux). PTHA results can be communicated using multiple typologies of outputs. Tsunami hazard curves relate tsunami *IM*s with their frequency (or probability) of being exceeded. They can be used for engineering design purposes (Chock, 2016) (see Chapter 4). Tsunami hazard maps are geographical representations of the tsunami *IM* at different scales (e.g., micro/meso/macro corresponding to city/region/nation) and can be used as key information to develop (1) evacuation plans and mitigation strategies (e.g., breakwaters) at the city level, (2) assess critical infrastructure at the regional level, and (3) allocate resources at the national level. Tsunami hazard disaggregation provides useful information on the relative importance of the different tsunamigenic sources to the hazard level for a specific site. It can be used as a tool to identify the most concerning tsunami scenarios

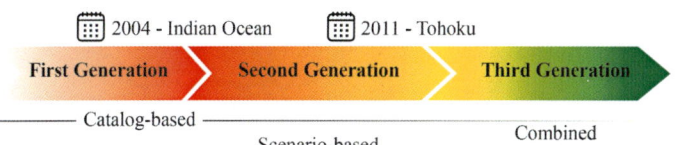

FIGURE 5.2 Evolution of probabilistic tsunami hazard analysis with time.

(i.e., near- versus far-field) and can be used to define the worst case scenario. Therefore, PTHA results are useful for making essential risk management decisions, such as designating safe evacuation areas and land-use planning, without resorting to PTRA (Muhammad et al., 2017).

On the other hand, PTRA requires additional components, such as vulnerability and consequence models, to convert the hazard into an impact. Similarly to PTHA, the main objective of PTRA is to estimate the probability of exceeding a specific threshold of loss in a particular time interval at a given location (or for an entire geographical region). Loss metrics are usually called loss (L) or, sometimes, decision variables (DV), as they are eventually used to inform decisions (e.g., economic losses, casualties, and injuries). PTRA results are informative for financial tsunami risk management, for example, tsunami insurance (Song & Goda, 2019) and tsunami catastrophe bonds (Goda et al., 2019). PTRA can be communicated using multiple outputs similar to PTHA. Tsunami loss curves relate tsunami losses to their frequency (or probability) of being exceeded. They can be used for decision-making purposes. For example, the impact associated with the current situation and the impact after applying mitigation strategies can be compared. Tsunami loss maps are geographical representations of the tsunami-induced loss at different scales (e.g., micro/meso/macro corresponding to city/region/nation) and can be used as the key information to prioritize interventions at the regional scale. Finally, tsunami expected losses are scalar numbers, allowing an easier comparison with respect to the loss curves, and can be integrated into benefit−cost or return-on-investment analyses (De Risi et al., 2018).

5.2.4 General probabilistic formulation

This section introduces an analytical formulation for PTHA. For illustrative purposes and to keep the framework as general as possible, let us consider a specific number of sources N_S that can trigger a tsunami affecting the site of interest. There are multiple potential tsunami sources: earthquakes, submarine landslides, subaerial landslides, volcanic eruptions, and meteorological conditions (Fig. 5.3 and Chapter 2).

Assuming that the set of tsunamigenic sources can be treated independently, the abovementioned PEER framework allows computing the mean annual rate of a tsunami event affecting a specific location with an IM larger or equal to a particular threshold im as

$$v_{IM}(IM \geq im) = \sum_{i=1}^{N_S} \lambda_{M\min,i} \int P_i(IM \geq im|\boldsymbol{\theta}) f_{i,\boldsymbol{\theta}|M}(\boldsymbol{\theta}|m) f_{i,M}(m) |d\boldsymbol{\theta}||dm| \tag{5.1}$$

where $\lambda_{M\min,i}$ is the mean annual rate of occurrence of triggering events with earthquake magnitude M exceeding the minimum value $M_{\min}$ associated with the ith source, $f_{i,M}(m)$ is the conditional probability density function for magnitude values M larger than the minimum value $M_{\min}$, and $f_{i,\boldsymbol{\theta}|M}(\boldsymbol{\theta}|m)$ is the conditional probability of the parameters $\boldsymbol{\theta}$ characterizing the tsunamigenic source for a specific value of the magnitude m. The term magnitude is used as a general indicator of the size/extent of the triggering event (i.e., it is not necessarily a seismic magnitude), and for the case of seismically triggered tsunamis, M_w can be used for M. The set of parameters $\boldsymbol{\theta}$ is denoted in bold because it is not a scalar but represents a vector of relevant parameters (e.g., maximum slip and fault dimensions). Finally, $P_i(IM \geq im|\boldsymbol{\theta})$ is the complementary cumulative distribution function (CCDF) of the IM conditioned on the source parameters $\boldsymbol{\theta}$. Graphically, Eq. (5.1) can be illustrated in Fig. 5.4, and it reflects the first module of the PEER framework shown in Fig. 5.1.

The convolution over all possible random variables (i.e., magnitude and all the other tsunamigenic parameters) and summation over all possible tsunamigenic sources produce the estimates of $v_{IM}(IM \geq im)$ which is also known as a hazard curve (Fig. 5.5). The hazard curves can be represented in linear space (Fig. 5.5A) and log−log space (Fig. 5.5B). The latter approach effortlessly displays the entire range of mean annual rates, spanning from low to high values. A computer code for the representation of hazard curves is provided in Example 5-1 (Appendix).

The rate $v_{IM}(IM \geq im)$ can be converted to a probability in a given time window, selecting an appropriate underlying probabilistic recurrence model. A classical occurrence model is the memory-less Poisson process (McGuire, 2004),

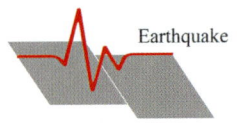

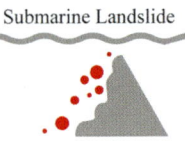

FIGURE 5.3 Potential tsunami sources.

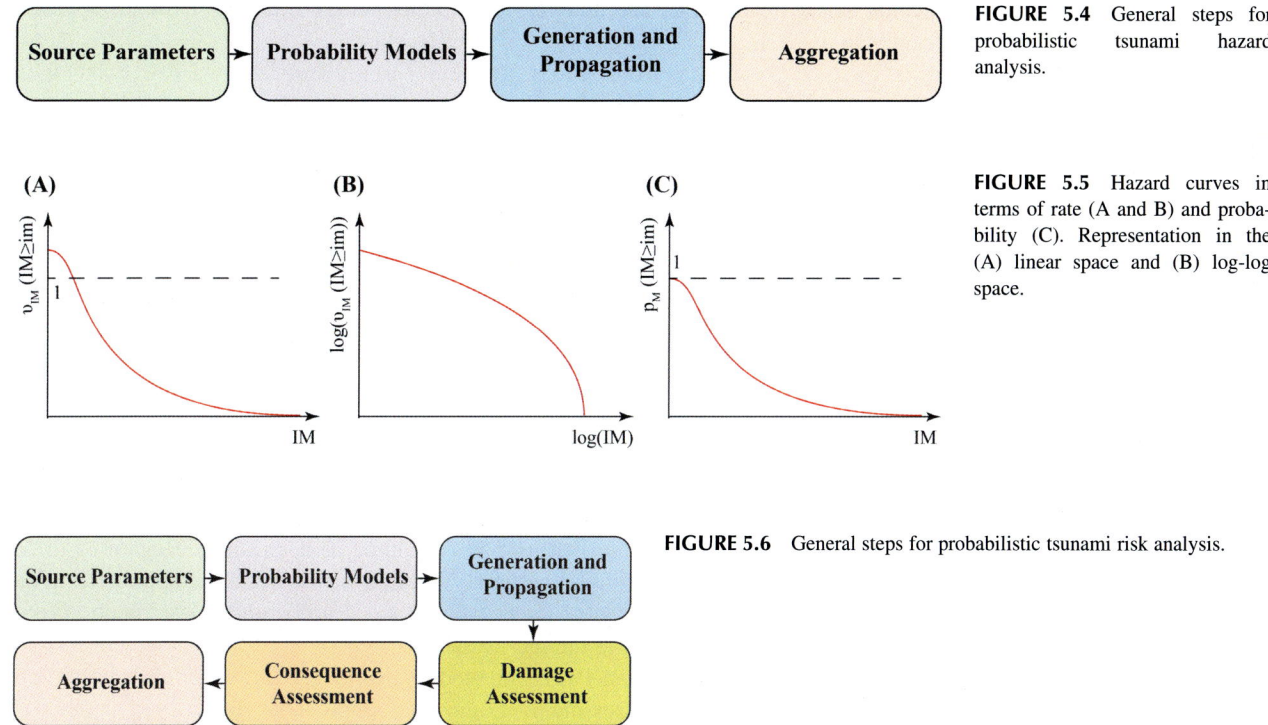

FIGURE 5.4 General steps for probabilistic tsunami hazard analysis.

FIGURE 5.5 Hazard curves in terms of rate (A and B) and probability (C). Representation in the (A) linear space and (B) log-log space.

FIGURE 5.6 General steps for probabilistic tsunami risk analysis.

generally adopted for long-term hazard assessments. In contrast, a renewal model (Goda, 2019; Matthews et al., 2002) may be applied for time-dependent hazard assessments by considering recent triggering activities (e.g., major earthquakes). Section 2.3.2 provides an overview of the available options related to underlying probabilistic models spanning from a simple Poisson process to a renewal occurrence process. In the following parts of this chapter, a Poisson process is adopted for the sake of simplicity. With this assumption, the rate can be converted to a probability of exceedance in a specific time window t as follows

$$P(IM \geq im) = 1 - \exp[-\upsilon_{IM}(IM \geq im)t] \tag{5.2}$$

It is important to emphasize that the hazard curve in terms of mean annual rate (Eq. 5.1) can have rate values larger than 1 (Fig. 5.5A), which means that it is possible to observe events with a specific intensity im more than once per year (although this is rare for tsunamis). On the other hand, the hazard curve in terms of probability (Eq. 5.2) has as an upper bound of 1 (Fig. 5.5C) by definition.

Building upon Eq. (5.1), the general formulation for PTRA can be expressed as follows

$$\upsilon_L(L \geq l) = \sum_{i=1}^{N_S} \lambda_{M\min,i} \int P_i(L \geq l|ds) f_{i,DS|IM}(ds|im) f_{i,IM|\theta}(im|\theta) f_{i,\theta|M}(\theta|m) f_{i,M}(m) |dds||dim||d\theta||dm| \tag{5.3}$$

In Eq. (5.3), $f_{i,IM}(im|\theta)$ is the probability of the *IM* conditioned on the source parameters θ, $f_{i,DS|IM}(ds|im)$ is the probability of the damage state (*DS*) conditioned on the *IM*, and $P_i(L \geq l|ds)$ is the CCDF of the loss conditioned on the experienced *DS*. The *DS* is the damage condition that is experienced by the asset due to the tsunami (e.g., building and bridge). Such damage states are usually accompanied by physical descriptions and are discretized in categories ranging from no damage (*ds*1) to collapse (*ds*6) or washed away (*ds*7) (Chapter 4). The term $P_i(L \geq l|ds)$ is known as the consequence model, associating each *DS* with a potential loss. For example, *ds*1 is associated with no loss, whereas *ds*6 is associated with the loss of the entire value of the asset plus the costs needed to remove the debris. Graphically, Eq. (5.3) can be illustrated in Fig. 5.6, reflecting the first four steps of the PEER approach shown in Fig. 5.1.

The convolution over all possible random variables (i.e., magnitude, tsunamigenic parameters θ, *IM*s, and *DS*s) and summation over all possible sources produce the estimates of $\upsilon_L(L \geq l)$ which is also known as a tsunami loss curve. As per the hazard curves, loss curves can be represented in linear space and log−log space, and they resemble the shapes of the hazard curves presented in Fig. 5.5. The main difference between the loss curve and the hazard curve is that losses (*L*) or *DV* are on the horizontal axis instead of *IM*. The integration of the loss curve (i.e., geometrically speaking, the area under the curve) corresponds to the AAL.

As per the hazard curve, the rate $v_L(L \geq l)$ can be converted to a probability in a given time window, selecting an appropriate underlying probabilistic recurrence model. Assuming a Poisson occurrence model, the probability is given by

$$P(L \geq l) = 1 - \exp[-v_L(L \geq l)t] \tag{5.4}$$

5.2.5 Logic tree approach

All uncertainties (both epistemic and aleatory) must be propagated appropriately. In Eq. (5.1), aleatory uncertainties are mainly represented by the term $f_{i,\theta|M}(\theta|m)$ describing the probabilistic prediction model related to the specific source of interest. Epistemic uncertainties, mainly consisting of different hypotheses for uncertain models and parameters, can be incorporated in Eq. (5.1) by building a logic tree (Fig. 5.7) where each option is associated with a weight (*w*) representative of the degree of confidence in the respective option. Computationally, Eq. (5.1) can be evaluated for each branch of the logic tree. Eventually, the hazard curve obtained for each set of option branches can be combined together using an ensemble approach (Marzocchi et al., 2015). The easiest approach is the weighted average of the hazard curves.

Weights associated with each option (i.e., branches and leaves) must be computed or determined. Regarding different hypotheses or models, to avoid subjectivism in their definition, expert elicitation methods can be applied (Aspinall & Cooke, 2013). If the different options correspond to a set of discrete values, and a probability mass function (*pmf*) describing their likelihood exists, then the weights can be set to the *pmf* values.

The logic tree approach, at the core of the first-generation PTHA methodologies (Fig. 5.2), has long been the preferred methodological framework for tsunami hazard assessment. Logic tree approaches have been applied to earthquake-induced (Annaka et al., 2007) and landslide-induced (Geist & Lynett, 2014) tsunamis. Typical "leaves" of the logic-tree approach are (1) source identification, (2) source features, (3) magnitude range, (4) triggering mechanisms features, (5) recurrence model, and (6) modification of tsunami *IM* based on mismatches between empirical observations and modeled results (Fig. 5.8).

The first-level leaf consists of the identification of the source. In a logic tree, similar to the first summation in Eq. (5.1), it is possible to consider all potential tsunamigenic sources of interest (e.g., close or distant subduction zones, nearby volcanos, and landslide-prone coastal areas). The second-level leaf generally captures a feature related to the constraints of the tsunamigenic source (e.g., the potential area where a rupture can occur in a subduction zone and the maximum volume of the mass that a landslide can mobilize). The third-level leaf corresponds to a measure of energy associated with the event; the moment magnitude and the sliding mass volume are often used for seismic events and landslides, respectively. The fourth-level leaf consists of the characterization of the triggering mechanisms. For example, in the case of subduction zones, it is essential to consider different possibilities for the areas where the largest slip will occur (i.e., asperity areas). The fifth-level leaf is usually related to the recurrence model of the failure. In theory, according to Eq. (5.1), the recurrence model and the magnitude range should be linked. However, historically, it was not always possible to specify these elements jointly; therefore, the two elements were defined separately. Consequently, a set of possible return periods is defined as possible alternatives. The sixth-level leaf of the logic tree consists of modifying the generic *IM* to account for empirical observations. Specifically, Aida (1978) suggested considering additional variability (i.e., a logarithmic standard deviation known as "Aida's κ") to account for the uncertainties

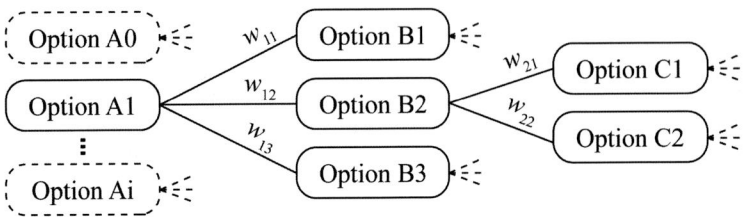

FIGURE 5.7 Example of a generic logic tree (*w* = weight).

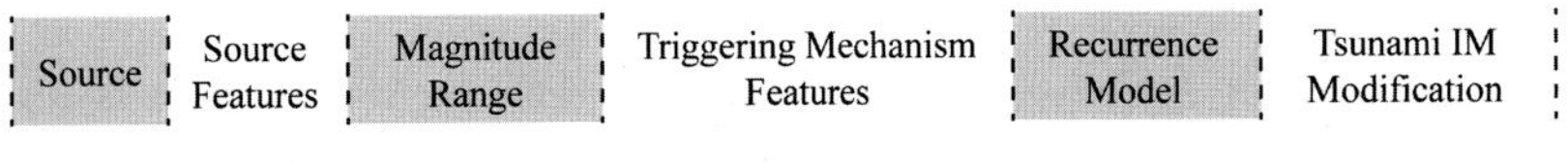

FIGURE 5.8 Leaf levels of typical PTHA logic tree.

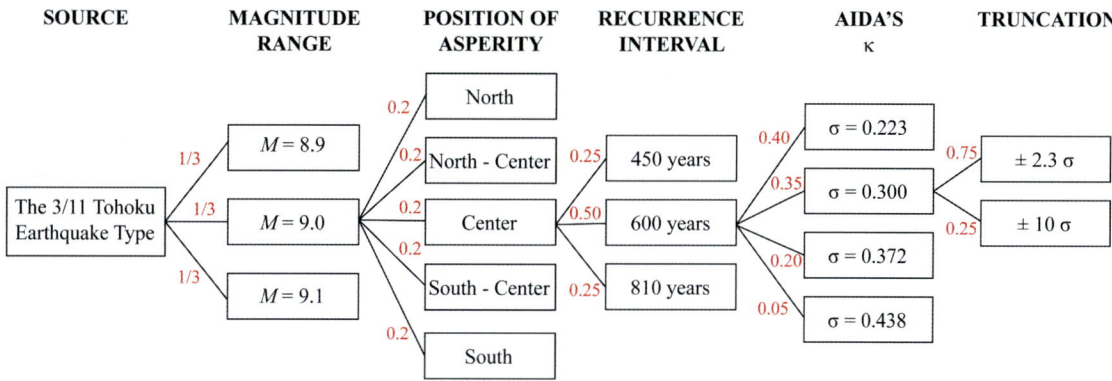

FIGURE 5.9 Logic tree. *Adapted from Fukutani, Y., Suppasri, A., & Imamura F. (2015). Stochastic analysis and uncertainty assessment of tsunami wave height using a random source parameter model that targets a Tohoku-type earthquake fault.* Stochastic Environmental Research and Risk Assessment, 29(7), 1763–1779. http://link.springer-ny.com/link/service/journals/00477/index.htm. https://doi.org/10.1007/s00477-014-0966-4.

due to the mismatch between the numerical simulations and the historical empirical data. For example, in the case of the Tohoku-type earthquakes, this factor has been calibrated by comparing observed runup and numerically derived runup for 11 different historical tsunami events (Fukutani et al., 2015). Since this logarithmic standard deviation σ depends on how the models are set up, they considered four possibilities of σ values varying from 22% to 44%. Regarding Aida's κ, a truncation of the distribution of the values may be considered to avoid unrealistically large runup values; therefore, a further logic-tree leaf is added at the very end to truncate extreme values. Fig. 5.9 shows an example of a logic tree developed for the seismically triggered tsunami case in the Tohoku region of Japan.

The logic tree reported in Fig. 5.9 generates 360 possible combinations. Each combination of parameters is associated with a different hazard curve. The 360 hazard curves can be either combined using weights (i.e., the product of the weights of the branches) or can be used to define percentiles and then the ensemble method. A computer code for the numerical implementation of the logic tree presented in Fig. 5.9 is provided in Examples 5-2 and 5-3 (Appendix). Moreover, a computer code of the ensemble approach is provided in Example 5-4 (Appendix).

The logic tree presented above for PTHA can be expanded for PTRA. Specifically, additional leaves and branches can be added to consider different vulnerability and consequence models. However, since not many tsunami vulnerability and consequence models are available in the literature, the logic tree approach is currently less utilized for PTRA applications. Notwithstanding such limitations, the methodology is still applicable and provides a transparent approach to propagate uncertainties related to the exposed built environment and vulnerability.

5.3 Probabilistic seismic tsunami hazard analysis

Earthquakes are the most common source of tsunamis. Around 80% of tsunamis worldwide are triggered by seismic events (https://www.ngdc.noaa.gov/hazard/tsu_db.shtml). Moreover, tsunamis triggered by landslides, volcanic eruptions, and meteorological phenomena have large uncertainties and knowledge gaps mainly due to the lack of complete understanding of phenomena, such as mobilized landslide and volcanic volumes, limited availability of data and benchmark cases, and difficulty in constraining the recurrence rates (Behrens et al., 2021). Therefore, in the following, the focus will be mainly on tsunamis triggered by seismic events. For the landslide-triggered tsunamis, useful information can be found in Chapter 10.

PTHA has many common features with PSHA. On the other hand, there are significant differences:

- In conventional PSHA, scaling relationships relating magnitude to seismic source characteristics are not explicitly considered because ground motion models (GMMs) used in PSHA do not account for such features, while details of the earthquake rupture process (e.g., geometry and slip distribution) have a major influence on tsunami simulations. See Chapter 2.
- Tsunami *IM*s are obtained through numerical inundation simulations and not using empirical statistical relationships, such as GMMs. See Chapter 3.
- Through numerical inundation simulations, PTHA automatically considers the spatial correlation among tsunami hazard estimates at different locations.

Eq. (5.1) applies to independent tsunamigenic sources; however, it is not suitable for cascading hazards that can magnify the effect of a tsunami (e.g., a seismically triggered landslide that exacerbates the tsunami that has already been triggered by the earthquake). For such cascading scenarios, more complicated tools must be used (De Risi et al., 2022). Nevertheless, Eq. (5.1) can be extended to model the seismic and tsunami hazards that are considered as compound hazards.

De Risi and Goda (2017) developed a simulation-based PTHA. The key features of their methodology comprise the following:

- A wide range of magnitude scenarios (e.g., M_w 7 −9) is considered by characterizing the regional seismicity of the target region in terms of the occurrence rate of major earthquakes and their relative frequency.
- The slip distribution of earthquake rupture is considered in the assessment. Some approaches consider uniform slip on the subduction plane; others use stochastic heterogeneous slip distributions with asperity areas having large slip concentrations. When the slip is modeled as uniform, uncertainties are usually propagated using a logic tree approach considering different geometry and possible uniform slip values (stress drop variations) together with Aida's κ factor. On the contrary, heterogeneous slips intrinsically capture several elements of the abovementioned aleatory uncertainties.
- Uncertainties of earthquake source parameters, such as fault width, fault length, and mean slip, are characterized using statistical prediction models (Chapter 2), accounting for their variability and dependency.
- Inland inundations of incoming and receding tsunami waves are simulated rather than stopping along the coast, producing more accurate and realistic estimates of tsunami hazard parameters.

Adopting realistic heterogeneous slip distributions in a stochastic simulation framework is a key point for rigorous results (Li et al., 2016). It is essential to highlight that the computational requirements for the simulation-based method are high compared to conventional methods. De Risi and Goda (2017) suggested using a bootstrap analysis to achieve an optimal trade-off between the number of simulations and the variability of the results; this approach will be explained later in this chapter. For the case of seismic tsunami hazard analysis, Eq. (5.1) can be written as follows

$$v_{IM}(IM \geq im) = \sum_{i=1}^{N_S} \lambda_{M\min,i} \int P_i(IM \geq im|\theta) S_i(\theta|m) f_{i,M}(m) |d\theta| |dm| \quad (5.5)$$

where N_S is the number of subduction seismic sources that are capable of generating a tsunami considered in the analysis. $\lambda_{M\min}$ is the mean annual rate of occurrence of seismic events having magnitudes greater than the minimum magnitude $M_{\min}$ for the ith source, whereas $f_{i,M}(m)$ is the magnitude—frequency distribution characterizing the ith source. In this case, with respect to Eq. (5.1), the magnitude is specific to the earthquake, and the preferred magnitude is the moment magnitude. The term $S_i(\theta|m)$ represents the functional distribution of the uncertain source parameters conditioned on the earthquake magnitude. It is important to emphasize that $S_i(\theta|m)$ is a special form of the term $f_{i,\theta|M}(\theta|m)$ in Eq. (5.1) as it represents the scaling relationships of the uncertain earthquake source parameters θ conditioned on the magnitude; this will be clearer in Section 5.3.2. $P_i(IM \geq im|\theta)$ is the probability that the tsunami IM produced by the ith source will exceed a prescribed value im at a given coastal location for a given set of tsunami source parameters θ.

To obtain the terms on the left-hand side of Eq. (5.5), four phases are defined, representing the specialization of the general concept shown in Fig. 5.4:

- Definition of input data (i.e., geometrical characteristics of each seismic source and magnitude-frequency distribution).
- Stochastic source simulations based on earthquake source scaling relationships.
- Tsunami inundation simulations.
- Statistical analysis of simulated tsunami results and final convolution.

Fig. 5.10 shows the first three phases of the general framework, whereas Fig. 5.11 shows schematically the statistical analyses consisting of the fourth phase. The description for each of these phases is presented in the following.

5.3.1 Source characterization and magnitude-frequency distribution

The first step is identifying all seismic sources capable of producing damaging tsunami inundation at a site. The sources that must be considered for PTHA are either crustal faults (Gusman et al., 2018, 2019) or faults located in subduction zones at convergent plate boundaries, which are known from seismological studies. Detailed geometrical information

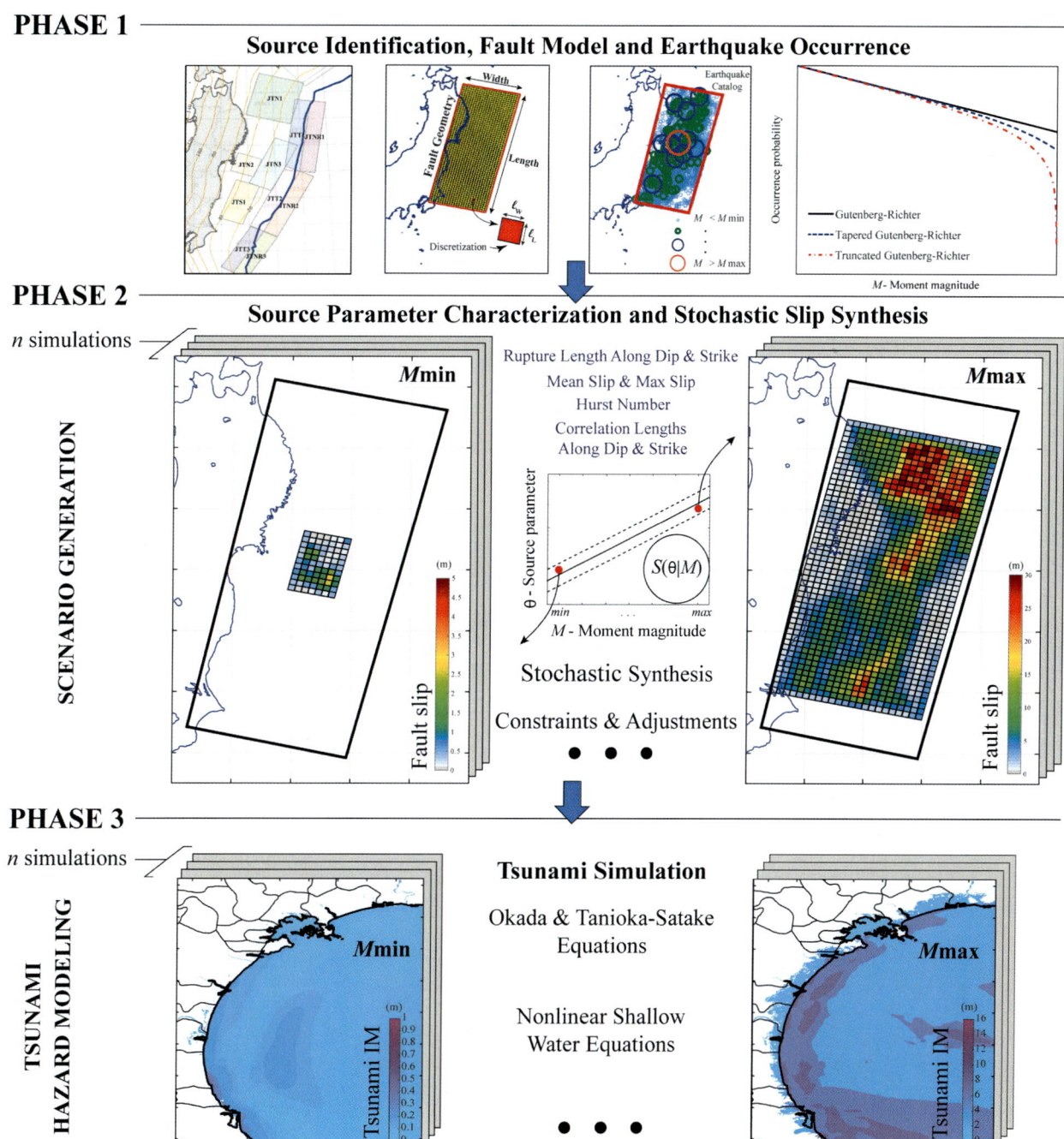

FIGURE 5.10 First three phases of the simulation-based stochastic tsunami hazard framework.

on subduction zones is available publicly (Hayes et al., 2018). As scientific knowledge progresses and new surveying techniques are invented, such data will be enriched and refined. Once the location of interest for which PTHA is performed is decided (e.g., the star in Fig. 5.12A), tsunamigenic sources can be divided into near-field and far-field (Fig. 5.12A). The main difference between the two source types is the tsunami arrival time, ranging from a few minutes to several hours for near-field and far-field sources. Tsunamis triggered by near-field seismic sources can be regarded as the main contributors to the tsunami impact, and they should be studied in detail. Compared to local tsunamis, a simpler parameterization is usually sufficient for far-field tsunamis because the seismic moment, source mechanism, fault geometry, and tsunami propagation pattern are more influential than slip distribution within a rupture plane (Geist & Parsons, 2006, 2016).

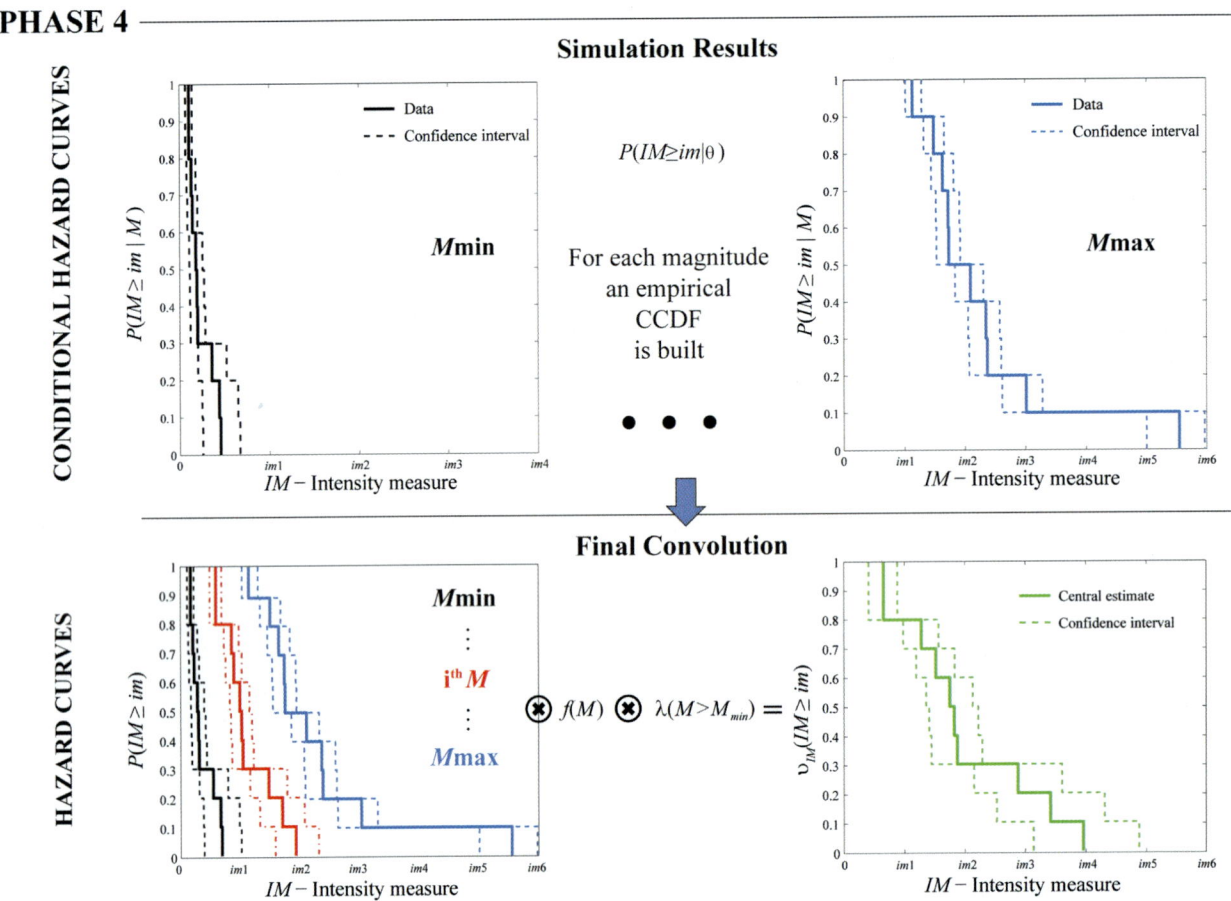

FIGURE 5.11 Statistical treatment of the stochastic tsunami simulation results.

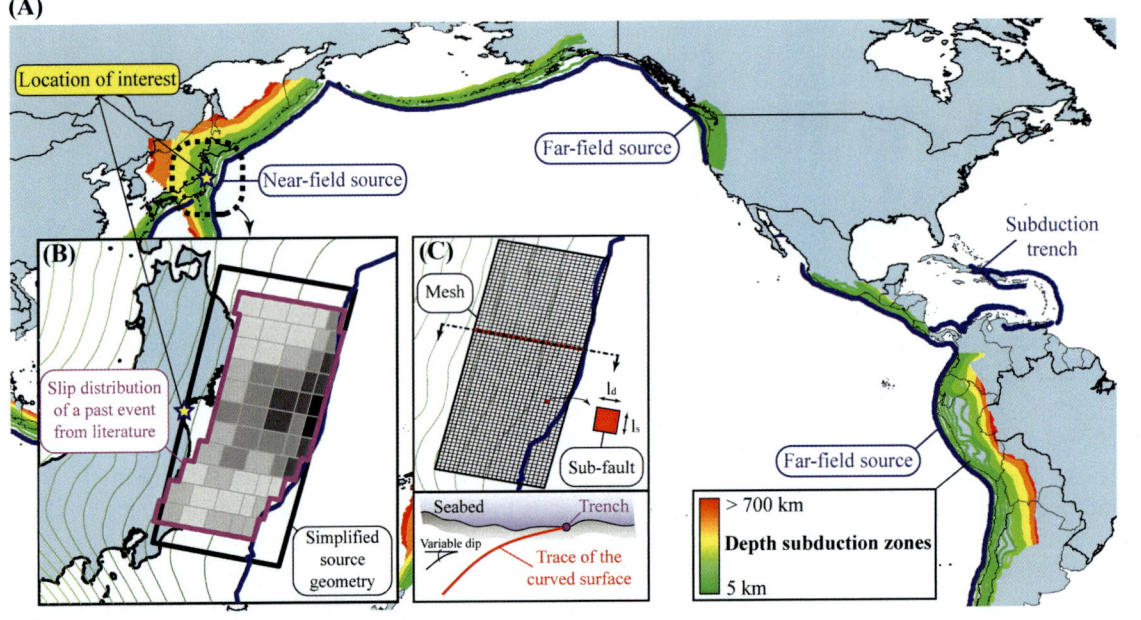

FIGURE 5.12 (A) Subduction zones, near- and far-field sources. (B) Simplified source geometry. (C) Source discretization.

The geometry of the tsunamigenic sources is simplified by defining one or more curved surfaces having rectangular shapes (Fig. 5.12B). Slip distributions of past events from literature can also be used to determine the source geometry (Fig. 5.12A). The seismic source must be able to accommodate the maximum magnitude (M_{max}) and should be consistent with the magnitude-frequency distribution. Extreme earthquake magnitudes (e.g., larger than 9) should be considered carefully because such large earthquakes are rare and may span multiple rupture segments. On the other hand, ignoring such large events results in underestimation of tsunami hazards, which has been proven to be deficient as decision-support tools for future tsunami hazards. These events may well be modeled by the characteristic magnitude model (Youngs & Coppersmith, 1985) (Section 2.2.3).

A stochastic synthesis of simulated slip distributions representative of realistic seismic events can be performed (Section 2.5.2). Therefore, discretizing the fault plane into many subfaults (Fig. 5.12C), generally having variable dip angles, is required. The dimension of the subfaults must allow accurate modeling of the slip distribution corresponding to the minimum magnitude (M_{min}) that is considered in the magnitude–frequency distribution. This minimum magnitude value is usually determined by considering that moderate earthquakes (magnitudes between 6 and 7) rarely generate significant tsunamis, and their contributions to the tsunami hazard are negligible. To describe the earthquake size in a target region, that is, the term $f_{i,M}(m)$ in Eq. (5.5), a truncated Gutenberg–Richter (G–R) relationship (Gutenberg & Richter, 1956) can be adopted (Section 5.2.2)

$$G_M(m) = \frac{1 - 10^{-b(m-M_{min})}}{1 - 10^{-b(M_{max}-M_{min})}} \tag{5.6}$$

where the b-value is calibrated based on historical events available from earthquake catalogs (e.g., http://earthquake.usgs.gov/earthquakes/search/, Fig. 5.13A). Subsequently, the mean annual rate of occurrence of seismic events having a magnitude greater than or equal to M_{min} falling in that area can be calculated from the fitted G–R relationship (Fig. 5.13B; Section 2.3.1).

For the simulation procedure, it is convenient to convert the continuous distribution of magnitudes into a discrete set of values ($M_{min}, \ldots, m_i, \ldots, M_{max}$), by adopting a specific discretization interval Δm. The discrete probability can be calculated as follows

$$p_M(m_i) = G_M(m_i + 0.5\Delta m) - G(m_i - 0.5\Delta m) \tag{5.7}$$

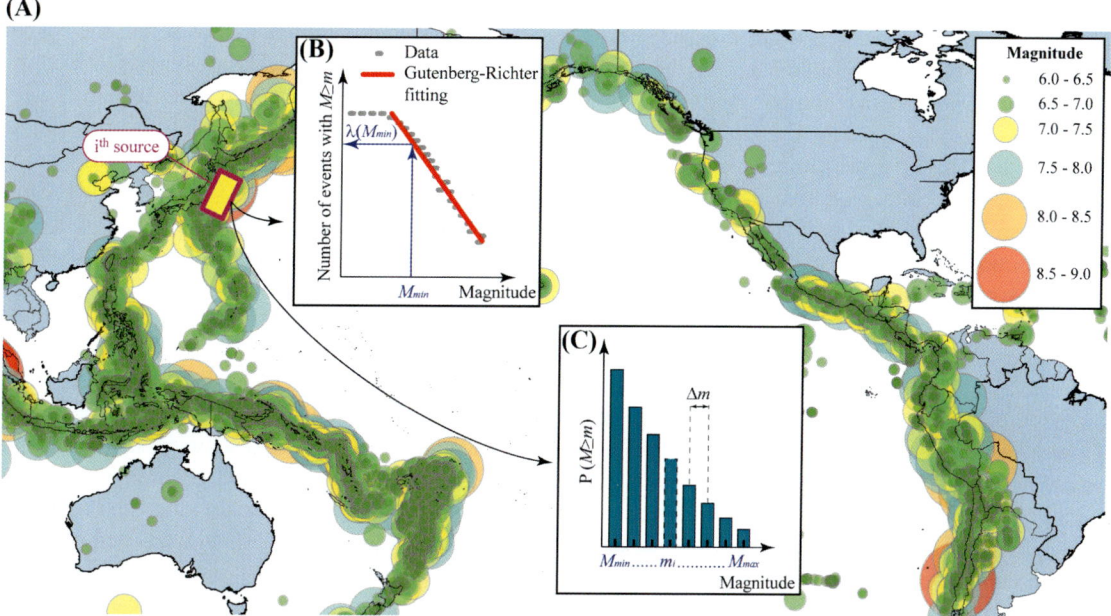

FIGURE 5.13 (A) NEIC earthquake catalog, (B) Gutenberg–Richter relationship, and (C) Gutenberg-Richter probability mass function. *NEIC*, National Earthquake Information Center.

The *pmf* (Fig. 5.13C) for the discrete values of magnitude presented in Eq. (5.7) is normalized with respect to the occurrence rate for the minimum magnitude event. Since the discrete magnitude is considered, the integral in Eq. (5.5) can be replaced by a summation.

$$\upsilon_{IM}(IM \geq im) = \sum_{i=1}^{N_S} \lambda_{M\min,i} \sum_{j=1}^{N_M} P_i(IM \geq im|\boldsymbol{\theta}) S_i(\boldsymbol{\theta}|m_j) p_{i,M}(m_j) \tag{5.8}$$

It is noteworthy that the adoption of the G–R model with a Poisson occurrence process and estimating model parameters based on short earthquake catalogs may not produce a reliable estimate of the long-term recurrence rate for large earthquakes (M 8.5+) given the lack of major historical events in modern instrumental catalogs (e.g., 869 Jogan Earthquake for the Tohoku case; Sawai et al., 2012). The extrapolation of the fitted magnitude–recurrence model should be considered carefully (Pisarenko & Rodkin, 2010).

5.3.2 Scaling relationships of earthquake source parameters and stochastic source models

The simulation-based method generates a certain number of stochastic source models (randomly generated nonuniform fault slip models) to consider uncertainty related to the rupture process. The simulation can be performed based on the statistical models of earthquake source parameters (Goda et al., 2016) and the spectral synthesis method (Goda et al., 2014), characterizing the earthquake slip distribution by wavenumber spectra (Mai & Beroza, 2002). Specifically, for each discrete value of magnitude m_i (also known as "target magnitude"), many samples of the source parameters are necessary to define a slip distribution on the fault plane. For example, the scaling relationships evaluating the source parameters (e.g., rupture size and spectral characteristics of the slip) as a function of moment magnitude can be used for stochastic source generation. Such scaling relationships are obtained from the statistical analysis of inverted source models (Section 2.4.3). Although specific approaches for characterizing earthquake rupture models are presented in the following, there are other applicable approaches (Sections 5.2.4 and 5.2.5).

The two geometrical parameters, that is, fault width W and length L, are used to create the rupture area randomly located inside the predefined subduction fault plane. Subsequently, a slip distribution is represented as a constrained random field based on desirable seismological features, which are characterized by anisotropic wavenumber spectra (Mai & Beroza, 2002), and a realization of such an earthquake slip distribution is obtained using a stochastic synthesis method (Goda et al., 2014) (Section 2.5.2). The simulation of the random slip distribution is carried out using a Fourier integral method (Pardo-Iguzquiza & Chica-Olmo, 1993). The amplitude spectrum of the target slip distribution is specified by a theoretical power spectrum, while a random phase matrix represents the phase spectrum. For the amplitude spectrum, the von Kármán model is adopted (Mai & Beroza, 2002). According to the von Kármán model, the correlation lengths (CL_z along the dip and CL_x along the strike) are important source parameters that define the spatial heterogeneity of small wavenumber components in the spectrum. On the other hand, the Hurst number H determines the spectral decay in the large wavenumber range. All three parameters that describe the slip heterogeneity can be simulated using the scaling relationships by Goda et al. (2016). The obtained complex Fourier coefficients are transformed into the spatial domain via a two-dimensional inverse Fast Fourier Transform. The synthesized slip distribution is then scaled nonlinearly to achieve suitable upper tail characteristics, in agreement with those observed in the finite-fault models, using the Box and Cox's (1964) parameter λ_{BC} that is modeled as a normal random variable. Finally, the generated slip distribution is further adjusted to have a mean slip (S_a) and maximum slip (S_m), which also scale with the magnitude; the slip parameters can also be simulated by using the scaling relationships by Goda et al. (2016).

The error terms of the source parameters W, L, CL_z, CL_x, S_a, and S_m (defined in the logarithmic space) mentioned above are considered to follow a multivariate normal distribution (Goda et al., 2016); therefore, values of these source parameters can be simulated jointly in the stochastic source generation. A joint random sampling of the source parameters ensures overall consistency in the simulated parameters. Nevertheless, due to uncertainty in the source parameters, sampled values of W, L, and S_a may result in a seismic moment different from the target magnitude M_i. To avoid sampled values of W, L, and S_a that do not match the target magnitude, consistency of the simulated magnitude with the target is ensured in determining an acceptable source model; in case the calculated moment magnitude does not fall within a certain range, the simulated combination of W, L, and S_a is discarded, and the sampling is repeated. A tolerance band of $\pm \delta M$ around each magnitude value can be used to define the acceptance criterion in this regard; as an acceptable value δM can be set to 0.05. An example of simulation with the central estimates and the confidence interval (16th and 84th percentiles) of the scaling relationships is shown in Fig. 5.14. The same figure also shows simulated data (gray dots) and associated statistics (colored circles), which are obtained from the stochastic source modeling.

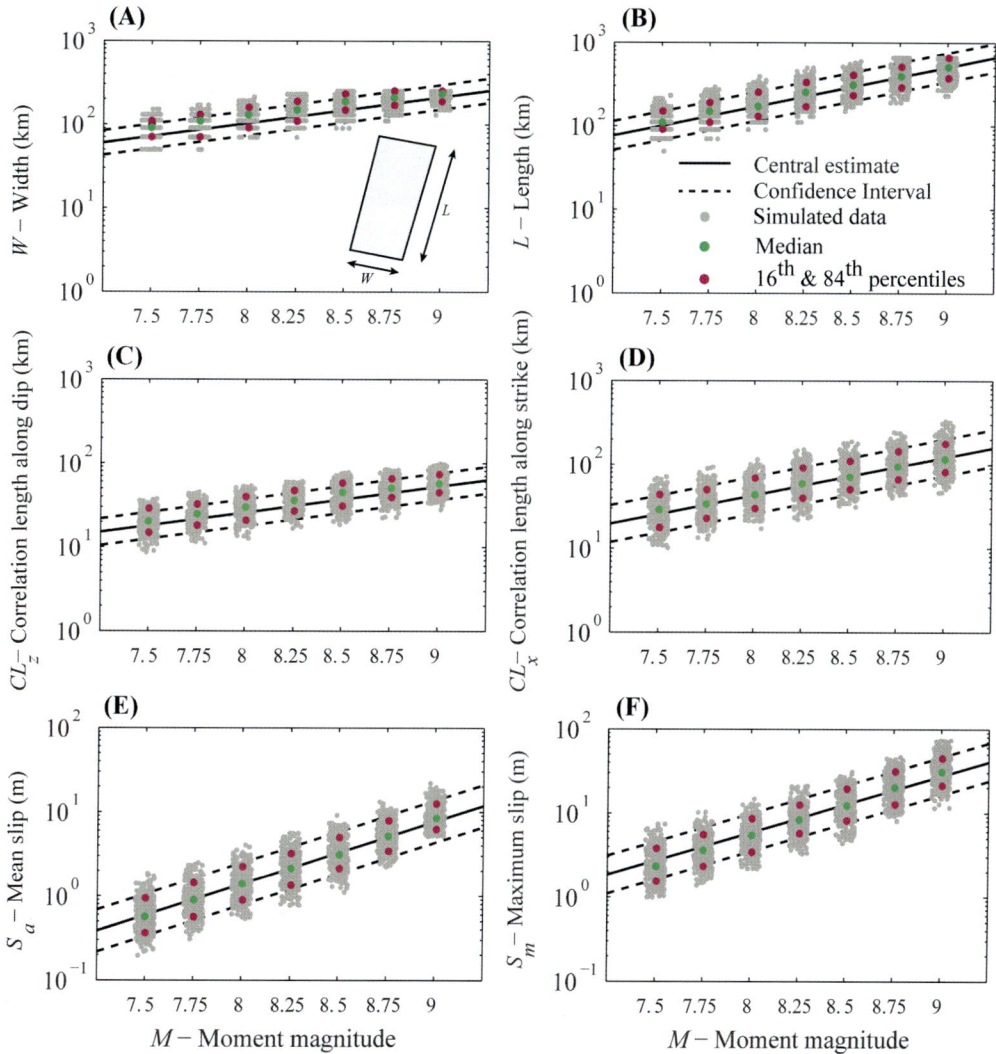

FIGURE 5.14 Scaling relationships for tsunamigenic earthquakes. (A) Rupture width versus moment magnitude. (B) Rupture length versus moment magnitude. (C) Correlation length along dip versus moment magnitude. (D) Correlation length along strike versus moment magnitude. (E) Mean slip versus moment magnitude. (F) Maximum slip versus moment magnitude. The simulated values (gray dots) and the corresponding percentiles (colored circles) are also shown.

5.3.3 Tsunami modeling

Tsunami *IM*s are not usually assessed using empirical prediction equations. Instead, for each stochastic event, the maximum inundation *IM*s for a specific location are computed through numerical tsunami simulations (Chapter 3).

The first step is computing the initial water surface elevation for a given earthquake slip scenario (Section 2.5.3). Specifically, assuming incompressibility of water, initial water surface elevation from the mean sea level can be considered equal to the seabed displacement field induced by the slip on the fault plane. Such a displacement field can be evaluated using analytical formulae for elastic dislocation (Okada, 1985; Tanioka & Satake, 1996). The calculated seafloor displacement may be filtered to account for the hydrodynamic response of the water. To optimize the computation of seafloor dislocation, the seafloor displacement field induced by a unity slip for each subfault can be computed in advance by creating a database of seabed displacement fields. To obtain the total effects of the *i*th slip distribution, each displacement field is scaled based on the slip in the sub-fault and summed.

Tsunami modeling is then carried out using a suitable numerical code capable of generating offshore tsunami propagation and inundation profiles by evaluating nonlinear shallow water equations with runup (Chapter 3). See Dutykh et al. (2011) for a summary of available computer codes in literature specific or adaptable to tsunami analysis (e.g., COMCOT, FUNWAVE, MOHID, TIDAL, TUNAMI, and SWAN). A comprehensive database of bathymetry and elevation, coastal/riverside structures (e.g., breakwater and levees), and surface roughness is required to run tsunami simulations.

5.3.4 Empirical tsunami hazard curve based on tsunami simulation results

A conceptual representation of the calculation of the empirical tsunami hazard curve is presented in Fig. 5.11. For each seismic source and for each magnitude, simulated *IM*s are used to evaluate the term $P_i(IM \geq im|m) = P_i(IM \geq im|\theta)S_i(\theta|m)$ for the locations of interest (Fig. 5.11A). Such probability is represented by the CCDF of the resulting *IM*. Specifically, *IM* is represented by the Kaplan–Meier estimator (Kaplan & Meier, 1958), being the central estimate

$$P(IM > im|m) = \prod_{IM < im} \frac{N_{\text{sim}}(IM \geq im|m) - N_{\text{sim}}(IM = im|m)}{N_{\text{sim}}(IM \geq im|m)} \tag{5.9}$$

where N_{sim} is the number of simulations per magnitude range. In addition, a confidence interval around the central estimate can be obtained by calculating the variance of the data through Greenwood's formula (Greenwood, 1926)

$$\text{Var}[P(IM > im|m)] = P(IM > im|m)^2 \sum_{IM < im} \frac{N_{\text{sim}}(IM = im|m)}{N_{\text{sim}}(IM \geq im|m)[N_{\text{sim}}(IM \geq im|m) - N_{\text{sim}}(IM = im|m)]} \tag{5.10}$$

A 95% confidence interval is generally deemed suitable to represent most of the tsunami scenarios in the hazard. A computer code for this analysis is provided in Example 5-5 (Appendix).

The hazard curves obtained in the previous step for each magnitude (i.e., conditional hazard curves) are then multiplied by the probability corresponding to the related magnitude (Eq. 5.7) and eventually are summed up (Eq. 5.8 and Fig. 5.11). Moreover, in this phase, three curves are obtained: one corresponding to the central value and two for the confidence interval.

Fig. 5.15 shows the results obtained from a case study in Sendai, Japan. Specifically, the CCDFs in terms of tsunami wave height are shown for all the magnitude values analyzed, spanning from M_w 7.5 to 9.0. Fig. 5.15B shows the CCDF weighted by the probability values obtained from the discretized G–R relationship (Fig. 5.13C). The final tsunami hazard curve, and its confidence interval, in terms of tsunami wave height, are presented in Fig. 5.16. A computer code for this analysis is provided in Example 5-6 (Appendix).

5.3.5 Hazard maps

The hazard computation procedure can be extended to obtain tsunami uniform hazard maps. The term "uniform" refers to the same annual probability of exceedance of the tsunami intensity values for all points in a given geographical region. Fig. 5.17 shows the tsunami uniform hazard maps for four values of return period: 30, 50, 475, and 2475 years

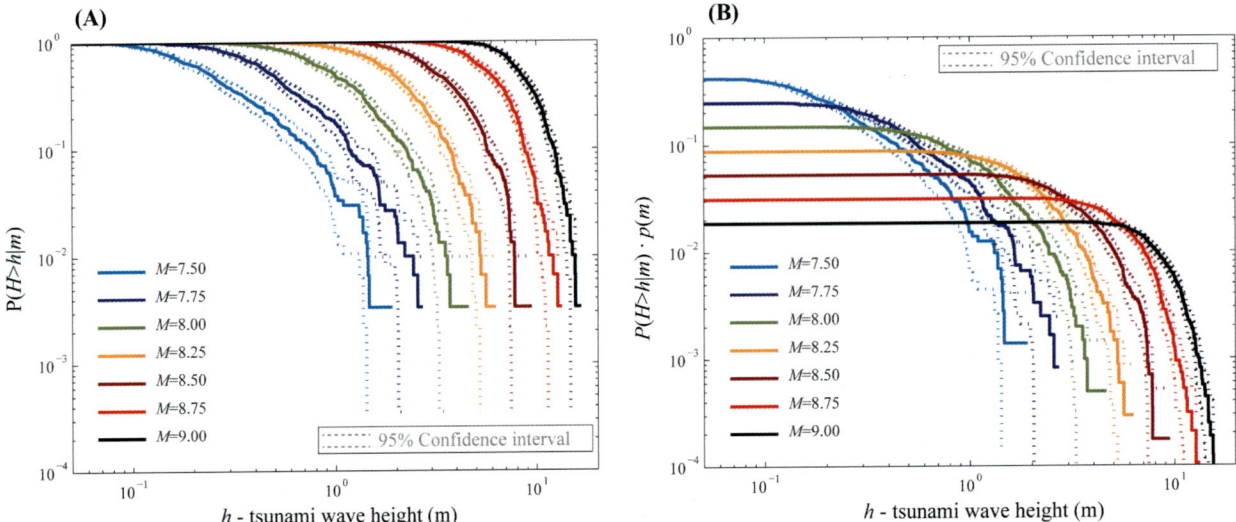

FIGURE 5.15 (A) Conditional hazard curves for tsunami wave height. (B) Weighted conditional hazard curves for tsunami wave height.

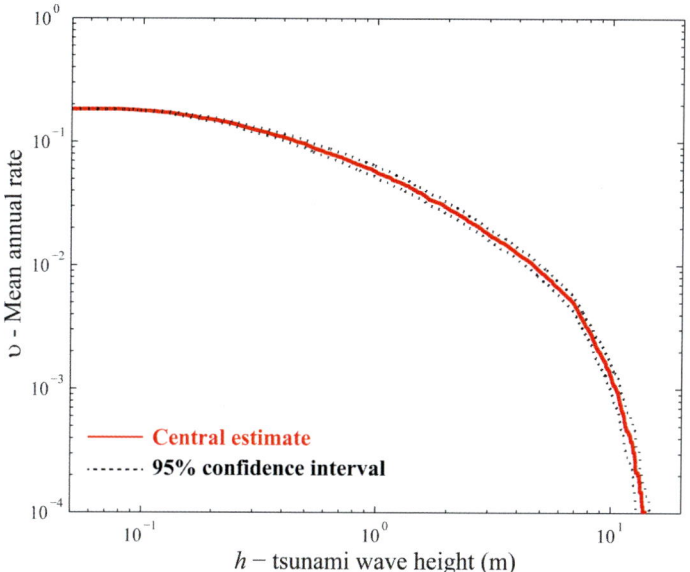

FIGURE 5.16 Final tsunami hazard curves.

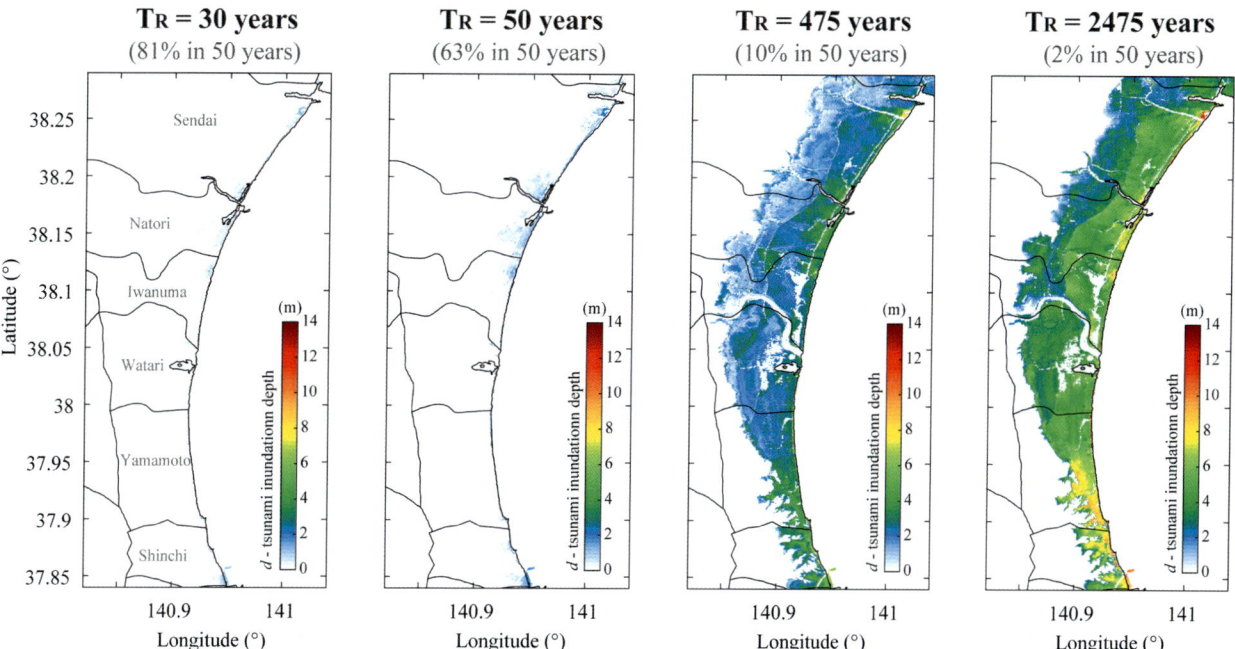

FIGURE 5.17 Tsunami uniform hazard maps for the plain coast of Miyagi Prefecture, Japan, in terms of tsunami inundation depth for four return periods.

(corresponding to 81.6%, 63.6%, 10%, and 2% probability of exceedance in 50 years) obtained for the coastal plain of Miyagi Prefecture in Japan. A computer code for this analysis is provided in Example 5-7 (Appendix). These maps are obtained by repeating the simulations shown for a single point for a set of grid points covering the area of interest. As tsunami *IM*s, maps show tsunami inundation depth. It is possible to observe that values of tsunami inundation depth increase with the reduction of the probability of occurrence, that is, with the decreasing mean annual rate. As expected, the tsunami depth maps show a decrease in the inundation depth with the distance from the shoreline.

It is important to stress that these maps are different from scenario-based maps (e.g., the worst case scenario). In fact, these maps do not reflect a specific rupture configuration for a specific source or magnitude. These maps are consistent with the classical seismic hazard maps that are used for seismic structural design.

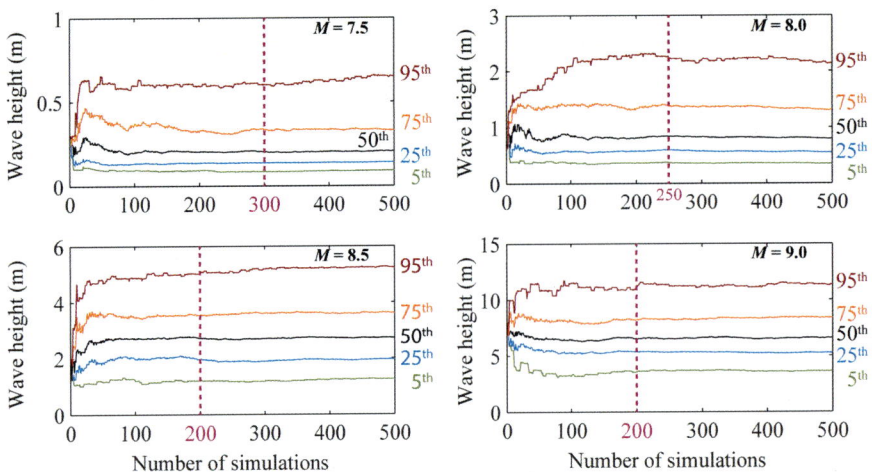

FIGURE 5.18 Bootstrap analysis for four values of magnitude.

Looking at the international building code scenarios, such maps represent the maximum considered tsunami, similar to what is expected by the American ASCE 7-22 (American Society of Civil Engineers, 2021) and the New Zealand MBIE 2020 (Ministry of Business Innovation and Employment, 2020). The Japanese AIJ 2015 (Architectural Institute of Japan, 2015) adopts the worst case scenario, whereas the Japanese JSCE 2016 for nuclear power plants (Japan Society of Civil Engineering, 2016) is more inclined to a probabilistic implementation that is presented in this book.

5.3.6 Optimal number of simulations

Short or incomplete data can lead to a biased estimation of the hazard parameters when conventional statistical methods are adopted (Lamarre et al., 1992). A bootstrap procedure can be carried out to study the effect of the number of simulations on the final hazard estimation. Through a Monte Carlo simulation, the bootstrap procedure provides the subsampling of a pool of m independent and identically distributed random variables from an initial sample of n elements (with $m \leq n$), having a distribution function identical to the empirical distribution function of the original sample. For each generated sample containing m elements, mean, median, and different percentiles can be computed. Such estimates can be used to quantify the uncertainty associated with these parameter values. A computer code for this analysis is provided in Example 5-8 (Appendix).

Fig. 5.18 shows an example of the Sendai coast in Japan. Five percentiles (i.e., 5th, 25th, 50th, 75th, and 95th) of the wave height are presented as a function of the number of simulations for different magnitude values (i.e., M_w 7.5, 8.0, 8.5, and 9.0). A fixed original sample of $n = 500$ simulations is considered for the analysis. The bootstrap procedure is then applied by considering the number of simulations m varying between 1 and 500; for each trial, a number of simulations m, 1000 Monte Carlo samples are realized.

Fig. 5.18 shows that the central estimate (i.e., the 50th percentile, represented by the black line) is stable after 200 simulations for all considered magnitude values. To obtain stable high percentiles, a larger number of simulations are needed. In particular, 300 simulations are necessary for M_w 7.5, 250 simulations for M_w 8.0, and 200 simulations for M_w 8.5 and M_w 9.0. Such a decreasing trend with the magnitude is consistent with the physical process. When the magnitude is relatively small, the rupture area can move more freely over the fault plane, resulting in the increased variability of the inundation IMs. When the magnitude is large, the fluctuation of the rupture area is more constrained (i.e., the major slip area tends to occupy the entire subduction plane).

5.3.7 Disaggregation

Disaggregation shows the relative contributions of dominant seismic scenarios to the specified hazard levels and is represented in terms of distance and magnitude (Bazzurro & Allin Cornell, 1999; McGuire, 2004). As for the seismic hazard case, the tsunami hazard disaggregation can be obtained as a byproduct of the probabilistic hazard procedure.

The disaggregation of the hazard for the same mean annual rate of occurrence can be computed according to the following equation

$$p(m, r | IM > im) = \frac{\sum_{i=1}^{N_S} \lambda_{Mmin,i} P_i(IM > im | \theta) S_i(\theta | m) f_{i,M}(m)}{v_{IM}(IM > im)} \tag{5.11}$$

In Eq. (5.11), the source-to-site distance r does not appear in the equation (like in the seismic hazard case), as this is obtained as a consequence of the slip simulation on the subduction plane, which is contained in the term $S_i(\theta|m)$. The simulation-based approach presented above to solve the hazard problem simplifies the computation of Eq. (5.11) because the problem can be transformed into counts of how many times a given IM is exceeded (or is observed) for specific magnitude values. Therefore, the disaggregation can be obtained as a byproduct of PTHA. Later in this chapter, more information will be provided regarding the computation of the distance.

Fig. 5.19 shows the disaggregation results for Sendai by considering tsunami inundation depth corresponding to two values of the mean annual rate of occurrence (i.e., 63% and 10% in 50 years). The 10% in 50 years hazard level (corresponding to an event with the 475-year return period) is commonly used to describe the life safety limit state, whereas the 63% in 50 years hazard level (corresponding to an event with the 50-year return period) corresponds to the damage control limit state. It is worth noting that only large-magnitude events contribute to higher values of IMs. Moreover, as observed before, the larger the magnitude is, the less the distance is influential.

5.4 Probabilistic seismic tsunami risk analysis

PTRA builds on PTHA. Eq. (5.3) is specialized for the earthquake-triggered tsunami, and the mean annual rate of having a loss L larger than or equal to a specific value l is given by

$$v_L(L \geq l) = \sum_{i=1}^{N_S} \lambda_{Mmin,i} \int P(L \geq l | ds) f_{DS|IM}(ds|im) f_{i,IM|\theta}(im|\theta) S_{i,\theta|M}(\theta|m) f_{i,M}(m) |dds||dim||d\theta||dm| \tag{5.12}$$

where $v_L(L \geq l)$ is the mean annual occurrence rate that the tsunami loss L for the exposed asset (e.g., a portfolio of buildings and infrastructure) exceeds a certain loss threshold l. The variables M, θ, IM, and DS are the earthquake magnitude, earthquake source parameters, tsunami IMs, and tsunami DSs, respectively. The integration should be performed over all random variables that are considered in Eq. (5.12). The key model components in Eq. (5.12) (terms that have been presented before, such as λ_{Mmin} and $S_{\theta|M}$, are omitted) are as follows:

- $f_{IM|\theta}$ is the probability density function of IM given θ and can be evaluated through tsunami simulations by solving the nonlinear shallow water equations (Goto et al., 1997) for initial boundary conditions of sea surface caused by earthquake rupture (Okada, 1985). IM can be obtained for a single location (e.g., water depth and flow velocity) as well as for some extended areas (e.g., inundated areas in a city).

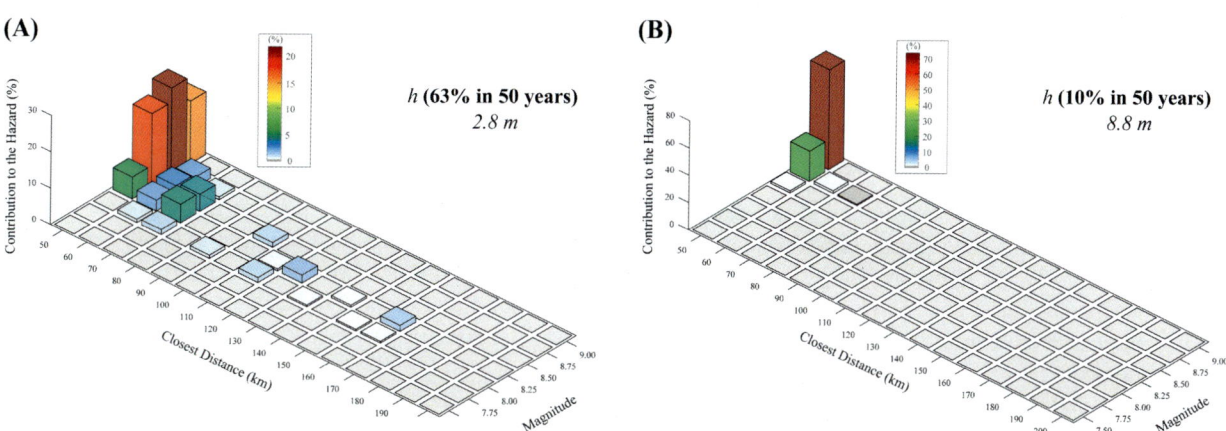

FIGURE 5.19 Disaggregation results for Sendai, Japan. (A) Tsunami hazard disaggregation for wave height corresponding to the hazard level of 63% in 50 years. (B) Tsunami hazard disaggregation for wave height corresponding to the hazard level of 10% in 50 years.

- $f_{DS|IM}$ is the tsunami fragility function, which predicts the probability of incurring a particular *DS* (e.g., collapse and complete damage) for a given *IM*. The fragility functions can be derived empirically (De Risi, Goda, Mori, et al., 2017; De Risi, Goda, Yasuda, et al., 2017) as well as analytically (Attary et al., 2017; Petrone et al., 2017). When adopting a bivariate fragility model, *IM* consists of a vector of multiple intensity values (e.g., depth and velocity). As an alternative to $f_{DS|IM}$, tsunami fragility analysis can be divided into two steps by introducing engineering demand parameters (*EDP*), that is, $f_{DS|IM} = f_{DS|EDP} \times f_{EDP|IM}$. Such refinements are particularly relevant for analytical fragility models, as often considered for performance-based earthquake engineering methods (Goulet et al., 2007; Porter et al., 2006).
- $P(L \geq l|ds)$ is the tsunami loss function given *DS*, which can be represented by the damage–loss function and the building cost model or the consequence function (when combined). A typical tsunami damage–loss function may be specified as a range of loss ratios for a given *DS* (Ministry of Land Infrastructure and Transportation, 2014). For instance, a complete *DS* may correspond to loss ratios between 0.5 and 1.0, expressed as a fraction of the total building replacement cost.

With respect to the information provided previously in this chapter and in this book, the critical elements that are needed for assessing the risk are the exposure and the vulnerability models. Two components and the terms $f_{DS|IM}$ and $P(L \geq l|ds)$ are presented in the following sections.

5.4.1 Exposure characterization

In the risk integral, the exposure identifies the assets under threat due to natural hazards. More specifically, an exposure model quantifies the geographical distribution of affected structures, movable property, infrastructure, and population and estimates the costs associated with the interruption of normal economic activity. Examples of exposure are portfolios of residential buildings, population, cultural heritage, critical buildings (e.g., hospitals, barracks, and schools), energy facilities (e.g., nuclear power plants and refineries), bridges, airports, and harbors. In general, components of the built and natural environments can be at risk and should be considered for the description of the exposure (Murnane et al., 2019).

The characterization of the exposure is paramount for two reasons. First, quantifying what is at risk allows defining (and creating, if they are lacking) vulnerability models, correlating the hazard with consequences. Second, a clear classification allows applying surveying methods tailored to quantify the economic value of the asset at risk, which is essential for quantifying the losses. Exposure characterization is not a trivial task and can be very time-consuming and resource-intensive.

Several techniques exist for acquiring exposure data, each leading to a different resolution and precision (Pittore et al., 2017). Data provided by governmental, national, and international agencies are the most commonly used resources for exposure characterization, as they are often openly available, comprehensive, and regularly updated for asset management (e.g., national technical maps) and fiscal reasons (e.g., cadastral data). Such data allow the characterization of exposure over large geographical areas and at the level of individual buildings; however, data are not collected for tsunami risk assessments, and essential information may be missing. Common issues in using such data are as follows:

- Cadastral data are often collected at the residential unit level, meaning that assumptions need to be made about how many units are in a building for the tsunami risk assessment.
- Specific details, such as construction material, are often indicated, but other features, such as structural system and foundation system, are not provided.

To account for such general issues, it is common to define some taxonomy properties for the exposure known as modifiers. Specifically, the primary and secondary modifiers are general parameters that are used to (1) differentiate the potential damage to the exposure and (2) differentiate them in terms of the peril of interest. The primary modifiers include occupancy class, construction type, construction year, and building height. The secondary modifiers are peril-specific and can include roof type and anchoring (wind), secondary defenses of property's contents (flood), and retrofitting (earthquake).

Governmental data, sometimes, may not be available. This applies especially to developing countries, where cities are under rapid urbanization and planning is not consistently enforced. In such cases, it is possible to collect exposure data from high-resolution satellite imagery, aerial orthophotos, and data made available from international projects (e.g., NASA's Earth Observing System Data and Information System) or using data coming from Volunteered Geographic Information systems (e.g., OpenStreetMap).

It is, therefore, clear that the geographical characterization and the location of each exposed asset are of primary importance. The main reason for this is that exposure features may be geographically correlated (e.g., structural typologies in the same geographical area may be similar). If they are nearly identical, it is possible to aggregate the exposure to different geographical resolutions. A classical geographical resolution is as follows (from detailed to coarse): (1) building, (2) street, (3) postal code, (4) city/town, (5) district, (6) state/municipality, and (7) country. On the other hand, the importance of geographical resolution can be understood with a simple example. When geographical coordinates are expressed with six decimal places, the resolution is 10 cm; moving to four decimal places, the resolution increases to 10 m.

Several international initiatives exist to create exposure models with a taxonomy that is as general as possible. Examples are the LandScan database, which provides the population with a lattice resolution of 1 km; the GED4GEM database, which provides the building stock worldwide; the GED-13 database that provides the residential building stock with a resolution of 5 km; the WorldPop database that provides the population for the entire globe at 1 km and 100 m resolutions; and the METEOR database that provides the current and the future building stock with a resolution of the 15-arcsecond grid for some countries. By all means, this list is not exhaustive; other resources exist, and future initiatives will provide more data.

Fig. 5.20 illustrates the exposure of Onagawa, Miyagi Prefecture, Japan. There are about 3400 buildings (Fig. 5.20A) and a population of about 6000 people. WorldPop provides the population distribution at 1-km and 100-m resolution (Fig. 5.20B and C, respectively). The data are well correlated, indicating that the population is concentrated in areas with buildings. In the area, the population may be distributed among public and private buildings. Moreover, buildings can have different structural typologies, for example, reinforced concrete, masonry, and timber. Since it is well known that the predominant structural typology consists of timber buildings in the geographical area of interest, more attention is given to low-rise wooden structures in this example. In such a case, it is possible to retrieve the mean unit construction cost and the coefficient of variation from Construction Research Institute (2011).

To study the consequence of any potential damage to the structures, it is common practice to use the definition of DS. The DS defines the damage and failure mechanisms of an engineering system (or of a single component) and their consequences (e.g., states associated with consequences in terms of costs, loss of lives, and impact on the environment). The definition of DSs provides an easy way to categorize the damage and eventually compute an expected loss associated with the damage. The repair/replacement costs associated with DSs are known as damage ratios (DRs). An example of mapping between DR and DS is 0%, 5%, 20%, 40%, 60%, and 100% for no damage (DS1), minor damage (DS2), moderate damage (DS3), major damage (DS4), complete damage (DS5), and collapse-wash-away (DS6), respectively (Chapter 4). Fig. 5.21 shows the distribution of the normalized construction cost and the repair/replacement costs associated with this case study. The availability of the geographical location of the exposed asset, and the financial and human capital at stake, consists of a complete definition of the consequence function associated with the exposure of

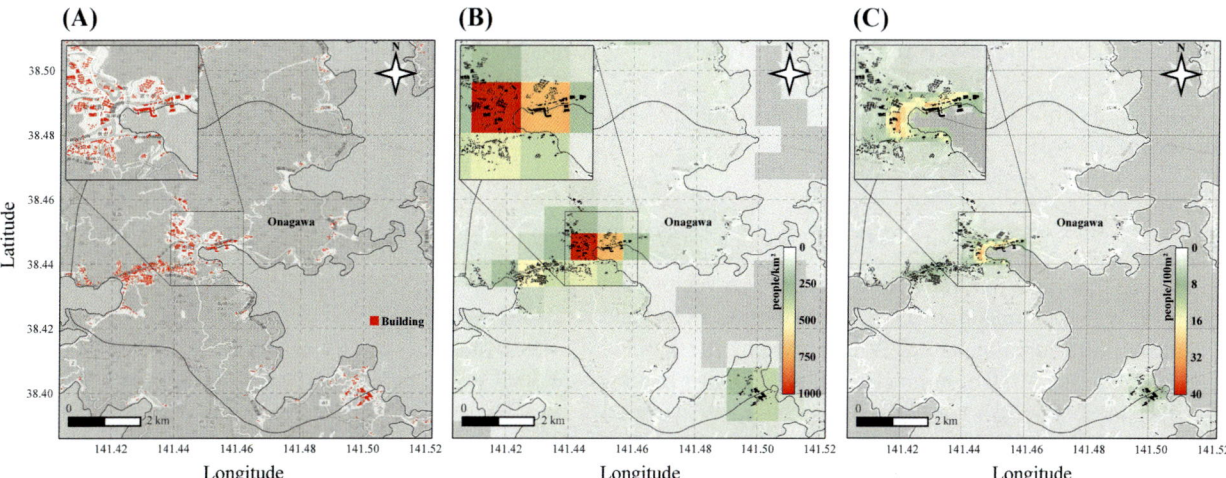

FIGURE 5.20 (A) Building distribution from OpenStreetMap in Onagawa, Japan, (B) 1-km resolution population counts in Onagawa in 2020 according to WorldPop, and (C) 100-m resolution population counts in Onagawa in 2020 according to WorldPop.

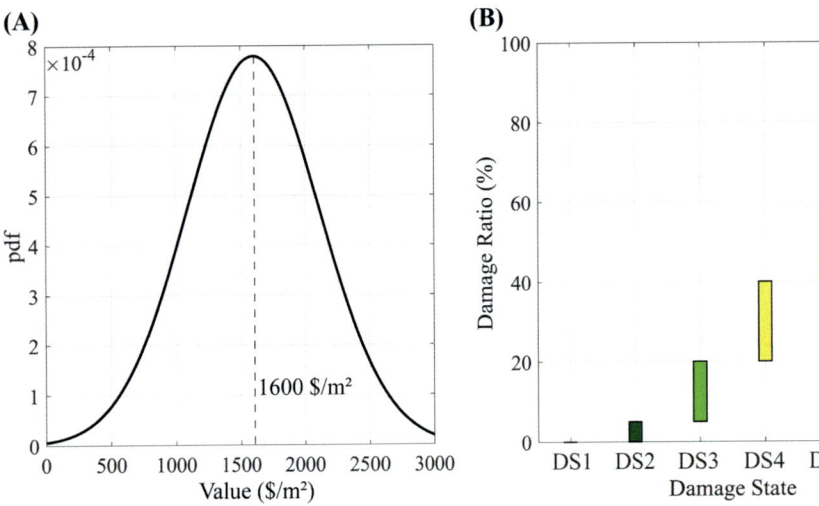

FIGURE 5.21 (A) Construction cost distribution and (B) damage ratios for different damage states.

interest that can now be used as one of the components of the risk convolution. A computer code for this analysis is provided in Example 5-9 (Appendix). Moreover, how to use the distribution in Fig. 5.21 in a simulation-based approach is described in the computer code of Example 5-10 (Appendix).

Data quality is paramount for developing an exposure database. Therefore, several checking operations are needed. First, standard mapping tables are critical to transforming the attributes of the original data into the standardized inputs of the final exposure database. Moreover, it is important to avoid duplicated data and erroneous/nonsensical data. Once the database of the exposed asset is compiled, a validation phase is needed to ensure that the data are homogeneous and up-to-date. Such validation can be performed by comparing data at multiple scales or with bespoke surveys, which can be very costly and time-consuming if not automated with modern image recognition techniques.

5.4.2 Vulnerability characterization

The vulnerability characterization connects tsunami hazard intensity to tsunami damage. Such a link involves both epistemic (e.g., material strength and details of the primary and secondary modifiers) and aleatory (e.g., wave-loading features) uncertainties. As already introduced in Chapter 4, a conventional way of discretizing the *DS* consists of defining a set of descriptive indicators, such as those defined by the Japanese Ministry of Land, Infrastructure, and Transportation (MLIT; Ministry of Land Infrastructure and Transportation, 2014), which adopts the tsunami damage scale with seven discrete states, namely, no damage (*DS1*) to wash-away (*DS7*). In several studies, the last two *DS*s (i.e., *DS6* and *DS7*) are combined as it is recognized that both collapse and wash-away are two different descriptions of a collapse mode (Charvet et al., 2014; De Risi, Goda, Mori, et al., 2017; De Risi, Goda, Yasuda, et al., 2017); in such a case, the *DS* numbering goes from 1 to 6. It is important to recognize that different studies have adopted different definitions of the *DS*s.

Vulnerability models come in the forms of fragility functions (or simply fragilities, Eq. (5.13)) and vulnerability functions (Eq. 5.14)

$$P(DS \geq ds|\boldsymbol{im}) \tag{5.13}$$

$$P(DR \geq dr|\boldsymbol{im}) \tag{5.14}$$

It is worth noting that **im** is represented as a vector (bold font) because these models can be functions of multiple tsunami *IM*s (e.g., inundation depth, inundation velocity, and momentum flux).

Vulnerability functions are CCDFs providing the probability of reaching or exceeding a specific *DR* conditioned on a specific level of *IM*. These functions combine information contained in fragility functions and consequence functions presented in Section 5.4.1 as *DR*s for specific *DS*s. The term presented in Eq. (5.14), multiplied by the value of the asset, can replace the product $P(L \geq l|ds) f_{DS|IM}(ds|im)$ in Eq. (5.12).

Tsunami vulnerability and fragility models are not as developed as for the seismic case; the available models are few and are mainly built after the empirical observations following past tsunami events. As discussed in Chapter 4, both fragility and vulnerability functions can be derived using (1) analytical, (2) empirical, and (3) hybrid methods

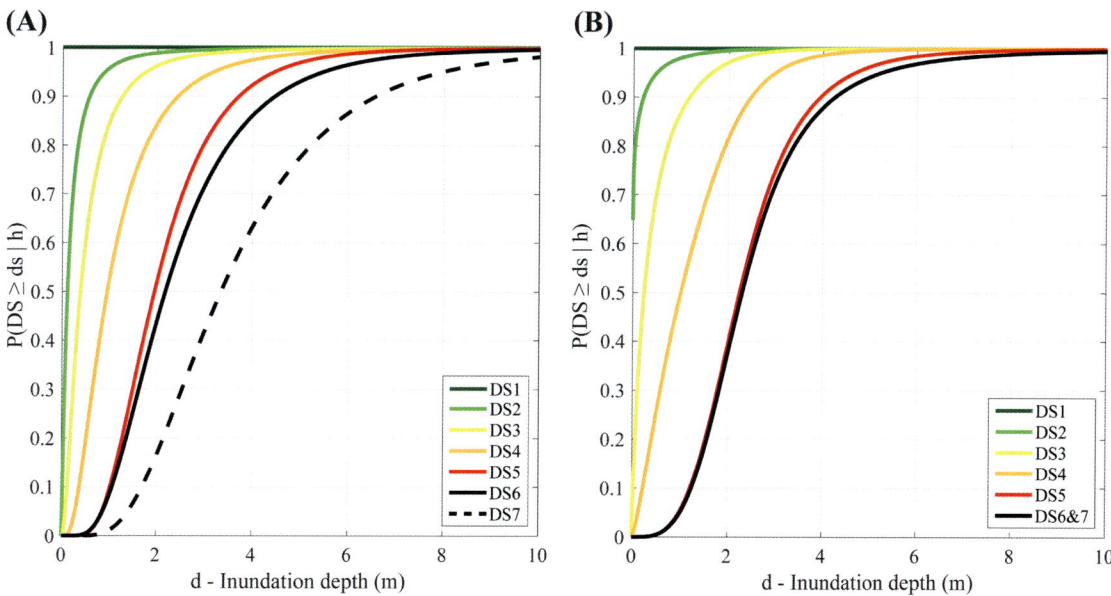

FIGURE 5.22 (A) Lognormal fragilities (Suppasri et al., 2013). (B) Multinomial fragilities (De Risi, Goda, Yasuda, et al., 2017).

combining the first two. Fig. 5.22 shows two fragility models for low-rise timber buildings derived from the damage observations collected in the aftermath of the 2011 Tohoku Earthquake and Tsunami. Specifically, Fig. 5.22A shows a lognormal model, and Fig. 5.22B shows a more sophisticated multinomial logistic model. It is possible to observe that the lognormal fragilities obtained according to Suppasri et al. (2013) are reported for the seven DSs distinction proposed by the MLIT. On the contrary, the multinomial fragilities are derived for the case where the last two DSs are combined (De Risi, Goda, Yasuda, et al., 2017). In addition to that, the fragility curves reported in Fig. 5.22B are obtained considering a flow velocity of 1 m/second; in fact, the multinomial case can incorporate multiple IMs in the equation. A computer code for this analysis is provided in Example 5-11 (Appendix). Moreover, how to use fragility curves in a simulation-based approach is provided in the computer code reported in Example 5-12 (Appendix).

5.4.3 Numerical evaluation of tsunami risk equation

The PTRA presented above can be performed (i.e., evaluation of Eq. (5.12)) using Monte Carlo tsunami simulations and tsunami damage—loss analyses. The calculation steps are illustrated in Fig. 5.23.

- The first step is the identification of the exposure (i.e., the portfolio of buildings at risk in a selected urban area, Fig. 5.23A).
- The second step is the computation of the IMs at the portfolio scale ($f_{IM|\theta}$, Fig. 5.23B); this consists of the computation of the tsunami hazard for each value of magnitude used to discretize the G—R relationship. Eventually, IMs can be simulated at each building location (Fig. 5.23C).
- The third step consists of calculating, for each building, the probability of attaining a specific DS for each simulated IM ($f_{DS|IM}$, Fig. 5.23D). To evaluate the tsunami damage (i.e., $f_{DS|IM}$), for each tsunami simulation, tsunami fragility models can be used (De Risi, Goda, Mori, et al., 2017; De Risi, Goda, Yasuda, et al., 2017). For each structure, a random number from the standard uniform distribution is compared with the DS probabilities (Fig. 5.23D). This determines the realized DS for this structure during the considered tsunami event. A computer code for this analysis is provided in Example 5-12 (Appendix).
- The fourth step consists of assigning a value that is sampled from a range of DRs to each realized DS. Each DS is related to a range of loss ratios (Fig. 5.23E). The loss ratio ranges are associated with minor, moderate, extensive, complete, and collapse DSs. The uniform distribution is assumed for the loss ratios. Note that the loss ratios are applied to the total cost of a building, including structural and nonstructural elements. By sampling the entire building cost of stores/offices and houses (which is usually lognormally distributed), the tsunami damage cost can be estimated (i.e., $P(L \geq l|ds)$). The building cost models are typically estimated based on two sources of information: unit building cost statistics and floor area statistics. A computer code for this analysis is provided in Example 5-10 (Appendix).

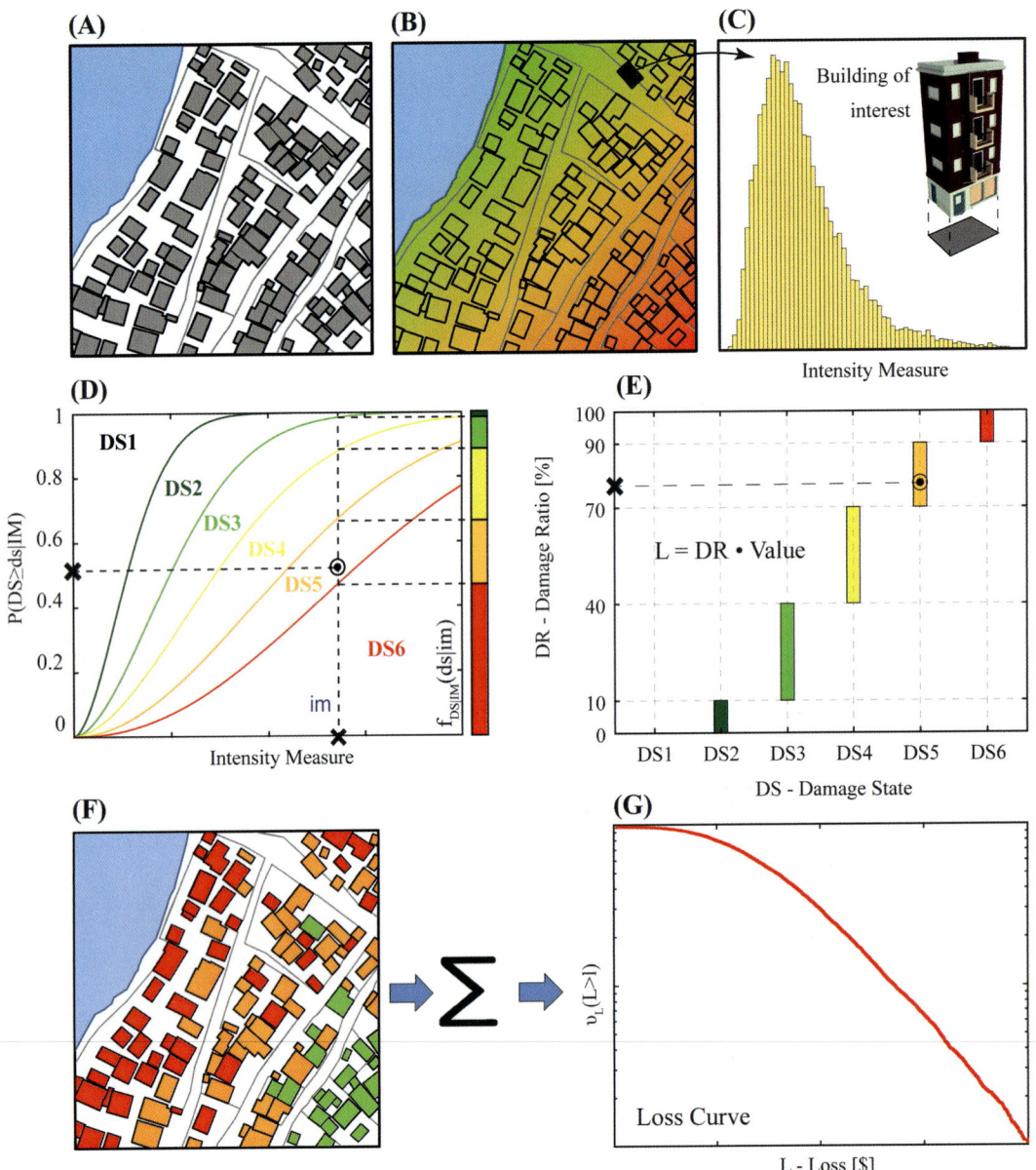

FIGURE 5.23 Tsunami risk assessment framework: (A) portfolio of buildings of interest, (B) hazard map for a simulated scenario, (C) intensity measures for each building, (D) fragility curves, (E) damage ratios, (F) risk maps, and (G) loss curve.

The Monte Carlo sampling is repeated for all structures and earthquake scenarios. Averaging the losses building by building, it is possible to obtain risk maps (Fig. 5.23F). Convoluting the losses together at the portfolio level, it is possible to obtain the whole loss curve (Fig. 5.23G). As in the hazard case, the optimal number of simulations can be assessed using bootstrap analysis, as indicated in Section 5.3.6. Computer codes for this analysis are provided in Examples 5-13 and 5-14 (Appendix).

5.4.4 Risk metrics

The loss curve is the main output from which it is possible to compute several risk metrics that are useful for different purposes. For example, it is possible to select loss values corresponding to specific return periods ($T_R = 1/\lambda$); the loss value at a particular quantile of the relevant loss distribution is known as the value at risk. Another metric is to compute the area under the loss curve expressed in terms of mean annual rate; such an area is known as the AAL or expected annual loss. This computation is easy when only economic aspects are accounted for. The problem becomes challenging when injuries and casualties are considered. The U.S. Federal Emergency Management Agency monetizes the population's human injury—fatality costs by using a value of statistical life approach where each fatality or 10 injuries is treated as $11.6 million in terms of economic loss.

5.5 Extension to multihazard risks

5.5.1 Compound earthquake—tsunami hazard and risk assessment

Let IM represent the vector of IMs of interest, such as spectral acceleration, inundation depth (h), flow velocity (v), flux momentum, and tsunami force. Eq. (5.1) can now be expressed as

$$\upsilon_{IM}(\boldsymbol{IM} \geq \boldsymbol{im}) = \sum_{i=1}^{N_S} \lambda_{M\min,i} \int P_i(\boldsymbol{IM} \geq \boldsymbol{im}|\theta) f_{i,\theta|M}(\theta|m) f_{i,M}(m) |d\theta| |dm| \tag{5.15}$$

where $P_i(\boldsymbol{IM} \geq \boldsymbol{im}|\theta)$ is the joint probability that the $\boldsymbol{IM}$ will exceed prescribed values $\boldsymbol{im}$ at a given coastal location for a given set of source parameters θ. With respect to Section 5.3, a further step (phase 3) is needed in addition to the four steps identified before for the PTHA:

- Definition of input data (i.e., geometrical characteristics of each seismic source and magnitude-frequency distribution).
- Stochastic source simulation based on earthquake source scaling relationships.
- Earthquake simulation.
- Tsunami inundation simulation.
- Development of joint earthquake—tsunami hazard curves.

Fig. 5.24 shows the phases 3 and 4 together. Phase 5 remains identical to the phase 4, illustrated in Fig. 5.11.

5.5.2 Earthquake simulation

GMMs are extensively used to predict seismic IMs for a given earthquake scenario (Wald et al., 2005). To account for seismic intensities at multiple locations for a given event, GMMs together with spatial correlations in the regression residuals, can be treated as statistical prediction models (Goda & Atkinson, 2010; Goda, 2011). This feature is particularly important in extending the seismic hazard assessment into a risk assessment of a portfolio of buildings/infrastructures. To simplify the simulation procedure, only the intraevent standard deviation can be propagated through the

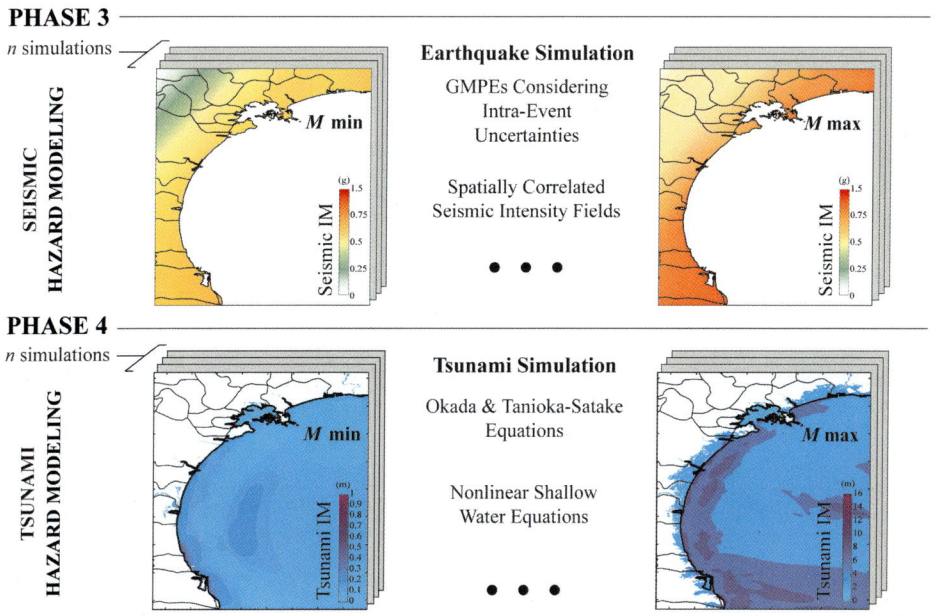

FIGURE 5.24 Phases 3 and 4 of the multihazard framework.

simulation procedure; such a choice is consistent with the simulation scenario of a single fault plane. GMMs developed for subduction zones are generally used for seismic simulations. An example of GMM is the model developed by Abrahamson et al. (2016); in this case, a global dataset of earthquakes in subduction zones was adopted and subsequently modified to consider the 2010 Maule Chile Earthquake and the 2011 Tohoku Japan Earthquake, which were not in the initial database. A second example of GMM is the model developed by Morikawa and Fujiwara (2013), which is suitable for M_w 9 earthquakes in Japan.

For seismic simulations, three main inputs are required: event magnitude, distance from the rupture, and near-surface shear wave velocity for the considered site. Regarding the distance from the rupture, the GMMs mentioned above are both based on the closest distance between the location of interest and the rupture area (Fig. 5.25). To optimize the computation of the shortest distance, the distances between the coastline location and each discretized element of a fault (e.g., the 2011 Tohoku-type fault) can be precomputed and stored (Fig. 5.25B). As an example, Fig. 5.25C shows the distances computed for 500 stochastic scenarios. It is worth noting that, for the specific case study reported in the figure, the minimum distance is approximately 50 km, corresponding to the depth of the fault plane under the considered location.

Goda and Atkinson (2010) observed that the source-to-site distance is influenced by both the location and size of the fault plane, which, in turn, are determined by the event magnitude. In Fig. 5.25C, it can be observed that the greater magnitude value results in a smaller variability of the closest distance (i.e., the distribution function has a steeper slope). This is because the rupture plane can move more freely within the overall fault plane when the earthquake magnitude is small. For the shear wave velocity, the U.S. Geological Survey global V_{S30} map server can be used when more detailed microzonation data are not available (Wald & Allen, 2007).

Finally, the multivariate lognormal distribution can be employed to generate shake maps of *IM*s. The median values of *IM* at sites of interest are calculated from the GMMs, whereas their variances are based on the intraevent components. The prediction errors in the GMMs are spatially correlated, and the correlation coefficient matrix has diagonal elements equal to one and off-diagonal elements equal to the correlation coefficient. The correlation coefficient can be calculated using equations found in the existing literature (Goda & Atkinson, 2010).

5.5.3 Development of joint earthquake–tsunami hazard curves

This procedure of developing a joint earthquake–tsunami hazard curve is identical to the one presented in Section 5.3.4 but is conducted at a vectorial level. For each value of magnitude, the simulations are used to evaluate the term

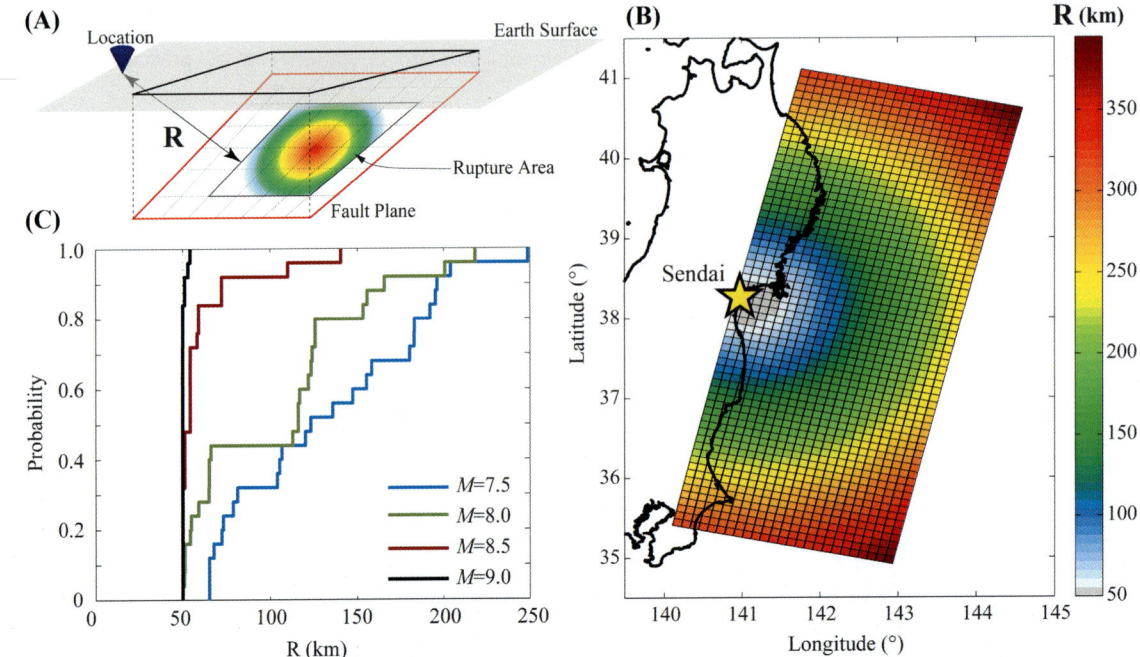

FIGURE 5.25 (A) Sketch of the shortest distance. (B) Precomputed distances between a location and the discretized fault plane. (C) Empirical distribution of 500 stochastic simulations for four increasing magnitude values.

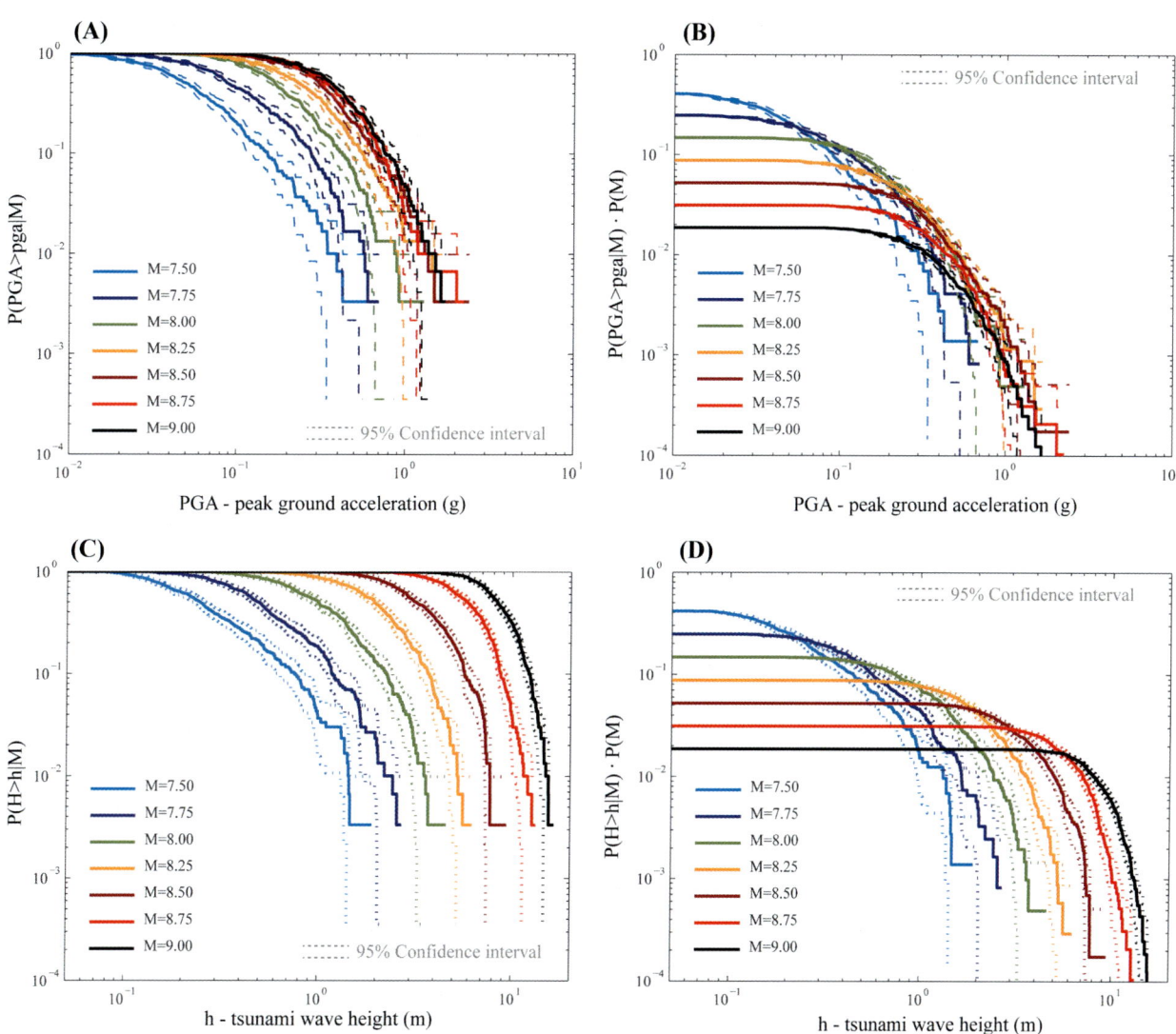

FIGURE 5.26 (A) Conditional hazard curves for PGA. (B) Weighted conditional hazard curves for PGA. (C) Conditional hazard curves for tsunami wave height. (D) Weighted conditional hazard curves for tsunami wave height. *PGA*, Peak ground acceleration.

$P(IM \geq im|m)$ for the location of interest. Such probability is represented by the CCDF of IM. Specifically, the CCDF of IM (i.e., spectral acceleration or tsunami inundation) is obtained as the Kaplan–Meier estimator (Kaplan & Meier, 1958), for which the variance can be calculated through Greenwood's formula (Greenwood, 1926). Therefore, a confidence interval around the central estimate can be obtained. For example, the 95% confidence interval can be considered.

The curves obtained in the previous step for each magnitude are then multiplied by the probabilities corresponding to the related magnitude and eventually summed up. Moreover, in this case, three curves are obtained, one corresponding to the central value and two for the confidence interval. The final joint hazard curves, representing the mean annual rate of occurrence of specific values of earthquake–tsunami IMs, are obtained by multiplying the previous three conditional curves (for each hazard) by the occurrence rate of events with magnitudes greater than the minimum magnitude considered in the magnitude–frequency distribution.

Fig. 5.26 shows the results of a case study in Sendai, Japan. Specifically, the conditional CCDFs in terms of peak ground acceleration (PGA) and tsunami wave height are shown in Fig. 5.26A and Fig. 5.26C, respectively, for all magnitude values analyzed. Figs. 5.26B and D show the CCDF, weighted by the probability values obtained from the discretized G–R relationship (Fig. 5.13C). The final hazard curves for PGA and tsunami wave height are presented in Figs. 5.27A and B. Fig. 5.27A shows the final hazard curves and the 95% confidence interval for PGA obtained using

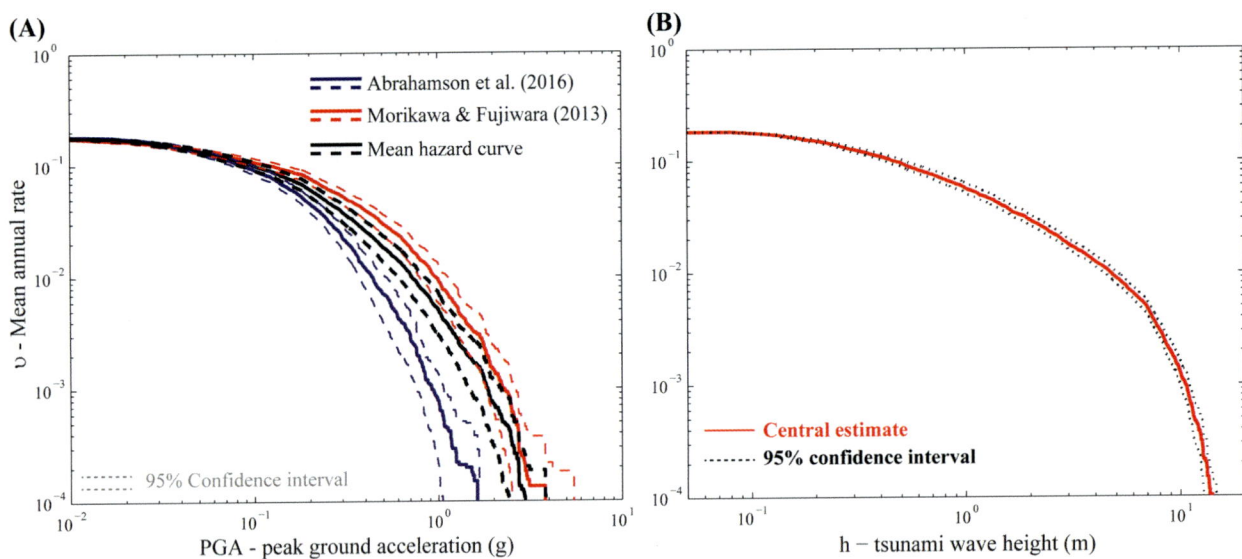

FIGURE 5.27 (A) Final seismic hazard curves for PGA. (B) Final tsunami hazard curves. *PGA*, Peak ground acceleration.

the two GMMs (blue and red curves). The same figure represents the weighted-average seismic hazard curves with black lines. Similarly, Fig. 5.27B shows the final tsunami hazard curve and its 95% confidence interval, noting that the confidence interval for the tsunami height is tight around the central estimate curve.

5.5.4 Earthquake–tsunami uniform hazard maps

To create a more resilient community, the knowledge of the extreme earthquake–tsunami intensities at municipality levels, associated with a specific mean annual rate of occurrence, is important and useful since it allows for preparing effective risk management plans in advance. Potential management plans include the design of evacuation routes and tsunami defense locations as well as upgrading/retrofitting existing buildings.

The procedure explained in Section 5.3.5 can be extended to obtain earthquake–tsunami uniform hazard maps for a coastal community. The term "uniform" refers to the same annual probability of exceedance of the earthquake and tsunami intensity values. As an example, a coastal neighborhood in Sendai is focused upon (Fig. 5.28); the area was devastated by the 2011 Tohoku Tsunami. According to the MLIT damage database, almost all buildings had collapsed or washed away (orange and red in Fig. 5.28A). Fig. 5.28A also shows a gradual decrease in damage severity when moving from the coast toward the inland areas.

The case study area (Fig. 5.28B) is located about 200 m from the coastline and has a size of 680 m × 180 m. In this area, there were 318 buildings, more than 90% of which were timber structures and were all destroyed by the 2011 Tsunami. Fig. 5.29 shows the earthquake–tsunami uniform hazard maps for three values of mean annual rate: (A,B) 2%, (C,D) 5%, and (E,F) 10% in 50 years. These maps are obtained by repeating the simulations shown for a single point, for a set of grid points covering the area of interest; in this case, a grid of 5 m × 5 m is adopted. PGA is adopted as the seismic *IM*; alternatively, different seismic *IM*s (e.g., spectral acceleration) can be considered. As a tsunami *IM*, maps show tsunami flow depths (i.e., inland tsunami wave heights corrected for the local topography/elevations). It is possible to observe that values of PGA and tsunami inundation depth decrease with the reduction of the probability of occurrence, that is, with the increasing mean annual rate. The PGA map for a given mean annual rate is more or less uniform because the source-to-site distance for a given scenario is nearly constant; the fluctuation of the PGA values is caused by the intraevent variability of ground motions. On the other hand, the tsunami flow depth map shows the decrease in the inundation depth with the distance from the shoreline.

5.5.5 Earthquake–tsunami disaggregation

The earthquake–tsunami hazard disaggregation can be obtained as a byproduct of the earthquake–tsunami hazard assessment. As explained in Section 5.3.7, the disaggregation provides information on the relative contributions of dominant seismic scenarios to specified hazard levels. Seismic scenarios are represented in terms of source-to-site distance

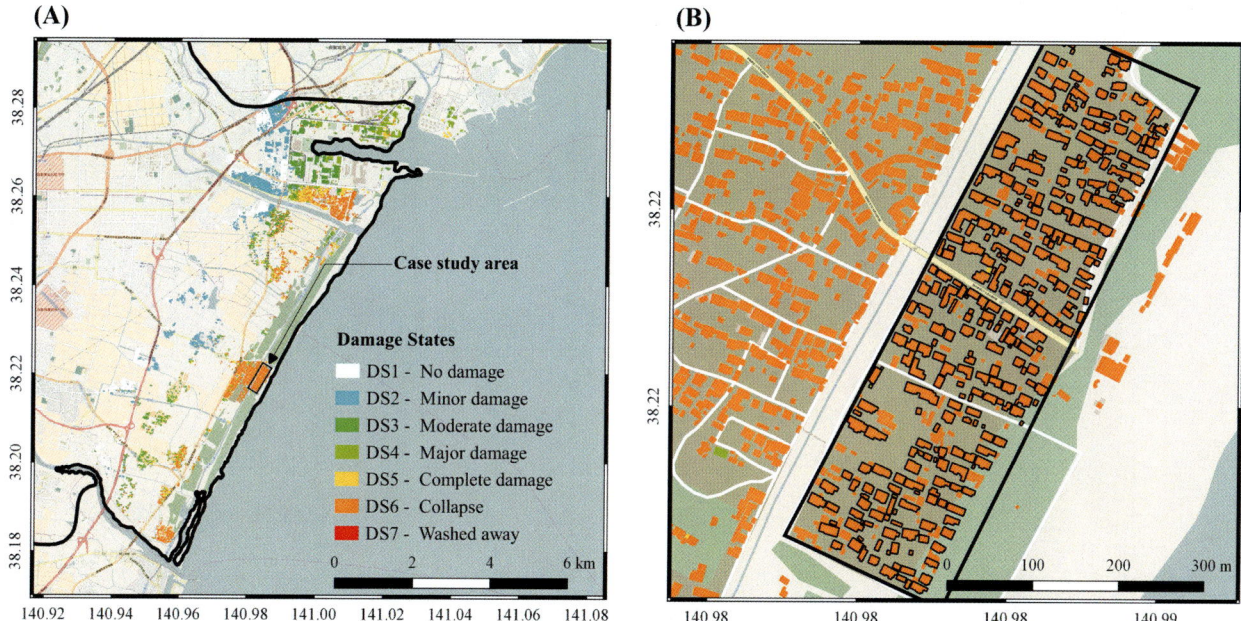

FIGURE 5.28 (A) Damage to the buildings in Sendai according to the MLIT database. (B) Zoom up of the case study area. *MLIT*, Ministry of Land, Infrastructure, and Transportation.

and magnitude. Fig. 5.30 shows the disaggregation results for Sendai by considering PGA (Fig. 5.30A) and tsunami wave height (Fig. 5.30B), corresponding to a mean annual rate of occurrence of 50% in 50 years. It is worth noting that only large-magnitude events contribute to higher values of *IM*s. Moreover, as observed in Section 5.3.7, the larger the magnitude is, the less the distance is influential. Finally, it is worth noting that the combinations of magnitude and distance significantly affecting the seismic and tsunami hazards are different.

This tool makes clear that although the two hazards are compound, their intensity for the same source can be significantly different. Such a result is even more evident by looking at the earthquake−tsunami risk assessment presented in the next subsection.

5.5.6 Earthquake−tsunami risk

Several recent studies have proposed multirisk approaches (Liu et al., 2015; Marzocchi et al., 2012; Mignan et al., 2014; Ming et al., 2015; Selva, 2013) by integrating the PEER formulation with new comprehensive methods that try to homogenize multiple risk components in different ways. Three main aspects of the multirisk framework that have been investigated in literature are (1) joint probability of relevant hazards, (2) vulnerability models that account for time-variant multihazard, and (3) combination of loss for each hazard in a coherent manner.

In this context, taking advantage of the multihazard procedure described in the previous section, Goda and De Risi (2017, 2018) have developed a multihazard loss estimation methodology for earthquakes and tsunamis. This formulation generalizes the formulation provided in Eq. (5.3). The developed multihazard loss estimation method can be regarded as a tool for implementing the performance-based earthquake−tsunami engineering methodology, particularly applicable to multiple buildings in coastal regions. The main equation for the loss assessment of a portfolio of buildings is given by

$$\upsilon_L(L \geq l) = \sum_{i=1}^{N_S} \lambda_{M\min,i} \int P_i(L \geq l|\boldsymbol{ds}) f_{i,DS|IM}(\boldsymbol{ds}|\boldsymbol{IM}) f_{i,IM|\theta}(\boldsymbol{im}|\boldsymbol{\theta}) S_{i,\theta|M}(\boldsymbol{\theta}|m) f_{i,M}(m) |d\boldsymbol{ds}||d\boldsymbol{im}||d\boldsymbol{\theta}||dm| \quad (5.16)$$

where, in addition to the terms presented in Eq. (5.3), $f(\boldsymbol{im}|\boldsymbol{\theta})$ is the joint probability density function of *IM* given $\boldsymbol{\theta}$, $f(\boldsymbol{ds}|\boldsymbol{im})$ is the earthquake−tsunami fragility function, which predicts the probability of incurring a particular *DS* for given *IM*, $P(L \geq l|\boldsymbol{ds})$ is the damage−loss function given *DS*, and $\nu_L(L \geq l)$ is the mean annual occurrence rate that the total loss *L* for a portfolio of buildings caused by earthquake and tsunami exceeds a certain loss threshold *l*. It is worth noting that in this case, the *DS*s are considered as vectorial as they are associated with both the earthquake and the tsunami.

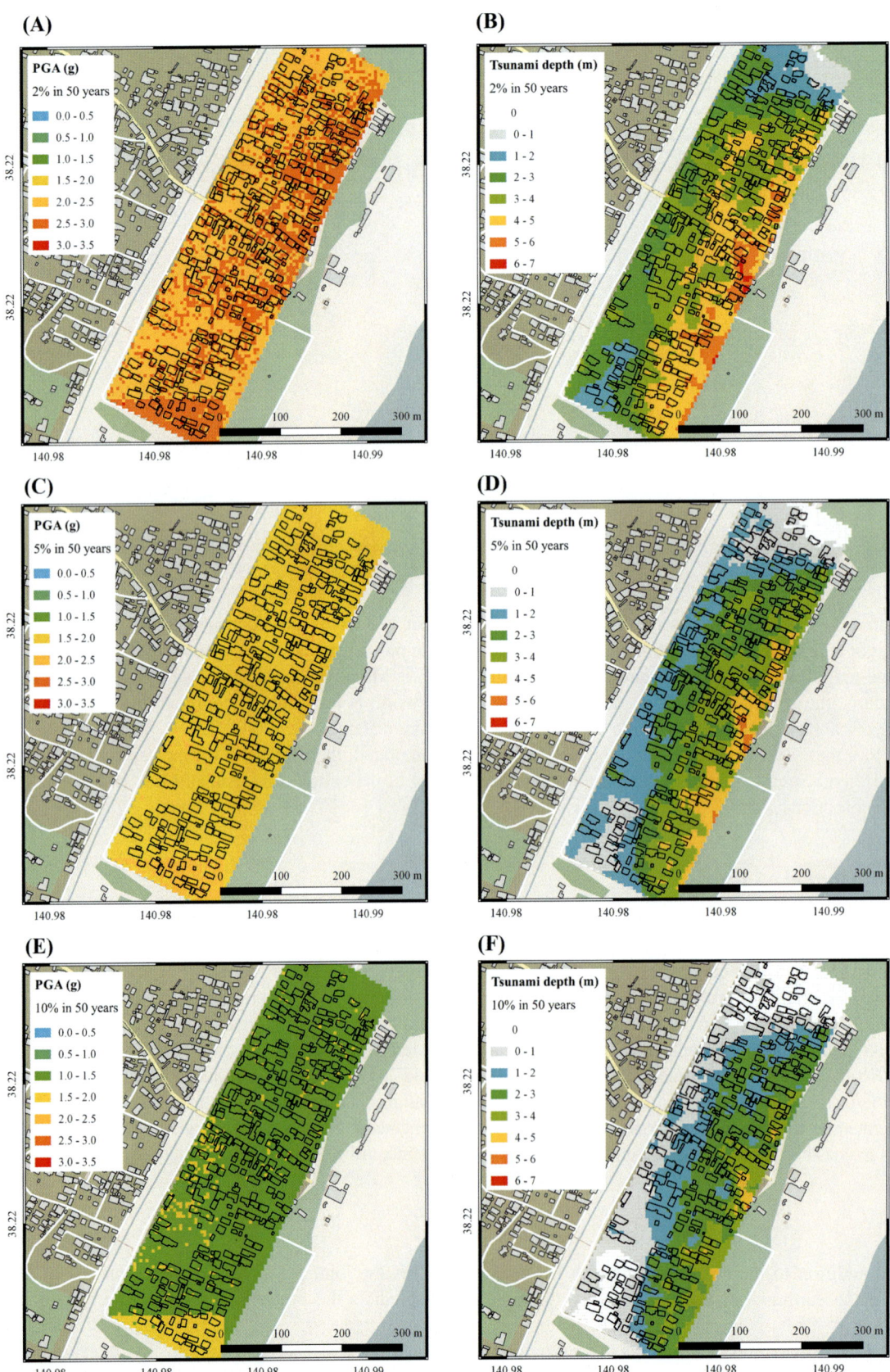

FIGURE 5.29 Earthquake–tsunami uniform hazard maps (PGA and inundation depth) for the mean annual rate of (A,B) 2% in 50 years, (C,D) 5% in 50 years, and (E,F) 10% in 50 years. *PGA*, Peak ground acceleration.

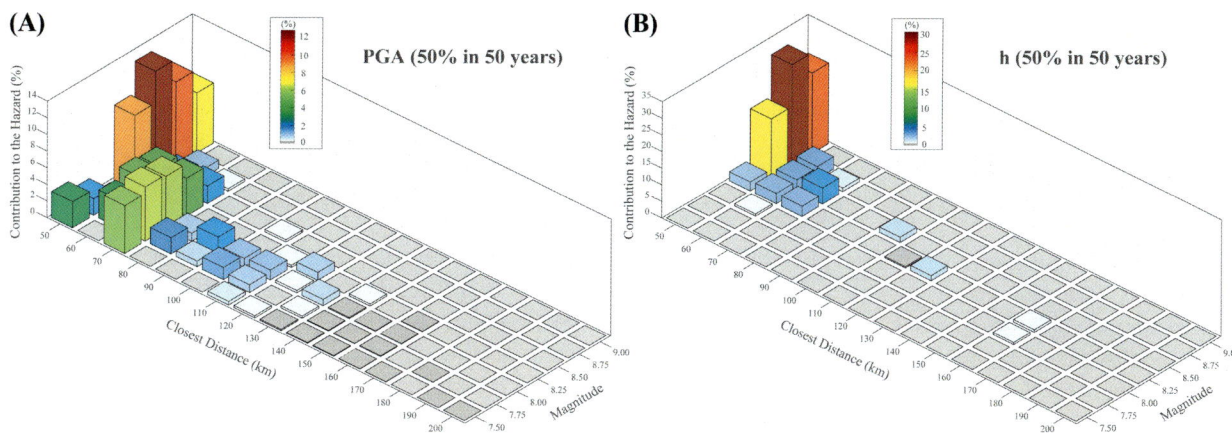

FIGURE 5.30 Disaggregation results for Sendai, Japan. (A) Peak ground acceleration, and (B) tsunami wave height corresponding to the hazard level of 50% in 50 years.

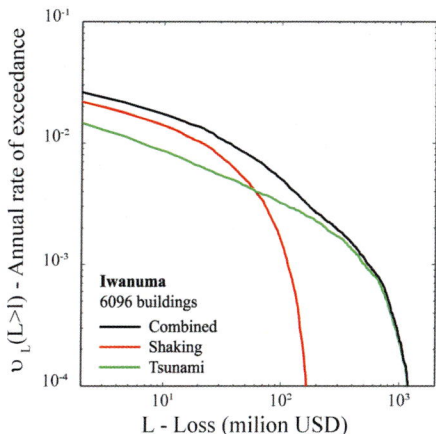

FIGURE 5.31 Comparison of loss exceedance curves for the combined earthquake–tsunami hazards and single hazards for Iwanuma, Miyagi Prefecture, Japan.

As before, the integral can be evaluated by Monte Carlo simulation. It is important to emphasize that there are not many multihazard earthquake–tsunami fragility models that apply to a wide range of buildings and infrastructures (Park et al., 2019). Therefore, conventional separate fragility models need to be adopted; the final loss due to earthquake and tsunami can be calculated as the dominant contribution from earthquake and tsunami damage. According to this specific procedure, after the joint earthquake–tsunami hazard is assessed, the hazard-specific fragility curves are used to assess the expected *DS*; subsequently, the *DR*s are simulated for both earthquakes and tsunamis (using hazard-specific consequence functions), and the maximum between the two can be considered. The rationale for adopting the larger loss value is that physical shaking always precedes the tsunami.

Similar to the hazard assessment, the main outcome of the multirisk procedure is a loss curve that represents the mean annual rate of exceedance of a given loss. Fig. 5.31 shows the results of a multihazard risk assessment for Iwanuma, Miyagi Prefecture, Japan (Goda & De Risi, 2018). Fig. 5.31 also compares the loss exceedance curves for multi- and single-hazard cases, offering a breakdown of the contributions of the two hazards toward the final result. It is possible to observe that shaking loss curves intersect with tsunami loss curves. At shorter return periods (e.g., $T_R <$ 200 years), shaking loss curves are generally positioned on the right-hand side with respect to tsunami loss curves, whereas at longer return periods (e.g., $T_R >$ 1000 years), their relative positions are reversed, and loss contributions, due to tsunami damage, become dominant (i.e., the combined loss curves and the tsunami loss curves nearly coincide). At intermediate return periods, the crossing of the two single-hazard curves occurs. In other words, earthquake damage tends to occur more frequently, but the overall impact is somewhat limited locally. On the other hand, tsunami damage is relatively rare but could be devastating. It is important to emphasize that the relative loss contributions of the earthquake and tsunami damage to total loss depend on various factors. For the earthquake loss curves, the proximity of the area to the fault rupture plane and the surface soil types at the building sites are crucial because these two factors control the severity of strong shaking in the area. For the tsunami as well as combined loss curves, the proportions of buildings that are located within low-elevation areas are the decisive factor because buildings at relatively high elevations or far from the coast will not be inundated by the tsunami. Iwanuma is an extreme situation because the majority of the buildings could be washed away completely (as they are built on a low-elevation flat plain).

5.6 Research needs

Existing methods for assessing tsunami risks still have deficiencies in each of their three main aspects: hazard, exposure, and vulnerability. Tsunami hazard and risk assessments depend on regional characteristics, and various examples of up-to-date tsunami hazard studies can be found in Chapters 13, 14, 15, and 16. A significant issue lies in the inherent lack of data for developing earthquake occurrence models and their uncertainties in evaluating the tsunami hazards and risks quantitatively (Chapter 19). To address this, a solution could be the integration of a time-dependent renewal model for earthquake occurrence. This would require supplementary details, such as the assessment time window, the probability distribution governing the time between earthquakes, the duration since the previous event, and the recurrence model (Goda, 2019).

Furthermore, PTRA lacks a precise characterization of the specific categories of losses that necessitate computation. Monetary losses can be easily approximated with a satisfactory level of certainty using established techniques. On the other hand, indirect losses usually constitute a vital element. Yet, evaluating them is more complicated due to the unavailability of consequence functions that are broadly applicable. When appraising the resilience of coastal communities, it is crucial to encompass the social, political, and economic facets in the evaluation. However, established risk approaches do not account for the effects of tsunamis on social, political, and economic dimensions.

PTRA methodologies, as demonstrated in this chapter, primarily concentrate on residential structures or buildings designated for tsunami evacuation. Nevertheless, coastal regions' resilience hinges on traditional and strategic infrastructure elements (Akiyama et al., 2020). Traditional infrastructure elements encompass a range of critical components, including roads, bridges, railways, power and communication networks, water distribution systems, wastewater facilities, and gas distribution grids. These elements form the foundation of economic and social endeavors within urban environments. Schools and hospitals are also pivotal in delivering essential education and healthcare services, integral to the recovery process. In coastal regions, critical infrastructure extends to include key features like harbors, pivotal for maritime industries, nuclear power plants requiring substantial water resources for reactor cooling, storage facilities for gas and oil housing fossil fuels transported via tankers, and early warning systems like tidal buoys and ocean bottom pressure gauges designed to detect hazardous occurrences in the open sea. Many of these infrastructures are complex, geographically distributed, and involve multiple sectors; the level of complexity is significantly increased when multiple infrastructures are interconnected and work as interacting systems (Dueñas-Osorio & Vemuru, 2009).

References

Abrahamson, N., Gregor, N., & Addo, K. (2016). BC hydro ground motion prediction equations for subduction earthquakes. *Earthquake Spectra, 32*(1), 23–44. Available from https://doi.org/10.1193/051712EQS188MR, http://earthquakespectra.org/doi/pdf/10.1193/051712EQS188MR.

Aida, I. (1978). Reliability of a tsunami source model derived from fault parameters. *Journal of Physics of the Earth, 26*(1), 57–73. Available from https://doi.org/10.4294/jpe1952.26.57.

Akiyama, M., Frangopol, D. M., & Ishibashi, H. (2020). Toward life-cycle reliability-, risk- and resilience-based design and assessment of bridges and bridge networks under independent and interacting hazards: Emphasis on earthquake, tsunami and corrosion. Taylor and Francis Ltd., Japan. *Structure and Infrastructure Engineering, 16*(1), 26–50. Available from https://doi.org/10.1080/15732479.2019.1604770, http://www.tandf.co.uk/journals/titles/15732479.asp.

American Society of Civil Engineers (2021). *Minimum design loads and associated criteria for buildings and other structures*. ASCE/SEI (pp. 7–22). Available from https://doi.org/10.1061/9780784415788.

Annaka, T., Satake, K., Sakakiyama, T., Yanagisawa, K., & Shuto, N. (2007). Logic-tree approach for probabilistic tsunami hazard analysis and its applications to the Japanese coasts. *Pure and Applied Geophysics, 164*(2-3), 577–592. Available from https://doi.org/10.1007/s00024-006-0174-3.

Architectural Institute of Japan. (2015). Architectural Institute of Japan Recommendations for Loads on Buildings. https://www.aij.or.jp/jpn/ppv/pdf/aij_recommendations_for_loads_on_buildings_2015.pdf.

Aspinall, W. P., & Cooke, R. M. (2013). *Risk and uncertainty assessment for natural hazard quantifying scientific uncertainty from expert judgement elicitation* (pp. 64–99). Cambridge University Press. Available from https://doi.org/10.1017/cbo9781139047562.005.

Attary, N., Unnikrishnan, V. U., van de Lindt, J. W., Cox, D. T., & Barbosa, A. R. (2017). Performance-based tsunami engineering methodology for risk assessment of structures. *Engineering Structures, 141*, 676–686. Available from https://doi.org/10.1016/j.engstruct.2017.03.071, http://www.journals.elsevier.com/engineering-structures/.

Bazzurro, P., & Allin Cornell, C. (1999). Disaggregation of seismic hazard. *Bulletin of the Seismological Society of America*, 89(2), 501–520. Available from https://doi.org/10.1785/bssa0890020501.

Behrens, J., Løvholt, F., Jalayer, F., Lorito, S., Salgado-Gálvez, M. A., Sørensen, M., Abadie, S., Aguirre-Ayerbe, I., Aniel-Quiroga, I., Babeyko, A., Baiguera, M., Basili, R., Belliazzi, S., Grezio, A., Johnson, K., Murphy, S., Paris, R., Rafliana, I., De Risi, R., Rossetto, T., Selva, J., Taroni, M., Del Zoppo, M., Armigliato, A., Bureš, V., Cech, P., Cecioni, C., Christodoulides, P., Davies, G., Dias, F., Bayraktar, H. B., González, M., Gritsevich, M., Guillas, S., Harbitz, C. B., Kânoğlu, U., Macías, J., Papadopoulos, G. A., Polet, J., Romano, F., Salamon, A., Scala, A., Stepinac, M., Tappin, D. R., Thio, H. K., Tonini, R., Triantafyllou, I., Ulrich, T., Varini, E., Volpe, M., & Vyhmeister, E. (2021). *Probabilistic Tsunami Hazard and Risk Analysis: A Review of Research Gaps*. Frontiers in Earth Science, 9, 628772. Available from https://doi.org/10.3389/feart.2021.628772.

Beven, K. J., Almeida, S., Aspinall, W. P., Bates, P. D., Blazkova, S., Borgomeo, E., Freer, J., Goda, K., Hall, J. W., Phillips, J. C., Simpson, M., Smith, P. J., Stephenson, D. B., Wagener, T., Watson, M., & Wilkins, K. L. (2018). Epistemic uncertainties and natural hazard risk assessment—Part 1: A review of different natural hazard areas. *Natural Hazards and Earth System Sciences*, 18(10), 2741–2768. Available from https://doi.org/10.5194/nhess-18-2741-2018, http://www.nat-hazards-earth-syst-sci.net/volumes_and_issues.html.

Beven, K. J., Aspinall, W. P., Bates, P. D., Borgomeo, E., Goda, K., Hall, J. W., Page, T., Phillips, J. C., Simpson, M., Smith, P. J., Wagener, T., & Watson, M. (2018). Epistemic uncertainties and natural hazard risk assessment—Part 2: What should constitute good practice? *Natural Hazards and Earth System Sciences*, 18(10), 2769–2783. Available from https://doi.org/10.5194/nhess-18-2769-2018, http://www.nat-hazards-earth-syst-sci.net/volumes_and_issues.html.

Box, G. E. P., & Cox, D. R. (1964). An analysis of transformations. *Journal of the Royal Statistical Society: Series B (Methodological)*, 26(2), 211–243. Available from https://doi.org/10.1111/j.2517-6161.1964.tb00553.x.

Burbidge, D., Cummins, P. R., Mleczko, R., & Thio, H. K. (2008). A probabilistic tsunami hazard assessment for Western Australia. *Pure and Applied Geophysics*, 165(11-12), 2059–2088. Available from https://doi.org/10.1007/s00024-008-0421-x.

Burroughs, S. M., & Tebbens, S. F. (2005). Power-law scaling and probabilistic forecasting of tsunami runup heights. *Pure and Applied Geophysics*, 162(2), 331–342. Available from https://doi.org/10.1007/s00024-004-2603-5.

Charvet, I., Suppasri, A., & Imamura, F. (2014). Empirical fragility analysis of building damage caused by the 2011 Great East Japan tsunami in Ishinomaki city using ordinal regression, and influence of key geographical features. *Stochastic Environmental Research and Risk Assessment*, 28(7), 1853–1867. Available from https://doi.org/10.1007/s00477-014-0850-2, http://link.springer-ny.com/link/service/journals/00477/index.htm.

Chock, G. Y. K. (2016). Design for tsunami loads and effects in the ASCE 7-16 Standard. *Journal of Structural Engineering*, 142(11). Available from https://doi.org/10.1061/(asce)st.1943-541x.0001565.

Construction Research Institute. (2011). *Japan building cost information*.

Cornell, C. A. (1968). Engineering seismic risk analysis. *Bulletin of the Seismological Society of America*, 58(5), 1583–1606. Available from https://doi.org/10.1785/bssa0580051583.

Cornell, C. A., & Krawinkler, H. (2000). Progress and challenges in seismic performance assessment. *PEER Center News*, 3, 1–3.

De Risi, R., & Goda, K. (2017). Simulation-based probabilistic tsunami hazard analysis: Empirical and robust hazard predictions. *Pure and Applied Geophysics*, 174(8), 3083–3106. Available from https://doi.org/10.1007/s00024-017-1588-9, http://www.springer.com/birkhauser/geo+science/journal/24.

De Risi, R., Goda, K., Mori, N., & Yasuda, T. (2017). Bayesian tsunami fragility modeling considering input data uncertainty. *Stochastic Environmental Research and Risk Assessment*, 31(5), 1253–1269. Available from https://doi.org/10.1007/s00477-016-1230-x, http://link.springer-ny.com/link/service/journals/00477/index.htm.

De Risi, R., Goda, K., Yasuda, T., & Mori, N. (2017). Is flow velocity important in tsunami empirical fragility modeling? *Earth-Science Reviews*, 166, 64–82. Available from https://doi.org/10.1016/j.earscirev.2016.12.015, http://www.sciencedirect.com/science/journal/00128252.

De Risi, R., Muhammad, A., De Luca, F., Goda, K., & Mori, N. (2022). Dynamic risk framework for cascading compounding climate-geological hazards: A perspective on coastal communities in subduction zones. *Frontiers in Earth Science*, 10. Available from https://doi.org/10.3389/feart.2022.1023018, https://www.frontiersin.org/journals/earth-science.

De Risi, R., De Paola, F., Turpie, J., & Kroeger, T. (2018). Life cycle cost and return on investment as complementary decision variables for urban flood risk management in developing countries. *International Journal of Disaster Risk Reduction*, 28, 88–106. Available from https://doi.org/10.1016/j.ijdrr.2018.02.026, http://www.journals.elsevier.com/international-journal-of-disaster-risk-reduction/.

Downes, G.L., & Stirling, M.W. (2001). *Groundwork for development of a probabilistic tsunami hazard model for New Zealand*. In Proceedings of International Tsunami Symposium (pp. 293–301).

Dueñas-Osorio, L., & Vemuru, S. M. (2009). Cascading failures in complex infrastructure systems. *Structural Safety*, 31(2), 157–167. Available from https://doi.org/10.1016/j.strusafe.2008.06.007.

Dutykh, D., Poncet, R., & Dias, F. (2011). The VOLNA code for the numerical modeling of tsunami waves: Generation, propagation and inundation. *European Journal of Mechanics, B/Fluids*, 30(6), 598–615. Available from https://doi.org/10.1016/j.euromechflu.2011.05.005.

Farreras, S., Ortiz, M., & Gonzalez, J. I. (2007). Steps towards the implementation of a tsunami detection, warning, mitigation and preparedness program for Southwestern coastal areas of Mexico. *Pure and Applied Geophysics*, 164(2-3), 605–616. Available from https://doi.org/10.1007/s00024-006-0175-2.

Fukutani, Y., Suppasri, A., & Imamura, F. (2015). Stochastic analysis and uncertainty assessment of tsunami wave height using a random source parameter model that targets a Tohoku-type earthquake fault. *Stochastic Environmental Research and Risk Assessment*, 29(7), 1763–1779. Available from https://doi.org/10.1007/s00477-014-0966-4, http://link.springer-ny.com/link/service/journals/00477/index.htm.

Geist, E.L. (2005). *Local tsunami hazards in the Pacific Northwest from Cascadia subduction zone earthquakes*. US Geological Survey Professional Paper (1661 B) (pp. 1–17).

Geist, E. L., & Dmowska, R. (1999). Local tsunamis and distributed slip at the source. *Pure and Applied Geophysics*, 154(3–4), 485–512. Available from https://doi.org/10.1007/s000240050241, http://www.springer.com/birkhauser/geo+science/journal/24.

Geist, E. L., & Lynett, P. J. (2014). Source processes for the probabilistic assessment of tsunami hazards. *Oceanography, 27*(2), 86–93. Available from https://doi.org/10.5670/oceanog.2014.43, http://www.tos.org/oceanography/archive/27-2_geist.pdf.

Geist, E. L., & Parsons, T. (2006). Probabilistic analysis of tsunami hazards. *Natural Hazards, 37*(3), 277–314. Available from https://doi.org/10.1007/s11069-005-4646-z.

Geist, E. L., & Parsons, T. (2016). Reconstruction of far-field tsunami amplitude distributions from earthquake sources. *Pure and Applied Geophysics, 173*(12), 3703–3717. Available from https://doi.org/10.1007/s00024-016-1288-x, http://www.springer.com/birkhauser/geo + science/journal/24.

Goda, K. (2011). Interevent variability of spatial correlation of peak ground motions and response spectra. *Bulletin of the Seismological Society of America, 101*(5), 2522–2531. Available from http://www.bssaonline.org/cgi/reprint/101/5/2522, https://doi.org/10.1785/0120110092, United Kingdom.

Goda, K. (2019). Time-dependent probabilistic tsunami hazard analysis using stochastic rupture sources. *Stochastic Environmental Research and Risk Assessment, 33*(2), 341–358. Available from https://doi.org/10.1007/s00477-018-1634-x, http://link.springer-ny.com/link/service/journals/00477/index.htm.

Goda, K., & Atkinson, G. M. (2010). Intraevent spatial correlation of ground-motion parameters using SK-net data. *Bulletin of the Seismological Society of America, 100*(6), 3055–3067. Available from https://doi.org/10.1785/0120100031, http://www.bssaonline.org/cgi/reprint/100/6/3055.pdf, United Kingdom.

Goda, K., Mai, P. M., Yasuda, T., & Mori, N. (2014). Sensitivity of tsunami wave profiles and inundation simulations to earthquake slip and fault geometry for the 2011 Tohoku earthquake. *Earth, Planets and Space, 66*(1). Available from https://doi.org/10.1186/1880-5981-66-105, http://rd.springer.com/journal/40623.

Goda, K., Mori, N., & Yasuda, T. (2019). Rapid tsunami loss estimation using regional inundation hazard metrics derived from stochastic tsunami simulation. *International Journal of Disaster Risk Reduction, 40*, 101152. Available from https://doi.org/10.1016/j.ijdrr.2019.101152.

Goda, K., & De Risi, R. (2017). Probabilistic tsunami loss estimation methodology: Stochastic earthquake scenario approach. *Earthquake Spectra, 33*(4), 1301–1323. Available from https://doi.org/10.1193/012617EQS019M, http://earthquakespectra.org/doi/pdf/10.1193/012617EQS019M.

Goda, K., & De Risi, R. (2018). Multi-hazard loss estimation for shaking and tsunami using stochastic rupture sources. *International Journal of Disaster Risk Reduction, 28*, 539–554. Available from https://doi.org/10.1016/j.ijdrr.2018.01.002, http://www.journals.elsevier.com/international-journal-of-disaster-risk-reduction/.

Goda, K., Yasuda, T., Mori, N., & Maruyama, T. (2016). New scaling relationships of earthquake source parameters for stochastic tsunami simulation. *Coastal Engineering Journal, 58*(3). Available from https://doi.org/10.1142/S0578563416500108, https://www.tandfonline.com/loi/tcej20.

González, F. I., Geist, E. L., Jaffe, B., Kânoğlu, U., Mofjeld, H., Synolakis, C. E., Titov, V. V., Arcas, D., Bellomo, D., Carlton, D., Horning, T., Johnson, J., Newman, J., Parsons, T., Peters, R., Peterson, C., Priest, G., Venturato, A., Weber, J., ... Yalciner, A. (2009). Probabilistic tsunami hazard assessment at Seaside, Oregon, for near- and far-field seismic sources. *Journal of Geophysical Research, 114*(C11). Available from https://doi.org/10.1029/2008jc005132.

Goto,C., Ogawa,Y., Shuto,N., & Imamura,F. (1997). *Numerical method of tsunami simulation with the leap-frog scheme (IUGG/IOC Time project)*. IOC Manual.

Goulet, C. A., Haselton, C. B., Mitrani-Reiser, J., Beck, J. L., Deierlein, G. G., Porter, K. A., & Stewart, J. P. (2007). Evaluation of the seismic performance of a code-conforming reinforced-concrete frame building—From seismic hazard to collapse safety and economic losses. *Earthquake Engineering and Structural Dynamics, 36*(13), 1973–1997. Available from https://doi.org/10.1002/eqe.694, http://onlinelibrary.wiley.com/journal/10.1002/(ISSN)1096-9845.

Greenwood, M. (1926). *The natural duration of cancer* (33, pp. 1–26). Her Majesty's Stationery Office.

Grezio, A., Babeyko, A., Baptista, M. A., Behrens, J., Costa, A., Davies, G., Geist, E. L., Glimsdal, S., González, F. I., Griffin, J., Harbitz, C. B., LeVeque, R. J., Lorito, S., Løvholt, F., Omira, R., Mueller, C., Paris, R., Parsons, T., Polet, J., ... Thio, H. K. (2017). Probabilistic tsunami hazard analysis: Multiple sources and global applications. *Reviews of Geophysics, 55*(4), 1158–1198. Available from https://doi.org/10.1002/2017RG000579, http://onlinelibrary.wiley.com/journal/10.1002/(ISSN)1944-9208.

Grezio, A., Gasparini, P., Marzocchi, W., Patera, A., & Tinti, S. (2012). Tsunami risk assessments in Messina, Sicily—Italy. *Natural Hazards and Earth System Science, 12*(1), 151–163. Available from https://doi.org/10.5194/nhess-12-151-2012.

Grezio, A., Marzocchi, W., Sandri, L., & Gasparini, P. (2010). A Bayesian procedure for probabilistic tsunami hazard assessment. *Natural Hazards, 53*(1), 159–174. Available from https://doi.org/10.1007/s11069-009-9418-8.

Gusman, A. R., Satake, K., Gunawan, E., Hamling, I., & Power, W. (2018). Contribution from multiple fault ruptures to tsunami generation during the 2016 Kaikoura earthquake. *Pure and Applied Geophysics, 175*(8), 2557–2574. Available from https://doi.org/10.1007/s00024-018-1949-z, http://www.springer.com/birkhauser/geo + science/journal/24.

Gusman, A. R., Supendi, P., Nugraha, A. D., Power, W., Latief, H., Sunendar, H., Widiyantoro, S., Daryono., Wiyono, S. H., Hakim, A., Muhari, A., Wang, X., Burbidge, D., Palgunadi, K., Hamling, I., & Daryono, M. R. (2019). Source model for the tsunami inside Palu Bay following the 2018 Palu earthquake, Indonesia. *Geophysical Research Letters, 46*(15), 8721–8730. Available from https://doi.org/10.1029/2019GL082717, http://agupubs.onlinelibrary.wiley.com/hub/journal/10.1002/(ISSN)1944-8007/.

Gutenberg, B., & Richter, C. F. (1956). Magnitude and energy of earthquakes. *Annals of Geophysics, 9*(1), 1–15.

Hayes, G. P., Moore, G. L., Portner, D. E., Hearne, M., Flamme, H., Furtney, M., & Smoczyk, G. M. (2018). Slab2, a comprehensive subduction zone geometry model. *Science, 362*(6410), 58–61. Available from https://doi.org/10.1126/science.aat4723, http://science.sciencemag.org/content/362/6410/58/tab-pdf.

Horspool, N., Pranantyo, I., Griffin, J., Latief, H., Natawidjaja, D. H., Kongko, W., Cipta, A., Bustaman, B., Anugrah, S. D., & Thio, H. K. (2014). A probabilistic tsunami hazard assessment for Indonesia. *Natural Hazards and Earth System Sciences, 14*(11), 3105–3122. Available from https://doi.org/10.5194/nhess-14-3105-2014, http://www.nat-hazards-earth-syst-sci.net/volumes_and_issues.html.

Japan Society of Civil Engineering. (2016). *Tsunami assessment method for nuclear power plants in Japan*. https://committees.jsce.or.jp/ceofnp/node/.

Jaynes, E. T. (2003). *Probability theory: The logic of science*. Cambridge University Press.

Kaplan, E. L., & Meier, P. (1958). Nonparametric estimation from incomplete observations. *Journal of the American Statistical Association*, *53*(282), 457−481. Available from https://doi.org/10.1080/01621459.1958.10501452.

Kulikov, E. A., Rabinovich, A. B., & Thomson, R. E. (2005). Estimation of tsunami risk for the coasts of Peru and Northern Chile. *Natural Hazards*, *35*(2), 185−209. Available from https://doi.org/10.1007/s11069-004-4809-3.

Lamarre, M., Townshend, B., & Shah, H. C. (1992). Application of the bootstrap method to quantify uncertainty in seismic hazard estimates. *Bulletin of the Seismological Society of America*, *82*(1), 104−119. Available from https://doi.org/10.1785/bssa0820010104.

Li, L., Switzer, A. D., Chan, C. H., Wang, Y., Weiss, R., & Qiu, Q. (2016). How heterogeneous coseismic slip affects regional probabilistic tsunami hazard assessment: A case study in the South China Sea. *Journal of Geophysical Research: Solid Earth*, *121*(8), 6250−6272. Available from https://doi.org/10.1002/2016JB013111, http://onlinelibrary.wiley.com/journal/10.1002/(ISSN)2169-9356.

Liu, Y., Santos, A., Wang, S. M., Shi, Y., Liu, H., & Yuen, D. A. (2007). Tsunami hazards along Chinese coast from potential earthquakes in South China Sea. *Physics of the Earth and Planetary Interiors*, *163*(1-4), 233−244. Available from https://doi.org/10.1016/j.pepi.2007.02.012.

Liu, Z., Nadim, F., Garcia-Aristizabal, A., Mignan, A., Fleming, K., & Luna, B. Q. (2015). A three-level framework for multi-risk assessment. *Georisk: Assessment and Management of Risk for Engineered Systems and Geohazards*, *9*(2), 59−74. Available from https://doi.org/10.1080/17499518.2015.1041989.

Løvholt, F., Glimsdal, S., Harbitz, C. B., Zamora, N., Nadim, F., Peduzzi, P., Dao, H., & Smebye, H. (2012). Tsunami hazard and exposure on the global scale. *Earth-Science Reviews*, *110*(1-4), 58−73. Available from https://doi.org/10.1016/j.earscirev.2011.10.002.

Mai, M. P., & Beroza, G. C. (2002). A spatial random field model to characterize complexity in earthquake slip. *Journal of Geophysical Research: Solid Earth*, *107*(B11). Available from https://doi.org/10.1029/2001jb000588, ESE 10-1-ESE 10-21.

Marzocchi, W., Garcia-Aristizabal, A., Gasparini, P., Mastellone, M. L., & Ruocco, A. D. (2012). Basic principles of multi-risk assessment: A case study in Italy. *Natural Hazards*, *62*(2), 551−573. Available from https://doi.org/10.1007/s11069-012-0092-x.

Marzocchi, W., Taroni, M., & Selva, J. (2015). Accounting for epistemic uncertainty in PSHA: Logic tree and ensemble modeling. *Bulletin of the Seismological Society of America*, *105*(4), 2151−2159. Available from https://doi.org/10.1785/0120140131, http://www.bssaonline.org/content/105/4/2151.full.pdf.

Matthews, M. V., Ellsworth, W. L., & Reasenberg, P. A. (2002). A Brownian model for recurrent earthquakes. *Bulletin of the Seismological Society of America*, *92*(6), 2233−2250. Available from https://doi.org/10.1785/0120010267.

McGuire, R.K. (2004). *Seismic hazard and risk analysis*. 240.

Mignan, A., Wiemer, S., & Giardini, D. (2014). The quantification of low-probability-high-consequences events: Part I. A generic multi-risk approach. *Natural Hazards*, *73*(3), 1999−2022. Available from https://doi.org/10.1007/s11069-014-1178-4, http://www.wkap.nl/journalhome.htm/0921-030X.

Ming, X., Xu, W., Li, Y., Du, J., Liu, B., & Shi, P. (2015). Quantitative multi-hazard risk assessment with vulnerability surface and hazard joint return period. *Stochastic Environmental Research and Risk Assessment*, *29*(1), 35−44. Available from https://doi.org/10.1007/s00477-014-0935-y, http://link.springer-ny.com/link/service/journals/00477/index.htm.

Ministry of Land Infrastructure and Transportation. (2014). *Survey of tsunami damage condition*. http://www.mlit.go.jp/toshi/toshi-hukkou-arkaibu.html.

Ministry of Business Innovation and Employment. (2020). *Tsunami loads and effects on vertical evacuation structures: Technical information*. Available from https://www.building.govt.nz/assets/Uploads/building-code-compliance/geotechnical-education/tsunami-vertical-evacuation-structures.pdf

Mitchell-Wallace, K., Jones, M., Hillier, J., & Foote, M. (2017). *Natural catastrophe risk management and modelling: A practitioner's guide* (536). Wiley-Blackwell.

Miyashita, T., Mori, N., & Goda, K. (2020). Uncertainty of probabilistic tsunami hazard assessment of Zihuatanejo (Mexico) due to the representation of tsunami variability. *Coastal Engineering Journal*, *62*(3), 413−428. Available from https://doi.org/10.1080/21664250.2020.1780676.

Morikawa, N., & Fujiwara, H. (2013). A new ground motion prediction equation for Japan applicable up to M9 mega-earthquake. *Journal of Disaster Research*, *8*(5), 878−888. Available from https://doi.org/10.20965/jdr.2013.p0878, http://www.fujipress.jp/finder/access_check.php?pdf_filename = DSSTR000800050004.pdf&frompage = abst_page&errormode = Login&pid = &lang = English.

Mueller, C., Power, W., Fraser, S., & Wang, X. (2015). Effects of rupture complexity on local tsunami inundation: Implications for probabilistic tsunami hazard assessment by example. *Journal of Geophysical Research: Solid Earth*, *120*(1), 488−502. Available from https://doi.org/10.1002/2014JB011301, http://onlinelibrary.wiley.com/journal/10.1002/(ISSN)2169-9356.

Muhammad, A., Goda, K., Alexander, N. A., Kongko, W., & Muhari, A. (2017). Tsunami evacuation plans for future megathrust earthquakes in Padang, Indonesia, considering stochastic earthquake scenarios. *Natural Hazards and Earth System Sciences*, *17*(12), 2245−2270. Available from https://doi.org/10.5194/nhess-17-2245-2017, http://www.nat-hazards-earth-syst-sci.net/volumes_and_issues.html.

Murnane, R. J., Allegri, G., Bushi, A., Dabbeek, J., de Moel, H., Duncan, M., Fraser, S., Galasso, C., Giovando, C., Henshaw, P., Horsburgh, K., Huyck, C., Jenkins, S., Johnson, C., Kamihanda, G., Kijazi, J., Kikwasi, W., Kombe, W., Loughlin, S., ... Verrucci, E. (2019). Data schemas for multiple hazards, exposure and vulnerability. *Disaster Prevention and Management: An International Journal*, *28*(6), 752−763. Available from https://doi.org/10.1108/DPM-09-2019-0293, http://www.emeraldinsight.com/info/journals/dpm/dpm.jsp.

Natawidjaja, D. H., Sieh, K., Chlieh, M., Galetzka, J., Suwargadi, B. W., Cheng, H., Edwards, R. L., Avouac, J. P., & Ward, S. N. (2006). Source parameters of the great Sumatran megathrust earthquakes of 1797 and 1833 inferred from coral microatolls. *Journal of Geophysical Research: Solid Earth*, *111*(6). Available from https://doi.org/10.1029/2005JB004025, http://onlinelibrary.wiley.com/journal/10.1002/(ISSN)2169-9356.

Okada, Y. (1985). Surface deformation due to shear and tensile faults in a half-space. *Bulletin of the Seismological Society of America*, *75*(4), 1135−1154. Available from https://doi.org/10.1785/bssa0750041135.

Orfanogiannaki, K., & Papadopoulos, G. A. (2007). Conditional probability approach of the assessment of tsunami potential: Application in three tsunamigenic Regions of the Pacific Ocean. *Pure and Applied Geophysics, 164*(2-3), 593−603. Available from https://doi.org/10.1007/s00024-006-0170-7.

Pardo-Iguzquiza, E., & Chica-Olmo, M. (1993). The Fourier integral method: An efficient spectral method for simulation of random fields. *Mathematical Geology, 25*(2), 177−217. Available from https://doi.org/10.1007/bf00893272.

Park, H., Alam, M. S., Cox, D. T., Barbosa, A. R., & van de Lindt, J. W. (2019). Probabilistic seismic and tsunami damage analysis (PSTDA) of the Cascadia Subduction Zone applied to Seaside, Oregon. *International Journal of Disaster Risk Reduction, 35*. Available from https://doi.org/10.1016/j.ijdrr.2019.101076, http://www.journals.elsevier.com/international-journal-of-disaster-risk-reduction/.

Park, H., & Cox, D. T. (2016). Probabilistic assessment of near-field tsunami hazards: Inundation depth, velocity, momentum flux, arrival time, and duration applied to Seaside, Oregon. *Coastal Engineering, 117*, 79−96. Available from https://doi.org/10.1016/j.coastaleng.2016.07.011, http://www.elsevier.com/inca/publications/store/5/0/3/3/2/5/.

Parsons, T., & Geist, E. L. (2008). Tsunami probability in the Caribbean region. *Pure and Applied Geophysics, 165*(11-12), 2089−2116. Available from https://doi.org/10.1007/s00024-008-0416-7.

Petrone, C., Rossetto, T., & Goda, K. (2017). Fragility assessment of a RC structure under tsunami actions via nonlinear static and dynamic analyses. *Engineering Structures, 136*, 36−53. Available from https://doi.org/10.1016/j.engstruct.2017.01.013, http://www.journals.elsevier.com/engineering-structures/.

Pianosi, F., Beven, K., Freer, J., Hall, J. W., Rougier, J., Stephenson, D. B., & Wagener, T. (2016). Sensitivity analysis of environmental models: A systematic review with practical workflow. *Environmental Modelling and Software, 79*, 214−232. Available from https://doi.org/10.1016/j.envsoft.2016.02.008, http://www.elsevier.com/inca/publications/store/4/2/2/9/2/1.

Pisarenko, V., & Rodkin, M. (2010). Heavy-tailed distributions in disaster analysis. *Advances in Natural and Technological Hazards Research, 30*. Available from https://doi.org/10.1007/978-90-481-9171-0.

Pittore, M., Wieland, M., & Fleming, K. (2017). Perspectives on global dynamic exposure modelling for geo-risk assessment. *Natural Hazards, 86*, 7−30. Available from https://doi.org/10.1007/s11069-016-2437-3, http://www.wkap.nl/journalhome.htm/0921-030X.

Porter, K. A., Scawthorn, C. R., & Beck, J. L. (2006). Cost-effectiveness of stronger woodframe buildings. *Earthquake Spectra, 22*(1), 239−266. Available from https://doi.org/10.1193/1.2162567, https://journals.sagepub.com/home/eqs.

Power, W., Downes, G., & Stirling, M. (2007). Estimation of tsunami hazard in New Zealand due to South American earthquakes. *Pure and Applied Geophysics, 164*(2-3), 547−564. Available from https://doi.org/10.1007/s00024-006-0166-3.

Rikitake, T., & Aida, I. (1988). Tsunami hazard probability in Japan. *Bulletin Seismological Society of America, 78*(3), 1268−1278.

Sawai, Y., Namegaya, Y., Okamura, Y., Satake, K., & Shishikura, M. (2012). Challenges of anticipating the 2011 Tohoku earthquake and tsunami using coastal geology. *Geophysical Research Letters, 39*(21). Available from https://doi.org/10.1029/2012GL053692, http://onlinelibrary.wiley.com/journal/10.1002/(ISSN)1944-8007/issues?year=2012.

Schlurmann, T., Kongko, W., Goseberg, N., Natawidjaja, D. H., & Sieh, K. (2010). *Near-field tsunami hazard map Padang, West Sumatra: Utilizing high resolution geospatial data and reseasonable source scenarios*. In Proceedings of the Coastal Engineering Conference.

Selva, J. (2013). Long-term multi-risk assessment: Statistical treatment of interaction among risks. *Natural Hazards, 67*(2), 701−722. Available from https://doi.org/10.1007/s11069-013-0599-9, http://www.wkap.nl/journalhome.htm/0921-030X.

Song, J., & Goda, K. (2019). Influence of elevation data resolution on tsunami loss estimation and insurance rate-making. *Frontiers in Earth Science, 7*. Available from https://doi.org/10.3389/feart.2019.00246, https://www.frontiersin.org/journals/earth-science.

Suppasri, A., Mas, E., Charvet, I., Gunasekera, R., Imai, K., Fukutani, Y., Abe, Y., & Imamura, F. (2013). Building damage characteristics based on surveyed data and fragility curves of the 2011 Great East Japan tsunami. *Natural Hazards, 66*(2), 319−341. Available from https://doi.org/10.1007/s11069-012-0487-8, http://www.wkap.nl/journalhome.htm/0921-030X.

Tang, L., Chamberlin, C., Tolkova, E., Spillane, M., Titov, V. V., Bernard, E. N., & Mofjeld, H. O. (2006). *Assessment of potential tsunami impact for Pearl Harbor*. NOAA Technical Memorandum OAR PMEL-131.

Tanioka, Y., & Satake, K. (1996). Tsunami generation by horizontal displacement of ocean bottom. *Geophysical Research Letters, 23*(8), 861−864. Available from https://doi.org/10.1029/96GL00736, http://onlinelibrary.wiley.com/journal/10.1002/(ISSN)1944-8007/issues?year=2012.

Thio, H. K., Somerville, P., & Ichinose, G. (2012). Probabilistic analysis of strong ground motion and tsunami hazards in SouthEast Asia. *Journal of Earthquake and Tsunami, 01*(02), 119−137. Available from https://doi.org/10.1142/s1793431107000080.

Tinti, S., Armigliaio, A., Tonini, R., Maramai, A., & Graziani, L. (2005). Assessing the hazard related to tsunamis of tectonic origin: A hybrid statistical-deterministic method applied to Southern Italy coasts. *ISET Journal of Earthquake Technology, 42*(4), 189−201.

Wald, D. J., & Allen, T. I. (2007). Topographic slope as a proxy for seismic site conditions and amplification. *Bulletin of the Seismological Society of America, 97*(5), 1379−1395. Available from https://doi.org/10.1785/0120060267.

Wald, D. J., Worden, B. C., Quitoriano, V., & Pankow, K. L. (2005). *ShakeMap manual: Technical manual, user's guide, and software guide*.

Yanagisawa, K., Imamura, F., Sakakiyama, T., Annaka, T., Takeda, T., & Shuto, N. (2007). Tsunami assessment for risk management at nuclear power facilities in Japan. *Pure and Applied Geophysics, 164*(2-3)), 565−576. Available from https://doi.org/10.1007/s00024-006-0176-1.

Youngs, R. R., & Coppersmith, K. J. (1985). Implications of fault slip rates and earthquake recurrence models to probabilistic seismic hazard estimates. *Bulletin of the Seismological Society of America, 75*(4), 939−964. Available from https://doi.org/10.1785/BSSA0750040939.

Chapter 6

Tsunami disaster risk reduction and management

Katsuichiro Goda[1], Raffaele De Risi[2], Ioan Nistor[3] and Aditya Gusman[4]

[1]*Department of Earth Sciences, Western University, London, ON, Canada,* [2]*School of Civil, Aerospace and Design Engineering, University of Bristol, Bristol, United Kingdom,* [3]*Department of Civil Engineering, University of Ottawa, Ottawa, ON, Canada,* [4]*GNS Science, Lower Hutt, Wellington, New Zealand*

6.1 Introduction

Disaster risk management (DRM) against catastrophic tsunami threats is critical to achieving disaster preparedness and resilience for coastal communities. The United Nations Office for Disaster Risk Reduction defines DRM as follows: *"the application of disaster risk reduction (DRR) policies and strategies to prevent new disaster risk, reduce existing disaster risk, and manage residual risk, contributing to the strengthening of resilience and reduction of disaster losses."* To promote and coordinate global efforts, the Sendai Framework for Disaster Risk Reduction 2015−2030 sets out four priorities for action:

- Understanding disaster risk.
- Strengthening disaster risk governance to manage disaster risk.
- Investing in disaster reduction for resilience.
- Enhancing disaster preparedness for effective response and *Building Back Better* in recovery, rehabilitation, and reconstruction.

Disaster resilience is the capacity of a community or society exposed to hazards to adapt to disturbances resulting from hazards by persevering, recuperating, or changing to reach and maintain an acceptable level of functioning. Developmental considerations contribute to all aspects of the disaster management cycle, that is, mitigation−preparedness−response−recovery (Chapter 1). Structural and nonstructural measures can reduce the loss of life and assets significantly. Long-term and comprehensive disaster prevention planning is essential to maximize the effectiveness of tsunami countermeasures. Ultimately, resilient capacity can be built by empowering citizens, responders, organizations, communities, governments, systems, and society to share the responsibility to keep hazards from becoming disasters. More broadly, linkages to sustainable development and climate change adaptation plans should be achieved. Effective DRR policies should be developed by targeting marginalized communities vulnerable to natural catastrophes (CATs) and climate risks. To inform DRR policies and to develop cost-effective mitigation solutions, qualitative and quantitative risk assessments serve as decision support tools. This chapter presents how these tsunami hazards and risk assessments can be used to enhance coastal community's capacities to cope with future tsunami hazards and risks.

Section 6.2 summarizes tsunami hazard−risk mapping and disaster preparedness planning for coastal communities that are under threat from tsunamis. By implementing improved tsunami design methods for coastal structures and buildings, tsunami risks can be mitigated physically (Chapter 4). Results from probabilistic tsunami hazard analysis (PTHA) and probabilistic tsunami risk analysis (PTRA) should inform prevalent tsunami perils in the communities (Chapter 5). PTHA and PTRA outputs can be displayed as tsunami hazard and risk maps and should reveal more exposed areas of the communities. Subsequently, these potential perils must be communicated with community stakeholders, and risk mitigation actions shall be developed and implemented. Where tsunami risk avoidance is possible, land-use planning and zoning can effectively achieve this goal.

Section 6.3 summarizes emergency response measures, such as early warning and evacuation. The importance of tsunami awareness/education and evacuation has been widely recognized after the 2004 and 2011 tsunami disasters in Indian Oceans and Japan (Chapter 1). Recently, more dense monitoring systems for near-source tsunamis have been deployed in

Japan for early warning purposes, while advanced evacuation modeling allows the consideration of behavioral aspects of evacuees after a tsunami and the integration of the tsunami early warning and evacuation. The key aspects of the new tsunami monitoring system and the development of effective evacuation protocols are explained in this section.

Section 6.4 summarizes approaches to quantify the financial risks to communities and their assets and to facilitate the decisions regarding tsunami risk mitigation. A concept of benefit–cost analysis is introduced to formalize the decision-making process regarding DRR mitigation actions. Moreover, risk financing of catastrophic tsunamis is critically important for prompt recovery. For such purposes, the main aspects of financial risk transfer instruments, such as tsunami insurance for homeowners and alternative risk transfer products for insurers, reinsurers, and governments, are mentioned.

Section 6.5 presents research gaps and needs related to tsunami risk reduction and resilience. Some of these research topics are further explained in the later chapters of the book.

6.2 Tsunami hazard–risk mapping for disaster preparedness planning

In the mitigation–preparedness phases of the disaster cycle, the primary objectives are to identify people and assets at risk from future catastrophic tsunamis, to communicate with stakeholders in the communities about the potential tsunami impacts in terms of the exposed population, tsunami loss, and infrastructure damage, and to create a plan for enhancing disaster resilience of the communities. For this purpose, outputs from PTHA and PTRA are useful. Tsunami hazard and risk maps provide information on the expected tsunami hazard and damage extent for the selected probability levels. The maps can be used for disaster preparedness planning. A robust tsunami planning strategy requires high-level governmental entities to develop and support legislation, policies, and guidance that can be implemented harmoniously at local community levels.

Effective planning requires multiple specialties at a local level, including personnel from (1) local building and public works departments responsible for technical review and implementation of the mitigation measures, (2) land-use planning departments responsible for the community planning and development approval processes, and (3) emergency management and response departments responsible for long-term implementation of tsunami safety practices. After the 2011 Tohoku Earthquake and Tsunami, the Japan Reconstruction Agency was established to coordinate the mitigation and land-use planning activities more efficiently (Japan Reconstruction Agency, 2016). A two-tiered approach has been implemented for residential and commercial development in coastal areas of Japan (Cosson, 2020; Onoda et al., 2018). *Level 1 tsunami* is caused by an earthquake magnitude (M) 8 class earthquake with a mean recurrence period of approximately 100 years, whereas *Level 2 tsunami* is caused by an M 9 class earthquake with a mean recurrence period of approximately 1000 years. Hazard areas at lower elevations (below the 1000-year return period inundation boundary, that is, below the *Level 2 flood line*) require land-use, mitigation, and evacuation strategies (Fig. 6.1). Some of the houses in low-lying areas may be relocated to higher lands. On the other hand, hazard areas at higher elevations mainly focus on evacuation strategies. It is important to recognize that the development and implementation of hard and soft measures must be integrated to achieve the intended goal of tsunami disaster preparedness and resilience.

In Section 6.2, three topics are focused upon. Section 6.2.1 introduces a general assessment procedure for disaster risks and a relative risk ranking method. Risk matrix approaches that are adopted from established emergency risk assessment guidelines (e.g., Australian Institute for Disaster Resilience (2015)) allow the translations of likelihoods and consequences of disaster events into descriptive expressions, such as "likely" and "unlikely" for the likelihood and "catastrophic" and "major" for the consequence, thereby facilitating the risk communication among stakeholders and policymakers. Section 6.2.2 summarizes approaches to tsunami hazard–risk mapping for tsunami disaster planning purposes, utilizing outputs from PTHA and PTRA. Section 6.2.3 describes tsunami life safety (i.e., fatality risks) from the viewpoint of tolerable societal risk. To provide concrete examples of the tsunami hazard–risk mapping and tsunami fatality estimation based on PTHA and PTRA, case study results from Tofino in British Columbia, Canada, subject to the Cascadia subduction megathrust events are adopted (Goda et al., 2023; Goda, 2022).

6.2.1 Risk assessment procedure and relative risk scoring

A risk assessment is a structured process of gathering information about risks and forming a judgment about them. Its main goal is to produce science-based objective information to make decisions related to risk management. It generally involves three types of questions to be addressed as part of the risk assessment process:

- What can go wrong to cause adverse consequences? (risk identification)

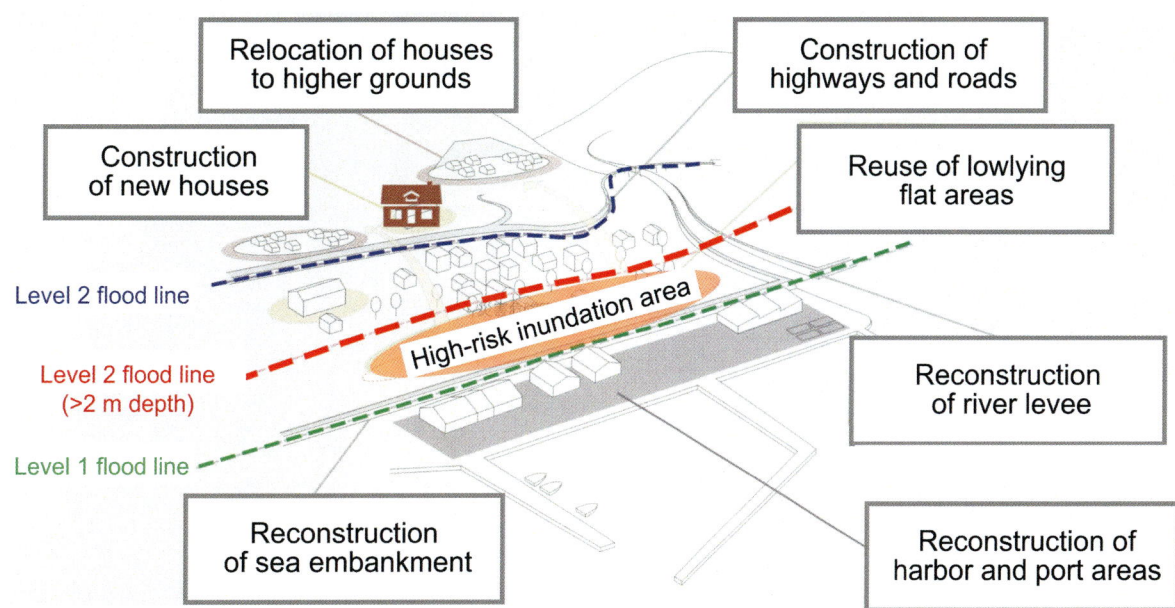

FIGURE 6.1 Reconstruction and redevelopment planning scheme of coastal areas in Japan. *From Cosson, C. (2020). "Build Back Better": Between public policy and local implementation, the challenges in Tohoku's reconstruction. Architecture and Urban Planning, 16(1), 1–4. https://doi.org/10.2478/aup-2020-0001.*

- What is the probability of occurrence of different severities of adverse consequences? (risk characterization and quantification)
- What can be done, and at what cost, to manage and reduce unacceptable adverse consequences? (risk mitigation and management)
  Fig. 6.2 shows a risk assessment process comprising six steps:
- The *description of intention* involves screening to decide whether further risk assessments are necessary and scoping to define the practical limits of the risk assessment.
- The *hazard assessment* typically consists of hazard identification and characterization. The key characteristics of the hazard include the magnitude–frequency pattern, time to onset, and speed of onset of hazard events. The latter two characteristics are important in determining the appropriate hazard monitoring system and establishing emergency response plans, including the development of warning systems and evacuation plans. The hazard assessment should encompass postevent hazards, secondary hazards, and follow-on hazards.
- The *consequence assessment* requires the characterization of the magnitude and severity of the hazardous events by taking into account a cascade of events triggered by an initiating event. It is crucial to recognize that the possible patterns of unfolding events are unlimited, unlike past historical catastrophic events. A wide range of possible scenarios must be accounted for in defining critical scenarios that are used for disaster preparedness and resilience planning. In this step, extreme events resulting in catastrophic consequences should be included (Dunant et al., 2021).
- *Risk estimation* involves an integration of the magnitude and probability of the hazard(s) with the magnitude and probability of the identified potential adverse consequences. Depending on how rigorously these preceding assessments are performed, the estimated risks can be expressed qualitatively, semiquantitatively, or quantitatively. The qualitative risk estimation can be formulated using a risk matrix, which combines the qualitative scoring of adverse consequences and the qualitative scoring of likelihood/probability. On the other hand, the quantitative risk estimation is often expressed by the exceedance probability (EP) curves in terms of aggregate tsunami damage and loss (Chapter 5).
- The *risk evaluation/assessment* is a judgment process to ascertain how significant the estimated risks are and to inform the subsequent *risk management* step. This step aims to get all losses expressed in the same units, for example, monetary units. However, it is difficult to assign a value to a scenic view, the environment, an amenity, or intangibles, such as artifacts, monuments, and buildings of cultural history. Similarly, a difficulty exists in assigning a value to human life, and the concept of a value of statistical life is often adopted.

In the *risk evaluation/assessment* step, the risk perception of stakeholders is taken into account (Fig. 6.2). Perceived risk is a subjective risk estimation by an individual having an imperfect view of probable outcomes, biased by belief, experience,

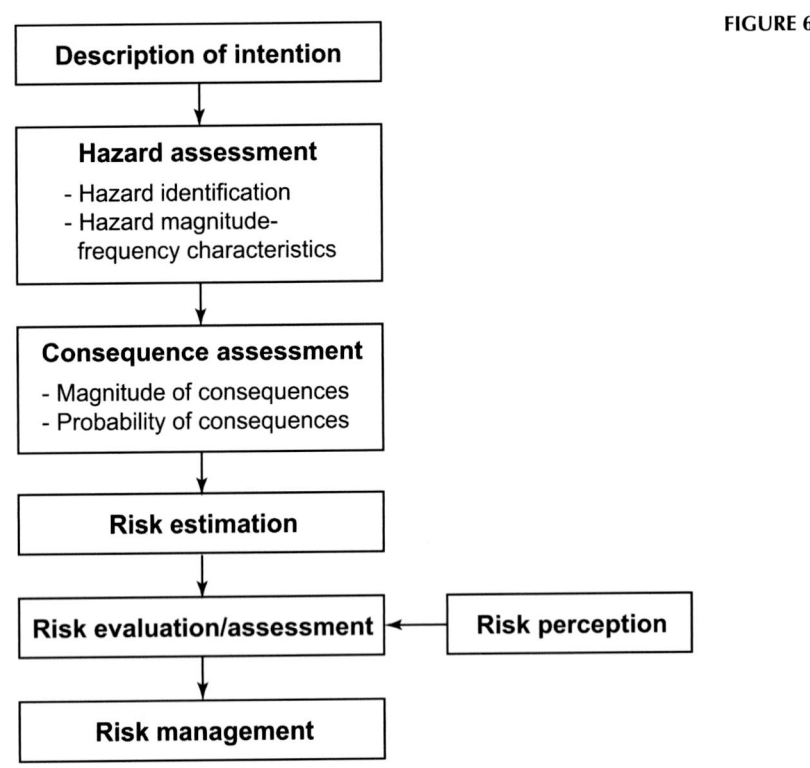

FIGURE 6.2 Risk assessment process.

TABLE 6.1 Qualitative risk matrix.

Likelihood level	Consequence level				
	Insignificant	Minor	Moderate	Major	Catastrophic
Almost certain	Medium	Medium	High	Extreme	Extreme
Likely	Low	Medium	High	Extreme	Extreme
Unlikely	Low	Low	Medium	High	Extreme
Rare	Very low	Low	Medium	High	High
Very rare	Very low	Very low	Low	Medium	High
Extremely rare	Very low	Very low	Low	Medium	High

Source: From Australian Institute for Disaster Resilience. (2015). *National emergency risk assessment guidelines*. https://knowledge.aidr.org.au/media/2030/handbook-10-national-emergency-risk-assessment-guidelines.pdf.

and personal disposition toward risk. Some of the influential factors for individual risk decisions include (1) the horror of the hazard and its outcomes, the feeling of lack of control, the possibility of fatal consequences, and CAT potential; (2) the unknown nature of the hazard and unobservable, unknown, new, delayed aspect of its manifestation; (3) experience of the activities and hazards involved; (4) environmental philosophy and world view; (5) race, gender, and socioeconomic status; (6) voluntariness; and (7) equity and nature of the threat to human generations. Good risk communication with stakeholders and the public is essential to reduce the discrepancy between objective and subjective risks. Risk communication can be facilitated by limiting technical jargon to a minimum, explaining any scientific terms properly, and explaining the basis of the conclusions and the uncertainties involved fully. Moreover, it is also essential to appreciate that individuals' perceptions of risk and resulting concerns are genuine and should be carefully considered in the risk assessment process.

Relative risk scoring is a practical approach to conducting the tsunami risk assessment. To illustrate the relative risk ranking method, a qualitative risk matrix proposed by Australian Institute for Disaster Resilience (2015) is shown in Table 6.1. The risk matrix has two main elements of the risk, that is, likelihood (rows) and consequence (columns). The likelihood level is divided into six categories, that is, "almost certain," "likely," "unlikely," "rare," "very rare," and "extremely rare." In contrast, the consequence level is divided into five categories, that is, "insignificant," "minor," "moderate," "major," and "catastrophic." The combinations of low probability and low consequence result in very low or low risk, whereas the combinations of high probability and high consequence are classified as high or extreme risk. The risk matrix captures a trade-off between likelihood and consequence qualitatively. When the likelihood level of a hazard is expressed in terms of annual EP, average recurrence interval, or frequency, a correspondence table like Table 6.2 can be used. This table is also functional when the quantitative estimates of hazard probability are available. To assign the consequence levels to hazard events, a correspondence table like Table 6.3 can be adopted. Table 6.3 is developed for classifying consequence levels in terms of the number of fatalities and injuries. Other consequences can be considered by creating correspondence tables in terms of economic, environmental, public administration, and social-setting consequences. Finally, it is important to evaluate the confidence of the likelihood and consequence

TABLE 6.2 Likelihood levels and criteria.

Likelihood	Annual exceedance probability	Average recurrence interval (indicative)	Frequency (indicative)
Almost certain	63% per year or more	Less than 1 year	Once or more per year
Likely	10% to <63% per year	1 to <10 years	Once per 10 years
Unlikely	1% to <10% per year	10 to <100 years	Once per 100 years
Rare	0.1% to <1% per year	100 to <1000 years	Once per 1000 years
Very rare	0.01% to <0.1% per year	1000 to <10,000 years	Once per 10,000 years
Extremely rare	Less than 0.01% per year	10,000 years or more	Once per 100,000 years

Source: From Australian Institute for Disaster Resilience. (2015). *National emergency risk assessment guidelines*. https://knowledge.aidr.org.au/media/2030/handbook-10-national-emergency-risk-assessment-guidelines.pdf.

TABLE 6.3 Casualty and fatality-related consequence levels and criteria.

Level	Death	Injury or illness
Catastrophic	Deaths directly from emergencies greater than 1 in 10,000 people for a population of interest	Critical injuries with long-term or permanent incapacitation greater than 1 in 10,000 people for a population of interest
Major	Deaths directly from emergencies greater than 1 in 100,000 people for a population of interest	Critical injuries with long-term or permanent incapacitation greater than 1 in 100,000 people for a population of interest or serious injuries greater than 1 in 10,000 people for a population of interest
Moderate	Deaths directly from emergency greater than 1 in 1,000,000 people for a population of interest	Critical injuries with long-term or permanent incapacitation greater than 1 in 1,000,000 people for a population of interest or serious injuries greater than 1 in 100,000 people for a population of interest
Minor	Deaths directly from emergencies greater than 1 in 10,000,000 people for a population of interest	Critical injuries with long-term or permanent incapacitation greater than 1 in 10,000,000 people for a population of interest or serious injuries greater than 1 in 1,000,000 people for a population of interest
Insignificant	Deaths directly from emergencies less than 1 in 10,000,000 people for a population of interest	Critical injuries less than 1 in 10,000,000 people for a population of interest or serious injuries less than 1 in 1,000,000 people for a population of interest or minor injuries to any number of people

Source: From Australian Institute for Disaster Resilience. (2015). *National emergency risk assessment guidelines*. https://knowledge.aidr.org.au/media/2030/handbook-10-national-emergency-risk-assessment-guidelines.pdf.

TABLE 6.4 Likelihood–consequence confidence matrix.

Confidence in likelihood	Confidence in consequence				
	Lowest	Low	Moderate	High	Highest
Highest	Moderate	Moderate	High	Highest	Highest
High	Moderate	Moderate	Moderate	High	Highest
Moderate	Low	Moderate	Moderate	Moderate	High
Low	Lowest	Low	Moderate	Moderate	Moderate
Lowest	Lowest	Lowest	Low	Moderate	Moderate

Source: From Australian Institute for Disaster Resilience. (2015). *National emergency risk assessment guidelines*. https://knowledge.aidr.org.au/media/2030/handbook-10-national-emergency-risk-assessment-guidelines.pdf.

assessments as a qualitative indicator for the robustness of the risk assessment. For this purpose, the level of confidence in the risk assessment process is used to identify and communicate uncertainty. An example of the confidence matrix is shown in Table 6.4.

6.2.2 Identification of critical tsunami scenarios and hazard–risk mapping

Critical tsunami scenarios for risk management purposes can be identified by performing tsunami hazard and risk analyses. A conventional, but limited, approach is to examine the list of historical tsunami events that occurred in the region of interest and select the most significant event as a critical tsunami scenario for disaster planning. Challenges faced with this approach are quantifying the uncertainty associated with historical events (magnitudes, locations, and earthquake rupture characteristics) and accounting for possibilities that extreme events may be missed in the available history of major tsunami events in the region of interest. Another difficulty is to assign the probability of occurrence to these historical events without a formal assessment of tsunami hazards from various sources. Such scenarios can be selected on the basis of the current tsunami threat in a region. For instance, for the Nankai–Tonankai region of Japan, the future occurrence of a megathrust subduction earthquake from the Nankai Trough has been identified for tsunami design and community evacuation purposes (Central Disaster Management Council, 2012).

Modern PTHA and PTRA frameworks, described in Chapter 5, can provide the basis for mitigating and controlling disaster risk exposures effectively in coastal areas. The key requirements are that the main uncertainties in earthquake occurrence, rupture process, and tsunami generation and propagation are quantified and incorporated into the analyses. In addition, epistemic uncertainty associated with PTHA elements should be accounted for by considering alternative models (Miyashita et al., 2020). Outputs from such hazard analyses include site-specific tsunami hazard curves, which can be used for engineering design (Chock, 2016), and tsunami inundation maps at different return period levels, which can serve as the fundamental input to develop local and regional risk mitigation plans (Zamora et al., 2021).

A standard method to define critical scenarios based on PTHA results is to adopt tsunami hazard scenarios at selected probability levels. The probability levels can be typically represented by mean return periods, such as 500 years, 1000 years, and 2500 years, approximately corresponding to 10%, 5%, and 2% in 50 years probability of exceedance, respectively. A tsunami hazard curve provides this information, and tsunami inundation simulation results at specified return period levels can be extracted to examine whether population, buildings, and infrastructure are affected by the tsunamis. More specifically, overlaying the building footprints and infrastructure networks with the inundation maps using a Geospatial Information System will permit a simple tsunami impact assessment for coastal communities. Furthermore, the inundation simulation results can be used to conduct tsunami risk assessments by integrating them with the building exposure model and tsunami vulnerability model. The tsunami damage and loss data can be shown on maps to display which populations and assets are at risk. To account for the effects of tidal levels and other ocean conditions, such as future sea-level rise from climate change, different baseline tidal conditions can be considered, and the effects due to these changing ocean conditions can be quantified in terms of tsunami risk metrics. In the following part of this subsection, examples for Tofino, British Columbia, Canada, under the Cascadia megathrust events are used to visualize the PTHA and PTRA results for disaster preparedness planning.

6.2.2.1 Tsunami hazard and exposure

The Cascadia subduction zone is one of the major seismic and tsunami sources along the Pacific coast of British Columbia, Canada. It involves the thrusting movements of the Juan de Fuca, Gorda, and Explorer Plates, which subduct underneath the North American Plate (Fig. 6.3A). A scenario of particular concern is the future occurrence of the moment magnitude (M_w) 9 megathrust earthquakes in Cascadia. To predict the future rupture patterns of the Cascadia events, geological and geophysical evidence and data have been collected, including onshore subsidence records and offshore marine turbidites (Goldfinger et al., 2012). Coastal municipalities and communities on Vancouver Island face an imminent need for accurate performance assessments of buildings and infrastructures under catastrophic tsunami threat.

Tofino is located at the tip of the Esowista Peninsula on Vancouver Island and is exposed to the Pacific Ocean to its west. Tofino is famous for its natural landscape and is a popular tourist destination. A map of Tofino, including buildings and roads, is shown in Fig. 6.3B. The elevations in Mackenzie Beach, Chesterman Beach, and Cox Bay are low, directly facing the Pacific Ocean. Hence, tsunami risks to people and buildings in the beach areas are high, and the particular concern for emergency managers and responders in Tofino is to communicate tsunami risks with tourists. The District of Tofino conducted comprehensive surveys to develop high-quality building and population exposure models for disaster risk assessment. The building dataset includes 1789 buildings, mostly 1- to 2-story wooden residential buildings constructed in the 1960s or afterward. The total asset value of the 1789 buildings is C$ 2.27 billion. On the other hand, the population dataset estimates the peak-summer populations at 2001 locations in the district (including marinas, camping sites, and touristic facilities). The peak-summer daytime and nighttime populations are 9600 and 11,590, respectively. These data can be used to develop a reliable population exposure model for tsunami fatality estimation (Section 6.2.3).

6.2.2.2 Tsunami hazard and risk mapping

PTHA for Tofino is conducted by implementing methods based on stochastic source models for the Cascadia subduction events. For this purpose, 5000 stochastic tsunami scenarios for the comprehensive magnitude range between M_w 8.1 and M_w 9.1 are considered by reflecting different rupture patterns along the Cascadia subduction zone (i.e., southern segment, southern-to-central segment, and full-margin ruptures). To derive a tsunami hazard curve for Tofino, the representative offshore point shown in Fig. 6.3B is adopted. The obtained tsunami hazard curve is shown in Fig. 6.4A. The maximum tsunami wave height, determined with respect to the mean sea level (i.e., no tidal effect), is 3.84, 5.37, 6.93, and 7.80 m at the return periods of 500, 1000, 2500, and 5000 years, respectively. These hazard levels can be used to define critical tsunami scenarios for tsunami hazard and risk mapping. In addition, the corresponding tsunami source models for the return periods of 500 and 2500 years are shown in Fig. 6.4B and C, respectively. Although both scenarios have similar earthquake magnitudes (i.e., M_w 8.86 vs M_w 8.96), their earthquake rupture patterns are significantly

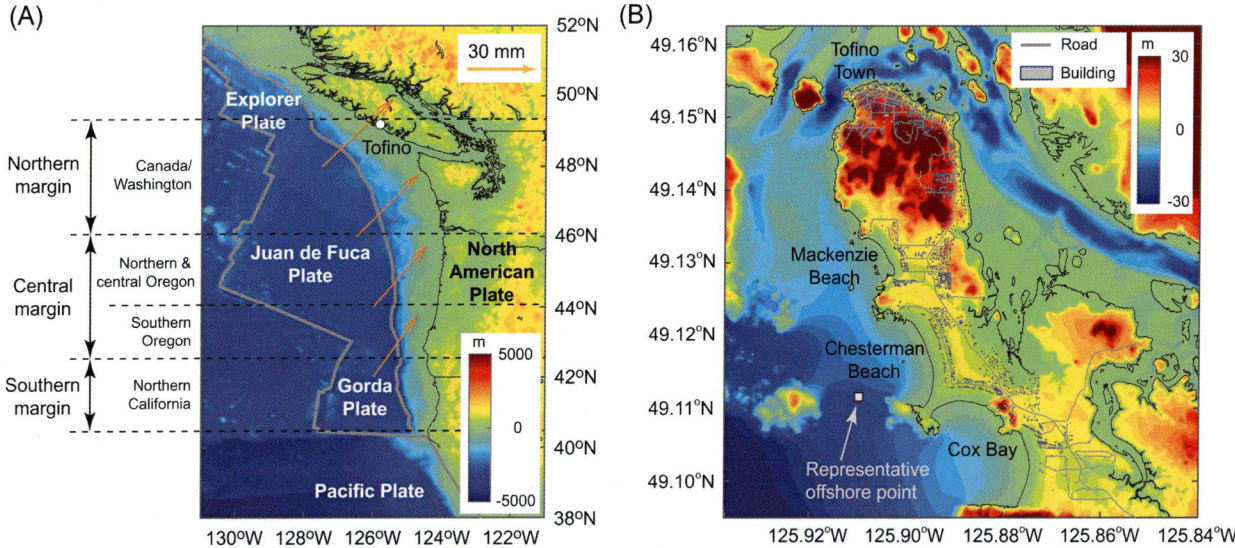

FIGURE 6.3 (A) Map of the Pacific Northwest and Cascadia subduction zone and (B) map of Tofino, British Columbia. Map lines delineate study areas and do not necessarily depict accepted national boundaries.

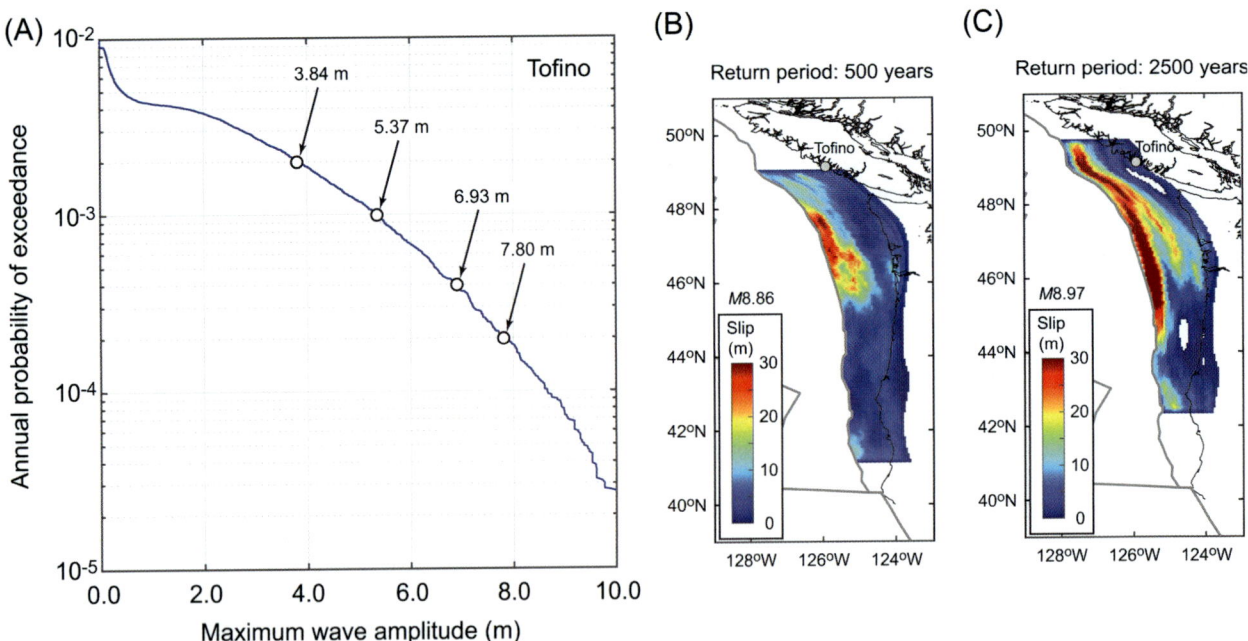

FIGURE 6.4 (A) Tsunami hazard curve at the offshore representative location off Tofino from the Cascadia subduction earthquakes. (B) Tsunami source model for the 500-year return period level. (C) Tsunami source model for the 2500-year return period level. Map lines delineate study areas and do not necessarily depict accepted national boundaries.

different. The tsunami source for the 500-year return period does not extend toward Tofino, whereas the tsunami source for the 2500-year return period has a large slip concentration in front of Tofino.

Multiple stochastic source scenarios achieve the maximum tsunami wave heights at each return period level. The tsunami inundation depths for the selected tsunami hazard levels can be visualized to examine the tsunami hazard impacts on buildings and roads. Fig. 6.5A and C shows tsunami inundation maps for Tofino at the return periods of 500 and 2500 years, respectively. At the return period of 500 years, tsunami inundations are limited to low-lying nearshore areas along Mackenzie Beach, Chesterman Beach, and Cox Bay. Under this scenario, the tsunami inundation depths are typically 2 m or less; therefore the tsunami damage will be limited to the first row of houses along Chesterman Beach. The tsunami inundation becomes significant when a more extreme scenario is considered. In fact, the low-lying area of the Esowista Peninsula will be completely inundated, and severe damage and loss will occur to buildings and infrastructure. Moreover, the comparison of the two inundation maps demonstrates the significant influence of earthquake slip distribution on the tsunami hazard results.

The tsunami inundation results for the two return period levels shown in Fig. 6.5A and C are for the average tidal condition in Tofino. However, tsunami waves may arrive during high tide periods. To investigate the effects of tidal conditions on the tsunami inundation results, a 2-m high tidal level is considered when simulating the tsunami inundation in Tofino under the same tsunami scenarios (i.e., Fig. 6.4B and C). The results are shown in Fig. 6.5B and D. Although the influence of the tide level on the maximum tsunami inundation depth is nearly commensurate (i.e., 2-m high tide will increase the tsunami inundation depth by approximately 2 m), the extent of the inundated areas is increased nonlinearly, depending on local topography. For the 500-year return period level (i.e., Fig. 6.5A and B), the tsunami wave can flood the low-lying part of the Esowista Peninsula. When the tsunami source becomes more extreme (i.e., Fig. 6.5C and D), the low-lying area will experience higher tsunami inundation depths, exposing more buildings to tsunami damage and loss. Paying attention to the road network in Tofino, a single main road connects Tofino Town (i.e., the tip of the Esowista Peninsula) and other parts of cities and towns on Vancouver Island (e.g., Ucluelet and Port Alberni). Noteworthily, the electricity grids in Tofino run parallel with the roads. At the return period of 500 years and the tidal level of 0 m, the passage of the traffic will not be significantly disrupted (if neither major building collapses nor liquefaction-related ground failures occur along the roads). At the return period of 500 years and the tidal level of 2 m, the main road will be submerged by the tsunami, but the disruption may not be severe given the relatively shallow inundation depths along the main road (0–2 m). In contrast, under the 2500-year return period scenario, the main road will be inundated over a relatively long distance, and the inundation depth becomes high, exceeding 4 m at several

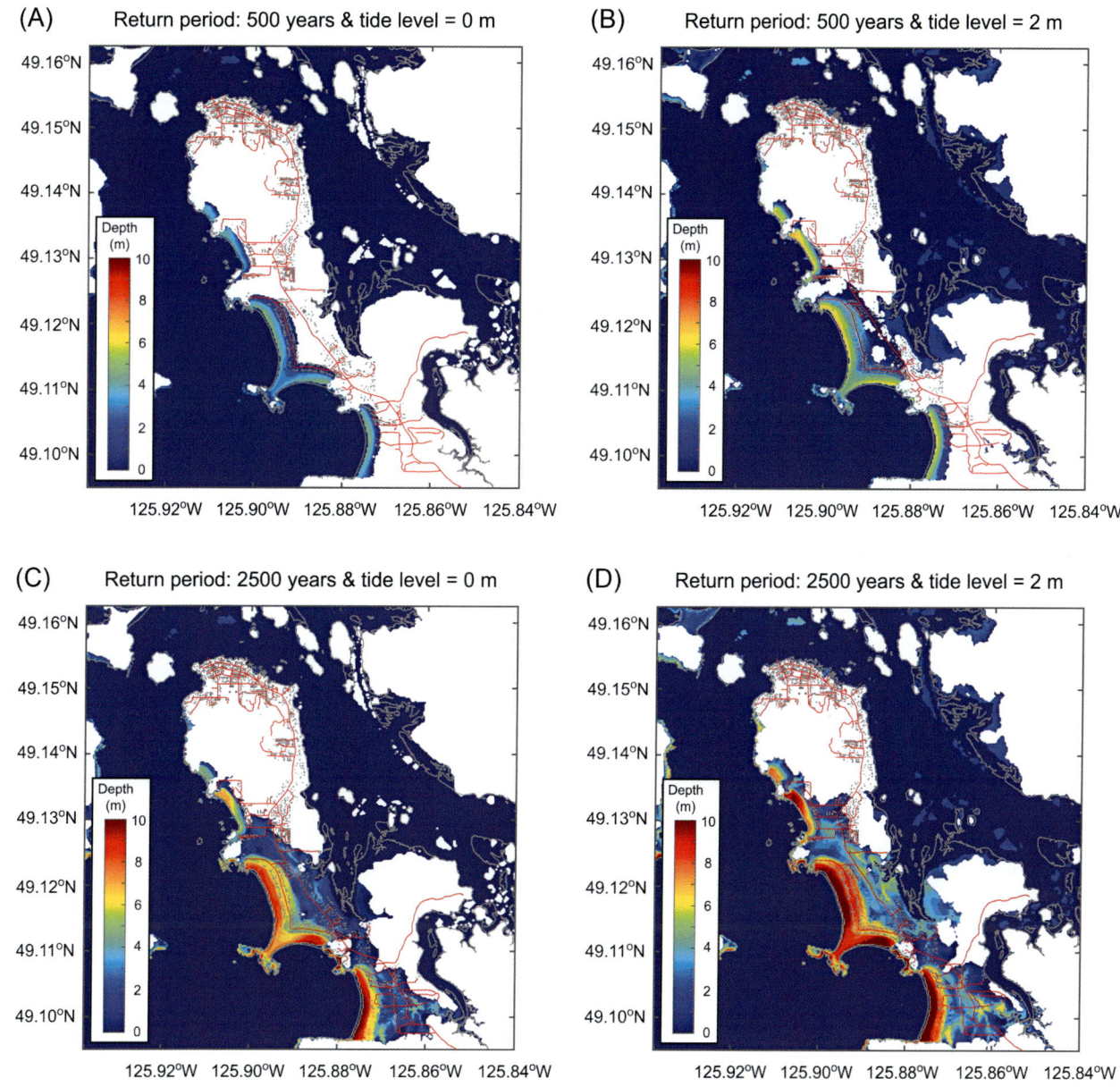

FIGURE 6.5 Tsunami inundation maps for Tofino considering different scenarios of the Cascadia subduction earthquakes: (A) return period of 500 years and tidal level of 0 m, (B) return period of 500 years and tidal level of 2 m, (C) return period of 2500 years and tidal level of 0 m, and (D) return period of 2500 years and tidal level of 2 m. The tidal level of 0 m corresponds to the mean sea level. In the maps, building locations (gray polygons) and roads (*red* lines) are shown. Map lines delineate study areas and do not necessarily depict accepted national boundaries.

locations. In such conditions, it is likely that roads may become impassable due to tsunami debris and may not be repairable over a short duration. In such worst-case scenarios, Tofino Town may be isolated from the rest of Vancouver Island, and residents in Tofino should be prepared for the prolonged disruptions of infrastructure.

A superposition of exposed buildings and infrastructure over tsunami inundation results will provide valuable information on possible tsunami disaster frequency and spatial extent. A simple tsunami planning map can be developed, which defines safe and unsafe areas based on land elevations in a binary manner under different critical scenarios and tidal conditions. From the maximum runup heights (similar to Fig. 6.5, but the tsunami heights are measured with respect to the reference sea level, rather than the ground elevation after accounting for coseismic deformation), the maximum runup heights for the return periods of 500 and 2500 years are obtained as 9.0 and 17.2 m by considering the tidal condition of 1 to 2 m. In such a case, two threshold values of 10 and 20 m can be considered for developing a tsunami planning map for Tofino. Such a tsunami planning map is illustrated in Fig. 6.6.

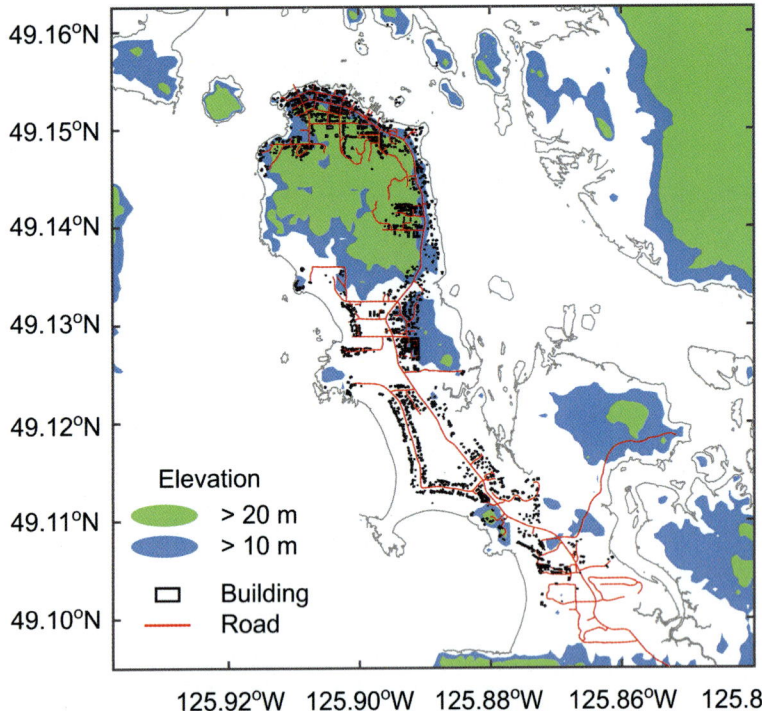

FIGURE 6.6 Tsunami planning map for Tofino by considering two elevation thresholds. The 10-m threshold approximately corresponds to the critical tsunami scenario for the 500-year return period, whereas the 20-m threshold approximately corresponds to the critical tsunami scenario for the 2500-year return period. Areas without colors may be regarded to be risky areas under possible Cascadia tsunami scenarios. Map lines delineate study areas and do not necessarily depict accepted national boundaries.

Beyond the tsunami hazard mapping, detailed tsunami inundation results can be used to develop tsunami risk maps for buildings and infrastructure. For tsunami damage and loss to buildings, tsunami fragility functions can be applied to individual buildings, and tsunami damage severity can be converted to tsunami loss estimates when building cost information is available. Fig. 6.7 shows tsunami loss maps for buildings in Tofino, where the color represents the building tsunami loss. Large economic losses are concentrated in the beach areas, where many houses are located, and tsunami inundation depths can be high. Because buildings in Tofino Town are at higher elevations above 20 m (Fig. 6.6), the tsunami risk to these buildings is not significant (but they may suffer from shaking damage in the case of the Cascadia megathrust event).

6.2.2.3 Risk-based critical tsunami scenarios

It is essential to recognize that in the tsunami hazard and risk mapping for Tofino presented in Section 6.2.2.2, critical tsunami scenarios are identified on the basis of the tsunami hazard curve at the offshore location (Fig. 6.4A). The underlying premise of using the hazard-based critical scenarios (e.g., Fig. 6.4B and C) to develop tsunami risk maps (e.g., Fig. 6.7) as a substitute for the risk-based critical scenarios is that the correlation between tsunami hazard parameter and tsunami building loss is strong and the hazard- and risk-based tsunami scenarios correspond well. To examine the correlation between the tsunami hazard and risk parameters, Fig. 6.8 shows a relationship between the maximum wave height and the tsunami building loss. The tsunami hazard parameter at the representative offshore point is correlated well with the tsunami building loss, but the correlation is not perfect.

A more direct approach to defining critical tsunami scenarios is to derive a tsunami loss curve for the building portfolio in Tofino and then choose the corresponding tsunami source models at the selected probability levels. The tsunami loss curve for the buildings in Tofino is shown in Fig. 6.9. Although defining risk-based critical tsunami scenarios is straightforward, the computational costs can be substantial and thus may be prohibitive in practice because numerous high-resolution tsunami inundation simulations (e.g., using 5-m grids) are required for several thousand tsunami scenarios, noting that each high-resolution tsunami simulation may take a few days.

To avoid numerous runs of high-resolution tsunami inundation simulations but to estimate the tsunami risk reliably, a two-step method can be implemented:

- In the first step, a full tsunami simulation is carried out by comprehensively considering several thousands of tsunami scenarios (e.g., 5000 Cascadia stochastic source models) and performing a full PTHA at the representative site (Fig. 6.4A). Because the representative site is offshore (e.g., Fig. 6.3B), a relatively coarse tsunami simulation grid

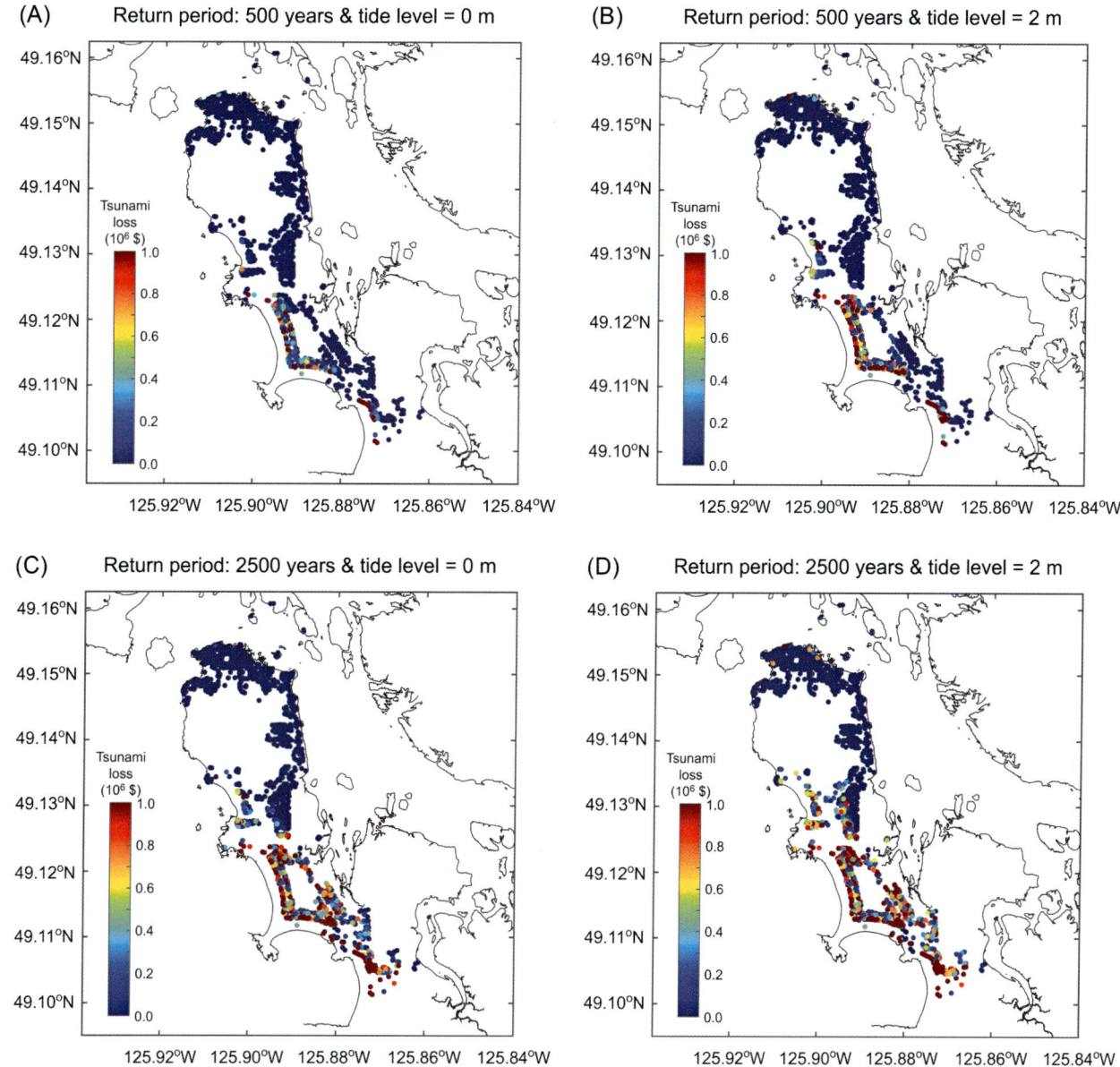

FIGURE 6.7 Tsunami loss maps for Tofino considering different scenarios of the Cascadia subduction earthquakes: (A) return period of 500 years and tidal level of 0 m, (B) return period of 500 years and tidal level of 2 m, (C) return period of 2500 years and tidal level of 0 m, and (D) return period of 2500 years and tidal level of 2 m. The tidal level of 0 m corresponds to the mean sea level. Map lines delineate study areas and do not necessarily depict accepted national boundaries.

spacing of several hundred meters is sufficient. This numerical setup will permit numerous runs of tsunami simulations with affordable computational requirements.
- In the second step, to estimate the tsunami risk for a building portfolio, 10−20 tsunami sources can be selected for a given return period. Subsequently, high-resolution tsunami inundation simulations can be carried out for these 10−20 tsunami sources, and the tsunami loss for the buildings can be evaluated for this smaller set of critical tsunami scenarios. Although the individual tsunami risk estimates based on hazard-based critical scenarios can vary significantly (e.g., Fig. 6.8), the median or mean of the estimated tsunami loss based on 10−20 high-resolution tsunami inundation simulations is more stable.

The overall computational costs can be reduced significantly by adopting the two-step approximate approach. The robustness of the tsunami risk estimation is demonstrated in Fig. 6.9 for the Tofino case study. Each of the orange

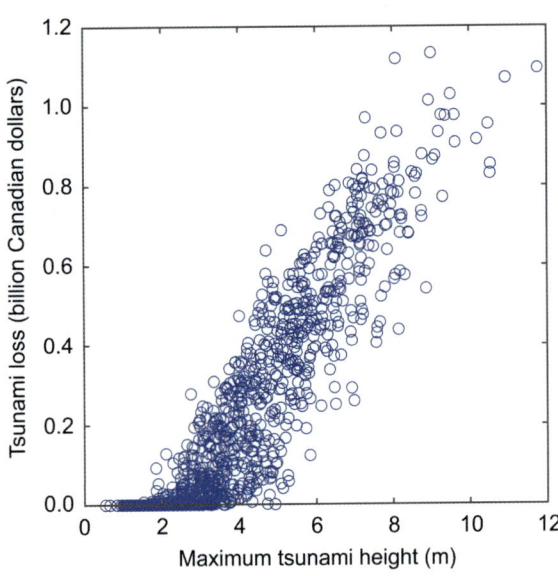

FIGURE 6.8 Scatter plot of the maximum tsunami height at the representative offshore point versus tsunami building loss for Tofino.

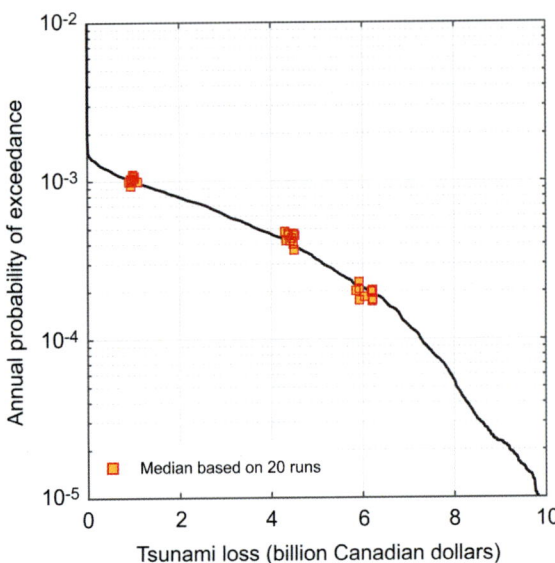

FIGURE 6.9 Tsunami loss curve and estimated median estimates based on 20 tsunami simulation runs. The tsunami loss curve (black curve) is derived on the basis of the 2000 stochastic source models between M_w 8.7 and M_w 9.1.

square markers is obtained on the basis of 20 high-resolution tsunami inundation simulations whose tsunami sources are identified from the full PTHA based on low-resolution tsunami simulations. For the three probability levels (i.e., return period levels of 1000, 2500, and 5000 years), the approximate method can estimate the tsunami loss values reliably.

6.2.3 Tsunami fatality risk and societal risk tolerability

Life safety remains the highest priority in mitigating the tsunami disaster due to its catastrophic nature. The life safety issue is severe for near-field tsunamis for several reasons. First, in nearshore areas, there is a short time between the seismic event and the resulting inundation, typically several minutes to 1 hour, compared to several hours of warning for far-field tsunamis. Second, evacuations will be self-initiated, relying on the individual's perception of risk and knowledge of the correct course of action. This can be problematic in areas where current generations of residents have not experienced major tsunamis, and many tourists are unfamiliar with local environments, including evacuation points. Third, some areas in a coastal community has a disproportionately larger population at risk due to age and socioeconomic status. Finally, life safety can be increased through a number of means, including structural measures, such as vertical evacuation facilities. Advances in evacuation modeling can help individuals better understand their risks due to near-field tsunamis and determine the best travel routes (Mas et al., 2015; Muhammad et al., 2021; Taubenböck et al., 2009; Wood et al., 2016).

Tsunami risk exposures are involuntary risks. The socially acceptable risk level is often controlled by authorized agencies; therefore, to control involuntary risk adequately, some guidelines need to be established by reflecting public value systems as well as public psychological effects (Starr, 1969). The ALARP (As Low As Reasonably Practicable) guideline, proposed by Health and Safety Executive (1992), is commonly represented as an $F-N$ curve that depicts the annual frequency F of an incident causing a loss of N or more lives. A typical $F-N$ curve presented by Health and Safety Executive (1992) is illustrated in Fig. 6.10. The ALARP region is bounded by the local tolerability and the negligibility lines, and the regions above the former and below the latter are considered "intolerable" and "negligible," respectively. The slopes of the limiting lines shown in Fig. 6.10 equal -1 on a logarithmic scale; such $F-N$ curves are considered a risk-neutral approach since the expected value on a limiting line (i.e., $F \times N$) is the same. In the literature, a quantitative fatality risk level has been discussed in evaluating the societal risk tolerability based on empirical and professional consensus. These studies suggest that an annual fatality risk of less than 10^{-6} is deemed to be negligible, whereas a value greater than 10^{-3} is significant. The range of the annual fatality frequency (i.e., $10^{-6}-10^{-3}$) is a useful yardstick for DRM.

To estimate the tsunami fatality risk, a population exposure model and an empirical fatality rate model can be integrated. The computational framework is similar to the tsunami building risk estimation. The first important step is to develop a realistic population exposure model. For Tofino, comprehensive building and facility inventories have been developed, which are accompanied by the peak-summer daytime and nighttime population distributions. In addition, various visitor surveys are available to determine monthly variations of the local populations and visitors. Based on the available information, 21 population profiles are defined. There are seven varied cases for the total population, ranging from 4000 to 10,000. This variation in the total population is shown in Fig. 6.11A. Regarding the daily variations of exposed people, the daytime and nighttime spatial distributions of the population are adopted, while an additional distribution is derived by averaging these spatial distributions of the population as a transition. The weights of 8/24, 10/24, and 6/24 can be assigned to the daytime, nighttime, and transition spatial distributions, respectively.

Due to the lack of direct observations of tsunami fatalities in British Columbia, a reliable, although approximate and potentially biased, approach is to adopt an empirical tsunami fatality rate model from recent catastrophic tsunamis. More specifically, the tsunami fatality rate model by Suppasri et al. (2016), which was developed using the extensive tsunami death database from the 2011 Tohoku Tsunami, is adopted. The fatality rate model developed by Suppasri et al. (2016) predicts the fatality rate as a function of tsunami flow depth, and their parameters involve statistical variability. Therefore, Monte Carlo sampling can be carried out for each tsunami event. The tsunami fatality ratio functions are shown in Fig. 6.11B.

The tsunami fatality risk curve in Tofino is developed by modifying simulation-based PTRA approaches for buildings. More specifically, the building exposure model is replaced by the population exposure model (i.e., 21 population profiles), whereas the tsunami fragility functions are replaced by the tsunami fatality rate model with variable model

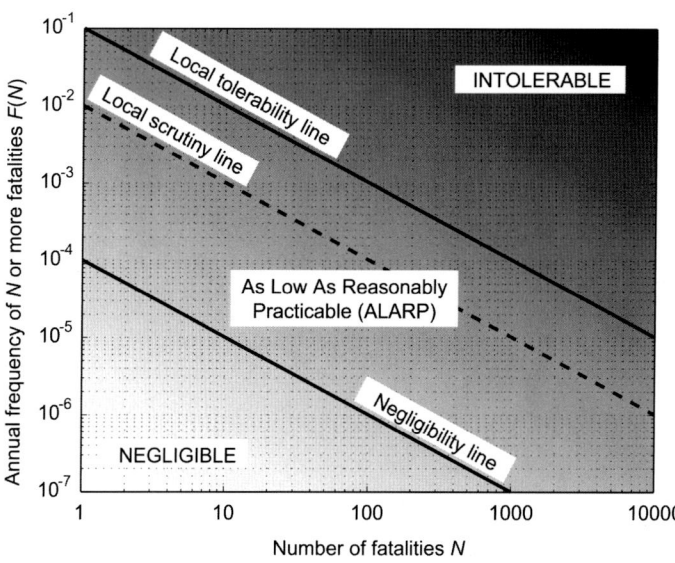

FIGURE 6.10 $F-N$ curves and risk criteria developed for major hazards of transport study.

204 SECTION | 1 Fundamentals of probabilistic tsunami hazard and risk analysis

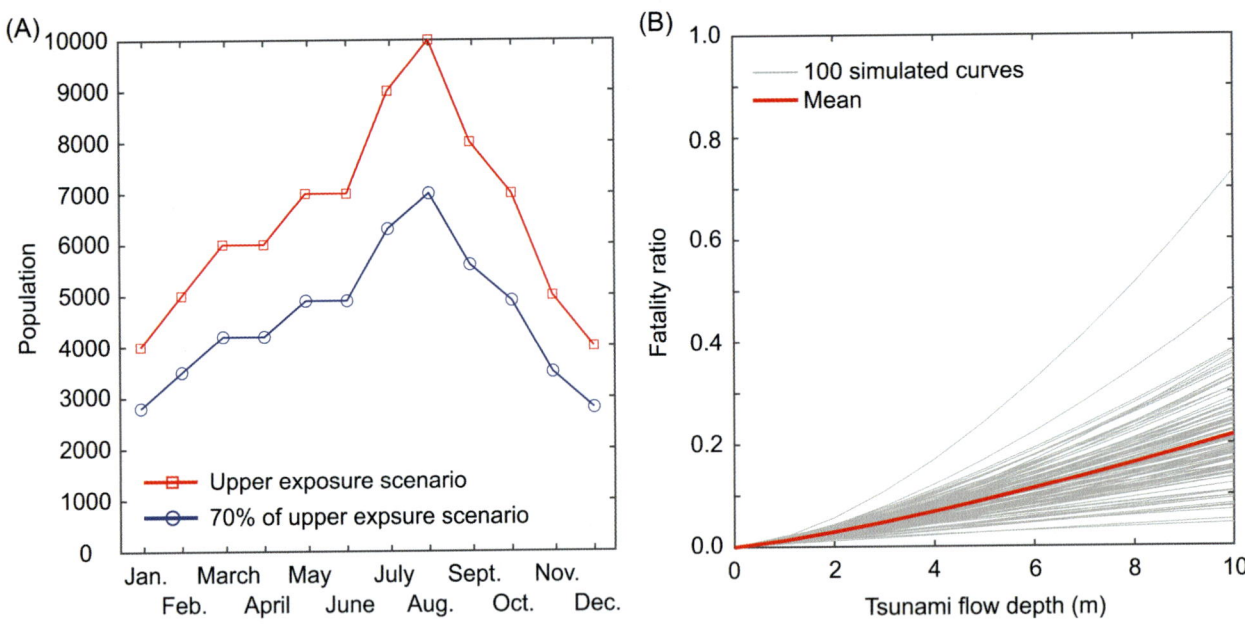

FIGURE 6.11 (A) Monthly variation of average population in Tofino. (B) Tsunami fatality ratio function.

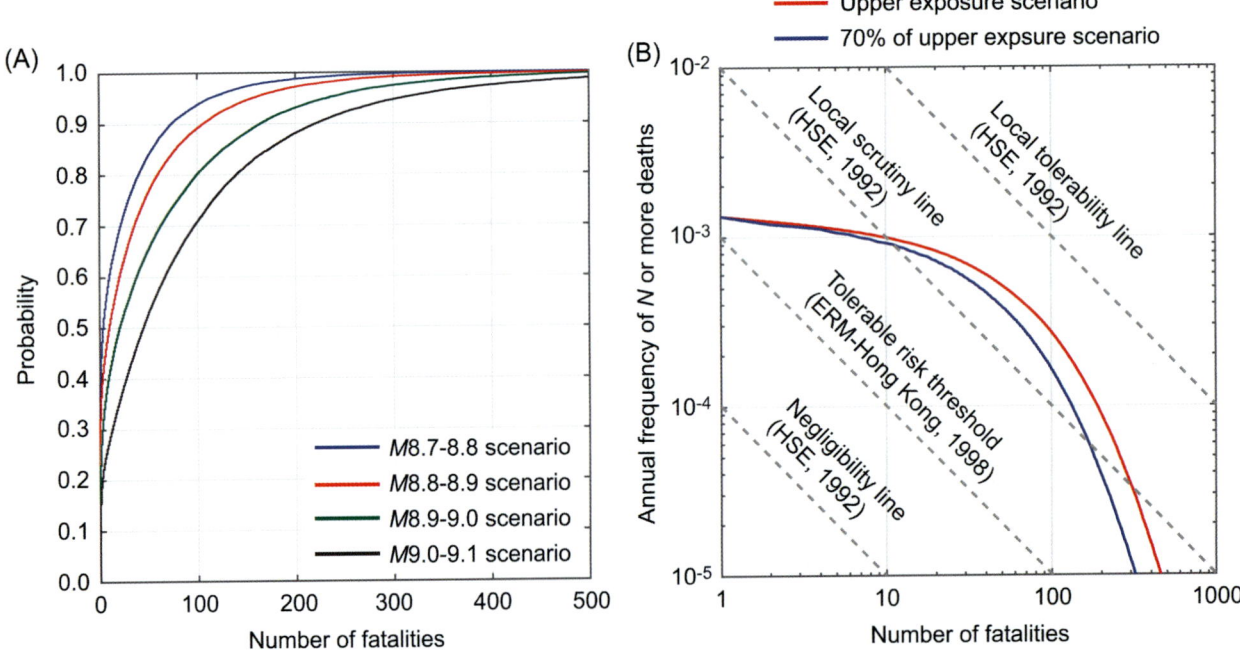

FIGURE 6.12 (A) Conditional cumulative distribution functions of tsunami fatalities for different earthquake magnitude scenarios. (B) Annual frequency of N or more fatalities due to the Cascadia megathrust events.

parameters. Using the tsunami inundation results, the tsunami fatalities at individual occupied locations in Tofino are evaluated as a Binomial variable (i.e., each person has a probability of survival for a given tsunami flow depth). By repeating this process for all events in the stochastic event catalogs, the EP curve can be obtained. This curve represents the tsunami $F-N$ curve for Tofino.

To show the tsunami fatality risk estimation results for Tofino, the cumulative distribution functions of tsunami fatalities in Tofino for four earthquake magnitudes are shown in Fig. 6.12A. With the increase of the earthquake magnitude, the curve shifts toward the right, causing greater tsunami fatalities. In the worst cases, the number of tsunami fatalities can exceed 200. In Fig. 6.12B, the obtained $F-N$ curves for Tofino are shown. The results are obtained for two exposure cases; one is with the upper exposure scenario (i.e., the maximum exposed population is 10,000), whereas the other is with the reduced exposure scenario (i.e., all population values for the upper exposure scenario are multiplied by 0.7). Both curves start at $N = 1$ and $F = 1/750$, and N gradually increases with decreasing F. The curves eventually reach some limit values; for these scenarios, low-lying parts of Tofino are completely inundated, while approximately 50% of the population is not directly affected by the tsunami risks because of high elevations (Fig. 6.5). In Fig. 6.12B, several societal risk criteria suggested by Health and Safety Executive (1992) and ERM-Hong Kong, Ltd (1998) are included. Both $F-N$ curves intersect with the local scrutiny line by Health and Safety Executive (1992), falling in the ALARP region (i.e., an area between the negligibility line and the local tolerability line). Compared to the tolerable risk threshold set by ERM-Hong Kong, Ltd (1998), the tsunami $F-N$ curves for Tofino are located above this criterion. The results indicate that the tsunami risks for Tofino residents and visitors are concerning, and when effective risk mitigation measures can be implemented, such actions should be pursued.

6.3 Emergency response

A destructive tsunami may be triggered after a large earthquake or a violent volcanic eruption and arrive at coastal areas as early as several minutes after the trigger event. When a long/strong shaking is felt, or a tsunami warning is heard from local siren systems or warning notification systems, the best way to protect them and their families from tsunamis is to evacuate to high grounds or vertical evacuation buildings/shelters. In responding to emergencies caused by tsunamis effectively, it is important to be familiar with high-risk areas in the neighborhood of home, school, and workplace. This step is aided by tsunami hazard and risk maps prepared by local municipalities (Section 6.2). It is also essential to know the evacuation and alerting systems in the community and be prepared to act when such an alarm is activated. For instance, a personalized preparedness emergency kit can be put together and placed at home, containing practical items, such as drinking water, unperishable food, cash, and a portable radio. Planning evacuation routes is also important. Local municipalities should specify destinations for tsunami evacuation and should be indicated on local tsunami hazard−risk maps. These locations are typically designated to areas a few tens of meters above sea level or a few kilometers inland (note: the maximum vertical runup height and horizontal runup distance depend on local topography and tsunami scenario). However, not all evacuees will know routes to these places in advance (e.g., tourists). In such cases, visible evacuation signages are particularly important. Evacuees must also be aware of fires after the earthquake and secondary hazards along the evacuation routes, such as liquefaction and landslides. These routes may be blocked by earthquake-shaking debris. During the nighttime evacuation, evacuation speeds tend to slow down. Due to these obstacles and congestions, evacuation from potential inundation areas may take longer times than initially planned. Some specialized systems and tools help municipality officers and residents plan their emergency response actions more effectively.

In Section 6.3, two topics are focused upon. Section 6.3.1 introduces a tsunami early warning system. Globally and nationally, such a system has been created and operated over the last several decades. A recent tsunami monitoring system in Japan will be used to highlight the key aspects of the tsunami early warning system to be effective in issuing accurate warning messages to evacuees. Section 6.3.2 summarizes approaches to tsunami evacuation planning. Utilizing detailed information on exposed populations and infrastructure, more sophisticated modeling of evacuees' behavior in responding to tsunamis can be performed.

6.3.1 Tsunami early warning system

A tsunami early warning system detects tsunamis offshore and issues suitable warning messages to mitigate their severe impacts on people and assets (Fig. 6.13). A typical system comprises a tsunami detection system (i.e., buoys, ocean bottom sensors, and seismic sensors) and a forecast-warning system (i.e., broadcast and early warning), assisting the exposed population to escape from the inundation zone in a timely and safe manner. Existing systems, such as the Japan Meteorological Agency (JMA) (Tatehata, 1997), the Deep-ocean Assessment and Reporting of Tsunami (Gonzalez et al., 1998), and the German-Indonesian system (Rudloff et al., 2010), operate/monitor/maintain observational networks, analyze earthquake information, and evaluate wave height information to issue accurate warning messages. Additional information on existing tsunami early warning systems can be found in Chapters 7 and 13. It is important for coastal residents to be aware of the potential threat of tsunamis and to know what to do in the event of a tsunami warning.

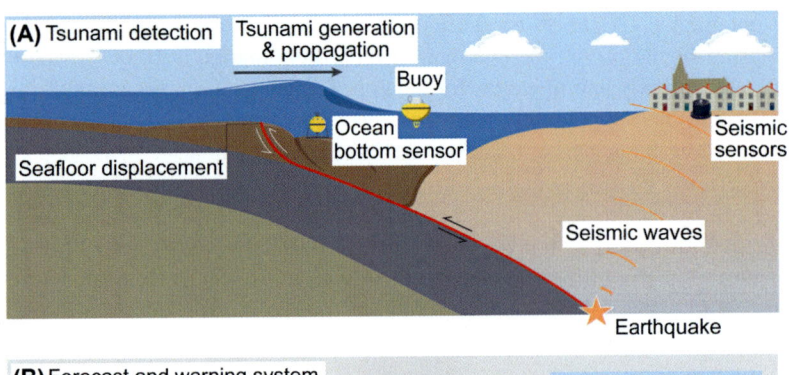

FIGURE 6.13 (A) Tsunami detection system. (B) Tsunami forecast and warning system.

Issuing accurate and prompt tsunami warnings to residents in coastal areas is critically important for megathrust tsunamigenic earthquakes. During the initial phase, it requires reliable estimation of key earthquake source characteristics, such as magnitude and location. The estimation of earthquake information is usually accurate and prompt; however, for large earthquakes, satisfactory performance may not be achieved during the early evacuation phase. This was exemplified by the 2011 Tohoku Earthquake and Tsunami (Hoshiba & Ozaki, 2014). The first estimate of the JMA magnitude (M_J) was 7.9 (3 minutes after the earthquake) and later was updated to M_J 8.4 (74 minutes after the earthquake). The significant underestimation was caused by the saturation of M_J. A correct estimate of the moment magnitude equal to M_w 8.8 (and eventually to M_w 9.0) was reached 134 minutes after the earthquake. It took a considerably long time to get the correct earthquake magnitude value because seismograms recorded at Japanese broadband stations exceeded the instruments' maximum amplitudes. On the other hand, overseas agencies, such as the U.S. Geological Survey, obtained correct estimates of the moment magnitude about 20 minutes after the earthquake using teleseismic signals recorded outside of Japan. Consequently, tsunami warnings issued by the JMA underestimated the observed tsunamis significantly (3–6 m vs 10 + m; Cyranoski (2011)). Different estimates of the earthquake source parameters had a significant influence on the predicted wave heights and inundation depths because seismic events of M_w 8.0 and M_w 9.0, for instance, correspond to very different tsunami hazard scenarios in terms of size and earthquake slip (Chapter 2).

After the 2011 Tohoku Earthquake and Tsunami, the tsunami wave monitoring system off the Kanto, Tohoku, and Hokkaido regions of Japan was expanded, and a new S-net was established (Chapter 11). The S-net project was initiated to improve the JMA's tsunami early warning system, which did not perform satisfactorily during the 2011 event. The current JMA system combines seismic networks (280 seismometers) with the S-net system, which consists of six cabled line sections and five on-land data management centers and extends over a 1000 km × 300 km area from off-Hokkaido to off-Kanto (Fig. 6.14). The S-net monitors real-time wave heights using 150 ocean bottom sensors to enhance the accuracy of tsunami forecasting by providing quick updates to early warning systems (Kanazawa, 2013; Sato et al., 2011) and has consistently demonstrated its ability to detect tsunamis, including tsunamis generated by the 2016 Fukushima Earthquake (Wang & Satake, 2021), the 2021 Miyagi Earthquake (Yoshida et al., 2022), and the 2022 Tonga Volcano eruption (Kubo et al., 2022).

An effective tsunami early warning system should consider not only early detection but also encompass information, such as the maximum onshore tsunami wave amplitude, inundation depth, and potential tsunami loss of coastal assets (Maeda et al., 2015; Tanioka & Gusman, 2018). Different statistical methods have been implemented to develop algorithms for tsunami early warning, including multiple linear regression (Kamiya et al., 2022; Li & Goda, 2022), random forest (Li & Goda, 2023b), and deep learning (Makinoshima et al., 2021; Mulia et al., 2022). In developing new methods for tsunami early warning systems, four perspectives are worthy of consideration: (1) use of the synthesized tsunami data for calibrating tsunami early warning algorithms, (2) data-driven statistical methods, (3) network system design

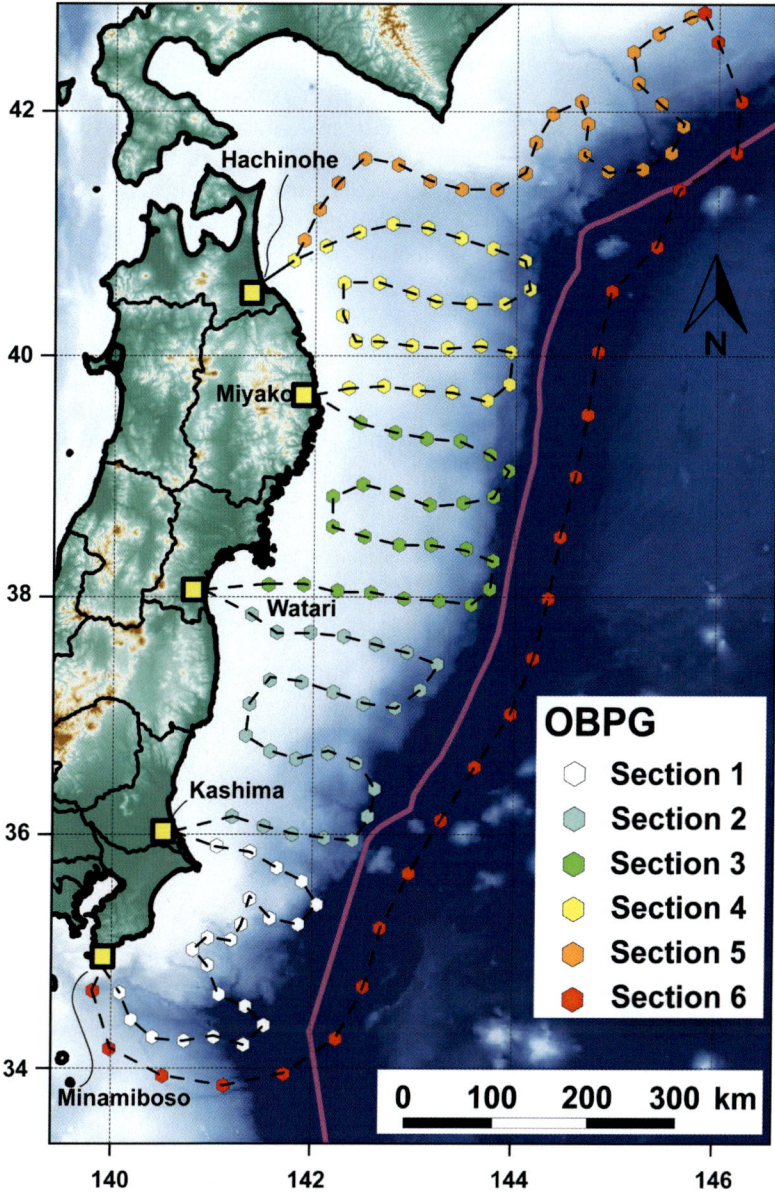

FIGURE 6.14 Layout of the Japanese early warning system S-net, consisting of 150 ocean bottom pressure gauges (OBPGs) connected by optic fiber cables. Map lines delineate study areas and do not necessarily depict accepted national boundaries.

and optimization, and (4) risk-based tsunami forecasting. These four aspects are interrelated. For instance, data-driven statistical approaches, such as machine learning and deep learning, require the availability of synthetic datasets for developing and testing new warning algorithms. These aspects are discussed in the following.

6.3.1.1 Use of synthetic tsunami datasets for developing tsunami early warning models

In calibrating a robust tsunami early warning system, the system must be tested to work well for a variety of possible earthquake ruptures. However, there is no abundant historical tsunami data that can be directly used for developing such a warning system. As it is uncertain how future tsunami events may unfold, using simulated synthetic data to represent future tsunami scenarios is a viable strategy (An et al., 2018; Wang et al., 2020). It is sensible to use synthetic tsunami wave data, generated from stochastic tsunami simulation models (Chapter 2), to calibrate a tsunami early warning model based on conventional earthquake information (e.g., JMA's magnitude estimate and epicenter location) as well as S-net wave data (Chapter 11). The stochastic tsunami simulations facilitate the consideration of realistic heterogeneous earthquake slips, which are seismologically and statistically compatible with inferred earthquake slips via source inversion analyses. Once a large set of synthetic earthquake source models is simulated, wave profiles at off-shore and onshore locations can be generated by solving physics-based governing equations (Chapter 3), which capture both spatial and temporal variations of tsunami waves. Subsequently, coastal tsunami inundation can be simulated, and the corresponding regional tsunami loss for buildings can be calculated by using suitable exposure and fragility models (Chapter 5). This numerical setup is particularly useful because it facilitates the development of tsunami early warning algorithms using data-driven statistical methods, such as machine and deep learning techniques (Section 6.3.1.2). The simulated data and the early warning algorithms can also be utilized for identifying the most effective configurations of the tsunami wave monitoring systems (Section 6.3.1.3). The use of synthetic tsunami data allows us to develop tsunami early warning systems for not only conventional hazard metrics, that is, the probability of occurrence of the tsunami event, which is quantified by wave amplitudes at shallow coastal locations near a town/city, but also risk metrics, that is, the product of hazard, exposure, and vulnerability, which is quantified by the total tsunami loss of damaged buildings in coastal communities (Section 6.3.1.4).

To illustrate synthetic tsunami data in the context of calibrating a tsunami early warning model, hypothetical rupture scenarios for two M_w 8.9 to 9.0 earthquakes in the Tohoku region are adopted, and the tsunami simulation results are shown in Fig. 6.15. The left panels show the synthesized earthquake slip distributions using the stochastic source modeling method (Chapter 2). The middle panels show the simulated wave profiles at three S-net stations, which are indicated in the left panels. The S-net stations close to the earthquake slip locations detect tsunami waves early, whereas

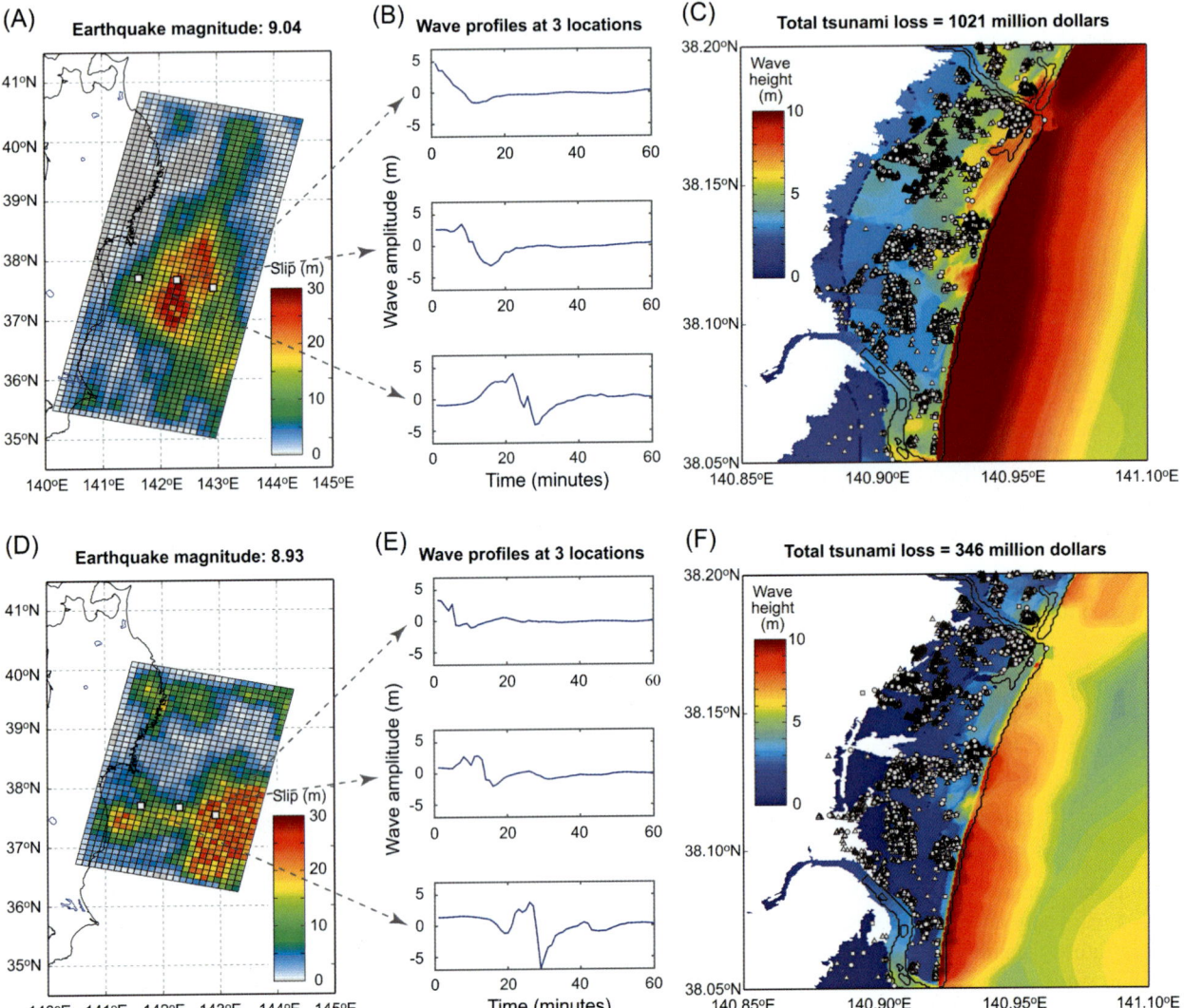

FIGURE 6.15 Earthquake ruptures, wave profiles, and coastal buildings affected. (A) to (C) present an M_w 9.0 earthquake with a tsunami loss of $1021 million. (D) to (F) present an M_w 8.9 earthquake with a tsunami loss of $346 million. (B) and (E) present the wave amplitude of three selected locations.

those close to the land detect their arrival later. For these synthetic rupture cases, detailed tsunami inundation simulations can also be performed, and inundation depths on land can be simulated, as displayed in the right panels. By applying tsunami fragility functions to 6152 individual buildings located in the inundation zone (see gray markers in the right panels), tsunami damage and loss to individual buildings can be evaluated. By aggregating tsunami losses from the 6152 buildings, the total loss is estimated as $1021 and $346 million for the two rupture scenarios. In summary, for this dataset, the following information can be generated: earthquake source information (earthquake magnitude and location), wave time-series profile data at S-net locations and other near-shore or on-shore locations, tsunami inundation areas, and tsunami losses to the building portfolio. By repeating these scenarios for several thousands of stochastic rupture scenarios, a large dataset can be established and used for developing a robust tsunami early warning model (Section 6.3.1.2). It is important to recognize that the synthetic data, although significant efforts have been directed to validate these models and their predictions, are still hypothetical. Therefore, there is always uncertainty regarding the fidelity of the underlying tsunami hazard and risk model.

6.3.1.2 Statistical approaches for developing tsunami early warning models

Using a synthetic tsunami hazard and risk simulator, rich information on earthquake sources, wave profiles, inundation profiles, and building damage and loss can be created for numerous hypothetical rupture scenarios (Section 6.3.1.1). Using these data, various tsunami early warning models can be developed. A tsunami early warning model can be formulated as a generic prediction equation for tsunami response variables of interest as a function of explanatory variables related to earthquake characteristics and tsunami wave characteristics. A general expression for this prediction model can be written as follows

$$y = f(x) \tag{6.1}$$

where y is the response variable (as a vector), x is the explanatory variable (as a vector), and f is the prediction function that relates x and y. In the context of tsunami early warning, the response variable y can represent tsunami wave heights or time-series profiles at shorelines and inland locations, tsunami inundation areas, tsunami damage/loss, or tsunami fatality (Section 6.3.2). On the other hand, the explanatory variable x may include earthquake source information (magnitude, location, and earthquake slip), ground deformation profiles at global positioning system (GPS) stations, or tsunami wave profiles at offshore locations (e.g., S-net). Mathematically, the problem of developing an effective tsunami early warning system is to find a suitable functional form together with input and output variables. A unique aspect of the tsunami early warning model, compared to other statistical problems, is that the optimal function needs to be developed by taking into account the trade-off between the prediction accuracy and the waiting time. Generally, with more data (i.e., wait longer before issuing a tsunami warning message), a tsunami warning model will perform better. On the other hand, if the warning message is issued late, evacuees will have less time to escape from the tsunami. The permitted maximum waiting time depends on the demographic and topographic characteristics of the communities in relation to tsunami characteristics. In particular, the locations of evacuation points in local communities are critical factors in developing a suitable tsunami early warning and evacuation system.

Returning to the general formulation shown in Eq. (6.1), one of the simplest forms for the function f is to use a multiple linear regression model. The functional form can be expressed as follows

$$y = \beta_0 + x_{\text{mag}}\beta_{\text{mag}} + x_{\text{lat}}\beta_{\text{lat}} + x_{\text{lon}}\beta_{\text{lon}} + \sum_{i=1}^{150} x_i \beta_i \tag{6.2}$$

where β with subscripts 0, mag, lat, lon, and $i = 1,\ldots,150$ (i is for the offshore sensor index) are the model coefficients, whereas x_{mag} is the earthquake magnitude, x_{lat} is the earthquake latitude, x_{lon} is the earthquake longitude, and x_i is the tsunami wave characteristic at the ith offshore sensor. It is noted that x_i is a tsunami wave quantity that is available up to the waiting time t_{wait}. For instance, x_i can be taken as the maximum tsunami wave height up to t_{wait}. To determine the model coefficients and suitable network configuration for early warning purposes, many trial-and-error investigations or systematic analyses, such as a sequential feature selection algorithm (Li & Goda, 2022), can be carried out. For a given value of the waiting time t_{wait}, the model variable selection criterion, such as Akaike (1973) information criterion (AIC), can be adopted. AIC is defined as follows

$$\text{AIC} = -2\log(L) + 2k \tag{6.3}$$

where L is the loglikelihood function, and k is the number of parameters in the model (i.e., the number of β coefficients). AIC rewards goodness of fit but discourages overfitting by penalizing the model when many explanatory variables are used. A smaller AIC value represents a superior model. The model variable selection refers to the determination of informative offshore sensors to be included in the early warning model.

According to the results presented by Li and Goda (2022), there is a general U-shape trend when the model performance (i.e., AIC) is plotted against the number of offshore sensors. When additional offshore sensors are included in the network, these sensors bring new information, thereby increasing the likelihood value (i.e., smaller AIC value). When enough offshore sensors are included in the model, newly added sensors do not improve the model performance significantly, while the model complexity penalty term increases the AIC value. Eventually, there will be no benefit by including additional offshore sensors in the early warning model. On the other hand, with a longer waiting time, the model performance tends to improve. An example of the model performance curve in terms of network configuration (horizontal axis) and waiting time (different curves) is shown in Fig. 6.16A, whereas part (B) of the figure shows the system configuration that achieves the optimal AIC value for the waiting time of 5 minutes based on the knee-point method (Thomas & Sheldon, 1999). In the figure, the knee-point-based optimal configurations are indicated by square markers, while the minimum AIC-based optimal configurations are indicated by circle markers. Six S-net sensors shown in Fig. 6.16B can achieve a good balance between the number of S-net stations and the model performance (Li & Goda, 2022).

Although the multiple linear regression approaches provide intuitive and transparent prediction models, methods have limitations in restricting the linear functional relationship between the tsunami response metrics and the earthquake/wave information, resulting in difficulties in accurately predicting the tsunami loss that is highly nonlinear. This limitation can be overcome and improved by adopting more advanced statistical methods. For instance, the use of the random forest model in predicting tsunami loss is advantageous because it utilizes nonparametric regression and ensemble learning (Breiman, 2001; Ho, 1998). These features make the random forest model well suited for predicting the nonlinear and scattered nature of tsunami loss (Li & Goda, 2023b). Due to the complex, ensemble, and nonparametric characteristics of random forest, the prediction equation cannot be explicitly expressed like multiple linear regression models shown in Eq. (6.2). The random forest model can effectively capture the complex relationships between offshore wave amplitudes at different offshore locations and tsunami loss in coastal areas and relies less on earthquake information. In other words, the random forest serves as an online forecasting method rather than the forward simulation that requires the estimated earthquake source parameters.

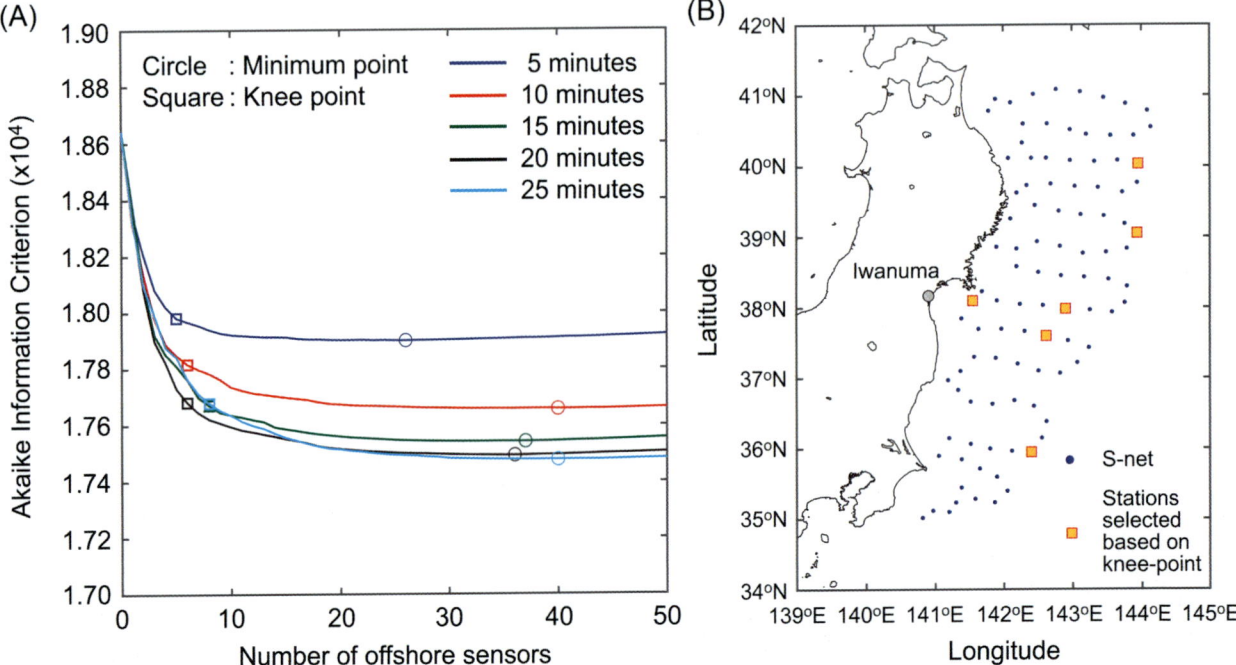

FIGURE 6.16 (A) AIC-based model performance curves in terms of network configuration (i.e., number of offshore sensors) and waiting time. (B) Tsunami early warning system configuration that achieves the optimal Akaike information criterion value based on the knee-point method for the waiting time of 5 minutes (i.e., blue square in (A)).

An example of the random forest-based tsunami loss forecasting model is presented in Fig. 6.17. Fig. 6.17A shows the trade-off of the model performance as a function of waiting time. The model performance is evaluated by the mean squared errors (MSE) between the original data and the random forest prediction. Four results are included in the figure by considering different system configurations (i.e., 6 selected sensors, as shown in Fig. 6.16B, vs all 99 sensors in the region) and by considering or ignoring earthquake information (i.e., earthquake magnitude, latitude, and longitude). The comparison of the blue and red curves shows the benefit of having earthquake information when there are only six offshore sensors available for tsunami risk forecasting. By utilizing all available sensors, the model performance is improved by a factor of two (i.e., MSE is halved), while the benefit of including the earthquake information in the forecasting model has been lost. In other words, monitoring tsunami waves in the target offshore region will provide accurate risk forecasting without earthquake information. Fig. 6.17B shows the comparison of the original synthetic tsunami loss data and the predicted loss data from the calibrated random forest models. The data points are scattered around the diagonal line, and the scatter of the data becomes less when more offshore sensors are included in the model. Similarly, the improved performance of the tsunami risk forecasting model is observed when the earthquake information is included or the waiting time becomes longer (Li & Goda, 2023b). These trends are consistent with the MSE-waiting time plot shown in Fig. 6.17A.

Deep learning approaches have also been successfully applied to tsunami hazard forecasting. Makinoshima et al. (2021) trained a convolutional neural network using a deep learning algorithm to forecast tsunami inundation from real-time observational data. A unique aspect of this research is that the response variable to be predicted is the tsunami inundation waveform at a single onshore location, which is different from the maximum tsunami wave height at an onshore site. On the other hand, Mulia et al. (2022) developed a nonparametric regression model using a fully connected neural network to predict coastal inundations. The response variable vector contained more than 200,000 grid points, covering a large region, and therefore their models can predict the spatial distribution of coastal inundations at a regional scale. Overall, the deep learning-based tsunami forecasting methods overcome the bottlenecks of existing simulation-based tsunami forecasting approaches by avoiding tsunami source estimations and high computational costs for solving nonlinear tsunami propagation and inundation processes in real time, thereby enabling real applications of the developed tsunami forecasting methods. Training a neural network requires significant computational times and trial-and-error investigations, but such a model calibration exercise can be performed offline, and using the calibrated neural network model for forecasting is computationally fast. It is also important to point out that the above-mentioned deep learning-based tsunami inundation forecasting models were developed on the basis of synthesized data generated from stochastic tsunami simulations (Section 6.3.1.1). Furthermore, Chapter 12 presents the latest application of deep learning algorithms for tsunami early warning.

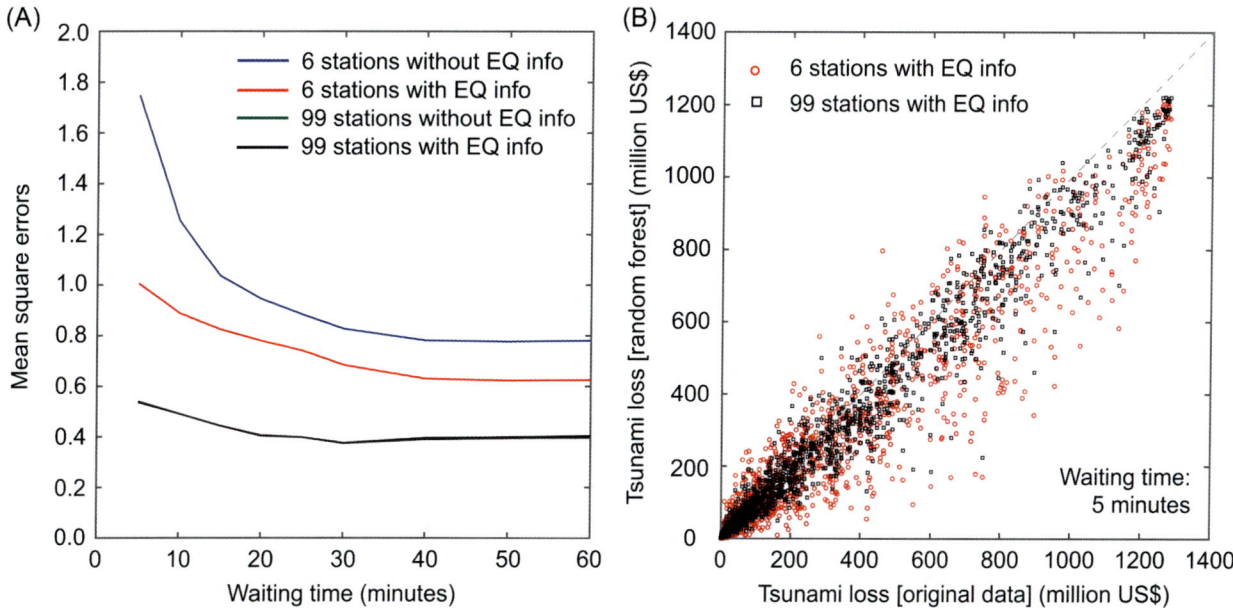

FIGURE 6.17 (A) MSE-waiting time plot from 5 to 60 min. (B) Plot of simulated tsunami loss (original data) with predicted tsunami loss (random forest) when the waiting time is 5 min. *MSE*, Mean squared errors.

6.3.1.3 Network design and optimization for developing tsunami early warning models

The existing tsunami early warning studies, for example, Li and Goda (2023b), Makinoshima et al. (2021), and Mulia et al. (2022), focused on the S-net due to its unprecedented density of ocean bottom pressure sensors. The real-time as well as synthetic data for the S-net promote data-driven statistical approaches to achieve high statistical correlations between input and output data. The consideration of the S-net data also permits the parametric investigations of network design and optimization. In many parts of the world, deploying 150 ocean bottom sensors is not economically viable. By using a dense network like S-net, various numerical investigations can be carried out to answer practical questions that will help refine and optimize the design of ocean bottom sensor networks for tsunami early warning. For instance, numerical investigations can be designed to answer how many stations will be necessary and where they should be placed to achieve a target forecasting performance (Li & Goda, 2022). A network with more dense sensors and with a broad spatial coverage will cost more. Therefore, a trade-off between the required accuracy and the installation and operational costs of the tsunami warning system can be examined during the design phase of the tsunami early warning system to make more informed decisions.

6.3.1.4 Risk-based tsunami early warning and rapid tsunami loss estimation

Conventional tsunami early warning approaches are focused on tsunami hazard metrics, such as the maximum tsunami height at the shoreline. When tsunami early warning messages are issued, these forecasted quantities are often communicated as categorical information, such as a 1–3 m tsunami or a 5 + m tsunami. The advanced and quantitative tsunami risk models allow tsunami forecasting to be based on tsunami risk quantities. The tsunami risk metrics can be aggregate tsunami loss for coastal communities or larger administrative units (e.g., Fig. 6.15). Alternatively, the number of trapped people or tsunami fatalities could be adopted as tsunami risk metrics in the forecasting models. The new trends and methods in developing tsunami forecasting models will enable the rapid tsunami impact forecasting tools (Chapter 24). Currently, there are no widely applicable rapid tsunami impact forecasting tools.

Moreover, the role of remote-sensing technology will become increasingly more important in tsunami forecasting and emergency response. Although the remote-sensing methods may not be able to acquire real-time data within the time frame of tsunami early warning (several minutes after the trigger event), remote-sensing data will enable more accurate estimation of the tsunami (and earthquake) damage to buildings and infrastructure regionally. Similar to the ideas put forward by Goda, Mori, et al. (2019) and Mulia et al. (2022), actual observations of tsunami inundation extent are particularly useful for evaluating the economic loss due to tsunamis and will facilitate prompt responses from various stakeholders, not only from local municipalities and governmental officers but also from insurers and reinsurers.

6.3.2 Tsunami evacuation

Stochastic earthquake–tsunami hazard maps can aid in developing robust and effective tsunami evacuation plans. Such plans should include horizontal routes to high grounds and vertical evacuation to earthquake and tsunami-resistant shelters (Muhammad et al., 2017; Wood et al., 2014). It should be emphasized that buildings are often struck by seismic waves within seconds, making evacuation before shaking impossible. Consequently, tsunami evacuation shelters must withstand ground shaking without suffering critical damage, such as immediate collapse, as stated by Federal Emergency Management Agency (2019).

Evacuation time maps are essential for residents in horizontal evacuation scenarios, as they help calculate the total time required to evacuate an area. The initial reaction time, which includes institutional decision, institutional notification, and community reaction times, is a key factor in determining the total evacuation time, along with the physical capacity of transport infrastructure and evacuation speed. The initial reaction time typically ranges from 10 to 30 minutes, depending on the level of social preparedness (Post et al., 2009). Evacuation time is affected by various factors, such as the mode of transportation, physical conditions of evacuees, and travel speed, which can range from 1 to 4 m/second (Muhammad et al., 2017). By comparing total evacuation time maps with the arrival time of the first major waves, it is possible to determine whether transportation infrastructure needs improvements. A possible solution is constructing tsunami evacuation buildings in areas at risk of inundation. Horizontal evacuation, vertical evacuation, and combined horizontal–vertical evacuation maps are illustrated in Fig. 6.18A–C, respectively. Fig. 6.18D shows a tsunami shelter used as a vertical evacuation structure. The integrated approach can significantly reduce evacuation time, particularly for those near the coastline.

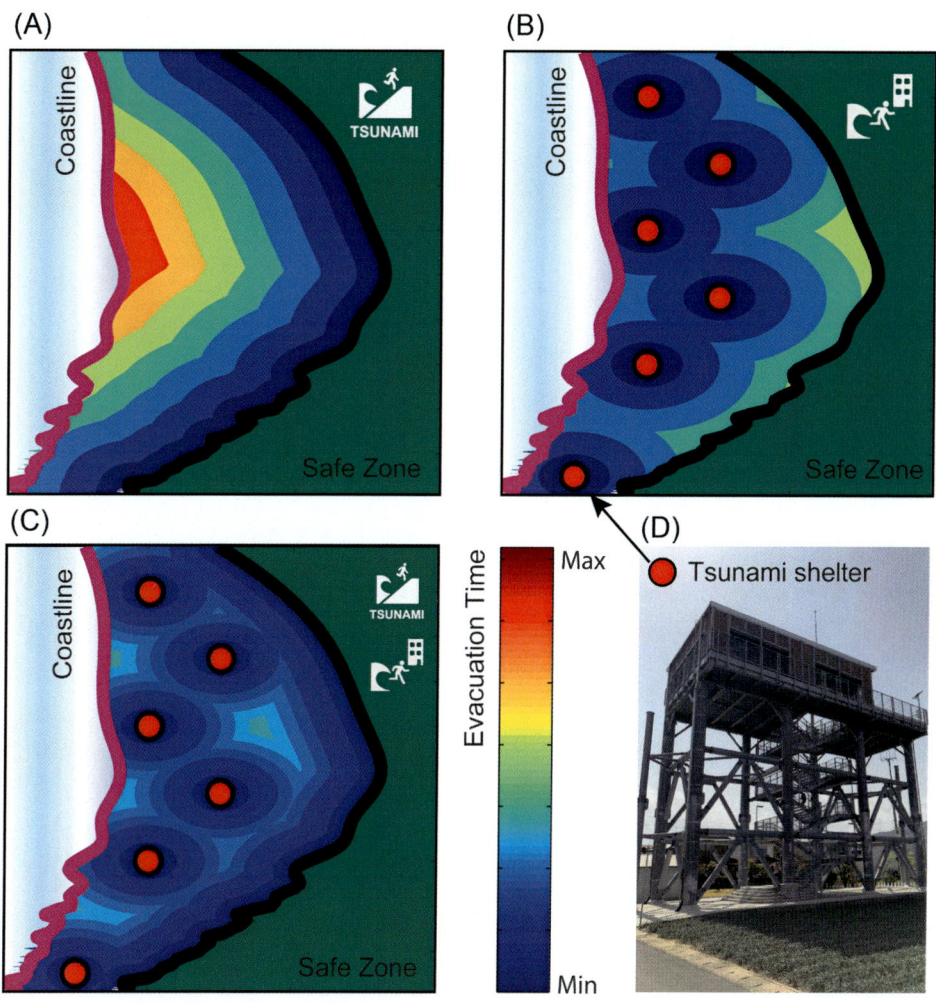

FIGURE 6.18 Illustration of evacuation time maps: (A) horizontal evacuation, (B) vertical evacuation, and (C) combined horizontal–vertical evacuation. (D) Vertical evacuation structure in Kuroshio, Kōchi Prefecture, Japan.

6.3.2.1 Tsunami evacuation model setup

An important subduction zone in Japan is the Nankai Trough. This zone in southwestern Japan has not experienced major ruptures since 1946 (Ando, 1975); however, it hosted M_w 8 + earthquakes that triggered significant tsunamis in the nearby regions (Ando, 1975), and it can generate massive tsunamis due to earthquakes of up to M_w 9.1 (Central Disaster Management Council, 2012; Goda, Yasuda et al., 2018).

The Saga district in Kuroshio, Kōchi Prefecture, Japan, faces the Nankai Trough directly, making it one of the coastal communities in Japan most susceptible to tsunamis caused by tsunamigenic earthquakes originating from the Nankai subduction zone (Fig. 6.19A). In the coastal areas of Saga, the maximum tsunami depth of 20 + m is predicted, and the tsunami waves could run up along the Iyoki River and overflow upstream areas in Saga (Kuroshio Town, 2019b). Several high-ground evacuation locations have been identified, and a vertical tsunami evacuation shelter was constructed at the center of the residential area (Fig. 6.19B).

In this section, an evacuation modeling case study is presented for Saga. Specifically, an agent-based tsunami evacuation modeling is used to evaluate the current existing evacuation plan (i.e., local high grounds and vertical shelter) in safeguarding the coastal communities. The agent-based modeling software package MATSim is used to simulate the evacuation process (Horni et al., 2016; Lämmel et al., 2009). A practical procedure regarding how to implement the evacuation scenarios in MATSim is provided by Muhammad et al. (2022). More specifically, MATSim requires the location of agents (evacuees) and the definition of links and nodes within a transportation network (Fig. 6.20A). The evacuation nodes represent the final destinations. Those evacuation nodes are connected to virtual evacuation links and nodes that have an infinite capacity to accommodate all evacuees.

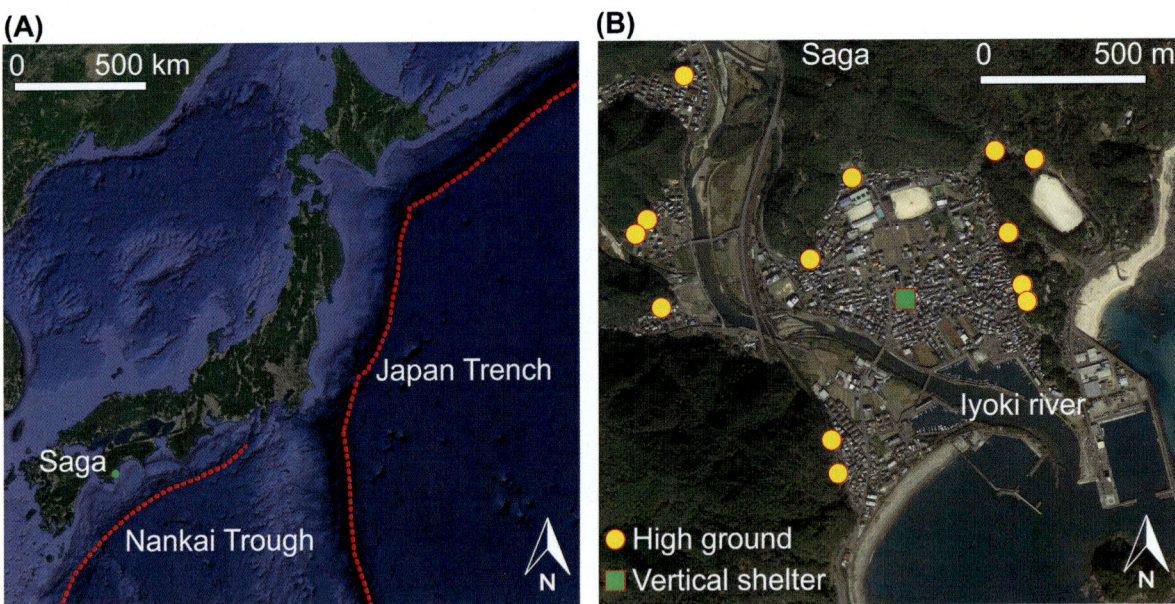

FIGURE 6.19 (A) Nankai Trough region and (B) zoom-in of the Saga district with the identification of the predesigned evacuation points. Map lines delineate study areas and do not necessarily depict accepted national boundaries.

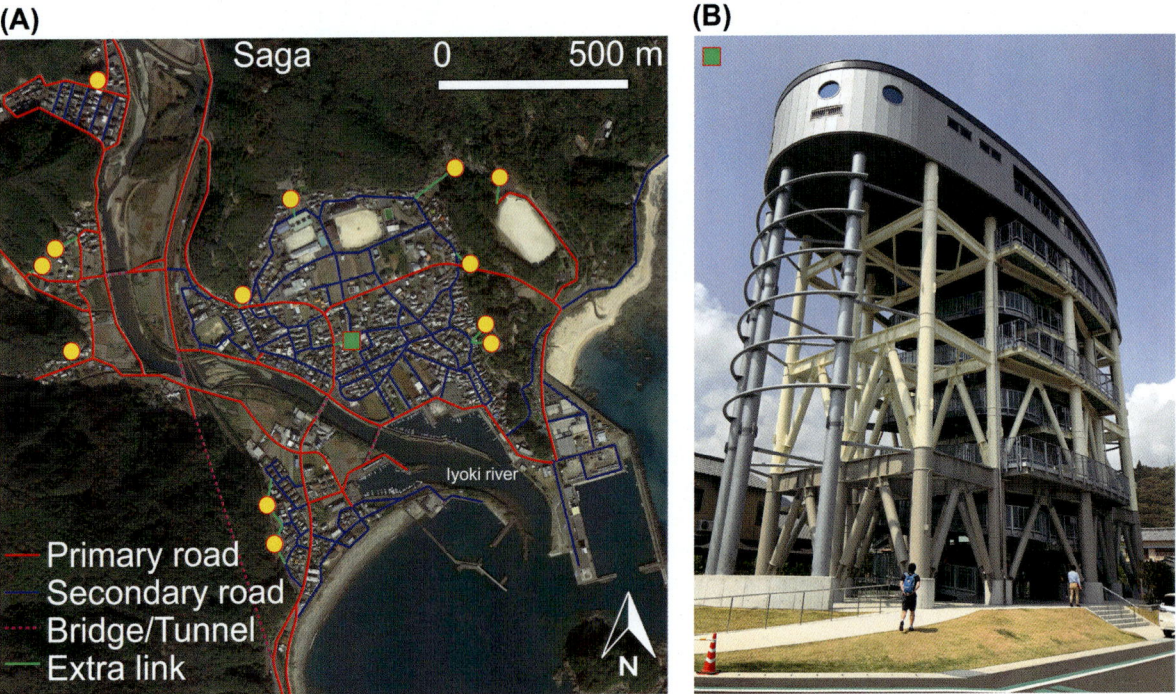

FIGURE 6.20 (A) Transportation network and (B) vertical evacuation shelter in Saga.

A simple pedestrian mode is used to evaluate the existing tsunami evacuation plan considering two scenarios: (1) including and (2) excluding the vertical tsunami evacuation shelter (Fig. 6.20B). Such a comparison demonstrates the importance of a vertical evacuation structure in saving people's lives. The single pedestrian mode is adopted because this is a recommended mode for evacuation (Yun & Hamada, 2015). Typically, the pedestrian speed ranges from 0.4 to 1.5 m/second (Sugimoto et al., 2003), and hence, three speeds are considered: 0.5, 1.0, and 1.5 m/second. A total of 6 cases (2 scenarios × 3 speeds) are simulated.

Network data are gathered by combining governmental data and freely available information from OpenStreetMap (http://www.openstreetmap.org). The road length, the road width, and the number of lanes are three essential parameters necessary to run the agent-based simulation in MATSim. This study categorizes the network into two types: primary roads (red lines in Fig. 6.20A) and secondary roads (blue lines in Fig. 6.20A). Primary roads are the main network in Saga with a width of 6 m and two lanes (i.e., 3 m width per lane), whereas the secondary roads are residential paths with a typical width of about 5 m. Secondary roads are not suitable for cars but can be accessed in two directions. Hence, they are assumed to have two lanes with a width of 2.5 m each. Moreover, since the location of the evacuation points is not connected with the existing network (i.e., links), extra pedestrian networks are added to the network system to access the evacuation areas (the green lines in Fig. 6.20A).

To determine the road capacities in the context of evacuation, first, the maximum flow speed is set to one of the considered pedestrian speeds (i.e., 0.5, 1.0, and 1.5 m/second). Second, the maximum flow capacity (FC) of each link is computed as follows

$$FC = w \times C_{max} \tag{6.4}$$

where w is the width of the link, and C_{max} is the maximum FC per unit width. The C_{max} value is assumed 1.3 persons/m/second (Lämmel et al., 2009; Weidmann, 1993). Finally, the storage capacity (SC) of a link (persons/m^2) is defined to limit the number of agents on the link and is calculated on the basis of the area (A) and the maximum density per unit area (D_{max}), where D_{max} is set to 5.4 persons/m^2 (Lämmel et al., 2009)

$$SC = A \times D_{max} \tag{6.5}$$

The speed in the simulation depends on the slope of the corresponding links; thus Tobler (1993) hiking function is adopted to change the speed of the link. MATSim also adopts a complex adaptive system, the coevoluntionary algorithm (Horni et al., 2016). With this algorithm, each agent iteratively simulates their activities and competes with other agents for space−time slots on the existing route. Therefore, a total of 250 iterations are adopted in MATSim simulation. Such a number is sufficient to produce a stable evacuation score for determining the best plan for each agent movement. Two movement strategies representing a motion plan of each agent are implemented: ReRoute (10%) and BestScore (90%). The ReRoute strategy generates a new plan in each iteration with a new evacuation route based on the information on experienced travel times from the last run. BestScore (90%) produces a new plan based on the best score in each iteration. Moreover, the score representing distance and time is also adopted to define effective evacuation routes. The results in terms of evacuation time from the agent-based evacuation modeling can eventually be used to evaluate the current tsunami evacuation plans.

The location of agents is distributed using three types of information: (1) the building footprints, (2) the total number of people living and working in the area, and (3) the average household size. The starting evacuation time is set to 10:00 a.m. (daytime). The building footprints are taken from the local government and OpenStreetMap, as shown in Fig. 6.21A. Buildings are divided into several categories: houses, district offices, fish companies, harbors, schools (i.e., elementary and junior high schools), markets, and nonresidential buildings. The classification is essential because each building type contains a different number of agents (e.g., the number of people in the market is different from the number of people in school). Other building types, for example, postoffice, community service, and supermarket, are not distinguished due to the lack of detailed information for specific buildings and are included in the nonresidential building category.

The total number of residential buildings is 1041. However, 315 buildings smaller than 30 m^2 are excluded since they are smaller than traditional wooden houses in Japan, and hence, only 726 houses are used to accommodate the agents. The population data for the Saga district in Kuroshio Town (2019c) are taken from governmental sources, which contain the detailed number of the population up to a subdistrict level. There are 28 subdistricts in Saga (which is larger than the study area), with a total population of 3457. However, the study area consists of five subdistricts only. Consequently, it is difficult to determine the number of agents in evacuation simulation precisely based on the total population in those five subdistricts (i.e., the entire population is ∼1500 people). Therefore, the average household size is considered to define the number of agents in each house (Fig. 6.21B). Two people are placed in each house, following the average household size in Saga (Kuroshio Town, 2019a).

Since the evacuation is set to start at 10:00 a.m., only 50% of local people are assumed to stay at home (i.e., 600 people); hence, the agents inside houses are in the range of 0−2 people. The percentage of schooling ages in Kuroshio Town (2019c) is 8.44%, leading to about 200 agents at school consisting of 150 students and 50 staff. Moreover, about 47.5% of the population is working (∼715), which may come from areas outside Saga. Eventually, the following setup for the agents' distribution is considered: 600 people in houses, 200 people at school, and 1400 people at factories, markets, and nonresidential buildings leading to a total of ∼2200 agents (Fig. 6.21B).

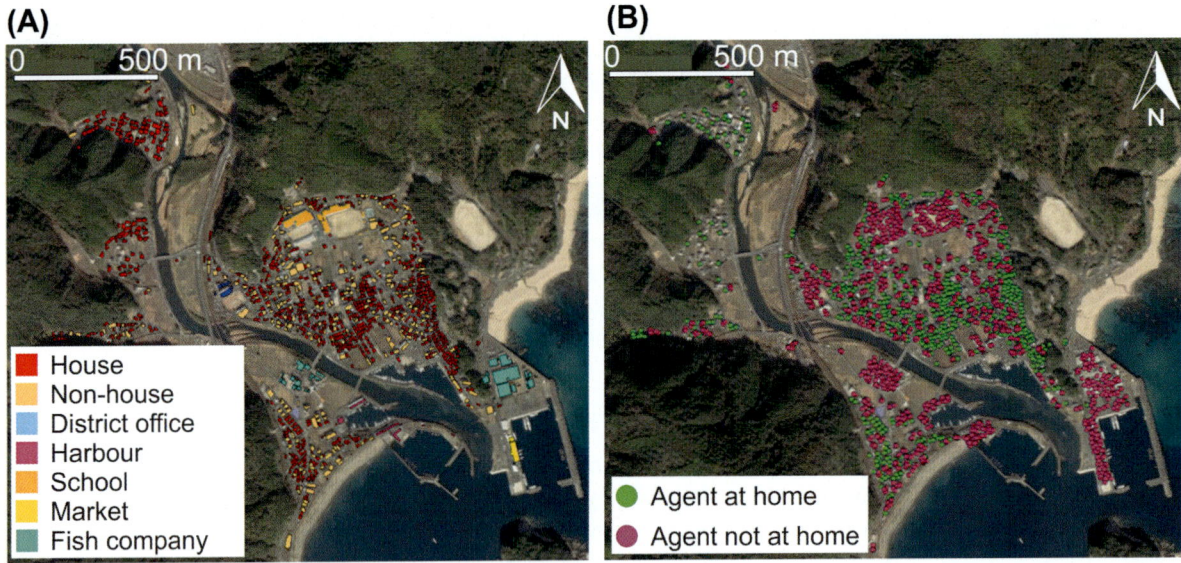

FIGURE 6.21 (A) Building categories and (B) spatial distribution of evacuees (agents).

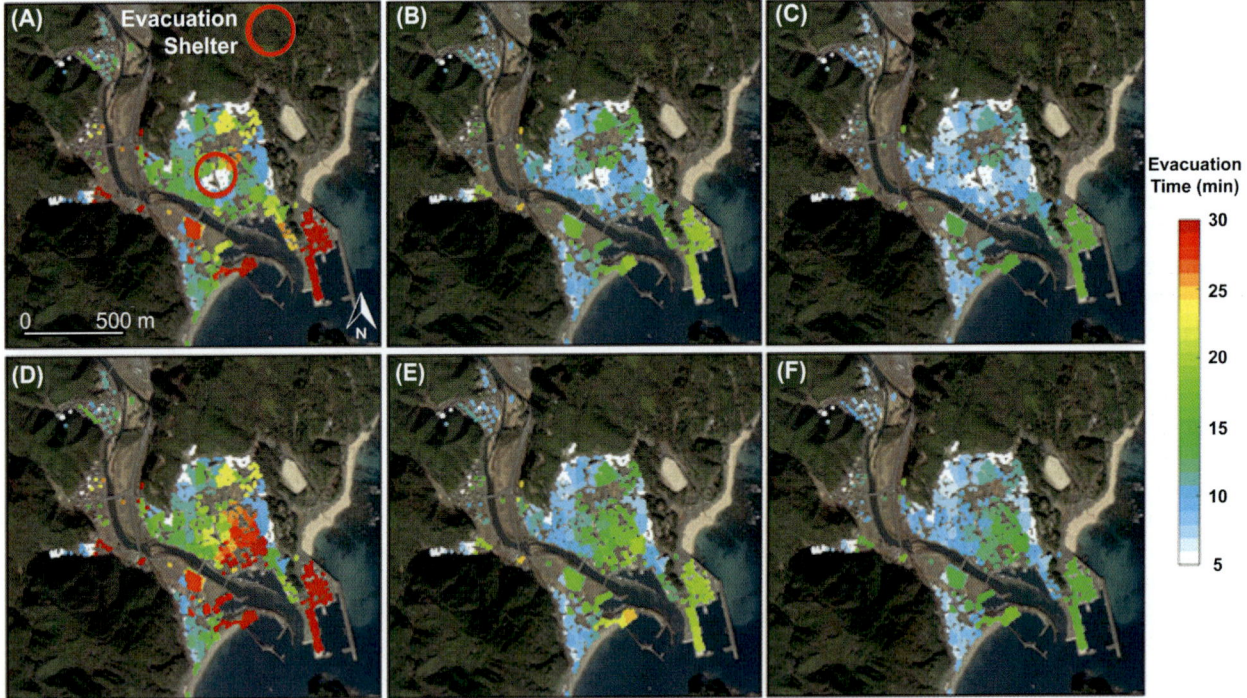

FIGURE 6.22 Maps of tsunami evacuation time: (A–C) considering and (D–F) neglecting the tsunami evacuation shelter with different speeds: (A, D) 0.5 m/second, (B, E) 1.0 m/second, and (C, F) 1.5 m/second. Map lines delineate study areas and do not necessarily depict accepted national boundaries.

6.3.2.2 Tsunami evacuation simulation results

The tsunami evacuation time is defined as the sum of the initial reaction time and the simulated evacuation time of the agents. The initial reaction time represents the starting time of the evacuee's response, that is, typically between 5 and 15 minutes (Yun & Hamada, 2015). In this case study, the initial reaction time of 5 minutes is adopted because communities in coastal environments of Japan are familiar with the tsunami evacuation. Fig. 6.22 shows tsunami evacuation time maps for the six cases considering two evacuation scenarios: including (top three subplots) and excluding (bottom

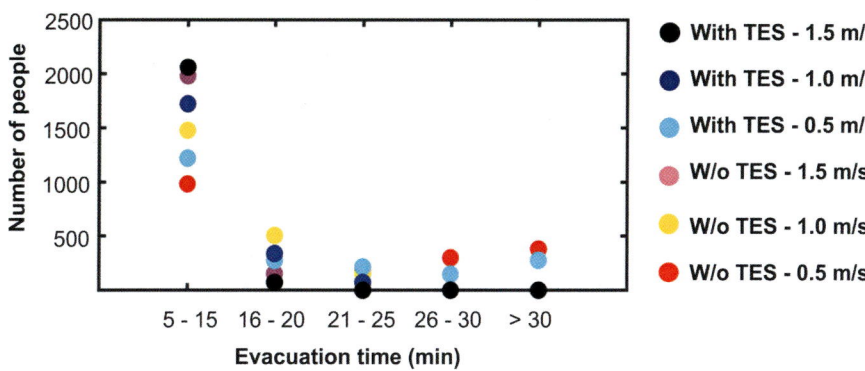

FIGURE 6.23 Number of evacuees in terms of different tsunami evacuation times with and without the tsunami evacuation shelter (TES).

three subplots) the vertical evacuation shelter. The presence of a vertical evacuation structure is vital as it reduces the tsunami evacuation time of the evacuees. The agents located close to the vertical evacuation structure have significantly shorter tsunami evacuation times (<25 minutes) in comparison to the case where such an infrastructure does not exist (>30 minutes).

Fig. 6.23 exemplifies the importance of having a tsunami evacuation shelter in areas prone to tsunamis, as it illustrates the distribution of individuals within various evacuation time ranges. The presence of a tsunami shelter leads to a greater number of people evacuating within a time frame of less than 15 minutes. The speed at which individuals evacuate significantly impacts the overall evacuation time in both scenarios. When the evacuation speed is 1.5 m/second, all agents achieve quicker evacuation times, with approximately 2000 individuals (around 90% of the agents) evacuating in less than 15 minutes. On the other hand, in the 0.5 m/second case, a higher percentage of agents (approximately 650 individuals, about 30% of the agents) encounter longer evacuation times (>20 minutes).

These findings emphasize several crucial points regarding tsunami evacuation. First, the presence of the tsunami evacuation shelter is essential for saving lives. With a predicted tsunami arrival time of around 16 minutes (Kuroshio Town, 2019b), the number of affected individuals could be reduced if the shelter is strategically located within residential areas. Second, providing access to higher ground evacuation points is crucial to enable prompt evacuation for people residing in coastal areas. Currently, individuals in the market and fish company areas can only rely on the primary network leading to residential areas, lacking direct access to evacuation points. Consequently, they experience longer evacuation times (>25 minutes). Third, regular tsunami evacuation drills are necessary to ensure that people can evacuate quickly once the shaking ceases. Lastly, the existing tsunami evacuation plan in Saga demonstrates relative effectiveness in safeguarding coastal communities. The established tsunami evacuation shelter has significantly reduced evacuation times for residents.

6.4 Tsunami risk financing

Recent earthquake CATs in Chile and Japan have revealed an insufficiency of conventional risk financing tools and systems, such as self-insurance, private insurance, and public support. In addition to life safety (Section 6.2.3), financial risk management is important for protecting livelihood in communities and for accelerating recovery after disasters. Muir-Wood (2011) suggested that for Chile, micro-insurance, targeted for low-income homeowners and small commercial enterprises, will be beneficial, and a new risk pooling system shall be set up to cover earthquake risks for government-owned buildings. The former issue is compounded by anticipation of ex-post subsidies from the governments by the Chilean people. However, it has been demonstrated that DRR measures are far more cost-effective than no DRR actions (Michel-Kerjan et al., 2013; Multi-Hazard Mitigation Council, 2020). In Japan, increasing the relatively low penetration rates of earthquake insurance (about 27% in 2013) is desirable to achieve a nationwide risk-sharing mechanism with stable incomes (Kajitani et al., 2013); this situation is in contrast with the earthquake insurance situation in New Zealand after the 2010−2011 Christchurch seismic sequences (King et al., 2014). Implementing risk-based insurance premiums for different structural types and seismic regions (e.g., California, Japan, and Turkey) and introducing financial incentive schemes (e.g., Japan) are important to promote DRR investments. Such proactive earthquake risk management is the key to achieving sustainable and resilient communities against earthquake disasters.

Quantitative tsunami hazard and risk assessments are instrumental in facilitating risk-based decision-making and management for earthquake and tsunami disasters (Wald & Franco, 2016). Outputs from PTRA can be obtained as an EP curve, which typically displays tsunami losses to individuals or institutions as a function of the probability of exceedance. From the EP curve, various risk metrics, such as average annual loss (AAL) and value at risk (VaR), can be derived (Yoshikawa & Goda, 2014) and used for making informed DRM decisions by insurers and reinsurers and for regulatory purposes by governmental agencies. The quantitative risk analysis also facilitates the benefit—cost analysis related to DRR investments (Multi-Hazard Mitigation Council, 2020). Due to the low-probability high-consequence nature of earthquake and tsunami risks, the benefits of conventional insurance products may not be appreciated by home and business owners, creating insurance gaps for financial risk protection. To resolve these issues, the insurance system needs to be designed by accounting for the behavioral aspects of stakeholders (Kunreuther et al., 2024; Meyer & Kunreuther, 2017). Existing (anticipated) governmental public support, after a disaster, can undermine the financial risk diversification mechanism in private markets. From the financial risk capacity perspective, natural CAT risks from megathrust earthquakes and tsunamis can present major difficulties in absorbing the financial impacts solely within insurance—reinsurance systems. In such cases, alternative risk transfer instruments, such as insurance-linked securities (ILS) and CAT bonds, can be utilized by insurers and reinsurers as well as governments to secure additional funds for disaster risk recovery (Cummins, 2008; Mitchell-Wallace et al., 2017).

In Section 6.4, three topics are focused upon. Section 6.4.1 introduces financial risk metrics for natural CAT risk management, which can be derived from PTRA. These metrics facilitate the concise representation of quantified tsunami risks to people and assets and the risk-informed decisions related to tsunami risk reduction. Section 6.4.2 summarizes an insurance-reinsurance system for tsunami disaster risks. The roles of different stakeholders in disaster risk financing are explained, and the challenges that are faced with achieving high coverage and take-up rates against tsunami risks are mentioned. Section 6.4.3 presents a summary of alternative risk transfer, utilizing financial markets. Different types of ILS and CAT bonds are mentioned to highlight how new PTHA and PTRA approaches can improve the performance of nontraditional disaster risk financing tools. Since tsunami risks are often regarded as secondary perils of earthquake risks, both types of risks are discussed together.

6.4.1 Financial risk metrics for natural catastrophe risk management

The outputs from PTHA and PTRA (Chapter 5) can be succinctly presented as an EP curve, that is, a plot of a risk variable in terms of annual probability of exceedance. An example of EP curves is shown in Fig. 6.9 for the monetary tsunami loss of a building portfolio or in Fig. 6.12B for the number of fatalities. Since the EP curve is a probability distribution of the adopted risk variable, various risk metrics can be derived. Other standard forms of quantitative risk outputs include the event loss table (ELT) and yearly loss table (YLT). These tables retain the granularity of the risk outputs on an event-by-event basis; the loss for each event is expressed by the mean and standard deviation of the event loss. Utilizing analytical formulae, an EP curve can be created from ELT and YLT (Mitchell-Wallace et al., 2017).

Fig. 6.24 shows an illustrative EP curve for an insurance company that underwrites tsunami risks. A standard approach to display an EP curve is to plot a risk variable on the horizontal axis and an EP on the vertical axis. As the

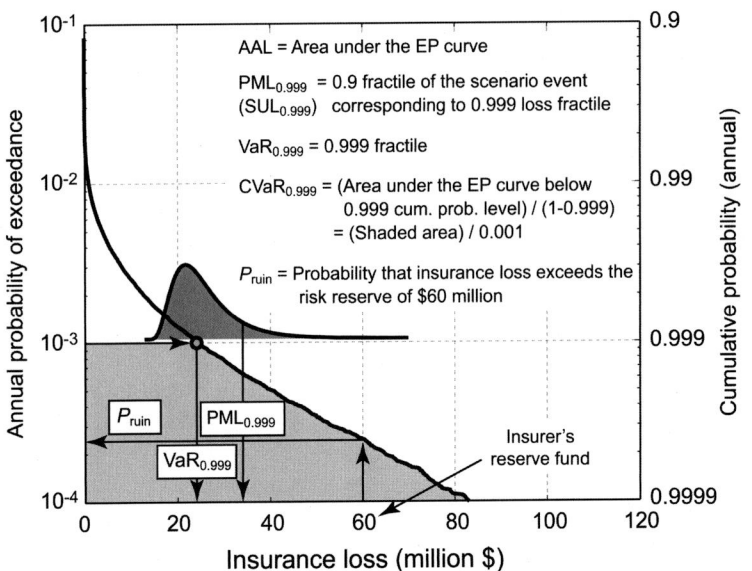

FIGURE 6.24 Exceedance probability curve and risk metrics.

TABLE 6.5 Summary of risk metrics for insurance purposes.

Risk metric	Equation	Relationship with the EP curve
Average annual loss: AAL	$AAL = E[L]$ where $E[\bullet]$ is the expectation	The area under the EP curve
Value at risk: VaR_α	$VaR_\alpha = \inf\{l : P(L > l) \leq 1 - \alpha\}$ where $\inf\{\bullet\}$ is the infimum and $P(L > l)$ is the probability of loss L exceeding a certain loss level l	Fractile value at a selected cumulative probability level α
Conditional value at risk: $CVaR_\alpha$	$CVaR_\alpha = \frac{1}{1-\alpha} \int_\alpha^1 VaR_u \, du$	The area under the EP curve below a selected cumulative probability level α, normalized by the exceedance probability $1-\alpha$
Ruin probability	$P_{ruin} = P(L > R_{insurer})$ where $R_{insurer}$ is the insurer's risk reserve	Probability when the insurer's reserve fund is depleted

AAL, Average annual loss; *EP*, exceedance probability.

EP becomes smaller (i.e., more extreme events), the consequences associated with these events become greater. Risk metrics that are widely used include AAL, VaR, probable maximum loss (PML), conditional VaR (CVaR), and ruin probability. In the literature, CVaR is sometimes referred to as tail VaR or expected shortfall. A graphical representation of these metrics is shown in Fig. 6.24, and the equations to compute these measures (except for PML) are summarized in Table 6.5. AAL is a useful metric for ordinary risks and is adopted to calculate insurance premiums; however, it is not suitable for fully characterizing catastrophic risks as it fails to capture the extent of devastating consequences due to rare events. VaR is the fractile value on an EP curve corresponding to a selected probability level (or mean return period), while CVaR accounts for rare events in terms of their frequency and severity by taking the conditional expectation of the EP curve. One of the crucial factors is "which probability level or return period" to focus upon in evaluating VaR and CVaR (also applicable to PML). In the context of earthquake−tsunami risks, the return periods of interest often lie between 500 years and 2500 years (which correspond to the ranges considered for earthquake and tsunami hazard mapping and for structural design provisions). For financial DRM, the 500-year return period (with a variation from 100 years to 1000 years) is often adopted. The ruin probability represents the chance of insurer's (or other risk-bearing entities') insolvency (i.e., all reserve funds are depleted) and is often used for insurance regulatory purposes. For instance, in the European Solvency II framework, financial institutions that operate in European Union countries must meet the solvency capital requirement, which is set at the 200-year return period economic loss outcome (Mitchell-Wallace et al., 2017). PML is one of the most popular metrics in financial risk management, and there are several definitions. Conventionally, it was defined as the fractile of the loss corresponding to the return period of 475 years (i.e., essentially the same as VaR). However, in different industries and countries, variants of the PML have been adopted. For example, in Japan, PML is defined as the (conditional) 0.9-fractile value for a scenario that corresponds to a selected probability level (typically, the return period of 475 years), and currently, specific nomenclatures are in use: scenario expected loss corresponds to the original PML definition (i.e., VaR), and scenario upper loss corresponds to the PML definition (i.e., 0.9-fractile).

The quantitative tsunami risk assessments are based on a so-called performance-based engineering methodology, which was originally developed for quantifying the seismic risk of structures and infrastructure (Cornell & Krawinkler, 2000; Goulet et al., 2007) and has been later extended to tsunamis (Attary et al., 2017; Goda & De Risi, 2018); see Chapter 5 for more details. The performance of a building or infrastructure asset is measured in terms of repair costs, life-safety impacts, or loss of function (e.g., dollars, deaths, and downtime), propagating all major sources of uncertainty in hazard, exposure, vulnerability, and loss components. Moreover, these quantitative methods promote the benefit−cost analysis of tsunami risk mitigation measures. For instance, in protecting coastal communities, coastal embankments may be considered physical tsunami risk mitigation countermeasures. Also, the implementation cost of the embankment system is denoted by C in monetary units. To evaluate the benefit B of the proposed mitigation measures, PTRA can be performed for two situations without and with the planned coastal embankment system. The coastal embankments will alter the tsunami loss distribution in terms of frequency and consequence of tsunami loss events. Assuming a constant discounting factor γ over the planned service period of the coastal embankment system τ (e.g., 50 years), the benefit of the mitigation measure can be quantified as follows

$$B = \left(EAL_{\text{status quo}} - EAL_{\text{mitigation}}\right) \frac{1 - \exp(-\gamma\tau)}{\gamma} \quad (6.6)$$

where $EAL_{\text{status quo}}$ and $EAL_{\text{mitigation}}$ represent the expected annual loss for the no mitigation and mitigation situations, respectively. Note that the benefit is discounted to the present value of benefits that will be realized in the future (i.e., loss events occur). When the benefit−cost ratio exceeds unity, that is

$$BCR = \frac{B}{C} \gg 1 \quad (6.7)$$

the risk mitigation countermeasure is considered cost-effective and therefore should be implemented. There are various practical evidences that prevention and mitigation measures are cost-effective (Multi-Hazard Mitigation Council, 2020). It is important to recognize that selecting the discounting factor and planning period for disaster risk mitigation measures is not always obvious (Ditlevsen & Friis-Hansen, 2009). A typical choice is to adopt a long-term financial interest rate. However, when the benefit includes life safety reduction and environmental improvement, a discounting factor smaller than the financial interest rate may be argued and accepted. Adopting a lower discounting rate increases the (future) benefit of the risk mitigation measures, thereby increasing the benefit−cost ratio.

6.4.2 Insurance-reinsurance system for tsunami risk

A traditional financial disaster risk diversification mechanism is achieved through an insurance-reinsurance system. A generic representation of such an insurance-reinsurance system for managing disaster risks is shown in Fig. 6.25. The key stakeholders are policyholders (e.g., owners of properties and enterprises), insurers operating at local/regional/national levels, reinsurers dealing with risk diversification at an international scale, insurance commissioners/governments, rating agencies as auditors, and capital markets where investors gather for financial transactions.

6.4.2.1 Policyholders, insurers, and governments

In the event of a major destructive earthquake and tsunami, policyholders may suffer from damage to their properties and assets and incur financial loss. To mitigate the potential financial impact, they may purchase earthquake−tsunami insurance coverage. Alternatively, they have options to mitigate the damage by adopting physical mitigation measures. Acceptance of earthquake−tsunami insurance and/or physical risk mitigation measures depends on various factors, including the price of insurance coverage (affordability), the type of properties (e.g., home vs business), the risk perception and awareness of property owners, and the expectation of postdisaster government assistance or compensation (Goda et al., 2020; Organisation for Economic Co-operation and Development, 2018).

Insurers offer earthquake risk coverage, usually as part of other insurance (e.g., fire insurance and multiperil coverage) (Goda, 2015; Mitchell-Wallace et al., 2017). Depending on country and property type, the purchase of earthquake coverage is compulsory or voluntary. A primary insurer collects insurance premiums from policyholders and reimburses

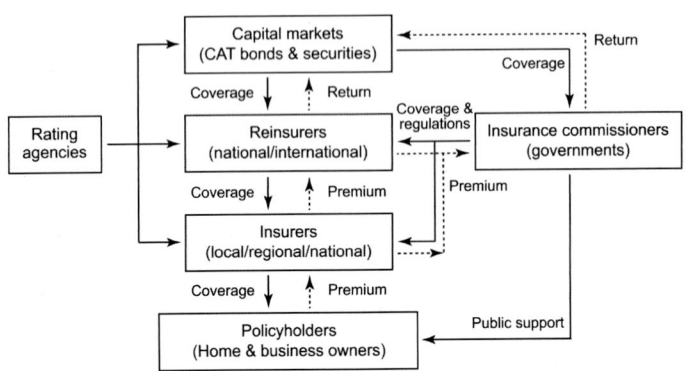

FIGURE 6.25 Overview of an insurance-reinsurance system.

incurred loss based on a preagreed policy upon the occurrence of a damaging hazardous event, such as earthquakes and tsunamis. The insurance premium P_I (also called technical price) typically consists of three components

$$P_I = P_{\text{pure}} + P_{\text{expense}} + P_{\text{risk}} \qquad (6.8)$$

where P_{pure} is the pure premium and is the expected cost of the ceded risks (i.e., fair price for the risk coverage); P_{expense} is the cost related to commission/brokerage, fees/levies, taxes, loss adjustment expenses associated with handling claims (e.g., allocated loss adjustment expenses, ALAE), reinsurance and retrocession costs, internal expenses, monitoring, and claim settlement; and P_{risk} is the risk premium (or insurer's profit) and is determined on the basis of various factors (e.g., insurer's capital reserve, reinsurance price/availability, and regulatory requirements). By expressing P_{expense} and P_{risk} as a function of P_{pure}, the technical price can be given by

$$P = (FE + P_{\text{pure}} \times (1 + ALAE\%))/(1 - VE\% - RoP\%) \qquad (6.9)$$

where FE is the flat expense per policy, $ALAE\%$ is the allocated expenses as a percentage of P_{pure}, $VE\%$ is the expense as a percentage of P_{pure}, and $RoP\%$ is the profit as a percentage of P_{pure}.

As both P_{expense} and P_{risk} are not negligible (particularly for earthquake and tsunami coverage, P_{risk} can be large, in comparison with P_{pure}), risk-neutral owners, who make decisions based on the expected values without considering the variability of the consequences, regard insurance coverage as overpriced and unfavorable economically. Thus only risk-averse owners, who avoid uncertain negative consequences, may seek insurance coverage. In cases where owners are affected/biased due to their risk perception (e.g., overestimation of potential consequences), earthquake–tsunami insurance coverage may or may not be attractive for them. Another important aspect to note is that insurance has the additional benefit of providing policyholders with financial liquidity in postdisaster situations. For example, a policyholder who has purchased a property with a mortgage may face liquidity constraints after a major earthquake and tsunami that damages the property severely; the policyholder may not be able to borrow money to recover the loss immediately after the event, because the collateral (i.e., property) may be damaged/lost and there may be an outstanding mortgage repayment. In such cases, insurance coverage is beneficial in enhancing financial resilience against major seismic events.

The policy (pay-out) function of an insurance contract F_{payout} for the earthquake–tsunami damage loss L_{ET} takes the following form

$$F_{\text{payout}}(L_{ET}) = \begin{cases} 0 & C_{EQ} \leq D \\ \varphi(L_{ET} - D) & D < L_{ET} < C \\ \varphi(C - D) & L_{ET} \geq C \end{cases} \qquad (6.10)$$

where D, C, and φ are the deductible, cap, and coinsurance factor, respectively. The deductible is the starting loss value of the insurance payout and is usually specified in terms of total insured value. The limit is the maximum loss value of the insurance payout, expressed in terms of total insured value. From insurers' viewpoint, the deductible is useful for reducing expense costs related to small claims, while the upper limit is effective in avoiding very large claims (critical for their solvency). The coinsurance factor specifies the proportional share of an incurred seismic loss between a policyholder and an insurer; this helps suppress the problem related to moral hazard.

The sales of earthquake risk coverage are often regulated by statutory bodies, such as insurance commissioners. The main reasons are that earthquake–tsunami insurance, unlike other liability insurance, tends to be public (i.e., protection against involuntary risks), and after major seismic disasters, governments and states/provinces may intervene to provide public support to the affected people (e.g., monetary gift and special loan programs). Although in reality, such disaster relief activities are inevitable (which are funded by tax and borrowing), it is desirable to minimize postdisaster intervention (ex-ante subsidies) as this discourages DRR efforts before earthquakes (Kunreuther, 1996). This is referred to as charity hazard.

6.4.2.2 Financial risk diversification

The basic principle of insurance portfolio management is the law of large numbers; with the increase in the number of policies, the variability of aggregate claims of a portfolio becomes small, and thus actual total loss approaches the expected loss value (which increases with the number of policies). In short, the financial risk becomes more predictable. Although this theory applies to independent risks (e.g., car insurance), earthquake–tsunami risks are spatially and temporally correlated. Insurers may experience a devastating surge of earthquake–tsunami loss claims in a disaster-hit region and face the possibility of insolvency. Therefore, insurers need to diversify their financial risks geographically.

A conventional approach for CAT risk transfer is reinsurance by global reinsurers (e.g., Munich Re and Swiss Re) and governments (Fig. 6.25). A notable functionality of governments as reinsurer is their ability to borrow money by issuing bonds; this achieves temporal diversification of earthquake−tsunami risks. Although reinsurers have greater risk-bearing capacities than insurers and achieve geographical risk diversification through their national/global portfolios, the potential size of the catastrophic earthquakes and tsunamis can be significant (in the order of tens to hundreds of billion dollars), and their default risk is not zero. Because of this, reinsurers charge high risk premiums to insurers and seek alternative means for risk transfer (Cummins, 2008), which in turn affects the risk premiums for primary insurance coverage. Reinsurers also utilize retrocession, that is, risk transfer from one reinsurer to another reinsurer to reduce the impact of peak risks. Recently, financial market instruments have become available through capital markets and have been used more frequently by insurers, reinsurers, and governments/states (Mitchell-Wallace et al., 2017). The emerging tools take advantage of the much greater financial capacity of capital markets, covering high-loss layers of insurance/reinsurance portfolios. Since earthquake−tsunami insurance is an integrated part of CAT risk management, insurance commissioners, on behalf of governments, regulate the solvency status and insurance policy/pricing of insurers. A high insolvency risk is not acceptable for public earthquake−tsunami insurance systems. For monitoring the financial status of the earthquake−tsunami risk underwriters, credit rating agencies (e.g., Standard & Poor's and Moody's) send signals to involved parties on their CAT exposures.

A useful tool for visualizing the composition of CAT risk diversification by different stakeholders is the loss diagram. It displays how the total loss is divided and covered by the involved parties. Fig. 6.26 shows a hypothetical loss diagram for aggregate earthquake−tsunami loss, together with a corresponding EP curve. In the diagram, policyholders retain loss up to $10 million (i.e., deductible for the entire portfolio) and loss exceeding $50 million (i.e., insurable limit). The loss layer between $10 million and $40 million is covered by an insurer and a reinsurer; the first $10 million layer is solely covered by the primary insurer, while the next $20 million layer is shared by the excess-of-loss reinsurance with a proportional rate of 0.2 (i.e., $20 million and $40 million are the attachment and exhaustion points of the reinsurance contract, respectively). The high loss layer between $40 million and $50 million is ceded to capital markets via CAT bonds/securities.

6.4.2.3 Risk perception and behavioral issues

Risk perception has a major influence on decision-making processes regarding the adoption of earthquake−tsunami insurance (Kunreuther, 1996). Facing catastrophic risks, home and business owners may have diverse opinions on the potential magnitude of earthquake−tsunami consequences. Therefore, the appreciation of the benefits of insurance coverage varies significantly, depending on personal characteristics (e.g., optimistic or pessimistic), proximity to physical hazards (e.g., near the shoreline), previous experience with earthquakes and tsunamis, and socioeconomic/demographic profiles (e.g., age, gender, income, and education). Another important aspect influencing insurance-related decisions is the cognitive limitation of human beings. Biases proven effective in supporting sound decisions for most people in most circumstances are found to be contributing to poor decisions with respect to low-probability/high-consequence hazards. Meyer and Kunreuther (2017) suggest addressing the following six biases:

- Myopia bias—Focus on short-time horizons in comparing upfront costs with expected benefits.

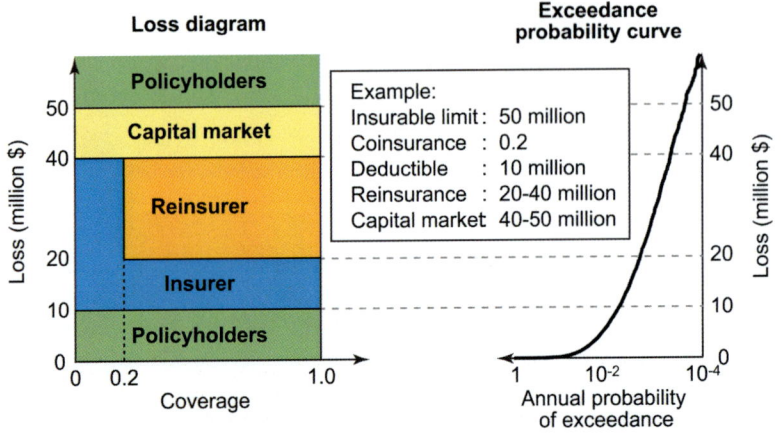

FIGURE 6.26 Loss diagram for stakeholders, insurer, reinsurer, and capital market.

- Amnesia bias—Forget the lessons of the past and decide not to undertake protective measures.
- Optimism bias—Underestimate the likelihood of hazards until below the threshold level of concern.
- Inertia bias—Maintain the status quo to avoid the effort to change uncertain outcomes.
- Simplification bias—Simplify complex situations and attend to only a few of the key factors.
- Herding bias—Choices based on the actions of others with little thought about the implications.

It is also well-known that people are affected by framing and have difficulties in estimating very low probabilities. An example of framing is that a factory owner who is considering earthquake insurance for financial protection is likely to take insurance if he/she is told that the chance of experiencing major earthquake damage is 1 in 5 over the entire period (20 years), rather than the chance of such an event is 1 in 100 per year. Even when the overall premium is reasonably priced, not many stakeholders voluntarily purchase earthquake−tsunami risk coverage.

Two known issues in offering insurance coverage are moral hazard and adverse selection (note: they apply to not only CAT earthquake−tsunami risks but other more common risk types). A moral hazard is a potential for change to imprudent behavior when the policyholder no longer needs to face the consequences of a risk. An example of moral hazard is that a policyholder, after a damaging earthquake and tsunami, may cause additional damage to his/her property to receive insurance payment. On the other hand, adverse selection is the tendency for more demand for insurance from high-risk compared to low-risk individuals. For instance, if an insurer charges a flat rate for all homeowners in a specific geographical region, earthquake−tsunami insurance is more attractive for those who live in seismically more vulnerable houses close to the sea. This leads to higher risk exposure for the insurer than originally expected and may result in increasing the premium rate (i.e., an adverse selection spiral can occur). One of the main causes of moral hazard and adverse selection is the information asymmetry between an insurer and a policyholder. The private information that the stakeholder has on his/her property is not entirely accessible to the insurer; in such cases, the stakeholder may have an incentive to behave inappropriately from the insurer's perspective.

6.4.2.4 Risk-based insurance rate-making for tsunamis

To alleviate the influence of moral hazard and adverse selection, introducing pro rata share of the risks and differentiating the rates according to physical parameters related to seismic−tsunami hazard and vulnerability of the covered properties (e.g., proximity to shoreline, geographical region, structural type, built year, and number of stories) are useful (Song & Goda, 2019). Risk-based insurance rate-making and effective incentive schemes can mitigate adverse selection and encourage physical DRR measures.

Song and Goda (2019) investigated the effects of building locations and tsunami simulation grid resolutions on insurance premium rate-making by performing detailed PTRA for Sendai and Onagawa in the Tohoku region of Japan. Their research interest was to compare EP curves and pure insurance premium rates (i.e., AAL for insurers) for buildings made of different materials (reinforced concrete, steel, wood, and masonry) and at different elevations and distances from shorelines. Intuitively, reinforced concrete and steel buildings are less vulnerable than wood and masonry (De Risi et al., 2017), and tsunami losses are expected to become smaller as the distance from the shoreline increases. In addition, since tsunami inundation simulations are often carried out using various grid resolutions, the use of a coarse digital elevation model (DEM) (e.g., 100-m resolution) may result in significant bias in the estimated tsunami loss and insurance premium rate. Fig. 6.27 shows PTRA results that are extracted from Song and Goda (2019). Fig. 6.27A and B shows the maps of the Tohoku region (indicating the location of Sendai) and the building portfolio in Sendai, respectively. In the zoomed area of Sendai, three building locations are selected: P1, P2, and P3. With the high-resolution DEM (10-m resolution) of the Sendai area, three locations are approximately at 2 m above the mean sea level (Table 6.6), but their distances from the shoreline are 0.5, 1.5, and 2 km for P1, P2, and P3, respectively (Fig. 6.27C). With the coarser DEM of 50- and 150-m resolutions, the elevations at P1 to P3 vary from 0.5 to 3.0 m (Fig. 6.27D and E). Due to the changes in the local topography represented in the different DEM data, EP curves for a wooden house at P1, P2, and P3 become significantly different (Fig. 6.27F−H). For instance, with the most accurate 10-m resolution DEM, the EP curves decrease from P1 to P3 due to the longer distance from the shoreline. This order is not maintained for the 50- and 150-m resolution DEM cases. Moreover, at the specific sites, differences in the EP curves are large, highlighting the importance of using high-resolution DEM data for inundation simulation and tsunami loss estimation. Finally, the effects of different locations and elevation data resolutions on the insurance pure premium rates are summarized in Table 6.6. The insurance premium rates are expressed as AAL per 1000 insured values. The trends of the insurance premium rates are similar to those of the EP curves shown in Fig. 6.27. When the coarse-resolution data are used, substantial errors can be introduced in the calculated insurance premium rates. Therefore, it is of paramount importance to use high-resolution elevation data in calculating risk-based tsunami insurance premium rates.

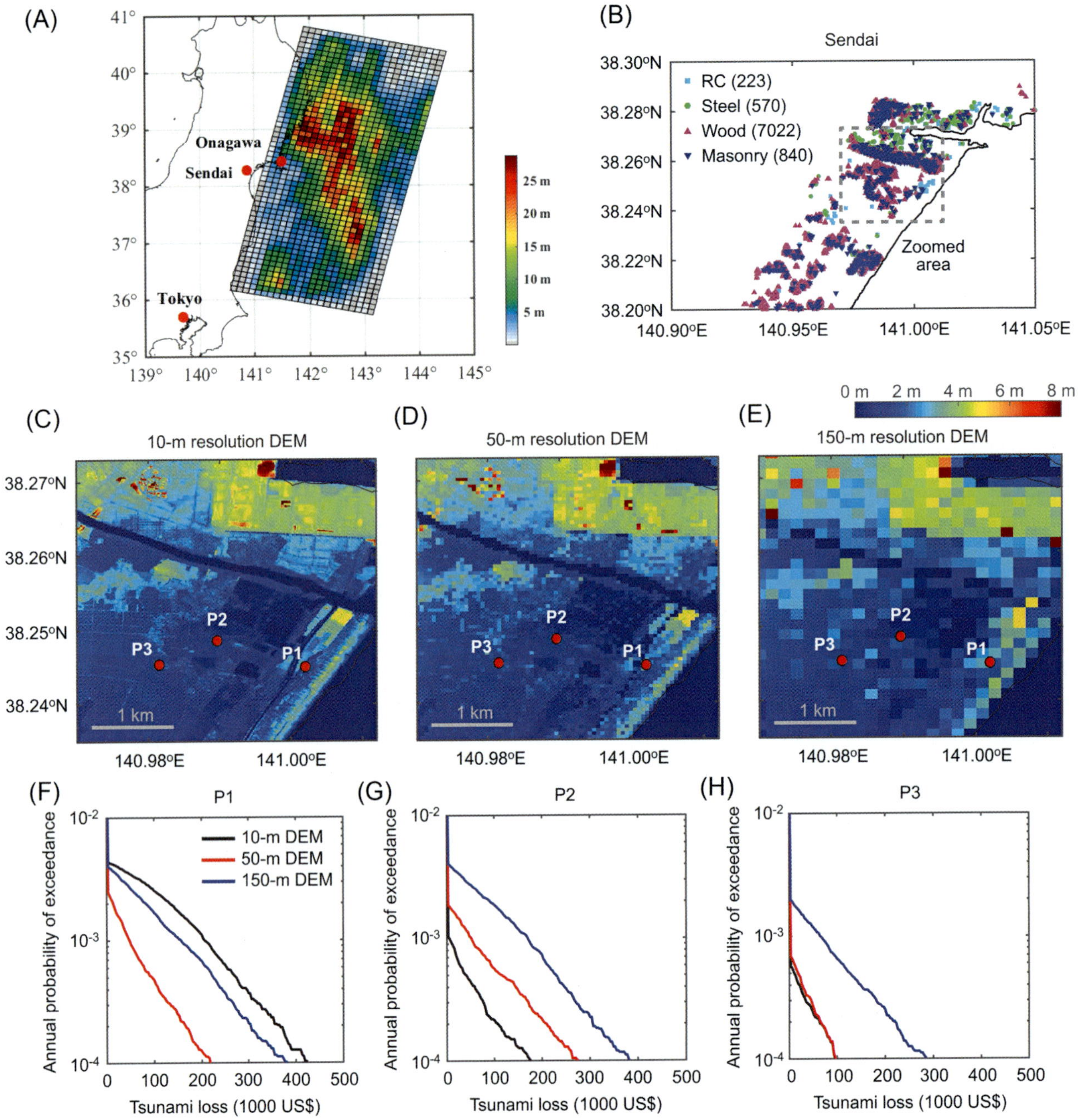

FIGURE 6.27 Effects of building locations and elevation data resolutions on tsunami insurance premium. (A) Map of the Tohoku region of Japan. (B) Building distribution in Sendai. (C) 10-m resolution DEM. (D) 50-m resolution DEM. (E) 150-m resolution DEM. (F) Tsunami loss curves for a wooden house at P1. (G) Tsunami loss curves for a wooden house at P2. (H) Tsunami loss curves for a wooden house at P3. The locations of P1 to P3 are indicated in (C–E). *DEM*, Digital elevation model. Map lines delineate study areas and do not necessarily depict accepted national boundaries.

6.4.3 Risk transfer for tsunami

In securing the financial coverage of upper loss layers, ILS serves as an effective risk financing tool (Fig. 6.24). The CAT bond is one of the most representative types of ILS to transfer catastrophic risks to investors in global capital markets. A typical structure of CAT bonds is shown in Fig. 6.28. A single-purpose vehicle collects funds (principal) from investors and issues CAT bonds. In cases where preagreed/specified trigger conditions are met, the single-purpose vehicle releases the principal to the sponsor; otherwise, the principal together with investment return, which is higher than typical securities, is paid back to the investors when the bonds mature (i.e., no trigger).

TABLE 6.6 Elevations of three selected locations using digital elevation models (DEMs) of different resolutions and tsunami insurance pure premium rates at different locations (per 1000 insured values).

Location	DEM resolution			Insurance premium rate		
	10 m	50 m	150 m	10 m DEM	50 m DEM	150 m DEM
P1	2.00	3.11	3.03	2.946	0.606	2.099
P2	1.99	1.39	0.50	0.284	0.758	2.115
P3	1.98	2.38	1.99	0.132	0.138	0.828

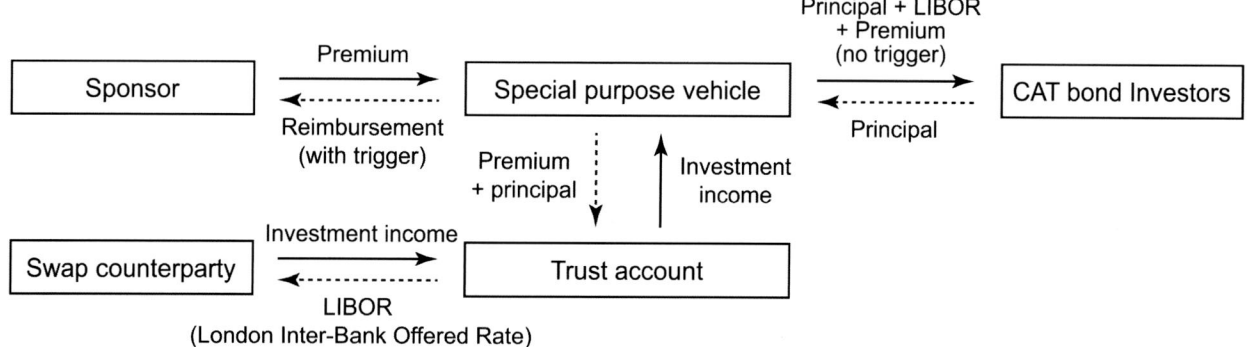

FIGURE 6.28 Structure of CAT bond. *CAT*, Catastrophe.

CAT bonds take different forms, depending on the bond triggers. Four popular types of CAT bond triggers are

- Indemnity trigger uses actual losses sustained by the sponsor; therefore, it is similar to insurance contracts. This trigger type involves the issue of moral hazard as the sponsor's actual loss is not transparent to the investors.
- Industry loss trigger uses the industry loss index issued by a third party, such as Property Claim Services.
- Modeled trigger uses an independent third party's estimate of the sponsor's loss, which is supplied by major CAT modeling firms, such as AIR Worldwide and RMS.
- Parametric trigger uses the physical measured characteristics of catastrophic events, such as earthquake magnitude and location, which are provided by third-party organizations, such as the U.S. Geological Survey.

The industry loss, modeled loss, and parametric triggers do not involve the loss settlement process, allowing the fast payment of the reimbursement for the sponsor upon the trigger event. The settlement process can be costly and take a long time because the insurance claims made by the insured need to be evaluated and verified by loss adjusters as well as engineers who are knowledgeable about earthquake—tsunami damage. Investors without expertise in the loss settlement process prefer a parametric approach because it saves them the effort associated with adjusting losses and provides transparency. Sponsors also like the transparency of parametric tools in addition to the prompt execution of payment. This aspect is particularly beneficial for mobilizing resources in emergency response and early recovery phases of critical importance as evidenced in the Christchurch earthquake scenario in New Zealand (King et al., 2014). On the other hand, the industry loss, modeled loss, and parametric triggers are influenced by the basis risk, which stems from an imperfect correlation between the actual losses experienced and the payments received. Basis risk arises because the parameters involved in determining the payment are insufficient to characterize the full consequences of the event. Since these risk transfer products are calibrated with numerical risk models, basis risk can also suffer from shortcomings in the models used. To refer to the limitations imposed by the parametric formulation, the terms of "trigger error" or "index error" are often used. On the other hand, "model risk" is used to refer to those imposed by the model.

Parametric approaches may be based on a few fundamental event characteristics (e.g., earthquake location and magnitude) or on a large number of intensity measurements, usually packaged into an index formulation. The former approach is referred to in the industry as a "CAT-in-a-Box" or "first-generation," whereas the latter approach is typically known as an "intensity-based index" or "second-generation" solution (Wald & Franco, 2016). Both approaches are useful, and their optimal application setting depends on several factors, such as the expertise of the parties using them, the complexity of the portfolio at risk, and the availability and disposition of the instruments providing the required measurements. CAT-in-a-Box solutions make sense when simplicity and transparency are important and when the potential loss is difficult to assess (e.g., business interruption) or when the measurements required for second-generation index solutions are not available. In seismic regions where reliable observation/monitoring networks are in place to provide reliable intensity measurements over a large portion of the geography (e.g., California and Japan), intensity-based indices can be considered (Goda, 2015; Pucciano et al., 2017). Local intensities near the sites of interest (e.g., tsunami wave intensity parameters recorded at offshore sensors; Section 6.3.1) are, in general, good predictors of earthquake–tsunami damage and loss and therefore typically increase the correlation between payment outcomes and actual loss (Goda, Franco, et al., 2019). New machine learning methods can be used to develop improved correlation models between tsunami intensity measures and tsunami loss (Li & Goda, 2023a). In other words, they have the potential to minimize basis risk and therefore constitute an appealing alternative.

6.5 Research needs

The ultimate goal of earthquake and tsunami DRM is achieving resilient and sustainable coastal communities around the world. A holistic risk management strategy that integrates all phases of a disaster cycle (i.e., mitigation, preparedness, response, and recovery) is necessary. This requires concerted efforts by citizens, emergency officers, policymakers, scientists, and professionals/practitioners in the public and private sectors. The research subjects for attaining improved disaster community resilience range across a variety of disciplines, including urban planning and land-use policy that are informed by sound hazard and risk assessments, physical risk mitigations through more robust earthquake–tsunami design methods and construction practices, emergency response approaches that integrate tsunami early warning systems and tsunami evacuation procedures, and recovery preparedness and strategies from financial viewpoints. In addition to conventional geological hazards, climate change and related perils will have a significant influence on communities' capacities to cope with future disasters in the long run.

In the later part of this book, several chapters that focus on the abovementioned aspects of effective tsunami DRM. The installation of dense tsunami monitoring systems in tandem with advanced machine learning algorithms can drastically improve the performance of tsunami early warning systems (Chapters 11 and 12). More advanced tsunami fragility methods will improve the accuracy of tsunami risk assessments (Chapters 17 and 18), thereby enhancing the effectiveness of risk-based management decisions. Further investigations on multihazard risk assessments should be explored in the future to consider the seismic–tsunami disaster risks more comprehensively at different time and spatial scales (Chapter 20). Related to the tsunami early warning, improved approaches for tsunami evacuation simulations by accounting for behavioral aspects of evacuees will be critically important in implementing effective tsunami risk mitigation procedures in the communities (Chapter 21). Climate change, in particular sea-level rise, will have a direct impact on the tsunami inundation risks for coastal communities, because more of existing buildings and infrastructure will be exposed to tsunami hazards. The effects of sea-level rise on tsunami risks to people and the built environment are nonlinear and depend on the extent of future scenarios as well as the topography and exposure (Chapter 22). Quantifying these effects for possible future climate scenarios is useful for long-term risk management strategies (Chapter 23). Finally, quantified disaster risks must be communicated effectively to those who are exposed. Future disaster impacts can be represented and understood better by utilizing digital-twin models for coastal disasters, including earthquakes, tsunamis, storm surges, and flooding (Chapter 24).

References

Akaike, H. (1973). Maximum likelihood identification of Gaussian autoregressive moving average models. *Biometrika, 60*(2), 255–265. Available from https://doi.org/10.1093/biomet/60.2.255.

An, C., Liu, H., Ren, Z., & Yuan, Y. (2018). Prediction of tsunami waves by uniform slip models. *Journal of Geophysical Research: Oceans, 123*(11), 8366–8382. Available from https://doi.org/10.1029/2018JC014363, http://agupubs.onlinelibrary.wiley.com/agu/jgr/journal/10.1002/(ISSN)2169-9291/.

Ando, M. (1975). Source mechanisms and tectonic significance of historical earthquakes along the Nankai trough, Japan. *Tectonophysics, 27*(2), 119–140. Available from https://doi.org/10.1016/0040-1951(75)90102-X.

Attary, N., Unnikrishnan, V. U., van de Lindt, J. W., Cox, D. T., & Barbosa, A. R. (2017). Performance-based tsunami engineering methodology for risk assessment of structures. *Engineering Structures, 141*, 676–686. Available from http://www.journals.elsevier.com/engineering-structures/, 10.1016/j.engstruct.2017.03.071.

Australian Institute for Disaster Resilience. (2015). *National Emergency Risk Assessment Guidelines*. https://knowledge.aidr.org.au/media/2030/handbook-10-national-emergency-risk-assessment-guidelines.pdf.

Breiman, L. (2001). Random forests. *Machine Learning*, *45*(1), 5−32. Available from https://doi.org/10.1023/A:1010933404324.

Central Disaster Management Council. (2012). *Working group report on mega-thrust earthquake models for the Nankai Trough, Japan*. Cabinet Office of the Japanese Government (unpublished content), http://www.bousai.go.jp/jishin/nankai/taisaku/pdf/20120829_2nd_report01.pdf.

Chock, G. Y. K. (2016). Design for tsunami loads and effects in the ASCE 7-16 Standard. *Journal of Structural Engineering*, *142*(11). Available from https://doi.org/10.1061/(asce)st.1943-541x.0001565.

Cornell, C. A., & Krawinkler, H. (2000). Progress and challenges in seismic performance assessment. *PEER Center News*, *3*, 1−3.

Cosson, C. (2020). Build Back Better": Between public policy and local implementation, the challenges in Tohoku's reconstruction. *Architecture and Urban Planning*, *16*(1), 1−4. Available from https://doi.org/10.2478/aup-2020-0001, http://content.sciendo.com/view/journals/aup/aup-overview.xml.

Cummins, J. D. (2008). CAT bonds and other risk-linked securities: State of the market and recent developments. *Risk Management and Insurance Review*, *11*(1), 23−47. Available from https://doi.org/10.1111/j.1540-6296.2008.00127.x.

Cyranoski, D. (2011). Japan's tsunami warning system retreats. *Nature*. Available from https://doi.org/10.1038/news.2011.477, 0028-0836.

De Risi, R., Goda, K., Yasuda, T., & Mori, N. (2017). Is flow velocity important in tsunami empirical fragility modeling? *Earth-Science Reviews*, *166*, 64−82. Available from https://doi.org/10.1016/j.earscirev.2016.12.015, http://www.sciencedirect.com/science/journal/00128252.

Ditlevsen, O., & Friis-Hansen, P. (2009). Cost and benefit including value of life, health and environmental damage measured in time units. *Structural Safety*, *31*(2), 136−142. Available from https://doi.org/10.1016/j.strusafe.2008.06.010.

Dunant, A., Bebbington, M., & Davies, T. (2021). Probabilistic cascading multi-hazard risk assessment methodology using graph theory, a New Zealand trial. *International Journal of Disaster Risk Reduction*, *54*, 102018. Available from https://doi.org/10.1016/j.ijdrr.2020.102018.

ERM-Hong Kong, Ltd. (1998). *Landslides and boulder falls from natural terrain: Interim risk guidelines GEO Report No. 75*. https://www.cedd.gov.hk/eng/publications/geo/geo-reports/geo_rpt075/index.html.

Federal Emergency Management Agency. (2019). *Guidelines for design of structures for vertical evacuation from tsunamis* (P−646). Applied Technology Council, https://www.fema.gov/sites/default/files/2020-08/fema_earthquakes_guidelines-for-design-of-structures-for-vertical-evacuation-from-tsunamis-fema-p-646.pdf.

Goda, K. (2015). Seismic risk management of insurance portfolio using catastrophe bonds. *Computer-Aided Civil and Infrastructure Engineering*, *30*(7), 570−582. Available from https://doi.org/10.1111/mice.12093United, http://onlinelibrary.wiley.com/journal/10.1111/(ISSN)1467-8667.

Goda, K. (2022). Stochastic source modeling and tsunami simulations of Cascadia subduction earthquakes for Canadian Pacific coast. *Coastal Engineering Journal*, *64*(4), 575−596. Available from https://doi.org/10.1080/21664250.2022.2139918, https://www.tandfonline.com/loi/tcej20.

Goda, K., & De Risi, R. (2018). Multi-hazard loss estimation for shaking and tsunami using stochastic rupture sources. *International Journal of Disaster Risk Reduction*, *28*, 539−554. Available from https://doi.org/10.1016/j.ijdrr.2018.01.002, http://www.journals.elsevier.com/international-journal-of-disaster-risk-reduction/.

Goda, K., Franco, G., Song, J., & Radu, A. (2019). Parametric catastrophe bonds for tsunamis: Cat-in-a-box trigger and intensity-based index trigger methods. *Earthquake Spectra*, *55*(1), 113−136. Available from https://doi.org/10.1193/030918EQS052M, https://earthquakespectra.org/doi/pdf/10.1193/030918EQS052M.

Goda, K., Mori, N., & Yasuda, T. (2019). Rapid tsunami loss estimation using regional inundation hazard metrics derived from stochastic tsunami simulation. *International Journal of Disaster Risk Reduction*, *40*, 101152. Available from https://doi.org/10.1016/j.ijdrr.2019.101152.

Goda, K., Orchiston, K., Borozan, J., Novakovic, M., & Yenier, E. (2023). Evaluation of reduced computational approaches to assessment of tsunami hazard and loss using stochastic source models: Case study for Tofino, British Columbia, Canada, subjected to Cascadia megathrust earthquakes. *Earthquake Spectra*, *39*(3), 1303−1327. Available from https://doi.org/10.1177/87552930231187407.

Goda, K., Wilhelm, K., & Ren, J. (2020). Relationships between earthquake insurance take-up rates and seismic risk indicators for Canadian households. *International Journal of Disaster Risk Reduction*, *50*, 101754. Available from https://doi.org/10.1016/j.ijdrr.2020.101754.

Goda, K., Yasuda, T., Mai, P. M., Maruyama, T., & Mori, N. (2018). Tsunami simulations of mega-thrust earthquakes in the Nankai−Tonankai Trough (Japan) based on stochastic rupture scenarios. *Geological Society, London, Special Publications*, *456*(1), 55−74. Available from https://doi.org/10.1144/sp456.1.

Goldfinger, C., Nelson, C. H., Morey, A. E., Johnson, J. E., Patton, J., Karabanov, E., Gutierrez-Pastor, J., Eriksson, A. T., Gracia, E., Dunhill, G., Enkin, R. J., Dallimore, A., & Vallier, T. (2012). *Turbidite event history: Methods and implications for Holocene paleoseismicity of the Cascadia subduction zone*. United States Geological Survey Professional Paper 1661. https://pubs.usgs.gov/publication/pp1661F.

Gonzalez, F. I., Milburn, H. B., Bernard, E. N., & Newman, J. (1998). *Deep-ocean assessment and reporting of tsunamis (DART): Brief overview and status report*. https://www.ndbc.noaa.gov/dart/brief.shtml.

Goulet, C. A., Haselton, C. B., Mitrani-Reiser, J., Beck, J. L., Deierlein, G. G., Porter, K. A., & Stewart, J. P. (2007). Evaluation of the seismic performance of a code-conforming reinforced-concrete frame building—From seismic hazard to collapse safety and economic losses. *Earthquake Engineering and Structural Dynamics*, *36*(13), 1973−1997. Available from https://doi.org/10.1002/eqe.694, http://onlinelibrary.wiley.com/journal/10.1002/(ISSN)1096-9845.

Health and Safety Executive. (1992). *The tolerability of risks from nuclear power stations*. https://www.onr.org.uk/documents/tolerability.pdf.

Ho, T. K. (1998). The random subspace method for constructing decision forests. *IEEE Transactions on Pattern Analysis and Machine Intelligence*, *20*(8), 832−844. Available from https://doi.org/10.1109/34.709601.

Horni, A., Nagel, K., & Axhausen, K. W. (2016). *The multi-agent transport simulation MATSim*. Ubiquity Press. Available from https://www.ubiquity-press.com/site/books/e/10.5334/baw/, 10.5334/baw.

Hoshiba, M., & Ozaki, T. (2014). *Early warning for geological disasters earthquake early warning and tsunami warning of the Japan Meteorological Agency, and their performance in the 2011 off the Pacific coast of Tohoku earthquake (Mw 9.0)* (pp. 1–28). Springer Science and Business Media LLC. Available from https://doi.org/10.1007/978-3-642-12233-0_1.

Japan Reconstruction Agency. (2016). *Basic guidelines for reconstruction in response to the Great East Japan Earthquake in the reconstruction and revitalization period.* https://www.reconstruction.go.jp/english/topics/Laws_etc/20160527_basic-guidelines.pdf.

Kajitani, Y., Chang, S. E., & Tatano, H. (2013). Economic impacts of the 2011 Tohoku-oki earthquake and tsunami. *Earthquake Spectra*, 29(1), S457–S478. Available from https://doi.org/10.1193/1.4000108, http://earthquakespectra.org/doi/pdf/10.1193/1.4000108.

Kamiya, M., Igarashi, Y., Okada, M., & Baba, T. (2022). Numerical experiments on tsunami flow depth prediction for clustered areas using regression and machine learning models. *Earth, Planets and Space*, 74(1). Available from https://doi.org/10.1186/s40623-022-01680-9, http://rd.springer.com/journal/40623.

Kanazawa, T. (2013). *Japan Trench earthquake and tsunami monitoring network of cable-linked 150 ocean bottom observatories and its impact to earth disaster science.* 2013 IEEE International Underwater Technology Symposium. Available from https://doi.org/10.1109/UT.2013.6519911, https://ieeexplore.ieee.org/document/6519911.

King, A., Middleton, D., Brown, C., Johnston, D., & Johal, S. (2014). Insurance: Its role in recovery from the 2010–2011 Canterbury earthquake sequence. *Earthquake Spectra*, 30(1), 475–491. Available from http://earthquakespectra.org/doi/pdf/10.1193/022813EQS058M, 10.1193/022813EQS058M.

Kubo, H., Kubota, T., Suzuki, W., Aoi, S., Sandanbata, O., Chikasada, N., & Ueda, H. (2022). Ocean-wave phenomenon around Japan due to the 2022 Tonga eruption observed by the wide and dense ocean-bottom pressure gauge networks. *Earth, Planets and Space*, 74(1). Available from https://doi.org/10.1186/s40623-022-01663-w, http://rd.springer.com/journal/40623.

Kunreuther, H. (1996). Mitigating disaster losses through insurance. *Journal of Risk and Uncertainty*, 12(2-3), 171–187. Available from https://doi.org/10.1007/BF00055792, http://www.klueronline.com/issn/0895-5646.

Kunreuther, H., Conell-Price, L., Li, B., Kovacs, P., & Goda, K. (2024). Influence of a private–public risk pool and an opt-out framing on earthquake protection demand for Canadian homeowners in Quebec and British Columbia. *Risk Analysis*. Available from https://doi.org/10.1111/risa.14285, http://onlinelibrary.wiley.com/journal/10.1111/(ISSN)1539-6924.

Kuroshio Town Council. (2019a). *Average household size in Saga district.* Kuroshio Town Council. https://www.town.kuroshio.lg.jp/pb/cont/machi-data.

Kuroshio TownCouncil. (2019b). *Earthquake-tsunami hazard map of Kuroshio, Kuroshio Town Data.* Kuroshio Town Council. https://www.town.kuroshio.lg.jp/img/files/pv/bousai/docs/map07.pdf.

Kuroshio Town Council. (2019c). *Population data in Saga.* Kuroshio Town Council. https://www.town.kuroshio.lg.jp/pb/cont/machi-data/20743.

Lämmel, G., Klüpfel, H., & Nagel, K. (2009). The MATSim network flow model for traffic simulation adapted to large-scale emergency egress and an application to the evacuation of the Indonesian city of Padang in case of a tsunami warning. In H. Timmermans (Ed.), *Pedestrian behavior* (pp. 245–265). Emerald Group Publishing Limited. Available from https://doi.org/10.1108/9781848557512-011.

Li, Y., & Goda, K. (2022). Hazard and risk-based tsunami early warning algorithms for ocean bottom sensor S-net system in Tohoku, Japan, using sequential multiple linear regression. *Geosciences*, 12(9), 350. Available from https://doi.org/10.3390/geosciences12090350.

Li, Y., & Goda, K. (2023a). Random forest-based multi-hazard loss estimation using hypothetical data at seismic and tsunami monitoring networks. *Geomatics, Natural Hazards and Risk*, 14(1). Available from https://doi.org/10.1080/19475705.2023.2275538, http://www.tandfonline.com/toc/tgnh20/current.

Li, Y., & Goda, K. (2023b). Risk-based tsunami early warning using random forest. *Computers & Geosciences*, 179105423. Available from https://doi.org/10.1016/j.cageo.2023.105423.

Maeda, T., Obara, K., Shinohara, M., Kanazawa, T., & Uehira, K. (2015). Successive estimation of a tsunami wavefield without earthquake source data: A data assimilation approach toward real-time tsunami forecasting. *Geophysical Research Letters*, 42(19), 7923–7932. Available from https://doi.org/10.1002/2015GL065588, http://onlinelibrary.wiley.com/journal/10.1002/(ISSN)1944-8007/issues?year = 2012.

Makinoshima, F., Oishi, Y., Yamazaki, T., Furumura, T., & Imamura, F. (2021). Early forecasting of tsunami inundation from tsunami and geodetic observation data with convolutional neural networks. *Nature Communications*, 12(1). Available from https://doi.org/10.1038/s41467-021-22348-0, http://www.nature.com/ncomms/index.html.

Mas, E., Koshimura, S., Imamura, F., Suppasri, A., Muhari, A., & Adriano, B. (2015). Recent advances in agent-based tsunami evacuation simulations: Case studies in Indonesia, Thailand. *Pure and Applied Geophysics*, 172(12), 3409–3424. Available from https://doi.org/10.1007/s00024-015-1105-y, http://www.springer.com/birkhauser/geo + science/journal/24.

Meyer, R., & Kunreuther, H. (2017). *The ostrich paradox.* Digital Press.

Michel-Kerjan, E., Hochrainer-Stigler, S., Kunreuther, H., Linnerooth-Bayer, J., Mechler, R., Muir-Wood, R., Ranger, N., Vaziri, P., & Young, M. (2013). Catastrophe risk models for evaluating disaster risk reduction investments in developing countries. *Risk Analysis*, 33(6), 984–999. Available from https://doi.org/10.1111/j.1539-6924.2012.01928.x.

Mitchell-Wallace, K., Jones, M., Hillier, J., & Foote, M. (2017). *Natural catastrophe risk management and modelling: A practitioner's guide.* Wiley-Blackwell.

Miyashita, T., Mori, N., & Goda, K. (2020). Uncertainty of probabilistic tsunami hazard assessment of Zihuatanejo (Mexico) due to the representation of tsunami variability. *Coastal Engineering Journal*, 413–428. Available from https://doi.org/10.1080/21664250.2020.1780676, https://www.tandfonline.com/loi/tcej20.

Muhammad, A., De Risi, R., De Luca, F., Mori, N., Yasuda, T., & Goda, K. (2021). Are current tsunami evacuation approaches safe enough. *Stochastic Environmental Research and Risk Assessment*, 35(4), 759–779. Available from https://doi.org/10.1007/s00477-021-02000-5, http://link.springer-ny.com/link/service/journals/00477/index.htm.

Muhammad, A., De Risi, R., De Luca, F., Mori, N., Yasuda, T., & Goda, K. (2022). *Manual of agent-based tsunami evacuation modelling using MATSim*. Bristol University. Available from https://data.bris.ac.uk/datasets/333uc5aebpzfz25mhmd83yt3yk/Manual_Agent_Based%20MATSim.pdf.

Muhammad, A., Goda, K., Alexander, N. A., Kongko, W., & Muhari, A. (2017). Tsunami evacuation plans for future megathrust earthquakes in Padang, Indonesia, considering stochastic earthquake scenarios. *Natural Hazards and Earth System Sciences*, 17(12), 2245−2270. Available from https://doi.org/10.5194/nhess-17-2245-2017, http://www.nat-hazards-earth-syst-sci.net/volumes_and_issues.html.

Muir-Wood, R. (2011). *Designing optimal risk mitigation and risk transfer mechanisms to improve the management of earthquake risk in Chile*. OECD Publishing.

Mulia, I. E., Ueda, N., Miyoshi, T., Gusman, A. R., & Satake, K. (2022). Machine learning-based tsunami inundation prediction derived from offshore observations. *Nature Communications*, 13(1). Available from https://doi.org/10.1038/s41467-022-33253-5, http://www.nature.com/ncomms/index.html.

Multi-Hazard Mitigation Council. (2020). A roadmap to resilience incentivization. *National Institute of Building Sciences*, 33. Available from https://www.nibs.org/files/pdfs/NIBS_MMC_RoadmapResilience_082020.pdf.

Onoda, Y., Tsukuda, H., & Suzuki, S. (2018). Complexities and difficulties behind the implementation of reconstruction plans after the Great East Japan earthquake and tsunami of March 2011. *Advances in Natural and Technological Hazards Research*, 47. Available from https://doi.org/10.1007/978-3-319-58691-5_1, http://www.springer.com/series/6362?detailsPage = titles.

Organisation for Economic Co-operation and Development. (2018). *Financial management of earthquake risk*. Organisation for Economic Co-operation and Development. Available from https://www.oecd-ilibrary.org/finance-and-investment/financial-management-of-earthquake-risk_eebded10-en. (2018).

Post, J., Wegscheider, S., Mück, M., Zosseder, K., Kiefl, R., Steinmetz, T., & Strunz, G. (2009). Assessment of human immediate response capability related to tsunami threats in Indonesia at a sub-national scale. *Natural Hazards and Earth System Science*, 9(4), 1075−1086. Available from https://doi.org/10.5194/nhess-9-1075-2009, http://www.nat-hazards-earth-syst-sci.net/volumes_and_issues.html.

Pucciano, S., Franco, G., & Bazzurro, P. (2017). Loss predictive power of strong motion networks for usage in parametric risk transfer: Istanbul as a case study. *Earthquake Spectra*, 33(4), 1513−1531. Available from https://doi.org/10.1193/021517EQS032M, http://earthquakespectra.org/doi/pdf/10.1193/021517EQS032M.

Rudloff, A., Lauterjung, J., & Münch, U. (2010). The GITEWS project (German-Indonesian Tsunami Early Warning System). *Natural Hazards and Earth System Sciences*, 9, 1381−1382. Available from https://doi.org/10.5194/nhess-9-1381-2009.

Sato, M., Ishikawa, T., Ujihara, N., Yoshida, S., Fujita, M., Mochizuki, M., & Asada, A. (2011). Displacement above the hypocenter of the 2011 Tohoku-oki earthquake. *Science*, 332(6036), 1395. Available from https://doi.org/10.1126/science.1207401Japan, http://www.sciencemag.org/content/332/6036/1395.full.pdf.

Song, J., & Goda, K. (2019). Influence of elevation data resolution on tsunami loss estimation and insurance rate-making. *Frontiers in Earth Science*, 7. Available from https://doi.org/10.3389/feart.2019.00246, https://www.frontiersin.org/journals/earth-science.

Starr, C. (1969). Social benefit versus technological risk. *Science*, 165(3899), 1232−1238. Available from https://doi.org/10.1126/science.165.3899.1232.

Sugimoto, T., Murakami, H., Kozuki, Y., Nishikawa, K., & Shimada, T. (2003). A human damage prediction method for tsunami disasters incorporating evacuation activities. *Natural Hazards*, 29(3), 585−600.

Suppasri, A., Hasegawa, N., Makinoshima, F., Imamura, F., Latcharote, P., & Day, S. (2016). An analysis of fatality ratios and the factors that affected human fatalities in the 2011 Great East Japan tsunami. *Frontiers in Built Environment*, 2. Available from https://doi.org/10.3389/fbuil.2016.00032, https://www.frontiersin.org/articles/10.3389/fbuil.2016.00032/full.

Tanioka, Y., & Gusman, A. R. (2018). Near-field tsunami inundation forecast method assimilating ocean bottom pressure data: A synthetic test for the 2011 Tohoku-oki tsunami. *Physics of the Earth and Planetary Interiors*, 283, 82−91. Available from https://doi.org/10.1016/j.pepi.2018.08.006, http://www.elsevier.com/inca/publications/store/5/0/3/3/5/6/.

Tatehata, H. (1997). *Perspectives on tsunami hazard reduction The new tsunami warning system of the Japan Meteorological Agency* (pp. 175−188). Springer Science and Business Media LLC. Available from https://doi.org/10.1007/978-94-015-8859-1_12.

Taubenböck, H., Goseberg, N., Setiadi, N., Lämmel, G., Moder, F., Oczipka, M., Klüpfel, H., Wahl, R., Schlurmann, T., Strunz, G., Birkmann, J., Nagel, K., Siegert, F., Lehmann, F., Dech, S., Gress, A., & Klein, R. (2009). Last-mile preparation for a potential disaster: Interdisciplinary approach towards tsunami early warning and an evacuation information system for the coastal city of Padang, Indonesia. *Natural Hazards and Earth System Science*, 9(4), 1509−1528. Available from https://doi.org/10.5194/nhess-9-1509-2009, http://www.nat-hazards-earth-syst-sci.net/volumes_and_issues.html.

Thomas, C., & Sheldon, B. (1999). The knee of a curve useful clue but incomplete support. *Military Operations Research*, 4(2), 17−24. Available from https://doi.org/10.5711/morj.4.2.17.

Wald, D. J., & Franco, G. (2016). Money matters: Rapid post-earthquake financial decision-making. *Natural Hazards Observer.*, 7, 24−27.

Wang, Y., Heidarzadeh, M., Satake, K., Mulia, I. E., & Yamada, M. (2020). A tsunami warning system based on offshore bottom pressure gauges and data assimilation for Crete Island in the eastern Mediterranean basin. *Journal of Geophysical Research: Solid Earth*, 125(10). Available from https://doi.org/10.1029/2020JB020293, http://agupubs.onlinelibrary.wiley.com/hub/jgr/journal/10.1002/(ISSN)2169-9356/.

Wang, Y., & Satake, K. (2021). Real-time tsunami data assimilation of S-net pressure gauge records during the 2016 Fukushima earthquake. *Seismological Research Letters*, 92(4), 2145−2155. Available from https://doi.org/10.1785/0220200447, https://watermark.silverchair.com/srl-2020447.1.pdf.

Tobler, W. (1993). Three presentations on geographical analysis and modeling: Nonisotropic geographic modeling. In *Speculations on the geometry of geography; and global spatial analysis* (pp. 93–1). National Center for Geographic Information and Analysis (NCGIA). https://escholarship.org/uc/item/05r820mz.

Weidmann, U. (1993). *Transporttechnik der fußgänger: Transporttechnische eigenschaften des Fußgängerverkehrs, Literaturauswertung. IVT Schriftenreihe*. Zürich: ETH. Available from https://www.research-collection.ethz.ch/handle/20.500.11850/242008.

Wood, N., Jones, J., Schelling, J., & Schmidtlein, M. (2014). Tsunami vertical-evacuation planning in the U.S. Pacific Northwest as a geospatial, multi-criteria decision problem. *International Journal of Disaster Risk Reduction, 9*, 68–83. Available from https://doi.org/10.1016/j.ijdrr.2014.04.009, http://www.journals.elsevier.com/international-journal-of-disaster-risk-reduction/.

Wood, N., Jones, J., Schmidtlein, M., Schelling, J., & Frazier, T. (2016). Pedestrian flow-path modeling to support tsunami evacuation and disaster relief planning in the U.S. Pacific Northwest. *International Journal of Disaster Risk Reduction, 18*, 41–55. Available from https://doi.org/10.1016/j.ijdrr.2016.05.010, http://www.journals.elsevier.com/international-journal-of-disaster-risk-reduction/.

Yoshida, K., Matsuzawa, T., & Uchida, N. (2022). The 2021 Mw7.0 and Mw6.7 Miyagi-oki earthquakes nucleated in a deep seismic/aseismic transition zone: possible effects of transient instability due to the 2011 Tohoku earthquake. *Journal of Geophysical Research: Solid Earth, 127*(8) e2022JB024887. Available from https://agupubs.onlinelibrary.wiley.com/doi/10.1029/2022JB024887.

Yoshikawa, H., & Goda, K. (2014). Financial seismic risk analysis of building portfolios. *Natural Hazards Review, 15*(2), 112–120. Available from https://doi.org/10.1061/(ASCE)NH.1527-6996.0000129.

Yun, N. Y., & Hamada, M. (2015). Evacuation behavior and fatality rate during the 2011 Tohoku-oki earthquake and tsunami. *Earthquake Spectra, 31*(3), 1237–1265. Available from https://doi.org/10.1193/082013EQS234M, http://earthquakespectra.org/doi/pdf/10.1193/082013EQS234M.

Zamora, N., Catalán, P. A., Gubler, A., & Carvajal, M. (2021). Microzoning tsunami hazard by combining flow depths and arrival times. *Frontiers Media S.A., Spain Frontiers in Earth Science, 8*. Available from https://doi.org/10.3389/feart.2020.591514, https://www.frontiersin.org/journals/earth-science.

Section 2

Advanced topics and applications related to probabilistic tsunami hazard and risk analysis

Chapter 7

Historical tsunami records and paleotsunamis

Kenji Satake
Earthquake Research Institute, The University of Tokyo, Tokyo, Japan

7.1 Introduction

Tsunamis are generated by submarine geological processes, such as earthquakes and landslides, and a typical way to model or compute tsunamis is to start from a given initial condition, simulate its propagation in the ocean, and calculate tsunami arrival times and water heights on coasts, or inundation on land (see Chapters 2 and 3). Such an approach is called forward modeling (Fig. 7.1). Once the initial condition is provided, tsunamis' propagation and coastal behavior can be numerically computed on actual bathymetry. Recent technological developments make it possible to carry out tsunami forward modeling with speed and accuracy usable for early warning and detailed hazard assessments (Musa et al., 2018).

The initial condition, or the tsunami generation process, is still poorly known because large tsunamis are rare and direct observations of tsunami generation are difficult. Indirect estimation of tsunami sources, mostly based on seismological analyses, is used as the initial condition of tsunami forward modeling, including an early warning. More direct estimation of tsunami sources is essential to better understand the tsunami generation process and to more accurately forecast the tsunami behavior on coasts.

Inverse modeling of tsunami sources starts from observed tsunami data (Fig. 7.1). The propagation process can be evaluated by using numerical simulation, in the same way as in the forward modeling. As the observed tsunami data, tsunami arrival times, heights, or waveforms recorded on instruments are used. For historical tsunamis, tsunami heights can be estimated from damage descriptions in historical documents. For prehistoric tsunamis, tsunami heights or flooding areas can be estimated from geological studies of tsunami deposits.

The inversion of tsunami waveforms recorded by tide gauges was first proposed by Satake (1987) to estimate the slip distribution on the fault. In this method (Fig. 7.2), the fault plane is first divided into several subfaults, and the

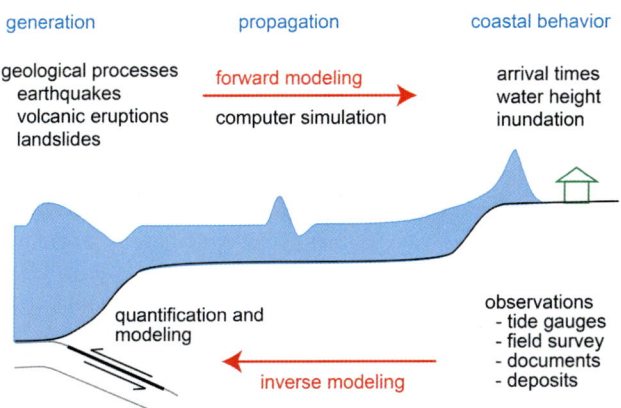

FIGURE 7.1 Tsunami forward and inverse problems.

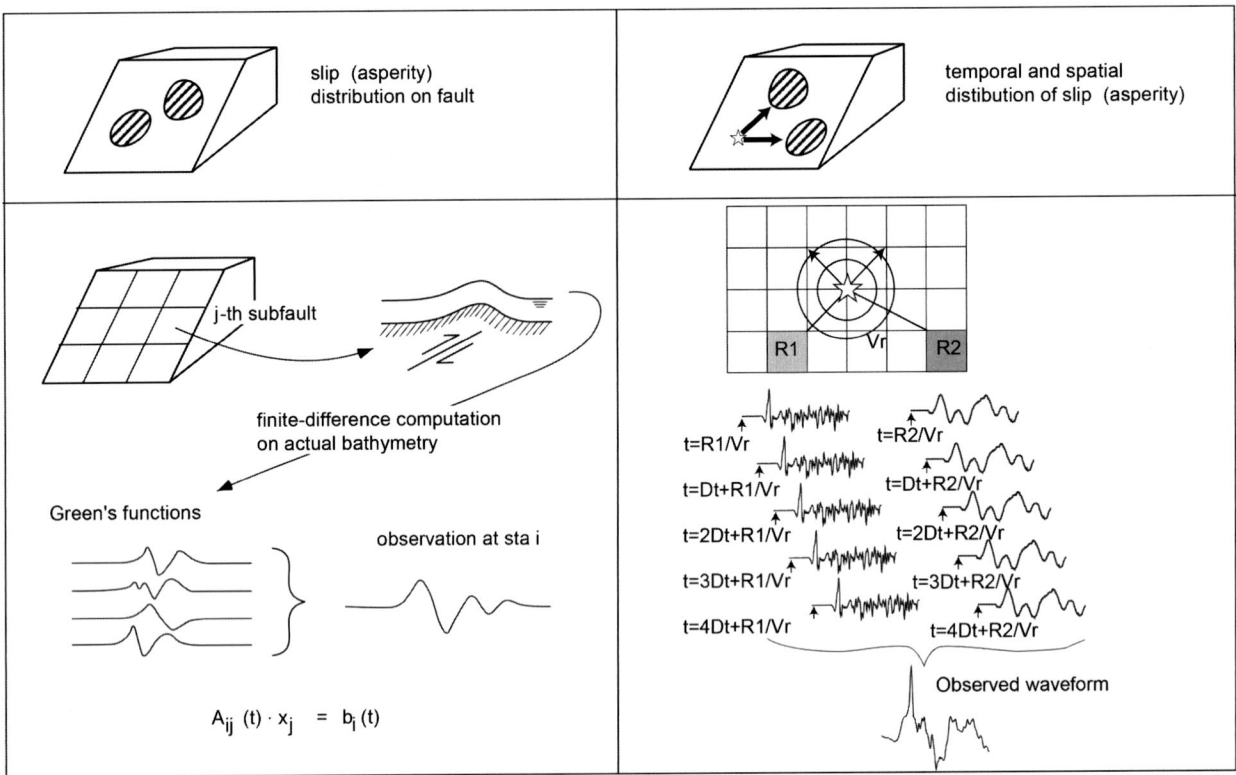

FIGURE 7.2 Inversion for earthquake slip distribution (left) and temporal and spatial distribution (right). *Modified from Satake, K., Fujii, Y., Harada, T., & Namegaya, Y. (2013). Time and space distribution of coseismic slip of the 2011 Tohoku earthquake as inferred from Tsunami waveform data.* Bulletin of the Seismological Society of America, 103(2B), 1473–1492. https://doi.org/10.1785/0120120122; Satake, K. (1987). *Inversion of tsunami waveforms for the estimation of a fault heterogeneity: Method and numerical experiments.* Journal of Physics of the Earth, 35(3), 241–254, https://doi.org/10.4294/jpe1952.35.241.

seafloor displacement is computed for each subfault with a unit amount of slip. Using this as an initial condition, tsunami waveforms are numerically computed for actual bathymetry. The observed tsunami waveforms are expressed as a superposition of waveforms computed for each subfault as

$$A_{ij}(t) \times x_j = b_i(t) \tag{7.1}$$

where $A_{ij}(t)$ is the computed waveform as a function of time t, or the Green function, at the ith station from the jth subfault, x_j is the amount of slip on the jth subfault, and $b_i(t)$ is the observed tsunami waveform at the ith station. The slip x_j on each subfault can be estimated by the least-squares inversion of the aforementioned set of equations. For the 2011 Tohoku Earthquake (Satake et al., 2013), because high-quality and high-sampling offshore tsunami waveforms were available, the temporal change and spatial distribution of the slips on subfaults were estimated by multiple time window inversion (Fig. 7.2).

Inverse modeling of tsunami waveforms has been further developed. Gusman et al. (2015) applied a multiple tide window inversion to the tsunami and GPS data from the 2014 Iquique Earthquake in Chile and compared the results with the teleseismic inversion. By varying the assumed rupture velocity, they found that the estimated moment rate functions are stable for the teleseismic inversion, while the estimated slip distributions are stable for the joint inversion of the tsunami and GPS data. In other words, the inversion of seismic waves provides a better temporal resolution, while the joint inversion of the tsunami and geodetic data provides a better spatial resolution. Gusman, Mulia, et al. (2018) proposed a two-step inversion of tsunami waveforms; in the first step, optimum sea surface displacement is estimated from observed tsunami waveforms, and then the fault slip distribution is estimated from the optimum sea surface displacement for various fault models.

Hossen et al. (2015) proposed time reversal imaging to reconstruct the tsunami source, in which the tsunami simulation is made from time-reversed tsunami waveforms. Mulia et al. (2018) further extended this to the adaptive tsunami source inversion method, which consists of two steps. In the first step, the reciprocity principle is incorporated for the computations of tsunami Green's functions, saving computational efforts for optimizing unit source locations. In the second step, an optimization scheme to search for the appropriate fault parameters is introduced.

7.2 Tsunami waveforms recorded on instruments

In this section, various modern instrumental measurements of tsunamis (Fig. 7.3) are described with examples of the 2011 Tohoku Tsunami.

7.2.1 Coastal sea-level measurements

Coastal sea-level monitoring has been conducted for more than a century for various research purposes in oceanography, climate, geophysics, and geodesy (Intergovernmental Oceanographic Commission, 2006). The Global Sea Level Observation System (GLOSS) was established by the UNESCO Intergovernmental Oceanographic Commission (IOC) and is currently formed by over 90 nations across the globe, consisting of approximately 300 tide gauge stations.

The station details (i.e., location, type of instruments, and datum) of the GLOSS stations are given at https://gloss-sealevel.org/gloss-station-handbook. The real-time sea-level data, or stored tsunami event data, are available at the UNESCO IOC site (http://www.ioc-sealevelmonitoring.org/). Locations of tide gauge stations are given as longitude and latitude with variable accuracy. For modeling tsunamis recorded on coastal tide gauges, the detailed location of the gauge, in relation to port or harbor shapes or coastal structures, such as breakwater, is very important because tsunami waveforms would be different, depending on whether the location is inside or outside the port or harbor.

The conventional tide gauge is a mechanical device that detects the sea-level change by a float set in a stilling well. The movements of the float had been recorded on paper charts until digital recording became popular in the last decades of the 20th century. More recent gauges include pressure, acoustic, and radar measurements. Because of low cost and ease of maintenance, radar gauges became popular in the 21st century (Intergovernmental Oceanographic Commission, 2016). In Japan, nearly 200 coastal tide gauge stations are operated by the Japan Meteorological Agency (JMA), Japan Coast Guard (JCG), Geospatial Information Authority of Japan (GSI), and Ministry of Land, Infrastructure, Transport and Tourism (MLIT) (Fig. 7.4).

The sampling interval of tide gauges varies, depending on the purposes of sea-level monitoring. Among various targets, the tsunami observation requires a short sampling interval (typically 1 minute or less), while the sampling interval for other phenomena can be longer, 1 month for sea-level change or crustal deformation, 1 day for ocean surface currents, and 15 minutes for tidal processes or storm surges.

The tide gauge data contain various components other than tsunami signals. The longer and tidal components can be removed by subtracting theoretically computed ocean tides or applying a high-pass filter. A simpler way is by fitting the tidal component with a polynomial function and then subtracting it from the original record. Shorter period noises, for example, seismic waves, can also be removed by applying a low-pass filter, or a simple moving average.

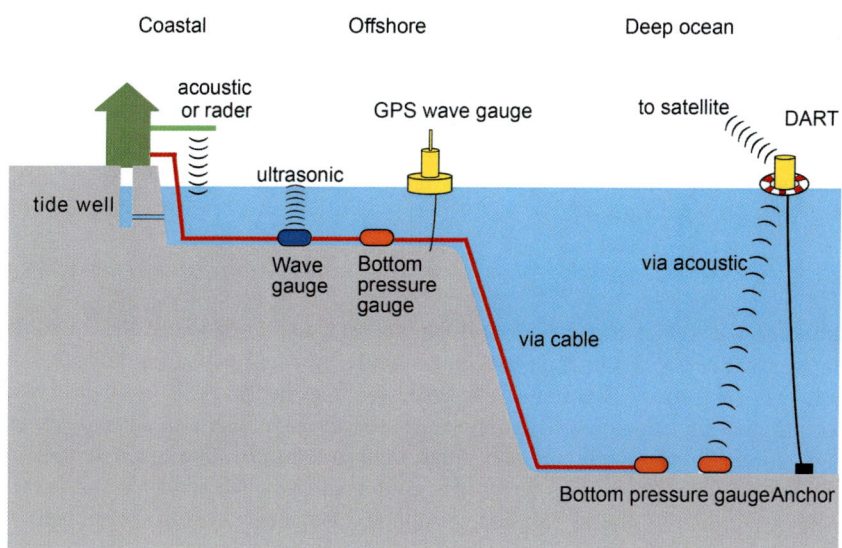

FIGURE 7.3 Tsunami measurement systems (Satake, 2014). *From Satake, K. (2014). Advances in earthquake and tsunami sciences and disaster risk reduction since the 2004 Indian Ocean tsunami. Geoscience Letters, 1(1). https://doi.org/10.1186/s40562–014-0015–7.*

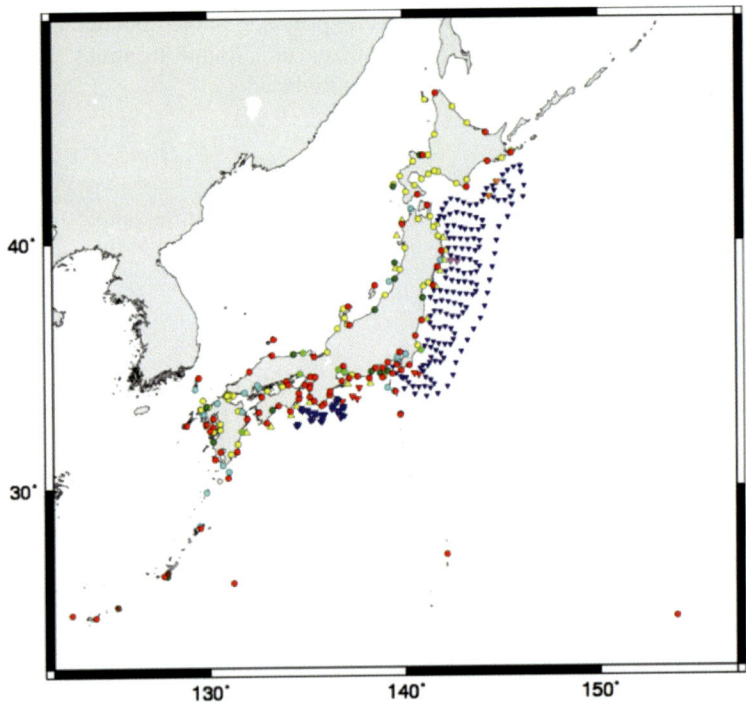

FIGURE 7.4 Tsunami observation stations in Japan, consisting of coastal tide gauges (circles: colors are different for different operating organizations), offshore wave gauges (triangles), and cabled bottom pressure gauges (inverse triangles). Map lines delineate study areas and do not necessarily depict accepted national boundaries. *https://www.data.jma.go.jp/eqev/data/tsunamimap/index.html.*

Fig. 7.5 (left) shows the 2011 Tohoku Tsunami recorded on three coastal tide gauges. The tsunami arrived at these coastal gauges at around 30 minutes after the earthquake, but the large tsunami destroyed the system, hence only the beginning parts of the tsunami waveforms were recorded, and the maximum tsunami amplitudes were not recorded. The 2011 Tohoku Tsunami was also recorded on coastal tide gauges across the Pacific Ocean (Heidarzadeh & Satake, 2013).

7.2.2 Offshore tsunami measurements

As a tsunami becomes slow but large and complex along coasts, it is ideal to monitor and detect its arrival offshore. Offshore wave gauges use ultrasonic waves to measure water height variation and its period. Around Japan, approximately 60 wave gauges are installed outside harbors at water depths of tens of meters as a part of the NOWPHAS system (https://nowphas.mlit.go.jp/). The NOWPHAS also includes 15 GPS wave gauges that have been installed typically 10 km offshore at water depths of about 100–200 m (Kato et al., 2000). These gauges use the Real-Time Kinematic GPS technique to estimate the location and altitude of the water surface. The sampling interval is 1 second with an accuracy of a few centimeters. Because it measures water surface movements, it is free from seismic waves, unlike bottom pressure gauges, and can also be used to monitor wave heights during storms. The 2011 Tohoku Tsunami was recorded on nearshore GPS wave gauges around Japan. Fig. 7.5 shows the record off Kamaishi about 10 km at a water depth of 200 m, approximately 12 minutes after the earthquake before its arrival on the coast.

7.2.3 Cabled bottom pressure gauges

In the deep ocean, tsunami waveforms are expected to be simpler (i.e., free from coastal topographic effects), although their amplitudes are smaller. Deep ocean measurements of tsunamis have been made by using ocean bottom pressure (OBP) gauges for early detection and warnings of a tsunami (Rabinovich & Eblé, 2015). A quartz crystal sensor is typically used as a pressure sensor. The sensor provides an accuracy of 1 cm in sea level at 6000 m depth. Because the pressure transducer is very sensitive to temperature, the measurements and corrections for temperature are essential. In addition, because the sensor is located at the ocean bottom, it also records ground motions and works as an ocean-bottom seismograph. Although the frequencies are often different between seismic and tsunami waves, it is necessary to separate them.

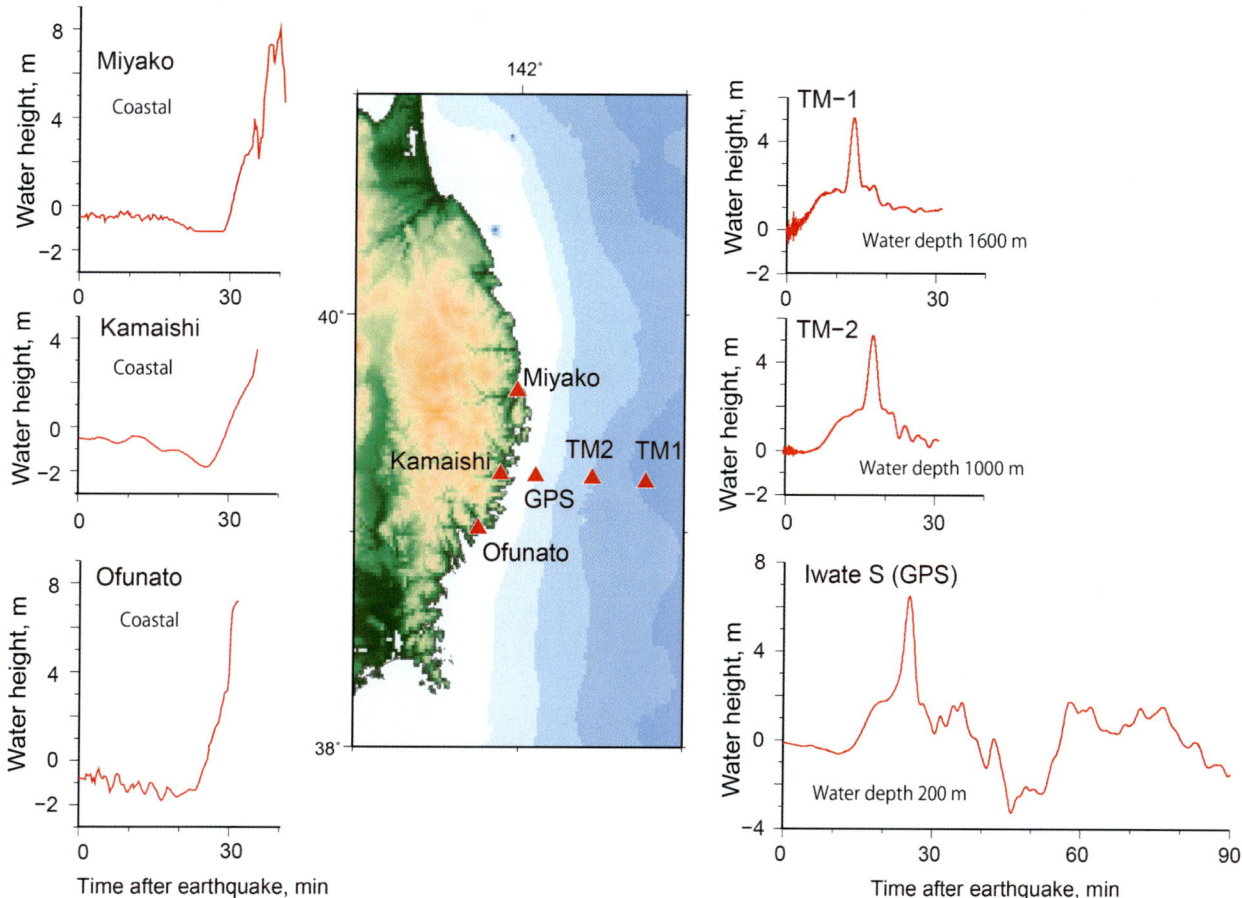

FIGURE 7.5 The 2011 Tohoku Tsunami recorded on cabled bottom pressure gauges (TM1 and TM2), GPS buoy (Iwate S), and three coastal gauges (Myako, Kamaishi, and Ofunato). TM1 and TM2 are operated by ERI, the University of Tokyo, GPS buoy belongs to NOWPHAS, Miyako and Ofunato tide gauges are by JMA, and Kamaishi tide gauge is by JCG. *ERI*, Earthquake Research Institute; *JMA*, Japan Meteorological Agency; *JCG*, Japan Coast Guard. Map lines delineate study areas and do not necessarily depict accepted national boundaries. *From Satake, K., Fujii, Y., Harada, T., & Namegaya, Y. (2013). Time and space distribution of coseismic slip of the 2011 Tohoku earthquake as inferred from Tsunami waveform data. Bulletin of the Seismological Society of America, 103(2B), 1473–1492. https://doi.org/10.1785/0120120122.*

Around Japan, several cabled OBP gauge networks have been constructed since the 1970s (Mulia & Satake, 2020; Rabinovich & Eblé, 2015). They are in operation by the JMA, the Earthquake Research Institute (ERI) of the University of Tokyo, the National Research Institute for Earth Science and Disaster Resilience (NIED), and the Japan Agency for Marine-Earth Science and Technology.

Off the Sanriku Coast, two cabled gauges (TM1 and TM2) were installed by the ERI in 1998. The 2011 Tohoku Tsunami was first recorded by offshore gauges (Fig. 7.5). On cabled OBP TM1, located at about 76 km off the coast of Sanriku at a water depth of 1600 m, a water rise of ~2 m was recorded immediately after the earthquake, followed by an impulsive wave that caused an additional 3 m rise within 2 minutes (Fujii et al., 2011). A very similar two-stage tsunami waveform was recorded on TM2, located 40 km offshore at a water depth of 1000 m, with about a 5-minute delay.

Following the 2011 Tohoku Earthquake Tsunami, the NIED installed the seafloor observation network for earthquakes and tsunamis along the Japan Trench (S-net), consisting of 150 observatories equipped with both OBP gauges and ocean bottom seismometers (OBSs) (Fig. 7.4). S-net recorded tsunamis from the 2016 Fukushima Earthquakes (Wang et al., 2022) as well as the 2022 Tonga Eruption (Kubota et al., 2022).

Along the Nankai Trough, the Deep Ocean-floor Network system for Earthquakes and Tsunamis (DONET) consists of 51 OBPs as well as OBSs. To the west of DONET, off the Shikoku area, a new N-net is under construction. The DONET detected tsunamis from the 2015 Torishima Volcanic Earthquake (Sandanbata et al., 2022), the 2011 Tohoku Earthquake (Nosov et al., 2018), and the 2022 Tonga Eruption (Wang et al., 2022). The waveform data of DONET and S-net are available from https://www.seafloor.bosai.go.jp.

7.2.4 Deep ocean measurements: offline and deep-ocean assessment and reporting of tsunamis

Marine seismological campaigns with OBSs typically deploy absolute or differential pressure gauges that record tsunamis. The array of pressure gauges recorded tsunamis from the 2009 Dusky Sound Earthquake in New Zealand (Sheehan et al., 2019) and the 2012 Haida Gwaii Earthquake (Fig. 7.6; Sheehan et al., 2015) and provided valuable data to test the tsunami data assimilation (Gusman et al., 2016).

The deep-ocean assessment and reporting of tsunamis (DART), or simply tsunameter, records water levels using bottom pressure gauges and sends signals to a surface buoy via acoustic telemetry in the ocean, then via satellites to a land station in real time (Fig. 7.3). The first generation DART system started measurements in 2000, and six sensors were deployed in the northeast Pacific Ocean off Alaska (González et al., 2005). Following the 2004 Indian Ocean Tsunami, the new DART systems have been developed, and the 4th generation DART system, consisting of approximately 40 stations, is now in operation. The real-time data can be accessed at the National Data Buoy Center (https://www.ndbc.noaa.gov/). The DART system is now exported to other countries. The DART stations consist of networks operated by Australia, New Zealand, Chile, Thailand, and India (Fig. 7.7).

In the DART system, the sea surface heights (SSHs) are estimated from pressure and temperature at 15-second intervals, but the system sends the data at 15-minute intervals in its standard transmission mode. When a tsunami event is detected, the system is switched to an event mode, and 15-second values are transmitted during the initial few minutes, followed by 1-minute averages. The system returns to the standard transmission after 4 hours of event mode transmission.

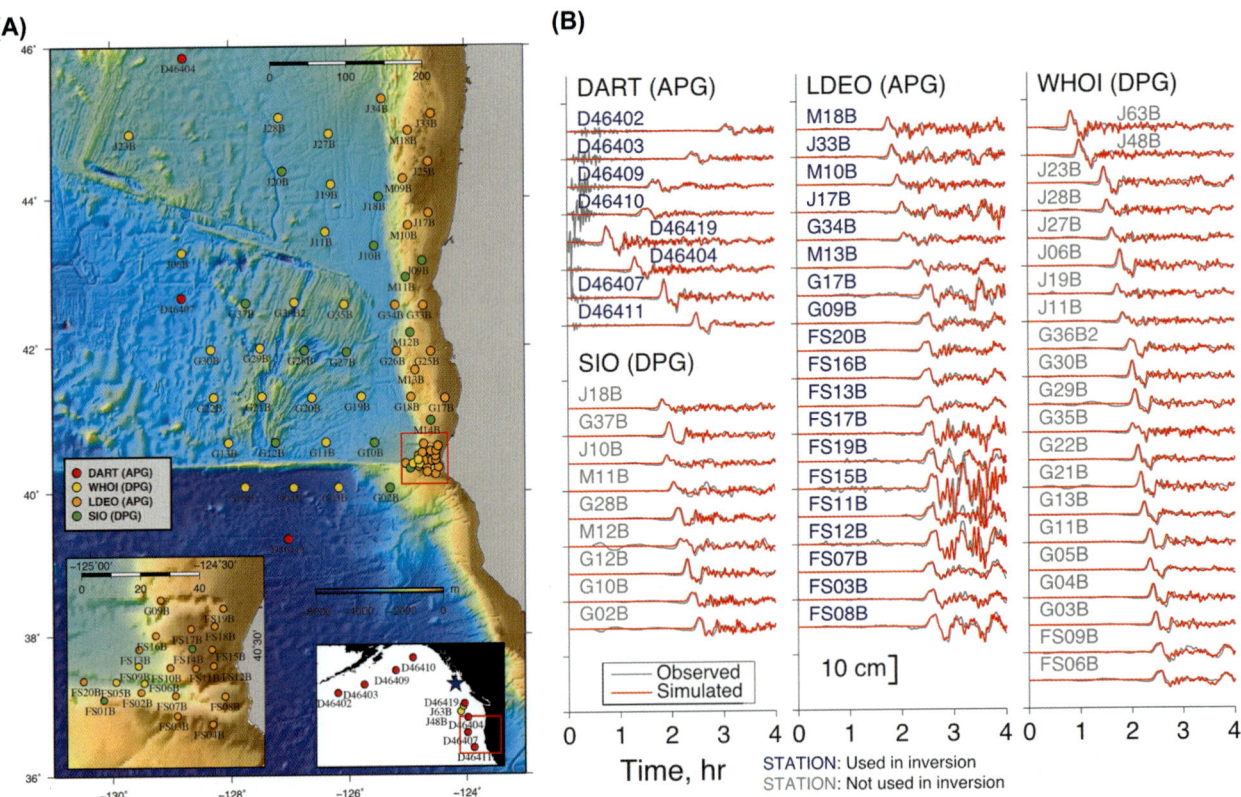

FIGURE 7.6 Tsunami observation and simulation of the 2012 Haida Gwaii Earthquake at OBSs and DART stations (Sheehan et al., 2015). The OBSs were equipped with APGs or DPGs. (A) Distribution of DARTs, APGs, and DPGs around the Cascadia and Alaska subduction zones. The blue star represents the earthquake epicenter. (B) Comparison of observed (gray lines) and simulated (red curves) tsunami waveforms from the slip distribution of the earthquake estimated by Gusman et al. (2016). Station names at which tsunami waveforms are used to estimate the earthquake source in the study are written in blue. *APGs*, Absolute pressure gauges; *DART*, deep-ocean assessment and reporting of tsunamis; *DPGs*, differential pressure gauges; *OBSs*, ocean bottom seismometers. Map lines delineate study areas and do not necessarily depict accepted national boundaries. *From Gusman, A. R., Sheehan, A. F., Satake, K., Heidarzadeh, M., Mulia, I. E., & Maeda, T. (2016). Tsunami data assimilation of Cascadia seafloor pressure gauge records from the 2012 Haida Gwaii earthquake. Geophysical Research Letters, 43(9), 4189–4196. https://doi.org/10.1002/2016GL068368.*

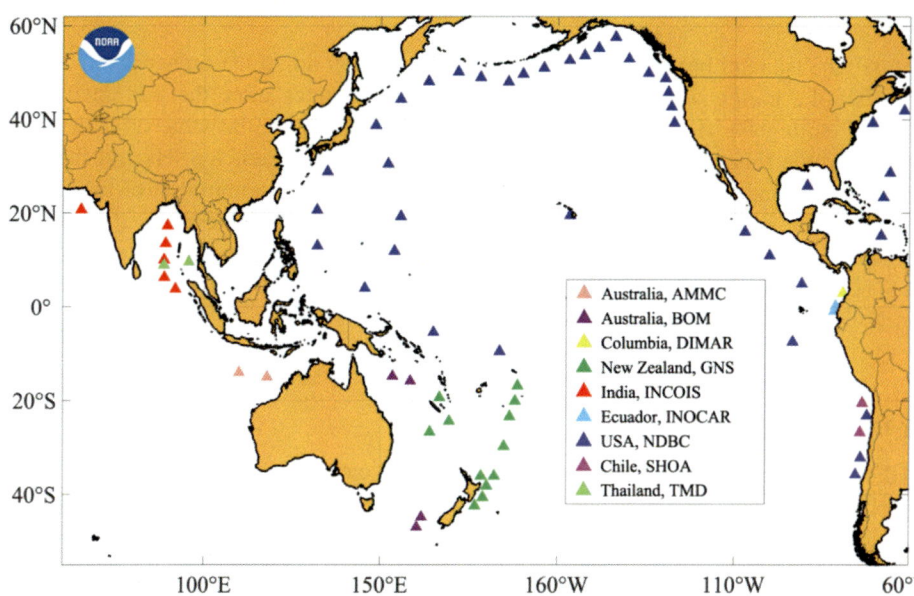

FIGURE 7.7 Location of DART gauges with operation agencies. DART, Deep-ocean assessment and reporting of tsunamis. Map lines delineate study areas and do not necessarily depict accepted national boundaries. *https://nctr.pmel.noaa.gov/Dart/*.

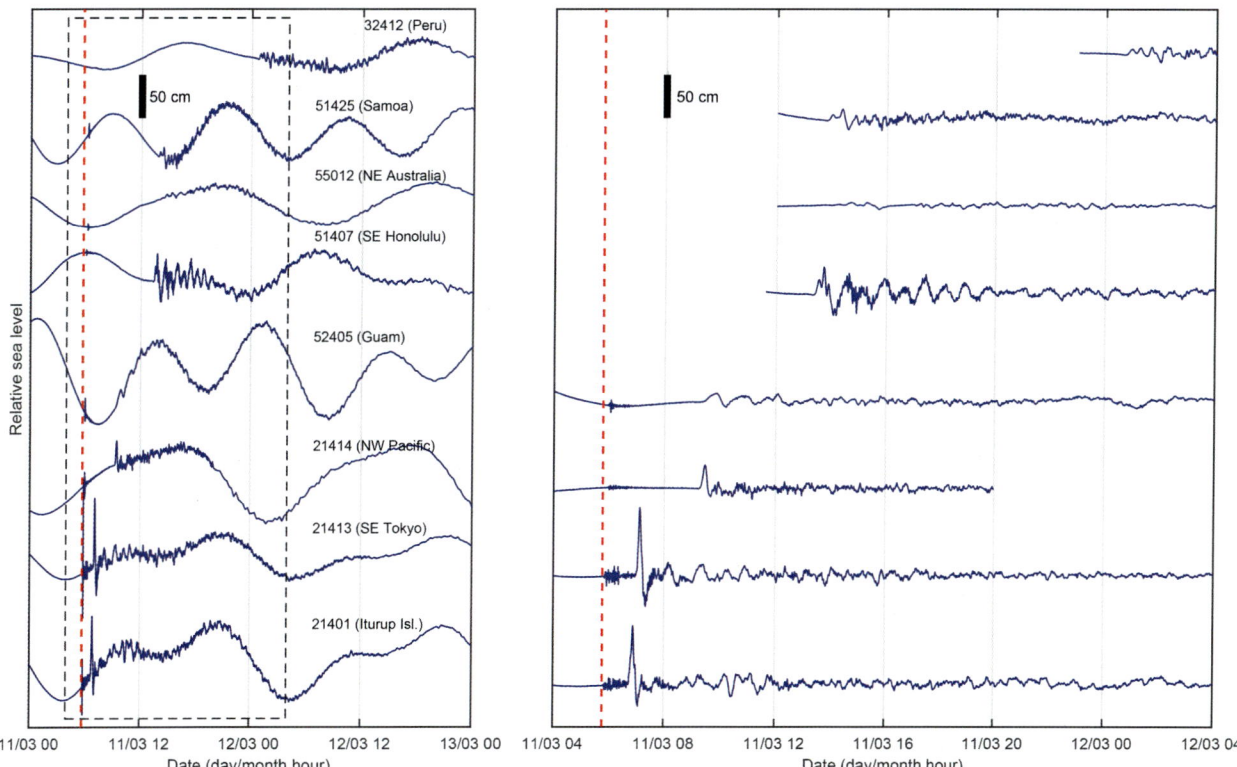

FIGURE 7.8 Original and filtered signals of the 2011 Tohoku Tsunami recorded on DART stations across the Pacific Ocean (Heidarzadeh & Satake, 2013). The red-vertical line represents the time of the earthquake. The dashed rectangle shows part of the waveform enlarged in the right panel. *From Heidarzadeh, M., & Satake, K. (2013). Waveform and spectral analyses of the 2011 Japan tsunami records on tide gauge and DART stations across the Pacific Ocean. Pure and Applied Geophysics, 170(6–8), 1275–1293. https://doi.org/10.1007/s00024–012-0558–5.*

The 2011 Tohoku Tsunami was recorded on DART gauges around the Pacific Ocean (Fig. 7.8). Saito et al. (2013) detected a signal of tsunami reflected on the Chilean coast at around 48 hours after the earthquake. They also estimated the decay of tsunami energy in the Pacific Basin.

7.2.5 Satellite observations

Four satellites with altimeters captured the tsunami propagation across the Indian Ocean on 26 December 2004; Jason-1 and TOPEX/Poseidon at 2 hours after the earthquake, ENVISAT at 3 hours 15 minutes after the earthquake, and Geostat Follow-On at 7 hours after the earthquake (Smith et al., 2005). These satellites fly in the same orbit in 10–35 days. Hayashi (2008) applied a multisatellite time-spatial interpolation method to define reference SSHs and then estimated the tsunami height profiles by subtracting the references from observed data. The maximum SSH of the tsunami was estimated as 1.1 m with 4–5 cm of root mean square errors and a spatial resolution of 5 km. For the 2011 Tohoku Tsunami, Jason-1 detected a propagating tsunami at about 7.5 hours after the earthquake, Jason-2 at about 8.3 hours, and ENVISAT at about 5.3 hours. The maximum amplitudes were approximately 0.2 m in the Pacific Ocean (Ho et al., 2017) (Fig. 7.9).

7.2.6 Oceanographic radar measurements

Oceanographic high-frequency (HF) radar, installed on land, can measure ocean currents over a large region of the coastal ocean, from a few kilometers up to about 200 km offshore, under any weather conditions. A unique aspect of radar measurements is that they can cover a wide range, thus spatial distribution of sea level, or tsunami wave, distribution can be captured. In addition, land-based station requires less maintenance compared to instruments in the ocean.

An HF (24.5 MHz) ocean surface radar installed on the eastern coast of Kii Channel detected the 2011 Tohoku Tsunami entering the channel, about 1.5 hours after the earthquake (Hinata et al., 2011). Temporal changes in radial components of surface current velocities at 12 km offshore from the radar are compared with tsunami waveforms recorded at GPS buoy, offshore wave gauge, and coastal tide gauge station (Fig. 7.10), indicating that the radar measurements can be used to monitor and detect tsunamis before its coastal arrival.

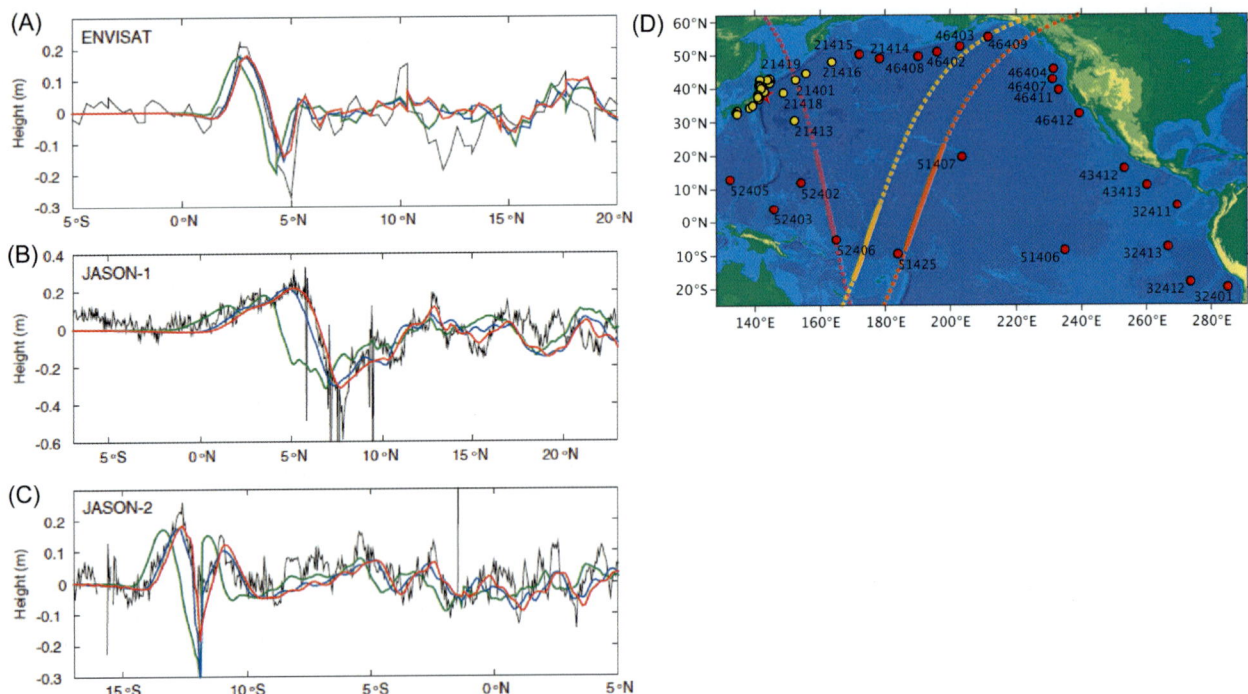

FIGURE 7.9 (A) The 2011 Tohoku Tsunami observed as sea surface height (black line) by ENVISAT with synthetic ones using different models (*red, cyan,* and *green* lines). (B) The same for Jason-1. (C) The same for Jason-2. Locations of DART stations (*red* circles), coastal tide stations (*yellow* circles), and paths of three satellites (ENVISAT (*magenta*), Jason-1(*orange*), and Jason-2(*yellow*)). DART, Deep-ocean assessment and reporting of tsunamis. (D) The magenta, orange, and yellow dashed paths indicate the satellite tracks of ENVISAT, Jason-1, and Jason-2, respectively. Map lines delineate study areas and do not necessarily depict accepted national boundaries. *From Ho, T.-C., Satake, K., & Watada, S. (2017). Improved phase corrections for transoceanic tsunami data in spatial and temporal source estimation: Application to the 2011 Tohoku earthquake. Journal of Geophysical Research: Solid Earth, 122(12), 10,155–10,175. https://doi.org/10.1002/2017jb015070.*

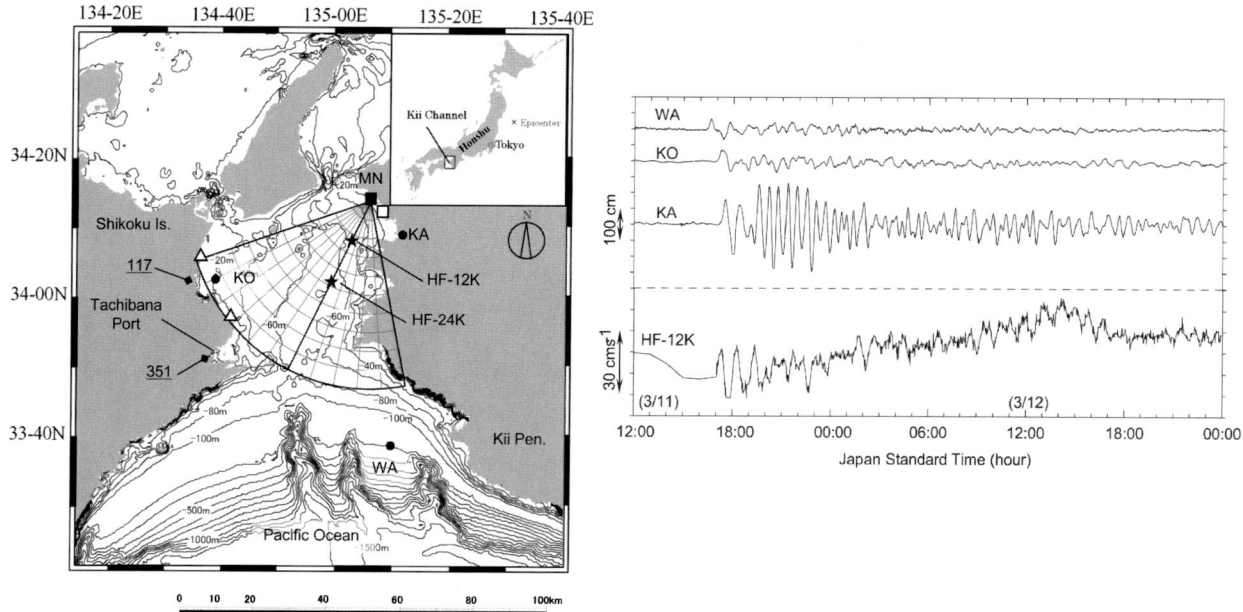

FIGURE 7.10 Map of high-frequency radar systems (squares and triangles) in the Kii Channel (Hinata et al., 2011). Locations of GPS buoy (WA), offshore wave gauge (KO), and coastal tide gauge station (KA) are also shown. Detided sea surface heights at WA, KO, KA, and radial velocity at HF-12K (12 km offshore from the radar) for the 2011 Tohoku Tsunami. Map lines delineate study areas and do not necessarily depict accepted national boundaries. *From Hinata, H., Fujii, S., Furukawa, K., Kataoka, T., Miyata, M., Kobayashi, T., Mizutani, M., Kokai, T., & Kanatsu, N. (2011). Propagating tsunami wave and subsequent resonant response signals detected by HF radar in the Kii Channel, Japan. Estuarine, Coastal and Shelf Science, 95(1), 268–273. https://doi.org/10.1016/j.ecss.2011.08.009.*

7.3 Source models of the 2011 Tohoku Earthquake based on the inversion of tsunami data

Numerous source models of the 2011 Tohoku Earthquake have been proposed, based on seismic, geodetic, and tsunami data. Maeda et al. (2011) used tsunami waveforms recorded on bottom pressure gauges to model the tsunami source. Fujii et al. (2011) inverted tsunami waveforms recorded on coastal gauges, GPS buoys, and bottom pressure gauges. Saito et al. (2011) inverted waveforms on GPS buoys and bottom pressure gauges. Gusman et al. (2012) inverted tsunami waveforms recorded on coastal tide gauges, GPS buoys, and cabled bottom pressure gauges, jointly with crustal deformation data recorded on GPS stations, and estimated the slip distribution with the largest slip of more than 40 m. These source models indicate the largest slip near the epicenter at ~38°N, while the measured tsunami height distribution showed a peak near Miyako at ~39.6°N (Fig. 7.11).

While the above models assumed instantaneous rupture on an earthquake fault, Satake et al. (2013) considered temporal change as well as the spatial distribution of fault slip in the source inversion and showed that the northern part of the source along the Japan Trench ruptured more than 3 minutes after the earthquake origin time. The delayed tsunami generation was required to better explain offshore tsunami waveforms and the maximum coastal tsunami heights observed near Miyako. Because the earthquake source models based on seismic and geodetic data do not require such a source in the north, the discrepancy was controversial (Lay et al., 2013; Satake & Fujii, 2014).

To resolve the controversy, Yamazaki et al. (2018) started with a slip model based on seismic waves, modified it to satisfy the tsunami data (waveforms and coastal heights), and obtained a model with a slip extending north to 39.5°N. They showed that this model, similar to that of Satake et al. (2013), can explain the seismic waves and crustal deformation data. In other words, the slip near the trench axis at 39.5°N cannot be resolved from seismic or inland geodetic data.

Ho et al. (2017) inverted near- and far-field (trans-Pacific) tsunami waveforms as well as SSH data obtained by satellites (Fig. 7.9). Simulation of trans-Pacific tsunami propagation has been improved to include the effects of elastic Earth and seawater (Watada et al., 2014). The inversion result from only far-field data showed a similar but smoother sea surface displacement than that from near-field data and all stations, including a large sea surface rise increasing toward the trench followed by a northward propagation of displacement along the trench.

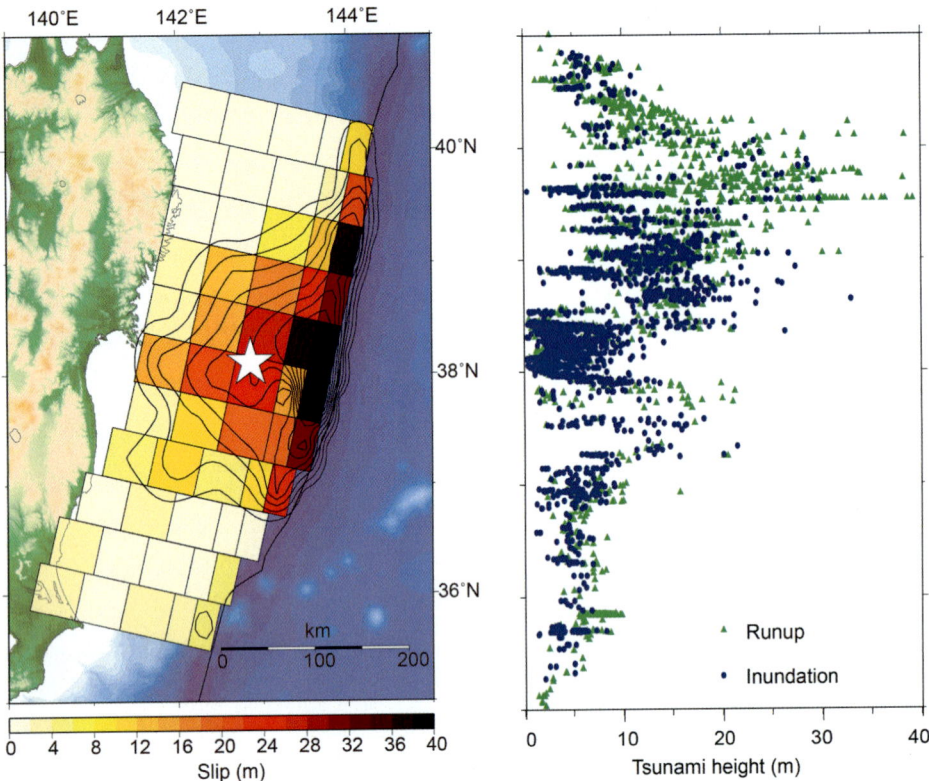

FIGURE 7.11 (Left) Slip distribution on subfaults of the 2011 Tohoku Earthquake based on inversion of tsunami waveforms (Satake et al., 2013). (Right) Coastal tsunami height, runup, and inundation height distributions by the Joint Survey Group (Mori et al., 2011). Map lines delineate study areas and do not necessarily depict accepted national boundaries. *From Satake, K., Fujii, Y., Harada, T., & Namegaya, Y. (2013). Time and space distribution of coseismic slip of the 2011 Tohoku earthquake as inferred from Tsunami waveform data. Bulletin of the Seismological Society of America, 103(2B), 1473–1492. https://doi.org/10.1785/0120120122.*

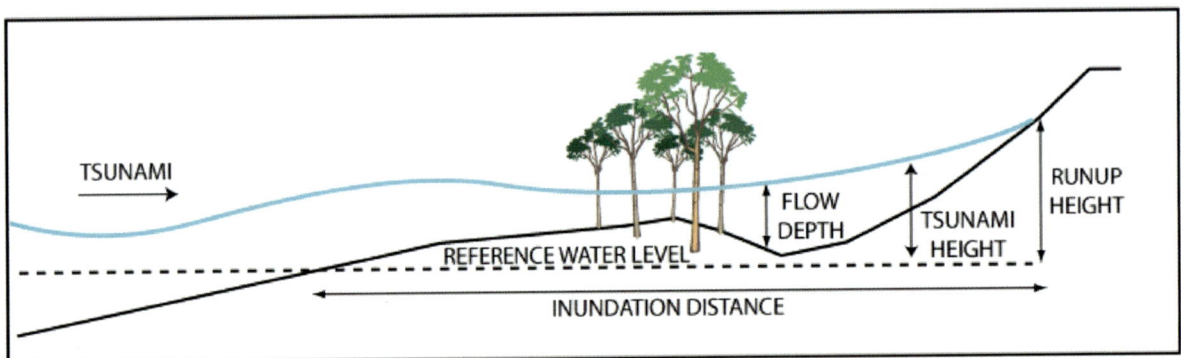

FIGURE 7.12 Definitions of tsunami heights and inundation distances on land. *From Intergovernmental Oceanographic Commission (2014). Post-tsunami Survey Field Guide (2nd ed.). UNESCO. https://unesdoc.unesco.org/ark:/48223/pf0000229456.*

A high-resolution slip distribution of the 2011 Tohoku Earthquake was acquired through an inversion analysis using tsunami waveforms. These waveforms were captured by temporarily deployed OBP sensors positioned both near the source area and at distant monitoring stations (Kubota et al., 2022). While the overall picture (very large coseismic slip near the trench axis) is similar to previous studies, the high-resolution distribution allowed them to calculate the stress drop distribution, which shows that high stress drop occurred at the deeper (between 10 and 20 km) region.

7.4 Field surveys to measure tsunami heights

Posttsunami field surveys are often conducted after the occurrence of large tsunamis (Intergovernmental Oceanographic Commission, 2014). The primary purpose of these surveys is to measure the tsunami heights along the coast. Several terms are defined to describe the tsunami heights; hence it is essential to use the appropriate term for the types of measurements consistently (Fig. 7.12).

Tsunami height, or inundation height, is measured from the sea level at the time of the tsunami's arrival or a reference datum, whereas flow depth is the depth of the tsunami flood over the local terrain height. Because the tide levels at the time of measurement and at the arrival of a tsunami are usually different, corrections are required for the survey data. Runup is the maximum tsunami height on a sloping shoreline. In the simplest case, the runup value is recorded at the maximum inundation distance measured from the coast, flooded by the wave. In cases where the ground topography is flat, large tsunamis can penetrate for hundreds of meters or even several kilometers inland.

During field surveys, inundation heights, runup heights, or flow depths are measured based on damage to buildings or other structures, watermarks left on walls, vegetative markers that include trim lines and salt-water discoloration or die-off, and lines of debris deposited on beaches. Locations are usually identified by portable GPS measurements and the heights are measured by traditional leveling or laser surveying. When no physical evidence is found, the tsunami heights are estimated based on eyewitness accounts but are considered less reliable. Interviews with local people also include the degree of earthquake ground shaking, tsunami arrival times, the number of waves with periods, and the duration.

Significant subsidence or uplift in the aftermath of a tsunami often prompts measurements as part of tsunami survey efforts. It is crucial to measure significant coseismic uplift or subsidence because it can have a significant impact on the magnitude and extent of the tsunami. An illustration of this is the Indian Ocean Earthquake and Tsunami in 2004. The earthquake that caused the tsunami resulted in significant coseismic subsidence along the Aceh coast in Indonesia (Borrero, 2005). The subsidence of the coastal topography allowed the tsunami waves to travel further inland than they would have otherwise. On the other hand, the Kaikoura Earthquake, which occurred in New Zealand in 2016, resulted in significant uplift of the land along the coast (Gusman, Satake, et al., 2018). This uplift helped reduce the impact of the subsequent tsunami that was generated by the earthquake.

The tsunami heights of the 2011 Tohoku Tsunami on the Japanese coast were measured by the Joint Survey Group, which consisted of ~300 researchers from more than 60 organizations (Mori et al., 2011). The total number of measurement points exceeded 5900 (Fig. 7.11). The surveys showed that the tsunami heights varied locally, particularly along the sawtooth-shaped Sanriku Coast where the tsunami reached about 30 minutes after the earthquake with the maximum runup height of almost 40 m. The measured tsunami heights were compared with those of old surveys of the 1896 and 1933 Sanriku Tsunamis (Tsuji et al., 2014).

The inland extent of tsunami inundation was mapped by field surveys as well as aerial photographs. The total inundation area of the 2011 Tsunami was estimated to be 561 km^2 by the GSI. The tsunami reached the Sendai Plain about 1 hour after the earthquake and inundated areas as far as 5 km from the coast.

Fig. 7.13 shows the inundation area of the 2011 Tohoku Tsunami in the Sendai Plain, compared with the distribution of tsunami deposits from the 869 Jogan Tsunami. Mapping the tsunami inundation area provides important information

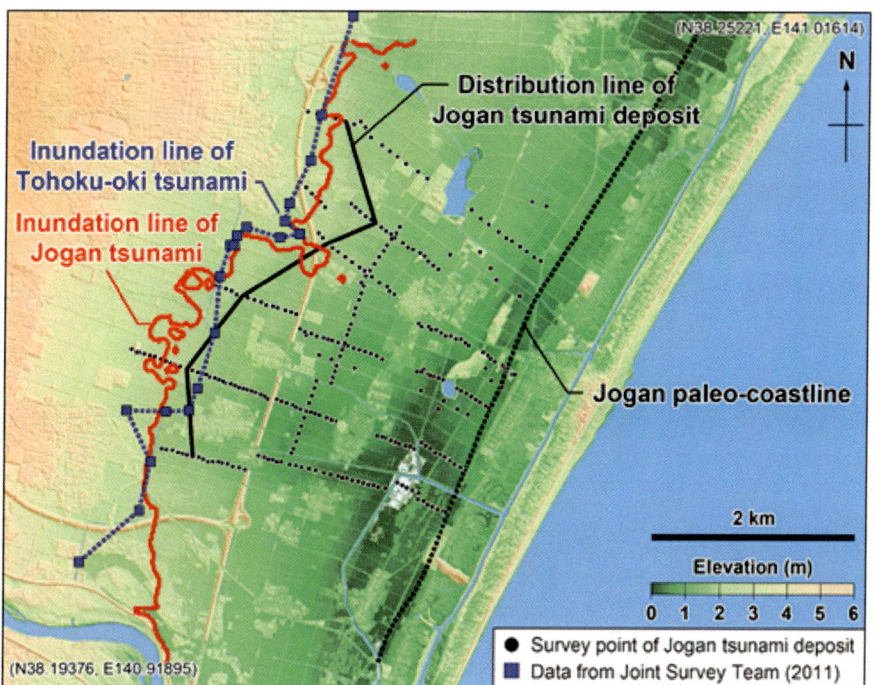

FIGURE 7.13 Surveyed inundation limit of the 2011 Tohoku Tsunami in the Sendai Plain, and distribution of tsunamis deposits of the 869 Jogan Tsunami. Map lines delineate study areas and do not necessarily depict accepted national boundaries. *From Sugawara, D., Goto, K., Imamura, F., Matsumoto, H., & Minoura, K. (2012). Assessing the magnitude of the 869 Jogan tsunami using sedimentary deposits: Prediction and consequence of the 2011 Tohoku-oki tsunami. Sedimentary Geology, 282, 14–26. https://doi.org/10.1016/j.sedgeo.2012.08.001.*

not only for modern tsunamis but also for past tsunamis. Moreover, tsunami inundation maps are often used as tsunami hazard maps for the preparation of future tsunamis.

For the damage caused by tsunamis, the MLIT of Japan collected building damage data, such as damage level, material of structure, and number of stories from more than 250,000 structures. The building damage data were grouped into six levels: (1) minor damage, (2) moderate damage, (3) major damage, (4) complete damage, (5) collapsed, and (6) washed away and compared with inundation depths.

Fig. 7.14 shows that about half of the buildings were washed away when the inundation (flow) depth was 3 m, and this ratio reached 80% for 5 m depth. The 250,000 structures were further grouped by the type of buildings (reinforced concrete, steel, wood, and masonry) or the number of stories, and the relations between inundation depth and damage probability, called tsunami fragility curve, are obtained (Suppasri et al., 2013). Such a fragility curve can also be used for both past and future tsunamis. For past tsunamis, tsunami heights can be estimated from descriptions of damage (assuming similar construction quality of the buildings). For future tsunamis, possible damage can be estimated once the tsunami height is calculated.

7.5 Historical data

Old tide gauge data, or marigrams, containing tsunami waveforms were preserved in microfiche in the 1970s, and the NOAA National Centers for Environmental Information (NCEI) converted tsunami marigrams from analog (microfiche) to high-resolution digital images. The NCEI collection (https://www.ngdc.noaa.gov/hazard/tide.shtml) includes records from 1854 to 1981. The tsunami waveforms, or time series data, further need to be extracted from the digital image. The oldest tsunami waveforms were recorded in San Francisco and San Diego from earthquakes along the Nankai Trough, Japan, in December 1854.

Tsunamis and their effects documented in historical records are also used to study old earthquakes and tsunamis for which no instrumental data are available. Some countries in Europe, South America, and Asia have a long history of written records that provide accurate dates of past earthquakes and tsunamis. For example, tsunami heights from the 1755 Lisbon Earthquake have been estimated from historical documents as 6 m in Lisbon, more than 15 m in southwestern Portugal, and more than 10 m in Spain and Morocco (Baptista et al., 1998).

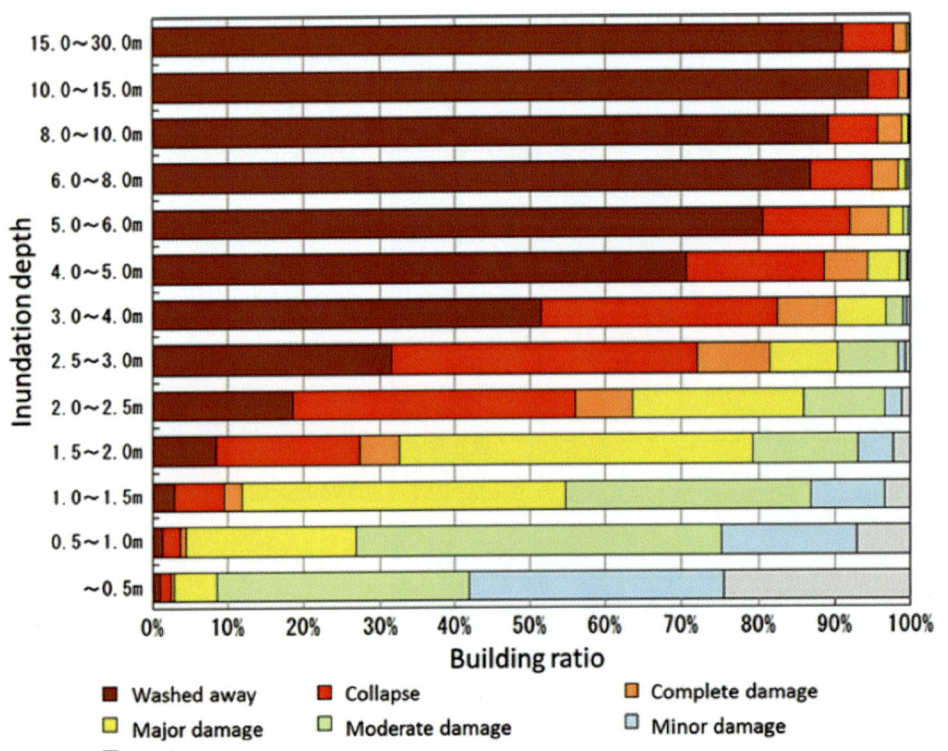

FIGURE 7.14 Distribution of the total 251,301 building data damaged during the 2011 Tohoku Tsunami surveyed by the MLIT and compiled by Suppasri et al. (2013). *MLIT*, Ministry of Land, Infrastructure, Transport and Tourism. *From Suppasri, A., Mas, E., Charvet, I., Gunasekera, R., Imai, K., Fukutani, Y., Abe, Y., & Imamura, F. (2013). Building damage characteristics based on surveyed data and fragility curves of the 2011 Great East Japan tsunami. Natural Hazards, 66(2), 319−341. https://doi.org/10.1007/s11069-012-0487-8.*

To estimate tsunami heights from historical damage descriptions, various assumptions need to be made. For example, if the building's strength against tsunamis is the same as modern buildings, the tsunami fragility curve described in the previous subsection can be used to estimate the inundation depth, and then the tsunami heights. In this section, historical data and the studies using them are described for four tsunami events that occurred in the southwest Japan and Tohoku regions before the 2011 Tsunami.

7.5.1 The 1854 Ansei Earthquakes and Tsunamis along Nankai Trough

The 1854 Ansei-Tokai Earthquake (M_w of 8.4) occurred on December 23 along the Nankai Trough off the coast of Japan. It was followed by the Ansei−Nankai Earthquake (M_w of 8.4) on the next day to the west of the first event. These earthquakes triggered large tsunamis that caused extensive damage and loss of life in coastal regions of Japan as described in many Japanese historical documents. It is also notable that the tsunami caused damage to a Russian frigate, which happened to be in Japan, and also was recorded on tide gauges on the west coast of the United States.

The origin time of the first earthquake was estimated between 9:00 and 10:00 a.m. (local time) on December 23 (Sugimori et al., 2022) based on descriptions in multiple documents. The timekeeping system used in Japan at that time had a resolution of 2 hours. At the time of the earthquake, a Russian frigate, Diana, was anchored in Shimoda Bay in the Izu Peninsula and experienced the earthquake and tsunami (Fig. 7.15). The frigate was damaged and eventually wrecked. The earthquake time as well as the tsunami arrival time and behavior was described in the logbook of the frigate, which had a time resolution of 15 minutes. According to the logbook, the earthquake occurred at 9:45 local time, and the tsunami arrived at 10:00 a.m. (Sugimori et al., 2022).

The origin time of the 1854 Tokai Earthquake was also estimated as 9:46 a.m. from the cross-correlation between the observed tide gauge record at San Francisco (Fig. 7.16) and synthetic tsunami waveforms (Kusumoto et al., 2020). The tsunami waveforms recorded in San Francisco also showed that the origin time difference between the 1854 Ansei−Tokai Earthquake and the 1854 Ansei−Nankai Earthquake was 30.9 hours (Kusumoto et al., 2022).

7.5.2 The 1896 Sanriku Tsunami

The 1896 Sanriku Tsunami occurred on June 15 and caused more than 20,000 casualties along the Sanriku Coast. This tsunami is sometimes called "Meiji Sanriku Tsunami," in contrast to the 1933 "Showa Sanriku Tsunami" which was also devastating with ∼3000 casualties.

FIGURE 7.15 Russian frigate Diana during the 1854 Tsunami. *"Shokoku umibe jishin tsunami gaki (Record on the damage caused by the 1854 Ansei Tokai and Nankai Earthquakes and Tsunamis)." Earthquake Research Institute, The University of Tokyo.*

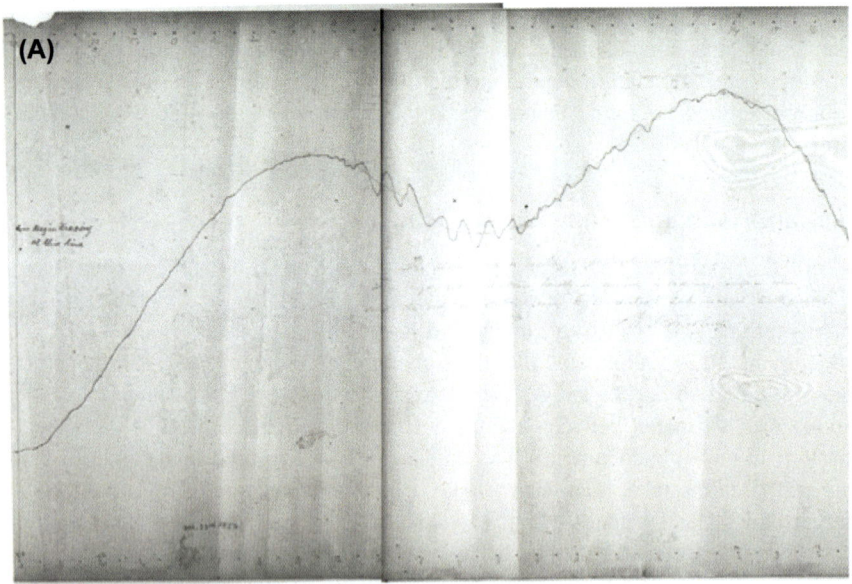
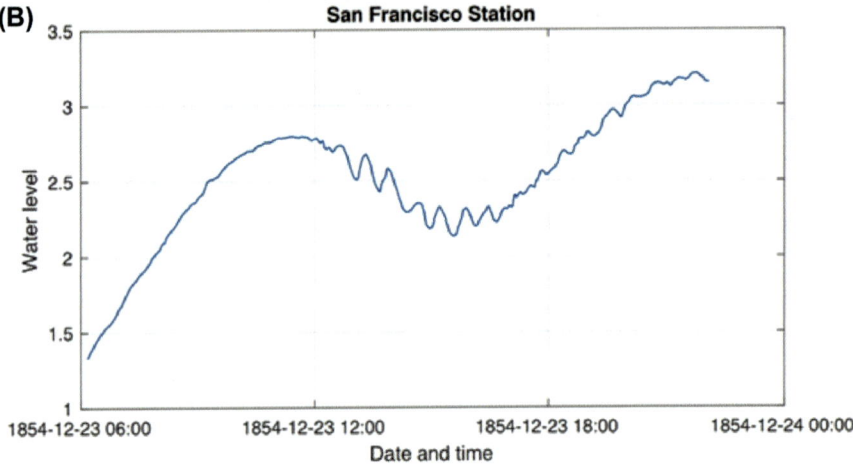

FIGURE 7.16 Tsunami record at San Francisco, U.S. tide gauge station on December 23, 1854, generated by the 1854 Tokai Earthquake in Japan. (A) Image of marigram of San Francisco tide gauge station from microfiche (created in the 1970s) of the original paper records. (B) The digitized water level of the tide gauge record. *The image and digitized data were made available by the National Centers for Environmental Information − National Oceanic and Atmospheric Administration.*

The 1896 Sanriku Tsunami was instrumentally recorded on three tide gauge stations at regional distances in Japan: Hanasaki, Ayukawa, and Choshi. Tanioka and Satake (1996) examined these waveforms, estimated the clock timing errors as large as 5 minutes, and modeled the waveforms from earthquake source models. A comparison of the 1896 Sanriku Tsunami and 2011 Tohoku Tsunami waveforms at these stations (Fig. 7.17) indicates that both periods and amplitudes of the 2011 waveforms are larger than those of the 1896 waveforms, reflecting the different sizes of the tsunami sources (Satake et al., 2017).

Field surveys to measure the tsunami heights and damage of the 1896 Tsunami were made by three groups (Tsuji et al., 2014). Yamana conducted a tsunami field survey from July through September of 1896 in all the villages along the Sanriku Coast, but the reliability of his measurements is very variable and questionable. Iki also conducted field surveys in June and July of 1896 along the Sanriku Coast and measured tsunami heights based on various kinds of traces and eyewitness accounts and assigned different reliability depending on the type of data. Following the 1933 Sanriku Tsunami, Matsuo conducted a field survey to measure the heights of both the 1896 and 1933 Tsunamis. The tsunami heights from the 1896 Sanriku Tsunami were measured 37 years after the event mostly based on eyewitness accounts. The coastal tsunami heights from the 1896 and 2011 events were similar along the northern Sanriku Coast (north of 39°N), while the 2011 heights were larger along the southern Sanriku Coast (Fig. 7.17).

The differences between the 1896 and 2011 Tsunami's heights, that is, similar on the Sanriku Coast but different at the regional distances, can be explained by different depths of the fault slips. While the 1896 Sanriku Tsunami was known as a "tsunami earthquake" and only a shallow part (3.5−7 km) of the plate interface slipped (Tanioka & Satake, 1996), the northern slip of the 2011 Tohoku Tsunami was even shallower at 0−3.5 km beneath seafloor (Satake et al., 2017).

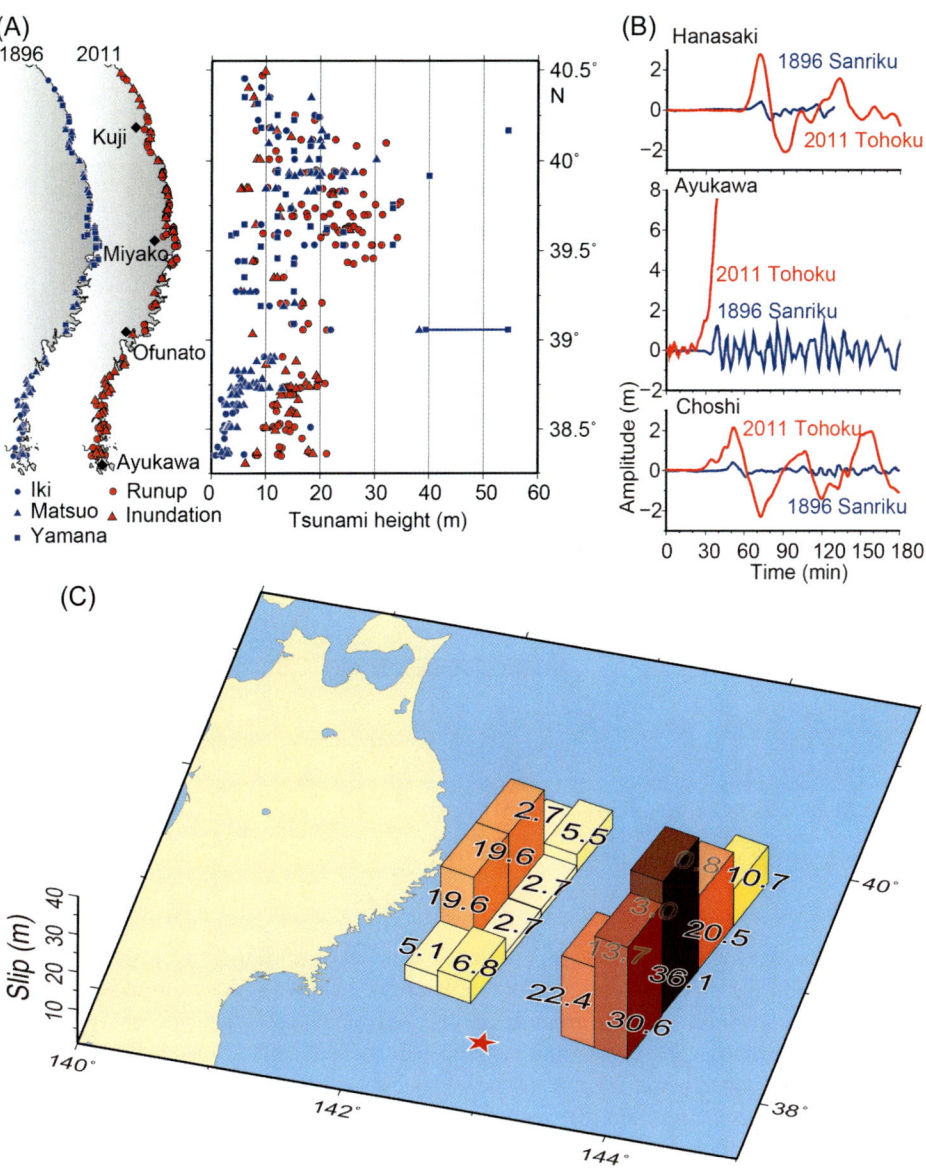

FIGURE 7.17 (A) Comparison of the 1896 and 2011 Tsunami heights on the Sanriku Coast (Satake et al., 2017). (B) Comparison of tsunami waveforms recorded at regional distances: Hanasaki, Ayukawa, and Choshi. (C) Slip distribution of the 1896 (left) and 2011 (right) Tsunamis. *From Satake, K., Fujii, Y., & Yamaki, S. (2017). Different depths of near-trench slips of the 1896 Sanriku and 2011 Tohoku earthquakes.* Geoscience Letters, 4 (1). https://doi.org/10.1186/s40562-017-0099-y.

7.5.3 The 1611 Keicho Tohoku Tsunami

On December 2, 1611 (October 28 in Keicho 11th year on the Japanese calendar), a large earthquake and tsunami occurred in the Tohoku region, and the tsunami damage was described in various historical documents (Ebina & Imai, 2014). The time of the earthquake and tsunami varied from 8:00 a.m. to 5:00 p.m. even in contemporary documents. Oral legends of this tsunami, such as inundation extent or tsunami heights, were also recorded in the survey reports of the 1896 Sanriku Tsunami. Ebina and Imai (2014) examined these documents, both contemporary and later, identified the locations, and measured the flow depths. Their estimated tsunami heights (Fig. 7.18) reached as high as 29 m at Koyadori. Yamanaka and Tanioka (2022) attributed this extreme runup height to a short-period component of local resonance in Yamada Bay. Based on this extreme height, they modeled the tsunami source as two separate faults along the Japan Trench (Fig. 7.18) with a total moment magnitude M_w of 8.5.

7.5.4 The 869 Jogan Tsunami

Another predecessor of the 2011 Tohoku Earthquake occurred on July 9, 869 (May 26 in Jogan 11th year on the Japanese calendar) and has been known as the Jogan Earthquake. The Japanese historical document, "Nihon-Sandai-Jitsuroku

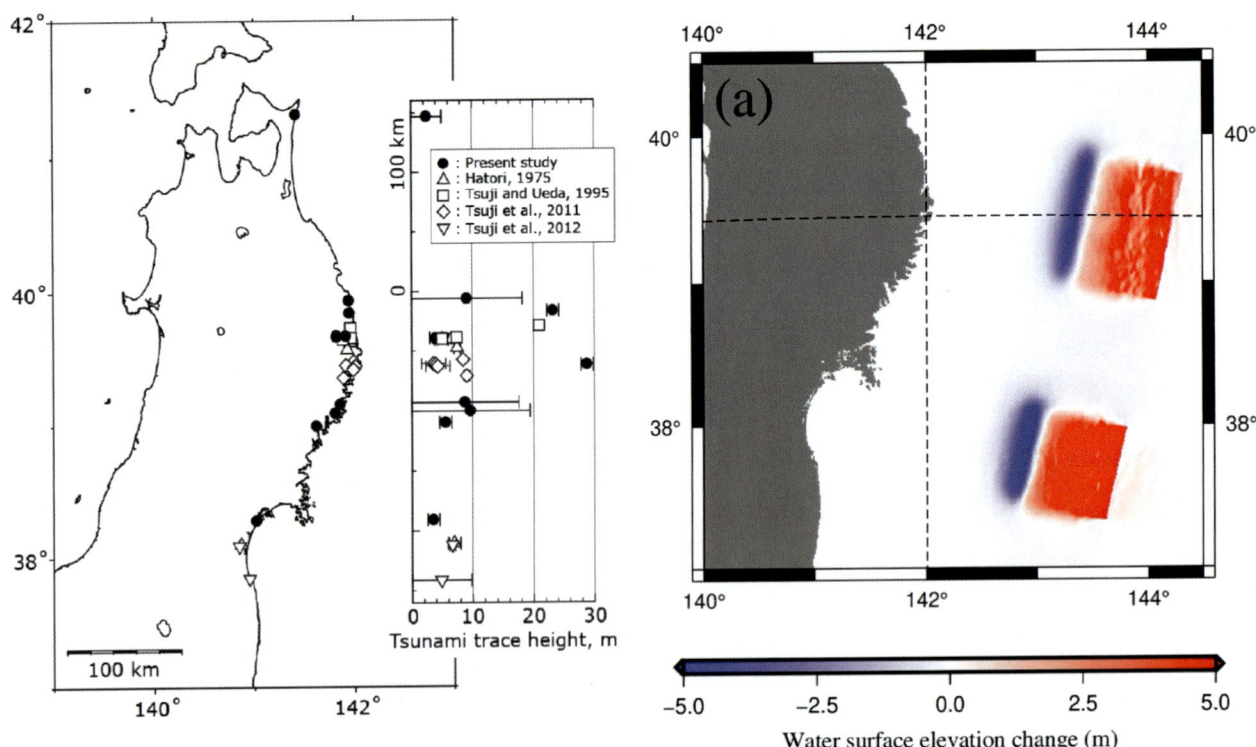

FIGURE 7.18 (Left) Estimated tsunami heights from damage descriptions in historical documents and modern surveys for the 1611 Keicho Tohoku Tsunami (Ebina and Imai, 2014). (Right) Slip distribution of the fault model of the 1611 Keicho Tohoku Tsunami (Yamanaka and Tanioka, 2022). *From Yamanaka, Y., & Tanioka, Y. (2022). Short-wave run-ups of the 1611 Keicho tsunami along the Sanriku Coast.* Progress in Earth and Planetary Science, *9(1). https://doi.org/10.1186/s40645−022−00496−1.*

(Chronicle of Japan)," described the tsunami as follows. "Roaring like thunder was heard from the sea. The tsunami soon rushed into the castle and converted the fields and roads into the sea for a few tens of hundred miles from the coast. There was no time for escape, to get onto boats or climb to high grounds. In this way, about 1000 people were drowned. Hundreds of hamlets and villages were left in ruins." These descriptions indicated that the tsunami inundated Sendai Plain up to Tagajo Castle, the capital of those days located about 5 km from the present coast. Because the coastline in 869 was estimated to be about 1 km inland from the present position, the tsunami inundation is considered to be approximately 4 km or more, which is similar to the inland extent of the 2011 Tsunami in the Sendai Plain (Fig. 7.13).

Minoura et al. (2001) estimated the source of the 869 Jogan Tsunami off the Sendai Coast with a moment magnitude M_w of 8.3. Sawai et al. (2012) considered various fault models, including plate-boundary ruptures with various lengths and widths, "tsunami earthquake" model, and outer-rise normal faults and compared the tsunami simulation results with inferred inundation limits from the historical documents and the distribution of tsunami deposits. The location of the coastline at the time of the 869 Earthquake was considered in the tsunami simulation. The best model for the 869 Earthquake is an interplate fault, 200 km long, 100 km wide, and 7 m slip with a moment magnitude M_w of 8.4 (Fig. 7.19).

7.6 Geological data

Geologists have studied the deposits of prehistoric tsunamis. In Japan, tsunami deposits or boulders in various environments, such as lagoons, marshes, or beaches, have been studied for modern, historic, and prehistoric tsunamis (Goto et al., 2011; Minoura & Nakaya, 1991; Nanayama et al., 2003).

In sedimentology, tsunami deposits are classified as event deposits, resulting from relatively short and rare episodes compared to the usual geological depositional processes. Events include floods, storm surges due to hurricanes and cyclones, and tsunamis. Tsunamis produce long waves, while storms produce short waves; hence some differences in their sedimentological characteristics are expected. In general, tsunami deposits are transported as a suspended load by

Historical tsunami records and paleotsunamis **Chapter | 7** 249

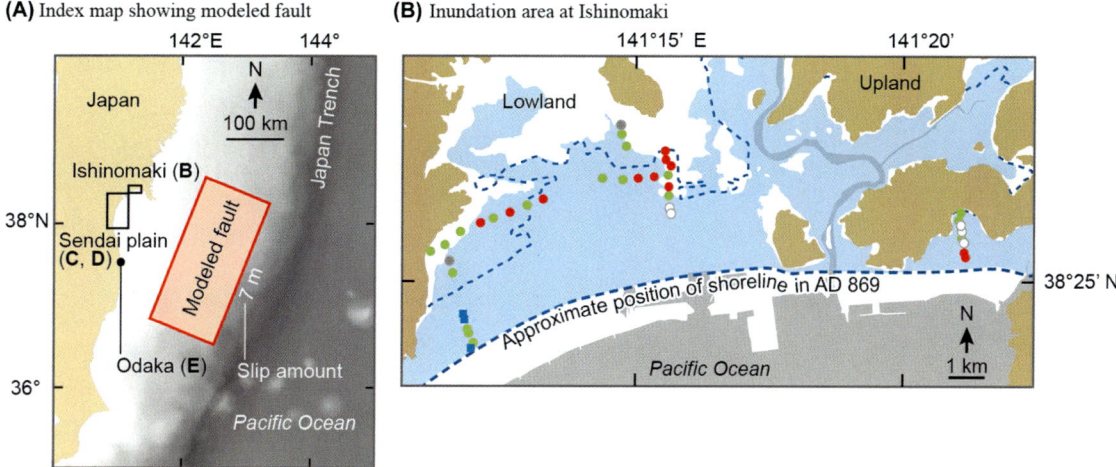
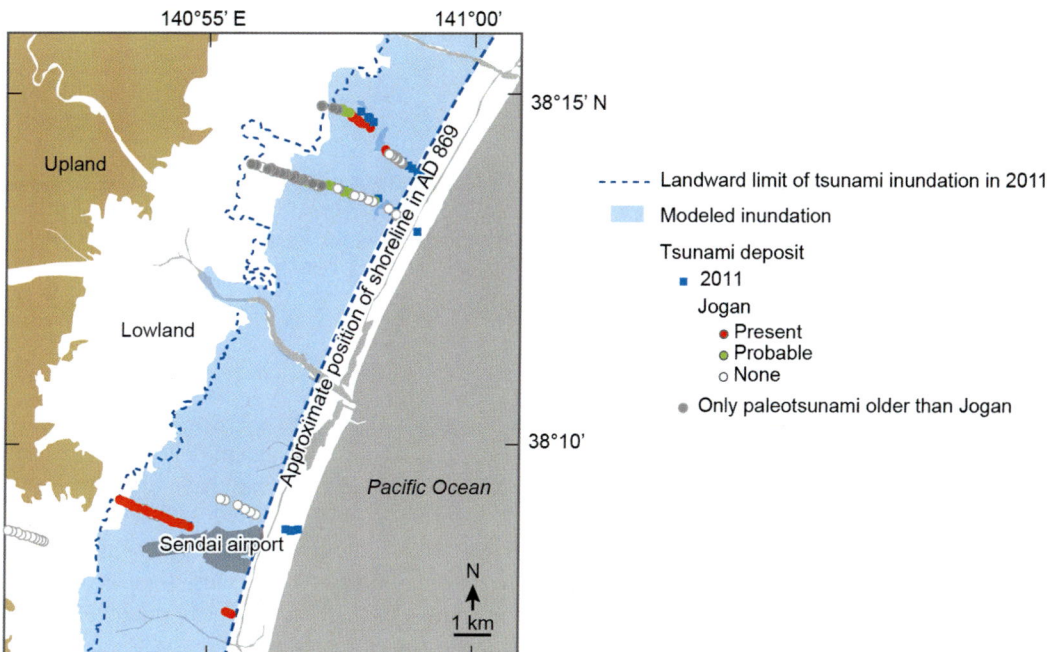

FIGURE 7.19 (A) Map of the 869 Jogan Tsunami. (B,C) Simulated tsunami inundation areas in Ishinomaki and in Sendai Plain from Model 10, the best model of the 869 Jogan Tsunami (Sawai et al., 2012). Map lines delineate study areas and do not necessarily depict accepted national boundaries. *From Sawai, Y., Namegaya, Y., Okamura, Y., Satake, K., & Shishikura, M. (2012). Challenges of anticipating the 2011 Tohoku earthquake and tsunami using coastal geology. Geophysical Research Letters, 39(21). https://doi.org/10.1029/2012GL053692.*

multiple waves, are deposited rapidly in sheets, and show graded bedding, whereas storm deposits are transported by bed load, and are continuously deposited forming cross-bedding (Morton et al., 2007; Nanayama et al., 2000). Once the deposit is attributed to tsunami origin, the source and date need to be estimated for paleoseismological purposes. If a tsunami is from a near-field earthquake, other phenomena, such as coseismic uplift/subsidence and liquefaction, may also be recorded as a geological description (Nelson et al., 1996; Sawai et al., 2012). Studies of deposits from modern tsunamis are important; hence, careful descriptions have been made for recent tsunamis.

7.6.1 Tsunami deposits in Sendai Plain

The 2011 Tohoku Tsunami left sand and mud deposits in the Sendai Plain and provided a valuable opportunity to study the sedimentological characteristics of modern tsunami deposits. While the 2011 Tsunami inundated nearly 5 km from

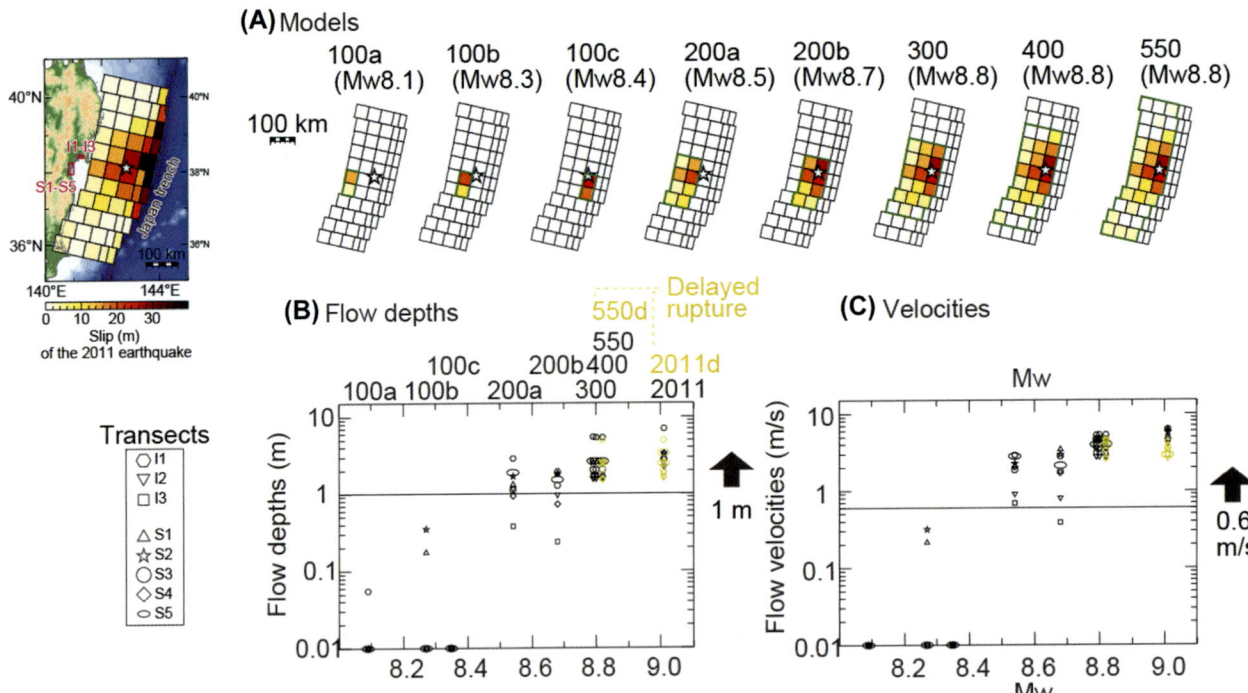

FIGURE 7.20 (A) Assumed fault models for the 869 Jogan Earthquake based on the slip distribution of the 2011 Tohoku Earthquake (Satake et al., 2013). Slip amounts outside the green rectangles were set to zero. (B) Calculated flow depths at the most landward sandy deposits of the Jogan Tsunami. (C) Calculated depth-averaged velocities at the most landward sandy deposits of the Jogan Tsunami. *From Namegaya, Y., & Satake, K. (2014). Reexamination of the A.D. 869 Jogan earthquake size from tsunami deposit distribution, simulated flow depth, and velocity. Geophysical Research Letters, 41(7), 2297–2303. https://doi.org/10.1002/2013GL058678.*

the coast, sand deposits with thicknesses of more than 0.5 cm could be traced up to 3 km, and only mud was deposited between 3 and 5 km (Goto et al., 2011). This indicates that previous studies on tsunami deposits and earthquake modeling, assuming that the extent of a tsunami deposit is the same as the inundation, may have underestimated the sizes of past earthquakes.

Namegaya and Satake (2014) first carried out a numerical simulation of the 2011 Tsunami and found that flow depths and velocities at most inland tsunami deposit sites were approximately 1 m and 0.6 m/second, respectively, on the Ishinomaki and Sendai Plains. They used these values to compare tsunami deposits and the inundation simulation of the 869 Jogan Earthquake. For the variable slip model similar to the 2011 slip distribution (Fig. 7.20), the minimum size of the 869 Jogan Earthquake was 300 km long with a moment magnitude M_w of 8.8.

In the Sendai Plain, another tsunami deposit layer between those from the 2011 Tohoku Earthquake and the 869 Jogan Earthquake was also found (Fig. 7.21). This tsunami layer is considered to be either from the 1454 Kyotoku Earthquake or the 1611 Keicho Tohoku Earthquake (Sawai et al., 2012). Thus the predecessors of the 2011 Tohoku Tsunami include at least one of them, as well as the Jogan Tsunami, making the recurrence interval several hundred years, rather than 1000 years.

7.6.2 Sediment transport modeling

The grain size distribution of tsunami deposits may correlate with the tsunami inundation process, and further with the tsunami source characteristics. Gusman, Goto, et al. (2018) developed and validated a tsunami sediment transport model that can simulate deposit thickness and grain size distribution of the 2011 Tohoku Tsunami deposit on the Numanohama coast (Goto et al., 2019) (Fig. 7.22). The computed net erosion and deposition from five tsunami scenarios suggest that it is possible to estimate tsunami wave amplitude and wave period from the sediment deposit thickness and grain size distribution data.

FIGURE 7.21 Tsunami deposit at archeological site. Section through the uppermost ~1.5 m of the deposits at the Takaose archeological site on the Sendai Plain showing three major sand layers thought to be related to the 869 Jogan, 1611 Keicho, and 2011 Tohoku Tsunamis, and the 915 Towada—a tephra layer. Identification of the older deposits is provisional and verification will require more complete documentation of their ages and origins. *From Wallis, S. R., Fujiwara, O., & Goto, K. (2018). Geological studies in tsunami research since the 2011 Tohoku earthquake.* Geological Society, London, Special Publications, *456(1), 39–53. https://doi.org/10.1144/sp456.12.*

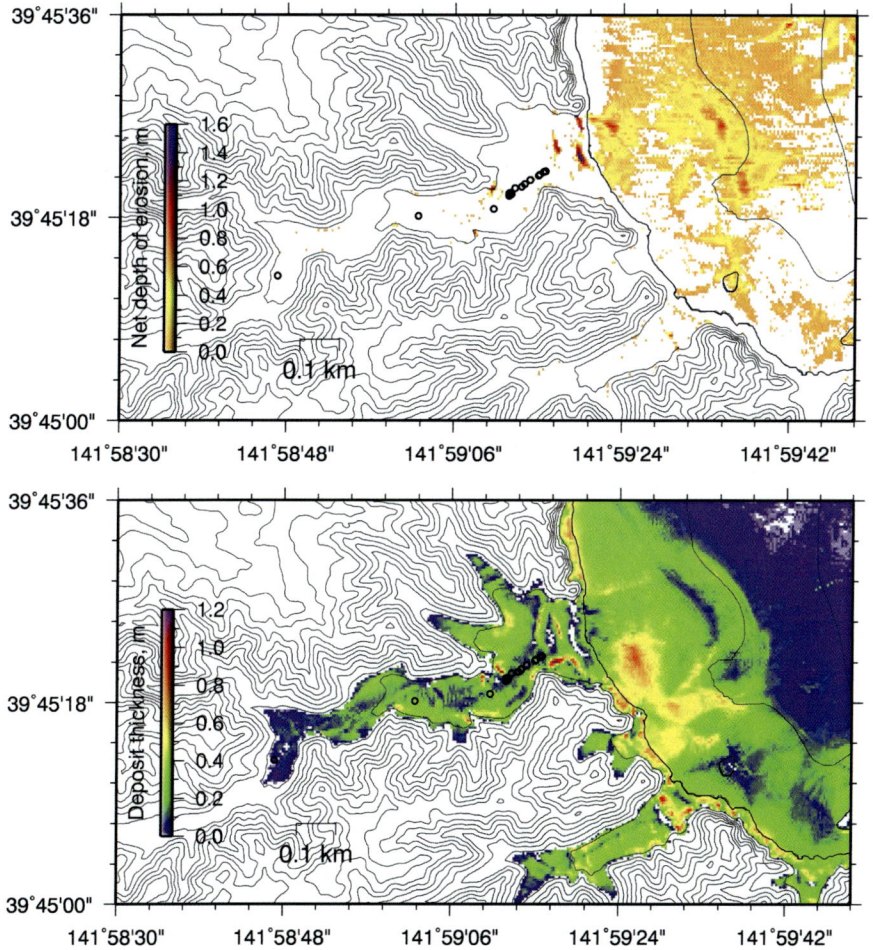

FIGURE 7.22 Simulated net depth of erosion (top) and tsunami deposit thickness (bottom) on Numanohama coast from the 2011 Tohoku Tsunami (Gusman, Goto, et al., 2018). Circles indicate locations for observed tsunami deposits (Goto et al., 2019). Map lines delineate study areas and do not necessarily depict accepted national boundaries. *From Gusman, A. R., Goto, T., Satake, K., Takahashi, T., & Ishibe, T. (2018). Sediment transport modeling of multiple grain sizes for the 2011 Tohoku tsunami on a steep coastal valley of Numanohama, northeast Japan.* Marine Geology, *405, 77–91. https://doi.org/10.1016/j.margeo.2018.08.003.*

Naruse and Abe (2017) proposed a tsunami inverse model from spatial variation of thickness and grain size distribution of deposits along a 1-D transect and applied it to the 2011 Tohoku Tsunami deposit along a 4-km long transect on the Sendai Plain. Their inversion model produced runup flow velocity, inundation depth, and concentration of suspended sediment. The result of inversion fits well with the observations from aerial videos and field surveys.

7.7 Conclusions

Tsunami sources can be estimated from observed tsunami data by inverse modeling. For modern tsunamis, various instrumental data are used, and waveform inversion methods have been developed. Recent offshore tsunami observations make it possible to use the results of inverse modeling for early warning (Wang et al., 2021). Tsunami field surveys provide detailed tsunami heights, which is particularly important for the sawtooth-shaped Sanriku Coast in Japan. The tsunami fragility curve, a relation between tsunami height and building damage, is also constructed from modern surveys and can be used to estimate past tsunami heights from damage descriptions in historical documents. For historical tsunamis, archived tsunami waveforms, field survey heights, and the tsunami flow depth or inundation areas estimated from documented damage observations are used to estimate the tsunami source. For prehistoric tsunamis, the sole available information comes from tsunami deposits. Distributions, inland extent, and altitude of tsunami deposits can be used as the tsunami data. With sediment transport modeling, the thickness and grain size distribution can also be used to estimate the local tsunami parameters, such as flow depth and velocity.

References

Baptista, M. A., Heitor, S., Miranda, J. M., Miranda, P., & Mendes Victor, L. (1998). The 1755 Lisbon Tsunami: Evaluation of the tsunami parameters. *Journal of Geodynamics, 25*. Available from https://doi.org/10.1016/S0264-3707(97.

Borrero, J. C. (2005). Field data and satellite imagery of tsunami effects in Banda Aceh. *Science, 308*(5728), 1596. Available from https://doi.org/10.1126/science.1110957.

Ebina, Y., & Imai, K. (2014). Tsunami trances survey of the 1611 Keicho Ohsyu earthquake tsunami based on historical documents and traditions. *Report of Tsunami Engineering, 31*, 139–148.

Fujii, Y., Satake, K., Sakai, S., Shinohara, M., & Kanazawa, T. (2011). Tsunami source of the 2011 off the Pacific coast of Tohoku Earthquake. *Earth, Planets and Space, 63*(7), 815–820. Available from https://doi.org/10.5047/eps.2011.06.010, http://rd.springer.com/journal/40623.

González, F. I., Bernard, E. N., Meinig, C., Eble, M. C., Mofjeld, H. O., & Stalin, S. (2005). *The NTHMP tsunameter network developing tsunami-resilient communities: The National Tsunami Hazard Mitigation Program* (pp. 25–39). Springer. Available from http://www.springerlink.com/openurl.asp?genre = book&isbn = 978-1-4020-3353-7, http://doi.org/10.1007/1-4020-3607-8_2.

Goto, K., Chagué-Goff, C., Fujino, S., Goff, J., Jaffe, B., Nishimura, Y., Richmond, B., Sugawara, D., Szczuciński, W., Tappin, D. R., Witter, R. C., & Yulianto, E. (2011). New insights of tsunami hazard from the 2011 Tohoku-Oki event. *Marine Geology, 290*(1-4), 46–50. Available from https://doi.org/10.1016/j.margeo.2011.10.004.

Goto, T., Satake, K., Sugai, T., Ishibe, T., Harada, T., & Gusman, A. R. (2019). Tsunami history over the past 2000 years on the Sanriku Coast, Japan, determined using gravel deposits to estimate tsunami inundation behavior. *Sedimentary Geology, 382*, 85–102. Available from https://doi.org/10.1016/j.sedgeo.2019.01.001, http://www.sciencedirect.com/science/journal/00370738.

Gusman, A. R., Goto, T., Satake, K., Takahashi, T., & Ishibe, T. (2018). Sediment transport modeling of multiple grain sizes for the 2011 Tohoku tsunami on a steep coastal valley of Numanohama, northeast Japan. *Marine Geology, 405*, 77–91. Available from https://doi.org/10.1016/j.margeo.2018.08.003, http://www.sciencedirect.com/science/journal/00253227.

Gusman, A. R., Mulia, I. E., & Satake, K. (2018). Optimum sea surface displacement and fault slip distribution of the 2017 Tehuantepec earthquake (Mw 8.2) in Mexico estimated from tsunami waveforms. *Geophysical Research Letters, 45*(2), 646–653. Available from https://doi.org/10.1002/2017GL076070, http://onlinelibrary.wiley.com/journal/10.1002/(ISSN)1944-8007/issues?year = 2012.

Gusman, A. R., Murotani, S., Satake, K., Heidarzadeh, M., Gunawan, E., Watada, S., & Schurr, B. (2015). Fault slip distribution of the 2014 Iquique, Chile, earthquake estimated from ocean-wide tsunami waveforms and GPS data. *Geophysical Research Letters, 42*(4), 1053–1060. Available from https://doi.org/10.1002/2014GL062604, http://onlinelibrary.wiley.com/journal/10.1002/(ISSN)1944-8007/issues?year = 2012.

Gusman, A. R., Satake, K., Gunawan, E., Hamling, I., & Power, W. (2018). Contribution from multiple fault ruptures to tsunami generation during the 2016 Kaikoura earthquake. *Pure and Applied Geophysics, 175*(8), 2557–2574. Available from https://doi.org/10.1007/s00024-018-1949-z, https://doi.org/10.1007/s00024-018-1949-z.

Gusman, A. R., Sheehan, A. F., Satake, K., Heidarzadeh, M., Mulia, I. E., & Maeda, T. (2016). Tsunami data assimilation of Cascadia seafloor pressure gauge records from the 2012 Haida Gwaii earthquake. *Geophysical Research Letters, 43*(9), 4189–4196. Available from https://doi.org/10.1002/2016GL068368, http://onlinelibrary.wiley.com/journal/10.1002/(ISSN)1944-8007/issues?year = 2012.

Gusman, A. R., Tanioka, Y., Sakai, S., & Tsushima, H. (2012). Source model of the great 2011 Tohoku earthquake estimated from tsunami waveforms and crustal deformation data. *Earth and Planetary Science Letters, 341-344*, 234–242. Available from https://doi.org/10.1016/j.epsl.2012.06.006.

Hayashi, Y. (2008). Extracting the 2004 Indian Ocean tsunami signals from sea surface height data observed by satellite altimetry. *Journal of Geophysical Research: Oceans, 113*(1). Available from https://doi.org/10.1029/2007JC004177, http://onlinelibrary.wiley.com/journal/10.1002/(ISSN)2169-9291.

Heidarzadeh, M., & Satake, K. (2013). Waveform and spectral analyses of the 2011 Japan tsunami records on tide gauge and DART stations across the Pacific Ocean. *Pure and Applied Geophysics, 170*(6-8), 1275–1293. Available from https://doi.org/10.1007/s00024-012-0558-5, http://www.springer.com/birkhauser/geo + science/journal/24.

Hinata, H., Fujii, S., Furukawa, K., Kataoka, T., Miyata, M., Kobayashi, T., Mizutani, M., Kokai, T., & Kanatsu, N. (2011). Propagating tsunami wave and subsequent resonant response signals detected by HF radar in the Kii Channel, Japan. *Estuarine, Coastal and Shelf Science*, *95*(1), 268−273. Available from https://doi.org/10.1016/j.ecss.2011.08.009.

Ho, T.-C., Satake, K., & Watada, S. (2017). Improved phase corrections for transoceanic tsunami data in spatial and temporal source estimation: Application to the 2011 Tohoku earthquake. *Journal of Geophysical Research: Solid Earth*, *122*(12), 10,155−10,715. Available from https://doi.org/10.1002/2017jb015070.

Hossen, M. J., Cummins, P. R., Roberts, S. G., & Allgeyer, S. (2015). Time reversal imaging of the tsunami source. *Pure and Applied Geophysics*, *172*(3-4), 969−984. Available from https://doi.org/10.1007/s00024-014-1014-5, http://www.springer.com/birkhauser/geo + science/journal/24.

Intergovernmental Oceanographic Commission (2006) Manual on sea level measurement and interpretation. Volume IV - An update to 2006. Paris, France, UNESCO, 87pp. (Intergovernmental Oceanographic Commission Manuals and Guides: 14, Vol. 4), (JCOMM Technical Report: 31), (WMO/TD: 1339). Available from https://doi.org/10.25607/OBP-1398.

Intergovernmental Oceanographic Commission. (2014). *Post-tsunami survey field guide* (2nd ed.). UNESCO. Available from: https://unesdoc.unesco.org/ark:/48223/pf0000229456.

Intergovernmental Oceanographic Commission. (2016). Manual on sea level measurement and interpretation. Volume V: Radar gauges. In *Manuals and guides* 14. UNESCO, Available from: https://unesdoc.unesco.org/ark:/48223/pf0000246981.

Kato, T., Terada, Y., Kinoshita, M., Kakimoto, H., Isshiki, H., Matsuishi, M., Yokoyama, A., & Tanno, T. (2000). Real-time observation of tsunami by RTK-GPS. *Earth, Planets and Space*, *52*(10), 841−845. Available from https://doi.org/10.1186/BF03352292, http://rd.springer.com/journal/40623.

Kubota, T., Saito, T., & Nishida, K. (2022). Global fast-traveling tsunamis driven by atmospheric Lamb waves on the 2022 Tonga eruption. *Science*, *377*(6601), 91−94. Available from https://doi.org/10.1126/science.abo4364, https://www.science.org/doi/10.1126/science.abo4364.

Kusumoto, S., Imai, K., & Hori, T. (2022). Time difference between the 1854 CE Ansei−Tokai and Ansei−Nankai earthquakes estimated from distant tsunami waveforms on the west coast of North America. *Progress in Earth and Planetary Science*, *9*(1). Available from https://doi.org/10.1186/s40645-021-00458-z, https://progearthplanetsci.springeropen.com/.

Kusumoto, S., Imai, K., Obayashi, R., Hori, T., Takahashi, N., Ho, T. C., Uno, K., Tanioka, Y., & Satake, K. (2020). Origin time of the 1854 Ansei-Tokai Tsunami estimated from tide gauge records on the West Coast of North America. *Seismological Research Letters*, *91*(5), 2624−2630. Available from https://doi.org/10.1785/0220200068, https://watermark.silverchair.com/srl-2020068.1.pdf?.

Lay, T., Fujii, Y., Geist, E., Koketsu, K., Rubinstein, J., Sagiya, T., & Simons, M. (2013). Introduction to the special issue on the 2011 Tohoku earthquake and tsunami. *Bulletin of the Seismological Society of America*, *103*(2B), 1165−1170. Available from https://doi.org/10.1785/0120130001, http://www.bssaonline.org/content/103/2B/1165.full.pdf + html.

Maeda, T., Furumura, T., Sakai, S., & Shinohara, M. (2011). Significant tsunami observed at ocean-bottom pressure gauges during the 2011 off the Pacific coast of Tohoku Earthquake. *Earth, Planets and Space*, *63*(7), 803−808. Available from https://doi.org/10.5047/eps.2011.06.005, http://rd.springer.com/journal/40623.

Minoura, K., Imamura., & Sugawara. (2001). The 869 Jogan tsunami deposit and recurrence interval of large-scale tsunami on the Pacific coast of northeast Japan. *Journal of Natural Disaster Science*, *23*, 83−88.

Minoura, K., & Nakaya, S. (1991). Traces of tsunami preserved in inter-tidal lacustrine and marsh deposits: Some examples from Northeast Japan. *The Journal of Geology*, *99*(2), 265−287. Available from https://doi.org/10.1086/629488.

Mori, N., Takahashi, T., Yasuda, T., & Yanagisawa, H. (2011). Survey of 2011 Tohoku earthquake tsunami inundation and run-up. *Geophysical Research Letters*, *38*(18). Available from https://doi.org/10.1029/2011GL049210, http://onlinelibrary.wiley.com/journal/10.1002/(ISSN)1944-8007/issues?year = 2012.

Morton, R. A., Gelfenbaum, G., & Jaffe, B. E. (2007). Physical criteria for distinguishing sandy tsunami and storm deposits using modern examples. *Sedimentary Geology*, *200*(3-4), 184−207. Available from https://doi.org/10.1016/j.sedgeo.2007.01.003.

Mulia, I. E., Gusman, A. R., Jakir Hossen, M., & Satake, K. (2018). Adaptive tsunami source inversion using optimizations and the reciprocity principle. *Journal of Geophysical Research: Solid Earth*, *123*(12), 10−760. Available from https://doi.org/10.1029/2018JB016439, http://agupubs.onlinelibrary.wiley.com/hub/jgr/journal/10.1002/(ISSN)2169-9356/.

Mulia, I. E., & Satake, K. (2020). Developments of tsunami observing systems in Japan. *Frontiers in Earth Science*, *8*. Available from https://doi.org/10.3389/feart.2020.00145, https://www.frontiersin.org/journals/earth-science.

Musa, A., Watanabe, O., Matsuoka, H., Hokari, H., Inoue, T., Murashima, Y., Ohta, Y., Hino, R., Koshimura, S., & Kobayashi, H. (2018). Real-time tsunami inundation forecast system for tsunami disaster prevention and mitigation. *Journal of Supercomputing*, *74*(7), 3093−3113. Available from https://doi.org/10.1007/s11227-018-2363-0, http://www.springerlink.com/content/0920-8542.

Namegaya, Y., & Satake, K. (2014). Reexamination of the A.D. 869 Jogan earthquake size from tsunami deposit distribution, simulated flow depth, and velocity. *Geophysical Research Letters*, *41*(7), 2297−2303. Available from https://doi.org/10.1002/2013GL058678, http://www.agu.org/journals/gl/.

Nanayama, F., Satake, K., Furukawa, R., Shimokawa, K., Atwater, B. F., Shigeno, K., & Yamaki, S. (2003). Unusually large earthquakes inferred from tsunami deposits along the Kuril trench. *Nature*, *424*(6949), 660−663. Available from https://doi.org/10.1038/nature01864.

Nanayama, F., Shigeno, K., Satake, K., Shimokawa, K., Koitabashi, S., Miyasaka, S., & Ishii, M. (2000). Sedimentary differences between the 1993 Hokkaido-Nansei-Oki Tsunami and the 1959 Miyakojima Typhoon at Taisei, southwestern Hokkaido, northern Japan. *Sedimentary Geology*, *135*(1-4), 255−264. Available from https://doi.org/10.1016/S0037-0738(00)00076-2.

Naruse, H., & Abe, T. (2017). Inverse tsunami flow modeling including nonequilibrium sediment transport, with application to deposits from the 2011 Tohoku-Oki tsunami. *Journal of Geophysical Research: Earth Surface*, *122*(11), 2159−2182. Available from https://doi.org/10.1002/2017JF004226, http://onlinelibrary.wiley.com/journal/10.1002/(ISSN)2169-9011.

Nelson, A. R., Shennan, I., & Long, A. J. (1996). Identifying coseismic subsidence in tidal-wetland stratigraphic sequences at the Cascadia subduction zone of western North America. *Journal of Geophysical Research: Solid Earth*, *101*(B3), 6115−6135. Available from https://doi.org/10.1029/95jb01051.

Nosov, M., Karpov, V., Kolesov, S., Sementsov, K., Matsumoto, H., & Kaneda, Y. (2018). Relationship between pressure variations at the ocean bottom and the acceleration of its motion during a submarine earthquake. *Federation Earth, Planets and Space*, *70*(1). Available from https://doi.org/10.1186/s40623-018-0874-9, http://rd.springer.com/journal/40623.

Rabinovich, A. B., & Eblé, M. C. (2015). Deep-ocean measurements of tsunami waves. *Pure and Applied Geophysics*, *172*(12), 3281−3312. Available from https://doi.org/10.1007/s00024-015-1058-1, http://www.springer.com/birkhauser/geo + science/journal/24.

Saito, T., Inazu, D., Tanaka, S., & Miyoshi, T. (2013). Tsunami coda across the Pacific Ocean following the 2011 Tohoku-Oki earthquake. *Bulletin of the Seismological Society of America*, *103*(2 B), 1429−1443. Available from https://doi.org/10.1785/0120120183Japan, http://www.bssaonline.org/content/103/2B/1429.full.pdf + html.

Saito, T., Ito, Y., Inazu, D., & Hino, R. (2011). Tsunami source of the 2011 Tohoku-Oki earthquake, Japan: Inversion analysis based on dispersive tsunami simulations. *Geophysical Research Letters*, *38*(21). Available from https://doi.org/10.1029/2011GL049089, http://onlinelibrary.wiley.com/journal/10.1002/(ISSN)1944-8007/issues?year = 2012.

Sandanbata, O., Watada, S., Satake, K., Kanamori, H., Rivera, L., & Zhan, Z. (2022). Sub-decadal volcanic tsunamis due to submarine trapdoor faulting at Sumisu Caldera in the Izu−Bonin Arc. *Journal of Geophysical Research: Solid Earth*, *127*(9). Available from https://doi.org/10.1029/2022JB024213, http://agupubs.onlinelibrary.wiley.com/hub/jgr/journal/10.1002/(ISSN)2169-9356/.

Satake, K. (1987). Inversion of tsunami waveforms for the estimation of a fault heterogeneity: Method and numerical experiments. *Journal of Physics of the Earth*, *35*(3), 241−254. Available from https://doi.org/10.4294/jpe1952.35.241.

Satake, K. (2014). Advances in earthquake and tsunami sciences and disaster risk reduction since the 2004 Indian Ocean Tsunami. *Geoscience Letters*, *1*(1). Available from https://doi.org/10.1186/s40562-014-0015-7, http://geoscienceletters.springeropen.com/.

Satake, K., & Fujii, Y. (2014). Review: Source models of the 2011 Tohoku earthquake and long-term forecast of large earthquakes. *Journal of Disaster Research*, *9*(3), 272−280. Available from https://doi.org/10.20965/jdr.2014.p0272, http://www.fujipress.jp/finder/access_check.php?pdf_filename = DSSTR000900030004.pdf&frompage = abst_page&errormode = Login&pid = 5148&lang = English.

Satake, K., Fujii, Y., Harada, T., & Namegaya, Y. (2013). Time and space distribution of coseismic slip of the 2011 Tohoku earthquake as inferred from Tsunami waveform data. *Bulletin of the Seismological Society of America*, *103*(2B), 1473−1492. Available from https://doi.org/10.1785/0120120122Japan, http://www.bssaonline.org/content/103/2B/1473.full.pdf + html.

Satake, K., Fujii, Y., & Yamaki, S. (2017). Different depths of near-trench slips of the 1896 Sanriku and 2011 Tohoku earthquakes. *Geoscience Letters*, *4*(1). Available from https://doi.org/10.1186/s40562-017-0099-y, http://geoscienceletters.springeropen.com/.

Sawai, Y., Namegaya, Y., Okamura, Y., Satake, K., & Shishikura, M. (2012). Challenges of anticipating the 2011 Tohoku earthquake and tsunami using coastal geology. *Geophysical Research Letters*, *39*(21). Available from https://doi.org/10.1029/2012GL053692, http://onlinelibrary.wiley.com/journal/10.1002/(ISSN)1944-8007/issues?year = 2012.

Sheehan, A. F., Gusman, A. R., Heidarzadeh, M., & Satake, K. (2015). Array observations of the 2012 Haida Gwaii tsunami using Cascadia initiative absolute and differential seafloor pressure gauges. *Seismological Research Letters*, *86*(5), 1278−1286. Available from https://doi.org/10.1785/0220150108, http://srl.geoscienceworld.org/content/86/5/1278.full.pdf + html.

Sheehan, A. F., Gusman, A. R., & Satake, K. (2019). Improving forecast accuracy with tsunami data assimilation: The 2009 Dusky Sound, New Zealand, Tsunami. *Journal of Geophysical Research: Solid Earth*, *124*(1), 566−577. Available from https://doi.org/10.1029/2018JB016575, http://agupubs.onlinelibrary.wiley.com/hub/jgr/journal/10.1002/(ISSN)2169-9356/.

Smith, W., Scharroo, R., Titov, V., Arcas, D., & Arbic, B. (2005). Satellite altimeters measure tsunami: Early model estimates confirmed. *Oceanography*, *18*(2), 11−13. Available from https://doi.org/10.5670/oceanog.2005.62.

Sugimori, R., Ariizumi, K., & Satake, K. (2022). Origin time of the 1854 Tokai earthquake recorded in the logbook of the Russian frigate Diana. *Journal of Disaster Research*, *17*(3), 409−419. Available from https://doi.org/10.20965/jdr.2022.p0409.

Suppasri, A., Mas, E., Charvet, I., Gunasekera, R., Imai, K., Fukutani, Y., Abe, Y., & Imamura, F. (2013). Building damage characteristics based on surveyed data and fragility curves of the 2011 Great East Japan tsunami. *Natural Hazards*, *66*(2), 319−341. Available from https://doi.org/10.1007/s11069-012-0487-8, http://www.wkap.nl/journalhome.htm/0921-030X.

Tanioka, Y., & Satake, K. (1996). Fault parameters of the 1896 Sanriku Tsunami earthquake estimated from tsunami numerical modeling. *Geophysical Research Letters*, *23*(13), 1549−1552. Available from https://doi.org/10.1029/96GL01479, http://onlinelibrary.wiley.com/journal/10.1002/(ISSN)1944-8007/issues?year = 2012.

Tsuji, Y., Satake, K., Ishibe, T., Harada, T., Nishiyama, A., & Kusumoto, S. (2014). Tsunami heights along the Pacific coast of northern Honshu recorded from the 2011 Tohoku and previous great earthquakes. *Pure and Applied Geophysics*, *171*(12), 3183−3215. Available from https://doi.org/10.1007/s00024-014-0779-x, http://www.springer.com/birkhauser/geo + science/journal/24.

Wang, Y., Imai, K., Kusumoto, S., & Takahashi, N. (2022). Tsunami early warning of the Hunga volcanic eruption using an ocean floor observation network off the Japanese islands. *Seismological Research Letters*, *94*(2A), 567−577. Available from https://doi.org/10.1785/0220220098.

Wang, Y., Tsushima, H., Satake, K., & Navarrete, P. (2021). Review on recent progress in near-field tsunami forecasting using offshore tsunami measurements: Source inversion and data assimilation. *Pure and Applied Geophysics*, *178*(12), 5109−5128. Available from https://doi.org/10.1007/s00024-021-02910-z, http://www.springer.com/birkhauser/geo + science/journal/24.

Watada, S., Kusumoto, S., & Satake, K. (2014). Traveltime delay and initial phase reversal of distant tsunamis coupled with the self-gravitating elastic Earth. *Journal of Geophysical Research: Solid Earth*, *119*(5), 4287−4310. Available from https://doi.org/10.1002/2013JB010841, http://onlinelibrary.wiley.com/journal/10.1002/(ISSN)2169-9356.

Yamanaka, Y., & Tanioka, Y. (2022). Short-wave run-ups of the 1611 Keicho Tsunami along the Sanriku Coast. *Progress in Earth and Planetary Science*, *9*(1). Available from https://doi.org/10.1186/s40645-022-00496-1, https://progearthplanetsci.springeropen.com/.

Yamazaki, Y., Cheung, K. F., & Lay, T. (2018). A self-consistent fault slip model for the 2011 Tohoku earthquake and tsunami. *Journal of Geophysical Research: Solid Earth*, *123*(2), 1435−1458. Available from https://doi.org/10.1002/2017JB014749, http://onlinelibrary.wiley.com/journal/10.1002/(ISSN)2169-9356.

Chapter 8

Informing megathrust tsunami source models with knowledge of tectonics and fault mechanics

Kelin Wang[1,2], Matías Carvajal[3], Yijie Zhu[2], Tianhaozhe Sun[1,2], Jiangheng He[1] and Matthew Sypus[2]

[1]Geological Survey of Canada, Sidney, BC, Canada, [2]School of Earth and Ocean Sciences, University of Victoria, Victoria, BC, Canada, [3]Instituto de Geografía, Pontificia Universidad Católica de Valparaíso, Valparaíso, Valparaiso, Chile

8.1 Introduction

This chapter discusses models of tsunami sources due to subduction earthquakes for tsunami hazard analysis in coastal areas facing megathrust rupture zones. In near-field tsunamis, the wave height and runup at coastal locations are sensitive to details of fault slip, but fault slip will remain the largest source of uncertainty for the foreseeable future. Even for the best studied events, such as the 2011 M_w 9 Tohoku, Japan, Earthquake and 2010 M_w 8.8 Maule, Chile, Earthquake, the fault slip is far from unambiguously constrained (Wang et al., 2018, 2020). Even if the slip in a past earthquake could be well constrained, it would not provide adequate information for future events because of the spatio-temporal complexities of real faults. Tsunami hazard assessment requires predictive source models that mathematically and numerically describe a wide range of plausible rupture scenarios. In these models, fault geometry, offshore and onshore topography, and rock properties are, or eventually will be, observationally constrained, but fault slip must be assumed or theoretically predicted.

Seismic moment (a product of rigidity, fault slip, and rupture area) is the best measure of earthquake size, but fault slip is a more fundamental parameter in near-field tsunami hazard analysis. One reason is the trade-off between the slip and rigidity. For the same moment and rupture area, the slip is much larger if the rigidity is much lower, and it is the slip that dictates tsunamigenic seafloor deformation. Another reason is the trade-off between the slip and rupture area, particularly in the strike dimension. If the rupture is very long along strike, the tsunami hazard at a coastal site is dominated by fault slip directly offshore, and the total moment is less important.

The design of fault slip in predictive source models should be informed by the knowledge of tectonics, which enables sensible estimates of the potential of seismic versus aseismic slip, updip and downdip limits of earthquake rupture, along-strike segmentation, and recurrence pattern. Hazard assessment commonly relies on kinematically assigned fault slip, but the slip assignment should be guided by fault mechanics principles, including those derived from dynamic rupture models. Application of these principles to a real subduction zone setting, combined with the knowledge of local tectonics and observations of past earthquakes and present megathrust locking state, allows physically reasonable judgment to be made regarding coseismic slip distribution, off-fault permanent deformation, and the likelihood of shallow or even trench-breaching slip.

To facilitate cross-disciplinary dialogs, we try to explain source modeling issues in a way that can be understood by those who do not normally deal with tectonics or solid mechanics. We first discuss general issues of tectonics and fault mechanics that are important to understanding tsunami sources (Section 8.2), and we then discuss issues more directly relevant to the design of predictive models (Section 8.3). We only focus on source characteristics. Recurrence behavior is another important but difficult subject in tsunami hazard analysis, and is beyond the scope of this chapter. Interested readers are referred to Chapter 2.

8.2 Subduction megathrust and tsunamigenic earthquakes

To the first order, a large coseismic slip causes a large tsunami runup (Geist, 1998), such that understanding tsunamigenesis hinges on understanding seismogenesis. Earthquake history, current seismicity, and the locking state of the megathrust provide the most direct information on the seismogenesis of a subduction zone. However, an understanding of the relevant tectonic processes enables us not only to explain these observations but also to postulate seismogenic behavior where these observations are absent or inadequate.

8.2.1 Fault zone structure affected by seafloor morphology and sediment subduction

Fault zone structure provides the basic geological condition for earthquakes. Seismic slip requires the localization of shear into a thin band of millimeter to centimeter scale thickness (Shipton et al., 2006; Sibson, 2003), and rupture propagation requires a certain degree of smoothness. A smooth megathrust can, although not always, exhibit interseismic locking of large fault patches and long-range propagation of rupture. The internal structure of the megathrust is strongly controlled by the morphology of the subducting seafloor. Various global syntheses suggest a strong tendency for large earthquakes to occur where the incoming seafloor is smooth (Brizzi et al., 2018; Lallemand et al., 2018; Scholl et al., 2015; van Rijsingen et al., 2018) but creep to prevail where the incoming seafloor is rugged (Bassett & Watts, 2015; Wang & Bilek, 2014). Similar control on the seismogenic behavior by fault roughness/smoothness is also observed at smaller scales in laboratory experiments (Xu et al., 2023).

Topographic features, such as seamounts, aseismic ridges, and large fracture zones, subduct by fracturing themselves and surrounding rocks, resulting in a fault zone with complex structure and extremely heterogeneous stresses that hinder coherent locking (Fig. 8.1A). There are smooth surfaces in this complex system, but they are usually not well connected

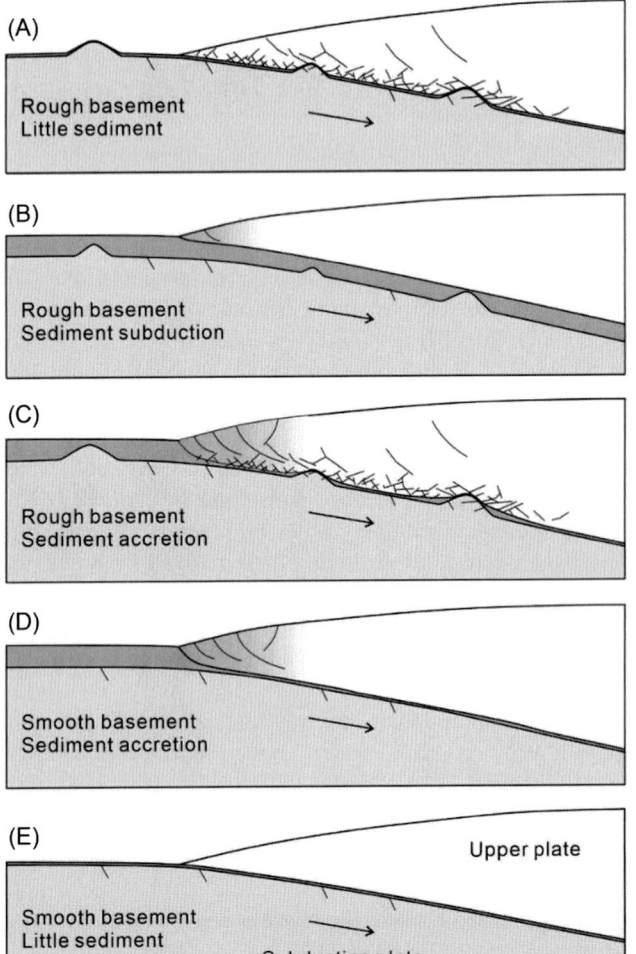

FIGURE 8.1 Megathrust structure controlled by the roughness of the subducting plate and sediment subduction or accretion. (A) Rough igneous basement with little sediment. (B or C) Rough basement with ample sediment that is mostly subducted or accreted, respectively. (D or E) Relatively smooth basement with sediment accretion or little sediment, respectively. Mild roughness of the basement may or may not lead to mild roughness of the megathrust, depending on sediment subduction. *Modified and expanded from Tan, H., Gao, X., Wang, K., Gao, J., & He, J. (2022). Hidden roughness of subducting seafloor and implications for megathrust seismogenesis: Example from northern Manila trench. Geophysical Research Letters, 49(17). http://agupubs.onlinelibrary.wiley.com/hub/journal/10.1002/(ISSN)1944-8007/. https://doi.org/10.1029/2022GL100146.*

and oriented to facilitate rupture propagation. The shear of this complex zone appears as creep accompanied by many small and medium-size earthquakes (Wang & Bilek, 2011, 2014). Fracturing and diminution may mechanically smooth the fault and localize shear farther downdip. At some margins, such as Costa Rica (Protti et al., 2014), downdip smoothing may be responsible for a tendency for large megathrust ruptures to occur beneath land (>25 km depths) and generate little tsunami. If the mechanical smoothing is very effective, shear localization may begin at shallower depths.

Some megathrusts are naturally smooth owing to the subduction of sediment (Ruff, 1989; Scholl et al., 2015) (Fig. 8.1B). The 1700 M_w 9 Cascadia, 1946 M_w 8.2 Nankai, 1960 M_w 9.5 Chile, 1964 M_w 9.2 Alaska, 2004 M_w 9.2 Sumatra, and 2010 M_w 8.8 Maule Earthquakes occurred in areas where the incoming plate is covered by thick sediment, and their rupture zones are considered examples of sediment-smoothed megathrusts. However, the presence of thick sediment on the incoming plate is not always a necessary or sufficient condition for a smooth megathrust.

If there is thick sediment on the incoming plate but little of it is subducted, the roughness of the igneous crust beneath the sediment cover can affect fault zone structure. At the northern Manila Trench, the igneous crust beneath a 1.5–2 km sediment cover is very rough, with numerous seamounts nearly 1 km tall. This hidden roughness is exposed as the sediment is scraped off at the trench, resulting in a rough megathrust (Tan et al., 2022) (Fig. 8.1C). At the Cascadia subduction zone, the igneous crust beneath the sediment cover is relatively smooth except for ancient ridge propagators and rare seamounts (Han et al., 2016). In this situation, the subduction of even a tiny fraction of the sediment may be adequate to result in a smooth megathrust (Fig. 8.1D), which may explain the prevalence of very large earthquakes along the Cascadia margin (Goldfinger et al., 2017).

In some situations, most notably in the area of the 2011 M_w 9.0 Tohoku Earthquake, there is little sediment on the incoming plate, but the seafloor is devoid of large topographic features (Fig. 8.1E). However, off Japan Trench, the incoming plate has mild roughness due to plate-bending normal faults (Fujie et al., 2020; Scholl et al., 2015). We envision the megathrust there to be borderline smooth with low-amplitude roughness. Limited sediment may fill some seafloor grabens to enhance smoothness, but some of the normal faults may increase offset during subduction to enhance roughness. The resultant complexity may explain observed variations in the slip behavior along strike (Nishikawa et al., 2019) and the occasional occurrence of large ruptures extending to the trench (Kodaira et al., 2021; Sun et al., 2017).

In summary, assessing seismogenesis requires knowing fault zone structure, along with earthquake history and the current state of megathrust locking. Bathymetry mapping provides the most basic information. Very rugged incoming seafloor tends to reduce the potential of large earthquakes but may lead to more frequent small and medium-size earthquakes. Where the seafloor is smoothed by a thick sediment cover, seismic imaging may constrain the degree of sediment subduction or accretion. If the sediment cover is scaped off upon subduction, large geometrical irregularities may be exposed to roughen the megathrust.

8.2.2 Fault zone rheology and fault friction

Fault zone rheology provides another basic geological condition for earthquakes. The creep of very rugged faults discussed in the preceding section involves distributed deformation of a highly irregular cataclastic shear zone. Because it occurs against strong resistance, it is called strong creep (Gao & Wang, 2014). Creep can also occur along smooth faults, but for a different reason; it is facilitated by a weak gouge and is called weak creep. For the scope of this chapter, we consider various slow slip events (Burgmann, 2018; Wallace, 2020) collectively a form of creep and do not separately address their special mechanisms. Whether a smooth megathrust exhibits stick–slip to produce large earthquakes or undergoes weak creep is controlled by the rheology of the fault zone material. If the smooth fault is thin enough to be imagined as a contact surface, its bulk rheology is parameterized as the frictional behavior of the contact. The friction law usually takes the following form

$$\tau_s = \mu(\sigma_n - P_f) = \mu\overline{\sigma}_n \tag{8.1}$$

where τ_s is shear strength (i.e., the shear stress that causes slip), σ_n is normal stress that is predominantly the weight of the overlying rock column, μ is the coefficient of friction, and P_f is pore fluid pressure along the fault which offsets the effect of σ_n to give rise to the effective normal stress $\overline{\sigma}_n$. The seismogenic capacity of a fault is controlled not by the absolute value of τ_s but by how it changes with slip or slip rate. There are two categories of friction law that describe these changes.

One category is rate- and state-dependent (R–S) friction in which μ changes with slip rate (Marone, 1998; Scholz, 1998). It is laboratory-derived at slip rates with the order of magnitude lower than that of seismic slip and therefore pertains mainly to rupture initiation. In simple terms, if μ (and hence τ_s) becomes smaller when the fault slips faster, the behavior is called rate weakening (or velocity weakening), which is the necessary condition to initiate a rupture. The opposite behavior is rate strengthening (or velocity strengthening), analogous to viscous flow in which shear resistance increases with shear rate. A rate-strengthening fault patch cannot initiate an earthquake, although it may not always be able to stop rupture propagation. If not pinned by neighboring locked patches, a rate-strengthening fault patch exhibits weak creep, usually accompanied by small earthquakes.

The other category consists of laws of dynamic weakening in which τ_s (through changes in μ and/or P_f) rapidly decreases with slip or slip rate (Di Toro et al., 2011; Tullis, 2015). These laws govern seismic slip (at rates around 1 m/second) and rupture propagation, thereby controlling whether a rupture can evolve into a large earthquake after initiation. It is important to recognize that a fault patch that exhibits rate strengthening at low slip rates can exhibit dynamic weakening at high rates. This behavior is thought to be responsible for the large slip of the shallowest part of the megathrust during the 2011 M_w 9.0 Tohoku Earthquake (Noda & Lapusta, 2013; Sun et al., 2017). A smooth fault may exhibit low- and high-rate frictional heterogeneity at multiple wavelengths. Some patches may facilitate but some resist seismic rupture, and the overall behavior depends on which type is more dominant.

Changes in P_f with slip can play a critical role in both low- and high-rate friction. During a slow slip event, a sudden decrease in P_f due to dilation of pore spaces may instantly increase $\bar{\sigma}_n$ to halt the slip (Liu & Rubin, 2010). During seismic rupture, a sudden increase in P_f due to thermal expansion of pore fluids as a result of frictional heating can instantly decrease $\bar{\sigma}_n$ (see Eq. 8.1) to promote rupture propagation (Noda & Lapusta, 2013). Without the sudden motion of the fault, the background P_f in the megathrust is associated with fluids brought down by the slab and metamorphic reactions within and around the fault zone (Saffer & Tobin, 2011). Experimental reports on how the background P_f may enhance rate-weakening or rate-strengthening are mixed (den Hartog et al., 2023). Nonetheless, everything else being equal, a higher P_f brings a locked fault closer to failure.

Unlike textbook explanations that involve the interaction of asperities of two surfaces in direct contact, the frictional behavior of a real fault is controlled by the composition of the fault gouge. For example, many phyllosilicate minerals, such as serpentine minerals chrysotile and lizardite and clay minerals talc, smectite, and illite, are rather weak (slippery). A gouge rich in these minerals promotes rate strengthening and thus weak creep (Ikari et al., 2011). The gouge behavior is affected by temperature that can be estimated using thermal modeling (Abers et al., 2020; Gao & Wang, 2014). For example, many clay minerals in the subducted sediments break down as temperature increases with depth, retarding rate-strengthening behavior.

The megathrust fault gouge can include the sediment brought down from the trench, material from the underside of the upper plate, and material from the igneous subducting crust. It is difficult to know in detail how the various materials are entrained into the fault zone and metamorphically modified. Direct sampling with drilling is possible only for very shallow depths. Rarely seen exhumed ancient megathrust fault zones provide valuable information, but deducing the original state of the megathrust and addressing the effects of later tectonic modification require careful analyses and various assumptions (Agard et al., 2018; Wang et al., 2023). Geophysical imaging using seismic and electromagnetic techniques can provide important information, especially on fluid contents and P_f, but the limited spatial resolution and the nonuniqueness of the interpretation present major challenges.

In summary, besides internal structure, fault frictional behavior governs seismogenesis and thus tsunamigenesis. In practice, the frictional behavior of a megathrust fault is deduced from its earthquake history and current state of locking or creep. Knowledge of the petrology of the subducting sediment and wall rocks and the thermal state of the megathrust allows us to postulate fault gouge composition and thus explain the deduced frictional behavior. The shallow part of the megathrust far offshore is usually difficult to monitor. However, the good knowledge of the sediments on the incoming plate may offer some help because the frictional behavior of the shallow megathrust is likely strongly influenced by subducting sediments.

8.2.3 Updip and downdip limits of megathrust rupture

A critically important issue in tsunami source modeling is the spatial extent of seismic rupture in the dip direction. Despite the huge body of literature on this subject, there are still many unknowns.

Some minerals in the subducted sediment cause the megathrust gouge to exhibit rate strengthening. Traditionally, it was thought that subsequent diagenetic processes would cause a change to rate weakening typically around a temperature of 150°C, defining the updip limit (Hyndman & Wang, 1993; Moore & Saffer, 2001). However, the process is now recognized to be more complicated for three reasons. (1) The transition from rate strengthening to rate weakening is not as clearly defined in nature as previously thought. For example, the transformation of clay mineral smectite to illite with increasing temperature was thought to trigger this transition (Hyndman & Wang, 1993), but it is now known that illite also promotes rate strengthening (Stanislowski et al., 2022). (2) New understanding of rupture dynamics blurs the concept of the updip limit. For example, although a rate-strengthening shallow segment cannot initiate seismic rupture, it may still participate in the rupture either passively while trying to resist rupture propagating from greater depths (Wang & He, 2008) or actively due to dynamic weakening as described in Section 8.2.2 (Noda & Lapusta, 2013). (3) Ruggedness of the subducting seafloor can impede large ruptures at the shallow end of the megathrust (Section 8.2.1). The downdip transition to seismogenic behavior depends on mechanical smoothing as well as temperature increase.

In an influential conceptual framework proposed in the 1990s (Hyndman et al., 1997; Oleskevich et al., 1999), the downdip limit is defined by either the ~350°C temperature or the mantle wedge corner (MWC, Fig. 8.2), whichever is shallower. The ~350°C temperature is thought to trigger a change from rate weakening to rate strengthening of quartz-rich fault gouge (Blanpied et al., 1991). Fluid released by the dehydrating slab causes serpentinization of the mantle wedge, at least along its base (Hyndman & Peacock, 2003), and serpentine was thought to promote rate strengthening. Today, the temperature limit still appears to be sound for very warm subduction zones, such as Cascadia, Nankai, and Mexico, where the temperature of 300°C−400°C is reached well before the megathrust intersects the MWC where episodic tremor and slip (ETS) occurs (Fig. 8.2A). Gao and Wang (2017) propose that the seismogenic zone and the ETS zone are separated by a zone of viscous or semi-frictional behavior.

In contrast, the MWC limit no longer holds. Many megathrust earthquakes over the past two decades have ruptured to depths much greater than the MWC. These events were recorded by modern seismic and geodetic networks, and their downdip limit is much better constrained than for earlier events. For example, at the Japan Trench (Fig. 8.2B) where the MWC is less than 30 km deep (Uchida et al., 2020), the 2011 M_w 9.0 Tohoku Earthquake ruptured to >50 km depth (Wang et al., 2018). In the area of the 2010 M_w 8.8 Maule Earthquake where the MWC is in the depth range of 32−40 km, the rupture extended to at least 55 km (Wang et al., 2020). Studies over the past three decades indicate that the frictional behavior of serpentine minerals is rather complex and to some degree controversial. While the weaker serpentine species chrysotile and lizardite are well known to exhibit rate strengthening, the "strong" species antigorite may exhibit rate weakening under some conditions (Wang et al., 2020). Besides, similar to the shallow end of the megathrust, a rupture may propagate into a rate-strengthening region. Where it stops depends on many known and unknown factors. In any case, it is rare to see reported megathrust ruptures extend deeper than 60 km. Below the depth of 70−80 km, the base of the hot mantle wedge is fully coupled with the slab, and the megathrust probably transitions to a zone of viscous shear (Wada & Wang, 2009).

In summary, multiple factors affect the updip and downdip limits of seismic rupture. In a smooth megathrust, although the shallow segment tends not to initiate a rupture, it may not be able to stop rupture propagation in a large earthquake, indicating that there may not be a fixed updip limit for a given subduction zone. For very warm subduction zones, the downdip limit is more likely to be confined to a depth shallower than the MWC. For other subduction zones, the definition of the downdip limit is unclear, although unlikely deeper than 60 km.

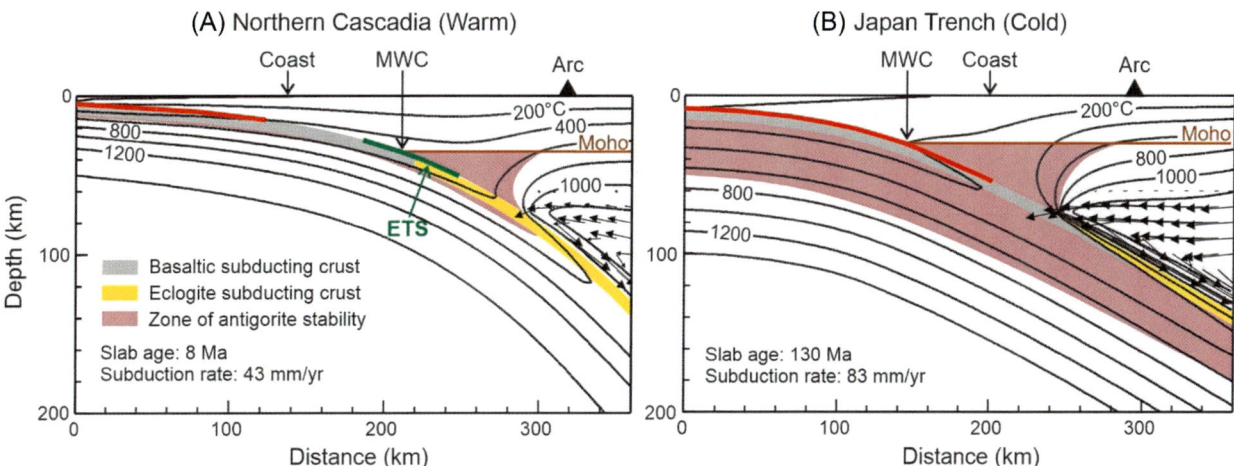

FIGURE 8.2 Downdip rupture limits in end-member warm and cold subduction zones. Model temperature and petrological zoning are based on Wada and Wang (2009). (A) In subduction zones with very young and thus warm subducting plate, the rupture zone (*red*) is shallower than the MWC and appears to be limited by a temperature of 350°C−400°C. Between the rupture zone and the zone of ETS (*green*), the fault exhibits viscous or semi-frictional behavior (Gao & Wang, 2017). (B) In colder subduction zones, the rupture zone extends deeper than the MWC, and the MWC does not harbor ETS. What controls the depth limit is not fully understood, but observed megathrust ruptures rarely exceed 60 km. *ETS*, Episodic tremor and slip; *MWC*, mantle wedge corner.

8.2.4 Intriguing strike dimension

Large tsunamigenic megathrust earthquakes have various rupture lengths. There certainly is much randomness in it because of spatiotemporal heterogeneities of the fault zone, but persistent segmentation boundaries can be recognized from a relatively good record of past earthquakes (Philibosian & Meltzner, 2020), such as along the Nankai subduction zone (Ando, 1975; Kumagai, 1996), much of the Chile margin (Molina et al., 2021; Sparkes et al., 2010), and the boundary between the 2004 M_w 9.2 and 2005 M_w 8.6 Sumatra Earthquakes (Meltzner et al., 2012). These boundaries are often associated with a rough fault as inferred from topographic features on the subducting plate (Section 8.2.1), although association with upper plate properties is also made (Bassett et al., 2016).

The striking difference between the 2004 M_w 9.2 Sumatra and 2011 M_w 9.0 Tohoku Earthquakes illustrates the importance of tectonic control on the strike dimension (Fig. 8.3). The Japan Trench megathrust exhibits large variations in slip behavior along strike (Nishikawa et al., 2019). The fast subduction rate (8–9 cm/year) combined with a long interval of locking enabled the relatively smooth central segment to produce the giant earthquake in 2011 with exceedingly large (>50 m) maximum slip but limited strike length. The segment to the north frequently produces M_w 6–8 earthquakes (Yamanaka & Kikuchi, 2004), and the segment to the south undergoes strong creep (Wang & Bilek, 2014). The Sumatra megathrust does not exhibit this degree of along-strike variability. The less heterogeneous megathrust produced the giant earthquake in 2004 with an exceedingly large strike length (~1300 km) but limited maximum slip. Nonetheless, there is no reason to believe that the next great tsunamigenic rupture will also be this long. In many situations, it is not clear to what degree the along-strike variability in earthquake history or within each earthquake reflects tectonic control or natural randomness. It continues to be an intriguing question deserving active research. However, as will be discussed in Section 8.3.4, the strike dimension may not be as critical as the dip dimension in near-field tsunami hazard analyses.

8.3 Slip distribution in predictive source scenarios

A megathrust rupture raises water to cause a tsunami through two primary mechanisms (Fig. 8.4). One is shortening of the upper plate to create a seafloor bulge, and the other is seaward translation of the sloping seafloor. A real earthquake usually shows a combination of the two mechanisms. A buried rupture intensifies the first mechanism (Fig. 8.4A), with greater shortening and thus bulging if the slip tapers more sharply toward the trench. The shallowest part that undergoes

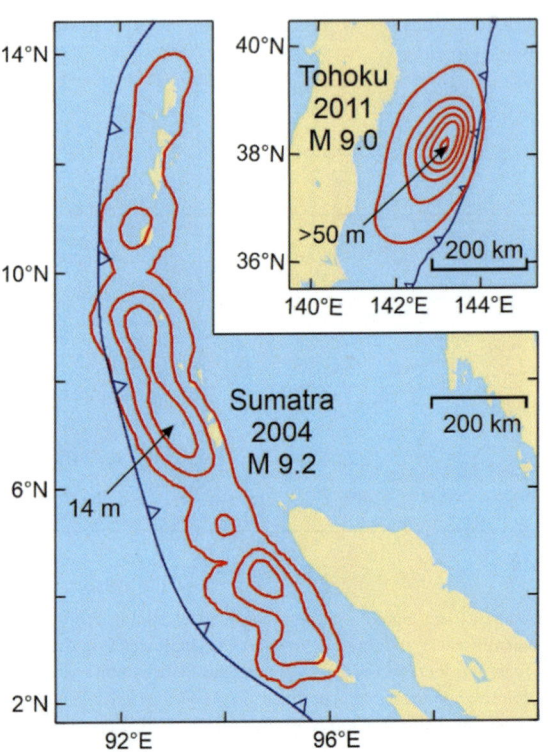

FIGURE 8.3 Contrasting rupture styles of two giant megathrust earthquakes. Their rupture areas are compared at the same spatial scale. The coseismic slip for the 2004 Sumatra Earthquake (contoured at 5 m) is a modification of Chlieh et al. (2007) provided by Hu and Wang (2012). The slip for the 2011 Tohoku Earthquake (contoured at 10 m) is from the model of Wang and Bilek (2014), but almost any published slip model for this earthquake (Brown et al., 2015) would convey the same message of large slip over a compact rupture area. *From Wang, K., Sun, T., Brown, L., Hino, R., Tomita, F., Kido, M., Iinuma, T., Kodaira, S., & Fujiwara, T. (2018). Learning from crustal deformation associated with the M9 2011 Tohoku-oki earthquake. Geosphere, 14(2), 552–571. https://doi.org/10.1130/GES01531.1.*

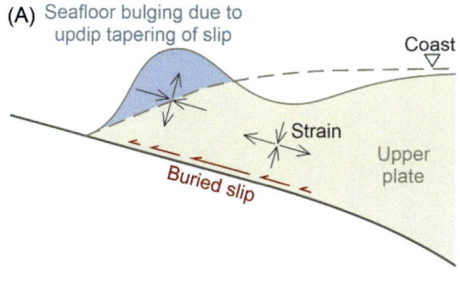
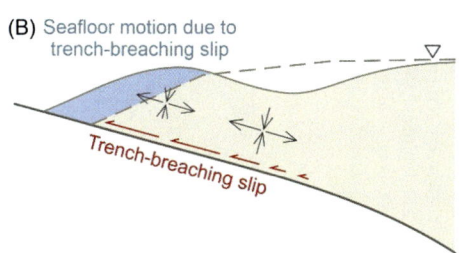

FIGURE 8.4 Two primary mechanisms of tsunamigenic seafloor uplift (blue-shaded area) in a megathrust earthquake illustrated using end-member buried and trench-breaching rupture scenarios. (A) Updip decrease in fault slip leads to upper plate shortening and hence seafloor bulging. (B) Slip of the shallow megathrust leads to the seaward motion of the sloping seafloor at the cost of seafloor bulging. A real earthquake usually shows a combination of the two mechanisms. *Modified from Carvajal, M., Sun, T., Wang, K., Luo, H., & Zhu, Y. (2022). Evaluating the tsunamigenic potential of buried versus trench-breaching megathrust slip.* Journal of Geophysical Research: Solid Earth, 127(8). https://doi.org/10.1029/2021jb023722.

no or less coseismic slip will catch up by having afterslip, interseismic creep, or rupture in a later earthquake. A trench-breaching rupture intensifies the second mechanism (Fig. 8.4B). There are many other mechanisms in the literature, but they are all variations of these two primary mechanisms.

8.3.1 Static kinematic models informed by fault mechanics

Fault slip in source models consists of two types: kinematically assigned or driven by a model stress field. Each type can be either static, featuring instantaneous rupture, or time-dependent, featuring rupture propagation. The simplest static kinematic model is that of a uniform-slip rectangular fault (Fig. 8.5A). The simplest static stress-driven model is that of a crack with prescribed uniform stress drop (Fig. 8.5B). Static stress-driven models can also be derived by prescribing a change in the frictional strength of the fault representing coseismic weakening or strengthening (Fig. 8.5C).

Dynamic rupture models are sophisticated time-dependent stress-driven models, in which fault strength changes with slip or slip rate following a friction law, and neither strength change nor stress drop needs to be prescribed (Geist & Oglesby, 2014). These models are valuable tools for investigating fundamental physics but are not suitable for routine use in predictive source modeling. In these models, preearthquake stress, location of rupture initiation, pore fluid distribution, and heterogeneous fault properties all sensitively control the rupture process, yet in reality, accurate definition of these parameters is intractable. Therefore, in predictive source modeling, it is common to use kinematic slip models that benefit from knowledge learned from the dynamic models. Because strength change is the fundamental cause of fault slip, the net slip and deformation produced by a dynamic model can also be produced by a static model of the type shown in Fig. 8.5C with appropriate friction parameters.

In near-field tsunami hazard analysis, the effect of rupture propagation and resultant temporal evolution of seafloor deformation is usually unimportant (Geist & Oglesby, 2014; Williamson et al., 2019). Megathrust rupture propagates (typically ∼2.5 km/second) an order of magnitude faster than tsunami waves propagating toward the coast and can be represented by instantaneous slip. Unless unrealistically slow rupture propagation is assumed (Riquelme et al., 2020), errors due to the neglect of rupture propagation are overshadowed by errors associated with many other parameters in these models. On some occasions, the instantaneous seafloor deformation is considered too extreme, and a fictitious rise time of tens of seconds is introduced to blunt the effect (Geist, 1998). However, the finite rise time mainly leads to smoother sea-surface uplift by allowing enough time for jagged features to attenuate. A similar role is played by the widely used Kajiura's (1963) transfer function, discussed in Chapter 2 of this book.

Although there is no technical obstacle to including real surface and fault geometry in numerical models (e.g., finite element), it is the most common to use analytical solutions of dislocation in an elastic half-space. It is convenient to integrate a point-source dislocation solution (Green's function) to simulate any slip distribution over a 3-D curved fault. However, the dislocation models feature a flat upper surface, but real subduction zones have a sloping seafloor. There are three options to handle the discrepancy as illustrated in Fig. 8.6. When using option A or B for tsunami modeling, some authors directly use model-predicted surface uplift, and some also incorporate the mechanism shown in Fig. 8.4B

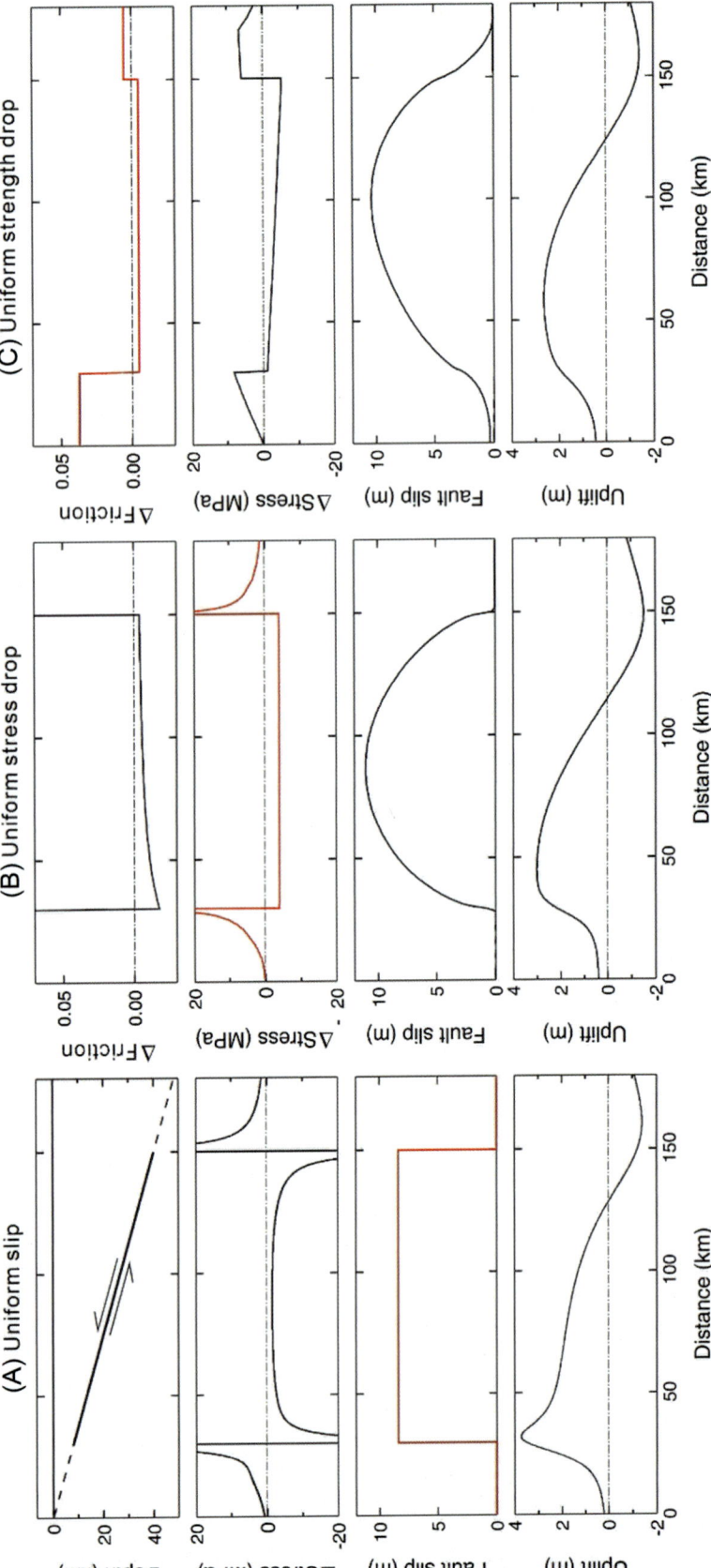

FIGURE 8.5 Simplest 2-D static deformation models in an elastic half-space with kinematically assigned or stress-driven fault slip. The models are designed to have similar seismic moments. In each model, prescribed variables are shown in red. (A) Dislocation model with uniform slip (Okada, 1985, 1992). The fault geometry shown in the top panel also applies to the other two models. (B) Crack model with uniform stress drop. The slip distribution is slightly skewed updip where the fault is shallower (lower stiffness). The top panel shows how the fault friction would need to change to obtain the uniform stress drop. (B) Static friction model with uniform weakening of the main rupture zone and associated strengthening of neighboring zones. The linear change in stress drop is due to the linear dependence of shear strength on normal stress. The slip distribution is skewed downdip because of greater stress drop. Based on Hu and Wang (2008), which explains details of the finite element modeling for (B) and (C).

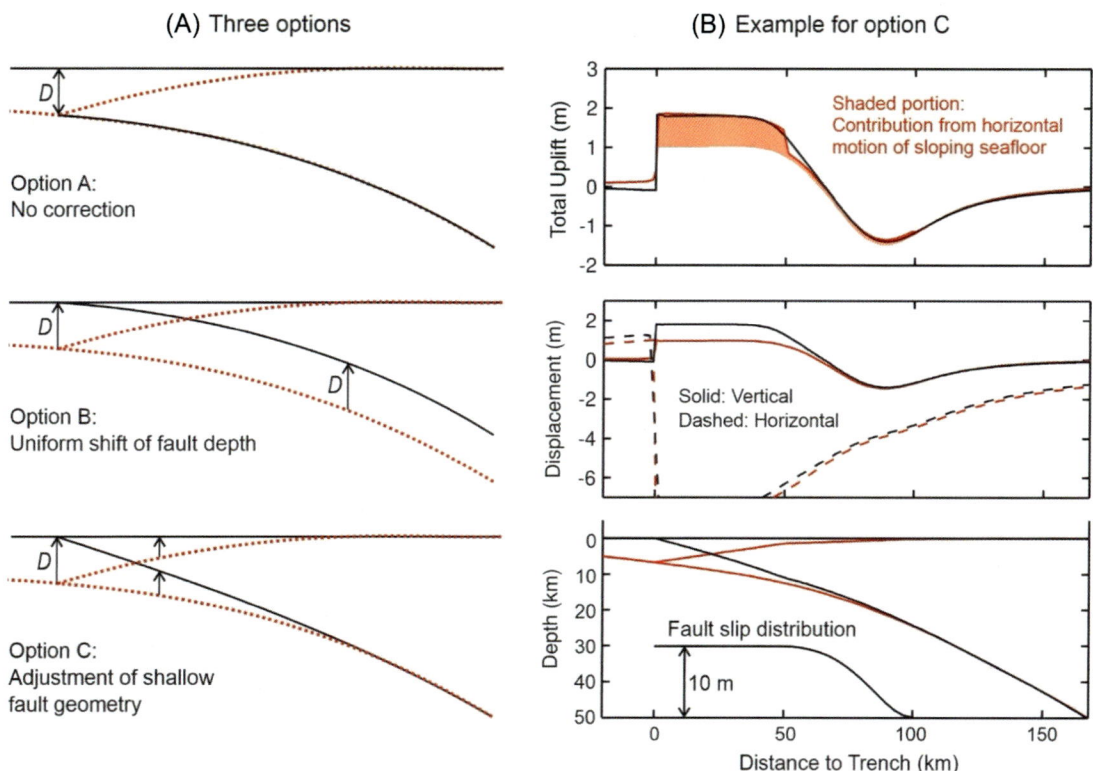

FIGURE 8.6 Options of handling real surface and fault geometry (*red*) in a flat-surface model (*black*). (A) The three options reviewed by Wang et al. (2018), where D is the water depth at the trench. (B) A trench-breaching example to show the effectiveness of option C. Bottom panel: Surface and fault geometry with (*black*) and without (*red*) the geometrical adjustment and the identical fault slip distribution assigned to both (elastic) models. Middle panel: Surface displacements predicted by both models, with uniform rigidity (any value) and Poisson's ratio (0.25). Results for the reality model (*red*) are obtained using the finite element method (Carvajal et al., 2022), and those for the flat-surface model (*black*) by integrating the point-source dislocation solution (Okada, 1985). Top: Tsunamigenic seafloor uplift from both models. For the flat-surface model, the uplift is simply the vertical displacement in the middle panel. For the reality model, the total uplift is the vertical displacement plus the uplift due to the horizontal motion of the sloping seafloor (shaded portion). The steps in the total uplift at 50 and 100 km are associated with the topographic kinks in the model geometry.

by adding uplift calculated from model-predicted horizontal displacement and the actual bathymetric slope using the method of Tanioka and Satake (1996). As discussed in Wang et al. (2018), the incorrect near-trench fault depth in option A or B causes errors in both the vertical and horizontal components of near-trench deformation, and option A cannot handle trench-breaching slip. In option C, an adjustment is made to the dip of the shallow fault so that the fault depth in the model is similar to the fault depth beneath the seafloor in reality. The greater near-trench fault dip gives rise to greater vertical displacements. Wang et al. (2018) reason that this additional uplift compensates for the missing sloping seafloor. Fig. 8.6B uses a simple model to illustrate how this compensation works. Here, the finite element model with a seafloor slope represents reality, and its total (tsunamigenic) seafloor uplift is a combination of the vertical displacement and the uplift due to the seaward motion of the sloping seafloor. In the flat-surface dislocation model, a dip adjustment is made to the fault using the seafloor bathymetry, and its predicted vertical displacement of the seafloor matches the total uplift of the reality model very well.

8.3.2 Slip distribution in the dip direction

8.3.2.1 Kinematic function of slip distribution

Basic physics requires that coseismic fault slip distribution is free of singularities in stress or displacement (i.e., infinite value or discontinuity) except where the fault breaks the surface. In Fig. 8.5A, both slip and stress are singular at rupture edges. The seafloor "spike" caused by the sudden termination of slip at the buried upper edge is a common artifact in subduction-zone tsunami modeling and is usually spikier than shown here because of a shallower depth of the edge. In Fig. 8.5B, stress is singular at both edges. The model in Fig. 8.5C is physically reasonable, and its general bell shape agrees with the slip distribution of real earthquakes. The following function proposed by Wang and He (2008), with a typographic error corrected by Wang et al. (2013), depicts a similar bell shape distribution (Fig. 8.7A and B)

$$s(x') = s_o \delta(x')[1 + \sin(\pi[\delta(x')]^b)] \tag{8.2a}$$

where $x' = x/W$ is downdip distance x from the upper edge of the rupture zone normalized by its downdip width W, s_o is peak slip, and b is a broadness parameter ranging from 0 to 0.3. The function $\delta(x')$ was proposed by Freund and Barnett (1976) (originally in derivative form)

$$\delta(x') = \begin{cases} \dfrac{6x'^2}{q^3}\left(\dfrac{q}{2} - \dfrac{x'}{3}\right) & 0 \leq x' \leq q \\ \dfrac{6(1-x')^2}{(1-q)^3}\left(\dfrac{1-q}{2} - \dfrac{1-x'}{3}\right) & q \leq x' \leq 1 \end{cases} \tag{8.2b}$$

where q is a skewness parameter ranging from 0 to 1. The slip peaks at the location $x' = q$. A lower or higher value of q causes the bell shape to be skewed updip or downdip, respectively. The skewness can strongly affect tsunami runup at the local coast because it determines the shape of the deformed seafloor that controls tsunami generation and propagation (Carvajal et al., 2022).

One consideration is the average stress drop. Stress drop (or increase) can exhibit large variations within the rupture zone, but seismological observations indicate that the stress drop averaged over the rupture area of large megathrust earthquakes is typically 1–5 MPa (Ye et al., 2016). The average stress drop scales with average slip $\bar{s}$ and characteristic rupture dimension L (Scholz, 2002):

$$\overline{\Delta\sigma} = CG\dfrac{\bar{s}}{L} \tag{8.3}$$

where C is a constant depending on rupture geometry and type (Wu et al., 2023), and G is the rigidity of the wall rock. Fault zone properties play no role in this scaling relationship. For large megathrust earthquakes, L is measured in the dip direction, and strike length does not play a role. Similar $\overline{\Delta\sigma}$ means that $\bar{s}$ and L increase or decrease more or less together between different earthquake sizes. Although there is no strict rule, a megathrust source model based on

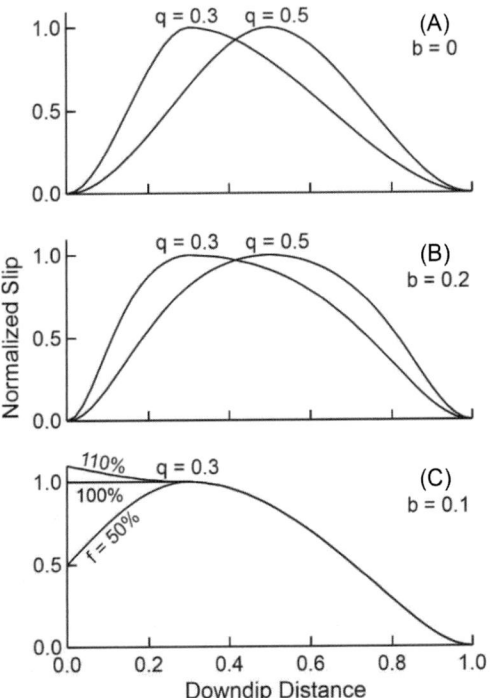

FIGURE 8.7 Examples of slip distribution based on Eqs. (8.2) and (8.4). (A and B) Distribution from Eq. (8.2) showing the effects of the skewness parameter q and broadness parameter b. (C) Distributions from Eq. (8.4) showing different degrees of trench breaching.

Eq. (8.2) with a small W but large s_o (and hence $\bar{s}$) so that $\overline{\Delta\sigma}$ greatly exceeds 5 MP is deemed less compatible with observations.

8.3.2.2 Trench-breaching slip

There can be different ways of assigning trench-breaching slip. Gao et al. (2018) and Carvajal et al. (2022) used the function below to replace the $x' \leq q$ part of Eq. (8.2A) (Fig. 8.7C):

$$s(x') = s_o \left[f + (1-f)\sin\left(\frac{\pi x'}{2q}\right) \right] \quad 0 \leq x' \leq q \tag{8.4}$$

where f is a constant describing the degree of trench breaching as a percentage of s_o and can be greater than 100%. Slip at the trench is $s(0) = fs_o$. Three important issues about trench-breaching slip should be emphasized:

1. *Tsunamigenic effects*. Of importance to hazard assessment, it is necessary to clarify a misconception that a trench-breaching rupture is much more tsunamigenic than a buried rupture. Through systematic model investigation, Carvajal et al. (2022) demonstrated a trade-off between the two mechanisms shown in Fig. 8.4. Greater trench breaching is accompanied by less or no seafloor bulging. Their results (Fig. 8.8) show that, given maximum slip, the degree of trench breaching has little control on tsunami runup. On the basis of this understanding, Carvajal et al. (2022) postulated that a "tsunami earthquake" generates a disproportionately large tsunami not simply by rupturing a very shallow segment of the megathrust but by having very large slip.
2. *Propensity*. Currently, it remains unclear whether trench-breaching rupture is common or rare. In the Tohoku Earthquake, repeated bathymetry surveys (Fujiwara et al., 2011; Sun et al., 2017) and repeated seismic surveys

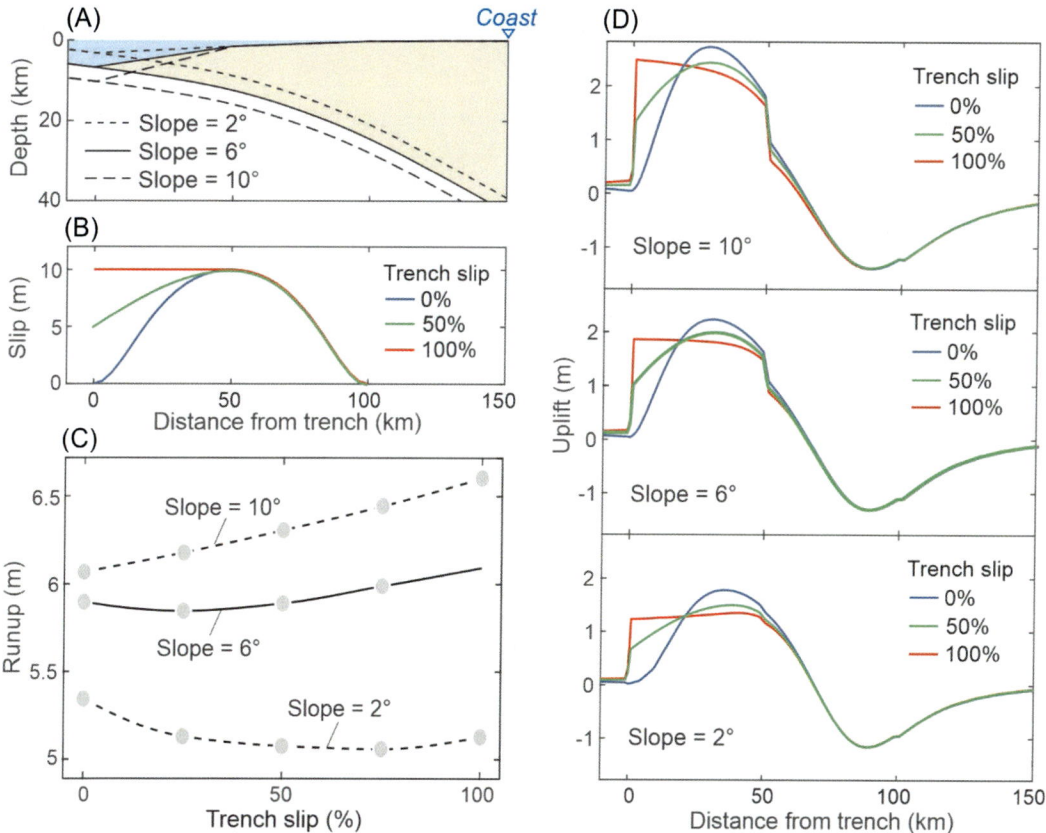

FIGURE 8.8 Model results of Carvajal et al. (2022) to illustrate the trade-off between the two mechanisms shown in Fig. 8.4 in causing tsunami runup. (A) Model geometry. (B) Slip distribution based on Eqs. (8.2) and (8.4) featuring different amounts of trench slip in the percentage of peak slip. (C) Runup as a function of trench slip for the three different slopes shown in (A). Each circle represents a model test. (D) Seafloor uplift for the three slopes in (A) and the three trench slip values in (B). Greater trench breaching is accompanied by smaller seafloor bulging, explaining the insensitivity of runup to trench slip shown in (C). The model with 6-degree slope and 100% trench slip is also shown in Figure 6B. See Carvajal et al. (2022) for details of the deformation and tsunami modeling.

(Kodaira et al., 2012) showed trench breaching to have occurred in a limited area (Kodaira et al., 2021). Similar observations are not available for other events. Some recent papers assume that the 1960 M_w 9.5 Chile, 1964 M_w 9.2 Alaska, and 2004 M_w 9.2 Sumatra Earthquakes featured trench-breaching slip (Qiu & Barbot, 2022), but there is no direct evidence for it, and numerous models can explain their tsunamis without trench breaching.

3. *Mechanism.* Because of the abundance of clay minerals in the seafloor sediment entering the subduction zone (Chester et al., 2013), the shallow megathrust is generally expected to exhibit rate strengthening, but it still can participate in a trench-breaching rupture if one or both of the two things happen. (1) It does not gain intense enough coseismic strengthening to resist the large slip from downdip (Kozdon & Dunham, 2013; Wang & He, 2008); or (2) it exhibits dynamic weakening if it acquires high enough slip rate in response to the large slip (see Section 8.2.2) (Noda & Lapusta, 2013; Ujiie et al., 2013). In dynamic rupture models, some factors, such as the free surface (seafloor) and low rigidity of the frontal wedge of the upper plate, tend to promote trench-breaching rupture (Kozdon & Dunham, 2013; Prada et al., 2021), but the predicted tendency can be readily suppressed by adjusting friction parameters to enhance coseismic strengthening of the shallow megathrust (Kuo et al., 2022; Ma & Nie, 2019; Prada et al., 2021; van Zelst et al., 2019).

8.3.2.3 Rigidity variations in the upper plate

Rock rigidity of the frontal wedge of the upper plate is very low due to sediment accretion and/or pervasive fracturing (Geist & Bilek, 2001; Sallarès & Ranero, 2019; Satake, 1994). However, with kinematically assigned slip distributions, the effect of heterogeneous rock rigidity on tsunamigenic seafloor deformation is usually very small so that a uniform half-space (Okada, 1985, 1992) is adequate for modeling. Fig. 8.9 compares deformation predicted by two models with identical fault slip (Fig. 8.9B), but one with uniform rigidity and the other with a low-rigidity wedge (Fig. 8.9A). In these flat-surface models, the effect of surface topography is assumed to have been accounted for by a fault dip adjustment following option C in Fig. 8.6. For either a buried rupture or a 100% trench-breaching rupture, the vastly different

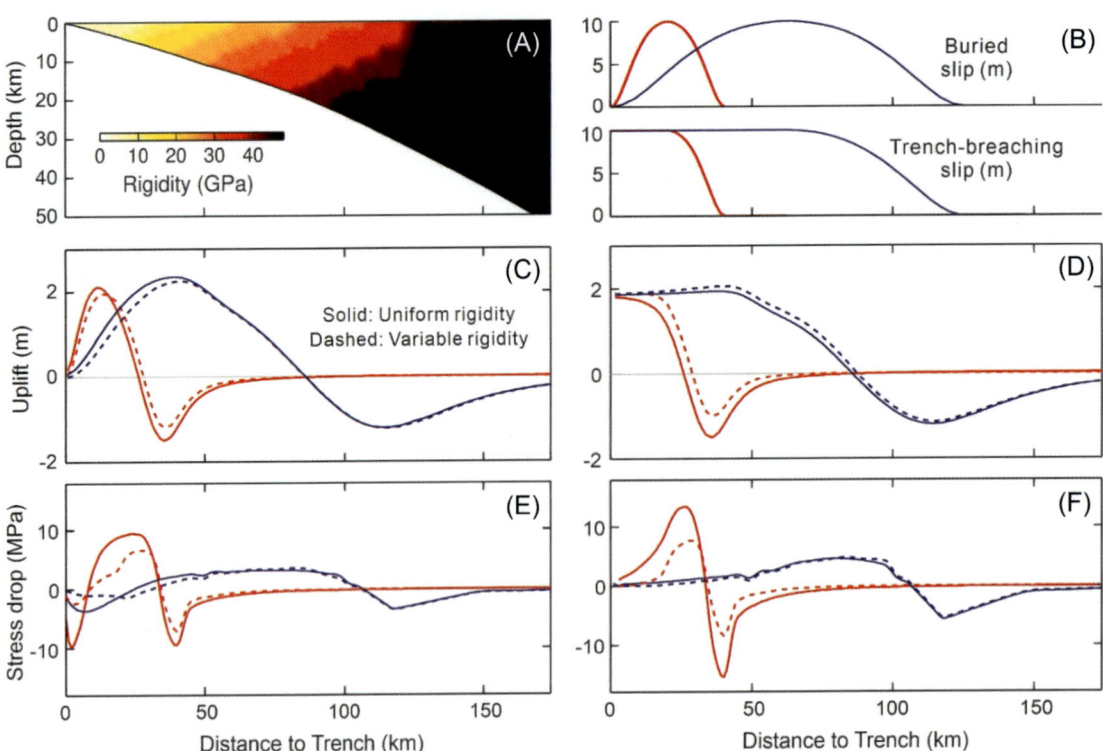

FIGURE 8.9 Comparison of deformation and stress drop predicted by models with uniform or variable rigidity. (A) The model of variable rigidity features a low-rigidity forearc wedge based on Sallarès and Ranero (2019); footwall rigidity is uniformly 48 GPa (not colored). The model of uniform rigidity (48 GPa) is not shown. Poisson's ratio is 0.25 in both models. (B) Two rupture modes and two downdip widths for each mode used for this comparison. (C and D) Surface uplift predicted by the buried and trench-breaching models, respectively, showing similar deformation for the two rigidity models. (E and F) Fault stress drop predicted by the buried and trench-breaching models, respectively, showing large differences for some slip distributions.

rigidity structures lead to very small differences in tsunamigenic surface deformation (Fig. 8.9C and D), although rather different stress drops if fault slip occurs mostly along the base of the low-rigidity wedge (Fig. 8.9E and F).

These examples also serve to reemphasize the important message that tsunamigenic seafloor deformation depends directly on slip regardless of seismic moment. Individual pairs of models in Fig. 8.9 with contrasting rigidities but identical slip have very different seismic moments but produce similar seafloor deformation. Conversely, for a fixed moment, a lower rigidity implies a larger slip (Section 8.1). In tsunami earthquakes, a very large shallow slip combined with a low-rigidity frontal wedge can explain the larger tsunamis than expected from their moment magnitude (Satake, 1994). However, the shallow slip over many earthquakes cannot exceed the long-term slip budget (Satake, 2015; Scala et al., 2020). See the discussion at the end of Section 8.3.2.2 for how rigidity may affect trench-breaching shallow slip if the slip is not kinematically assigned.

8.3.3 Off-megathrust permanent deformation

The vast majority of tsunami models for real earthquakes in the literature can explain observations with the two primary mechanisms shown in Fig. 8.4. If the assumed slip is large enough, it can also explain tsunami earthquakes (Cheung et al., 2022). However, permanent deformation of the upper plate, especially the frontal area, may take place during megathrust earthquakes. Fig. 8.10 shows several proposed modes of tsunamigenic permanent deformation seen in the literature. These scenarios can be viewed as variations of the two primary mechanisms, and the trade-off principle discussed in Section 8.3.2.2 and illustrated in Fig. 8.8 still applies. Mega-splay is a variation of megathrust rupture with the shallow part of the fault having a much greater dip. Frontal thrust and popup are trench-breaching ruptures in the presence of thick trench sediment. Multiple faulting and diffuse permanent deformation are variations of buried rupture with some of the seafloor bulging being permanent deformation.

Mega-splay. The potentially most tsunamigenic mode of off-megathrust permanent deformation is probably the activation of a mega-splay. Here, mega-splay refers to a mature thrust fault that soles into the megathrust, is located at some distance from the trench, and marks a distinct major structural boundary (Fig. 8.10A). Because of its deep rooting

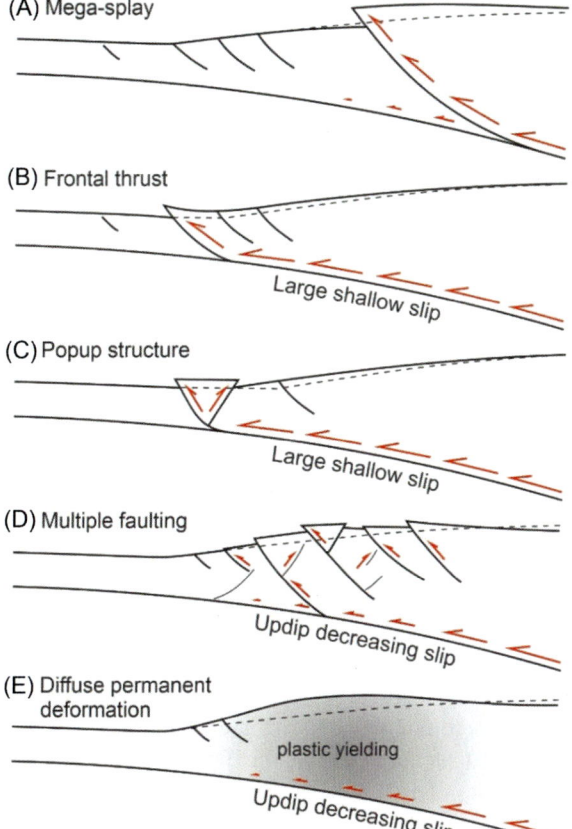

FIGURE 8.10 Different modes of off-megathrust permanent coseismic deformation proposed in the literature. (A through E) Schematic illustrations of the different modes as labeled in each panel.

and large strike length, its coseismic activation can enhance seafloor uplift over a large area. Such a mega-splay is clearly seen in seismic imaging at the Nankai subduction zone and was speculated to be responsible for enhancing tsunami generation in the 1946 M_w 8.2 Earthquake (Baba & Cummins, 2005).

Frontal thrust. A single frontal thrust is a greatly downsized version of the mega-splay (Fig. 8.10B). At sediment-rich margins, simple trench-breaching slip as shown in Figs. 8.4B and 8.8 cannot occur. The large slip propagating to the trench must break the sediment cover, and the activation of a single frontal fault is the simplest way to do it. Frontal thrusts are shallowly rooted, and in real subduction zones, they are short along strike. In the hypothetical situation of a very large strike length, a single frontal thrust could cause a greater tsunami than simple trench breaching, especially if the sediment cover is very thick (Carvajal et al., 2022). A slight modification of the frontal thrust model is the frontal fault ramp (Felix et al., 2022). It causes local uplift at the top of the ramp. In Felix et al. (2022), the narrow uplift is damped by the water column and makes little contribution to tsunami generation.

Popup structure. In another scenario of trench-breaching slip breaking thick sediment, a seaward vergent thrust and a landward vergent back thrust are activated together in one earthquake, causing the wedge-shaped volume between them to popup (Fig. 8.10C) (Hananto et al., 2020; Kuncoro et al., 2015). Because of the shallow rooting of the thrust faults and thus the short wavelength of the popup structure, the effect on enhancing tsunami generation is not expected to be very different from that of a single frontal thrust (or frontal fault ramp).

Multiple faulting. In this scenario, multiple faults are activated in one earthquake over a broad area (Fig. 8.10D). Qiu and Barbot (2022) envisioned the activation of many small faults within 15–50 km of the trench. Using a 2-D numerical stress-driven model based on van Zelst et al. (2019), van Zelst et al. (2022) simulated the activation of several seaward vergent thrust faults in a single earthquake over distances 50–200 km from the trench, with the most landward one akin to a mega-splay. Although the latter work portrays unrealistic geology, it quantitatively illustrates an important point, that is, the horizontal shortening through multiple faulting requires an updip decrease in megathrust slip, similar to how the seafloor bulge is generated in the elastic buried-rupture models shown in Figs. 8.4A and 8.8. The model of van Zelst et al. (2022) features strong coseismic strengthening of the shallow megathrust, with a factor of 2.5 increases in friction coefficient (0.35–0.875), such that the slip tapers to zero at the trench.

Diffuse permanent deformation. A continuum version of multiple faulting is diffuse plastic yielding during the dynamic rupture process (Fig. 8.10E) (Ma, 2012). It also results from horizontal shortening associated with an updip decrease in megathrust slip. In different numerical models of this scenario (Ma & Nie, 2019; Ma, 2012; Wilson & Ma, 2021), the updip slip decrease is a result of coseismic strengthening of the shallow megathrust, feedback with upper plate plastic yielding, and/or kinematically preventing the slip from breaking the sediment at the trench. Similar to the elastic buried-rupture model (Fig. 8.4A), a sharper slip gradient leads to greater shortening of the upper plate, which in turn leads to greater thickening and hence seafloor uplift.

For constructing predictive tsunami source scenarios using a static elastic model with prescribed fault slip, some of these off-megathrust deformations can be readily included, such as mega-splay and frontal thrust. Others, such as popup and multiple faulting, may be accomplished by summing models of individual faults, with kinematically compatible total slip along the megathrust. However, there are a few practical considerations.

1. There are questions regarding the necessity and validity of explicitly invoking some of these deformation modes. Numerous faults form in offshore forearc in the process of sediment accretion or tectonic erosion (Clift & Vannucchi, 2004), but their maturity and along-strike continuity tend to be poor. The limited strike dimension also limits coseismic displacement in the strike-normal direction. If a single short fault is activated, the effect on tsunami hazard is minimal. If numerous small faults are simultaneously activated, the process is better described as diffuse permanent deformation (Fig. 8.10E).
2. For a buried rupture scenario (Figs. 8.4A and 8.10E), given the same megathrust slip gradient, whether the seafloor bulging is due to elastic deformation or diffuse permanent deformation is of second-order importance. Different rheologies will affect details of the predicted seafloor deformation, but the difference is undoubtedly much smaller than errors due to overarching assumptions in either type of model.
3. It cannot be overemphasized that when introducing the added effects shown in Fig. 8.10, the model should use the relevant primary processes that drive the permanent deformation (Figs. 8.4 and 8.8). Although frontal thrust and popup produce locally high uplift, the large slip of the shallow megathrust that triggers them retards seafloor bulging farther landward, as shown in Fig. 8.8. Overlooking this trade-off leads to double counting and hence overprediction of tsunami generation. For example, in the popup model for the 2010 Mentawai Tsunami Earthquake by Hananto et al. (2020), the broad seafloor bulging predicted by the buried rupture of Hill et al. (2012) and a huge (8 m) local popup due to trench-breaching rupture were both included.

8.3.4 Along-strike variability of rupture behavior

It is challenging to know to what degree the along-strike variability of megathrust rupture behavior is influenced by tectonic control and to what degree by natural randomness (Section 8.2.4). In the terminology of probabilistic hazard analysis, they respectively contribute to epistemic and aleatoric uncertainties. The distinction between them is not sharp, though, since statistical characteristics of the randomness may have tectonic reasons. There are two end-member approaches in kinematically constructing slip scenarios.

One end member emphasizes tectonic control and is represented by the currently used tsunami hazard model of the State of Oregon (Priest et al., 2010; Witter et al., 2011). In this model, along-strike variability is based on Holocene megathrust earthquake history along the Cascadia margin inferred from offshore turbidity records (Goldfinger et al., 2017), assuming that persistent patterns seen in this history reflect local tectonics. The history includes both less frequent full-margin ruptures and more frequent shorter ruptures, with the latter tending to occur toward the southern end of the margin. Accordingly, the Oregon hazard model includes several strike segments with different recurrence intervals, and the peak coseismic slip for each segment is assigned to be consistent with the recurrence interval and the rate of subduction (Witter et al., 2011).

The other end member emphasizes natural randomness and assumes coseismic slip distributions to be realizations of a random field (i.e., spatial stochastic process) (Geist & Bilek, 2001; Geist et al., 2022). The random field applies to all directions along the fault, but here, we only focus on its ability to describe along-strike variability. Numerous slip distributions can be quickly generated using a numerical random-field generator. Parameters chosen for the random-field generator necessarily reflect the researchers' understanding or assumption of the natural process, such as parameters that control the smoothness of the random field. The stochastic approach not only brings operational convenience but also allows tectonic processes that are difficult to constrain, such as those discussed in Section 8.2.2, to be treated as natural randomness. However, there is a risk of losing relevance if tectonic processes that often can be constrained to some degree, such as those discussed in Section 8.2.1, are ignored.

Obviously, some middle ground between the two end members is more appropriate. For example, Goda (2022) was able to use the pattern of along-strike variations inferred from the turbidity records to guide the generation of stochastic slip scenarios for Cascadia, essentially combining the Oregon approach and the purely stochastic approach. Using thousands of stochastically generated Cascadia megathrust slip scenarios, Goda (2022) also quantitatively verified what appears to be an obvious expectation: for a given coastal area, the slip directly offshore dominates the tsunami hazard, and contributions from slip elsewhere along strike diminish with increasing distance.

There is another obvious expectation: for a given coastal area, a tsunami is very large if the slip directly offshore is very large. This was verified by Melgar et al. (2019) using stochastically generated models. However, contrary to what these authors argued in their paper, this point holds regardless of whether slip distribution is heterogeneous or uniform along strike. Again, slip instead of moment is the more fundamental parameter in tsunami hazard analysis (Section 8.1). For this reason, a single along-strike uniform slip model may serve the purpose of numerous heterogeneous models. This is why tsunami hazard models recently developed by U.S. states along the Cascadia margin used piecewise uniform (in the strike dimension) slip models to calculate tsunami waves. Here, "along-strike uniform" refers to either uniform slip or, if the subduction rate varies along strike, uniform recovery of slip deficit accrued over a specified time window (Witter et al., 2011). A long rupture may exhibit very large slip in one fault patch but smaller slip elsewhere. If the occurrence of very large slip is deemed uniformly rare along strike, a uniform model with very large slip and very long recurrence is adequate, although rigorous determination of their probabilistic recurrence interval deserves further research. In other words, along-strike variations (space) are cast into recurrence behavior (time) at each location. Examples are the XL or XXL scenarios in Witter et al. (2011).

8.4 Concluding remarks

To constrain near-field megathrust source scenarios, the most direct information comes from local earthquake and tsunami histories and the current state of megathrust locking. General characteristics of global megathrust earthquakes provide average scaling relationships regarding rupture dimension and slip size and average parameters for stochastic generation of rupture scenarios. However, megathrust seismogenic behavior persistently differs between different subduction zones (e.g., Figs. 8.1–8.3), which cannot be adequately represented by global averages. It is important to study how local tectonics affect earthquake and tsunami processes.

To inform predictive source scenarios, useful tectonic information includes the structure of the megathrust fault zone that is strongly influenced by the roughness of the incoming seafloor and the process of sediment accretion or subduction (Section 8.2.1), the rheology or frictional behavior of the megathrust (Section 8.2.2), and the thermo-petrologic and mechanical constraints of the updip and downdip rupture limits (Section 8.2.3). Many of the parameters are still poorly known, but the accuracy of the source scenarios will improve with advancing scientific research.

For hazard assessment, it is feasible to use source models with prescribed fault slip, but the slip assignment should be informed by fault mechanics, including knowledge learned from dynamic rupture models (Section 8.3.1). A uniform elastic half-space (Okada, 1985, 1992) is generally adequate. The missing seafloor slope can be compensated by a fault dip adjustment to approximately preserve fault depth below the seafloor (Fig. 8.6). Fault slip instead of seismic moment directly controls tsunamigenic seafloor deformation. If fault slip is prescribed, heterogenous rock rigidity plays a minimal role in controlling surface deformation (Fig. 8.9).

The two primary mechanisms of tsunamigenic deformation are (1) seafloor bulging due to shortening and (2) seaward motion of the sloping seafloor (Fig. 8.4). It is important to recognize the trade-off between them. The dominance of one mechanism is always at the cost of the other, so that, for the same maximum slip, trench-breaching rupture is generally not more tsunamigenic than buried rupture (Fig. 8.8). Each primary mechanism may be accompanied by additional off-megathrust permanent deformation (Fig. 8.10). It is neither necessary nor valid to incorporate complex permanent deformation in kinematic source modeling. Simplifying approaches should be sought, such as using a frontal thrust to accommodate trench-breaching rupture in the presence of thick trench sediment and using elastic deformation to approximate diffuse permanent deformation.

Along-strike variations in seismogenic behavior reflect both tectonics and natural randomness. However, for a given coastal region, tsunami hazard is dominated by large slip directly offshore. Along-strike uniform slip models can serve the purpose of heterogeneous models if recurrence behavior in individual parts of the margin is established.

References

Abers, G. A., van Keken, P. E., & Wilson, C. R. (2020). Deep decoupling in subduction zones: Observations and temperature limits. *Geosphere*, *16*(6), 1408−1424. Available from https://doi.org/10.1130/GES02278.1, http://geosphere.gsapubs.org/content/by/year.

Agard, P., Plunder, A., Angiboust, S., Bonnet, G., & Ruh, J. (2018). The subduction plate interface: Rock record and mechanical coupling (from long to short timescales). *Lithos*, *320-321*, 537−566. Available from https://doi.org/10.1016/j.lithos.2018.09.029.

Ando, M. (1975). Source mechanisms and tectonic significance of historical earthquakes along the Nankai Trough, Japan. *Tectonophysics*, *27*(2), 119−140. Available from https://doi.org/10.1016/0040-1951(75)90102-x.

Baba, T., & Cummins, P. R. (2005). Contiguous rupture areas of two Nankai Trough earthquakes revealed by high-resolution tsunami waveform inversion. *Geophysical Research Letters*, *32*(8), 1−4. Available from https://doi.org/10.1029/2004GL022320.

Bassett, D., Sandwell, D. T., Fialko, Y., & Watts, A. B. (2016). Upper-plate controls on co-seismic slip in the 2011 magnitude 9.0 Tohoku-oki earthquake. *Nature*, *531*(7592), 92−96. Available from https://doi.org/10.1038/nature16945, http://www.nature.com/nature/index.html.

Bassett, D., & Watts, A. B. (2015). Gravity anomalies, crustal structure, and seismicity at subduction zones: 1. Seafloor roughness and subducting relief. *Geochemistry, Geophysics, Geosystems*, *16*(5), 1508−1540. Available from https://doi.org/10.1002/2014GC005684, http://onlinelibrary.wiley.com/journal/10.1002/(ISSN)1525-2027.

Blanpied, M. L., Lockner, D. A., & Byerlee, J. D. (1991). Fault stability inferred from granite sliding experiments at hydrothermal conditions. *Geophysical Research Letters*, *18*(4), 609−612. Available from https://doi.org/10.1029/91gl00469.

Brizzi, S., Sandri, L., Funiciello, F., Corbi, F., Piromallo, C., & Heuret, A. (2018). Multivariate statistical analysis to investigate the subduction zone parameters favoring the occurrence of giant megathrust earthquakes. *Tectonophysics*, *728-729*, 92−103. Available from https://doi.org/10.1016/j.tecto.2018.01.027.

Brown, L., Wang, K., & Sun, T. (2015). Static stress drop in the Mw 9 Tohoku-oki earthquake: Heterogeneous distribution and low average value. *Geophysical Research Letters*, *42*(24), 10,595−10,600. Available from https://doi.org/10.1002/2015gl066361.

Burgmann, R. (2018). The geophysics, geology, and mechanics of slow fault slip. *Earth and Planetary Science Letters*, *495*, 112−134.

Carvajal, M., Sun, T., Wang, K., Luo, H., & Zhu, Y. (2022). Evaluating the tsunamigenic potential of buried versus trench-breaching megathrust slip. *Journal of Geophysical Research: Solid Earth*, *127*(8). Available from https://doi.org/10.1029/2021jb023722.

Chester, F. M., Rowe, C., Ujiie, K., Kirkpatrick, J., Regalla, C., Remitti, F., Moore, J. C., Toy, V., Wolfson-Schwehr, M., Bose, S., Kameda, J., Mori, J. J., Brodsky, E. E., Eguchi, N., & Toczko, S. (2013). Structure and composition of the plate-boundary slip zone for the 2011 Tohoku-oki earthquake. *Science*, *342*(6163), 1208−1211. Available from https://doi.org/10.1126/science.1243719, http://www.sciencemag.org/content/342/6163/1208.full.pdf.

Cheung, K. F., Lay, T., Sun, L., & Yamazaki, Y. (2022). Tsunami size variability with rupture depth. *Nature Geoscience*, *15*(1), 33−36. Available from https://doi.org/10.1038/s41561-021-00869-z, http://www.nature.com/ngeo/index.html.

Chlieh, M., Avouac, J.-P., Hjorleifsdottir, V., Song, T.-R. A., Ji, C., Sieh, K., Sladen, A., Hebert, H., Prawirodirdjo, L., Bock, Y., & Galetzka, J. (2007). Coseismic slip and afterslip of the Great Mw 9.15 Sumatra−Andaman earthquake of 2004. *Bulletin of the Seismological Society of America*, *97*(1A), S152−S173. Available from https://doi.org/10.1785/0120050631.

Clift, P., & Vannucchi, P. (2004). Controls on tectonic accretion versus erosion in subduction zones: Implications for the origin and recycling of the continental crust. *Reviews of Geophysics, 42*(2). Available from https://doi.org/10.1029/2003rg000127.

Felix, R. P., Hubbard, J. A., Moore, J. D. P., & Switzer, A. D. (2022). The role of frontal thrusts in tsunami earthquake generation. *Bulletin of the Seismological Society of America, 112*(2), 680−694. Available from https://doi.org/10.1785/0120210154, http://www.bssaonline.org/.

Freund, L. B., & Barnett, D. M. (1976). A two-dimensional analysis of surface deformation due to dip-slip faulting. *Bulletin of the Seismological Society of America, 66*(3), 667−675.

Fujie, G., Kodaira, S., Nakamura, Y., Morgan, J. P., Dannowski, A., Thorwart, M., Grevemeyer, I., & Miura, S. (2020). Spatial variations of incoming sediments at the northeastern Japan arc and their implications for megathrust earthquakes. *Geology, 48*(6), 614−619. Available from https://doi.org/10.1130/g46757.1.

Fujiwara, T., Kodaira, S., No, T., Kaiho, Y., Takahashi, N., & Kaneda, Y. (2011). The 2011 Tohoku-oki earthquake: Displacement reaching the trench axis. *Science (New York, N.Y.), 334*(6060). Available from https://doi.org/10.1126/science.1211554, http://www.sciencemag.org/content/334/6060/1240.full.pdf.

Gao, D., Wang, K., Insua, T. L., Sypus, M., Riedel, M., & Sun, T. (2018). Defining megathrust tsunami source scenarios for northernmost Cascadia. *Natural Hazards, 94*(1), 445−469. Available from https://doi.org/10.1007/s11069-018-3397-6.

Gao, X., & Wang, K. (2014). Strength of stick-slip and creeping subduction megathrusts from heat flow observations. *Science, 345*(6200), 1038−1041. Available from https://doi.org/10.1126/science.1255487.

Gao, X., & Wang, K. (2017). Rheological separation of the megathrust seismogenic zone and episodic tremor and slip. *Nature, 543*(7645), 416−419. Available from https://doi.org/10.1038/nature21389.

Geist, E. L. (1998). Local tsunamis and earthquake source parameters. *Advances in Geophysics, 39*, 117−209. Available from https://doi.org/10.1016/S0065-2687(08)60276-9.

Geist, E. L., & Bilek, S. L. (2001). Effect of depth-dependent shear modulus on tsunami generation along subduction zones. *Geophysical Research Letters, 28*(7), 1315−1318. Available from https://doi.org/10.1029/2000GL012385.

Geist, E. L., & Oglesby, D. D. (2014). Encyclopedia of earthquake engineering. Earthquake mechanism and seafloor deformation for tsunami generationIn M. Beer, I. Kougioumtzoglou, E. Patelli, & I. K. Au (Eds.), Springer. Available from 10.1007/978-3-642-36197-5_296-1.

Geist, E. L., Oglesby, D. D., & Ryan, K. J. (2022). Complexity in Tsunamis, Volcanoes, and their Hazards. In R. I. Tilling (Ed.), *Tsunamis: Stochastic models of occurrence and generation mechanisms*. Springer. Available from 10.1007/978-1-0716-1705-2_595.

Goda, K. (2022). Stochastic source modeling and tsunami simulations of cascadia subduction earthquakes for Canadian Pacific coast. *Coastal Engineering Journal, 64*(4), 575−596. Available from https://doi.org/10.1080/21664250.2022.2139918.

Goldfinger, C., Galer, S., Beeson, J., Hamilton, T., Black, B., Romsos, C., Patton, J., Hans Nelson, C., Hausmann, R., & Morey, A. (2017). The importance of site selection, sediment supply, and hydrodynamics: A case study of submarine paleoseismology on the northern Cascadia margin, Washington USA. *Marine Geology, 384*, 4−46. Available from https://doi.org/10.1016/j.margeo.2016.06.008.

Han, S., Carbotte, S. M., Canales, J. P., Nedimovic, M. R., Carton, H., Gibson, J. C., & Horning, G. W. (2016). Seismic reflection imaging of the Juan de Fuca plate from ridge to trench: New constraints on the distribution of faulting and evolution of the crust prior to subduction. *Journal of Geophysical Research: Solid Earth, 121*(3), 1849−1872. Available from https://doi.org/10.1002/2015JB012416, http://onlinelibrary.wiley.com/journal/10.1002/(ISSN)2169-9356.

Hananto, N. D., Leclerc, F., Li, L., Etchebes, M., Carton, H., Tapponnier, P., Qin, Y., Avianto, P., Singh, S. C., & Wei, S. (2020). Tsunami earthquakes: Vertical pop-up expulsion at the forefront of subduction megathrust. *Earth and Planetary Science Letters, 538*. Available from https://doi.org/10.1016/j.epsl.2020.116197.

den Hartog, S. A. M., Marone, C., & Saffer, D. M. (2023). Frictional behavior downdip along the subduction megathrust: Insights from laboratory experiments on exhumed samples at in situ conditions. *Journal of Geophysical Research: Solid Earth, 128*(1). Available from https://doi.org/10.1029/2022jb024435.

Hill, E. M., Borrero, J. C., Huang, Z., Qiu, Q., Banerjee, P., Natawidjaja, D. H., Elosegui, P., Fritz, H. M., Suwargadi, B. W., Pranantyo, I. R., Li, L. L., Macpherson, K. A., Skanavis, V., Synolakis, C. E., & Sieh, K. (2012). The 2010 Mw 7.8 Mentawai earthquake: Very shallow source of a rare tsunami earthquake determined from tsunami field survey and near-field GPS data. Blackwell Publishing Ltd, Singapore. *Journal of Geophysical Research: Solid Earth, 117*(6). Available from https://doi.org/10.1029/2012JB009159, http://onlinelibrary.wiley.com/journal/10.1002/(ISSN)2169-9356.

Hu, Y., & Wang, K. (2008). Coseismic strengthening of the shallow portion of the subduction fault and its effects on wedge taper. *Journal of Geophysical Research, 113*(B12). Available from https://doi.org/10.1029/2008jb005724.

Hu, Y., & Wang, K. (2012). Spherical-Earth finite element model of short-term postseismic deformation following the 2004 Sumatra earthquake. *Journal of Geophysical Research: Solid Earth, 117*(B5). Available from https://doi.org/10.1029/2012jb009153.

Hyndman, R. D., & Peacock, S. M. (2003). Serpentinization of the forearc mantle. *Earth and Planetary Science Letters, 212*(3-4), 417−432. Available from https://doi.org/10.1016/S0012-821X(03)00263-2, http://www.sciencedirect.com/science/journal/0012821X/321-322.

Hyndman, R. D., & Wang, K. (1993). Thermal constraints on the zone of major thrust earthquake failure: The Cascadia Subduction Zone. *Journal of Geophysical Research: Solid Earth, 98*(B2), 2039−2060. Available from https://doi.org/10.1029/92jb02279.

Hyndman, R. D., Yamano, M., & Oleskevich, D. A. (1997). The seismogenic zone of subduction thrust faults. *The Island Arc, 6*(3), 244−260. Available from https://doi.org/10.1111/j.1440-1738.1997.tb00175.x.

Ikari, M. J., Marone, C., & Saffer, D. M. (2011). On the relation between fault strength and frictional stability. *Geology, 39*(1), 83−86. Available from https://doi.org/10.1130/G31416.1, http://geology.gsapubs.org/content/39/1/83.full.pdf.

Kajiura, K. (1963). The leading wave of a tsunami. *Bulletin of the Earthquake Research Institute, 41*(3), 535–571.

Kodaira, S., Iinuma, T., & Imai, K. (2021). Investigating a tsunamigenic megathrust earthquake in the Japan Trench. *Science, 371*(6534). Available from https://doi.org/10.1126/science.abe1169.

Kodaira, S., No, T., Nakamura, Y., Fujiwara, T., Kaiho, Y., Miura, S., Takahashi, N., Kaneda, Y., & Taira, A. (2012). Coseismic fault rupture at the trench axis during the 2011 Tohoku-oki earthquake. *Nature Geoscience, 5*(9), 646–650. Available from https://doi.org/10.1038/ngeo1547.

Kozdon, J. E., & Dunham, E. M. (2013). Rupture to the trench: Dynamic rupture simulations of the 11 March 2011 Tohoku earthquake. *Bulletin of the Seismological Society of America, 103*(2B), 1275–1289. Available from https://doi.org/10.1785/0120120136.

Kumagai, H. (1996). Time sequence and the recurrence models for large earthquakes along the Nankai Trough revisited. *Geophysical Research Letters, 23*(10), 1139–1142. Available from https://doi.org/10.1029/96gl01037.

Kuncoro, A. K., Cubas, N., Singh, S. C., Etchebes, M., & Tapponnier, P. (2015). Tsunamigenic potential due to frontal rupturing in the Sumatra locked zone. *Earth and Planetary Science Letters, 432*, 311–322. Available from https://doi.org/10.1016/j.epsl.2015.10.007, http://www.sciencedirect.com/science/journal/0012821X/321-322.

Kuo, S.-T., Duan, B., & Meng, Q. (2022). Comparing roles of fault friction and upper-plate rigidity in depth-dependent rupture characteristics of megathrust earthquakes. *ESS Open Archive*. Available from https://doi.org/10.1002/essoar.10511621.1.

Lallemand, S., Peyret, M., van Rijsingen, E., Arcay, D., & Heuret, A. (2018). Roughness characteristics of oceanic seafloor prior to subduction in relation to the seismogenic potential of subduction zones. *Geochemistry, Geophysics, Geosystems, 19*(7), 2121–2146. Available from https://doi.org/10.1029/2018GC007434, http://agupubs.onlinelibrary.wiley.com/agu/journal/10.1002/(ISSN)1525-2027/.

Liu, Y., & Rubin, A. M. (2010). Role of fault gouge dilatancy on aseismic deformation transients. *Journal of Geophysical Research: Solid Earth, 115*(10). Available from https://doi.org/10.1029/2010JB007522, http://onlinelibrary.wiley.com/journal/10.1002/(ISSN)2169-9356.

Ma, S. (2012). A self-consistent mechanism for slow dynamic deformation and tsunami generation for earthquakes in the shallow subduction zone. *Geophysical Research Letters, 39*(11). Available from https://doi.org/10.1029/2012gl051854.

Ma, S., & Nie, S. (2019). Dynamic wedge failure and along-arc variations of tsunamigenesis in the Japan Trench margin. *Geophysical Research Letters, 46*(15), 8782–8790. Available from https://doi.org/10.1029/2019gl083148.

Marone, C. (1998). Laboratory-derived friction laws and their application to seismic faulting. *Annual Review of Earth and Planetary Sciences, 26*(1), 643–696. Available from https://doi.org/10.1146/annurev.earth.26.1.643.

Melgar, D., Williamson, A. L., & Salazar-Monroy, E. F. (2019). Differences between heterogenous and homogenous slip in regional tsunami hazards modelling. *Geophysical Journal International, 219*(1), 553–562. Available from https://doi.org/10.1093/gji/ggz299, http://gji.oxfordjournals.org/.

Meltzner, A. J., Sieh, K., Chiang, H. W., Shen, C. C., Suwargadi, B. W., Natawidjaja, D. H., Philibosian, B., & Briggs, R. W. (2012). Persistent termini of 2004- and 2005-like ruptures of the Sunda megathrust. *Journal of Geophysical Research: Solid Earth, 117*(4). Available from https://doi.org/10.1029/2011JB008888, http://onlinelibrary.wiley.com/journal/10.1002/(ISSN)2169-9356.

Molina, D., Tassara, A., Abarca, R., Melnick, D., & Madella, A. (2021). Frictional segmentation of the Chilean megathrust from a multivariate analysis of geophysical, geological, and geodetic data. *Journal of Geophysical Research: Solid Earth, 126*(6). Available from https://doi.org/10.1029/2020jb020647.

Moore, J. C., & Saffer, D. (2001). Updip limit of the seismogenic zone beneath the accretionary prism of Southwest Japan: An effect of diagenetic to low-grade metamorphic processes and increasing effective stress. *Geology, 29*(2), 183–186. Available from https://doi.org/10.1130/0091-7613, https://pubs.geoscienceworld.org/geology.

Nishikawa, T., Matsuzawa, T., Ohta, K., Uchida, N., Nishimura, T., & Ide, S. (2019). The slow earthquake spectrum in the Japan Trench illuminated by the S-net seafloor observatories. *Science, 365*(6455), 808–813. Available from https://doi.org/10.1126/science.aax5618, https://science.sciencemag.org/content/365/6455/808/tab-pdf.

Noda, H., & Lapusta, N. (2013). Stable creeping fault segments can become destructive as a result of dynamic weakening. *Nature, 493*(7433), 518–521. Available from https://doi.org/10.1038/nature11703.

Okada, Y. (1985). Surface deformation due to shear and tensile faults in a half-space. *Bulletin of the Seismological Society of America, 75*(4), 1135–1154. Available from https://doi.org/10.1785/bssa0750041135.

Okada, Y. (1992). Internal deformation due to shear and tensile faults in a half-space. *Bulletin of the Seismological Society of America, 82*(2), 1018–1040. Available from https://doi.org/10.1785/bssa0820021018.

Oleskevich, D. A., Hyndman, R. D., & Wang, K. (1999). The updip and downdip limits to great subduction earthquakes: Thermal and structural models of Cascadia, south Alaska, SW Japan, and Chile. *Journal of Geophysical Research: Solid Earth, 104*(B7), 14965–14991. Available from https://doi.org/10.1029/1999jb900060.

Philibosian, B., & Meltzner, A. J. (2020). Segmentation and supercycles: A catalog of earthquake rupture patterns from the Sumatran Sunda Megathrust and other well-studied faults worldwide. *Quaternary Science Reviews, 241*. Available from https://doi.org/10.1016/j.quascirev.2020.106390.

Prada, M., Galvez, P., Ampuero, J. P., Sallarès, V., Sánchez-Linares, C., Macías, J., & Peter, D. (2021). The influence of depth-varying elastic properties of the upper plate on megathrust earthquake rupture dynamics and tsunamigenesis. *Journal of Geophysical Research: Solid Earth, 126*(11). Available from https://doi.org/10.1029/2021JB022328, http://agupubs.onlinelibrary.wiley.com/hub/jgr/journal/10.1002/(ISSN)2169-9356/.

Priest, G. R., Goldfinger, C., Wang, K., Witter, R. C., Zhang, Y., & Baptista, A. M. (2010). Confidence levels for tsunami-inundation limits in northern Oregon inferred from a 10,000-year history of great earthquakes at the Cascadia subduction zone. *Natural Hazards, 54*(1), 27–73. Available from https://doi.org/10.1007/s11069-009-9453-5.

Protti, M., González, V., Newman, A. V., Dixon, T. H., Schwartz, S. Y., Marshall, J. S., Feng, L., Walter, J. I., Malservisi, R., & Owen, S. E. (2014). Nicoya earthquake rupture anticipated by geodetic measurement of the locked plate interface. *Nature Geoscience*, 7(2), 117−121. Available from https://doi.org/10.1038/ngeo2038.

Qiu, Q., & Barbot, S. (2022). Tsunami excitation in the outer wedge of global subduction zones. *Earth-Science Reviews*, 230. Available from https://doi.org/10.1016/j.earscirev.2022.104054.

van Rijsingen, E., Lallemand, S., Peyret, M., Arcay, D., Heuret, A., Funiciello, F., & Corbi, F. (2018). How subduction interface roughness influences the occurrence of large interplate earthquakes. *Geochemistry, Geophysics, Geosystems*, 19(8), 2342−2370. Available from https://doi.org/10.1029/2018gc007618.

Riquelme, S., Schwarze, H., Fuentes, M., & Campos, J. (2020). Near-field effects of earthquake rupture velocity into tsunami runup heights. *Journal of Geophysical Research: Solid Earth*, 125(6). Available from https://doi.org/10.1029/2019JB018946, http://agupubs.onlinelibrary.wiley.com/hub/jgr/journal/10.1002/(ISSN)2169-9356/.

Ruff, L. J. (1989). Do trench sediments affect great earthquake occurrence in subduction zones? *Pure and Applied Geophysics*, 129(1-2), 263−282. Available from https://doi.org/10.1007/BF00874629.

Saffer, D. M., & Tobin, H. J. (2011). Hydrogeology and mechanics of subduction zone forearcs: Fluid flow and pore pressure. *Annual Review of Earth and Planetary Sciences*, 39(1), 157−186. Available from https://doi.org/10.1146/annurev-earth-040610-133408.

Sallarès, V., & Ranero, C. R. (2019). Upper-plate rigidity determines depth-varying rupture behaviour of megathrust earthquakes. *Nature Research, Spain Nature*, 576(7785), 96−101. Available from https://doi.org/10.1038/s41586-019-1784-0, http://www.nature.com/nature/index.html.

Satake, K. (1994). Mechanism of the 1992 Nicaragua tsunami earthquake. *Geophysical Research Letters*, 21(23), 2519−2522. Available from https://doi.org/10.1029/94gl02338.

Satake, K. (2015). Tsunamis. In *Treatise on geophysics*, (2nd ed.). Japan: Elsevier Inc. Available from http://www.sciencedirect.com/science/book/9780444538031.

Scala, A., Lorito, S., Romano, F., Murphy, S., Selva, J., Basili, R., Babeyko, A., Herrero, A., Hoechner, A., Løvholt, F., Maesano, F. E., Perfetti, P., Tiberti, M. M., Tonini, R., Volpe, M., Davies, G., Festa, G., Power, W., Piatanesi, A., & Cirella, A. (2020). Effect of shallow slip amplification uncertainty on probabilistic tsunami hazard analysis in subduction zones: Use of long-term balanced stochastic slip models. *Pure and Applied Geophysics*, 177(3), 1497−1520. Available from https://doi.org/10.1007/s00024-019-02260-x.

Scholl, D. W., Kirby, S. H., von Huene, R., Ryan, H., Wells, R. E., & Geist, E. L. (2015). Great (≥Mw8.0) megathrust earthquakes and the subduction of excess sediment and bathymetrically smooth seafloor. *Geosphere*, 11(2), 236−265. Available from https://doi.org/10.1130/GES01079.1, http://geosphere.gsapubs.org/content/11/2/236.full.pdf.

Scholz, C. H. (1998). Earthquakes and friction laws. *Nature*, 391(6662), 37−42. Available from https://doi.org/10.1038/34097.

Scholz, C. H. (2002). *The mechanics of earthquakes and faulting* (471). Cambridge University Press.

Shipton, Z. K., Evans, J. P., Abercrombie, R. E., & Brodsky, E. E. (2006). The missing sinks: Slip localization in faults, damage zones, and the seismic energy budget. *Geophysical Monograph Series*, 170. Available from https://doi.org/10.1029/170GM22, http://as.wiley.com/WileyCDA/Section/id-815707.html.

Sibson, R. H. (2003). Thickness of the seismic slip zone. *Bulletin of the Seismological Society of America*, 93(3), 1169−1178. Available from https://doi.org/10.1785/0120020061.

Sparkes, R., Tilmann, F., Hovius, N., & Hillier, J. (2010). Subducted seafloor relief stops rupture in South American great earthquakes: Implications for rupture behaviour in the 2010 Maule, Chile earthquake. *Earth and Planetary Science Letters*, 298(1-2), 89−94. Available from https://doi.org/10.1016/j.epsl.2010.07.029.

Stanislowski, K., Roesner, A., & Ikari, M. J. (2022). Implications for megathrust slip behavior and pore pressure at the shallow northern Cascadia subduction zone from laboratory friction experiments. *Earth and Planetary Science Letters*, 578. Available from https://doi.org/10.1016/j.epsl.2021.117297, http://www.sciencedirect.com/science/journal/0012821X/321-322.

Sun, T., Wang, K., Fujiwara, T., Kodaira, S., & He, J. (2017). Large fault slip peaking at trench in the 2011 Tohoku-oki earthquake. *Nature Communications*, 8(1). Available from https://doi.org/10.1038/ncomms14044.

Tan, H., Gao, X., Wang, K., Gao, J., & He, J. (2022). Hidden roughness of subducting seafloor and implications for megathrust seismogenesis: Example from northern Manila trench. *Geophysical Research Letters*, 49(17). Available from https://doi.org/10.1029/2022GL100146, http://agupubs.onlinelibrary.wiley.com/hub/journal/10.1002/(ISSN)1944-8007/.

Tanioka, Y., & Satake, K. (1996). Tsunami generation by horizontal displacement of ocean bottom. *Geophysical Research Letters*, 23(8), 861−864. Available from https://doi.org/10.1029/96gl00736.

Di Toro, G., Han, R., Hirose, T., De Paola, N., Nielsen, S., Mizoguchi, K., Ferri, F., Cocco, M., & Shimamoto, T. (2011). Fault lubrication during earthquakes. *Nature*, 471(7339), 494−498. Available from https://doi.org/10.1038/nature09838.

Tullis, T. E. (2015). *Mechanisms for friction of rock at earthquake slip rates*, . (2nd ed.). *Treatise on Geophysics*, (4). Elsevier Inc. Available from http://www.sciencedirect.com/science/book/9780444538031, 10.1016/B978-0-444-53802-4.00073-7.

Uchida, N., Nakajima, J., Wang, K., Takagi, R., Yoshida, K., Nakayama, T., Hino, R., Okada, T., & Asano, Y. (2020). Stagnant forearc mantle wedge inferred from mapping of shear-wave anisotropy using S-net seafloor seismometers. *Nature Communications*, 11(1). Available from https://doi.org/10.1038/s41467-020-19541-y.

Ujiie, K., Tanaka, H., Saito, T., Tsutsumi, A., Mori, J. J., Kameda, J., Brodsky, E. E., Chester, F. M., Eguchi, N., & Toczko, S. (2013). Low coseismic shear stress on the Tohoku-oki megathrust determined from laboratory experiments. *Science*, 342(6163), 1211−1214. Available from https://doi.org/10.1126/science.1243485.

Wada, I., & Wang, K. (2009). Common depth of slab-mantle decoupling: Reconciling diversity and uniformity of subduction zones. *Geochemistry, Geophysics, Geosystems*, *10*(10). Available from https://doi.org/10.1029/2009gc002570.

Wallace, L. M. (2020). Slow slip events in New Zealand. *Annual Review of Earth and Planetary Sciences*, *48*(1), 175−203. Available from https://doi.org/10.1146/annurev-earth-071719-055104.

Wang, K., & Bilek, S. L. (2011). Do subducting seamounts generate or stop large earthquakes? *Geology*, *39*(9), 819−822. Available from https://doi.org/10.1130/G31856.1.

Wang, K., & Bilek, S. L. (2014). Fault creep caused by subduction of rough seafloor relief. *Tectonophysics*, *610*, 1−24. Available from https://doi.org/10.1016/j.tecto.2013.11.024.

Wang, K., & He, J. (2008). Effects of frictional behavior and geometry of subduction fault on coseismic seafloor deformation. *Bulletin of the Seismological Society of America*, *98*(2), 571−579. Available from https://doi.org/10.1785/0120070097.

Wang, K., Huang, T., Tilmann, F., Peacock, S. M., & Lange, D. (2020). Role of serpentinized mantle wedge in affecting megathrust seismogenic behavior in the area of the 2010 M = 8.8 Maule earthquake. *Geophysical Research Letters*, *47*(22). Available from https://doi.org/10.1029/2020GL090482, http://agupubs.onlinelibrary.wiley.com/hub/journal/10.1002/(ISSN)1944-8007/.

Wang, K., Sun, T., Brown, L., Hino, R., Tomita, F., Kido, M., Iinuma, T., Kodaira, S., & Fujiwara, T. (2018). Learning from crustal deformation associated with the M9 2011 Tohoku-oki earthquake. *Geosphere*, *14*(2), 552−571. Available from https://doi.org/10.1130/GES01531.1.

Wang, P. L., Engelhart, S. E., Wang, K., Hawkes, A. D., Horton, B. P., Nelson, A. R., & Witter, R. C. (2013). Heterogeneous rupture in the great Cascadia earthquake of 1700 inferred from coastal subsidence estimates. *Journal of Geophysical Research: Solid Earth*, *118*(5), 2460−2473. Available from https://doi.org/10.1002/jgrb.50101, http://onlinelibrary.wiley.com/journal/10.1002/(ISSN)2169-9356.

Wang, Y., Wang, K., He, J., & Zhang, L. (2023). On unusual conditions for the exhumation of subducted oceanic crustal rocks: How to make rocks hotter than models. *Earth and Planetary Science Letters*, *615*. Available from https://doi.org/10.1016/j.epsl.2023.118213.

Williamson, A., Melgar, D., & Rim, D. (2019). The effect of earthquake kinematics on tsunami propagation. *Journal of Geophysical Research: Solid Earth*, *124*(11), 11639−11650. Available from https://doi.org/10.1029/2019jb017522.

Wilson, A., & Ma, S. (2021). Wedge plasticity and fully coupled simulations of dynamic rupture and tsunami in the Cascadia subduction zone. *Journal of Geophysical Research: Solid Earth*, *126*(7). Available from https://doi.org/10.1029/2020jb021627.

Witter, R. C., Zhang, Y., Wang, K., Priest, G. R., & Goldfinger, C. (2011). Using hypothetical Cascadia and Alaska earthquake scenarios. *Oregon Department of Geology and Mineral Industries Special Paper*, *43*.

Wu, B., Kyriakopoulos, C., Oglesby, D. D., & Ryan, K. J. (2023). Variation of proportionality between stress drop and slip, with implications for megathrust earthquakes. *Geophysical Research Letters*, *50*(4). Available from https://doi.org/10.1029/2022GL100568, http://agupubs.onlinelibrary.wiley.com/hub/journal/10.1002/(ISSN)1944-8007/.

Xu, S., Fukuyama, E., Yamashita, F., Kawakata, H., Mizoguchi, K., & Takizawa, S. (2023). Fault strength and rupture process controlled by fault surface topography. *Nature Geoscience*, *16*(1), 94−100. Available from https://doi.org/10.1038/s41561-022-01093-z.

Yamanaka, Y., & Kikuchi, M. (2004). Asperity map along the subduction zone in northeastern Japan inferred from regional seismic data. *Journal of Geophysical Research: Solid Earth*, *109*(B7). Available from https://doi.org/10.1029/2003jb002683.

Ye, L., Lay, T., Kanamori, H., & Rivera, L. (2016). Rupture characteristics of major and great (Mw ≥ 7.0) megathrust earthquakes from 1990 to 2015: 2. Depth dependence. *Journal of Geophysical Research: Solid Earth*, *121*(2), 845−863. Available from https://doi.org/10.1002/2015jb012427.

van Zelst, I., Rannabauer, L., Gabriel, A. -A., & Van Dinther, Y. (2022). Earthquake rupture on multiple splay faults and its effect on tsunamis. *Journal of Geophysical Research: Solid Earth*, *127*(8). Available from https://doi.org/10.1029/2022jb024300.

van Zelst, I., Wollherr, S., Gabriel, A. A., Madden, E. H., & van Dinther, Y. (2019). Modeling megathrust earthquakes across scales: One-way coupling from geodynamics and seismic cycles to dynamic rupture. Blackwell Publishing Ltd, Switzerland. *Journal of Geophysical Research: Solid Earth*, *124*(11), 11414−11446. Available from https://doi.org/10.1029/2019JB017539, http://agupubs.onlinelibrary.wiley.com/hub/jgr/journal/10.1002/(ISSN)2169-9356/.

Chapter 9

Tsunamis triggered by splay faulting

Mohammad Mokhtari[1] and Katsuichiro Goda[2]

[1]International Institute of Earthquake Engineering and Seismology, Tehran, Iran, [2]Department of Earth Sciences, Western University, London, ON, Canada

9.1 Introduction

Subduction zones are geological areas where one tectonic plate moves beneath another (McCaffrey, 2009). These zones often feature wedge-shaped sedimentary complexes, known as accretionary prisms, that form when sediments are scraped off from the subducting plate and added to the overriding plate (Wang & Hu, 2006). Within these prisms, large thrust faults can be found, cutting through the layers of sediment. These faults are known as splay faults and can have steep dip angles (Ranero et al., 2003), leading to significant seafloor deformation when they rupture. Consequently, tsunami runup heights due to splay faults can be increased significantly in the near field (Hsu et al., 2006).

The effect of splay faulting on tsunami wave heights in the near field has been well-documented in various studies (Kanamori, 1972), particularly during some of the largest tsunamis in history. The 1946 Nankai Earthquake and Tsunami (Cummins & Kaneda, 2000), the 1960 Chilean and the 1964 Alaskan Earthquakes and Tsunamis (Kanamori, 1972), and the 2004 Indian Ocean Earthquake and Tsunami (Lay et al., 2005) all demonstrated the significant impacts of splay faulting on tsunami wave heights in the near field. Recently, Heidarzadeh et al. (2021) reported a high potential for splay faulting in the Molucca Sea, Indonesia.

The significance of splay faults in generating tsunamis has become increasingly apparent (Gao et al., 2018; Momeni et al., 2023). These faults, which branch off from the main fault plane, can play a crucial role in the generation and propagation of tsunamis. Since splay faults are relatively weak zones (faulted), they can be chosen as pathways for rupture propagation of earthquakes from the main subduction interface. In light of this, it is imperative that coastal communities and regions that are susceptible to tsunamis factor in the potential impact of splay faults when assessing the risks associated with these events. Careful consideration of splay faults can aid in developing improved evacuation plans and building infrastructure that can better withstand the effects of tsunamis.

In this chapter, a comprehensive overview of splay faults is provided, including their definition, seismic expression, and specific interpretations in two different margins. The potential of splay faults is examined to generate tsunamis as both primary and secondary sources. Subsequently, numerical modeling of tsunamis is used to demonstrate the effect of splay faulting on tsunami wave heights in the near field. Through an in-depth discussion, the hazardous nature of these faults is highlighted, especially when the combined ruptures from primary and secondary sources affect coastal areas that could potentially be impacted by tsunamis.

9.2 Tectonic background

A splay fault is a secondary fault that branches off from a larger fault when it becomes significantly misaligned with the direction of the maximum principal stress. It is often found in subduction zones (e.g., Japan, Chile, Iran-Pakistan [Makran], and USA-Canada [Cascadia]) where one tectonic plate slides beneath another. In particular, a first-order splay fault is defined as a secondary fault that develops at acute angles to the primary fault with a slip rate similar to that of the primary fault. Since splay faults are usually steeper than the main fault (e.g., 20–40 degrees for splay faults in comparison with 5 to 15 degrees for the main fault/subduction interface), they cause significant vertical displacements of the seafloor compared to the main fault rupture for the same amount of slip. As a result, these faults can play a crucial role in the generation of tsunamis. Splay faults can develop at varying depths and angles in relation to the primary fault, and their orientation and geometry can differ depending on the specific geological context. Fig. 9.1 illustrates a splay fault that branches from the

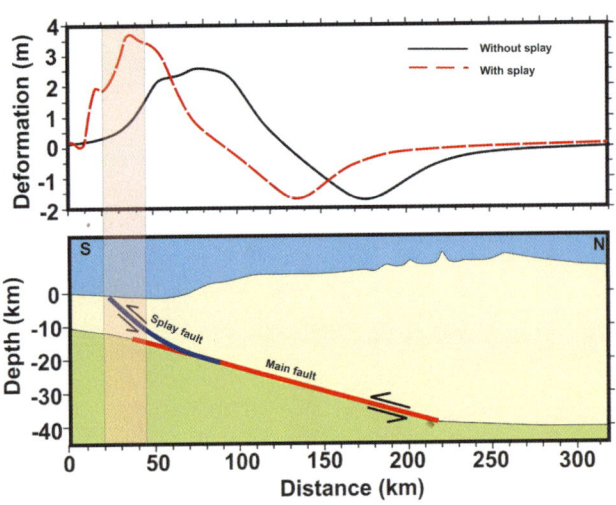

FIGURE 9.1 Sketch of a splay fault that branches off upward from a subduction zone plate boundary (bottom) along with the variation of seafloor uplift due to the rupture (top).

plate boundary. The fault's location (shaded area in Fig. 9.1) is characterized by a sudden increase in seafloor uplift, resulting in higher tsunami wave amplitudes in the immediate areas. Although the rupture of splay faults alone may be insufficient to cause large tsunamis, when triggered by the megathrust earthquake, it will increase the potential consequences due to tsunamis (e.g., loss of life and economic loss) along the coastal line due to their capacity to generate powerful tsunamis.

New insights into the behavior and characteristics of splay faults have emerged from recent studies that use seismic reflection data. For instance, the 2010 Chile Earthquake and Tsunami, which caused extensive damage and loss of life in Chile and neighboring countries, was triggered by a subduction zone fault that included splay faults along with the main thrust fault (Quezada et al., 2020).

Recent studies have investigated the possibility of multiple splay faults breaking during a single earthquake and the resulting impact on tsunamis. van Zelst et al. (2022) used numerical simulations to examine this phenomenon. Their investigation demonstrated that multiple splay faults could rupture during a single earthquake, with stress changes from trapped seismic waves playing a significant role. The rupture of splay faults causes greater seafloor displacements with shorter wavelengths, resulting in more complex and larger tsunamis. The findings suggest that the inclusion of splay faults is critical for accurately assessing tsunami hazards.

9.3 Seismic expression of offshore splay faults

Among the subduction zones that have splay faults in the offshore areas, two subduction zones are focused upon to illustrate the geological characteristics of the splay faults. The selected subduction zones are the Makran region off Iran and Pakistan and the Nankai Trough region in southwestern Japan.

9.3.1 Makran subduction zone

Taking the Makran subduction zone off Iran and Pakistan as an example, the eastern and western Makran are characterized by numerous splay faults (Fig. 9.2), as identified through seismic reflection data used by Mokhtari (2015), Pajang et al. (2021), and Smith et al. (2012). These splay faults are predominantly thrust faults located near the deformation front. The activation of these splay thrust faults may be triggered by rupture along the Makran plate boundary (Kame & Yamashita, 1999).

Seismic reflection lines are used to illustrate the imbricate thrusts (splay faults), which are mainly landward dipping with some secondary back thrusts. The presence of occasional secondary back-thrusts suggests that the deformation is not simple. Using the 2D seismic profile data for western Makran (Fig. 9.3) and eastern Makran (Fig. 9.4), the spacing between the thrust faults varies along the margin, ranging from 2.2 to 18.6 km. The same spacing for another example in the Makran region was reported to be in the range of 7 to 9 km, while they also dip landward and are distributed over 70 km width. The results indicate that the distribution of fault spacing is not random but likely reflects variations in the local stress regime and material properties. A more detailed analysis of fault spacing patterns could provide insights into the tectonic processes controlling the deformation of the accretionary prism (Mokhtari, 2015; Pajang et al., 2021; Smith et al., 2012).

Importantly, the proximity of these identified splay faults to the coastline varies. While some are located closer to the coastline, others are further away. The location of these splay faults is significant as their impact is localized, and splay

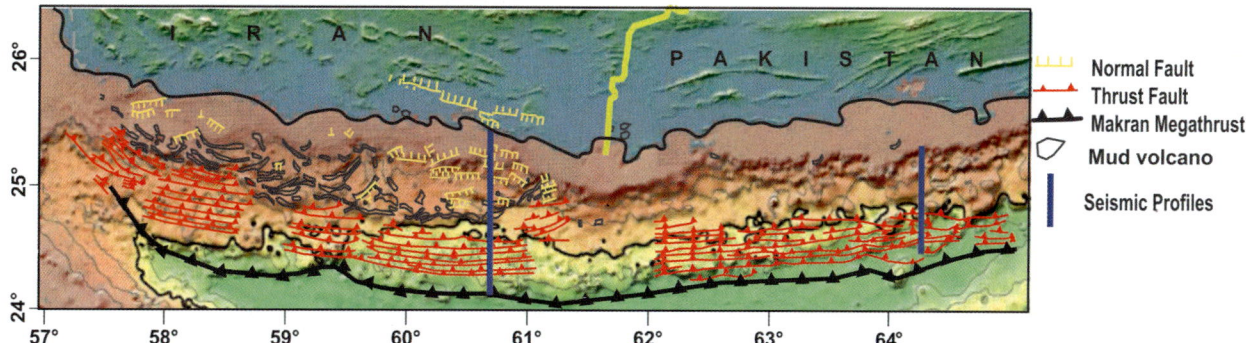

FIGURE 9.2 Map of the locations of the two seismic reflection profiles in western and eastern Makran. In addition, the geographical distribution of the identified splay faults in the Makran subduction zone area is indicated. Map lines delineate study areas and do not necessarily depict accepted national boundaries. *Modified from Pajang, S., Cubas, N., Letouzey, J., Le Pourhiet, L., Seyedali, S., Fournier, M., Agard, P., Khatib, M. M., Heyhat, M., & Mokhtari, M. (2021). Seismic hazard of the western Makran subduction zone: Insight from mechanical modelling and inferred frictional properties. Earth and Planetary Science Letters, 562, 116789. https://doi.org/10.1016/j.epsl.2021.116789.*

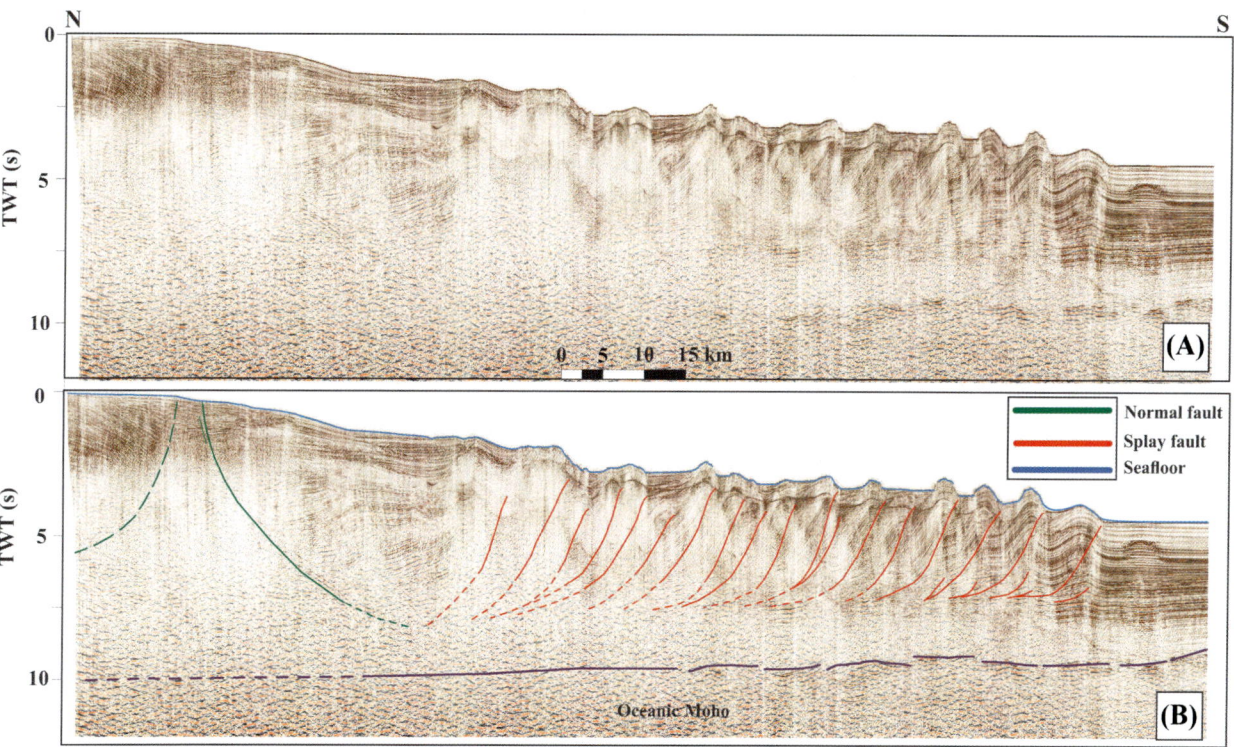

FIGURE 9.3 (A) Uninterpreted and (B) interpreted 2D seismic line in time in western Makran, showing imbricated structures (splay faults) and out-of-sequence faults. In (B), the color identifies different horizons within the cross-section, whereas the red line indicates the splay faults (see Fig. 9.2 for the location).

faults located closer to the coastline may pose a greater tsunami risk compared to those farther away (Momeni et al., 2023). The dip angle of these splay faults also plays a crucial role in assessing the tsunami hazard, as steeper faults are more efficient in causing greater vertical deformation on the seabed, consequently leading to larger tsunami amplitudes.

9.3.2 Nankai Trough zone

The Nankai Trough is another area that has been well documented for its mega-splay thrust system in the Kumano Basin. With the use of 3D seismic data, the splay fault geometry and its lateral continuity have been successfully mapped. The Kumano mega-splay fault can be traced from the main plate interface fault up to the seafloor, where it

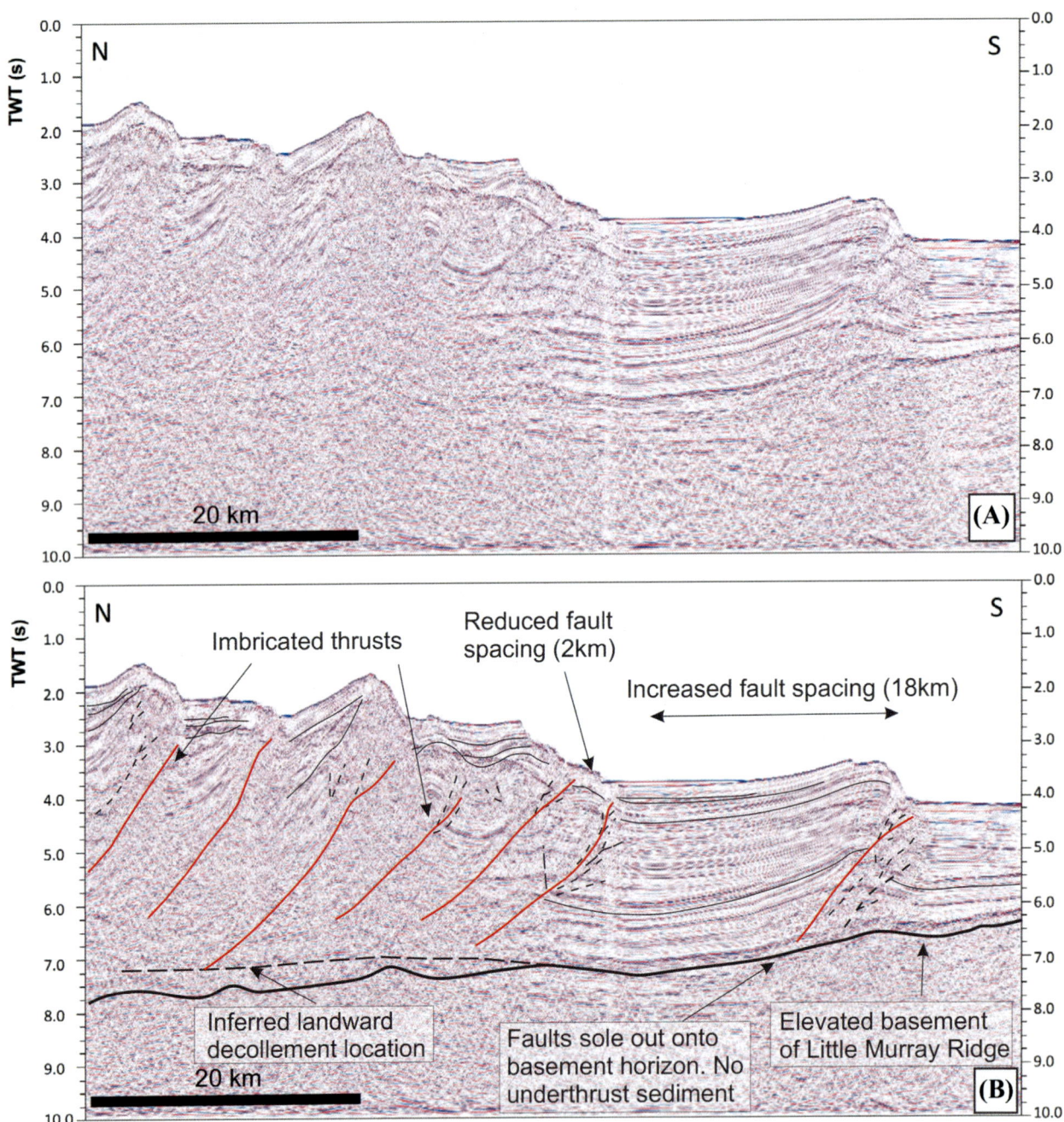

FIGURE 9.4 (A) Uninterpreted and (B) interpreted 2D seismic line in time in eastern Makran, showing imbricated structures (splay faults). The dotted line is the décollement zone (Smith et al., 2013). See Fig. 9.2 for the location.

intersects with older thrust slices of the frontal accretionary prism (Fig. 9.5). The thrust geometry and evidence of large-scale slumping of surficial sediments provide evidence of the active fault, which has evolved toward the landward direction over time, contrary to the usual seaward progression of accretionary thrusts. The steepening of the mega-splay fault has substantially increased the potential for vertical uplift of the seafloor with slip. The study conducted by Moore et al. (2007) concluded that slip on the mega-splay fault most likely contributed to generating historic tsunamis, such as the 1944 Tonankai Earthquake with a magnitude of 8.1. Details of 3D seismic inline in the Nankai Trough zone show the sediment accumulation on top of older thrusts (blue) and the overriding of young slope sediment by a block moving along the mega-splay fault (black) (Fig. 9.6) (Moore et al., 2007). This unique geometry of the mega-splay fault makes this margin particularly prone to tsunami genesis.

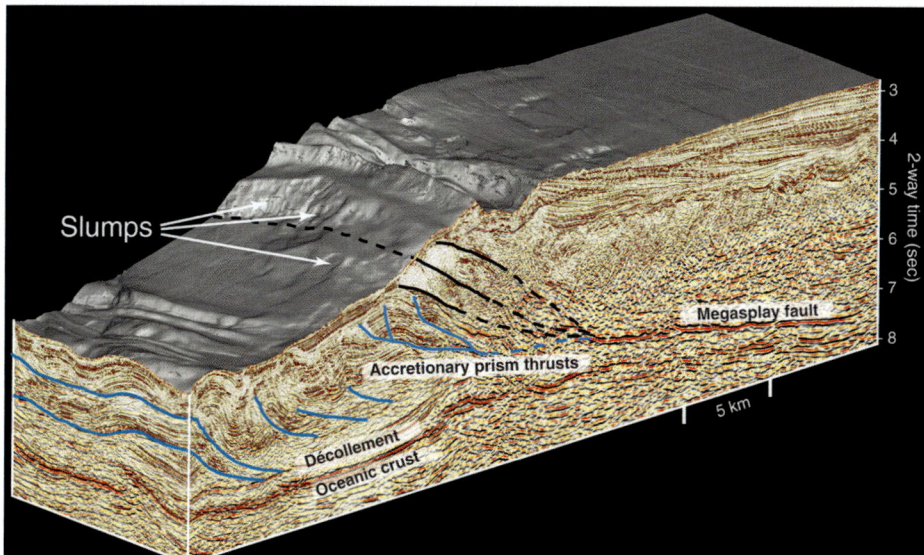

FIGURE 9.5 3D seismic data volume depicting the location of the mega-splay fault (*black* lines) and its relationship to older in-sequence thrusts of the frontal accretionary prism (*blue* lines). Steep sea-floor topography and numerous slumps above the splay fault are shown. *From Moore, G. F., Bangs, N. L., Taira, A., Kuramoto, S., Pangborn, E., & Tobin, H. J. (2007). Three-dimensional splay fault geometry and implications for tsunami generation. Science, 318(5853), 1128−1131, https://doi.org/10.1126/science.1147195.*

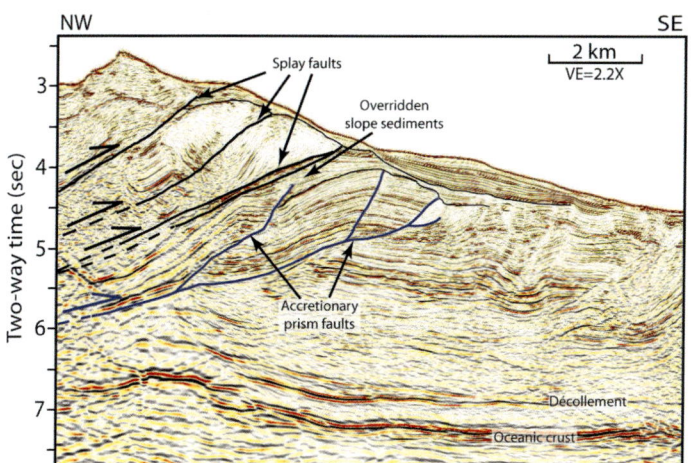

FIGURE 9.6 Detail of 3D seismic inline in the Nankai Trough zone showing the sediment accumulation on top of older thrusts (*blue*) and the overriding of young slope sediment by a block moving along the mega-splay fault (*black*). *From Moore, G. F., Bangs, N. L., Taira, A., Kuramoto, S., Pangborn, E., & Tobin, H. J. (2007). Three-dimensional splay fault geometry and implications for tsunami generation. Science, 318(5853), 1128−1131, https://doi.org/10.1126/science.1147195.*

9.4 Splay fault and tsunami generation

Splay faults can be classified as the primary and secondary sources for tsunami generation. One of the important parameters for tsunami modeling is the location of the identified splay fault because the effect of splay faults closer to the coastline tends to pose a greater tsunami risk as compared to the farther ones. Moreover, the dip angle of the splay faults plays a crucial role in assessing tsunami hazards because steeper faults cause greater vertical deformations on the seabed and consequently lead to larger tsunami amplitudes (Wendt et al., 2009).

Splay faults branching upward from plate boundary megathrust and extending to the seafloor have been observed on seismic and bathymetric images in various subduction margins, such as Cascadia (McCaffrey & Goldfinger, 1995), south-central Chile (Melnick et al., 2006), Sumatra (Singh et al., 2011), Nankai (Park et al., 2002), and Makran subduction zone (Mokhtari, 2015; Smith et al., 2012). Some of these margins have been shown in the previous section. Slip along these faults may occur independently or during major megathrust earthquakes.

Fliss et al. (2005) investigated the correlation between fault branching and rupture directivity. Building upon observations from various geological evidence, Wang and Hu (2006) proposed a novel theory to explain the formation of accretionary prisms in subduction earthquake cycles. Their theory provided a valuable framework for studying the history of accretionary prisms and forearc basins as well as related phenomena, such as splay faulting during major earthquakes. Understanding the role of splay-faulting in accommodating deformation and generating tsunamis is crucial in predicting the risks associated with these disasters.

Past studies have provided important insight into the partitioning of slip during great megathrust earthquakes and the role of splay faults in this process. Based on surface ruptures on land, Plafker (1965) observed that slip can be distributed between the subduction interface and splay faults. Additionally, modeling studies have supported these observations, using geophysical and historical data to infer patterns of slip partitioning (Cummins & Kaneda, 2000; Park et al., 2002). Recent research has focused on using advanced technologies and techniques to better understand the behavior of splay faults during earthquakes. For example, high-resolution imaging of the seafloor has revealed previously unknown details about the geometry and activity of splay faults (Jia et al., 2020).

The steeper dip angles for splay faults can significantly affect the tsunami hazard for the nearby regions. The location relative to the main fault and the dip angle of the splay faults are two main factors for increasing the strengthening effect on the generation of higher amplitude wave height and tsunami hazard at the shoreline. Momeni et al. (2023) used the splay fault dip angles in the range of 30–40 degrees to include the splay faults in the Makran subduction zone. For a specific cross-section profile in the Makran region, splay faults with sharper dip (41 degrees) are located at the front of the accretionary prism, and inner splay faults show gentler angles (32 degrees). The result indicated a large increase in the wave height at the location of the splay faults. Moreover, the geometry of fault systems has been recognized as a critical factor in tsunami generation, and dynamic processes play a vital role in determining the path of rupture in complex fault systems.

Despite these advances, the specific parameters that control slip activation in splay faults and the exact patterns of slip partitioning between the megathrust and splay faults remain poorly understood. One major factor contributing to this lack of understanding is the scarcity of direct observations during earthquakes as well as the offshore locations of the splay faults.

Recent studies have updated information on the 2004 Indian Ocean Tsunami arrival in Sri Lanka, challenging the previous theoretical model based on standard earthquake source and shallow-water gravity wave models. While the previous models suggested that the tsunami should have arrived as a single large wave followed by a depression, eyewitness accounts of the tsunami arrival indicated that there were multiple wave arrivals separated by a recession (Liu et al., 2005). This was confirmed by altimetry data that showed a double-peaked lead wave. Further research has suggested that the observed double-peaked wave may be the result of two areas of vertical uplift in the source region caused by slips on the plate interface and on a steeper splay fault branching from this detachment. Only a small amount of slip is needed on the steep splay to vertically displace the seafloor enough to create a double-peaked wave (Geist et al., 2007). These findings highlight the importance of considering complex fault geometry and the potential for multiple sources of vertical uplift when modeling tsunami generation and propagation.

Cooke (1997) investigated the effect of frictional strength variations near fault tips on the pattern of splay fault branching. His work showed that gradual changes in frictional strength could produce broad zones of splay fractures, while abrupt changes could result in a single splay fracture. In another study, Kame et al. (2003) explored the effects of prestress state, rupture velocity, and branch angle on dynamic fault branching. Their research demonstrated that the prestress state significantly influences the most favored direction for dynamic branching. Additionally, the study revealed that the enhanced dynamic stressing of a rapidly propagating rupture could cause failure on a fault that may not be the most favorably oriented based on the prestress state. In addition, numerical modeling studies have been used to simulate earthquake scenarios and investigate the behavior of splay faults under different conditions (Wang et al., 2018).

In summary, recent studies have emphasized the importance of considering the role of splay faults in understanding tsunami hazards and improving risk assessments. Newly acquired seismic reflection data and dynamic earthquake modeling have provided a better understanding of the fault system's geometry and its influence on tsunami generation.

9.5 Numerical example

A numerical example of the effects of the splay fault is useful for grasping the importance of considering splay fault ruptures as part of critical tsunami scenarios for tsunami hazard mapping and tsunami risk mitigation purposes. Although the consideration of the splay fault rupture scenarios is becoming more popular as research investigations, this has not yet been taken up in the national tsunami hazard assessment and mapping for coastal areas. An exception is the Nankai–Tonankai Trough rupture scenarios generated by the Central Disaster Management Council of Japan. Learning the lessons from the 2011 Tohoku Earthquake and Tsunami, Central Disaster Management Council (2012) generated 11 characteristic tsunami scenarios having moment magnitudes between 9.0 and 9.2 for tsunami preparedness planning for coastal communities in southwestern Japan. The 11 scenarios included the following:

- Five single asperity rupture cases having the slip concentrations in (Case 1) Suruga Bay to Kii Peninsula, (Case 2) Kii Peninsula, (Case 3) Kii Peninsula to Shikoku, (Case 4) Shikoku, or (Case 5) Shikoku to Kyushu;
- Two splay fault sources cases in Kumano Basin with asperities in either (Case 6) western or (Case 7) eastern side of Kumano Basin; and
- Four two-asperity rupture cases having slip concentrations in (Case 8) Suruga Bay to Kii Peninsula and Kii Peninsula to Shikoku, (Case 9) Suruga Bay to Kii Peninsula and Shikoku, (Case 10) Kii Peninsula to Shikoku and Shikoku to Kyushu, or (Case 11) Shikoku and Shikoku to Kyushu.

Details of the tsunami source models by Central Disaster Management Council (2012) can be found in Goda et al. (2018).

In this chapter, it is interesting to compare the two tsunami source models, Case 1 (asperities span from Suruga Bay to Kii Peninsula) and Case 6 (asperities are concentrated off Suruga Bay to Enshu Nada with the splay fault in western Kumano Basin), by focusing on coastal areas that directly face the Kumano splay fault. Fig. 9.7 shows these two source models. Both source models assign the majority of earthquake slips to the eastern side of Kii Peninsula to Suruga Bay, focusing on the Tonankai earthquake scenario. In Case 1 (Fig. 9.7A), the large asperity areas (exceeding 30-m slip) span across the eastern Kii Peninsula to Suruga Bay along the trench line. In Case 6 (Fig. 9.7B), the large asperity off the eastern Kii Peninsula is replaced by the Kumano splay fault. The moment magnitude for Case 1 is 9.16 (=seismic moment of 6.12×10^{22} N m), which is calculated by using the rock rigidity of 40.9 GPa (Central Disaster Management

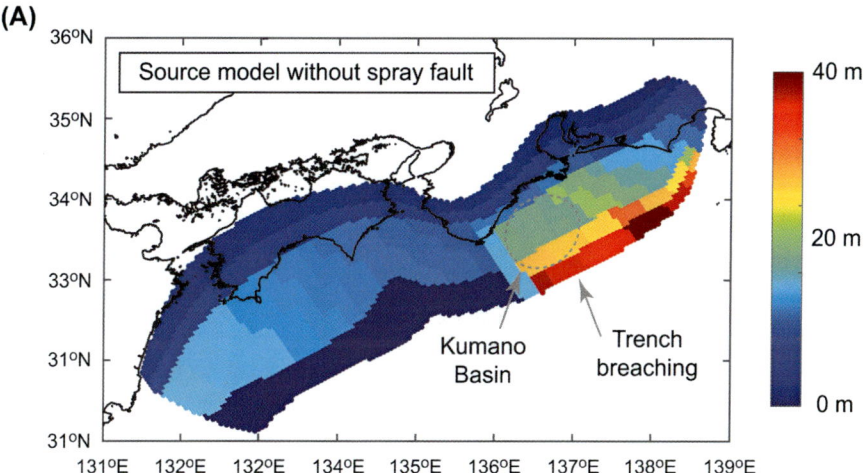
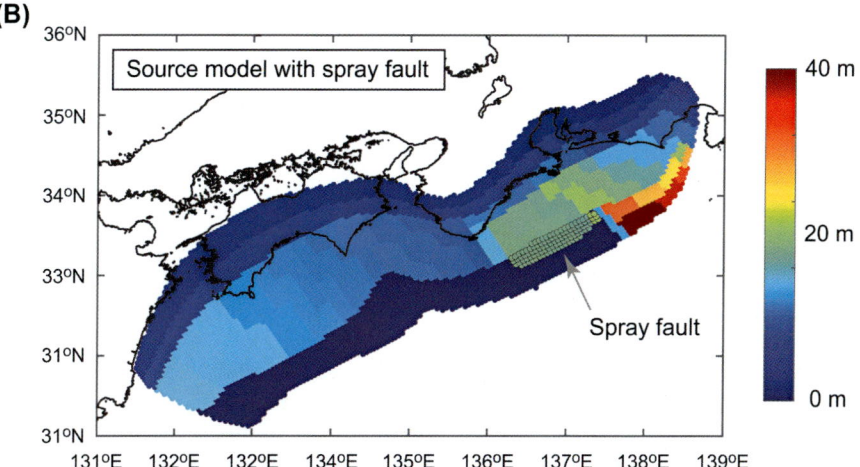

FIGURE 9.7 Tsunami source models by Central Disaster Management Council (2012) for the future Nankai-Tonankai Trough Earthquakes. (A) Source model without the splay fault (Case 1) and (B) source model with the splay fault (Case 6).

Council, 2012). On the other hand, the moment magnitude for Case 6 is 9.12 (=seismic moment of 5.37×10^{22} N m), and this released energy is split into the main fault rupture (moment magnitude of 9.11 and seismic moment of 5.16×10^{22} N m) and the splay fault rupture (moment magnitude of 8.17 and seismic moment of 2.05×10^{21} N m). In other words, only 4% of the total seismic moment was allocated to the splay fault rupture. Note that the splay fault plane has an area of 2600 km^2, compared to the main fault plane (141,725 km^2). In the Central Disaster Management Council's model, the average dip angle of 20 degrees and the mean slip of 18.1 m are assigned to the Kumano splay fault (note: there are only minor variations for the earthquake slip values on the splay fault plane; Fig. 9.7B).

Tsunami propagation and runup simulations for Cases 1 and 6 are conducted to investigate the effects of the splay fault as implemented by Central Disaster Management Council (2012). The ground deformation data (i.e., input initial conditions for tsunami simulations) that are used for this investigation are provided by Central Disaster Management Council (2012). The bathymetry and elevation data are also directly obtained from Central Disaster Management Council (2012). The bathymetry-elevation data are based on the nested grid system by connecting grid data at 2430-m resolution gradually down to 10-m resolution with a nesting grid factor of 3. See Goda et al. (2018) for details of the tsunami simulation setup.

Fig. 9.8 shows tsunami propagation and runup simulations for Cases 1 and 6 by focusing on three different spatial scales. The top-row contour maps show the results for the Tonankai region; the middle-row contour maps show results near the Kumano Basin; and the bottom-row contour maps show local results in Kihoku Town in Mie Prefecture. At the regional levels (Fig. 9.8A and B), the influence of the splay fault can be clearly seen; there is a stripe of yellow-colored maximum wave heights off Kumano Basin. The comparison of Fig. 9.8A and B indicates that despite less extensive earthquake slips on the splay fault than the main subduction fault, the splay fault can generate vertical seafloor deformation effectively. By looking at the results at the intermediate spatial scale (Fig. 9.8C and D), the influence of the splay fault along the coastline that faces Kumano Basin directly can be observed (green to yellow colors in Fig. 9.8C versus yellow to red colors in Fig. 9.8D). The differences between the nonsplay versus splay tsunami cases in terms of tsunami runup and inundation can be seen more clearly in Fig. 9.8E and F. In Kihoku Town, the inundated areas due to the Nankai-Tonankai Trough tsunami have been expanded significantly, and the maximum tsunami heights have been increased significantly (up to 5 m) due to the presence of the splay fault rupture. These results demonstrate the importance of considering the splay fault ruptures as part of critical tsunami scenarios for tsunami preparedness and mitigation purposes.

9.6 Discussions

Observations of accretionary wedges in subduction zones reveal the presence of multiple splay faults with varying sizes and dips. While not all splay faults are expected to be seismically active simultaneously, earthquake ruptures originating from the megathrust can potentially activate these faults. This may complicate rupture dynamics and increase the efficiency of tsunami generation, as rupture on splay faults could lead to significant ramifications for tsunami genesis (Fabbri et al., 2020; Fukao, 1979; Kimura et al., 2007; Kopp, 2013; Moore & Saffer, 2001; Waldhauser et al., 2012). Several studies have suggested that splay fault ruptures played an essential role in large tsunamigenic megathrust earthquakes, such as the 2004 M_w 9.1–9.3 Indian Ocean Earthquake (DeDontney & Rice, 2012) and the 2010 M_w 8.8 Maule Earthquake (Melnick et al., 2012).

Accretionary wedges have been imaged in detail, revealing splay fault networks that could pose a significant tsunami hazard. However, the dynamics of multiple splay fault activation during megathrust earthquakes and the consequent effects on tsunami generation are not well understood, indicating a critical knowledge gap in our understanding of tsunami hazards. According to the study by van Zelst et al. (2022), understanding the behavior and mechanics of splay faults during earthquake events is essential in accurately assessing the tsunami risk posed by accretionary wedges.

Priest et al. (2009) conducted a study on the tsunami hazard of local earthquakes in the Cascadia subduction zone and considered the effect of splay faulting. They developed a logic-tree approach and assigned weights to different scenarios, including splay faulting. In the splay fault scenario, the slip is distributed between the megathrust and the splay fault. They found that splay faulting can amplify tsunami wave height and runup by 6%–31% and 2%–20%, respectively. Further research by Gao et al. (2018) on the northernmost Cascadia region used numerical simulations to compare the effects of different faulting models (buried, splay, and trench-breaching). Their results showed that the splay faulting model produced higher tsunami waves (50%–100%) than the commonly used model of buried rupture. Heidarzadeh et al. (2009) studied the effect of splay faulting on tsunami amplitudes in Makran. They considered a combined tsunami source from the eastern Makran segment and a splay fault and found that this scenario caused an increase

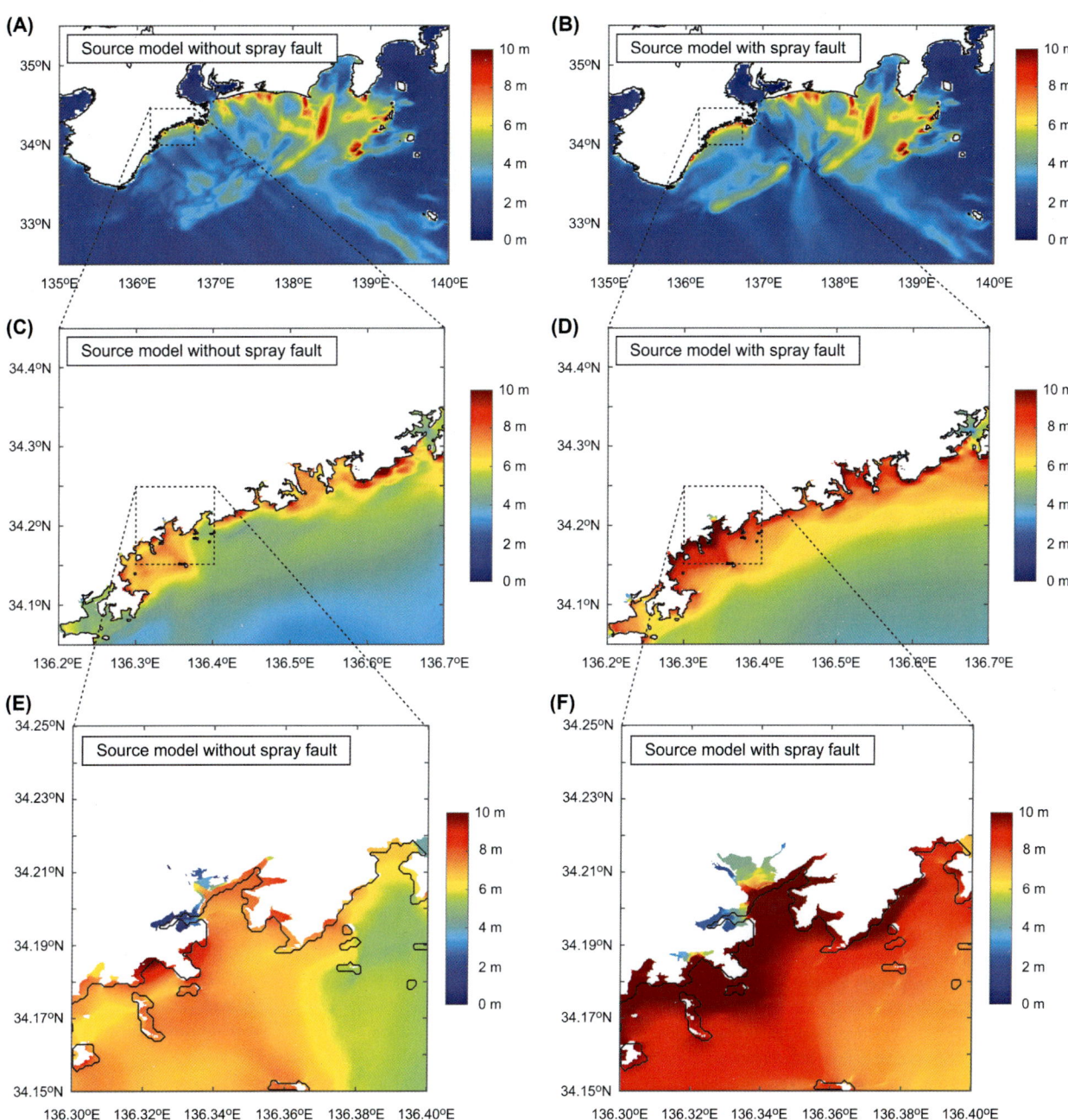

FIGURE 9.8 Tsunami propagation and runup simulations for the two tsunami source models without the splay fault (A, C, E) and with the splay fault (B, D, F). The contour maps show the maximum tsunami wave heights (offshore locations) or runup heights (onshore locations). The maps in (A) and (B) show results for the Tonankai region; the maps in (C) and (D) show results along the Kumano Basin; and the maps in (E) and (F) show local results in Kihoku Town.

in seabed displacement by a factor of 1.5 and an increase in tsunami amplitude by a factor of 2 at some locations. A systematic study by Momeni et al. (2023) showed that even with the presence of heterogeneous earthquake slip on the main fault plane, the splay fault rupture generates significantly larger tsunami waves along the shoreline directly facing the splay fault. For M_w 8.6 earthquakes in western Makran, splay faults can amplify tsunami heights up to 3−6 times as compared to ruptures exclusively on the plate boundary.

In summary, the studies discussed above highlight the significant impact of splay faulting on a tsunami hazard assessment. The findings underscore the importance of incorporating splay faulting scenarios in assessing the tsunami hazard potential of subduction zones.

9.7 Conclusions

In recent years, there have been significant advancements in our understanding of how megathrust slip triggers motions along splay faults. Such advancement is the use of numerical modeling techniques, such as the finite element method (FEM), to investigate the relationship between slip on a megathrust and the behavior of splay faults. Recent studies using FEM have shown that the amplitude and spatial distribution of seafloor deformation caused by slip on a megathrust can have a significant influence on splay faults. This includes the possibility of splay faults being activated by dynamic stress changes during megathrust rupture, which can lead to complex slip patterns and significant variations in seismic radiation. Further studies using FEM and other modeling techniques are likely to continue to shed light on the complex interaction between megathrusts and splay faults and help improve our ability to predict and mitigate earthquake and tsunami hazards.

In addition to numerical modeling, recent research has also focused on improving our understanding of splay fault geometry and its role in subduction zones. For example, studies have identified splay faults that have both landward and seaward dip directions, with different implications for seafloor deformation and tsunami generation. Furthermore, recent studies have also investigated the effects of changes in coseismic static stress resulting from megathrust earthquakes on splay faults, using a range of earthquake scenarios (Lay et al., 2012).

While significant progress has been made in recent years, the understanding of splay faults and their interaction with megathrust earthquakes remains an area of active research. The current understanding of slip partitioning between subduction zones and splay faults is still limited, with no clear patterns established (Baba et al., 2006; Park et al., 2002). Moreover, it remains unclear whether splay faults can rupture independently or not (Sykes & Menke, 2006). Despite these challenges, researchers have made significant efforts to study the complex phenomenon of splay fault branching during large subduction earthquakes. It is worth noting that ongoing research and analysis are continuously revealing new phenomena on the mechanics of splay faulting and its relationship to tsunami generation. As a result, our understanding of these phenomena is constantly evolving, and it is important to stay informed about the latest findings and insights.

In light of recent studies that highlight the importance of splay faults in generating tsunamis, it is crucial to integrate this information into risk assessment and mitigation plans for coastal areas at risk due to these catastrophic events. It has become increasingly clear that splay faults can play a significant role in the creation of tsunamis in subduction zones. Thus, scientists and policymakers need to consider the potential hazards associated with these types of faults when evaluating tsunami risks.

References

Baba, T., Cummins, P. R., Hori, T., & Kaneda, Y. (2006). High precision slip distribution of the 1944 Tonankai earthquake inferred from tsunami waveforms: Possible slip on a splay fault. *Tectonophysics, 426*(1-2), 119–134. Available from https://doi.org/10.1016/j.tecto.2006.02.015.

Central Disaster Management Council. (2012). *Working group report on mega-thrust earthquake models for the Nankai Trough*, Japan (unpublished content). Cabinet Office of the Japanese Government. http://www.bousai.go.jp/jishin/nankai/nankaitrough_info.html

Cooke, M. L. (1997). Fracture localization along faults with spatially varying friction. *Journal of Geophysical Research: Solid Earth, 102*(10), 22425–22434. Available from http://agupubs.onlinelibrary.wiley.com/hub/jgr/journal/10.1002/(ISSN)2169-9356/.

Cummins, P. R., & Kaneda, Y. (2000). Possible splay fault slip during the 1946 Nankai Earthquake. *Geophysical Research Letters, 27*(17), 2725–2728. Available from https://doi.org/10.1029/1999GL011139, http://onlinelibrary.wiley.com/journal/10.1002/(ISSN)1944-8007/issues?year=2012.

DeDontney, N., & Rice, J. R. (2012). Tsunami wave analysis and possibility of splay fault rupture during the 2004 Indian Ocean Earthquake. *Pure and Applied Geophysics, 169*(10), 1707–1735. Available from https://doi.org/10.1007/s00024-011-0438-4, http://www.springer.com/birkhauser/geo+science/journal/24.

Fabbri, O., Goldsby, D. L., Chester, F., Karpoff, A. M., Morvan, G., Ujiie, K., Yamaguchi, A., Sakaguchi, A., Li, C. F., Kimura, G., Tsutsumi, A., Screaton, E., & Curewitz, D. (2020). Deformation structures from splay and décollement faults in the Nankai Accretionary Prism, SW Japan (IODP NanTroSEIZE Expedition 316): Evidence for slow and rapid slip in Fault Rocks. *Geochemistry, Geophysics, Geosystems, 21*(6). Available from https://doi.org/10.1029/2019GC008786, http://agupubs.onlinelibrary.wiley.com/agu/journal/10.1002/(ISSN)1525-2027/.

Fliss, S., Bhat, H. S., Dmowska, R., & Rice, J. R. (2005). Fault branching and rupture directivity. *Journal of Geophysical Research: Solid Earth, 110*(6), 1–22. Available from https://doi.org/10.1029/2004JB003368, http://onlinelibrary.wiley.com/journal/10.1002/(ISSN)2169-9356.

Fukao, Y. (1979). Tsunami earthquakes and subduction processes near deep-sea trenches. *Journal of Geophysical Research, 84*(B5), 2303. Available from https://doi.org/10.1029/jb084ib05p02303.

Gao, D., Wang, K., Insua, T. L., Sypus, M., Riedel, M., & Sun, T. (2018). Defining megathrust tsunami source scenarios for northernmost Cascadia. *Natural Hazards, 94*(1), 445–469. Available from https://doi.org/10.1007/s11069-018-3397-6, http://www.wkap.nl/journalhome.htm/0921-030X.

Geist, E. L., Titov, V. V., Arcas, D., Pollitz, F. F., & Bilek, S. L. (2007). Implications of the 26 December 2004 Sumatra–Andaman earthquake on tsunami forecast and assessment models for great subduction-zone earthquakes. *Bulletin of the Seismological Society of America*, *97*(1), S249–S270. Available from https://doi.org/10.1785/0120050619.

Goda, K., Yasuda, T., Martin Mai, P., Maruyama, T., & Mori, N. (2018). Tsunami simulations of mega-thrust earthquakes in the Nankai–Tonankai Trough (Japan) based on stochastic rupture scenarios. *Geological Society, London, Special Publications*, *456*(1), 55–74. Available from https://doi.org/10.1144/sp456.1.

Heidarzadeh, M., Ishibe, T., Harada, T., Natawidjaja, D. H., Pranantyo, I. R., & Widyantoro, B. T. (2021). High potential for splay faulting in the Molucca Sea, Indonesia: November 2019 Mw 7.2 earthquake and tsunami. *Seismological Research Letters*, *92*(5), 2915–2926. Available from https://doi.org/10.1785/0220200442, https://pubs.geoscienceworld.org/srl/article/92/5/2915/596106/High-Potential-for-Splay-Faulting-in-the-Molucca.

Heidarzadeh, M., Pirooz, M. D., & Zaker, N. H. (2009). Modeling the near-field effects of the worst-case tsunami in the Makran subduction zone. *Ocean Engineering*, *36*(5), 368–376. Available from https://doi.org/10.1016/j.oceaneng.2009.01.004.

Hsu, Y. J., Simons, M., Avouac, J. P., Galeteka, J., Sieh, K., Chlieh, M., Natawidjaja, D., Prawirodirdjo, L., & Bock, Y. (2006). Frictional afterslip following the 2005 Nias–Simeulue earthquake, Sumatra. *Science*, *312*(5782), 1921–1926. Available from https://doi.org/10.1126/science.1126960.

Jia, D., Li, Y., Yan, B., Li, Z., Wang, M., Chen, Z., & Zhang, Y. (2020). The Cenozoic thrusting sequence of the Longmen Shan fold-and-thrust belt, eastern margin of the Tibetan plateau: Insights from low-temperature thermochronology. *Journal of Asian Earth Sciences*, *198*, 104381. Available from https://doi.org/10.1016/j.jseaes.2020.104381.

Kame, N., Rice, J. R., & Dmowska, R. (2003). Effects of prestress state and rupture velocity on dynamic fault branching. *Journal of Geophysical Research: Solid Earth*, *108*(5). Available from http://onlinelibrary.wiley.com/journal/10.1002/(ISSN)2169-9356.

Kame, N., & Yamashita, T. (1999). A new light on arresting mechanism of dynamic earthquake faulting. *Geophysical Research Letters*, *26*(13), 1997–2000. Available from https://doi.org/10.1029/1999GL900410, https://agupubs.onlinelibrary.wiley.com/doi/abs/10.1029/1999GL900410.

Kanamori, H. (1972). Mechanism of tsunami earthquakes. *Physics of the Earth and Planetary Interiors*, *6*(5), 346–359. Available from https://doi.org/10.1016/0031-9201(72)90058-1.

Kimura, G., Kitamura, Y., Hashimoto, Y., Yamaguchi, A., Shibata, T., Ujiie, K., & Okamoto, S. (2007). Transition of accretionary wedge structures around the up-dip limit of the seismogenic subduction zone. *Earth and Planetary Science Letters*, *255*(3-4), 471–484. Available from https://doi.org/10.1016/j.epsl.2007.01.005.

Kopp, H. (2013). Invited review paper: The control of subduction zone structural complexity and geometry on margin segmentation and seismicity. *Tectonophysics*, *589*, 1–16. Available from https://doi.org/10.1016/j.tecto.2012.12.037.

Lay, T., Kanamori, H., Ammon, C. J., Koper, K. D., Hutko, A. R., Ye, L., Yue, H., & Rushing, T. M. (2012). Depth-varying rupture properties of subduction zone megathrust faults. *Journal of Geophysical Research: Solid Earth*, *117*(4). Available from https://doi.org/10.1029/2011JB009133, http://onlinelibrary.wiley.com/journal/10.1002/(ISSN)2169-9356.

Lay, T., Kanamori, H., Ammon, C. J., Nettles, M., Ward, S. N., Aster, R. C., Beck, S. L., Bilek, S. L., Brudzinski, M. R., Butler, R., Deshon, H. R., Ekström, G., Satake, K., & Sipkin, S. (2005). The great Sumatra–Andaman earthquake of 26 December 2004. *Science*, *308*(5725), 1127–1133. Available from https://doi.org/10.1126/science.1112250.

Liu, P. L. F., Lynett, P., Fernando, H., Jaffe, B. E., Fritz, H., Higman, B., Morton, R., Goff, J., & Synolakis, C. (2005). Observations by the International Tsunami Survey Team in Sri Lanka. *Science*, *308*(5728), 1595. Available from https://doi.org/10.1126/science.1110730.

McCaffrey, R. (2009). The tectonic framework of the Sumatran subduction zone. *Annual Review of Earth and Planetary Sciences*, *37*, 345–366. Available from https://doi.org/10.1146/annurev.earth.031208.100212, http://arjournals.annualreviews.org/doi/pdf/10.1146/annurev.earth.031208.100212.

McCaffrey, R., & Goldfinger, C. (1995). Forearc deformation and great subduction earthquakes: Implications for Cascadia offshore earthquake potential. *Science*, *267*(5199), 856–859. Available from https://doi.org/10.1126/science.267.5199.856.

Melnick, D., Bookhagen, B., Echtler, H. P., & Strecker, M. R. (2006). Coastal deformation and great subduction earthquakes, Isla Santa María, Chile (37°S). *Bulletin of the Geological Society of America*, *118*(11-12), 1463–1480. Available from https://doi.org/10.1130/B25865.1.

Melnick, D., Moreno, M., Motagh, M., Cisternas, M., & Wesson, R. L. (2012). Splay fault slip during the Mw 8.8 2010 Maule Chile earthquake. *Geology*, *40*(3), 251–254. Available from https://doi.org/10.1130/G32712.1, http://geology.gsapubs.org/content/40/3/251.full.pdf.

Mokhtari, M. (2015). The role of splay faulting in increasing the devastation effect of tsunami hazard in Makran, Oman Sea. *Arabian Journal of Geosciences*, *8*(7), 4291–4298. Available from https://doi.org/10.1007/s12517-014-1375-1, http://www.springer.com/geosciences/journal/12517?cm_mmc = AD-_-enews-_-PSE1892-_-0.

Momeni, P., Goda, K., Mokhtari, M., & Heidarzadeh, M. (2023). A new tsunami hazard assessment for eastern Makran subduction zone by considering splay faults and applying stochastic modeling. *Coastal Engineering Journal*, *65*(1), 67–96. Available from https://doi.org/10.1080/21664250.2022.2117585, https://www.tandfonline.com/loi/tcej20.

Moore, G. F., Bangs, N. L., Taira, A., Kuramoto, S., Pangborn, E., & Tobin, H. J. (2007). Three-dimensional splay fault geometry and implications for tsunami generation. *Science*, *318*(5853), 1128–1131. Available from https://doi.org/10.1126/science.1147195.

Moore, J. C., & Saffer, D. (2001). Updip limit of the seismogenic zone beneath the accretionary prism of Southwest Japan: An effect of diagenetic to low-grade metamorphic processes and increasing effective stress. *Geology*, *29*(2), 183–186. Available from https://pubs.geoscienceworld.org/geology.

Pajang, S., Cubas, N., Letouzey, J., Le Pourhiet, L., Seyedali, S., Fournier, M., Agard, P., Khatib, M. M., Heyhat, M., & Mokhtari, M. (2021). Seismic hazard of the western Makran subduction zone: Insight from mechanical modelling and inferred frictional properties. *Earth and Planetary Science Letters*, *562*, 116789. Available from https://doi.org/10.1016/j.epsl.2021.116789.

Park, J. O., Tsuru, T., Kodaira, S., Cummins, P. R., & Kaneda, Y. (2002). Splay fault branching along the Nankai subduction zone. *Science*, *297* (5584), 1157−1160. Available from https://doi.org/10.1126/science.1074111, http://www.sciencemag.org.

Plafker, G. (1965). Tectonic deformation associated with the 1964 Alaska Earthquake. *Science*, *148*(3678), 1675−1687. Available from https://doi.org/10.1126/science.148.3678.1675.

Priest, G.R., Goldfinger, C., Wang, K., Witter, R. C., Zhang, Y., Baptista, A. M. (2009). *Tsunami hazard assessment of the northern Oregon coast: A multi-deterministic approach tested at Cannon Beach.* Special Paper 41.

Quezada, J., Jaque, E., Catalán, N., Belmonte, A., Fernández, A., & Isla, F. (2020). Unexpected coseismic surface uplift at Tirúa-Mocha island area of South Chile before and during the Mw 8.8 Maule 2010 earthquake: A possible upper plate splay fault. *Andean Geology*, *47*(2), 295−315. Available from https://doi.org/10.5027/andgeov47n2-3057, https://scielo.conicyt.cl/pdf/andgeol/v47n2/0718-7106-andgeol-47-02-0295.pdf.

Ranero, C. R., Phipps Morgan, J., McIntosh, K., & Relchert, C. (2003). Bending-related faulting and mantle serpentinization at the Middle America trench. *Nature*, *425*(6956), 367−373. Available from https://doi.org/10.1038/nature01961.

Singh, S. C., Hananto, N. D., & Chauhan, A. P. S. (2011). Enhanced reflectivity of backthrusts in the recent great Sumatran earthquake rupture zones. *Geophysical Research Letters*, *38*(4). Available from https://doi.org/10.1029/2010GL046227, http://onlinelibrary.wiley.com/journal/10.1002/(ISSN)1944-8007/issues?year = 2012.

Smith, G., McNeill, L., Henstock, I. J., & Bull, J. (2012). The structure and fault activity of the Makran accretionary prism. *Journal of Geophysical Research: Solid Earth*, *117*(7). Available from https://doi.org/10.1029/2012JB009312, http://onlinelibrary.wiley.com/journal/10.1002/(ISSN)2169-9356.

Smith, G. L., McNeill, L. C., Wang, K., He, J., & Henstock, T. J. (2013). Thermal structure and megathrust seismogenic potential of the Makran subduction zone. *Geophysical Research Letters*, *40*(8), 1528−1533. Available from https://doi.org/10.1002/grl.50374.

Sykes, L. R., & Menke, W. (2006). Repeat times of large earthquakes: Implications for earthquake mechanics and long-term prediction. *Bulletin of the Seismological Society of America*, *96*(5), 1569−1596. Available from https://doi.org/10.1785/0120050083.

Waldhauser, F., Schaff, D. P., Diehl, T., & Engdahl, E. R. (2012). Splay faults imaged by fluid-driven aftershocks of the 2004 Mw 9.2 Sumatra−Andaman earthquake. *Geology*, *40*(3), 243−246. Available from https://doi.org/10.1130/G32420.1, http://geology.gsapubs.org/content/40/3/243.full.pdf.

Wang, K., & Hu, Y. (2006). Accretionary prisms in subduction earthquake cycles: The theory of dynamic Coulomb wedge. *Journal of Geophysical Research: Solid Earth*, *111*(B6). Available from https://doi.org/10.1029/2005jb004094, n/a-n/a.

Wang, X., Bradley, K. E., Wei, S., & Wu, W. (2018). Active backstop faults in the Mentawai region of Sumatra, Indonesia, revealed by teleseismic broadband waveform modeling. *Earth and Planetary Science Letters*, *483*, 29−38. Available from https://doi.org/10.1016/j.epsl.2017.11.049, http://www.sciencedirect.com/science/journal/0012821X/321-322.

Wendt, J., Oglesby, D. D., & Geist, E. L. (2009). Tsunamis and splay fault dynamics. *Geophysical Research Letters,*, *36*(15). Available from https://doi.org/10.1029/2009GL038295, http://onlinelibrary.wiley.com/journal/10.1002/(ISSN)1944-8007/issues?year = 2012.

van Zelst, I., Rannabauer, L., Gabriel, A. A., & van Dinther, Y. (2022). Earthquake rupture on multiple splay faults and its effect on tsunamis. *Journal of Geophysical Research: Solid Earth*, *127*(8). Available from https://doi.org/10.1029/2022JB024300, http://agupubs.onlinelibrary.wiley.com/hub/jgr/journal/10.1002/(ISSN)2169-9356/.

Chapter 10

Tsunami hazard from subaerial landslides

Finn Løvholt[1], Sylfest Glimsdal[2] and Carl Bonnevie Harbitz[2]

[1]*Offshore Energy Department, Norwegian Geotechnical Institute, Oslo, Norway,* [2]*Natural Hazards Department, Norwegian Geotechnical Institute, Oslo, Norway*

10.1 Introduction

Landslide tsunamis represent a severe threat to coastlines and lakes-shores (Harbitz, Løvholt, et al., 2014; Heller & Ruffini, 2023; Løvholt et al., 2015). Systematic hazard analysis methods for landslide tsunamis remain relatively unexplored, despite landslides being the second-most frequent source of tsunamis globally (Harbitz, Løvholt, et al., 2014). Earthquakes are the most frequent source of tsunamis. Nevertheless, with better documentation of tsunamis in recent years, it is evident that landslide tsunamis, and in particular those of subaerial origin, also take place relatively frequently. As shown in Table 10.1, a significant number of these are of high intensity often involving large local runup heights; a more exhaustive list of events can be found in Heller and Ruffini (2023). More than 10 tsunamis have reached a runup height above 10 m since the year 2000, implying that such high-intensity tsunamis have taken place globally about every second year or so. Although most of these events are relatively local, several have propagated significant distances and impacted distant locations with considerable runup heights (e.g., 2014 Karratfjord, 2014 Icy Bay, and 2018 Anak Krakatau).

Traditionally, landslide tsunamis have been treated using scenario-based methods (Grilli et al., 2015; Harbitz, Løvholt, et al., 2014; Løvholt et al., 2005; Tehranirad et al., 2015; Zaniboni et al., 2014), while for earthquake tsunamis, probabilistic tsunami hazard analyses (PTHA) have been the accepted standard for at least 10 years or so (Geist & Lynett, 2014; Grezio et al., 2017). The main reason for this is partly that past landslide and tsunami data for estimating landslide tsunami frequency are lacking (Behrens et al., 2021), but also that landslide tsunamis necessitate a higher degree of model sophistication. The latter holds in particular for subaerial landslide tsunamis involving impact dynamics that cannot be straightforwardly treated by traditional shallow-water tsunami solvers. Yet, a few examples of landslide PTHA (LPTHA) have been conducted (Lane et al., 2016; Løvholt et al., 2020). Moreover, recent recipes provide background for conducting more elaborate hazard analysis (Løvholt et al., 2022; Zengaffinen-Morris et al., 2022), and the preparation of landslide data statistics that can be used as input for LPTHA (Geist & ten Brink, 2019) indicates that LPTHA is an emerging area for research and development.

Landslide tsunamis can be separated into two main types, submarine landslide tsunamis (Harbitz, Løvholt, et al., 2014; Løvholt et al., 2015) and subaerial landslide tsunamis (including volcano flank collapses), also called impact waves (Harbitz, Glimsdal, et al., 2014). Tsunami genesis and propagation of submerged submarine landslides can most often be simulated using dispersive (Glimsdal et al., 2013; Løvholt et al., 2015) or sometimes even shallow-water (Løvholt et al., 2005) long-wave models. Simulating subaerial landslide tsunamis resulting from rock or soil plunging into the water, involving nonlinearity, breaking, and mixing, requires consideration of more complex dynamics, and acquiring accurate and efficient numerical methods is more challenging. Hence, a range of methods exist; see the comprehensive review of Yavari-Ramshe and Ataie-Ashtiani (2016). In principle, a complete modeling of impact waves, including all main elements of the physics, should involve a fully three-dimensional (3D) model with advanced landslide rheology.

Rauter et al. (2022) exemplified that a fully vertically resolved 3D coupled landslide-tsunami model with an advanced rheological landslide model can provide accurate predictions of both laboratory and field-scale observations for the Lake Askja event in 2014 (Gylfadóttir et al., 2017). Somewhat less sophisticated models, yet with significant rigor, include layered nonhydrostatic models (Esposti Ongaro et al., 2021; Ma et al., 2012; Macías et al., 2021).

The layered models cannot take into account cratering, rotational flow, and overturning waves, but are still able to hindcast past events relatively accurately (Esposti Ongaro et al., 2021). On the other hand, 3D models are complex to set up and computationally too expensive to use for almost any type of hazard analysis, let alone probabilistic analysis that

TABLE 10.1 Selected high-impact subaerial landslide tsunamis since the year 2000.

Year	Event	Country	Volume (Mm^3)	Maximum runup (m)	References
2000	Paatuut	Greenland	~50	>50	Dahl-Jensen et al. (2004)
2002	Stromboli (two slides)	Italy	4–9 and 20	>10	Tinti et al. (2006)
2003	Qianjiangping	China	24	39	Yin et al. (2015)
2007	Dayantang	China	3	~50	Roberts et al. (2013)
2007	Aysen Fjord	Chile	High number of slides	~50	Sepúlveda et al. (2010)
2008	Chehalis Lake	Canada	3	38	Roberts et al. (2013)
2008	Three Gorges	China	0.38	13	Huang et al. (2012)
2014	Lake Askja	Iceland	~20	80	Gylfadóttir et al. (2017)
2015	Icy Bay	USA	76	193	Higman et al. (2018)
2017	Karrat Fjord	Greenland	~40	90	Paris et al. (2019)
2018	Bureya	Russia	25	90	Gusiakov and Makhinov (2020)
2018	Baige	China	>10	54	Hu et al. (2020)
2018	Anak Krakatau	Indonesia	~210	85	Grilli et al. (2019)
2020	Elliot Creek	Canada	13.5	120	Geertsema et al. (2022)

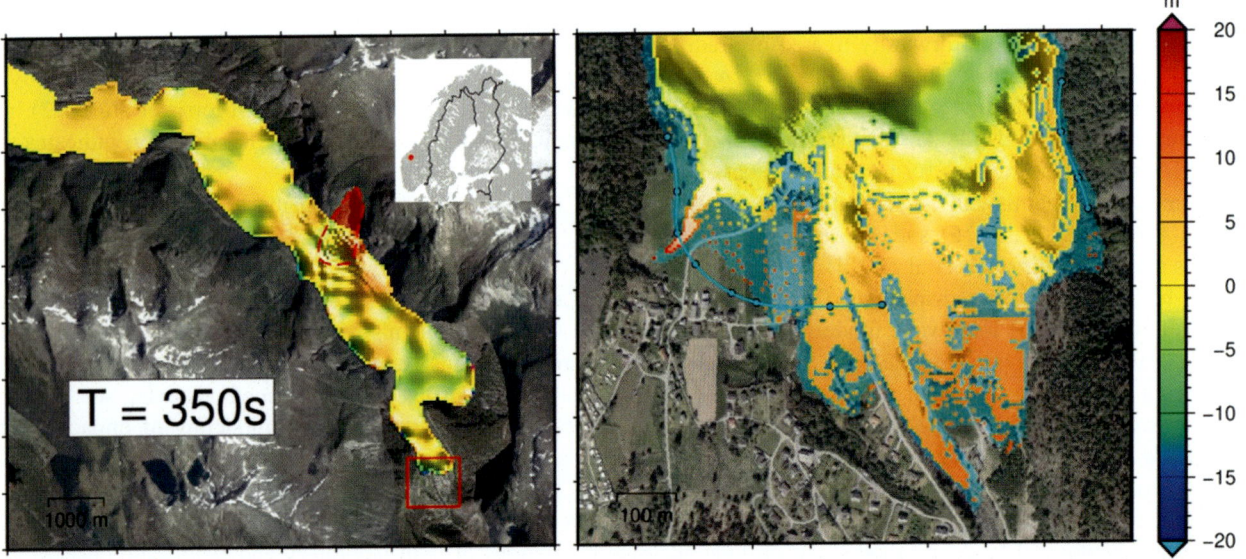

FIGURE 10.1 Simulation of the 1934 Tafjord Tsunami in Norway. The snapshot shows the tsunami propagation (left panel) and the inundation in Tafjord (right panel) 350 seconds after the rockslide plunged into the fjord. The figure shows the withdrawal of the tsunami after the impact of the highest wave.

requires simulations of a large number of scenarios. Hence, it is necessary to explore how computationally simpler methods, for exapmle, single-layer long-wave models (Kirby, 2016; Løvholt et al., 2013), can be used to assess the tsunami hazard.

In today's numerical models used in LPTHA applications, possible uncertainties related to either model simplifications or unknowns are not systematically analyzed. These effects can be framed as epistemic uncertainties, which we define as uncertainties due to a lack of knowledge. To better constrain these uncertainties, a methodology based on model calibration that uses hindcast of past events is proposed. To this end, we will discuss in this chapter how the long-wave models, reminiscent of those already being used for earthquake and submarine landslide tsunamis, can be used for simulating subaerial landslide tsunamis. However, as these models are more simplified, they will at the same time exhibit a larger uncertainty, mainly related to the tsunami generation. We will compare simulations with field observations to discuss how this uncertainty can be addressed by a source-calibration process to improve present LPTHA approaches for landslide tsunamis. This chapter is organized as follows: In Section 10.2, we discuss previous LPTHA approaches and outline ways of performing LPTHA calibration. In Section 10.3, results from such a calibration exercise are presented. For the calibration of the LPTHA, we have focused on the 1934 Tafjord Tsunami (Harbitz et al., 1993), Fig. 10.1, and the 1983 Årdalstangen Tsunami, which were caused by the rockslides. The results are discussed further in Section 10.4, and the main findings of this chapter are summarized in Section 10.5.

10.2 Method for subaerial landslide probabilistic hazard analysis

We follow the previous LPTHA method by Løvholt et al. (2020). Below, we outline different steps in their analyses and discuss possible measures to improve LPTHA by analyzing past events more systematically. To this end, the method analyzes the tsunami hazard from a limited number of unstable slopes that pose a threat to local communities in a fjord system. For each potential landslide volume, different landslide scenarios are associated with different probabilities, followed by modeling the impact in several localities. Finally, the hazard is aggregated for all scenario outcomes. As there is often a need to perform thousands of inundation simulations, some model simplifications are necessary.

10.2.1 Numerical modeling

The following steps are undertaken for modeling each landslide tsunami (Løvholt et al., 2020).

First, the landslide dynamics are modeled by using a prescribed landslide kinematic model combined with a box-shaped landslide, where the rounded box moves downslope like a flexible blanket with a fixed aspect ratio (length, width, and thickness). Typical dimensions of the length, width, and thickness for rockslide scenarios can vary over several orders of magnitude, yet examples can be found in Harbitz, Glimsdal, et al. (2014). Estimates of the frontal area of this source for the box-shaped landslide were determined through more sophisticated "pre-simulations" using the depth-averaged frictional–collisional landslide model VoellmyClaw (Kim, 2014) for each of the unstable landslide volumes that were included in the probabilistic analysis. Second, this box-shaped landslide acts as a tsunami generator, displacing water through linear flux source terms in the tsunami model GloBouss (Løvholt et al., 2008), implying that only the volume displacement is considered for the wave generation.

The tsunami propagation model GloBouss is a Boussinesq-type long-wave model that is used in a linear dispersive mode (i.e., omitting the nonlinear terms) for propagating the tsunami from the source, into the fjord, and to far-field locations within the fjord system. A typical grid resolution of 25–50 m is used for the propagation modeling, before computing the inundation. Nonlinearities are neglected in the offshore propagation, but these are often important near the slide impact, and the linear wave assumption thus holds in the far field. In turn, the wave propagation model is coupled with the shallow water type MOST model (Titov et al., 2011) for conducting high-resolution inundation simulations for several tens of coastal locations using a local inner resolution of 10 m or less (see Løvholt et al., 2010 for details regarding the model coupling). A set of local high-resolution grids covering all different sites of interest for simulating the inundation is coupled with the coarser "global" grid covering the entire fjord.

10.2.2 Probabilistic treatment

Several landslide volumes are included in the analysis (Løvholt et al., 2020). A key element in enabling LPTHA is to associate each of these scenario volumes with annualized release probabilities, below referred to as scenario rates. The scenario rates for the unstable rock-slope volumes in the analysis were established using the method of Hermanns et al. (2013), which takes into account the stability of each slope. To convert slope-stability observations to rates, regional or

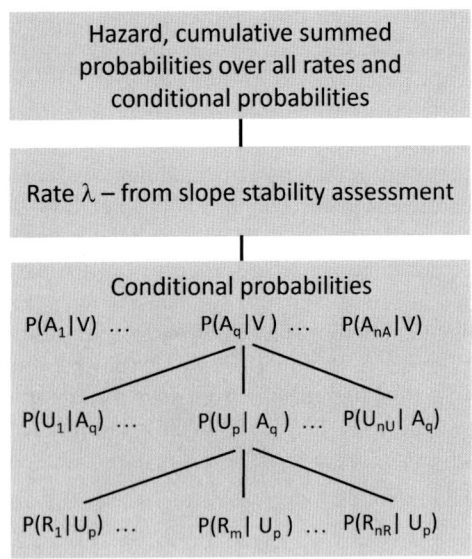

FIGURE 10.2 Probabilistic tsunami hazard event tree (Løvholt et al., 2020). Simplified drawing of the probabilistic event tree used for aggregating landslide tsunami hazard based on the method of Løvholt et al. (2020). The annual rate is denoted by λ, while conditional probabilities are represented by the landslide frontal area A_q with nA branches, the velocity U_p up to nU branches, and runout distance Rm up to nR branches.

national magnitude–frequency characteristics of past landslides for all of Norway have been used to tune the parameter in the model describing the rates as a function of the rock-slope instability. On the other hand, as the inferred rates reflect the instability, they are not directly linked to past landslide frequency as done in classical earthquake PTHA (Geist & Parsons, 2006; Grezio et al., 2017). The description of establishing the rates is beyond the scope of this chapter, and the reader is referred to Hermanns et al. (2013) for further details.

Each scenario represents a realization of possible outcomes in the probabilistic analysis and is given a discrete probability. The hazard is aggregated using an event tree, illustrated in Fig. 10.2, taking the products of the probabilities of three different landslide kinematics parameter values representing the run-out distance (horizontal travel distance of the landslide), the impact velocity (defined as the velocity when the landslide hits the water), and the frontal area of the landslide. These parameters thus represent the epistemic uncertainty related to the landslide kinematics.

For each point in the inundation area, we sum the annualized scenario probabilities for all source scenarios. Using a contouring algorithm, exceedance probability thresholds are placed into the topography. For the study of Løvholt et al. (2020), 1/1000 year and 1/5000 year exceedance rates adapted to the national legislation for landuse were used.

10.2.3 Shortcomings

The simplicity of the tsunami generation model of Løvholt et al. (2020) (e.g., omission of nonlinear effects in the generation phase) implies that some physics is lacking in the depth-averaged numerical model used so far in LPTHA. This can in principle partly be mitigated by including more physics to gain more accurate landslide–tsunami models, for example, by running 3D coupled landslide–tsunami models. However, as discussed above, the resources needed for running many realizations of these models render them unfeasible for use in most practical LPTHA, because the high number of simulations needed to resolve landslide parameter uncertainty would take up too many computational resources unless abundant high-performance computing resources are available.

In addition, there are several aspects related to subaerial landslide dynamics and tsunami generation that are presently not well understood or documented, which affects the uncertainty. The material behavior of the sliding rock mass can be described by more advanced rheologies, such as $\mu(I)$ type (Jop et al., 2006), and if subaqueous also with pore fluid pressure (Guazzelli & Pouliquen, 2018), where μ represents the friction coefficient and I is the inertia number. On the other hand, complex features, such as rock disintegration and friction, are not fully understood. For example, it has been shown that the effective landslide friction scales inversely with the volume (Lucas et al., 2014), which implies that rockslide mass cannot be represented by a single material–based frictional factor. Furthermore, factors, such as erosion, entrainment, and deposition, which imply that the landslide mass and momentum change due to interaction with the bed, represent significant epistemic uncertainties that are not presently built into the advanced 3D models used for simulating landslide tsunamis. Finally, the mixing of the landslide bulk material with air or water implies that the effective frontal area that directly determines the landslide tsunamigenic strength (Fritz et al., 2003; Rauter et al., 2021) is increased. The so-called bulking factor, which is the ratio of fluid to bulk material in the flowing landslide, can typically increase the landslide volume. Modeling these phenomena involves multiphase aspects that are most often not embedded in landslide dynamics or coupled landslide–tsunami models in use today.

Consequently, there are a substantial number of effects that cannot be factored into an LPTHA analysis due to the high computational cost, lack of understanding, or simply lack of inclusion of the physics in the models. The typical characteristic is that the simpler the model, the larger the epistemic uncertainty. In addition to the abovementioned imperfections, it is stressed that the modeling of the source rates clearly includes substantial uncertainty as it includes a calibration based on the frequency of past events regionally mixed with expert judgment-based techniques applied for each slope. However, possible improvements in the source rate estimation are beyond the scope of this chapter.

10.2.4 Suggested improved methodology using hindcasting to calibrate parameter probabilities

The basis for the methodology is that we will use observed runup heights from past events to calibrate the outputs of the numerical model. On the other hand, these past events are rare in the vicinity of presently unstable rock slopes, if any, and possible previous events may have changed the water depths and thus the conditions for tsunami generation considerably. Hence, we propose to use observations of subaerial landslide tsunamis stemming from different geographical areas, but still having similar landslide or material characteristics, to calibrate the outputs.

Once the volume is determined, the other key mechanism to the tsunami generation is the landslide dynamics. The simulations carried out by Løvholt et al. (2020) showed that the tsunami generation is more sensitive to the frontal area of the box landslide than to the runout distance and the impact velocity. We stress that this is partly in contradiction with laboratory-based scaling relationships (Fritz et al., 2003; Heller & Ruffini, 2023; Rauter et al., 2021) and that generally, tsunami genesis is more sensitive to the impact velocity. The discrepancy is likely due to the linearization of the tsunami generation process in the simplified modeling scheme adopted.

In the following, we revise the methodology proposed by Løvholt et al. (2020) outlined in Section 10.2.1, combining box-shaped landslide models with prescribed kinematics for tsunami generation, a linear dispersive model for tsunami propagation, and a nonlinear shallow-water model for inundation. First, we attribute the uncertainty of statistical observations of past landslide runout distances and can hence set the corresponding probabilities based on runout versus volume statistics (Løvholt et al., 2020). The runout statistics shown in Fig. 10.3 are used for selecting conditional probabilities for a given landslide volume. Second, we have observed that to more closely reproduce past tsunami runup events, relatively large frontal areas are needed, and this gives rise to unrealistically short landslide configurations. Hence, it was further assumed that the tail of the landslides was infinitely long, hence avoiding too short-crested waves.

Furthermore, we choose a set of landslide frontal area factors based on depth-averaged numerical simulations of the studied landslide volumes, in this case using a depth-averaged landslide Voellmy-type model (Kim, 2014). The landslide model is then run with a large number of combinations (N) of the friction parameters. For the i-th simulation, we extract the total integrated area of the landslide intersecting the shoreline at all simulation times, $a_{t,i}$. A_T is then defined

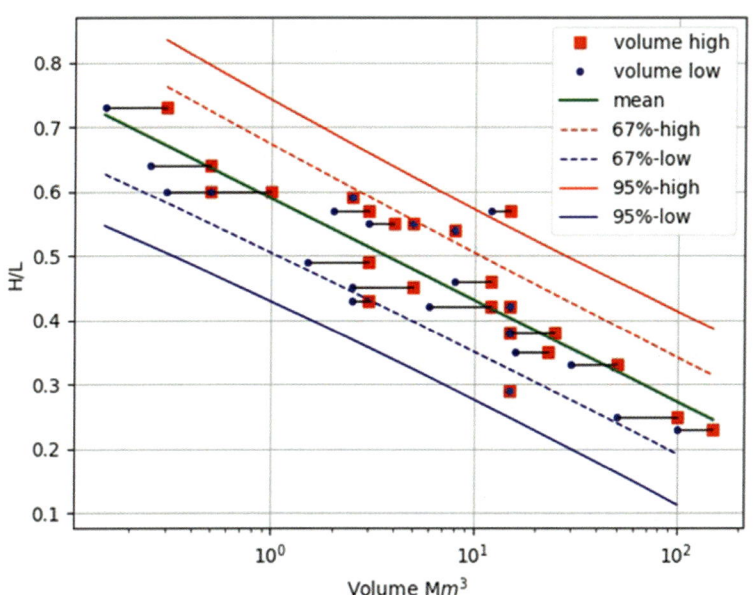

FIGURE 10.3 Landslide runout statistics from Norway. The figure shows the relationship between landslide H/L ratio (vertical drop height over horizontal runout distance ratio) and landslide volume. *From Løvholt, F., Glimsdal, S., & Harbitz, C. B. (2020). On the landslide tsunami uncertainty and hazard. Landslides, 17(10), 2301–2315. https://www.springer.com/journal/10346, https://doi.org/10.1007/s10346-020-01429-z.*

TABLE 10.2 Different epistemic uncertainty representation sets combining frontal areas with conditional probabilities considered in the present study. A is the area factor multiplied with model result A_T, and p_A is the respective conditional probability for the event tree analysis.

Realization		A1	A2	A3	A4	A5
$p01$	A	3.0	2.0	1.0	0.8	0.6
	p_A	0.2	0.45	0.2	0.1	0.05
$p02$	A	3.0	2.0	1.0	0.8	0.6
	p_A	0.5	0.28	0.1	0.07	0.05
$p03$	A	1.5	1.0	0.8	0.6	0.4
	p_A	0.05	0.2	0.35	0.25	0.15
$p04$	A	2.0	1.5	1.0	0.8	0.6
	p_A	0.1	0.4	0.3	0.15	0.05
$p05$	A	1.5	1.0	0.8	0.6	0.4
	p_A	0.1	0.3	0.4	0.15	0.05

as the largest area of all simulations with areas in the set $[a_{t,1}, a_{t,i}, ..., a_{t,N}]$. To consider tsunami generation effects not incorporated in the linear generation model, such as bulking factors, tsunami cratering, tsunami buildup to the critical landslide speed, and entrainment, a set of area factors A (each given different conditional probabilities p_A) is multiplied to the model results A_T to obtain the frontal area for the box model used in the LPTHA by capturing the uncertainty in the tsunami generation (see Table 10.2). The area factors then represent the major source of epistemic uncertainty in the tsunami generation model. As the Voellmy-type landslide model tends to make (smear out) the landslide thin, the area factors also provide a bias correction. The area factor times A_T should therefore be expected to significantly surpass the numerically modeled frontal areas for some of the realizations.

To test how sensitive the different sets of conditional probabilities are to the hazard, we carry out tsunami simulations for alternative full sets of conditional probabilities for past hazard analyses (some of them unpublished) for different rockslides at several different inundation sites. Without providing a stringent statistical analysis, we attempt to select sets of frontal areas that give mean hazard results comparing as closely as possible with the observed runup height trim lines (contour that defines the innermost wet points). In the next sections, we show results from the analysis following this new schema before we discuss the results.

10.3 Results—model calibration for constraining uncertainty analysis

The calibration process following the procedure outlined in Section 10.2.4 is undertaken by investigating a large number of landslide simulations for two previous well-documented rockslide tsunamis in Norwegian fjords, the 1934 Tafjord Tsunami (Harbitz et al., 1993) with an estimated volume of 1.5 Mm^3, rising to 3 Mm^3 when entrainment is included, and the 1983 Årdal Tsunami with a volume of about 200,000 m^3 (Harbitz, Glimsdal, et al., 2014). The Tafjord Tsunami resulted in runup heights of several tens of meters in the vicinity of the rockslide impact and caused 40 fatalities, while the Årdal Tsunami was more localized with a maximum runup height of 5 m and with no fatalities. In the following, we will use observations of the inundation from the 1934 Tafjord event at three different locations, that is, Tafjord, Fjøra, and Sødalsvika, and from the 1983 Årdal event at the location Årdalstangen (see Fig. 10.4).

For simulating the events, we ran VoellmyClaw simulations using the failed rock-slope volume from the release area, and then extracted the maximum velocity of the landslide as well as the total landslide intersection area over all simulation output times (A_T) at the fjord impact as inputs to creating the box-shaped landslides in the LPTHA. We then defined a set of different area factors A that are multiplied by either value of A_T to provide the frontal area of the box landslide. Each of these area factors represents a realization of the epistemic uncertainty in the tsunami generation related to the effective frontal impact area and is hence given a set of conditional probabilities that are collectively exhaustive. We applied the combinations of area factors and conditional probabilities from past hazard analyses (apart

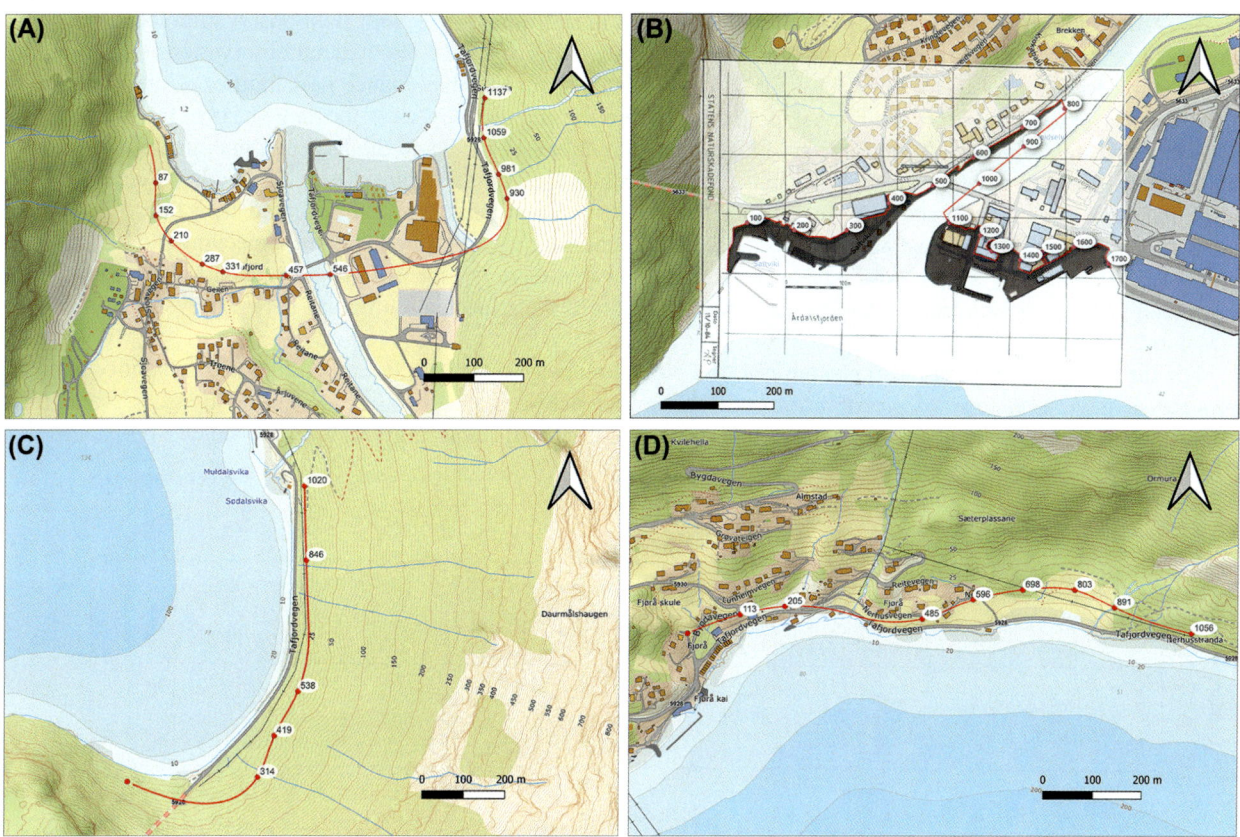

FIGURE 10.4 Maps showing the study sites with observed trim lines. Sites with recorded inundation: (A) Tafjord, (B) Årdalstangen, (C) Sødalsvika, and (D) Fjøra. Map lines delineate study areas and do not necessarily depict accepted national boundaries.

TABLE 10.3 Model parameters used for runout distance and frontal velocity for the 1934 Tafjord and the 1983 Årdal events. The conditional probabilities are indicated as p_R and p_U for runout and velocity, respectively.

Runout distance	Tafjord (m)	Årdal (m)	p_R	Impact velocity	Tafjord (m/s)	Årdal (m/s)	p_U
R1	1146	611	0.07	U1	70	79	0.2
R2	1146	421	0.24	U2	50	61	0.6
R3	1002	240	0.38	U3	30	41	0.2
R4	857	40	0.24				
R5	704	0	0.07				

from Lyngen described by Løvholt et al. (2020), these factors are all unpublished), and in addition, we created a set of synthetic factor and area combinations. The respective values are listed in Table 10.2.

The next step of the analysis was carrying out all the simulations in the event tree for the parameters given in Tables 10.2 and 10.3, providing high-resolution inundation maps for each of the four inundation locations. Then the exceedance probabilities using the PTHA event tree analysis for all the conditional probabilities were aggregated. Compared to the previous LPTHA application by Løvholt et al. (2020), we apply the probability aggregation over the event tree for past events, hence excluding the annualized rates of landside release from the analysis. This implies that the probability of all simulations for each volume in this investigation sums up to unity. In the following, we will compare the average exceedance probabilities (relative probability plots with 0.5 exceedance probability) with the observed trim lines obtained from the field investigations.

Fig. 10.5 shows the aggregated probability plots for the four different sites with field observation locations defining the trim line using the full ensemble simulation sets. We initially remark on a few general observations. First, the different probability and areal factor sets ($p01$–$p05$) provide a broad range of exceedance probabilities at the trim line location. This is expected as the conditional probabilities for the simulations with the largest area factors are largely different, and in particular, $p01$ and $p02$ give higher conditional probabilities to the larger frontal areas. As linear models are used, we also expect a linear dependence of the frontal area on the induced tsunami heights. Second, we see a large variation from site to site in terms of either over- or underestimation of the observed trim lines. For example, simulations for the Tafjord location clearly provide more observational trim line locations exceeding the 0.5 probability compared to the nearby location of Sødalsvika. This implies that there will be differences related to how well the models match at a given inundation site. This comes in addition to the expected variability between events (e.g., between Årdal and Tafjord).

For the Tafjord and Fjøra locations, the $p01$ and $p02$ parameter sets generally exceed the 0.5 probability, while $p04$ provides a more even mix of probabilities both above and below 0.5 probability. A similar trend is found for the Årdalstangen location, although here in particular the $p04$ and $p01$ sets tend to give somewhat smaller probabilities. For

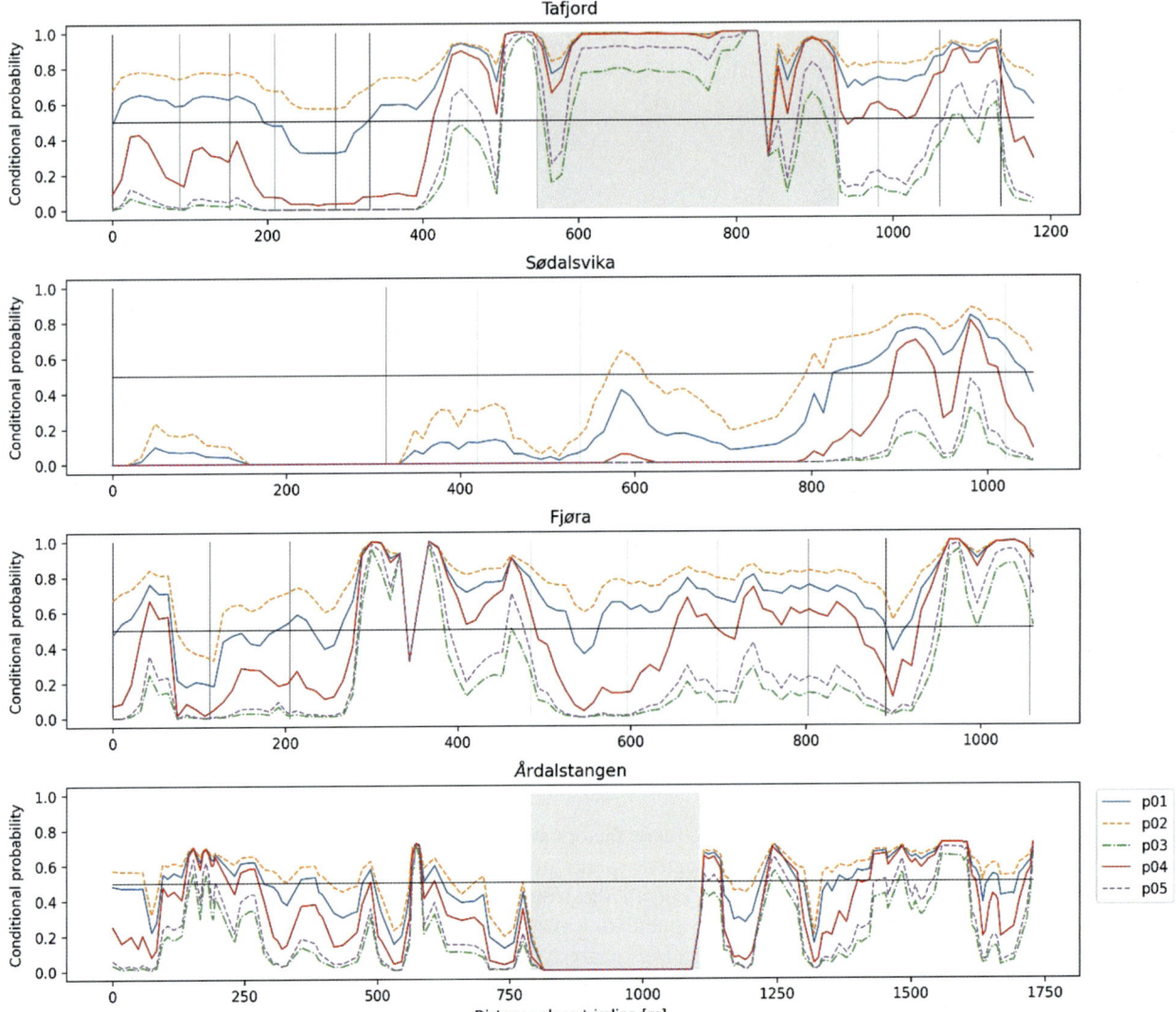

FIGURE 10.5 Overall 50% exceedance probability for the different epistemic probability sets extracted along the observed trim lines for the four different sites. Upper panel: Tafjord; upper mid: Sødalsvika; lower mid: Fjøra; and lower panel: Årdalstangen. The gray rectangles correspond to areas without observation points (also river mouths).

the Sødalsvika location, all combinations of area factors and probability sets tend to give probabilities less than 0.5, except for $p02$ and partly $p01$ in the east part of the domain. The two other parameter sets, $p03$ and $p05$, give probabilities that are most often well below 0.5 for all the locations.

TABLE 10.4 Comparison of $L2$ norms for the probability values shown in Fig. 10.5 with the 0.5 probability. The $L2$ norm for Tafjord, Sødalsvika, and Fjøra is calculated only at the observation points (vertical lines in Fig. 10.5).

Scenario	Site	$p01$	$p02$	$p03$	$p04$	$p05$
Tafjord	Tafjord	0.29	0.30	0.52	0.49	0.53
	Sødalsvika	0.38	0.37	0.49	0.45	0.49
	Fjøra	0.30	0.31	0.42	0.43	0.45
Årdalstangen		0.23	0.22	0.39	0.32	0.38
Average $L2$ norm		0.30	0.30	0.45	0.42	0.46

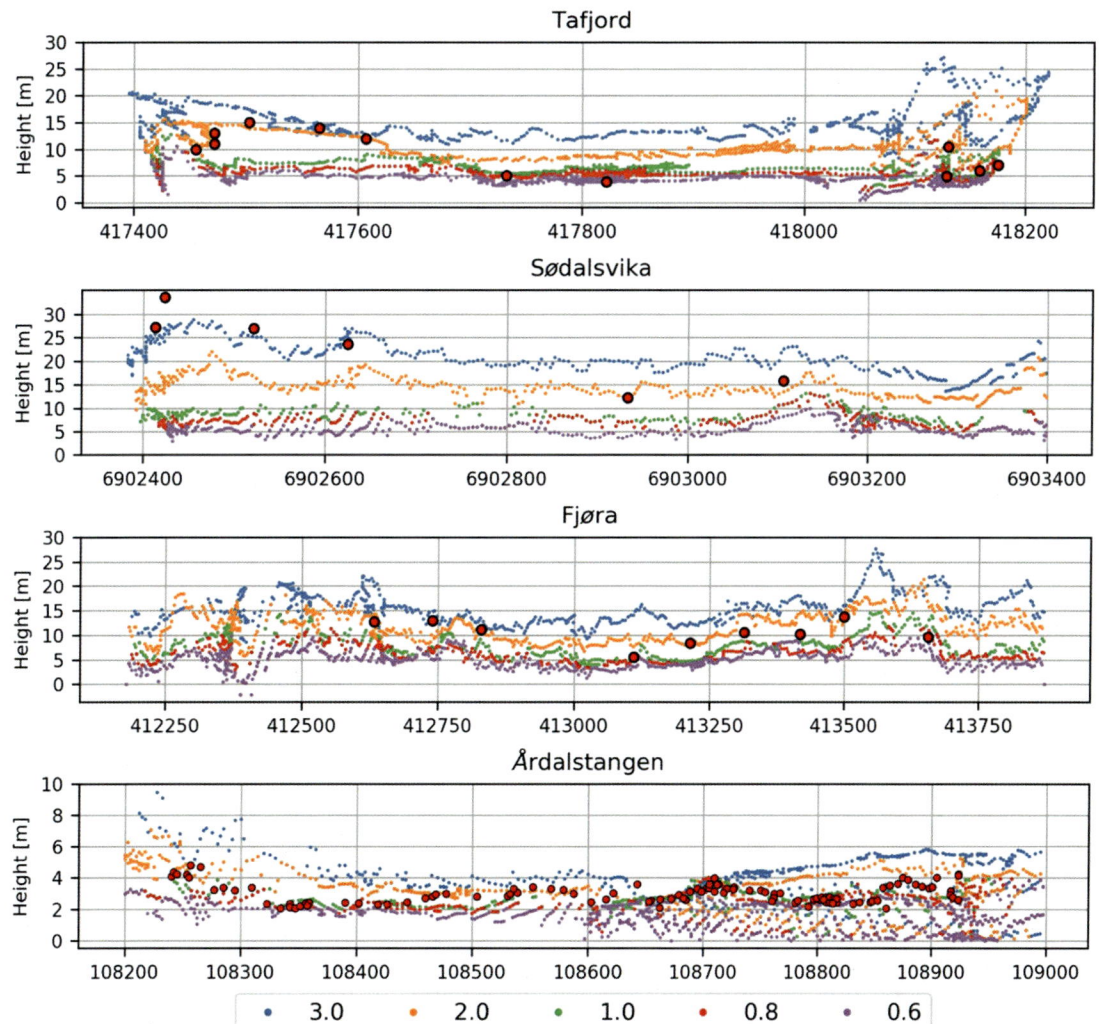

FIGURE 10.6 Height of the simulated trim lines with different factors A for the frontal area. Upper panel, Tafjord; upper mid, Sødalsvika; lower mid, Fjøra; and lower panel, Årdalstangen. Red dots are observed runup heights. The horizontal axis shows the UTM easting (meters) except for UTM northing (meters) at Sødalsvika.

Our visual inspection of the plots would indicate that the *p*01 and *p*02 parameter sets provide the closest agreement with observations, while *p*04 also provides a relatively good agreement. A more formal analysis of the simulations includes computing *L*2 error norms comparing the probabilities with a 0.5 value for all the observation points. The result of this analysis is shown in Table 10.4, with individual *L*2 norms normalized by the number of runup observation points. This verifies that the *p*01 and *p*02 parameters and probability sets give the lowest average error norm. The other parameter sets give significantly higher error norms, in particular *p*03 and *p*05. It should be stressed that the Sødalsvika location is relatively anomalous compared to the three other sites, and this is expected to influence the results.

Fig. 10.6 compares individual simulations with the observed runup heights for the Tafjord and Årdalstangen simulations, using five different area factors *A*. The modeled landslide runout distance for the Tafjord simulations is close to the observed deposits on the seafloor, while for Årdal, we have chosen the mean runout distance ($R3$ from the national statistics) since the runout distance is not known. For both cases, we have chosen the average impact velocity from the VoellmyClaw test simulations, for example, $U1$ for Tafjord and $U2$ for Årdal. The figure shows that the $A = 2$ simulations tend to give the closest agreement with the observations for Tafjord, while the simulations at Årdalstangen show that a factor between $A = 1$ and $A = 2$ gives closer agreement. This advocates that the highest conditional probability in the *p*01 and *p*02 sets for exactly $A = 2$ makes a good choice based on the hindcast. The figure further shows that the highest runup in Sødalsvika can only be reached with $A = 3$, while for many of the other runup locations, much smaller frontal areas are needed to get agreement with the observations. This highlights the need for using a probabilistic representation of the factors to cover the observation range. It is further noted that the topography around the river mouth in Tafjord has changed since 1934 and that this might be the reason for the lower observations compared to the simulations at this point.

10.4 Discussion—implications for future methods

As shown in Section 10.3, the parameter sets *p*01 and *p*02 tended to give the closest match between the 0.5 probability curves and the inundation observations. The same was to some extent also the case with *p*03. In all of these parameter sets, there are relatively high conditional probabilities on the largest area factors that are needed to match the observed inundations. This implies that for this method, we find that the frontal area footprint from the numerical simulation is not enough to provide sufficient tsunami generation. There are various possible reasons for this, and we propose that some or all of the following possible explanations can contribute:

- The landslide model used in this study tend to reduce the effective frontal area of the landslide too much, implying that the landslide spreads its material more than a real landslide would do, both azimuthally (perpendicular to the flow direction) and along the flow. This likely has to do with both the rheology that is included and the depth averaging of the landslide, and potentially because of not including centrifugal and curvature effects. The Voellmy rheological model (Kim, 2014) does not allow the landslide to change frictional properties during the flow, such as more advanced models like $\mu(I)$ with pore pressure. In addition, the Voellmy rheological model has a simplified one-phase treatment of the quadratic resistance due to ambient fluids, and curvature effects are not included.
- Bulking factors expected to increase the effective frontal area of the landslide are not included. In addition, we do not include erosion and entrainment but rather lump the volume at the start of the simulation.
- The applied linear tsunami model implies that the buildup of the wave generation does not include the strong nonlinearities expected during impact. At this point, the landslide speed is highly supercritical, but nonlinearities may potentially give rise to generation effects that are unmodeled in our method. Moreover, the possible strong retardation of the landslide at the moment it hits the water (Rauter et al., 2022) is not embedded, and such effects may give rise to increased frontal areas. We cannot rule out that these higher order effects that are not modeled with our setup can give rise to an additional increase in the tsunamigenic strength of the landslide.
- There is also epistemic uncertainty related to the initial conditions of the slide configuration that will enter the modeling, and this may affect both the landslide dynamics, the aspect ratio of the landslide during impact, and in turn the tsunamigenesis. This uncertainty can only be modeled through probabilistic scenario realizations. Other explanations can be inaccuracies in the topobathymetric representation.

Hence, the present hindcasts of past events show that with the simplified models we presently use in LPTHA, it is necessary to weigh unmodeled effects into the uncertainty in the LPTHA. We treat this through the area factor, and we must typically use area factors significantly larger than unity.

It is emphasized that the present analysis is run for several events and locations only and should ideally be supplemented by hindcasting several additional landslide tsunamis. In general, the number of events modeled, and the sites considered are too few to provide a solid basis for the calibration. However, there are only a few events available, in particular in the region of interest (western Norway), and the modeling provides a valuable link between understanding past events and LPTHA. Nevertheless, including several additional analyses would be necessary for better constraining the uncertainty treatment.

The present analysis provides a preliminary scheme on how to use past events to better calibrate the conditional probabilities used for modeling alternative parameter realizations in LPTHA. With the present limitations of the employed models in mind, the range of parameter values needed to cover the uncertainty is relatively large. On the other hand, the methodology used could in principle be used for any combination of models, and with more accurate models, we generally expect reduced epistemic uncertainties.

The present method only investigates the matching of past events in a semi-quantitative way without statistical rigor. We select the best set of parameter values based on the lowest $L2$ norm (identifying the set of conditional probabilities that most closely matches the mean runup observations). We foresee improving the method to involve a more stringent statistical analysis of the epistemic uncertainty factors. We should also include an analysis of the variability of the applied parameter set, as we only consider the mean values.

10.5 Conclusions

We reviewed a recently developed LPTHA method (Løvholt et al., 2020) and proposed a new method for calibrating the treatment of epistemic uncertainties due to landslide tsunami generation. This method consists of four main steps:

1. Assigning the annualized rates for the landslide sources.
2. Setting a set of landslide dynamics parameter values (including the conditional probabilities for each parameter value) representing the epistemic uncertainty in the tsunami generation process.
3. Running the tsunami generation, propagation, and inundation simulations.
4. Aggregating the uncertainties through an event tree.

In this chapter, we focused on step (2) and discussed methods for constraining the epistemic uncertainties. By using the event tree in a prognostic fashion, we simulated two past rockslide tsunami events in Norway, the 1983 Årdal and 1934 Tafjord Tsunamis. Respective inundation simulations were carried out for four sites. These events took place at locations that are different from expected future LPTHA study areas but are yet expected to be representative of future rockslide tsunamis in Norwegian fjords. The parameter sets used for the event trees were a mix of parameter sets used in past analyses and synthetic parameter combinations. The employed tsunami generation and propagation model was linear, and the most sensitive generation parameter was the frontal landslide area. Hence, we represented the epistemic uncertainty mainly by varying the frontal area in the tsunami generation. A relatively wide range of frontal areas was used in the analysis to reflect the epistemic uncertainty and thus capture the range of inundation observations for the two events.

The epistemic uncertainty treatment indirectly includes several processes not embedded in the numerical models, such as landslide bulking factors, landslide entrainment, and nonlinear generation processes. In principle, the same scheme could be implemented based on more advanced numerical models and physics in the generation process, potentially reducing uncertainties. These include fully nonlinear Boussinesq models (Shi et al., 2012); layered nonhydrostatic models (Esposti Ongaro et al., 2021; Ma et al., 2012; Macías et al., 2021); or Navier−Stokes type models (Rauter et al., 2022). On the other hand, we also expect the presence of aleatory uncertainties related to the variability from event to event that cannot be removed by increased model sophistication.

There are several aspects of the outlined method that can be improved as it does not involve a formal uncertainty analysis nor a model parameter optimization. Furthermore, we only attempted to match the mean value of the model results with the observed runup height and not the variability. It is also emphasized that the number of test cases simulated so far was small, and this should ideally be expanded to better reflect the natural variability of past rockslide tsunami events. On the other hand, this is the first systematic step to constrain landslide tsunami uncertainty in LPTHA, which is necessary for a more solid understanding of this phenomenon in the future.

References

Behrens, J., Løvholt, F., Jalayer, F., Lorito, S., Salgado-Gálvez, M. A., Sørensen, M., Abadie, S., Aguirre-Ayerbe, I., Aniel-Quiroga, I., Babeyko, A., Baiguera, M., Basili, R., Belliazzi, S., Grezio, A., Johnson, K., Murphy, S., Paris, R., Rafliana, I., De Risi, R., ... Vyhmeister, E. (2021). Probabilistic tsunami hazard and risk analysis: A review of research gaps. *Frontiers in Earth Science*, *9*. Available from https://doi.org/10.3389/feart.2021.628772, https://www.frontiersin.org/journals/earth-science.

Dahl-Jensen, T., Larsen, L. M., Pedersen, S. A. S., Pedersen, J., Jepsen, H. F., Pedersen, G. K., Nielsen, T., Pedersen, A. K., Von Platen-Hallermund, F., & Weng, W. (2004). Landslide and tsunami 21 November 2000 in Paatuut, West Greenland. *Natural Hazards*, *31*(1), 277−287. Available from https://doi.org/10.1023/B:NHAZ.0000020264.70048.95.

Esposti Ongaro, T., Vitturi, M., Cerminara, M., Fornaciai, A., Nannipieri, L., & Favalli, M. (2021). Escalante, modeling tsunamis generated by submarine landslides at Stromboli Volcano (Aeolian Islands, Italy): A numerical benchmark study. *Frontiers in Earth Science, 9*.

Fritz, H. M., Hager, W. H., & Minor, H. E. (2003). Landslide generated impulse waves. 1. Instantaneous flow fields. *Experiments in Fluids, 35*(6), 505−519. Available from https://doi.org/10.1007/s00348-003-0659-0.

Geertsema, M., Menounos, B., Bullard, G., Carrivick, J. L., Clague, J. J., Dai, C., Donati, D., Ekstrom, G., Jackson, J. M., Lynett, P., Pichierri, M., Pon, A., Shugar, D. H., Stead, D., Del Bel Belluz, J., Friele, P., Giesbrecht, I., Heathfield, D., Millard, T., ... Sharp, M. A. (2022). The 28 November 2020 landslide, tsunami, and outburst flood—A hazard cascade associated with rapid deglaciation at Elliot Creek, British Columbia, Canada. *Geophysical Research Letters, 49*(6). Available from https://doi.org/10.1029/2021gl096716.

Geist, E. L., & Lynett, P. J. (2014). Source processes for the probabilistic assessment of tsunami hazards. *Oceanography, 27*(2), 86−93. Available from https://doi.org/10.5670/oceanog.2014.43, http://www.tos.org/oceanography/archive/27-2_geist.pdf.

Geist, E. L., & Parsons, T. (2006). Probabilistic analysis of tsunami hazards. *Natural Hazards, 37*(3), 277−314. Available from https://doi.org/10.1007/s11069-005-4646-z.

Geist, E. L., & ten Brink, U. S. (2019). Offshore landslide hazard curves from mapped landslide size distributions. *Journal of Geophysical Research: Solid Earth, 124*(4), 3320−3334. Available from https://doi.org/10.1029/2018JB017236, http://agupubs.onlinelibrary.wiley.com/hub/jgr/journal/10.1002/(ISSN)2169-9356/.

Glimsdal, S., Pedersen, G., Carl, H., & Løvholt, F. (2013). Dispersion of tsunamis: Does it really matter. *Natural Hazards and Earth System Sciences, 13*, 1507−1526. Available from https://doi.org/10.5194/nhess-13-1507-2013.

Grezio, A., Babeyko, A., Baptista, M. A., Behrens, J., Costa, A., Davies, G., Geist, E. L., Glimsdal, S., González, F. I., Griffin, J., Harbitz, C. B., LeVeque, R. J., Lorito, S., Løvholt, F., Omira, R., Mueller, C., Paris, R., Parsons, T., Polet, J., ... Thio, H. K. (2017). Probabilistic tsunami hazard analysis: Multiple sources and global applications. *Reviews of Geophysics, 55*(4), 1158−1198. Available from https://doi.org/10.1002/2017RG000579, http://onlinelibrary.wiley.com/journal/10.1002/(ISSN)1944-9208.

Grilli, S. T., O'Reilly, C., Harris, J. C., Bakhsh, T. T., Tehranirad, B., Banihashemi, S., Kirby, J. T., Baxter, C. D. P., Eggeling, T., Ma, G., & Shi, F. (2015). Modeling of SMF tsunami hazard along the upper US East Coast: Detailed impact around Ocean City, MD. *Natural Hazards, 76*(2), 705−746. Available from https://doi.org/10.1007/s11069-014-1522-8, http://www.wkap.nl/journalhome.htm/0921-030X.

Grilli, S. T., Tappin, D. R., Carey, S., Watt, S. F. L., Ward, S. N., Grilli, A. R., Engwell, S. L., Zhang, C., Kirby, J. T., Schambach, L., & Muin, M. (2019). Modelling of the tsunami from the December 22, 2018 lateral collapse of Anak Krakatau volcano in the Sunda Straits, Indonesia. *Scientific Reports, 9*(1). Available from https://doi.org/10.1038/s41598-019-48327-6, http://www.nature.com/srep/index.html.

Guazzelli, É., & Pouliquen, O. (2018). Rheology of dense granular suspensions. *Journal of Fluid Mechanics, 852*, P11−P173. Available from https://doi.org/10.1017/jfm.2018.548, http://journals.cambridge.org/action/displayJournal?jid = FLM.

Gusiakov, V., & Makhinov, A. (2020). *Understanding and reducing landslide disaster risk December 11, 2018 landslide and 90-m icy tsunami in the Bureya water reservoir* (pp. 351−360). Springer Science and Business Media LLC. Available from http://doi.org/10.1007/978-3-030-60196-6_25.

Gylfadóttir, S. S., Kim, J., Helgason, J. K., Brynjólfsson, S., Höskuldsson, Á., Jóhannesson, T., Harbitz, C. B., & Løvholt, F. (2017). The 2014 Lake Askja rockslide-induced tsunami: Optimization of numerical tsunami model using observed data. *Journal of Geophysical Research: Oceans, 122*(5), 4110−4122. Available from https://doi.org/10.1002/2016JC012496, http://onlinelibrary.wiley.com/journal/10.1002/(ISSN)2169-9291.

Harbitz, C. B., Glimsdal, S., Løvholt, F., Kveldsvik, V., Pedersen, G. K., & Jensen, A. (2014). Rockslide tsunamis in complex fjords: From an unstable rock slope at Åkerneset to tsunami risk in western Norway. *Coastal Engineering, 88*, 101−122. Available from https://doi.org/10.1016/j.coastaleng.2014.02.003.

Harbitz, C. B., Løvholt, F., & Bungum, H. (2014). Submarine landslide tsunamis: How extreme and how likely. *Natural Hazards, 72*(3), 1341−1374. Available from https://doi.org/10.1007/s11069-013-0681-3, http://www.wkap.nl/journalhome.htm/0921-030X.

Harbitz, C. B., Pedersen, G., & Gjevik, B. (1993). Numerical simulations of large water waves due to landslides. *Journal of Hydraulic Engineering, 119*(12), 1325−1342. Available from https://doi.org/10.1061/(asce)0733-9429(1993)119:12(1325).

Heller, V., & Ruffini, G. (2023). A critical review about generic subaerial landslide-tsunami experiments and options for a needed step change. *Earth-Science Reviews, 242*, 104459. Available from https://doi.org/10.1016/j.earscirev.2023.104459, https://www.sciencedirect.com/science/article/pii/S0012825223001484.

Hermanns, R. L., Oppikofer, T., Anda, E., Blikra, L. H., Böhme, M., Bunkholt, H., Crosta, G. B., Dahle, H., Devoli, G., Fischer, L., Jaboyedoff, M., Loew, Si, Sætre, S., & Molina, F. X. Y. (2013). Hazard and risk classification for large unstable rock slopes in Norway. *Italian Journal of Engineering Geology and Environment, 2013*(2), 245−254. Available from https://doi.org/10.4408/IJEGE.2013-06.B-22, http://www.ijege.uniroma1.it.

Higman, B., Shugar, D. H., Stark, C. P., Ekström, G., Koppes, M. N., Lynett, P., Dufresne, A., Haeussler, P. J., Geertsema, M., Gulick, S., Mattox, A., Venditti, J. G., Walton, M. A. L., McCall, N., Mckittrick, E., MacInnes, B., Bilderback, E. L., Tang, H., Willis, M. J., ... Loso, M. (2018). The 2015 landslide and tsunami in Taan Fiord, Alaska. *Scientific Reports, 8*(1). Available from https://doi.org/10.1038/s41598-018-30475-w, http://www.nature.com/srep/index.html.

Hu, Yx, Yu, Zy, & Zhou, Jw (2020). Numerical simulation of landslide-generated waves during the 11 October 2018 Baige landslide at the Jinsha River. *Landslides, 17*(10), 2317−2328. Available from https://doi.org/10.1007/s10346-020-01382-x, https://www.springer.com/journal/10346.

Huang, B., Yin, Y., Liu, G., Wang, S., Chen, X., & Huo, Z. (2012). Analysis of waves generated by Gongjiafang landslide in Wu Gorge, three Gorges reservoir, on November 23, 2008. *Landslides, 9*(3), 395−405. Available from https://doi.org/10.1007/s10346-012-0331-y.

Jop, P., Forterre, Y., & Pouliquen, O. (2006). A constitutive law for dense granular flows. *Nature, 441*(7094), 727−730. Available from https://doi.org/10.1038/nature04801, http://www.nature.com/nature/index.html.

Kim, J. (2014). *Finite volume methods for tsunamis generated by submarine landslides*. University of Washington.

Kirby, J. T. (2016). Boussinesq models and their application to coastal processes across a wide range of scales. *Journal of Waterway, Port, Coastal and Ocean Engineering*, *142*(6). Available from https://doi.org/10.1061/(ASCE)WW.1943-5460.0000350, http://ojps.aip.org/wwo/.

Lane, E. M., Mountjoy, J. J., Power, W. L., & Mueller, C. (2016). Probabilistic hazard of tsunamis generated by submarine landslides in the Cook Strait canyon (New Zealand). *Pure and Applied Geophysics*, *173*(12), 3757−3774. Available from https://doi.org/10.1007/s00024-016-1410-0, http://www.springer.com/birkhauser/geo + science/journal/24.

Løvholt, F., Glimsdal, S., & Harbitz, C. B. (2020). On the landslide tsunami uncertainty and hazard. *Landslides*, *17*(10), 2301−2315. Available from https://doi.org/10.1007/s10346-020-01429-z, https://www.springer.com/journal/10346.

Løvholt, F., Harbitz, C. B., & Haugen, K. B. (2005). A parametric study of tsunamis generated by submarine slides in the Ormen Lange/Storegga area off western Norway. *Marine and Petroleum Geology*, *22*(1-2), 219−231. Available from https://doi.org/10.1016/j.marpetgeo.2004.10.017.

Løvholt, F., Lynett, P., & Pedersen, G. (2013). Simulating run-up on steep slopes with operational Boussinesq models; capabilities, spurious effects and instabilities. *Nonlinear Processes in Geophysics*, *20*(3), 379−395. Available from https://doi.org/10.5194/npg-20-379-2013.

Løvholt, F., Pedersen, G., & Gisler, G. (2008). Oceanic propagation of a potential tsunami from the La Palma Island. *Journal of Geophysical Research: Oceans*, *113*(9). Available from https://doi.org/10.1029/2007JC004603, http://onlinelibrary.wiley.com/journal/10.1002/(ISSN)2169-9291.

Løvholt, F., Pedersen, G., & Glimsdal, S. (2010). Coupling of dispersive tsunami propagation and shallow water coastal response. *The Open Oceanography Journal*, *4*(1), 71−82. Available from https://doi.org/10.2174/1874252101004010071.

Løvholt, F., Pedersen, G., Harbitz, C. B., Glimsdal, S., & Kim, J. (2015). On the characteristics of landslide tsunamis. *Philosophical Transactions of the Royal Society A: Mathematical, Physical and Engineering Sciences*, *373*(2053). Available from https://doi.org/10.1098/rsta.2014.0376.

Løvholt, F., Urgeles, R., Harbitz, C. B., Vanneste, M., & Carlton, B. (2022). *Submarine landslides. Treatise on Geomorphology* (pp. 919−959). Elsevier. Available from https://www.sciencedirect.com/book/9780128182352, http://doi.org/10.1016/B978-0-12-818234-5.00139-5.

Lucas, A., Mangeney, A., & Ampuero, J. P. (2014). Frictional velocity-weakening in landslides on Earth and on other planetary bodies. *Nature Communications*, *5*. Available from https://doi.org/10.1038/ncomms4417, http://www.nature.com/ncomms/index.html.

Ma, G., Shi, F., & Kirby, J. T. (2012). Shock-capturing non-hydrostatic model for fully dispersive surface wave processes. *Ocean Modelling*, *43-44*, 22−35. Available from https://doi.org/10.1016/j.ocemod.2011.12.002.

Macías, J., Escalante, C., & Castro, M. J. (2021). Multilayer-HySEA model validation for landslide-generated tsunamis-Part 1: Rigid slides. *Natural Hazards and Earth System Sciences*, *21*(2), 775−789. Available from https://doi.org/10.5194/nhess-21-775-2021, http://www.nat-hazards-earth-syst-sci.net/volumes_and_issues.html.

Paris, A., Okal, E. A., Guérin, C., Heinrich, P., Schindelé, F., & Hébert, H. (2019). Numerical modeling of the June 17, 2017 landslide and tsunami events in Karrat Fjord, West Greenland. *Pure and Applied Geophysics*, *176*(7), 3035−3057. Available from https://doi.org/10.1007/s00024-019-02123-5, http://www.springer.com/birkhauser/geo + science/journal/24.

Rauter, M., Hoße, L., Mulligan, R. P., Take, W. A., & Løvholt, F. (2021). Numerical simulation of impulse wave generation by idealized landslides with OpenFOAM. *Coastal Engineering*, *165*. Available from https://doi.org/10.1016/j.coastaleng.2020.103815.

Rauter, M., Viroulet, S., Gylfadóttir, S. S., Fellin, W., & Løvholt, F. (2022). Granular porous landslide tsunami modelling—The 2014 Lake Askja flank collapse. *Nature Communications*, *13*(1). Available from https://doi.org/10.1038/s41467-022-28296-7, http://www.nature.com/ncomms/index.html.

Roberts, N. J., McKillop, R. J., Lawrence, M. S., Psutka, J. F., Clague, J. J., Brideau, M. A., & Ward, B. C. (2013). Landslide science and practice: impacts of the 2007 landslide-generated tsunami in Chehalis Lake, Canada. *Landslide Science and Practice: Risk Assessment, Management and Mitigation*, *6*. Available from https://doi.org/10.1007/978-3-642-31319-6_19.

Sepúlveda, S. A., Serey, A., Lara, M., Pavez, A., & Rebolledo, S. (2010). Landslides induced by the April 2007 Aysén fjord earthquake, Chilean Patagonia. *Landslides*, *7*(4), 483−492. Available from https://doi.org/10.1007/s10346-010-0203-2.

Shi, F., Kirby, J. T., Harris, J. C., Geiman, J. D., & Grilli, S. T. (2012). A high-order adaptive time-stepping TVD solver for Boussinesq modeling of breaking waves and coastal inundation. *Ocean Modelling*, *43-44*, 36−51. Available from https://doi.org/10.1016/j.ocemod.2011.12.004.

Tehranirad, B., Harris, J. C., Grilli, A. R., Grilli, S. T., Abadie, S., Kirby, J. T., & Shi, F. (2015). Far-field tsunami impact in the North Atlantic Basin from large scale flank collapses of the Cumbre Vieja volcano, La Palma. *Pure and Applied Geophysics*, *172*(12), 3589−3616. Available from https://doi.org/10.1007/s00024-015-1135-5, http://www.springer.com/birkhauser/geo + science/journal/24.

Tinti, S., Pagnoni, G., & Zaniboni, F. (2006). The landslides and tsunamis of the 30th of December 2002 in Stromboli analysed through numerical simulations. *Bulletin of Volcanology*, *68*(5), 462−479. Available from https://doi.org/10.1007/s00445-005-0022-9.

Titov, V. V., Moore, C. W., Greenslade, D. J. M., Pattiaratchi, C., Badal, R., Synolakis, C. E., & Kânoğlu, U. (2011). A new tool for inundation modeling: Community modeling interface for tsunamis (ComMIT). *Pure and Applied Geophysics*, *168*(11), 2121−2131. Available from https://doi.org/10.1007/s00024-011-0292-4.

Yavari-Ramshe, S., & Ataie-Ashtiani, B. (2016). Numerical modeling of subaerial and submarine landslide-generated tsunami waves—Recent advances and future challenges. *Landslides*, *13*(6), 1325−1368. Available from https://doi.org/10.1007/s10346-016-0734-2.

Yin, Y.-p, Huang, B., Chen, X., Liu, G., & Wang, S. (2015). Numerical analysis on wave generated by the Qianjiangping landslide in Three Gorges Reservoir, China. *Landslides*, *12*(2), 355−364. Available from https://doi.org/10.1007/s10346-015-0564-7.

Zaniboni, F., Armigliato, A., Pagnoni, G., & Tinti, S. (2014). Continental margins as a source of tsunami hazard: The 1977 Gioia Tauro (Italy) landslide-tsunami investigated through numerical modeling. *Marine Geology*, *357*, 210−217. Available from https://doi.org/10.1016/j.margeo.2014.08.011, http://www.sciencedirect.com/science/journal/00253227.

Zengaffinen-Morris, T., Urgeles, R., & Løvholt, F. (2022). On the inference of tsunami uncertainties from landslide run-out observations. *Journal of Geophysical Research: Oceans*, *127*(4). Available from https://doi.org/10.1029/2021JC018033, http://agupubs.onlinelibrary.wiley.com/agu/jgr/journal/10.1002/(ISSN)2169-9291/.

Chapter 11

Dense tsunami monitoring system

Yuichiro Tanioka

Institute of Seismology and Volcanology, Faculty of Science, Hokkaido University, Sapporo, Hokkaido, Japan

11.1 Introduction

The 2011 Tohoku-Oki Earthquake (moment magnitude M_w 9.0) generated a huge tsunami, maximum runup height of about 40 m, along the Pacific coast of northern Japan (Mori & Takahashi, 2011). Although the Japan Meteorological Agency (JMA) issued a major tsunami warning along the Pacific coast of Japan following the earthquake (Ozaki, 2011), the tsunami caused a catastrophic disaster with approximately 19,000 fatalities. After this event, it became clear that a near-field tsunami warning based on earthquake magnitude estimation using seismometers on land had a critical limitation in terms of the accuracy of tsunami height estimation within an evacuation time before the tsunami hits the coast. As a result, the construction of seafloor observation networks has become an urgent issue in Japan.

The Dense Ocean floor Network system for Earthquakes and Tsunamis (DONET) is an ocean-bottom observation network consisting of 51 stations that monitor earthquakes and tsunamis along the Nankai Trough in southwestern Japan (Fig. 11.1). These stations cover the eastern half of the Nankai subduction zone, one of the regions where the next megathrust earthquake is anticipated. At the time of the 2011 Tohoku Earthquake and Tsunami, DONET1 was under construction by the Japan Agency for Marine-Earth Science Technology and 10 stations were in operation. Then, 20 stations of DONET1 were in operation in July 2011, and all 51 stations of DONET1 and DONET2 were completed

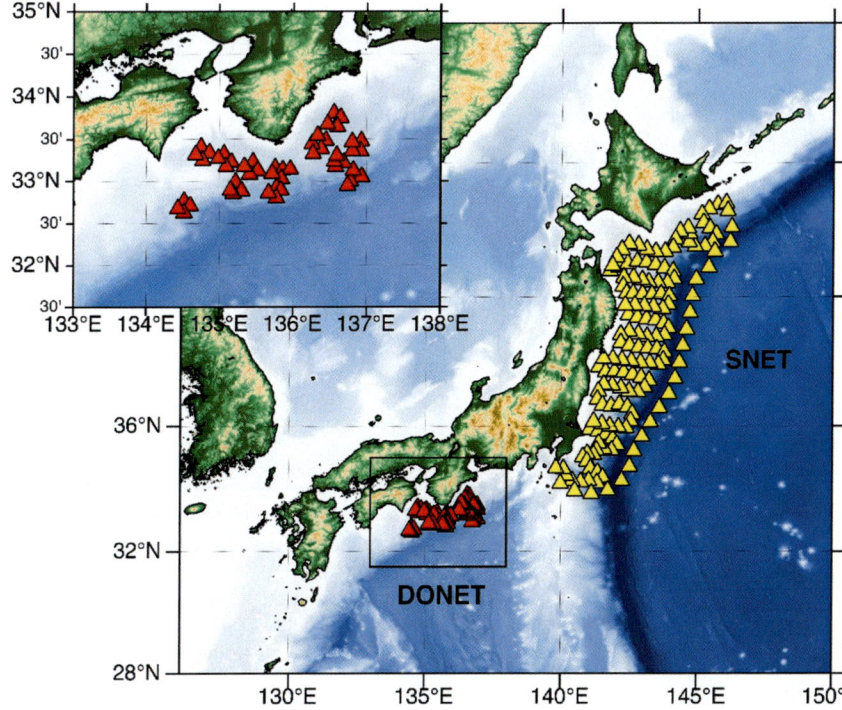

FIGURE 11.1 Station distribution of dense tsunami observation networks, S-net (*yellow triangles*), and DONET (*red triangles*). *DONET*, Dense Ocean floor Network system for Earthquakes and Tsunamis.

in March 2016. The DONET consists of three major components, a backbone cable, nodes, and observatories. A backbone cable is a cable connected from a station in land through each node to that station in land. The nodes serve as the connectors that can link several observatories. They also enable the addition of further observatories to the system and also the repair of malfunctioning observatories. Each observatory consists of several sensors, a broadband seismometer, a strong motion accelerometer, a pressure gauge, a differential pressure gauge, a hydrophone, and a thermometer (Kawaguchi et al., 2011).

After the 2011 Tohoku Earthquake and Tsunami, the Japanese Government decided to construct the Seafloor observation network for earthquakes and tsunamis (S-net) along the Japan Trench off the Pacific coast of eastern Japan (Fig. 11.1). The National Research Institute for Earth Science and Disaster Resilience (NIED) completed the construction of the S-net in March 2017 and began operating the observation network for real-time direct measurements of earthquakes and tsunamis. S-net currently consists of 150 observatories equipped with seismometers and pressure gauges linked together by ocean-bottom fiber optic cables. S-net provides prompt and accurate earthquake and tsunami observations to increase the lead time of tsunami and earthquake early warnings for offshore large earthquakes. The S-net consists of six cables and four stations on land. Each observatory is directly connected to one of these cables in this network. Each observatory consists of four different types of seismometers and two pressure sensors (Uehira et al., 2012).

For the first time in the world, such dense tsunami observation systems were built in Japan (Aoi et al., 2020). As a result, research using tsunami waveforms observed by these dense tsunami observation networks has progressed rapidly. This chapter reviews and explains the recent development of accurate source estimation methods, tsunami assimilation methods, and tsunami forecasting methods using dense tsunami observation data.

11.2 Accurate earthquake source estimation

After the installation of the S-net, the 2016 Fukushima Earthquake (M_w 7.1) occurred beneath the region of the S-net sensors (Fig. 11.1). The tsunami caused by the earthquake was observed by the dense S-net sensors for the first time. A detailed slip distribution of the M_w 7.1 event could be estimated from the tsunami waveform inversion because the tsunami source was surrounded by approximately 50 ocean-bottom pressure sensors (Kubota et al., 2021). The fault model of the 2016 Fukushima Earthquake was estimated using tsunami waveforms observed by five ocean-bottom pressure sensors off Kamaishi, and tide gauges along the coast (Gusman et al., 2017). Kubota et al. (2021) estimated a more accurate and detailed slip distribution of the earthquake (Fig. 11.2) using data from the

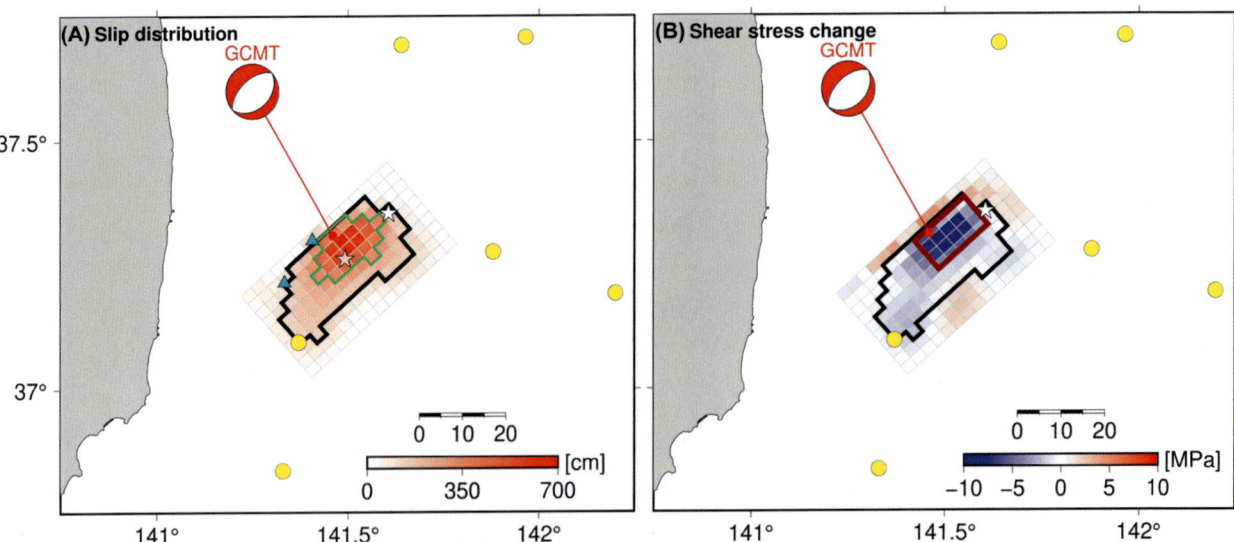

FIGURE 11.2 Result of the slip inversion. (A) Slip distribution. The pink and white stars indicate the slip-weighted averaged centroid and the JMA epicenter, respectively. Subfaults with slip amounts larger than $0.2 \times D_{max}$ (the main rupture area) and larger than $0.5 \times D_{max}$ are marked by thick black lines and green lines, respectively. (B) Shear stress change along the fault. The dark red rectangle denotes the optimum rectangular fault obtained by the grid-search analysis. *JMA*, Japan Meteorological Agency. *From Kubota, T., Kubo, H., Yoshida, K., Chikasada, N.Y., Suzuki, W., Nakamura, T., & Tsushima, H. (2021). Improving the Constraint on the Mw 7.1 2016 Off-Fukushima Shallow Normal-Faulting Earthquake With the High Azimuthal Coverage Tsunami Data From the S-Net Wide and Dense Network: Implication for the Stress Regime in the Tohoku Overriding Plate. Journal of Geophysical Research: Solid Earth, 126(10). https://doi.org/10.1029/2021JB022223.*

S-net, and other smaller networks installed by the University of Tokyo and Tohoku University. After the detailed slip distribution was estimated, the distribution of the shear stress change along the fault (i.e., stress drop) was obtained using the equation of Okada (1992), as implemented by Kubota, et al. (2021) (Fig. 11.2). Additionally, the energy-based stress drop, or the slip-weighted average stress drop, $\Delta\sigma_E$ (Noda et al., 2013) can be calculated using the following equation

$$\Delta\sigma_E = \frac{\sum_i D_i \Delta\sigma_i}{\sum_i D_i} \tag{11.1}$$

where D_i is the slip amount at the ith subfault, and $\Delta\sigma_i$ is the stress drop at the ith subfault. The authors further discussed the stress regime around the source area to understand the generation mechanism of the earthquake. This is one of the advancements of source studies using dense tsunami observation data.

The back-projection method can be used to estimate earthquake sources using dense tsunami observation data. The back-projection method has been developed as a powerful array-based analysis method for imaging the source processes of large earthquakes using seismological data (Ishii et al., 2005; Yagi et al., 2012). This approach uses the seismograms recorded in a dense seismic network. The advantages of this method over conventional and popular waveform inversions (Kiser et al., 2011) are that we do not require prior information, such as the geometry and location of a finite fault plane, and that the operation of the method is only to stack the seismic records shifted by each theoretical travel time. Therefore, calculations for solving the inverted matrices in inversion methods are not required. Ishii et al. (2005) indicated that the back-projection image represents the seismic energy release on a fault plane.

Mizutani and Yomogida (2022) applied the back-projection method to dense tsunami observation data for the 2016 Fukushima event. The back-projection method for tsunami data is developed from that used for seismic data. A back-projection image of the seismic data is expressed as follows (Ishii et al., 2005)

$$s_i(t) = \sum_{k=1}^{N} w_k d_k(t + t_{ki}^{travel}) \tag{11.2}$$

where s_i represents the stacked waveform at the ith potential source grid, d_k is the seismic or tsunami waveform observed at the kth station, w_k is the weighting factor for the kth station, and t_{ki}^{travel} is the theoretical travel time between the ith source grid and the kth station, respectively. The weighting factor, w_k, is defined to normalize each waveform by the maximum absolute value. Each theoretical travel time t_{ki}^{travel} is calculated using the fast marching method (FMM; Sethian, 1999). The FMM numerically solves the eikonal equation in space, and a stable travel time connecting any source−station pair can be obtained. The phase speed map for the FMM is defined as $\sqrt{gh}$, the travel time under the linear long-wave approximation, where h is the depth of the ocean and g is the gravitational acceleration. Then, the tsunami back-projection image is given by

$$BP_l(t) = \frac{1}{\max_l\{BP_l\}} \int_{t-\alpha}^{t+\alpha} \{s_l(\tau)\}^2 d\tau \tag{11.3}$$

where $BP_l(t)$ represents the back-projection image at the lth grid and α is the time window for integrating the stacked waveform $s_l(t)$ of Eq. (11.2) (Mizutani & Yomogida, 2022). The image is normalized by the maximum of all the grids at each time step, $BP_l(t)$. The earthquake origin time was set to $t = 0$. For the back-projection method, stacking coherent waveforms among stations is essential; therefore, cluster analysis needs to be conducted to group stations. Mizutani and Yomogida (2022) used a hierarchical cluster analysis (Romesburg, 2004) with the correlation coefficients estimated by the unweighted pair-group method using arithmetic averages for the normalized waveforms. Clusters were obtained with a correlation coefficient limit of 0.6, that is, the correlation coefficients of the waveforms belonging to each group were larger than 0.6. They only used the waveforms in the largest cluster for the back-projection analysis. Because the back-projection image is the normalized image of the initial ocean surface deformation, the image needs to be multiplied by a constant, C, to obtain the initial ocean surface deformation. Therefore, a tsunami numerical simulation from the back-projection image is necessary to define the constant, C, by comparing the observed and computed tsunami waveforms. It is important to notice that the essence of the back-projection analysis is only to stack many coherent tsunami waveforms without any inversion methods. Therefore, the initial ocean surface deformation image due to a large earthquake can easily be obtained using dense tsunami observation data. The results of the back-projection analysis for the 2016 Fukushima Earthquake (Fig. 11.3) show that the back-projection analysis worked very well by using the S-net data.

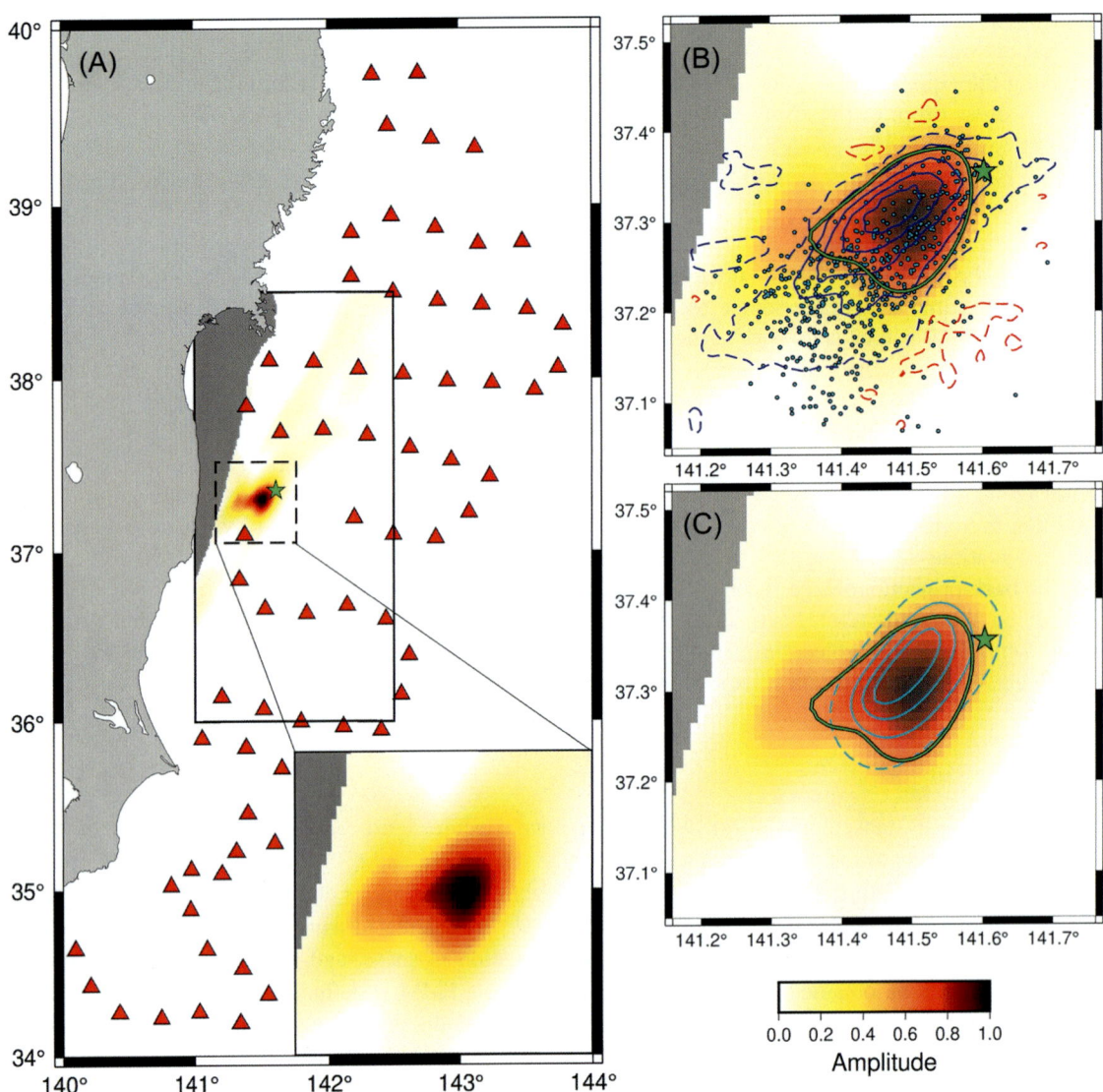

FIGURE 11.3 Back-projection image at $t = 0$ or the origin time. The red triangles are the stations used in the analysis. The green star represents the epicenter of the 2016 Off-Fukushima Earthquake. The black solid rectangle represents the target area of the back-projection imaging, and the dashed one corresponds to the enlarged area of the right bottom, (B) or (C). (B) Enlarged image of the black dashed rectangle of (A). The green line represents the area with an amplitude larger than 0.6. The cyan dots represent the aftershock epicenters. The red and blue contour lines represent the positive and negative amplitudes of the initial tsunami height estimated by Kubota, Kubo, et al. (2021) with the solid for 0.5 m interval and the dashed for 0.1 m. There are no solid red contours as the maximum uplift was less than 0.5 m. (C) Same as (B) except that the cyan contour lines represent the subsidence for the single uniform fault slip model of (Kubota, Kubo, et al., 2021). *From Mizutani, A., & Yomogida, K. (2022). Back-projection imaging of a tsunami excitation area with ocean-bottom pressure gauge array data.* Journal of Geophysical Research: Oceans, *127(7). https://doi.org/10.1029/2022JC018480.*

11.3 Data assimilation

11.3.1 Tsunami data assimilation

Data assimilation methods have been used for operational weather forecasts in atmospheric science (Lynch, 2008). The utilization of the dense observation data along the Japan and Nankai subduction zones has advanced the development of tsunami predictions by using the assimilation method for observed tsunami waveforms. Tsunami numerical simulations typically start from the initial ocean surface deformation calculated from the source model of the earthquake or obtained directly by inversion of the tsunami waveforms. Recently, significant advancements have been made in the field of tsunami data assimilation, particularly through the utilization of dense observation data. This approach becomes a powerful technique for mitigating the uncertainties associated with the initial estimation of tsunami sources. Maeda et al. (2015) developed an assimilation method for tsunami wavefields using real-time tsunami observation data. This assimilation method proposed by Maeda et al. (2015) is explained in the following.

Tsunami data assimilation consists of two steps: tsunami propagation and data assimilation. These two steps are performed at each time step in the simulation. The tsunami propagation step is a traditional tsunami simulation step and is expressed as follows

$$\mathbf{x}_n^f = \mathbf{F}\mathbf{x}_{n-1}^a \tag{11.4}$$

$$\mathbf{x}_n = (h(x,y,n\Delta t), M(x,y,n\Delta t), N(x,y,n\Delta t))^T \tag{11.5}$$

where h is the tsunami height, M and N are the velocities in the horizontal directions, $\mathbf{x}_n^f$ and $\mathbf{x}_n^a$ are the column vectors representing the tsunami wavefield at the nth time step after the propagation and assimilation steps, respectively, and $\mathbf{F}$ is the tsunami propagation matrix corresponding to the tsunami numerical propagation model. For example, the matrix, $\mathbf{F}$, consists of the coefficients of the finite difference equations of the linear long-wave tsunami. When the total grid number of the tsunami simulation is L, $\mathbf{x}_n^f$ or $\mathbf{x}_n^a$ has $3L$ components.

The data assimilation step is expressed as

$$\mathbf{x}_n^a = \mathbf{x}_n^f + \mathbf{W}(\mathbf{y}_n - \mathbf{H}\mathbf{x}_n^f) \tag{11.6}$$

where $\mathbf{W}$ ($3L \times m$) is the weight matrix, $\mathbf{y}_n$ (m) is a column vector for the observed tsunami heights, and $\mathbf{H}$ ($m \times 3L$) is the observation matrix of a sparse linear matrix with a value of 1 at the grids of the observation stations and 0 at other grids. The number of observation stations is m. In this second step, the residual between the observed and the simulated tsunami wavefields, $\mathbf{y}_n - \mathbf{H}\mathbf{x}_n^f$, is used to produce the assimilated tsunami wavefield using an appropriate weight matrix $\mathbf{W}$. This weight matrix, $\mathbf{W}$, is a key controlling matrix for the assimilation. The matrix defines the weights to correct the tsunami wavefield at all grid points in the tsunami numerical simulation from the residual at the observation grids. The weight matrix, $\mathbf{W}$, can be estimated by minimizing the covariance matrix of $\langle \varepsilon^a \varepsilon^{aT} \rangle = \langle (\mathbf{x}_n^a - \mathbf{x}_n^{true})(\mathbf{x}_n^a - \mathbf{x}_n^{true})^T \rangle$ as a solution of the following equation

$$\mathbf{W}(\mathbf{R} + \mathbf{H}\mathbf{P}^f\mathbf{H}^T) = \mathbf{P}^f\mathbf{H}^T \tag{11.7}$$

where $\mathbf{P}^f = \langle \varepsilon^f \varepsilon^{fT} \rangle$ and $\mathbf{R} = \langle \varepsilon^o \varepsilon^{oT} \rangle$ are the covariance matrices of the numerical simulation error and the observation error, respectively. Because the observation error is not correlated among stations, $\mathbf{R}$ is simplified as a diagonal matrix in which the diagonal components are the variance of the observation errors at the stations. By assuming that the standard errors between the grids of the numerical simulation are the same in space, Eq. (11.7) can be simplified to

$$\sum_{j=1}^{m} w_{gj}\left(\delta_{ij}\rho_i\rho_j + \mu_{ij}\right) = \mu_{gj} \tag{11.8}$$

where w_{gj} represents components of the weight matrix $\mathbf{W}$, μ_{ij} and μ_{gj} are the correlation functions of errors in the numerical simulation between two observation points (i and j) and between computational grid (g) and observation point (j), respectively, and ρ_i is the observational error relative to the numerical simulation error. Maeda et al. (2015) assumed that the correlation functions, μ_{ij} and μ_{gj}, are the Gaussian functions with the characteristic distance of 10 km and ρ_i is 1 by trial and error. Then, w_{gj} can be obtained by numerically solving Eq. (11.8). The characteristic distance of the Gaussian function and the value of ρ_i can be changed for the other observation systems.

Gusman et al. (2016) applied this tsunami data assimilation method to real observation data of a tsunami generated by the 2012 Haida Gwaii Earthquake. Dense ocean-bottom stations equipped with seismometers and pressure sensors were deployed offshore in Oregon and California (Fig. 11.4) during the earthquake event (Sheehan et al., 2015). Gusman et al. (2016) used those observed data to test the tsunami data assimilation method by Maeda et al. (2015) for the first time. In this data assimilation, the Gaussian function with the characteristic distance of 20 km was used for the correlation function, μ_{ij} and μ_{gj} in Eq. (11.8). Fig. 11.5 compares the tsunami data assimilation result with the tsunami simulation result computed from the estimated source model. At 79 minutes after the earthquake, the tsunami was observed at one station in the north, so the data assimilation started. At that time, the observed wave at one station was used to correct the simulated tsunami for the grids surrounding the station grid by employing the weight matrix $\mathbf{W}$. Because the correlation function is a Gaussian function with a characteristic distance of 20 km, the pattern of the tsunami height distribution in Fig. 11.5 is similar to that of the Gaussian function. At 140 minutes after the earthquake,

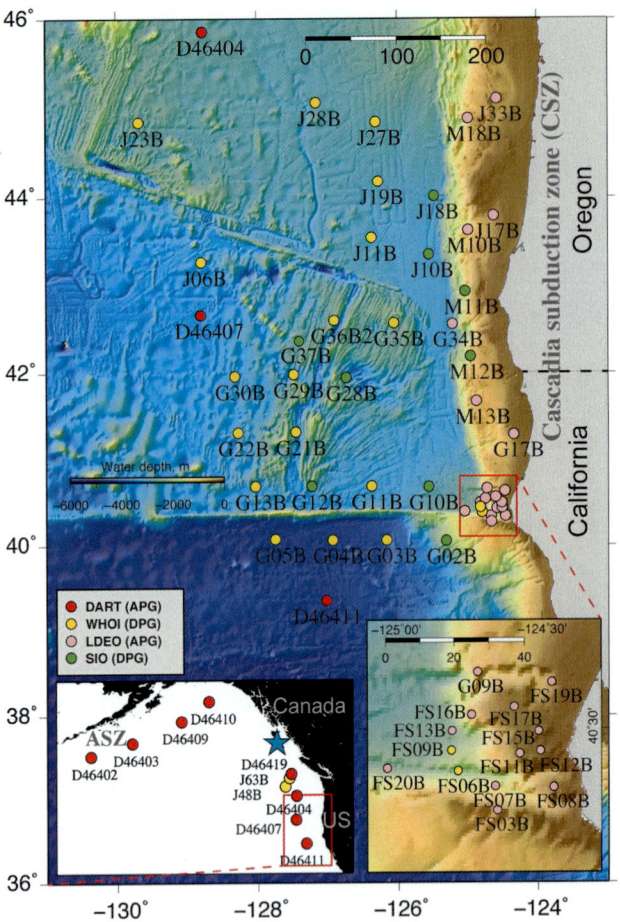

FIGURE 11.4 Station map of the tsunami dense array. Distribution of Deep-ocean Assessment and Reporting of Tsunamis (DART), Absolute Pressure Gauge (APG), and Differential Pressure Gauge (DPG) stations in the CSZ and ASZ. The light blue star represents the earthquake's epicenter. *ASZ*, Aleutian subduction zones; *CSZ*, Cascadia subduction zones. Map lines delineate study areas and do not necessarily depict accepted national boundaries. *From Gusman, A. R., Sheehan, A. F., Satake, K., Heidarzadeh, M., Mulia, I. E., & Maeda, T. (2016). Tsunami data assimilation of Cascadia seafloor pressure gauge records from the 2012 Haida Gwaii earthquake. Geophysical Research Letters, 43(9), 4189–4196. https://doi.org/10.1002/2016GL068368.*

the tsunami had already been observed by many stations, tsunami waveforms were well simulated by the data assimilation. This demonstrated that the tsunami data assimilation method developed by Maeda et al. (2015) worked well.

11.3.2 Green's function-based tsunami data assimilation

Wang et al. (2017) improved the assimilation method to reduce the time required for real-time tsunami simulation for early warning purposes. They introduced Green's function-based tsunami data assimilation method. In this method, Green's functions are calculated in advance. No tsunami simulation was necessary for real-time tsunami forecasts. The forecasted tsunami waveforms were produced using a simple matrix calculation (Fig. 11.5). This method is explained as follows.

The residual vector in Eq. (11.6) can be written as $\mathbf{x}_n^r = \mathbf{W}(\mathbf{y}_n - \mathbf{H}\mathbf{x}_n^f)$. This can be expressed using a linear combination of the unit column vector $\mathbf{e}_i^T$ and the corresponding residual of the tsunami height at the ith station r_n^i

$$\mathbf{x}_n^r = \mathbf{W} \sum_i r_n^i \mathbf{e}_i^T \tag{11.9}$$

The assimilation correction of the ith station is represented as $r_n^i \mathbf{W} \mathbf{e}_i^T$. Because the propagation model is linear, the assimilation corrections of different steps and stations can be superimposed. Then, the overall data assimilation steps can be described as follows

$$\mathbf{x}_n^a = \mathbf{x}_n^f + \mathbf{W}(\mathbf{y}_n - \mathbf{H}\mathbf{x}_n^f) = \mathbf{x}_n^f + \mathbf{x}_n^r = \mathbf{F}\mathbf{x}_{n-1}^a + \mathbf{x}_n^r = \mathbf{F}\left(\mathbf{x}_{n-1}^f + \mathbf{x}_{n-1}^r\right) + \mathbf{x}_n^r = \sum_{t=0}^n \mathbf{F}^{n-t} \mathbf{x}_t^r$$

$$\sum_{t=0}^n \mathbf{F}^{n-t} \mathbf{x}_t^r = \sum_{t=0}^n \mathbf{F}^{n-t} \left(\sum_{i=1}^m r_n^i \mathbf{W} \mathbf{e}_i^T\right) = \sum_{i=1}^m \left(\sum_{t=0}^n r_n^i \mathbf{F}^{n-t} \mathbf{W} \mathbf{e}_i^T\right) \tag{11.10}$$

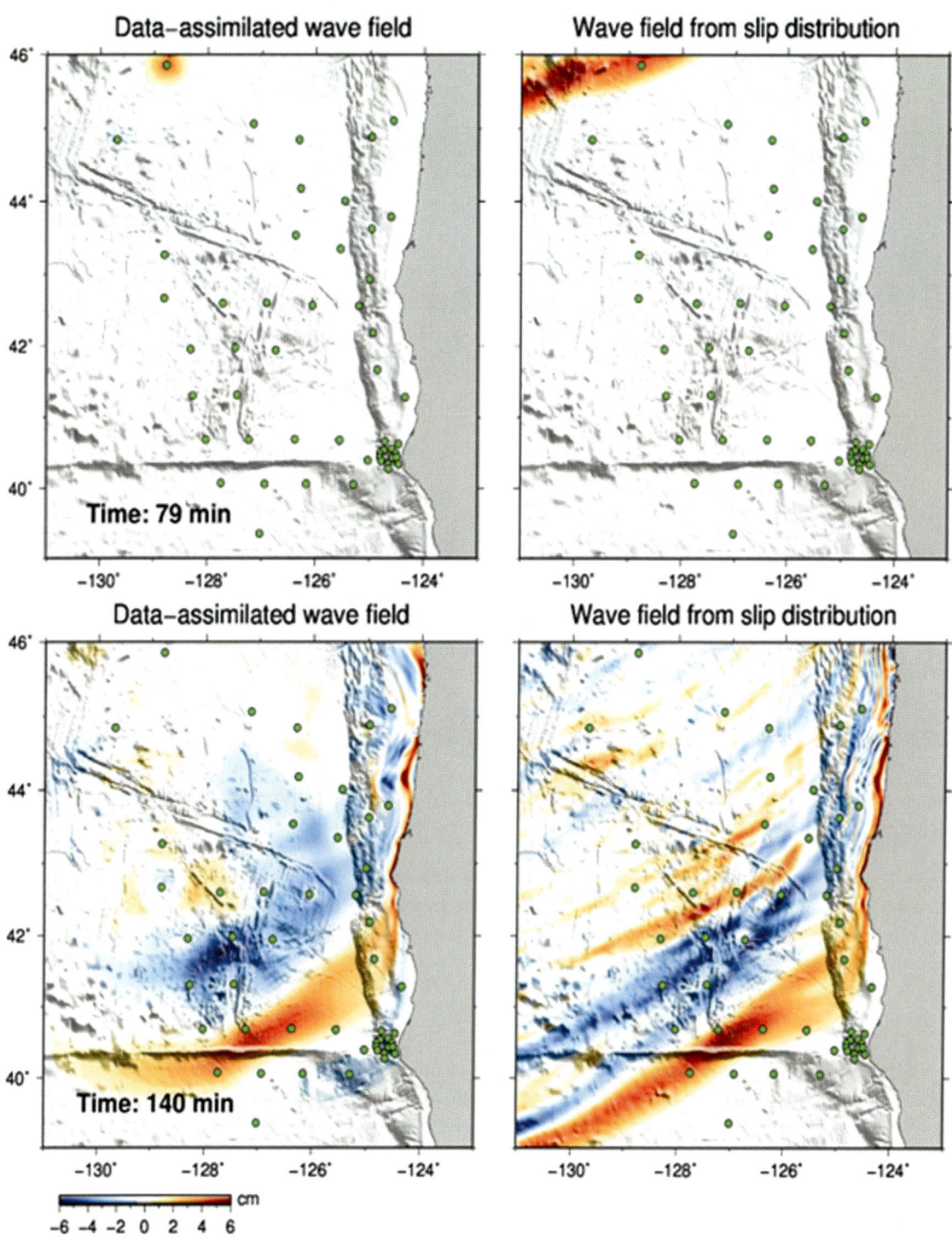

FIGURE 11.5 Tsunami propagation forecast offshore the Cascadia subduction zone. Left: wavefields at 79 and 140 minutes after the earthquake's origin time produced by tsunami data assimilation. Green circles show the station distribution. Right: wavefields at 79 and 140 minutes simulated from the estimated slip distribution of the 2012 Haida Gwaii Earthquake by Gusman et al. (2016).

where $\mathbf{We}_i^T$ is the unit assimilation correction for the ith station. Here, $\mathbf{F}^{n-t}\mathbf{We}_i^T$ represents the propagation computation of the assimilation correction, that is the Green's function between the ith station and the other grid points. When the tsunami waveform at a particular grid point needs to be forecasted, the Green's functions at that point from all ocean-bottom stations should be computed and stored in a database. When the boundary conditions for the tsunami inundation

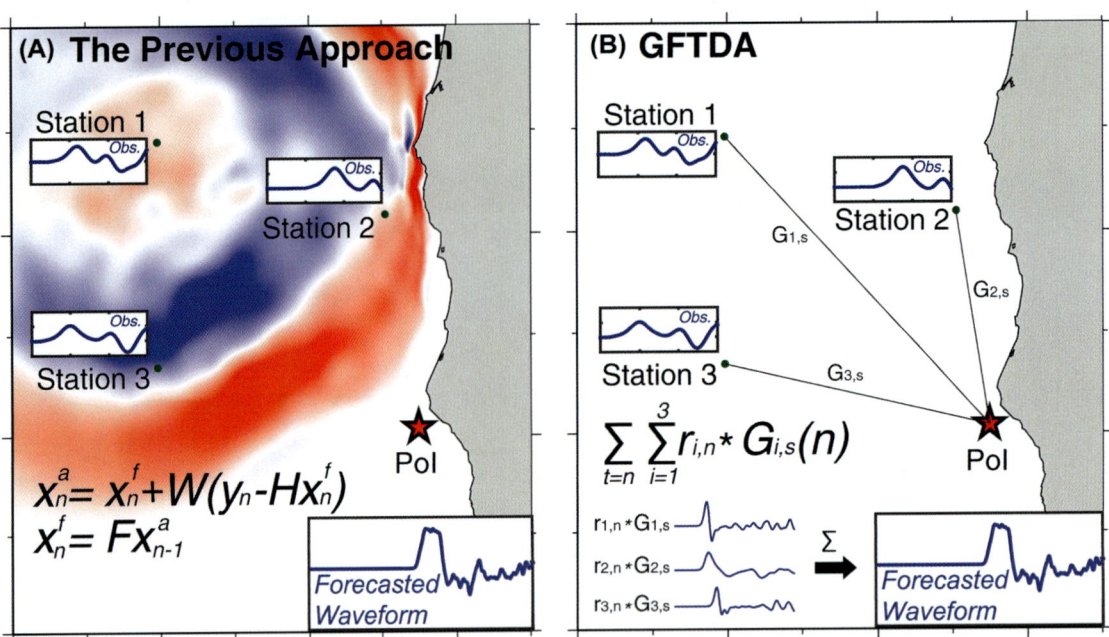

FIGURE 11.6 Illustration of (A) the previous data assimilation approach and (B) the GFTDA approach. For simplification, three stations are considered to forecast the waveform of the PoI (*red* star). In the previous approach, the observed tsunami height of stations is assimilated, and the wavefield for the entire modeling domain is computed. Then the forecasted waveform of the PoI can be recorded in the wavefield. In the GFTDA approach, Green's functions are computed in advance, which represents the waveform of the PoI resulting from the propagation of each station's assimilation response. The forecasted waveform is directly synthesized with the precomputed Green's functions by matrix manipulation. Both approaches give the same result for the forecasted waveform. *GFTDA*, Green's Function-based Tsunami Data Assimilation method. *From Wang, Y., Satake, K., Maeda, T., & Gusman, A.R. (2017). Green's function-based tsunami data assimilation: A fast data assimilation approach toward tsunami early warning. Geophysical Research Letters, 44(20), 10−289. https://doi.org/10.1002/2017GL075307.*

computation are needed for real-time tsunami inundation forecasts, Green's functions at the boundary grids from all stations should be computed and stored in the database. Therefore, no real-time tsunami numerical computation is necessary to save time for tsunami early warnings (Fig. 11.6).

Because a dense observation system at the ocean bottom, such as S-net and DONET, uses pressure sensors to observe tsunami waveforms, the ocean surface displacement above the source area immediately after the earthquake cannot be observed by the ocean-bottom pressure sensors within the source area. The data assimilation method needs to be improved further to forecast near-field tsunamis. In addition, nontsunami components, such as sea-bottom acceleration changes due to the fault motion of the earthquake, should be removed from the ocean-bottom pressure data. Saito and Tsushima (2016) theoretically showed that the first 2 minutes of ocean-bottom pressure data are highly affected by acoustic waves and seismic waves. Subsequently, Mizutani et al. (2020) developed an early tsunami detection method for near-fault ocean-bottom pressure data by comparing the seismic data and pressure data at the same site to eliminate nontsunami components. The effectiveness of this approach is supported by the comprehensive instrumentation of seismometers and pressure sensors at all stations within the S-net and DONET dense observation systems.

11.3.3 Near-field tsunami data assimilation

For near-field tsunami, instead of using the pressure data directly, Tanioka (2018) and Tanioka and Gusman (2018) developed a near-field tsunami forecast method using the time derivative of the pressure waveforms observed at the ocean-bottom pressure sensors near the source area. In this method, tsunami computation by assimilating data can be performed without any tsunami source information as soon as an earthquake or tsunami generation is completed. Tanioka (2020) improved their method by applying it to an actual station distribution of S-net, noting that Tanioka and Gusman (2018) considered ocean-bottom pressure sensors equally distributed at 15-arcminute intervals or approximately 30 km apart. This method is explained as follows.

Tanioka (2018) developed a tsunami simulation method that can assimilate dense ocean-bottom pressure data near the tsunami source region and overcome the limitation in the data assimilation method proposed by Maeda et al.

(2015). In the method (Tanioka, 2018), the tsunami height distribution is estimated from the pressure changes for each time step observed at dense ocean-bottom pressure sensors. Then, this estimated tsunami height distribution at a given time step is used to substitute the tsunami heights in the tsunami numerical simulation.

After nontsunami components are eliminated by the method developed by Mizutani et al. (2020), the pressure changes at the ocean bottom, $\Delta p(x, y, t)$ is expressed as

$$\Delta p(x, y, t) = \rho g [\eta(x, y, t) - B(x, y, t)] \tag{11.11}$$

where $\eta(x, y, t)$ is the tsunami height, $B(x, y, t)$ is the sea floor deformation due to faulting of a large earthquake, ρ is the density of water, and g is the gravitational acceleration. Then, the water-depth fluctuation, $h_f(x, y, t)$, is expressed as

$$h_f(x, y, t) = \frac{1}{\rho g} \Delta p(x, y, t) = \eta(x, y, t) - B(x, y, t) \tag{11.12}$$

After the faulting of the earthquake is completed, the wave equation of the linear shallow water approximation is as follows

$$\frac{\partial^2 h_f(x, y, t)}{\partial t^2} = \frac{\partial^2 \eta(x, y, t)}{\partial t^2} = gd\left(\frac{\partial^2 \eta(x, y, t)}{\partial x^2} + \frac{\partial^2 \eta(x, y, t)}{\partial y^2}\right) \tag{11.13}$$

where d is the ocean depth. The finite difference equation of the wave propagation (Eq. 11.13) is as follows

$$\frac{h_{f_{i,j}}^{k+1} - 2h_{f_{i,j}}^{k} + h_{f_{i,j}}^{k-1}}{\Delta t^2} = gd_{i,j}\left(\frac{\eta_{i+1,j}^{k} - 2\eta_{i,j}^{k} + \eta_{i-1,j}^{k}}{\Delta x^2} + \frac{\eta_{i,j+1}^{k} - 2\eta_{i,j}^{k} + \eta_{i,j-1}^{k}}{\Delta y^2}\right) \tag{11.14}$$

where i and j are indices of points for the x- and y-directions in the computed domain, k is the index for time steps, Δx and Δy are the spacing intervals in the x and y directions, respectively, and Δt is the time interval. When the ocean-bottom pressure data are available at equally spaced intervals, such as 10 arcminutes or 18 km, the left-hand side of Eq. (11.14) is calculated from those pressure data. The unknown parameters are tsunami height field, $\eta_{i,j}^{k}$, at a particular time, k. The tsunami heights outside the computational domain are assumed to be 0. Eq. (11.14) for all grid points is combined to obtain the following

$$\mathbf{A}\boldsymbol{\eta}^k = \mathbf{d}^k \tag{11.15}$$

$$\mathbf{d}^k = \left(\frac{h_{f_{i,j}}^{k+1} - 2h_{f_{i,j}}^{k} + h_{f_{i,j}}^{k-1}}{\Delta t^2}\right)^T \tag{11.16}$$

where $\boldsymbol{\eta}^k$ is an unknown column vector representing tsunami height field at the time of k, the matrix, $\mathbf{A}$, includes coefficients representing the right side of Eq. (11.14), and $\mathbf{d}^k$ is a column vector for the left side of Eq. (11.14) which is obtained from the pressure data. We can solve Eq. (11.15) numerically to obtain the tsunami height field, $\boldsymbol{\eta}^k$. Tanioka and Gusman (2018) demonstrated that the tsunami data assimilation method can be combined with the tsunami inundation forecasting method (Gusman et al., 2014) to predict the tsunami inundation in coastal towns along the Tohoku coast for the 2011 event. They assumed that the ocean-bottom pressure sensors are available at the equally spacing grid of 15 arcminutes or 30 km.

Tanioka (2020) improved the method by using the exact locations of the S-net sensors. The improved method was tested for two large earthquakes expected to occur off the Pacific coast of Hokkaido, Japan. To use Eq. (11.14) or (11.15), equal spacing data were generated using the appropriate interpolation method to match the pressure data with the exact locations of the S-net sensor. First, the left side of Eq. (11.14) at each sensor position $p(x_p, y_p)$, $d_p^k = \frac{h_{f_p}^{k+1} - 2h_{f_p}^{k} + h_{f_p}^{k-1}}{\Delta t^2}$, is calculated from observed pressure data. The d_{ij}^k at equally spaced grids, such as 10 arcminutes, is obtained from the weighted interpolation equation as follows

$$d_{ij}^k = \sum_{p=1}^{n} w_{pij} d_p^k \bigg/ \sum_{p=1}^{n} w_{pij} \tag{11.17}$$

$$w_{pij} = \frac{1}{\left(\sqrt{(x_p - x_{ij})^2 + (y_p - y_{ij})^2}\right)^{\alpha}} \tag{11.18}$$

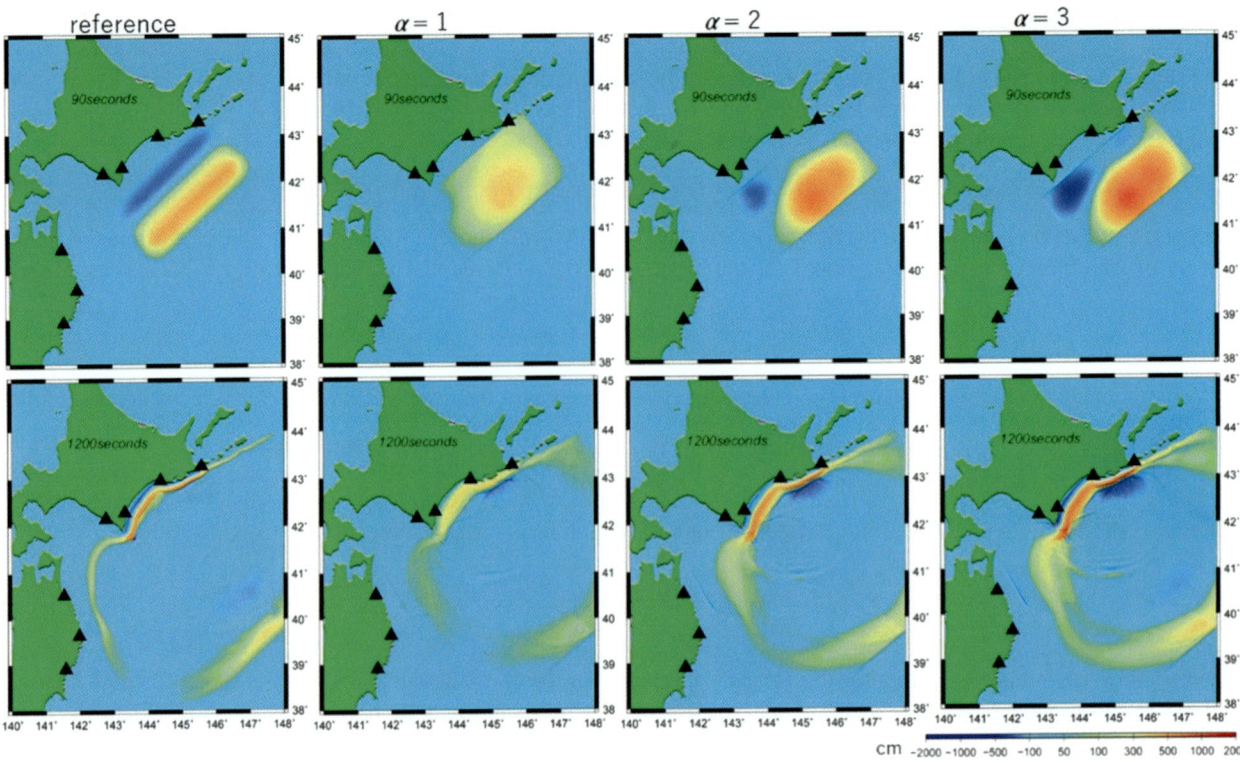

FIGURE 11.7 Comparisons of the tsunami height fields at 90 and 1200 seconds after the initiation of the earthquake for the reference tsunami of the giant earthquake and three assimilation tsunamis as computed by the interpolation of the simulated observation data at the S-net sensors using Eq. (11.17), where control factor α from Eq. (11.18) ranges from 1 to 3. *From Tanioka, Y. (2020). Improvement of near-field tsunami forecasting method using ocean-bottom pressure sensor network (S-net).* Earth, Planets and Space, 72(1). https://doi.org/10.1186/s40623-020-01268-1.

where w_{pij} is the weight factor for each observation point $p(x_p, y_p)$ to calculate at equally spaced grids and α is the control factor. When the control factor α equals 1, the weights are inversely proportional to the distance between the observed points in the S-net and hypothetical points for assimilation. Tanioka (2020) showed that $\alpha = 2$ produced the best interpolation solution for the station distribution of S-net off the Hokkaido coast (Fig. 11.7). However, this factor should be adjusted for a different station distribution. Although the method worked for numerical tests, Tanioka (2020) indicated that several problems still needed to be solved for practical use. First, because of the station distribution of S-net (the distance between the closest two stations is approximately 30 km), the small wavelength of the ocean surface displacement cannot be resolved by the interpolation method. Second, the end of the earthquake is unknown for practical purposes. It may be necessary to determine the ending time of an earthquake using seismic records at the same sites as the pressure sensors to complete the replacement of the assimilation. Finally, after pressure sensors experience strong motions due to a large earthquake, mechanical malfunction of the sensors may occur, as suggested by Kubota et al. (2018). Those malfunctioned sensors need to be identified as soon as possible, and the data from those sensors should not be used in the forecasting. It may be important to combine this method with the data assimilation method developed by Maeda et al. (2015) which is previously explained in this chapter.

11.4 Tsunami early warning

11.4.1 Tsunami forecasting method based on inversion for initial sea-surface height

The tsunami early warning system has significantly improved after the installation of dense ocean-bottom observation systems. Tsushima et al. (2009) developed the tsunami forecasting method based on inversion for initial sea-surface height (tFISH). In this method, tsunami waveforms observed at ocean-bottom pressure sensors are inverted to estimate the initial sea-surface height for the tsunami forecast simulation. Because dense ocean-bottom pressure sensors were available after the 2011 Tohoku Earthquake, the effect of the seafloor deformation due to a large earthquake cannot be ignored. Tsushima et al. (2012) improved the tFISH to include the effect of coseismic seafloor deformation in Green's functions of the inversion method. The tFISH was installed in the tsunami early warning system of the JMA in 2019 and has been operating since then. This method is explained as follows.

The observed tsunami waveforms between the origin time of the earthquake ($T = 0$) and the time duration used in the inversion ($T = T_0$) are described by the superposition of Green's functions as follows

$$\mathbf{d}(T) = \mathbf{G}(T)\mathbf{m} \quad \text{for} \quad 0 \leq T \leq T_0 \tag{11.19}$$

where $\mathbf{d}(T)$ is a vector of the observed tsunami waveforms, $\mathbf{G}(T)$ is a matrix of Green's functions, which are computed tsunami waveforms at observation points from a unit displacement function at the ocean surface, and $\mathbf{m}$ is a vector representing the initial ocean surface displacement.

For a tsunami early warning system, the time required for forecasting is critical. To maintain the accuracy of the forecast with a shorter time T_0, a constraint is applied to the spatial distribution of the ocean surface displacement, which is set to zero if the epicenter of the earthquake is sufficiently far away. The constraint is formulated as follows

$$w(r_i)m_i = 0 \tag{11.20}$$

or

$$\mathbf{Wm} = 0 \tag{11.21}$$

where $w(r_i)$ is a weighted function of the horizontal distance r_i from the epicenter to ith point on the ocean surface, and $\mathbf{W}$ is a diagonal matrix whose components are the weighted functions, $\delta_{ij}w(r_i)$ with the Kronecker delta δ_{ij}. Tsushima et al. (2009) used $w(r_i) = r_i/r_{max}$ where r_{max} is the horizontal distance from the epicenter to the cutoff distance, which is the maximum distance of the influence area. The influence area at $T = T_0$ can be defined by drawing back-propagated waveforms from the epicenter. As the influence area increases with time, r_{max} increases. Therefore, the effect of this damping constraint is weakened when T_0 increases.

A smoothness constraint on the spatial distribution of the ocean surface displacement is used under the following conditions

$$m_{i+1,j} + m_{i-1,j} + m_{i,j+1} + m_{i,j-1} - 4m_{i,j} = 0 \tag{11.22}$$

or

$$\mathbf{Sm} = 0 \tag{11.23}$$

where $m_{i,j}$ is the amount of the initial ocean surface displacement at the points defined by i and j on the east and north axes, respectively, and $\mathbf{S}$ is a matrix representing the smoothness operator.

From Eqs. (11.19), (11.21), and (11.23), the combined equation to be solved in the inversion can be expressed as follows

$$\begin{pmatrix} \mathbf{d}(T) \\ 0 \\ 0 \end{pmatrix} = \begin{pmatrix} \mathbf{G}(T) \\ \alpha \mathbf{S} \\ \beta \mathbf{W} \end{pmatrix} \mathbf{m} \quad \text{for} \quad 0 \leq T \leq T_0 \tag{11.24}$$

where α and β are weighting factors or hyperparameters of the smoothness and the damping constraints, respectively. The equation can be solved by using the singular value decomposition method (Press et al., 1992). Tsushima et al. (2009) used $\alpha = 10$ and $\beta = 100$, determined by trial and error. Those weighting factors are needed to be defined for the other observation system.

As a unit ocean surface displacement function, $I(\mathbf{x})$, Tsushima et al. (2009) used the function shown in Fig. 11.8. The unit function was represented by a square of 700×700 arcseconds. The spacing between the unit functions is 540 arcseconds. Therefore, the neighboring unit functions overlapped at their margins (Fig. 11.8). The Green's functions at each observation point from the unit function $I(\mathbf{x})$ were numerically computed by solving the linear long-wave equations or the linear Boussinesq equations using the finite difference scheme.

Finally, Tsushima et al. (2012) improved the Green's functions, used in Tsushima et al. (2009), by correcting the ocean-bottom deformation caused by earthquakes. Because the water depth fluctuation $h_f(\mathbf{x};t)$ is calculated from the pressure change $dp(\mathbf{x};t)$ observed by the pressure sensor at the ocean bottom, $h_f(\mathbf{x};t)$ in the coseismic deformation area is expressed as

$$h_f(\mathbf{x};t) = \frac{1}{\rho g}dp(\mathbf{x};t) = \eta(\mathbf{x};t) - B(\mathbf{x};t) \tag{11.25}$$

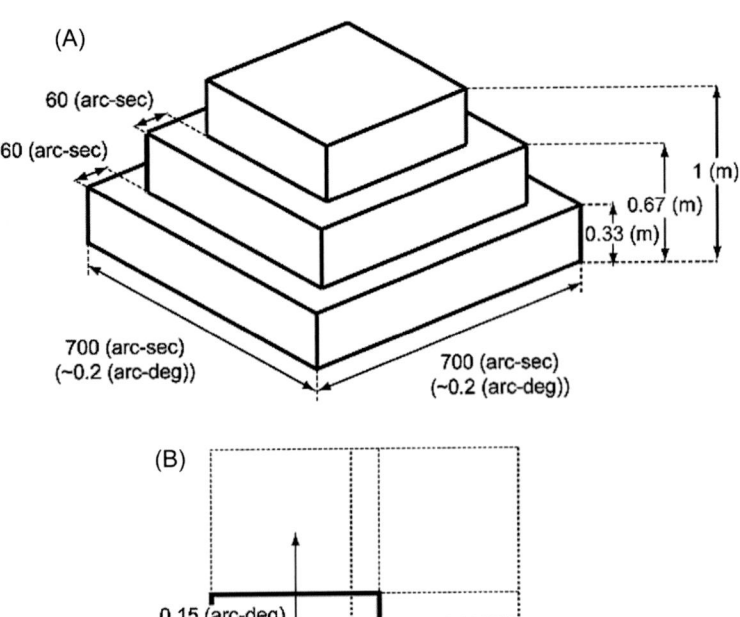

FIGURE 11.8 Diagram illustrating unit elements of the tsunami source used to compute Green's function. (A) Geometry and distribution of sea-surface displacement within an element. (B) Overlapping alignment of adjoining source elements. *From Tsushima, H., Hino, R., Fujimoto, H., Tanioka, Y., & Imamura, F. (2009). Near-field tsunami forecasting from cabled ocean bottom pressure data.* Journal of Geophysical Research: Solid Earth, *114(6). https://doi.org/10.1029/2008JB005988.*

where ρ is the density of seawater, g is the gravitational acceleration, $\eta(\mathbf{x}; t)$ is the ocean surface displacement or tsunami height, and $B(\mathbf{x}; t)$ is the coseismic ocean-bottom deformation. By assuming that the ocean bottom deforms with a constant speed during the source duration of the earthquake $0 < t \leq T_b$, the coseismic ocean-bottom deformation is expressed as

$$B(\mathbf{x};t) = \begin{cases} B(\mathbf{x};T_b)\dfrac{t}{T_b} & \text{for } 0 < t \leq T_b \\ B(\mathbf{x};T_b) & \text{for } t > T_b \end{cases} \tag{11.26}$$

Then, the observed water depth fluctuation $h_f(\mathbf{x}_j; t)$ for the sensor located at $\mathbf{x}_j$ becomes

$$h_f(\mathbf{x}_j; t) = \sum_i G'_{ij}(t) m_i \tag{11.27}$$

$$G'_{ij}(t) = \begin{cases} G_{ij}(t) - I_i(\mathbf{x}_j)\dfrac{t}{T_b} & \text{for } 0 < t \leq T_b \\ G_{ij}(t) - I_i(\mathbf{x}_j) & \text{for } t > T_b \end{cases} \tag{11.28}$$

where $G'_{ij}(t)$ is the improved Green's function at the jth ocean-bottom sensor computed from the ith unit displacement function, $G_{ij}(t)$ is the previous Green's function numerically computed from the ith unit displacement function. Eq. (11.28) includes the correction of coseismic ocean-bottom deformation by assuming that the unit ocean-surface deformation function $I(\mathbf{x})$ is the same as the unit ocean-bottom deformation function $B(\mathbf{x}; T_b)$. As shown in Eq. (11.28), for the sensors located in a unit ocean-surface function, the improved Green's function is different from the previous Green's function. Then Eq. (11.24) with the improved Green's function can be used for the inversion to forecast the tsunamis. The tFISH has been used in JMA's tsunami early warning system since 2019. Furthermore, Tsushima et al. (2014) improved the method by incorporating it into a real-time automatic detection method for permanent displacement (RAPiD) developed by Ohta et al. (2012). The RAPiD uses global navigation satellite system data on land and can estimate a fault model 6 minutes after a large earthquake. Tsushima et al. (2014) then developed a method in which the initial ocean-surface deformation estimated by the RAPiD was used as the initial condition for the tFISH, called tFISH/RAPiD.

11.4.2 Tsunami forecast using multiindex method for a scenario database

Another real-time tsunami forecasting system using dense ocean-bottom observation systems was developed by the NIED (Aoi et al., 2019). The method uses a database that includes a large number of source models and finds the scenarios to fit the precomputed tsunami waveforms with the tsunami waveforms observed at dense ocean-bottom sensors. More than 60,000 source models of large interplate earthquakes on the Pacific Plate along the Kuril, Japan, and Izu-Bonin Trenches, and on the Philippine Sea Plate along the Sagami and Nankai Trough are used. Source models of outer-rise earthquakes that occur outside of the trench at shallow depths and of a normal fault type are also included. The intraplate earthquakes that occurred in the upper plate at shallow depths are also included. The multiindex method developed by Yamamoto et al. (2016) was used to select source models from the database by fitting the observed and precomputed tsunamis. This tsunami forecasting system has been in operation in the NIED since 2017. The multiindex method is explained as follows.

Three indices, the correlation coefficient and two kinds of variance reductions to compare the observed waveform $O(\mathbf{x}_i, t)$ and the precomputed waveform $C(\mathbf{x}_i, t)$ are used to select the source models. First, the correlation coefficient $R(t)$ is defined as follows

$$R(t) = \frac{\sum_{i=1}^{n} O(\mathbf{x}_i, t) C(\mathbf{x}_i, t)}{\sqrt{\sum_{i=1}^{n} O^2(\mathbf{x}_i, t)} \sqrt{\sum_{i=1}^{n} C^2(\mathbf{x}_i, t)}} \quad (11.29)$$

where n is the number of observation stations. $R(t)$ has a maximum value of 1 when $O(\mathbf{x}_i, t)$ is identical to $C(\mathbf{x}_i, t)$ and a minimum value of -1 when $O(\mathbf{x}_i, t)$ and $C(\mathbf{x}_i, t)$ are opposite. $R(t)$ strongly depends on the distribution patterns of $O(\mathbf{x}_i, t)$ and $C(\mathbf{x}_i, t)$ but not on the amplitudes of $O(\mathbf{x}_i, t)$ and $C(\mathbf{x}_i, t)$. Next, the variance reduction VRO(t) normalized by the L2-norm of the observed waveform $O(\mathbf{x}_i, t)$ is expressed as follows

$$\text{VRO}(t) = 1 - \frac{\sum_{i=1}^{n} (O(\mathbf{x}_i, t) - C(\mathbf{x}_i, t))^2}{\sum_{i=1}^{n} O^2(\mathbf{x}_i, t)} \quad (11.30)$$

Finally, the variance reduction VRC(t) normalized by the L2-norm of the observed waveform $C(\mathbf{x}_i, t)$ is expressed as follows

$$\text{VRC}(t) = 1 - \frac{\sum_{i=1}^{n} (O(\mathbf{x}_i, t) - C(\mathbf{x}_i, t))^2}{\sum_{i=1}^{n} C^2(\mathbf{x}_i, t)} \quad (11.31)$$

From Eqs. (11.30) and (11.31), Yamamoto et al. (2016) showed that VRO(t) was sensitive to the overestimation of the precomputed waveform $C(\mathbf{x}_i, t)$ with respect to the observed waveform $O(\mathbf{x}_i, t)$ and that VRC(t) was sensitive to the underestimation of $C(\mathbf{x}_i, t)$ with respect to $O(\mathbf{x}_i, t)$. Therefore, three indices, $R(t)$, VRO(t), and VRC(t), are needed to select appropriate scenarios from the database. To avoid the underestimation of tsunami heights for the tsunami forecast, the maximum value of the absolute values of the waveforms is used for the calculation of the three indices. The observed maximum $O_{\max}(\mathbf{x}_i, t)$ and precomputed maximum $C_{\max}(\mathbf{x}_i, t)$ are expressed as follows

$$O_{\max}(\mathbf{x}_i, t) = \max_{t' \leq t} |O(\mathbf{x}_i, t')| \quad (11.32)$$

$$C_{\max}(\mathbf{x}_i, t) = \max_{t' \leq t} |C(\mathbf{x}_i, t')| \quad (11.33)$$

By selecting appropriate thresholds for three indices, dozens of scenarios are selected from the database. Fig. 11.9 demonstrates one example of the selection procedure shown by Yamamoto et al. (2016). Then, the ranges of tsunami heights are forecasted along the coast by using this multiindex method.

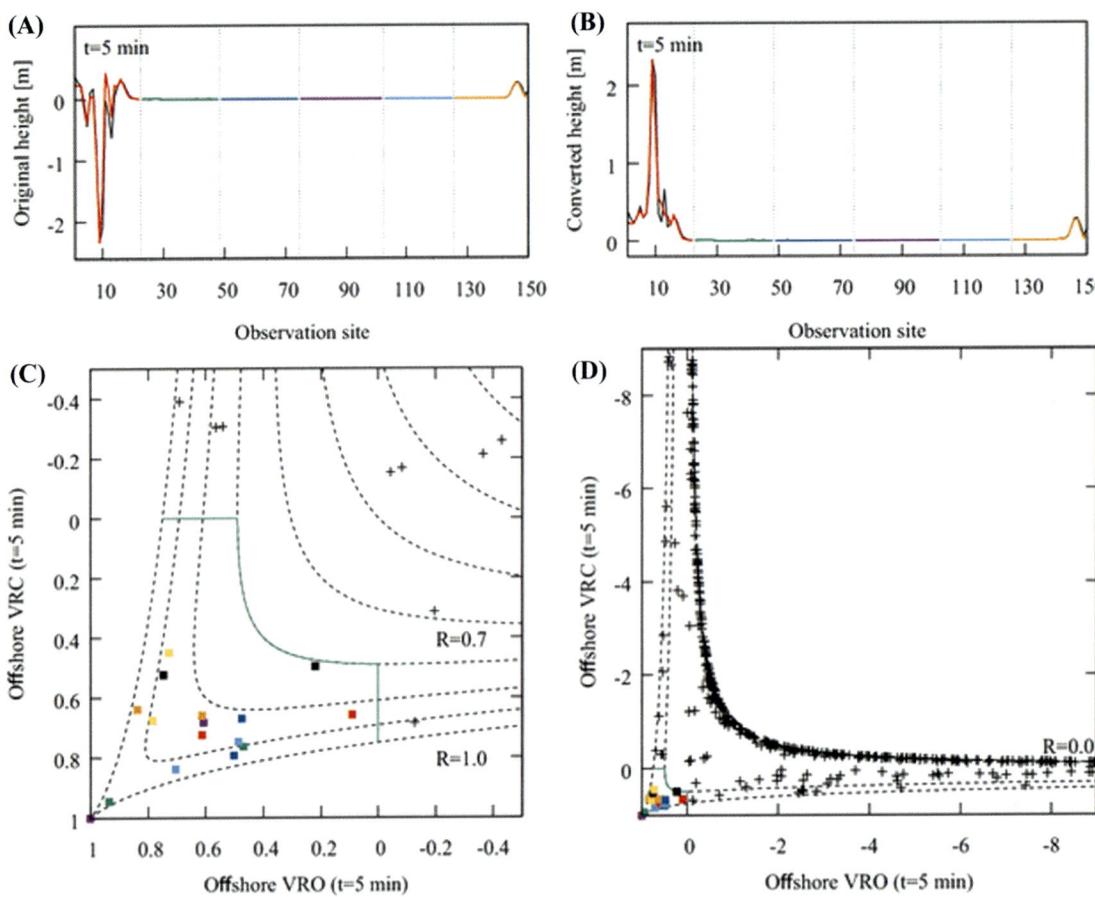

FIGURE 11.9 Offshore comparisons at 5 minutes after the occurrence of an earthquake for Model I. (A) Comparison of the tsunami height distributions of Model I (*colored lines*) with those of scenario A (*gray line*). (B) Comparison of the distributions in the converted wave height changes in Eqs. (11.32) and (11.33). (C,D) A close-up plot around the criteria and a wide view of the VRO–VRC diagram, respectively. The area surrounded by the green curves indicates the applied criteria, VRO ($t = 5$ min) ≥ 0.0, VRC ($t = 5$ min) ≥ 0.0, and R ($t = 5$ min) ≥ 0.7. The colored squares represent the selected tsunami scenarios and the gray crosses represent the unselected tsunami scenarios. The pseudo-observation scenario and scenario A are represented by the purple square at (1, 1) and the green square at (0.94, 0.95) in (C), respectively. *From Yamamoto, N., Aoi, S., Hirata, K., Suzuki, W., Kunugi, T., & Nakamura, H. (2016). Multi-index method using offshore ocean-bottom pressure data for real-time tsunami forecast 4. Earth, Planets and Space, 68(1). https://doi.org/10.1186/s40623-016-0500-7.*

11.5 Observed tsunamis from other sources

11.5.1 Meteotsunamis

Tsunami-like ocean waves are sometimes generated by meteorological phenomena and are referred to as meteorological tsunamis or meteotsunamis (Rabinovich, 2020). The generation mechanisms of the meteotsunamis have been extensively studied by Hibiya and Kajiura (1982), Proudman (1929) and Saito et al. (2021). Actual meteotsunamis were observed by coastal tide gauges and studied by Fukuzawa and Hibiya (2020), Rabinovich et al. (2021), and Šepic et al. (2015). Those studies indicated that meteotsunamis became sufficiently large to be observed at those coastal tide gauges when the enhancement mechanisms of tsunami-like waves were available. Recently, Kubota, Saito, et al. (2021) reported that the S-net could observe a small meteotsunami. By showing the waveforms of dense stations in the deep ocean as shown in Fig. 11.10 (Kubota, Saito, et al., 2021), the coherent tsunami-like waveforms were identified. This small tsunami-like wave could not be recognized if only a few observations were available.

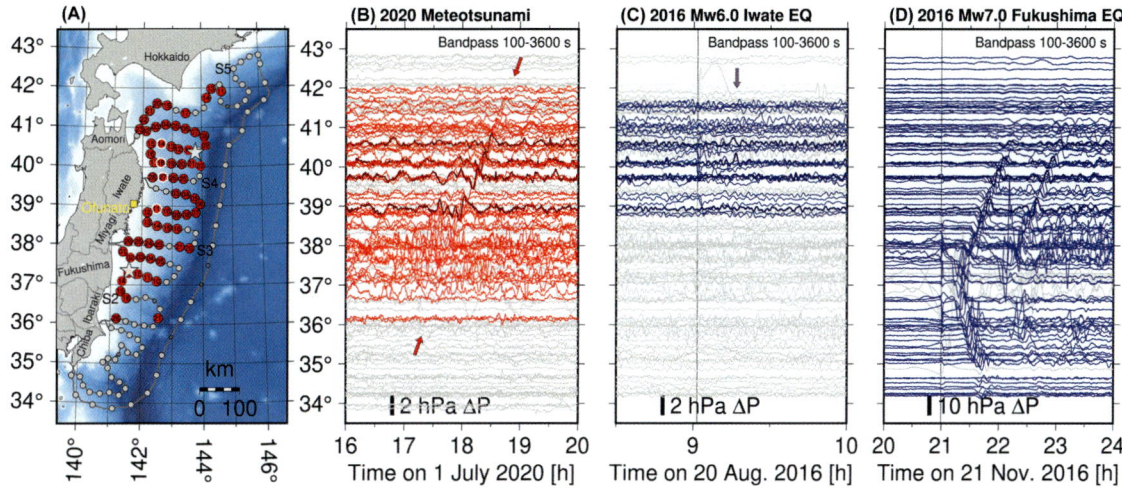

FIGURE 11.10 (A) Station map of this study and (B) pressure waveforms recorded on July 1, 2020. The horizontal axis is the time on 1 July (Universal Time Coordinated) and the vertical axis corresponds to the station latitude. The waveforms from the ocean-bottom pressure gauges marked by white circles in panel (A) are shown by thick lines. Data with low quality are plotted using gray lines. The tsunami-like pressure changes are denoted by red arrows. Pressure waveforms for (C) the 2016 Off-Iwate Earthquake and (D) the 2016 Off-Fukushima Earthquake. Epicenters for each earthquake are shown by white stars in panel (A). The horizontal scale in panel (C) is different from other panels. Map lines delineate study areas and do not necessarily depict accepted national boundaries. *From Kubota, T., Saito, T., Chikasada, N.Y., & Sandanbata, O. (2021). Meteotsunami observed by the deep-ocean seafloor pressure gauge network off Northeastern Japan.* Geophysical Research Letters, *48(21). https://doi.org/10.1029/2021GL094255.*

The equations for meteotsunami propagation are obtained by adding an external force term related to the atmospheric pressure, P_{atm}, to the conventional linear long-wave equations (An et al., 2012; Hibiya & Kajiura, 1982) as follows

$$\frac{\partial \eta}{\partial t} + \frac{\partial M}{\partial x} + \frac{\partial N}{\partial y} = 0 \tag{11.34}$$

$$\rho_0 \frac{1}{d} \frac{\partial M}{\partial t} + \rho_0 g_0 \frac{\partial \eta}{\partial x} = -\frac{\partial P_{atm}}{\partial x} \tag{11.35}$$

$$\rho_0 \frac{1}{d} \frac{\partial N}{\partial t} + \rho_0 g_0 \frac{\partial \eta}{\partial y} = -\frac{\partial P_{atm}}{\partial y} \tag{11.36}$$

where d is the sea depth, g_0 is the gravity acceleration, ρ_0 is the density of the seawater, $\eta(x,y,t)$ is the sea-surface height change, and $M(x,y,t)$ and $N(x,y,t)$ are the horizontal velocity in the x and y directions, respectively. Kubota, Saito, et al. (2021) numerically modeled the tsunami-like waves observed by the S-net by solving Eqs. (11.34)–(11.36) using the input from the line atmospheric pressure source propagated at a constant speed. Then, the ocean-bottom pressure changes, P_{bot}, were calculated as the sum of the pressure changes due to waves, P_η, and the atmospheric pressure disturbance, P_{atm}, and were compared with those observed by the S-net.

They found that a line source with a peak amplitude of pressure change of −0.5 hPa, a speed of 45 m/second, and a source azimuth of 70 degrees, explained the observed waves well. It clearly showed the small meteotsunami caused by such a small pressure change of −0.5 hPa was able to be identified by the dense ocean-bottom observation network (S-net).

11.5.2 Analysis of the 2022 Tonga Tsunami

A large eruption of the Hunga Tonga–Hunga Ha'apai Volcano in Tonga on January 15, 2022, caused devastating disasters in nearby areas. The eruption generated air waves coupled with seawater waves that propagated through the Pacific Ocean. Tsunamis, which are long waves in the ocean, generated by the airwaves from the eruption were observed by the S-net in Japan (Kubo et al., 2022). Because dense observation data were available, a record section (Fig. 11.11) or snapshots of the interpolated data (Fig. 11.12) clearly show that several linear waves are propagated in the northwest direction (Kubo et al., 2022). They estimated the velocities and the directions of propagation for each phase using a semblance analysis, commonly used in the array observation data of seismic waves (Honda et al., 2008; Neidell & Turhan Taner, 1971).

For these dense tsunami observation data, Mizutani and Yomogida (2023) applied the Vespa analysis, velocity spectral analysis, which is a widely used array-based method in seismology to estimate the incident angles and arrival times

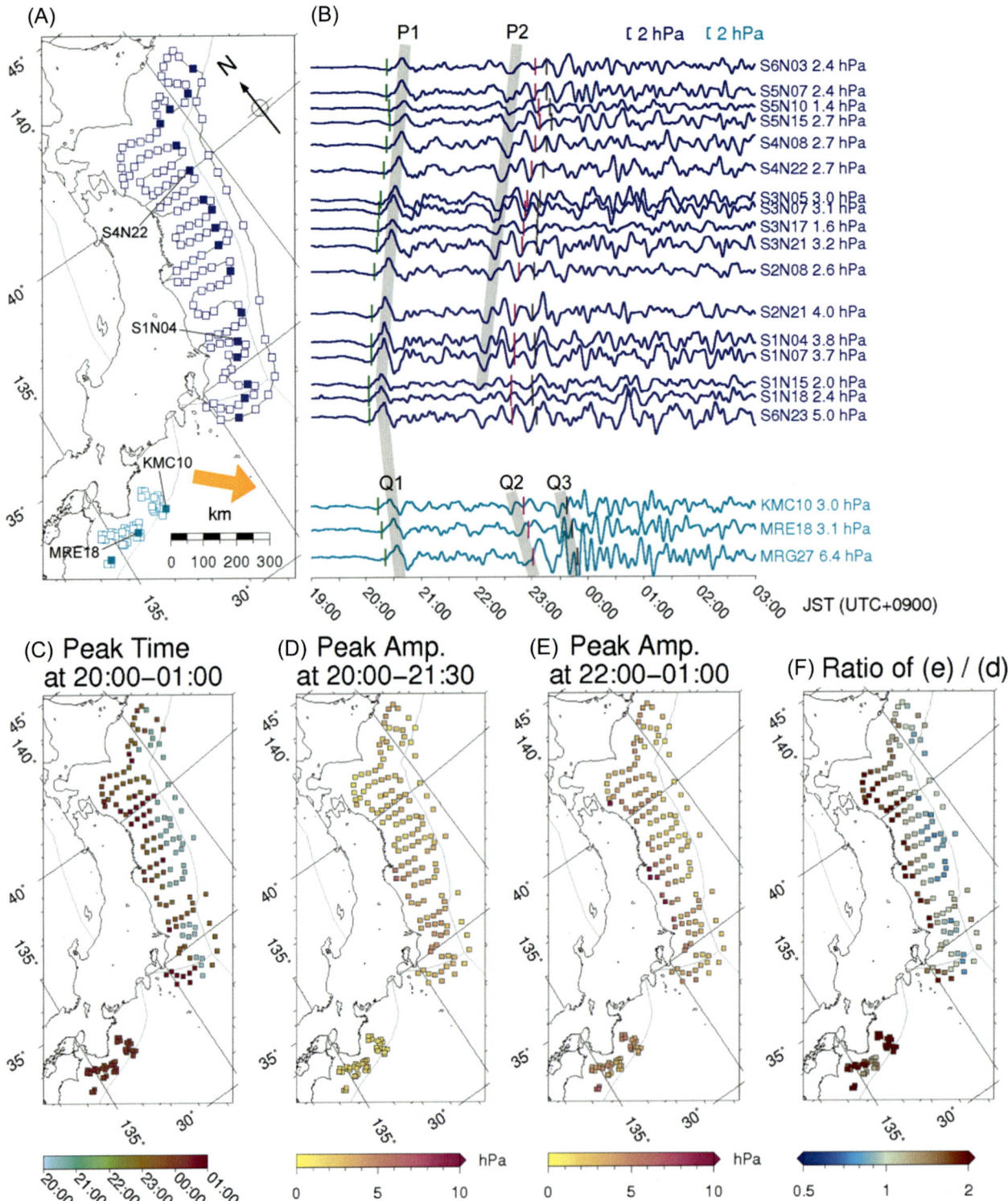

FIGURE 11.11 (A) Map of the study area. Blue and cyan squares indicate the S-net and DONET stations used in this study, respectively. Solid squares denote stations for which waveform records are shown in (B). The orange arrow indicates a direction that is 140 degrees clockwise from north, which roughly corresponds to the direction of the short great circle between Tonga and Japan. (B) Record section of ocean-bottom pressure from 20:00 JST (UTC + 0900) on January 15, 2022, to 03:00 JST on January 16, 2022. Blue and cyan lines denote waveform records of ocean-bottom pressure gauges at the S-net and DONET stations, respectively. Significant phases (P1, P2, Q1, Q2, and Q3) are shaded in gray. Black bars indicate the theoretical arrival times of direct tsunamis. The origin time of the tsunami travel time was assumed at 13:00. Green and pink bars indicate the theoretical travel times assuming propagation along the short great circle path from the Hunga Tonga–Hunga Ha'apai Volcano at velocities of 300 and 220 m/second, respectively. (C) Distribution of peak time of ocean-bottom pressure change between 20:00 and 01:00. (D) Distribution of peak amplitude of ocean-bottom pressure change at 20:00–21:30. (E) The same as (C) but at 22:00–01:00. (F) Distribution of the ratio of (D) the peak amplitude at 20:00–21:30 to (E) the peak amplitude at 22:00–01:00. *DONET*, Dense Ocean floor Network system for Earthquakes and Tsunamis. Map lines delineate study areas and do not necessarily depict accepted national boundaries. *From Kubo, H., Kubota, T., Suzuki, W., Aoi, S., Sandanbata, O., Chikasada, N., & Ueda, H. (2022). Ocean-wave phenomenon around Japan due to the 2022 Tonga eruption observed by the wide and dense ocean-bottom pressure gauge networks.* Earth, Planets and Space, 74(1). https://doi.org/10.1186/s40623-022-01663-w.

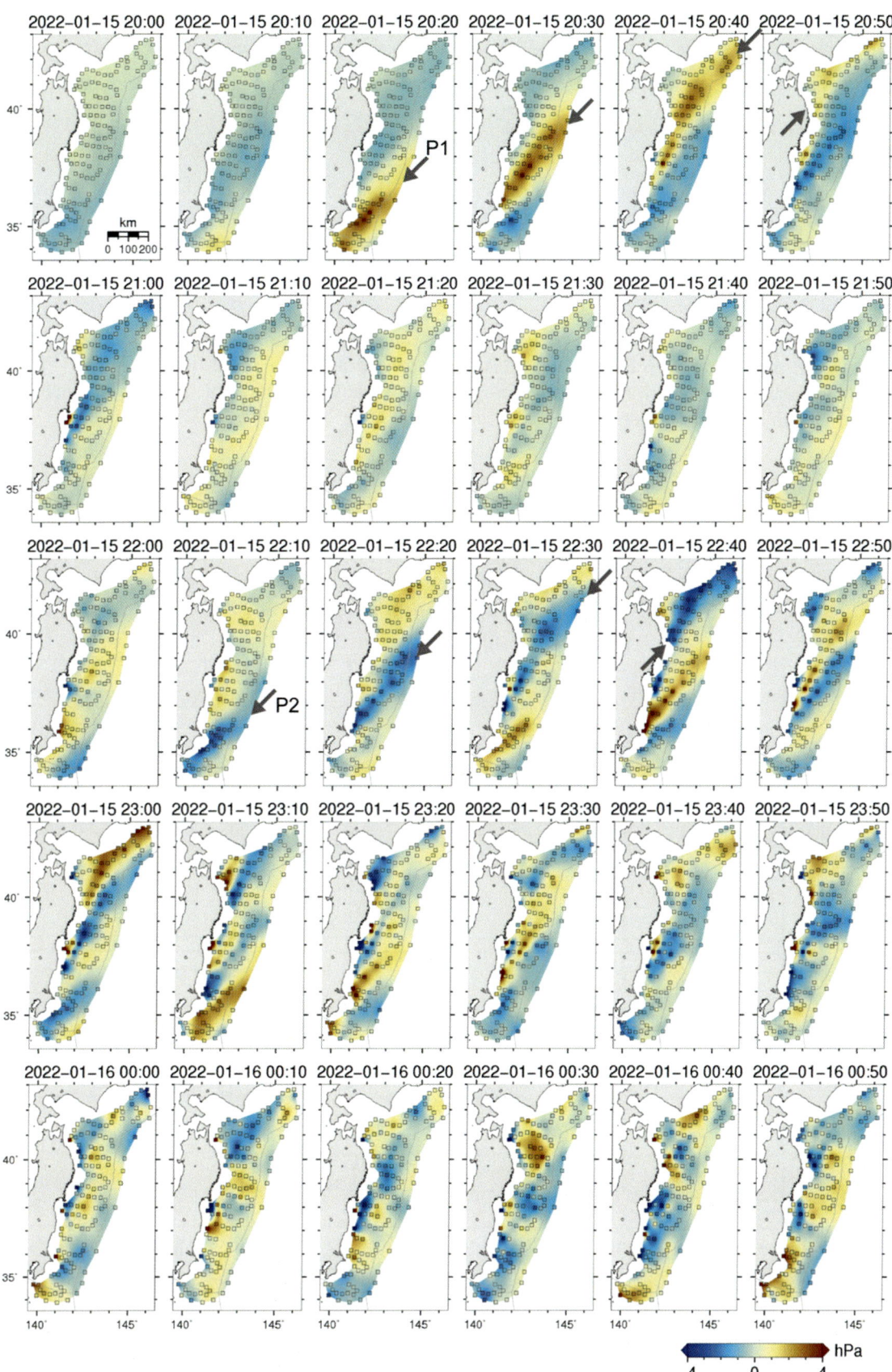

FIGURE 11.12 Snapshot of the ocean-bottom pressure wavefield derived by interpolating S-net observations. Squares indicate S-net observation stations. Map lines delineate study areas and do not necessarily depict accepted national boundaries. *From Kubo, H., Kubota, T., Suzuki, W., Aoi, S., Sandanbata, O., Chikasada, N., & Ueda, H. (2022). Ocean-wave phenomenon around Japan due to the 2022 Tonga eruption observed by the wide and dense ocean-bottom pressure gauge networks.* Earth, Planets and Space, *74(1). https://doi.org/10.1186/s40623-022-01663-w.*

of tsunami phases (Rost & Thomas, 2002). The results of the Vespa analysis are shown in a diagram of the energy of the incoming signals as a function of slowness or back-azimuth and time. In this analysis, the observed waveforms are stacked using the following Nth-root stacking

$$\tilde{S}_{N,\phi}(t) = \frac{1}{M} \sum_{k=1}^{M} \left| d_k\left(t - t_{k,\phi}^{travel}\right) \right|^{1/N} \text{sgn}[d_k(t)] \quad (11.37)$$

where $\tilde{S}_{N,\phi}(t)$ is the Nth-root beam trace of each incident angle ϕ, $d_k(t)$ is the observed waveforms at the kth station ($k = 1,\ldots,M$), $t_{k,\phi}^{travel}$ is the relative travel time from the reference point to station k with incident angle ϕ, and sgn is the sign function. Then, the beam $S_{N,\phi}(t)$ is calculated by taking Nth power with keeping sign

$$S_{N,\phi}(t) = \left| \tilde{S}_{N,\phi}(t) \right|^N \text{sgn}\left[\tilde{S}_{N,\phi}(t)\right] \quad (11.38)$$

Mizutani and Yomogida (2023) selected $N = 4$ for their tsunami Vespa analysis using a trial-and-error approach. The relative travel time, $t_{k,\phi}^{travel}$, with each incident angle is computed using the FMM (Sethian, 1999). The tsunami is assumed to be a plane wave before reaching the array. Then, the moving average of the squared amplitude of each beam is calculated to estimate the energy

$$E_\phi(t) = \int_{t-\alpha}^{t+\alpha} [S_{N,\phi}(\tau)]^2 d\tau \quad (11.39)$$

where $E_\phi(t)$ is the energy at each azimuth and α is a half-length of the time window. Mizutani and Yomogida (2023) set α to be 150 seconds. Fig. 11.13 shows the vespagrams for DONET data and S-net data (Mizutani & Yomogida, 2023).

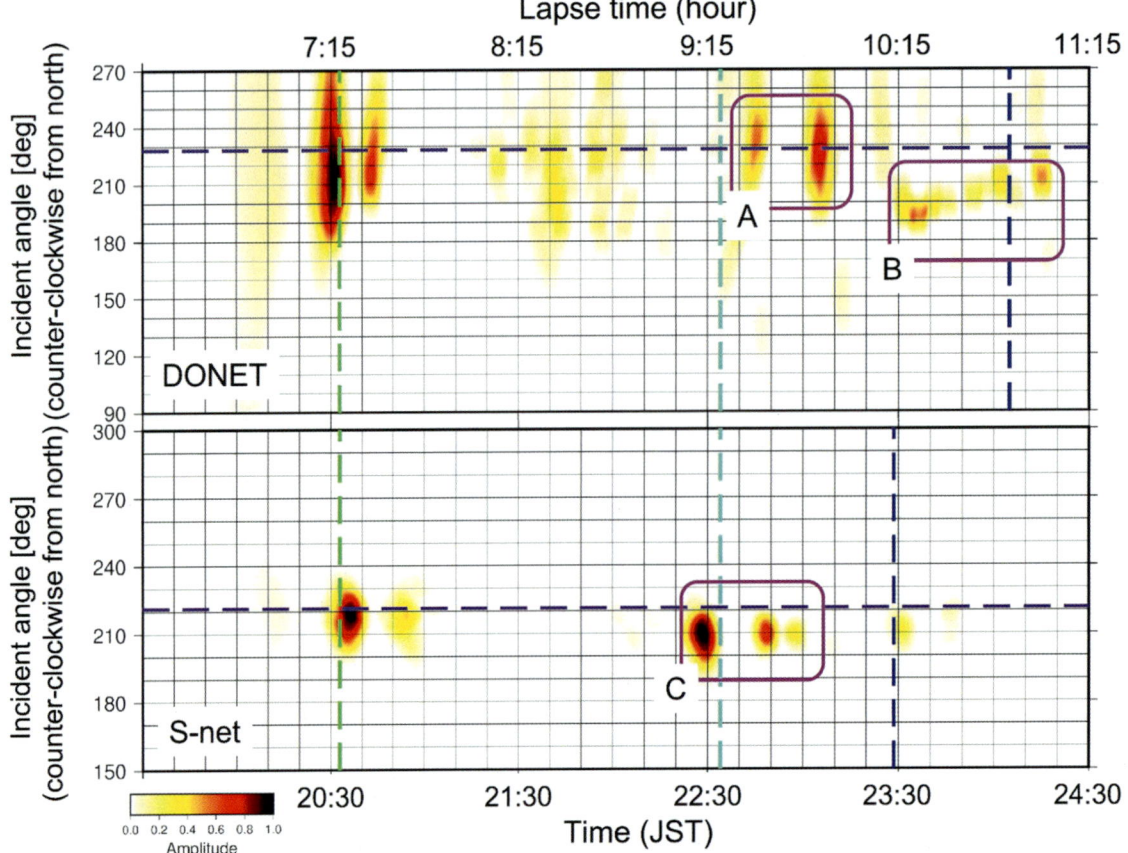

FIGURE 11.13 Vespagrams for DONET (upper panel) and S-net (lower panel). Amplitudes are normalized by the maximum value of each plot. The purple line represents the incident angle of the great circle path from Tonga to the reference points of each network. The green and cyan lines give the arrival times of the waves along the great circle with speeds of 300 and 235 m/second, respectively. The blue line represents the theoretical tsunami arrival time from Tonga. The lapse time is from the origin time of the eruption, 13:15 (JST) or 4:15 (UTC). *DONET*, Dense Ocean floor Network system for Earthquakes and Tsunamis. *From Mizutani, A., & Yomogida, K. (2023). Source estimation of the tsunami later phases associated with the 2022 Hunga Tonga volcanic eruption. Geophysical Journal International, 234(3), 1885–1902. https://doi.org/10.1093/gji/ggad174.*

The green line represents the arrival time of the waves along the great circle path from Tonga with a speed of 300 m/second which should be the arrival time of the Lamb wave. The blue line represents the theoretical tsunami arrival time from Tonga. Fig. 11.13 shows that the tsunami energy arrived with the Lamb wave generated by the Tonga eruption and the other large energies from Tonga were also found after the Lamb wave arrival time but before the theoretical tsunami arrival time. This analysis was only possible by using the dense tsunami observation networks, such as S-net and DONET. Mizutani and Yomogida (2023) discussed the tsunami generation mechanism of those later arrival phases from Tonga. Tanioka et al. (2022) numerically modeled these dense tsunami data generated by the Lamb wave from Tonga using Eq. (11.36) and discussed the characteristics of tsunami propagation from the deep ocean to the coast.

11.6 Summary

In this chapter, advancements in tsunami research using dense tsunami observation systems, such as S-net in which 150 sensors were installed approximately 30–50 km apart, were reviewed and discussed. Geophysical researchers, such as seismologists, volcanologists, and meteorologists, have been analyzing such dense observation data or array data using advanced methods and techniques. The dense tsunami array data enable the precise and reliable estimation of tsunami source models, whether they result from earthquakes or other mechanisms. Their methods and techniques can be adjusted to analyze dense tsunami observation data. The methods and techniques described in this chapter are useful for advanced tsunami forecasts and also for understanding tsunami generation mechanisms or propagation phenomena.

The bottom pressure gauges of S-net recorded the tsunami generated by the 2016 Fukushima Earthquake, and the collected data have been instrumental in estimating a detailed earthquake source model. Furthermore, it has facilitated the development of a back-projection method utilizing tsunami waveforms.

The S-net, installed after the 2011 Tohoku Earthquake and Tsunami, has significantly enhanced Japan's tsunami early warning capabilities. The tsunami forecasting method that is based on tsunami waveform inversion developed by the JMA has been greatly improved by the dense network.

Another tsunami forecasting method that harnesses the capabilities of the S-net was developed by the NIED. The multiindex method relies on a database containing numerous source models, aiming to identify scenarios that match the precomputed tsunami waveforms with those observed by the dense ocean-bottom sensors.

The S-net data have also proven to be highly valuable in studying a relatively uncommon type of tsunamis resulting from the coupling of air and seawater waves, known as meteotsunamis. The network recorded the tsunami generated by the 2022 Hunga Tonga - Hunga Ha'apai Volcano eruption, which occurred approximately 8000 km away from Japan.

References

An, C., Liu, P. L. F., & Seo, S. N. (2012). Large-scale edge waves generated by a moving atmospheric pressure. *Theoretical and Applied Mechanics Letters*, *2*(4). Available from https://doi.org/10.1063/2.1204201, http://www.journals.elsevier.com/theoretical-and-applied-mechanics-letters/.

Aoi, S., Asano, Y., Kunugi, T., Kimura, T., Uehira, K., Takahashi, N., Ueda, H., Shiomi, K., Matsumoto, T., & Fujiwara, H. (2020). MOWLAS: NIED observation network for earthquake, tsunami and volcano. *Earth, Planets and Space*, *72*(1). Available from https://doi.org/10.1186/s40623-020-01250-x.

Aoi, S., Suzuki, W., Chikasada, N. Y., Miyoshi, T., Arikawa, T., & Seki, K. (2019). Development and utilization of real-time tsunami inundation forecast system using S-net data. *Fuji Technology Press, Japan Journal of Disaster Research*, *14*(2), 212−224. Available from https://doi.org/10.20965/jdr.2019.p0212, https://www.fujipress.jp/main/wp-content/themes/Fujipress/pdf_subscribed.php.

Fukuzawa, K., & Hibiya, T. (2020). The amplification mechanism of a meteo-tsunami originating off the western coast of Kyushu Island of Japan in the winter of 2010. *Journal of Oceanography*, *76*(3), 169−182. Available from https://doi.org/10.1007/s10872-019-00536-3.

Gusman, A. R., Satake, K., Shinohara, M., Sakai, S., & Tanioka, Y. (2017). Fault slip distribution of the 2016 Fukushima earthquake estimated from tsunami waveforms. *Pure and Applied Geophysics*, *174*(8), 2925−2943. Available from https://doi.org/10.1007/s00024-017-1590-2, http://www.springer.com/birkhauser/geo+science/journal/24.

Gusman, A. R., Sheehan, A. F., Satake, K., Heidarzadeh, M., Mulia, I. E., & Maeda, T. (2016). Tsunami data assimilation of Cascadia seafloor pressure gauge records from the 2012 Haida Gwaii earthquake. *Geophysical Research Letters*, *43*(9), 4189−4196. Available from https://doi.org/10.1002/2016GL068368, http://onlinelibrary.wiley.com/journal/10.1002/(ISSN)1944-8007/issues?year=2012.

Gusman, A. R., Tanioka, Y., MacInnes, B. T., & Tsushima, H. (2014). A methodology for near-field tsunami inundation forecasting: Application. *Journal of Geophysical Research: Solid Earth*, *119*(11), 8186−8206.

Hibiya, T., & Kajiura, K. (1982). Origin of the Abiki phenomenon (a kind of seiche) in Nagasaki Bay. *Journal of the Oceanographical Society of Japan*, *38*(3), 172−182. Available from https://doi.org/10.1007/BF02110288.

Honda, R., Aoi, S., Sekiguchi, H., & Fujiwara, H. (2008). Imaging an asperity of the 2003 Tokachi-oki earthquake using a dense strong-motion seismograph network. *Geophysical Journal International*, *172*(3), 1104−1116. Available from https://doi.org/10.1111/j.1365-246X.2007.03702.x.

Ishii, M., Shearer, P. M., Houston, H., & Vidale, J. E. (2005). Extent, duration and speed of the 2004 Sumatra-Andaman earthquake imaged by the Hi-net array. *Nature*, *435*(7044), 933–936. Available from https://doi.org/10.1038/nature03675.

Kawaguchi, K., Araki, E., & Kaneda, Y. (2011). Establishment for a method for real-time and long-term seafloor monitoring. *Journal of Advanced Marine Science and Technology Society*, *17*, 125–135. Available from https://doi.org/10.14928/amstec.17.2_125.

Kiser, E., Ishii, M., Langmuir, C. H., Shearer, P. M., & Hirose, H. (2011). Insights into the mechanism of intermediate-depth earthquakes from source properties as imaged by back projection of multiple seismic phases. *Journal of Geophysical Research*, *116*(B6). Available from https://doi.org/10.1029/2010jb007831.

Kubo, H., Kubota, T., Suzuki, W., Aoi, S., Sandanbata, O., Chikasada, N., & Ueda, H. (2022). Ocean-wave phenomenon around Japan due to the 2022 Tonga eruption observed by the wide and dense ocean-bottom pressure gauge networks. *Earth, Planets and Space*, *74*(1). Available from https://doi.org/10.1186/s40623-022-01663-w, http://rd.springer.com/journal/40623.

Kubota, T., Kubo, H., Yoshida, K., Chikasada, N. Y., Suzuki, W., Nakamura, T., & Tsushima, H. (2021). Improving the constraint on the Mw 7.1 2016 off-Fukushima shallow normal-faulting earthquake with the high azimuthal coverage tsunami data from the S-net wide and dense network: Implication for the stress regime in the Tohoku overriding plate. *Journal of Geophysical Research: Solid Earth*, *126*(10). Available from https://doi.org/10.1029/2021JB022223, http://agupubs.onlinelibrary.wiley.com/hub/jgr/journal/10.1002/(ISSN)2169-9356/.

Kubota, T., Saito, T., Chikasada, N. Y., & Sandanbata, O. (2021). Meteotsunami observed by the deep-ocean seafloor pressure gauge network off Northeastern Japan. *Geophysical Research Letters*, *48*(21). Available from https://doi.org/10.1029/2021GL094255, http://agupubs.onlinelibrary.wiley.com/hub/journal/10.1002/(ISSN)1944-8007/.

Kubota, T., Suzuki, W., Nakamura, T., Chikasada, N. Y., Aoi, S., Takahashi, N., & Hino, R. (2018). Tsunami source inversion using time-derivative waveform of offshore pressure records to reduce effects of non-tsunami components. *Geophysical Journal International*, *215*(2), 1200–1214. Available from https://doi.org/10.1093/GJI/GGY345, http://gji.oxfordjournals.org/.

Lynch, P. (2008). The origins of computer weather prediction and climate modeling. *Journal of Computational Physics*, *227*(7), 3431–3444. Available from https://doi.org/10.1016/j.jcp.2007.02.034.

Maeda, T., Obara, K., Shinohara, M., Kanazawa, T., & Uehira, K. (2015). Successive estimation of a tsunami wavefield without earthquake source data: A data assimilation approach toward real-time tsunami forecasting. *Geophysical Research Letters*, *42*(19), 7923–7932. Available from https://doi.org/10.1002/2015GL065588, http://onlinelibrary.wiley.com/journal/10.1002/(ISSN)1944-8007/issues?year = 2012.

Mizutani, A., & Yomogida, K. (2022). Back-projection imaging of a tsunami excitation area with ocean-bottom pressure gauge array data. *Journal of Geophysical Research: Oceans*, *127*(7). Available from https://doi.org/10.1029/2022JC018480, http://agupubs.onlinelibrary.wiley.com/agu/jgr/journal/10.1002/(ISSN)2169-9291/.

Mizutani, A., & Yomogida, K. (2023). Source estimation of the tsunami later phases associated with the 2022 Hunga Tonga volcanic eruption. *Journal International*, *234*(3), 1885–1902. Available from https://doi.org/10.1093/gji/ggad174, http://gji.oxfordjournals.org/.

Mizutani, A., Yomogida, K., & Tanioka, Y. (2020). Early tsunami detection with near-fault ocean-bottom pressure gauge records based on the comparison with seismic data. *Journal of Geophysical Research: Oceans*, *125*(9). Available from https://doi.org/10.1029/2020jc016275.

Mori, N., & Takahashi, T. (2011). The 2011 Tohoku Earthquake Tsunami Joint Survey Group, Nationwide post event survey and analysis of the 2011 Tohoku earthquake tsunami. *Coastal Engineering Journal*, *54*(1). Available from https://doi.org/10.1142/S0578563412500015.

Neidell, N. S., & Turhan Taner, M. (1971). Semblance and other coherency measures for multichannel data. *Geophysics*, *36*(3), 482–497. Available from https://doi.org/10.1190/1.1440186.

Noda, H., Lapusta, N., & Kanamori, H. (2013). Comparison of average stress drop measures for ruptures with heterogeneous stress change and implications for earthquake physics. *Geophysical Journal International*, *193*(3), 1691–1712. Available from https://doi.org/10.1093/gji/ggt074.

Ohta, Y., Kobayashi, T., Tsushima, H., Miura, S., Hino, R., Takasu, T., Fujimoto, H., Iinuma, T., Tachibana, K., Demachi, T., Sato, T., Ohzono, M., & Umino, N. (2012). Quasi real-time fault model estimation for near-field tsunami forecasting based on RTK-GPS analysis: Application to the 2011 Tohoku-Oki earthquake (Mw 9.0). *Journal of Geophysical Research: Solid Earth*, *117*(B2). Available from https://doi.org/10.1029/2011JB008750.

Okada, Y. (1992). Internal deformation due to shear and tensile faults in a half-space. *Bulletin of the Seismological Society of America*, *82*(2), 1018–1040. Available from https://doi.org/10.1785/bssa0820021018.

Ozaki, T. (2011). Outline of the 2011 off the Pacific coast of Tohoku Earthquake (Mw 9.0): Tsunami warnings/advisories and observations. *Earth, Planets and Space*, *63*(7), 827–830. Available from https://doi.org/10.5047/eps.2011.06.029, http://rd.springer.com/journal/40623.

Press, W. H., Teukolsky, S. A., Vetterling, W. T., & Flannery, B. P. (1992). *Numerical recipes in Fortran 77: The art of scientific computing*. Cambridge University Press.

Proudman, J. (1929). The effects on the sea of changes in atmospheric pressure. *Geophysical Journal International*, *2*, 197–209. Available from https://doi.org/10.1111/j.1365-246X.1929.tb05408.x.

Rabinovich, A. B. (2020). Twenty-seven years of progress in the science of meteorological tsunamis following the 1992 Daytona Beach event. *Pure and Applied Geophysics*, *177*(3), 1193–1230. Available from https://doi.org/10.1007/s00024-019-02349-3, http://www.springer.com/birkhauser/geo + science/journal/24.

Rabinovich, A. B., Šepić, J., & Thomson, R. E. (2021). The meteorological tsunami of 1 November 2010 in the southern Strait of Georgia: A case study. *Natural Hazards*, *106*(2), 1503–1544. Available from https://doi.org/10.1007/s11069-020-04203-5, http://www.wkap.nl/journalhome.htm/0921-030X.

Romesburg,C. (2004). *Cluster analysis for researchers*.

Rost, S., & Thomas, C. (2002). Array seismology: Methods and applications. *Reviews of Geophysics*, *40*(3). Available from https://doi.org/10.1029/2000RG000100, http://agupubs.onlinelibrary.wiley.com/agu/journal/10.1002/(ISSN)1944-9208/.

Saito, T., Kubota, T., Chikasada, N. Y., Tanaka, Y., & Sandanbata, O. (2021). Meteorological tsunami generation due to sea-surface pressure change: Three-dimensional theory and synthetics of ocean-bottom pressure change. *Journal of Geophysical Research: Oceans*, *126*(5). Available from https://doi.org/10.1029/2020JC017011, http://agupubs.onlinelibrary.wiley.com/agu/jgr/journal/10.1002/(ISSN)2169-9291/.

Saito, T., & Tsushima, H. (2016). Synthesizing ocean bottom pressure records including seismic wave and tsunami contributions: Toward realistic tests of monitoring systems. *Journal of Geophysical Research: Solid Earth*, *121*(11), 8175−8195. Available from https://doi.org/10.1002/2016JB013195, http://onlinelibrary.wiley.com/journal/10.1002/(ISSN)2169-9356.

Sethian, J. (1999). *Level set methods and Fast Marching methods*. Cambridge Unversity Press.

Šepic, J., Vilibic, I., Rabinovich, A. B., & Monserrat, S. (2015). Widespread tsunami-like waves of 23-27 June in the Mediterranean and Black Seas generated by high-altitude atmospheric forcing. *Scientific Reports*, *5*. Available from https://doi.org/10.1038/srep11682, http://www.nature.com/srep/index.html.

Sheehan, A. F., Gusman, A. R., Heidarzadeh, M., & Satake, K. (2015). Array observations of the 2012 Haida Gwaii tsunami using Cascadia initiative absolute and differential seafloor pressure gauges. *Seismological Society of America, United States Seismological Research Letters*, *86*(5), 1278−1286. Available from https://doi.org/10.1785/0220150108, http://srl.geoscienceworld.org/content/86/5/1278.full.pdf + html.

Tanioka, Y. (2018). Tsunami simulation method assimilating ocean bottom pressure fata near a tsunami source region. *Pure and Applied Geophysics*, *175*(2), 721−729. Available from https://doi.org/10.1007/s00024-017-1697-5.

Tanioka, Y. (2020). Improvement of near-field tsunami forecasting method using ocean-bottom pressure sensor network (S-net). *Earth, Planets and Space*, *72*(1). Available from https://doi.org/10.1186/s40623-020-01268-1.

Tanioka, Y., & Gusman, A. R. (2018). Near-field tsunami inundation forecast method assimilating ocean bottom pressure data: A synthetic test for the 2011 Tohoku-Oki tsunami. *Physics of the Earth and Planetary Interiors*, *283*, 82−91. Available from https://doi.org/10.1016/j.pepi.2018.08.006.

Tanioka, Y., Yamanaka, Y., & Nakagaki, T. (2022). Characteristics of tsunamis observed in Japan due to the Air Wave from the 2022 Tonga Eruption. *Research Square*. Available from https://doi.org/10.21203/rs.3.rs-1320093/v1, https://www.researchsquare.com/browse.

Tsushima, H., Hino, R., Fujimoto, H., Tanioka, Y., & Imamura, F. (2009). Near-field tsunami forecasting from cabled ocean bottom pressure data. *Journal of Geophysical Research: Solid Earth*, *114*(6). Available from https://doi.org/10.1029/2008JB005988, http://onlinelibrary.wiley.com/journal/10.1002/(ISSN)2169-9356.

Tsushima, H., Hino, R., Ohta, Y., Iinuma, T., & Miura, S. (2014). tFISH/RAPiD: Rapid improvement of near-field tsunami forecasting based on offshore tsunami data by incorporating onshore GNSS data. *Geophysical Research Letters*, *41*(10), 3390−3397. Available from https://doi.org/10.1002/2014gl059863.

Tsushima, H., Hino, R., Tanioka, Y., Imamura, F., & Fujimoto, H. (2012). Tsunami waveform inversion incorporating permanent seafloor deformation and its application to tsunami forecasting. *Journal of Geophysical Research: Solid Earth*, *117*(B3). Available from https://doi.org/10.1029/2011jb008877.

Uehira,K., Kanazawa,T., Noguchi,S., Aoi,S., Kunugi,T., Matsumoto,T., Okada,Y., Sekiguchi,S., Shiomi,K., Shinohara,M., & Yamada,T. (2012). *Ocean bottom seismic and tsunami network along the Japan Trench*. In American Geophysical Union, Fall Meeting 2012. Available from: https://ui.adsabs.harvard.edu/abs/2012AGUFMOS41C1736U/abstract.

Wang, Y., Satake, K., Maeda, T., & Gusman, A. R. (2017). Green's function-based tsunami data assimilation: A fast data assimilation approach toward tsunami early warning. *Geophysical Research Letters*, *44*(20), 10−289. Available from https://doi.org/10.1002/2017GL075307, http://onlinelibrary.wiley.com/journal/10.1002/(ISSN)1944-8007/issues?year = 2012.

Yagi, Y., Nakao, A., & Kasahara, A. (2012). Smooth and rapid slip near the Japan Trench during the 2011 Tohoku-oki earthquake revealed by a hybrid back-projection method. *Earth and Planetary Science Letters*, *355-356*, 94−101. Available from https://doi.org/10.1016/j.epsl.2012.08.018.

Yamamoto, N., Aoi, S., Hirata, K., Suzuki, W., Kunugi, T., & Nakamura, H. (2016). Multi-index method using offshore ocean-bottom pressure data for real-time tsunami forecast 4. *Earth, Planets and Space*, *68*(1). Available from https://doi.org/10.1186/s40623-016-0500-7, http://rd.springer.com/journal/40623.

Chapter 12

Machine learning approaches for tsunami early warning

Iyan E. Mulia[1,2]

[1]Hydrography Research Group, Faculty of Earth Sciences and Technology, Bandung Institute of Technology, Bandung, Indonesia, [2]Research Center for Disaster Mitigation, Bandung Institute of Technology, Bandung, Indonesia

12.1 Introduction

Since the advent of the digital era, data volume and complexity have increased immensely. This circumstance has led to the development of intelligent ways to process and analyze those data for solving diverse real-world problems, such as classification, regression, clustering, or dimensionality reduction tasks (Sarker, 2021). Machine learning is one of the most widely used methods, which can learn large amounts of data and draw inferences from it. In the past few decades, machine learning has considerably transformed our way of utilizing information from the ever-growing volume of data. Within a relatively short period, the rapid advancement of machine learning capabilities is evident (Emmert-Streib et al., 2020), contributing to the exponential growth of data across fields. In addition to gaining insight from data, machine learning can facilitate fast decision-making owing to its massively parallel processing system and information storage. This feature is often leveraged to complement or even supersede the conventional physics-based modeling approach. The trend of machine learning applications will continue to grow and become more mature, along with ceaseless data collection.

In geophysics, the increasing amount of seismic, geodetic, and tsunami instrumentation enables the accumulation of a wealth of geophysical information. Besides enhancing our understanding of related natural occurrences, the numerous observational systems can also be viewed as supporting apparatus in the context of early warning. For example, the dense tsunami observation networks in Japan (Aoi et al., 2020; Kaneda et al., 2015) and the global networks over the world's oceans (Bernard & Titov, 2015; González et al., 2005) have greatly improved our preparedness against the threat of tsunamis. Apart from the early detection of tsunamis, to be more meaningful, the data registered at observational sensors must be used to forecast tsunami impacts along the coastline, which is an integral component of a tsunami early warning system. Several tsunami forecasting methods are already integrated into operational warning systems (Titov et al., 2005; Tsushima et al., 2009), which have been tested in past notable events (Tang et al., 2016; Tsushima et al., 2011). The methods typically involve traditional forward and inverse modeling configured to operate globally and on regional or local scales.

Applying machine learning for tsunami forecasting is an innovative strategy to optimally exploit the existing observational systems' potential that may have yet to be fully reached using the conventional physics-based modeling approach. More data will likely improve the accuracy of full-physics models through a data assimilation scheme (Gusman et al., 2016; Hossen et al., 2021; Maeda et al., 2015) or provide a better initial condition obtained from an inversion analysis (Mulia et al., 2017; Wang et al., 2021). However, a numerical solution to the interrelated variables governing the complex tsunami hydrodynamics is expensive. The computational cost increases manifold for nearshore or overland flow simulations where the nonlinear effect prevails. Thus it limits the implementation of the traditional simulation setup, particularly when the model is needed in the operational mode. Machine learning is a prospective technology to address the issue, considering its computational efficiency compared to physics-based models. It has recently gained more attention and has been implemented in numerous geophysical applications (Espeholt, Agrawal, et al., 2022; Espeholt, Palma, et al., 2022; Lee et al., 2021; Mousavi et al., 2020).

Although there have been many viable applications, developing a machine learning method for predicting extreme and infrequent events is rather challenging, as it is, by definition, a data-driven model. Despite the growing number of geophysical observations, the lack of data associated with rare natural events, such as tsunamis, hinders the standalone application of machine learning. A physics-based model is needed to produce realistic pseudo data, which is reasonable considering physical understandings of tsunamis—from source to impact—have significantly matured, as reflected by several well-validated software packages (Marras & Mandli, 2021; Sugawara, 2021). Therefore, machine learning is commonly trained on many hypothetical scenarios resulting from physics-based simulations. The fully trained machine learning model is expected to have a comparable predictive skill to the physics-based models with a substantial computational cost reduction. In this sense, machine learning is regarded as an alternative surrogate for the ordinary tsunami numerical models evolving from the prior scenario-based selection or interpolation methods (Gusman et al., 2014; Mulia et al., 2018).

Unlike the traditional scenario-based selection or interpolation, by which the real-time operational forecast is made via reading the precalculated database, machine learning forecasts rely on its model parameters. Therefore, although it depends on the complexity of the model structure, machine learning models generally consume less computer memory and data storage than their predecessor methods. This relatively small memory footprint is favorable for efficient data communications, facilitating modern internet-based computational infrastructures like cloud or edge computing (Chen & Ran, 2019). The efficacy of such computing technology in conjunction with the utilization of Internet of Things (IoT) sensors for tsunami early warning support systems has been investigated and proposed (Behrens et al., 2022; Esposito et al., 2022). It is of additional interest to explore how machine learning would fit such information technology-oriented solutions and contribute to tsunami early warning systems.

This chapter discusses the applicability of machine learning for tsunami forecasting and its implications in tsunami early warning. Tsunami forecasting methods using machine learning with diverse settings and architectural building blocks can be found in the recent literature (Fauzi & Mizutani, 2020a, 2020b; Giles et al., 2022; Kamiya et al., 2022; Liu et al., 2021; Makinoshima et al., 2021; Mulia et al., 2022; Mulia, Gusman, et al., 2020; Rim et al., 2022; Rodríguez et al., 2022). A general theoretical basis and simulation workflow relevant to all currently available models are provided to elaborate various applications. Furthermore, assessments of existing studies provide deeper insight into the development of machine learning in tsunami forecasting, which is still in its infancy. An analysis of the advantages and limitations of machine learning is also required to clarify its position within the global and regional tsunami early warning frameworks. Based on the analysis, future directions will be suggested to improve the present stage and overcome commonly identified obstacles, thus expediting the implementation of machine learning-based tsunami forecasting in the operational systems.

12.2 Theoretical framework

Along with other subsets (e.g., expert systems, fuzzy logic, and robotics), machine learning is a branch of artificial intelligence, a system mimicking human-like cognitive capacity for advanced problem-solving (Janiesch et al., 2021). The algorithm of machine learning is inspired by the structure and function of the brain called artificial neural networks. Initially, machine learning was developed with a simple architecture or shallow neural networks. More recently, adapting to the vast amount and complex structure of data, deep learning with various topologies stems from shallow machine learning models (Goodfellow et al., 2016). For conciseness, this chapter refers to any type of such computer algorithms as machine learning as defined in Janiesch et al. (2021). Machine learning operates by iteratively capturing the patterns on the problem-specific training data and using the acquired knowledge to make predictions based on a new dataset. In general, a model's learning process or training is categorized into supervised, unsupervised, and reinforcement learning.

Here, the description focuses on the supervised learning paradigm used by most tsunami-related studies, meaning that the dataset contains input features and the corresponding output labels. As most studies account for a regression problem, the label is a real or continuous value. In tsunami forecasting cases, the input features usually compose tsunami waveforms (Liu et al., 2021; Makinoshima et al., 2021), including their statistics at observation points (Mulia et al., 2022) or wavefields (Fauzi & Mizutani, 2020b) at present and previous states. Additionally, auxiliary information from geodetic or seismic data can be incorporated (Makinoshima et al., 2021; Rim et al., 2022; Rodríguez et al., 2022). The outputs are crucial tsunami properties representing coastal impacts, such as maximum heights and arrival times along the coasts (Rodríguez et al., 2022), inundation coverage or flow depth (Fauzi & Mizutani, 2020a; Kamiya et al., 2022; Mulia et al., 2022; Mulia, Gusman, et al., 2020), and near-shore or inland waveforms (Liu et al., 2021; Makinoshima et al., 2021; Rim et al., 2022) at specified locations of interest and forecast horizons.

Machine learning generally works as a nonlinear function that maps the input data $\mathbf{x}_n \in R^{in}$ to the desired output $\mathbf{y}_n \in R^{out}$, where *in* and *out* denote the input and output feature dimensions, respectively. The N data points on a training set in the supervised learning setting consist of pairs $\{(\mathbf{x}_n, \mathbf{y}_n)\}_{n=1}^{N}$, assumed to be identically distributed. Another common assumption is that the data points are statistically independent of each other. The input–output mapping can be expressed by $f(\cdot, \theta): R^{in} \to R^{out}$, where f indicates machine learning methods of any type parameterized by $\theta \in R^{par}$. The number of model parameters *par* depends on the selected architectural building blocks. A nonnegative loss function $\ell(\mathbf{y}_n, \hat{\mathbf{y}}_n)$ is defined to fit the data, where $\hat{\mathbf{y}}_n = f(\mathbf{x}_n, \theta)$ is the predicted output. Several formulations of loss functions are commonly used in a regression problem, one of which is the classical squared error, $\ell(\mathbf{y}_n, \hat{\mathbf{y}}_n) = (\mathbf{y}_n - \hat{\mathbf{y}}_n)^2$. The objective of the training or learning process is to minimize the loss function, such that

$$\min_{\theta \in \mathbb{R}^{par}} \ell(\mathbf{y}_n, \hat{\mathbf{y}}_n). \tag{12.1}$$

Note that this chapter is meant to provide a general idea of how machine learning works; readers are referred to a specific textbook (Goodfellow et al., 2016) for more details on the methodology.

The dataset for machine learning is prepared in a similar manner to the traditional precalculated tsunami database approach (Gusman et al., 2014; Mulia et al., 2018). Such methodologies rely on creating a wide range of plausible tsunami scenarios representing the anticipated hazard. Owing to the advanced understanding of tsunamis' dynamics and physical characteristics, state-of-the-art tsunami propagation models can simulate tsunamis realistically (Marras & Mandli, 2021). Similarly, high-fidelity tsunami source mechanisms can be simulated using stochastic earthquake slip models considering the heterogeneous slip distribution (Goda, 2022; Herrero & Murphy, 2018; Mai & Beroza, 2002; Melgar et al., 2016). These simulated data will be captured as knowledge stored in machine learning model parameters upon completion of training. In the testing stage, the model can then predict or forecast tsunami impacts per the intended design using new inputs previously unseen in the training set. The training time varies with the class of machine learning, the amount of data, and computer specifications. On the other hand, the computing time for testing the fully trained model on a single tsunami event usually takes seconds on a standard computer, significantly faster than any physics-based tsunami model. The general workflow of tsunami forecasting using machine learning is illustrated in Fig. 12.1.

12.3 Existing studies

Earlier machine learning-based tsunami forecasting studies used simpler model configurations and architectures (Barman et al., 2006; Hadihardaja et al., 2011). Nevertheless, the basic principle of the methods remains the same as the latest applications, which the above fundamental theoretical framework can explain. In addition to the outdated techniques, these studies typically did not leverage more sophisticated physics-based tsunami models. For instance, Barman et al. (2006) developed a model trained on simulated tsunami travel times using the Huygens principle instead

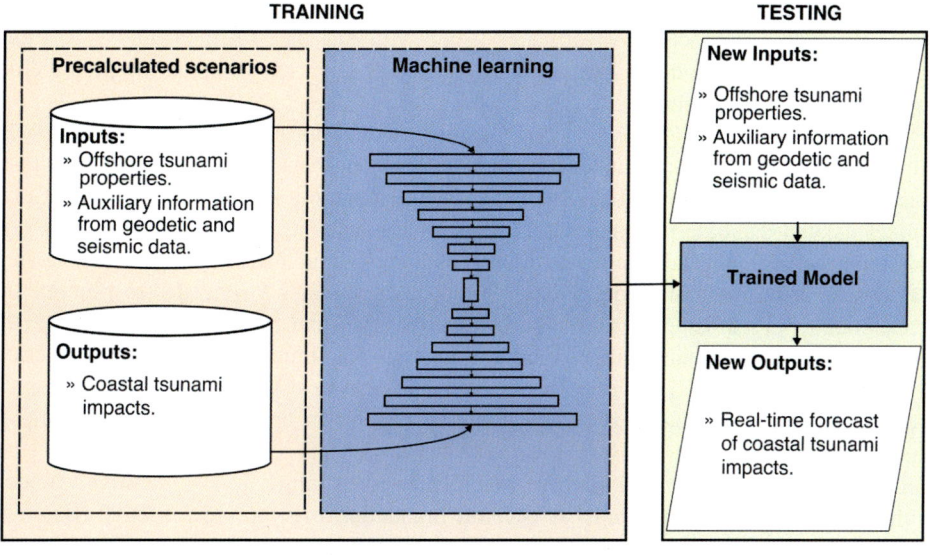

FIGURE 12.1 General workflow of machine learning-based tsunami forecasting. The training stage is conducted *offline* in advance prior to the *online* testing stage in the operational forecasting system.

of a full-physics model. Due to the flexibility of machine learning, the method could be extended to account for the coastal tsunami heights in addition to the travel times if the dataset were generated using a standard tsunami model. Later on, machine learning-based forecasting for both coastal tsunami heights and arrival times was proposed by Hadihardaja et al. (2011). Their method utilized earthquake fault parameters as predictors similar to the first generation of tsunami forecasting based on interpolating precalculated libraries (Tatehata, 1997). However, this approach was subjected to large uncertainties; thus an additional statistical correction was needed in their final result. Recently, Rodríguez et al. (2022) applied the same approach but with a more advanced machine learning model and extensive database.

Other studies provided a better alternate modeling design and dataset quality incorporating the nonlinear characteristics of the tsunami in their machine learning models (Mulia et al., 2016; Namekar et al., 2009). Namekar et al. (2009) used machine learning to translate observed near-source tsunami waveforms into far-field waveforms and runups. They run multiple tsunami simulations derived from nonlinear long-wave equations to generate data for their machine learning. The nonlinear dynamics of the tsunami were well captured by the machine learning known as a universal nonlinear function approximator. However, the method is less compelling now because the current generation of even standard personal computers fairly runs a physics-based tsunami model for far-field events. An application of machine learning-based tsunami forecasting for a near-field event was introduced by Mulia et al. (2016), where the predictions of coastal impacts should be made under stricter time constraints. Their method is, in principle, similar to the inverse approach for tsunami forecasting (Tsushima et al., 2009), but it provides advantages over the conventional linear inversion analysis, imposing the nonlinearity of machine learning. One limitation of this approach is that a large dataset of tsunami Green's functions needs to be loaded into the memory when making on-the-fly predictions.

Recent machine learning-based tsunami forecasting applications employed more advanced algorithms and modeling environments. Using synthetic experiments with more than 900 hypothetical tsunami scenarios, Liu et al. (2021) demonstrated different types of machine learning for forecasting tsunami waveforms at one end of a strait from tsunami data recorded at the strait entrance. Such a coastal geometry allows them to build a forecasting algorithm with a limited observation point. Their machine learning models can forecast 5-hours tsunami waveforms at the specified points using 30- to 60-minutes data from a single observation station. Furthermore, the models were configured to account for the forecast uncertainty, an essential feature in a forecasting system. Subsequently, a follow-up version of the machine learning model was introduced, with inputs based on displacement waveforms measured at global navigation satellite system (GNSS) stations (Rim et al., 2022). The GNSS data from multiple observation points improved the forecast accuracy and significantly shortened the observation window to only 9 minutes. An example of forecast results by Rim et al. (2022) and Liu et al. (2021) is shown in Fig. 12.2A.

A combined use of GNSS and tsunami data for machine learning inputs was proposed by Makinoshima et al. (2021). Their method requires more tsunami observation points due to the nature of the considered coastal areas being directly exposed to the sea, which is not an issue considering the existence of the Seafloor Observation Network for Earthquakes and Tsunamis along the Japan Trench (S-net) in the study area (Aoi et al., 2020). 12,000 hypothetical tsunami scenarios were used to train, validate, and test their machine learning model. The model aimed to forecast tsunami waveforms at a single site inland using offshore tsunami data from 49 S-net stations in conjunction with five onshore GNSS observatories. As the model attained a satisfactory performance using synthetic experiments (Fig. 12.2B), they extended its application to real events of the 2011 Tohoku Tsunami with limited observation points at that time before the S-net establishment, resulting in a reasonable accuracy.

Another tsunami inundation waveform forecasting using machine learning was proposed by Núñez et al. (2022). However, due to the unavailability of dense offshore tsunami observation networks in their study area, virtual observation points were introduced to facilitate the modeling. Previous studies have implemented this technique with traditional best scenario matching from a precalculated database (Gusman et al., 2014; Setiyono et al., 2017). Time series of tsunami waveforms at these virtual stations distributed at specific water depths were considered inputs for their machine learning model. Consequently, a physics-based tsunami model, including a source estimate, was needed to generate the time series at the predefined virtual stations in real-time operational computation. The total computing time to make the prediction is expected to be within a relatively acceptable range because the inputs are simulated on a coarse numerical model grid representation. Overall, the proposed model achieved good accuracy for both synthetic and real event applications. While the additional computing time is probably not ideal, their approach can be beneficial, particularly in regions with limited observational resources.

A tsunami waveform indicates important arrival time and wave height properties, but it is less intuitively understandable to the general public than flood or inundation maps. A tsunami inundation map provides spatial awareness and guidance for evacuees in finding the safest route hence is suited for a warning system. Several studies proposed

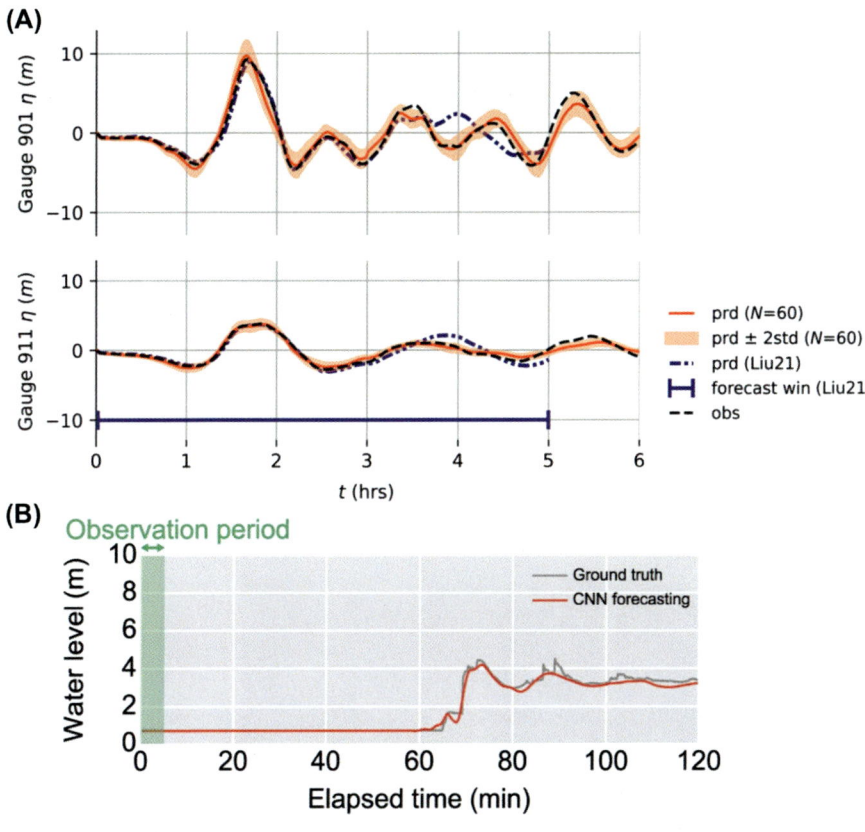

FIGURE 12.2 Tsunami waveforms forecasts by Liu et al. (2021), Rim et al. (2022), and Makinoshima et al. (2021). (A) Tsunami forecasts at gauges 901 and 911 using inputs from 60 GNSS stations by Rim et al. (2022). The figure also compares with the results of Liu et al. (2021). (B) Tsunami forecast at an inland point by Makinoshima et al. (2021) using inputs from the offshore tsunami and onshore GNSS data based on CNN, a type of deep learning algorithm. *CNN*, Convolutional neural networks; *GNSS*, global navigation satellite system. *Modified from Liu et al. (2021), Makinoshima et al. (2021), and Rim et al. (2022).*

using machine learning to produce a real-time tsunami inundation map (Fauzi & Mizutani, 2020a; Mulia, Gusman, et al., 2020). The main premise of their machine learning models is to translate maximum tsunami heights from a low-resolution linear model into a nonlinear high-resolution inundation map. The same idea was used by Giles et al. (2022) to predict local tsunami variability from a computationally cheap regional forecast. Therefore, similar to Núñez et al. (2022), the methods require real-time computation of a linear physics-based model on relatively coarse grids, albeit no virtual observation stations are needed. Fauzi and Mizutani (2020a) and Mulia, Gusman, et al. (2020) trained their machine learning models using 328 and 532 hypothetical tsunami scenarios, respectively. The testing result by Fauzi and Mizutani (2020a) for a hypothetical tsunami event is shown in Fig. 12.3A, and that by Mulia, Gusman, et al. (2020) for a real event of the 2011 Tohoku Tsunami is shown in Fig. 12.3B.

Improving the previous studies (Fauzi & Mizutani, 2020a; Mulia, Gusman, et al., 2020), Mulia et al. (2022) proposed a machine learning-based tsunami inundation prediction derived directly from offshore tsunami data to circumvent the need to infer the tsunami source and numerically simulate the tsunami propagation in real time. This approach is favorable as it can considerably increase the lead time and preclude the uncertainties associated with source estimates, which is also advocated by the previous studies (Liu et al., 2021; Makinoshima et al., 2021; Rim et al., 2022). However, instead of utilizing full tsunami waveforms, they used statistical properties in terms of maximum or mean wave amplitudes at S-net stations as inputs for their machine learning model, similar to Kamiya et al. (2022) but for different observation networks. The single representative value at each observation point is efficient but will likely be effective only with a dense or large number of observation stations. Mulia et al. (2022) considered the training set of more than 3000 hypothetical scenarios originating from megathrust and outer-rise earthquakes. The extensive inundation map coverage is one of the attractive features of their model. Fig. 12.4 shows the predicted inundation map compared to a physics-based simulation reference for seven coastal cities stretching approximately 100 km.

A unique application for forecasting sequences of tsunami wavefields was introduced by Fauzi and Mizutani (2020b), which demonstrates the adaptability of machine learning on various tasks. Their machine learning model takes inputs from assimilated tsunami wavefields at the previous time steps to forecast the future state of wavefields at several forecast horizons. Such an approach can speed up the computation of the common tsunami data assimilation scheme (Gusman et al., 2016; Maeda et al., 2015) without substantially compromising the accuracy. Using only 90 hypothetical

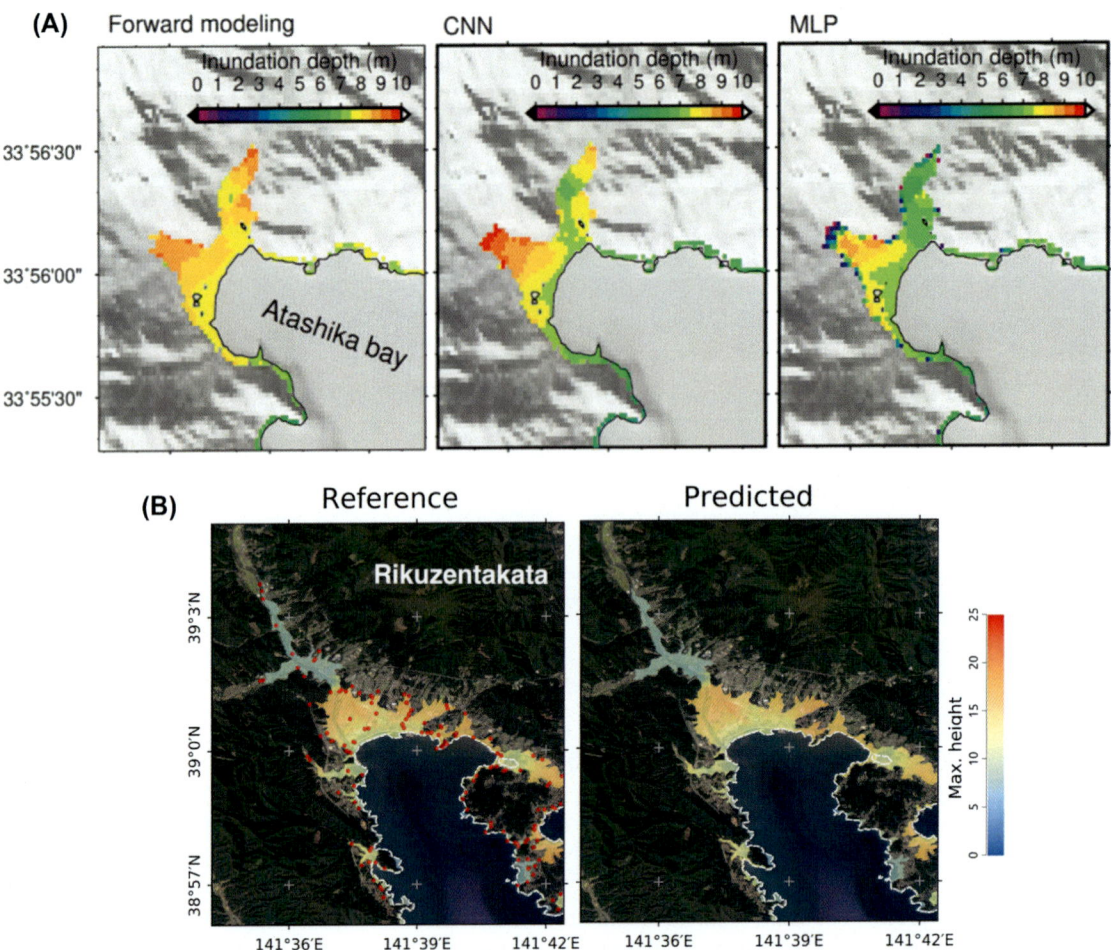

FIGURE 12.3 Tsunami inundation forecasts by Fauzi and Mizutani (2020a) and Mulia, Gusman, et al. (2020), Mulia, Hirobe, et al. (2020), Mulia, Watada, et al. (2020). (A) Comparison of tsunami inundation maps produced using a physics-based (forward modeling), CNN, and MLP by Fauzi and Mizutani (2020a). (B) Comparison of tsunami inundation map produced using a physics-based tsunami model (reference) and machine learning (predicted) by Mulia, Gusman, et al. (2020), Mulia, Hirobe, et al. (2020), Mulia, Watada, et al. (2020). Red dots in the left panel mark locations of inundation measurements. *MLP*, Multilayer perceptron. *Modified from Fauzi and Mizutani (2020a) and Mulia, Gusman, et al. (2020), Mulia, Hirobe, et al. (2020), Mulia, Watada, et al. (2020).*

scenarios as a training dataset, the proposed machine learning model shows comparable performance to the standard data assimilation method and forward tsunami model for a single tsunami event (Fig. 12.5). Notwithstanding the satisfactory performance, the model is currently applicable only for forecasting linear tsunami characteristics in the open ocean. Further enhancements are needed to apply the method to a more complex coastal environment.

12.4 Advantages, limitations, and future direction

Tsunami forecasting studies advocating machine learning have significantly grown in recent years. Typically, their performance refinements align with the expansion of state-of-the-art machine learning algorithms invented under the proliferation of data science. Machine learning-based tsunami forecasting offers several advantages, mainly of practical relevance, but it is subjected to limitations. Here, the advantages and limitations of machine learning-based tsunami forecasting are discussed, aiming to define future research directions and developments toward the actual implementation in the operational systems.

Computational efficiency is the main attractive feature of machine learning, facilitating rapid real-time tsunami forecasts and a cost-effective early warning. The remarkable computing speed of machine learning is accredited to its ability to model various forms of tsunami properties (e.g., wavefields, waveforms, or runup and inundation) without calculating the entire state variables of the tsunami dynamical system, as that is requisite for physics-based models.

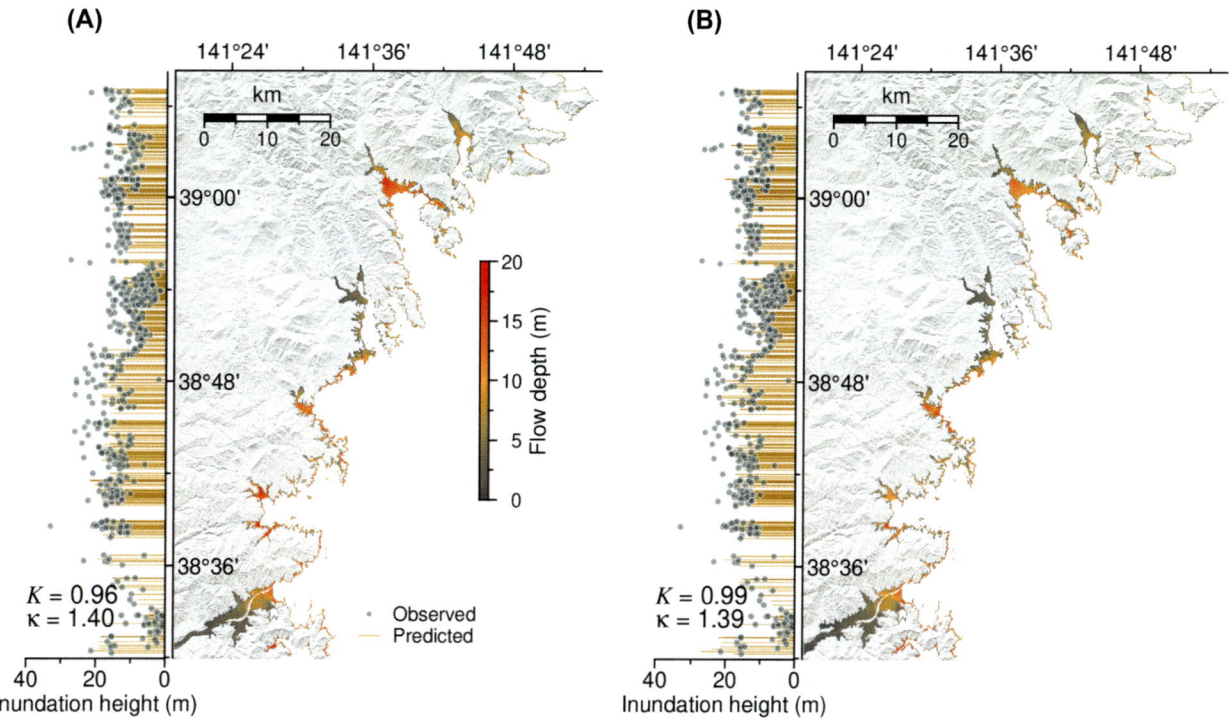

FIGURE 12.4 The forecasted inundation flow depth map is shown in the right panels, and the comparison between observed and predicted inundation heights is shown in the left panels, including accuracy measures of K and κ (Mulia et al., 2022). (A) Physics-based model. (B) Machine learning-based model. *Modified from Mulia, I. E., Ueda, N., Miyoshi, T., Gusman, A. R., & Satake, K. (2022). Machine learning-based tsunami inundation prediction derived from offshore observations.* Nature Communications, *13(1). http://www.nature.com/ncomms/index.html. https://doi.org/10.1038/s41467-022-33253-5.*

This notion is attainable because a machine learning model is constructed merely upon statistical correlations between input and output regardless of causality. The distribution of samples on the dataset used to train and test a machine learning model plays a more important role than their physical significance. Furthermore, the easily parallelizable algorithm, as machine learning consists of a parallel-distributed information repository and processing system, is an essential attribute that expedites information processing tasks. This trait can be well optimized by accelerated solutions using graphical processing units (Steinkraus et al., 2005), a relatively inexpensive computational framework suitable for parallel computing compared to its central processing units counterpart.

Another advantage of machine learning is the versatility of its architecture, allowing for the inclusion of multiple datasets by which the inference is made straightforwardly. Machine learning models can easily account for fusing seismic, geodetic, and tsunami data considered primary predictors for tsunami characteristics. Even though, generally, machine learning models have no physical meaning, incorporating more physically relevant variables as inputs improves their prediction accuracy (Makinoshima et al., 2021). Current physical theory and historical accounts suggest that tsunami occurrences can be traced from those multiple observational data sources. However, assimilating all the related datasets or developing a comprehensive full-physics simulation via coupled modeling can be costly. Also, while scientifically pertinent, such a model will incur extra computational efforts diminishing its practicality. For that reason, machine learning can be a worthy substitute for physics-based models as a practical operational tsunami forecasting system.

Machine learning can also be advantageous for uncertainty analysis. Real-time uncertainty quantifications should be provided explicitly as part of tsunami forecast products (Selva et al., 2021). The estimated uncertainty can then be translated into more reliable alert levels, augmenting decision-making processes. One way to achieve such quantifications is through ensemble modeling typically conducted in numerical weather prediction (Hawcroft et al., 2021; Mamgain et al., 2020). The computational efficiency of machine learning allows for tractable computation of ensemble modeling in an operational tsunami forecasting system. Mulia et al. (2022) demonstrated that 100 ensemble predictions of tsunami inundation from the perturbed input variables could be made using a standard personal computer within seconds. Furthermore, particular machine learning models possess a specific feature for ease of treatment of uncertainty, which

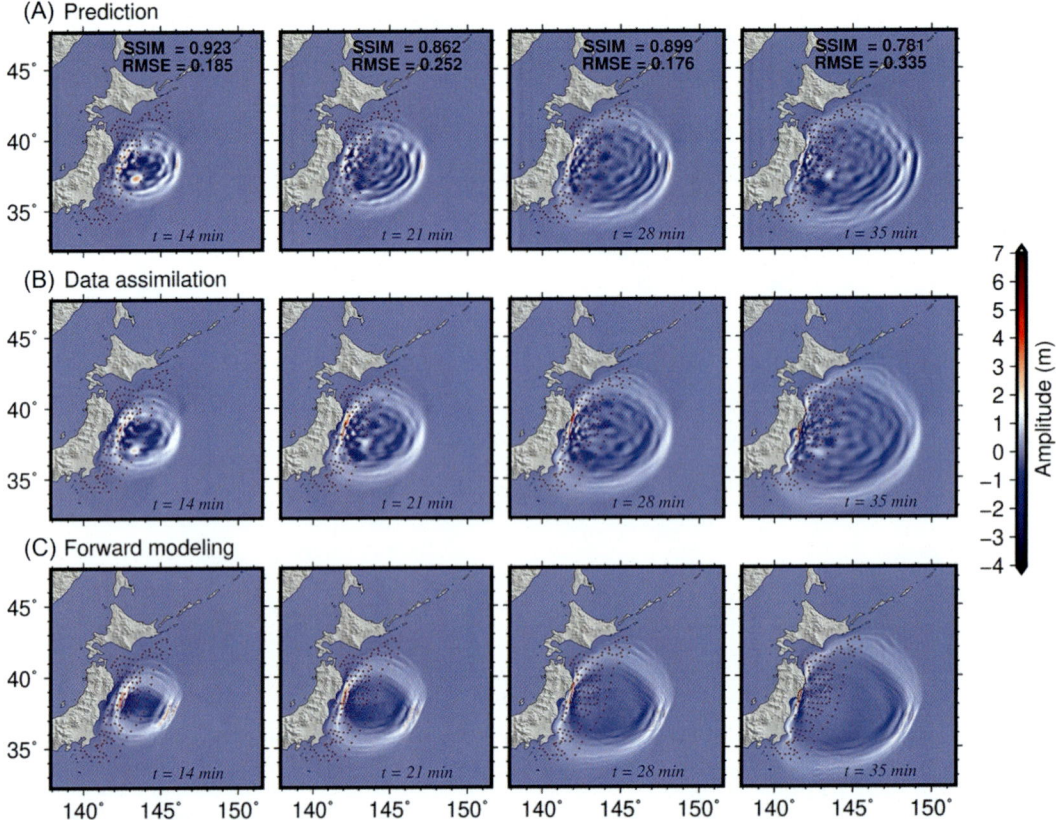

FIGURE 12.5 Comparison of estimated wavefields from the proposed machine learning method (A), data assimilation (B), and forward modeling (C) by Fauzi and Mizutani (2020b). The predictions are evaluated relative to the data assimilation results using a SSIM and a RMSE. *RMSE*, Root mean square error; *SSIM*, structural similarity index measurement. *From Fauzi, A., & Mizutani, N. (2020b). Potential of deep predictive coding networks for spatiotemporal tsunami wavefield prediction. Geoscience Letters, 7(1). http://geoscienceletters.springeropen.com/. https://doi.org/10.1186/s40562-020-00169-1.*

has been implemented in the tsunami case (Liu et al., 2021). Therefore, machine learning enables straightforward and efficient uncertainty quantification in tsunami forecasting, which can be extended further to account for the total uncertainty attributed to tsunami predictions.

Despite the above advantages, a longstanding issue regarding the black-box nature of machine learning remains, which renders the model interpretability difficult, especially for those with more complex architectures. This limitation can, to some extent, undermine the reliability of machine learning predictions. Efforts have been made to uncover factors underlying model outcomes through explainable machine learning algorithms (Ras et al., 2022; Subel et al., 2023). A machine learning model can also be trained while conforming to the law of physics governed by nonlinear partial differential equations (Raissi et al., 2019). The innovative strategy to combine mathematical models with data in the form of physics-informed machine learning addresses the interpretability problem considerably. A recent study has successfully applied such machine learning for modeling crustal deformation caused by earthquakes (Okazaki et al., 2022). The same approach can be adopted for tsunami modeling to make the prediction more discernible. However, the contribution may not be as significant from a practical application standpoint.

Machine learning limitation that can affect its practical use is related to the susceptibility of the algorithm to data errors. For instance, Mulia et al. (2022) showed that prediction accuracy is degraded when a considerable number of input data becomes corrupted due to observational errors or unavailable caused by device malfunctioning. Such a condition causes the inconsistency between training and test set distributions, possibly resulting in incoherent results. Machine learning-based tsunami forecasting that utilizes observation data directly as inputs (Liu et al., 2021; Makinoshima et al., 2021; Mulia et al., 2022; Rim et al., 2022) can suffer from data errors. Therefore, proper data quality assurance is needed, which can be implemented using automated control as in seismic (Aoi et al., 2020) or atmospheric (Bäserud et al., 2020) observations. Contrarily, machine learning models with inputs from simulations (Fauzi & Mizutani, 2020a; Mulia, Gusman, et al., 2020; Núñez et al., 2022) are less likely influenced by the same issue.

Observational errors may affect the source model in the standard tsunami simulation without a data assimilation scheme, thus eventually affecting the machine learning prediction due to the propagated errors. However, they can be relatively easier to resolve than in machine learning models with observation-based input.

Although the number of tsunami observation networks has increased following several major events in the past decades, they are only well-distributed in a few developed countries. Therefore, machine learning models designed to work under limited observational resources serve the remaining tsunami-prone regions. Further developments of this type of machine learning methods can lead to a more global impact. The only drawback of these models is the need to run the numerical simulation in the operational mode incurring additional real-time computation. One potential solution is to employ the linear superposition of tsunami Green's functions (Mulia & Satake, 2021) to characterize the offshore tsunami as input features for forecasting inland or coastal impacts by machine learning. As discussed above, this approach requires extra storage for the precomputed synthetic waveforms. Nonetheless, investing in storage is probably more affordable than computing power. Furthermore, additional data from geodetic, seismic, or other related observations are necessary to compensate for the limited offshore tsunami observing systems. Thus developing machine learning models with combined inputs from various data sources should be regarded as one of the essential directions for future studies.

Despite the scarcity of observatories, machine learning models with input features from real-time tsunami observations are preferable, as tsunami data give viable information on the occurrence of tsunamis. In particular, tsunami data provide a more accurate reading to estimate the exacerbated impacts caused by anomalous events, such as tsunami earthquakes (Kanamori, 1972). Several studies have proposed the deployment of offshore tsunami observation networks in many tsunami-prone countries, for example, south of Java of Indonesia (Mulia et al., 2019), north of Chile (Meza et al., 2020; Navarrete et al., 2020), and the Mediterranean Sea (Heidarzadeh et al., 2019; Wang et al., 2020). Machine learning-based tsunami forecasting can leverage these prospective tsunami observing systems. Additionally, unconventional platforms for observing tsunamis have been proposed using: undersea telecommunications cables (Howe et al., 2019; Salaree et al., 2022), radar on aircraft (Hirobe et al., 2019; Mulia, Hirobe, et al., 2020) or coastlines (Dzvonkovskaya et al., 2018; Mulia, Watada, et al., 2020; Wang et al., 2023), and GNSS receiver on vessels (Inazu et al., 2016; Mulia et al., 2017; Mulia, Watada, et al., 2020). As demonstrated by the existing studies, the machine learning model versatility can take advantage of these atypical tsunami observing platforms, thus leading toward a global and sustainable tsunami early warning system.

Furthermore, an on-demand computing environment has been proposed for cost-effective tsunami early warning (Behrens et al., 2022). The study comprehensively reviewed the benefits of cloud computing against traditional on-premise systems. The on-demand scheme is convenient for tsunami early warning operated occasionally during infrequent events, and the online nature of the systems equips for warning dissemination. Moreover, major cloud computing providers, such as Amazon, Google Cloud, and Microsoft Azure, offer specific infrastructure for training and deploying machine learning models at reasonable rates. These facilities will ensure that machine learning applications for tsunami forecasting become accessible worldwide. However, a potential issue associated with cloud computing is often caused by data scalability. Sending a large amount of data to the centralized cloud can lead to a bottleneck, inflicting latency. To circumvent the problem, edge computing technology can be a viable option (Chen & Ran, 2019). Operational machine learning models can be executed on edge devices or servers closer to the data source. Nowadays, tsunami observing systems are based on IoT frameworks (Esposito et al., 2022). Being an IoT sensor or edge device, most tsunami observation networks are compatible with the edge computing ecosystem.

As reviewed in this chapter, recent developments in machine learning-based tsunami forecasting have shown promising results. Although still in the early stage, current results on machine learning applications in tsunami forecasting are encouraging, as sustaining an accurate and timely tsunami early warning system remains challenging for scientific communities and practitioners. While some recent developments are arguably incremental, underpinned by modern geophysical instrumentations, computational advances, and fast communication networks, machine learning can be a reliable tool. However, as with all tools, machine learning is also subjected to limitations commonly encountered in their applications. Further developments in machine learning are envisaged to bring considerable advantages that can outweigh their intrinsic constraints. Addressing this issue will push machine learning closer to being readily implemented in operational tsunami forecasting systems, thus eventually enhancing tsunami early warning systems' capability.

References

Aoi, S., Asano, Y., Kunugi, T., Kimura, T., Uehira, K., Takahashi, N., Ueda, H., Shiomi, K., Matsumoto, T., & Fujiwara, H. (2020). MOWLAS: NIED observation network for earthquake, tsunami and volcano. *Earth, Planets and Space*, 72(1). Available from https://doi.org/10.1186/s40623-020-01250-x, http://rd.springer.com/journal/40623.

Barman, R., Prasad Kumar, B., Pandey, P. C., & Dube, S. K. (2006). Tsunami travel time prediction using neural networks. *Geophysical Research Letters*, *33*(16). Available from https://doi.org/10.1029/2006GL026688, http://onlinelibrary.wiley.com/journal/10.1002/(ISSN)1944-8007/issues?year = 2012.

Behrens, J., Schulz, A., & Simon, K. (2022). Performance assessment of the cloud for prototypical instant computing approaches in geoscientific hazard simulations. *Frontiers in Earth Science*, *10*. Available from https://doi.org/10.3389/feart.2022.762768, https://www.frontiersin.org/journals/earth-science.

Bernard, E., & Titov, V. (2015). Evolution of tsunami warning systems and products. *Philosophical Transactions of the Royal Society A: Mathematical, Physical and Engineering Sciences*, *373*(2053), 20140371. Available from https://doi.org/10.1098/rsta.2014.0371.

Bäserud, L., Lussana, C., Nipen, T. N., Seierstad, I. A., Oram, L., & Aspelien, T. (2020). TITAN automatic spatial quality control of meteorological in-situ observations. *Advances in Science and Research*, *17*, 153−163. Available from https://doi.org/10.5194/asr-17-153-2020, http://www.advances-in-science-and-research.net/index.html.

Chen, J., & Ran, X. (2019). Deep learning with edge computing: A review. *Proceedings of the IEEE*, *107*(8), 1655−1674. Available from https://doi.org/10.1109/jproc.2019.2921977.

Dzvonkovskaya, A., Petersen, L., Helzel, T., & Kniephoff, M. (2018). High-frequency ocean radar support for Tsunami Early Warning Systems. *Geoscience Letters*, *5*(1). Available from https://doi.org/10.1186/s40562-018-0128-5, http://geoscienceletters.springeropen.com/.

Emmert-Streib, F., Yang, Z., Feng, H., Tripathi, S., & Dehmer, M. (2020). An introductory review of deep learning for prediction models with big data. *Frontiers in Artificial Intelligence*, *3*. Available from https://doi.org/10.3389/frai.2020.00004, http://www.frontiersin.org/journals/artificial-intelligence#.

Espeholt, L., Agrawal, S., Sønderby, C., Kumar, M., Heek, J., Bromberg, C., Gazen, C., Carver, R., Andrychowicz, M., Hickey, J., Bell, A., & Kalchbrenner, N. (2022). Deep learning for twelve hour precipitation forecasts. *Nature Communications*, *13*(1). Available from https://doi.org/10.1038/s41467-022-32483-x, http://www.nature.com/ncomms/index.html.

Esposito, M., Palma, L., Belli, A., Sabbatini, L., & Pierleoni, P. (2022). Recent advances in internet of things solutions for early warning systems: A review. *Sensors*, *22*(6), 2124. Available from https://doi.org/10.3390/s22062124.

Fauzi, A., & Mizutani, N. (2020a). Machine learning algorithms for real-time tsunami inundation forecasting: A case study in Nankai Region. *Pure and Applied Geophysics*, *177*(3), 1437−1450. Available from https://doi.org/10.1007/s00024-019-02364-4, http://www.springer.com/birkhauser/geo + science/journal/24.

Fauzi, A., & Mizutani, N. (2020b). Potential of deep predictive coding networks for spatiotemporal tsunami wavefield prediction. *Geoscience Letters*, *7*(1). Available from https://doi.org/10.1186/s40562-020-00169-1, http://geoscienceletters.springeropen.com/.

Giles, D., Gailler, A., & Dias, F. (2022). Automated approaches for capturing localized tsunami response—Application to the French Coastlines. *Journal of Geophysical Research: Oceans*, *127*(6). Available from https://doi.org/10.1029/2022JC018467, http://agupubs.onlinelibrary.wiley.com/agu/jgr/journal/10.1002/(ISSN)2169-9291/.

Goda, K. (2022). Stochastic source modeling and tsunami simulations of cascadia subduction earthquakes for Canadian Pacific coast. *Coastal Engineering Journal*, *64*(4), 575−596. Available from https://doi.org/10.1080/21664250.2022.2139918, https://www.tandfonline.com/loi/tcej20.

González, F. I., Bernard, E. N., Meinig, C., Eble, M. C., Mofjeld, H. O., & Stalin, S. (2005). The NTHMP tsunameter network. *Natural Hazards*, *35*(1), 25−39. Available from https://doi.org/10.1007/s11069-004-2402-4.

Goodfellow, I., Bengio, Y., & Courville, A. (2016). *Deep learning*. MIT Press.

Gusman, A. R., Sheehan, A. F., Satake, K., Heidarzadeh, M., Mulia, I. E., & Maeda, T. (2016). Tsunami data assimilation of Cascadia seafloor pressure gauge records from the 2012 Haida Gwaii earthquake. *Geophysical Research Letters*, *43*(9), 4189−4196. Available from https://doi.org/10.1002/2016GL068368, http://onlinelibrary.wiley.com/journal/10.1002/(ISSN)1944-8007/issues?year = 2012.

Gusman, A. R., Tanioka, Y., Macinnes, B. T., & Tsushima, H. (2014). A methodology for near-field tsunami inundation forecasting: Application to the 2011 Tohoku tsunami. *Journal of Geophysical Research: Solid Earth*, *119*(11), 8186−8206. Available from https://doi.org/10.1002/2014JB010958, http://onlinelibrary.wiley.com/journal/10.1002/(ISSN)2169-9356.

Hadihardaja, I. K., Latief, H., & Mulia, I. E. (2011). Decision support system for predicting tsunami characteristics along coastline areas based on database modelling development. *Journal of Hydroinformatics*, *13*(1), 96−109. Available from https://doi.org/10.2166/hydro.2010.001, http://www.iwaponline.com/jh/013/0096/0130096.pdf.

Hawcroft, M., Lavender, S., Copsey, D., Milton, S., Rodríguez, J., Tennant, W., Webster, S., & Cowan, T. (2021). The benefits of Ensemble prediction for forecasting an extreme event: The Queensland Floods of February 2019. *Monthly Weather Review*, *149*(7), 2391−2408. Available from https://doi.org/10.1175/MWR-D-20-0330.1, https://journals.ametsoc.org/view/journals/mwre/149/7/MWR-D-20-0330.1.xml.

Heidarzadeh, M., Wang, Y., Satake, K., & Mulia, I. E. (2019). Potential deployment of offshore bottom pressure gauges and adoption of data assimilation for tsunami warning system in the western Mediterranean Sea. *Geoscience Letters*, *6*(1). Available from https://doi.org/10.1186/s40562-019-0149-8, http://geoscienceletters.springeropen.com/.

Herrero, A., & Murphy, S. (2018). Self-similar slip distributions on irregular shaped faults. *Geophysical Journal International*, *213*(3), 2060−2070. Available from https://doi.org/10.1093/gji/ggy104, http://gji.oxfordjournals.org/.

Hirobe, T., Niwa, Y., Endoh, T., Mulia, I. E., Inazu, D., Yoshida, T., Tatehata, H., Nadai, A., Waseda, T., & Hibiya, T. (2019). Observation of sea surface height using airborne radar altimetry: A new approach for large offshore tsunami detection. *Journal of Oceanography*, *75*(6), 541−558. Available from https://doi.org/10.1007/s10872-019-00521-w, http://www.springer.com/earth + sciences + and + geography/oceanography/journal/10872.

Hossen, M. J., Mulia, I. E., Mencin, D., & Sheehan, A. F. (2021). Data assimilation for tsunami forecast with ship-borne GNSS data in the Cascadia subduction zone. *Earth and Space Science*, *8*(3). Available from https://doi.org/10.1029/2020EA001390, http://agupubs.onlinelibrary.wiley.com/hub/journal/10.1002/(ISSN)2333-5084/.

Howe, B. M., Arbic, B. K., Aucan, J., Barnes, C., Bayliff, N., Becker, N., Butler, R., Doyle, L., Elipot, S., Johnson, G. C., Landerer, F., Lentz, S., Luther, D. S., Müller, M., Mariano, J., Panayotou, K., Rowe, C., Scholl, R., Ota, H., ... Weinstein, S. (2019). Smart cables for observing the global ocean: Science and implementation. *Frontiers in Marine Science*, *6*. Available from https://doi.org/10.3389/fmars.2019.00424, https://www.frontiersin.org/articles/10.3389/fmars.2019.00424.

Inazu, D., Waseda, T., Hibiya, T., & Ohta, Y. (2016). Assessment of GNSS-based height data of multiple ships for measuring and forecasting great tsunamis. *Geoscience Letters*, *3*(1). Available from https://doi.org/10.1186/s40562-016-0059-y, http://geoscienceletters.springeropen.com/.

Janiesch, C., Zschech, P., & Heinrich, K. (2021). Machine learning and deep learning. *Electronic Markets*, *31*(3), 685–695. Available from https://doi.org/10.1007/s12525-021-00475-2, http://www.springer.com/business/business+information+systems/journal/12525?detailsPage=aimsAndScopes.

Kamiya, M., Igarashi, Y., Okada, M., & Baba, T. (2022). Numerical experiments on tsunami flow depth prediction for clustered areas using regression and machine learning models. *Earth, Planets and Space*, *74*(1). Available from https://doi.org/10.1186/s40623-022-01680-9, http://rd.springer.com/journal/40623.

Kanamori, H. (1972). Mechanism of tsunami earthquakes. *Physics of the Earth and Planetary Interiors*, *6*(5), 346–359. Available from https://doi.org/10.1016/0031-9201(72)90058-1.

Kaneda, Y., Kawaguchi, K., Araki, E., Matsumoto, H., Nakamura, T., Kamiya, S., Ariyoshi, K., Hori, T., Baba, T., & Takahashi, N. (2015). Development and application of an advanced ocean floor network system for megathrust earthquakes and tsunamis. *Seafloor Observatories: A New Vision of the Earth from the Abyss* (pp. 643–662). Springer. Available from http://www.dx.doi.org/10.1007/978-3-642-11374-1, 10.1007/978-3-642-11374-1_25.

Lee, J. W., Irish, J. L., Bensi, M. T., & Marcy, D. C. (2021). Rapid prediction of peak storm surge from tropical cyclone track time series using machine learning. *Coastal Engineering*, *170*. Available from https://doi.org/10.1016/j.coastaleng.2021.104024, http://www.elsevier.com/inca/publications/store/5/0/3/3/2/5/.

Liu, C. M., Rim, D., Baraldi, R., & LeVeque, R. J. (2021). Comparison of machine learning approaches for tsunami forecasting from sparse observations. *Pure and Applied Geophysics*, *178*(12), 5129–5153. Available from https://doi.org/10.1007/s00024-021-02841-9, http://www.springer.com/birkhauser/geo+science/journal/24.

Maeda, T., Obara, K., Shinohara, M., Kanazawa, T., & Uehira, K. (2015). Successive estimation of a tsunami wavefield without earthquake source data: A data assimilation approach toward real-time tsunami forecasting. *Geophysical Research Letters*, *42*(19), 7923–7932. Available from https://doi.org/10.1002/2015GL065588, http://onlinelibrary.wiley.com/journal/10.1002/(ISSN)1944-8007/issues?year=2012.

Mai, P. M., & Beroza, G. C. (2002). A spatial random field model to characterize complexity in earthquake slip. *Journal of Geophysical Research: Solid Earth*, *107*(B11). Available from https://doi.org/10.1029/2001jb000588, ESE 10-1-ESE 10-21.

Makinoshima, F., Oishi, Y., Yamazaki, T., Furumura, T., & Imamura, F. (2021). Early forecasting of tsunami inundation from tsunami and geodetic observation data with convolutional neural networks. *Nature Communications*, *12*(1). Available from https://doi.org/10.1038/s41467-021-22348-0, http://www.nature.com/ncomms/index.html.

Mamgain, A., Sarkar, A., & Rajagopal, E. N. (2020). Medium-range global ensemble prediction system at 12 km horizontal resolution and its preliminary validation. *Meteorological Applications*, *27*(1). Available from https://doi.org/10.1002/met.1867, http://onlinelibrary.wiley.com/journal/10.1002/(ISSN)1469-8080.

Marras, S., & Mandli, K. T. (2021). Modeling and simulation of tsunami impact: A short review of recent advances and future challenges. *Geosciences*, *11*(1), 5. Available from https://doi.org/10.3390/geosciences11010005.

Melgar, D., LeVeque, R. J., Dreger, D. S., & Allen, R. M. (2016). Kinematic rupture scenarios and synthetic displacement data: An example application to the Cascadia subduction zone. *Journal of Geophysical Research: Solid Earth*, *121*(9), 6658–6674. Available from https://doi.org/10.1002/2016JB013314, http://onlinelibrary.wiley.com/journal/10.1002/(ISSN)2169-9356.

Meza, J., Catalán, P. A., & Tsushima, H. (2020). A multiple-parameter methodology for placement of tsunami sensor networks. *Pure and Applied Geophysics*, *177*(3), 1451–1470. Available from https://doi.org/10.1007/s00024-019-02381-3, http://www.springer.com/birkhauser/geo+science/journal/24.

Mousavi, S. M., Ellsworth, W. L., Zhu, W., Chuang, L. Y., & Beroza, G. C. (2020). Earthquake transformer—An attentive deep-learning model for simultaneous earthquake detection and phase picking. *Nature Communications*, *11*(1). Available from https://doi.org/10.1038/s41467-020-17591-w, http://www.nature.com/ncomms/index.html.

Mulia, I. E., Asano, T., & Nagayama, A. (2016). Real-time forecasting of near-field tsunami waveforms at coastal areas using a regularized extreme learning machine. *Coastal Engineering*, *109*, 1–8. Available from https://doi.org/10.1016/j.coastaleng.2015.11.010, http://www.elsevier.com/inca/publications/store/5/0/3/3/2/5/.

Mulia, I. E., Gusman, A. R., & Satake, K. (2018). Alternative to non-linear model for simulating tsunami inundation in real-time. *Geophysical Journal International*, *214*(3), 2002–2013. Available from https://doi.org/10.1093/gji/ggy238.

Mulia, I. E., Gusman, A. R., & Satake, K. (2020). Applying a deep learning algorithm to tsunami inundation database of megathrust earthquakes. *Journal of Geophysical Research: Solid Earth*, *125*(9). Available from https://doi.org/10.1029/2020JB019690, http://agupubs.onlinelibrary.wiley.com/hub/jgr/journal/10.1002/(ISSN)2169-9356/.

Mulia, I. E., Gusman, A. R., Williamson, A. L., & Satake, K. (2019). An optimized array configuration of tsunami observation network off Southern Java, Indonesia. *Journal of Geophysical Research: Solid Earth*, *124*(9), 9622–9637. Available from https://doi.org/10.1029/2019JB017600, http://agupubs.onlinelibrary.wiley.com/hub/jgr/journal/10.1002/(ISSN)2169-9356/.

Mulia, I. E., Hirobe, T., Inazu, D., Endoh, T., Niwa, Y., Gusman, A. R., Tatehata, H., Waseda, T., & Hibiya, T. (2020). Advanced tsunami detection and forecasting by radar on unconventional airborne observing platforms. *Scientific Reports*, *10*(1). Available from https://doi.org/10.1038/s41598-020-59239-1, http://www.nature.com/srep/index.html.

Mulia, I. E., Inazu, D., Waseda, T., & Gusman, A. R. (2017). Preparing for the future Nankai Trough Tsunami: A data assimilation and inversion analysis from various observational systems. *Journal of Geophysical Research: Oceans*, *122*(10), 7924−7937. Available from https://doi.org/10.1002/2017JC012695, http://onlinelibrary.wiley.com/journal/10.1002/(ISSN)2169-9291.

Mulia, I. E., & Satake, K. (2021). Synthetic analysis of the efficacy of the S-net system in tsunami forecasting. *Earth, Planets and Space*, *73*(1). Available from https://doi.org/10.1186/s40623-021-01368-6, http://rd.springer.com/journal/40623.

Mulia, I. E., Ueda, N., Miyoshi, T., Gusman, A. R., & Satake, K. (2022). Machine learning-based tsunami inundation prediction derived from offshore observations. *Nature Communications*, *13*(1). Available from https://doi.org/10.1038/s41467-022-33253-5, http://www.nature.com/ncomms/index.html.

Mulia, I. E., Watada, S., Ho, T. C., Satake, K., Wang, Y., & Aditiya, A. (2020). Simulation of the 2018 tsunami due to the flank failure of Anak Krakatau Volcano and implication for future observing systems. *Geophysical Research Letters*, *47*(14). Available from https://doi.org/10.1029/2020GL087334, http://agupubs.onlinelibrary.wiley.com/hub/journal/10.1002/(ISSN)1944-8007/.

Namekar, S., Yamazaki, Y., & Cheung, K. F. (2009). Neural network for tsunami and runup forecast. *Geophysical Research Letters*, *36*(8). Available from https://doi.org/10.1029/2009GL037184.

Navarrete, P., Cienfuegos, R., Satake, K., Wang, Y., Urrutia, A., Benavente, R., Catalán, P. A., Crempien, J., & Mulia, I. (2020). Sea surface network optimization for tsunami forecasting in the near field: Application to the 2015 Illapel earthquake. *Geophysical Journal International*, *221*(3), 1640−1650. Available from https://doi.org/10.1093/gji/ggaa098, http://gji.oxfordjournals.org/.

Núñez, J., Catalán, P. A., Valle, C., Zamora, N., & Valderrama, A. (2022). Discriminating the occurrence of inundation in tsunami early warning with one-dimensional convolutional neural networks. *Scientific Reports*, *12*(1). Available from https://doi.org/10.1038/s41598-022-13788-9, http://www.nature.com/srep/index.html.

Okazaki, T., Ito, T., Hirahara, K., & Ueda, N. (2022). Physics-informed deep learning approach for modeling crustal deformation. *Nature Communications*, *13*(1). Available from https://doi.org/10.1038/s41467-022-34922-1, https://www.nature.com/ncomms/.

Raissi, M., Perdikaris, P., & Karniadakis, G. E. (2019). Physics-informed neural networks: A deep learning framework for solving forward and inverse problems involving nonlinear partial differential equations. *Journal of Computational Physics*, *378*, 686−707. Available from https://doi.org/10.1016/j.jcp.2018.10.045, http://www.journals.elsevier.com/journal-of-computational-physics/.

Ras, G., Xie, N., Van Gerven, M., & Doran, D. (2022). Explainable deep learning: A field guide for the uninitiated. *Journal of Artificial Intelligence Research*, *73*, 329−397. Available from https://doi.org/10.1613/jair.1.13200.

Rim, D., Baraldi, R., Liu, C. M., LeVeque, R. J., & Terada, K. (2022). Tsunami early warning from global navigation satellite system data using convolutional neural networks. *Geophysical Research Letters*, *49*(20). Available from https://doi.org/10.1029/2022GL099511, http://agupubs.onlinelibrary.wiley.com/hub/journal/10.1002/(ISSN)1944-8007/.

Rodríguez, J. F., Macías, J., Castro, M. J., de la Asunción, M., & Sánchez-Linares, C. (2022). Use of neural networks for tsunami maximum height and arrival time predictions. *GeoHazards*, *3*(2), 323−344. Available from https://doi.org/10.3390/geohazards3020017.

Salaree, A., Howe, B. M., Huang, Y., Weinstein, S. A., & Sakya, A. E. (2022). A numerical study of SMART cables potential in marine hazard early warning for the Sumatra and Java Regions. *Pure and Applied Geophysics*. Available from https://doi.org/10.1007/s00024-022-03004-0, http://www.springer.com/birkhauser/geo + science/journal/24.

Sarker, I. H. (2021). Machine learning: Algorithms, real-world applications and research directions. *SN Computer Science*, *2*(3). Available from https://doi.org/10.1007/s42979-021-00592-x, https://www.springer.com/journal/42979.

Selva, J., Lorito, S., Volpe, M., Romano, F., Tonini, R., Perfetti, P., Bernardi, F., Taroni, M., Scala, A., Babeyko, A., Løvholt, F., Gibbons, S. J., Macías, J., Castro, M. J., González-Vida, J. M., Sánchez-Linares, C., Bayraktar, H. B., Basili, R., Maesano, F. E., ... Amato, A. (2021). Probabilistic tsunami forecasting for early warning. *Nature Communications*, *12*(1). Available from https://doi.org/10.1038/s41467-021-25815-w, http://www.nature.com/ncomms/index.html.

Setiyono, U., Gusman, A. R., Satake, K., & Fujii, Y. (2017). Pre-computed tsunami inundation database and forecast simulation in Pelabuhan Ratu, Indonesia. *Pure and Applied Geophysics*, *174*(8), 3219−3235. Available from https://doi.org/10.1007/s00024-017-1633-8, http://www.springer.com/birkhauser/geo + science/journal/24.

Steinkraus, D., & Buck, I. (2005). Using GPUs for machine learning algorithms. In *Proceedings of the International Conference on Document Analysis and Recognition, ICDAR*, (pp. 1115−1120). Available from https://doi.org/10.1109/ICDAR.2005.251.

Subel, A., Guan, Y., Chattopadhyay, A., Hassanzadeh, P., & Yortsos, Y. (2023). Explaining the physics of transfer learning in data-driven turbulence modeling. *PNAS Nexus*, *2*(3). Available from https://doi.org/10.1093/pnasnexus/pgad015.

Sugawara, D. (2021). Numerical modeling of tsunami: Advances and future challenges after the 2011 Tohoku earthquake and tsunami. *Earth-Science Reviews*, *214*, 103498. Available from https://doi.org/10.1016/j.earscirev.2020.103498.

Tang, L., Titov, V. V., Moore, C., & Wei, Y. (2016). Real-time assessment of the 16 September 2015 Chile tsunami and implications for near-field forecast. *Pure and Applied Geophysics*, *173*(2), 369−387. Available from https://doi.org/10.1007/s00024-015-1226-3, http://www.springer.com/birkhauser/geo + science/journal/24.

Tatehata, H. (1997). *The new tsunami warning system of the Japan Meteorological Agency* (pp. 175−188). Springer Science and Business Media LLC. Available from 10.1007/978-94-015-8859-1_12.

Titov, V. V., González, F. I., Bernard, E. N., Eble, M. C., Mofjeld, H. O., Newman, J. C., & Venturato, A. J. (2005). Real-time tsunami forecasting: Challenges and solutions. *Natural Hazards*, *35*(1), 41−58. Available from https://doi.org/10.1007/s11069-004-2403-3.

Tsushima, H., Hino, R., Fujimoto, H., Tanioka, Y., & Imamura, F. (2009). Near-field tsunami forecasting from cabled ocean bottom pressure data. *Journal of Geophysical Research: Solid Earth*, *114*(6). Available from https://doi.org/10.1029/2008JB005988, http://onlinelibrary.wiley.com/journal/10.1002/(ISSN)2169-9356.

Tsushima, H., Hirata, K., Hayashi, Y., Tanioka, Y., Kimura, K., Sakai, S., Shinohara, M., Kanazawa, T., Hino, R., & Maeda, K. (2011). Near-field tsunami forecasting using offshore tsunami data from the 2011 off the Pacific coast of Tohoku Earthquake. *Earth, Planets and Space*, *63*(7), 821–826. Available from https://doi.org/10.5047/eps.2011.06.052, http://rd.springer.com/journal/40623.

Wang, Y., Heidarzadeh, M., Satake, K., Mulia, I. E., & Yamada, M. (2020). A tsunami warning system based on offshore bottom pressure gauges and data assimilation for Crete Island in the Eastern Mediterranean basin. *Journal of Geophysical Research: Solid Earth*, *125*(10). Available from https://doi.org/10.1029/2020JB020293, http://agupubs.onlinelibrary.wiley.com/hub/jgr/journal/10.1002/(ISSN)2169-9356/.

Wang, Y., Imai, K., Mulia, I. E., Ariyoshi, K., Takahashi, N., Sasaki, K., Kaneko, H., Abe, H., & Sato, Y. (2023). Data assimilation using high-frequency radar for tsunami early warning: A case study of the 2022 Tonga Volcanic Tsunami. *Journal of Geophysical Research: Solid Earth*, *128*(2). Available from https://doi.org/10.1029/2022JB025153, http://agupubs.onlinelibrary.wiley.com/hub/jgr/journal/10.1002/(ISSN)2169-9356/.

Wang, Y., Tsushima, H., Satake, K., & Navarrete, P. (2021). Review on recent progress in near-field tsunami forecasting using offshore tsunami measurements: Source inversion and data assimilation. *Pure and Applied Geophysics*, *178*(12), 5109–5128. Available from https://doi.org/10.1007/s00024-021-02910-z, http://www.springer.com/birkhauser/geo + science/journal/24.

Chapter 13

Global tsunami hazards and risks

Yong Wei[1,2]

[1]*Cooperative Institute for Climate, Ocean, & Ecosystem Studies, University of Washington, Seattle, WA, United States,* [2]*National Oceanic & Atmospheric Administration Pacific Marine Environmental Laboratory, Seattle, WA, United States*

13.1 Introduction

The horrific December 26, 2004, Indian Ocean Tsunami, which killed over 230,000 people and displaced 1.7 million across 14 countries, prompted governments worldwide to address tsunami hazards. During the following years, the U.S. National Oceanic and Atmospheric Administration (NOAA) responded by deploying an array of 39 Deep-ocean Assessment and Reporting of Tsunamis (DART) stations, which were designed and patented at NOAA's Pacific Marine Environmental Laboratory (PMEL), along the world's major subduction zones in the Pacific, Atlantic, and Indian Oceans (Bernard et al., 2023). Built on data assimilation utilizing DART measurements, the Short-term Inundation Forecasting of Tsunamis (SIFT) emerged in NOAA's Tsunami Warning System (TWS) to support NOAA's mission in issuing real-time and near-real-time tsunami flooding forecasts for the U.S. coastlines and territories (Titov et al., 2005). This short-term tsunami hazard assessment method proved highly effective in tsunami forecasts during the 2011 Japan Tsunami, providing modeling accuracies of 70% in a basin-wide forecast (Tang et al., 2012) and 86% in near-field inundation assessment (Wei et al., 2013), respectively. It has become an operational tool at NOAA's Tsunami Warning Centers (TWCs) since 2013 to provide real-time inundation forecasts for more than 70 ports in the United States (Fig. 13.1). The time frame of short-term assessments for an unfolding tsunami event typically ranges from

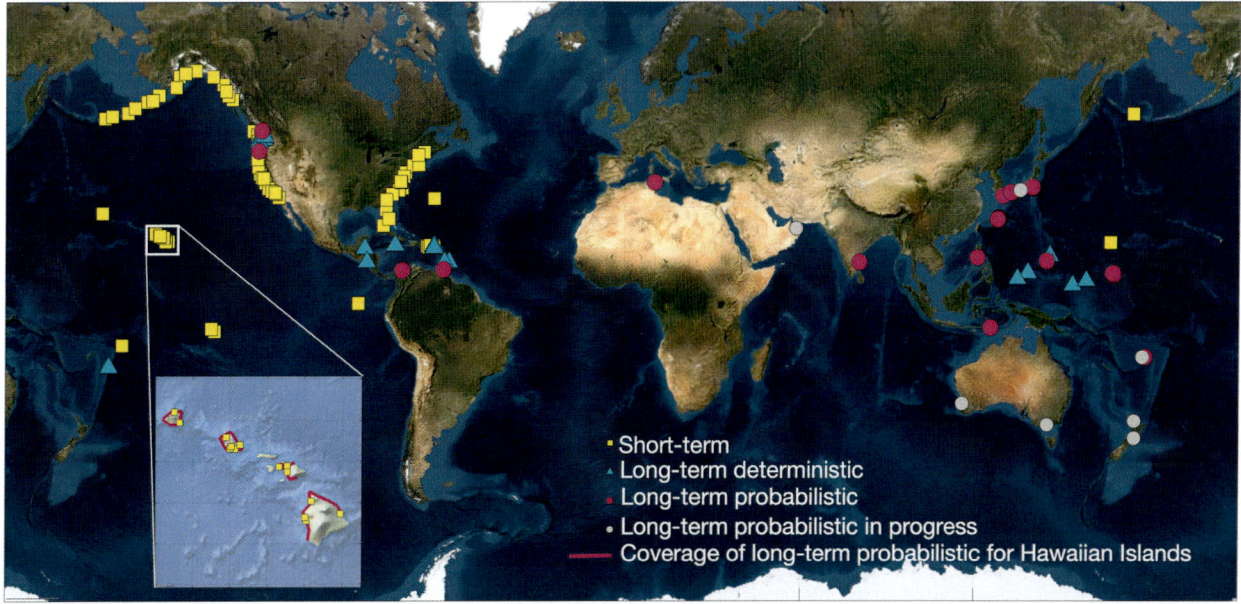

FIGURE 13.1 Location coverage of PMEL's short-term and long-term tsunami inundation hazard assessment models. *PMEL*, Pacific Marine Environmental Laboratory. Map lines delineate study areas and do not necessarily depict accepted national boundaries. *From Bernard, E., Meinig, C., Titov, V. V., & Wei, Y. (2023). 50 Years of PMEL tsunami research and development. Oceanography, 36(2−3), 175−185. https://doi.org/10.5670/oceanog.2023.208.*

minutes to hours and may expand to days under circumstances, such as long-lasting resonance or global reach. The short-term assessments provide prompt forecast information to support early warning for the public, emergency management decision-making in evacuation, and rescue missions during and after the event.

A long-term tsunami hazard assessment applies modeling technologies to identify the potential impact of tsunamis on coastal communities at risk. It involves the comprehensive evaluation of potential tsunami sources, historical records, geological evidence, and numerical modeling to predict the likelihood and impact of future tsunamis. Such assessments aid in developing effective risk-reduction strategies and land-use planning. They also inform civil and infrastructure building codes and emergency preparedness, ultimately enhancing community resilience to future events. For decades, long-term tsunami hazard assessments have primarily relied on deterministic approaches. These methods aim to identify and analyze the potential impacts of specific worst-case tsunami events, factoring in the characteristics of the tsunami source, as well as the resulting wave propagation and inundation patterns. However, the inherent limitation in the deterministic practice that did not explicitly account for aleatory uncertainty has been revealed by unsustainable losses in the last two decades due to "unexpected" tsunamis. The probabilistic tsunami hazard assessment (PTHA) evaluates tsunami perils by considering the uncertainty and variability of the seismic events (Geist & Parsons, 2006). PTHA has become an increasingly important tool to understand and mitigate tsunami risk. It has been widely used on regional and global scales, including the United States (Cheung et al., 2011; Chock et al., 2018; González et al., 2009; Wei et al., 2017), Japan (Annaka et al., 2007), China (Liu et al., 2007; Yuan et al., 2021), Australia (Burbidge et al., 2008; Davies & Griffin, 2018), New Zealand (Lane et al., 2013; Power & Gale, 2011), Mediterranean Sea (Sørensen et al., 2012), Makran region (Hoechner et al., 2016), and the globe (Davies et al., 2018). PTHA has been employed to inform coastal planning, risk management strategies, and the design of tsunami evacuation plans. The return periods of the hazards assessed using long-term methods typically range from tens to 10,000 years.

This chapter gives an overview of short-term and long-term assessments of tsunami hazards and risks globally that have been primarily conducted at PMEL (Fig. 13.1). Section 13.2 will discuss the methods and recent enhancements in the short-term assessment. Section 13.3 will discuss the methods and main challenges of long-term assessment using both deterministic and probabilistic approaches. Section 13.4 provides concluding remarks in terms of global hazards and risks.

13.2 Short-term hazard assessment

13.2.1 Methods

The devastating 2004 Indian Ocean Tsunami highlighted the need for accurate and timely tsunami forecasts. In response, NOAA began integrating PMEL's flooding forecast capabilities into operational TWSs. This integration eventually led to the development of NOAA's SIFT system, which consists of three core components: the DART buoy network, source inversion processes that assimilate DART observations, and inundation forecast models.

13.2.1.1 DART buoy network

The DART system is comprised of three parts: the bottom pressure recorder (BPR), the surface buoy with related electronics, and an anchored mooring chain (Fig. 13.2A). The DART system was first developed at PMEL by a joint effort of engineers and modelers (Titov et al., 2023), and the prototype was deployed offshore of the Aleutians in the late 1980s and early 1990s (Eble & Gonzalez, 1991; Gonzalez et al., 1991). The first-generation DART design featured an automatic detection and reporting algorithm triggered by a threshold wave-height value. The second-generation DART incorporated two-way communications that enabled tsunami data transmission on demand and independence of the automatic algorithm. This improvement ensured the measurement and reporting of tsunamis with amplitudes below the auto-reporting threshold (Fig. 13.2A). The third-generation tsunami detection buoy featured an easy-to-deploy (ETD) capability. The small-sized ETD can be transported utilizing small vessels and minimally trained staff. Its deployment required less than 30 seconds, changing the way how deep-water oceanographic moorings were deployed. The fourth generation (4G) is an enhanced version of ETD that incorporated advanced sensors, updated software, and power management to measure tsunamis at a sampling rate of 1 Hz (data reported every 15 seconds). The increased number of measurements allowed earthquake signals to be separated from tsunami waveforms. This way, the 4Gs can be sited closer to the seismic source zone than any predecessor DART for early detection of an ongoing tsunami.

During an earthquake-generated tsunami, the DART BPR is primarily triggered by seismic waves, though it can also be remotely triggered by the TWC staff. As shown in Fig. 13.2B, once triggered, the DART station transitions into an event mode. In this mode, the BPR data are transmitted to the surface buoy via an acoustic link and then relayed to

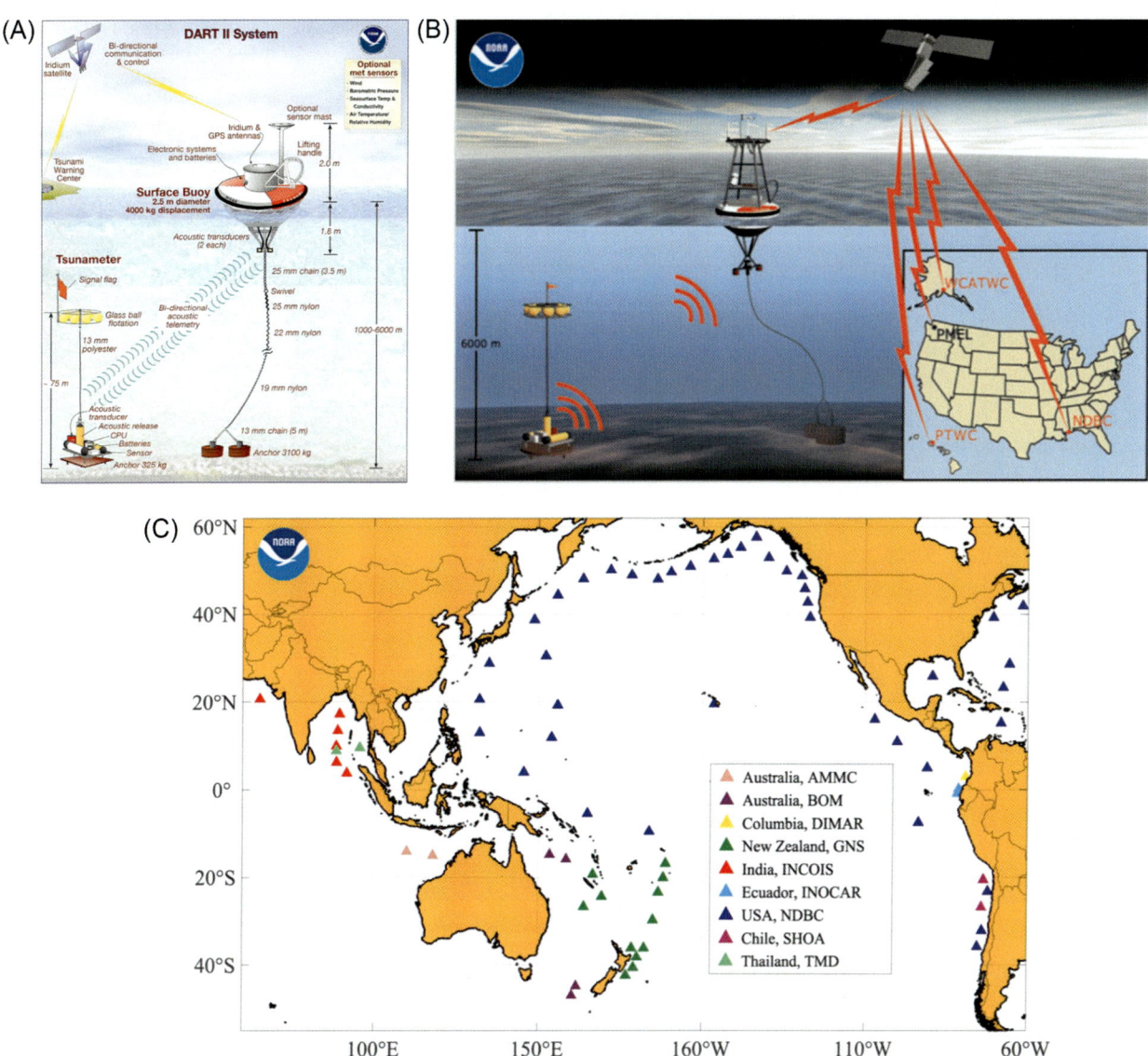

FIGURE 13.2 (A) Components of a DART system, (B) DART schematic diagram, and (C) up-to-date (March 2024) DART buoy locations across the world. *DART*, Deep-ocean Assessment and Reporting of Tsunamis. Map lines delineate study areas and do not necessarily depict accepted national boundaries.

ground stations via satellite for immediate dissemination to TWCs. Presently, the DART network has been expanded to a total of 74 stations (Fig. 13.2C) with many additional deployments by other countries worldwide beyond the United States. This includes New Zealand's recent deployments of 12 advanced 4Gs in the southwest Pacific (Fry et al., 2020). Real-time measurements from all DART stations can be accessed through NOAA's National Data Buoy Center (NDBC).

13.2.1.2 Tsunami source inversion using precomputed propagation database

The second core component of SIFT is an inversion process to constrain the tsunami source based on a data assimilation method utilizing DART observations and a precomputed database of tsunami propagation scenarios. The tsunami propagation database stores simulation results of global propagation of more than 2000 tsunami unit sources that are placed in all major subduction zones across the Pacific Ocean, Indian Ocean, and Atlantic Ocean (Fig. 13.3). Each unit source, aligned to fit fault geometry, features a magnitude 7.5 earthquake source sized at 100 km along strike and 50 km downdip with a unit slip. The Method of Splitting Tsunami (MOST) model, based on shallow water equations, computes up to 48 hours of tsunami propagation for each unit source at a grid resolution of 4 arcminutes (~7.2 km at the equator).

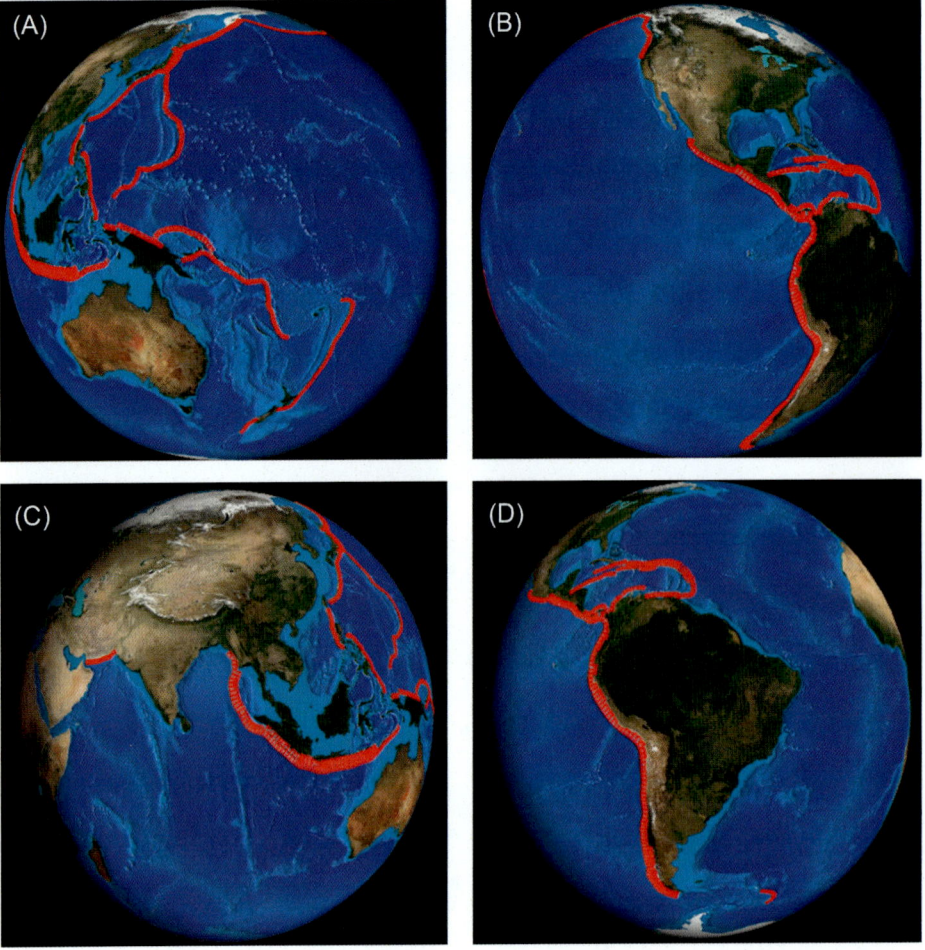

FIGURE 13.3 PMEL's tsunami unit sources, indicated by red rectangles, used to develop the precomputed tsunami propagation database. (A) West Pacific Ocean, (B) East Pacific Ocean, (C) Indian Ocean, and (D) Atlantic Ocean. *PMEL*, Pacific Marine Environmental Laboratory. Map lines delineate study areas and do not necessarily depict accepted national boundaries. *From NCTR/PMEL/NOAA.*

The model results at every four grid nodes (16 arcminutes, ~28.8 km) are stored in the database (Gica et al., 2008). Utilizing the linearity of tsunami wave characteristics in the deep ocean, these unit-source propagation scenarios can be superimposed to provide a best match with observations from selected DART stations. This source-constraint process is performed using an inversion algorithm to adjust precomputed models to minimize residuals between the adjusted models and the DART data in the least squares sense (Percival et al., 2011).

13.2.1.3 Inundation forecast models

An advancement of SIFT is its capability to provide real-time assessments of tsunami inundation hazards for more than 70 ports and communities along the U.S. coastlines and territories. The development of inundation forecast models was a major modeling effort in SIFT along with the growth of DART observations. The yellow squares in Fig. 13.1 indicate the ports equipped with such a short-term capability. All forecast models are based on the MOST model computation using nonlinear shallow water equations. The model grids were created from the best-available digital elevation models (DEMs) developed by the National Center for Environmental Information (NCEI) at the time of development. To speed up model computation, each model employs three telescoped grids, with the finest grid resolution of 30–90 m, to resolve the wave dynamics from deep-water propagation to onshore inundation. To ensure the model quality and accuracy, all models have undergone a strict model validation process using historical records (Tang et al., 2009). A high-resolution (10-m) reference model was developed in parallel to ensure the quality of the forecast results.

Once the tsunami source is determined from the data assimilation mentioned in Sections 13.2.1.1 and 13.2.1.2, a tsunami propagation scenario in the deep ocean can be rapidly assembled to provide a forecast that furnishes initial and boundary conditions for the forecast models to perform real-time inundation hazard assessments at various ports and communities. For distant sites, the SIFT assessments can be completed with a few hours of lead time before the tsunami arrives. Data assimilation that integrates observations and numerical models is a major advancement compared to traditional assessments solely based on seismic sources. This short-term assessment contributes significantly to improving forecast accuracy and reducing false alarms.

The SIFT short-term assessment was first applied to the real-time forecast of the November 17, 2003 Rat Island Tsunami and successfully predicted 30-cm high tsunami waves ~ 45 minutes before they arrived at Hilo, Hawaii (Titov et al., 2005). SIFT became an operational tool at NOAA's TWCs in 2013 after numerous experimental forecasts for nearly all detectable tsunamis since 2005. The notable events were the November 15, 2006 M_w 8.3 and January 13, 2007 M_w 8.1 Kuril Islands, the August 15, 2007 M_w 8.0 Peru (Wei et al., 2008), September 29, 2009 M_w 8.1 Samoa (Zhou et al., 2012), February 27, 2010 M_w 8.8 Chile, March 11, 2011 M_w 9.1 Japan (Tang et al., 2012; Wei et al., 2013), October 27, 2012 M_w 7.8 Haida Gwaii, and December 7, 2012 M_w 7.2 Japan (Bernard et al., 2014).

13.2.2 Short-term hazard assessment: the March 11, 2011 Japan Tsunami

The earthquake that caused the March 11, 2011 Japan Tsunami was significantly underestimated at the early stage of the tsunami evolution. The initial magnitude estimate was only M_w 7.9 when it ruptured at 05:46:24 UTC. The strength of SIFT lies in its ability to generate forecast results grounded in DART observations, thereby eliminating bias from source magnitude estimates. However, it is worth noting that SIFT does offer a preliminary forecast based on the initial source magnitude estimate, prior to the availability of DART data. The DART station 21418, located about 500 km east of the epicenter (38.322°N, 142.369°E), registered a 1.8-m high wave crest 56 minutes after the earthquake, SIFT's first estimate of the tsunami source showed the source energy was equivalent to an M_w 8.8 earthquake (Tang et al., 2012). SIFT produced a refined source estimate after the second DART D20401 registered a 0.5-m high half wave. The second source estimation was obtained from a joint inversion using measurements up to 1.5 hours from D21418 and D21401. This DART-constrained source indicated a rupture area of ~400 km × 100 km with >20 m slips near the trench and the largest slip of 26 m in a deeper portion of the fault (Fig. 13.4A). The model predictions showed excellent matches with observations at D21428 and D20401 (Fig. 13.4B), which were used for inversion. This model prediction was further confirmed consistent with the measurements from two additional DART buoys, D21419 and D21413, which were not used for the source inversion (Fig. 13.4B).

During the 2011 Japan Tsunami, the SIFT inundation hazard assessments along the U.S. coastlines were all produced using the source inversion obtained about 1.5 hours after the earthquake. Model forecast results were provided for 32 ports in Hawaii, Alaska, American Samoa, and the U.S. West Coast, including the successful forecast of tsunami inundation at Kahului Harbor, Maui (Fig. 13.5). The model forecasts were obtained 5 hours before the tsunami waves reached Hawaii and 9 hours before the waves started impacting the U.S. West Coast. The forecast accuracy across all 32 forecast sites was 68% and a 0.2 m average error (Tang et al., 2012). For eight sites with observed amplitudes larger than 1.0 m, the accuracy was 74%. It was further improved to 87% for four sites with observations greater than 1.5 m. The forecast accuracy was measured by the difference between the model forecast and the measurement at tide gauges, $|h_{obs} - h_{forecast}|/h_{obs}$, where h_{obs} and $h_{forecast}$ represent the measured and model-predicted maximum tsunami amplitudes, respectively.

The same SIFT tsunami source estimate constrained from DART measurements also proved valid for the near-field coasts that were reached by tsunami waves within 1 hour after the earthquake. The model results closely reproduced the wave series recorded at the Global Positioning System (GPS) buoys and wave gauges deployed offshore of Japan (Fig. 13.6A). To further demonstrate the validity of the forecast source, Wei et al. (2013) developed multiple inundation models at the grid resolution of 60 m that cover the entire coastlines in East Japan impacted by the tsunami, and model accuracy reached 86% in terms of the total inundated area. As demonstrated in Fig. 13.6B, the modeled inundation agreed very well with the measured inundation limit in Sendai, Japan.

13.2.3 Enhancements to short-term hazard assessments

On the basis of SIFT, PMEL has also developed two prototype web tools: (1) the Community Model Interface for Tsunamis (ComMIT), which enables the development, use, and sharing of tsunami modeling results (Titov et al., 2011), and (2) Tsunami Web (Tweb), which is capable of sharing short-term assessment results for different coastlines via a graphical web client (Bernard & Titov, 2015). ComMIT and Tweb have opened up opportunities for the fast development of the tsunami forecast capability for specific locations. Fig. 13.7 illustrates the snapshots of the graphic user interface of the three short-term assessment tools.

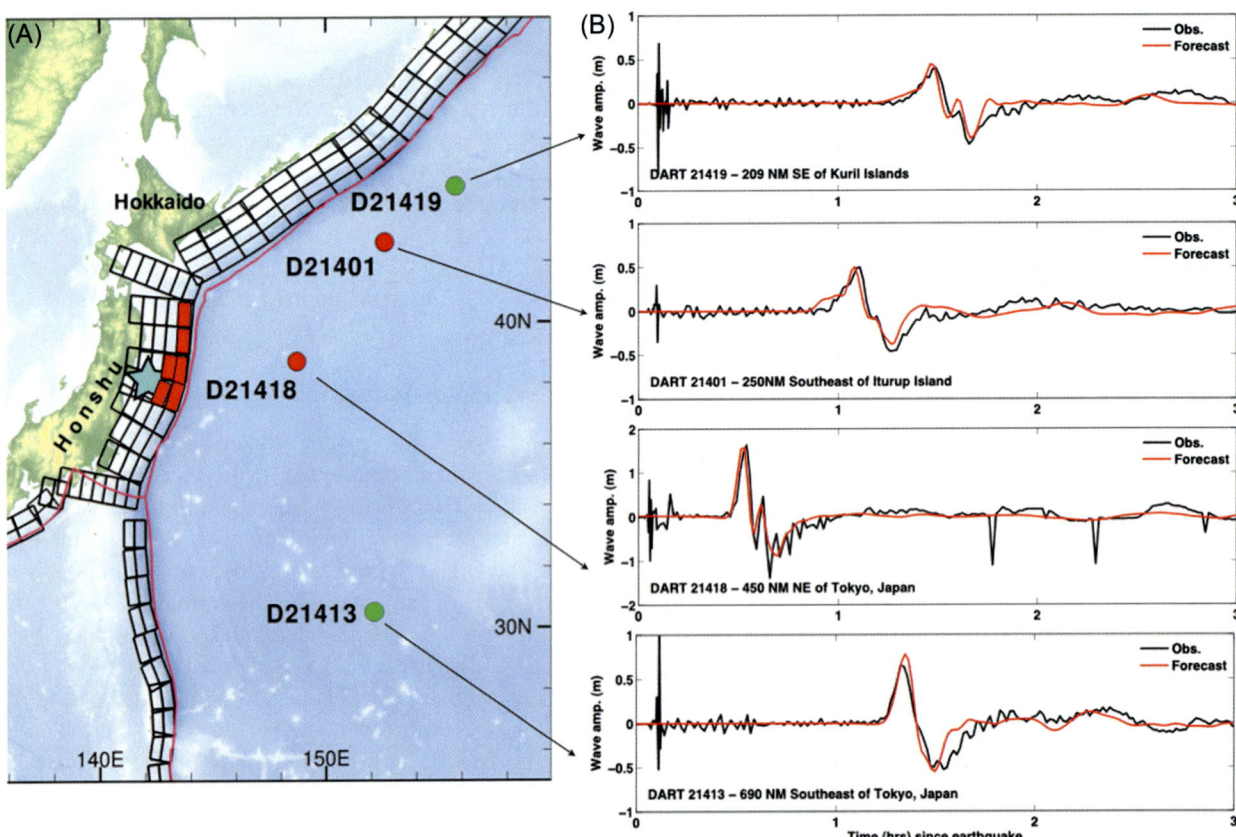

FIGURE 13.4 SIFT source inversion using DART measurements during the March 11, 2011 Japan Tsunami. (A) Estimated tsunami source from inversion of DART measurements, where the red blocks represent the estimated slip zone of the rupture area, the black blocks represent the tsunami unit sources used to precompute tsunami propagation scenarios, the cyan star is the epicenter location, the red dots are the two DARTs used in the source inversion, and the green dots are the two DARTs not used in the source inversion but used for model forecast confirmation. (B) Comparison of time-series between the model forecast and observations at nearby DARTs. *DART*, Deep-ocean Assessment and Reporting of Tsunamis; *SIFT*, Short-Term Inundation Forecasting of Tsunamis.

Although SIFT proved to work well for short-term model assessments of tsunami hazards generated by large earthquakes at distant shores, challenges remain in terms of the speed and robustness of the existing SIFT system. First, the type, location, and size of the earthquake/tsunami sources are largely limited by the existing tsunami propagation database and their predetermined parameters, making the short-term assessments less robust. Second, the latency of older-generation DART detection, which usually are not available until 30 to 90 minutes after the earthquake, may result in delayed response to provide timely forecasts for the near field. In recent years, there have been multiple efforts to improve SIFT performance and capabilities toward more rapid and robust hazard assessments. These enhancements include (1) more rapid model computation, (2) modeling a broader range of earthquake sources, (3) more rapid tsunami detection, and (4) the development of a global tsunami propagation database.

13.2.3.1 Enhancement in model computational speed

Boosting of the model's computational speed is essential in improving the speed of short-term model assessments. Utilizing the graphic processor unit (GPU) technology, the MOST model computation has been improved by more than 60 times compared to a model run using the central processing unit (CPU). Table 13.1 compares the computational time between CPU and GPU runs of MOST for basin-wide and regional domains. A single GPU model run for the entire Pacific Basin (~7.4 million grid nodes) can be completed within several minutes if the model saves outputs frequently, and the model time can be shortened to less than 2 minutes if the model only saves the output at the end of the model run. A four-GPU basin-wide run can further reduce the computing time by two to five times. For a regional model with about one million grid nodes, a 6-hour model run only needs a few seconds. This is particularly efficient for rapid source inversion since the tsunami propagation only needs to be modeled from the source to a few nearest

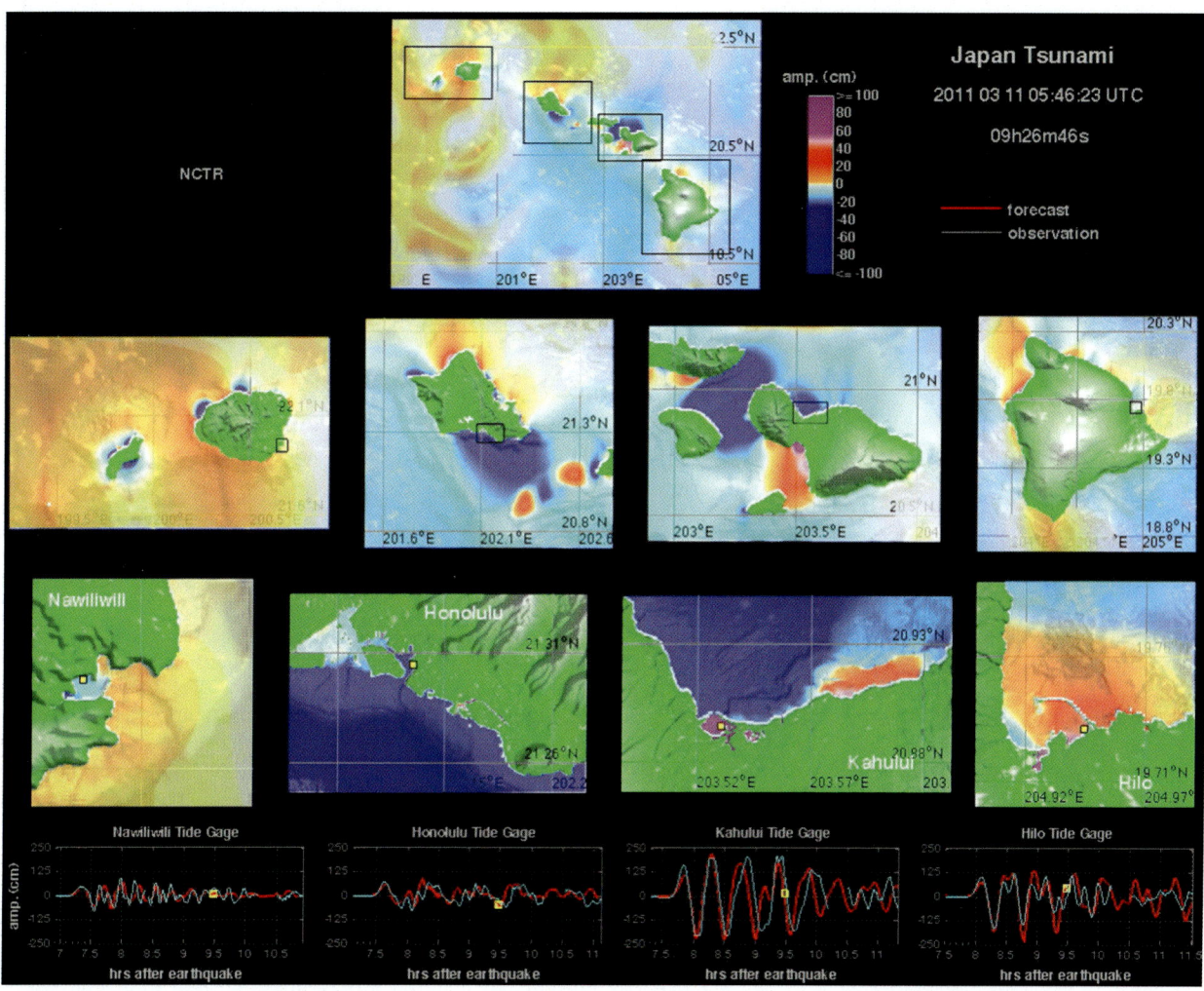

FIGURE 13.5 SIFT model forecast of the tsunami flooding in Hawaii from the 2011 Japan Tsunami 5 hours before arrival using DART data. The top, middle, and bottom panels show the telescoped grids at the regional, islands, and harbor levels, respectively, used in the MOST model to simulate tsunami propagation and inundation. *DART*, Deep-ocean Assessment and Reporting of Tsunamis; *SIFT*, Short-Term Inundation Forecasting of Tsunamis. Map lines delineate study areas and do not necessarily depict accepted national boundaries. *From Bernard, E., Meinig, C., Titov, V. V., & Wei, Y. (2023). 50 YEARS OF PMEL TSUNAMI RESEARCH AND DEVELOPMENT. Oceanography, 36(2−3), 175−185. https://doi.org/10.5670/oceanog.2023.208.*

DART locations. The high-performance GPU capability of MOST is a game-changer as it now enables us to perform "on-the-fly" computation of tsunami propagation for arbitrary source characteristics obtained from different methods, such as the Centroid Moment Tensor (CMT), W-Phase, or Global Navigation Satellite System (GNSS). The "on-the-fly" creation of unit sources fitting the seismic source characteristics has also become feasible. The tsunami propagation simulations from these new "on-the-fly" unit sources can be modeled in real time and then used in SIFT to refine the source inversion. The MOST GPU capability allows the inundation forecast to be completed more rapidly, a major improvement towards timely forecast for the near field.

13.2.3.2 Enhancement in modeling a broader range of earthquake sources

The tsunami unit sources stored in the propagation database are determined from preset source characteristics and therefore limit the range of earthquake sources to be modeled. To provide valid and accurate model forecasts, the epicenter location needs to be within the extents covered by the unit sources and the source mechanism needs to be consistent with the predetermined source characteristics. Due to these constraints, three types of earthquake sources cannot be well presented by the existing tsunami source database: (1) the earthquake source mechanism is not well represented by the source mechanism of the precomputed unit sources; (2) the strong intraplate earthquake rupture is outside the major

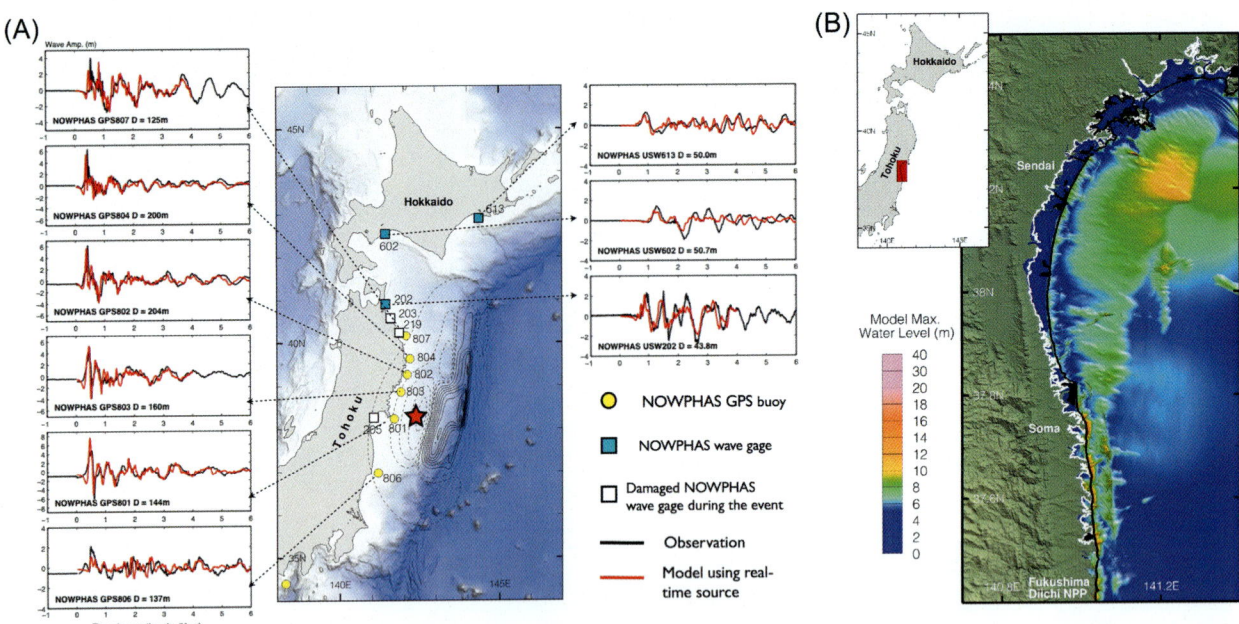

FIGURE 13.6 SIFT model results in the near field using the DART-inverted tsunami source for the 2011 Japan Earthquake. (A) Model results compared to measurements from GPS buoys and wave gauges along Japan's eastern coast. (B) Modeled tsunami inundation compared to the measured inundation limit from the 2011 Japan Tsunami, where the white solid line indicates the measured inundation limit. *DART*, Deep-ocean Assessment and Reporting of Tsunamis; *GPS*, Global Positioning System; *SIFT*, Short-Term Inundation Forecasting of Tsunamis. Map lines delineate study areas and do not necessarily depict accepted national boundaries. *From Wei, Y., Chamberlin, C., Titov, V. V., Tang, L., & Bernard, E. N. (2013). Modeling of the 2011 Japan Tsunami: Lessons for near-field forecast. Pure and Applied Geophysics, 170(6–8), 1309–1331. https://doi.org/10.1007/s00024-012-0519-z.*

fault structures and thus is not covered by the unit sources; and (3) earthquakes within the major fault structures with a magnitude <7.5 cannot be properly represented by the magnitude 7.5 unit sources. The high-performance computational capability is lifting these restrictions and making it possible to address these source challenges on the fly, as shown by the following case studies.

The source of the September 8, 2017 M_w 8.1 Chiapas Earthquake featured a high-angle normal fault that was not well represented by the predetermined thrust-fault unit sources across the source region. Based on the U.S. Geological Survey (USGS) CMT solution (epicenter dip 73 degrees, rake −96 degrees, and strike 315 degrees), a total of 40 high-angle normal-fault unit sources were created at the size of 50 km × 50 km on the fly (Fig. 13.8A). The model computation of tsunami propagation to the nearest DARTs for these new unit sources was completed before the two closest DARTs, D43413 and D32411, registered the 8-cm high tsunami signals. Fig. 13.8A shows the DART-constrained inversion indicated five slip patches, out of 40, with slip amounts ranging from 2.2 to 4.3 m. The coseismic vertical deformation resulting from the inverted source compared well with the deformation derived by Gusman et al. (2018) and Ye et al. (2017). The model forecasts compared well with observations at the nearby DARTs (Fig. 13.8A) and tide gauges in Central America and Galapagos (Figs. 13.8B and C).

The second example is the January 23, 2018 M_w 7.9 Alaska Tsunami. This event represents one of the most challenging situations of SIFT in which the earthquake source falls outside the coverage of the existing unit source database. The initial solution using the existing database predicted misfits in wave amplitudes and arrival times at five nearby DART buoys (Fig. 13.9A). Five strike-slip unit sources were created based on the USGS W-phase solution on the fly (https://earthquake.usgs.gov/earthquakes/eventpage/us7000asvb/moment-tensor) and the tsunami propagation from each of these new unit sources to the five nearby DART stations was computed. A quick source inversion using measurements from DART D46410 and D46403 offered a more reconcilable interpretation between DART and seismic observations (Figs. 13.9B and C). The inverted slips range from 2.2 to 5.8 m, with a magnitude of 7.96. Fig. 13.9A shows that the coseismic vertical displacement obtained from the inverted source fitted well with the aftershocks-affected area. The model results agreed well with observations at all five DARTs, especially DART D46409, while the USGS finite fault solution gave overpredictions of the wave amplitudes (Fig. 13.9C). The earthquake ruptured only a small portion (∼600 km²) dipping at 20 degrees along the subducting Pacific Plate underneath the North American

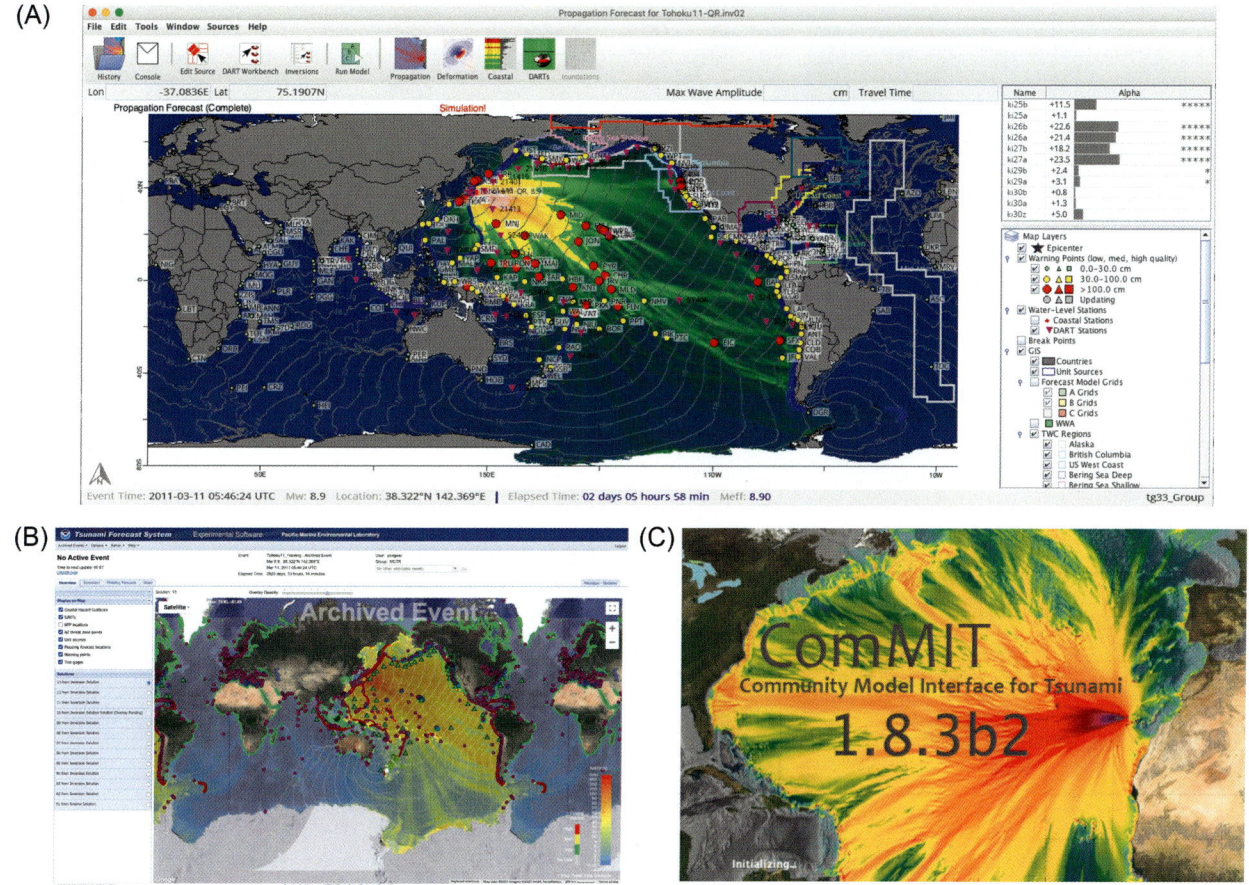

FIGURE 13.7 Snapshots of graphic user interfaces of the short-term hazard assessment tools developed at PMEL: (A) SIFT, (B) Tsunami Web, and (C) ComMIT. *ComMIT*, Community Model Interface for Tsunamis; *PMEL*, Pacific Marine Environmental Laboratory; *SIFT*, Short-Term Inundation Forecasting of Tsunamis. Map lines delineate study areas and do not necessarily depict accepted national boundaries. *From NCTR/PMEL/NOAA.*

TABLE 13.1 Sensitivity tests of the MOST computational time using the central processing unit (CPU) versus the graphic processor unit (GPU).

Domain size	No. of processors	No. of nodes (in million)	Length of simulation	Computing time (frequent model output)	Computing time (model output only at the end)
Basin wide	1 CPU	~7.4	24 hours	~136 minutes	–
Basin wide	1 GPU	~7.4	24 hours	~10 minutes	~100 seconds
Basin wide	4 GPU	~7.4	24 hours	~105 seconds	~30 seconds
Regional	1 GPU	~1	6 hours	<8 seconds	~1.5 seconds

Plate. The energy released from the small rupture area was probably dissipated quickly when propagating away from the source. Multiple uninhabited islands blocked the tsunami energy from direct impact on coastal communities in the Alaska Peninsula (Fig. 13.9B). This powerful earthquake also ruptured into the Shumagin Gap (Crowell & Melgar, 2020), an area that has been considered immune from large earthquakes due to the constant release of fault pressure (Fig. 13.9B).

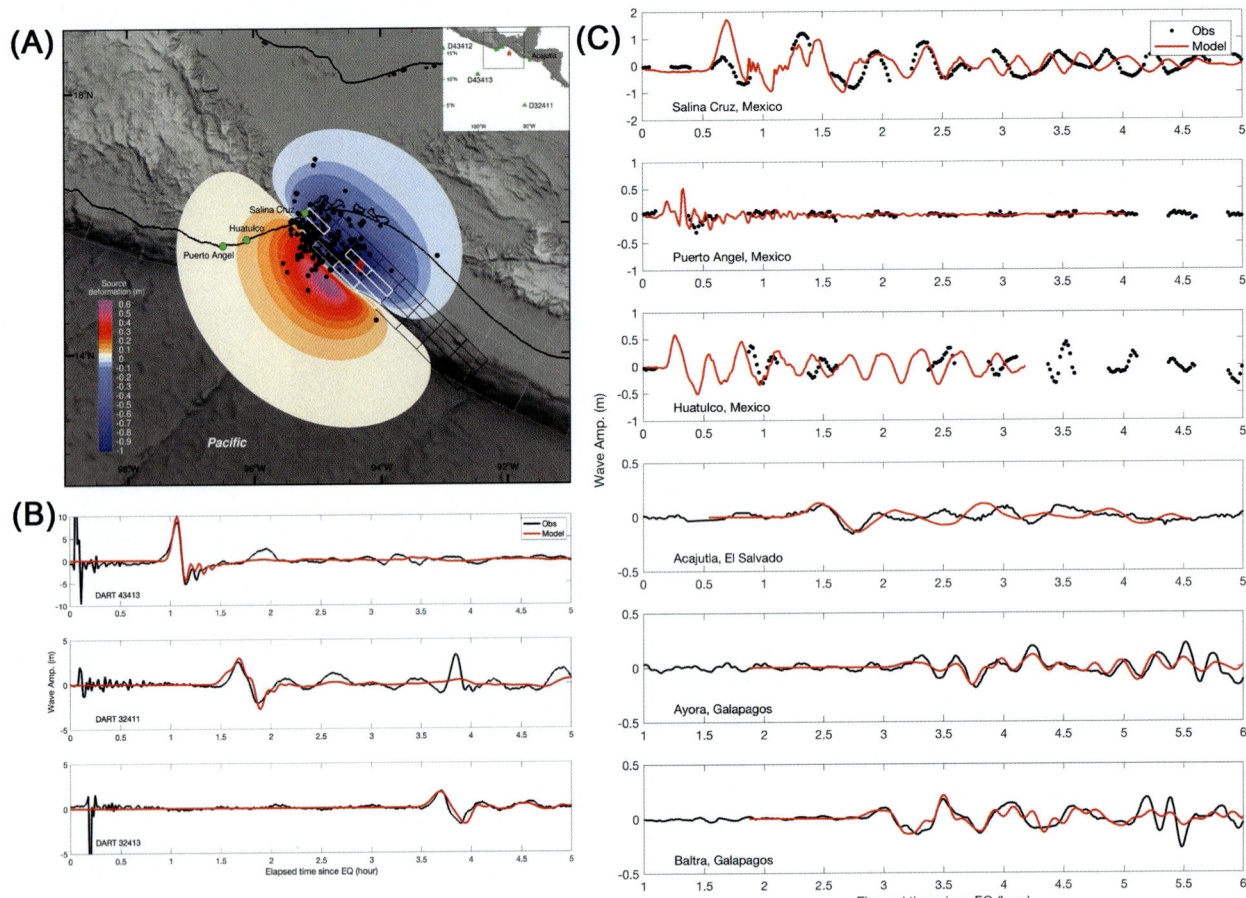

FIGURE 13.8 Source inversion and model/data comparisons of the September 8, 2017 M_w 8.1 Chiapas Tsunami utilizing the "on-the-fly" capability. (A) Initial vertical displacement estimated by the source inversion. The red star indicates the epicenter location. The thick black blocks represent the tsunami unit sources developed on the fly based on the USGS CMT solution. The white blocks are the slip patches obtained from the DART-constrained source. The color-coded contours represent the coseismic vertical displacement resulting from the DART-constrained source. (B) Comparison between model and observations at the three nearest DARTs. (C) Comparisons between model and observations at multiple tide gauges. *DART*, Deep-ocean Assessment and Reporting of Tsunamis.

13.2.3.3 Enhancements toward more rapid tsunami detection

More rapid tsunami detection leads to more rapid forecasts. The latency in tsunami detection at many existing DART stations comes from the difficulty in the separation of tsunami signals from seismic waves. When a DART system is placed too close to the earthquake source, the registered tsunami signals are usually contaminated by the seismic Rayleigh waves. The DART 4Gs emerged as the key solution to this problem. Due to the high sampling rate at 1 Hz, a DART 4G possesses an exceptional ability to separate earthquake vibrations and other noises from tsunami waveforms. New Zealand has recently deployed 12 DART 4Gs offshore (Fig. 13.2C), and they are playing critical roles in tsunami early detection and source determination in the southwest Pacific Ocean (Fry et al., 2020). For example, the March 4, 2021 Kermadec Islands Earthquake sequence was the first event that the DARTs recorded three consecutive earthquake/tsunami events within a few hours (Fig. 13.10A). The first two earthquakes with M_w 7.3 and 7.4 were about 4 hours apart. They were followed by a much stronger M_w 8.1 event 2 hours later, which generated visible tsunami signals at nearby New Zealand's DART 4G stations (Romano et al., 2021). The estimated tsunami source was preliminarily determined using DART 4G D5401001 and then refined using D5401000 (Fig. 13.10B). The source inversion produced a reasonable agreement between model forecasts and observations at tide gauge stations in New Zealand and across the Pacific Ocean (Fig. 13.10C).

The GNSS devices are advantageous in directly measuring ground displacements without saturation issues for the largest earthquakes. GNSS can rapidly detect the peak ground displacement of land deformation within a few minutes of the earthquake. The GNSS network has great potential to provide accurate estimates of earthquake magnitudes for

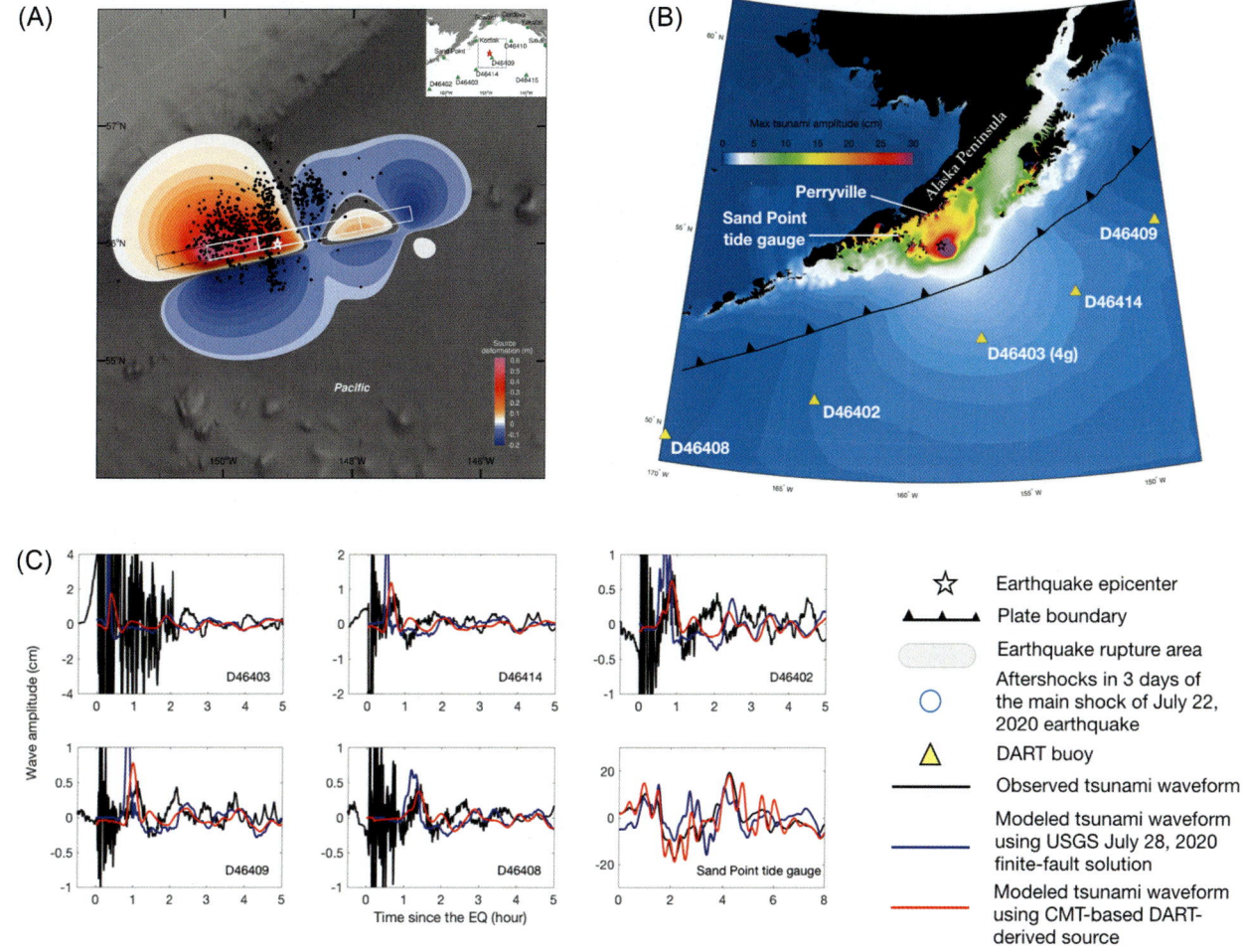

FIGURE 13.9 Summary of the July 22, 2020 M_w 7.8 Alaska Peninsula Earthquake and Tsunami. (A) Coseismic deformation computed from the CMT-based DART-derived tsunami source of the earthquake, where the thin black blocks are the new unit sources developed on the fly, and the white blocks are the slip patches obtained from the DART-constrained source; (B) max tsunami wave amplitude computed using the CMT-based DART-derived tsunami source of the earthquake; and (C) model results versus observations at five nearby DART buoys and the tide gauge at Sand Point. *DART*, Deep-ocean Assessment and Reporting of Tsunamis.

early warning of tsunamis. A GNSS-based analysis tool, named GFAST (Geodetic First Approximation of Size and Time), utilizing these rapid GNSS measurements (Crowell et al., 2018), is currently being implemented in the tsunami forecast system for testing at both TWCs. This capability is expected to produce a preliminary source magnitude estimation within 90–120 seconds of an earthquake followed by a finite fault solution within 3 minutes of the event. GFAST is expected to enhance tsunami early detection to warn the near field before tsunami measurements become available. On the other hand, the GNSS measurements are restricted almost entirely to land and contain little information on the offshore rupture process where the largest earthquake deformation usually occurs during a great seismic event (Newman, 2011). This limitation may lead to a biased estimation of the tsunami source. Integration of the GNSS measurements with observations from deep-ocean instruments, such as the offshore GPS/acoustic station (Newman, 2011) or DART network (Wei et al., 2014), proved critical in reducing the error of the source estimation.

13.2.3.4 Enhancement in developing a global tsunami propagation database

A global tsunami propagation database has been recently included in the upgraded SIFT system. The existing database in SIFT was developed separately for three ocean basins, the Pacific, Atlantic, and Indian Oceans (Fig. 13.11A). Every propagation model run was done for 30-hour propagation at 4 arcminutes resolution for about ~2000 unit sources. Previously, the domain separation led to an artificial boundary effect for locations close to the subdomain boundaries and therefore was not able to fully address the wave reflection from continents at the other end of the ocean basin. The

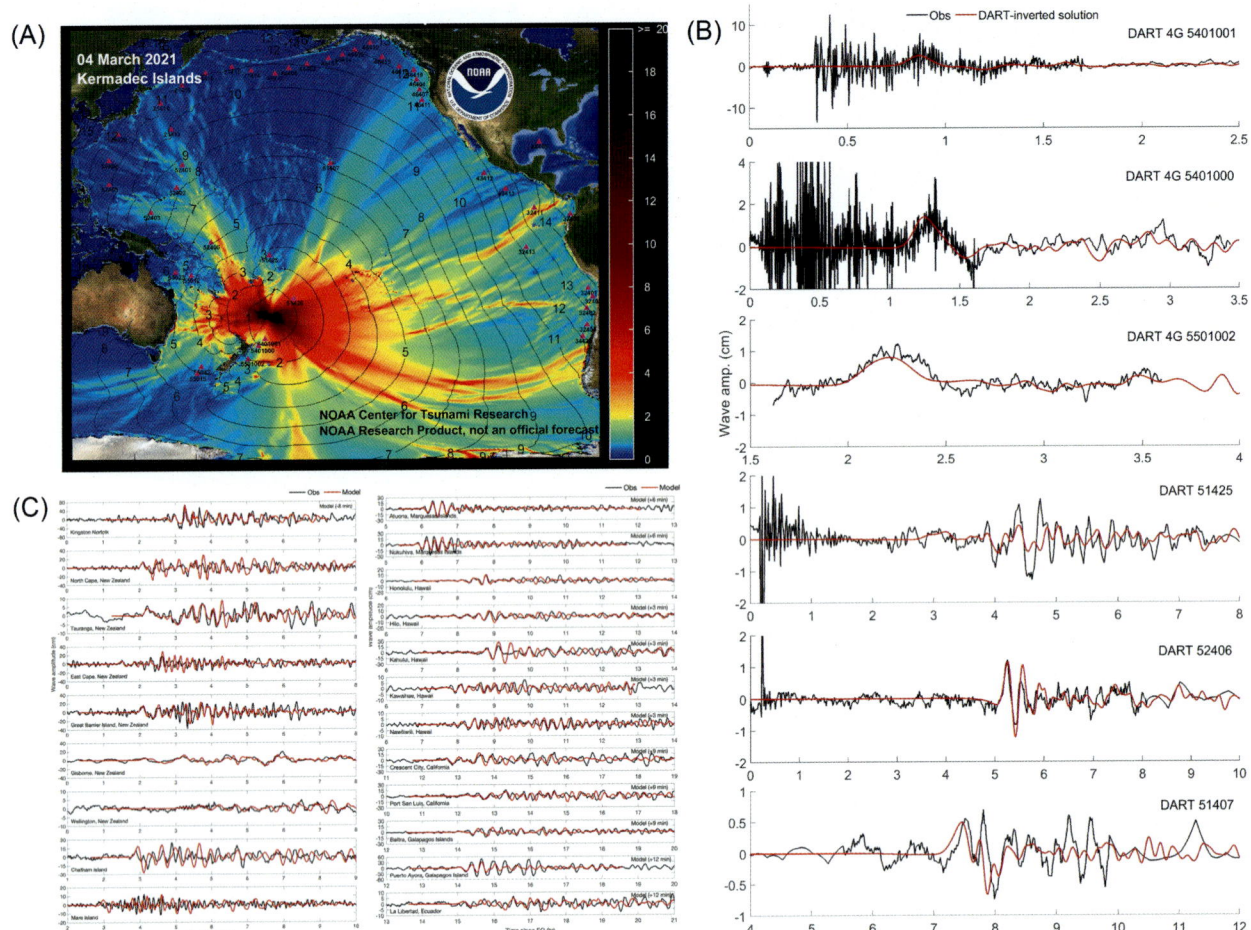

FIGURE 13.10 Model forecast results of the March 4, 2021 Kermadec Islands Earthquake sequence. (A) Maximum tsunami amplitudes calculated with the MOST forecast model, (B) comparison between model forecast and observations at New Zealand's DART 4Gs and U.S. DART stations, and (C) comparison between model forecast and observations at multiple tide gauges across the Pacific Ocean. *DART*, Deep-ocean Assessment and Reporting of Tsunamis. *From NCTR/PMEL/NOAA.*

newly developed global database addresses these issues utilizing a 48-hour model run for each unit source with a smooth transition of model parameters across the model boundaries (Fig. 13.11B).

13.2.4 Short-term tsunami hazard assessment systems in other countries and regions

Presently, there are four regional tsunami warning and mitigation systems for the Pacific Ocean (PTWS), the Indian Ocean Tsunami Warning and Mitigation System (IOTWMS), the Caribbean and adjacent regions (CARIBE-EWS), and the North-Eastern Atlantic, Mediterranean, and connected seas (NEAMTWS) to provide short-term hazards assessments for their member states across the world (Fig. 13.12). The United Nations Educational, Scientific, and Cultural Organiztion (UNESCO) Intergovernmental Oceanographic Commission (IOC) facilitates the coordination of these regional systems in both short-term assessment and long-term hazard mitigation and planning, including training and capacity development, preparedness, and awareness.

The U.S. tsunami forecast system utilizes the SIFT tool for inundation forecast at both NOAA's TWCs. Meanwhile, each TWC also operates its independent short-term assessment tools to provide forecasts of tsunami impact, particularly the tsunami heights, along the coastlines of their responsibilities. The Pacific TWC (PTWC), located in Honolulu, Hawaii, utilizes the Real-Time Forecast of tsunamis tool (RIFT) to compute the maximum tsunami heights offshore, which are then used to derive the maximum tsunami heights at the coastline based on Green's Law. The National TWC employs the Alaska Tsunami Forecast Model (ATFM) to provide forecasts of tsunami heights at the coasts through data assimilation of hundreds of precomputed hypothetical scenarios and real-time observations.

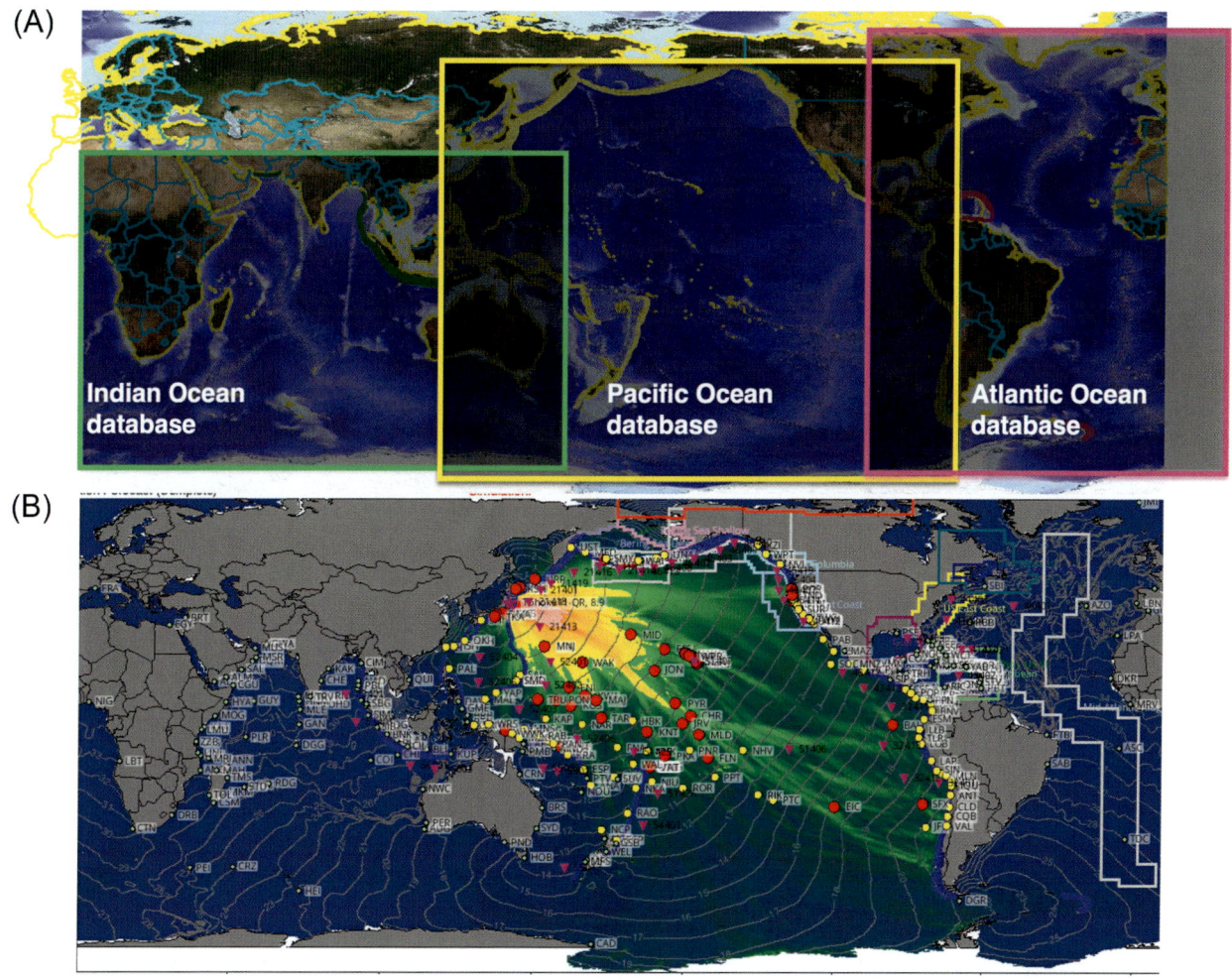

FIGURE 13.11 (A) Coverage of the three basin-wide models in the previous SIFT tsunami propagation database. (B) New global tsunami propagation database demonstrated using the maximum tsunami wave amplitude of the 2011 Japan Tsunami. *SIFT*, Short-Term Inundation Forecasting of Tsunamis. Map lines delineate study areas and do not necessarily depict accepted national boundaries. *From NCTR/PMEL/NOAA.*

The New Zealand Tsunami Monitoring Network receives advice from PTWC or a national geological hazard monitoring system (GeoNET) maintained by the Institute of Geological and Nuclear Science Limited. An on-call duty team will then provide scientific advice and convene a tsunami expert panel for more detailed advice if time permits (Power & Gale, 2011). The New Zealand forecasting system embraced a major upgrade in 2019 after their government deployed 12 DART 4Gs offshore to monitor tsunamis generated by large undersea earthquakes or volcanic eruptions. These new deep-ocean tsunami observations also proved vital for real-time assessment of tsunami impacts generated by nonseismic origins, for example, the January 15, 2022 Tonga Volcanic Tsunami (Borrero et al., 2023; Gusman et al., 2022; Lynett et al., 2022; Ren et al., 2023).

Japan's tsunami warnings are primarily provided by the Japan Meteorological Agency (JMA) utilizing a comprehensive network that consists of seismological observations, GNSS devices on land, GNSS buoys deployed at water depths of 100–400 m, ocean-bottom pressure gauges connected by seafloor cables (DONET, DONET2, and S-net). For the model forecast, JMA utilizes tsunami models combined with a real-time data assimilation method named tFISH/RAPiD (tsunami Forecasting based on Inversion for initial sea-Surface Height/Real-time Automatic detection method for Permanent Displacement) (Mori et al., 2022). Similar to SIFT, the JMA forecasting system is now being migrated to include real-time inundation forecasts for near-field coastlines.

The Joint Australian TWC (JATWC) is an independent tsunami warning service operated by the Australian Government to provide tsunami warnings to Australia and its offshore territories across the Pacific and Indian Oceans. The tsunami warning forecasts are provided by JATWC seismologists utilizing a network of coastal sea-level gauges

FIGURE 13.12 Global tsunami warning and mitigation systems that provide tsunami advisory and information services. Map lines delineate study areas and do not necessarily depict accepted national boundaries. *From International Tsunami Information Center (ITIC) (http://itic.ioc-unesco.org/index.php?option = com_content&view = category&layout = blog&id = 2005&Itemid = 2005).*

and deep-ocean tsunami buoys combined with tsunami numerical models to generate a first estimate of the tsunami size, arrival time, and potential impact locations.

The South China Sea Tsunami Advisory Center provides timely advice on potentially destructive tsunamis to its member states bordering the South China Sea region. It relies on a combination of real-time processing of seismic and sea-level data for an early decision on the earthquake/tsunami source. The real-time tsunami impact assessment is then carried out utilizing model capabilities in both the precomputed scenario database and the high-performance GPU-accelerated computation.

In the Indian Ocean, the IOTWMS has established a reliable earthquake/tsunami instrumental network comprised of seismographic devices, sea-level observation stations, and deep-sea tsunami buoys. Like other regional systems, IOTWMS is also equipped with precomputed ocean-wide tsunami characteristics for a large number of earthquake scenarios. The short-term tsunami impacts are provided to individual coastal forecasting zones (approximately 100 km long) along the coastlines in the Indian Ocean after confirming with the real-time sea-level observations (Hettiarachchi, 2018).

In the Mediterranean and north-eastern Atlantic, the NEAMTWS provides rapid tsunami hazard assessments for its 39 member states by starting with tsunami source determination using their core seismic network. It is then followed by rapid model estimates of the arrival time and amplitudes of the tsunami at all forecasting points, which will be confirmed by sea-level data collected by tide gauges and deep-ocean buoys throughout the region. NEAMTWS has taken steps to implement probabilistic methodologies to quantify forecast uncertainties in real time based on rapid probabilistic analyses (Selva et al., 2021).

13.3 Long-term hazard assessment

13.3.1 Challenges for long-term assessment

Long-term tsunami hazard assessments present multiple challenges that require careful consideration and ongoing research efforts to address. The key challenges include uncertainties in tsunami recurrence intervals due to limited historical records, potential changes in geological processes, and the complexities of subduction zones and fault systems.

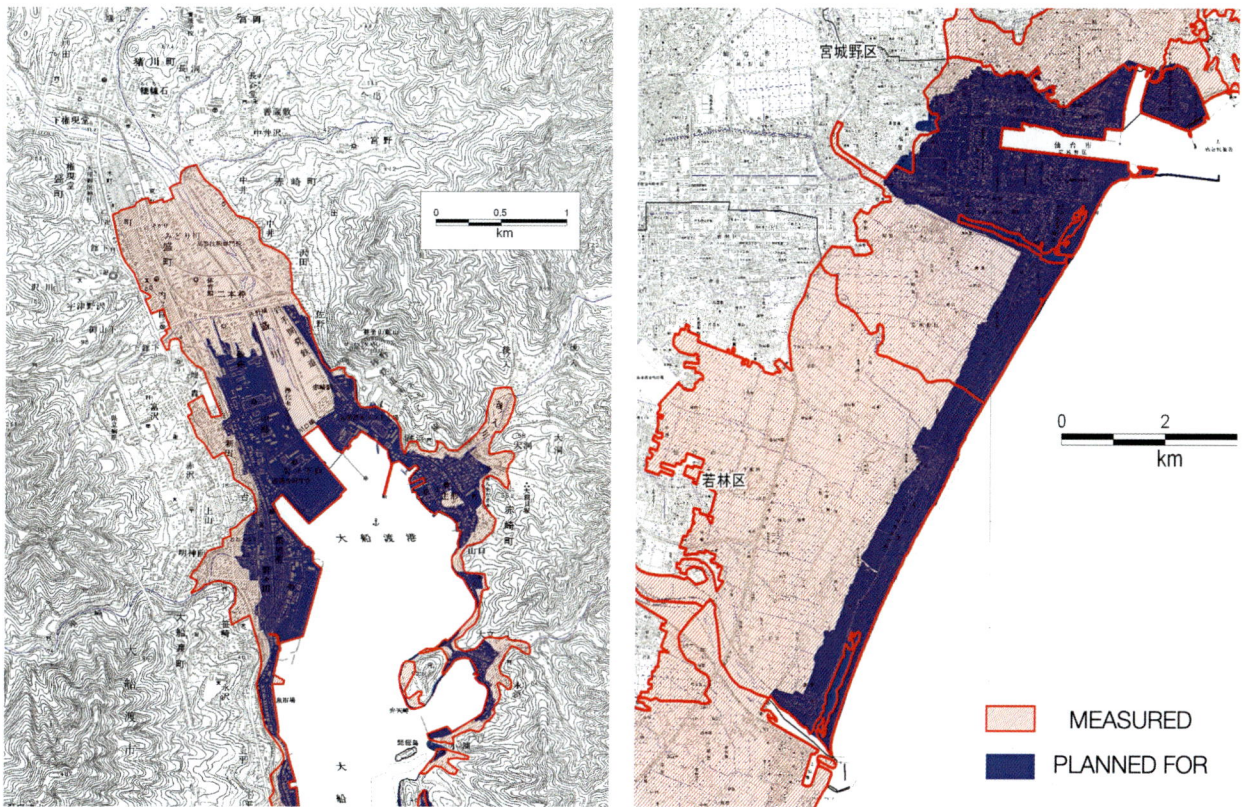

FIGURE 13.13 Examples of pre-2011 tsunami evacuation maps of Ofunato City, Iwate Prefecture (left) and Sendai City, Miyagi Prefecture (right) versus the measured inundation resulting from the March 11, 2011 Japan Tsunami, where the blue polygon represents the planned evacuation zone, and the red polygon represents the flood zone from the 2011 Japan Tsunami. Map lines delineate study areas and do not necessarily depict accepted national boundaries. *From Japanese Cabinet Office.*

Incomplete historical records, especially in regions with limited historical documentation or where tsunamis occur infrequently, make it difficult to assess the full extent of potential tsunami hazards and understand the long-term recurrence patterns. Fig. 13.13 shows the tsunami hazards reflected from inundation maps developed before 2011 versus the measured inundation extent from the 2011 Japan Tsunami for Ofunato, Iwate Prefecture, and Sendai, Miyagi Prefecture of Japan. In both areas, the extent of the flooding damage exceeded far beyond what was anticipated from the prior tsunami hazard mapping. Rapid coastal development and urbanization lead to changes in land use, infrastructure, and population density, affecting the vulnerability of coastal communities to tsunamis. Assessing the long-term hazards in the context of changing coastal landscapes presents a significant challenge for planners and researchers. Climate change and sea-level rise also influence the long-term tsunami hazard by altering coastal geomorphology and potentially increasing the frequency and impact of tsunamis. Communicating the results of long-term tsunami hazard assessments to at-risk communities and fostering public awareness and preparedness pose significant challenges. Overcoming cultural, linguistic, and educational barriers is essential for ensuring that communities understand and respond effectively to long-term tsunami risks. Provided in the next sections are methods and examples of deterministic and probabilistic long-term tsunami hazard assessment approaches conducted at PMEL to address some of these challenges.

13.3.2 Methods

13.3.2.1 Literature review and data collection

The long-term tsunami hazard assessment starts with a literature review to gain knowledge from different publications about a study site, including historical, present, and potential tsunami impact on the location. This will be followed by data collection focusing on historical records of water level, tide gauge data, inundation, and runup from witnesses and publications. The NCEI historical tsunami database presents a reliable resource to obtain many of these historical data worldwide (https://data.noaa.gov/metaview/page?xml = NOAA/NESDIS/NGDC/MGG/Hazards/iso/xml/G02151.xml&view = getDataView).

13.3.2.2 Global and regional digital elevation models and model grids

For a modeling study, high-quality and high-resolution DEMs for both bathymetry and topography are critical for accurate modeling. NCEI provides access to a variety of DEMs at different scales ranging from the globe to a city or county. The NCEI global relief ETOPO models are provided at three different resolutions, 15 arcseconds (~450 m), 30 arcseconds (~900 m), and 60 arcseconds (~1800 m) (NOAA National Centers for Environmental Information, 2022). Alternatively, the global relief data can be obtained through the General Bathymetric Chart of the Oceans (GEBCO) that is supported by the International Hydrographic Organization (IHO) and UNESCO. GEBCO is a global terrain model for ocean and land, providing global coverage of elevation data on grid resolution of 15 arcseconds (~450 m). GEBCO is traditionally more accurate at depths of 200 m or deeper, although high-quality bathymetric data in shallower water have been significantly improved over the past years due to the inclusion of the Electronic Nautical Charts contributed by IHO member states. Higher-quality DEMs are needed to model the tsunami wave shoaling process in the nearshore region of a study site, where a grid resolution of 3–6 arcseconds (90–180 m) is normally applied in PMEL's short-term modeling efforts. The NCEI coastal elevation model tool featured a bathymetric data viewer (https://www.ncei.noaa.gov/maps/bathymetry/) that gives users access to high-resolution bathymetric and topographic data developed along the U.S. coastlines. The DEMs supporting regional model studies are provided at a cell size of 3–24 arcseconds (90–720 m), while those supporting tsunami inundation computation have a minimum grid resolution of 1/3 arcseconds (~10 m). The finest grid resolution reaches 1/9 arcseconds (~3 m) to support high-quality coastal inundation modeling (Amante et al., 2023). Fig. 13.14 shows that the NCEI's coastal DEMs cover the U.S. coastlines and territories comprehensively, with exceptions along the uninhabited coastlines in Alaska and the Aleutians.

For the world's coastlines not covered by NCEI's DEM database, high-resolution DEMs, particularly the land topography, can be acquired from LiDAR surveys, satellite-derived datasets, nautical charts, and local authorities' data. One useful satellite-derived dataset is the elevation map FABDEM (Forest And Buildings removed Copernicus DEM) developed at the University of Bristol, United Kingdom. FABDEM provides global coverage of land elevations at the grid resolution of 1 arcsecond (~30 m) that removes building and tree height from the Copernicus GLO 30 DEM (Hawker et al., 2022). The global elevation dataset TanDEM-X (TerraSAR-X add-on for Digital Elevation Measurements, https://tandemx-science.dlr.de/) is derived from an earth observation radar mission by the German Aerospace Center (DLR) to create a precise 3D map of the Earth's land surfaces that is homogeneous in quality and unprecedented in accuracy. TanDEM-X has a global coverage of land elevation at 12 m resolution. Unlike FABDEM, the TanDEM-X data have not been processed to remove buildings, vegetation, and other structures to provide the "bare-earth" topography for inundation modeling.

DEMs are used to develop model grids at different resolutions to propagate tsunami waves from an ocean-basin (or even a global) scale to the site of interest. Nested grids are the most used strategy in tsunami inundation modeling to compensate this scale variation. For instance, the MOST model utilizes one-way nested computational grids to telescope down to the high-resolution area of interest for inundation computation. The numerical coupling between all nested grids in MOST is unidirectional from the outer grid. That is, the inner grid has no effect on the outer grid, and the outer grid provides the inner grid with computed wave amplitudes and flow velocities at all four boundaries by linear interpolation from coarser resolution. MOST usually employs a grid resolution ranging from 30 arcseconds (~900 m) to 4 arcminutes (~7.2 km) to model tsunami propagation at the levels of global, ocean basin, or regional sea simulations, depending on the water depth and the size of the model domain. A grid resolution of 3–15 arcseconds (90–450 m) is commonly used in the transition zone from offshore (1000–2000 m) to nearshore (100–500 m). A grid resolution of 1–2 arcseconds (30–60 m) is appropriate for nearshore areas, where water depths are less than hundreds of meters, to capture the shoaling effect of the tsunami waves. For long-term assessment, the tsunami inundation hazards are modeled at a grid resolution of 10 m or finer to conform with the requirements of the United States National Tsunami Hazard Mitigation Program (NTHMP).

13.3.2.3 Tsunami model benchmarking

To ensure the accuracy and quality of tsunami hazard assessments, the tsunami model used for inundation mapping is strongly recommended to be benchmarked following the NTHMP standards and validated for the modeling study. NTHMP has developed a number of benchmarking exercises that include analytical, laboratory, and real-world

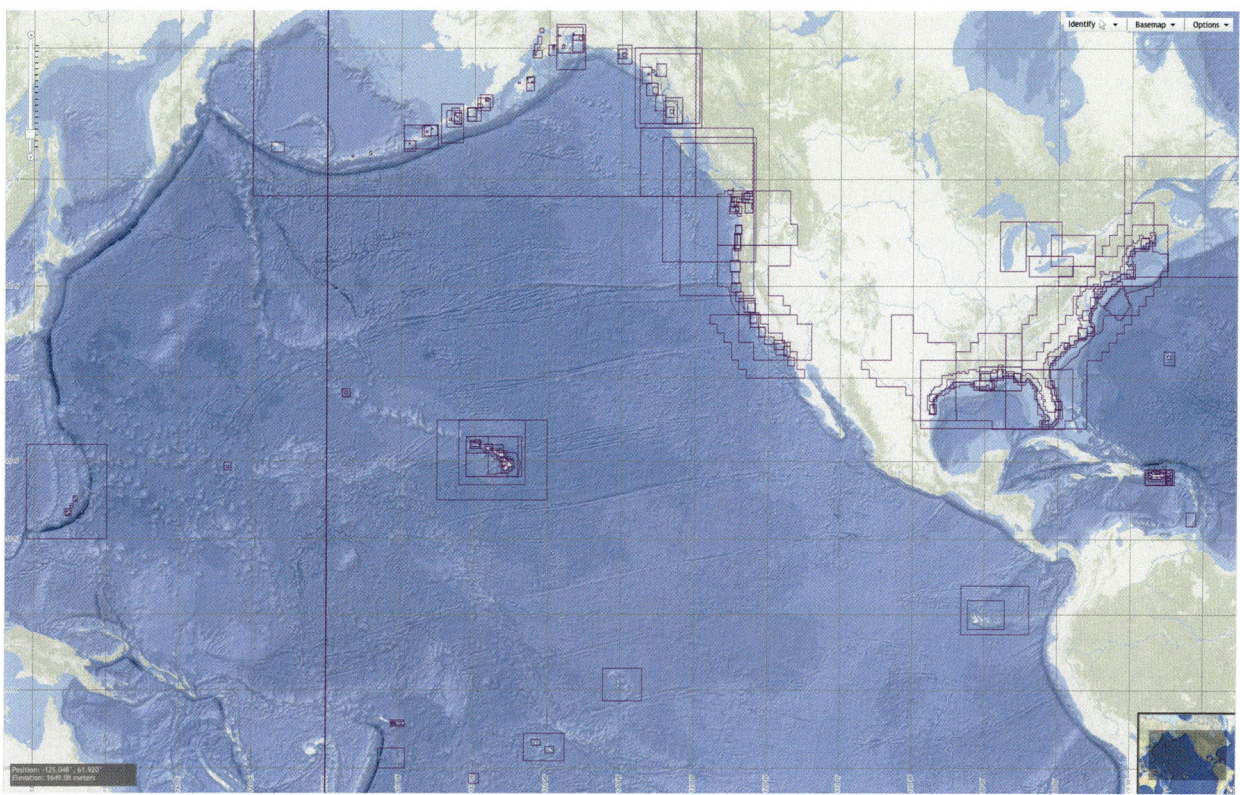

FIGURE 13.14 Coverage of NCEI's coastal digital elevation models. *NCEI*, National Center for Environmental Information. Map lines delineate study areas and do not necessarily depict accepted national boundaries. *From NCEI.*

simulations, available in the NOAA special report (https://nws.weather.gov/nthmp/documents/nthmpWorkshopProcMerged.pdf) and (Horrillo et al., 2015). For example, the MOST model has been extensively tested against a number of laboratory experiments and benchmarks and proved successful for several historical tsunami events (Synolakis et al., 2008; Tang et al., 2008, 2009, 2012; Titov et al., 2005; Titov, 2009; Uslu et al., 2011; Wei et al., 2008; Zhou et al., 2012). PMEL's NOAA Center for Tsunami Research (NCTR) event website (https://nctr.pmel.noaa.gov/database_devel.html) provides model forecast results for many tsunami events over the past two decades, fully demonstrating the quality and accuracy of the MOST model.

13.3.2.4 Model validation using historical data

Model validation using historical data is a required step at NCTR to ensure the model parameters, including the constructed model grids, are appropriate for the study site. The historical data include tide gauge records, field surveys of tsunami runup, height, or inundation from past events at the study site or vicinities. Tang et al. (2009) discussed how the model validation was conducted to develop the tsunami inundation models for real-time forecasting for four sites in Hawaii. Sensitivity tests of nearshore tsunami wave characteristics were conducted for a variety of model grid setups, resolutions, and parameters. Each of the four sites was validated using 14 historical events. The modeled inundation at Hilo was also examined with field survey results from the 1946 Unimak, Alaska Tsunami. These testing and validation procedures were adopted in most of the NCTR's short- and long-term site-specific hazard assessments.

The NCEI tsunami database archives a rich set of historical data, such as runup heights, inundation distance, witness reports, publications, and paper marigrams for historical events. Tide gauge records in the United States can be extracted from NOAA's tides and currents website (https://tidesandcurrents.noaa.gov/). The tsunami-capable stations provide data at a sampling rate of 1 minute. As shown in Fig. 13.2C, all DART buoy data are archived at the NDBC (https://www.ndbc.noaa.gov/). The IOC sea-level station monitoring facility provides access to real-time water level measurements at numerous stations contributed by 175 countries and regions globally (https://www.ioc-sealevelmonitoring.org/index.php).

Tsunami scenarios to be modeled for a site-specific modeling assessment can be obtained through three different approaches: a deterministic method, a probabilistic method, or a hybrid of both, which will be discussed in the next two sections. Once these model inputs are ready, the inundation model runs can be carried out for all scenarios to develop map products and draw assessment conclusions.

13.3.3 Deterministic tsunami hazard assessment

Scenario-based deterministic methods are commonly used in long-term tsunami hazard assessments, particularly for developing tsunami inundation maps. It involves the prediction and assessment of potential tsunami impacts based on specific scenarios. Deterministic assessments are a vital component of tsunami risk management and mitigation efforts. At NCTR, these deterministic assessments have been conducted utilizing different approaches based on: (1) a historically worst-case event, (2) single or multiple worst-credible scenario(s) derived from historical data and source characteristics, and (3) a sensitivity study with synthetic tsunamis generated by earthquakes of single or multiple magnitudes along major subduction zones to identify probable maximum tsunamis. In this section, case studies are discussed to illustrate how these deterministic studies are performed and how these modeling assessments provide valuable information for emergency management, land-use planning, and evacuation plans.

13.3.3.1 Worst-case scenarios of crustal faults in the Pacific Northwest

Crustal faults across the Salish Sea in the Pacific Northwest of the United States are capable of $M7+$ earthquakes and have the potential to generate hazardous tsunamis in the Puget Sound and Strait of Georgia. Historically, the tsunami inundation mapping for the Puget Sound started with the model assessment of tsunami impact caused by the Seattle Fault. Based on historical accounts of the source magnitude and field evidence of seismic deformation, Titov et al. (2003) derived a worst-case source model of the Seattle Fault using six subfaults with a source magnitude of 7.3 (Fig. 13.15A), which appeared to agree well with the surface displacements produced by the 900 A.D. Seattle Fault Earthquake (Bucknam et al., 1999; ten Brink et al., 2002) at multiple locations, including a maximal 7-m uplift at the Restoration Point. The model results show significant tsunami flooding in Elliott Bay (Fig. 13.15B). An updated study by Venturato et al. (2007) modeled the tsunami inundation in Tacoma, Washington using a similar Seattle Fault geometry but a greater downdip width (35 km) and a couple of source configurations for the Tacoma Fault (Fig. 13.15A). All three scenarios caused inundation along the Port of Tacoma and Puyallup River but presented different wave impacts—the Seattle Fault produced inundation during the initial wave (Fig. 13.15C) but the Tacoma Fault created inundation over time. Recently, adopting the source configuration of Venturato et al. (2007), the Washington Geological Survey (WGS) conducted a newly updated modeling assessment of the tsunami inundation, current speeds, and arrival times generated in the Puget Sound from the Seattle Fault (Dolcimascolo et al., 2022). Two tsunami models, MOST and the Geophysical Conservation Laws Package (GeoClaw), were used to model different areas in the Puget Sound. A comparative study between the two models was conducted for Bainbridge Island and its vicinity, where the larger value from the two models was used to obtain the final map. The new study utilized the high-resolution 1/3 arcseconds (~ 10 m) bare-earth DEM developed by NCEI in 2014. Figs. 13.15D and E show the final inundation maps for the Eastern South Sound and the East Passage published by the WGS, respectively. It is worth noting that a more recent study by Black et al. (2023), using dendrochronological dating and a cosmogenic radiation pulse to constrain the death dates of earthquake-killed trees, indicated that the Seattle Fault may have occurred either as a single composite M_w 7.8 earthquake or a double earthquake sequence of M_w 7.5 and M_w 7.3. Whether this new study will trigger future modification of the newly developed WGS inundation mapping for the Puget Sound remains unknown at this point.

13.3.3.2 Worst-credible scenarios for Hawaii and the Caribbean Sea

A worst-credible scenario is the scenario considered plausible and reasonably believable among experts and communities to produce the most severe consequences. Unlike the worst-case scenario, there is usually none or very little historical evidence and records to prove the worst-credible scenario(s) have occurred or have caused a severe impact on the study site. The worst-credible scenarios are mostly derived from a combination of historical data and source characteristics based on hypotheses.

The tsunami evacuation maps in Hawaii State have two tiers. The tier-one standard evacuation maps are based on historical tsunamis, while the tier-two extreme maps represent hazards from a worst-credible scenario of a magnitude 9.2 earthquake in the eastern Aleutians (Butler et al., 2017), which is believed to far exceed the flooding observed in Hawaii from past historical tsunamis. Geological evidence discovered in the Makauwahi Sinkhole on the Island of

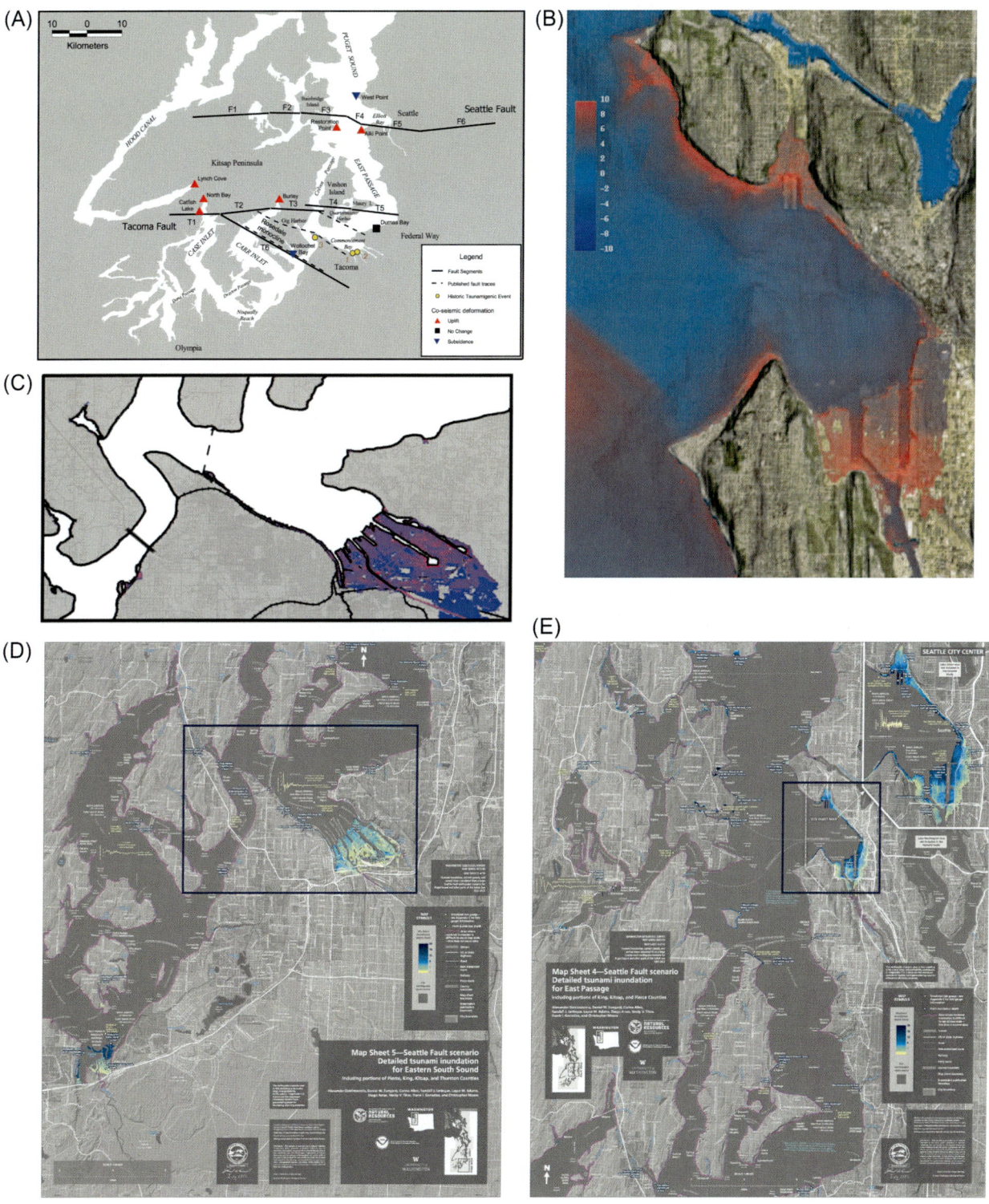

FIGURE 13.15 Deterministic assessments of tsunami hazards in the Puget Sound from the worst-case Seattle Fault. (A) Source configurations of the Seattle Fault, Tacoma Fault, and the Rosedale-Tacoma Fault (Venturato et al., 2007), (B) modeled tsunami inundation at the Tacoma Port (Venturato et al., 2007), (C) modeled tsunami inundation at the Elliot Bay (Titov et al., 2003), (D) updated tsunami inundation maps for the Easter South Sound caused by the Seattle Fault (Dolcimascolo et al., 2022), and (E) updated tsunami inundation maps for the East Passage of Puget Sound caused by the Seattle Fault (Dolcimascolo et al., 2022). Map lines delineate study areas and do not necessarily depict accepted national boundaries. *Modified from (A and B) Venturato et al., 2007; (C) Titov et al., 2003; (D and E) Dolcimascolo et al., 2022.*

358 SECTION | 2 Advanced topics and applications related to probabilistic tsunami hazard and risk analysis

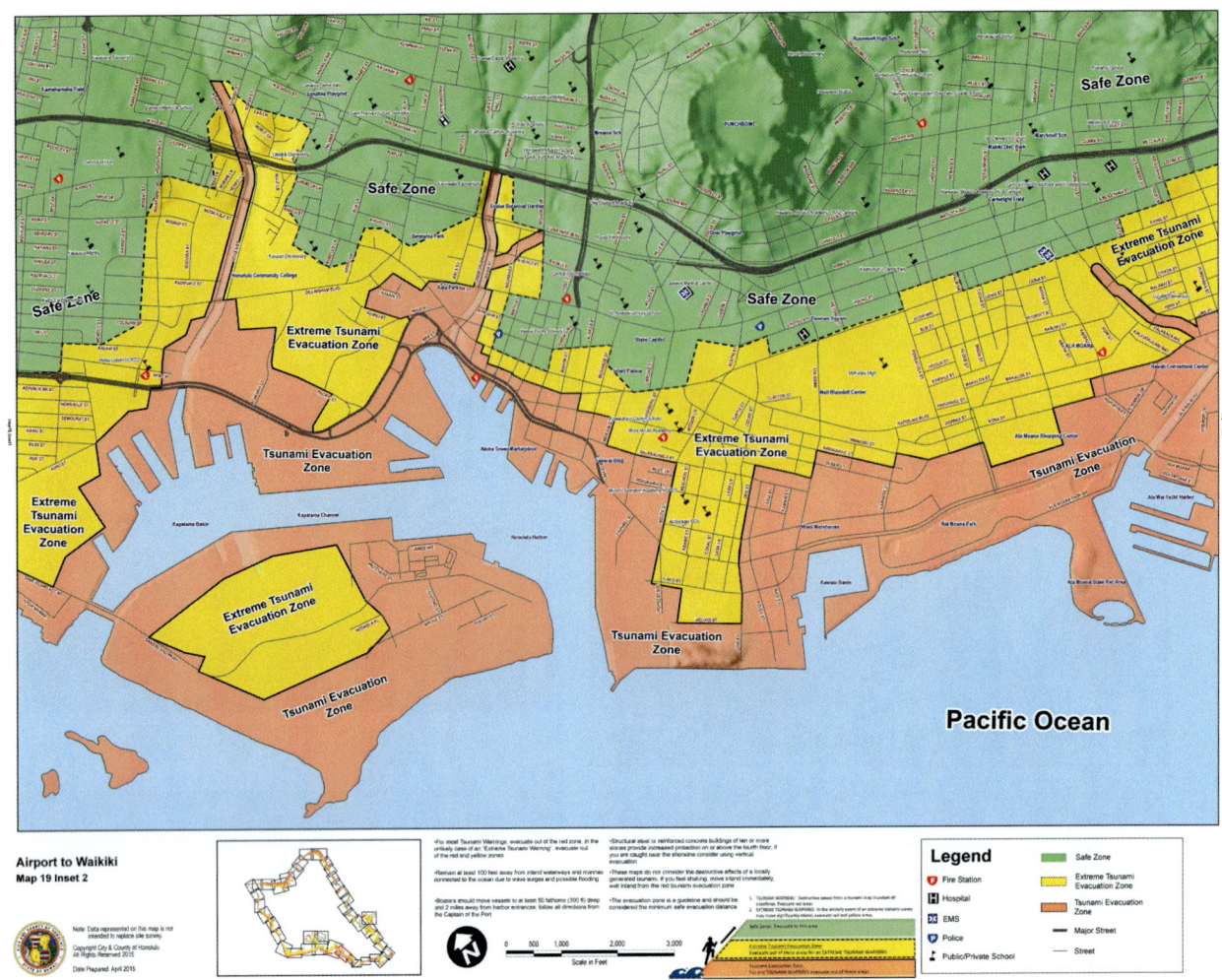

FIGURE 13.16 Two-tier tsunami evacuation maps for the coastlines between the Honolulu International Airport and Waikiki, Oahu, Hawaii, where the brown-colored area indicates the tier-one standard tsunami evacuation zone based on historical events, and the yellow-colored area represents the tier-two extreme tsunami evacuation zone based on the worst-credible scenario. Map lines delineate study areas and do not necessarily depict accepted national boundaries. *From City and County of Honolulu, Department of Planning and Permitting.*

Kauai may support this hypothesis (https://manoa.hawaii.edu/news/article.php?aId = 8635). Fig. 13.16 gives an example of such two-tier evacuation maps developed for the coastlines between the Honolulu International Airport and Waikiki, Oahu, Hawaii. The tier-two extreme evacuation map features a much broader flooding zone than the tier-one map, leading to more conservative tsunami hazard mitigation planning in the State of Hawaii.

For the Caribbean Sea, NCEI and the Universidad Nacional, Costa Rica, developed a map viewer named the "Caribbean and Adjacent regions Tsunami Sources And Models (CATSAM)" that presents a collection of worst-credible source scenarios for model assessment of tsunami hazards in the Caribbean and adjacent regions. The CATSAM effort aims at defining tsunami potential within the region. One can obtain the detailed parameters of each earthquake source from the NCEI-hosted CATSAM interface (https://www.ncei.noaa.gov/maps/CATSAM/). Fig. 13.17 is a snapshot of the CATSAM interface including an example of the source parameters of a hypothetical M_w 8.71 earthquake in the Puerto Rico Trench determined at an IOC-led expert meeting in 2016.

13.3.3.3 Deterministic hazards assessment based on sensitivity tests using synthetic tsunamis

The deterministic sensitivity test of synthetic tsunami scenarios represents a unique hazard assessment method at NCTR fully utilizing NOAA's tsunami propagation database, which features ~2000 unit tsunami sources covering major subductions around the world. It presents an effective way to identify the most threatening tsunami sources and their impacts on the study area.

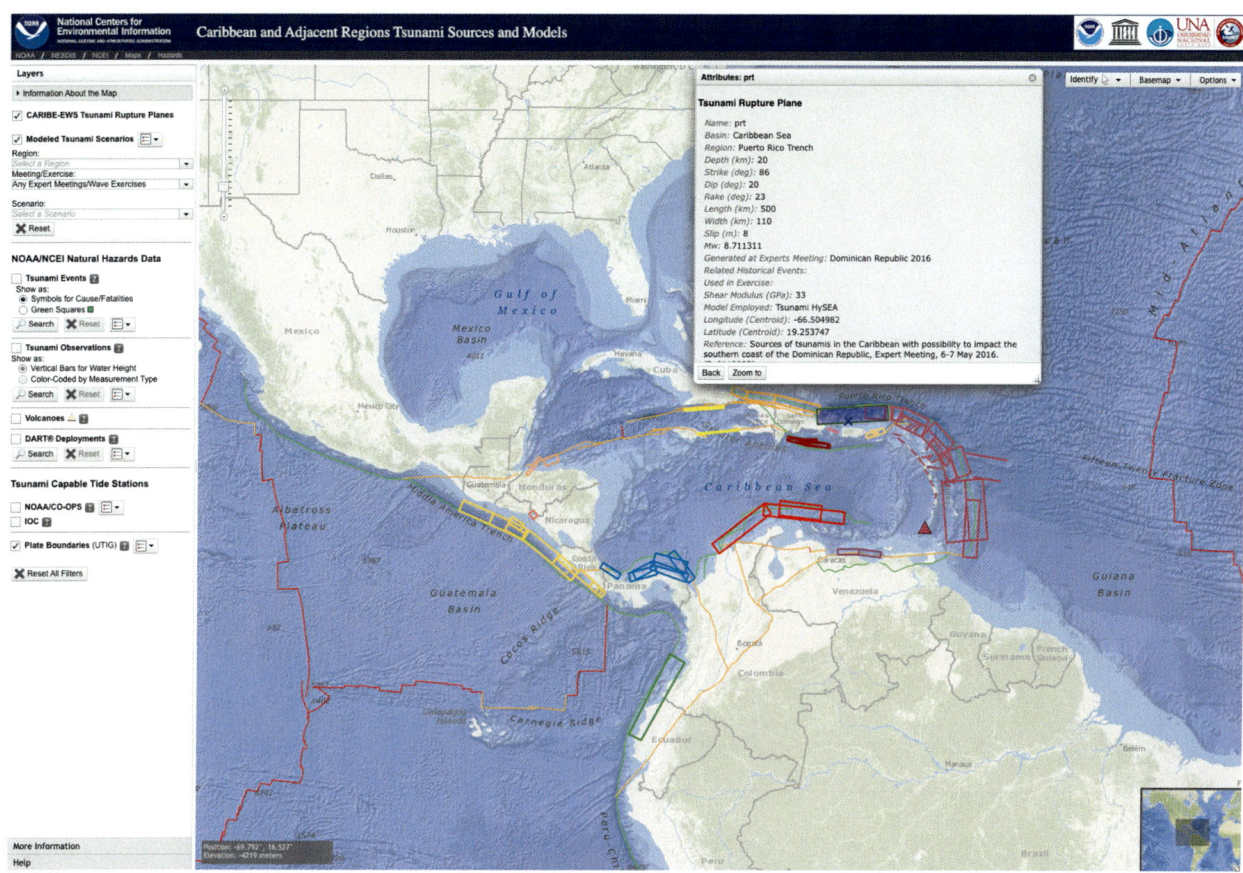

FIGURE 13.17 The NCEI-hosted CATSAM interface. The inset shows the source parameters and other information about a hypothetical M_w 8.71 earthquake in the Puerto Rico Trench, with its epicenter and fault area represented by the blue cross and the transparent blue box, respectively. *CATSAM*, Caribbean and Adjacent regions Tsunami Sources And Models; *NCEI*, National Center for Environmental Information. Map lines delineate study areas and do not necessarily depict accepted national boundaries. *From NCEI (https://www.ncei.noaa.gov/maps/CATSAM/).*

Using the tsunami unit sources, Tang et al. (2006) configured a set of 18 synthetic tsunamis of M_w 9.3 from all major subduction zones in the Pacific Ocean to identify the most threatening source for the Ford Island in Pearl Harbor, Hawaii. Each of these M_w 9.3 scenarios is composed of 20 of the 100 km × 50 km unit sources (a total rupture area of 1000 km × 100 km) with a uniform slip of 29 m. Fig. 13.18A shows the configuration of the 18 scenarios utilizing grouped unit sources. The sensitivity tests indicated that the M_w 9.3 Kamchatka scenario produced the most severe impact in Pearl Harbor among the 18 scenarios (Figs. 13.18B and C). The test results also indicated that for M_w 9.3 events hazardous waves are likely to be created by sources from the Kamchatka, East Philippines, Japan, Alaska-Aleutian, South America, and Cascadia subduction zone (CSZ).

Adopting the same practice, Uslu et al. (2010, 2013) performed similar sensitivity tests for hazard assessments at locations in Guam and the Commonwealth of the Northern Mariana Islands (CNMI). These studies investigated the hazards produced from multiple source magnitudes and from more source combinations. Specifically, Uslu et al. (2010) conducted the modeling test for M_w 8.5 and 9.0 and a total of 725 source scenarios, all composed of precomputed unit source propagation database (Figs. 13.19A and B). In the CNMI study, Uslu et al. (2013) used 349 M_w 9.0 earthquake sources to study inundation impact on those islands, and the results showed 26 potential earthquake scenarios from West Aleutians, CSZ, Philippines, Japan Trench, Marianas, Manus Trench, New Guinea, and Ryukyu-Nankai are capable of producing hazardous tsunami impact to the CNMI. Model results predicted that an M_w 9.0 earthquake originating from a source south of Japan could result in wave amplitudes exceeding 11 m in Saipan, and an M_w 9.0 earthquake occurring in the East Philippines could trigger tsunami wave amplitudes exceeding 3 m at Rota and 4 m at Saipan and Tinian (Figs. 13.19C, D, and E).

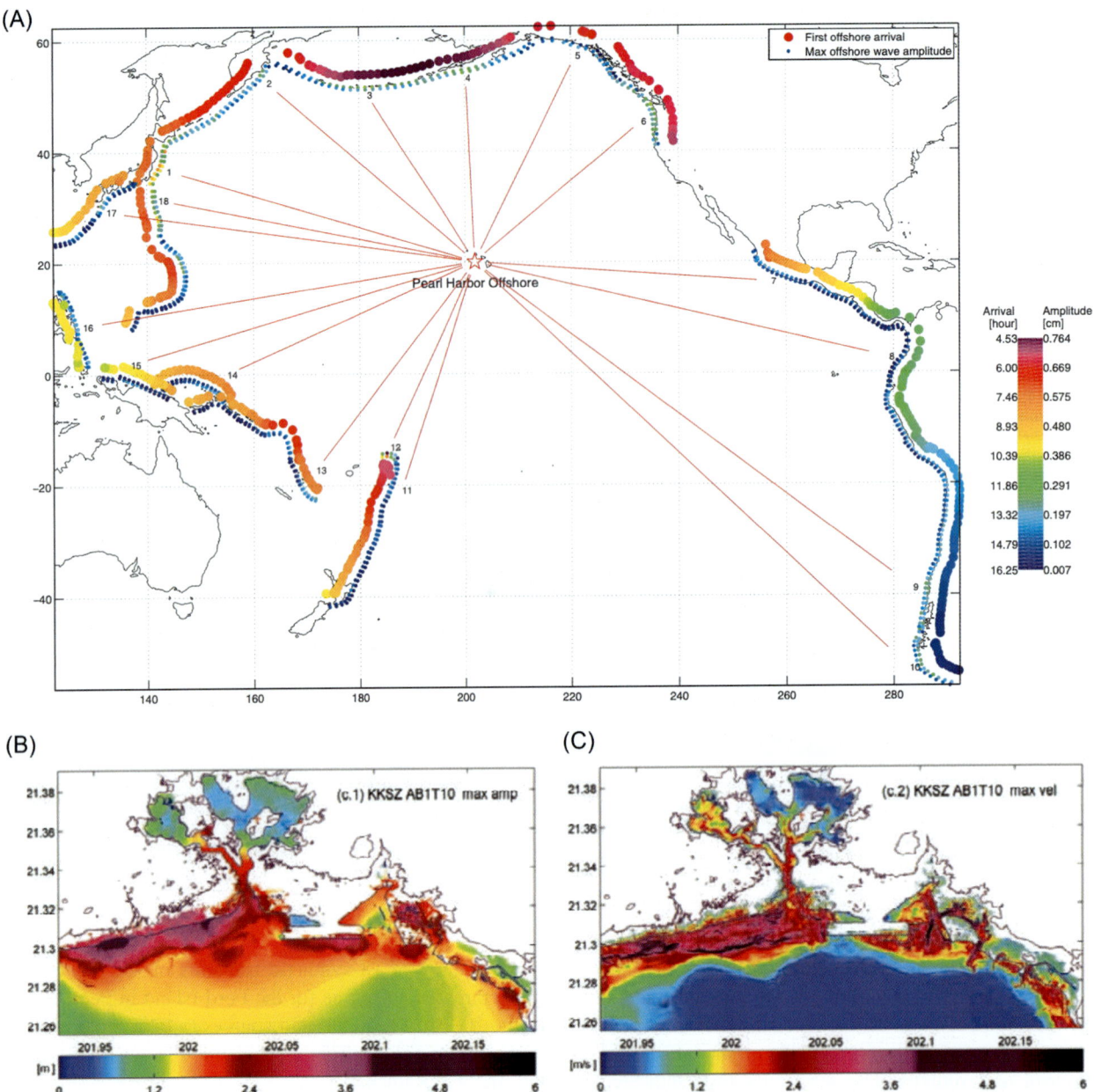

FIGURE 13.18 Model assessment of tsunami impact at Pearl Harbor, Honolulu, Hawaii using synthetic M_w 9.3 earthquakes around the Pacific Rim. (A) First arrival and maximum tsunami amplitude offshore Pearl Harbor. The dots indicate the centers of the unit sources. The color on the large-size dots indicates the arrival time from that unit source to Pearl Harbor, while the color on the small dots shows the tsunami amplitude offshore Pearl Harbor produced by each unit source scenario. The red line represents the 18 scenarios tested in Tang et al. (2006). (B) The maximum tsunami wave amplitude in Pearl Harbor and its vicinity produced by the Kuril-Kamchatka M_w 9.3 scenario, the most threatening source out of the 18 scenarios. (C) The maximum tsunami current speed produced by the Kuril-Kamchatka M_w 9.3 scenario. *From Tang, L., Chamberlin, C., Tolkova, E., Spillane, M., Titov, V. V., Bernard, E. N., & Mofjeld, H. O. (2006). Assessment of potential tsunami impact for Pearl Harbor, Hawaii. NOAA Tech. Memo. OAR PMEL.*

13.3.4 Probabilistic tsunami hazard assessment—the American Society of Civil Engineers framework of tsunami design zone

Although deterministic methods inform authorities and emergency management of high-risk areas and extreme potential hazards, they rely only on specific scenarios and may not account for less predictable events. Probabilistic methods, on the other hand, have several advantages over deterministic methods: (1) PTHA considers numerous (tens to thousands

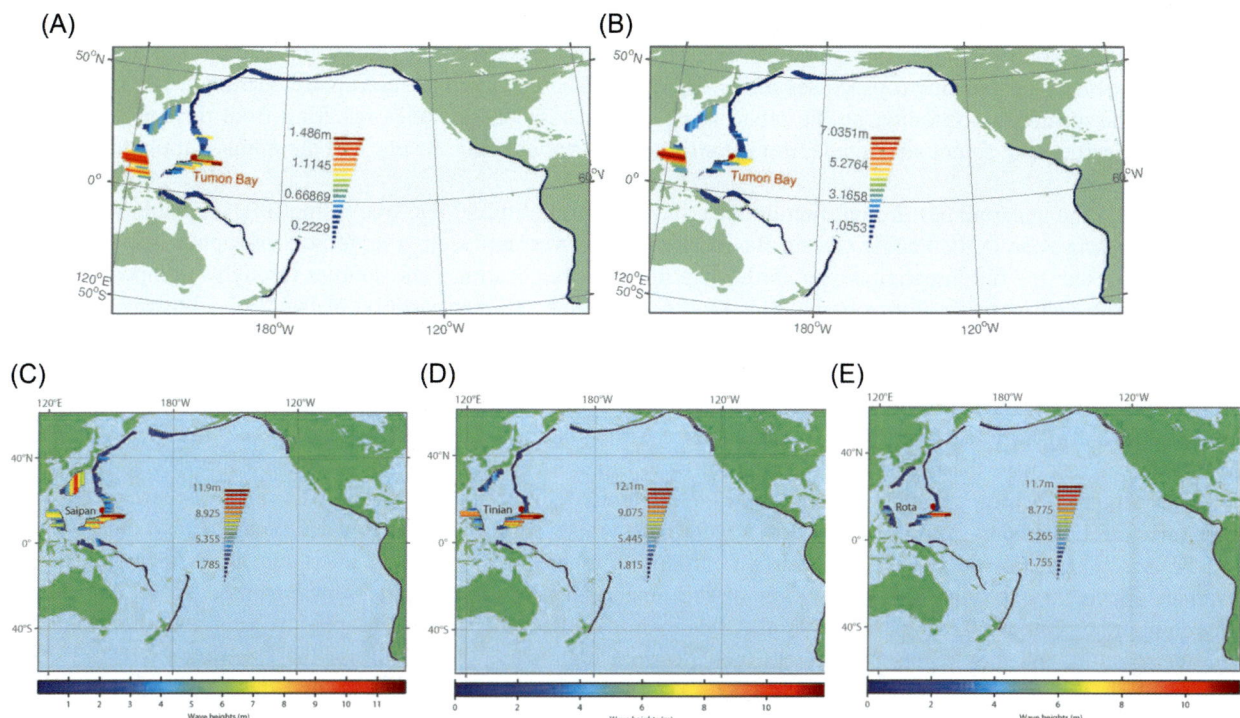

FIGURE 13.19 Deterministic model assessment for Guam (Uslu et al., 2010) and the Commonwealth of the Northern Mariana Islands (Uslu et al., 2013). (A) and (B) show the maximum tsunami amplitude computed for Tumon Bay, Guam from tsunamis triggered by synthetic M_w 8.5 and 9.0 earthquakes, respectively, along the Pacific Basin subduction zones. (C), (D), and (E) show the maximum tsunami amplitude computed for Saipan, Titian, and Rota, respectively, from tsunamis triggered by synthetic M_w 9.0 earthquakes along the Pacific Basin subduction zones. Map lines delineate study areas and do not necessarily depict accepted national boundaries. *From Uslu, B., Eble, M., Arcas, D., & Titov, V. (2013). Tsunami hazard assessment for the Commonwealth of the northern Mariana Islands. NOAA OAR Special Report, 3. https://nctr.pmel.noaa.gov/hazard_assessment_reports/H03_CNMI_3949_lowres.pdf; Uslu, B., Titov, V. V., Eble, M., & Chamberlin, C. (2010). Tsunami hazard assessment for Guam. NOAA OAR Special Report, Tsunami Hazard Assessment Special Series (Vol. 1, pp. 186).*

or more) scenarios and provides a more complete understanding of the overall hazards including rare but potentially catastrophic events; (2) PTHA can be used to quantify uncertainties by integrating a range of potential scenarios and associated probabilities, which the TWCs are keen to include in their real-time forecast; (3) PTHA provides a nuanced view of potential tsunami hazards, such as the annual probability of exceedance, which supports clearer communication with stakeholders, policymakers, and the general public; (4) PTHA naturally aligns with broader risk analysis frameworks, facilitating the integration of tsunami hazards into comprehensive risk management strategies; and (5) PTHA is well suited for informing building codes, land-use planning, and infrastructure design standards, supporting the development of more resilient coastal communities. With these merits, probabilistic methods are becoming widely used in long-term assessments and engineering design across the world.

In the United States, a groundbreaking achievement in probability-based tsunami hazard assessment is the establishment of the American Society of Civil Engineers (ASCE) 7-16 Tsunami Design Provisions in 2016. It is the first U.S. national, consensus-based standard for engineering design for tsunami loads and effects on buildings for the States of Alaska, Hawaii, California, Oregon, and Washington (American Society of Civil Engineers, 2017; Chock et al., 2018). This standard was also adopted by the International Building Code to address global hazards and risks of tsunami impact on building design. This section focuses on the model development of the ASCE tsunami design zone (TDZ), which is defined as the area vulnerable to being inundated by the Maximum Considered Tsunami that corresponds to a hazard with a 2% probability of being exceeded in 50 years (Chock et al., 2018).

13.3.4.1 Derivation of probabilistic 2475-year offshore tsunami amplitudes

The ASCE 7 tsunami provisions strictly require the TDZ maps to be developed by integrating generation, propagation, and inundation tsunami models that replicate the given offshore tsunami amplitude and period obtained from PTHA. In the ASCE tsunami provisions, the method by Thio (2019) is used to obtain the PTHA 2475-year offshore amplitudes for the five Pacific states of the United States, including Alaska, California, Hawaii, Oregon, and Washington. Thio (2019) considered both the epistemic and aleatory uncertainties. The epistemic uncertainty addresses human's incomplete understanding of the natural processes of the earthquake sources, such as the rupture characteristics in dip, strike, and rake. The aleatory uncertainty, on the other hand, accounts for uncertainties resulting from the random nature of modeling in earthquake source magnitude, rupture slip amount, variable tide levels, and the numerical model of tsunami waves.

To address the uncertainties in fault rupture area and slip variability, a logic tree framework is established for every major subducting zone in the Pacific Ocean. Based on the logic tree approach, a large pool of rupture scenarios is generated to represent the full integration over earthquake magnitudes, locations, and sources through a composition of predetermined subfault partition. For example, Fig. 13.20A shows the subfault partition of the CSZ following the Slab 1.0 geometry (Hayes et al., 2012). Although each subfault patch is 1 km $\times$ 1 km to accommodate the geometrical complexity, the actual analysis was conducted such that the slip on every 30 km $\times$ 10 km subfault is uniform (Thio, 2019).

A linear shallow water model COMCOT (Liu et al., 1995) was used to obtain a database of Green's functions at the strategic PTHA offshore locations at 100 m depths for every subfault. These fundamental Green's functions can be linearly combined to obtain the tsunami waveforms for each rupture scenario that is associated with a probability defined by the earthquake recurrence model, which defines the magnitude of earthquakes with their rate of occurrence. The aleatory uncertainties arising from the model errors were included based on a misfit between the model and data.

The ASCE probabilistic analysis utilized two earthquake recurrence models the maximum magnitude and the characteristic model to determine the upper limit of the magnitudes for large faults. Three earthquake scaling relations (Murotani et al., 2008, 2013; Papazachos et al., 2004; Strasser et al., 2010) were adopted to determine the rupture area and average slip, and they represent different branches in a logic tree (an example of the CSZ logic tree is shown in Fig. 13.20B). The recurrence models for the CSZ, Alaska-Aleutian subduction zone, and the Japan-Kuril-Kamchatka subduction zone adapted from existing studies (Fujiwara et al., 2006; Witter et al., 2016; Petersen, et al., 2014) were extended to include multisegment ruptures. For the remaining subduction zones around the Pacific Rim, the recurrence models were generic: a full-rupture geometry following Slab 1.0 (Hayes et al., 2012) and their recurrence rates based on plate convergence rates.

Slip variability of the earthquake source plays a key role in local tsunami impact. To address the aleatory uncertainty caused by the slip variability, the ASCE PTHA utilized variable slip rupture models with one-third of the rupture as an asperity with twice the average slip and the other two-thirds of the rupture at half the average slip. For each event, three scenarios are computed where the asperity occupies every part of the rupture once to avoid over or underestimation of the tsunami hazards due to incomplete or overlapping asperity coverage offshore (Thio, 2019). Fig. 13.20C illustrates two examples of variable slip rupture scenarios with different asperity locations for a full rupture and a partial rupture in the south of the CSZ.

Using Green's functions, more than 10,000 events from all the source regions in the Pacific Ocean were integrated to obtain the probabilistic results considering both epistemic (source characterization) and aleatory variability (modeling error). The wave amplitude exceedance rates at every 100-m offshore location were computed from the hazard curves reflecting all earthquake source compositions.

The PTHA source disaggregation analysis stores information on the relative contribution of every source to the offshore tsunami amplitudes, and it enables the users to identify the dominant source regions and magnitudes for a site at risk. For example, Fig. 13.20D shows that the PTHA disaggregation analysis finds the dominant tsunami hazards at a point (124.653°W, 47.003°N) offshore Westport, Washington, are mostly attributed from the CSZ earthquake sources, as indicated by the tallest bars (i.e., the largest source contribution). The source disaggregation analysis emphasizes dominant source regions and disregards less dominant sources, reducing the number of probabilistic scenarios to be considered for inundation modeling.

13.3.4.2 Derivation of offshore wave amplitudes

Based on the PTHA source disaggregation, the NOAA precomputed tsunami propagation database can be utilized to reconstruct tsunami sources from the most significant contributing source regions. The reconstructed tsunami scenarios need to provide a good approximation of the PTHA offshore amplitudes compliant with the ASCE requirements. These source scenarios can then be used as input to a verified tsunami inundation model to compute the TDZ for a site of interest. Wei et al. (2017) described the procedure used to produce the ASCE 7-16 TDZ maps based on the PTHA offshore wave amplitudes. The process includes three steps: (1) generating a database of tsunami waveforms, that is, Green's functions; (2) an inversion process to derive 2500-year tsunami sources in a system of delineated and

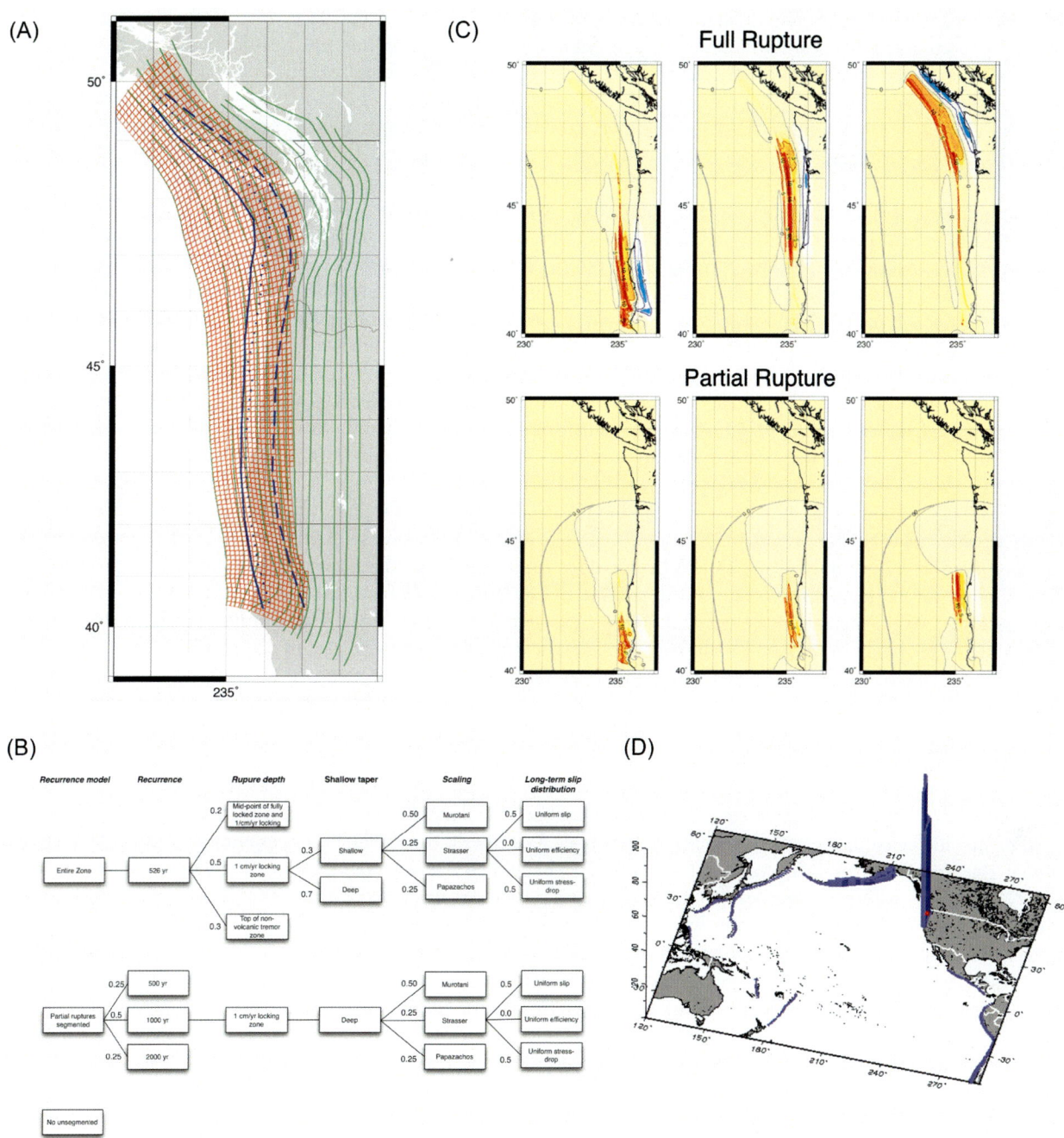

FIGURE 13.20 ASCE's probabilistic tsunami hazard assessment method and results for the CSZ. (A) 1 km × 1 km subfault partition based on Slab 1.0 (Hayes et al., 2012) for the CSZ, (B) logic tree extended from Petersen et al. (2014) for the CSZ to account for multisegment ruptures, (C) slip distributions for two events, fall rupture and partial rupture, illustrating different asperity occupancies for each event, and (D) probabilistic offshore exceedance amplitudes for the return period of 2475 years based on the CSZ sources. *ASCE*, American Society of Civil Engineers; *CSZ*, Cascadia subduction zone. *From Thio, H. K., & Li, W. (2015). Probabilistic tsunami hazard analysis of the Cascadia Subduction Zone and the role of epistemic uncertainties and aleatory variability. In Proceedings of the 11th Canadian Conference on Earthquake Engineering.*

discretized rectangular subfaults; and (3) adjustment of the inverted tsunami sources to satisfy the ASCE 7 requirements for matching offshore wave amplitudes.

The first step of the tsunami source reconstruction involves computing a database of Green's functions for every offshore location (100-m (328-ft) deep) where the tsunami amplitude is provided by the ASCE 7 Tsunami Geodatabase. It is worth pointing out that Green's functions are mostly computed in the tsunami model at a grid resolution suitable for basin-wide propagation. Since the offshore tsunami amplitudes in the NOAA database were obtained using a grid resolution of 4 arcminutes (7.2 km), which may not be fine enough to fully address the 100-m contour features. A nested grid level at a spatial resolution of 24–30 arcseconds (~720–900 m) is employed to obtain Green's functions at all 100-m offshore points.

In the second step, a nonlinear least squares inversion method is used to reconstruct the tsunami sources. Based on the PTHA source disaggregation, a group of unit sources is selected in the dominating rupture zones. The inversion method then adjusts the combination of the slip amount of each unit source until the model results match the PTHA offshore amplitudes so that the reconstructed sources are compliant with the ASCE requirements: (1) for the results obtained from the hydrostatic model, the mean value of the model-generated offshore wave amplitudes must be greater than 100% of the mean value of the corresponding ASCE 7 Tsunami Geodatabase values and (2) all individual model generated offshore wave amplitudes are greater than 80% of the corresponding ASCE 7 Tsunami Geodatabase values. Additional criteria have been amended in the ASCE 7-22 version for the results obtained from higher-order nonhydrostatic models. In these cases, the mean value of the model-generated offshore wave amplitudes must be greater than 85% of the mean value of the corresponding ASCE 7 Tsunami Geodatabase values, and all individual model-generated offshore wave amplitudes must be greater than 75% of the corresponding ASCE 7 Tsunami Geodatabase values. As a result, the final adjusted solution of slip combination across the selected subfaults represents a tsunami source that satisfies the ASCE 7-22 criteria for offshore tsunami amplitudes. It can then be used for site-specific inundation computation to develop the probabilistic TDZ maps.

For the TDZ computation, we mostly utilized the NCEI high-resolution DEMs developed at the spatial resolution of 1/3 arcseconds (~10 m) for selected U.S. coastal regions. These DEMs have as follows: (1) a global, geographic coordinate system; (2) a mean high water (MHW) vertical datum for modeling the maximum flooding; (3) a grid file format of the Environmental Systems Research Institute (ESRI) ArcGIS; and (4) bare earth with buildings and trees excluded from the DEM. Most regions along the U.S. West Coast are covered by these high-resolution DEMs. In Hawaii, 1/3 arcseconds DEMs are developed for most of the islands. The exceptions are east of Maui, which has a coarser grid resolution at 1 arcsecond (~30 m), and western and central Molokai and the southern tip of the Island of Hawaii have a grid resolution of 6 arcseconds (~180 m). In Alaska, high-resolution DEMs at a spatial resolution of 1 arcsecond (30 m) or coarser are only available for populated areas, and the remaining coastlines have much lower resolutions and are not suitable for inundation computation.

The ASCE 7-16 TDZ maps are computed using the MOST model. Due to the large extent of the coastlines and limited resources, it was not practical to carry out all inundation computations at a grid resolution of 1/3 arcseconds (~10 m) for the coastlines of all five states at that time. As a result, the inundation computation was carried out at an optimal grid resolution of 2 arcseconds (~60 m). The bathymetry and topography of all model grids were derived from NCEI DEMs based on their best available data at that time. As discussed earlier, MOST uses telescoped grids (A, B, and C grids) to account for tsunami wave transformation from deep water to onshore flooding. For a coastline of interest, the 24-arcsecond grid resolution was used in the A grid. A smaller B grid, with a grid resolution of 6 arcseconds (~180 m), was nested within the A grid to further capture tsunami wave characteristics at water depths of hundreds of meters. The 2-arcsecond (~60 m) grid resolution was used at the innermost C grid to compute tsunami inundation. Fig. 13.21 illustrates all C grids used to develop the ASCE TDZ for the U.S. West Coast. A total of 43 models were used to provide full coverage for the coastal regions of the U.S. West Coast, Hawaii, and Alaska. The vertical datum for all inundation computation is the MHW. A constant Manning's coefficient of 0.03 was applied to all inundation computations.

13.3.4.3 Updates of tsunami design zone maps

Accompanying the publication of the ASCE 7-16 tsunami provision, an ASCE Tsunami Hazard Tool (https://asce7tsunami.online/) was also developed to integrate all geocoded reference points of the PTHA offshore tsunami amplitude and period, runup elevation, and associated inundation limit of the TDZ. Fig. 13.22A shows the final coverage of coastlines in the ASCE 7-16 maps (American Society of Civil Engineers, 2017).

Since 2017, there have been continuous modeling efforts supported by state and federal governments to update the ASCE 7-16 low-resolution maps using up-to-date high-resolution (10 m) model grids. These high-resolution model grids provide improved descriptions of small terrain features, such as levees, roads, river channels, and sand dunes, that potentially affect the inundation hazards at many coastal communities. Figs. 13.22B and C demonstrate the differences between the ASCE 7-16 and 7-22 TDZs for coastlines between Los Angeles Harbor and Huntington Beach, California. The 10-m high-resolution TDZ maps in Hawaii State were completed in 2023 for the entire Island of Oahu and most of

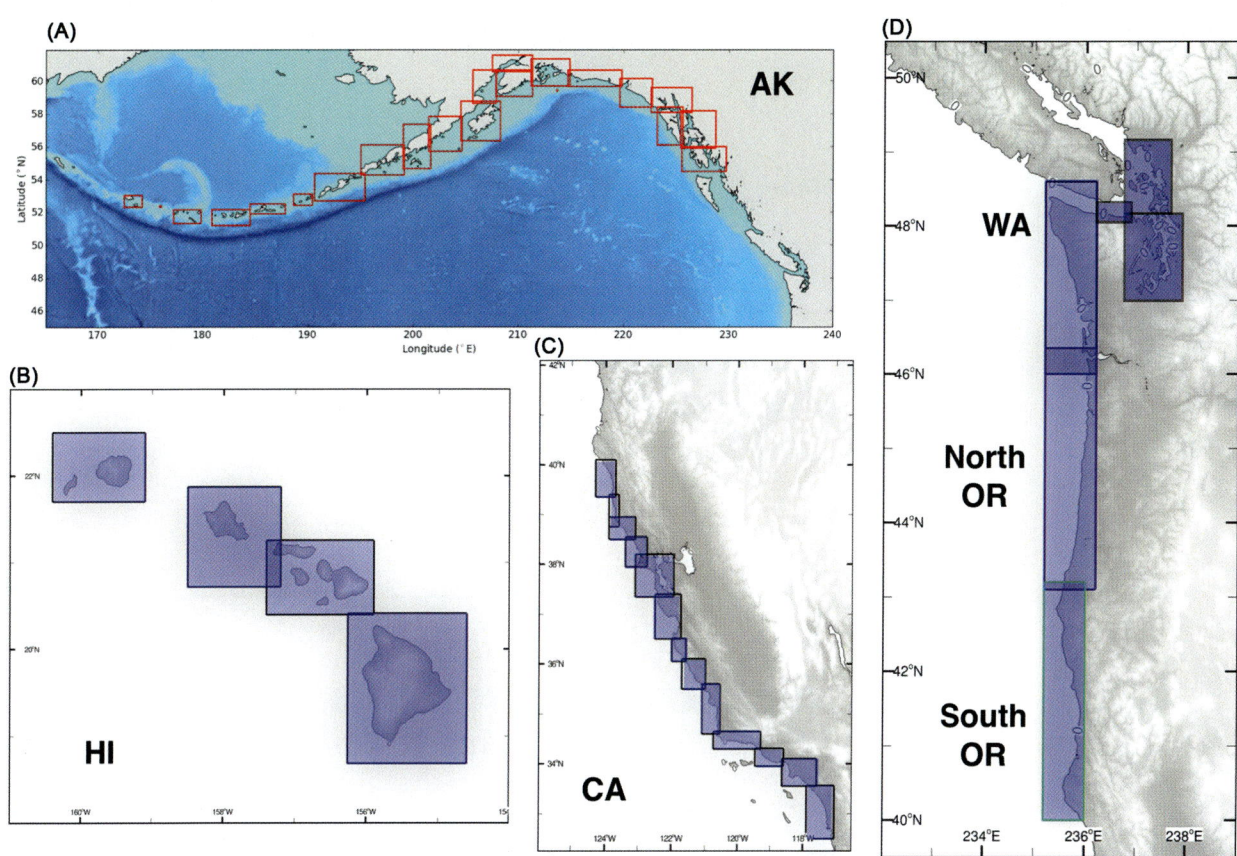

FIGURE 13.21 Model coverage of all C grids for MOST inundation computation along the U.S. West Coast for ASCE 7-16: (a) Alaska; (b) Hawaii; (c) California; and (d) Washington and Oregon. *ASCE*, American Society of Civil Engineers.

the populated coastlines on Maui and Kauai using a nonhydrostatic inundation model NEOWAVE (Nonhydrostatic Evolution of Ocean Wave). Maps for Hilo and Kona on the Hawaii Island were also updated using the 10-m high-resolution MOST results from a modeling study conducted in 2016 (Wei, 2016). These updated Hawaii TDZ maps are expected to be included in the ASCE 7-28 Tsunami Hazard Tool. There are continuous efforts to update the ASCE 7-16 TDZ maps for other regions utilizing the next-generation logic trees of the Pacific subduction-zone sources.

13.3.4.4 Examples of developing tsunami design zone maps for building design

The ASCE tsunami design standards are now adopted by the state building codes in California and Hawaii. Tsunami Risk Category II, III, and IV buildings and structures (American Society of Civil Engineers, 2017) in the tsunami hazard zone are required to be designed following the ASCE criteria to ensure their compliance with tsunami resilience. These TDZ maps are currently valid for the five Pacific states of the United States, whereas the ASCE standards can be applied to develop TDZ maps for building and structure design in other U.S. territories and overseas facilities. Following the development of ASCE 7-16 TDZ maps, the NCTR has continued probabilistic inundation modeling studies for sites globally as shown in Fig. 13.1.

The increasing potential tsunami hazards pose great challenges for infrastructures along the coastlines of the U.S. Pacific Northwest. Tsunami impact at a coastal site is usually assessed from deterministic scenarios based on 10,000 years of geological records in the CSZ. In 2016–17, the Oregon State University (OSU) acquired a site-specific modeling for the new Hatfield Marine Science Center to be built within the tsunami hazard zone in Newport, Oregon (Fig. 13.23A). It was also to be designated as a tsunami vertical evacuation shelter for local communities. A site-specific modeling study was performed to obtain the inundation depth, current speed, and momentum flux at the building site using not only deterministic worst-credible CSZ tsunami scenarios but also probabilistic methods to ensure the building design is compliant with the ASCE 7-16 standards. Three Cascadia scenarios, two deterministic scenarios, XXL1 and L1 (Witter et al., 2011), and a 2475-year probabilistic scenario compliant with the new ASCE 7-16 standard

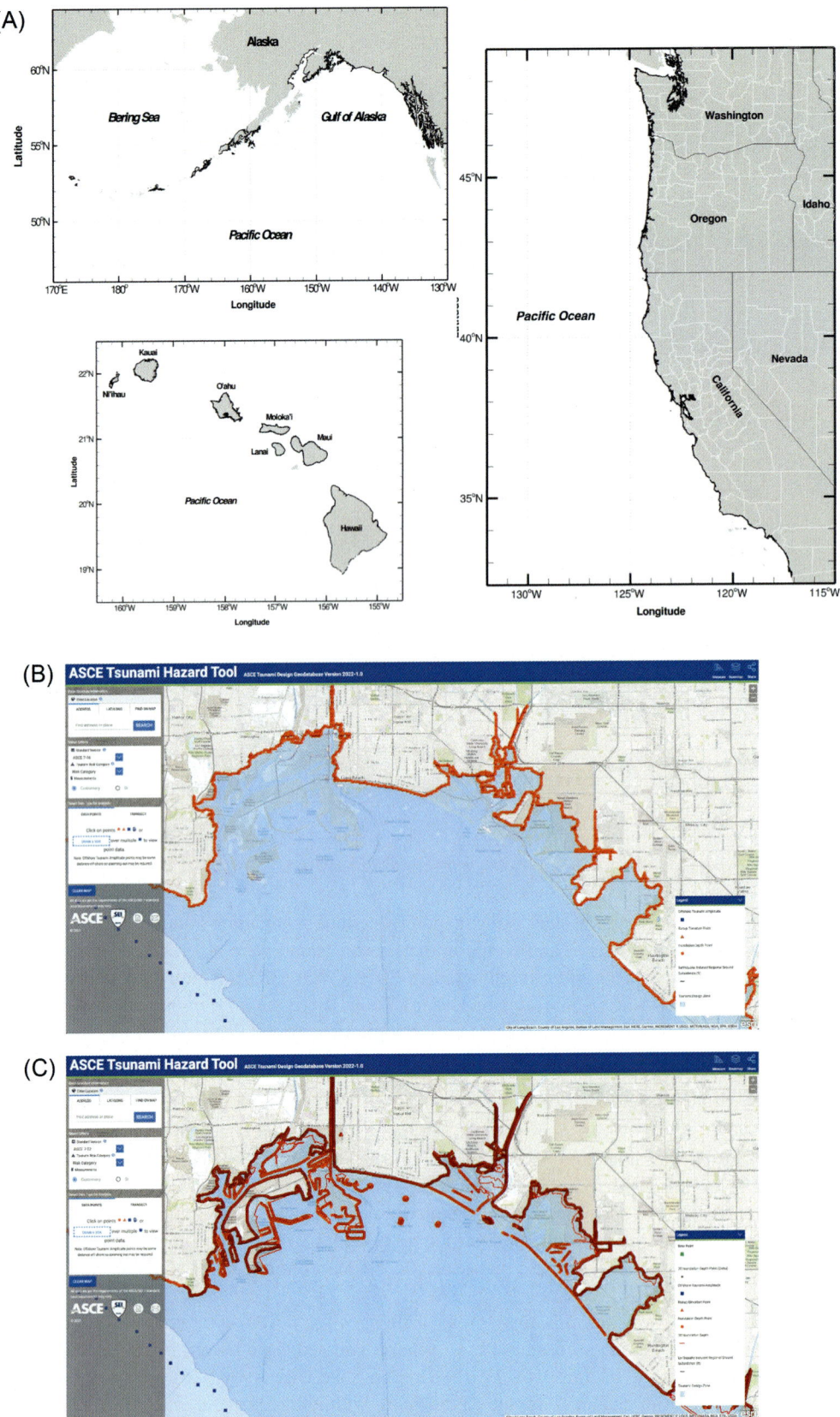

FIGURE 13.22 ASCE TDZ maps. (A) Coverage of ASCE 7-16 TDZ maps (American Society of Civil Engineers, 2017), where the black solid line indicates the coastlines where TDZ maps are available, (B) ASCE 7-16 TDZ maps, based on 60-m grid-resolution model results, for coastlines between Los Angeles Harbor and Huntington Beach, California, and (C) updated ASCE 7-22 TDZ maps based on 10-m grid-resolution model results for coastlines between Los Angeles Harbor and Huntington Beach, California. (B) and (C) are extracted from the ASCE Tsunami Hazard Tool, where the darker blue indicates the TDZ and the orange triangle represents the runup elevation point. *ASCE*, American Society of Civil Engineers; *TDZ*, tsunami design zone. Map lines delineate study areas and do not necessarily depict accepted national boundaries. *(A) From ASCE (2017); (B) and (C) are extracted from the ASCE Tsunami Hazard Tool.*

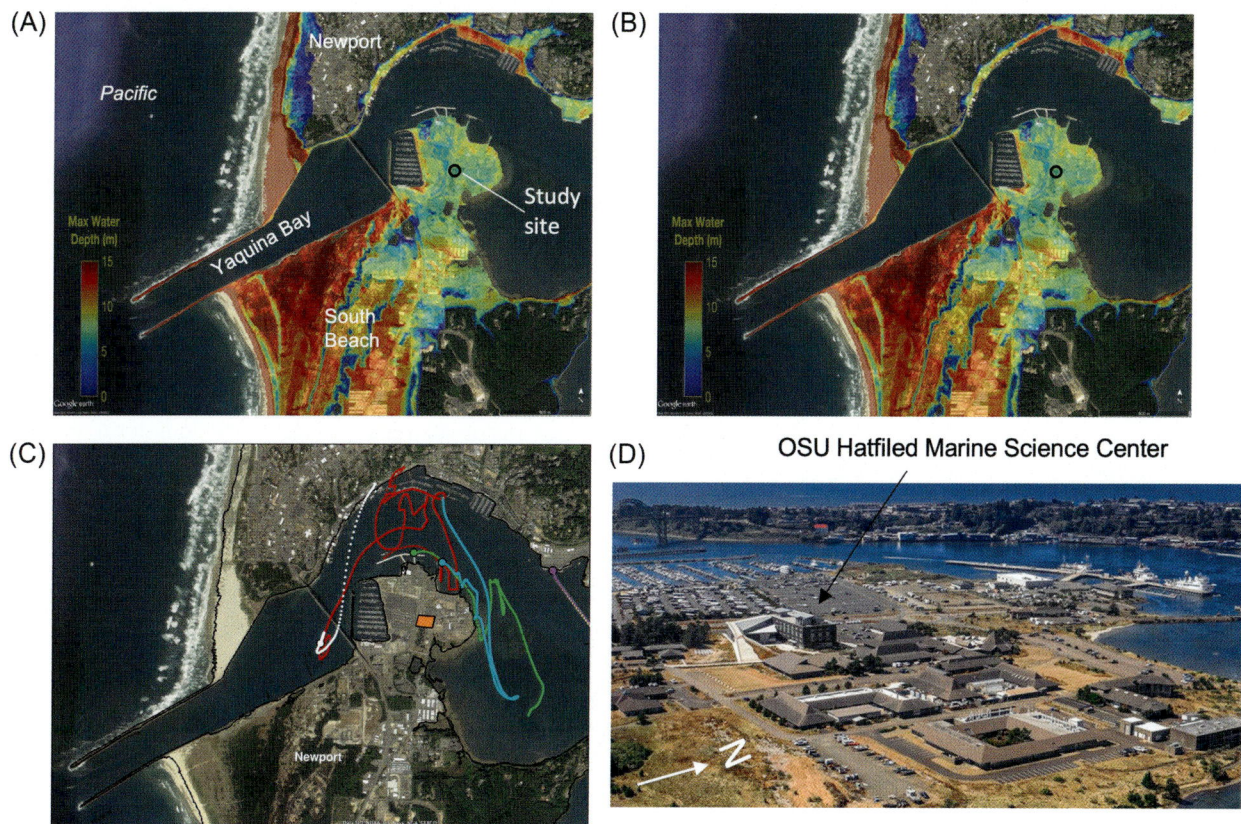

FIGURE 13.23 Site-specific modeling assessment of ASCE-compliant tsunami hazards for the OSU Hatfield Marine Science Center and vicinity. (A) Maximum inundation depth, (B) maximum current speed, and (C) simulated tracks of ship vessels (with mooring) for the ASCE-compliant scenario. Red = track of ship NOAA ship 1; cyan = track of ship NOAA ship 2; white = track of NOAA ship 3; green = track of the OSU research vessel; yellow = track of Cargo ship 1; and purple = track of Cargo ship 2. (D) A photograph of the constructed new OSU Hatfield Marine Science Center. *ASCE*, American Society of Civil Engineers; *OSU*, Oregon State University. *From https://hmsc.oregonstate.edu/.*

were simulated using a combination of the MOST model for offshore propagation and a Boussinesq-type model for onshore inundation. In particular, the deterministic XXL1 and L1 CSZ scenarios involving the activation of local splay faults were estimated to have recurrence intervals of 2500 years and 10,000 years, respectively, based on geological evidence in the past 10,000 years. The model results show that the ASCE 7-16 consistent hazards fall between those obtained from deterministic L1 and XXL1 scenarios, and the greatest impact on the building may come from later waves. Around the building site, the model-estimated maximum inundation depth and current speed from the ASCE-compliant source were 8.3 m and 11.9 m/second (Figs. 13.23A and B), larger than 4.9 m and 9.3 m/second produced by L1, but less than 10.4 m and 12.7 m/second produced by XXL1. As a further step, the inundation model results were utilized to numerically track the movement of large vessels in the vicinity of the building site and estimate how likely these vessels might impact the building site during the extreme XXL1 and ASCE 7-16 hazard-consistent scenarios (Fig. 13.23C). The simulation results showed that the possibility of the large vessels impacting the building site is low in all simulated scenarios. With the rooftop about 14.3 m above the ground, the finished OSU building was constructed to survive an M_w 9.0 + earthquake and resulting tsunami. It was designated as a vertical evacuation structure for more than 900 people after the earthquake (Fig. 13.23D).

13.4 Conclusions

Tsunami hazard assessments are usually conducted for different time scales. The short-term (minutes to days) assessment focuses on providing prompt forecast information to support early warnings and near real-time decisions in evacuation and rescue for an ongoing tsunami event. The long-term assessments, however, aim at identifying the potential impact of tsunamis of return periods ranging from tens to 10,000 years on coastal communities at risk for planning and preparedness.

PMEL's short-term hazard assessments feature three core components: real-time observations of tsunami waves especially in the deep ocean from the global tsunami buoy network, source inversion utilizing precomputed tsunami propagation dataset, and inundation forecast models for major harbors and coastal communities. NOAA's SIFT system integrates these components to provide a forecast of the tsunami hazards in real time before the tsunami impacts distant coasts at risk. This chapter provided an overview of the multiple enhancements of the existing short-term tsunami hazard assessments, including (1) the implementation of high-performance GPU computational capability, (2) modeling capability to compute tsunami propagation for arbitrary earthquake sources, (3) early detection of tsunami waves using the fourth-generation DART systems, (4) GNSS-based early determination of the earthquake magnitude and source, and (5) expansion of the tsunami propagation database from ocean basins to the globe. These enhancements have proved valuable in improving short-term hazard assessment for near-field coastlines.

The long-term assessments can be done using deterministic or probabilistic approaches. A deterministic approach involves the consideration of worst-case or worst-credible scenarios based on historical data and the best knowledge of source characteristics. For long-term hazard map development and mitigation planning, a tsunami model needs to be rigorously benchmarked using the NTHMP model standards and historical records, and these long-term modeling assessments need to be conducted at a grid resolution of 10 m or finer based on the best available DEMs. Case studies across the world are provided to show the long-term modeling assessments utilizing worse-case scenarios, worst-credible regional scenarios, and sensitivity tests of synthetical scenarios of different magnitudes to identify the most probable threatening hazards. PTHA is adopted in the framework of ASCE 7 tsunami effects and loads provisions to ensure the design of critical coastal infrastrucures in the tsunami hazard zones of the United States compliant with the 2500-year return period criteria.

References

Amante, C. J., Love, M., Carignan, K., Sutherland, M. G., MacFerrin, M., & Lim, E. (2023). Continuously Updated Digital Elevation Models (CUDEMs) to support coastal inundation modeling. *Remote Sensing, 15*(6). Available from https://doi.org/10.3390/rs15061702, https://www.mdpi.com/2072-4292/15/6/1702.

American Society of Civil Engineers. (2017). *Minimum design loads and associated criteria for buildings and other structures.* ASCE/SEI (pp. 7–16).

Annaka, T., Satake, K., Sakakiyama, T., Yanagisawa, K., & Shuto, N. (2007). Logic-tree approach for Probabilistic Tsunami Hazard Analysis and its applications to the Japanese coasts. *Pure and Applied Geophysics, 164*(2-3), 577–592. Available from https://doi.org/10.1007/s00024-006-0174-3.

Bernard, E., Meinig, C., Titov, V. V., & Wei, Y. (2023). 50 Years of PMEL tsunami research and development. *Oceanography, 36*(2-3), 175–185. Available from https://doi.org/10.5670/oceanog.2023.208, https://tos.org/oceanography/assets/docs/36-2-3-bernard.pdf.

Bernard, E., & Titov, V. (2015). Evolution of tsunami warning systems and products. *Philosophical Transactions of the Royal Society A: Mathematical, Physical and Engineering Sciences, 373*(2053), 2014.0371. Available from https://doi.org/10.1098/rsta.2014.0371.

Bernard, E., Wei, Y., Tang, L., & Titov, V. (2014). Impact of near-field, deep-ocean tsunami observations on forecasting the 7 December 2012 Japanese tsunami. *Pure and Applied Geophysics, 171*(12), 3483–3491. Available from https://doi.org/10.1007/s00024-013-0720-8, http://www.springer.com/birkhauser/geo + science/journal/24.

Black, B. A., Pearl, J. K., Pearson, C. L., Pringle, P. T., Frank, D. C., Page, M. T., Buckley, B. M., Cook, E. R., Harley, G. L., King, K. J., Hughes, J. F., Reynolds, D. J., & Sherrod, B. L. (2023). A multifault earthquake threat for the Seattle metropolitan region revealed by mass tree mortality. *Science Advances, 9*(39). Available from https://doi.org/10.1126/sciadv.adh4973, http://www.sciencemag.org/journals.

Borrero, J. C., Cronin, S. J., Latu'ila, F. H., Tukuafu, P., Heni, N., Tupou, A. M., Kula, T., Fa'anunu, O., Bosserelle, C., Lynett, P., & Kong, L. (2023). Tsunami runup and inundation in Tonga from the January 2022 eruption of Hunga Volcano. *Pure and Applied Geophysics, 180*, 1–22. Available from https://doi.org/10.1007/s00024-022-03215-5, https://link.springer.com/article/10.1007/s00024-022-03215-5#citeas.

ten Brink, U. S., Molzer, P. C., Fisher, M. A., Blakely, R. J., Bucknam, R. C., Parsons, T., Crosson, R. S., & Creager, K. C. (2002). Subsurface geometry and evolution of the Seattle fault zone and the Seattle Basin, Washington. *Bulletin of the Seismological Society of America, 92*(5), 1737–1753. Available from https://doi.org/10.1785/0120010229.

Bucknam, R. C., Sherrod, B. L., & Elfendahl, G. (1999). A fault scarp of probable Holocene age in the Seattle Fault Zone. *Seismological Research Letters, 70*(2).

Burbidge, D., Cummins, P. R., Mleczko, R., & Thio, H. K. (2008). A probabilistic tsunami hazard assessment for Western Australia. *Pure and Applied Geophysics, 165*(11-12), 2059–2088. Available from https://doi.org/10.1007/s00024-008-0421-x.

Butler, R., Walsh, D., & Richards, K. (2017). Extreme tsunami inundation in Hawai'i from Aleutian–Alaska subduction zone earthquakes. *Natural Hazards, 85*(3), 1591–1619. Available from https://doi.org/10.1007/s11069-016-2650-0, http://www.wkap.nl/journalhome.htm/0921-030X.

Cheung, K. F., Wei, Y., Yamazaki, Y., & Yim, S. C. S. (2011). Modeling of 500-year tsunamis for probabilistic design of coastal infrastructure in the Pacific Northwest. *Coastal Engineering, 58*(10), 970–985. Available from https://doi.org/10.1016/j.coastaleng.2011.05.003.

Chock, G. Y. K., Carden, L., Robertson, I., Wei, Y., Wilson, R., & Hooper, J. (2018). Tsunami-resilient building design considerations for coastal communities of Washington, Oregon, and California. *Journal of Structural Engineering*, *144*(8). Available from https://doi.org/10.1061/(ASCE)ST.1943-541X.0002068, https://ascelibrary.org/doi/10.1061/%28ASCE%29ST.1943-541X.0002068.

Crowell, B. W., & Melgar, D. (2020). Slipping the Shumagin gap: A kinematic coseismic and early afterslip model of the M_w 7.8 Simeonof Island, Alaska, earthquake. *Geophysical Research Letters*, *47*(19). Available from https://doi.org/10.1029/2020GL090308, http://agupubs.onlinelibrary.wiley.com/hub/journal/10.1002/(ISSN)1944-8007/.

Crowell, B. W., Schmidt, D. A., Bodin, P., Vidale, J. E., Baker, B., Barrientos, S., & Geng, J. (2018). G-FAST earthquake early warning potential for great earthquakes in Chile. *Seismological Research Letters*, *89*(2A), 542−556. Available from https://doi.org/10.1785/0220170180, https://pubs.geoscienceworld.org/ssa/srl/article-pdf/89/2A/542/4087552/srl-2017180.1.pdf.

Davies, G., & Griffin, J. (2018). *The 2018 Australian Probabilistic Tsunami Hazard Assessment: Hazards from earthquake generated tsunamis*. Geoscience Australia Record. https://ecat.ga.gov.au/geonetwork/static/api/records/fd1533d2-b176-45b1-9d40-5936aaef7d6f. doi:10.11636/Record.2018.041.

Davies, G., Griffin, J., Løvholt, F., Glimsdal, S., Harbitz, C., Thio, H. K., Lorito, S., Basili, R., Selva, J., Geist, E., & Baptista, M. A. (2018). A global probabilistic tsunami hazard assessment from earthquake sources. *Geological Society London, Special Publications*, *456*(1), 219−244. Available from https://doi.org/10.1144/sp456.5.

Dolcimascolo, A., Eungard, D.W., Allen, C., LeVeque, R.J., Adams, L.M., Arcas, D., Titov, V.V., González, F.I., & Moore, C. (2022). *Tsunami inundation, current speeds, and arrival times simulated from a large Seattle Fault earthquake scenario for Puget Sound and other parts of the Salish Sea*. Washington Geological Survey Map Series 2022-03, 16 sheets.

Eble, M. C., & Gonzalez, F. I. (1991). Deep-ocean bottom pressure measurements in the Northeast Pacific. *Journal of Atmospheric and Oceanic Technology*, *8*(2), 221−233, 10.1175/1520-0426(1991)008 < 0221:dobpmi > 2.0.co;2.

Fry, B., McCurrach, S.-J., Gledhill, K., Power, W., Williams, M., Angove, M., Arcas, D., & Moore, C. (2020). Sensor network warns of stealth tsunamis. *EOS*, *101*. Available from https://doi.org/10.1029/2020eo144274, https://eos.org/science-updates/sensor-network-warns-of-stealth-tsunamis.

Fujiwara, H., Kawai, S., Aoi, S., Morikawa, N., Senna, S., Kobayashi, K., Ishii, T., Okumura, T., & Hayakawa, V. (2006). National seismic hazard maps of Japan. *Bulletin of Earthquake Research Institute, University of Tokyo*, *81*, 221−232.

Geist, E. L., & Parsons, T. (2006). Probabilistic analysis of tsunami hazards. *Natural Hazards*, *37*(3), 277−314. Available from https://doi.org/10.1007/s11069-005-4646-z.

Gica, E., Spillane, M., Titov, V. V., Chamberlin, C., & Newman, J. C. (2008). *Development of the forecast propagation database for NOAA's Short-term Inundation Forecast for Tsunamis (SIFT)*. NOAA Technical Memorandum OAR PMEL-139.

Gonzalez, F. I., Mader, C. L., Eble, M. C., & Bernard, E. N. (1991). The 1987-88 Alaskan Bight tsunamis: Deep ocean data and model comparisons. *Natural Hazards*, *4*(2-3), 119−139. Available from https://doi.org/10.1007/BF00162783.

González, F. I., Geist, E. L., Jaffe, B., Kânoğlu, U., Mofjeld, H., Synolakis, C. E., Titov, V. V., Arcas, D., Bellomo, D., Carlton, D., Horning, T., Johnson, J., Newman, J., Parsons, T., Peters, R., Peterson, C., Priest, G., Venturato, A., Weber, J., ... Yalciner, A. (2009). Probabilistic tsunami hazard assessment at Seaside, Oregon, for near- and far-field seismic sources. *Journal of Geophysical Research: Oceans*, *114*(C11). Available from https://doi.org/10.1029/2008jc005132.

Gusman, A. R., Mulia, I. E., & Satake, K. (2018). Optimum sea surface displacement and fault slip distribution of the 2017 Tehuantepec earthquake (M_w 8.2) in Mexico estimated from tsunami waveforms. *Geophysical Research Letters*, *45*(2), 646−653. Available from https://doi.org/10.1002/2017GL076070, http://onlinelibrary.wiley.com/journal/10.1002/(ISSN)1944-8007/issues?year = 2012.

Gusman, A. R., Roger, J., Noble, C., Wang, X., Power, W., & Burbidge, D. (2022). The 2022 Hunga Tonga-Hunga Ha'apai Volcano air-wave generated tsunami. *Pure and Applied Geophysics*, *179*(10), 3511−3525. Available from https://doi.org/10.1007/s00024-022-03154-1.

Hawker, L., Uhe, P., Paulo, L., Sosa, J., Savage, J., Sampson, C., & Neal, J. (2022). A 30 m global map of elevation with forests and buildings removed. *Environmental Research Letters*, *17*(2), 024016. Available from https://doi.org/10.1088/1748-9326/ac4d4f.

Hayes, G. P., Wald, D. J., & Johnson, R. L. (2012). Slab1.0: A three-dimensional model of global subduction zone geometries. *Journal of Geophysical Research: Solid Earth*, *117*(1). Available from https://doi.org/10.1029/2011JB008524, http://onlinelibrary.wiley.com/journal/10.1002/(ISSN)2169-9356.

Hettiarachchi, S. (2018). Establishing the Indian Ocean Tsunami Warning and Mitigation System for human and environmental security. *Procedia Engineering*, *212*, 1339−1346. Available from https://doi.org/10.1016/j.proeng.2018.01.173, http://www.sciencedirect.com/science/journal/18777058.

Hoechner, A., Babeyko, A. Y., & Zamora, N. (2016). Probabilistic tsunami hazard assessment for the Makran region with focus on maximum magnitude assumption. *Natural Hazards and Earth System Sciences*, *16*(6), 1339−1350. Available from https://doi.org/10.5194/nhess-16-1339-2016, http://www.nat-hazards-earth-syst-sci.net/volumes_and_issues.html.

Horrillo, J., Grilli, S. T., Nicolsky, D., Roeber, V., & Zhang, J. (2015). Performance benchmarking tsunami models for NTHMP's inundation mapping activities. *Pure and Applied Geophysics*, *172*(3-4), 869−884. Available from https://doi.org/10.1007/s00024-014-0891-y, http://www.springer.com/birkhauser/geo + science/journal/24.

Lane, E. M., Gillibrand, P. A., Wang, X., & Power, W. (2013). A probabilistic tsunami hazard study of the Auckland region, Part II: Inundation modelling and hazard assessment. *Pure and Applied Geophysics*, *170*(9-10), 1635−1646. Available from https://doi.org/10.1007/s00024-012-0538-9.

Liu, P. L. F., Cho, Y. S., Briggs, M. J., Kanoglu, U., & Synolakis, C. E. (1995). Runup of solitary waves on a circular Island. *Journal of Fluid Mechanics*, *302*(37), 259−285. Available from https://doi.org/10.1017/S0022112095004095.

Liu, Y., Santos, A., Wang, S. M., Shi, Y., Liu, H., & Yuen, D. A. (2007). Tsunami hazards along Chinese coast from potential earthquakes in South China Sea. *Physics of the Earth and Planetary Interiors*, *163*(1-4), 233–244. Available from https://doi.org/10.1016/j.pepi.2007.02.012.

Lynett, P., McCann, M., Zhou, Z., Renteria, W., Borrero, J., Greer, D., Fa'anunu, O., Bosserelle, C., Jaffe, B., La Selle, S. P., Ritchie, A., Snyder, A., Nasr, B., Bott, J., Graehl, N., Synolakis, C., Ebrahimi, B., & Cinar, G. E. (2022). Diverse tsunamigenesis triggered by the Hunga Tonga-Hunga Ha'apai eruption. *Nature*, *609*(7928), 728–733. Available from https://doi.org/10.1038/s41586-022-05170-6, https://www.nature.com/nature/.

Mori, N., Satake, K., Cox, D., Goda, K., Catalan, P. A., Ho, T. C., Imamura, F., Tomiczek, T., Lynett, P., Miyashita, T., Muhari, A., Titov, V., & Wilson, R. (2022). Giant tsunami monitoring, early warning and hazard assessment. *Nature Reviews Earth and Environment*, *3*(9), 557–572. Available from https://doi.org/10.1038/s43017-022-00327-3, http://nature.com/natrevearthenviron/.

Murotani, S., Miyake, H., & Koketsu, K. (2008). Scaling of characterized slip models for plate-boundary earthquakes. *Earth, Planets and Space*, *60*(9), 987–991. Available from https://doi.org/10.1186/BF03352855, http://rd.springer.com/journal/40623.

Murotani, S., Satake, K., & Fujii, Y. (2013). Scaling relations of seismic moment, rupture area, average slip, and asperity size for $M \sim 9$ subduction-zone earthquakes. *Geophysical Research Letters*, *40*(19), 5070–5074. Available from https://doi.org/10.1002/grl.50976.

Newman, A. V. (2011). Hidden depths. *Nature*, *474*(7352), 441–443. Available from https://doi.org/10.1038/474441a.

NOAA National Centers for Environmental Information. (2022). *ETOPO 2022 15 arc-second global relief model*. https://www.ncei.noaa.gov/products/etopo-global-relief-model. 10.25921/fd45-gt74.

Papazachos, B. C., Scordilis, E. M., Panagiotopoulos, D. G., Papazachos, C. B., & Karakaisis, G. F. (2004). Global relations between seismic fault parameters and moment magnitude of earthquakes. *Bulletin of the Geological Society of Greece*, *36*(3), 1482. Available from https://doi.org/10.12681/bgsg.16538.

Percival, D. B., Denbo, D. W., Eblé, M. C., Gica, E., Mofjeld, H. O., Spillane, M. C., Tang, L., & Titov, V. V. (2011). Extraction of tsunami source coefficients via inversion of DART® buoy data. *Natural Hazards*, *58*(1), 567–590. Available from https://doi.org/10.1007/s11069-010-9688-1.

Petersen, M. D., Moschetti, M. P., Powers, P. M., Mueller, C. S., Haller, K. M., Frankel, A. D., Zeng, Y., Rezaeian, S., Harmsen, S. C., Boyd, O. S., Field, N., Chen, R., Rukstales, K. S., Luco, N., Wheeler, R. L., Williams, R. A., & Olsen, A. H. (2014). *Documentation for the 2014 Update of the United States National Seismic Hazard Maps Open-File* (Report 2014–1091). US Geological Survey. https://pubs.usgs.gov/of/2014/1091/. doi:https://doi.org/10.3133/ofr20141091.

Power, W., & Gale, N. (2011). Tsunami forecasting and monitoring in New Zealand. *Pure and Applied Geophysics*, *168*(6-7), 1125–1136. Available from https://doi.org/10.1007/s00024-010-0223-9, http://www.springer.com/birkhauser/geo+science/journal/24.

Ren, Z., Higuera, P., & Li-Fan Liu, P. (2023). On tsunami waves induced by atmospheric pressure shock waves after the 2022 Hunga Tonga-Hunga Ha'apai Volcano eruption. *Journal of Geophysical Research: Oceans*, *128*(4). Available from https://doi.org/10.1029/2022JC019166, http://agupubs.onlinelibrary.wiley.com/agu/jgr/journal/10.1002/(ISSN)2169-9291/.

Romano, F., Gusman, A. R., Power, W., Piatanesi, A., Volpe, M., Scala, A., & Lorito, S. (2021). Tsunami source of the 2021 M_w 8.1 Raoul Island earthquake from DART and tide-gauge data inversion. *Geophysical Research Letters*, *48*(17). Available from https://doi.org/10.1029/2021GL094449, http://agupubs.onlinelibrary.wiley.com/hub/journal/10.1002/(ISSN)1944-8007/.

Selva, J., Lorito, S., Volpe, M., Romano, F., Tonini, R., Perfetti, P., Bernardi, F., Taroni, M., Scala, A., Babeyko, A., Løvholt, F., Gibbons, S. J., Macías, J., Castro, M. J., González-Vida, J. M., Sánchez-Linares, C., Bayraktar, H. B., Basili, R., Maesano, F. E., ... Amato, A. (2021). Probabilistic tsunami forecasting for early warning. *Nature Communications*, *12*(1). Available from https://doi.org/10.1038/s41467-021-25815-w, http://www.nature.com/ncomms/index.html.

Strasser, F. O., Arango, M. C., & Bommer, J. J. (2010). Scaling of the source dimensions of interface and intraslab subduction-zone earthquakes with moment magnitude. *Seismological Research Letters*, *81*(6), 941–950. Available from https://doi.org/10.1785/gssrl.81.6.941, http://srl.geoscienceworld.org/cgi/reprint/81/6/941, South Africa.

Synolakis, C. E., Bernard, E. N., Titov, V. V., Kânoğlu, U., & González, F. I. (2008). Validation and verification of tsunami numerical models. *Pure and Applied Geophysics*, *165*(11-12), 2197–2228. Available from https://doi.org/10.1007/s00024-004-0427-y.

Sørensen, M. B., Spada, M., Babeyko, A., Wiemer, S., & Grünthal, G. (2012). Probabilistic tsunami hazard in the Mediterranean Sea. *Journal of Geophysical Research: Solid Earth*, *117*(B1). Available from https://doi.org/10.1029/2010jb008169.

Tang, L., Chamberlin, C., Tolkova, E., Spillane, M., Titov, V. V., Bernard, E. N., Mofjeld, H. O. (2006). *Assessment of potential tsunami impact for Pearl Harbor, Hawaii*. NOAA Technical Memorandum OAR PMEL-131. https://nctr.pmel.noaa.gov/Pdf/tang2984_low_res.pdf.

Tang, L., Titov, V. V., Bernard, E. N., Wei, Y., Chamberlin, C. D., Newman, J. C., Mofjeld, H. O., Arcas, D., Eble, M. C., Moore, C., Uslu, B., Pells, C., Spillane, M., Wright, L., & Gica, E. (2012). Direct energy estimation of the 2011 Japan tsunami using deep-ocean pressure measurements. *Journal of Geophysical Research: Oceans*, *117*(8). Available from https://doi.org/10.1029/2011JC007635, http://onlinelibrary.wiley.com/journal/10.1002/(ISSN)2169-9291.

Tang, L., Titov, V. V., & Chamberlin, C. D. (2009). Development, testing, and applications of site-specific tsunami inundation models for real-time forecasting. *Journal of Geophysical Research: Oceans*, *114*(C12). Available from https://doi.org/10.1029/2009jc005476.

Tang, L., Titov, V. V., Wei, Y., Mofjeld, H. O., Spillane, M., Arcas, D., Bernard, E. N., Chamberlin, C., Gica, E., & Newman, J. (2008). Tsunami forecast analysis for the May 2006 Tonga tsunami. *Journal of Geophysical Research: Oceans*, *113*(12). Available from https://doi.org/10.1029/2008JC004922, http://onlinelibrary.wiley.com/journal/10.1002/(ISSN)2169-9291.

Thio, H. K. (2019). *Probabilistic tsunami hazard maps for the State of California (Phase 2)*. California Geological Survey. Available from https://www.conservation.ca.gov/cgs/tsunami/reports.

Titov, V. V. (2009). *The Sea Tsunami forecasting*. Harvard University Press. Available from https://www.nhbs.com/the-sea-volume-15-tsunamis-book.

Titov, V. V., González, F. I., Bernard, E. N., Eble, M. C., Mofjeld, H. O., Newman, J. C., & Venturato, A. J. (2005). Real-time tsunami forecasting: Challenges and solutions. *Natural Hazards*, *35*(1), 41–58. Available from https://doi.org/10.1007/s11069-004-2403-3.

Titov, V. V., González, F. I., Mofjeld, H. O., & Venturato, A. J. (2003). NOAA TIME Seattle Tsunami Mapping Project: Procedures, data sources, and products. *NOAA Technical Memorandum OAR PMEL-124*, 2004–101635. Available from https://repository.library.noaa.gov/view/noaa/11033.

Titov, V. V., Meinig, C., Stalin, S., Wei, Y., Moore, C., & Bernard, E. (2023). Technology transfer of PMEL tsunami research protects populations and expands the New Blue Economy. *Oceanography*, *36*(2-3), 186–195. Available from https://doi.org/10.5670/oceanog.2023.205, https://tos.org/oceanography/assets/docs/36-2-3-titov.pdf.

Titov, V. V., Moore, C. W., Greenslade, D. J. M., Pattiaratchi, C., Badal, R., Synolakis, C. E., & Kânoğlu, U. (2011). A new tool for inundation modeling: Community Modeling Interface for Tsunamis (ComMIT). *Pure and Applied Geophysics*, *168*(11), 2121–2131. Available from https://doi.org/10.1007/s00024-011-0292-4.

Uslu, B., Eble, M., Arcas, D., & Titov, V. (2013). Tsunami hazard assessment for the Commonwealth of the northern Mariana Islands. *NOAA OAR Special Report*, *3*. Available from https://nctr.pmel.noaa.gov/hazard_assessment_reports/H03_CNMI_3949_lowres.pdf.

Uslu, B., Power, W., Greenslade, D., Eblé, M., & Titov, V. (2011). The July 15, 2009 Fiordland, New Zealand tsunami: Real-time assessment. *Pure and Applied Geophysics*, *168*(11), 1963–1972. Available from https://doi.org/10.1007/s00024-011-0281-7.

Uslu, B., Titov, V. V., Eble, M., & Chamberlin, C. (2010). Tsunami hazard assessment for Guam. *NOAA OAR Special Report*, *1*. Available from https://nctr.pmel.noaa.gov/hazard_assessment_reports/01_Guam_3528_web.pdf.

Venturato, A. J., Arcas, D., Titov, V. V., Mofjeld, H. O., Chamberlin, C. C., & Gonzalez, F. I. (2007). Tacoma, Washington, tsunami hazard mapping project: Modeling tsunami inundation from Tacoma and Seattle fault earthquakes. *NOAA Technical Memorandum OAR PMEL-132*. Available from https://www.pmel.noaa.gov/pubs/PDF/vent2981/vent2981.pdf.

Wei, Y. (2016). *Tsunami probabilistic reference maps for benchmarking Hawaii tsunami design zone maps per the ASCE 7-16 standard*. Hawaii State Emergency Management Agency.

Wei, Y., Bernard, E. N., Tang, L., Weiss, R., Titov, V. V., Moore, C., Spillane, M., Hopkins, M., & Kânoğlu, U. (2008). Real-time experimental forecast of the Peruvian tsunami of August 2007 for US coastlines. *Geophysical Research Letters*, *35*(4). Available from https://doi.org/10.1029/2007GL032250.

Wei, Y., Chamberlin, C., Titov, V. V., Tang, L., & Bernard, E. N. (2013). Modeling of the 2011 Japan tsunami: Lessons for near-field forecast. *Pure and Applied Geophysics*, *170*(6-8), 1309–1331. Available from https://doi.org/10.1007/s00024-012-0519-z, http://www.springer.com/birkhauser/geo + science/journal/24.

Wei, Y., Newman, A. V., Hayes, G. P., Titov, V. V., & Tang, L. (2014). Tsunami forecast by joint inversion of real-time tsunami waveforms and seismic or GPS data: Application to the Tohoku 2011 tsunami. *Pure and Applied Geophysics*, *171*(12), 3281–3305. Available from https://doi.org/10.1007/s00024-014-0777-z, http://www.springer.com/birkhauser/geo + science/journal/24.

Witter, R. C., Carver, G. A., Briggs, R. W., Gelfenbaum, G., Koehler, R. D., La Selle, S., Bender, A. M., Engelhart, S. E., Hemphill-Haley, E., & Hill, T. D. (2016). Unusually large tsunamis frequent a currently creeping part of the Aleutian megathrust. *Geophysical Research Letters*, *43*(1), 76–84. Available from https://doi.org/10.1002/2015GL066083, http://onlinelibrary.wiley.com/journal/10.1002/(ISSN)1944-8007/issues?year = 2012.

Witter, R. C., Zhang, Y., Wang, K., Priest, G. R., Goldfinger, C., Stimely, L. L., English, J. T., & Ferro, P. (2011). Simulating tsunami inundation at Bandon, Coos County, Oregon, using hypothetical Cascadia and Alaska earthquake scenarios. *DOGAMI Special Paper*, *43*. Available from https://pubs.oregon.gov/dogami/sp/p-SP-43.htm.

Ye, L., Lay, T., Bai, Y., Cheung, K. F., & Kanamori, H. (2017). The 2017 M_w 8.2 Chiapas, Mexico, earthquake: Energetic slab detachment. *Geophysical Research Letters*, *44*(23), 11–832. Available from https://doi.org/10.1002/2017GL076085, http://onlinelibrary.wiley.com/journal/10.1002/(ISSN)1944-8007/issues?year = 2012.

Yuan, Y., Li, H., Wei, Y., Shi, F., Wang, Z., Hou, J., Wang, P., & Xu, Z. (2021). Probabilistic Tsunami Hazard Assessment (PTHA) for southeast coast of Chinese Mainland and Taiwan Island. *Journal of Geophysical Research: Solid Earth*, *126*(2). Available from https://doi.org/10.1029/2020JB020344, http://agupubs.onlinelibrary.wiley.com/hub/jgr/journal/10.1002/(ISSN)2169-9356/.

Zhou, H., Wei, Y., & Titov, V. V. (2012). Dispersive modeling of the 2009 Samoa tsunami. *Geophysical Research Letters*, *39*(16). Available from https://doi.org/10.1029/2012GL053068, http://onlinelibrary.wiley.com/journal/10.1002/(ISSN)1944-8007/issues?year = 2012.

Wei, Y., Thio, H. K., Titov, V., Chock, G., Zhou, H., Tang, L., & Moore, C. (2017). *Inundation modeling to create 2500-year return period tsunami design zone maps for the ASCE 7-16 standard*. Paper No. 450 In Proceedings of the 16th World Conference on Earthquake Engineering, Santiago, Chile.

Chapter 14

Probabilistic tsunami hazard assessment for New Zealand

William Power, Aditya Gusman, David Burbidge and Xiaoming Wang
GNS Science, Lower Hutt, Wellington, New Zealand

14.1 Introduction

New Zealand is a tectonically active country that sits astride the boundary between the Pacific and Australian Plates. As a consequence, several local earthquakes have caused tsunamis over the past 200 years, and the geological record shows evidence of larger earthquakes and tsunamis further back in time. New Zealand's position in the southwest corner of the Pacific Ocean also makes it vulnerable to tsunamis created at greater distances, both within the southwestern Pacific region and from the Pacific Basin as a whole. Given the potential for tsunamis, it is useful to take steps to mitigate the risks that they pose, for example, by developing evacuation zones, incorporating tsunami hazards into land-use planning, and applying tsunami-resilient engineering techniques to building design. To apply these mitigation techniques appropriately, it is necessary to understand and quantify the hazard that needs to be mitigated, and this is the motivation for New Zealand's National Tsunami Hazard Model (NTHM).

Although the written history of New Zealand is relatively short, less than 200 years, it contains records of many tsunamis from a wide variety of sources. The most notable events from distant sources have typically come from South America, with tsunamis caused by large earthquakes in 1868 (Peru), 1877 (Chile), and 1960 (Chile). Of these, the 1868 tsunami, caused by an earthquake of around moment magnitude (M_w) 9 (Okal et al., 2006), appears to have been the largest; the greatest impacts of this tsunami in New Zealand were observed on the Chatham Islands, where inundation was extensive. There was at least one fatality officially recorded from this event, though Māori oral history indicates that the true number of casualties is likely to be substantially higher (Goff, 2021; Thomas et al., 2020).

For tsunamis caused by local earthquakes, the most notable historical events occurred in 1855, 1947, and 2016. The 1855 Wairarapa Fault Earthquake, of magnitude M_w 8.0–8.3, was one of the largest primarily strike-slip earthquakes to have been recorded (Grapes & Downes, 1997). Though largely on land, the rupture extended into Cook Strait and demonstrated large vertical deformation at the coast driven by movement on the Wharekauhau thrust. The subsequent tsunami reached runup heights of around 11 m in Palliser Bay and inundated land around the southern coast of Wellington. The two tsunami-causing earthquakes in March and May 1947 (both of around M_w 7) are notable for having ruptured on the Hikurangi subduction interface and for having the characteristics of "tsunami earthquakes." The largest of these tsunamis followed the March event and had a maximum recorded runup of about 10 m (Bell et al., 2014). It was fortunate that these tsunamis primarily affected a coast that was very sparsely populated at that time, and consequently they caused only a few injuries and no fatalities. Of recent tsunamis caused by faulting close to New Zealand, the M_w 7.8 Kaikoura Earthquake and Tsunami are the most remarkable (Gusman et al., 2018). The earthquake itself was composed of fault movement on at least 12 different faults, possibly including a section of plate interface (Hamling et al., 2017). Maximum runup heights on nearby coastlines reached up to about 7 m (Power et al., 2017).

Not all historical tsunamis in New Zealand originated from earthquakes. Tsunamis caused by subaerial landslides have been recorded around the coast and in sounds and lakes. The 1929 Buller Earthquake (M_w 7.3) caused a cliff collapse on the coast near Karamea on the western coast of the South Island that is believed to have been responsible for a 2.5-m high tsunami (Downes et al., 2017). On two occasions, in 1846 and 1910, landslides resulting from dam breaks caused by geothermal cliff collapses are thought to have caused tsunamis in Lake Taupo (Massey et al., 2009). More recently, a landslide in Charles Sound, triggered by the M_w 7.2 Fiordland Earthquake of 2003, produced a tsunami with runup of 4–5 m and localized damage within the Sound (Power et al., 2005). Of tsunamis caused by volcanic events, the most prominent examples are both recent: the 2022 eruption of Hunga Tonga—Hunga-Ha'apai caused a tsunami, with the longer range effects primarily the result of the blast wave, which was widely observed and mildly damaging in New Zealand (Gusman et al., 2022; Lawson et al., 2022); also in 2022, a volcanic event in Lake Taupo, associated with an M_w 5.7 earthquake, caused inundation and minor damage around the lake (https://www.geonet.org.nz/news/LuzOzDmQcQUUmdeiL67oX).

While the short historical record of tsunamis in New Zealand contains a wide variety of events, it is also important to note that the paleotsunami and paleoseismic records (corroborated in some instances with Māori oral history) indicate that larger earthquakes and their subsequent tsunamis have occurred before the period of written history, in particular on the Hikurangi subduction plate interface (Clark et al., 2019; Pizer et al., 2021), where it is thought that as many as 10 such earthquakes have occurred in the past 7000 years.

The development of probabilistic tsunami hazard models in New Zealand has taken place in several stages over the past two decades. An initial assessment of tsunami hazards and risks to New Zealand's coastal cities was made in 2005 (Berryman, 2005). That assessment combined earthquake statistical modeling techniques from seismic hazard modeling with empirical and semiempirical methods for tsunami height estimation. Power et al. (2007) developed a hazard model for tsunamis originating on the coast of South America that estimated the subsequent hazard along the whole of New Zealand's coastline. Following studies to better understand the potential for tsunamis originating on the Kermadec and southern New Hebrides Trenches (Power et al., 2012), tsunami hazard models were developed to evaluate the hazard that these sources posed to New Zealand (Power et al., 2013), including the inundation hazard to the Auckland region (Lane et al., 2013).

The original NTHM, which was the first hazard model to include the whole coast of New Zealand and to include tsunamis from all identified seismic sources, was completed in 2013 (Power et al., 2018; Power, 2013), and an update was made in 2021 (Power et al., 2022). The NTHM provides estimates of coastal tsunami heights at the New Zealand shoreline for return periods of up to 2500 years, and subsequent studies have built upon the NTHM to include onshore inundation hazards for particular locations (Burbidge et al., 2021, 2022; Gusman, Power et al., 2019). The NTHM has been used as a basis for many studies across New Zealand to develop evacuation zones and provide tsunami inundation assessments for land-use planning and quantitative risk assessments.

14.2 Overview of statistical modeling approach

The NTHM uses a synthetic catalog approach to estimate tsunami hazards. This involves creating artificial catalogs of tsunami-causing events that are generated to be consistent with our understanding of their likelihoods, estimating the heights of the subsequent tsunamis on New Zealand's coast, and then processing the resulting set of tsunami height estimates to evaluate the tsunami hazard. This section provides an overview of the process and explains how magnitude-frequency information about the tsunami sources is used to create the synthetic catalogs. Subsequent sections will detail how the tsunami heights are calculated for the events in the catalogs.

The NTHM outputs estimates of tsunami hazards along the coast, which, for this purpose, has been divided into 268 coastal sections, each of which covers approximately 20 km of coastline (Fig. 14.1). Currently, the NTHM quantifies the hazard from tsunamis caused by earthquakes, which are thought to be the source of about 80% of tsunamis globally. In reality, other tsunami sources, such as landslides and volcanic eruptions, also contribute to the total tsunami hazard, but these tend to have more localized consequences and are less likely to cause large tsunami heights far from the source. However, there are some exceptions to this, such as the 2022 Hunga Tonga—Hunga Ha'apai volcanic tsunami (Gusman et al., 2022; Lynett et al., 2022; Omira et al., 2022).

To calculate the tsunami hazard in the NTHM, one of the requirements is to determine suitable parameters for describing the magnitude-frequency distributions of the earthquake sources. The process of estimating magnitude–frequency distributions for earthquake sources is a major endeavor requiring multidisciplinary research (including paleoseismology, seismology, geodesy, and paleotsunami). Fortunately, much work has been done by the seismic-hazard research community, which can be drawn on. For the 2021 version of the NTHM, three types of sources were quantified based as far as possible on existing research:

- Subduction zone sources—these were based on the 2015 Global Earthquake Model Faulted Earth Subduction Zone Characterization Project model (Berryman et al., 2015). Additional assumptions regarding the segmentation of subduction zones were carried over from the previous 2013 NTHM (Power, 2013). These sources were characterized using the truncated Guttenberg-Richter distributions.

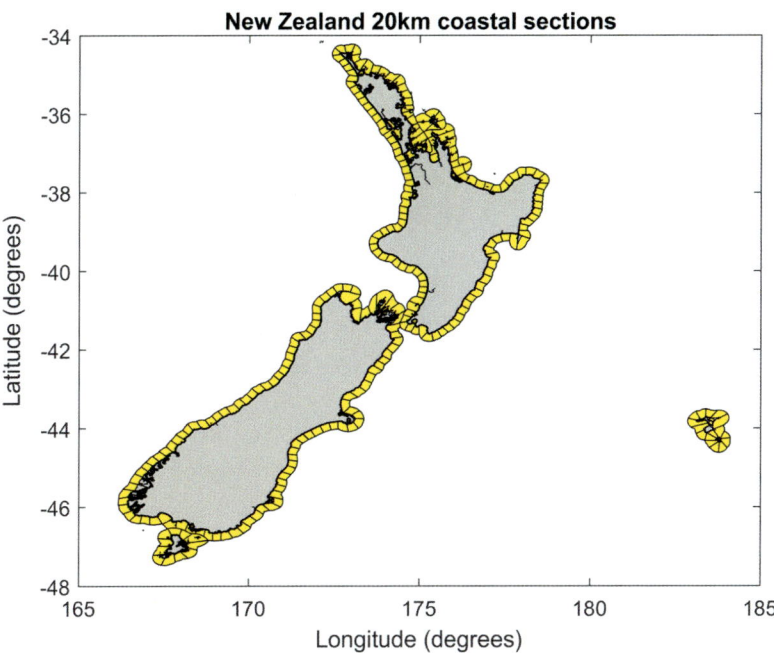

FIGURE 14.1 New Zealand coastal sections. The 268 coastal sections used for defining the tsunami hazard. Each section is approximately 20 km in length. Map lines delineate study areas and do not necessarily depict accepted national boundaries.

- Identified local crustal faults—these were based on fault sources that were characterized in the 2010 update to the National Seismic Hazard Model (NSHM) - (Stirling et al., 2012). Following the NSHM, these sources were assumed to have characteristic magnitude-frequency distributions.
- Tentatively identified local crustal faults—these are faults that were not characterized or not identified at the time of the 2010 update to the NSHM but are considered to pose a tsunami hazard. These were also assumed to have characteristic magnitude-frequency distributions. The parameters of these distributions were evaluated through discussions with researchers studying those sources.

The approach used to estimate tsunami hazards is based on a Monte Carlo modeling process. The method aims to estimate the maximum tsunami height expected over a specified interval of time within each of the 20 km sections of New Zealand's coast. There are many uncertainties in tsunami hazard assessment, and some of these are quite large; consequently, an estimate of tsunami hazard is of little value without an assessment of the associated uncertainty, and the estimation of uncertainties plays a major role in the analysis.

It is useful to clearly distinguish between variability and uncertainty. Variability refers to the natural variations that occur between different events. For instance, the magnitude of earthquakes on a fault naturally varies from one earthquake to the next. Uncertainty, on the other hand, is a measure of our lack of knowledge about things that are constant in time. For example, while the shape of a fault is fixed (within the timeframes we are interested in), its shape is not known exactly, and the uncertainty is a measure of how well it is known. In probabilistic seismic hazard analysis, these concepts are also sometimes called aleatory and epistemic uncertainty.

The Monte Carlo analysis used in the NTHM operates on two levels (Fig. 14.2). On the inner level, it is assumed that we have perfect knowledge of the uncertain parameters and carry out a hazard assessment using Monte Carlo sampling of those properties that naturally vary between events. On the outer level, we perform Monte Carlo sampling of the uncertain parameters and use this to build up a set of different hazard estimates. The spread of these estimates represents the uncertainty in the hazard.

A more detailed representation of the method is shown in Fig. 14.3. In this figure, each row going across the chart describes the steps used to construct one tsunami hazard curve. These steps are repeated many times using different samples of the uncertain parameters, and from these catalogs it is possible to assign "error bars" to the tsunami hazard curves. At this stage in our overview, the process by which the coastal tsunami heights are calculated for the events in the catalogs has not been explained; this will be covered in detail in Sections 14.3–14.5. Each hazard curve describes the maximum tsunami height reached within a coastal section as a function of return period. By sampling from the uncertain parameters and creating multiple hazard curves, it is possible to estimate the uncertainty in the tsunami hazard (Fig. 14.4).

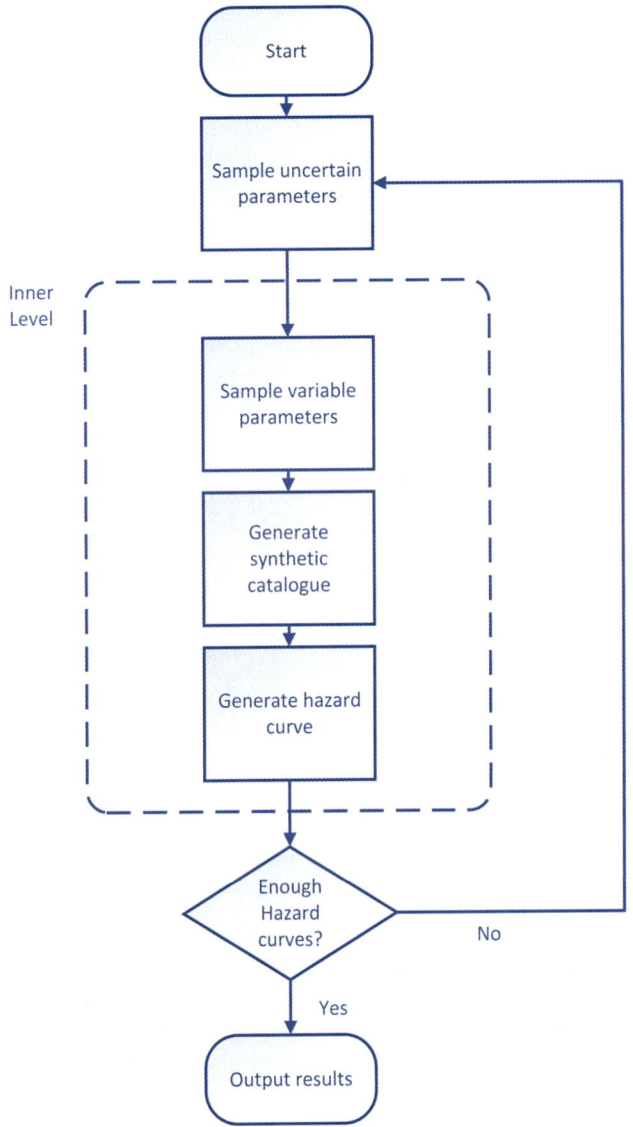

FIGURE 14.2 Simplified flowchart representation of the Monte Carlo modeling scheme. *Adapted from Power, W. L. (2013). Review of tsunami hazard in New Zealand (2013 update). GNS Science, 131. Available from: https://www.gns.cri.nz/assets/Data-and-Resources/Download-files/Tsunami-Report-2013.pdf and Power, W. L., Burbidge, D. R., & Gusman, A. R. (2022). The 2021 update to New Zealand's National Tsunami Hazard Model. GNS Science, 06. https://doi.org/10.21420/X2XQ-HT52.*

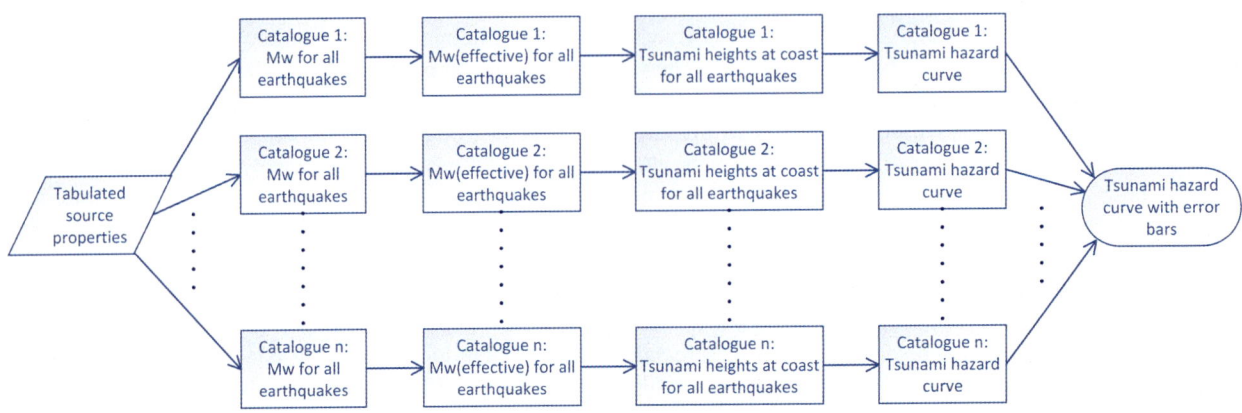

FIGURE 14.3 Representation of the Monte Carlo modeling scheme. *Adapted from Power, W. L. (2013). Review of tsunami hazard in New Zealand (2013 update). GNS Science, 131. Available from: https://www.gns.cri.nz/assets/Data-and-Resources/Download-files/Tsunami-Report-2013.pdf and Power, W. L., Burbidge, D. R., & Gusman, A. R. (2022). The 2021 update to New Zealand's National Tsunami Hazard Model. GNS Science, 06. https://doi.org/10.21420/X2XQ-HT52.*

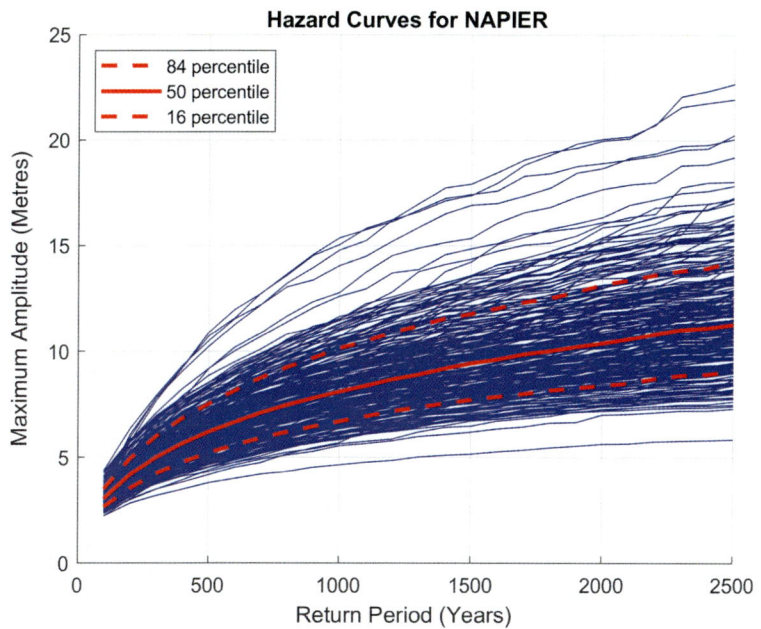

FIGURE 14.4 Example of uncertainty in tsunami hazard curves. Hazard curves for 600 samples of the uncertain parameters, illustrating how the 16th, 50th, and 84th percentiles of uncertainty are calculated for one coastal section. *Adapted from Power, W. L. (2013). Review of tsunami hazard in New Zealand (2013 update). GNS Science, 131. Available from: https://www.gns.cri.nz/assets/Data-and-Resources/Download-files/Tsunami-Report-2013.pdf and Power, W. L., Burbidge, D. R., & Gusman, A. R. (2022). The 2021 update to New Zealand's National Tsunami Hazard Model. GNS Science, 06. https://doi.org/10.21420/X2XQ-HT52.*

The "effective magnitude" in Fig. 14.3 allows for uncertainties and variabilities that are not directly included in the tsunami modeling, for example, the variation in tsunami height that results from variations in the distribution of slip (Mueller et al., 2015), to be represented as having equivalent effects as small changes in magnitude. Several other sources of uncertainty and variability can also be approximated in this way. See Power (2013) for further details. This is particularly useful when using scenarios selected from the NTHM to estimate tsunami inundation for purposes of evacuation zone design or coastal planning (see Section 14.6.1 for more details about how this is done). The inundation maps resulting from using a slightly higher "effective magnitude" approximate the envelope of many possible tsunami extents caused by, for example, variability in slip without requiring explicit modeling of the inundation extents from many possible slip variations.

The uncertainties and variabilities fall into two broad categories—those associated with the earthquake source and those associated with the modeling process. For earthquakes, the primary uncertainty is in the true form of the magnitude-frequency distribution of the faults (i.e., knowing how often earthquakes of varying magnitudes occur along a fault), though it also encompasses such things as uncertainty in the geometry of the faults. The earthquake variabilities represent the variation in magnitude from event to event on a particular fault and also the variation in the distribution of slip (even among earthquakes of the same magnitude). Modeling uncertainty, on the other hand, reflects the inability of the model to fully capture the physics of tsunami generation and propagation and uncertainties in bathymetric data.

An essential input to the probabilistic NTHM is a definition of the physical and statistical properties of the various tsunami sources. The scope of the NTHM is to estimate the tsunami hazard within return periods of up to 2500 years. On these return periods, the major contributions to tsunami hazards come from both distant and local earthquakes.

The definition of tsunami sources from subduction-zone earthquakes, which constitute all distant earthquake sources and the most important local ones, drew heavily on work that has been done for the Global Earthquake Model. The assumed parameters used to generate the magnitude-frequency distributions for the subduction-zone earthquakes in this study are given in Power et al. (2022). Characterization of the geometrical properties of the sources as used as inputs to the tsunami propagation modeling is described in Section 14.3.

As stated earlier, the starting point for defining tsunami sources for local nonsubduction zone earthquakes was the NSHM (Stirling et al., 2012). The faults in the seismic hazard model were filtered to exclude those with characteristic magnitudes below M_w 6.5 (which are too small to generate enough displacement to cause a tsunami), those with strike-slip mechanisms, and those that are entirely on-shore. Additional fault sources were added in the outer-rise areas, the Taranaki Basin, and along the western coast of the South Island; these fault sources are only tentatively identified in geophysical data. More details on the local crustal fault models used for the NTHM can be seen in Section 14.3.2. The creation of synthetic earthquake catalogs from the tabulated fault and subduction zone properties is illustrated in Fig. 14.5.

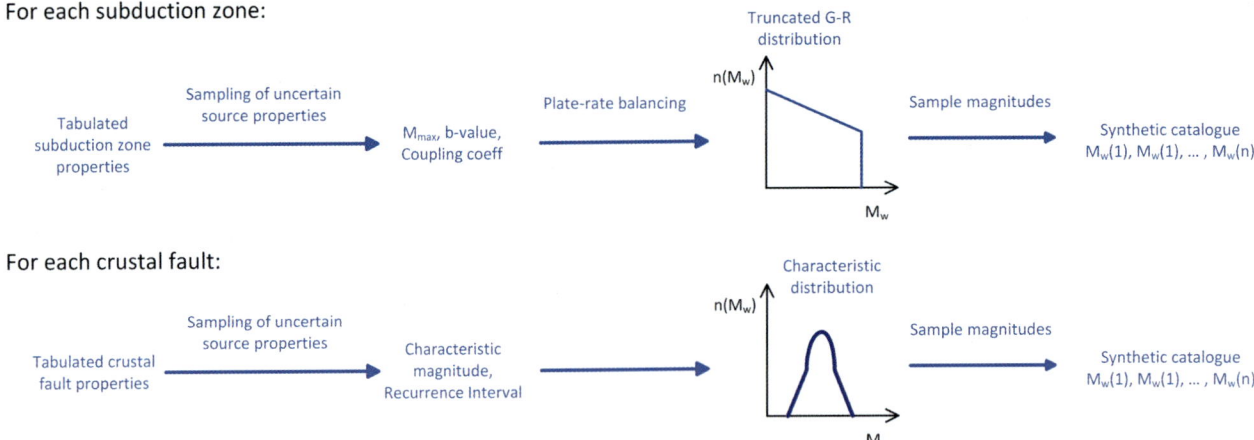

FIGURE 14.5 Magnitude-frequency derivation. Illustration of the steps by which the tabulated fault properties are used to create synthetic earthquake catalogs. *Adapted from Power, W. L. (2013). Review of tsunami hazard in New Zealand (2013 update). GNS Science, 131. Available from: https://www.gns.cri.nz/assets/Data-and-Resources/Download-files/Tsunami-Report-2013.pdf and Power, W. L., Burbidge, D. R., & Gusman, A. R. (2022). The 2021 update to New Zealand's National Tsunami Hazard Model. GNS Science, 06. https://doi.org/10.21420/X2XQ-HT52.*

14.3 Earthquake source models

The calculation of tsunami heights for the events in the synthetic catalogs is the focus of this and the following two sections. The height estimation method is informed by a wide set of numerical models of tsunami scenarios, which are used to parameterize semiempirical functions (Section 14.5). It is these semiempirical functions that are applied in the calculation of tsunami hazard, which allows for generalization from the finite set of numerical models to estimates of tsunami height for the broader range of events in the synthetic catalogs (e.g., estimates of tsunami heights for events that are intermediate in magnitude from those numerically modeled), and to do so in a way that is fast to calculate. This section describes the geometrical models of the fault sources used for the numerical modeling, the scaling relations used to generate earthquake scenarios, and the way that scenarios are distributed across the source regions.

The regional and distant source tsunami scenarios were developed by Gusman et al. (2020) and Gusman, Wang et al. (2019), originally to inform real-time tsunami forecasts. In all (including some local scenarios not used in the NTHM), there are 998 scenarios in the forecast database. An additional set of scenarios was developed for the local subduction zones and the local crustal faults. In the tsunami simulations, the ground surface and seafloor displacement in an earthquake event are calculated using the elastic theory documented in Okada (1985). The ruptures in the scenarios are assumed to be instantaneous.

14.3.1 Plate interface earthquake models

The earthquake scenarios have been categorized into three groups: distant, regional, and local sources. Distant source regions encompass South America, Central America, Cascadia, Alaska-Aleutian, Kamchatka, Kuril, Japan, Izu, Nankai, Mariana, Yap, East Philippines, Papua, and Manus subduction zones. Regional sources include New Britain, Solomon, Vanuatu (New Hebrides), Tonga, and Hjort subduction zones, while local source regions consist of Hikurangi, Puysegur, and Kermadec subduction zones.

The earthquake magnitude interval for all groups is 0.2, but the magnitude ranges differ among distant, regional, and local sources. Magnitudes for the scenarios range from 6.9 to 9.3, depending on the source region. Reference points were distributed along the subduction zones, which serve as the centers for the earthquake scenarios.

In the distant source regions, the minimum magnitude for each reference point is sufficiently small to ensure that all 43 of New Zealand's tsunami warning regions have tsunami amplitudes less than 1.0 m (Gusman et al., 2019). For the regional sources, the scenario database was created using minimum magnitudes such that at least one scenario at each reference point results in a "No Threat" warning level (tsunami amplitude <0.3 m) for all warning regions.

The spatial distance between reference points for distant earthquake sources is 300 km. For regional sources, the spatial distances between reference points vary, being either 100, 150, or 300 km depending on the magnitude range. For local sources, such as Hikurangi, Kermadec, and Puysegur, the distances are either 100 or 150 km. A comprehensive listing of magnitude and reference point setups for the scenarios is provided in Table 14.1.

TABLE 14.1 Patch size, distance between reference points, number of reference points, earthquake magnitudes for the scenarios, and number of earthquake scenarios for each subduction zone group.

Subduction zone	Patch size (km × km)	Distance between reference points (km)	Number of reference points	Earthquake magnitudes	Number of scenarios
Central and South America	100 × 50	300	37	8.7, 8.9, 9.1, 9.3	148
Cascadia	100 × 50	300	9	8.7, 8.9, 9.1, 9.3	36
Alaska-Aleutian	100 × 50	300	12	8.7, 8.9, 9.1, 9.3	48
Kamchatka, Kuril, Japan, Izu, Mariana and Yap	100 × 50	300	25	8.7, 8.9, 9.1, 9.3	100
Nankai	100 × 50	300	6	8.7, 8.9, 9.1, 9.3	24
East Philippines	100 × 50	300	6	8.7, 8.9, 9.1, 9.3	24
Papua	100 × 50	300	4	8.7, 8.9, 9.1, 9.3	16
Manus	100 × 50	300	4	8.7, 8.9, 9.1, 9.3	16
New Britain, Solomon and Vanuatu	100 × 50	300	12	8.1, 8.3, 8.5, 8.7, 8.9, 9.1, 9.3	84
New Britain, Solomon and Vanuatu	50 × 25	150	10	7.5, 7.7, 7.9	30
New Britain, Solomon and Vanuatu	50 × 25	100	15	6.9, 7.1, 7.3	25
Tonga and North Kermadec	50 × 25	300	10	8.1, 8.3, 8.5, 8.7, 8.9, 9.1, 9.3	70
Tonga and North Kermadec	50 × 25	150	19	7.5, 7.7, 7.9	57
Tonga and North Kermadec	50 × 25	100	28	6.9, 7.1, 7.3	50
Tonga and North Kermadec (deeper depth)	50 × 25	100	11	6.9, 7.1, 7.3	33
South Kermadec and Hikurangi	50 × 25	150	5	7.5, 7.7, 7.9, 8.1, 8.3, 8.5, 8.7, 8.9, 9.1, 9.3	50
South Kermadec and Hikurangi	50 × 25	100	8	6.9, 7.1, 7.3	24
South Kermadec and Hikurangi (deeper depth)	50 × 25	100	8	6.9, 7.1, 7.3	24
Puysegur	50 × 25	150	3	7.5, 7.7, 7.9, 8.1, 8.3, 8.5, 8.7, 8.9, 9.1, 9.3	30
Puysegur	50 × 25	100	5	6.9, 7.1, 7.3	15
Hjort	100 × 50	300	2	8.1, 8.3, 8.5, 8.7, 8.9, 9.1, 9.3	14
Hjort	50 × 25	150	3	7.5, 7.7, 7.9	9
Hjort	50 × 25	100	4	7.1, 7.3	8
Additional scenarios	100 × 50	300	N/A	7.9, 8.3, 8.5	63

Source: Data from Gusman, A. R., Power, W. L., & Mueller, C. (2019). Tsunami modelling for Porirua City: The methodology to inform land use planning response. GNS Science, 80. https://doi.org/10.21420/SEG2-Q850; Gusman, A. R., Wang, X., Power, W. L., Lukovic, B., Mueller, C., & Burbidge, D. R. (2019). Tsunami threat level database update. GNS Science, 67. https://doi.org/10.21420/QM31-NA61.

Earthquakes in the subduction zones of the Pacific Ocean are the main sources of tsunami risks to New Zealand. For simulating tsunamis, the National Oceanic and Atmospheric Administration's (NOAA) Center for Tsunami Research has created a set of fault patches that cover the subduction zones throughout the Pacific Ocean. Each fault patch has a fault length of 100 km and a width of 50 km, equivalent to an M_w 7.7 earthquake. The strike and dip angles were determined by the plate interface geometry, and the rake angle was assumed to be 90 degrees. The number of rows of these patches (extending downdip) can be two or more. For the NTHM, the original NOAA fault patches were used for all subduction zones except for Puysegur, Hikurangi, Kermadec, and Tonga. For these four subduction zones, NOAA's and other available fault geometries (Power et al., 2012; Williams et al., 2013) were used to build smaller fault patches with a fault length of 50 km and a fault width of 25 km, equivalent to typical dimensions of an M_w 7.1 earthquake. The NOAA's fault patches were further subdivided into patches of 50 km by 25 km for the New Britain, Solomon, Vanuatu, and Hjort subduction zones. This subdivision enables the creation of scenarios with magnitudes smaller than 8.1 in these regions. The fault patches used in this work are shown in Fig. 14.6.

To make earthquake scenarios, scaling relations are needed to derive rupture lengths and widths from the earthquake magnitudes. For the subduction zone earthquake scenarios, the scaling relations of Blaser et al. (2010) were chosen. This scaling relation is based on a large dataset of thrust faulting events on worldwide subduction zones. Here, we use the scaling relations of Blaser et al. (2010) that were fitted with orthogonal regression. The magnitude to fault length (L) scaling relation that is used for these thrust fault events is $\log(L) = -2.37 + 0.57 \times M_w$, while the fault width (W) is calculated using $\log(W) = -1.86 + 0.46 \times M_w$. The length and width are both in kilometers.

Reference points for earthquake scenarios were distributed for the scenarios along all the subduction zones. The distance between the two reference points for the local sources in the Hikurangi and Puysegur subduction zones is 150 km, while the distance between the two reference points for the regional and distant sources is 300 km. The scenarios along each zone are centered on the reference points. The fault length and width for the scenarios were calculated using the above scaling relations. These dimensions were then used to select the fault patches. There are two patch sizes, one 100 km long and 50 km wide, and the other one 50 km long and 25 km wide. The number of 100 × 50 fault patches in the downdip direction is limited to four, or 200 km in total width. Although not all subduction zones have four patches down the dip. For the smaller 50 × 25 fault patches, which is the size used for the Tonga, Kermadec, Hikurangi, and Puysegur subduction zones, the number of available patches down the dip is four or 100 km in total width. The number of patches for each scenario along the strike is based on the scaling relation above but also depends on the local plate interface geometry.

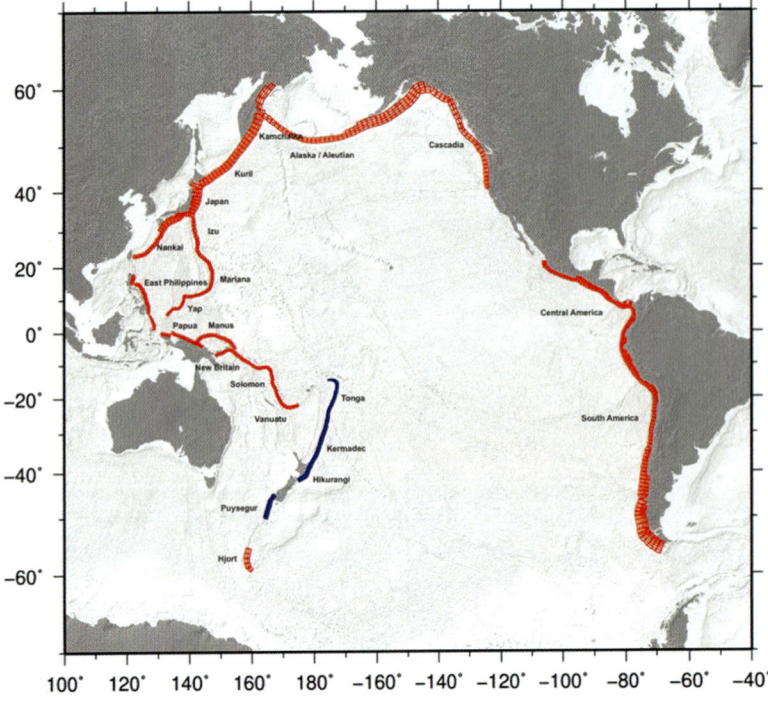

FIGURE 14.6 Plate interface patches used in the earthquake scenarios for the tsunami simulations. The NOAA's fault patches with a fault length of 100 km and a width of 50 km are shown in red. Fault patches for Tonga, Kermadec, Hikurangi, and Puysegur with a fault length of 50 km and a width of 25 km are shown in blue. *NOAA's*, National Oceanic and Atmospheric Administration's. Map lines delineate study areas and do not necessarily depict accepted national boundaries. *From Gusman, A. R., Power, W. L., & Mueller, C. (2019). Tsunami modelling for Porirua City: The methodology to inform land use planning response.* GNS Science, 80. *https://doi.org/10.21420/SEG2-Q850; Gusman, A. R., Wang, X., Power, W. L., Lukovic, B., Mueller, C., & Burbidge, D. R. (2019). Tsunami threat level database update.* GNS Science, 67. *https://doi.org/10.21420/QM31-NA61.*

Uniform slip models were used for the simulations. Depending on which fault patches were selected, the resulting total fault area (A) of scenarios with the same magnitude can be different. This is because the geometry of the plate interface at each location is unique. The slip amount (S) of each scenario is calculated using $M_0 = \mu S A$ where M_0 is the desired seismic moment and μ is the rigidity. The rigidity is assumed to be 4×10^{10} N/m^2, and the seismic moment is converted from the earthquake moment magnitude using the formula $M_w = \frac{2}{3}(\log(M_0) - 9.1)$ (Kanamori, 1977).

14.3.2 Crustal fault earthquake models

Crustal fault earthquakes have previously triggered tsunamis. In 2016, New Zealand experienced a powerful earthquake during the Kaikoura event, which involved the rupture of several crustal faults, including those located both onshore and offshore. This earthquake-induced tsunami reached heights of up to 7 m in an uninhabited area at Goose Bay, south of Kaikoura, and caused minimal damage, affecting only one house at the end of the V-shaped Little Pigeon Bay on the Banks Peninsula (Gusman et al., 2018; Lane et al., 2017).

In previous studies, crustal fault source models have been developed for the NSHM (Stirling et al., 2002, 2012). The dimensions and orientations of these source models were created after considering onshore and offshore geological mapping and their interpretations in terms of fault sources. The fault model magnitudes of the NSHM were estimated from the fault dimensions using equations for global strike-slip faults (Hanks & Bakun, 2002) and equations for New Zealand earthquakes (Stirling et al., 2008; Villamor et al., 2007). Just like the plate interface earthquake models, uniform slip models were used. To estimate the slip amount from the earthquake magnitude, a rigidity of 3.43×10^{10} N/m^2 was assumed, which is commonly used for crustal fault earthquakes (Satake et al., 2022).

Local crustal sources that are offshore and capable of producing an earthquake above M_w 6.5 were selected based on Stirling et al. (2012). Additionally, any onshore local sources were included if they were capable of generating an M_w 7.5 earthquake or greater, or M_w 8.0 if the mechanism is strike-slip (the rake angle for strike-slip scenarios was assumed to be 165 degrees). The reason for including onshore earthquakes with these characteristics was that they may result in partial offshore vertical displacement capable of generating tsunamis larger than 30 cm. The locations of the 250 local crustal fault sources used in the NSHM are shown in Fig. 14.7.

14.4 Tsunami numerical simulation

In this section, we present the numerical modeling that is applied to the earthquake source scenarios developed in Section 14.3 and used to estimate tsunami heights at the New Zealand coast for each of the scenario events. The numerical models that form the scenario database were made using the Cornell Multi-grid Coupled Tsunami (COMCOT) modeling software (Wang & Power, 2011). The nested bathymetric grid scheme was kept consistent for the local,

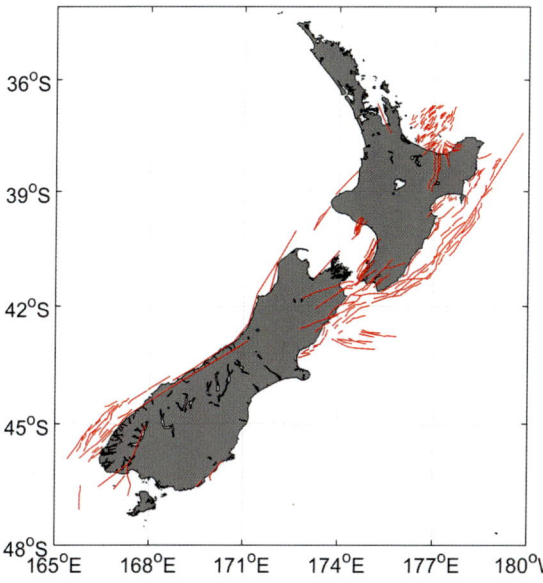

FIGURE 14.7 New Zealand tsunamigenic crustal fault map, showing the local crustal fault sources used in the NTHM 2021. *NTHM*, National Tsunami Hazard Model. Map lines delineate study areas and do not necessarily depict accepted national boundaries. *From Power, W. L., Burbidge, D. R., & Gusman, A. R. (2022). The 2021 update to New Zealand's National Tsunami Hazard Model. GNS Science, 06. https://doi.org/10.21420/X2XQ-HT52.*

regional, and distant sources (Table 14.2 and Fig. 14.8). Modeling grids at three levels of grid spacing refinement were used to simulate tsunami generation, propagation, and coastal interaction to satisfy the resolution requirements for tsunami evolutions across different regimes of ocean bathymetry and achieve sufficient accuracy for tsunami threat level determination along the New Zealand coasts.

TABLE 14.2 Boundaries of the nested grids in degrees of longitude and latitude.

Grid layer	Grid size (arcseconds)	West (degrees)	East (degrees)	South (degrees)	North (degrees)
1	240	150	200	−55	−25
2	60	160	190	−50	−30
3	15	166	179	−48	−34
4	15	182.5	184.5	−45	−43

Source: From Gusman, A. R., Power, W. L., & Mueller, C. (2019). Tsunami modelling for Porirua City: The methodology to inform land use planning response. *GNS Science, 80*. https://doi.org/10.21420/SEG2-Q850; Gusman, A. R., Wang, X., Power, W. L., Lukovic, B., Mueller, C., & Burbidge, D. R. (2019). Tsunami threat level database update. *GNS Science, 67*. https://doi.org/10.21420/QM31-NA61; Gusman, A. R., Lukovic B., & Peng, B. (2020). Tsunami threat level database update: Regional sources and tsunami warning text. *GNS Science, 83*. Available from: http://natlib.govt.nz/records/42967871.

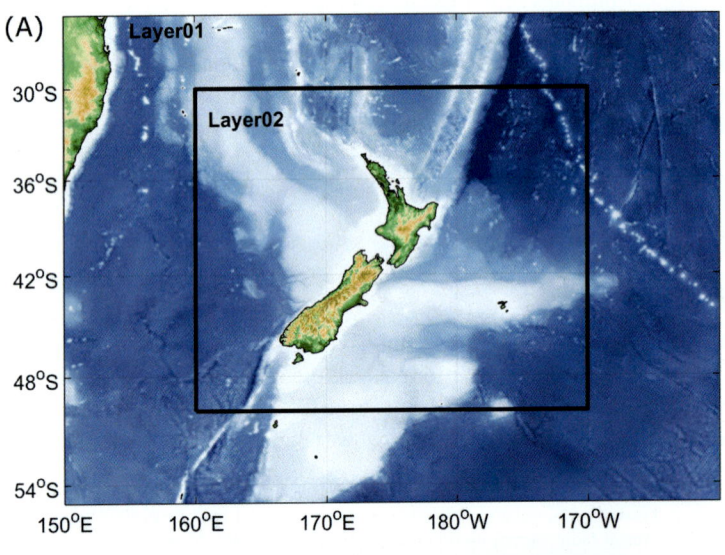

FIGURE 14.8 Nested grid modeling set-up used for the NTHM 2021 update (A) Southwest Pacific and (B) New Zealand. *NTHM*, National Tsunami Hazard Model. Map lines delineate study areas and do not necessarily depict accepted national boundaries.

The digital elevation model (DEM, a combination of topography and bathymetry) data for the first level grids, Layer01 (Fig. 14.8A), was extracted from the National Centers for Environmental Information Gridded Global Relief Data (ETOPO2v2, https://www.ngdc.noaa.gov/mgg/global/etopo2.html), which covers the whole Pacific to simulate tsunami generations and propagations from local, regional, and distant sources, modeled at a grid spacing of 4.0 arcminutes. The same 2.0 arcminutes DEM data were also used for the second level grids, Layer02 (Fig. 14.8B) but were interpolated at 1.0 arcminute spacing (1.2–1.6 km in New Zealand) to cover the entirety of New Zealand and its offshore regions.

A national scale DEM was derived from LINZ (Land Information New Zealand) charts and elevation datasets, the Seabed Mapping C-MAP, and the General Bathymetric Chart of the Oceans (GEBCO) datasets, covering New Zealand and its offshore regions at a spatial resolution of 10.0 arcseconds (~640–740 m grid spacing in New Zealand). Within this 10.0-arcsecond DEM, higher resolution local DEM datasets from other studies were incorporated to update the topographic and bathymetric data in Southland (Prasetya et al., 2010), the Bay of Plenty (Beban et al., 2012, 2013), Poverty Bay, Hawke's Bay (Fraser et al., 2014), and Greater Wellington Region (Mueller et al., 2014) for better accuracy. The third level modeling grids include Layer03, which covers the two main islands of New Zealand, and Layer04 for the Chatham Islands and their nearshore areas. Their spatial resolution of 15.0 arcseconds (310–380 m) was resampled from the 10.0-arcsecond DEM data.

The linear shallow water equations were used to simulate the tsunami on Layer01 and Layer02, while the nonlinear shallow water equations were employed for Layer03 and Layer04. Manning's formula was used to model seafloor friction effects on the tsunami, with a roughness value of $n = 0.015$. Vertical wall boundaries were implemented at the 10-cm water depth contours. This setup was chosen to make the simulated coastal tsunami heights on the finest grids as realistic as possible.

14.5 Tsunami height estimation

For each earthquake in a synthetic catalog (Section 14.2), an estimate of the subsequent tsunami height at the coast needs to be made for each of the 268 coastal sections (Fig. 14.1). For the NTHM, we define the tsunami height as the maximum positive amplitude, which is the highest elevation that the tsunami reaches relative to the background sea level. As heights must be estimated for many thousands of events, a computationally efficient approach is required. For this, we make use of the database of tsunami scenarios created by numerical modeling (Sections 14.3 and 14.4) and use additional techniques of interpolation and extrapolation to estimate the tsunami heights for events that differ from those in the database. This is accomplished differently for crustal faults, local subduction zones, and regional and distant subduction zones, as will be explained below.

14.5.1 Regional and distant sources

For regional and distant sources, the NTHM makes use of the precalculated scenarios to estimate the coefficients in a semiempirical scaling relationship based on Abe (1979, 1995):

$$Ht_{ij} = 10^{M_w - B_{ij}} \tag{14.1}$$

where Ht_{ij} is the tsunami height at the coastal section j due to an earthquake of moment magnitude M_w in the source region i (for distant and regional sources, the "source regions" are the subduction zones) and B_{ij} is a coefficient specific to tsunami traveling from the source region i to the coastal section j.

Analysis of the modeled scenarios indicated that the B_{ij} coefficients were not entirely independent of magnitude; consequently, the average B_{ij} and their standard deviations were calculated at the specific magnitudes of the scenarios by rearranging Eq. (14.1) as

$$B_{ij}(M_w) = M_w - \log_{10} Ht_{ij} \tag{14.2}$$

Specifically, $B_{ij}(M_w)$ was calculated for each of the scenarios that originate in the source region i (i.e., for all of the scenarios whose central reference point occurs within the source region), and these were evaluated to produce a mean and standard deviation of $B_{ij}(M_w)$ for the source region i as a whole.

When used to estimate tsunami heights in the hazard model, interpolation was used to find a value of $B_{ij}(M_w)$ for the specific magnitude of each event in the synthetic catalogs. To allow for the effects of different locations within the source regions, the standard deviation $\sigma B_{ij}(M_w)$ was also estimated from the set of modeled scenario data. The tsunami height is then calculated using the following equation:

$$Ht_{ij} = 10^{M_w - \left(B_{ij}(M_w) + \overline{\sigma B_{ij}(M_w)}\right)} \tag{14.3}$$

where $\overline{\sigma B_{ij}(M_w)}$ is sampled from a normal distribution with mean zero and standard deviation $\sigma B_{ij}(M_w)$ and the normal distribution is truncated at $\pm 2\sigma$.

14.5.2 Local subduction zones

For the local subduction zones within 1 hour travel time to the New Zealand coast (i.e., the Hikurangi, Puysegur, and Kermadec subduction zones), a similar method was developed that was used for the distant and regional sources. However, for these nearby sources, there is a greater sensitivity of tsunami heights to the effects of the earthquake location, and, rather than treating this sensitivity as a random input (via the standard deviation in B_{ij}), a different approach was used to find $B_{ij}(M_w)$ that assumed a random position for the earthquake within the subduction zone.

The scenarios used for this modeling were similar to those in the 2019 scenario database (Gusman et al., 2020, 2019), except that the scenario magnitudes and locations were chosen such that for each magnitude a whole number of rupture areas would fit within the subduction zones without gaps or overlap (as in Fig. 14.9), giving equal coverage to all parts.

14.5.3 Local crustal faults

For the local crustal sources, hydrodynamic modeling was used to calculate the tsunami heights at the characteristic magnitudes of each fault (see Sections 14.3 and 14.4), and then Abe's equation (Eq. 14.1) was used with the B_{ij} estimated from the characteristic magnitude scenario (using Eq. 14.2) to calculate the tsunami shoreline wave heights for other earthquake magnitudes. As described in Section 14.3, local crustal sources were selected that were offshore and capable of producing an earthquake above M_w 6.5 according to Stirling et al. (2012). Also included were any onshore local source capable of generating an M_w 7.5 earthquake or greater, or M_w 8.0 if the mechanism was strike-slip. Fig. 14.7 shows the location of the 250 local crustal fault sources used in this study.

14.6 Tsunami hazard model results

With the methods for estimating the tsunami heights presented in Sections 14.3–14.5, all the components are present to run the nested Monte Carlo analysis described in Section 14.2. The implementation of the algorithm is in MATLAB® and can be run on a personal computer (PC) or Unix workstation. The calculation is relatively quick (typically a few

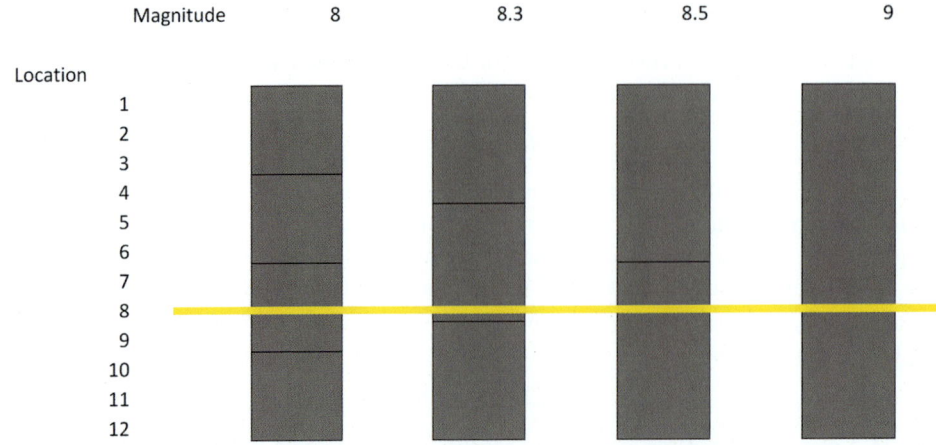

FIGURE 14.9 Schematic illustration of how scenarios are used to obtain B_{ij} values for the Hikurangi and Puysegur subduction zones (the method for Kermadec Trench is similar but uses different magnitudes and more locations). A random position 1–12 is selected (yellow bar), then B_{ij} values from scenarios whose magnitudes bracket the event magnitude are selected (gray boxes represent scenario rupture areas) and a magnitude-dependent $B_{ij}(M_w)$ is calculated from them by interpolation (Power et al., 2022). *From Power, W. L., Burbidge, D. R., & Gusman, A. R. (2022). The 2021 update to New Zealand's National Tsunami Hazard Model. GNS Science, 06. https://doi.org/10.21420/X2XQ-HT52.*

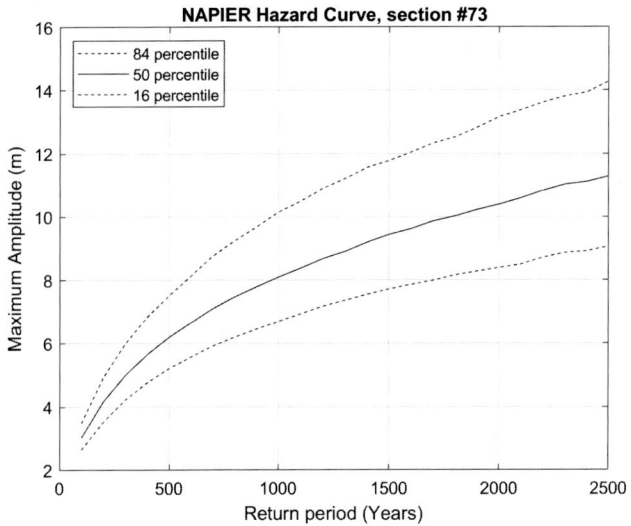

FIGURE 14.10 Tsunami hazard curve for Napier from the NTHM 2021. *NTHM*, National Tsunami Hazard Model. *From Power, W. L., Burbidge, D. R., & Gusman, A. R. (2022). The 2021 update to New Zealand's National Tsunami Hazard Model.* GNS Science, 06. https://doi.org/10.21420/X2XQ-HT52.

hours) as the numerically intensive numerical modeling required to create the scenario database was previously performed and does not take place within the hazard calculation.

14.6.1 Tsunami hazard curves and disaggregation results

For each of the 268 coastal sections, the NTHM produces a hazard curve with "error bars" at the 16% and 84% levels of confidence. Fig. 14.10 shows one example of the city of Napier on the east coast of the North Island. Napier is located on the eastern coast of New Zealand's North Island and is a city known for its Art Deco architecture, beautiful beaches, and vibrant tourism industry. While the city's proximity to the Pacific Ocean has contributed to its appeal, it also means that Napier faces significant exposure to tsunami hazards.

The dashed lines in Fig. 14.10 represent the 16% and 84% confidence intervals, while the solid black line shows the median hazard as a function of the return period, determined by the highest shoreline amplitude, also known as the tsunami height. The confidence curves at 16% and 84% indicate the level of uncertainty in the hazard estimates for a specific return time. Statistically, there is an 84% chance that the true hazard will be below the upper dotted line and there is a 16% chance that the true hazard will be below the lower dotted line. As we extend the return periods, the level of uncertainty increases as one might expect.

Additionally, the synthetic event catalogs can be examined to produce a disaggregation plot for a given coastal section. Fig. 14.11 shows such a plot for Napier at the 500-year return period. The disaggregation shows which tsunami sources contribute the most to the tsunami hazard at a specified return period and is useful for choosing scenarios for later inundation modeling. In addition to location-specific hazard curves, it is possible to plot the tsunami heights in a map view showing the expected maximum tsunami height to be exceeded at a specific return period and level of confidence. Figs. 14.12 and 14.13 show the 500- and 2500-year tsunami hazards at the median (50th percentile) of confidence. As we would expect, the estimated hazard is greatest along those coasts that are most exposed to tsunamis from the three local subduction zones: the eastern coast of the North Island (Hikurangi), the far-northeast of the North Island (Kermadec), and the southwest of the South Island (Puysegur).

14.6.2 Tsunami inundation hazard modeling

We provide an example of how to apply the NTHM to develop tsunami inundation hazard maps. Probabilistic tsunami inundation maps for 100, 500, and 1000-year return periods have been developed for Wellington City (Burbidge et al., 2021). The maps were made to inform Wellington City Council's Urban Growth Plan to improve Wellington's resilience to tsunamis. To produce these maps, the NTHM was used to determine which earthquake scenarios contributed most to the shoreline tsunami hazard at the three return periods for the coastal Section 91 encompassing Wellington City (Fig. 14.14). For Section 91, median, 16%, and 84% tsunami hazard curves from the 2021 NTHM are shown in Fig. 14.15 (Power et al., 2022). In this study, the 50% confidence level (i.e., median) was used as the basis for the inundation modeling to give an unbiased estimate of the hazard for land-use planning.

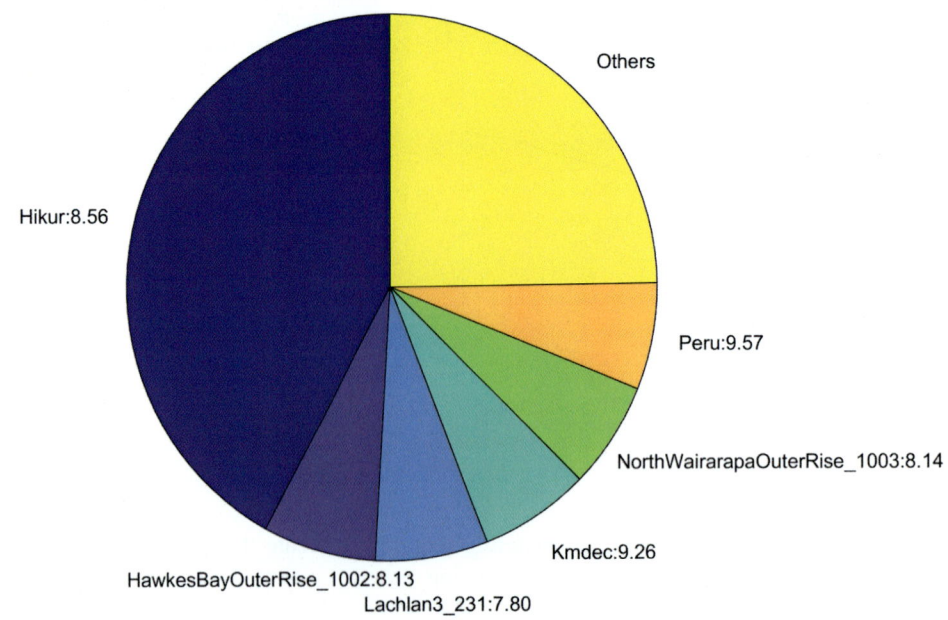

FIGURE 14.11 Disaggregation of tsunami sources for Napier at the 500-year return period. *Modified from Power, W. L., Burbidge, D. R., & Gusman, A. R. (2022). The 2021 update to New Zealand's National Tsunami Hazard Model. GNS Science, 06. https://doi.org/10.21420/X2XQ-HT52.*

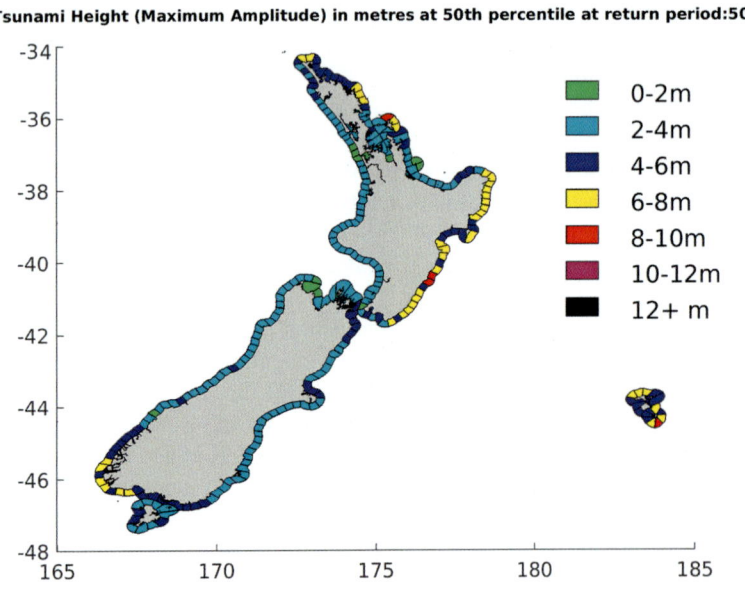

FIGURE 14.12 Expected maximum tsunami height in meters at the 500-year return period, shown at median (50th) percentile of epistemic uncertainty. Map lines delineate study areas and do not necessarily depict accepted national boundaries. *From Power, W. L., Burbidge, D. R., & Gusman, A. R. (2022). The 2021 update to New Zealand's National Tsunami Hazard Model. GNS Science, 06. https://doi.org/10.21420/X2XQ-HT52.*

To determine the earthquake source models for the tsunami inundation modeling, the disaggregation process was used to determine which sources contribute most to the hazard at the three return periods, which for the purpose of inundation modeling are expressed in terms of annual probability of exceedance. In Tables 14.3—14.5, the top six hazard source scenarios for each of the three annual probabilities of exceedances are shown for Section 91, as determined by the disaggregation. This process gives the effective magnitude and the contribution to the tsunami hazard for each selected scenario, and also the tsunami height at Section 91 for a given annual probability of exceedance.

The majority of the hazards in this coastal zone have been contributed by the Hikurangi subduction zone for all return periods examined. The majority of the hazard is attributed to an M_w 8.1 earthquake in this subduction zone at the 1:100 annual probability of exceedance, or 100-year return period. The magnitude of the Hikurangi earthquake increases to M_w 8.6 for the 1:500 annual probability of exceedance and M_w 8.7 for the 1:1000 annual probability of

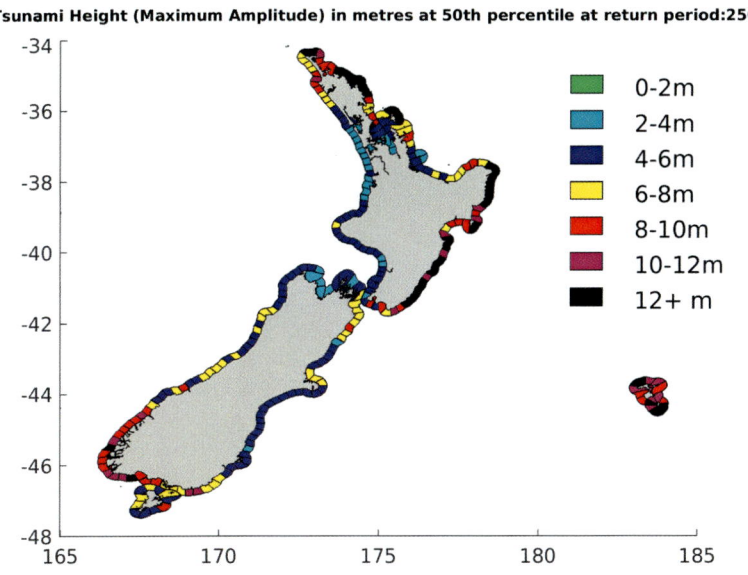

FIGURE 14.13 Expected maximum tsunami height in meters at the 2500-year return period, shown at median (50th) percentile of epistemic uncertainty. Map lines delineate study areas and do not necessarily depict accepted national boundaries. *From Power, W. L., Burbidge, D. R., & Gusman, A. R. (2022). The 2021 update to New Zealand's National Tsunami Hazard Model. GNS Science, 06. https://doi.org/10.21420/X2XQ-HT52.*

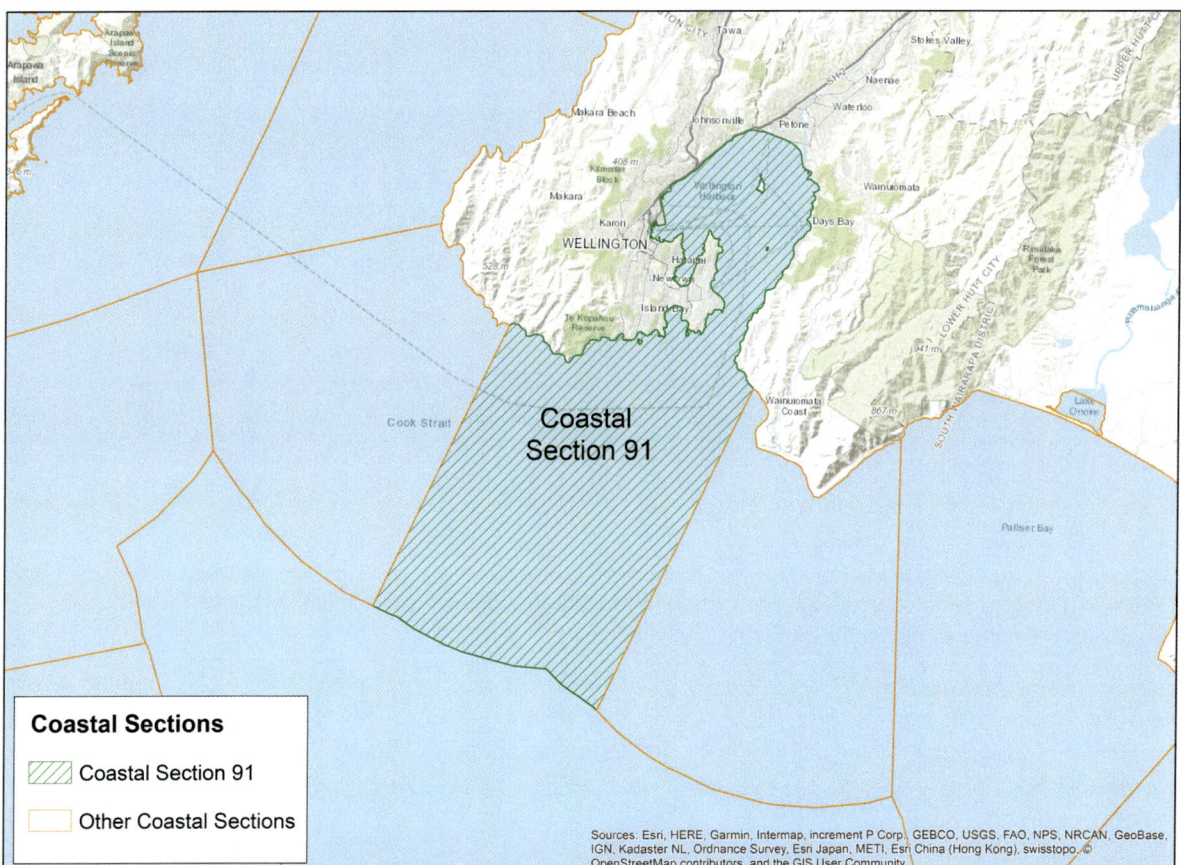

FIGURE 14.14 Tsunami coastal sections around the Wellington region, highlighting Section 91 of the NTHM, which was used to calculate the tsunami hazard for Wellington. Map lines delineate study areas and do not necessarily depict accepted national boundaries.

exceedance. Most of the remaining threats for the 1:100 annual probability of exceedance originate from distant subduction zones, like Peru. However, the majority of the remaining hazards at the 1:1000 annual probability of exceedance are caused by large (M_w 7.9 to 8.4) magnitude earthquakes on local crustal faults that have a fault section offshore of New Zealand (e.g., the offshore portion of the Wairarapa Fault).

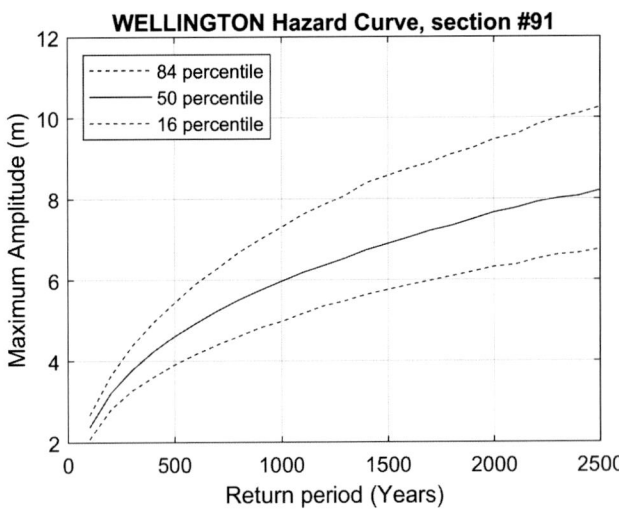

FIGURE 14.15 Tsunami hazard curves for Section 91 in the updated NTHM (Power et al., 2023). The shoreline tsunami height (i.e., amplitude in the figure) above mean sea level is a function of the return period. *NTHM*, National Tsunami Hazard Model. *From Power, W. L., Burbidge, D. R., & Gusman, A. R. (2022). The 2021 update to New Zealand's National Tsunami Hazard Model. GNS Science, 06. https://doi.org/10.21420/X2XQ-HT52.*

TABLE 14.3 Disaggregated source scenarios for the inundation modeling for the 1:100 annual probability of exceedance in Section 91 for Wellington.

Tsunami height (m)	Source name	Disaggregation magnitude	Contribution (%)	Contribution of top six (normalized) (%)
2.37	Hikurangi	8.09	39.33	55.65
2.37	Peru	9.34	9.00	12.73
2.37	Jordan, Kekerengu, and Needles	7.65	7.37	10.42
2.37	Northern Chile	9.20	5.20	7.36
2.37	Central Chile	9.52	5.02	7.10
2.37	Kermadec	9.07	4.77	6.74
2.37	Other sources	N/A	29.32	N/A

Source: Data from Burbidge, D. R., Gusman, A. R., Power, W. L., Wang, X., & Lukovic, B. (2021). Wellington City probabilistic tsunami hazard maps. *GNS Science, 91*. Available from: https://wellington.govt.nz/-/media/your-council/plans-policies-and-bylaws/plans-and-policies/a-to-z/spatial-plan/tsunami-hazards-report--september-2021.pdf.

TABLE 14.4 Disaggregated source scenarios for the inundation modeling for the 1:500 annual probability of exceedance in Section 91 for Wellington.

Tsunami height (m)	Source name	Disaggregation magnitude	Contribution (%)	Contribution of top six (normalized) (%)
4.61	Hikurangi	8.57	36.03	49.98
4.61	Jordan, Kekerengu, and Needles	7.94	16.90	23.44
4.61	Wairarapa and Wharekauhau	8.25	5.50	7.63
4.61	Hope and Te Rapa	7.76	5.20	7.21
4.61	Wairarapa	8.29	4.97	6.89
4.61	South Wairarapa Outer Rise	8.33	3.50	4.85
4.61	Other sources	N/A	27.90	N/A

Source: Data from Burbidge, D. R., Gusman, A. R., Power, W. L., Wang, X., & Lukovic, B. (2021). Wellington City probabilistic tsunami hazard maps. *GNS Science, 91*. Available from: https://wellington.govt.nz/-/media/your-council/plans-policies-and-bylaws/plans-and-policies/a-to-z/spatial-plan/tsunami-hazards-report--september-2021.pdf.

TABLE 14.5 Disaggregated source scenarios for the inundation modeling for the 1:1000 annual probability of exceedance in Section 91 for Wellington.

Tsunami height (m)	Source name	Disaggregation magnitude	Contribution (%)	Contribution of top six (normalized) (%)
5.96	Hikurangi	8.69	33.82	43.41
5.96	Jordan, Kekerengu, and Needles	8.05	18.55	23.81
5.96	Wairarapa and Wharekauhau	8.37	8.17	10.48
5.96	Wairarapa	8.4	7.25	9.31
5.96	Hope and Te Rapa	7.87	5.22	6.70
5.96	South Wairarapa Outer Rise	8.44	4.90	6.29
5.96	Other sources	N/A	22.10	N/A

Source: Data from Burbidge, D. R., Gusman, A. R., Power, W. L., Wang, X., & Lukovic, B. (2021). Wellington City probabilistic tsunami hazard maps. GNS Science, 91. Available from: https://wellington.govt.nz/-/media/your-council/plans-policies-and-bylaws/plans-and-policies/a-to-z/spatial-plan/tsunami-hazards-report--september-2021.pdf.

TABLE 14.6 Roughness values for different land-cover groups for the tsunami modeling.

Land-cover group	Manning's n (roughness coefficient)
Built-up area (e.g., urban/residential/industrial/Central Business District)	0.060
Tall vegetation (e.g., forest)	0.040
Scrub (e.g., low trees/bushes)	0.040
Low vegetation (e.g., grass)	0.030
Urban open area (e.g., paved/smoothed)	0.025
Bare land (e.g., farmland)	0.025
Water area (e.g., riverbed/seabed)	0.011

Source: Data from Wang, X., Lukovic, B., Power, W. L., & Mueller, C. (2017). High-resolution inundation modelling with explicit buildings. GNS Science, 13. .

To model tsunami inundation, COMCOT was used with a set of nested elevation grids of increasing resolution as we approached the area of interest. To create the highest resolution grid, we incorporated light detection and ranging (LiDAR) data for the tsunami simulation DEM. High-resolution DEM data (10-m resolution) were used to cover the Wellington Harbor and its surrounding suburbs. The data were created for hydrodynamic inundation modeling and delineation of tsunami evacuation zones for Wellington Harbor (Mueller et al., 2014) and were derived from a combination of LiDAR topographic data provided by Wellington Regional Council and multibeam bathymetric survey data from the National Institute of Water & Atmospheric Research (Pallentin et al., 2009), covering the interior of the harbor. Outside the harbor, the bathymetric data were derived from LINZ nautical charts. For the nested grids, we used five grid levels with grid resolutions of 2.7 km, 675 m, 135 m, 34 m, and 11 m. The coarsest grid covers the whole Pacific Ocean, while the finest grid covers the Wellington Harbor. The tsunami was simulated by solving the nonlinear shallow water equations for the finest grid while for the coarser grids, the linear shallow water equations were solved.

We utilized a series of Manning's roughness coefficient (n) for a simplified set of land-cover groups as proposed in Wang et al. (2017) to account for resistance effects of land-cover features on tsunami flow through Manning's formula in the COMCOT simulation model (Wang & Power, 2011). The land-cover groups and their corresponding roughness values can be found in Table 14.6. These values were derived by comparing, grouping, and averaging the roughness values found in the literature (Acrement & Schneider, 1989; Bricker et al., 2015; Gayer et al., 2010; Kaiser et al., 2011; Power et al., 2016; Prasetya & Wang, 2011; Wang et al., 2009; Wang & Liu, 2007; Fujima, 2001; Imamura et al., 2006). With the use of constant roughness values for each land-cover group, this approach cannot capture well the detailed tsunami flow dynamics due to spatial variations of ground surface features within a land-cover group, for example, variations in building densities and types in built-up area, which require advanced modeling approaches (Muhari et al., 2011; Wang et al., 2017, 2020).

In all the simulations presented here, it is assumed that the tsunami occurs at the current mean high water springs (MHWS). MHWS was modeled as a static level, set at 0.69 m above local mean sea level, as determined in the 2019 edition of the New Zealand Nautical Almanac (https://www.linz.govt.nz/sites/default/files/2023-06/hydro_202324-almanac_full-nautical-almanac_pdf.pdf). This level remains constant and does not vary with tide fluctuations over time.

For each of the scenarios indicated above, the models were run over the set of nested grids through to inundation on the innermost grid level. The contribution of each of the inundation scenarios was then weighted and normalized according to their contribution to the hazard and used to find the weighted median of flow depths across the area of interest. The "contribution" percentage describes how large a proportion that source represents in the total disaggregation. These percentages were normalized by including only the top six sources and rescaling the percentages so they make a total of 100%.

The resulting maps of median inland tsunami flow depths and shoreline heights are shown in Figs. 14.16–14.18. As expected, at the 1:100-year annual probability of exceedance (Fig. 14.16), the extent of inundation is small at the current MHWS. However, the tsunami inundation extents in the populated coastal areas of Lyall Bay, Evans Bay, and Lambton Harbor are significant at 1:500 (Fig. 14.17) and 1:1000 (Fig. 14.18) annual probabilities of exceedance. At the 1:500 and 1:1000 annual probabilities of exceedance, the expected tsunami inundation also covers much of the runways and the surrounding terminal area of Wellington Airport, which is located between the shores of Lyall and Evans Bays.

FIGURE 14.16 Probabilistic tsunami inundation map for Wellington City showing median onshore flow depths and offshore tsunami heights, with a 1:100 annual probability of being exceeded at the current MHWS. Onshore values refer to maximum flow depths, while offshore values refer to tsunami heights. *MHWS*, Mean high water springs. Map lines delineate study areas and do not necessarily depict accepted national boundaries. *Data from Burbidge, D. R., Gusman, A. R., Power, W. L., Wang, X., & Lukovic, B. (2021). Wellington City probabilistic tsunami hazard maps. GNS Science, 91. Available from: https://wellington.govt.nz/-/media/your-council/plans-policies-and-bylaws/plans-and-policies/a-to-z/spatial-plan/tsunami-hazards-report--september-2021.pdf.*

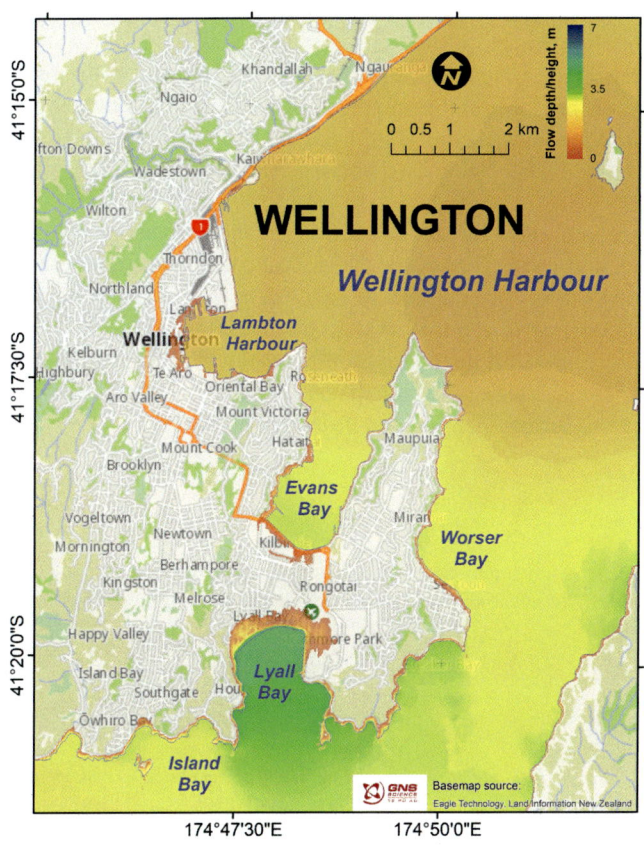

FIGURE 14.17 Probabilistic tsunami inundation map for Wellington City showing median onshore flow depths and offshore tsunami heights, with a 1:500 annual probability of being exceeded at the current MHWS. Onshore values refer to maximum flow depths, while offshore values refer to tsunami heights. *MHWS*, Mean high water springs. Map lines delineate study areas and do not necessarily depict accepted national boundaries. *Data from Burbidge, D. R., Gusman, A. R., Power, W. L., Wang, X., & Lukovic, B. (2021). Wellington City probabilistic tsunami hazard maps. GNS Science, 91. Available from: https://wellington.govt.nz/-/media/your-council/plans-policies-and-bylaws/plans-and-policies/a-to-z/spatial-plan/tsunami-hazards-report--september-2021.pdf.*

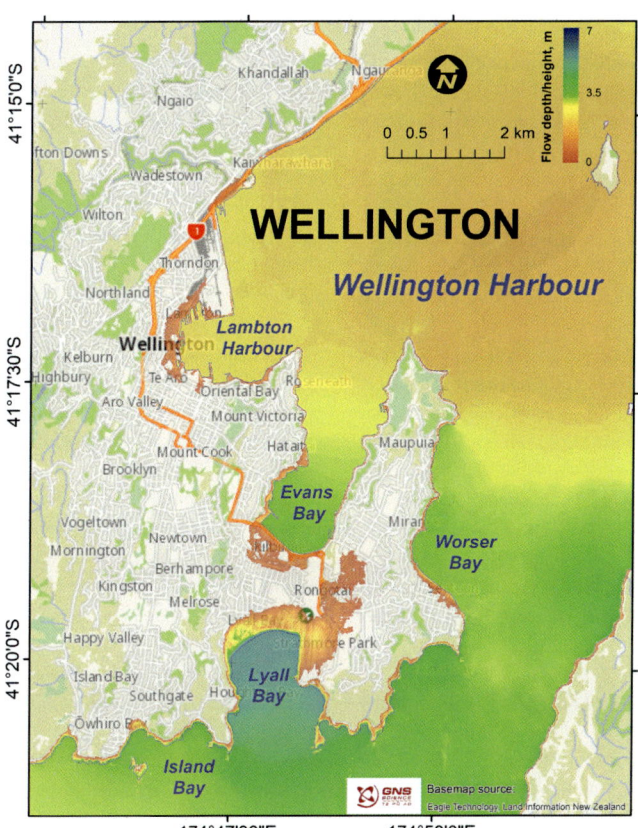

FIGURE 14.18 Probabilistic tsunami inundation map for Wellington City showing median onshore flow depths and offshore tsunami heights, with a 1:1000 annual probability of being exceeded at current MHWS. Onshore values refer to maximum flow depths, while offshore values refer to tsunami heights. *MHWS*, Mean high water springs. Map lines delineate study areas and do not necessarily depict accepted national boundaries. *Data from Burbidge, D. R., Gusman, A. R., Power, W. L., Wang, X., & Lukovic, B. (2021). Wellington City probabilistic tsunami hazard maps. GNS Science, 91. Available from: https://wellington.govt.nz/-/media/your-council/plans-policies-and-bylaws/plans-and-policies/a-to-z/spatial-plan/tsunami-hazards-report--september-2021.pdf.*

14.7 Conclusions

This chapter described the process that has been used to develop the New Zealand NTHM. This model covers return periods from 100 to 2500 years and includes a quantification of the uncertainties involved. Results from the model have been published online via the website https://www.gns.cri.nz/data-and-resources/2021-national-tsunami-hazard-model/, which includes hazard curves and disaggregation plots for each of the 268 coastal sections (Power et al., 2023) and a downloadable Geographic Information System (GIS) layer.

In Section 14.6, it was demonstrated how the NTHM can be used to inform tsunami evacuation modeling for land-use planning. It has also been used to define tsunami evacuation zones for several New Zealand regions, as these are required to encompass at least the 2500-year tsunami at the 84% level of confidence. This has been done using the disaggregations to choose appropriate scenarios for inundation modeling in a similar way to that described in Section 14.6. Often this has been supplemented with explicit modeling of variations in the distribution of coseismic slip for the most important sources to provide additional conservatism (Wang et al., 2022). It is anticipated that tsunami inundation hazard maps derived from the NTHM will also be used in the future for engineering and insurance purposes.

Acknowledgments

This project was supported by the New Zealand Ministry of Business, Innovation, and Employment through the Hazards and Risk Management program (Strategic Science Investment Fund, contract C05X1702). The probabilistic inundation modeling of Wellington City was funded by Wellington City Council.

References

Abe, K. (1995). *Tsunami: Progress in prediction, disaster prevention and warning. Estimate of tsunami run-up heights from earthquake magnitudes. Advances in natural and technological hazards research* (pp. 21–35). Springer. Available from :10.1007/978-94-015-8565-1_2.

Abe, K. (1979). Size of great earthquakes of 1837–1974 inferred from tsunami data. *Journal of Geophysical Research: Solid Earth*, *84*(B4), 1561–1568. Available from https://doi.org/10.1029/jb084ib04p01561.

Acrement, G. J., & Schneider, V. R. (1989). *Guide for selecting Manning's roughness coefficients for natural channels and flood plains. Water supply paper*. Geological Survey. Available from https://ton.sdsu.edu/usgs_report_2339.pdf, 10.3133/wsp2339.

Beban, J. G., Cousins, W. J., Wang, X., & Becker, J. S. (2012). Modelling of the tsunami risk to Papamoa, Wairakei and Te Tumu assuming an altered ground level due to development of Wairakei and Te Tumu, and the implications for the SmartGrowth Strategy. *GNS Science*, *54*.

Beban, J. G., Wang, X., & Cousins, W. J. (2013). Understanding the tsunami hazard and potential life safety risk in Matata, Bay of Plenty from the variation to the Southern Kermadec scenario. *GNS Science*, *113*.

Bell, R., Holden, C., Power, W. L., Wang, X., & Downes, G. (2014). Hikurangi margin tsunami earthquake generated by slow seismic rupture over a subducted seamount. *Earth and Planetary Science Letters*, *397*, 1–9.

Berryman, K. (2005). Review of tsunami hazard and risk in New Zealand review of tsunami hazard and risk in New Zealand. *GNS Science*, *104*.

Berryman, K., Wallace, L., Hayes, G., Bird, P., Wang, K., Basili, R., Lay, T., Pagani, M., Stein, R., Sagiya, T., Rubin, C., Barreintos, S., Kreemer, C., Litchfield, N., Stirling, M., Gledhill, K., Haller, K., & Costa, C. (2015). *The GEM Faulted Earth Subduction Interface Characterisation Project*. Available from http://www.nexus.globalquakemodel.org/gem-faulted-earth/posts.

Blaser, L., Krüger, F., Ohrnberger, M., & Scherbaum, F. (2010). Scaling relations of earthquake source parameter estimates with special. *Bulletin of the Seismological Society of America*, *100*(6), 2914–2926. Available from https://doi.org/10.1785/0120100111.

Bricker, J. D., Gibson, S., Takagi, H., & Imamura, F. (2015). On the need for larger Manning's roughness coefficients in depth-integrated tsunami inundation models. *Coastal Engineering Journal*, *57*(2). Available from https://doi.org/10.1142/S0578563415500059, https://www.tandfonline.com/loi/tcej20.

Burbidge, D. R., Gusman, A. R., Power, W. L., Wang, X., & Lukovic, B. (2021). Wellington City probabilistic tsunami hazard maps. *GNS Science*, *91*. Available from https://wellington.govt.nz/-/media/your-council/plans-policies-and-bylaws/plans-and-policies/a-to-z/spatial-plan/tsunami-hazards-report--september-2021.pdf.

Burbidge, D. R., Roger, J. H. M., Wang, X., Lukovic, B., & Power, W. L. (2022). Level 3 tsunami modelling in Hawke's Bay final report. *GNS Science*, *58*. Available from https://www.eqc.govt.nz/assets/Publications-Resources/3166-Final-Report-Level-3-Tsunami-Modelling-for-Hawkes-Bay.pdf.

Clark, K., Howarth, J., Litchfield, N. J., Cochran, U., Turnbull, J., Dowling, L., Howell, A., Berryman, K., & Wolfe, F. (2019). Geological evidence for past large earthquakes and tsunamis along the Hikurangi subduction margin, New Zealand. *Marine Geology*, *412*, 139–172. Available from https://doi.org/10.1016/j.margeo.2019.03.004.

Downes, G., Barberopoulou, A., Cochran, U., Clark, K., & Scheele, F. (2017). The New Zealand tsunami database: Historical and modern records. *Seismological Research Letters*, *88*(2), 342–353. Available from https://doi.org/10.1785/0220160135, https://www.gns.cri.nz/data-and-resources/new-zealand-tsunami-database-historical-and-modern-records/.

Fraser, S. A., Power, W. L., Wang, X., Wallace, L. M., Mueller, C., & Johnston, D. M. (2014). Tsunami inundation in Napier, New Zealand, due to local earthquake sources. *Natural Hazards, 70*, 415−445. Available from https://doi.org/10.1007/s11069-013-0820-x.

Fujima,K. (2001). *Long wave propagation on large roughness*. In Proceedings of the ITS 2001 (pp. 7−22).

Gayer, G., Leschka, S., Nöhren, I., Larsen, O., & Günther, H. (2010). Tsunami inundation modelling based on detailed roughness maps of densely populated areas. *Natural Hazards and Earth System Sciences, 10*(8), 1679−1687. Available from https://doi.org/10.5194/nhess-10-1679-2010.

Goff, J. (2021). New Zealand's tsunami death toll rises. *Natural Hazards, 107*(2), 1925−1934. Available from https://doi.org/10.1007/s11069-021-04665-1.

Grapes, R., & Downes, G. (1997). The 1855 Wairarapa, New Zealand, earthquake: Analysis of historical data. *Bulletin of the New Zealand Society for Earthquake Engineering, 30*(4), 271−368. Available from https://www.eqc.govt.nz/resilience-and-research/research/search-all-research-reports/the-1855-wairarapa-new-zealand-earthquake-analysis-of-historical-data/.

Gusman, A. R., Lukovic, B., & Peng, B. (2020). Tsunami threat level database update: Regional sources and tsunami warning text. *GNS Science, 83*. Available from http://natlib.govt.nz/records/42967871.

Gusman, A. R., Power, W. L., & Mueller, C. (2019). Tsunami modelling for Porirua City: The methodology to inform land use planning response. *GNS Science, 80*. Available from https://doi.org/10.21420/SEG2-Q850.

Gusman, A. R., Roger, J., Noble, C., Wang, X., Power, W. L., & Burbidge, D. (2022). The 2022 Hunga Tonga-Hunga Ha'apai volcano air-wave generated tsunami. *Pure and Applied Geophysics, 179*(10), 3511−3525. Available from https://doi.org/10.1007/s00024-022-03154-1, https://www.springer.com/journal/24.

Gusman, A. R., Satake, K., Gunawan, E., Hamling, I., & Power, W. L. (2018). Contribution from multiple fault ruptures to tsunami generation during the 2016 Kaikoura earthquake. *Pure and Applied Geophysics, 175*, 2557−2574. Available from https://doi.org/10.1007/s00024-018-1949-z.

Gusman, A. R., Wang, X., Power, W. L., Lukovic, B., Mueller, C., & Burbidge, D. R. (2019). Tsunami threat level database update. *GNS Science, 67*. Available from https://doi.org/10.21420/QM31-NA61.

Hamling, I. J., Hreinsdóttir, S., Clark, K., Elliott, J., Liang, C., Fielding, E., Litchfield, N., Villamor, P., Wallace, L., & Wright, T. J. (2017). Complex multifault rupture during the 2016 M w 7.8 Kaikōura earthquake, New Zealand. *Science, 356*(6334), eaam7194. Available from https://doi.org/10.1126/science.aam71.

Hanks, T. C., & Bakun, W. H. (2002). A bilinear source-scaling model for M-log A observations of continental earthquakes. *Bulletin of the Seismological Society of America, 92*(5), 1841−1846. Available from https://doi.org/10.1785/0120010148.

Imamura, F., Yalciner, A. C., & Ozyurt, G. (2006). Tsunami modelling manual. *UNESCO IOC international training course on Tsunami Numerical Modelling*, 137-209. Available from: https://www.tsunami.irides.tohoku.ac.jp/media/files/_u/project/manual-ver-3_1.pdf.

Kaiser, G., Scheele, L., Kortenhaus, A., Løvholt, F., Römer, H., & Leschka, S. (2011). The influence of land cover roughness on the results of high resolution tsunami inundation modeling. *Natural Hazards and Earth System Sciences, 11*(9), 2521−2540. Available from https://doi.org/10.5194/nhess-11-2521-2011.

Kanamori, H. (1977). The energy release in great earthquakes. *Journal of Geophysical Research, 82*(20), 2981−2987. Available from https://doi.org/10.1029/JB082i020p02981.

Lane, E. M., Borrero, J., Whittaker, C. N., Bind, J., Chagué-Goff, C., Goff, J., Goring, D., Hoyle, J., Mueller, C., Power, W. L., Reid, C. M., Williams, J. H., & Williams, S. P. (2017). Effects of inundation by the 14th November, 2016 Kaikōura tsunami on Banks Peninsula, Canterbury, New Zealand. *Pure and Applied Geophysics, 174*(5), 1855−1874. Available from https://doi.org/10.1007/s00024-017-1534-x, https://doi.org/10.1007/s00024-017-1534-x.

Lane, E. M., Gillibrand, P. A., Wang, X., & Power, W. L. (2013). A probabilistic tsunami hazard study of the Auckland Region, Part II: Inundation modelling and hazard assessment. *Pure and Applied Geophysics, 170*, 1635−1646. Available from https://doi.org/10.1007/s00024-012-0538-9.

Lawson, R. V., Potter, S. H., & Clark, K. J. (2022). Crowdsourcing tsunami observations in Aotearoa following the Tonga-Hunga Ha'apai eruption. *Coastal News, 79*, 6−8. Available from https://www.coastalsociety.org.nz/media/view/publications/crowdsourcing-tsunami-observations/.

Lynett, P., McCann, M., Zhou, Z., Renteria, W., Borrero, J., Greer, D., Fa'anunu, O., Bosserelle, C., Jaffe, B., La Selle, S. P., Ritchie, A., Snyder, A., Nasr, B., Bott, J., Graehl, N., Synolakis, C., Ebrahimi, B., & Cinar, G. E. (2022). Diverse tsunamigenesis triggered by the Hunga Tonga-Hunga Ha'apai eruption. *Nature, 609*(7928), 728−733. Available from https://doi.org/10.1038/s41586-022-05170-6, https://www.nature.com/nature/.

Massey, C. I., Beetham, R., Severne, C., Archibald, G., Hancox, G. T. H., & Power, W. L. (2009). Field investigations at Waihi landslide, Taupo 30 June & 1 July 2009. *GNS Science, 34*. Available from https://static.geonet.org.nz/info/reports/landslide/SR_2009-034.pdf.

Mueller, C., Power, W. L., Fraser, S., & Wang, X. (2015). Effects of rupture complexity on local tsunami inundation: Implications for probabilistic tsunami hazard assessment by example. *Journal of Geophysical Research: Solid Earth, 120*(1), 488−502. Available from https://doi.org/10.1002/2014JB011301.

Mueller, C., Wang, X., & Power, W. L. (2014). Investigation of the effects of earthquake complexity on tsunami inundation hazard in Wellington Harbour. *GNS Science, 198*.

Muhari, A., Imamura, F., Koshimura, S., & Post, J. (2011). Examination of three practical run-up models for assessing tsunami impact on highly populated areas. *Natural Hazards and Earth System Sciences, 11*(12), 3107−3123. Available from https://doi.org/10.5194/nhess-11-3107-2011.

Okada, Y. (1985). Surface deformation due to shear and tensile faults in a half-space. *Bulletin of the Seismological Society of America, 75*(4), 1135−1154. Available from https://doi.org/10.1785/BSSA0750041135.

Okal, E. A., Borrero, J. C., & Synolakis, C. E. (2006). Evaluation of tsunami risk from regional earthquakes at Pisco, Peru. *Bulletin of the Seismological Society of America*, 96(5), 1634–1648. Available from https://doi.org/10.1785/0120050158.

Omira, R., Ramalho, R. S., Kim, J., González, P. J., Kadri, U., Miranda, J. M., Carrilho, F., & Baptista, M. A. (2022). Global Tonga tsunami explained by a fast-moving atmospheric source. *Nature*, 609(7928), 734–740. Available from https://doi.org/10.1038/s41586-022-04926-4, https://www.nature.com/nature/.

Pallentin, A., Verdier, A.L., & Mitchell, J.S. (2009). *NIWA miscellaneous chart series 87 Beneath the waves: Wellington Harbour*. https://niwa.co.nz/media-gallery/detail/109673/27262.

Pizer, C., Clark, K., Howarth, J., Garrett, E., Wang, X., Rhoades, D., & Woodroffe, S. (2021). Paleotsunamis on the southern Hikurangi subduction zone, New Zealand, show regular recurrence of large subduction earthquakes. *The Seismic Record*, 1(2), 75–84. Available from https://doi.org/10.1785/0320210012.

Power, W. L. (2013). Review of tsunami hazard in New Zealand (2013 update). *GNS Science*, 131. Available from https://www.gns.cri.nz/assets/Data-and-Resources/Download-files/Tsunami-Report-2013.pdf.

Power, W. L., Burbidge, D. R., & Gusman, A. R. (2022). The 2021 update to New Zealand's National Tsunami Hazard Model. *GNS Science*, 06. Available from https://doi.org/10.21420/X2XQ-HT52.

Power, W. L., Burbidge, D. R., & Gusman, A. R. (2023). Tsunami hazard curves and deaggregation plots for 20 km coastal sections, derived from the 2021 National Tsunami Hazard Model. *GNS Science*, 61. Available from https://doi.org/10.21420/XPA4-VD47.

Power, W. L., Clark, K., King, D. N., Borrero, J., Howarth, J., Lane, E. M., Goring, D., Goff, J., Chagué-Goff, C., & Williams, J. (2017). Tsunami runup and tide-gauge observations from the 14 November 2016M7. 8 Kaikōura earthquake, New Zealand. *Pure and Applied Geophysics*, 174, 2457–2473.

Power, W. L., Downes, G., McSaveney, M., Beavan, J., & Hancox, G. (2005). The Fiordland earthquake and tsunami New Zealand, 21 August 2003. In K. Satake (Ed.), *Tsunamis. Advances in natural and technological hazards research* (pp. 31–42). Springer. Available from https://doi.org/10.1007/1-4020-3331-1_2.

Power, W. L., Downes, G., & Stirling, M. (2007). *Tsunami and Its Hazards in the Indian and Pacific Oceans. Estimation of tsunami hazard in New Zealand due to South American earthquakes* (pp. 547–564). Springer. Available from 10.1007/978-3-7643-8364-0_15.

Power, W. L., Horspool, N. A., Wang, X., & Mueller, C. (2016). Probabilistic mapping of tsunami hazard and risk for Gisborne City and Wainui Beach. *GNS Science*, 219. Available from https://ref.coastalrestorationtrust.org.nz/site/assets/files/6261/probabilistic-mapping-of-tsunami-hazard-and-risk-gns-report-2015-.pdf.

Power, W. L., Wallace, L., Wang, X., & Reyners, M. (2012). Tsunami hazard posed to New Zealand by the Kermadec and southern New Hebrides subduction margins: An assessment based on plate boundary kinematics, interseismic coupling, and historical seismicity. *Pure and Applied Geophysics*, 169, 1–36. Available from https://doi.org/10.1007/s00024-011-0299-x.

Power, W. L., Wang, X., Lane, E., & Gillibrand, P. (2013). A probabilistic tsunami hazard study of the auckland region, part I: Propagation modelling and tsunami hazard assessment at the shoreline. *Pure and Applied Geophysics*, 170, 1621–1634. Available from https://doi.org/10.1007/s00024-012-0543-z.

Power, W. L., Wang, X., Wallace, L., Clark, K., & Mueller, C. (2018). The New Zealand probabilistic tsunami hazard model: Development and implementation of a methodology for estimating tsunami hazard nationwide. *Geological Society*, 456(1), 199–217. Available from https://doi.org/10.1144/SP456.6.

Prasetya, G., & Wang, X. (2011). Tsunami inundation modelling for Tauranga and Mount Maunganui. *GNS Science*, 35.

Prasetya, G., Wang, X., & Palmer, N. G. (2010). Tsunami inundation modelling for Tiwai Point. *GNS Science*, 293.

Satake, K., Ishibe, T., Murotani, S., Mulia, I. E., & Gusman, A. R. (2022). Effects of uncertainty in fault parameters on deterministic tsunami hazard. *Earth, Planets and Space*, 74(1), 36. Available from https://doi.org/10.1186/s40623-022-01594-6.

Stirling, M., Gerstenberger, M., Litchfield, N., McVerry, G., Smith, W., Pettinga, J., & Barnes, P. (2008). Seismic hazard of the Canterbury Region, New Zealand: New earthquake source model and methodology. *Bulletin of the New Zealand Society for Earthquake Engineering*, 41(2), 51–67. Available from https://doi.org/10.5459/bnzsee.41.2.51-67.

Stirling, M., McVerry, G., & Berryman, K. (2002). A new seismic hazard model for New Zealand. *Bulletin of the Seismological Society of America*, 92(5), 1878–1903. Available from https://doi.org/10.1785/0120010156.

Stirling, M., McVerry, G., Gerstenberger, M., Litchfield, N., Van Dissen, R., Berryman, K., Barnes, P., Wallace, L., Villamor, P., Langridge, R., Lamarche, G., Nodder, S., Reyners, M., Bradley, B., Rhoades, D., Smith, W., Nicol, A., Pettinga, J., Clark, K., ... Jacobs, K. (2012). National Seismic Hazard Model for New Zealand: 2010 update. *Bulletin of the Seismological Society of America*, 102(4), 1514–1542. Available from https://doi.org/10.1785/0120110170NewZealand, http://www.bssaonline.org/content/102/4/1514.full.pdf + html.

Thomas, K. L., Kaiser, L., Campbell, E., Johnston, D., Campbell, H., Solomon, R., Jack, H., Borrero, J., & Northern, A. (2020). Disaster memorial events for increasing awareness and preparedness: 150 years since the Arica tsunami in Aotearoa-New Zealand. *Australian Journal of Emergency Management*, 35(3), 71–78.

Villamor, P., Van Dissen, R., Alloway, B. V., Palmer, A. S., & Litchfield, N. (2007). The Rangipo fault, Taupo rift, New Zealand: An example of temporal slip-rate and single-event displacement variability in a volcanic environment. *Geological Society of America Bulletin*, 119(5-6), 529–547. Available from https://doi.org/10.1130/B26000.1.

Wang, X., Gusman, A., Lukovic, B., & Power, W. L. (2022). Multiple scenario tsunami modelling for mid- and south Canterbury. *GNS Science*, 45. Available from https://www.ecan.govt.nz/document/download/?uri = 4675747.

Wang, X., & Liu, P. (2007). Numerical simulations of the 2004 Indian Ocean tsunamis—Coastal effects. *Journal of Earthquake and Tsunami*, *1*(03), 273–297. Available from https://doi.org/10.1142/S179343110700016X.

Wang, X., Lukovic, B., Power, W. L., & Mueller, C. (2017). High-resolution inundation modelling with explicit buildings. *GNS Science*, *13*. Available from https://doi.org/10.21420/G2RW2N.

Wang, X., & Power, W. L. (2011). COMCOT: A tsunami generation, propagation and run-up model. *GNS Science*, *43*.

Wang, X., Power, W. L., Lukovic, B., & Mueller, C. (2017). Effect of explicitly representing building on tsunami inundation : a pilot study of Wellington CBD. paper O3C.3. In Seismic Isolation, Energy Dissipation and Active Vibration Control of Structures : next generation of low damage and resilient structures : conference handbook & book of abstracts, Wellington: New: Zealand Society for Earthquake Engineering.

Wang, X., Power, W. L., Lukovic, B., Mueller, C., & Liu, Y. (2020). A pilot study on effectiveness of flow depth as sole intensity measure of tsunami damage potential. paper 136. In *Valuing societal benefits of earthquake engineering excellence NZSEE Annual Conference 2020, 22-24 April 2020*, Wellington. Wellington, N.Z.: New: Zealand Society for Earthquake Engineering.

Wang, X., Prasetya, G., Power, W. L., Lukovic, B., Brackley, H., & Berryman, K. R. (2009). Gisborne District Council tsunami inundation study. *GNS Science*, *233*.

Williams, C. A., Eberhart-Phillips, D., Bannister, S., Barker, D., Henrys, S., Reyners, M., & Sutherland, R. (2013). Revised interface geometry for the Hikurangi subduction zone, New Zealand. *Seismological Research Letters*, *84*(6), 1066–1073. Available from https://doi.org/10.1785/0220130035.

Chapter 15

Tsunami hazard and risk in the Mediterranean Sea

Anita Grezio[1], Marco Anzidei[2], Alberto Armigliato[3], Enrico Baglione[1], Alessandra Maramai[4], Jacopo Selva[5], Matteo Taroni[4], Antonio Vecchio[6] and Filippo Zaniboni[3]

[1]*Sezione di Bologna, Istituto Nazionale di Geofisica e Vulcanologia, Bologna, Italy,* [2]*Osservatorio Nazionale Terremoti, Istituto Nazionale di Geofisica e Vulcanologia, Rome, Italy,* [3]*Dipartimento di Fisica e Astronomia "Augusto Righi", Alma Mater Studiorum—Università di Bologna, Bologna, Italy,* [4]*Sezione Roma1, Istituto Nazionale di Geofisica e Vulcanologia, Rome, Italy,* [5]*Dipartimento di Scienze della Terra, dell'Ambiente e delle Risorse, Università di Napoli—Federico II, Napoli, Italy,* [6]*Department of Astrophysics/IMAPP, Radboud University, Nijmegen, The Netherlands*

15.1 Introduction

The Mediterranean countries are densely populated regions with about 150 million people living on the coasts in 2005, reaching 200 million by 2030. The coastal population increases dramatically during the summer because of tourism increasing human exposure and the associated tsunami risk (UNEP, 2017). In this region, tsunami early warning is a challenge (Amato et al., 2021). In fact, in the case of a submarine earthquake, it has been demonstrated that tsunami inundation could occur in a time range spanning from a few minutes to about 3 hours, depending on whether the tsunami source is local or at the Hellenic Arc considering the case of annual probability of 0.001 (Sørensen et al., 2012; Vecchio et al., 2014).

Submarine earthquakes are considered the primary tsunamigenic sources in the Mediterranean Sea (Basili et al., 2021), and they are about 80% of the total events in the European historical catalog (Maramai et al., 2019). However, other local nonseismic sources should be considered for tsunami hazards (Favalli et al., 2009). Landslide-triggered tsunamis have only recently received the deserved attention, as they are the main potential source of nonseismic tsunami hazards (Sassa et al., 2022). The continuously growing availability of computational power has made it possible to adopt more sophisticated models by accounting for the main aspects of the phenomenon, which presents a wide spectrum of uncertainties due to the complex and highly nonlinear features of the slide motion and mass-wave interactions. Volcanic activity may generate pyroclastic density currents (PDCs) from Somma-Vesuvius (Grezio et al., 2020) in the Gulf of Naples and Stromboli (Maramai et al., 2005) or underwater explosions (UWEs) in the Campi Flegrei caldera (Paris et al., 2019) and Santorini (Ulvrova et al., 2016) representing the potential causes of tsunamis in the area with varying probabilities. On the other hand, meteotsunamis in the Mediterranean Sea have been observed and named locally as *Rissaga* (in the Balearic Islands), *Marrobbio* (in the Strait of Sicily), *Milghuba* (in the Maltese Islands), and *Šćiga* (in the Adriatic Sea) (Vilibić, Denamiel, et al., 2021). By cataloging (Maramai et al., 2022) and modeling (Denamiel et al., 2019) recently observed meteotsunamis and historical events, tsunami hazard estimates can be improved at specific coastal locations.

Coastal variability caused by sea-level rise (SLR) due to global warming (IPCC, 2022), in addition to local vertical land movements (VLMs), such as subsidence (Anzidei et al., 2014, 2021; Vecchio et al., 2023), constitutes another uncertain component in tsunami hazard evaluations for the Mediterranean Sea. In fact, probabilistic tsunami hazard procedures are usually based on time-independent analysis, but time-dependent factors in the tsunami impact and inundation phases should be considered (De Risi et al., 2022).

Historically, tsunami scientists have used two main approaches (among those available, see Chapter 5) in the Mediterranean Sea to assess the tsunami hazard for a given coastal area:

- "Scenario-Based Tsunami Hazard Analysis" (SBTHA), based on a relatively small number of selected scenarios. In this way, source mechanisms and inundation representations may be delineated with high resolutions and details and with the additional scope to understand the tsunami features (Tinti et al., 2005).
- "Probabilistic Tsunami Hazard Analysis" (PTHA), consisting of a large number of scenarios with a large variability of the source parameters and with event trees and/or other analyses to explore epistemic and aleatory uncertainties. The primary scope is to provide probabilistic hazard estimates with the relative range of uncertainties (Basili et al., 2021; Pampell-Manis et al., 2016; Selva et al., 2022).

Probabilistic Tsunami Risk Analysis (PTRA) has not been carried out for the Mediterranean Sea at the regional scale yet, and a detailed analysis of the tsunami risk for the entire Mediterranean coastline is beyond the scope of this chapter. The available tsunami risk assessments in the Mediterranean Sea are at the local scale with a high level of detail in a few selected cities (and only a few of them are based on a full probabilistic approach):

- Alexandria (El-Barmelgy, 2014; El-Hattab et al., 2018; Pagnoni et al., 2015; Jelínek et al., 2009) in Egypt;
- Rhodes (Flouri et al., 2012), Santorini (Nomikou et al., 2014), and Crete (Papadopoulos & Dermentzopoulos, 1998; Triantafyllou et al., 2019) in Greece;
- Malta (Camilleri, 2006);
- Göcek Peninsula (Şalap et al., 2001) and Gulf of Fethiye (Dilmen et al., 2015) in Turkey;
- Messina (Grezio et al., 2012) and Augusta (Pagnoni et al., 2021) in Italy.

In this chapter, various approaches are illustrated for tsunami hazard analyses considering the different types of sources in the region (submarine earthquakes, subaerial/submarine slides, volcanic activity, and meteotsunamis). Specifically, the Mediterranean seismic PTHA is focused upon by presenting the projections for the next 50 years of the SLR. Subsequently, exposure elements are introduced for tsunami risk analyses at the regional scale as further components of the tsunami risk from a multidisciplinary perspective.

15.2 Regional tsunami hazard

Tsunami hazard studies, which are aimed at facilitating the implementation of prevention measures to reduce the tsunami risk, generally follow both SBTHA methods and PTHA procedures. Fig. 15.1 shows the different types of sources that are considered major potential tsunamigenic sources in the Mediterranean Sea: submarine earthquakes (http://neamtic.ioc-unesco.org/index.php?option=com_content&view=article&id=211&Itemid=444; https://doi.org/10.3390/geosciences10110447), subaerial/submarine slides (Tinti et al., 2003; https://ls3gp.icm.csic.es/?page_id=753; https://ls3gp.icm.csic.es/?page_id=553), volcanic activity (https://www.ingv.it/organizzazione/dipartimenti-di-ricerca/dipartimento-vulcani), and meteotsunamis (Vilibić, Denamiel et al., 2021; Vilibić, Rabinovich, et al., 2021).

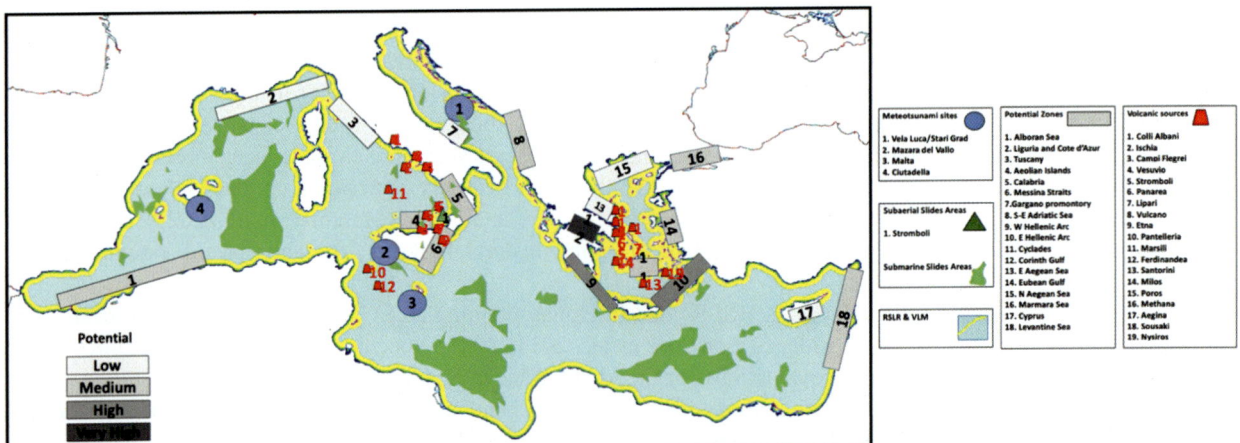

FIGURE 15.1 Potential tsunamigenic sources in the Mediterranean Sea. Gray rectangles indicate the tsunamigenic zones, the red cones are the active volcanoes, the green portions are the major slide areas, and the blue circles are the meteotsunami sites. Map lines delineate study areas and do not necessarily depict accepted national boundaries. *Modified from: http://neamtic.ioc-unesco.org/index.php?option=com_content&view=article&id=211&Itemid=444; https://ls3gp.icm.csic.es/?page_id=753; https://ls3gp.icm.csic.es/?page_id=553; https://www.ingv.it/it/organizzazione/dipartimenti-di-ricerca/dipartimento-vulcani.*

15.2.1 Seismic tsunami hazard

Assessing the tsunami hazard in a given region involves several steps, some of which are independent of the specific mechanisms of tsunami generation, while others are highly dependent on the mechanism and/or the studied area. The first step of the methodology is to gather information and data on historical (or, more generally, past) tsunamis in the region. Historical catalogs and databases of observed tsunami deposits (coastal and/or offshore) are key tools in this respect. For the Euro-Mediterranean area, the available reference catalogs are the EMTC-v2 (Maramai et al., 2019) and the ASTARTE Paleotsunami Deposits Database (Pantosti et al., 2017). Among the information available in the catalogs, the indication of the generating cause has a prominent role in tsunami hazard assessment. For instance, in the EMTC-v2 catalog, approximately 75% of the events were generated by coastal or offshore earthquakes, and a similar percentage is deducible at global scale (NCEI/WDS, 2023). Because of this, methodologies for assessing tsunami hazards have traditionally focused on earthquake sources. It is, hence, straightforward that a second fundamental ingredient of the assessment strategy involves as thorough as possible knowledge of the seismo-tectonics of the studied region together with the evaluation of the past and present seismicity. For the Euro-Mediterranean area, this information can be retrieved from regional fault source databases (EFSM20; Basili et al., 2022) or national-scale databases (e.g., for Italy; DISS Working Group, 2021), from earthquakes catalogs (Grünthal et al., 2013; Rovida et al., 2022), and seismicity (see the EMSC-CSEM online service https://www.emsc-csem.org/#2 as well as the national surveys services).

Since tsunami catalogs contain insufficient events to support "direct" statistical/probabilistic analyses, both SBTHA and PTHA methodologies adopt numerical modeling. In the case of tectonic sources, the modeling ranges from the computation of the coseismic displacement field of the seafloor (i.e., tsunami initial condition) to the simulation of the tsunami wave propagation in the open sea, and finally to the inundation of the coasts. The purpose of PTHA is to estimate the probability for a tsunami "intensity measure" (*IM*) to exceed a given threshold in a predefined time interval. In contrast to the use of the "tsunami intensity scale" in catalogs, which defines a tsunami "size" or its effects on inland areas, the term *IM* has a distinct meaning within the PTHA framework. An *IM* is a physical observable quantity that is strictly associated with the physics of the process. Common *IM*s are wave amplitude, flow depth, current velocity, momentum flux, and maximum inundation height (Grezio et al., 2017). PTHA has been carried out for the coastlines of the NEAM region (NE Atlantic, the Mediterranean, and connected seas) within the EU project TSUMAPS-NEAM (http://www.tsumaps-neam.eu/) considering the maximum inundation height as *IM*. The latter is estimated using the method of the amplification factors with the inclusion of the shoaling on the local bathymetry (Løvholt et al., 2012, 2015). The TSUMAPS-NEAM PTHA assesses the probability $P(H \geq h,t)$ that the tsunami height H exceeds a certain maximum inundation height h at specific coastal locations of the NEAM regions within a selected time window $t = 50$ years, providing hazard curves at selected tsunami points of interest, almost regularly spaced with a distance from each other of approximately 50 km.

15.2.2 Submarine and subaerial landslides tsunami hazard

Applying a probabilistic approach to landslide-generated tsunamis, that is, landslide PTHA (LPTHA), is a challenging task that is still under development (Chapter 10). Some attempts have been made to assess landslide-triggered tsunami features, starting from the landslide characteristics, by adopting probabilistic approaches. Most of the studies concentrated on submarine failures (Collico et al., 2020; Grilli et al., 2009; Lane et al., 2016; Pampell-Manis et al., 2016; Salmanidou et al., 2019; Watts, 2004; Zengaffinen-Morris et al., 2022). Subaerial failures concentrate on impulsive waves from subaerial landslides, typically involving more complicated processes than submarine failures (especially in the impact zone). Du et al. (2020) proposed a methodology of assessing the uncertainties of subaerial failures from different stages of the phenomenon: landslide mechanism, triggering scenario, and material parameters, with an event tree approach (i.e., a special case of a logic tree) combined with a Monte Carlo analysis to account for the uncertainty propagation to develop hazard and probability maps.

In general, the described approaches are data-driven, that is, deduce some correlations for areas that are prone to instability. The setup of such LPTHA procedures requires a detailed database of events, reconstructing the main characteristics of mass failures to parametrize the complex processes involved and build correlations with the generated waves. In this way, it is possible to quantify the tsunami hazard and the probability of exceeding a given amplitude threshold in a certain time period. The main drawback of this approach is evident. To provide a complete and reliable estimate of the hazard for a specific coastal stretch, a detailed characterization of the underwater potential or failure case histories should be completed, together with accurate knowledge of the tsunami effects on the coast. These two classes of datasets are not easily achieved or available everywhere, so another approach can be attempted, more focused on the physical characteristics of the process.

An alternative methodology for studying landslide-generated tsunamis is the definition of four different stages, each of which is characterized by its features, unknowns, and level of uncertainty (Table 15.1). The quantification of the probability of occurrence for each phase, the related uncertainty, and the computation of their propagation across the different phases can contribute to providing an estimate of the whole probability of tsunami impact from this class of events.

Trigger phase: The first elements to consider are the conditions for landslide onset. The mass destabilizing factors are linked to the geotechnical characteristics (e.g., cohesion, density, and internal friction) and the geometry (e.g., thickness, surface of rupture, and slope angle) of the mass, and to external factors affecting the stability of the mass, such as seismic shaking, pore-pressure, and change in water content (preparatory factors). A classic approach to this kind of problem, and still the most used, is the Limit Equilibrium Method (Bishop, 1955), aiming at calculating the safety factor F as the ratio between the resistive forces (i.e., capacity) and the stress acting on the sliding body (i.e., demand). When $F > 1$, the stabilizing factors prevail, and the mass does not move; when the demand is greater than the capacity, $F < 1$, and the mass is unstable. The probability of landslide destabilization can be quantified, in the first-order approximation, as the probability of exceedance of a given peak ground acceleration. Other factors that can contribute to instability are the geotechnical parameters of the ground: cohesion, friction angle, and soil saturation.

Dynamic phase: The influence of body geometry changes during the descent on tsunami generation is a well-known issue. For example, the use of rigid body rheology for the landslide causes overestimation of the tsunami effects (Harbitz et al., 2006). The factors influencing the slide motion and the shape changes during the descent are many. The most significant factors include: the basal friction coefficient, influencing the sliding motion dynamics, that is, acceleration, velocity, runout, and other characteristics that deeply affect the wave generation process; the rheology of the slide, accounting for the thickness change rate during the motion, then for the capability of the mass to perturb the whole water column; and the drag forces, that is, the stresses acting at the interface between the mass and the water, which can contribute to kinetic energy dissipation.

Tsunami generation phase: The wave generation process from a moving body is one of the most controversial and unknown characteristics. In the Storegga Tsunami, for example, the energy transferred from the slide to the water was less than 1% as computed by Harbitz et al. (2006) using the simulation by Bondevik et al. (2005). Conversely, in simulating the Cumbre Vieja (Canary Islands) collapse, Abadie et al. (2012) found that 30%−40% of the slide energy was transferred to the tsunami. In general, Ruff (2003) states that the energy conversion can range from zero to a maximum value of 50%. A frequently used method to assess the efficiency of the energy transfer from the mass to the water is the Froude number. This dimensionless quantity is computed as the ratio between the horizontal component of the slide velocity and the phase velocity of the tsunami ($Fr = vs/\sqrt{gh}$, with vs slide velocity, g gravity acceleration, and h water depth). When this ratio is smaller than 1 (subcritical condition), the wave travels faster than the slide and does not receive additional impulse. On the contrary, when $Fr > 1$ (supercritical range), the mass moves faster than the wave, and again, the wave build due to the energy transfer is limited. For Fr values around the critical value of 1, the mass and the tsunami move at the same speed, and the energy transfer is at the maximum. Typically, submarine landslides show a subcritical behavior, while subaerial landslides are initially found in supercritical conditions when impacting the water (Harbitz et al., 2014). The computation of Fr can provide useful indications of the capability of the moving mass to generate a tsunami.

TABLE 15.1 Primary landslide probabilistic tsunami hazard analysis factors to be quantified for each landslide tsunami process phase.

Landslide tsunami phase	Main factors for probabilistic assessment
Slope stability	Geotechnical parameters (cohesion and friction) and seismic shaking
Slide dynamics	Basal friction coefficient and rheology
Tsunamigenic impulse	Mass-wave energy transfer (Froude number)
Tsunami propagation	Dispersion and coastal effects

Tsunami propagation and impact on the coast: The final phase of the process is the most extensively studied, involving the propagation of tsunamis. A classic approach utilizes the shallow water approximation for Navier−Stokes equations, but recent advancements in computational capabilities and higher order hydrodynamics equations have greatly improved applications. The main source of uncertainty of the landslide tsunami propagation concerns wave dispersion, which becomes more pronounced as the wavelength decreases in relation to sea depth. In general, tsunamis generated by landslides are composed of shorter oscillations than earthquake-generated ones, which means that the long-wave approximation (that forms the basis of the shallow-water approach) is no longer valid. A quantification of the distance from the source where dispersion effects prevail is given by Glimsdal et al. (2013). Neglecting this effect can lead, in particular conditions, to overestimation of the impacting wave (Harbitz et al., 2006). Dispersion affects more subaerial landslides that generate shorter waves than submarine cases. A comprehensive review by Yavari-Ramshe and Ataie-Ashtiani (2016) outlines the advantages and drawbacks of the different approaches of landslide tsunami modeling and simulations.

The above four-stage procedure has undoubtedly a series of drawbacks. First, it considers the process as a sequence of separated phases, while there are pieces of evidence of interactions that can affect the phenomena involved in the process. As an example, it is known that in shallow water, the slide motion can be influenced by the water perturbation that itself is provoking, with an amount of about 10%−15% (Harbitz et al., 2014). In many cases, the landslide−water system is modeled as a two-phase fluid, each characterized by its rheology, to account for these mutual interactions during the slide motion. Other processes are also neglected, for example, the nonlinear coastal effects in shallow water and the resonance phenomena that can be excited by short waves entering coastal basins. However, this methodology could be tested on sites where much evidence is reported to define the most influencing factors on tsunami generation and impact on the coast.

15.2.3 Volcanic tsunami hazard

The necessary phases for volcanic tsunami hazard and risk assessments are, first, the selection of the potential volcanoes and/or volcanic areas and the selection of the type of sources; and second, the definition of the source variability (in space and size) to explore the natural variability and the kinematics of the selected source(s). In the case of probabilistic analysis, the third phase is the definition of a large set of potential tsunamigenic scenarios and their mean annual rates of occurrence in a selected exposure time.

Volcanic tsunamis may be caused by (1) eruptive dynamics (due to the eruption processes, e.g., UWEs, pyroclastic flows, collapse of an eruptive column, caldera collapse, and atmospheric acoustic-gravity waves) and (2) noneruptive sources (due to the structure of the volcanic edifice, for example, subaerial and submarine landslides, debris flows, lahars, and volcanic earthquakes), with different characteristics in terms of duration, volume, mass flux, and energy, which has to be pondered in the tsunami wave modeling (Paris, 2015). In the Mediterranean Sea, a probabilistic hazard assessment was carried out within the Gulf of Naples. Two different eruptive sources causing tsunamis have been studied: PDCs and UWEs.

Pyroclastic flows are hot mixtures of gas and particles and may generate tsunami waves when a dense basal component of the pyroclastic flow is present (Watts & Waythomas, 2003), or a high-velocity pyroclastic flow occurs with a bulk density near or even below that of water, whatever their temperature (Bougouin et al., 2020; Freundt, 2003). The controlling parameters of tsunami generation are the pyroclastic flow volume and mass flux, the flow density and permeability, the angle of incidence, and the transport distance from the eruptive vent (Bougouin et al., 2020; Grezio et al., 2020, 2015; Maeno & Imamura, 2011; Watts & Waythomas, 2003). In the Mediterranean Sea, pyroclastic flows potentially generating tsunamis are located, for example, at the Somma-Vesuvius and Stromboli. A PTHA study was carried out by Grezio et al. (2020) considering PDCs from the summit crater of Somma-Vesuvius with the probability of generation and frequency of the arrival at the coast based on the statistics and simulations provided by previous studies (Tierz et al., 2017). An extensive set of dense PDCs entering the sea and generating tsunami waves was considered and simulated in a multisource hazard approach. UWEs are characterized by the development of a water crater, depending on the water depth and energy of the explosion, which are the main input parameters necessary to delineate the initial wave. The subsequent expansion, rise, and gravitational collapse of the crater create two successive bores followed by many smaller undulations propagating radially from the source. In Paris et al. (2019), a Bayesian event tree was developed to explore the natural variability of potential submarine explosions based on past volcanic activity and to quantify the tsunami hazard conditional upon the occurrence of an underwater eruption at Campi Flegrei. The quantification of long-term conditional hazard is common in volcanic hazard studies and its extension to unconditional PTHA may be obtained by multiplying the conditional hazard by the mean annual rate or the probability of occurrence of volcanic eruption (Selva et al., 2022).

Studies of the noneruptive sources in the Mediterranean Sea were focused mainly on flank instability of volcanoes and landslides, such as Stromboli (Fornaciai et al., 2019), Ischia (Selva et al., 2019), and Santorini (Necmioglu et al., 2023). While their probability of occurrence may be strongly related to the dynamics of the volcano and its behavior during unrest episodes or eruptions, the debris avalanches at volcanoes are similar to nonvolcanic landslides in generating tsunamis. The main source parameters are the Froude number, the thickness, the mass flux and volume of the sliding mass, the initial acceleration, and the maximum velocity (Fritz et al., 2004; Grilli & Watts, 2005; Harbitz et al., 2006; Ward, 2001; Yavari-Ramshe & Ataie-Ashtiani, 2016). The high instability of the volcano flanks is the result of both endogenous (structural discontinuities, hydrothermal alteration, and magmatic intrusions inside the edifice, rapid growth by accumulation of tephra and lava flows) and exogenous factors (earthquake, tectonic uplift, weather and climatic events, and sea-level variations) (Paris et al., 2023) with a complex triggering process that may be related to volcanic phenomena (e.g., deformations and material alteration due to degassing) or nonvolcanic phenomena (e.g., regional earthquakes). These phenomena should, in principle, facilitate the probabilistic quantification of the related hazard.

15.2.4 Meteotsunami hazard

The meteotsunami phenomenon is a relatively recent scientific discovery, considering that the first description of tsunami-like effects produced by atmospheric disturbances appeared only in 1931 and that the term "meteotsunamis" was introduced only in 1961 (Pattiaratchi & Wijeratne, 2015). Meteotsunamis are significant sea-level oscillations, which can be destructive at the coast, originating from extreme meteorological phenomena. These waves have characteristics (wave period and height) comparable to tsunami waves generated by earthquakes, volcanic eruptions, or submarine landslides. They can impact the coastline similarly to the previously mentioned tsunamis, especially in specific bays and inlets. Meteotsunamis are caused by storm systems moving rapidly across the water. Their development depends on several factors, such as the intensity, direction, and speed of the disturbance as it travels over a water body with sufficient depth to increase wave magnification. Unlike tsunamis of seismic origin, the atmospheric disturbances that generate meteotsunamis have limited cross-propagation dimensions of up to a few tens of kilometers (Horvath et al., 2018; Horvath & Vilibić, 2015).

A common generation mechanism of meteotsunamis is a noticeable long-lasting (at least 2–3 hours) air pressure disturbance in the atmosphere, characterized by an abrupt air pressure change of at least 2–4 hPa over 5–10 minutes (Okal et al., 2014; Šepić & Vilibić, 2011; Thomson et al., 2009; Vilibić et al., 2004). Such a disturbance crosses the open sea and resonantly transfers its energy to long-period sea waves, traveling at the same speed as the overhead weather system. The first step of the meteotsunami generation mechanism is atmospherically conditioned, but the local bathymetry plays a critical role in the amplification of the long ocean waves approaching a coastline and within the harbors and bays (Monserrat et al., 1998, 2006; Rabinovich et al., 2009; Šepić et al., 2012; Vučetić et al., 2009). In addition, harbor oscillations are the last stage of the phenomenon (Hibiya & Kajiura, 1982; Ličer et al., 2017).

While most meteotsunamis occur unnoticed due to their small size, larger ones can cause devastating impacts on coastlines. In comparison with earthquake-triggered tsunamis, wave heights and spatial extent of meteotsunamis are usually smaller. Still, they can cause damaging waves, flooding, and strong currents that can last from a few hours to a day, causing damage and victims. Furthermore, although meteotsunamis are not as catastrophic as major seismically induced events, since the atmospheric disturbances that generate meteotsunamis are much more frequent than seismic tsunamis, the occurrence rate of meteotsunamis in time and space is greater. Recent research has shown that meteotsunamis are more common than previously thought and suggested that some past events may have been confused with other types of coastal flooding, such as storm surges or tidal surges, which tend to be caused by wind. There are some areas in the world, such as the northeastern Gulf of Mexico, the eastern coast of the United States, the southwestern coast of Japan, southern Britain, and the Mediterranean Sea, that are known to be particularly prone to meteotsunamis due to a combination of variables, such as geography, weather patterns, and bathymetry (size, shape, and depth of the water body).

Concerning the Mediterranean Sea, meteotsunamis mainly occur in three regions, namely, the northern and central coasts of the Adriatic Sea, the southwestern coast of Sicily, and the Balearic Islands. In each of these three areas, the phenomenon is called by a different local term: in Croatia, they are referred to as *Šćiga*, in Sicily as *Marrobbio*, and in the Balearic islands as *Rissaga*. Among these coasts, several locations are known as "meteotsunami hot spots": Vela Luka and Stari Grad in Croatia, Mazara del Vallo in southwestern Sicily, and Ciutadella harbor in the Balearic islands.

A number of devastating meteotsunamis with wave heights exceeding several meters occurred at these sites over time (Šepić et al., 2018). It has been shown that meteotsunamis in the Mediterranean Sea tend to occur during the warm half of the year, between April and October, when the atmospheric conditions are suitable. Despite calm conditions at ground level, fast winds of dry air from Africa in the atmosphere at 1500 m trigger atmospheric waves (Maramai et al., 2022). In late spring/early summer, the Mediterranean air is much colder than the African air, and warmer African air flows above colder Mediterranean air, generating relevant perturbation.

Along the central coast of the Adriatic Sea, the strongest meteotsunami ever recorded in the world occurred, strongly hitting the town of Vela Luka in Croatia, in June 1978. This event lasted several hours and had the maximum effects in Vela Luka, where the wave height was about 6 m. The flood reached the first floor of the houses located on the top of the bay, causing significant damage and the maximum inland inundation was about 650 m. The impact on the infrastructure was enormous, and the destruction spread over the city. Fortunately, no human fatalities and only a few minor injuries were reported. The meteotsunami involved about 30 localities along both the Croatian and the Italian Adriatic coasts with slight to moderate effects (Maramai et al., 2022; Orlić, 2015; Vučetić et al., 2009). Due to its complex topography, characterized by a large number of funnel-shaped bays and harbors with high amplification factors, the eastern coast of the Adriatic Sea is prone to meteotsunami. Since 1931, the Adriatic experienced 33 meteotsunami events (some with waves over 2 m high), observed or recorded at 54 places (Maramai et al., 2022) along both the eastern and western Adriatic coasts. Occurred mostly during summer time, these events often affected the lifestyles of the coastal communities, particularly in the Dalmatian islands, where severe flooding was observed. In Croatia, the towns of Vela Luka and Stari Grad were affected by the largest number of events, while many localities on the Italian Adriatic coast experienced the June 1978 meteotsunami. A few locations were also slightly affected by the June 2014 event.

Southwestern Sicily coast is another Mediterranean area hit by meteotsunami for centuries, early described by Admiral Smyth in his "Memoir descriptive of the resources, inhabitants, and hydrography of Sicily - London, 1824" (Honda et al., 1908; Rabinovich, 2020). The phenomenon, locally known as *Marrobbio*, is commonly related to the passage of large-scale low-pressure atmospheric systems over the region and it has been characterized as a regional gravitational phenomenon trapped in the local topography. Atmospheric pressure is the main forcing mechanism for the *Marrobbio* and produces a response of around an order of magnitude larger than the wind stress forcing (Candela et al., 1999; Šepić et al., 2018). In this region, the mainly affected town is Mazara del Vallo as the geometry of its port permits the amplification of certain oscillations within the port itself. In addition, Mazara del Vallo, the most important fishing harbor in Sicily, is located along the Mazaro River, and due to the river bathymetry, a unique phenomenon has been observed during extreme events. A bore rising to a meter or higher, driven by an incoming meteotsunami and propagating upstream the river channel, has been frequently concurrent with the *Marrobbio* (Candela et al., 1999; Colucci & Michelato, 1976). Generally, every year one *Marrobbio* occurs in this area, and in the last decades, several relevant events have taken place in this town, some of them producing severe damage to the local fishing fleet, and sea-level oscillations with periods ranging from a few minutes to an hour are also quite common in other ports on the southern coast of Sicily and in the Lampedusa island. The most recent strong *Marrobbio* hit Mazara del Vallo on June 25–26, 2014, with the formation of a bore propagating upstream along the harbor and the Mazaro River, initiating an SLR of about 1.5 m in less than 2 minutes. The bore produced some damage to boats and flooded the promenade in Mazara del Vallo on the river embankment. Sea-level oscillations lasted more than 1 day inside Mazara del Vallo harbor and the Mazaro River. Noticeable oscillations had also been observed in neighboring harbors and on beaches along the southwestern coast of Sicily; an initial sea retreat for more than 30 m was observed at the Tonnarella beach, followed in a few minutes by a sea inundation that reached the coastal road.

In the Balearic islands, meteotsunamis are locally named *Rissaga*. In this region, the meteotsunami "hot spot" is the harbor of Ciutadella, on the Menorca island, which experienced many relevant events, with extraordinary sea-level fluctuations that can reach 2 m in amplitude in periods of 10 minutes; this bay, 1 km long and 100 m wide, with a depth around 5 m, is perfectly suited to enhance the effects of the phenomenon. The *Rissaga* in the port of Ciutadella has been known for a long time and there are references from the 15th century reporting of ships sinking in the port due to extraordinary and sudden tides. The phenomenon is also described in detail by Riudavets as early as 1885 in his book: History of Menorca Island (Vilibić et al., 2021). More than 35 meteotsunamis are known to have occurred on the Balearic islands, most of them producing only some small floods in the area without significant negative consequences. Besides Ciutadella, other locations in Menorca (Mahon, Porto Colom, Santa Ponça, and Pollensa) and Maiorca (Palma and Sa Rapita) have been affected by meteotsunamis over time.

Mechanisms that originate *Rissaga* and the interaction between the atmosphere and the sea have been studied, showing that during late spring and summer meteorological conditions in the western Mediterranean, with the entrance of warm air from the Sahara at near-surface levels, are favorable for the formation of high-frequency atmospheric pressure disturbances capable of generating *Rissaga* (Alonso et al., 1989; Jansà, 1986; Šepić et al., 2009). For sea-level fluctuations of less than 50 cm, it is usually difficult to discriminate whether the cause is a *Rissaga* or another phenomenon, but more significant fluctuations are almost always caused by a *Rissaga*. In the port of Ciutadella, this phenomenon occurs about ten times a year, but significant oscillations (more than 1.5 m high) occur every 4–5 years. In the last 30 years, a very strong *Rissaga* occurred on June 20–21, 1984 with wave heights of more than 4 m, destroying more than 300 boats and yachts and causing major damage in the harbor (Jansà, 1986; Šepić et al., 2009). On June 15, 2006, an extraordinary *Rissaga* event was reported, the strongest in the last 20 years, with an initial catastrophic drying of a significant part of the harbor: Most boats broke their moorings and were dragged by the current only a few minutes later. More than 40 boats were sunk or severely damaged, with the total economic loss estimated to be in the order of tens of millions of euros. An impressive pressure rise, accompanied by a strong wind, caused by a squall line system, was the probable cause of that extraordinary event (Jansa et al., 2007). More recently, on July 16, 2018, a meteotsunami hit the coasts of Mallorca and Menorca, causing the death of a German tourist swept away by the current while on the beach and causing considerable damage all along the coast. The wave flooded numerous places and businesses, and some boats broke their moorings and were swept away due to the strong currents.

Meteotsunamis will occur in the future in the abovementioned regions with favorable conditions for this phenomenon. The Earth's rising temperature, climate change, global warming, and rising sea levels may increase the frequency and likelihood of these potentially destructive waves. Therefore, forecasting and warning systems for meteotsunamis are essential. Currently, meteotsunami warning systems already exist in Spain and Croatia. In Spain, the Meteorological Centre of the Spanish Instituto Nacional de Meteorologia at the Balearics has been forecasting *Rissaga* events since 1984, and the Balearic Rissaga Forecasting System has started, in a pilot phase, since 2010. It is based on the identification of favorable synoptic conditions. A meteotsunami forecast is given a few days ahead, but only at the qualitative level (Vilibić et al., 2016). For the Adriatic region, a similar system has been designed and implemented in Croatia since 2015, and it is based on real-time detection of intense air pressure disturbances (Vilibić et al., 2016). These systems must be further integrated and improved because they rely on statistical predictions based on atmospheric conditions or models that cannot yet provide precise forecasts of atmospheric waves locally .

15.2.5 Effects of climate change and vertical land movements on tsunamis hazard

The Sixth Assessment Report released by the Intergovernmental Panel on Climate Change (IPCC) in 2022 (http://www.ipcc.ch; IPCC, 2022) stated that the rate of global sea level (GSL) started to rise since the beginning of the industrial era and is rising faster than in the last two millennia. This is mainly due to global warming, which triggers the melting of ice sheets and mountain glaciers and the thermal expansion of the oceans. Measurements from ground-based and satellite sensors indicate that the GSL is rising at about 3.7 mm/year in the time interval between 2006 and 2018 (Fox-Kemper, 2021) due to a combination of different effects, such as eustatic, glacio-hydro-isostatic, and land-hydrology contributions (Lambeck & Purcell, 2005). In the framework of the IPCC, the different SSP (Shared Socioeconomic Pathways) scenarios, such as SSP1-1.9, SSP1-2.6, SSP2-4.5, SSP3-7.0, and SSP5-8.5, and the global warming levels (specified as 1.5°C, 2°C, 3°C, 4°C, and 5°C increase above the current mean global temperature) were developed, showing that if the global temperature continues to rise in the following decades without any mitigation, the sea level will rise even more than 1 m throughout the 21st century and likely several meters by 2300 and beyond.

Along the coasts, the sea-level changes can significantly differ from the mean GSL rise due to various solid earth and oceanographic changes driven by geophysical and climate processes acting at the regional scale. The following processes can exceed climatically-driven mean SLR at regional scales: oceanographic and climatic effects, glacial isostatic adjustment, and continuous and episodic VLM, especially coastal subsidence due to tectonics or anthropogenic factors. As a consequence, they cause a spatial pattern in the SLR variability. In the Mediterranean Sea, the SLR has been estimated as about 2 mm/year since 1880 (Anzidei et al., 2014; Fenoglio-Marc et al., 2013, 2012; Tsimplis et al., 2013; Wöppelmann & Marcos, 2012). The SLR is already affecting low-lying coastal plains, river deltas, and beaches with consequent coastal erosion, beach retreat, and salinization of the water table, representing a significant hazard factor for the coastal populations and infrastructures.

VLM, in particular land subsidence, mainly caused by active tectonics and volcanism, is a significant issue along the Mediterranean coasts. Coastal cities, like Venice in Italy and Istanbul in Turkey, are undergoing the combined effects of SLR and natural and anthropogenic land subsidence. The combination of the above factors is causing sea levels to swamp coastal areas, eroding dune belts, and retreating beaches, with social and economic consequences and exposing about a billion people living along the global coasts to this threat. SLR projections up to 2100 and beyond, combined with data about land subsidence, seismicity, potential tsunami occurrence, and extreme waves from storm surges, are all crucial factors for coastal hazards that need to be assessed to defend coastal infrastructures and prepare communities for the expected changes. This means that developing realistic SLR projections and flooding scenarios, which include local effects, such as the VLM, is crucial in both scientific and social frameworks to raise the awareness of people on coastal hazards and foster cognizant coastal management to mitigate the socioeconomic impacts of SLR.

Assessing the potential effects of SLR combined with increasing frequency and intensity, due to climate change, storm surges, and tsunamis, is a challenging task. However, assessing the above phenomena concerning potential future scenarios is critical since it may inform policymakers and planners of this threat looming over coastal areas and encourage them to take practical actions. To achieve this goal, high-resolution topographic data of specific coastal zones are needed to build up maps of coastal flooding extension due to the combined effects of SLR, tsunamis, and ordinary or extreme storm surge events.

The effect of VLM and relative SLR (RSLR) can be directly included in the tsunami hazard computation by modifying the hazard curves (Sepúlveda et al., 2021). This correction, called linear superposition (Sepúlveda et al., 2021), affects a hazard curve with a horizontal translation of the curve in the increasing direction. To compute this translation, future tsunamis that will arrive at the coastlines will need to be evaluated for a selected time window (e.g., 50 years) (Dura et al., 2021). To do so, the average of the VLM and RSLR effects in the considered time window can be used. Since both VLM and RSLR primarily affect the sites of the tsunami hazard analysis, the correction needed for the hazard curves is the same for all possible tsunamigenic sources: seismicity, volcanic activity, landslides, and meteotsunami. Therefore, we can compute the correction, which is site-dependent, and then apply this correction to all different hazard curves.

To show an example of this method, we applied the correction for VLM and RSLR (Anzidei et al., 2021) to the hazard curves of the EU project TSUMAPS-NEAM (Basili et al., 2021). Fig. 15.2A shows the probabilistic tsunami hazard, including the projection for the next 50 years taking into account local VLM and RSLR in the Mediterranean Sea due

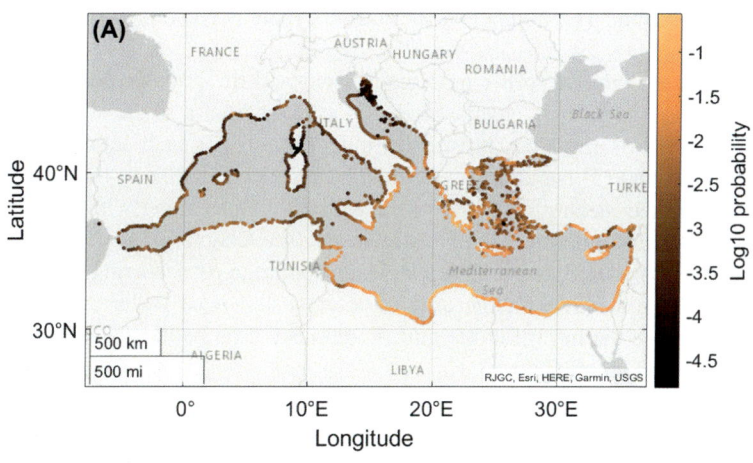

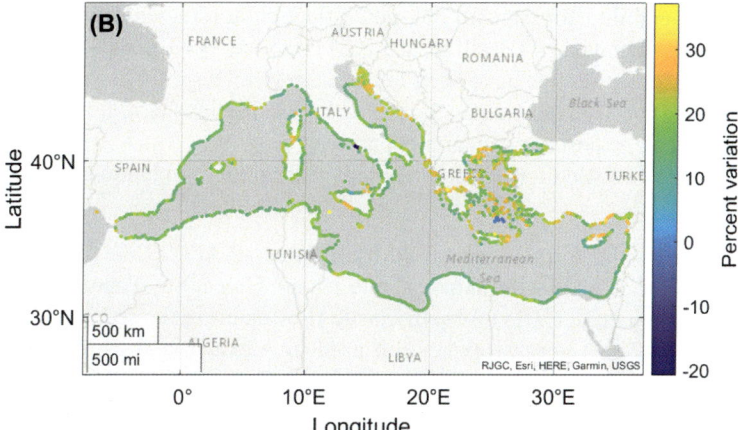

FIGURE 15.2 Tsunami hazard maps : (A) Map of the probability of exceeding the 2 m maximum inundation height (fraction of unit) in 50 years, including the relative sea-level rise in the Mediterranean Sea and (B) map of the relative percentage variations with respect to the current PTHA. *PTHA*, Probabilistic tsunami hazard analysis. Map lines delineate study areas and do not necessarily depict accepted national boundaries.

to global warming caused by anthropogenic greenhouse gasses. The percentual variations with respect to the current PTHA are shown in Fig. 15.2B. From these maps, it is possible to appreciate the effect of VLM and RSLR on PTHA. In some cases, the probability of exceeding 2 m maximum inundation height in 50 years is increased by up to about 30%. This example then demonstrates the importance of the inclusion of climate change effects and VLM in PTHA studies.

15.3 Multisource probabilistic tsunami hazard analysis in the Gulf of Naples

In the case of different types of sources, a probabilistic framework is an appropriate tool to estimate the tsunami hazard with the scope to prioritize the first-order different sources (Grezio et al., 2020): seismic sources (Ss), landslides (Ls), and volcanic sources (Vs). In the multisource case, given the ith tsunamigenic source type TS_i ($i = Ss, Ls, Vs$), the corresponding conditional hazard curve is determined by the probability P that IM exceeds a selected threshold im in the exposure interval t

$$P_i(IM \geq im, t) = P(IM > im | TS_i) \cdot P(TS_i, t) \tag{15.1}$$

If P defines the probability of occurrence of at least one tsunami event, then $(1 - P)$ indicates the generic probability that no tsunami occurs in the exposure interval t.

Assuming their statistical independence, the product over N_S different types of tsunamigenic source TS_i represents the probability that none of the TSs produces a value of the tsunami IM larger than the threshold im in the time window t, so that the comprehensive PTHA for the N_S tsunamigenic source types is calculated by the following equation

$$P(IM > im, t) = 1 - \prod_{i=1}^{N_S} [1 - P_i(IM \geq im, t)] \tag{15.2}$$

Each contribution to the total PTHA is evaluated separately by considering that the ith source type TS_i can be represented by a number N_S of independent physical source events. The number of the source events is conveniently large to explore the aleatoric uncertainties.

Assuming that the probabilistic hazard assessments are relative to the next main event and the possible subsequent events are not included because the characteristics of the system may change after the occurrence of a single significant event, the following statements are further required: (1) TSs must be statistically independent; (2) each source event must be independent of other events within the same TS; and (3) tsunami waves must be independent also in the case of almost simultaneous events. The final hazard curves are the mean probability of exceedance values along with the corresponding 16th and 84th percentiles, which provide the epistemic uncertainty estimations (see Chapter 5).

A case study for the multisource tsunami hazard is considered the Gulf of Naples. Submarine Ss occurring even far in the Mediterranean Basin may generate potential tsunami waves in the area of interest and need to be evaluated. Submarine Ls occurring in the Tyrrhenian Sea may increase the hazard in the gulf. Potential Vs can be identified mainly as PDCs at the Somma-Vesuvius and UWEs at the Campi Flegrei Caldera.

15.4 Regional tsunami risk index

In the Mediterranean Sea, studies on exposure and vulnerability for tsunami risk quantification at a regional scale have not been published yet. PTHA is present for submarine earthquakes (Basili et al., 2021) and has been mainly focused on tsunami alerts (Amato et al., 2021). Considering that other studies have been undertaken to manage the tsunami risk for Ls and Vs (Selva et al., 2021), a new perspective on the tsunami risk analysis for the Mediterranean region is presented by taking into account elements at the regional scale exposed to tsunami events.

Tsunami risk has been evaluated for a limited number of Mediterranean cities using different formulations for the risk assessments:

- Risk = Hazard × Vulnerability × Exposure. This is the formula, proposed by Fournier d'Albe (1982), and has been used in Messina (Italy) by Grezio et al. (2012) for Ss and in Heraklion (Crete Island, Greece) by Papadopoulos and Dermentzopoulos (1998) for Ss and by Triantafyllou et al. (2019) for pyroclastic flow entering the sea.
- Risk = Hazard × Vulnerability. This is the UNDP (2004) general expression used in Alexandria (Egypt) by El-Barmelgy (2014) and Jelínek et al. (2009), in Rhodes (Greece) by Flouri et al. (2012), in Malta by Camilleri (2006), in the Göcek Peninsula (Turkey) by Şalap et al. (2001), and Gulf of Fethiye (Turkey) by Dilmen et al. (2015), in all cases considering Ss.

- *Risk = Hazard × Vulnerability / Resilience*. This is the UK Aid 2015 approach used in Alexandria (Egypt) by El-Hattab et al. (2018) for seismic sources.

Although those formulations evaluate the tsunami risk in different ways, they are all consistent with definitions of the relevant terms and glossaries. In particular, vulnerability refers to the exposure or the degree of resistance to a hazard, even if the exposure term is not explicitly mentioned in the risk formulations. Tsunami vulnerability is related to a set of objects, conditions, variables, and processes resulting from physical, social, economic, and environmental factors, which increase the susceptibility of a community to the tsunami hazard (Flouri et al., 2012). Various data sources and techniques are used to identify the exposure of critical elements to tsunami hazards. Exposure elements incorporate population, buildings, roads, ports, and infrastructures and present both spatial variations due to the locations and temporal changes due to the modifications of population and the environment at daily/seasonal/yearly and even longer time scales. Resilience is recently indicated as an additional component to recuperate after disasters and it is strongly dependent on the social, cultural, and economic structure of the community under hazardous phenomena (Pushpalal et al., 2023).

Fig. 15.3 shows a synoptic view of the exposed elements in the Mediterranean Sea considering people (population and relative density in 2012 and tourists in 2015), environment (relevant marine conservation areas in 2017), and historical/cultural sites (in 2018). Information on the population density of the coastal regions, major coastal cities, and international tourists' arrivals is from the Barcelona Convention Mediterranean 2017 Quality Status Report (https://www.medqsr.org/tourism) (Fig. 15.3A). Marine conservation areas were extracted from Piante and Ody (2015) and the UNESCO World Heritage Sites are indicated by Reimann et al. (2018) (Fig. 15.3B).

Fig. 15.4 shows coastal transport infrastructures and economic activities exposed to tsunamis in the Mediterranean region. Airports and major roads (within 1 km from the coast) in 2012 and ports with the TEU (20-foot equivalent unit) volume × 1000 of commercial exchanges in 2015 were extracted from GRID-Arendal 2013 (https://www.grida.no/resources/5905). The containers throughout the main Mediterranean ports are indicated by Grifoll et al. (2018) (Fig. 15.4A). The aquaculture fish farm sites with total fishery production in terms of tons/year were extracted from economic indices (Piante & Ody, 2015). Activities related to energetic production and supplies are the offshore wind farms (operative installations or potential locations indicated by Piante and Ody (2015)) and the marine gas pipelines (routes activated or projected displayed on https://en.wikipedia.org/wiki/Trans-Saharan_gas_pipeline, https://en.wikipedia.org/wiki/Maghreb%E2%80%93Europe_Gas_Pipeline, https://it.wikipedia.org/wiki/Galsi, https://it.wikipedia.org/wiki/Transmed, https://it.wikipedia.org/wiki/Greenstream, https://depa-int.gr/en/poseidon-pipeline/, https://en.wikipedia.org/wiki/EastMed_pipeline) (Fig. 15.4B).

15.5 Discussion and final remarks

Assessing tsunami hazards and risks in the Mediterranean Sea is a complex task, compounded by the increasing rates of SLR due to climate change. At a regional scale, the major causes of tsunamis (earthquakes, submarine and subaerial failures, volcanic activity, and meteotsunamis) are present with different rates of occurrence. Regional hazard models are useful in providing a starting point for local analyses. The offshore PTHA may be insufficient to reproduce the tsunami effect, but its results can contribute to simplifying the onshore hazard treatment by reducing the number of simulations while keeping the errors small compared to the ideal case. For a local tsunami hazard, high-resolution onshore tsunami simulations are required to accurately capture fine-scale phenomena, but they require powerful computing resources and a large number of scenarios to explore aleatory and epistemic uncertainties. When high-performance computing resources are available, a way to proceed is to extend the numerical simulations to the entire set of scenarios. However, another way to tackle the problem is to reduce the number of simulations for the final calculation of the hazard. The first way is easier to implement since it does not require any kind of scientific effort more than a regional hazard, once the ensemble of starting sources and the numerical software to solve the equations are ready. On the other hand, the second way is more useful and accessible to the scientific community. Finding a way to reduce the number of simulations without losing the detail necessary for the correct quantification of the local hazard is, therefore, a highly relevant problem in the transition from a regional to a local description.

In the Mediterranean Sea, the tsunami risk is assessed mainly at a local scale at specific locations. At the regional scale, it is extremely important to take into account the exposure elements in the risk analysis. In fact, the hazard is just one part of the overall risk, and other components, for example, vulnerability and exposure, must be included for a comprehensive and interdisciplinary analysis. Although the exposure elements may be relatively far from the area of interest, they can still have significant cascading effects, which can lead to underestimation of the tsunami risk if they are

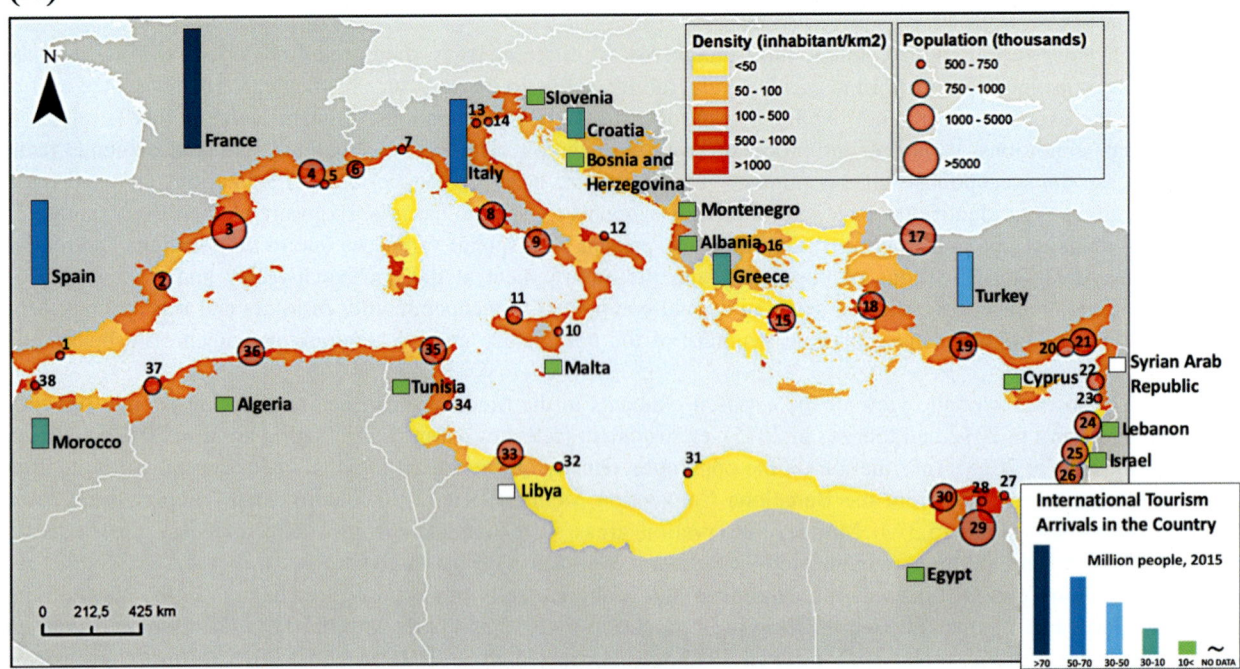

MAIN CITIES

1. Malaga, 2. Valencia, 3. Barcelona, 4. Marseille, 5. Toulon, 6. Nice, 7. Genoa, 8. Rome, 9. Naples, 10. Catania, 11. Palermo, 12. Bari, 13. Padua, 14. Venice, 15. Athens, 16. Thessaloniki, 17. Istanbul, 18. Izmir, 19. Antalya, 20. Mersin, 21. Adana, 22. Samandag, 23. Tripoli, 24. Beirut, 25. Haifa, 26. Tel-Aviv, 27. Port Said, 28. Mansoura, 29. Cairo, 30. Alexandria, 31. Benghazi, 32. Misrata, 33. Tripoli, 34. Sfax, 35. Tunis, 36. Algiers, 37. Oran, 38. Tetouan

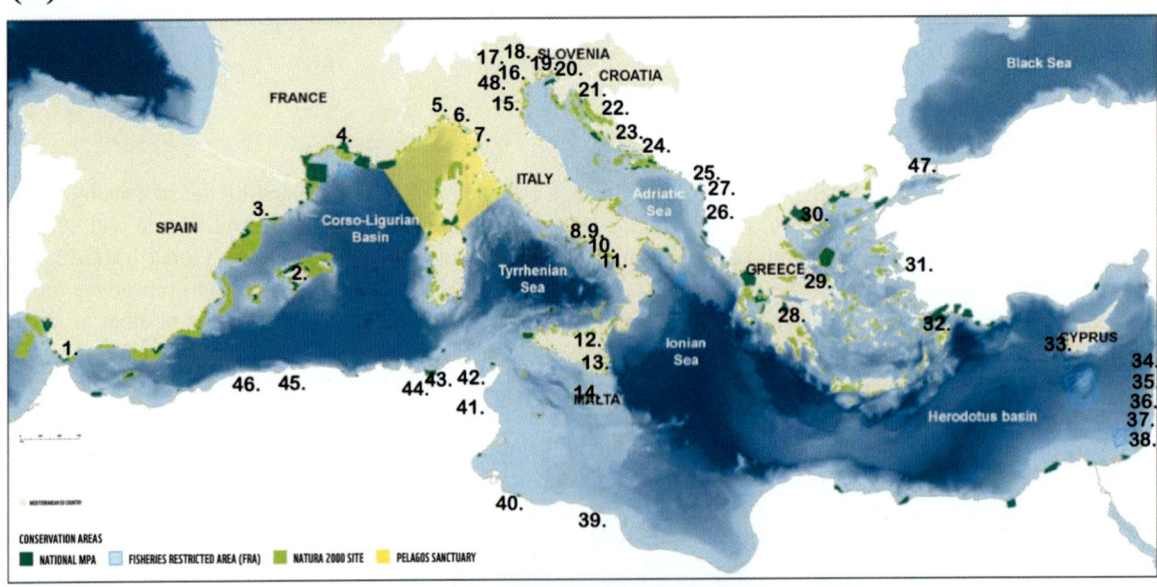

UNESCO WORLD HERITAGE SITES

1. Gorham's Cave, 2. Serra de Tramontana, 3. Arch. Site Tàrraco, 4. Arles, 5. Genoa, 6. Cinque Terre-Portovenere, 7. Pisa, Piazza Duomo, 8. Naples hist. city centre, 9. Pompei-Herculaneum-Torre Annunziata, 10. Costiera Amalfitana, 11. Paestum-Velia-Padula, 12. Val di Noto, 13. Syracusa-Pantalica, 14. Valletta, 15. Ravenna, 16. Venice, 17. Vicenza, Palladian Villas, 18. Aquileia, 19. Porec, 20. Sibenic, 21. Trogir, 22. Split, 23. Stari Grad, 24. Dubrovnik, 25. Kotor, 26. Corfu, 27. Butrint, 28. Delos, 29. Samos, 30. Ephesus, 31. Rhodes, 32. Xanrhos-Letoon, 33. Paphos, 34. Byblos, 35. Tyre, 36. Acre, 37. Haifa, 38. Tel-Aviv, 39. Leptis Magna, 40. Sabratha, 41. Medina of Sousse, 42. Kerkuane, 43. Tunis, 44. Carthage, 45. Algiers, 46. Tipasa, 47. Istanbul, 48. Ferrara

FIGURE 15.3 Exposure of population, cities, marine protected areas, and heritage/cultural sites. (A) Population and population density of the coastal regions, major coastal cities (more than 500,000 inhabitants) in 2012 and international tourists' arrivals in 2015 and (B) most relevant marine protected areas and heritage historical sites indicated by the UNESCO. Map lines delineate study areas and do not necessarily depict accepted national boundaries.

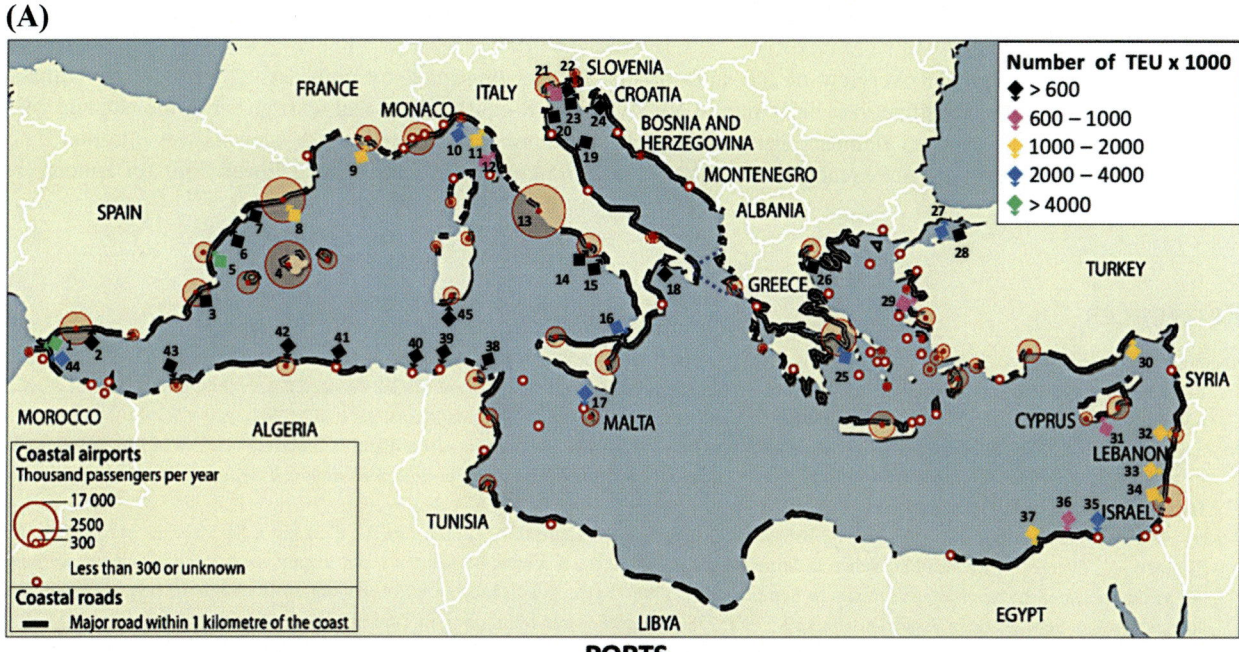

PORTS

1. Algeciras, 2. Malaga, 3. Alicante, 4. Maiorca, 5. Valencia, 6. Castellon, 7. Tarragona, 8. Barcelona, 9. Marseille, 10. Genoa, 11. La Spezia, 12. Livorno, 13. Civitavecchia, 14. Naples, 15. Salerno, 16. Gioia Tauro, 17. Marsaxlokk, 18. Taranto, 19. Ancona, 20. Ravenna, 21. Venezia, 22. Trieste, 23. Koper, 24. Rijeka, 25. Piraeus, 26. Thessaloniki, 27. Ambarli, 28. Haydarpasa, 29. Izimir, 30. Mersin, 31. Limasol, 32. Beirut, 33. Haifa, 34. Ashood, 35. Port Said, 36. Damietta, 37. Alexandria, 38. Tunisi, 39. Annaba, 40. Skikda, 41. Bejaia, 42. Algier, 43. Oran, 44. Tanger-Med, 45. Cagliari

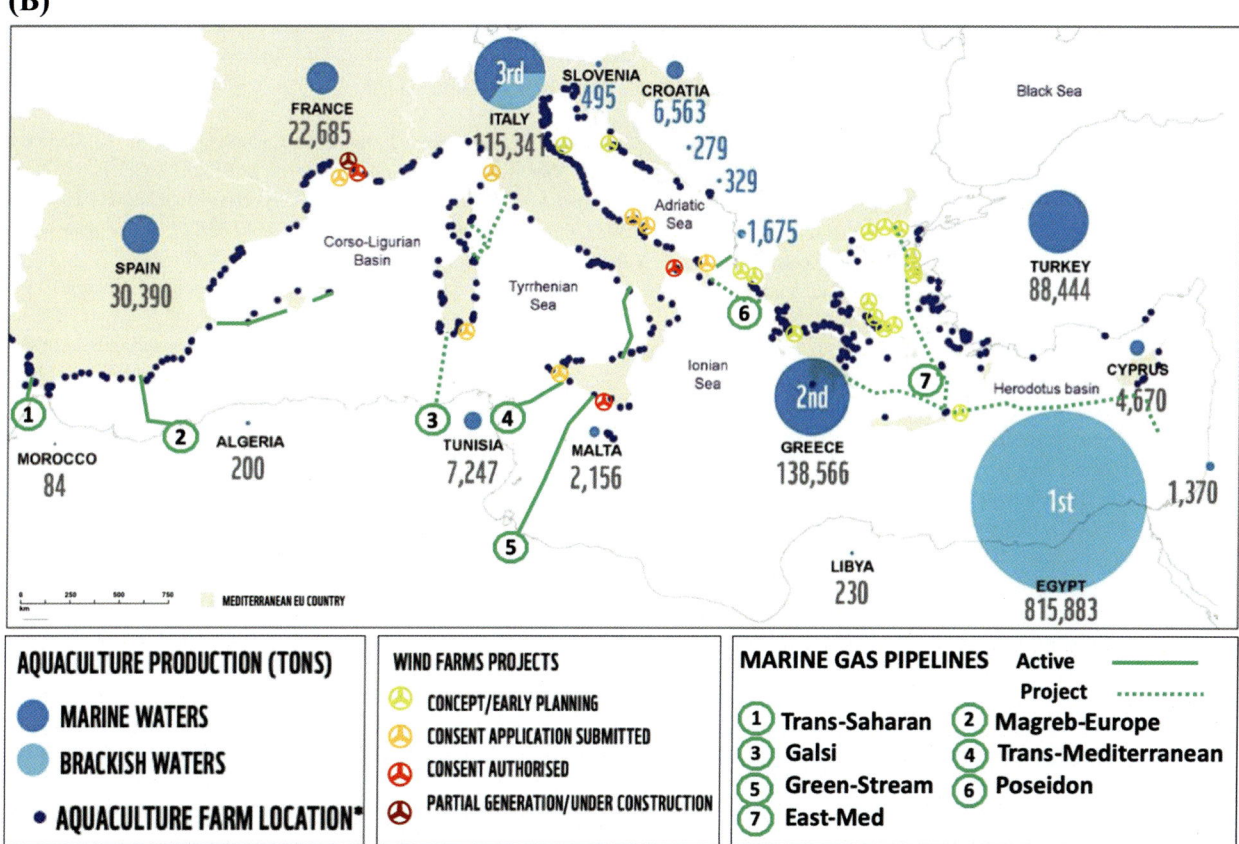

FIGURE 15.4 Exposure of infrastructures and economic activities. (A) Coastal infrastructures (airports and roads) and the number of exchanged containers throughout the main Mediterranean ports (with TEU > 200,000) and (B) aquaculture fish farms and total fishery production (big circle indicate tons/year), potential locations for offshore wind farms and marine gas pipelines. Map lines delineate study areas and do not necessarily depict accepted national boundaries.

overlooked. Therefore, it is essential to consider the relevant factors in assessing the tsunami risk in the Mediterranean Sea.

In this chapter, a regional overview of the state of the art was presented by considering different tsunamigenic sources and including the coastline variability due to VLMs and SLR caused by global warming. For the tsunami risk, a regional selection of the exposed elements that were useful for a more comprehensive risk analysis was introduced. The selected exposed elements are not homogeneous in space and time and may constitute a lower limit for tsunami risk estimations.

References

Abadie, S. M., Harris, J. C., Grilli, S. T., & Fabre, R. (2012). Numerical modeling of tsunami waves generated by the flank collapse of the Cumbre Vieja Volcano (La Palma, Canary Islands): Tsunami source and near field effects. *Journal of Geophysical Research: Oceans, 117*(5). Available from https://doi.org/10.1029/2011JC007646. Available from: http://onlinelibrary.wiley.com/journal/10.1002/(ISSN)2169-9291.

Alonso, S., Tintoré, J., & Di Bladé, I. G. (1989). Estudio Teórico sobre las Rissagues de Ciutadella. Parte I: Ondas de talud en la plataforma. In Tom II (Ed.), *Modos libres de oscilación del puerto. A: Les Rissagues de Ciutadella i altres oscil.lacions de nivell de la mar de gran amplitud a la Mediterrànea Occidental*. Institut Menorquí d'Estudis.

Amato, A., Avallone, A., Basili, R., Bernardi, F., Brizuela, B., Graziani, L., Herrero, A., Lorenzino, M. C., Lorito, S., Mele, F. M., Michelini, A., Piatanesi, A., Pintore, S., Romano, F., Selva, J., Stramondo, S., Tonini, R., & Volpe, M. (2021). From seismic monitoring to tsunami warning in the Mediterranean Sea. *Seismological Research Letters, 92*(3), 1796–1816. Available from https://doi.org/10.1785/0220200437. Available from: https://pubs.geoscienceworld.org/ssa/srl/article/92/3/1796/595804/From-Seismic-Monitoring-to-Tsunami-Warning-in-the.

Anzidei, M., Lambeck, K., Antonioli, F., Furlani, S., Mastronuzzi, G., Serpelloni, E., & Vannucci, G. (2014). Coastal structure, sea-level changes and vertical motion of the land in the Mediterranean. *Geological Society, London, Special Publications, 388*(1), 453–479. Available from https://doi.org/10.1144/sp388.20.

Anzidei, M., Scicchitano, G., Scardino, G., Bignami, C., Tolomei, C., Vecchio, A., Serpelloni, E., De Santis, V., Monaco, C., Milella, M., Piscitelli, A., & Mastronuzzi, G. (2021). Relative sea-level rise scenario for 2100 along the coast of southeastern Sicily (Italy) by InSAR data, satellite images and high-resolution topography. *Remote Sensing, 13*(6). Available from https://doi.org/10.3390/rs13061108. Available from: https://www.mdpi.com/2072-4292/13/6/1108/pdf.

Basili, R., Brizuela, B., Herrero, A., Iqbal, S., Lorito, S., Maesano, F. E., Murphy, S., Perfetti, P., Romano, F., Scala, A., Selva, J., Taroni, M., Tiberti, M. M., Thio, H. K., Tonini, R., Volpe, M., Glimsdal, S., Harbitz, C. B., Løvholt, F., ... Zaytsev, A. (2021). The making of the NEAM tsunami hazard model 2018 (NEAMTHM18). *Frontiers in Earth Science, 8*. Available from https://doi.org/10.3389/feart.2020.616594. Available from: https://www.frontiersin.org/journals/earth-science.

Basili, R., Danciu, L., Beauval, C., Sesetyan, K., Vilanova, S., Adamia, S., Arroucau, P., Atanackov, J., Baize, S., Canora, C., Caputo, R., Carafa, M., Cushing, M., Custódio, S., Demircioglu Tumsa, M., Duarte, J., Ganas, A., García-Mayordomo, J., Gómez de la Peña, L., Gràcia, E., Jamšek Rupnik, P., Jomard, H., Kastelic, V., Maesano, F., MartínBanda, R., Martínez-Loriente, S., Neres, M., Perea ,H., Sket-Motnikar, B., Tiberti ,M., Tsereteli, N., Tsironi, V., Vallone, R., Vanneste, K., Zupančič, P. (2022). *European Fault-Source Model 2020 (EFSM20): online data on fault geometry and activity parameters*. Istituto Nazionale di Geofisica e Vulcanologia (INGV). https://doi.org/10.13127/efsm20.

Bishop, A. W. (1955). The use of the slip circle in the stability analysis of slopes. *Geotechnique, 5*(1), 7–17. Available from https://doi.org/10.1680/geot.1955.5.1.7.

Bondevik, S., Løvholt, F., Harbitz, C., Mangerud, J., Dawson, A., & Inge Svendsen, J. (2005). The Storegga slide tsunami: Comparing field observations with numerical simulations. *Marine and Petroleum Geology, 22*(1–2), 195–208. Available from https://doi.org/10.1016/j.marpetgeo.2004.10.003.

Bougouin, A., Paris, R., & Roche, O. (2020). Impact of fluidized granular flows into water: Implications for tsunamis generated by pyroclastic flows. *Journal of Geophysical Research: Solid Earth, 125*(5). Available from https://doi.org/10.1029/2019JB018954. Available from: http://agupubs.onlinelibrary.wiley.com/hub/jgr/journal/10.1002/(ISSN)2169-9356/.

Camilleri, D. H. (2006). Tsunami construction risks in the Mediterranean—Outlining Malta's scenario. *Disaster Prevention and Management: An International Journal, 15*(1), 146–162. Available from https://doi.org/10.1108/09653560610654301. Available from: http://www.emeraldinsight.com/info/journals/dpm/dpm.jsp.

Candela, J., Mazzola, S., Sammari, C., Limeburner, R., Lozano, C. J., Patti, B., & Bonanno, A. (1999). The 'Mad Sea' phenomenon in the Strait of Sicily. *Journal of Physical Oceanography, 29*(9), 2210–2231. Available from http://journals.ametsoc.org/loi/phoc, doi:10.1175/1520-0485(1999)029<2210:TMSPIT>2.0.CO;2.

Collico, S., Arroyo, M., Urgeles, R., Gràcia, E., Devincenzi, M., & Peréz, N. (2020). Probabilistic mapping of earthquake-induced submarine landslide susceptibility in the South-West Iberian margin. *Marine Geology, 429*106296. Available from https://doi.org/10.1016/j.margeo.2020.106296.

Colucci, P., & Michelato, A. (1976). An approach to study ot the 'Marubbio' phenomenon. *Bollettino di Geofisica Teorica ed Applicata, 13*(69), 3–10.

Denamiel, C., Šepić, J., Ivanković, D., & Vilibić, I. (2019). The Adriatic sea and coast modelling suite: Evaluation of the meteotsunami forecast component. *Ocean Modelling, 135*, 71–93. Available from https://doi.org/10.1016/j.ocemod.2019.02.003. Available from: http://www.elsevier.com/inca/publications/store/6/0/1/3/7/6/index.htt.

De Risi, R., Muhammad, A., De Luca, F., Goda, K., & Mori, N. (2022). Dynamic risk framework for cascading compounding climate-geological hazards: A perspective on coastal communities in subduction zones. *Frontiers in Earth Science*, 10. Available from https://doi.org/10.3389/feart.2022.1023018. Available from: https://www.frontiersin.org/journals/earth-science.

DISS Working Group (2021). *DISS 3.3.0. Database of Individual Seismogenic Sources (DISS), Version 3.3.0: A compilation of potential sources for earthquakes larger than M 5.5 in Italy and surrounding areas. Istituto Nazionale di Geofisica e Vulcanologia (INGV)*. https://doi.org/10.13127/diss3.3.0

Dilmen, D. I., Kemec, S., Yalciner, A. C., Düzgün, S., & Zaytsev, A. (2015). Development of a tsunami inundation map in detecting tsunami risk in Gulf of Fethiye, Turkey. *Pure and Applied Geophysics*, 172(3−4), 921−929. Available from https://doi.org/10.1007/s00024-014-0936-2. Available from: http://www.springer.com/birkhauser/geo + science/journal/24.

Du, J., Yin, K., Glade, T., Woldai, T., Chai, B., Xiao, L., & Wang, Y. (2020). Probabilistic hazard analysis of impulse waves generated by multiple subaerial landslides and its application to Wu Gorge in Three Gorges Reservoir, China. *Engineering Geology*, 276105773. Available from https://doi.org/10.1016/j.enggeo.2020.105773.

Dura, T., Garner, A. J., Weiss, R., Kopp, R. E., Engelhart, S. E., Witter, R. C., Briggs, R. W., Mueller, C. S., Nelson, A. R., & Horton, B. P. (2021). Changing impacts of Alaska-Aleutian subduction zone tsunamis in California under future sea-level rise. *Nature Communications*, 12(1). Available from https://doi.org/10.1038/s41467-021-27445-8.

El-Barmelgy, H. M. (2014). Strategic tsunami hazard analysis and risk assessment planning model: A case study for the city of Alexandria. Egypt. *International Journal of Development and Sustainability*, 3(4), 784−809.

El-Hattab, M. M., Mohamed, S. A., & El Raey, M. (2018). Potential tsunami risk assessment to the city of Alexandria, Egypt. *Environmental Monitoring and Assessment*, 190(9). Available from https://doi.org/10.1007/s10661-018-6876-z. Available from: http://www.wkap.nl/journalhome.htm/0167-6369.

Favalli, M., Boschi, E., Mazzarini, F., & Pareschi, M. T. (2009). Seismic and landslide source of the 1908 Straits of Messina tsunami (Sicily, Italy). *Geophysical Research Letters*, 36(16). Available from https://doi.org/10.1029/2009GL039135. Available from: http://onlinelibrary.wiley.com/journal/10.1002/(ISSN)1944-8007/issues?year = 2012.

Fenoglio-Marc, L., Mariotti, A., Sannino, G., Meyssignac, B., Carillo, A., Struglia, M. V., & Rixen, M. (2013). Decadal variability of net water flux at the Mediterranean Sea Gibraltar Strait. *Global and Planetary Change*, 100, 1−10. Available from https://doi.org/10.1016/j.gloplacha.2012.08.007.

Fenoglio-Marc, L., Rietbroek, R., Grayek, S., Becker, M., Kusche, J., & Stanev, E. (2012). Water mass variation in the Mediterranean and Black Seas. *Journal of Geodynamics*, 59−60, 168−182. Available from https://doi.org/10.1016/j.jog.2012.04.001.

Flouri, E.T., Mitsoudis, D.A., Chrysoulakis, N., & Synolakis, C.E. (2012). *Tsunami risk and vulnerability analysis for the city of Rhodes*. In Proceedings of the International Offshore and Polar Engineering Conference, pp. 257−264.

Fornaciai, A., Favalli, M., & Nannipieri, L. (2019). Numerical simulation of the tsunamis generated by the Sciara del Fuoco landslides (Stromboli Island, Italy). *Scientific Reports*, 9(1). Available from https://doi.org/10.1038/s41598-019-54949-7. Available from: http://www.nature.com/srep/index.html.

Fournier d'Albe, E. M. (1982). An approach to earthquake risk management. *Engineering Structures*, 4(3), 147−152. Available from https://doi.org/10.1016/0141-0296(82)90002-5.

Fox-Kemper, B. (2021). *In climate change 2021: The physical science basis. Contribution of Working Group I to the sixth assessment report of the intergovernmental panel on climate change ocean, cryosphere and sea level change*. Cambridge University Press. Available from 10.1017/9781009157896.011.

Freundt, A. (2003). Entrance of hot pyroclastic flows into the sea: Experimental observations. *Bulletin of Volcanology*, 65(2−3), 144−164. Available from https://doi.org/10.1007/s00445-002-0250-1. Available from: https://rd.springer.com/journal/volumesAndIssues/445.

Fritz, H. M., Hager, W. H., & Minor, H. E. (2004). Near field characteristics of landslide generated impulse waves. *Journal of Waterway, Port, Coastal and Ocean Engineering*, 130(6), 287−302. Available from https://doi.org/10.1061/(ASCE)0733-950X(2004)130:6(287).

GRID-Arendal. (2013). *State of the Mediterranean Marine and coastal environment*. GRID_Arendal UNEP partner. Pubblication 192, 2013. https://www.grida.no/publications/192

Glimsdal, S., Pedersen, G. K., Harbitz, C. B., & Løvholt, F. (2013). Dispersion of tsunamis: Does it really matter? *Natural Hazards and Earth System Sciences*, 13, 1507−1526. Available from https://doi.org/10.5194/nhess-13-1507-2013. Available from: http://www.nat-hazards-earth-syst-sci.net/volumes_and_issues.html.

Grezio, A., Babeyko, A., Baptista, M. A., Behrens, J., Costa, A., Davies, G., Geist, E. L., Glimsdal, S., González, F. I., Griffin, J., Harbitz, C. B., LeVeque, R. J., Lorito, S., Løvholt, F., Omira, R., Mueller, C., Paris, R., Parsons, T., Polet, J., ... Thio, H. K. (2017). Probabilistic tsunami hazard analysis: Multiple sources and global applications. *Reviews of Geophysics*, 55(4), 1158−1198. Available from https://doi.org/10.1002/2017RG000579. Available from: http://onlinelibrary.wiley.com/journal/10.1002/(ISSN)1944-9208.

Grezio, A., Cinti, F. R., Costa, A., Faenza, L., Perfetti, P., Pierdominici, S., Pondrelli, S., Sandri, L., Tierz, P., Tonini, R., & Selva, J. (2020). Multisource Bayesian probabilistic tsunami hazard analysis for the Gulf of Naples (Italy). *Journal of Geophysical Research: Oceans*, 125(2). Available from https://doi.org/10.1029/2019JC015373. Available from: http://agupubs.onlinelibrary.wiley.com/agu/jgr/journal/10.1002/(ISSN)2169-9291/.

Grezio, A., Gasparini, P., Marzocchi, W., Patera, A., & Tinti, S. (2012). Tsunami risk assessments in Messina, Sicily—Italy. *Natural Hazards and Earth System Science*, 12(1), 151−163. Available from https://doi.org/10.5194/nhess-12-151-2012.

Grezio, A., Tonini, R., Sandri, L., Pierdominici, S., & Selva, J. (2015). A methodology for a comprehensive probabilistic tsunami hazard assessment: Multiple sources and short-term interactions. *Journal of Marine Science and Engineering*, *3*(1), 23–51. Available from https://doi.org/10.3390/jmse3010023. Available from: http://www.mdpi.com/2077-1312/3/1/23/pdf.

Grifoll, M., Karlis, T., & Ortego, M. (2018). Characterizing the evolution of the container traffic share in the Mediterranean Sea using hierarchical clustering. *Journal of Marine Science and Engineering*, *6*(4), 121. Available from https://doi.org/10.3390/jmse6040121.

Grilli, S. T., Taylor, O. D. S., Baxter, C. D. P., & Maretzki, S. (2009). A probabilistic approach for determining submarine landslide tsunami hazard along the upper east coast of the United States. *Marine Geology*, *264*(1–2), 74–97. Available from https://doi.org/10.1016/j.margeo.2009.02.010.

Grilli, S. T., & Watts, P. (2005). Tsunami generation by submarine mass failure. I: Modeling, experimental validation and sensitivity analyses. *Journal of Waterway, Port, Coastal and Ocean Engineering*, *131*(6), 283–297. Available from https://doi.org/10.1061/(ASCE)0733-950X(2005)131:6(283).

Grünthal, G., Wahlström, R., & Stromeyer, D. (2013). The SHARE European Earthquake catalogue (SHEEC) for the time period 1900–2006 and its comparison to the European-Mediterranean Earthquake catalogue (EMEC). *Journal of Seismology*, *17*(4), 1339–1344. Available from https://doi.org/10.1007/s10950-013-9379-y.

Harbitz, C. B., Løvholt, F., & Bungum, H. (2014). Submarine landslide tsunamis: How extreme and how likely? *Natural Hazards*, *72*(3), 1341–1374. Available from https://doi.org/10.1007/s11069-013-0681-3. Available from: http://www.wkap.nl/journalhome.htm/0921-030X.

Harbitz, C. B., Løvholt, F., Pedersen, G., & Masson, D. G. (2006). Mechanisms of tsunami generation by submarine landslides: A short review. *Norsk Geologisk Tidsskrift*, *86*(3), 255–264.

Hibiya, T., & Kajiura, K. (1982). Origin of the Abiki phenomenon (a kind of seiche) in Nagasaki Bay. *Journal of the Oceanographical Society of Japan*, *38*(3), 172–182. Available from https://doi.org/10.1007/BF02110288.

Honda, K., Terada, T., Yoshida, Y., & Isitani, D. (1908). Secondary undulations of oceanic tides. *Philosophical Magazine Series 6*, *15*(85), 88–126.

Horvath, K., Šepić, J., & Prtenjak, M. T. (2018). Atmospheric forcing conducive for the Adriatic 25 June 2014 Meteotsunami event. *Pure and Applied Geophysics*, *175*, 3817–3837. Available from https://doi.org/10.1007/978-3-030-11958-4_7.

Horvath, K., & Vilibić, I. (2015). *Meteorological tsunamis: The U.S. East Coast and other coastal regions. Atmospheric mesoscale conditions during the Boothbay meteotsunami: A numerical sensitivity study using a high resolution mesoscale model* (pp. 55–74). Springer International Publishing Available from: https://link.springer.com/chapter/10.1007/978-3-319-12712-5_4. Available from 10.1007/978-3-319-12712-5_4.

IPCC. (2022). *Intergovernmental Panel on Climate Change: Sixth Assessment Report*. IPCC. Available from https://www.ipcc.ch/assessment-report/ar6/.

Jansà, A. (1986). Respuesta marina a perturbaciones mesometeorológicas: la "rissaga" de 21 de junio de 1984 en Ciutadella (Menorca). *Revista de Meteorología, Junio*, *3*(7), 5–29.

Jansa, A., Monserrat, S., & Gomis, D. (2007). The rissaga of 15 June 2006 in Ciutadella (Menorca), a meteorological tsunami. *Advances in Geosciences*, *12*, 1–4. Available from https://doi.org/10.5194/adgeo-12-1-2007. Available from: http://www.adv-geosci.net/volumes.html.

Jelínek, R., Eckert, S., Zeug, G., & Krausmann, E. (2009). Tsunami vulnerability and risk analysis applied to the city of Alexandria, Egypt. EUR 23967 EN. Luxembourg (Luxembourg): OP. JRC53316, Available from https://publications.jrc.ec.europa.eu/repository/handle/JRC53316. (2009).

Lambeck, K., & Purcell, A. (2005). Sea-level change in the Mediterranean Sea since the LGM: Model predictions for tectonically stable areas. *Quaternary Science Reviews*, *24*(18–19), 1969–1988. Available from https://doi.org/10.1016/j.quascirev.2004.06.025.

Lane, E. M., Mountjoy, J. J., Power, W. L., & Popinet, S. (2016). Initialising landslide-generated tsunamis for probabilistic tsunami hazard assessment in Cook Strait. *The International Journal of Ocean and Climate Systems*, *7*(1), 4–13. Available from https://doi.org/10.1177/1759313115623162.

Ličer, M., Mourre, B., Troupin, C., Krietemeyer, A., Jansá, A., & Tintoré, J. (2017). Numerical study of Balearic meteotsunami generation and propagation under synthetic gravity wave forcing. *Ocean Modelling*, *111*, 38–45. Available from https://doi.org/10.1016/j.ocemod.2017.02.001. Available from: http://www.elsevier.com/inca/publications/store/6/0/1/3/7/6/index.htt.

Løvholt, F., Glimsdal, S., Harbitz, C. B., Zamora, N., Nadim, F., Peduzzi, P., Dao, H., & Smebye, H. (2012). Tsunami hazard and exposure on the global scale. *Earth-Science Reviews*, *110*(1–4), 58–73. Available from https://doi.org/10.1016/j.earscirev.2011.10.002.

Løvholt, F., Griffin, J., & Salgado-Gálvez, M. A. (2015). *Encyclopedia of complexity and systems science. Tsunami hazard and risk assessment on the global scale* (pp. 1–34). Springer. Available from 10.1007/978-3-642-27737-5_642-1.

Maeno, F., & Imamura, F. (2011). Tsunami generation by a rapid entrance of pyroclastic flow into the sea during the 1883 Krakatau eruption, Indonesia. *Journal of Geophysical Research: Solid Earth*, *116*(9). Available from https://doi.org/10.1029/2011JB008253. Available from: http://onlinelibrary.wiley.com/journal/10.1002/(ISSN)2169-9356.

Maramai, A., Brizuela, B., & Graziani, L. (2022). A database for tsunamis and meteotsunamis in the Adriatic Sea. *Applied Sciences*, *12*(11), 5577. Available from https://doi.org/10.3390/app12115577.

Maramai, A., Graziani, L., Alessio, G., Burrato, P., Colini, L., Cucci, L., Nappi, R., Nardi, A., & Vilardo, G. (2005). Near- and far-field survey report of the 30 December 2002 Stromboli (Southern Italy) tsunami. *Marine Geology*, *215*(1–2), 93–106. Available from https://doi.org/10.1016/j.margeo.2004.11.009.

Monserrat, S., Rabinovich, A. B., & Casas, B. (1998). On the reconstruction of the transfer function for atmospherically generated seiches. *Geophysical Research Letters*, *25*(12), 2197–2200. Available from https://doi.org/10.1029/98GL01506. Available from: http://onlinelibrary.wiley.com/journal/10.1002/(ISSN)1944-8007/issues?year = 2012.

Monserrat, S., Vilibić, I., & Rabinovich, A. B. (2006). Meteotsunamis: Atmospherically induced destructive ocean waves in the tsunami frequency band. *Natural Hazards and Earth System Science*, *6*(6), 1035–1051. Available from https://doi.org/10.5194/nhess-6-1035-2006. Available from: http://www.nat-hazards-earth-syst-sci.net/volumes_and_issues.html.

Maramai, A., Graziani, L., Brizuela, B. (2019). Euro-Mediterranean Tsunami Catalogue (EMTC), version 2.0 (Version 2.0) [Data set]. Istituto Nazionale di Geofisica e Vulcanologia (INGV). https://doi.org/10.13127/TSUNAMI/EMTC.2.0

NCEI/WDS (2023). *National Geophysical Data Center / World Data Service: Global Historical Tsunami Database*. NOAA National Centers for Environmental Information. Available from: https://doi.org/10.7289/V5PN93H7.

Necmioglu, O., Heidarzadeh, M., Vougioukalakis, G. E., & Selva, J. (2023). Landslide induced tsunami hazard at volcanoes: the case of Santorini. *Pure and Applied Geophysics*. Available from https://doi.org/10.1007/s00024-023-03252-8. Available from: https://www.springer.com/journal/24.

Nomikou, P., Carey, S., Bell, K. L. C., Papanikolaou, D., Bejelou, K., Cantner, K., Sakellariou, D., & Perros, I. (2014). Tsunami hazard risk of a future volcanic eruption of Kolumbo submarine volcano, NE of Santorini Caldera, Greece. *Natural Hazards*, 72(3), 1375–1390. Available from https://doi.org/10.1007/s11069-012-0405-0. Available from: http://www.wkap.nl/journalhome.htm/0921-030X.

Okal, E. A., Visser, J. N. J., & de Beer, C. H. (2014). The Dwarskersbos, South Africa local tsunami of August 27, 1969: Field survey and simulation as a meteorological event. *Natural Hazards*, 74(1), 251–268. Available from https://doi.org/10.1007/s11069-014-1205-5. Available from: http://www.wkap.nl/journalhome.htm/0921-030X.

Orlić, M. (2015). The first attempt at cataloguing tsunami-like waves of meteorological origin in Croatian coastal waters. *Acta Adriatica*, 56(1), 83–96. Available from: http://jadran.izor.hr/acta/pdf/56_1_pdf/56_1_4.pdf.

Pagnoni, G., Armigliato, A., & Tinti, S. (2015). Scenario-based assessment of buildings' damage and population exposure due to earthquake-induced tsunamis for the town of Alexandria, Egypt. *Natural Hazards and Earth System Sciences*, 15(12), 2669–2695. Available from https://doi.org/10.5194/nhess-15-2669-2015. Available from: http://www.nat-hazards-earth-syst-sci.net/volumes_and_issues.html.

Pagnoni, G., Armigliato, A., & Tinti, S. (2021). Estimation of human damage and economic loss of buildings related to tsunami inundation in the city of Augusta, Italy. *Geological Society, London, Special Publications*, 501(1), 327–342. Available from https://doi.org/10.1144/sp501-2019-134.

Pampell-Manis, A., Horrillo, J., Shigihara, Y., & Parambath, L. (2016). Probabilistic assessment of landslide tsunami hazard for the northern Gulf of Mexico. *Journal of Geophysical Research: Oceans*, 121(1), 1009–1027. Available from https://doi.org/10.1002/2015JC011261. Available from: http://onlinelibrary.wiley.com/journal/10.1002/(ISSN)2169-9291.

Pantosti, D., De Martini, D. P., Orefice, S., Smedile, S., Patera, A., Paris, R., Terrinha, P., Hunt, J., Papadopoulos, G. A., Noiva, J., Triantafyllou, I., & Yalciner, A. C. (2017). *The ASTARTE Paleotsunami Deposits data base—Web-based references for tsunami research in the NEAM region*. Available from http://hdl.handle.net/2122/11136.

Papadopoulos, G. A., & Dermentzopoulos, T. (1998). A tsunami risk management pilot study in Heraklion, Crete. *Natural Hazards*, 18(2), 91–118. Available from https://doi.org/10.1023/A:1008070306156. Available from: http://www.wkap.nl/journalhome.htm/0921-030X.

Paris, R. (2015). Source mechanisms of volcanic tsunamis. *Philosophical Transactions of the Royal Society A: Mathematical, Physical and Engineering Sciences*, 373(2053)20140380. Available from https://doi.org/10.1098/rsta.2014.0380.

Paris, R., Selva, J., & Grezio, A. (2023). *Mathematics of planet Earth source modeling*. Springer.

Paris, R., Ulvrova, M., Selva, J., Brizuela, B., Costa, A., Grezio, A., Lorito, S., & Tonini, R. (2019). Probabilistic hazard analysis for tsunamis generated by subaqueous volcanic explosions in the Campi Flegrei caldera, Italy. *Journal of Volcanology and Geothermal Research*, 379, 106–116. Available from https://doi.org/10.1016/j.jvolgeores.2019.05.010. Available from: http://www.sciencedirect.com/science/journal/03770273.

Pattiaratchi, C. B., & Wijeratne, E. M. S. (2015). Are meteotsunamis an underrated hazard? *Philosophical Transactions of the Royal Society A: Mathematical, Physical and Engineering Sciences*, 373(2053)20140377. Available from https://doi.org/10.1098/rsta.2014.0377.

Piante, C., & Ody, D. (2015). *Blue growth in the Mediterranean Sea: The challenge of good environmental status*. World Wildlife Fund. Available from: https://d2ouvy59p0dg6k.cloudfront.net/downloads/medtrends_regional_report.pdf.

Pushpalal, D., Wanner, P. J., & Pak, K. (2023). Notions of resilience and qualitative evaluation of tsunami resiliency using the theory of springs. *Journal of Safety Science and Resilience*, 4(1), 1–8. Available from https://doi.org/10.1016/j.jnlssr.2022.09.002.

Rabinovich, A. B. (2020). Twenty-seven years of progress in the science of meteorological tsunamis following the 1992 Daytona Beach event. *Pure and Applied Geophysics*, 177(3), 1193–1230. Available from https://doi.org/10.1007/s00024-019-02349-3. Available from: http://www.springer.com/birkhauser/geo + science/journal/24.

Rabinovich, A. B., Vilibić, I., & Tinti, S. (2009). Meteorological tsunamis: Atmospherically induced destructive ocean waves in the tsunami frequency band. *Physics and Chemistry of the Earth*, 34(17–18), 891–893. Available from https://doi.org/10.1016/j.pce.2009.10.006.

Reimann, L., Vafeidis, A. T., Brown, S., Hinkel, J., & Tol, R. S. J. (2018). Mediterranean UNESCO World Heritage at risk from coastal flooding and erosion due to sea-level rise. *Nature Communications*, 9(1). Available from https://doi.org/10.1038/s41467-018-06645-9.

Rovida, A., Antonucci, A., & Locati, M. (2022). The European Preinstrumental Earthquake catalogue EPICA, the 1000–1899 catalogue for the European Seismic Hazard Model 2020. *Earth System Science Data*, 14(12), 5213–5231. Available from https://doi.org/10.5194/essd-14-5213-2022. Available from: http://www.earth-system-science-data.net.

Ruff, L. J. (2003). Some aspects of energy balance and tsunami generation by earthquakes and landslides. *Pure and Applied Geophysics*, 160(10–11), 2155–2176. Available from https://doi.org/10.1007/s00024-003-2424-y.

Salmanidou, D. M., Heidarzadeh, M., & Guillas, S. (2019). Probabilistic landslide-generated tsunamis in the Indus Canyon, NW Indian Ocean, using statistical emulation. *Pure and Applied Geophysics*, 176(7), 3099–3114. Available from https://doi.org/10.1007/s00024-019-02187-3. Available from: http://www.springer.com/birkhauser/geo + science/journal/24.

Sassa, S., Grilli, S. T., Tappin, D. R., Sassa, K., Karnawati, D., Gusiakov, V. K., & Løvholt, F. (2022). Understanding and reducing the disaster risk of landslide-induced tsunamis: A short summary of the panel discussion in the World Tsunami Awareness Day Special Event of the Fifth World Landslide Forum. *Landslides*, 19(2), 533–535. Available from https://doi.org/10.1007/s10346-021-01819-x. Available from: https://www.springer.com/journal/10346.

Selva, J., Acocella, V., Bisson, M., Caliro, S., Costa, A., Della Seta, M., De Martino, P., De Vita, S., Federico, C., Giordano, G., Martino, S., & Cardaci, C. (2019). Multiple natural hazards at volcanic islands: A review for the Ischia volcano (Italy). *Journal of Applied Volcanology*, 8(1). Available from https://doi.org/10.1186/s13617-019-0086-4. Available from: https://appliedvolc.biomedcentral.com/.

Selva, J., Amato, A., Armigliato, A., Basili, R., Bernardi, F., Brizuela, B., Cerminara, M., de' Micheli Vitturi, M., Di Bucci, D., Di Manna, P., Esposti Ongaro, T., Lacanna, G., Lorito, S., Løvholt, F., Mangione, D., Panunzi, E., Piatanesi, A., Ricciardi, A., Ripepe, M., ... Zaniboni, F. (2021). Tsunami risk management for crustal earthquakes and non-seismic sources in Italy. *La Rivista del Nuovo Cimento*, 44, 69−144. Available from https://doi.org/10.1007/s40766-021-00016-9.

Selva, J., Sandri, L., Taroni, M., Sulpizio, R., Tierz, P., & Costa, A. (2022). A simple two-state model interprets temporal modulations in eruptive activity and enhances multi-volcano hazard quantification. *Science Advances*, 8(44). Available from https://doi.org/10.1126/sciadv.abq4415. Available from: https://www.science.org/doi/10.1126/sciadv.abq4415.

Šepić, J., & Vilibić, I. (2011). The development and implementation of a real-time meteotsunami warning network for the Adriatic Sea. *Natural Hazards and Earth System Science*, 11(1), 83−91. Available from https://doi.org/10.5194/nhess-11-83-2011.

Šepić, J., Vilibić, I., & Monserrat, S. (2009). Teleconnections between the Adriatic and the Balearic meteotsunamis. *Physics and Chemistry of the Earth, Parts A/B/C*, 34(17−18), 928−937. Available from https://doi.org/10.1016/j.pce.2009.08.007.

Šepić, J., Vilibić, I., Rabinovich, A., & Tinti, S. (2018). Meteotsunami ("Marrobbio") of 25−26 June 2014 on the southwestern coast of Sicily, Italy. *Pure and Applied Geophysics*, 175(4), 1573−1593. Available from https://doi.org/10.1007/s00024-018-1827-8.

Šepić, J., Vilibić, I., & Strelec Mahović, N. (2012). Northern Adriatic meteorological tsunamis: Observations, link to the atmosphere, and predictability. *Journal of Geophysical Research: Oceans*, 117(C2). Available from https://doi.org/10.1029/2011jc007608.

Sepúlveda, I., Haase, J. S., Liu, P. L. F., Grigoriu, M., & Winckler, P. (2021). Non-stationary probabilistic tsunami hazard assessments incorporating climate-change-driven sea level rise. *Earth's Future*, 9(6). Available from https://doi.org/10.1029/2021EF002007. Available from: http://onlinelibrary.wiley.com/journal/10.1002/(ISSN)2328-4277.

Sørensen, M. B., Spada, M., Babeyko, A., Wiemer, S., & Grünthal, G. (2012). Probabilistic tsunami hazard in the Mediterranean Sea. *Journal of Geophysical Research: Solid Earth*, 117(1). Available from https://doi.org/10.1029/2010JB008169. Available from: http://onlinelibrary.wiley.com/journal/10.1002/(ISSN)2169-9356.

Thomson, R. E., Rabinovich, A. B., Fine, I. V., Sinnott, D. C., McCarthy, A., Sutherland, N. A. S., & Neil, L. K. (2009). Meteorological tsunamis on the coasts of British Columbia and Washington. *Physics and Chemistry of the Earth*, 34(17−18), 971−988. Available from https://doi.org/10.1016/j.pce.2009.10.003.

Tierz, P., Woodhouse, M. J., Phillips, J. C., Sandri, L., Selva, J., Marzocchi, W., & Odbert, H. M. (2017). A framework for probabilistic multi-hazard assessment of rain-triggered lahars using Bayesian belief networks. *Frontiers in Earth Science*, 5. Available from https://doi.org/10.3389/feart.2017.00073. Available from: http://journal.frontiersin.org/article/10.3389/feart.2017.00073/full.

Tinti, S., Armigliato, A., Pagnoni, G., & Zaniboni, F. (2005). Scenarios of giant tsunamis of tectonic origin in the Mediterranean. *ISET Journal of Earthquake Technology*, 42(4), 171−188.

Tinti, S., Pagnoni, G., Zaniboni, F., & Bortolucci, E. (2003). Tsunami generation in Stromboli island and impact on the south-east Tyrrhenian coasts. *Natural Hazards and Earth System Science*, 3(5), 299−309. Available from https://doi.org/10.5194/nhess-3-299-2003. Available from: http://www.nat-hazards-earth-syst-sci.net/volumes_and_issues.html.

Triantafyllou, I., Novikova, T., Charalampakis, M., Fokaefs, A., & Papadopoulos, G. A. (2019). Quantitative tsunami risk assessment in terms of building replacement cost based on tsunami modelling and GIS methods: The case of Crete Island, Hellenic Arc. *Pure and Applied Geophysics*, 176(7), 3207−3225. Available from https://doi.org/10.1007/s00024-018-1984-9. Available from: http://www.springer.com/birkhauser/geo+science/journal/24.

Tsimplis, M. N., Calafat, F. M., Marcos, M., Jordà, G., Gomis, D., Fenoglio-Marc, L., Struglia, M. V., Josey, S. A., & Chambers, D. P. (2013). The effect of the NAO on sea level and on mass changes in the Mediterranean Sea. *Journal of Geophysical Research: Oceans*, 118(2), 944−952. Available from https://doi.org/10.1002/jgrc.20078. Available from: http://agupubs.onlinelibrary.wiley.com/agu/jgr/journal/10.1002/(ISSN)2169-9291/.

Ulvrova, M., Paris, R., Nomikou, P., Kelfoun, K., Leibrandt, S., Tappin, D. R., & McCoy, F. W. (2016). Source of the tsunami generated by the 1650 AD eruption of Kolumbo submarine volcano (Aegean Sea, Greece). *Journal of Volcanology and Geothermal Research*, 321, 125−139. Available from https://doi.org/10.1016/j.jvolgeores.2016.04.034. Available from: http://www.sciencedirect.com/science/journal/03770273.

Şalap,S., Ayça, A. Akyürek,Z., Yalçıner,A.C. (2001). *Tsunami risk analysis and disaster management by using GIS: A case study in Southwest Turkey*. Conference Paper: AGILE International Conference on Geographic Information ScienceAt: Utrect, The Netherlands.

UNEP. (2017). *Mediterranean action plan*. United Nations Environment Programme. Available from https://www.unep.org/unepmap/.

Vecchio, A., Anzidei, M., & Carbone, V. (2014). New insights on the tsunami recording of the May, 21, 2003, M_w 6.9 Boumerdès earthquake from tidal data analysis. *Journal of Geodynamics*, 79, 39−49. Available from https://doi.org/10.1016/j.jog.2014.05.001. Available from: http://www.sciencedirect.com/science/journal/02643707/68.

Vecchio, A., Anzidei, M., & Serpelloni, E. (2023). Sea level rise projections up to 2150 in the northern Mediterranean coasts. *Environmental Research Letters*, 19(1). Available from https://doi.org/10.1088/1748-9326/ad127e.

Vilibić, I., Denamiel, C., Zemunik, P., & Monserrat, S. (2021). The Mediterranean and Black Sea meteotsunamis: An overview. *Natural Hazards*, 106(2), 1223−1267. Available from https://doi.org/10.1007/s11069-020-04306-z. Available from: http://www.wkap.nl/journalhome.htm/0921-030X.

Vilibić, I., Domijan, N., Orlić, M., Leder, N., & Pasarić, M. (2004). Resonant coupling of a traveling air pressure disturbance with the east Adriatic coastal waters. *Journal of Geophysical Research: Oceans*, 109(10). Available from https://doi.org/10.1029/2004JC002279. C10001-12, Available from: http://agupubs.onlinelibrary.wiley.com/agu/jgr/journal/10.1002/(ISSN)2169-9291/.

Vilibić, I., Rabinovich, A. B., & Anderson, E. J. (2021). Special issue on the global perspective on meteotsunami science: Editorial. *Natural Hazards*, *106*(2), 1087−1104. Available from https://doi.org/10.1007/s11069-021-04679-9. Available from: http://www.wkap.nl/journalhome.htm/0921-030X.

Vilibić, I., Šepić, J., Rabinovich, A. B., & Monserrat, S. (2016). Modern approaches in meteotsunami research and early warning. *Frontiers in Marine Science*, *3*. Available from https://doi.org/10.3389/fmars.2016.00057.

Vučetić, T., Vilibić, I., Tinti, S., & Maramai, A. (2009). The Great Adriatic flood of 21 June 1978 revisited: An overview of the reports. *Physics and Chemistry of the Earth*, *34*(17−18), 894−903. Available from https://doi.org/10.1016/j.pce.2009.08.005.

Ward, S. N. (2001). Landslide tsunami. *Journal of Geophysical Research: Solid Earth*, *106*(6), 11201−11215. Available from https://doi.org/10.1029/2000jb900450. Available from: http://agupubs.onlinelibrary.wiley.com/hub/jgr/journal/10.1002/(ISSN)2169-9356/.

Watts, P. (2004). Probabilistic predictions of landslide tsunamis off Southern California. *Marine Geology*, *203*(3−4), 281−301. Available from https://doi.org/10.1016/S0025-3227(03)00311-6. Available from: http://www.sciencedirect.com/science/journal/00253227.

Watts, P., & Waythomas, C. F. (2003). Theoretical analysis of tsunami generation by pyroclastic flows. *Journal of Geophysical Research: Solid Earth*, *108*(B12). Available from https://doi.org/10.1029/2002JB002265.

Wöppelmann, G., & Marcos, M. (2012). Coastal sea level rise in southern Europe and the non-climate contribution of vertical land motion. *Journal of Geophysical Research: Oceans*, *117*(C1). Available from https://doi.org/10.1029/2011jc007469.

Yavari-Ramshe, S., & Ataie-Ashtiani, B. (2016). Numerical modeling of subaerial and submarine landslide-generated tsunami waves—Recent advances and future challenges. *Landslides*, *13*(6), 1325−1368. Available from https://doi.org/10.1007/s10346-016-0734-2. Available from: http://springerlink.metapress.com/app/home/journal.asp?wasp = e39xlqwvtg0yvw9n9h2m&referrer = parent&backto = browsepublicationsresults,328,541;.

Zengaffinen-Morris, T., Urgeles, R., & Løvholt, F. (2022). On the inference of tsunami uncertainties from landslide run-out observations. *Journal of Geophysical Research: Oceans*, *127*(4). Available from https://doi.org/10.1029/2021JC018033. Available from: http://agupubs.onlinelibrary.wiley.com/agu/jgr/journal/10.1002/(ISSN)2169-9291/.

Chapter 16

Tsunami hazard assessment in Chile

Patricio Andrés Catalán[1,2] and Natalia Zamora[3]

[1]*Centro Nacional de Investigación para la Gestión Integrada de Desastres (CIGIDEN), Santiago, Chile,* [2]*Departamento de Obras Civiles, Universidad Tecnica Federico Santa Maria, Valparaiso, Chile,* [3]*Wave Phenomena Group, Barcelona Supercomputing Center, Barcelona, Spain*

16.1 Introduction

The estimation and forecasting of hazards that tsunamis pose for a given region are a key task in the overarching goal of mitigating their effects. Hence, an accurate forecasting ability can be of significant societal relevance. Over the last few years, scientific advances in the related fundamental science, methods and techniques, and the increasing computing power, have enabled the possibility of conducting probabilistic tsunami hazard assessments (PTHA) as an alternative or complement to scenario-based tsunami hazard assessments (SBTHA). Despite these advances, several research gaps remain (Behrens et al., 2021). In this work, improvements of the current status and future challenges of the implementation of PTHA approaches in Chile are presented.

Chile lies in a tsunami-prone area with high potential for tsunami genesis and is exposed to teletsunamis (i.e., far-field). The convergence of several tectonic plates makes the Chilean margin one of the most active and complex seismotectonic settings in the Pacific Basin. It is located on South America's western coast and stretches for around 4000 km from northern Chile ($\sim 18°$S) to the southernmost point of the continent ($\sim 56°$S). The occurrence of earthquakes and tsunamis in Chile is well known, mostly in continental Chile (this understood as the territory comprised between $\sim 18°$S and 42°S), to differentiate from the coastline south of 42°S, which comprises the Chiloé, Guaitecas, and Chonos Archipelagos, and the inner sea fjords. The subduction of the Nazca Plate beneath the South American Plate hosted large earthquakes, about moment magnitude M_w 8.0 or above, on average, every 10 years based on recorded events along the Chilean subduction zone (NGDC/WDS, 2018).

Several of these events have triggered large tsunamis, causing casualties and huge economic losses, such as those from the 1960 M_w 9.5 Valdivia and 2010 M_w 8.8 Maule Earthquakes (Daniell et al., 2017; Soulé, 2014). The 1960 Valdivia Earthquake (Barrientos & Ward, 1990; Ho et al., 2019) remains the largest earthquake ever recorded worldwide and highlighted the large seismic potential of the Chilean margin. However, earthquakes and tsunamis also may occur in Patagonia as a result of the subducting Scotia Plate. Southward of the triple juncture of the Nazca, South American, and Antarctic Plates, seismicity is mostly due to local faults that triggered landslides and occasional local tsunamis (Costa et al., 2020; Easton et al., 2013; Lastras et al., 2013).

This Chilean tectonic context has influenced urban development and planning of coastal areas by accounting for tsunamis as relevant factors. The 2010 Maule Earthquake and Tsunami (Fritz et al., 2011) highlighted the need and urgency to assess earthquake and tsunami hazards with different methods and perspectives. Although the seismic hazard is well known and has led to robust seismic codes, engineering practice, and building procedures, until recently the tsunami design codes and practice did not have the same level of comprehension nationwide, in part affected by the 50-year gap between damaging tsunami events (Soulé, 2014). While many studies had been conducted for decades following the 1960 event, especially abroad, local advancement did not have the same rate of progress, aside from the efforts done by the Chilean Navy and its Hydrographic and Oceanic Service (SHOA, by its acronym in Spanish), the Office of Emergency of the Ministry of Interior (ONEMI), and a small group of experts and scientists. The SHOA is the office responsible for tsunami assessment in accordance with Chilean laws, whereas the ONEMI is responsible for postdisaster management. The ONEMI has recently transformed into the National Service for Disaster Prevention (SENAPRED), which highlights a shift in focus from emergency to preparedness. Only after the 2010 systemic failure, a reassessment of the tsunami early warning system began, with a strong focus on the understanding of tsunami propagation patterns and inundation processes that have prompted a rapid development of tsunami science and engineering in Chile. The 2014 Iquique (Catalán et al., 2015) and 2015 Illapel Earthquakes (Aránguiz et al., 2016; Contreras-López et al., 2016) as well as several other small-scale tsunamis, such as the 2016 Melinka Earthquake (Melgar et al., 2017), the 2017 Valparaiso sequence (Nealy et al., 2017), and other events in the range M_w 6.7–7.0 generated along the Chilean margin since 2015, have provided invaluable experience and learning opportunities, which fortunately reduced the cost of lives. A summary of the history of past tsunamigenic events is shown in Fig. 16.1.

Tsunami hazard assessments are complex owing to the multiple generation mechanisms that need to be accounted for and likewise for the eventually complex dependences resulting from the interaction of tsunamis and morphological features along their propagation paths. Besides, the time frame for the assessment also constrains the problem. In the case of early warning (or near real-time assessments), it is assumed that at least some information on the characteristics of the originating source is available, which reduces the level of uncertainty. In such a situation, the aleatory component of uncertainty, originating from the physical process itself and its natural variability, is reduced. However, epistemic uncertainty, related to inaccuracy or lack of knowledge, can remain significant, especially with regard to the tsunami. Cienfuegos et al. (2018) and MacInnes et al. (2013), among others, have shown that even with a wide range of methods and time to assess them, there are different solutions of the source of a given earthquake. Even small variability associated with the estimates at the source can have significant effects while modeling tsunami propagation and inundation. As it will be reviewed below, addressing this uncertainty within the time frame of the emergency is essential.

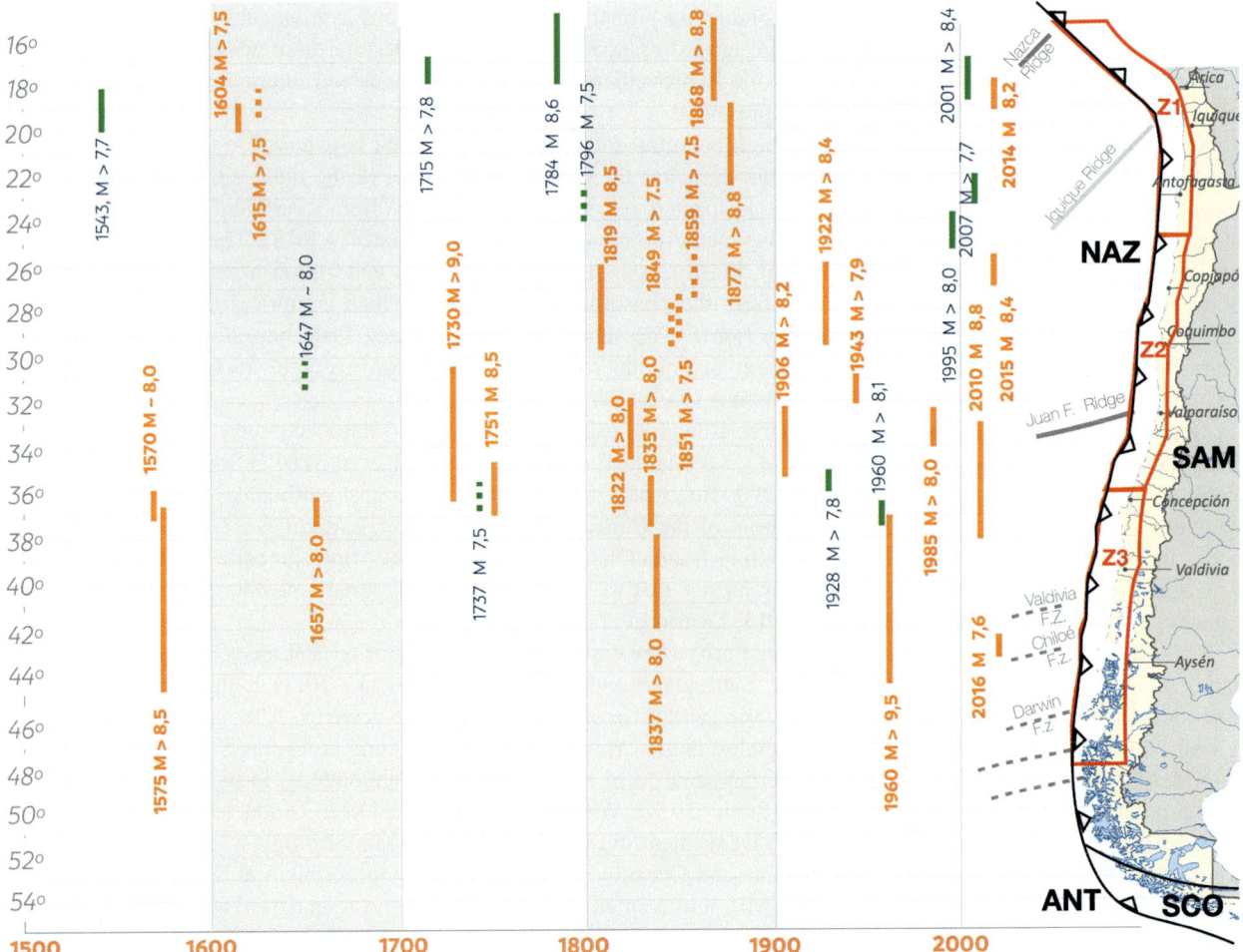

FIGURE 16.1 Timeline of historical earthquakes along the Chilean margin, based on Lomnitx (2004) and Carena (2011). In orange, events with destructive tsunamis. Lines represent estimated rupture length, and dashed lines indicate events with limited information. To the right, the tectonic setting is presented, highlighting the NAZ, SAM, ANT, and SCO Plates. Red zones denote the segmentation of Poulos et al. (2019). *ANT*, Antarctic; *NAZ*, Nazca; *SAM*, South American; *SCO*, Scotia.

Long-term studies, on the other hand, ought to consider the aleatory and epistemic uncertainty. Ideally, all possible sources should be considered by assessing local and distant sources, earthquakes of different types (subduction, local faults, outer rise, and tsunami earthquakes), landslides, and volcanoes. Most of these are poorly understood or lack robust evidence to sustain their proper inclusion in the assessment, or it can be done at the expense of considerable uncertainty (Behrens et al., 2021). Therefore, it is common to focus the effort on only the most relevant among them for a given target location. This situation has led to the focus on earthquake sources, typically in subduction zones, where earthquake and tsunami data are used to estimate target scenarios and their recurrences. Seismic events are the source of about 80% of the tsunamis worldwide (Harbitz et al., 2014) and have been the common focus of hazard studies, such as PTHA (González et al., 2009; Grezio et al., 2017; Thio et al., 2007), among many others. Grezio et al. (2017) and Geist and Parsons (2006) provided a thorough review of PTHA. Although Geist and Parsons (2006) referred to PTHA, this terminology has been revisited later, and a distinction is made when the focus is on tsunamis generated by coseismic floor displacements, which is termed seismic PTHA, that is, SPTHA (Selva et al., 2016; Volpe et al., 2019). In the Chilean case, there is little empirical evidence of events other than subduction zone tsunamis causing damage or devastation, with the exception of landslides in the southern region (Carena, 2011; Lomnitz, 2004). As such, seismic tsunamis will be the main focus of this review.

16.2 Tsunami occurrence in Chile

Along the Chilean subduction zone, three major tectonic plates interact. Most of the seismicity results from the Nazca Plate's subduction beneath the South American Plate, whose margin exhibits a wide range of geological features due to the varying plate convergence rates along its length. Remarkably, the Chilean margin has been characterized as highly coupled (Chlieh et al., 2011; Métois et al., 2013; Sippl et al., 2021; Yáñez-Cuadra et al., 2022). Toward the northern section (13°S), the subduction dip angle is relatively shallow, with a narrow continental margin (Bloch et al., 2014). The central Chilean margin is characterized as a flat slab (28°S–33°S) with a gap in the Andean volcanic arc in the Pampean section of the subduction zone in Chile (Contreras-López et al., 2016; Manea et al., 2012; Nikulin et al., 2019; Olsen et al., 2020). Further south, the subduction dip angle steepens, and more frequent and powerful earthquakes have occurred, especially in latitudes limiting the triple point where the two aforementioned plates meet with the Antarctic Plate (45.5°S–46°S). The tectonic complexity of the Chilean margin is further enhanced by the interaction of other microplates and faults in the region. For instance, the Juan Fernández Ridge and the Chile Rise interact with the subduction zone, creating additional seismic activity and geological heterogeneity (Barrientos, 1980; Comte et al., 1986; Ruiz et al., 2017).

Significant efforts have been dedicated to improving the monitoring of the seismicity of the region. Seismological studies (Barrientos, 2018; Sippl et al., 2023) as well as geodetic and geological studies (Molina et al., 2021; Ruiz et al., 2017; Saillard et al., 2017; Sippl et al., 2021) have contributed to characterize seismogenic zones. In this regard, the long margin in Chile exhibits variability in the expected recurrence rates and the maximum magnitudes that can rupture along the interplate seismogenic zone, which will determine the seismic tsunami hazard. The fact that segments can be differentiated along strike and dip has been a matter of discussion without consensus because of the multifactorial geophysical aspects needed to define segments. These segments should recognize common spatial patterns, in particular, attributing to similar seismological, physical, and mechanical aspects (Basili et al., 2013; González et al., 2020; Molina et al., 2021).

The Chilean margin is prone to mega-earthquakes, as described by paleotsunami records and historical data. At least 25 earthquakes with $M_w > 8.0$ have been registered since 1471 along the Chilean coasts. The 1960 M_w 9.5 Valdivia event (38.8°S) is the largest earthquake ever recorded worldwide. Moreover, according to the National Geophysical Data Center (NGDC) of the National Oceanic and Atmospheric Administration in the U.S., at least 168 tsunamis have been registered in Chile since 1562, causing more than 33,000 casualties and about $US 31 billion in economic loss. At least 10 tsunamis have struck the Chilean coasts with wave heights (most likely tsunami runup) between 10 and 60 m.

A series of paleoseismological studies show the recurrence of large events (León et al., 2023; Moernaut et al., 2018). Hence, it is clear that such earthquakes will occur again in the southern region, with recurrence rates of about 285 years on average (Cisternas et al., 2005). Similar behavior may be expected in northern and central Chile, where large earthquakes of $M_w > 9.0$ have occurred, as evidenced in the stratigraphic records in the Atacama and the Valparaíso regions (DePaolis et al., 2021; Dura et al., 2015; Goff et al., 2022; Salazar et al., 2022). This updated estimate will urge a reassessment of the tsunami hazard in northern Chile, as most analyses to date considered the events of 1868 and 1877 as the largest (M_w 8.8–9.0) (Comte & Pardo, 1991; Hayes et al., 2014; Monge & Mendoza, 1993; Okal et al., 2006). Revisiting the upper end of the catalog is not uncommon as methods and techniques evolve and new or more evidence is acquired.

The 1730 Valparaíso event could have ruptured with M_w 9.0−9.3 and generated a tsunami that was registered in several coastal areas of Chile and Japan (Carvajal, Cisternas, & Catalán, 2017; Dura et al., 2015; Ruiz et al., 2017;

Udias et al., 2012). However, it was not until 2012 that this event was included in the tsunami assessment (SHOA, 2012), as previous studies were based on the devastating earthquake that occurred near Valparaíso in 1906. Now, the 1906 event is understood to be relatively deep, thus explaining the large damage due to the earthquake but with a relatively small tsunami (Carvajal, Cisternas & Gubler, Catalán, et al., 2017). Besides, paleoseismic data suggest that mega-earthquakes like the 1730 event could have shorter recurrence times than expected from the historical records (NGDC/WDS, 2018), thereby the estimation of earthquake recurrence and the tsunami hazard for mid-to-long-term studies needs to be reassessed. This situation is also present along the Atacama region based on very recent paleotsunami findings (Goff et al., 2022; Salazar et al., 2022) and in the central-southern Chile (Hocking et al., 2021) where more frequent occurrence of tsunami events may result from combining geological and historical records in tsunami likelihood estimates.

Much less information exists for the knowledge of tsunamis generated by submarine landslides along the continental margin (north of latitude 42°S) (Harbitz et al., 2014; Völker et al., 2009). Völker et al. (2012) performed an analysis of bathymetry and identified 62 candidates for submarine landslides. Among those, the Reloca slide was massive, and it could have generated a large tsunami (Contreras-Reyes et al., 2016). In the south of 42°S, the fjords–archipelago region is susceptible to subaerial landslides that could trigger tsunamis, such as the Aysen event in 2007. This area is sparsely populated, with insufficient instrumentation to characterize recurrence rates. Finally, a few meteotsunamis have been recorded, but their amplitudes usually fall into the instrumental range, with no damaging effects on coastal settlements (Carvajal, Contreras-López, et al., 2017).

16.3 Tsunami hazard assessments in Chile

Despite its large concentration of population and economic relevance, and being one of the most tsunamigenic regions in the world, to date there is no integrated tsunami hazard assessment in Chile that accounts for the probabilities of occurrence of tsunamis. However, after the 2010 Maule Tsunami, tsunami science has evolved rapidly in Chile owing to the need to enhance preparedness. This was done first by updating the early warning system, followed by the need to improve coastal planning. However, these scientific advances have been adapted only partially in standard procedures, owing to various factors. The current status and challenges of the assessment for characterizing the tsunami hazard are reviewed in this section.

The tsunami problem is treated as an initial and boundary value problem. The initial condition is associated with the water surface deformation induced by a tsunamigenic event. Next, propagation is controlled by the characteristics of this initial condition and the bottom boundary, which are the bathymetry and topography. A proper characterization of both is required. Further details can be found in Chapters 2 and 3.

16.3.1 Boundary conditions and local hydrodynamics

Tsunami generation is directly affected by the type of earthquake, size, segmentation, and depth. This is in part related to the characteristics of the subduction zone along which earthquakes rupture. For instance, Bilek (2010) suggested a relation of mega-earthquake occurrence in regions where more materials can be added to the overriding plate. The 1960 M_w 9.5 Valdivia and the 2010 M_w 8.8 Maule Earthquakes ruptured along a region where thick layers of sediment prevail in the incoming Nazca Plate (Heuret et al., 2011). Seismicity relates to physical imprints of subducting plate and the upper plate, which affects the orogeny. The Andes and the coastal range are a consequence of the seismicity (Armijo et al., 2015; Charrier et al., 2007). The high elevation of the Andes and the narrow extent of the valleys at their feet yield large average slopes. In addition, the latitudinally varying climate results in enhanced river discharges toward southern regions, being influenced by the mean river slope (Pepin et al., 2010). This also affects the morphologic imprints, as these rivers transport significant amounts of sediments to the coast, resulting in typically shallow, sandy bays or estuaries. These conditions contribute to controlling exposure to tsunamis, as most of the relevant settlements along the Chilean coast are located at the southward end of bays, thus offering protection to the dominant waves coming from the southern Pacific (Lucero et al., 2017). However, these areas are usually shallower, thus enhancing tsunami shoaling and energy-trapping processes (Geist, 2013).

These rivers have also led to the development of several coastal canyons, which interrupt the continental shelf, defined as the 200 m deep contour isobath and the coastline, and could offer some level of control on the hydrodynamics (Aránguiz, 2012). The northern coastline (north of 25°S) is characterized by relatively narrow continental shelves, which can contribute to some degree of energy trapping (Catalán et al., 2015; Cortés et al., 2017), whereas in the south of 34°S, the continental shelf broadens and could reach up to 60 km (Charrier et al., 2007), resulting in a gently sloping shelf that promotes high levels of energy trapping (Aránguiz et al., 2019). Toward the southern latitudes (south of 42°S), the territory comprises archipelagos and fjords, which may yield different tsunami propagation characteristics.

Energy trapping can lead to standing and edge waves along the coast, as well as tsunami resonance within bays. While some efforts have been made to assess and characterize resonance along the Chilean coasts (Aránguiz et al., 2019; Catalán et al., 2015; Cortés et al., 2017; Zaytsev et al., 2016), the exact balance between the role of the source and the bathymetry is not fully understood. Resonance dominates the response at places, such as Arica, in northern Chile, as well as Coquimbo (Catalán et al., 2015; Cortés et al., 2017) and Talcahuano (Aránguiz et al., 2019; Farreras, 1978). However, the interaction of bay and shelf can lead to mesoscale resonance dominating the response of small bays at periods longer than those of the bay, as in Dichato (Aránguiz et al., 2019). Notably, Dichato was one of the locations more severely affected by the 2010 Maule Tsunami (Martínez et al., 2017; Mas et al., 2012).

The problem is compounded by energy trapping and the generation of propagating waves along the coast, in the form of edge waves, which have been observed to be ubiquitous in Chile (Geist, 2013, 2016, 2018). The occurrence of these can be sensitive to details of the initial free surface deformation leading to nonlinear coupling (Geist, 2013, 2016, 2018). Even though it could be difficult to accurately reproduce edge waves by numerical models (Geist, 2013; Liu et al., 1998), they could be used as a first approximation. Geist (2013) concluded that using simple earthquake characterizations, such as uniform slip models, may underestimate the hazard. However, its estimation would require a larger number of simulations to include the required variability of slip distributions, as well as modeling for an extended period of time. In addition, Aránguiz et al. (2014) showed that even earthquakes in northern Chile can trigger broadside waves that induce relatively large waves in Talcahuano Bay, more than 2000 km away, thus requiring long simulation times.

This poses an operational challenge for tsunami modeling, as it is required to run multiple simulations with distributed slip over a large spatial domain, at least latitude wise, as well as simulations of long duration. To address the details of resonance, high-resolution grids are needed. To a certain extent, these limitations have restricted tsunami hazard estimation efforts to only local sources.

16.3.2 Initial conditions: seismotectonic segmentation along Chile

The first step in the PTHA workflow is to establish the source recurrence rates and parameters. Since the seminal works of Cornell (1968), seismotectonic segmentation has been a key aspect to establish the maximum size of an earthquake and the derivation of recurrence rate models. Thus defining segments based on transdisciplinary criteria, such as geologic, geodetic, paleotsunami, and archeological, can influence hazard estimates. Besides characterizing the earthquake source recurrence, a crucial ingredient is to identify the maximum magnitude along the seismogenic zone. The limits of the maximum expected rupture may give rise to the definition of a segment that differentiates in at least four aspects: (1) maximum magnitude limited by geological barriers, (2) seismicity patterns, (3) asperity distribution, and (4) seismic coupling.

Multiple of these features are used in many seismotectonic segmentations for the Chilean margin. Studies in the early 1980s served as the foundation of the earliest segmentations that were used in probabilistic seismic hazard assessment. To estimate seismic recurrence models for both the interplate and inslab regimes, seismic recurrence models were proposed by Barrientos (1980) and then modified by Leyton et al. (2009). Poulos et al. (2019) used a model similar to the latter to update recurrence rates in Chile. More recently, Saillard et al. (2017) concluded that the spatial variations of frictional properties along the megathrust dictate the tectonic-geomorphological evolution of the coastal zone and limit the extent of seismic ruptures along strike. The latter may be limited by areas of creeping where mostly aseismic deformation occurs and could arrest seismic ruptures. Using a different approach based on a multivariate analysis of gravity anomalies, basal friction, and interplate locking from Global Navigation Satellite System (GNSS) velocities, Molina et al. (2021) presented a seismic segmentation model of the Chilean margin. While Saillard et al. (2017) and Molina et al. (2021) used similar geophysical data, the final models differ, which will, in turn, affect the estimation of the recurrence rates. A summary of the existing models for segmentation is shown in Fig. 16.2.

The characterization of the plausible maximum magnitude in a segment is a matter of debate. This is closely related to the definition of a fixed segment or the estimates of multiple-segment ruptures from models (Molina et al., 2021; Poulos et al., 2019). Molina et al. (2021) described 18 unitary segments along the interplate; Saillard et al. (2017) described five, whereas Poulos et al. (2019) used only three. Evidence of large events can challenge these segmentations. For instance, the results from Salazar et al. (2022) challenge the notion of a boundary near the Mejillones Peninsula (25°S).

One key aspect when conducting SPTHA is to characterize the earthquake occurrence. It is worth noting that most of these studies focused on assessing earthquake hazard and risk and not tsunami hazard (Das et al., 2020; Leyton et al., 2009; Poulos et al., 2019), hence they could put the emphasis on local or regional events. They also used different

422 SECTION | 2 Advanced topics and applications related to probabilistic tsunami hazard and risk analysis

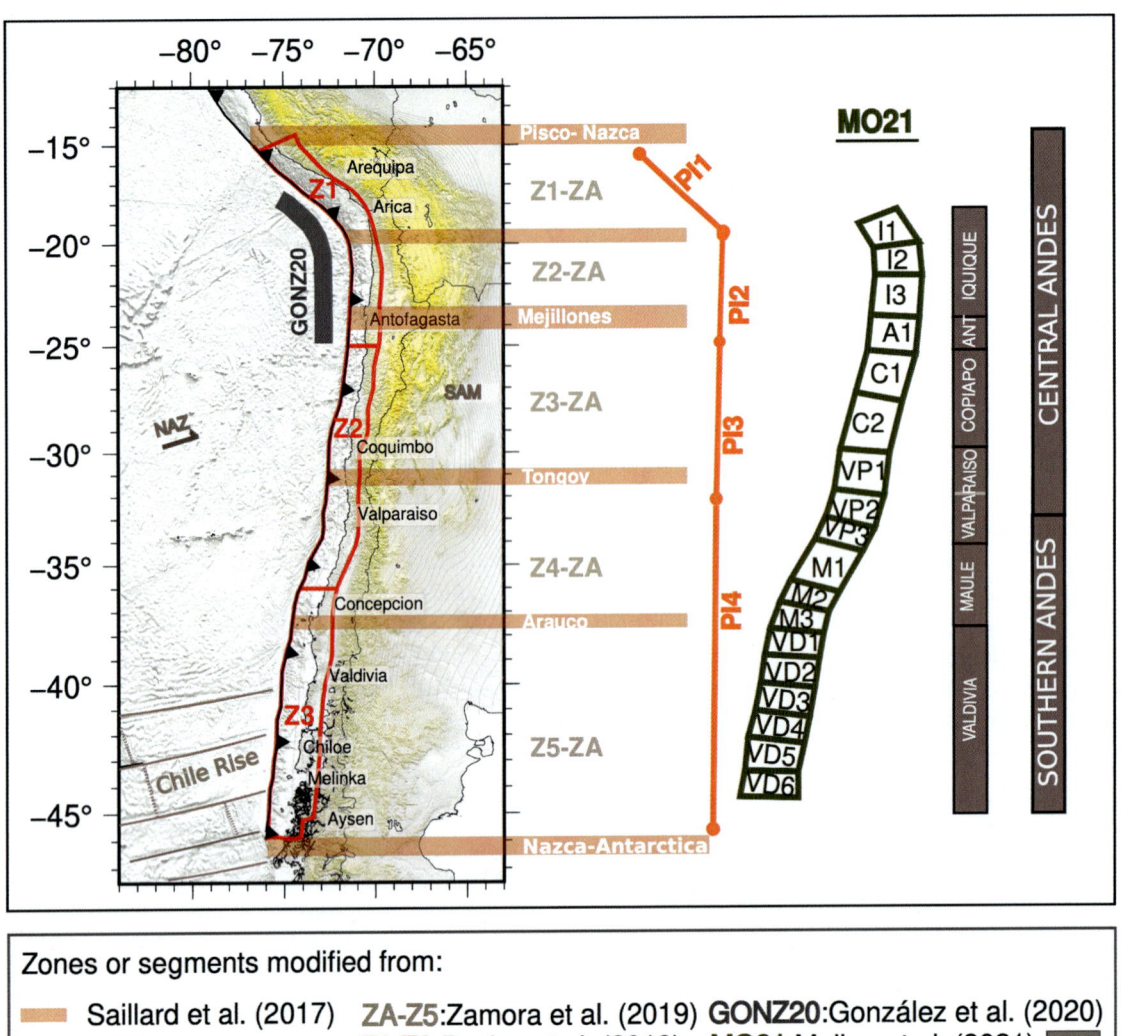

FIGURE 16.2 Summary of different zones, seismotectonic or frictional segmentation along the Chilean margin, up to the Nazca–Antarctica Plate boundary. Map lines delineate study areas and do not necessarily depict accepted national boundaries.

databases for building a catalog. This may omit aspects, such as catalog homogeneity and magnitude standardization (Das et al., 2013), zonation (Hoechner et al., 2016), and completeness (Wiemer & Wyss, 2000) which prompted Carvajal, Cisternas, and Catalán (2017) to further incorporate tsunami data as a way to improve tsunami studies.

Pinilla and Zamora (2019) established a unified catalog for Chile, which used the magnitude as the main descriptor owing to the large range of events in Chile. Orthogonal regressions were used to improve homogeneity following Das et al. (2013), as well as the global regressions from Scordilis (2006) for a sensitivity analysis. Several databases were considered. The first was the database provided by the National Earthquake Information Center. This database does not contain many events along the Chilean margin that ought to be retrieved from other sources. These include the Global Historical Earthquake Catalog (Albini et al., 2013), which was further complemented by the Global Instrumental Earthquake Catalog, originally published by Storchak et al. (2013), expanded with events up to 2017. Newer events were also obtained from the Global Centroid Moment Tensor (http://www.globalcmt.org/CMTsearch.html), considering only events from 2005 due to their more precise timing. To complement the low magnitude range, the catalog from the International Seismological Center (http://www.isc.ac.uk) was used. Finally, a regional catalog was considered to include events in the Andean region. This was elaborated by the Centro Regional de Sismología para América del Sur. The aggregated data were reviewed, with all magnitudes converted to M_w, and eventually updated by reassessing historical earthquakes in terms of location and/or magnitude (Pinilla & Zamora, 2019). To ensure independence among events, the methodology of Gardner and Knopoff (1974) was applied, and the completeness was estimated by the method of Stepp (1972).

The resulting catalog yielded 10,925 earthquakes, spanning from 1471 to 2017 (Pinilla & Zamora, 2019). The impact of catalog choice was assessed by comparing and merging the catalogs of Poulos et al. (2019) and from the South America Risk Assessment by the Global Earthquaque Model (GEM). Different zonations were considered to estimate the maximum possible magnitude and recurrence. First, Poulos et al. (2019) considered three latitudinal zones, and Saillard et al. (2017) defined five, which are shown as PI in Fig. 16.2. For each of these zones, b-values were estimated using OpenQuake algorithms and the Aki-Utsu maximum likelihood method (Aki, 1965; Kijko & Smit, 2012; Utsu, 1965).

Recently, using other statistical techniques, Morales-Yáñez et al. (2022) estimated b-values for central Chile, albeit for instrumental data only and with a smaller maximum magnitude. A Bayesian trans-dimensional approach was employed based on the seismic catalog, giving results that are consistent with the state of stress that was estimated from seismological and geodetic studies (Sippl et al., 2021). Besides, the paleotsunami data provide a larger window in time and space that can be incorporated into the recurrence rate assessment or can be used to validate statistical models. The latter was shown in recent studies to test how reliable paleoseismic records are in forecasting earthquake occurrence (Acuña et al., 2022; Parsons, 2012).

Fig. 16.3 shows the comparison among methods and data, in terms of the maximum estimated for the segment. The b-values show a relatively large variability depending on the database and the methods used. Using these values for forecasting can yield large differences in terms of the return periods for the maximum magnitudes.

16.3.3 Scenario-based tsunami hazard assessment

The main tsunamigenic sources considered correspond to large earthquakes generated along the subduction zone between the Nazca and South American Plates (see the previous sections). However, its location and orientation along the Pacific eastern seaboard make Chile susceptible to tsunamis generated by events of large magnitude along Japan, such as the case of the 2011 Tohoku Tsunami that generated Peak Coastal Tsunami Amplitude (PCTA) of up to 2.20 m at different locations (SHOA, http://www.shoa.cl/s3/servicios/tsunami/data/tsunamis_historico.pdf), or from New Zealand, where tsunami directivity can pose a threat to northern Chile and southern Perú (Goff et al., 2022). Moreover, this directivity could also explain large tsunami intensity metrics along Chile when extreme volcanic events are considered. For instance, the Hunga Tonga-Hunga Ha'Pai Tsunami reached its largest magnitude (with the exception of Tonga itself) along the coasts of Chile (Carvajal et al., 2022).

The SHOA began a systematic effort to prepare tsunami hazard maps in 1997, by using numerical models to estimate the tsunami inundation as posed by a single, reference historical event for each map. Similar deterministic approaches were used elsewhere, for instance, in Italy (Tinti & Armigliato, 2003). In the aftermath of the 2010 event, the SBTHA shifted toward a variant where several seismic sources are considered, and the hazard map is the envelope of the flow depth at any given point. This approach does not consider the recurrence nor likelihood of earthquakes; therefore, they can be treated as a susceptibility map. This is the current procedure, which may be limited by three main factors. First, there is reliance on historical data to constrain the magnitude of the events. For instance, the current hazard map for Viña del Mar and Valparaíso, two of the most populated coastal cities in central Chile, upgraded the previous hazard map that used (using the 1906 Earthquake as reference) to M_w 8.8 in 2012, following an earlier magnitude estimate of the 1730 Earthquake. However, it does not yet capture the current M_w 9.1–9.3 estimate. The second aspect is the reliance on regional events, that is, earthquakes whose rupture zone is in the vicinity of the area of interest. This could be appropriate to estimate the maximum flow depth and inundation as the focus is on evacuation and minimizing loss of life. It could affect the estimation of recurrence in areas that are prone to tsunami amplification even for distant sources. Last but not least, most of the models are based on finite area uniform slip (FAUS), following Davies (2019), where a uniform slip distribution is used, which is estimated using scaling laws relating magnitude to rupture length and width (Blaser et al., 2010; Papazachos et al., 2004).

16.3.4 Seismic probabilistic tsunami hazard assessment

The deterministic or quasi-deterministic SBTHA approach is suitable for determining "worst case" or "worst credible" situations that aim to minimize the loss of life or determine the design of evacuation routes as a fail-safe approach. However, it must be noted that the notion of the worst case is challenged by the use of FAUS, which is known to underestimate tsunami intensity metrics (Davies, 2019).

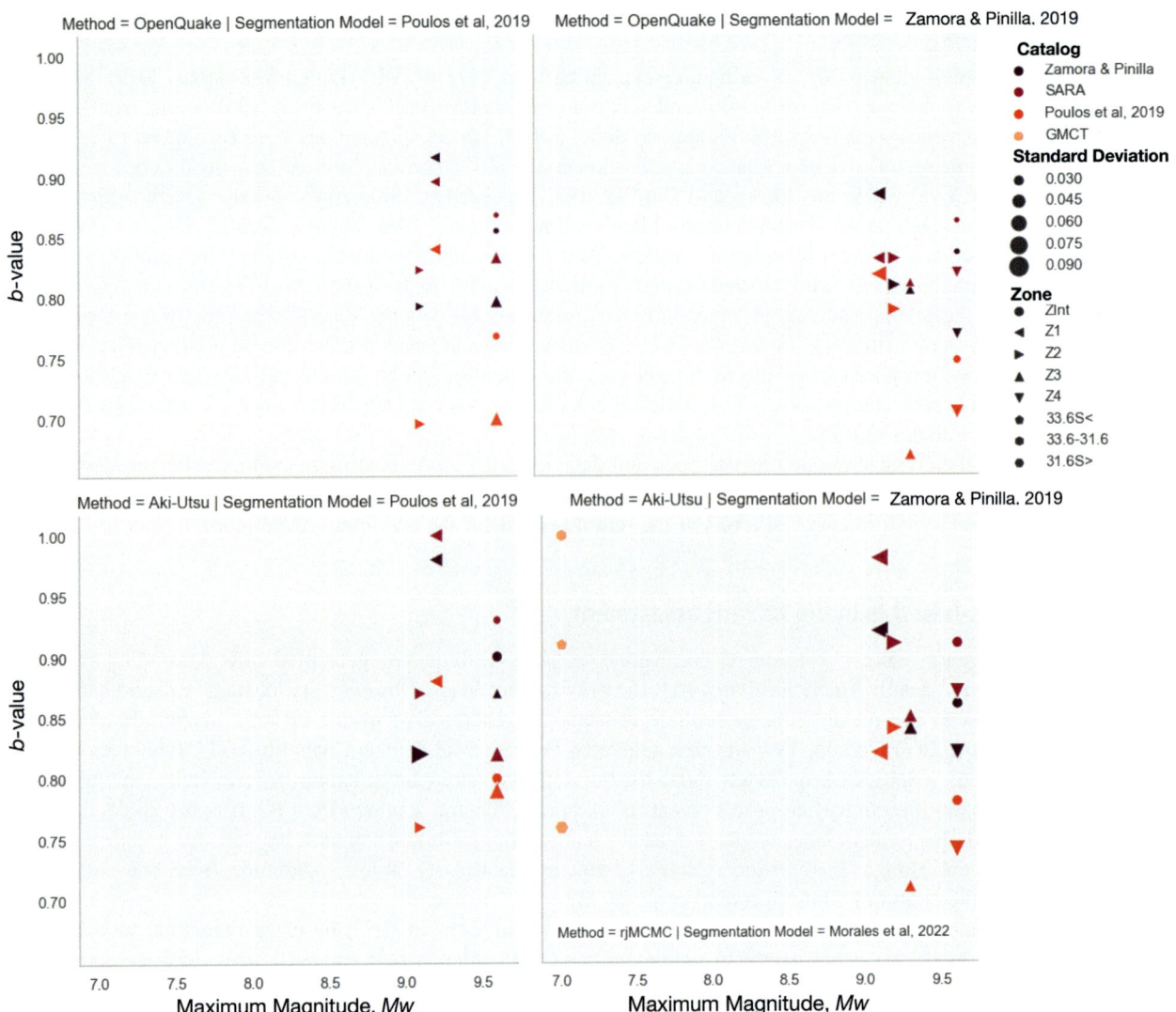

FIGURE 16.3 Comparison of different *b*-values as estimated by Pinilla and Zamora (2019). Plots differentiate between the method of estimation (rows, top to bottom: OpenQuake, Aki-Utsu) and the segmentation used (columns, left to right: Pinilla & Zamora, 2019; Poulos et al., 2019). The results of Morales-Yáñez et al. (2022) using the Markov Chain Monte Carlo trans-dimensional Bayesian approach are inset in the last panel (orange symbols). Colors differentiate among catalogs, and markers by the resulting zonation in latitude. The size of the markers corresponds to the standard deviation of the *b*-value.

Other objectives can benefit from probabilistic analyses. While methods and applications for seismic hazard assessments are ubiquitous, the case of tsunamis poses additional difficulties. Due to both the minimal attenuation of tsunamis along its propagation path and local amplification effects, a complete tsunami hazard assessment might require a large number of non-local sources (Geist & Parsons, 2006). It may also need to account for different source mechanisms. This is especially critical for a coastal zone as exposed as the Chilean one, which is potentially susceptible to tsunamis generated anywhere along the Pacific Ocean. This poses a problem not only due to the need to include events with vastly varying recurrence times, which may require the use of a logic tree, for which the appropriate weights might need to be determined by expert elicitation. In addition, it requires computation at very large spatial scales with enough resolution to capture propagation details both during transoceanic propagation and also when tsunamis generated by subduction earthquakes travel along the Chilean coast.

The key elements of SPTHA are the determination of the appropriate magnitude range, and recurrence rates of the earthquakes to account for the interevent probability. However, due to the large dependence of local hydrodynamics on the details of the earthquake source (Cienfuegos et al., 2018; Davies et al., 2015; Geist & Dmowska, 1999; Ruiz et al., 2015), the inclusion of variability in the slip distribution is required to account for intraevent variability, that is, the aleatory uncertainty remaining even if the magnitude and location of the earthquake are constrained (Li et al., 2016;

Mueller et al., 2015; Sepúlveda et al., 2017). However, the computational cost of this is high, and different approaches have been proposed to reduce its burden, including clustering or filtering (Lorito et al., 2015; Volpe et al., 2019), reduce order models (Sepúlveda et al., 2017), and efficient sampling (Davies et al., 2022).

Only a few examples of applications of SPTHA have been applied to Chile to date. Rara et al. (2010) used a set of 3600 earthquake scenarios where the orientation, dip, and depth of the earthquakes were randomized to generate initial conditions for a tsunami model, which considered inundation. However, no estimation of the recurrence was provided, nor a clear estimate of the susceptibility. Løvholt et al. (2012) provided a global tsunami hazard estimate based on oceanic propagation and amplification factors related to wave shoaling to estimate the inundated area using coarse grids. They considered only a few local events along the Perú—Chile trench (five scenarios with plane fault solutions), as well as regional and far-field events in the Pacific.

Sepúlveda et al. (2017) selected the northern city of Iquique as a study location and applied a stochastic reduced order model (SROM) to better quantify the probabilities at a reduced computational cost. The analysis was concentrated around a small area for the ruptures that could host an M_w 8.0 event. The scenarios were constructed using the Karhunen—Loève expansion (KL) as proposed by LeVeque et al. (2016) and Melgar et al. (2016) to estimate the intraevent variability without including the recurrence. To further reduce the computational cost, only tsunami propagation was considered. Becerra et al. (2020) built on this methodology and performed a local analysis at a single bay considering inundation. Events only in the central Chile segment similar to the along-strike boundaries suggested by Saillard et al. (2017) were included for a single magnitude bin spanning a large window M_w 8.8—9.2. The return period for a given flow depth $\hat{d}$ was estimated following Sepúlveda et al. (2019)

$$T_R = \left[\sum_j^{N_M} \sum_i^{N_s} \lambda_{M_{w,j,x_i}}^{EQ} P_d(d > \hat{d}|M_{w,j,x_i}) \right]^{-1} \quad (16.1)$$

$P_d(d > \hat{d}|M_{w,j,x_i})$ represents the exceedance probability of flow depths being larger than $\hat{d}$ for earthquakes in the jth interval of magnitudes $M_{w,j}$ within the ith zone; thus the intra-event variability. λ^{EQ} is the recurrence rate of earthquakes as obtained from the recurrence analysis (Fig. 16.4). This corresponds to the interevent variability, related to the variations in location and magnitude. These quantities are computed for each of N_M magnitude bins and Ns segmentation zones.

Despite Quintero being a small bay subject to resonance, which somewhat limits the variability, and that a single zone ($N_Z = 1$) was used with a single magnitude bin ($N_M = 1$), the results showed significant variability in the hazard curves depending on whether 50 or 100 samples were used. For example, at return periods longer than 100 years, differences larger than 1 m in flow depth (up to 25%) were common. No specific assessment of the convergence of the results was performed. Hence, the size of the simulation is not yet well constrained for such reduced-order models.

González et al. (2020) tackled the problem in a slightly different way. They also used the same KL expansion to generate 20,000 synthetic earthquakes but to reduce the number to a manageable number in performing the tsunami inundation modeling; they used Welch's t-test to find populations with the same mean value. This is a much simpler method than the SROM approach. While the b-value was also estimated from the catalog of Das et al. (2018) using the Aki-Utsu maximum likelihood method, the focus was given to finding the worst credible scenario rather than a complete SPTHA.

Two recent projects aim to assess the integrated risk of earthquakes and tsunamis. The estimation of the probabilities for both is based on the generation of random scenarios following the method of Crempien et al. (2020) to account for intraevent variability. For each of these, tsunami inundation at San Antonio was estimated using the GeoClaw model (Berger et al., 2011). A similar approach was used to estimate the return periods, using the recurrence rates (Poulos et al., 2019).

16.3.5 Other susceptibility estimation studies

Zamora et al. (2021) carried out a stochastic study aimed at determining the susceptibility of the hazard at two locations in central Chile, the densely populated cities of Valparaíso and Viña del Mar. Here, susceptibility is interpreted as a hazard

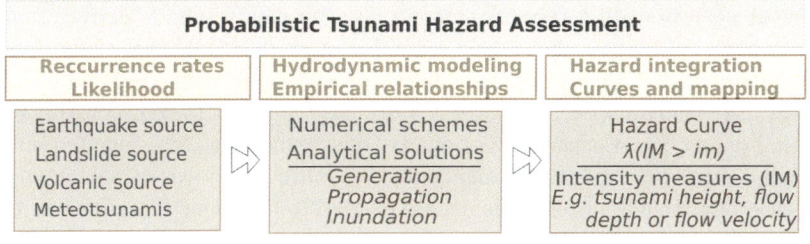

FIGURE 16.4 Probabilistic tsunami hazard assessment general workflow.

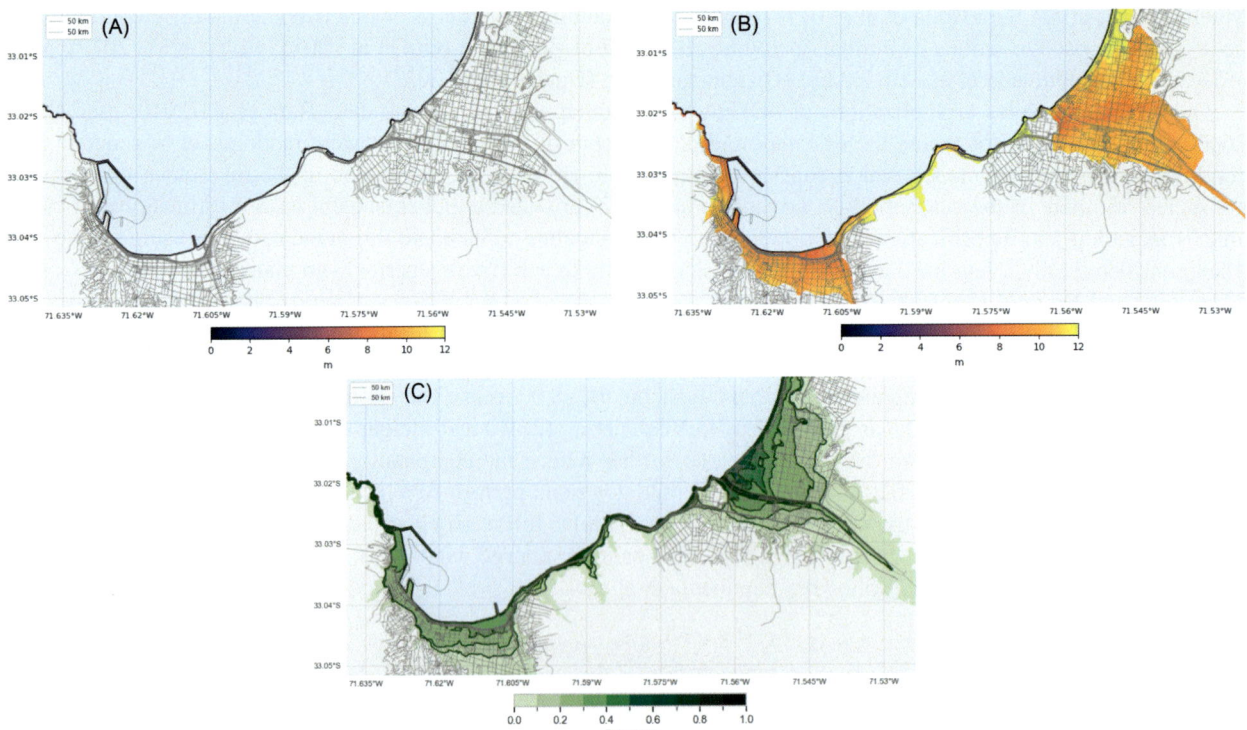

FIGURE 16.5 Variability on inundation simulations. Panels (A) and (B) show sample extreme inundation flow depth maps resulting from specific M_w 9.2 events. Panel (C) shows a spatial map of the percentage of scenarios that inundate any given grid point, as estimated from 400 M_w 9.2 events. Map lines delineate study areas and do not necessarily depict accepted national boundaries.

estimate without including recurrence time. The goal was to establish how likely tsunami inundation is in the area by adopting stochastic scenarios. They also used a single seismogenic zone, Zone 2 (Poulos et al., 2019) and modeled 2800 scenarios in the magnitude range of M_w 8.6–9.2. It was found that among these 2800 scenarios, about 220 (∼8%) induced inundation, the extent of which varied significantly. For instance, Fig. 16.5 shows sample flow depth maps of inundation, as well as the percentage of scenarios that inundate the area out of the 400 scenarios modeled for M_w 9.2. Up to 40 events are capable of inundating most of Viña del Mar (Panel B), but still there are nearly 40 scenarios that do not inundate at all (Panel A). This shows that details of the slip distribution, rather than magnitude, control inundation at this location. This finding is similar to that of González et al. (2020) at Iquique, who found a weak correlation between flow depths and earthquake magnitude or peak slip when comparing all their 400 simulated scenarios (M_w 8.3–8.9).

In addition, events of magnitude as low as M_w 8.6 were capable of inundating as much as events of M_w 9.2, if the slip distribution concentrated over a specific areal patch. Moreover, not only flow depth extremes but also the arrival time of the events varied. Inundating waves could arrive as early as 10 minutes, but also that they could arrive as late as 150 minutes. This is a relevant aspect that needs to be considered in evacuation procedures since the warning needs to determine not only when evacuation is required but also when returning is safe.

16.3.6 Short-term assessment: applications to early warning

While the concept of PTHA pertains to the long-term assessment, some of the methods and techniques can be adapted for tsunami early warning to include a certain degree of variability. The rationale for this stems from the observation that most source inversion approaches are greatly improved when tsunami data are considered. In other words, sources derived from such data are not sufficient to accurately forecast the tsunami, as shown by Cienfuegos et al. (2018). However, for early warning, the goal is to include an estimate of the uncertainty, mainly epistemic, that is present during the assessment.

This problem can be tackled in various ways. A standard approach is to use a database of precomputed scenarios for each of which the hazard is estimated at coastal points. Upon occurrence of an earthquake, the database is queried; ideally, the scenario that best matches the estimated earthquake magnitude and location is retrieved. However, this implicitly assumes that the source details are accurate. To include uncertainties, a set of closely matching scenarios can be considered, which are then treated as the parameter set. Next, the estimation of the hazard can be done in several ways.

The Chilean Tsunami Warning System uses the envelope of the hazard at each coastal point to estimate a hazard distribution that becomes decoupled from the earthquake source but allows for a conservative estimate. Recently, Selva et al. (2021) proposed to determine the probability of a given hazard level constructed from the set scenarios. They termed the approach probabilistic tsunami forecast (PTF), which allows for a formal procedure to establish hazard thresholds that might differ from the envelope, if the decision-makers consider it appropriate. In the case of the Chilean system (Catalan et al., 2020), the envelope is the preferred approach, assuming that source details are lacking in accuracy during the very early stages of the emergency phase (the initial 10 to 20 minutes). It is noted that the envelope approach matched well the 95th percentile obtained with the PTF (Selva et al., 2021).

The use of precomputed databases does not take advantage of the improvements in the estimation of the source that might result in longer observation times. Here, a probabilistic approach can be considered. De Risi and Goda (2017), Goda et al. (2014, 2015), Goda and Abilova (2016), and Mori et al. (2017) improve upon early developments by Mai and Beroza (2002) to characterize the observed source in terms of its spectral characteristics, namely, its radial wavenumber spectrum $\hat{S}(k)$, where $k^2 = k_s^2 + k_d^2$, and k_d and k_s, are the along dip and along strike wavenumbers of the observed source model. Once the nominal spectrum is estimated $\hat{S}$, stochastic realizations can be obtained by sampling the phase spectrum using a uniform distribution between $[-\pi, \pi]$. Next, scenarios departing from the baseline model can be discarded. In the Chilean Tsunami Warning System, a slight variation of the method has been included, whereby the observed phase spectrum is retained. This is then sampled using a unitary normal distribution

$$\phi(k_d, k_s) = \hat{\phi}(k_d, k_s) + \alpha N(0, 1), \tag{16.2}$$

where α is a free parameter that accounts for the level of variability to be considered, and $N(0,1)$ are Gaussian random values with zero mean and unit variance. Under the central limit theorem, the average of a large number of scenarios will resemble the observed source. The method has been tested against the observations from the 2010 Maule and 2015 Illapel Tsunamis, with $\alpha \approx \frac{\pi}{6}$ to $\frac{\pi}{5}$ offering an adequate balance between observed hazard metrics, and the observed variance among predictions from published earthquake sources.

The generation of stochastic scenarios is computationally efficient, allowing for multiple stochastic sources. The computational burden is then transferred to the tsunami modeling. Owing to the very short arrival times of tsunamis in Chile, which can be as low as 10 minutes (Williamson & Newman, 2019), only a limited number of scenarios can be considered. Offline testing showed that modeling tsunami propagation along the entire Chilean coast of 50 random realizations can be obtained in about 15 minutes. These results are then evaluated independently, and the hazard estimates are displayed for the user to estimate the validity of the results. For instance, the hazard categorization that each scenario produces and the hazard envelope among them are shown to operators. In addition, the exceedance probability of PCTA, derived from the simulations, can be displayed to assess the hazard. This allows the operator to establish whether one event drives the hazard estimate or not, for example.

The inclusion of this procedure is still under testing but offers a conservative step in the direction of near real-time assessments without neglecting the epistemic uncertainty present in the process. As an example, Fig. 16.6 shows the

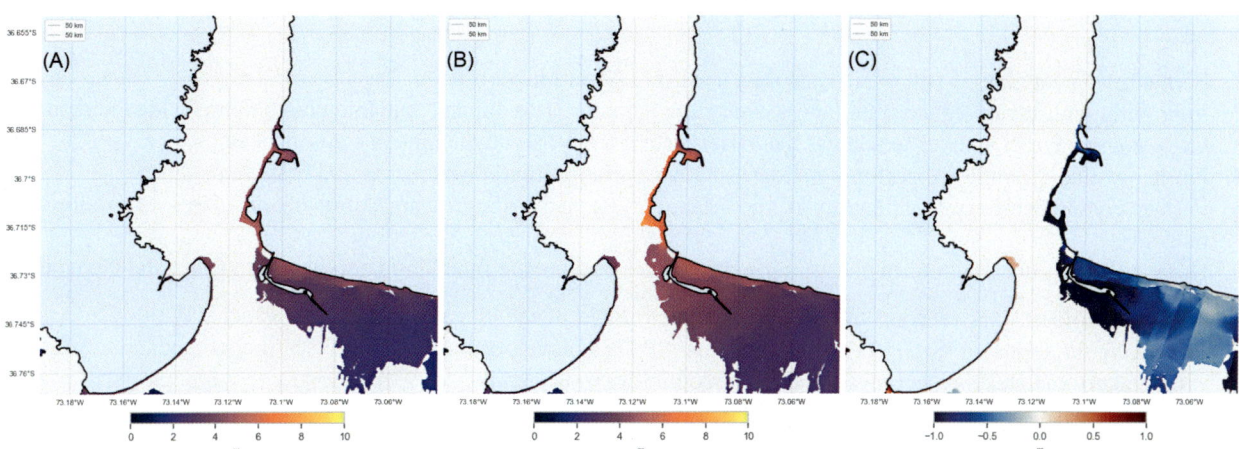

FIGURE 16.6 Comparison of inundation maps at Talcahuano. (A) Median inundation obtained from the 17 models simulated in Cienfuegos et al. (2018); (B) Median of 24 scenarios; (C) Difference between (A) and (B). Map lines delineate study areas and do not necessarily depict accepted national boundaries.

comparison between two inundation maps. On the left, the median inundation flow depth is shown as obtained from independent modeling of the 17 published sources of the 2010 Maule Earthquake used in Cienfuegos et al. (2018), modeled using Tsunami-HySEA (Macías et al., 2017) with nested grids of up to 1 arcsecond resolution. To derive the stochastic assessment, the median of the slip distributions is used as a seed to obtain 50 synthetic earthquakes. The middle panel shows the median of 24 simulations using $\alpha = \frac{\pi}{5}$. Although 24 scenarios are a small number to have adequate convergence, the difference between the target and simulated median inundations does not exceed ± 1 m, with negative values dominating, suggesting a conservative approach, as shown in the right panel.

16.4 Discussion

The review by Behrens et al. (2021) highlighted that there are several gaps in understanding and scarcity of data to constrain PTHA. In this regard, the Chilean situation is not unique. However, there might be some other limiting factors that need to be taken into account. For example, the study of the tsunami hazard at the national scale gained momentum over the last few years. Previously, most of the focus was set to the so-called seismic gaps, most notably in northern Chile (Métois et al., 2013). Scientific advances have brought forward to other areas, such as central Chile (Carvajal, Cisternas, & Catalán, 2017; Morales-Yáñez et al., 2022), the Atacama gap (Yáñez-Cuadra et al., 2022), and others (Lay & Nishenko, 2022).

The short written history of the country (roughly 500 years), and the uneven coverage and evolution of settlements along the coast affect the characterization of the recurrence of earthquakes, especially of very large magnitude events. Though paleosismicity offers a valuable complement, the environmental conditions along the coast have also affected more studies to be based in the southern areas. Regardless, these studies have contributed significantly to characterize the upper tail of the magnitude distribution.

On the other hand, there has been a significant growth of instrumentation both in terms of earthquakes and tsunamis. Critical in this aspect has been the growth of the GNSS network (Báez et al., 2018), which has allowed significant improvements in geodetic coupling analyses and the possible identification of asperities. However, the linear nature of the Chilean coast poses a hurdle to characterizing the shallower section of the megathrust, where the largest tsunamis are produced if the slip concentrates there (Carvajal & Gubler, 2016). Even in light of these advances, discrepancies among the results are common, which indicates that a large epistemic uncertainty remains.

There are other operational problems as well. Even if some of the uncertainty remaining can be tackled using logic trees, for instance, to include different zonations or b-values, the computational cost is large for the reasons explained above, length and number of locations to be detailed, regions with poor bathymetry, and long simulation runs. The needs for computing power are high, and even the use of high-performance computing and GPUs is constrained by a lack of hardware. The National Laboratory for High-Performance Computing (http://www.nlhpc.cl) has only four Graphics Processing Units (GPUs) NVIDIA Tesla V100 for all computing requirements.

It should be stressed that addressing these shortcomings will require a communal and integrated effort, where processes of elicitation, integration, and distributed computing might be required and the scientific community must focus on collaboration. Despite the advances over the last decade, much remains to be done. Among the key tasks, the following can be mentioned:

1. Further understanding of past events and data to inform hazard models.
2. Improving fault identification, fault and source zone parameterization, and tsunamigenic potential characterization.
3. Expand coverage of sources to address the variety, complexity, and dynamics of fault mechanics.
4. Further testing of empirical scaling relations, such as a relationship between area and maximum slip.
5. Further comprehension of stationarity of the seismic cycle and testing of nonstationary processes as an alternative model.
6. Other emerging uncertainties in constraining the rupture or dynamic modeling coupling earthquake and tsunami.

Last but not least, it is relevant to elicit the intended usage of the tsunami hazard assessment. While Chile is currently lacking in terms of SPTHA and PTHA, the SBTHA developed to date has set the grounds for informing decision-makers and stakeholders regarding the susceptibility to tsunamis.

16.5 Conclusions and perspectives

In this chapter, the current status of PTHA in Chile has been reviewed, as well as some of the challenges that remain. A wide range of magnitudes that can occur in Chile, yet the poorly constrained record, and still insufficient understanding of the seismicity with regards to segmentation present a significant challenge for characterizing the recurrence of the tsunamigenic events. Large magnitude events also impose the need to account for slip distributions, which increase significantly the computational efforts.

In addition to these, Chilean geography also poses some challenges. Its long, almost linear coast makes it prone to tsunamis originating from multiple locations. The typically wide continental shelf is efficient in energy trapping and along-shore propagation of tsunami energy, which can reach locations far from the source almost unabated. Next, coastal morphology is efficient in energy trapping and resonance, at specific locations, which requires longer tsunami simulations.

While it is tempting to apply some of the techniques developed elsewhere to the Chilean environment, this has to be done carefully. It is likely that certain elements of uncertainty, such as the seismic segmentation, recurrence rates, or the use of different catalogs ought to be included by utilizing logic trees. This, in turn, will require processes of elicitation as well as a combined effort to model and analyze synthetic tsunamis.

References

Acuña, F., Montalva, G. A., & Melnick, D. (2022). How good is a paleoseismic record of megathrust earthquakes for probabilistic forecasting. *Seismological Research Letters*, *93*(2A), 739–748. Available from https://doi.org/10.1785/0220210044, https://pubs.geoscienceworld.org/ssa/srl/article/93/2A/739/609528/How-Good-is-a-Paleoseismic-Record-of-Megathrust.

Aki, K. (1965). Maximum likelihood estimate of b in the formula $\log_{10} n = a - bm$ and its confidence limits. *Bulletin of Earthquake Research*, *43*, 237–239.

Albini, P., Musson, R. M. W., Capera, A. A., Locati, M., Rovida, A., Stucchi, M., Viganò, D. (2013). *Global historical earthquake archive and catalogue (1000-1903)*. GEM Foundation.

Armijo, R., Lacassin, R., Coudurier-Curveur, A., & Carrizo, D. (2015). Coupled tectonic evolution of Andean orogeny and global climate. *Earth-Science Reviews*, *143*, 1–35. Available from https://doi.org/10.1016/j.earscirev.2015.01.005.

Aránguiz, R. (2012). The effect of a submarine canyon on tsunami propagation in the Gulf of Arauco, Chile. *Proceedings of the Coastal Engineering Conference*. Available from https://doi.org/10.9753/icce.v33.currents.5, https://icce-ojs-tamu.tdl.org/icce/article/view/6727.

Aránguiz, R., Catalán, P. A., Cecioni, C., Bellotti, G., Henriquez, P., & González, J. (2019). Tsunami resonance and spatial pattern of natural oscillation modes with multiple resonators. *Journal of Geophysical Research: Oceans*, *124*(11), 7797–7816. Available from https://doi.org/10.1029/2019jc015206.

Aránguiz, R., González, G., González, J., Catalán, P. A., Cienfuegos, R., Yagi, Y., Okuwaki, R., Urra, L., Contreras, K., Del Rio, I., & Rojas, C. (2016). The 16 September 2015 Chile tsunami from the post-tsunami survey and numerical modeling perspectives. *Pure and Applied Geophysics*, *173*(2), 333–348. Available from https://doi.org/10.1007/s00024-015-1225-4, http://www.springer.com/birkhauser/geo+science/journal/24.

Aránguiz, R., Shibayama, T., & Yamazaki, Y. (2014). Tsunamis from the Arica–Tocopilla source region and their effects on ports of Central Chile. *Natural Hazards*, *71*(1), 175–202. Available from https://doi.org/10.1007/s11069-013-0906-5.

Barrientos, S. E. (1980). *Regionalización sísmica de Chile* (p. 72) Universidad de Chile.

Barrientos, S. E. (2018). National Seismological Center (CSN) Team; The Seismic Network of Chile. *Seismological Research Letters*, *89*(2A), 467–474. Available from https://doi.org/10.1785/0220160195.

Barrientos, S. E., & Ward, S. N. (1990). The 1960 Chile earthquake: Inversion for slip distribution from surface deformation. *Geophysical Journal International*, *103*(3), 589–598. Available from https://doi.org/10.1111/j.1365-246X.1990.tb05673.x.

Basili, R., Tiberti, M. M., Kastelic, V., Romano, F., Piatanesi, A., Selva, J., & Lorito, S. (2013). Integrating geologic fault data into tsunami hazard studies. *Natural Hazards and Earth System Sciences*, *13*(4), 1025–1050. Available from https://doi.org/10.5194/nhess-13-1025-2013.

Becerra, I., Aránguiz, R., González, J., & Benavente, R. (2020). An improvement of tsunami hazard analysis in Central Chile based on stochastic rupture scenarios. *Coastal Engineering Journal*, *62*(4), 473–488. Available from https://doi.org/10.1080/21664250.2020.1812943.

Behrens, J., Løvholt, F., Jalayer, F., Lorito, S., Salgado-Gálvez, M. A., Sørensen, M., Abadie, S., Aguirre-Ayerbe, I., Aniel-Quiroga, I., Babeyko, A., Baiguera, M., Basili, R., Belliazzi, S., Grezio, A., Johnson, K., Murphy, S., Paris, R., Rafliana, I., De Risi, R., ... Vyhmeister, E. (2021). Probabilistic tsunami hazard and risk analysis: A review of research gaps. *Frontiers in Earth Science*, *9*. Available from https://doi.org/10.3389/feart.2021.628772, https://www.frontiersin.org/journals/earth-science.

Berger, M. J., George, D. L., LeVeque, R. J., & Mandli, K. T. (2011). The GeoClaw software for depth-averaged flows with adaptive refinement. *Advances in Water Resources*, *34*(9), 1195–1206. Available from https://doi.org/10.1016/j.advwatres.2011.02.016.

Bilek, S. L. (2010). Invited review paper: Seismicity along the South American subduction zone: Review of large earthquakes, tsunamis, and subduction zone complexity. *Tectonophysics*, *495*(1-2), 2–14. Available from https://doi.org/10.1016/j.tecto.2009.02.037.

Blaser, L., Krüger, F., Ohrnberger, M., & Scherbaum, F. (2010). Scaling relations of earthquake source parameter estimates with special focus on subduction environment. *Bulletin of the Seismological Society of America*, *100*(6), 2914–2926. Available from https://doi.org/10.1785/0120100111Germany, http://www.bssaonline.org/cgi/reprint/100/6/2914.pdf.

Bloch, W., Kummerow, J., Salazar, P., Wigger, P., & Shapiro, S. A. (2014). High-resolution image of the North Chilean subduction zone: Seismicity, reflectivity and fluids. *Geophysical Journal International*, *197*(3), 1744–1749. Available from https://doi.org/10.1093/gji/ggu084, http://gji.oxfordjournals.org/.

Báez, J. C., Leyton, F., Troncoso, C., del Campo, F., Bevis, M., Vigny, C., Moreno, M., Simons, M., Kendrick, E., Parra, H., & Blume, F. (2018). The Chilean GNSS network: Current status and progress toward early warning applications. *Seismological Research Letters*, *89*(4), 1546−1554. Available from https://doi.org/10.1785/0220180011.

Carena, S. (2011). Subducting-plate topography and nucleation of great and Giant earthquakes along the South American Trench. *Seismological Research Letters*, *82*(5), 629−637. Available from https://doi.org/10.1785/gssrl.82.5.629.

Carvajal, M., Cisternas, M., & Catalán, P. A. (2017). Source of the 1730 Chilean earthquake from historical records: Implications for the future tsunami hazard on the coast of Metropolitan Chile. *Journal of Geophysical Research: Solid Earth*, *122*(5), 3648−3660. Available from https://doi.org/10.1002/2017JB014063, http://onlinelibrary.wiley.com/journal/10.1002/(ISSN)2169-9356.

Carvajal, M., Cisternas, M., Gubler, A., Catalán, P. A., Winckler, P., & Wesson, R. L. (2017). Reexamination of the magnitudes for the 1906 and 1922 Chilean earthquakes using Japanese tsunami amplitudes: Implications for source depth constraints. *Journal of Geophysical Research: Solid Earth*, *122*(1), 4−17. Available from https://doi.org/10.1002/2016jb013269.

Carvajal, M., Contreras-López, M., Winckler, P., & Sepúlveda, I. (2017). Meteotsunamis occurring along the southwest coast of South America during an intense storm. *Pure and Applied Geophysics*, *174*(8), 3313−3323. Available from https://doi.org/10.1007/s00024-017-1584-0.

Carvajal, M., & Gubler, A. (2016). The effects on tsunami hazard assessment in Chile of assuming earthquake scenarios with spatially uniform slip. *Pure and Applied Geophysics*, *173*(12), 3693−3702. Available from https://doi.org/10.1007/s00024-016-1332-x, http://www.springer.com/birkhauser/geo + science/journal/24.

Carvajal, M., Sepúlveda, I., Gubler, A., & Garreaud, R. (2022). Worldwide Signature of the 2022 Tonga Volcanic Tsunami. *Geophysical Research Letters*, *49*(6). Available from https://doi.org/10.1029/2022GL098153, http://agupubs.onlinelibrary.wiley.com/hub/journal/10.1002/(ISSN)1944-8007/.

Catalan, P. A., Gubler, A., Cañas, J., Zuñiga, C., Zelaya, C., Pizarro, L., Valdes, C., Mena, R., Toledo, E., & Cienfuegos, R. (2020). Design and operational implementation of the integrated tsunami forecast and warning system in Chile (SIPAT). *Coastal Engineering Journal*, *62*(3), 373−388. Available from https://doi.org/10.1080/21664250.2020.1727402, https://www.tandfonline.com/loi/tcej20.

Catalán, P. A., Aránguiz, R., González, G., Tomita, T., Cienfuegos, R., González, J., Shrivastava, M. N., Kumagai, K., Mokrani, C., Cortés, P., & Gubler, A. (2015). The 1 April 2014 Pisagua tsunami: Observations and modeling. *Geophysical Research Letters*, *42*(8), 2918−2925. Available from https://doi.org/10.1002/2015gl063333.

Charrier, R., Pinto, L., & Rodríguez, M. P. (2007). Tectonostratigraphic evolution of the Andean Orogen in Chile. *Geological Society of London, Chile Geological Society Special Publication*, 21−114. Available from https://doi.org/10.1144/goch.3, http://sp.lyellcollection.org/.

Chlieh, M., Perfettini, H., Tavera, H., Avouac, J.-P., Remy, D., Nocquet, J.-M., Rolandone, F., Bondoux, F., Gabalda, G., & Bonvalot, S. (2011). Interseismic coupling and seismic potential along the Central Andes subduction zone. *Journal of Geophysical Research*, *116*(B12). Available from https://doi.org/10.1029/2010jb008166.

Cienfuegos, R., Catalán, P. A., Urrutia, A., Benavente, R., Aránguiz, R., & González, G. (2018). What can we do to forecast tsunami hazards in the near field given large epistemic uncertainty in rapid seismic source inversions. *Geophysical Research Letters*, *45*(10), 4944−4955. Available from https://doi.org/10.1029/2018GL076998, http://agupubs.onlinelibrary.wiley.com/hub/journal/10.1002/(ISSN)1944-8007/.

Cisternas, M., Atwater, B. F., Torrejón, F., Sawai, Y., Machuca, G., Lagos, M., Eipert, A., Youlton, C., Salgado, I., Kamataki, T., Shishikura, M., Rajendran, C. P., Malik, J. K., Rizal, Y., & Husni, M. (2005). Predecessors of the Giant 1960 Chile earthquake. *Nature*, *437*(7057), 404−407. Available from https://doi.org/10.1038/nature03943, http://www.nature.com/nature/index.html.

Comte, D., Eisenberg, A., Lorca, E., Pardo, M., Ponce, L., Saragoni, R., Singh, S. K., & Suárez, G. (1986). The 1985 Central Chile earthquake: A repeat of previous great earthquakes in the region. *Science*, *233*(4762), 449−453. Available from https://doi.org/10.1126/science.233.4762.449.

Comte, D., & Pardo, M. (1991). Reappraisal of great historical earthquakes in the northern Chile and southern Peru seismic gaps. *Natural Hazards*, *4*(1), 23−44. Available from https://doi.org/10.1007/bf00126557.

Contreras-López, M., Winckler, P., Sepúlveda, I., Andaur-Álvarez, A., Cortés-Molina, F., Guerrero, C. J., Mizobe, C. E., Igualt, F., Breuer, W., Beyá, J. F., Vergara, H., & Figueroa-Sterquel, R. (2016). Field Survey of the 2015 Chile Tsunami with emphasis on coastal wetland and conservation areas. *Pure and Applied Geophysics*, *173*(2), 349−367. Available from https://doi.org/10.1007/s00024-015-1235-2, http://www.springer.com/birkhauser/geo + science/journal/24.

Contreras-Reyes, E., Völker, D., Bialas, J., Moscoso, E., & Grevemeyer, I. (2016). Reloca Slide: An ~ 24 km3 submarine mass-wasting event in response to over-steepening and failure of the central Chilean continental slope. *Terra Nova*, *28*(4), 257−264. Available from https://doi.org/10.1111/ter.12216, http://www.blacksci.co.uk/$\sim$cgilib/jnlpage.bin?Journal = terra&File = terra&Page = aims.

Cornell, C. A. (1968). Engineering seismic risk analysis. *Bulletin of the Seismological Society of America*, *58*(5), 1583−1606. Available from https://doi.org/10.1785/bssa0580051583.

Cortés, P., Catalán, P. A., Aránguiz, R., & Bellotti, G. (2017). Tsunami and shelf resonance on the northern Chile coast. *Journal of Geophysical Research: Oceans*, *122*(9), 7364−7379. Available from https://doi.org/10.1002/2017JC012922, http://onlinelibrary.wiley.com/journal/10.1002/(ISSN)2169-9291.

Costa, C., Alvarado, A., Audemard, F., Audin, L., Benavente, C., Bezerra, F. H., Cembrano, J., González, G., López, M., Minaya, E., Santibañez, I., Garcia, J., Arcila, M., Pagani, M., Pérez, I., Delgado, F., Paolini, M., & Garro, H. (2020). Hazardous faults of South America; compilation and overview. *Journal of South American Earth Sciences*, *104*. Available from https://doi.org/10.1016/j.jsames.2020.102837, http://www.sciencedirect.com/science/journal/08959811.

Crempien, J. G. F., Urrutia, A., Benavente, R., & Cienfuegos, R. (2020). Effects of earthquake spatial slip correlation on variability of tsunami potential energy and intensities. *Scientific Reports*, *10*(1). Available from https://doi.org/10.1038/s41598-020-65412-3, http://www.nature.com/srep/index.html.

Daniell, J. E., Schaefer, A. M., & Wenzel, F. (2017). Losses associated with secondary effects in earthquakes. *Frontiers in Built Environment*, *3*. Available from https://doi.org/10.3389/fbuil.2017.00030, https://www.frontiersin.org/articles/10.3389/fbuil.2017.00030/pdf.

Das, R., Gonzalez, G., de la Llera, J. C., Saez, E., Salazar, P., Gonzalez, J., & Meneses, C. (2020). A probabilistic seismic hazard assessment of southern Peru and Northern Chile. *Engineering Geology*, *271*, 105585. Available from https://doi.org/10.1016/j.enggeo.2020.105585.

Das, R., Wason, H. R., Gonzalez, G., Sharma, M. L., Choudhury, D., Lindholm, C., Roy, N., & Salazar, P. (2018). Earthquake magnitude conversion problem. *Bulletin of the Seismological Society of America*, *108*(4), 1995−2007. Available from https://doi.org/10.1785/0120170157.

Das, R., Wason, H. R., & Sharma, M. L. (2013). General orthogonal regression relations between body-wave and moment magnitudes. *Seismological Research Letters*, *84*(2), 219−224. Available from https://doi.org/10.1785/0220120125.

Davies, G. (2019). Tsunami variability from uncalibrated stochastic earthquake models: Tests against deep ocean observations 2006−2016. *Geophysical Journal International*, *218*(3), 1939−1960. Available from https://doi.org/10.1093/gji/ggz260.

Davies, G., Horspool, N., & Miller, V. (2015). Tsunami inundation from heterogeneous earthquake slip distributions: Evaluation of synthetic source models. *Journal of Geophysical Research: Solid Earth*, *120*(9), 6431−6451. Available from https://doi.org/10.1002/2015jb012272.

Davies, G., Weber, R., Wilson, K., & Cummins, P. (2022). From offshore to onshore probabilistic tsunami hazard assessment via efficient Monte Carlo sampling. *Geophysical Journal International*, *230*(3), 1630−1651. Available from https://doi.org/10.1093/gji/ggac140.

De Risi, R., & Goda, K. (2017). Simulation-based probabilistic tsunami hazard analysis: Empirical and robust hazard predictions. *Pure and Applied Geophysics*, *174*(8), 3083−3106. Available from https://doi.org/10.1007/s00024-017-1588-9.

DePaolis, J. M., Dura, T., MacInnes, B., Ely, L. L., Cisternas, M., Carvajal, M., Tang, H., Fritz, H. M., Mizobe, C., Wesson, R. L., Figueroa, G., Brennan, N., Horton, B. P., Pilarczyk, J. E., Corbett, D. R., Gill, B. C., & Weiss, R. (2021). Stratigraphic evidence of two historical tsunamis on the semi-arid coast of north-central Chile. *Quaternary Science Reviews*, *266*. Available from https://doi.org/10.1016/j.quascirev.2021.107052.

Dura, T., Cisternas, M., Horton, B. P., Ely, L. L., Nelson, A. R., Wesson, R. L., & Pilarczyk, J. E. (2015). Coastal evidence for Holocene subduction-zone earthquakes and tsunamis in central Chile. *Quaternary Science Reviews*, *113*, 93−111. Available from https://doi.org/10.1016/j.quascirev.2014.10.015, http://www.journals.elsevier.com/quaternary-science-reviews/.

Easton, G. V., Rebolledo, S., Sepúlveda, S. A., Lahsen, A., Thiele, R., Townley, B., Padilla, C., Rauld, R., Herrera, M. J., & Lara, M. (2013). Submarine earthquake rupture, active faulting and volcanism along the major Liquiñe−Ofqui Fault Zone and implications for seismic hazard assessment in the Patagonian Andes. *Andean Geology*, *40*(1). Available from https://doi.org/10.5027/andgeoV40n1-a07.

Farreras, S. F. (1978). Tsunami resonant conditions of conceptión bay (Chile). *Marine Geodesy*, *1*(4), 355−360. Available from https://doi.org/10.1080/01490417809387981.

Fritz, H. M., Petroff, C. M., Catalán, P. A., Cienfuegos, R., Winckler, P., Kalligeris, N., Weiss, R., Barrientos, S. E., Meneses, G., Valderas-Bermejo, C., Ebeling, C., Papadopoulos, A., Contreras, M., Almar, R., Dominguez, J. C., & Synolakis, C. E. (2011). Field survey of the 27 February 2010 Chile Tsunami. *Pure and Applied Geophysics*, *168*(11), 1989−2010. Available from https://doi.org/10.1007/s00024-011-0283-5.

Gardner, J. K., & Knopoff, L. (1974). Is the sequence of earthquakes in Southern California, with aftershocks removed, Poissonian. *Bulletin of the Seismological Society of America*, *64*(5), 1363−1367. Available from https://doi.org/10.1785/bssa0640051363.

Geist, E. L. (2013). Near-field tsunami edge waves and complex earthquake rupture. *Pure and Applied Geophysics*, *170*(9-10), 1475−1491. Available from https://doi.org/10.1007/s00024-012-0491-7.

Geist, E. L. (2016). Non-linear resonant coupling of tsunami edge waves using stochastic earthquake source models. *Geophysical Journal International*, *204*(2), 878−891. Available from https://doi.org/10.1093/gji/ggv489, http://gji.oxfordjournals.org/.

Geist, E. L. (2018). Effect of dynamical phase on the resonant interaction among tsunami edge wave modes. *Pure and Applied Geophysics*, *175*(4), 1341−1354. Available from https://doi.org/10.1007/s00024-018-1796-y, http://www.springer.com/birkhauser/geo+science/journal/24.

Geist, E. L., & Dmowska, R. (1999). Local tsunamis and distributed slip at the source. *Pure and Applied Geophysics*, *154*(3-4), 485−512. Available from https://doi.org/10.1007/s000240050241.

Geist, E. L., & Parsons, T. (2006). Probabilistic analysis of tsunami hazards. *Natural Hazards*, *37*(3), 277−314. Available from https://doi.org/10.1007/s11069-005-4646-z.

Goda, K., & Abilova, K. (2016). Tsunami hazard warning and risk prediction based on inaccurate earthquake source parameters. *Natural Hazards and Earth System Sciences*, *16*(2), 577−593. Available from https://doi.org/10.5194/nhess-16-577-2016.

Goda, K., Mai, P. M., Yasuda, T., & Mori, N. (2014). Sensitivity of tsunami wave profiles and inundation simulations to earthquake slip and fault geometry for the 2011 Tohoku earthquake. *Earth, Planets and Space*, *66*(1). Available from https://doi.org/10.1186/1880-5981-66-105, http://rd.springer.com/journal/40623.

Goda, K., Yasuda, T., Mori, N., & Mai, P. M. (2015). Variability of tsunami inundation footprints considering stochastic scenarios based on a single rupture model: Application to the 2011 Tohoku earthquake. *Journal of Geophysical Research: Oceans*, *120*(6), 4552−4575. Available from https://doi.org/10.1002/2014JC010626, http://onlinelibrary.wiley.com/journal/10.1002/(ISSN)2169-9291.

Goff, J., Borrero, J., & Easton, G. (2022). In search of Holocene trans-Pacific palaeotsunamis. *Earth-Science Reviews*, *233*. Available from https://doi.org/10.1016/j.earscirev.2022.104194.

González, F. I., Geist, E. L., Jaffe, B., Kânoğlu, U., Mofjeld, H., Synolakis, C. E., Titov, V. V., Arcas, D., Bellomo, D., Carlton, D., Horning, T., Johnson, J., Newman, J., Parsons, T., Peters, R., Peterson, C., Priest, G., Venturato, A., Weber, J., ... Yalciner, A. (2009). Probabilistic tsunami hazard assessment at Seaside, Oregon, for near- and far-field seismic sources. *Journal of Geophysical Research*, *114*(C11). Available from https://doi.org/10.1029/2008jc005132.

González, J., González, G., Aránguiz, R., Melgar, D., Zamora, N., Shrivastava, M. N., Das, R., Catalán, P. A., & Cienfuegos, R. (2020). A hybrid deterministic and stochastic approach for tsunami hazard assessment in Iquique, Chile. *Natural Hazards*, *100*(1), 231−254. Available from https://doi.org/10.1007/s11069-019-03809-8, http://www.wkap.nl/journalhome.htm/0921-030X.

Grezio, A., Babeyko, A., Baptista, M. A., Behrens, J., Costa, A., Davies, G., Geist, E. L., Glimsdal, S., González, F. I., Griffin, J., Harbitz, C. B., LeVeque, R. J., Lorito, S., Løvholt, F., Omira, R., Mueller, C., Paris, R., Parsons, T., Polet, J., ... Thio, H. K. (2017). Probabilistic tsunami hazard analysis: Multiple sources and global applications. *Reviews of Geophysics*, 55(4), 1158−1198. Available from https://doi.org/10.1002/2017RG000579, http://onlinelibrary.wiley.com/journal/10.1002/(ISSN)1944-9208.

Harbitz, C. B., Løvholt, F., & Bungum, H. (2014). Submarine landslide tsunamis: How extreme and how likely. *Natural Hazards*, 72(3), 1341−1374. Available from https://doi.org/10.1007/s11069-013-0681-3.

Hayes, G. P., Herman, M. W., Barnhart, W. D., Furlong, K. P., Riquelme, S., Benz, H. M., Bergman, E., Barrientos, S., Earle, P. S., & Samsonov, S. (2014). Continuing megathrust earthquake potential in Chile after the 2014 Iquique earthquake. *Nature,*, 512(7514), 295−298. Available from https://doi.org/10.1038/nature13677, http://www.nature.com/nature/index.html.

Heuret, A., Lallemand, S., Funiciello, F., Piromallo, C., & Faccenna, C. (2011). Physical characteristics of subduction interface type seismogenic zones revisited. *Geochemistry, Geophysics, Geosystems*, 12(1). Available from https://doi.org/10.1029/2010gc003230, n/a-n/a.

Ho, T. C., Satake, K., Watada, S., & Fujii, Y. (2019). Source estimate for the 1960 Chile earthquake from joint inversion of geodetic and transoceanic tsunami data. *Journal of Geophysical Research: Solid Earth*, 124(3), 2812−2828. Available from https://doi.org/10.1029/2018JB016996, http://agupubs.onlinelibrary.wiley.com/hub/jgr/journal/10.1002/(ISSN)2169-9356/.

Hocking, E. P., Garrett, Ed, Aedo, D., Carvajal, M., & Melnick, D. (2021). Geological evidence of an unreported historical Chilean tsunami reveals more frequent inundation. *Communications Earth & Environment*, 2(1). Available from https://doi.org/10.1038/s43247-021-00319-z.

Hoechner, A., Babeyko, A. Y., & Zamora, N. (2016). Probabilistic tsunami hazard assessment for the Makran region with focus on maximum magnitude assumption. *Natural Hazards and Earth System Sciences*, 16(6), 1339−1350. Available from https://doi.org/10.5194/nhess-16-1339-2016, http://www.nat-hazards-earth-syst-sci.net/volumes_and_issues.html.

Kijko, A., & Smit, A. (2012). Extension of the Aki-Utsu b-Value estimator for incomplete catalogs. *Bulletin of the Seismological Society of America*, 102(3), 1283−1287. Available from https://doi.org/10.1785/0120110226SouthAfrica, http://www.bssaonline.org/content/102/3/1283.full.pdf + html.

Lastras, G., Amblas, D., Calafat, A. M., Canals, M., Frigola, J., Hermanns, R. L., Lafuerza, S., Longva, O., Micallef, A., Sepúlveda, S. A., Vargas, G., De Batist, M., van Daele, M., Azpiroz, M., Bascuñán, I., Duhart, P., Iglesias, O., Kempf, P., & Rayo, X. (2013). Landslides cause tsunami waves: Insights from Aysén Fjord, Chile. *Eos, Transactions American Geophysical Union*, 94(34), 297−298. Available from https://doi.org/10.1002/2013EO340002.

Lay, T., & Nishenko, S. P. (2022). Updated concepts of seismic gaps and asperities to assess great earthquake hazard along South America. *Proceedings of the National Academy of Sciences*, 119(51). Available from https://doi.org/10.1073/pnas.2216843119.

LeVeque, R. J., Waagan, K., González, F. I., Rim, D., & Lin, G. (2016). Generating random earthquake events for probabilistic tsunami hazard assessment. *Pure and Applied Geophysics*, 173(12), 3671−3692. Available from https://doi.org/10.1007/s00024-016-1357-1, http://www.springer.com/birkhauser/geo + science/journal/24.

Leyton, F., Ruiz, S., & Sepúlveda, S. A. (2009). Preliminary re-evaluation of probabilistic seismic hazard assessment in Chile: From Arica to Taitao Peninsula. *Advances in Geosciences*, 22, 147−153. Available from https://doi.org/10.5194/adgeo-22-147-2009.

León, T., Lau, A. Y. A., Easton, G., & Goff, J. (2023). A comprehensive review of tsunami and palaeotsunami research in Chile. *Earth-Science Reviews*, 236. Available from https://doi.org/10.1016/j.earscirev.2022.104273, http://www.sciencedirect.com/science/journal/00128252.

Li, L., Switzer, A. D., Chan, C. H., Wang, Y., Weiss, R., & Qiu, Q. (2016). How heterogeneous coseismic slip affects regional probabilistic tsunami hazard assessment: A case study in the South China Sea. *Journal of Geophysical Research: Solid Earth*, 121(8), 6250−6272. Available from https://doi.org/10.1002/2016JB013111, http://onlinelibrary.wiley.com/journal/10.1002/(ISSN)2169-9356.

Liu, P. L.-F., Yeh, H., Lin, P., Chang, K.-T., & Cho, Y.-S. (1998). Generation and evolution of edge-wave packets. *Physics of Fluids*, 10(7), 1635−1657. Available from https://doi.org/10.1063/1.869682.

Lomnitz, C. (2004). Major earthquakes of Chile: A historical survey, 1535−1960. *Seismological Research Letters*, 75(3), 368−378. Available from https://doi.org/10.1785/gssrl.75.3.368.

Lorito, S., Selva, J., Basili, R., Romano, F., Tiberti, M. M., & Piatanesi, A. (2015). Probabilistic hazard for seismically induced tsunamis: Accuracy and feasibility of inundation maps. *Geophysical Journal International*, 200(1), 574−588. Available from https://doi.org/10.1093/gji/ggu408.

Lucero, F., Catalán, P. A., Ossandón, Á., Beyá, J., Puelma, A., & Zamorano, L. (2017). Wave energy assessment in the central-south coast of Chile. *Renewable Energy*, 114, 120−131. Available from https://doi.org/10.1016/j.renene.2017.03.076, http://www.journals.elsevier.com/renewable-and-sustainable-energy-reviews/.

Løvholt, F., Glimsdal, S., Harbitz, C. B., Zamora, N., Nadim, F., Peduzzi, P., Dao, H., & Smebye, H. (2012). Tsunami hazard and exposure on the global scale. *Earth-Science Reviews*, 110(1-4), 58−73. Available from https://doi.org/10.1016/j.earscirev.2011.10.002.

MacInnes, B. T., Gusman, A. R., LeVeque, R. J., & Tanioka, Y. (2013). Comparison of earthquake source models for the 2011 Tohoku event using tsunami simulations and near-field observations. *Bulletin of the Seismological Society of America*, 103(2B). Available from https://doi.org/10.1785/0120120121, 1256−1274-1256−1274.

Macías, J., Castro, M. J., Ortega, S., Escalante, C., & González-Vida, J. M. (2017). Performance benchmarking of tsunami-HySEA model for NTHMP's inundation mapping activities. *Pure and Applied Geophysics*, 174(8), 3147−3183. Available from https://doi.org/10.1007/s00024-017-1583-1, http://www.springer.com/birkhauser/geo + science/journal/24.

Mai, P. M., & Beroza, G. C. (2002). A spatial random field model to characterize complexity in earthquake slip. *Journal of Geophysical Research: Solid Earth*, 107(B11). Available from https://doi.org/10.1029/2001jb000588, ESE 10-1-ESE 10-21.

Manea, V. C., Pérez-Gussinyé, M., & Manea, M. (2012). Chilean flat slab subduction controlled by overriding plate thickness and trench rollback. *Geology*, 40(1), 35−38. Available from https://doi.org/10.1130/g32543.1.

Martínez, C., Rojas, O., Villagra, P., Aránguiz, R., & Sáez-Carrillo, K. (2017). Risk factors and perceived restoration in a town destroyed by the 2010 Chile tsunami. *Natural Hazards and Earth System Sciences*, *17*(5), 721–734. Available from https://doi.org/10.5194/nhess-17-721-2017.

Mas, E., Koshimura, S., Suppasri, A., Matsuoka, M., Matsuyama, M., Yoshii, T., Jimenez, C., Yamazaki, F., & Imamura, F. (2012). Developing tsunami fragility curves using remote sensing and survey data of the 2010 Chilean tsunami in Dichato. *Natural Hazards and Earth System Sciences*, *12*(8), 2689–2697. Available from https://doi.org/10.5194/nhess-12-2689-2012.

Melgar, D., LeVeque, R. J., Dreger, D. S., & Allen, R. M. (2016). Kinematic rupture scenarios and synthetic displacement data: An example application to the Cascadia subduction zone. *Journal of Geophysical Research: Solid Earth*, *121*(9), 6658–6674. Available from https://doi.org/10.1002/2016JB013314, http://onlinelibrary.wiley.com/journal/10.1002/(ISSN)2169-9356.

Melgar, D., Riquelme, S., Xu, X., Baez, J. C., Geng, J., & Moreno, M. (2017). The first since 1960: A large event in the Valdivia segment of the Chilean subduction zone, the 2016M7.6 Melinka earthquake. *Earth and Planetary Science Letters*, *474*, 68–75. Available from https://doi.org/10.1016/j.epsl.2017.06.026, http://www.sciencedirect.com/science/journal/0012821X/321-322.

Moernaut, J., Van Daele, M., Fontijn, K., Heirman, K., Kempf, P., Pino, M., Valdebenito, G., Urrutia, R., Strasser, M., & De Batist, M. (2018). Larger earthquakes recur more periodically: New insights in the megathrust earthquake cycle from lacustrine turbidite records in south-central Chile. *Earth and Planetary Science Letters*, *481*, 9–19. Available from https://doi.org/10.1016/j.epsl.2017.10.016.

Molina, D., Tassara, A., Abarca, R., Melnick, D., & Madella, A. (2021). Frictional segmentation of the Chilean megathrust from a multivariate analysis of geophysical, geological, and geodetic data. *Journal of Geophysical Research: Solid Earth*, *126*(6). Available from https://doi.org/10.1029/2020jb020647.

Monge, J., & Mendoza, J. (1993). Study of the effects of tsunami on the coastal cities of the region of Tarapacá, north Chile. *Tectonophysics*, *218*(1–3), 237–246. Available from https://doi.org/10.1016/0040-1951(93)90270-t.

Morales-Yáñez, C., Bustamante, L., Benavente, R., Sippl, C., & Moreno, M. (2022). *B*-Value variations in the Central Chile seismic gap assessed by a Bayesian transdimensional approach. *Scientific Reports*, *12*(1). Available from https://doi.org/10.1038/s41598-022-25338-4.

Mori, N., Mai, P. M., Goda, K., & Yasuda, T. (2017). Tsunami inundation variability from stochastic rupture scenarios: Application to multiple inversions of the 2011 Tohoku, Japan earthquake. *Coastal Engineering*, *127*, 88–105. Available from https://doi.org/10.1016/j.coastaleng.2017.06.013, http://www.elsevier.com/inca/publications/store/5/0/3/3/2/5/.

Mueller, C., Power, W., Fraser, S., & Wang, X. (2015). Effects of rupture complexity on local tsunami inundation: Implications for probabilistic tsunami hazard assessment by example. *Journal of Geophysical Research: Solid Earth*, *120*(1), 488–502. Available from https://doi.org/10.1002/2014jb011301.

Métois, M., Socquet, A., Vigny, C., Carrizo, D., Peyrat, S., Delorme, A., Maureira, E., Valderas-Bermejo, M.-C., & Ortega, I. (2013). Revisiting the North Chile seismic gap segmentation using GPS-derived interseismic coupling. *Geophysical Journal International*, *194*(3), 1283–1294. Available from https://doi.org/10.1093/gji/ggt183.

Nealy, J. L., Herman, M. W., Moore, G. L., Hayes, G. P., Benz, H. M., Bergman, E. A., & Barrientos, S. E. (2017). 2017 Valparaíso earthquake sequence and the megathrust patchwork of central Chile. *Geophysical Research Letters*, *44*(17), 8865–8872. Available from https://doi.org/10.1002/2017GL074767, http://onlinelibrary.wiley.com/journal/10.1002/(ISSN)1944-8007/issues?year = 2012.

NGDC/WDS. (2018). *National geophysical data center* https://www.ngdc.noaa.gov/.

Nikulin, A., Bourke, J. R., Domino, J. R., & Park, J. (2019). Tracing geophysical indicators of fluid-induced serpentinization in the Pampean flat slab of central Chile. *Geochemistry, Geophysics, Geosystems*, *20*(9), 4408–4425. Available from https://doi.org/10.1029/2019gc008491.

Okal, E. A., Borrero, J. C., & Synolakis, C. E. (2006). Evaluation of tsunami risk from regional earthquakes at Pisco, Peru. *Bulletin of the Seismological Society of America*, *96*(5), 1634–1648. Available from https://doi.org/10.1785/0120050158.

Olsen, K. M., Bangs, N. L., Tréhu, A. M., Han, S., Arnulf, A., & Contreras-Reyes, E. (2020). Thick, strong sediment subduction along south-central Chile and its role in great earthquakes. *Earth and Planetary Science Letters*, *538*. Available from https://doi.org/10.1016/j.epsl.2020.116195, http://www.sciencedirect.com/science/journal/0012821X/321-322.

Papazachos, B. C., Scordilis, E. M., Panagiotopoulos, D. G., Papazachos, C. B., & Karakaisis, G. F. (2004). Global relations between seismic fault parameters and moment magnitude of earthquakes. *Bulletin of the Geological Society of Greece*, *36*(3). Available from https://doi.org/10.12681/bgsg.16538, ISSN: 2529-1718.

Parsons, T. (2012). Paleoseismic interevent times interpreted for an unsegmented earthquake rupture forecast. *Geophysical Research Letters*, *39*(13). Available from https://doi.org/10.1029/2012gl052275.

Pepin, E., Carretier, S., Guyot, J. L., & Escobar, F. (2010). Specific suspended sediment yields of the Andean rivers of Chile and their relationship to climate, slope and vegetation. *Hydrological Sciences Journal*, *55*(7), 1190–1205. Available from https://doi.org/10.1080/02626667.2010.512868.

Pinilla, D., & Zamora, N. (2019). *Análisis de la influencia del catálogo sísmico en la estimación de la recurrencia de terremotos*. Civil Engineering Thesis, Universidad Técnica Federico Santa María, Valparaíso, Chile (in Spanish)

Poulos, A., Monsalve, M., Zamora, N., & de la Llera, J. C. (2019). An updated recurrence model for Chilean subduction seismicity and statistical validation of its Poisson nature. *Bulletin of the Seismological Society of America*, *109*(1), 66–74. Available from https://doi.org/10.1785/0120170160, https://pubs.geoscienceworld.org/ssa/bssa/article-pdf/109/1/66/4627425/bssa-2017160.1.pdf.

Rara, V., Arango, C., Puncochar, P., Trendafiloski, G., Ewing, C., Podlaha, A., Vatvani, D., & van Ormondt, M. (2010) Probabilistic tsunami model for Chile. In *Proceedings of the 11th international conference on hydroinformatics*.

Ruiz, J. A., Contreras-Reyes, E., Ortega-Culaciati, F., & Manríquez, P. (2017). Mw 6.9 Valparaíso earthquake from the joint inversion of teleseismic body waves and near-field data. *Physics of the Earth and Planetary Interiors*, *279*, 1–14. Available from https://doi.org/10.1016/j.pepi.2018.03.007.

Ruiz, J. A., Fuentes, M., Riquelme, S., Campos, J., & Cisternas, A. (2015). Numerical simulation of tsunami runup in northern Chile based on non-uniform $k-2$ slip distributions. *Natural Hazards*, *79*(2), 1177–1198. Available from https://doi.org/10.1007/s11069-015-1901-9.

Saillard, M., Audin, L., Rousset, B., Avouac, J.-P., Chlieh, M., Hall, S. R., Husson, L., & Farber, D. L. (2017). From the seismic cycle to long-term deformation: Linking seismic coupling and Quaternary coastal geomorphology along the Andean megathrust. *Tectonics*, *36*(2), 241–256. Available from https://doi.org/10.1002/2016tc004156.

Salazar, D., Easton, G., Goff, J., Guendon, J. L., González-Alfaro, J., Andrade, P., Villagrán, X., Fuentes, M., León, T., Abad, M., Izquierdo, T., Power, X., Sitzia, L., Álvarez, G., Villalobos, A., Olguín, L., Yrarrázaval, S., González, G., Flores, C., … Campos, J. (2022). Did a 3800-year-old $M_W \sim 9.5$ earthquake trigger major social disruption in the Atacama Desert? *Science Advances*, *8*(14). Available from https://doi.org/10.1126/sciadv.abm2996, https://www.science.org/doi/10.1126/sciadv.abm2996.

Scordilis, E. M. (2006). Empirical global relations converting MS and mb to moment magnitude. *Journal of Seismology*, *10*(2), 225–236. Available from https://doi.org/10.1007/s10950-006-9012-4.

Selva, J., Lorito, S., Volpe, M., Romano, F., Tonini, R., Perfetti, P., Bernardi, F., Taroni, M., Scala, A., Babeyko, A., Løvholt, F., Gibbons, S. J., Macías, J., Castro, M. J., González-Vida, J. M., Sánchez-Linares, C., Bayraktar, H. B., Basili, R., Maesano, F. E., … Amato, A. (2021). Probabilistic tsunami forecasting for early warning. *Nature Communications*, *12*(1). Available from https://doi.org/10.1038/s41467-021-25815-w.

Selva, J., Tonini, R., Molinari, I., Tiberti, M. M., Romano, F., Grezio, A., Melini, D., Piatanesi, A., Basili, R., & Lorito, S. (2016). Quantification of source uncertainties in seismic probabilistic tsunami hazard analysis (SPTHA). *Geophysical Journal International*, *205*(3), 1780–1803. Available from https://doi.org/10.1093/gji/ggw107.

Sepúlveda, I., Liu, P. L. F., & Grigoriu, M. (2019). Probabilistic Tsunami hazard assessment in South China sea with Consideration of uncertain earthquake characteristics. *Journal of Geophysical Research: Solid Earth*, *124*(1), 658–688. Available from https://doi.org/10.1029/2018JB016620, http://agupubs.onlinelibrary.wiley.com/hub/jgr/journal/10.1002/(ISSN)2169-9356/.

Sepúlveda, I., Liu, P. L. F., Grigoriu, M., & Pritchard, M. (2017). Tsunami hazard assessments with consideration of uncertain earthquake slip distribution and location. *Journal of Geophysical Research: Solid Earth*, *122*(9), 7252–7271. Available from https://doi.org/10.1002/2017JB014430, http://onlinelibrary.wiley.com/journal/10.1002/(ISSN)2169-9356.

SHOA (2012). *SHOA Carta de inundación por tsunami (CITSU)* [Unpublished content]. de Valparaíso y Viña del Mar. Available in https://www.shoa.cl/php/citsu.php.

Sippl, C., Moreno, M., & Benavente, R. (2021). Microseismicity appears to outline highly coupled regions on the Central Chile megathrust. *Republic Journal of Geophysical Research: Solid Earth*, *126*(11). Available from https://doi.org/10.1029/2021JB022252, http://agupubs.onlinelibrary.wiley.com/hub/jgr/journal/10.1002/(ISSN)2169-9356/.

Sippl, C., Schurr, B., Münchmeyer, J., Barrientos, S., & Oncken, O. (2023). The Northern Chile Forearc constrained by 15 years of permanent seismic monitoring. *Journal of South American Earth Sciences*, *126*, 104326. Available from https://doi.org/10.1016/j.jsames.2023.104326.

Soulé, B. (2014). Post-crisis analysis of an ineffective tsunami alert: The 2010 earthquake in Maule, Chile. *Disasters*, *38*(2), 375–397. Available from https://doi.org/10.1111/disa.12045.

Stepp, J. C. (1972). Analysis of completeness of the earthquake sample in the Puget Sound area and its effect on statistical estimates of earthquake hazard. In *Proceedings of the 1st International Conference on Microzonation* (Vol. 2, pp. 897–910).

Storchak, D. A., Di Giacomo, D., Bondar, I., Engdahl, E. R., Harris, J., Lee, W. H. K., Villasenor, A., & Bormann, P. (2013). Public release of the ISC-GEM global instrumental earthquake catalogue (1900-2009). *Seismological Research Letters*, *84*(5), 810–815. Available from https://doi.org/10.1785/0220130034.

Thio, H. K., Somerville, P., & Ichinose, G. (2007). Probabilistic analysis of strong ground motion and tsunami hazards in Southeast Asia. *Journal of Earthquake and Tsunami*, *02*, 1–19.

Tinti, S., & Armigliato, A. (2003). The use of scenarios to evaluate the tsunami impact in southern Italy. *Marine Geology*, *199*(3-4), 221–243. Available from https://doi.org/10.1016/s0025-3227(03)00192-0.

Udias, A., Madariaga, R., Buforn, E., Munoz, D., & Ros, M. (2012). The large Chilean historical earthquakes of 1647, 1657, 1730, and 1751 from contemporary documents. *Bulletin of the Seismological Society of America*, *102*(4), 1639–1653. Available from https://doi.org/10.1785/0120110289.

Utsu, K. (1965). A method for determining the value of b in a formula $\log n = a - bm$ showing the magnitude–frequency relation for earthquakes. *Geophysical Bulletin*, *13*, 99–103.

Volpe, M., Lorito, S., Selva, J., Tonini, R., Romano, F., & Brizuela, B. (2019). From regional to local SPTHA: Efficient computation of probabilistic tsunami inundation maps addressing near-field sources. *Natural Hazards and Earth System Sciences*, *19*(3), 455–469. Available from https://doi.org/10.5194/nhess-19-455-2019.

Völker, D., Geersen, J., Behrmann, J. H., & Weinrebe, W. R. (2012). Submarine mass wasting off southern central Chile: Distribution and possible mechanisms of slope failure at an active continental margin. *Submarine Mass Movements and Their Consequences - 5th International Symposium*, 379–389. Available from https://doi.org/10.1007/978-94-007-2162-3_34.

Völker, D., Weinrebe, W., Behrmann, J. H., Bialas, J., & Klaeschen, D. (2009). Mass wasting at the base of the south central Chilean continental margin: The Reloca slide. *Advances in Geosciences*, *22*, 155–167. Available from https://doi.org/10.5194/adgeo-22-155-2009.

Wiemer, S., & Wyss, M. (2000). Minimum magnitude of completeness in earthquake catalogs: Examples from Alaska, the Western United States, and Japan. *Bulletin of the Seismological Society of America*, *90*(4), 859–869. Available from https://doi.org/10.1785/0119990114, http://www.bssaonline.org/.

Williamson, A. L., & Newman, A. V. (2019). Suitability of open-ocean instrumentation for use in near-field tsunami early warning along seismically active subduction zones. *Pure and Applied Geophysics*, *176*(7), 3247–3262. Available from https://doi.org/10.1007/s00024-018-1898-6, http://www.springer.com/birkhauser/geo+science/journal/24.

Yáñez-Cuadra, V., Ortega-Culaciati, F., Moreno, M., Tassara, A., Krumm-Nualart, N., Ruiz, J., Maksymowicz, A., Manea, M., Manea, V. C., Geng, J., & Benavente, R. (2022). Interplate coupling and seismic potential in the Atacama seismic gap (Chile): Dismissing a rigid Andean sliver. *Geophysical Research Letters*, *49*(11). Available from https://doi.org/10.1029/2022GL098257, http://agupubs.onlinelibrary.wiley.com/hub/journal/10.1002/(ISSN)1944-8007/.

Zamora, N., Catalán, P. A., Gubler, A., & Carvajal, M. (2021). Microzoning tsunami hazard by combining flow depths and arrival times. *Frontiers in Earth Science*, *8*. Available from https://doi.org/10.3389/feart.2020.591514.

Zaytsev, O., Rabinovich, A. B., & Thomson, R. E. (2016). A comparative analysis of coastal and open-ocean records of the great Chilean tsunamis of 2010, 2014 and 2015 off the coast of Mexico. *Pure and Applied Geophysics*, *173*(12), 4139–4178. Available from https://doi.org/10.1007/s00024-016-1407-8, http://www.springer.com/birkhauser/geo + science/journal/24.

Chapter 17

Uncertainty in empirical tsunami fragility curves

Fatemeh Jalayer[1] and Hossein Ebrahimian[2]

[1]Department of Risk and Disaster Reduction, University College London, London, United Kingdom, [2]Department of Structures for Engineering and Architecture, University of Naples Federico II, Naples, Italy

17.1 Introduction

Tsunami fragility is a relatively new concept. It was introduced by Koshimura et al. (2009) to quantify the building damage due to tsunamis in the aftermath of the 2004 Indian Ocean Tsunami. In this seminal work, fragility is defined as "structural damage probability or fatality ratio with regard to the hydrodynamic features of tsunami inundation flow, such as inundation depth, current velocity, and hydrodynamic force." As this definition demonstrates, the fragility curve depicts "tsunami consequences" as a function of "tsunami intensity."

The fragility curves are divided into two main categories, namely, analytical and empirical fragility curves. The empirical fragility curves are obtained based on available data from damage and consequences observed in the aftermath of past tsunami events. These data can be obtained based on post-tsunami field surveys, remote sensing techniques, and machine learning (Mas et al., 2020; Tarbotton et al., 2015). The tsunami intensity data can be obtained through numerical modeling and propagation of tsunamis (Koshimura et al., 2009), field surveys (Suppasri et al., 2013), or a combination of both (Mas et al., 2012). Compared to earthquake engineering, the use of tsunami empirical curves is more common. This can be attributed to a few factors, such as the relatively smaller number of studies dedicated to numerical modeling of the effect of tsunami-induced actions on buildings and infrastructure. Other factors are the complexity of modeling the tsunami wave impact on the physical infrastructure and the fact that the tsunami waves hit buildings already subjected to the effects of the triggering event(s). For instance, in the case of seismically triggered tsunamis, the physical infrastructure might be subjected to the earthquake and the ensuing sequence of aftershocks before being hit by the tsunami waves. In fact, gathered information about the observed damage in the aftermath of tsunamis usually encompasses the cumulative damage due to the triggering events (e.g., the main earthquake event and the ensuing seismic sequence). Separating such effects to evaluate the damage caused only by the tsunami is quite complicated. Therefore, it should be kept in mind that the empirical fragility curves developed based on observed damage data can include other effects beyond those caused by the tsunami itself.

Empirical fragility curves are obtained based on data available from different buildings and infrastructure. Therefore, they are classified as fragility curves for a "class" of buildings or physical infrastructure. It should be noted, however, that although tsunami empirical fragility assessment is mostly done for buildings, the concept has also been used for deriving fragility curves for transportation infrastructure (Williams et al., 2020) and industrial facilities (Chua et al., 2021) at tsunami risk. In general, the concept of tsunami fragility can be used for any physical asset at risk, even areas covered by vegetation (e.g., forests, parks, meadows, and agricultural fields), coral reefs, and mangroves (Lahcene et al., 2023). Therefore, in this chapter, the general term "assets at risk" is used to reflect the wider application of the tsunami fragility concept. It is also worth mentioning that the empirical fragility concept itself has broader applications beyond tsunamis. It can be applied to define fragility to any phenomenon as long as its occurrence can be (roughly) described by a homogenous Poisson process, an ordinal damage scale can be defined to describe different levels of damage and a (usually scalar) measure of intensity can be assigned to the phenomenon (Jalayer et al., 2023).

The terms tsunami fragility and tsunami vulnerability are sometimes used interchangeably. According to the United Nations Office for Disaster Risk Reduction (UNDRR) terminology (https://www.undrr.org/drr-glossary/terminology), tsunami vulnerability is defined as "the conditions determined by physical, social, economic and environmental factors or processes which increase the susceptibility of an individual, a community, assets or systems to the impacts of hazards." However, a tsunami fragility curve is defined as a relationship showing the probability of exceeding a specific physical damage level as a function of a tsunami intensity measure (*IM*) (e.g., flow depth and momentum flux). It is evident that vulnerability is a more polyhedric concept and encompasses also the socioeconomic consequences, such as economic and social losses. More specifically, the tsunami vulnerability curve can be defined as the probability of exceeding a specific fatality or economic loss value (often normalized) as a function of a tsunami *IM*. The tsunami vulnerability curves can also be derived based on empirical data, such as the vulnerability curves derived by Koshimura et al. (2009).

There are different ways in which the input data for empirical tsunami assessment can be classified. One possible classification is to distinguish between the *aggregated* versus *point-wise* data. Aggregated data are usually provided for a specific areal extent, for example, an administrative unit or even a city. As the name suggests, the data are provided for a group of specific asset typologies located in a specific area assumed to be lumped together. Aggregated damage data are usually available in terms of percentages of exposed assets (e.g., buildings) located in that specific area exceeding certain damage levels. Aggregated intensity data are usually provided as an average over the areal extent of reference and are reported at a reference point, such as the centroid. The Japanese Ministry of Land, Infrastructure, Transport and Tourism (MLIT)[1] database for buildings damaged by the 2011 Tohoku Earthquake and Tsunami is an example of an extensive database providing aggregated damage information for different building typologies. The data can also be provided in a disaggregated manner, meaning that damage data are provided on a point-wise basis and the intensity data are provided at the location of each surveyed asset. An example of such a database is the one reported by Reese et al. (2011)[2].

The empirical fragility curves obtained based on the tsunamis that occurred in the past two decades show significant variability. Such variability can be attributed to the following (Tarbotton et al., 2015): (1) unclear definition of the damage levels; (2) unclear definition of the typology or class of the asset at risk; (3) post-tsunami survey (damage versus intensity) not being a proper representation of the damage incurred to the class or typology of interest; (4) improper statistical post-processing to obtain the fragility curves; and (5) improper designation of the tsunami-impacted buildings (assets). All these aspects will be discussed in this chapter.

This chapter focuses on the uncertainties in the empirical tsunami assessment. These uncertainties will be associated both with the assessment of damage and consequences and the assessment of the tsunami intensity. Moreover, we will discuss the epistemic uncertainties stemming from the choice of the fragility model, the uncertainty related to the parameters of the fragility model, and the limited available data. The chapter is organized as follows: (1) the definition of empirical fragility curves as fragility for a "class" of buildings is discussed; (2) various methods for the evaluation of empirical fragility curves are discussed; (3) the uncertainty in the evaluation of tsunami intensity is discussed; (4) the uncertainty in the evaluation of the tsunami damage is discussed; and (5) the uncertainties related to the fragility model and its parameters are discussed.

17.2 Definitions and limitations of empirical fragility curves

The tsunami fragility for an asset at risk can be defined as the probability of exceeding a specific damage level as a function of the intensity. For a fragility curve to be well-defined, three essential elements need to be specified:

- Asset at risk
- Measure of intensity
- Damage scale (the scale of the increasing damage levels)

The fragility curves for different damage levels are usually color-coded and represented on the same plot to show the damage progression. An example of fragility curves for various damage levels can be seen in Jalayer et al. (2023). Fragility curves for lower damage levels are associated with higher values (they are positioned to the left), and fragility curves for higher damage levels are associated with lower values (they are positioned to the right).

Sometimes, the terms fragility and vulnerability are used interchangeably. The vulnerability curves are usually used to indicate the socioeconomic consequences. For example, a vulnerability curve can depict the probability of exceeding a certain number of fatalities or the probability of exceeding a certain level of economic losses as a function of *IM*

---

1. https://www.mlit.go.jp/toshi/toshi-hukkou-arkaibu.html
2. https://github.com/eurotsunamirisk/etris_data_and_data_products/tree/main/etris_data

(e.g., see the vulnerability curves in the European Tsunami Risk Service, ETRIS, https://eurotsunamirisk.org/, also available on the EPOS Data Portal, https://www.ics-c.epos-eu.org/).

A fragility or vulnerability curve, in probabilistic terms, is a cumulative distribution function (CDF). This interpretation is common to both analytical and empirical fragility curves. That is, the distinction between the analytical and empirical fragility curves is mainly due to the method used for assessing/gathering the level of damage and the processing of the data and not in the underlying probabilistic model.

17.2.1 Class of asset at risk

The concept of fragility assumes inherently that the occurrence of the tsunami events of interest can be described by a homogenous Poisson process (Der Kiureghian, 2005; Jalayer & Ebrahimian, 2020). The Poisson process is a renewal process, which means that each time a new tsunami event occurs, it will impact the same intact structure. Strictly speaking, the concept of fragility curve applies to a single structure with time-invariant characteristics in the time scale of the interarrival times of tsunami events (of a certain impact).

However, we know that the empirical fragility curves are derived based on data from spatially distributed assets at risk. Therefore, the empirical fragility curves are characterized for a "class" or "typology" of the asset at risk. Applying the fragility curve concept to a class in substance means that different assets belonging to the same class are conceptually replaced by an "average" building representing intra-class or asset-to-asset variability, which is usually quite significant. Hence, data from spatially distributed assets inherently include the asset-to-asset variability within the same class (Miano et al., 2020), reflected in the dispersion in the empirical fragility curve.

It is important to distinguish fragility and vulnerability modeling for a single asset (building and infrastructure) from modeling at a spatial level (neighborhood, city, and regional) for risk mapping purposes. The first category needs to adhere to strict rules regarding data availability (the minimum criteria are defined in building codes and standards). The second one is used for decision-making at an urban or regional level. For example, if we perform safety-checking for a single building (e.g., for structural retrofit purposes), we tend to use the terminology *limit states* instead of damage levels. The limit states have very specific definitions, usually related to the local response of the building and the expected performance (e.g., serviceability, life safety, and near collapse). These definitions are described in the building codes. In summary, different risk modeling scales are associated with different definitions, procedures, and data availability criteria.

The empirical fragility curves are suitable for risk assessment at the city level or regional level. They are not suitable for risk assessment for a single asset. For instance, they are not suitable for safety checking of vertical evacuation structures. These structures need to be designed and assessed based on detailed numerical analysis and data available from in situ structural tests and investigations (Chock, 2016).

17.2.2 Tsunami intensity measure

The tsunami *IM* (or simply "intensity"; e.g., the tsunami flow depth) refers to a parameter (or even a vector of parameters) used to quantify information about the intensity of a tsunami. The *IM* plays the role of an intermediate variable, bridging from the hazard level to the fragility or vulnerability level. The output of a probabilistic tsunami hazard analysis is usually expressed as the annual rate of exceeding increasing *IM* levels. The tsunami intensity may be measured in various ways, such as runup height, flow depth, tsunami height at the coast, current velocity, and momentum flux.

Empirical fragility assessment is usually done based on scalar *IM*s. Flow depth, which is the height of the tsunami relative to the local topography at a specific location inland, is the most frequently used scalar measure of intensity for empirical fragility assessment. However, it might not be an effective representation of the destructive power of the tsunamis, and other parameters, such as the current velocity, debris impact, and scour, can become increasingly more important for higher damage levels. For instance, Charvet et al. (2015) and De Risi et al. (2017) developed bivariate tsunami fragilities, which account for the interaction between the two tsunami *IM*s, flow depth and current velocity (defined as the velocity with which the water particles move).

17.2.3 Tsunami damage scale

The damage scale defines how the range of possible damage is discretized. It usually implies a monotonic progress of damage from no damage to destruction. Sometimes, damage scales for tsunamis may have an extra damage level beyond collapse, known as being washed away, as used in the Japanese MLIT damage scale.

Formally, the damage scale provides information about the set of ordered and monotonically increasing physical damage levels. Normally, damage level zero denotes the "no-damage" threshold. In contrast, the last damage level defines the "total collapse" or "being totally washed away" threshold (e.g., the European Macroseismic Scale (EMS)-98 damage scale (Grünthal, 1998) for damage caused by an earthquake and Japanese MLIT for tsunami-induced damage). The damage parameter, D, is usually a discrete variable. The damage levels are expressed as ordered and increasing thresholds of damage. The vector of damage levels, known also as the *damage scale*, is expressed as $\{D_j, j = 0{:}N\}$, where D_j is the jth *damage level* (threshold) and N is the total number of damage levels considered (depending on the damage scale being used and on the type of hazard, e.g., earthquake, tsunami, and debris flow). Normally, D_0 denotes the *no-damage* threshold, while DS_N defines the total *collapse* or *being totally washed away*. The damage states, denoted by DS, indicate that the incurred damage is between two consecutive damage levels. The damage states are defined to be comprehensive, that is, they cover all the possibilities. They are also exclusive, meaning that a given structure cannot have two distinct damage states simultaneously. Fig. 17.1 illustrates a damage scale, the different damage levels, and damage states. Note that the damage levels are usually color-coded to convey the damage progression.

There are many damage scales available in the literature. The team that gathers or detects the tsunami damage in the aftermath of a destructive tsunami will need to establish a specific damage scale. Up to now, there are no standard damage scales for tsunami empirical fragility assessment. There are almost as many damage scales as the number of databases available for damage from tsunami events. It becomes complicated to compare fragility curves that are derived based on different damage scales, even for the same class of assets.

Ideally, the damage scale is set before commencing the survey. Survey forms should be prepared in advance following specific damage scales (e.g., EMS 98). This is quite important. The team doing the survey needs to receive sufficient training to ensure that they are familiar with the damage scales of choice.

It is possible to merge consecutive damage states; for example, the modified MLIT damage scale (De Risi, Goda, Yasuda, et al., 2017) is distinguished from the original Japanese MLIT damage scale by merging collapse and washed-away damage states together. See Table 17.1 for some examples of the damage scales in the literature.

17.2.4 Representation of empirical fragility curves for a damage scale

For a given damage scale, distinct damage states can be described by mutually exclusive events that span the entire range of possibilities (in terms of damage). Formally, this means that the probability of being in different damage states can be described by a discrete probability distribution (Chua et al., 2021; Jalayer et al., 2023). This representation of

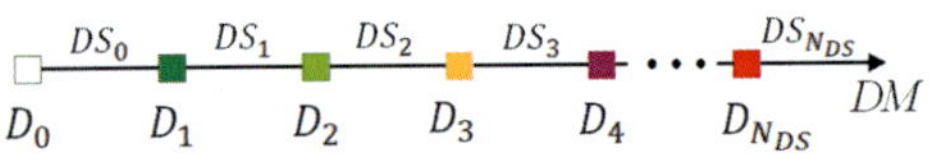

FIGURE 17.1 Graphical representation of damage levels D_j and damage states DS_j, where $j = 0{:}N$.

TABLE 17.1 Damage scales from the literature.

Damage scale	Number of damage levels (Tiers)	References
Peiris 2006	3 (D_0 to D_3)	Rossetto et al. (2007)
Reese 2011	5 (D_0 to D_5)	Reese et al. (2011)
EMS 98 masonry	5 (D_0 to D_5)	Grünthal (1998)
EMS 98 reinforced concrete (RC)	5 (D_0 to D_5)	Grünthal (1998)
MLIT	7 (D_0 to D_7)	https://www.mlit.go.jp/toshi/toshi-hukkou-arkaibu.html
Modified MLIT	6 (D_0 to D_6)	De Risi, Goda, Yasuda, et al. (2017)
Paulik 2019	3 (D_0 to D_3)	Paulik et al. (2019)

MLIT, Ministry of Land, Infrastructure, Transport and Tourism. *EMS*, European Macroseismic Scale.

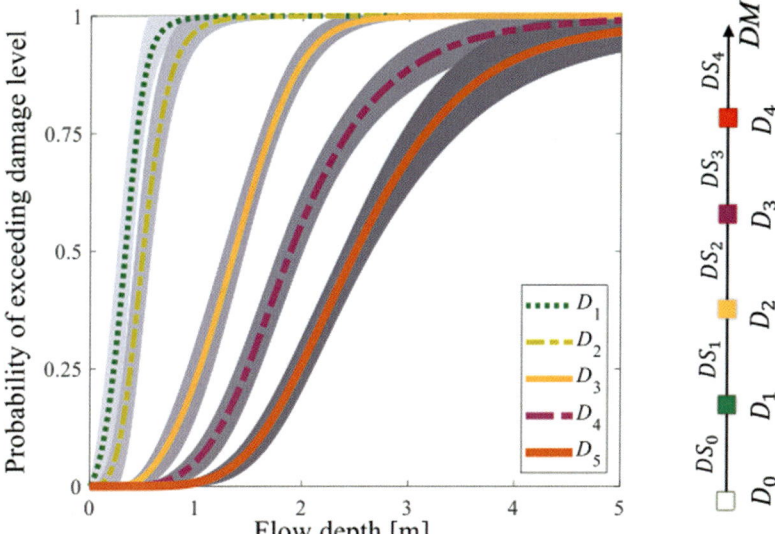

FIGURE 17.2 Fragility curves for different damage levels.

the fragility curves provides the possibility of evaluating the fragility curves for the entire damage scale as an ensemble, with the advantage of ensuring that they are ordered (i.e., do not cross). Fragility curves are usually plotted for different damage levels (for the entire damage scale) on the same figure (Fig. 17.2).

17.3 Methods for empirical tsunami fragility assessment

In the earlier days, the fragility curves were calculated based on simple linear regression methods, where the domain of the tsunami *IM* (the independent variable) was divided into bins. The center of each bin was used as the representative *IM*. For each bin, the ratio or percentage of assets that exceeded the prescribed damage level was estimated. The independent regression variable was usually calculated as the inverse Gaussian of the damage ratio per bin. This was then related to the dependent variable, defined as the natural logarithm of the intensity value representing the bin, through the least squares regression method (Mas et al., 2020; Suppasri et al., 2013). This method lends itself better to aggregated fragility assessment and is not suitable for point-wise assessment since the ratio/percentage of exceeding certain damage levels needs to be estimated.

The empirical fragility curves are, by definition, based on visual observations/surveys of damage. The information provided by the visual surveys usually is just (barely) enough to decide between a set of discrete damage states. However, linear regression is more suitable for numerically quantifiable dependent variables. Therefore, it is more straightforward to use a binary dependent variable for the regression, which assumes the value of one when the prescribed damage level is exceeded and zero otherwise. In this manner, the probability that this variable is equal to one can be related to the natural logarithm of *IM* value through linear regression. This probability can be expressed through alternative functional forms known as "links." Common "link" functions used in the literature are the "probit" link, which has the functional form of a lognormal distribution, the "logit" link, which has a logistic functional form, and the "cloglog," which has a double exponential form. A major advantage of using the generalized regression models is that the *IM* values observed or detected at the position of the damaged assets can be used directly (i.e., in a point-wise manner) without the need for using *IM* bins. The generalized regression models are increasingly used for empirical fragility assessment (Charvet et al., 2014, 2015; Jalayer et al., 2023; Lahcene et al., 2021).

Several issues related to the characterization of the empirical fragility curves include:

Crossing fragility curves. The easiest way of developing empirical fragility curves is to fit them one at a time to each damage level. This results in multiple fragility curves equal to the number of damage levels. One potential issue is to ensure that fragility curves for the different damage levels do not cross. This is important because the probability of being simultaneously in two distinct damage states is zero. This condition is not automatically satisfied when the curves are fitted separately to each damage level. One solution to this problem is to model the fragility curves as "parallel." This translates into using the same dispersion in the fragility curves. Another solution is to model the fragility curves as an ensemble of curves and condition the exceedance of each damage level to the fact that the immediately lower damage level is exceeded; examples are with binned *IM*s (De Risi, Goda, Mori, et al., 2017) and with point-wise *IM*s (Jalayer et al., 2023).

Interclass correlations. The fragility curves for different classes of assets at risk are usually derived for each class of assets at a time. An exception is Miano et al. (2020), where fragility curves for different classes of buildings are modeled together. Fragility modeling for each class at a time may lead to ignoring the possible correlations between the observed damage for different classes of buildings and the correlations between the *IM* values at each asset location for different classes (i.e., interclass correlations). Charvet et al. (2014) used a nested model to take into account the buildings' classes. Apart from possible correlations between damage across different classes, it is important to perform a fragility assessment for all the building classes together when different tsunami inundation scenarios (e.g., different realizations of the spatial distribution of *IM*) are considered. Ideally, the fragility curves for different classes of buildings need to be modeled together as an ensemble to allow for the modeling of possible correlations.

Reference set of assets. Another common issue in the modeling of empirical fragility curves is the definition of the portfolio of exposed assets. That is, how the areal extent of buildings to consider in the survey is defined. This will affect the total number of buildings in the portfolio (i.e., the number of buildings with damage greater than or equal to D_0). In other words, this number is the common denominator for the fragility assessment and the fragility curves may show significant sensitivity to how the spatial extent of the portfolio is defined. A natural way to define the reference set is to delimit the inundation extent and identify the assets that fall within such extent as the reference set.

Missing damage levels. Another common issue is when the damage survey misses some of the damage levels across the scale. This could be potentially problematic when an ensemble modeling approach is used (i.e., the fragility curves for all the damage levels are modeled together). This problem can be solved by creating a "new" damage scale through the union of the damage scales for which specific data are not available. For example, the survey might be lacking data about D_0. In such a case, the new damage scale could start from D_1 or the first damage level for which survey results are available. As a general rule, if the damage scale, for which data are available, consists of N damage levels, this will lead to $N-1$ fragility curves. A simple explanation is that the first damage level serves to define the reference set of assets (domain of assets).

17.3.1 Sources of uncertainties in tsunami intensity characterization

It is more common to think that the uncertainties associated with empirical fragility assessment are those related to damage assessment only. However, one of the main sources of uncertainty in the empirical fragility assessment is associated with the characterization of the tsunami *IM*. In general, one could consider tsunami inundation as an ungauged situation. That is, we do not have gauges and sensors that provide the value of the tsunami *IM*s at the position of each building or infrastructure for which we measure the damage. In other words, tsunami *IM* values at the position of the exposed asset may not be directly observable. Therefore, there can be a significant uncertainty associated with the value of the tsunami *IM* evaluated at the position of each asset/building. One can classify the different ways of estimating the tsunami *IM* at the position of each building into the following groups:

Using observed *IM*. This category applies mainly to measurements of flow depth at the position of each building within the reference set. This type of information can be gathered by measuring the signs of water on the buildings as part of in situ field surveys. In some cases, this information is not available for all the buildings, and spatial averaging or other smoothing techniques can be used to interpolate the flow depth for positions of interest for which observations are not available. Measuring the maximum flow depth from direct in situ observations provides reasonable means; however, such direct measurements are usually not possible for other *IM*s, such as velocity and moment flux. Examples include Lahcene et al. (2021), Paulik et al. (2019), and Reese et al. (2011).

Using simulated *IM*. This method is quite common for tsunami empirical fragility assessment. That is, the tsunami event is modeled from the initial stages of its genesis (most often, these are seismically induced tsunamis), estimation of the dislocation of the sea-bed, deep water propagation of the tsunami waves, and shallow water propagation and inundation of the tsunami waves. Tsunami *IM* estimation through numerical simulation is subjected to many sources of uncertainties, for example, in the tsunami magnitude, source parameters, Manning's constant (friction terms in the shallow water propagation), bathymetry, digital elevation models, presence of buildings and obstacles, and so on. The common method in the literature is to simulate various plausible scenarios and compare them with available observations (not necessarily at the position of the buildings) and identify a scenario that provides a better fit with the observed data. Examples include Aránguiz et al. (2017) and Koshimura et al. (2009).

Smoothing the available observations. Spatial averaging or kernel functional forms can be used for creating a smooth inundation surface based on available observed flow depth. Clearly, in this case, the estimation of flow depth at the positions for which direct observations are not available is subjected to approximations. Examples include Mas et al. (2012, 2020).

17.3.2 Sources of uncertainties related to tsunami damage evaluation

Observed damage is at the heart of empirical fragility assessment and consists of assigning relevant damage labels to buildings/exposed assets in the aftermath of a disaster. The classification of damage is usually done based on a specific damage scale. There are different methods for damage evaluation (Mas et al., 2020).

In situ surveys: Damage data acquisition based on in situ surveys is the most reliable method for damage state assignment. However, it may be complicated by the difficulties in accessing the areas affected by the tsunami and the potential risks to the survey team. Potential sources of uncertainties are related to the subjectiveness of damage level attribution by the survey team and their level of experience. Nevertheless, the margins of error are usually reasonable. Another advantage of in situ surveys is that the building class or typology is assigned based on visual observation of the building. One way to reduce the bias in in situ surveys is to have multiple surveyors assess the same asset. Performing the same survey by more people has various advantages: (1) calibrating the error and formally considering it in the fragility assessment and (2) reducing the dispersion in damage assessment by averaging the results.

Remote sensing: In recent years, it has become very common to use before- and after-tsunami satellite images for damage detection. For example, the Copernicus EMS damage grading maps visualize the different damage states assigned to buildings by comparing satellite images (Fig. 17.3). This method has several advantages: For example, it is the preferred method for damage acquisition in areas that are difficult to reach in post-disaster situations. Moreover,

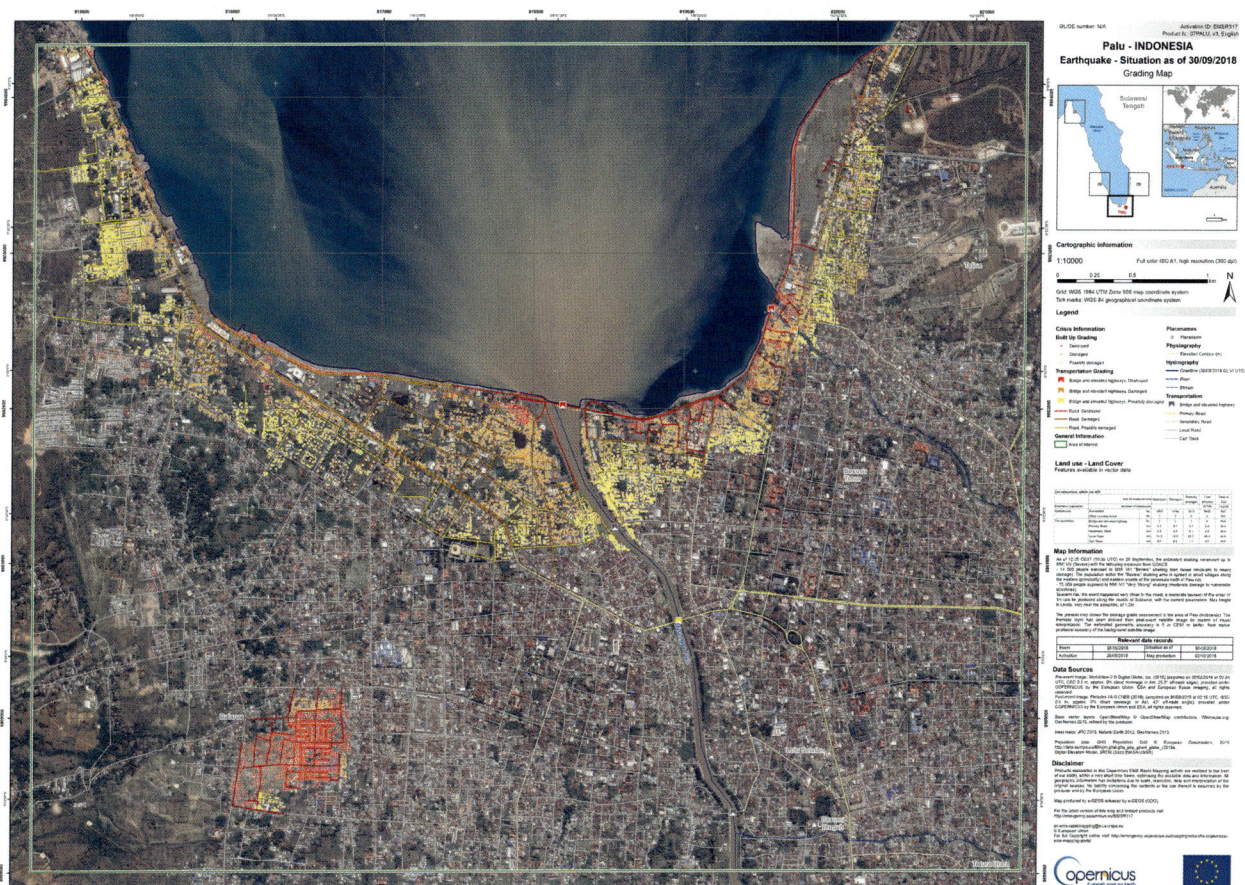

FIGURE 17.3 Damage grading map of Palu city where almost over 37,000 people, 10,000 buildings, and other infrastructure were affected. Map lines delineate study areas and do not necessarily depict accepted national boundaries. *From Copernicus EMS © 2018 EU, [EMSR317] Palu: Grading Map.*

damage acquisition by remote sensing does not jeopardize the safety of the survey team and does not cause potential interference with the rescue and recovery operations. It is, however, prone to errors. First of all, the satellite photos used for damage detection only provide information about the plan view of the buildings. They are more suitable for the ultimate damage levels, such as "collapse" and being "washed away." They are less accurate for lower damage levels as signs of nonstructural and minor damage are harder to detect from the top. Second, they are prone to a degree of subjectiveness and depend on the person who has done the damage level assignment. Last, it is not easy to detect the structural type from the satellite photos. Therefore, remote sensing usually needs to be complemented by other sources of information (e.g., exposure layers and urban morphology maps) to integrate such information.

Artificial intelligence (AI): This method has proven to be a reliable way of acquiring damage data. It uses "before" and "after" photos. It avoids some of the shortcomings mentioned above in in situ surveys and remote sensing. First, the machine assigns the damage states and not human beings, therefore, the "surveyor's bias" in assigning the damage states is significantly reduced. Second, the AI process is usually more versatile and can also examine lateral photos of the building, and this adds sources of information with respect to photos from the top. For the rest, it is similar to the remote sensing process (Mas et al., 2020).

17.3.3 Sources of uncertainties related to the choice of fragility models

The fragility curve is essentially a CDF. The choice of a fragility model signifies finding the best probability model conditioned on the observed data. In this chapter, we focus on the Bayesian methods for finding the best fragility model. However, in the literature, there are several examples of finding the best model through hypothesis testing (e.g., the likelihood test) and the use of the Akaike information criterion (Charvet et al., 2014, 2017).

When the Bayesian methods are used for empirical fragility assessment, usual tests of goodness of fit (for model validation) are not going to be very meaningful. This is because, in Bayesian inference, available data are used to update the model (in this case, the fragility model). Therefore, the equivalent of a test of goodness of fit in the Bayesian inference is to evaluate the evidence provided by the data in favor of a specific model. This procedure is called the Bayesian model class selection.

The Bayesian model class selection is based on the following characteristics. First, the alternative fragility models are assumed to be mutually exclusive and collectively exhaustive. Being mutually exclusive means that only one model will be suitable among the set of candidate models (due to incomplete knowledge, we do not know which one is with certainty). Being collectively exhaustive means that the set of candidate models builds the entire set of possible alternative fragility models. Consequently, the probability or the degree of beliefs assigned to each probability model will sum to unity. Second, it is shown that, roughly speaking, the "best" model is the one that provides the best fit to the observed data with the simplest model (e.g., less number of parameters in the fragility model). This is described in detail in Jalayer et al. (2023) based on an original derivation in Muto and Beck (2008). An application to tsunami hazard modeling is shown in De Risi and Goda (2017).

The best model in this approach is defined as the model that has the largest evidence (or log evidence) among the mutually exclusive and collectively exhaustive set of alternative models, where evidence is defined as the likelihood of observing the data given a specific fragility model. It is shown that this likelihood is equal to the expected value of likelihood over the domain of the fragility model parameters (goodness of fit term) minus the amount of information (on average) that is "gained" about fragility model parameters from the observed data (relative entropy term). Interestingly, the relative entropy term penalizes model complexity; that is, if the model extracts more information from data (which is a sign of being a complex model with more model parameters), the log evidence is reduced. Once the evidence is obtained, it can be used to find the probability/weight/degree of belief assigned to each model. These weights can be used to calculate the ensemble average of the alternative fragility models.

17.3.4 Sources of uncertainties related to fragility model parameters

The reliability of an empirical fragility curve depends also on the number of buildings (or assets at risk) surveyed. The more buildings are included in the survey dataset, the more confidence can be gained in the results. This is an "epistemic type" of uncertainty since it is related to the amount of information we have. It is common to visualize epistemic uncertainties as confidence intervals around the fragility curve. In this sense, the fragility curve can be treated as an average (mean) or median fragility curve. Representing the empirical fragility as "mean fragility" between all possible fragility curves, the confidence interval can be marked as the mean plus and minus a certain number of (e.g., one and two) standard deviations. Here, the standard deviation corresponds to the one for the fragility values; that is, the

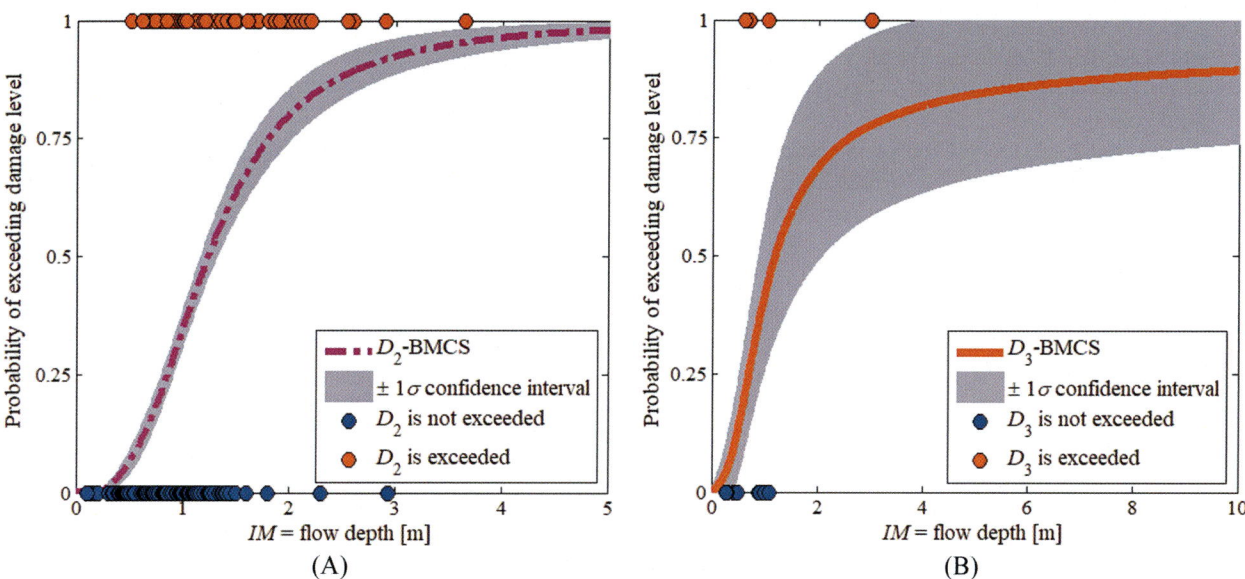

FIGURE 17.4 Fragility curve and the confidence band: (A) $N = 279$ data points and (B) $N = 14$ data points.

probability of exceeding certain damage levels for a given *IM*. The larger the number of standard deviations, the larger the confidence interval and the larger the probability that the "true" fragility curve is contained within the confidence interval. Alternatively, if the empirical fragility is represented as a median (50th percentile) of all possible fragility curves, the confidence interval can be represented as the distance between the two fragility curves corresponding to different percentiles equidistance from the median (e.g., 16th and 84th, 5th and 95th).

Fig. 17.4 shows an example of empirical fragility curves and their confidence interval for two different building classes with different numbers of data points (279 and 14) derived based on observed damage data from the 2018 Sulawesi Tsunami. The confidence intervals are visualized as the distance between the mean plus and minus one standard deviation fragility curves. The confidence interval is much wider in the case where only 14 buildings are surveyed (Fig. 17.4B) compared to the case where 279 buildings are surveyed (Fig. 17.4A).

In the literature, there are two methods for obtaining the confidence intervals: bootstrap and Bayesian inference.

Bootstrap method: In the Bootstrap method (Efron & Tibshirani, 1994), different subsets of the survey data are sampled with replacement and the empirical fragility is evaluated. This operation is repeated many times and the resulting fragility curves are plotted and their statistics (e.g., mean and standard deviation and percentiles) are calculated and visualized. The bootstrap simulates surveys with a smaller number of data points concerning the current sample. The bootstrap method is quite straightforward to apply. The main criticism of the bootstrap method is that it implicitly assumes that survey data (or the sample in general) build the entire range of possibilities.

Bayesian inference: Using the Bayesian inference to evaluate the confidence interval in the fragility curves mainly works around prescribed fragility models and evaluates the uncertainty in their parameters. This uncertainty is reported as the joint probability distribution for the fragility parameters conditioned on the available data. The updated joint probability distribution for the fragility model parameters is known as *the posterior joint PDF* of the fragility model parameters. It is then propagated to obtain the confidence interval of the fragility curve (see Jalayer et al. (2023) for more detail). The uncertainty propagation is done by calculating the expected value and the variance of the fragility curve over the entire domain of the fragility model parameters. The uncertainty in the fragility model parameters is related directly to the size of the survey data.

17.4 Conclusions

This chapter provided an overview of empirical tsunami fragility assessment, with a specific focus on the characterization and propagation of various sources of uncertainty. Here are some main gaps that emerge from the state-of-the-art of empirical tsunami fragility assessment:

- **Lack of standard damage scales**: Having standard damage scales will lead to harmonized fragility curves, to the possibility of updating fragility curves if new data are available, to the overall reduction of surveyors' bias, and to an increased possibility for preparation of standard damage assessment templates and preevent training of surveyors.
- **Surveyors' bias**: One of the main sources of errors in damage data acquisition is the subjectivity in assigning the damage labels. This can be reduced through the use of machine learning, standardizing the damage scale, having more surveyors assess the same asset, and formal consideration of such bias as a source of uncertainty in fragility assessment.
- **Uncertainty in *IM***: The uncertainty in the evaluation of the tsunami *IM* at the position of the damaged buildings is a significant contributor to the overall uncertainty in the fragility curves. Numerical tsunami simulation and calibrating to onshore observations can be a good way of characterizing and propagating such uncertainty.
- **Spatial and temporal gaps**: Since empirical fragility curves are derived based on observed events from past tsunamis, there are significant gaps associated with their spatial and temporal availability. One obvious solution is to dedicate more efforts to the development of the analytical fragility curves. Another solution is to dedicate efforts toward homogenization and harmonization of the derivation of the empirical fragility curves. This will facilitate making direct comparisons between fragility curves obtained based on different datasets.
- **Scale issues**: Empirical fragility curves are useful for spatial risk assessment purposes, such as impact forecasting and land-use planning. They are not useful for safety-checking and analysis of single infrastructure, such as vertical evacuation structures.
- **Correlations and dependences**: There could be potential sources of correlations or dependencies between fragility model parameters derived for various damage levels and for various classes based on the same suite of post-disaster data. Ideally, these sources of correlations need to be addressed and considered in fragility modeling.

As a final note, empirical fragility curves are data-driven products and can benefit from the emergence of machine learning and AI to make inferences from the available and alternative data sources to provide fragility models. To this end, Bayesian machine learning can provide a powerful tool not just for inference but also for thorough characterization and propagation of uncertainties.

References

Aránguiz, R., González, G., González, J., Catalán, P. A., Cienfuegos, R., Yagi, Y., Okuwaki, R., Urra, L., Contreras, K., Rio, I. D., & Rojas, C. (2017). The 16 September 2015 Chile Tsunami from the post-tsunami survey and numerical modeling perspectives. *Pure and Applied Geophysics*, *173*(2), 219–234. Available from https://doi.org/10.1007/978-3-319-57822-4_16, https://www.springer.com/journal/24.

Charvet, I., Ioannou, I., Rossetto, T., Suppasri, A., & Imamura, F. (2014). Empirical fragility assessment of buildings affected by the 2011 Great East Japan tsunami using improved statistical models. *Natural Hazards*, *73*(2), 951–973. Available from https://doi.org/10.1007/s11069-014-1118-3, http://www.wkap.nl/journalhome.htm/0921-030X.

Charvet, I., Macabuag, J., & Rossetto, T. (2017). Estimating tsunami-induced building damage through fragility functions: Critical review and research needs. *Frontiers in Built Environment*, *3*. Available from https://doi.org/10.3389/fbuil.2017.00036, https://www.frontiersin.org/articles/10.3389/fbuil.2017.00036/pdf.

Charvet, I., Suppasri, A., Kimura, H., Sugawara, D., & Imamura, F. (2015). A multivariate generalized linear tsunami fragility model for Kesennuma City based on maximum flow depths, velocities and debris impact, with evaluation of predictive accuracy. *Natural Hazards*, *79*(3), 2073–2099. Available from https://doi.org/10.1007/s11069-015-1947-8, http://www.wkap.nl/journalhome.htm/0921-030X.

Chock, G. Y. K. (2016). Design for tsunami loads and effects in the ASCE 7-16 Standard. *Journal of Structural Engineering*, *142*(11). Available from https://doi.org/10.1061/(ASCE)ST.1943-541X.0001565, http://ascelibrary.org/journal/jsendh.

Chua, C. T., Switzer, A. D., Suppasri, A., Li, L., Pakoksung, K., Lallemant, D., Jenkins, S. F., Charvet, I., Chua, T., Cheong, A., & Winspear, N. (2021). Tsunami damage to ports: Cataloguing damage to create fragility functions from the 2011 Tohoku Event. *Natural Hazards and Earth System Sciences*, *21*(6), 1887–1908. Available from https://doi.org/10.5194/nhess-21-1887-2021, http://www.nat-hazards-earth-syst-sci.net/volumes_and_issues.html.

De Risi, R., & Goda, K. (2017). Simulation-based probabilistic tsunami hazard analysis: Empirical and robust hazard predictions. *Pure and Applied Geophysics*, *174*(8), 3083–3106. Available from https://doi.org/10.1007/s00024-017-1588-9, http://www.springer.com/birkhauser/geo + science/journal/24.

De Risi, R., Goda, K., Mori, N., & Yasuda, T. (2017). Bayesian tsunami fragility modeling considering input data uncertainty. *Stochastic Environmental Research and Risk Assessment*, *31*(5), 1253–1269. Available from https://doi.org/10.1007/s00477-016-1230-x, http://link.springer-ny.com/link/service/journals/00477/index.htm.

De Risi, R., Goda, K., Yasuda, T., & Mori, N. (2017). Is flow velocity important in tsunami empirical fragility modeling? *Earth-Science Reviews*, *166*, 64–82. Available from https://doi.org/10.1016/j.earscirev.2016.12.015, http://www.sciencedirect.com/science/journal/00128252.

Der Kiureghian, A. (2005). Non-ergodicity and PEER's framework formula. *Earthquake Engineering and Structural Dynamics, 34*(13), 1643–1652. Available from https://doi.org/10.1002/eqe.504, http://onlinelibrary.wiley.com/journal/10.1002/(ISSN)1096-9845.

Efron, B., & Tibshirani, R. J. (1994). *An introduction to the bootstrap.* Chapman and Hall/CRC. Available from https://doi.org/10.1201/9780429246593.

Grünthal, G. (1998). *European macroseismic scale 1998.* European Seismological Commission, ESC. Available from https://gfzpublic.gfz-potsdam.de/rest/items/item_227033_2/component/file_227032/content.

Jalayer, F., & Ebrahimian, H. (2020). Seismic reliability assessment and the nonergodicity in the modelling parameter uncertainties. *Earthquake Engineering and Structural Dynamics, 49*(5), 434–457. Available from https://doi.org/10.1002/eqe.3247, http://onlinelibrary.wiley.com/journal/10.1002/(ISSN)1096-9845.

Jalayer, F., Ebrahimian, H., Trevlopoulos, K., & Bradley, B. (2023). Empirical tsunami fragility modelling for hierarchical damage levels. *Natural Hazards and Earth System Sciences, 23*(2), 909–931. Available from https://doi.org/10.5194/nhess-23-909-2023, http://www.nat-hazards-earth-syst-sci.net/volumes_and_issues.html.

Koshimura, S., Namegaya, Y., & Yanagisawa, H. (2009). Tsunami fragility—A new measure to identify tsunami damage. *Journal of Disaster Research, 4*(6), 479–488. Available from https://doi.org/10.20965/jdr.2009.p0479, https://doi.org/10.20965/jdr.2009.p0479.

Lahcene, E., Ioannou, I., Suppasri, A., Pakoksung, K., Paulik, R., Syamsidik, S., Bouchette, F., & Imamura, F. (2021). Characteristics of building fragility curves for seismic and non-seismic tsunamis: Case studies of the 2018 Sunda Strait, 2018 Sulawesi-Palu, and 2004 Indian Ocean tsunamis. *Natural Hazards and Earth System Sciences, 21*(8), 2313–2344. Available from https://doi.org/10.5194/nhess-21-2313-2021, http://www.nat-hazards-earth-syst-sci.net/volumes_and_issues.html.

Lahcene, E., Suppasri, A., Pakoksung, K., & Imamura, F. (2023). Coral reef response in the Maldives during the 2004 Indian Ocean tsunami. *International Journal of Disaster Risk Reduction,* 103952. Available from https://doi.org/10.1016/j.ijdrr.2023.103952.

Mas, E., Koshimura, S., Suppasri, A., Matsuoka, M., Matsuyama, M., Yoshii, T., Jimenez, C., Yamazaki, F., & Imamura, F. (2012). Developing tsunami fragility curves using remote sensing and survey data of the 2010 Chilean tsunami in Dichato. *Natural Hazards and Earth System Science, 12*(8), 2689–2697. Available from https://doi.org/10.5194/nhess-12-2689-2012.

Mas, E., Paulik, R., Pakoksung, K., Adriano, B., Moya, L., Suppasri, A., Muhari, A., Khomarudin, R., Yokoya, N., Matsuoka, M., & Koshimura, S. (2020). Characteristics of tsunami fragility functions developed using different sources of damage data from the 2018 Sulawesi earthquake and tsunami. *Pure and Applied Geophysics, 177*(6), 2437–2455. Available from https://doi.org/10.1007/s00024-020-02501-4, http://www.springer.com/birkhauser/geo+science/journal/24.

Miano, A., Jalayer, F., Forte, G., & Santo, A. (2020). Empirical fragility assessment using conditional GMPE-based ground shaking fields: Application to damage data for 2016 Amatrice Earthquake. *Bulletin of Earthquake Engineering, 18*(15), 6629–6659. Available from https://doi.org/10.1007/s10518-020-00945-6, https://rd.springer.com/journal/10518.

Muto, M., & Beck, J. L. (2008). Bayesian updating and model class selection for hysteretic structural models using stochastic simulation. *Journal of Vibration and Control, 14*(1-2), 7–34. Available from https://doi.org/10.1177/1077546307079400.

Paulik, R., Gusman, A., Williams, J. H., Pratama, G. M., Lin, Sl, Prawirabhakti, A., Sulendra, K., Zachari, M. Y., Fortuna, Z. E. D., Layuk, N. B. P., & Suwarni, N. W. I. (2019). Tsunami hazard and built environment damage observations from Palu City after the September 28 2018 Sulawesi earthquake and tsunami. *Pure and Applied Geophysics, 176*(8), 3305–3321. Available from https://doi.org/10.1007/s00024-019-02254-9, http://www.springer.com/birkhauser/geo+science/journal/24.

Reese, S., Bradley, B. A., Bind, J., Smart, G., Power, W., & Sturman, J. (2011). Empirical building fragilities from observed damage in the 2009 South Pacific Tsunami. *Earth-Science Reviews, 107*(1–2), 156–173. Available from https://doi.org/10.1016/j.earscirev.2011.01.009.

Rossetto, T., Peiris, N., Pomonis, A., Wilkinson, S. M., Del Re, D., Koo, R., & Gallocher, S. (2007). The Indian Ocean Tsunami of December 26, 2004: Observations in Sri Lanka and Thailand. *Natural Hazards, 42*(1), 105–124. Available from https://doi.org/10.1007/s11069-006-9064-3.

Suppasri, A., Mas, E., Charvet, I., Gunasekera, R., Imai, K., Fukutani, Y., Abe, Y., & Imamura, F. (2013). Building damage characteristics based on surveyed data and fragility curves of the 2011 Great East Japan tsunami. *Natural Hazards, 66*(2), 319–341. Available from https://doi.org/10.1007/s11069-012-0487-8, http://www.wkap.nl/journalhome.htm/0921-030X.

Tarbotton, C., Dall'Osso, F., Dominey-Howes, D., & Goff, J. (2015). The use of empirical vulnerability functions to assess the response of buildings to tsunami impact: Comparative review and summary of best practice. *Earth-Science Reviews, 142*, 120–134. Available from https://doi.org/10.1016/j.earscirev.2015.01.002, http://www.sciencedirect.com/science/journal/00128252.

Williams, J. H., Wilson, T. M., Horspool, N., Paulik, R., Wotherspoon, L., Lane, E. M., & Hughes, M. W. (2020). Assessing transportation vulnerability to tsunamis: Utilising post-event field data from the 2011 Tōhoku tsunami, Japan, and the 2015 Illapel Sunami, Chile. *Natural Hazards and Earth System Sciences, 20*(2), 451–470. Available from https://doi.org/10.5194/nhess-20-451-2020, http://www.nat-hazards-earth-syst-sci.net/volumes_and_issues.html.

Chapter 18

Analytical tsunami fragility curves

Tiziana Rossetto[1], Marta Del Zoppo[2], Marco Baiguera[3] and Jonas Cels[1]
[1]UCL EPICentre, University College London, London, United Kingdom, [2]Department of Structures for Engineering and Architecture, University of Naples Federico II, Naples, Italy, [3]School of Engineering, University of Southampton, Southampton, United Kingdom

18.1 Introduction

The ability to estimate what damage will occur in the built environment under different tsunami inundation scenarios is key to effective tsunami risk modeling, disaster management, evacuation, and response planning. A quantitative tool commonly used for this is fragility functions, which usually comprise a set of curves (f), each representing the damage state (DS) exceedance probability for a specific damage state (ds_i) conditioned on the expected tsunami intensity measure (IM), as given in the following equation

$$f = P(DS \geq ds_i | IM) \tag{18.1}$$

Other chapters (Chapters 4 and 17) have already described fragility functions and their form, as well as how fragility functions can be derived from asset damage statistics generated in different ways. In this chapter, we look at the analytical approach for developing fragility functions and focus on buildings as the asset of interest.

Analytical fragility curves adopt as their statistical basis damage distributions, which are simulated from the analyses of numerical models of buildings under the loading regimes that are created by tsunami inundations of increasing intensity. The analytical approach can present several advantages over the empirical approach:

- It can be used to develop a fragility function for any type of structure, as long as the structure and its response to tsunami loading can be reliably represented in a numerical model.
- Analytical fragility functions can be developed for a single specific building or a building class, which comprises buildings with similar expected responses but having varied geometric and material properties (i.e., using several building models).
- The analytical approach allows full control over the design and composition of the represented buildings, which is not possible in empirical fragility. In the latter, the building classes are usually defined coarsely based on observable characteristics of tsunami-affected buildings.
- A range of tsunami inundations can be used as input to the numerical analysis, and hence damage simulations can be run for a wide range of tsunami IMs.
- Damage simulations can be run for multiple inundation scenarios that produce the same IM value at the structure, allowing a better representation of tsunami inundation aleatory uncertainty.

Despite many advantages, relatively few analytical fragility curves for buildings affected by tsunamis have been developed in the past. This is partly due to the relatively recent establishment and wide recognition of tsunami engineering as a discipline. The devastating nature of tsunamis has been long known, and tsunami hazards have been studied for decades. However, the engineering research community was spurred into action following recent tsunami events that have resulted in massive human and monetary losses. In particular, the 2004 Indian Ocean Tsunami that devastated coastlines across the Indian Ocean and killed over 230,000 people across 14 countries (Rossetto et al., 2007) resulted in a significant worldwide effort toward a better characterization of tsunami-induced scour and forces on structures (Foster et al., 2017; Nistor et al., 2017; Qi et al., 2014). The subsequent 2010 Maule Earthquake and Tsunami in Chile and the 2011 Tohoku Earthquake and Tsunami in Japan, which damaged numerous modern engineered buildings, industrial facilities, bridges, and large coastal defense infrastructure (Pomonis et al., 2011), resulted in the revision of building codes for tsunami design and the development of analytical methods for assessing buildings under tsunami loads (Alam et al., 2018; ASCE/SEI, 2017; Petrone et al., 2017). This marked the beginning of analytical tsunami fragility function development.

Given its recent nature, tsunami analytical fragility is currently an evolving field of research that is influenced significantly by advancements in tsunami simulation capacity, knowledge of tsunami inundation and its interaction with structures, and developments of new approaches for simulating structural and nonstructural performance under the induced tsunami actions. The following sections present the analytical approach to fragility function development, highlighting currently used methods and their challenges.

18.2 Tsunami actions on buildings

A thorough understanding of the actions imposed on a structure from the tsunami inundation is required to simulate its response to a tsunami. The nature of the tsunami actions influences the choice of numerical modeling approach and structural analysis method and also determines the appropriateness of adopted simplifying assumptions. In this section, the estimation of tsunami inundation flow characteristics is first examined and then the estimation of tsunami actions on buildings is described.

18.2.1 Characterization of tsunami inundation flows

Tsunami can be considered extremely long waves. When these reach the coastline, they inundate the land, with the inundation flow characterized by a large velocity and duration (of the order of tens of minutes). The inundation flow transports soil and other debris and flows back to the sea following inundation. The process is highly complex and is affected by several features of the tsunami (e.g., wave height and wavelength), the bathymetry and coastal topography, and the presence of any defense structures.

Tsunami inundation flows interact with buildings onshore, applying pressures and loads on the building that are highly dependent on the parameters of inundation depth and velocity. As seen later, the estimation of building response using a nonlinear dynamic analysis method requires as input time-histories of tsunami inundation height and velocity at the location of the building. Numerical modeling of tsunami inundation through computational fluid dynamics can provide estimates of these time-histories. As discussed in Chapter 3, onshore inundation simulation can vary in complexity from detailed 3D models considering flow around individual buildings to simplified 2D models that represent built-up areas using Manning's roughness coefficient and making assumptions regarding the depth distribution of velocities and pressures (Charvet et al., 2017). The latter approach is the most commonly used in past analytical fragility studies. The numerical inundation flow simulation results are highly sensitive to the quality and resolution of the local coastal bathymetry and onshore topography, the choice of numerical methods, and the mesh sizes used (Song & Goda, 2019). Inundation flow velocity is particularly affected, while tsunami inundation depth and runup estimates are considered more robust, especially as numerical models can be calibrated/validated with observational data for these parameters from past tsunamis.

The full simulation of inundation from a single tsunami is complex and computationally very expensive. It creates data storage issues when inundation characteristics need to be retained for each analysis time step. The use of numerical simulations for multiple tsunami inundation simulations required to develop analytical fragility functions is often considered prohibitive. Hence, empirical and semiempirical approaches for estimating tsunami inundation flow parameters on inundated coastlines have been proposed as an alternative. Of particular note is the approach proposed in ASCE/SEI (2022), hereafter ASCE7-22, which uses the Energy Grade-line Analysis method (Kriebe et al., 2017) to estimate the maximum onshore depth and velocity fields from data on offshore tsunami height and resulting runup. These maximum values are then entered into normalized time-history curves also present in ASCE7-22 (Baiguera et al., 2022) to obtain the required inundation depth and velocity time-histories for structural analysis. It is highlighted that in the case of ASCE7-22, the tsunami height to runup relationships are mapped for the U.S. coastline and derived from numerous tsunami wave and inundation numerical simulations. An alternative, but less rigorous approach, can be to estimate runup heights for particular offshore tsunami heights using empirical equations based on observation or experiments (McGovern et al., 2018).

18.2.2 Characterization of tsunami actions on buildings

Tsunami inundation flows apply a number of sustained and impulsive loads on buildings (Chapter 4). Sustained lateral loads on buildings from tsunami inundation comprise unbalanced hydrostatic pressures on structural components and hydrodynamic (drag) forces. Sustained vertical loads are mainly related to hydrostatic buoyancy and hydrodynamic surge.

These are discussed in greater depth in this section, as different approaches have been used for the estimation of these loads, and different assumptions have been made as to how these are applied to building models in analytical fragility function development.

Impulsive lateral and vertical loads comprise the impact of the leading edge of the arriving water mass and impact forces from waterborne debris. Current analytical approaches to fragility largely ignore the influence of these impulsive loads on the response of buildings. For the case of the impact of the leading edge of the arriving water mass, this is justified both by the highly transient nature of this force and by the lack of high water impact forces being observed in experiments that have simulated tsunami inundation interaction with buildings (Foster et al., 2017; Qi et al., 2014). In the case of waterborne debris impact, this has been neglected from past analytical fragility studies due to the large uncertainties associated with the variable nature of the debris (that can range from a plank or brick to a car or shipping container) and the location where it might impact the structure. These impulsive forces are, therefore, not further described in this section, although debris accumulation at openings is accounted for in the calculation of the sustained hydrodynamic lateral forces.

It is also noted that tsunami flows can scour the soil around structure foundations, causing loss of bearing capacity and structure instability. Although empirical equations exist for estimating scour depth around structures (McGovern et al., 2019; Mehrzad et al., 2022), this mechanism of structural failure has not yet been included in analytical fragility functions, and hence scour is also not further discussed here.

18.2.2.1 Sustained lateral loads

Unbalanced lateral hydrostatic forces develop on walls and vertical structural elements when there is a difference in water height on either side of the element. ASCE7-22 assumes that walls with openings greater than 10% of their area will allow hydrostatic pressures to equalize on opposite sides of the wall. However, hydrostatic loads should be considered if an enclosure forms on one side of a vertical element, that is, hydrostatic forces act on one side of the wall only. Hydrostatic pressures follow a triangular distribution with water depth, and the size of the hydrostatic load on an element depends on at what height the vertical element is located with respect to the inundation depth. For the general case of a vertical element of width b, with the base at $z = h_1$ above grade, and top at $z = h_2$ above grade, and a tsunami inundation depth h that exceeds h_2, the lateral hydrostatic force, F_h, can be calculated from the following equation

$$F_h = -\gamma_s b \int_{z=h-h_1}^{z=h-h_2} z\,dz \tag{18.2}$$

where $\gamma_s = 1127.5$ kN/m^3 is the specific weight of seawater.

The largest sustained lateral loads on buildings typically result from hydrodynamic (drag) forces induced by the tsunami flow passing around and through the building. Two main approaches have been followed for calculating hydrodynamic forces on structures subjected to tsunami inundation in the analytical fragility literature.

The first of these approaches involves the use of the hydrodynamic load equations of ASCE7-22 for determining overall drag forces on the building and each vertical element within it. ASCE7-22 defines the drag forces acting on the building at each building story, F_{dx}, as

$$F_{dx} = \frac{1}{2}\rho_s I_{tsu} C_D C_{cx} B (hu^2) \tag{18.3}$$

where C_D is the drag coefficient given in Table 6.10-1 of ASCE7-22. ρ_s is the minimum mass density accounting for suspended solids and debris flow-embedded smaller objects. In most analytical fragility approaches that use the ASCE7 loads, ρ_s is simply assumed to be equal to the density of seawater. The closure coefficient, C_{cx}, represents the ratio between the vertical projected area of structural components and the projected area of the submerged portion of the building and can be calculated from the following equation

$$C_{cx} = \frac{\sum (A_{col} + A_{wall}) + 1.5 A_{beam}}{B h_{sx}} \tag{18.4}$$

where A_{col}, A_{wall}, and A_{beam} are the vertical projected areas at the level x of columns, walls, and beams, respectively. h_{sx} is the inundation depth ratio.

The equation to evaluate the drag force on individual building components takes the same form as Eq. (18.3) and is given by the following equation

$$F_d = \frac{1}{2}\rho_s I_{tsu} C_D b (h_e u^2) \tag{18.5}$$

The element drag forces are calculated as the resultant pressure on the projected inundated height, h_e, of all the inundated structural components. b is the width of the component perpendicular to the flow. The drag coefficient C_D is assigned based on a different set of values for the component level, given in Table 6.10-2 of ASCE7-22.

An alternative approach to tsunami hydrodynamic load calculation that has been used in analytical fragility studies is the force formulations proposed in Foster et al. (2017). In this approach, the tsunami inundation flow is assumed quasisteady, as its very long duration means that the temporal variation of the flow is small with respect to the length scale of a building. Foster et al. (2017) and Qi et al. (2014) demonstrated through experiments that either subcritical or supercritical (choked) flow conditions can occur in the case of quasisteady and steady flows passing rectangular objects, respectively. The flow condition is represented by the Froude number, Fr, which is calculated from the flow velocity (v) and inundation depth (h) as

$$Fr = \frac{v}{\sqrt{gh}} \tag{18.6}$$

A critical Froude number value, Fr_c, for the flow is defined, which signals the transition from subcritical to supercritical flow and is estimated using the following equation

$$Fr_c = \left(1 - \frac{0.58B}{w}\right)Fr_{d,c}^{-\frac{4}{3}} + Fr_{d,c}^{\frac{2}{3}} \tag{18.7}$$

where

$$Fr_{d,c} = \left(1 - \frac{0.58B}{w}\right)^{\frac{1}{2}} \tag{18.8}$$

In these equations, B/w is the blocking ratio and represents the ratio between the width of the building (B) and the width of the flume (w), wherein the tests were conducted to empirically derive these relationships. It is a key parameter in the load calculations. In situations of an actual building within an urban context, the blocking ratio can be considered to represent the urban density. A blocking ratio value of 0.1 can be considered to represent rural areas, whereas blocking ratios between 0.5 and 0.8 may represent densely populated areas.

The hydrodynamic force on a building is then calculated for the appropriate flow regime as per the following equations

$$F_t = B \times \text{sign}(v)0.5C_D\rho v^2 h \quad \text{if} \quad Fr < Fr_c \text{ (subcritical flow)} \tag{18.9}$$

and

$$F_t = B \times \text{sign}(v)\lambda\rho g^{1/3}v^{4/3}h^{4/3} \quad \text{if} \quad Fr \geq Fr_c \text{ (supercritical flow)} \tag{18.10}$$

where $\text{sign}(v)$ is the flow velocity sign function, C_D is the drag coefficient, ρ is the fluid density, and h is the inundation depth. λ is the leading coefficient for steady flows, which can be interpreted as the empirical closure for F_t. The drag and leading coefficients are both functions of the blocking ratio (B/w) and are represented by Eqs. (18.11) and (18.12), respectively.

$$C_D = 1.9\left(1 + \frac{1.9B}{2w}\right)^2 \tag{18.11}$$

$$\lambda = \frac{1}{2}C_D Fr_c^{2/3} + \frac{1}{2}0.58\left(Fr_c^{-2/3} - Fr_{d,c}^{-4/3}\right) \tag{18.12}$$

Neither ASCE7-22 nor Foster et al. (2017) explicitly state how to apply tsunami forces to a structure nor to structural elements in tsunami performance assessments. Baiguera et al. (2022) conducted a sensitivity analysis using different load discretization methods for systemic and component-level checks. The study applied point loads as concentrated loads to each story level and compared it to the structural response where loads are distributed along columns (with five equally distributed load application points per column). The study found that when forces are concentrated at the story level, the structural response is biased toward the flexural failure of columns and the shear demand is significantly underestimated. The authors, therefore, recommended that hydrodynamic loads are applied to vertical elements in a structure in a distributed way, using five or more load application points per vertical element of the building.

The magnitude of the hydrodynamic forces over the building height depends on what load pattern distribution is assumed. Currently, there is little consensus on what lateral load pattern should be used. If a subcritical flow condition is considered, the tsunami inundation height at the front and back of a building within the flow would be very similar. This would promote a uniform distribution of the drag forces, as hydrostatic components would effectively cancel out. However, if the tsunami flow around the building enters a supercritical regime, there would be a significant difference in inundation depth between the front and back of the building. This would promote a trapezoidal load distribution, given that there would be an imbalance of hydrostatic loads over the height difference. Trapezoidal lateral load distributions were adopted by Petrone et al. (2017). Rossetto, De la Barra, et al. (2019) compared the tsunami structural response for a structure assuming different lateral load patterns for hydrodynamic loading and showed that the approach recommended in ASCE7-22, where a uniform distribution of hydrodynamic loads is assumed to act up to the full inundation depth, is conservative. Examples of the load patterns used to represent global lateral loads imposed by tsunami inundations are shown in Fig. 18.1.

18.2.2.2 Sustained vertical loads

The leading cause of vertical sustained loads on buildings subjected to tsunami inundation is buoyancy. Buoyancy produces upward pressures on structures and acts mainly on horizontal structural and nonstructural elements (e.g., beams, slabs, and foundation slabs). Buoyancy pressures result from water being displaced, and hence in a tsunami-inundated building, the buoyancy pressures depend on the inundation flow depth and the arrangement of structural and nonstructural elements that displace the water and create air voids. Buoyancy is not dependent on other parameters of the tsunami inundation flow, such as velocity or Froude number. According to ASCE7-22, the vertical forces from tsunami-inundated buoyancy, F_b, can be evaluated from the following equation

$$F_b = \gamma_s V_w \tag{18.13}$$

where $\gamma_s = 1127.5$ kN/m^3 is the specific weight of seawater and V_w is the volume of the displaced water. ASCE7-22 also highlights three main mechanisms for buoyancy force development in buildings, shown in Fig. 18.2 and described as

- Buoyancy due to air pockets: Air pockets can form because air is trapped between the floor slab and beam and frame downstands as inundation rises above each story level. In this case, V_w is computed as the volume of air pockets, and the force acts vertically on the element above the trapped air pockets.
- Buoyancy due to submergence: A reduction of weight due to buoyancy is expected in submerged structural elements. The corresponding vertical force is computed by taking V_w as the submerged volume of the structural element and acting on the structural element.

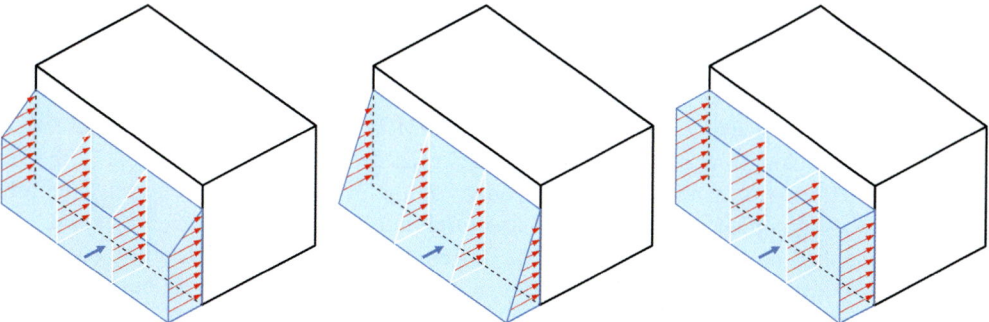

FIGURE 18.1 Global lateral load patterns under tsunami loading. From left to right, trapezoidal loading, triangular loading, and uniform loading.

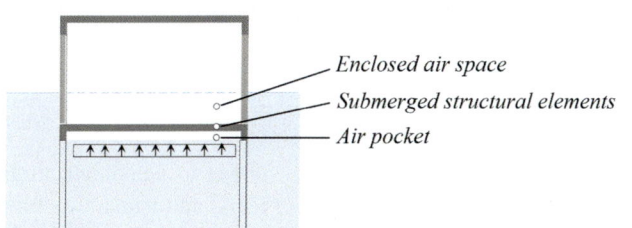

FIGURE 18.2 Depiction of the three main components of buoyancy loads in tsunami-inundated buildings. The shaded area represents the inundation level.

- Buoyancy due to enclosed air spaces: A buoyancy force is also created when an enclosed area (similar to an air pocket) is submerged. Such an enclosed area can arise in buildings between the slab and exterior walls above it when these remain intact as the inundation rises above a slab level. In this case, the vertical forces acting on the slab can be computed assuming V_w as the product of the floor area and the net distance between the top of the slab and the tsunami inundation depth. This can be a very significant force, which strictly depends on the out-of-plane capacity of exterior walls in perimeter frames. Indeed, after the failure of exterior walls, the water is allowed into the building relieving this buoyancy effect (Del Zoppo et al., 2023).

In structural analyses, buoyancy-related forces should be modeled to vary with inundation depth and can be applied with a uniform upward distribution to horizontal structural components and as point loads to vertical elements in buildings. A further sustained vertical force is caused by residual water surcharge. This arises when the inundation waters do not fully drain from elevated floors. ASCE7-22 specifies that the volume of retained water should be limited to the height of the continuous perimeter structural components that can survive the tsunami loads during tsunami surge and drawdown. This vertical force, which acts as a surcharge on building slabs, has not been included in analytical fragility functions as it applies after the tsunami inundation phase.

18.2.3 Tsunami intensity measures for fragility analysis

The tsunami *IM* (independent variable) defines the abscissa of tsunami analytical fragility functions. A good tsunami *IM* can represent the damaging potential of tsunami inundation flows on buildings and can be estimated by tsunami hazard analyses, that is, from the numerical simulation of tsunami inundation flows. Given the latter, the vast majority of past analytical fragility functions have adopted as the *IM* the maximum inundation depth (h_{max}) or the maximum inundation flow velocity (v_{max}) at the site of the building (Nanayakkara & Dias, 2016). As previously stated, h_{max} can be estimated by tsunami inundation numerical models with a reasonable degree of robustness, and as a consequence most tsunami hazard or inundation maps in existence present values of h_{max}. This parameter has, therefore, historically been the preferred *IM* due to its computability and ease of integration of developed fragility functions into existing risk models. However, it is highlighted that the same inundation depth can be observed for tsunami inundation flows with significantly different velocities (and vice versa). This means that the *IM*s of h_{max} and v_{max} provide, at best, a partial view of the damage potential of tsunami flows on buildings.

In recognition of this, the maximum momentum flux, that is, the product of inundation depth and the square of the velocity, max hv^2, has been proposed by some studies for use as *IM* (Alam et al., 2018; Attary et al., 2017). No tsunami inundation maps currently present momentum flux, and its determination requires a full simulation of the tsunami inundation, given that maximum values of velocity and inundation depth do not occur at the same time. Hence, the momentum flux should be evaluated for the entire tsunami inundation time-history to determine its maximum value. As stated, an alternative approach can be to use empirical approximations for tsunami inundation characteristics, as described in Section 18.2.1.

One of the main rationales for using momentum flux as an *IM* is the argument that it represents the drag force imposed by a tsunami inundation flow on a building. However, as described in Section 18.2.2, the size of the sustained loads on buildings from a tsunami is highly dependent on the geometry of the building, the urban context in which it is sited (i.e., blockage ratio), and the area of structural and nonstructural elements exposed to the flow (as the sustained loads are applied in the form of pressures). Therefore, recent analytical fragility studies have moved to using the maximum equivalent quasisteady lateral force (F_{max}) to partly account for these building–flow interactions. The equations adopted for the evaluation of F_{max} are the same as those presented in Eqs. (18.2)–(18.5) (ASCE7-22) or Eqs. (18.7)–(18.12) (Foster et al., 2017), assuming the lateral pressures are applied to the entire flow-facing façade of the building with 0% or 30% permeability, respectively. F_{max} is considered a better proxy than h_{max}, v_{max}, and max hv^2 for tsunami inundation damage potential (Macabuag et al., 2016) but has the same difficulties as max hv^2 associated with its determination from numerical modeling of tsunami inundations (or empirical approximations of these).

18.3 Tsunami structural assessment approaches

Different types of structural analysis approaches can be adopted to determine building response to tsunami actions. These can be grouped into two categories: tsunami nonlinear time-history dynamic analyses (TDA) and tsunami nonlinear static analyses (pushover analyses, PO). These approaches have been inspired by established practices in the

TABLE 18.1 Summary of key features of different tsunami structural analysis methods for use in fragility assessment.

Computation	Analysis method	Analysis type	Integrator	Required input for tsunami load calculation	EDPs achievable
Increasing computational expense ↑	TDA	Nonlinear dynamic time-history analysis	Newmark	Inundation flow depth and velocity time histories, blocking ratio	Transient and permanent EDPs, in the axis of applied loading. Energy-based EDPs
	VDPO-BI	Nonlinear static analysis	Force-based followed by displacement based	Froude number of inundation and blocking ratio	Permanent EDPs, in the direction of applied loading
	VDPO2	Nonlinear static analysis	Force-based followed by displacement based	Froude number of inundation and blocking ratio	Permanent EDPs, in the direction of applied loading
	VDPO	Nonlinear static analysis	Force-based	Froude number of inundation and blocking ratio	Permanent EDPs, in the direction of applied loading
	CDPO	Nonlinear static analysis	Displacement-based	Target inundation depth and blocking ratio	Permanent EDPs, in the direction of applied loading

CDPO, Constant depth pushover; *EDPs*, engineering demand parameters; *TDA*, tsunami nonlinear time-history dynamic analyses; *VDPO*, variable depth pushover.

earthquake response analyses of buildings. Although pushover analysis provides a crude and overconservative approximation of the highly dynamic seismic response of buildings when applied in the context of tsunami loading, pushover analyses can closely replicate the actual hydrodynamic drag on the overall structure and individual structural components due to the sustained nature of the loading (Rossetto et al., 2018).

The choice of structural analysis approaches has a direct correlation to the computational expense of the fragility assessment. Moreover, it has implications for the form of the required input (in terms of tsunami inundation characterization) and the engineering demand parameters (*EDP*) that can be extracted from the buildings' structural analyses to determine the building's tsunami performance (i.e., damage level). These are summarized in Table 18.1 for the most common tsunami structural analysis approaches, which are further described in this section.

18.3.1 Tsunami nonlinear time-history dynamic analysis

In TDA, time-histories of inundation depth $h(t)$ and velocity $v(t)$ are input to the analysis. These time-histories are obtained from numerical simulations of tsunami inundation. They are used to calculate corresponding lateral and vertical force time-histories $F(t)$ and $F_v(t)$, respectively, using equations like those presented in Eqs. (18.2)–(18.12). At each time step of the analysis, the tsunami forces are applied to the structure assuming a given load distribution up to the tsunami inundation depth at that time step, and the resulting structural response is computed. To do this, load time-histories need to be calculated for each structural node according to the assumed load distribution, level of submergence, and contributing surface area for that node. The performance of the structure under the applied tsunami time-history is determined through the interpretation of the *EDP* values measured during the analysis. Both transient and permanent values of *EDP* can be determined, for example, maximum interstory drift during the time-history as well as permanent interstory drift at the end of the analysis. TDA analyses of reinforced concrete frames have been conducted by Petrone et al. (2017), considering only lateral forces.

Although appropriate for the response assessment of tall, highly irregular, or critical infrastructure, TDA may not be ideal for developing analytical data to derive fragility functions. This is because TDA is very computationally expensive as the duration of the time history can span tens of minutes. Moreover, for each structural model, TDA must be rerun numerous times to capture the variability in tsunami inundation time-history associated with a given *IM* value, and then this process is repeated for multiple *IM* values.

18.3.2 Tsunami nonlinear static analyses

A number of tsunami nonlinear static analysis procedures have been developed for use in the assessment of building performance under tsunami loading. This is in recognition of the adequacy of pushover analyses to represent the long-duration loads induced by tsunami inundation flows. In each static pushover approach, the tsunami load is applied incrementally to the structure, and the structural response is calculated for each load increment until structural collapse is reached. Like in TDA, the loading applied to each node of the structure depends on the contributing surface area over which the tsunami inundation acts, the assumed pressure distribution, and the level of submergence of the node at the particular analysis step. Similarly to TDA, this requires the definition of force time-histories for each structural model node, with the difference being that the loading is monotonically increased at each analysis step rather than being a time variable. A tsunami pushover curve can be drawn from the results of the analysis by plotting the total horizontal base shear of the structure versus its roof drift, as shown in Fig. 18.3C.

The constant depth pushover (CDPO), also called Constant Height Pushover, was the first tsunami nonlinear static analysis procedure developed. In the CDPO, a value of tsunami inundation depth is first assumed. Then, equivalent lateral forces are calculated using formulae like those in Eqs. (18.2)–(18.9) for a small value of flow velocity. The force is then incrementally increased by recalculating the lateral loads for increasing values of the velocity while maintaining the same inundation depth. The force at each step is applied along the height of the structure up to the set inundation depth through an assumed load profile. Several studies have developed pushover curves or fragility functions for reinforced concrete buildings using the CDPO approach (Alam et al., 2018; Attary et al., 2017; Petrone et al., 2017). However, all have only considered the horizontal sustained loads applied by tsunamis.

The CDPO approach presents one main disadvantage: analysts should know the target tsunami inundation depth a priori. Nowadays, the CDPO has been largely superseded by the variable depth pushover (VDPO) approach, where instead of fixing the inundation depth value, the inundation depth is incremented during the analysis. At each step of the VDPO analysis, the flow velocity corresponding to the inundation depth at that step is calculated assuming a constant Froude number. The flow velocity and inundation height values are then used to calculate the equivalent forces, which are applied to the structure according to a set pressure profile up to the inundation height level for the analysis step. It is highlighted that, as the water depth increases through the analysis, new nodes are loaded along the height of the building model. A schematic comparison of the CDPO and VDPO loading approaches is shown in Fig. 18.3. Several versions of the VDPO follow the same concept of applying increasing loads to the structure by incrementally increasing the inundation depth while maintaining a constant Froude number. The first VDPO approach was proposed by Petrone et al. (2017) and was set up as a load-controlled analysis. This presented numerical instabilities when the structure neared its peak base-shear capacity and was not capable of capturing the postpeak degrading behavior of the structure.

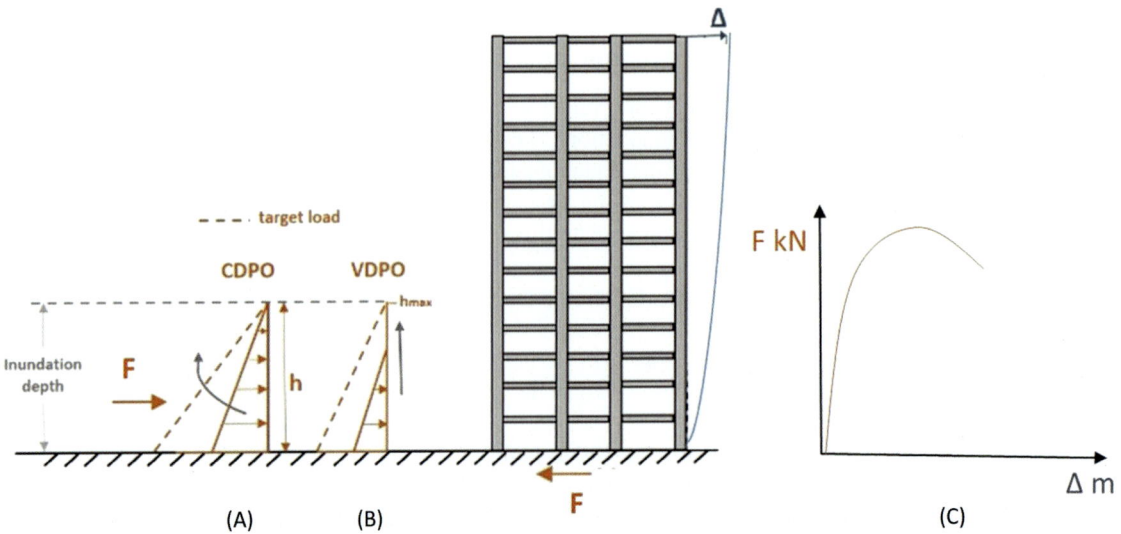

FIGURE 18.3 Schematic representation of the loading approach adopted in (A) CDPO and (B) VDPO. (C) An illustrative pushover curve developed from the building base shear (F) and roof displacement (Δ). *CDPO*, Constant depth pushover; *VDPO*, variable depth pushover.

To overcome this issue, Baiguera et al. (2019) proposed the VDPO2 approach, which consists of a two-phase nonlinear static analysis, where in phase one, a load-control pushover analysis is conducted assuming that the inundation depth and flow velocity increase incrementally, and in phase two, the analysis switches to a response-control pushover analysis, where the displacement is increased incrementally and the corresponding tsunami force is calculated assuming that there is no further increase in inundation depth. The switch from the first to the second phase occurs either when a predefined load level is reached or when the analysis encounters a numerical convergence issue, whichever occurs first.

To provide a tool that is consistent with the requirements for tsunami design in ASCE7-22, Baiguera et al. (2022) proposed a variant of the VDPO2 approach called ASCE-VDPO2, where instead of maintaining the Froude number constant in phase 1 of the analysis, loads are applied incrementally following the ASCE7-22 inundation depth and velocity time-histories up to a specific load case, after which the analysis switches to response control (phase 2). The ASCE-VDPO2 is currently referenced in ASCE7-22 as a recommended nonlinear static method for the tsunami design of buildings exposed to tsunami risk.

The VDPO was also modified by Del Zoppo, Di Ludovico, et al. (2021) to better assess infilled reinforced concrete structures. Called VDPO-BI, this version of VDPO explicitly considers the changes in horizontal and vertical loading that occur in the building when there is progressive failure of infill walls during tsunami inundation. Within VDPO-BI, the out-of-plane capacity of each infill wall is first calculated, accounting for support conditions. The inundation height at which the applied load on the walls will exceed their capacity is precalculated. Instead of explicitly modeling the infill walls in the numerical model of the structure, the influence of the infill walls is accounted for through modification of the loading history applied to the structural model nodes. In other words, for each analysis step where the inundation depth is lower than a level that will cause the infill collapse, the full infill wall surface area is assumed to contribute to the load calculation for the nodes in adjacent structural elements. When the inundation depth exceeds the value at which infill out-of-plane failure occurs, only the structural element surface area is adopted in the pressure to force calculation for the nodes in adjacent elements. Similarly, air voids created by the presence of intact infills can be accounted for in the vertical loading calculation up until the infill wall point of failure (Del Zoppo et al., 2023; Del Zoppo et al., 2021; Del Zoppo, Wijesundara, et al., 2021). The failure of boundary infill walls can also signal the start of tsunami inundation loading being applied to internal elements of the buildings. A schematic of the process is shown in Fig. 18.4.

In all cases of tsunami static pushover, the structure's performance for a given tsunami *IM* value is assessed by first calculating the equivalent lateral force associated with the *IM* (using Eqs. 18.2–18.12). This is regarded as the demand base shear, and the corresponding roof drift can be found through the interpretation of the pushover curve. Other *EDPs* can then be determined from the results of the structural analysis at that roof drift value. As there is no time-dependent (dynamic) component in the analysis, only permanent *EDPs* can be determined for the interpretation of structural damage.

The tsunami pushover approaches presented vary in complexity but present a significantly reduced computational run time compared to TDA. Like TDA tsunami, pushover analyses need to be rerun to account for the variation in the flow parameters associated with a given *IM*. This is because the applied forces and their distribution depend on the Froude number and blockage ratio assumed. Therefore, a variation in these parameters results in a difference in the structural response. However, each developed pushover curve can be used to assess the structural performance under a range of *IM* values, thus avoiding the rerun of analyses under increasing tsunami intensities.

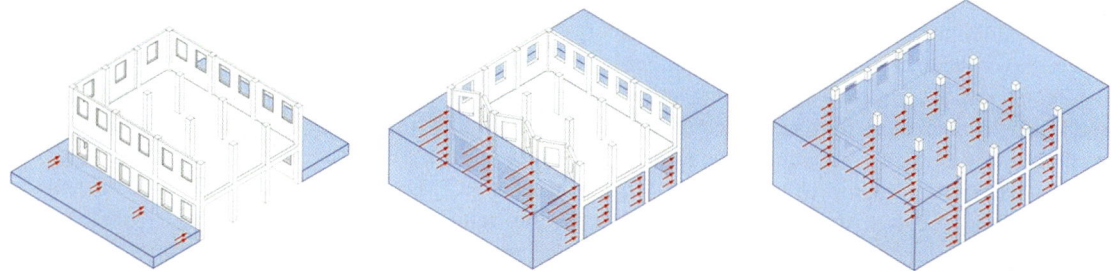

FIGURE 18.4 Depiction of how the tsunami inundation loading is affected by the integrity of external infill walls. The inundation flow is assumed to apply pressures to the external envelope of the structure (left) until the infill breaks away (middle), after which the water enters the building and applies loads to the internal structural elements (right).

18.4 Structural modeling for tsunami analysis

At the core of analytical fragility is the definition of a finite element model (FEM) of the structure that is capable of representing all potential local and global damage mechanisms that might occur in the building when it is subjected to a tsunami inundation flow. This requires the use of a relatively sophisticated FEM of the building, and models based on single-degree-of-freedom systems (Park et al., 2017) should be avoided as they are unable to capture damage mechanisms effectively.

At the basic level, the analyst should choose whether to model the structure in 2D (planar) or 3D. This depends on the plan asymmetry characteristics of the building and on whether the analyst wants to consider the effects of sequential damage to nonstructural elements. For example, a 3D model is most appropriate wherever significant eccentricity exists in the structure (e.g., causing enhanced torsional effects) and when the progressive damage to infills and the evolution of damage to external and internal structural elements need to be modeled. The type of structural analysis method adopted can also influence the modeling choice. For example, in the case of a time-history dynamic analysis, the use of a 3D model will result in a significantly higher computation time than a 2D model, which may impact the practicality of analytical fragility function development.

In the construction of the structural model, each structural element (e.g., beam, column, and load-bearing wall) and nonstructural element (e.g., masonry infill panel) should be represented by one or more finite elements if it is contributing to the lateral or vertical capacity of the structure. The exception is nonstructural infill walls with axes perpendicular to the flow. These do not provide a significant contribution to the lateral resistance of buildings to tsunami flow but do significantly contribute to loading on the structure. The latter can be considered by accounting for their contributing surface area in the calculation of tsunami actions on buildings (see VDPO-BI above), without the need to include them as elements in the structural model.

All permanent gravity loads (i.e., dead loads and live loads) must be defined and included in the model. According to existing guidelines for earthquake loading (European Committee for Standardization, 1998), it is possible to reduce the live loads acting on the building at the time of an earthquake. In the case of earthquake ground shaking, which excites the building mass, this reduction in live loads results in lower earthquake loading and less conservatism in structural assessment. In the case of tsunami loading, where structural mass does not affect the applied actions, a reduction in the effective mass is not necessarily beneficial. This is because the tsunami inundation can induce significant buoyancy forces in the structure, which are counteracted by the structural weight. It is also important to highlight that the approach to modeling dead and live loads may change according to the structural analysis method chosen. In TDA, the dead and live loads are set at the start of the analysis and remain constant throughout the analysis. In VDPO, the dead loads and live loads can be varied at each analysis step to consider the progressive failure of infills (reducing the dead loads) and inundation of floors (that may remove elements contributing to live loads).

Two main analytical modeling procedures for nonlinear analysis are adopted in the literature: fiber-based or plastic hinge-based structural modeling. Plastic-hinge-based structural modeling typically assumes plasticity is concentrated in predetermined zones within the structural element. In this way, "hinges" can be added to nodes in the model wherein the nonlinear behavior of the element is defined, with connecting elements assumed to remain elastic. In typical plastic hinge-based modeling, hinges are placed at the extremes of structural elements, and yielding is assumed to occur at the member ends. Readers are referred to Fragiadakis and Papadrakakis (2008) for further guidance. In the case of tsunami, element yielding and failure do not always occur at the ends of structural elements, and indeed, areas of failure are difficult to predict in advance of the structural analysis. Hence, the use of a plastic-hinge-based model is not recommended.

Fiber-based numerical techniques are most commonly adopted in past tsunami structural assessment studies, mainly to overcome the difficulty in predicting where elements will fail. In the fiber-based approach, each structural member is discretized into a number of elements with each boundary being linked to a discrete cross-section with a grid of fibers. Each fiber in the cross-section is associated with a uniaxial stress−strain relationship related to the material. The material stress−strain response in each fiber is integrated to get stress-resultant forces and stiffness terms, and from these, forces and stiffness over the length of the element are obtained through finite element interpolation functions, which must satisfy equilibrium and compatibility conditions (D'Ayala et al., 2014). The material constitutive models, therefore, drive the hysteretic response of the element. This numerical technique allows a more detailed characterization of the nonlinearity in elements with composite cross-sections (like reinforced concrete elements) by modeling separately the different behavior of the materials constituting the element cross-section (i.e., cover and core concrete and longitudinal steel). Moreover, using this approach, axial load-bending moment interaction can be modeled directly.

In fiber-based FEM, it is essential that the analyst ensures that the number of section fibers used for the equilibrium computations is sufficient to adequately represent the stress−strain distribution across the element's cross-section.

There is no set guidance for this, and sensitivity analyses may need to be carried out. Typically, this number lies between 100 and 150 fibers for reinforced concrete sections (Seismosoft, 2024). The number of integration sections also needs to be defined, with literature proposing between 4 and 7 integration sections for elements, with higher numbers required to represent element hardening response (D'Ayala et al., 2014). Again, a sensitivity check is advisable.

In general, both plastic hinge-based and fiber-based modeling approaches provide reliable simulations of flexural modes of failure and combined axial and flexural modes of failure. However, neither can simulate pure shear failures nor the reduction in flexural capacity due to combined shear and bending. This is particularly problematic for tsunami structural analysis, as tsunami loading applies large shear forces to structural elements, and many studies have reported shear failure to dominate structural failure under tsunami loading (Baiguera et al., 2022; Petrone et al., 2020). In all these studies, shear failure has been assessed through postanalysis checks with semiempirical equations for shear capacity evaluation obtained from codes of practice or the literature (Biskinis et al., 2004).

18.5 Tsunami damage assessment

Damage scales for assessing building damage to tsunamis have mainly been developed for use in posttsunami damage reconnaissance and provide descriptive definitions of *DS*s. A notable example is the Damage scale for Earthquakes and Tsunami, named (DET) scale (Rossetto, Raby, et al., 2019), developed for the assessment of earthquake and tsunami damage to buildings following the 2018 Sulawesi Earthquake and Tsunami, Indonesia, and adopted in an empirical fragility study (Baskaran et al., 2023). The scale is constituted by descriptions of observable damage (Table 18.2) but makes one crucial observation that nonstructural and structural damage is uniquely defined, irrespective of which hazard causes it. This is also recognized by Federal Emergency Management Agency (2017, 2022), which provides a damage scale containing three *DS*s (moderate, extensive, and collapse) that correspond with their HAZUS earthquake model.

This philosophy should underpin all hazard damage assessments and simply requires that all structural damage and failure mechanisms possible are captured by the damage scale used. To do this, a damage scale must be able to represent both local (section, element, and story-based) and global damage mechanisms and provide unambiguous guidance to evaluate *DS*s from these. This requires the damage scale to distinguish between buildings of different materials, load-resisting systems, and structural design. Moreover, descriptive *DS*s are not sufficient to be used for analytical fragility assessment, as structural performance is determined through the interpretation of *EDP*s that are output from the structural analysis.

It is common in the existing literature to find tsunami fragility functions being developed using only a single *EDP*, typically the maximum transient interstory drift. This is done either because a very simplified structural model is used or because the model response values for other *EDP*s are not retained due to the need to store large quantities of data output from the analyses (often the case with TDA). This practice is inherited from earthquake engineering, where nonlinear dynamic analysis is the gold standard for structural damage assessment and where the structural response is usually flexure-dominated. However, in the case of tsunamis, the representativeness of nonlinear static-based analysis enables the use of detailed structural models and the recording of many *EDP*s with fewer data storage issues. Moreover, it cannot be assumed that structural damage will be dominated by flexure-based mechanisms. It is, therefore, of critical importance to ensure that the damage scale used adopts *EDP*s that are defined at all of the following levels:

- Section level (e.g., strains, stresses, and curvature)
- Member level (e.g., rotations and local damage indices)
- Story level (e.g., interstory drift)
- Global level (e.g., top drift and maximum base shear)

Beyond the analysis type (see Section 18.3), the construction type and modeling approach may limit the choice and determine the appropriateness of *EDP*s used for the *DS* definition. Hence, it is difficult to define a single damage scale that can universally be used for all structure types, models, and analyses in analytical fragility. In practice, it is useful to start by plotting the occurrence of different *EDP*s onto the tsunami pushover curve to understand the evolution of damage and damage mechanisms in the structure being analyzed, as shown in Fig. 18.5.

The point of occurrence of different descriptive *DS*s can then be determined from the interpretation of the *DS* descriptions and damage evolution in the structure analyzed. Damage scales that are used in the fragility literature (e.g., DET) or widely recognized guidance (e.g., HAZUS) can be adopted for damage interpretation. The final damage scale will thus contain a tailored set of *EDP* types and values for damage thresholds that can appropriately represent the structure being assessed and all its possible ductile and brittle failure mechanisms.

The approach followed in Del Zoppo et al. (2023) to define a damage scale for the analytical assessment of the performance of reinforced concrete infilled frames illustrates this process. Del Zoppo et al. (2023) adopt as a basis the

TABLE 18.2 Overview of the damage scale for earthquakes and tsunami (DET).

DET damage state	Equivalent EMS-98 damage state	Usability level postdisaster	Reinforced concrete building damage	Masonry building damage [variation for confined masonry buildings in square brackets]	Timber building damage
No damage (DET0)	DS0	Immediate occupancy	No visible structural or nonstructural damage observed during the survey. Inundation with contents damage is possible.		
Slight damage (DET1)	DS1	Occupiable. Minor repair needed.	No structural damage. Damage to nonstructural elements only. Minor damage to infill walls and partitions. Damage to cladding, windows, doors, and fixtures.	No structural damage. Damage to nonstructural elements only. Damage limited to loss of plaster on walls, hairline cracking (<1 mm) visible in masonry walls [or in tie-beam/column joints], damage to cladding, windows, doors, and fixtures.	No frame damage. Light damage in wall panels (full or perforated). Damage in nonstructural elements like windows, doors and roof cover.
Moderate damage (DET2)	DS2 & 3	Suitable for occupancy after significant repair	No structural member failure. Fine cracking to spalling of concrete in structural elements. Repairable damage from debris impact to individual to few structural members, without compromising structural stability. Out-of-plane failure or collapse of parts of or whole sections of infill walls without compromising structural stability. Scouring at corners of the structures leaving foundations partly exposed but repairable by backfilling.	No structural component fails. Cracks up to 3 mm in masonry walls [in most of the tie beam/column joints and at the base of the tie-columns] without compromising structural integrity. Masonry wall can be repaired or rebuilt to restore integrity. Scouring at corners of the structures leaving foundations partly exposed but repairable by backfilling. Cracks caused by undermined foundations are visible on walls but not critical. Heavy damage to collapse of non-structural components (e.g., roof cover, gables, or parapets)	No frame component fails. Some frame connections damaged (e.g., pull-out of some nails or fixings) but structural stability maintained. Failure of some wall panels. Scouring at corners of the structures leaving foundations partly exposed but repairable by backfilling. Heavy damage to collapse of nonstructural components (e.g., roof cover, gables, or parapets)
Very heavy damage (DET3)	DS4	Structure is unsafe and will require demolition	The structure is standing but is very heavily damaged. Collapse of a few columns or of a single upper floor possible. Structural elements damaged or failed (e.g., large cracks in structural elements, compression failure of concrete, buckling/fracture of rebar). Roofs are damaged and have to be totally replaced or repaired.	The structure is standing but is very heavily damaged. Out-of-plane failure or collapse of masonry wall panels beyond repair [major cracks, > 5 mm, and/or rebar yielding in concrete elements and joints]. Structural integrity compromised. Roof structure damaged. Excessive foundation settlement and tilting beyond repair. Partial collapse or corner failure of walls due to scouring. Collapse of most (>70%) nonstructural components (e.g., roof cover, gables, or parapets).	The frame is standing but is very heavily damaged. Structural integrity compromised. Roof structure damaged. Excessive foundation settlement or frame tilting beyond repair. Collapse of most (>70%) nonstructural components (e.g., roof cover, gables, or parapets).
Collapse/washed away (DET4)	DS5	Structure requires demolition	Complete structural damage or collapse. Structure and tilting.	Complete structural damage or collapse. Structure might have been washed away, or structure highly unstable due to excessive foundation settlement	

Source: Adapted from Rossetto, T., Raby, A., Brennan, A., Lagesse, R., Robinson, D., Adhikari, R. K., Rezki-Hr, M., Meilianda, E., Idris, Y., Rusydy, Y., & Kumala, I. D. (2019). *The Central Sulawesi, Indonesia earthquake and tsunami of 28th September 2018—A field report by EEFIT-TDMRC.* Earthquake Engineering Field Investigation Team Report. Available from: https://www.istructe.org/IStructE/media/Public/Resources/report-eefit-mission-sulawesi-indonesia-20200214.pdf.

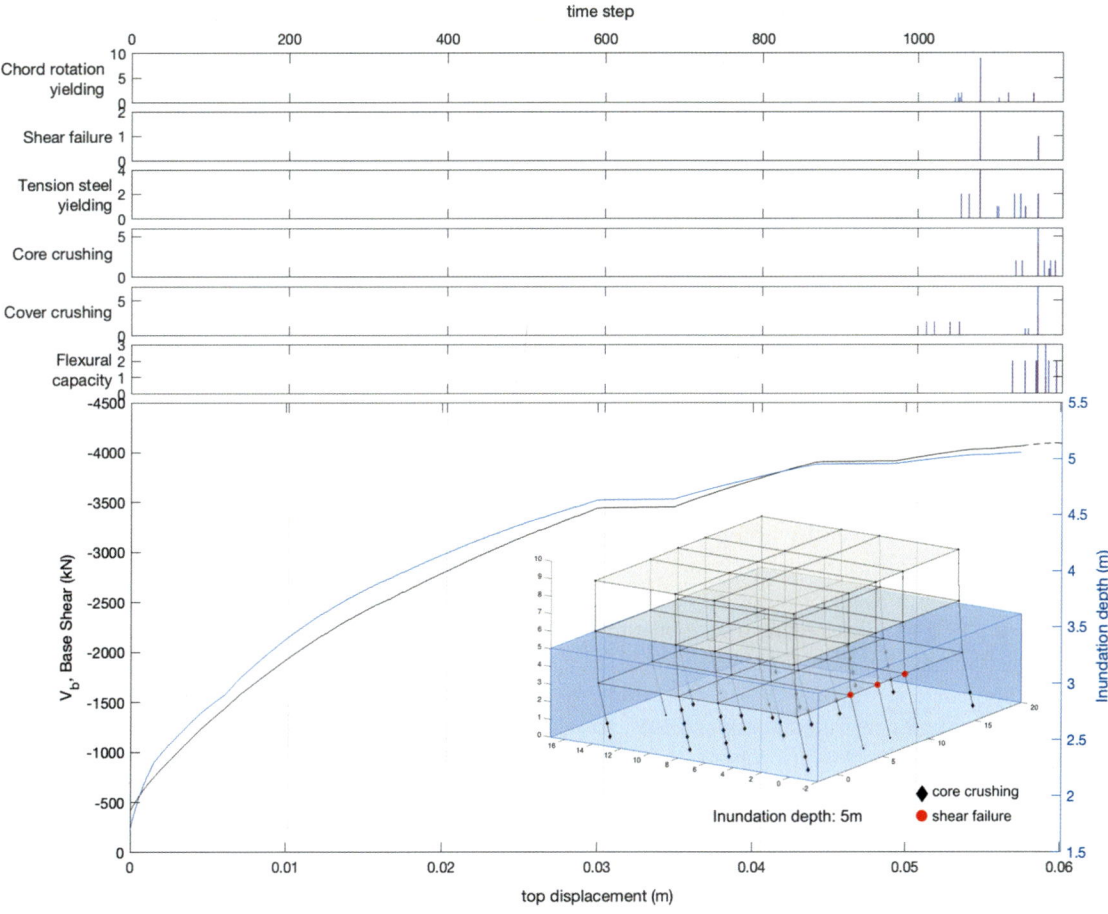

FIGURE 18.5 Depiction of damage evolution in an infilled frame analyzed through VDPO2. The inset shows the scaled-up displaced shape of the building at the point of collapse (inundation depth of 5 m). *VDPO*, Variable depth pushover.

TABLE 18.3 Summary of the damage scale used for interpreting structural damage states in Del Zoppo et al. (2023).

Damage state	Slight	Moderate	Extensive	Complete
Damage state definition	First achievement in any vertical member of concrete cracking	First achievement in any vertical member of ½ steel yield strain in the longitudinal steel Or First achievement of slab flexural capacity	First achievement in any vertical member of steel yield strain in the longitudinal steel Or First achievement in any vertical member of shear capacity	Peak base shear is reached Or Achievement of the shear capacity in two consecutive vertical members

damage descriptions in Federal Emergency Management Agency (2017), which, however, focus on the global damage to the structure rather than on the failure of individual elements and do not define damage to nonstructural elements sufficiently. The resulting damage scale reinterpretation by Del Zoppo et al. (2023) is shown in Table 18.3.

It is important to note that collapse is defined by the occurrence of either the maximum base shear capacity of the building being reached or the occurrence of column shear failure (in two columns in this case), whichever occurs first. The former metric is commonly adopted for the definition of structural collapse in existing analytical fragility functions in recognition that due to the sustained nature of tsunami inundation flow loads, structural ductility cannot be relied on to help withstand induced postpeak structural deformations (Rossetto et al., 2018). The latter *EDP* recognizes that shear capacity is not explicitly modeled in the structural modeling but the occurrence of shear failure in key structural elements can precipitate collapse. Moreover, the arrangement of elements has been considered in defining how many elements need to reach a certain performance level before the *DS* is reached.

18.6 Development of analytical fragility curves

Once the *DS*s have been appropriately defined, data in the form of *IM* values corresponding to the achievement of each *DS* are derived from the structural assessment procedures discussed in Section 18.3. For the nonlinear static analysis approach, the inundation depth at different levels of structural damage can be directly interpreted from the pushover curve. These velocity values corresponding to the inundation depth can then be interpreted through the adopted Froude number. To obtain a large set of *IM*-damage data, the analysis can be repeated for several Froude number values, different blockage ratios (both of these change the loading history), or different variations in the structural model. In the case of TDA, the *DS* is determined from the response of the structure during the time-history, and the maximum *IM* values are calculated from the tsunami time-history. For TDA, the analysis has to be rerun for a different (or scaled up) tsunami time-history to obtain *IM* data for other *DS*s.

Fragility functions are derived by fitting statistical models to the developed *IM*-damage data. The same procedures for model fitting and confidence bound definition are used in analytical and empirical fragility. Fig. 18.6 depicts an analytical fragility curve for a 3-story reinforced concrete residential structure and the collapse (DET 4) *DS*. The reader is referred to Chapter 17 for further details on the procedures.

Aleatory uncertainties should be considered for both capacity and demand parameters in the probabilistic framework adopted to develop tsunami fragility curves. Uncertainties on capacity are related to geometric and material properties adopted for structural and nonstructural components, as well as the capacity models used to assess threshold values for *EDP*s. Historical data or in situ inspections may be helpful in the characterization of uncertainties related to the mechanical properties of materials. Uncertainties on tsunami demand are mainly related to the velocity and the Froude number characterizing the tsunami flow at the building location. Uncertain parameters are considered independent random variables and are modeled through their statistical distributions. Latin Hypercube algorithm and Monte Carlo simulation can be used to generate random combinations of the selected aleatory variables.

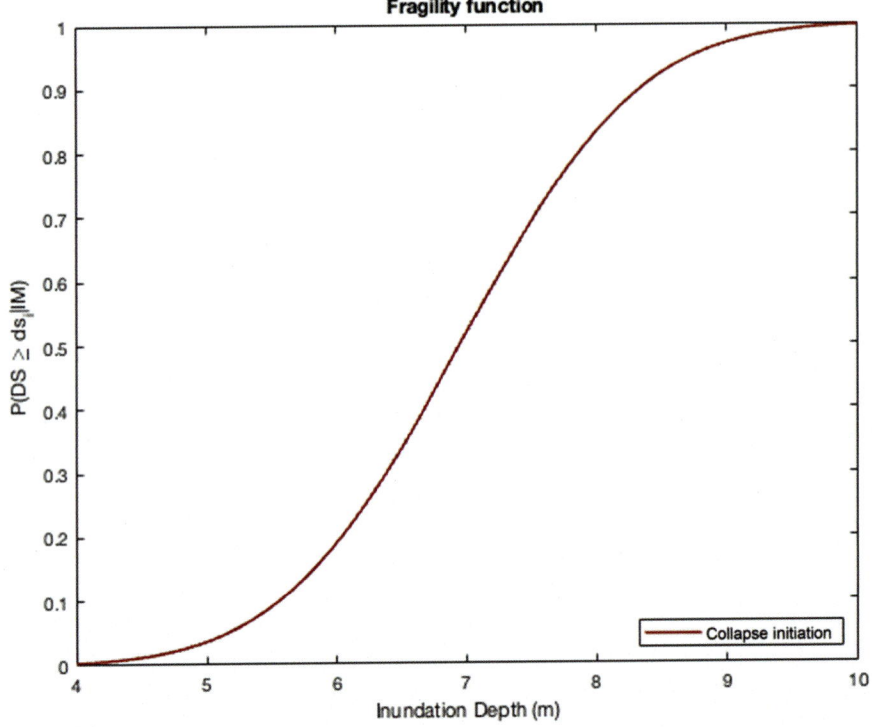

FIGURE 18.6 Tsunami fragility function for the collapse damage state of a three-story residential reinforced concrete building.

18.7 Conclusions

This chapter has provided insights into current approaches for developing analytical fragility functions. The fundamental components of tsunami loading characterization, structural response simulation, and damage definition have been described, and it is shown that challenges still exist in each of these components. The field of analytical fragility, and indeed of tsunami engineering, is relatively new. Advances in knowledge on tsunami−structure interaction and our ability to model tsunami-induced failures are in fast evolution. It is observed that a concerted effort by the research community is still needed to recognize and implement widely a hazard-agnostic approach to building damage characterization and *DS* tailoring to suit the characteristics of individual buildings. It is hoped that new research into tsunami scour and debris will see these damage mechanisms entering the analytical literature in the near future. It is also observed that structural analysis techniques are constantly being updated and improved and are gradually entering the realm of commercial software, accompanied by a wider offering of open analysis tools.

References

Alam, M. S., Barbosa, A. R., Scott, M. H., Cox, D. T., & van de Lindt, J. W. (2018). Development of physics-based tsunami fragility functions considering structural member failures. *Journal of Structural Engineering*, *144*(3). Available from https://doi.org/10.1061/(asce)st.1943-541x.0001953.

ASCE/SEI, American Society of Civil Engineers. (2017). *Minimum design loads and associated criteria for buildings and other structures*. ASCE/SEI. Available from: https://ascelibrary.org/doi/book/10.1061/9780784414248.

ASCE/SEI, American Society of Civil Engineers. (2022). *Minimum design loads and associated criteria for buildings and other structures*. ASCE/SEI. Available from: https://ascelibrary.org/doi/book/10.1061/9780784415788.

Attary, N., Unnikrishnan, V. U., van de Lindt, J. W., Cox, D. T., & Barbosa, A. R. (2017). Performance-based tsunami engineering methodology for risk assessment of structures. *Engineering Structures*, *141*, 676−686. Available from https://doi.org/10.1016/j.engstruct.2017.03.071, http://www.journals.elsevier.com/engineering-structures/.

Baiguera, M., Rossetto, T., Robertson, I. N., & Petrone, C. (2019). Towards a tsunami nonlinear static analysis procedure for the ASCE 7 standard. *In Proceedings of the International Conference on Natural Hazards and Infrastructure*, Chania, Greece. http://iconhic.com/2019/.

Baiguera, M., Rossetto, T., Robertson, I. N., & Petrone, C. (2022). A procedure for performing nonlinear pushover analysis for tsunami loading to ASCE 7. *Journal of Structural Engineering*, *148*(2). Available from https://doi.org/10.1061/(asce)st.1943-541x.0003256.

Baskaran, H., Salah, P., Sathurshan, M., Cels, J., Ioannou, I., Rossetto, T., Thamboo, J., Dias, P., Warder, S., & Piggott, M. (2023). *Empirical tsunami fragility assessment of masonry schools in Sri Lanka. Society for Earthquake and Civil Engineering Dynamics*. In Proceedings of the SECED2023 Conference: Earthquake Engineering and Dynamics for a Sustainable Future. Cambridge. Available from: https://registrations.hg3conferences.co.uk/hg3/frontend/reg/thome.csp?pageID = 89507&eventID = 237&traceRedir = 2.

Biskinis, D. E., Roupakias, G. K., & Fardis, M. N. (2004). Degradation of shear strength of reinforced concrete members with inelastic cyclic displacements. *ACI Structural Journal.*, *101*(6), 773−783. Available from https://doi.org/10.14359/13452.

Charvet, I., Macabuag, J., & Rossetto, T. (2017). Estimating tsunami-induced building damage through fragility functions: Critical review and research needs. *Frontiers in Built Environment*, *3*. Available from https://doi.org/10.3389/fbuil.2017.00036, https://www.frontiersin.org/articles/10.3389/fbuil.2017.00036/pdf.

D'Ayala, D., Meslem, A., Vamvastikos, D., Porter, K., Rossetto, T., Crowley, H., & Silva, V. (2014). *Guidelines for analytical vulnerability assessment of low/mid-rise Buildings—Methodology*. Pavia, Italy: GEM Foundation.

Del Zoppo, M., Di Ludovico, M., & Prota, A. (2021). Methodology for assessing the performance of RC structures with breakaway infill walls under tsunami inundation. *Journal of Structural Engineering.*, *147*(2).

Del Zoppo, M., Rossetto, T., Di Ludovico, M., & Prota, A. (2023). Effect of buoyancy loads on the tsunami fragility of existing reinforced concrete frames including consideration of blow-out slabs. *Scientific Reports*, *13*(1). Available from https://doi.org/10.1038/s41598-023-36237-7, https://www.nature.com/srep/.

Del Zoppo, M., Wijesundara, K., Rossetto, T., Dias, P., Baiguera, M., Di Ludovico, M., Thamboo, J., & Prota, A. (2021). Influence of exterior infill walls on the performance of RC frames under tsunami loads: Case study of school buildings in Sri Lanka. *Engineering Structures*, *234*, 111920. Available from https://doi.org/10.1016/j.engstruct.2021.111920.

European Committee for Standardization. (1998). *Eurocode 8: Design provisions for earthquake resistance of structures, Part 1.1: General rules, seismic actions and rules for buildings*.

Federal Emergency Management Agency. (2017). *HAZUS tsunami model technical guidance (Version4.0)*.

Federal Emergency Management Agency. (2022). *HAZUS tsunami model technical manual (HAZUS 5.1)*. Federal Emergency Management Agency. Available from: https://www.fema.gov/sites/default/files/documents/fema_hazus-tsunami-model-technical-manual-5-1.pdf.

Foster, A. S. J., Rossetto, T., & Allsop, W. (2017). An experimentally validated approach for evaluating tsunami inundation forces on rectangular buildings. *Coastal Engineering*, *128*, 44−57. Available from https://doi.org/10.1016/j.coastaleng.2017.07.006, http://www.elsevier.com/inca/publications/store/5/0/3/3/2/5/.

Fragiadakis, M., & Papadrakakis, M. (2008). Performance-based optimum seismic design of reinforced concrete structures. *Earthquake Engineering and Structural Dynamics*, *37*(6), 825−844. Available from https://doi.org/10.1002/eqe.786, http://onlinelibrary.wiley.com/journal/10.1002/(ISSN)1096-9845.

Grünthal, G. (1998). *European macroseismic scale 1998 (EMS-98)*. Cahiers de Centre Europèen de Géodynamique et de Séismologie. ISBN No2-87977-008-4.

Kriebe, D. L., Lynett, P. J., Cox, D. T., Petroff, C. M., Robertson, I. N., & Chock, G. Y. K. (2017). Energy method for approximating overland tsunami flows. *Journal of Waterway, Port, Coastal and Ocean Engineering*, *143*(5). Available from https://doi.org/10.1061/(ASCE)WW.1943-5460.0000393, http://ojps.aip.org/wwo/.

Macabuag, J., Rossetto, T., Ioannou, I., Suppasri, A., Sugawara, D., Adriano, B., Imamura, F., Eames, I., & Koshimura, S. (2016). A proposed methodology for deriving tsunami fragility functions for buildings using optimum intensity measures. *Natural Hazards*, *84*(2), 1257−1285. Available from https://doi.org/10.1007/s11069-016-2485-8, http://www.wkap.nl/journalhome.htm/0921-030X.

McGovern, D. J., Robinson, T., Chandler, I. D., Allsop, W., & Rossetto, T. (2018). Pneumatic long-wave generation of tsunami-length waveforms and their runup. *Coastal Engineering*, *138*, 80−97. Available from https://doi.org/10.1016/j.coastaleng.2018.04.006, http://www.elsevier.com/inca/publications/store/5/0/3/3/2/5/.

McGovern, D. J., Todd, D., Rossetto, T., Whitehouse, R. J. S., Monaghan, J., & Gomes, E. (2019). Experimental observations of tsunami induced scour at onshore structures. *Coastal Engineering*, *152*, 103505. Available from https://doi.org/10.1016/j.coastaleng.2019.103505.

Mehrzad, R., Nistor, I., & Rennie, C. (2022). Scour mechanics of a tsunami-like bore around a square structure. *Journal of Waterway, Port, Coastal and Ocean Engineering*, *148*(1). Available from https://doi.org/10.1061/(ASCE)WW.1943-5460.0000686, https://ascelibrary.org/journal/jwped5.

Nanayakkara, K. I. U., & Dias, W. P. S. (2016). Fragility curves for structures under tsunami loading. *Natural Hazards*, *80*(1), 471−486. Available from https://doi.org/10.1007/s11069-015-1978-1, http://www.wkap.nl/journalhome.htm/0921-030X.

Nistor, I., Palermo, D., Nouri, Y., Murty, T., & Saatcioglu, M. (2017). *Tsunami-induced forces on structures*, . Handbook of coastal and ocean engineering: Expanded edition (1−2). World Scientific. Available from http://www.worldscientific.com/worldscibooks/10.1142/10353, doi:10.1142/9789813204027_0018.

Park, H., Cox, D. T., & Barbosa, A. R. (2017). Comparison of inundation depth and momentum flux based fragilities for probabilistic tsunami damage assessment and uncertainty analysis. *Coastal Engineering*, *122*, 10−26. Available from https://doi.org/10.1016/j.coastaleng.2017.01.008, http://www.elsevier.com/inca/publications/store/5/0/3/3/2/5/.

Petrone, C., Rossetto, T., Baiguera, M., la Barra Bustamante, C. D., & Ioannou, I. (2020). Fragility functions for a reinforced concrete structure subjected to earthquake and tsunami in sequence. *Engineering Structures*, *205*. Available from https://doi.org/10.1016/j.engstruct.2019.110120, http://www.journals.elsevier.com/engineering-structures/.

Petrone, C., Rossetto, T., & Goda, K. (2017). Fragility assessment of a RC structure under tsunami actions via nonlinear static and dynamic analyses. *Engineering Structures*, *136*, 36−53. Available from https://doi.org/10.1016/j.engstruct.2017.01.013, http://www.journals.elsevier.com/engineering-structures/.

Pomonis, A., Saito, K., Fraser, S., Chian, S. C., Goda, K., Macabuag, J., Offord, M., Raby, A., & Sammonds, P. (2011). *The Mw9.0 Tohoku earthquake and tsunami of 11th March 2011—A field Report by EEFIT*. Available from: https://www.istructe.org/IStructE/media/Public/Resources/report-eefit-mission-japan-20111203.pdf.

Qi, Z. X., Eames, I., & Johnson, E. R. (2014). Force acting on a square cylinder fixed in a free-surface channel flow. *Journal of Fluid Mechanics*, *756*, 716−727. Available from https://doi.org/10.1017/jfm.2014.455, http://journals.cambridge.org/action/displayJournal?jid = FLM.

Rossetto, T., De la Barra, C., Petrone, C., De la Llera, J. C., Vásquez, J., & Baiguera, M. (2019). Comparative assessment of nonlinear static and dynamic methods for analysing building response under sequential earthquake and tsunami. *Earthquake Engineering and Structural Dynamics*, *48*(8), 867−887. Available from https://doi.org/10.1002/eqe.3167, http://onlinelibrary.wiley.com/journal/10.1002/(ISSN)1096-9845.

Rossetto, T., Peiris, N., Pomonis, A., Wilkinson, S. M., Del Re, D., Koo, R., & Gallocher, S. (2007). The Indian Ocean tsunami of December 26, 2004: Observations in Sri Lanka and Thailand. *Natural Hazards*, *42*(1), 105−124. Available from https://doi.org/10.1007/s11069-006-9064-3.

Rossetto, T., Petrone, C., Eames, I., De La Barra, C., Foster, A., & Macabuag, J. (2018). Advances in the assessment of buildings subjected to earthquakes and tsunami. *Geotechnical, Geological and Earthquake Engineering*, *46*. Available from https://doi.org/10.1007/978-3-319-75741-4_23, http://www.springerlink.com/content/1573-6059/.

Rossetto, T., Raby, A., Brennan, A., Lagesse, R., Robinson, D., Adhikari, R. K., Rezki-Hr, M., Meilianda, E., Idris, Y., Rusydy, Y., & Kumala, I. D. (2019). *The Central Sulawesi, Indonesia earthquake and tsunami of 28th September 2018—A field report by EEFIT-TDMRC*. Earthquake Engineering Field Investigation Team Report. Available from: https://www.istructe.org/IStructE/media/Public/Resources/report-eefit-mission-sulawesi-indonesia-20200214.pdf.

Seismosoft. (2024). *SeismoStruct 2024 − A computer program for static and dynamic nonlinear analysis of framed structures*. Available from https://seismosoft.com/.

Song, J., & Goda, K. (2019). Influence of elevation data resolution on tsunami loss estimation and insurance rate-making. *Frontiers in Earth Science*, *7*. Available from https://doi.org/10.3389/feart.2019.00246, https://www.frontiersin.org/journals/earth-science.

Chapter 19

Modeling and uncertainty in probabilistic tsunami hazard and risk assessment

Nobuhito Mori and Takuya Miyashita
Disaster Prevention Research Institute, Kyoto University, Uji, Kyoto, Japan

19.1 Introduction

Giant tsunamis have caused several catastrophic disasters, resulting in enormous casualties and economic losses in the last few decades. Furthermore, the 2004 Indian Ocean Earthquake Tsunami (Fujii & Satake, 2007; Hébert & Schindelé, 2015; McCloskey et al., 2008) and the 2011 Tohoku Earthquake Tsunami (Mori et al., 2012; Shimozono et al., 2012; Suppasri et al., 2013) revealed a particularly complex tsunami behavior that was not observed before. It is also apparent that the layout of the building significantly influences both tsunami inundation inflow and backflow (Moris et al., 2021; Prasetyo et al., 2019). Understanding the fluid-dynamic characteristics of tsunamis from offshore to inland is essential for accurate tsunami hazard and risk assessments for disaster risk reduction.

The assessment of tsunami behavior is also complex and challenging because of many epistemic and natural (i.e., aleatory) uncertainties associated with the source, inundation, propagation, and impact phenomena (Annaka et al., 2007; Fukutani et al., 2018). The randomness of earthquake phenomena, fluid behavior, prediction errors, and numerical model inaccuracies cause these uncertainties. Fig. 19.1 shows that, when focusing on seismogenic (earthquake-induced) tsunamis, there are uncertainties from tsunami generation and inundation to tsunami damage and loss. These uncertainties can be broadly classified into three categories: earthquake source, offshore tsunami propagation, and coastal and river inundation.

Incorporating these uncertainties into the design process of soft and hard countermeasures (i.e., mitigation strategies) requires appropriate uncertainty modeling and propagation. Moreover, modeling uncertainty (a special typology of epistemic uncertainty) should be appropriately accounted for, for example, using multiple models for the earthquake recurrence time, such as simple Poisson or renewal models (Goda, 2020). These assumptions significantly affect the tsunami intensity measures (IMs) (e.g., tsunami wave height, tsunami depth, and flow velocity), exposure damage states (e.g., minor and washed away), and losses (e.g., economic loss and casualties). Probabilistic tsunami hazard assessment (PTHA) and probabilistic tsunami risk assessment (PTRA) are methodologies that model and propagate uncertainties and quantitatively assess tsunami IMs and losses as a function of frequency or probability. These methodologies are structured to provide a comprehensive quantitative evaluation of tsunami hazards and risks, avoiding biased evaluations. Moreover, they allow easy implementation of model uncertainties using logic trees.

Fig. 19.2 shows a general flowchart for PTHA (Mori et al., 2018). The primary outcome of PTHA is the hazard curve for the target location (Geist & Parsons, 2006; Grezio et al., 2017). This hazard curve represents the relationship between a specific IM and the mean annual rate of occurrence. Typical IMs for tsunami hazards are tsunami wave height and inundation depth. Such IMs are equivalent to the peak ground acceleration (PGA) or peak ground velocity (PGV) for seismic hazard assessments (Cornell, 1968). Similarly, the risk curve is the main output of PTRA and represents the relationship between economic loss and the mean annual rate of occurrence (Goda & De Risi, 2018; Goda & Song, 2016).

Fig. 19.3 shows an example of tsunami hazard curves for the coast of the Tonankai region of Japan by two different PTHA methods (Miyashita et al., 2019). Fig. 19.3A shows an example of conditional tsunami hazard curves by the logic-tree method. The horizontal and vertical axes represent the inundation depth (IM) and annual exceedance probability, respectively. Different curves are computed for each value of magnitude, accounting for different assumptions and models. The curves are then integrated, as per Chapter 5, into a single hazard curve, which is the outcome of PTHA (Fig. 19.3B).

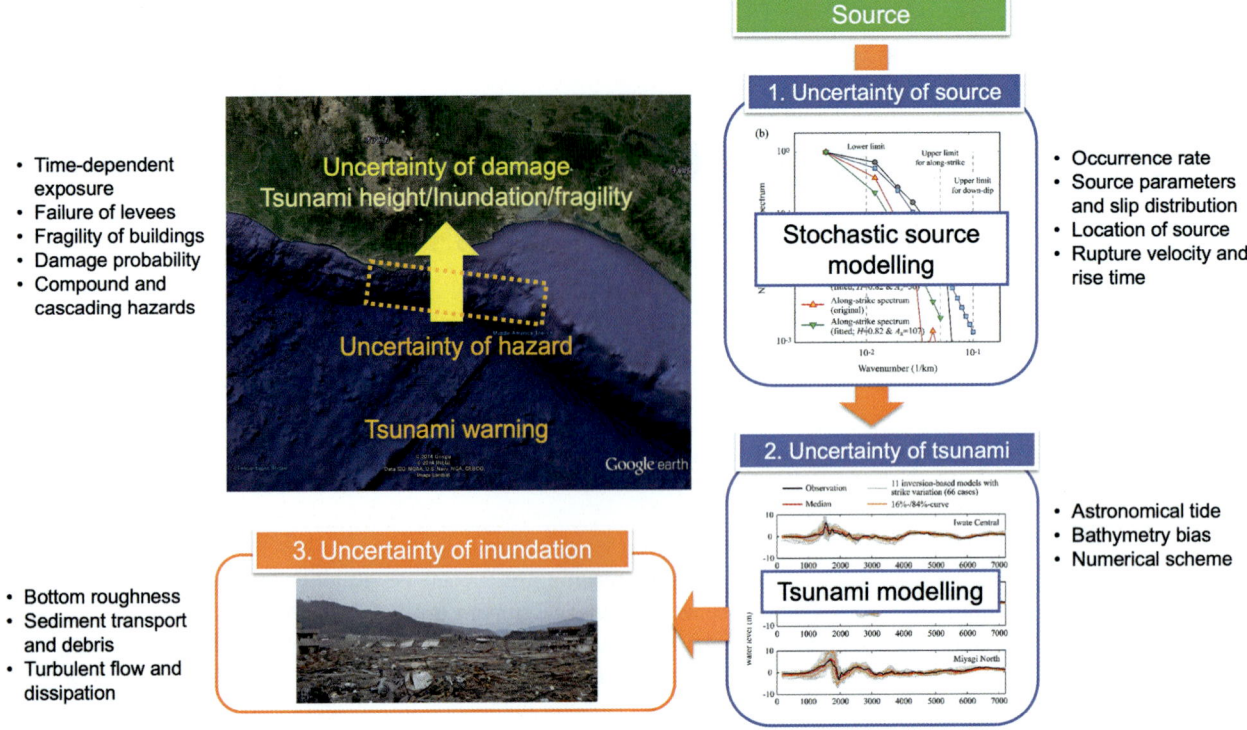

FIGURE 19.1 Individual uncertainties considered during probabilistic tsunami hazard analysis and probabilistic tsunami risk assessment.

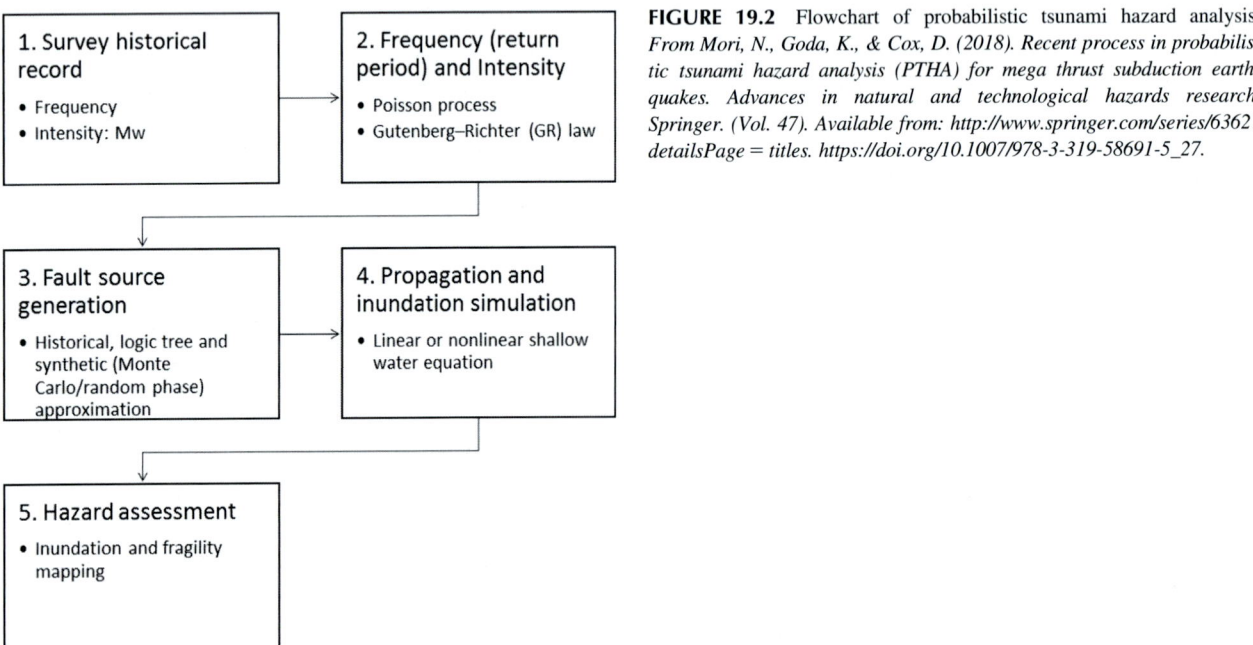

FIGURE 19.2 Flowchart of probabilistic tsunami hazard analysis. *From Mori, N., Goda, K., & Cox, D. (2018). Recent process in probabilistic tsunami hazard analysis (PTHA) for mega thrust subduction earthquakes. Advances in natural and technological hazards research. Springer. (Vol. 47). Available from: http://www.springer.com/series/6362?detailsPage = titles. https://doi.org/10.1007/978-3-319-58691-5_27.*

A generic equation for the tsunami hazard curve can be expressed as

$$\upsilon_{IM}(IM \geq im) = \sum_{i=1}^{N_S} \lambda_{M_{\min,i}} \int P_i(IM \geq im|\boldsymbol{\theta}) S_i(\boldsymbol{\theta}|m) f_{i,M}(m) |dim\|d\boldsymbol{\theta}\|dm| \quad (19.1)$$

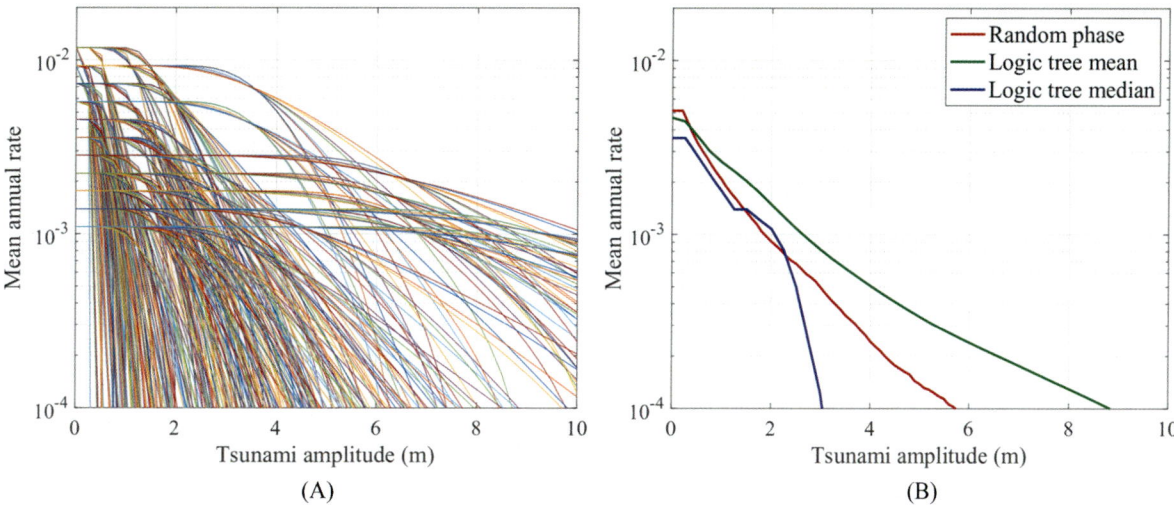

FIGURE 19.3 Examples of (A) conditional tsunami hazard curves in a logic tree model and (B) combined hazard curves. The horizontal axis represents the tsunami amplitude/inundation depth (typical intensity measure for probabilistic tsunami hazard analysis), and the vertical axis represents the mean annual rate. These figures were regenerated based on Miyashita et al. (2019) for the coast of the Tonankai region of Japan. *Modified from Miyashita, T., Mori, N., & Goda, K. (2019). Uncertainty quantification of tsunami height for future earthquakes in West Japan. International Conference on Applications of Statistics and Probability in Civil Engineering, ICASP 2019 Seoul National University, https://doi.org/10.22725/ICASP13.466.*

where $v_{IM}(IM \geq im)$ is the mean annual rate of a tsunami event with an *IM* larger than the threshold *im*, $\lambda_{M_{min,i}}$ is the mean annual rate of occurrence of tsunamigenic events from source i, $f_{i,M}(m)$ is the conditional probability density function for $M \geq M_{min}$, $S_i(\theta|m)$ is the probability density function of the set of source parameters *S* for a given magnitude *M*, and $P_i(IM \geq im|\theta)$ is the complementary cumulative distribution function of *IM* given the parameters θ. $P_i(IM \geq im|\theta)$ is evaluated through numerical tsunami simulations. The uncertainty associated with the variable source characteristics is captured by $S_i(\theta|m)$.

This chapter provides an overview of the uncertainties inherent in tsunami hazard and risk assessment methodologies and their applications. First, the main uncertainties are listed. Subsequently, the procedures to model these uncertainties are described. Finally, the method for integrating all the models, their applications, limitations, and issues to be resolved are discussed.

19.2 Uncertainty classification

19.2.1 Uncertainties related to tsunami source

This section describes the uncertainties associated with earthquake-induced tsunamis. Megathrust earthquakes are fault rupture phenomena caused by slips on the interplate surface. The rupture nucleates at a specific point and propagates on the interplate surface (Fan & Zhao, 2021). The rupture area and the overall slip determine the earthquake magnitude (Kanamori, 1978). Predicting when and where rupture starts, propagates, and ends is difficult. This is why the uncertainties regarding the seismic source contribute most significantly to the variation in inundation *IM*s (Goda et al., 2014; Mori et al., 2017). Therefore, a stochastic approach is essential in quantifying the variability of earthquakes and tsunamis. This section lists the uncertainties associated with the seismic source significantly influencing tsunami hazard and risk assessments.

First, the earthquake occurrence rate in the target seismic source area is essential for the probabilistic assessment, as shown in Fig. 19.2. Significant variations exist in the recurrence intervals and rates of these large earthquakes. For example, according to the historical earthquake catalog and rupture regions of the Nankai Trough, Japan, major earthquakes with fault ruptures longer than 500 km occur approximately once every 100–250 years. However, not all trough segments ruptured simultaneously; they ruptured partially, with adjacent segments sometimes rupturing within a relatively short time difference of hours to several years (Goda et al., 2020). This time interval varies, and seismic gap areas are known to exist. However, some seismic source areas experienced nearly reproducible slips of a specific magnitude at regular intervals, such as off the coast of the Sanriku area along the Japan Trench (Matsuzawa, 2002). In general, earthquake occurrence and slip amount are strongly related to strain accumulation caused by plate movement. They can be divided into areas where strain is either easy or difficult to release. While the relationship between strain accumulation and release on a plate interface governs recurrence intervals, many aspects remain unexplained, making it challenging to determine recurrence intervals quantitatively (Satake, 2015).

Earthquake slip distributions cannot be determined before the progression of the fault rupture. The physical processes of small to large earthquakes are similar during the initial rupture stages (Ide, 2019). Therefore, whether an earthquake will be large or small may not be known until the rupture ends. The amount of slip affects seafloor topography changes for tsunamigenic earthquakes (i.e., the initial water level of a tsunami), and where the rupture stops determines the fault length and width and, consequently, the earthquake magnitude. It follows that, for PTHA and PTRA, considering many possible magnitude values and slip patterns is essential.

The uncertainty of the tsunami source, owing to insufficient information on the three-dimensional structure of the plate boundary surface, is also significant. The interplate surface is estimated from past geophysical surveys (Nakanishi et al., 1998). However, its top-depth information may contain some bias. Tsunami inundation depth and damage are highly sensitive to the depth of the rupture fault. In addition, although the shear modulus of rocks is a crucial parameter that determines the moment magnitude, it is often assumed to be spatially uniform. Some studies have used a heterogeneous shear modulus (Geist & Bilek, 2001; Salazar-Monroy et al., 2021). This parameter is closely related to the rupture velocity, which is the speed at which the rupture propagates and is an important factor in the kinematics of a fault rupture. Tsunami waveforms are also sensitive to these parameters (Fukutani et al., 2016).

Various hypocenters of an earthquake and starting points of the rupture (nucleation point) are also possible for the target fault plane. According to the historical database, there is a low correlation between the nucleation point and the slip distribution, indicating that it can occur anywhere regardless of the slip amount (Chang & Ide, 2021; Mai et al., 2005). Moreover, determining the total area of the slips is not straightforward. Additionally, nonuniform slips are needed to consider the occurrence locations and slips probabilistically.

19.2.2 Uncertainty related to tsunami propagation

The tsunami propagation process involves further uncertainties. These uncertainties are independent of the seismic activity. First, the astronomical sea-level variability, known as the tide, depends on the location and time. Given the unpredictability of the earthquake occurrence time, the tidal level at the time of tsunami generation and propagation is uncertain. The tidal range cannot be ignored in most of the open ocean, and damage can depend significantly on whether the tide is high or low when a tsunami strikes. In particular, the high tide condition can dramatically increase the force acting on levees, potentially determining whether flooding occurs.

As water depth is a dominant factor for long-wave propagation, bathymetry significantly influences tsunami propagation. The bathymetry distribution includes measurement errors in both shallow and deep waters that significantly affect the results. In the middle- to long-term, the bathymetry and onshore topography can change due to storm waves, high tides during storm surges, or other coastal processes, and it affects the trend and variability of bathymetry distributions. Furthermore, coastal uplift and subsidence patterns can vary due to plate tectonic movements not accompanied by earthquakes (typically less than 1 m over 100 years).

Breaking waves occur when water particle velocity exceeds a certain threshold, and the wave energy is dissipated. Introducing the breaking energy dissipation term in inundation simulations is computationally demanding. In particular, three-dimensional fluid dynamic behavior due to breaking is necessary to determine the maximum fluid velocity in urban areas. However, such detailed inundation simulations for urban areas are not feasible practically due to current computational costs.

In addition to tidal and topographic conditions, uncertainty in propagation related to numerical models is also important. Tsunami inundation depths and flow speeds can differ slightly depending on numerical models. Fig. 19.4 compares a few numerical tsunami models (Lynett et al., 2017). The study compared the time series of tsunami amplitudes and velocities at specific points for a tsunami benchmark problem. As shown in the figure, even if the phases are almost the same, differences in amplitude are observed depending on the numerical models, particularly in the subsequent waves. Therefore, divergent assessments of tsunamis that reach their maximum amplitudes in successive waves are possible, depending on the numerical model used. This is owing to the differences in the governing equations and numerical schemes. The governing equations for tsunamis are generally nonlinear shallow-water equations; however, in some cases, considering dispersive waves is necessary to reproduce realistic behavior depending on the wavelength and water depth (Baba et al., 2017; Glimsdal et al., 2013; Watada, 2023). Recent studies have found that the impact of errors in

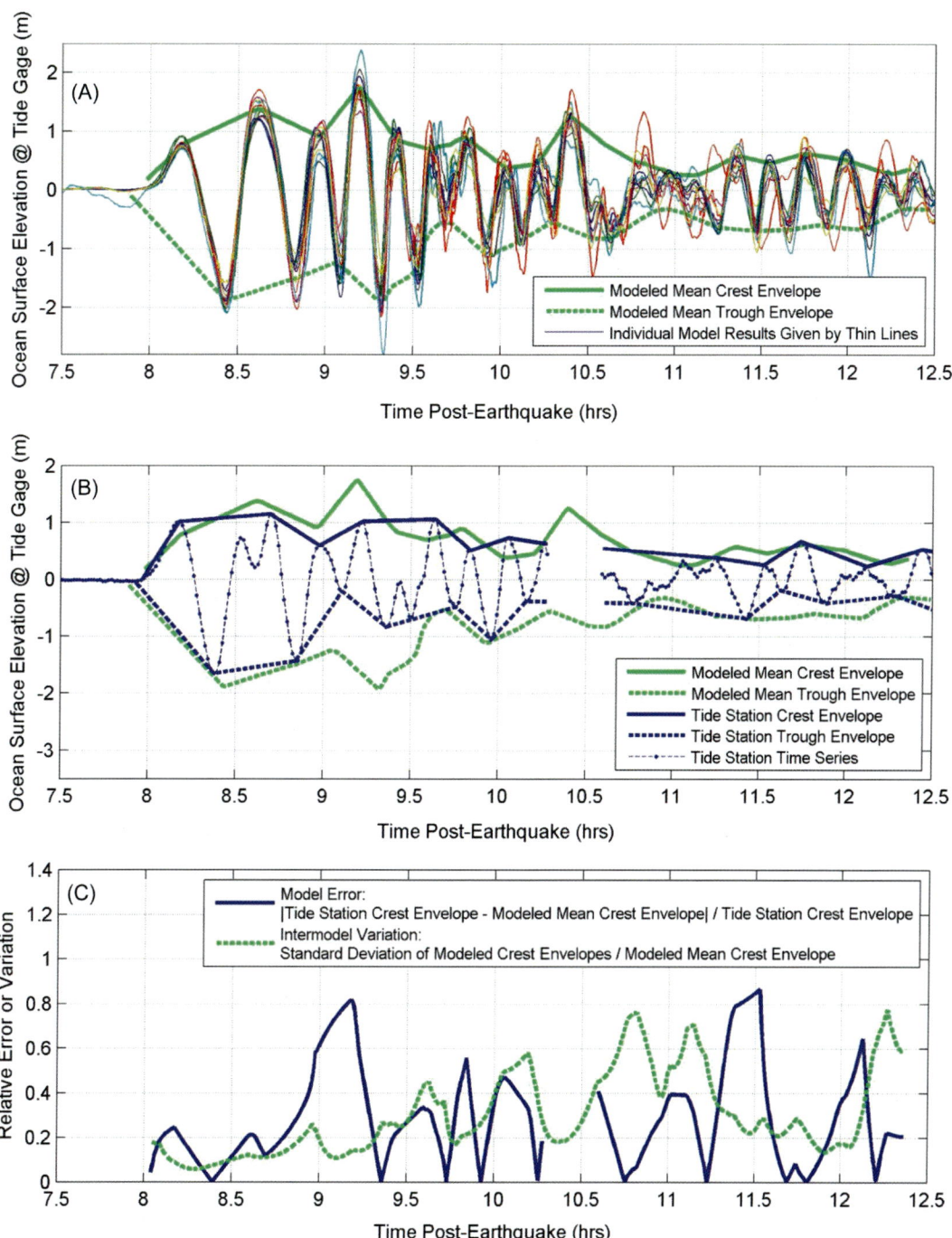

FIGURE 19.4 Intermodel comparison of numerical tsunami models (Lynett et al., 2017). Simulated results of several numerical models for a benchmarking experiment are compared. (A): predictions from all models (thin solid lines), inter-model mean crest envelope (thick solid line), and inter-model mean trough envelope (thick dashed line). (B) Comparison of inter-model envelope to measured tide station data envelope; also shown in the time series from the measured data. (C): Mean inter-model error (solid line) and intermodal standard deviation (dashed line) for the crest envelope. *From Lynett, P. J., Gately, K., Wilson, R., Montoya, L., Arcas, D., Aytore, B., Bai, Y., Bricker, J. D., Castro, M. J., Cheung, K. F., David, C. G., Dogan, G. G., Escalante, C., González-Vida, J. M., Grilli, S. T., Heitmann, T. W., Horrillo, J., Kânoğlu, U., Kian, R., ... Zhang, Y. J. (2017). Inter-model analysis of tsunami-induced coastal currents. Ocean Modelling, 114, 14–32. Available from: http://www.elsevier.com/inca/publications/store/6/0/1/3/7/6/index.htt, https://doi.org/10.1016/j.ocemod.2017.04.003.*

the model is small in offshore areas. However, the tsunami characteristics simulated using different numerical models can differ more significantly in coastal areas with complex shoreline shapes.

19.2.3 Uncertainty related to tsunami inundation

Additional uncertainties can arise during tsunami inundation. The inland behavior of a tsunami needs to be modeled accurately. Especially the tsunami inundation over an urban area becomes complex and involves many uncertain factors (Lynett, 2016). The major uncertainties are related to the differences between tsunami simulations and actual phenomena. This is due to a lack of observations and the need for simplified modeling approaches to reduce computational costs.

Although tsunamis behave mostly as long waves offshore, the nonhydrostatic behavior tends to govern as tsunamis run up. The spatial scale of tsunami behavior in coastal and land areas is significantly smaller than in offshore areas. An inundation simulation generally uses horizontal resolutions of less than 50 or 10 m for practical applications. The high-resolution simulation must consider the effects on streets and buildings in urban areas on tsunami hazard and risk assessments.

The bottom friction term is essential for considering the bottom boundary layer. Manning's roughness or friction coefficients have been calibrated. Manning's roughness coefficient is usually set to a constant value for the seafloor and is varied according to land use for onshore areas. However, the representation of the roughness coefficient in land use does not consider the building density. Moreover, boundary information, such as the building shape, is typically inexplicitly considered in tsunami simulations. Some studies have shown that the three-dimensional fluid characteristics around buildings are significant in determining building damage caused by tsunamis. Thus clarifying the fluid characteristics that reflect the building arrangement is necessary.

When simulating tsunami inundation of urban cities, the height of coastal structures, such as levees, is incorporated into the numerical model. The criteria for the failure of the levees are also modeled. The resilience of coastal systems sometimes makes a significant difference in the inundated area (Shimozono & Sato, 2016; Takahashi et al., 2019). Therefore, uncertainties regarding structural failures should be considered depending on the required modeling level.

19.2.4 Uncertainty related to exposure and damage

When considering long-term risk over tens of years, the exposure variation (e.g., the variation of population and built environment) and the consequent vulnerability variation raise further uncertainties to be addressed. Even if, for a specific city, the tsunami hazard remains more or less the same, the uncertainties on exposure and vulnerability and their variability over time can lead to significant damage scenarios. Even in the short term, populations in urban areas exhibit large differences between daytime and nighttime. In residential areas with high inundation potential, the risk of human casualties increases when a tsunami strikes at nighttime.

The strength of buildings and houses depends on their materials, structures, and age (Suppasri et al., 2018). A tsunami fragility curve provides the probability of attaining or exceeding a specific damage state conditioned on the *IM*. Existing fragility curves are mainly based on data coming from historical events; however, the available data are limited and nation-specific. There is uncertainty about the fragility curve from one country to another due to different structures and related regulations. Additionally, obtaining information on all components of buildings and houses is challenging. Moreover, even under the same tsunami force, older buildings are more likely to collapse due to aging. Quantifying such temporal changes in tsunami vulnerability is necessary because building damage is a major contributor to economic loss.

Several combinations of concurrent/compound damage can occur. For example, in an earthquake-induced tsunami, damage may be caused by seismic motion before the tsunami inundation. Whether dikes and structures are destroyed before a tsunami is a crucial factor affecting the extent of tsunami damage. Similarly, fires can occur due to earthquakes or tsunamis. The fires can diminish the function of tsunami evacuation towers. Therefore, compound damage cannot be overlooked in disaster risk assessments. However, the interrelationship between seismic motions and tsunamis for structures is currently unclear, and the location of fires during earthquakes and tsunamis is unpredictable.

19.3 Uncertainty propagation

The previous section discussed various uncertainties inherent in the tsunami hazard and risk, from the generation of a tsunami, its runup, and tsunami damage. This section describes incorporating these uncertainties into the probabilistic assessments, that is, PTHA and PTRA (hereafter denoted as probabilistic assessment). The uncertainties can vary depending on the target region and assumed conditions. It is noted that not all possible uncertainties will be modeled; only those expected to be significant will be selected. The significance of each uncertainty can be established a priori by performing a sensitivity analysis (see Chapter 5).

19.3.1 Earthquake source modeling

The probability of earthquake occurrence is a primary parameter for probabilistic assessment. It is modeled by defining the earthquake occurrence probability distribution. The mean occurrence interval is assumed to be stable in the regional areas associated with strain accumulation owing to the plate motions. However, it also includes variations due to the physical process of the stress loading and release (Matthews et al., 2002). The Brownian passage time (BPT) distribution explicitly considers these physical processes. The occurrence probability of the next earthquake is low immediately after an earthquake occurs; however, the occurrence probability increases as time approaches the mean occurrence interval. The occurrence period is determined randomly using a specific time-dependent probability density function. After setting the occurrence time to zero, the probability density of the occurrence of the next event is reset. This process is repeated in a Monte Carlo simulation to generate a stochastic event catalog (Goda, 2020). The renewal process is performed using random numbers. However, the Poisson distribution, which assumes a constant occurrence rate without considering physical constraints, has also been practically used. Only the mean occurrence interval parameter is required to determine the occurrence probability because considering previous earthquake events is unnecessary. The mean occurrence interval and other parameters are also uncertain. Thus, even for a specific distribution, the uncertainty of each parameter should be considered.

The regional occurrence rate characterizes the earthquake magnitude based on historical records. The Gutenberg–Richter (G–R)-type model is most frequently used. In the G–R model, the occurrence rate of each magnitude bin is assumed to follow a power law (Aki, 1965). The slope of the power law is estimated by the maximum likelihood method using regional seismicity (Nishikawa & Ide, 2014). The tapered G–R distribution, in which the occurrence rate is below the power law for larger magnitudes, is also used (Kagan, 2002). Other magnitude models include characteristic magnitude models with uniform or truncated distributions. The mean occurrence interval should be estimated from the regional seismicity. However, it is often difficult to evaluate from regional observation records alone due to the small sample size. For this reason, global seismicity data may be partially utilized, or the parameter range may be constrained from regional seismic moment release.

Many stochastic source catalogs are necessary to consider the earthquake source uncertainty. Several stochastic source models have been proposed to generate synthetic events (i.e., source model). A key of the stochastic source model is the random generation of the source but keeping several constraints based on regional characteristics. For example, Goda et al. (2014) proposed a stochastic source model for the megathrust subduction earthquake that generates random slip distributions having spatial correlations considering historical earthquakes. In that study, first, the wavenumber spectral shape of the slip distributions of historical earthquakes inferred by inversion analyses is approximated using von Karman-type spectra. Based on the slip spectral shape, the inverse Fourier transformation randomly generates a two-dimensional slip distribution (so-called random phase model; Mori et al., 2018). The random phase model characterizes spatial decay and correlation of slips in different wavenumbers (spatial scale). An arbitrary number of source models can be obtained by changing the combinations of phases randomly. To incorporate variable magnitudes, source parameters, such as fault width and length, can vary based on the scaling law as a function of earthquake magnitude (Goda et al., 2016). However, the uncertainties of these parameters are also randomly determined within certain ranges.

To consider kinematic rupture, the nucleation point of the fault is determined by a random number within statistical constraints. Mai et al. (2005) obtained the statistical relationship between slip distribution and the rupture starting point over 80 source fault models (M_w 4.1–8.1). The source fault models were classified into three regions: very large slip ($D > 2/3D_m$), large slip ($1/3D_m < D < 2/3D_m$), and low slip ($D < 1/3D_m$) based on the maximum slip amount D_m and region in which rupture nucleation occurred. They differ significantly in terms of the area ratio occupied by each region. This suggests that the nucleation point is not completely random and can be constrained based on the slip values. The rupture nucleation is more likely to occur near a large slip region. Some stochastic source models can consider a two-dimensional probability distribution of the rupture nucleation point from the distribution of the rupture fault plane determined using random numbers.

After the nucleation of the fault, the rupture gradually expands from the nucleation point spatially; therefore, the rupture velocity and rise time are also needed in the kinematic rupture model. These values cannot be determined unless the fault structure is well understood, a typical value is used, or they vary by random numbers from that value.

However, the rupture velocity should correspond when using nonuniform rock rigidity because it depends on this parameter. Rock rigidity generally increases with depth, affecting the seafloor deformation due to seismic rupture and the total moment magnitude (Bilek & Lay, 1999). Variable rock rigidity can significantly affect tsunami simulations (Geist & Bilek, 2001).

19.3.2 Tsunami modeling

Various uncertainties also arise in assumptions of physical parameters and modeling methods of the tsunami simulation. First, considering various astronomical tidal heights is possible due to the randomness of the earthquake occurrence times. The mean water level (MWL) and high water level (HWL) of the astronomical tide are widely used in engineering design. Thus HWL should be considered for the worst case scenarios. However, the HWL time and the arrival of a tsunami peak are unlikely to occur simultaneously. There are methods for simulating tsunamis under both MWL and HWL conditions or randomly determining the initial height of the astronomical tide. Besides, in reality, astronomical tides are spatially nonuniform over time; however, the spatially uniform tidal level is often considered for most simulations.

Tsunami simulation results contain numerical modeling- and bathymetry-related errors besides the errors related to seismic sources. To account for these errors, the predicted tsunami amplitude at the point of interest is often replaced by a probability density function based on the simulated amplitude. The actual amplitudes will vary according to this function, even if the same scenario occurs (Fukutani et al., 2018). A lognormal distribution is often used for the assumed representative probability distribution of the simulated tsunami amplitudes. In some studies, the logarithm of the standard deviation of the lognormal distribution is defined based on Aida's κ. However, it is essential to note that Aida (1978) κ was initially used to evaluate spatial variability. The influence of the representative probability distribution is significant in the upper tail of tsunami amplitude distribution (Fukutani et al., 2015; Miyashita et al., 2020). The probability of exceeding the maximum of all simulated tsunami amplitudes is zero without the representative probability distribution of tsunami amplitude (i.e., solely based on the simulated samples of the tsunami amplitude); however, the probability is not zero when considering the representative probability distribution.

Third, a parameterization of bottom friction is important when the tsunami inundates the land. The bottom roughness is widely expressed based on land use (e.g., cropland). However, this parameterization does not consider some factors, such as building or house shapes and layout (Moris et al., 2021; Park et al., 2013). The high-resolution simulation resolving buildings and house shapes is necessary to consider the drag forces in the numerical model in addition to high-resolution topography (Son et al., 2011). In recent years, new approaches have been developed to account for drag forces depending on individual building heights and shapes (Fukui et al., 2022).

Tsunami debris, such as ships and port containers, also causes another type of damage; however, predicting their behavior is difficult (Charvet et al., 2015; Chida & Mori, 2023). Many tsunami debris models incorporating object rolling, sliding, and landfall have been developed recently. Based on the comparative studies between the numerical models and physical experiments, the final position of the tsunami debris is sensitive to the initial position of the debris, topography, and building layout. The new result indicates the importance of considering perturbations in the initial position of the debris in the same scenario for more rigorous risk assessment. Incorporating the fluid and debris interaction into the modeling is challenging owing to computational costs.

19.3.3 Fragility modeling

PTRA quantifies damage or risk of exposure. The extent of damage and casualties is evaluated based on the tsunami characteristics. Such damage estimation involves uncertainty. The fragility function parameterizes the damage probability of a building based on the observed damage with modeled *IM*s, such as tsunami inundation depth and flow velocity, and is statistically calculated. The tsunami fragility function often assumes a lognormal distribution, similar to building damage caused by earthquake shaking (Wu et al., 2016; Yamaguchi & Yamazaki, 2001). In addition, the observed damage probability depends on the building structure (e.g., wooden and reinforced concrete), the number of stories, and age. Therefore, the fragility function incorporates detailed structural information. In particular, wooden and reinforced concrete structures and buildings with three or fewer and four or more stories have significantly different damage probabilities, even with the same value of tsunami intensity; therefore, the same fragility function should not be applied. Fig. 19.5 shows an example of the tsunami fragility curve for a building. Damage functions assuming a lognormal cumulative distribution function are constructed for each of the six damage states (see Chapters 4 and 5). The probability of damage increases with the *IM* (in this case, tsunami inundation depth); however, the shape of this curve varies

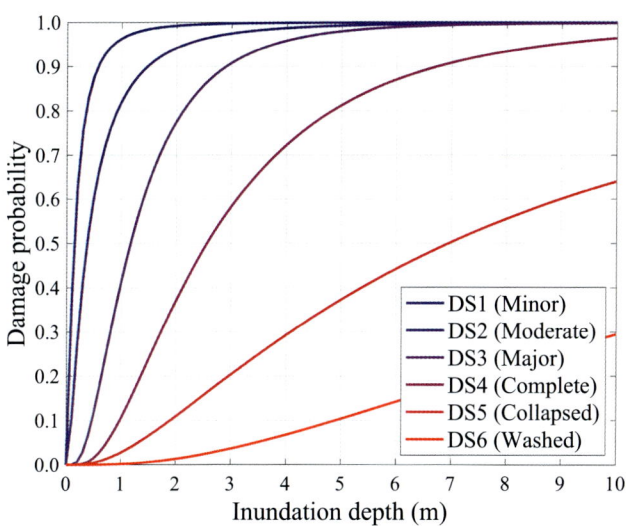

FIGURE 19.5 Tsunami fragility curves of a building for different damage states.

according to the damage state. Different probability distribution parameters are used depending on the main components of the building and the number of stories.

Fragility functions have been constructed based on the records of half-collapse and total collapse of the buildings. However, these damage probabilities primarily depend on the design standards of each country and region. Therefore, the applicable range is limited by country or region. Furthermore, building strength decreases with age. Some buildings are renovated or reconstructed, and the building regulations can also be revised. Thus recognizing the limitations of not considering these aspects in long-term assessments is necessary. Furthermore, surveys of past tsunami damage have shown that the level of damage can vary significantly among adjacent buildings. This indicates that damage varies according to local building information and topography, even with approximately the same inundation depth and duration. Some studies have formalized the statistics of tsunami inundation damage as damage probabilities and constructed fragility functions for building damage (Charvet et al., 2014, 2015; Suppasri et al., 2013). These fragility functions are commonly used to estimate the probability of building damage in urban areas through inundation simulations.

19.3.4 Compound and cascading hazard modeling

Seismic waves are faster than tsunami waves in near-field events. Even buildings safe from tsunamis can be damaged by preceding seismic motions in advance. In addition, embankments or other tsunami protection measures might collapse due to seismic ground motions before the arrival of the tsunami, and it potentially causes the expansion of the area inundated by tsunamis. These compound hazard assessments have been integrated with previously conducted Probabilistic Seismic Hazard Assessment (PSHA) after PTHA became widely used, and modeling has progressed in recent years (Goda et al., 2019; Goda et al., 2020; Ishibashi et al., 2021; Park et al., 2017). Earthquake shaking damage is often evaluated using a fragility function with the PGA or PGV as the *IM*.

The most straightforward probabilistic seismic and tsunami hazard assessment individually calculates earthquake shaking and tsunami damage. The larger probability is considered the damage probability for the corresponding building for near-field tsunamis. This method is simple and easy to implement. However, it does not consider a temporal sequence of hazards and damage accumulation over time, such as the reduction in the building strength due to seismic motion preceding the tsunami or the total destruction of a building by a tsunami after it partially collapsed by seismic motion. Furthermore, unifying the number and definitions/descriptions of damage states in PSHA and PTHA is challenging. In addition, local parameters of ground condition parameters, such as surface soil types, are essential to assess seismic hazards. However, a sufficient resolution of such parameters is not available in many areas and generally does not match the spatial resolution of tsunami hazard assessment.

During an earthquake or tsunami inundation, fire damage may also occur in addition to damage from seismic motions and tsunamis. During the 2011 Tohoku Earthquake, a fire broke out in Kesennuma Bay, Miyagi Prefecture, where a large volume of petrochemical products were stored. There was a possibility that the fire could have caused unexpected damage to critical facilities, such as evacuation buildings. Therefore, understanding high fire-risk locations is important for additional damage estimation or identifying appropriate evacuation routes and locations. However,

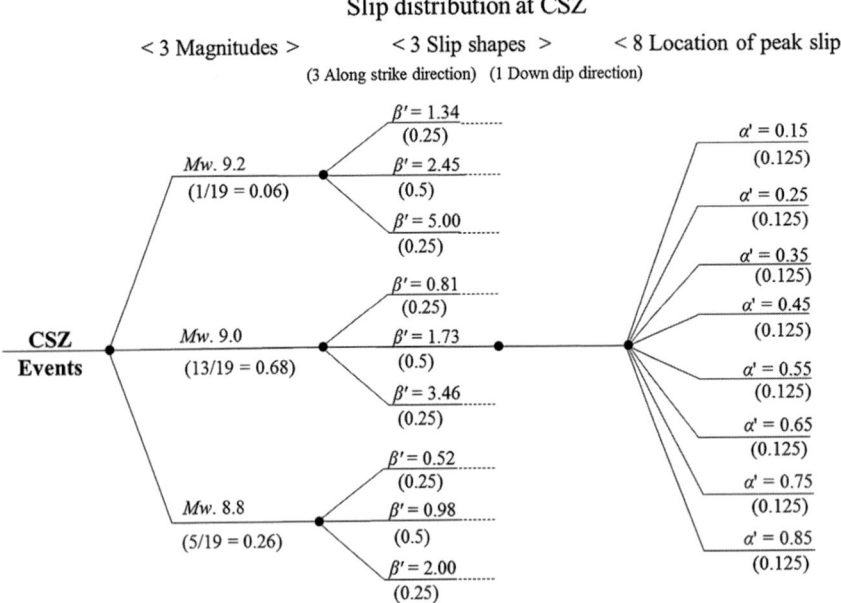

FIGURE 19.6 A generic logic tree model in probabilistic tsunami hazard analysis (Park & Cox, 2016). Branching represents different models, assumptions, or events. Each uncertainty has a branch, and each path from the beginning to the end of the branch constitutes a hazard curve. *From Park, H., & Cox, D. T. (2016). Probabilistic assessment of near-field tsunami hazards: Inundation depth, velocity, momentum flux, arrival time, and duration applied to Seaside, Oregon. Coastal Engineering, 117, 79–96. Available from: http://www.elsevier.com/inca/publications/store/5/0/3/3/2/5/, https://doi.org/10.1016/j.coastaleng.2016.07.011.*

predicting the ignition points of ignitable fuels and their spread is generally impossible. Therefore, it is common to randomly select ignition points from potential areas and conduct combustion simulations to estimate high fire-risk locations. These combustion simulations require a physical model of the behavior of oil that leaks from overturned oil tanks and in the tsunami flow field. Details on oil behavior modeling can be found in Nishino and Imazu (2018) and Nishino and Takagi (2020).

19.3.5 Incorporating different uncertainty models

This section discusses the uncertainties in each process and their modeling methods. Moreover, a methodology for aggregating these individual models to form significant outcomes, such as hazard and risk curves for PTHA and PTRA, is introduced. A logic tree is a method for representing the outcome of PTHA (Annaka et al., 2007; Park et al., 2017; Park & Cox, 2016). This is used to integrate different hypothetical models for the same type of uncertainty. For example, a model using a time-dependent function for the annual earthquake occurrence probability (e.g., BPT distribution) and a time-invariant model (e.g., Poisson process) can be assessed.

Fig. 19.6 shows an example of a logic tree in the PTHA (Park & Cox, 2016). Branches are provided for each uncertainty, and all branches are calculated from the tsunami simulation results. Each path from the beginning to the end branch corresponds to a calculation using Eq. (19.1) with weights assigned to each path. The sum of the weights of all paths is equal to one, and the percentiles of the hazard curve corresponding to the *IM* are plotted. The weighting of each branch is an epistemic uncertainty. Moreover, it is important to note that this can potentially cause considerable sensitivity to the outcome.

19.4 Summary and conclusions

This chapter introduced the uncertain factors in conducting PTHA and PTRA and their respective modeling approaches. The uncertainty of the tsunami hazards is multifaceted and can be broadly categorized into the source, tsunami propagation, inundation, and vulnerability or exposure uncertainty. Hazard curves are generally obtained by integrating various parameters and magnitude ranges while accounting for these uncertainties. However, the sources of uncertainty are diverse, and considering all uncertainties in a probabilistic assessment is impossible.

Research gaps exist because uncertainties cannot be considered without subjectivity (Behrens et al., 2021). For example, either or both global seismic and geodetic data can be used when estimating the frequency of a seismic source. Local source information alone is insufficient to estimate the seismic rates of large earthquakes; whether the seismic rate is calculated by referring to local or global seismic data, the range of estimates, and their impact on the final outcome should be discussed. Another example is that the resolution and level of details of exposure modeling data vary from study to study.

A sensitivity analysis of each outcome uncertainty is necessary to fill these research gaps. The sensitivity of the parameters in the area of interest can be clarified by setting scientifically acceptable ranges for specific parameters and quantifying changes in the lower and upper bounds of the hazard or risk curve. Parameters with relatively low sensitivity indicate that the modeling uncertainty in that domain may be insignificant. Rigorous PTHA and PTRA can be implemented by identifying uncertainties that should always be considered and those of lower importance.

Interpreting the results is essential for developing designs, evacuation plans, and disaster mitigation measures. The hazard curve includes an envelope with a specific range corresponding to the fractiles (or percentiles). Decision-makers must determine which fractile of this hazard curve is adopted as a reference. The mean hazard curve has heavier tails than the median (Fukutani et al., 2016), which is similar to the case of PSHA (Bommer & Abrahamson, 2006; McGuire et al., 2005; McGuire, 1995). Those conducting PTHA and PTRA must be clear about the characteristics of each and the variability of the dominant curve so that decision-makers can properly refer to the hazard curve.

The PTHA can be applied to other tsunami trigger mechanisms, such as nonsubduction earthquakes, landslides, volcanic eruptions, and meteorological tsunamis, although this chapter focuses on subduction earthquake tsunamis. Since these tsunamis differ only in their sources, the same methods can be applied except for source modeling.

References

Aida, I. (1978). Reliability of a tsunami source model derived from fault parameters. *Journal of Physics of the Earth*, 26(1), 57−73. Available from https://doi.org/10.4294/jpe1952.26.57.

Aki, K. (1965). Maximum likelihood estimate of b in the formula $\log N = a - bM$ and its confidence limits. *Bulletin of the Earthquake Research Institute*, 43, 237−239.

Annaka, T., Satake, K., Sakakiyama, T., Yanagisawa, K., & Shuto, N. (2007). Logic-tree approach for probabilistic tsunami hazard analysis and its applications to the Japanese coasts. *Pure and Applied Geophysics*, 164(2−3), 577−592. Available from https://doi.org/10.1007/s00024-006-0174-3.

Baba, T., Allgeyer, S., Hossen, J., Cummins, P. R., Tsushima, H., Imai, K., Yamashita, K., & Kato, T. (2017). Accurate numerical simulation of the far-field tsunami caused by the 2011 Tohoku earthquake, including the effects of Boussinesq dispersion, seawater density stratification, elastic loading, and gravitational potential change. *Ocean Modelling*, 111, 46−54. Available from https://doi.org/10.1016/j.ocemod.2017.01.002, http://www.elsevier.com/inca/publications/store/6/0/1/3/7/6/index.htt.

Behrens, J., Løvholt, F., Jalayer, F., Lorito, S., Salgado-Gálvez, M. A., Sørensen, M., Abadie, S., Aguirre-Ayerbe, I., Aniel-Quiroga, I., Babeyko, A., Baiguera, M., Basili, R., Belliazzi, S., Grezio, A., Johnson, K., Murphy, S., Paris, R., Rafliana, I., De Risi, R., ... Vyhmeister, E. (2021). Probabilistic tsunami hazard and risk analysis: A review of research gaps. *Frontiers in Earth Science*, 9. Available from https://doi.org/10.3389/feart.2021.628772, https://www.frontiersin.org/journals/earth-science.

Bilek, S. L., & Lay, T. (1999). Rigidity variations with depth along interplate megathrust faults in subduction zones. *Nature*, 400(6743), 443−446. Available from https://doi.org/10.1038/22739.

Bommer, J. J., & Abrahamson, N. A. (2006). Why do modern probabilistic seismic-hazard analyses often lead to increased hazard estimates. *Bulletin of the Seismological Society of America*, 96(6), 1967−1977. Available from https://doi.org/10.1785/0120060043.

Chang, T. W., & Ide, S. (2021). Hypocenter hotspots illuminated using a new cross-correlation-based hypocenter and centroid relocation method. *Journal of Geophysical Research: Solid Earth*, 126(9). Available from https://doi.org/10.1029/2021jb021991.

Charvet, I., Ioannou, I., Rossetto, T., Suppasri, A., & Imamura, F. (2014). Empirical fragility assessment of buildings affected by the 2011 Great East Japan tsunami using improved statistical models. *Natural Hazards*, 73(2), 951−973. Available from https://doi.org/10.1007/s11069-014-1118-3.

Charvet, I., Suppasri, A., Kimura, H., Sugawara, D., & Imamura, F. (2015). A multivariate generalized linear tsunami fragility model for Kesennuma City based on maximum flow depths, velocities and debris impact, with evaluation of predictive accuracy. *Natural Hazards*, 79(3), 2073−2099. Available from https://doi.org/10.1007/s11069-015-1947-8.

Chida, Y., & Mori, N. (2023). Numerical modeling of debris transport due to tsunami flow in a coastal urban area. *Coastal Engineering*, 179. Available from https://doi.org/10.1016/j.coastaleng.2022.104243.

Cornell, C. A. (1968). Engineering seismic risk analysis. *Bulletin of the Seismological Society of America*, 58(5), 1583−1606. Available from https://doi.org/10.1785/bssa0580051583.

Fan, J., & Zhao, D. (2021). Subslab heterogeneity and giant megathrust earthquakes. *Nature Geoscience*, 14(5), 349−353. Available from https://doi.org/10.1038/s41561-021-00728-x.

Fujii, Y., & Satake, K. (2007). Tsunami source of the 2004 Sumatra−Andaman earthquake inferred from tide gauge and satellite data. *Bulletin of the Seismological Society of America*, 97(1A), S192−S207. Available from https://doi.org/10.1785/0120050613.

Fukui, N., Mori, N., Miyashita, T., Shimura, T., & Goda, K. (2022). Subgrid-scale modeling of tsunami inundation in coastal urban areas. *Coastal Engineering*, 177. Available from https://doi.org/10.1016/j.coastaleng.2022.104175.

Fukutani, Y., Suppasri, A., & Imamura, F. (2015). Stochastic analysis and uncertainty assessment of tsunami wave height using a random source parameter model that targets a Tohoku-type earthquake fault. *Stochastic Environmental Research and Risk Assessment*, 29(7), 1763−1779. Available from https://doi.org/10.1007/s00477-014-0966-4.

Fukutani, Y., Anawat, S., & Imamura, F. (2016). Uncertainty in tsunami wave heights and arrival times caused by the rupture velocity in the strike direction of large earthquakes. *Natural Hazards*, 80(3), 1749–1782. Available from https://doi.org/10.1007/s11069-015-2030-1.

Fukutani, Y., Suppasri, A., & Imamura, F. (2018). Quantitative assessment of epistemic uncertainties in tsunami hazard effects on building risk assessments. *Geosciences*, 8(1). Available from https://doi.org/10.3390/geosciences8010017.

Geist, E. L., & Bilek, S. L. (2001). Effect of depth-dependent shear modulus on tsunami generation along subduction zones. *Geophysical Research Letters*, 28(7), 1315–1318. Available from https://doi.org/10.1029/2000GL012385.

Geist, E. L., & Parsons, T. (2006). Probabilistic analysis of tsunami hazards. *Natural Hazards*, 37(3), 277–314. Available from https://doi.org/10.1007/s11069-005-4646-z.

Glimsdal, S., Pedersen, G. K., Harbitz, C. B., & Løvholt, F. (2013). Dispersion of tsunamis: Does it really matter? *Natural Hazards and Earth System Sciences*, 13(6), 1507–1526. Available from https://doi.org/10.5194/nhess-13-1507-2013.

Goda, K. (2020). Multi-hazard portfolio loss estimation for time-dependent shaking and tsunami hazards. *Frontiers in Earth Science*, 8. Available from https://doi.org/10.3389/feart.2020.592444.

Goda, K., & De Risi, R. (2018). Multi-hazard loss estimation for shaking and tsunami using stochastic rupture sources. *International Journal of Disaster Risk Reduction*, 28, 539–554. Available from https://doi.org/10.1016/j.ijdrr.2018.01.002.

Goda, K., & Song, J. (2016). Uncertainty modeling and visualization for tsunami hazard and risk mapping: A case study for the 2011 Tohoku earthquake. *Stochastic Environmental Research and Risk Assessment*, 30(8), 2271–2285. Available from https://doi.org/10.1007/s00477-015-1146-x.

Goda, K., Mai, P. M., Yasuda, T., & Mori, N. (2014). Sensitivity of tsunami wave profiles and inundation simulations to earthquake slip and fault geometry for the 2011 Tohoku earthquake. *Earth, Planets and Space*, 66(1). Available from https://doi.org/10.1186/1880-5981-66-105.

Goda, K., Yasuda, T., Mori, N., & Maruyama, T. (2016). New scaling relationships of earthquake source parameters for stochastic tsunami simulation. *Coastal Engineering Journal*, 58(3). Available from https://doi.org/10.1142/S0578563416500108, https://www.tandfonline.com/loi/tcej20.

Goda, K., Mori, N., Yasuda, T., Prasetyo, A., Muhammad, A., & Tsujio, D. (2019). Cascading geological hazards and risks of the 2018 Sulawesi Indonesia earthquake and sensitivity analysis of tsunami inundation simulations. *Frontiers in Earth Science*, 7. Available from https://doi.org/10.3389/feart.2019.00261.

Goda, K., Yasuda, T., Mori, N., Muhammad, A., De Risi, R., & De Luca, F. (2020). Uncertainty quantification of tsunami inundation in Kuroshio, Kochi Prefecture, Japan, using the Nankai–Tonankai megathrust rupture scenarios. *Natural Hazards and Earth System Sciences*, 20(11), 3039–3056. Available from https://doi.org/10.5194/nhess-20-3039-2020.

Grezio, A., Babeyko, A., Baptista, M. A., Behrens, J., Costa, A., Davies, G., Geist, E. L., Glimsdal, S., González, F. I., Griffin, J., Harbitz, C. B., LeVeque, R. J., Lorito, S., Løvholt, F., Omira, R., Mueller, C., Paris, R., Parsons, T., Polet, J., ... Thio, H. K. (2017). Probabilistic tsunami hazard analysis: Multiple sources and global applications. *Reviews of Geophysics*, 55(4), 1158–1198. Available from https://doi.org/10.1002/2017RG000579, http://onlinelibrary.wiley.com/journal/10.1002/(ISSN)1944-9208.

Hébert, H., & Schindelé, F. (2015). Tsunami impact computed from offshore modeling and coastal amplification laws: Insights from the 2004 Indian Ocean tsunami. *Pure and Applied Geophysics*, 172(12), 3385–3407. Available from https://doi.org/10.1007/s00024-015-1136-4.

Ide, S. (2019). Frequent observations of identical onsets of large and small earthquakes. *Nature*, 573(7772), 112–116. Available from https://doi.org/10.1038/s41586-019-1508-5.

Ishibashi, H., Akiyama, M., Kojima, T., Aoki, K., Koshimura, S., & Frangopol, D. M. (2021). Risk estimation of the disaster waste generated by both ground motion and tsunami due to the anticipated Nankai Trough earthquake. *Earthquake Engineering and Structural Dynamics*, 50(8), 2134–2155. Available from https://doi.org/10.1002/eqe.3440, http://onlinelibrary.wiley.com/journal/10.1002/(ISSN)1096-9845.

Kagan, Y. Y. (2002). Seismic moment distribution revisited: I. Statistical results. *Geophysical Journal International*, 148(3), 520–541. Available from https://doi.org/10.1046/j.1365-246x.2002.01594.x.

Kanamori, H. (1978). Quantification of earthquakes. *Nature*, 271(5644), 411–414. Available from https://doi.org/10.1038/271411a0.

Lynett, P. J. (2016). Precise prediction of coastal and overland flow dynamics: A grand challenge or a fool's errand. *Journal of Disaster Research*, 11(4), 615–623. Available from https://doi.org/10.20965/jdr.2016.p0615, https://www.fujipress.jp/main/wp-content/themes/Fujipress/pdf_subscribed.php.

Lynett, P. J., Gately, K., Wilson, R., Montoya, L., Arcas, D., Aytore, B., Bai, Y., Bricker, J. D., Castro, M. J., Cheung, K. F., David, C. G., Dogan, G. G., Escalante, C., González-Vida, J. M., Grilli, S. T., Heitmann, T. W., Horrillo, J., Kânoğlu, U., Kian, R., ... Zhang, Y. J. (2017). Inter-model analysis of tsunami-induced coastal currents. *Ocean Modelling*, 114, 14–32. Available from http://www.elsevier.com/inca/publications/store/6/0/1/3/7/6/index.htt, 10.1016/j.ocemod.2017.04.003.

Mai, P. M., Spudich, P., & Boatwright, J. (2005). Hypocenter locations in finite-source rupture models. *Bulletin of the Seismological Society of America*, 95(3), 965–980. Available from https://doi.org/10.1785/0120040111.

Matsuzawa, T. (2002). Characteristic small-earthquake sequence off Sanriku, northeastern Honshu, Japan. *Geophysical Research Letters*, 29(11). Available from https://doi.org/10.1029/2001gl014632.

Matthews, M. V., Ellsworth, W. L., & Reasenberg, P. A. (2002). A Brownian model for recurrent earthquakes. *Bulletin of the Seismological Society of America*, 92(6), 2233–2250. Available from https://doi.org/10.1785/0120010267.

McCloskey, J., Antonioli, A., Piatanesi, A., Sieh, K., Steacy, S., Nalbant, S., Cocco, M., Giunchi, C., Huang, J. D., & Dunlop, P. (2008). Tsunami threat in the Indian Ocean from a future megathrust earthquake west of Sumatra. *Earth and Planetary Science Letters*, 265(1–2), 61–81. Available from https://doi.org/10.1016/j.epsl.2007.09.034.

McGuire, R. K. (1995). Probabilistic seismic hazard analysis and design earthquakes: Closing the loop. *Bulletin of the Seismological Society of America*, 85(5), 1275–1284. Available from https://doi.org/10.1785/bssa0850051275.

McGuire, R. K., Cornell, C. A., & Toro, G. R. (2005). The case for using mean seismic hazard. *Earthquake Spectra*, *21*(3), 879−886. Available from https://doi.org/10.1193/1.1985447.

Miyashita, T., Mori, N., & Goda, K. (2019). *Uncertainty quantification of tsunami height for future earthquakes in West Japan. International Conference on Applications of Statistics and Probability in Civil Engineering, ICASP 2019*. Seoul National University. Available from https://doi.org/10.22725/ICASP13.466.

Miyashita, T., Mori, N., & Goda, K. (2020). Uncertainty of probabilistic tsunami hazard assessment of Zihuatanejo (Mexico) due to the representation of tsunami variability. *Coastal Engineering Journal*, *62*(3), 413−428. Available from https://doi.org/10.1080/21664250.2020.1780676.

Mori, N., Takahashi, T., Hamaura, S. E., Miyakawa, K., Tanabe, K., Tanaka, K., Tanaka, M., Watanabe, T., Matsutomi, H., Naoe, K., Noumi, T., Yamaguchi, E., Ando, S., Fujii, Y., Kashima, T., Okuda, Y., Shibazaki, B., Sakakiyama, T., Matsuyama, M., . . . Suzuki, T. (2012). Nationwide post event survey and analysis of the 2011 Tohoku earthquake tsunami. *Coastal Engineering Journal*, *54*(1). Available from https://doi.org/10.1142/S0578563412500015, https://www.tandfonline.com/doi/abs/10.1142/S0578563412500015.

Mori, N., Mai, P. M., Goda, K., & Yasuda, T. (2017). Tsunami inundation variability from stochastic rupture scenarios: Application to multiple inversions of the 2011 Tohoku, Japan earthquake. *Coastal Engineering*, *127*, 88−105. Available from http://www.elsevier.com/inca/publications/store/5/0/3/3/2/5/, https://doi.org/10.1016/j.coastaleng.2017.06.013.

Mori, N., Goda, K., & Cox, D. (2018). *Recent process in probabilistic tsunami hazard analysis (PTHA) for mega thrust subduction earthquakes*, . The 2011 Japan earthquake and tsunami: Reconstruction and restoration (Advances in natural and technological hazards research) (47). Springer. Available from http://doi.org/10.1007/978-3-319-58691-5_27.

Moris, J. P., Kennedy, A. B., & Westerink, J. J. (2021). Tsunami wave run-up load reduction inside a building array. *Coastal Engineering*, *169*. Available from http://www.elsevier.com/inca/publications/store/5/0/3/3/2/5/, 10.1016/j.coastaleng.2021.103910.

Nakanishi, A., Shiobara, H., Hino, R., Kodaira, S., Kanazawa, T., & Shimamura, H. (1998). Detailed subduction structure across the eastern Nankai Trough obtained from ocean bottom seismographic profiles. *Journal of Geophysical Research: Solid Earth*, *103*(B11), 27151−27168. Available from https://doi.org/10.1029/98jb02344.

Nishikawa, T., & Ide, S. (2014). Earthquake size distribution in subduction zones linked to slab buoyancy. *Nature Geoscience*, *7*(12), 904−908. Available from https://doi.org/10.1038/ngeo2279.

Nishino, T., & Imazu, Y. (2018). A computational model for large-scale oil spill fires on water in tsunamis: Simulation of oil spill fires at Kesennuma Bay in the 2011 Great East Japan Earthquake and Tsunami. *Journal of Loss Prevention in the Process Industries*, *54*, 37−48. Available from https://doi.org/10.1016/j.jlp.2018.02.009.

Nishino, T., & Takagi, Y. (2020). Numerical analysis of tsunami-triggered oil spill fires from petrochemical industrial complexes in Osaka Bay, Japan, for thermal radiation hazard assessment. *International Journal of Disaster Risk Reduction*, *42*. Available from https://doi.org/10.1016/j.ijdrr.2019.101352.

Park, H., & Cox, D. T. (2016). Probabilistic assessment of near-field tsunami hazards: Inundation depth, velocity, momentum flux, arrival time, and duration applied to Seaside, Oregon. *Coastal Engineering*, *117*, 79−96. Available from http://www.elsevier.com/inca/publications/store/5/0/3/3/2/5/, 10.1016/j.coastaleng.2016.07.011.

Park, H., Cox, D. T., Lynett, P. J., Wiebe, D. M., & Shin, S. (2013). Tsunami inundation modeling in constructed environments: A physical and numerical comparison of free-surface elevation, velocity, and momentum flux. *Coastal Engineering*, *79*, 9−21. Available from https://doi.org/10.1016/j.coastaleng.2013.04.002.

Park, H., Cox, D. T., Alam, M. S., & Barbosa, A. R. (2017). Probabilistic seismic and tsunami hazard analysis conditioned on a megathrust rupture of the Cascadia subduction zone. *Frontiers in Built Environment*, *3*. Available from https://doi.org/10.3389/fbuil.2017.00032, https://www.frontiersin.org/articles/10.3389/fbuil.2017.00032/pdf.

Prasetyo, A., Yasuda, T., Miyashita, T., & Mori, N. (2019). Physical modeling and numerical analysis of tsunami inundation in a coastal city. *Frontiers in Built Environment*, *5*. Available from https://doi.org/10.3389/fbuil.2019.00046.

Salazar-Monroy, E. F., Melgar, D., Jaimes, M. A., & Ramirez-Guzman, L. (2021). Regional probabilistic tsunami hazard analysis for the Mexican subduction zone from stochastic slip models. *Journal of Geophysical Research: Solid Earth*, *126*(6). Available from https://doi.org/10.1029/2020jb020781.

Satake, K. (2015). Geological and historical evidence of irregular recurrent earthquakes in Japan. *Philosophical Transactions of the Royal Society A: Mathematical, Physical and Engineering Sciences*, *373*(2053). Available from https://doi.org/10.1098/rsta.2014.0375.

Shimozono, T., & Sato, S. (2016). Coastal vulnerability analysis during tsunami-induced levee overflow and breaching by a high-resolution flood model. *Coastal Engineering*, *107*, 116−126. Available from https://doi.org/10.1016/j.coastaleng.2015.10.007.

Shimozono, T., Sato, S., Okayasu, A., Tajima, Y., Fritz, H. M., Liu, H., & Takagawa, T. (2012). Propagation and inundation characteristics of the 2011 Tohoku tsunami on the central Sanriku coast. *Coastal Engineering Journal*, *54*(1). Available from https://doi.org/10.1142/S0578563412500040, 1250004-1-1250004-17.

Son, S., Lynett, P. J., & Kim, D. H. (2011). Nested and multi-physics modeling of tsunami evolution from generation to inundation. *Ocean Modelling*, *38*(1−2), 96−113. Available from https://doi.org/10.1016/j.ocemod.2011.02.007.

Suppasri, A., Mas, E., Charvet, I., Gunasekera, R., Imai, K., Fukutani, Y., Abe, Y., & Imamura, F. (2013). Building damage characteristics based on surveyed data and fragility curves of the 2011 Great East Japan tsunami. *Natural Hazards*, *66*(2), 319−341. Available from https://doi.org/10.1007/s11069-012-0487-8.

Suppasri, A., Fukui, K., Yamashita, K., Leelawat, N., Ohira, H., & Imamura, F. (2018). Developing fragility functions for aquaculture rafts and eelgrass in the case of the 2011 Great East Japan tsunami. *Natural Hazards and Earth System Sciences*, *18*(1), 145–155. Available from https://doi.org/10.5194/nhess-18-145-2018.

Takahashi, H., Morikawa, Y., Mori, N., & Yasuda, T. (2019). Collapse of concrete-covered levee under composite effect of overflow and seepage. *Soils and Foundations*, *59*(6), 1787–1799. Available from https://doi.org/10.1016/j.sandf.2019.08.008.

Watada, S. (2023). Progress and application of the synthesis of trans-oceanic tsunamis. *Progress in Earth and Planetary Science*, *10*(1). Available from https://doi.org/10.1186/s40645-023-00555-1.

Wu, H., Masaki, K., Irikura, K., & Kurahashi, S. (2016). Empirical fragility curves of buildings in northern Miyagi Prefecture during the 2011 off the Pacific coast of Tohoku earthquake. *Journal of Disaster Research*, *11*(6), 1253–1270. Available from https://doi.org/10.20965/jdr.2016.p1253.

Yamaguchi, N., & Yamazaki, F. (2001). Estimation of strong motion distribution in the 1995 Kobe earthquake based on building damage data. *Earthquake Engineering & Structural Dynamics*, *30*(6), 787–801. Available from https://doi.org/10.1002/eqe.33.

Chapter 20

Multihazard risk assessments

Hyoungsu Park
Department of Civil, Environmental, and Construction Engineering, University of Hawaii at Manoa, Honolulu, HI, United States

20.1 Introduction

The concept of multihazards refers to the circumstance where a primary hazard is accompanied by secondary hazards that interact or cascade, leading to additional damage and risk. Tsunamis are often generated by megathrust earthquakes, which not only cause strong ground motions but also inflict damage on coastal communities before the actual tsunami waves reach the shore. The initial damage caused by the earthquake can increase the vulnerability of infrastructure systems, including buildings, to further damage from subsequent tsunami-induced inundation. Therefore, it is essential to consider both seismic and tsunami effects when assessing tsunami hazards and conducting risk assessments in coastal regions. However, major hazard and risk assessments for earthquakes and tsunamis have traditionally been treated as separate entities. There have been limited studies on multihazard risk assessments because of the relatively short history of research about tsunami-related hazards, damage, and risks, after the historical tsunami events about two decades ago (e.g., the India Ocean Tsunami in 2004 and the Japan Tsunami in 2011).

Stochastic approaches to the hazard and risk assessments from earthquakes are widely adopted to quantify the damage and risk associated with both aleatory and epidemic uncertainties. A probabilistic seismic hazard analysis (PSHA) evaluates the frequency of exceedance of ground motion intensity of shaking. PSHA has become the basis for seismic assessment and design of new and existing engineered facilities ranging from civil structures, such as buildings and bridges, to critical facilities (Bazzurro, 1998). Recently, there has been an increasing interest in developing probabilistic tsunami hazard analysis (PTHA) for coastal communities in parallel to PSHA (González et al., 2009; Mori et al., 2018; Power et al., 2013; Priest et al., 2010; Thio & Somerville, 2009). One of the outputs from both PSHA and PTHA includes a hazard curve that provides the recurrence of specific intensity measures (*IM*s) that are generated from each natural hazard. These hazard curves are generally site-specific depending on the characteristics of geometry conditions and vary for each *IM*. Due to the simple format and easy application, *IM*s were utilized for engineering design purposes and for creating maps at different return period levels (e.g., 1000 years), which can serve as the fundamental input to developing local and regional hazard mitigation and evacuation plans.

Recently, there has been a growing focus on the multihazard risk assessments in earthquakes and tsunamis, but facing challenges due to the need for interdisciplinary collaboration between researchers specializing in earthquakes and tsunamis (Goda & De Risi, 2023). Moreover, the difficulties lie in quantifying and predicting the cascading or accumulated damage resulting from ground shaking (*GS*) and flow inundation, especially when confronted with limited field data and the inherent uncertainties involved. Another obstacle is the analysis of distinguishing damage states (*DS*s) that are predominantly caused by earthquakes or tsunamis. Consequently, only a limited number of studies are available that address the multihazard aspects of earthquakes and tsunamis, as well as their cascading hazards, damage, and associated risk assessments.

De Risi and Goda (2016) first assessed the combined earthquake and tsunami hazards in communities using the common earthquake and tsunami source models to evaluate regional hazards in Japan by performing separate hazard modeling for earthquake and tsunami. Goda and De Risi (2018) evaluated the regional damage and developed a multirisk exceedance curve to evaluate the direct loss on buildings utilizing empirical fragility functions that were developed using a dataset from Japan. This study highlighted the difference in levels of damage between single and multirisk at the regional scale. Recently, Goda et al. (2021) assessed the potential earthquake and tsunami risks from the Nankai-Tonankai megathrust subduction zone based on a stochastic earthquake source modeling approach and identified critical multihazard loss scenarios. Similarly, Park, Cox, Alam, et al. (2017) and Park et al. (2019) adopted a probabilistic approach to evaluate hazards and damage to built environments in coastal communities in the United States from multihazard conditions sharing the earthquake source model environments on a regional scale.

The primary objective of this chapter is to introduce a framework for evaluating multihazards and subsequent combined damage and risks in built environments. The following sections discuss the threat of multihazards in Seaside, Oregon and outline the framework, which comprises two phases: (1) probabilistic seismic and tsunami hazard analysis, PSTHA, and (2) probabilistic seismic and tsunami damage analysis, PSTDA. PSTHA evaluates seismic and tsunami-driven hazards, utilizing common source inputs for earthquakes and tsunamis while considering local geophysical characteristics and variations in rupture models. The probabilistic characterization of earthquake and tsunami *IM*s is assessed by adopting ground motion prediction equations (GMPEs) and simulating numerical models for tsunami wave propagation and inundation. Subsequently, PSTDA employs fragility analysis at the defined *IM*s from the PSTHA at specific exceedance intervals to evaluate damage probabilities in built environments resulting from earthquakes, tsunamis, and combined earthquake-tsunami events. Finally, this chapter presents an example application of risk assessment at Seaside, Oregon, utilizing the results of PSTDA, and concludes with a discussion and summary.

20.2 Multihazards, damage, and risk assessments at Seaside, Oregon

20.2.1 Study site

The coasts of the US Pacific Northwest (Washington, Oregon, and northern California) are under a significant threat from megathrust earthquakes and resulting tsunamis originating from the Cascadia subduction zone (CSZ) (Wood et al., 2010). The CSZ is located where the Juan de Fuca Plate and the North American Plate converge (Fig. 20.1A), and the Juan de Fuca Plate moves northeast and subducts beneath the North American Plate (Heaton & Hartzell, 1987). Historically, the accumulated elastic potential energy between these plates has been released in megathrust earthquakes, causing multihazards from *GS* and tsunami inundation at the coast. The most recent megathrust event in the CSZ occurred on January 26, 1700, with a full rupture across the entire zone, estimated to have a moment magnitude scale,

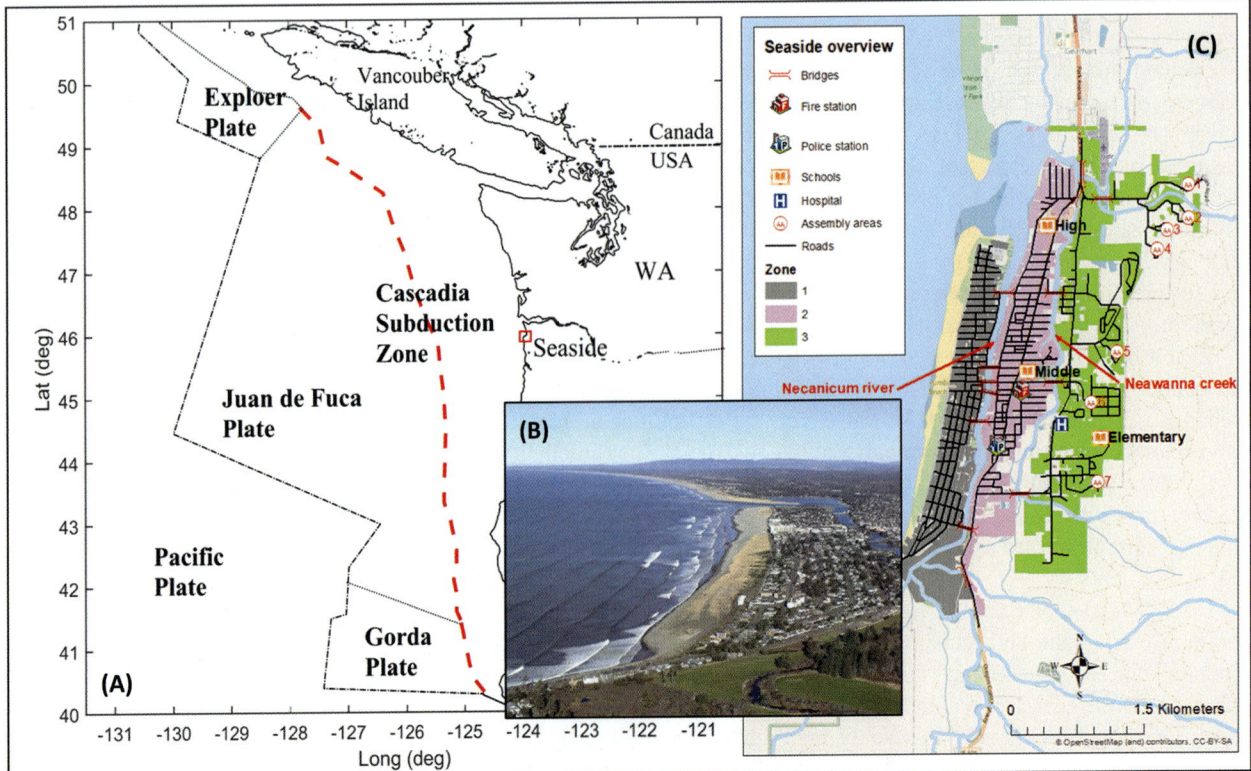

FIGURE 20.1 (A) Map of the Pacific Northwest, (B) aerial view of the study site, Seaside, Oregon, and (C) sketch of the study site with information. Map lines delineate study areas and do not necessarily depict accepted national boundaries.

M_w, ranging from 8.7 to 9.2 (Satake et al., 2003). The probability of the next M_w 9.0 exceeding event in the CSZ has been estimated to be 17% in the next 50 years and 25% in the next 100 years (Goldfinger et al., 2012).

The study site, Seaside, Oregon, is recognized as one of the most vulnerable coastal towns to future tsunamis originating from the CSZ due to its shallow and flat bathymetry, as shown in Fig. 20.1B. The city has a population of approximately 6500 residents, and the number of people in the city can increase to over 20,000 during the summer season (Park & Cox, 2016). There are over 5700 buildings in Seaside, with larger hotels, constructed of steel and concrete, in the city center and with wooden residential structures in the northern and southern parts of the city. In the event of a full-rupture CSZ scenario, approximately 87% of the town area is projected to be inundated by the tsunami. Fig. 20.1C shows the sketch of Seaside along with the locations of critical facilities (hospital, police station, and fire station), roads, bridges, and potential shelters for the tsunami. The city area could be divided into zones 1–3 by the Necanicum River and the Neawanna Creek (Fig. 20.1C). Zone 1 (gray region) and Zone 2 (pink region) are mostly flat areas, and most critical facilities are placed, while Zone 3 (green region) has assembly areas where people can take shelter during tsunami events. Existing research has extensively examined the city's vulnerability to potential earthquake and tsunami hazards stochastically, including assessments of damage to built environments, life safety, and infrastructure resilience, including potential debris impacts due to tsunamis or earthquakes (Amini et al., 2023; Kameshwar et al., 2021; Park et al., 2019; Park, Cox, & Barbosa, 2017; Wang et al., 2016).

20.2.2 Probabilistic seismic and tsunami hazard analysis

As the priority task in the assessment of risk from both earthquake and tsunami hazards, PSTHA is adopted. A key aspect of the PSTHA process is that the results from PSHA and PTHA share the same source event that triggers the shaking of the ground and the initial displacement of water along the fault. In this study, PSTHA is performed by the following three steps:

Step 1—Defining earthquake and tsunami source models,
Step 2—Modeling of earthquake and tsunami hazards, and
Step 3—Evaluating annual exceedance probability (*AEP*) and hazard map.

In Step 1, the geophysical characteristics and intensity of earthquakes at the sources (e.g., subduction zones) could be discretized for several scenarios through a logic tree model. We utilized a multihazard logic tree model for earthquake and tsunami hazard modeling, shown in Fig. 20.2. Initially, tsunamigenic earthquakes affecting the site are identified and classified as near- or far-field based on their source locations. Here, we only utilized the megathrust events in the CSZ as a case study to represent possible near-field sources of earthquakes that will generate significant damage and losses to the built environment in Seaside, Oregon. Other tsunami sources from different subduction zones or other types of seismic events are excluded in the current PSTHA because the large inundation events at the test site are

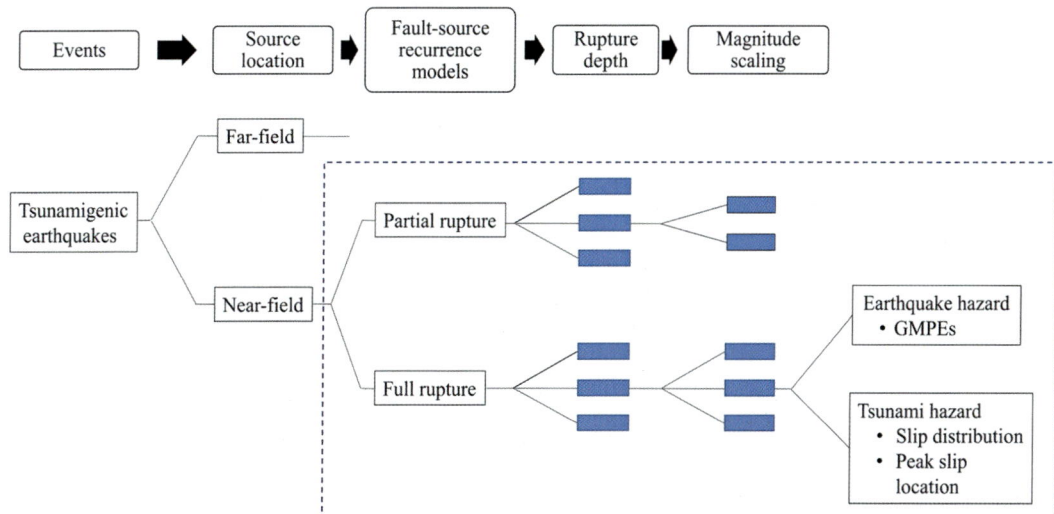

FIGURE 20.2 Multihazard logic tree model for the probabilistic seismic and tsunami hazard analysis.

mostly dominated by the CSZ events. Thus our PTHA analysis is conditioned on the CSZ events only (boxed area in Fig. 20.2).

Recurrence models and characteristic seismic moment–frequency distributions are required to determine the earthquake frequency in a logic tree model. In our model, the rupture depth is fixed as the given geophysical models utilized in tsunami generation, while it affects ground motion intensity since most GMPEs use the closest distance to rupture as a distance metric. For the CSZ, three seismological models were used in the 2014 National Seismic Hazard Map. Various magnitude scaling relationships exist for subduction earthquakes, including the scaling laws based on a global earthquake database (Skarlatoudis et al., 2015). Once the appropriate magnitude and scaling laws are determined, earthquake and tsunami simulations are performed along different logic tree branches. GMPEs are used for ground motion estimation, while tsunami simulations require the seismic moment and the earthquake slip distribution for a given rupture to assess tsunami hazard intensity at a specific site.

To describe the distribution of earthquake magnitudes in the region of interest, the Gutenberg–Richter (GR) relationship is adopted. The GR relationship is given by

$$\log(\lambda_m) = a - bm \tag{20.1}$$

where λ_m is the mean annual rate of exceedance of an earthquake of magnitude, m, that represents the overall rate of earthquakes in a region, a represents the overall seismic activity rate, and b represents the relative ratio of small and large magnitudes. For the simplification of notations in Eqs. (20.1–20.4), the variable m is used for earthquake magnitude; this can be interchanged with the moment magnitude scale, M_w. For the CSZ, a tapered GR (TGR) distribution is adopted (Rong et al., 2014). The TGR is expressed as a function of the seismic moment, M_0, instead of M_w, and an exponential taper is applied to the number of events with a very large seismic moment. The TGR complementary CDF (CCDF) is given by Kagan (2002)

$$F(M_0) = \left(\frac{M_{0t}}{M_0}\right)^\beta \exp\left(\frac{(M_{0t} - M_0)}{M_{0c}}\right) \quad \text{for} \quad M_{0t} \leq M_0 < \infty \tag{20.2}$$

where β is the index parameter of the distribution and $\beta = (\tfrac{2}{3})b$. M_{0c} and M_{0t} are the corner moment and the threshold moment above which the earthquake catalog is assumed to be complete. The CCDF of TGR can be rewritten in terms of m as

$$F(m) = \left(10^{1.5(m_t - m)}\right)^\beta \exp\left(10^{1.5(m_t - m_c)} - 10^{1.5(m - m_c)}\right) \tag{20.3}$$

For the CSZ, Rong et al. (2014) estimated $\beta = 0.59$ and $m_c = 9.02$ considering the 10,000-year paleoseismic record based on the turbidite studies (Goldfinger et al., 2012), in addition to the limited number of instrumental earthquake data, using the maximum likelihood method. In the performed implementation, it is convenient to convert the continuous distribution of moment magnitudes into a discrete set of possible moment magnitudes, which are given by

$$P(M = m_j) = G(m_j + 0.5\Delta m) - G(m_j - 0.5\Delta m) \tag{20.4}$$

where $G(m)$ is the cumulative distribution function and Δm is the adopted discretization interval. A discretization interval of $\Delta m = 0.2$ is adopted, resulting in 10 central magnitude values $m = 7.4–9.2$ for this study.

The TGR that was recommended by Rong et al. (2014) is shown in Fig. 20.3. According to the TGR distribution, M_w 8.8 + earthquakes are expected with a return period of 500 years ($\lambda_m = 0.002$), while M_w 9.0 + earthquakes are expected with a return period of 1000 years ($\lambda_m = 0.001$). Goldfinger et al. (2012) reconstructed the large earthquake history of the CSZ for approximately 10,000 years based on earthquake-induced turbidite deposits in marine sediments and onshore paleoseismic records. They suggested four types of earthquake rupture along the CSZ based on the interpretation of the turbidite data during the past 10,000 years: (1) 19–20 full-margin or nearly full-margin ruptures; (2) 3–4 ruptures along the 50%–70% of the southern margins; (3) 10–12 southern ruptures from central Oregon southward; and (4) 7–8 southern Oregon/northern California ruptures. Though the turbidite data do not provide a direct indication of the probable earthquake magnitudes, Goldfinger et al. (2012) estimated the earthquake magnitudes of different rupture events based on the relations observed among the rupture length (distance between offshore core sites containing turbidites from the same event), turbidite thickness, and turbidite mass. They estimated that full-rupture events constituted M_w 8.7–9.3. Considering 19 full-ruptures over the past 10,000 years, $\lambda_{fullrupture} \approx 19/10,000 = 0.0019$, which is consistent with the $\lambda_{M \geq 8.8} = 0.0019$ estimated the TGR distribution in Fig. 20.3. Similarly, we utilize the TGR to calculate for partial rupture events constituted $M_w < 8.7$.

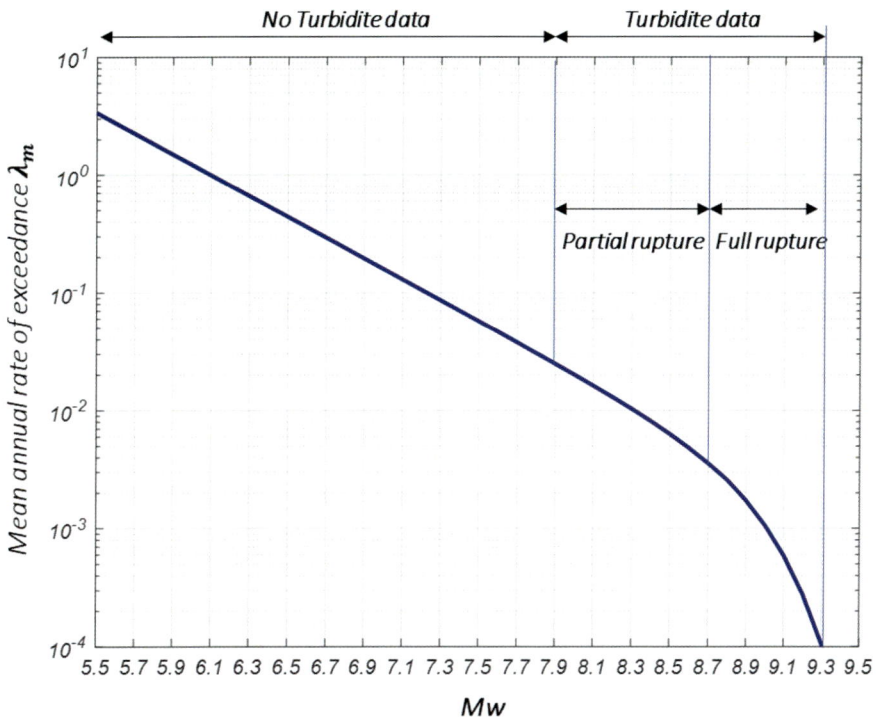

FIGURE 20.3 Tapered Gutenberg–Richter relationship for the Cascadia subduction zone.

20.2.2.1 Tsunami hazard model

In general terms, tsunami hazard characterization at a particular site requires three steps of modeling: (1) tsunami generation, (2) propagation, and (3) inundation. For most modeling efforts, the tsunami generation is given as an initial surface water displacement along the fault. The Okada model (Okada, 1985), which is based on the linear coseismic dislocation of fault slips, is widely used for simplicity, and the initial displacement is assumed to occur simultaneously along the fault. This study adopts this simplified model to generate initial tsunami displacements for each scenario defined from the logic tree model.

For the CSZ tsunami hazard model, the tsunami logic tree model considers both partial rupture and full rupture events. Based on historical turbidite data in the CSZ (Fig. 20.3), smaller moment magnitudes ($M_w < 8.7$) have been associated with a partial rupture, whereas, larger moment magnitudes ($M_w \geq 8.7$) have been linked to full rupture events. The model discretized moment magnitude ranges, such as M_w 8.1 and 8.5 for partial rupture, and M_w 8.8, 9.0, and 9.2 for the full-rupture event. The logic tree model excluded the event that the moment magnitude is less than 8.1, assuming no significant impacts from smaller magnitude events at Seaside.

The fault slip model suggested by Park and Cox (2016) characterizes the randomness of fault slip distribution at the CSZ as a Gaussian shape, parameterized in terms of the moment magnitude, peak slip location, and a fault slip shape as

$$f\left(Y'/dL | \alpha_1, \alpha_2\right) = \frac{1}{\sqrt{2\pi}\alpha_2} \exp\left(\frac{-\left(Y'/dL - \alpha_1\right)^2}{2\alpha_2^2}\right) \tag{20.5}$$

where α_1 and α_2 are the slip distribution parameters along rupture strike direction (Y), and dL is the unit length of the subfault utilized in the slip model. α_1 and α_2 control the location of the peak slip and the shape of the slips along the strike direction (narrow, medium, and wide) with empirically determined weighting factors (w_S). A total of 120 scenarios are considered along the CSZ, having five representative seismic moments, three slip shapes, and eight peak slip locations.

The mean recurrence interval (T_m) for a full-rupture event with magnitudes M_w 8.8, 9.0, and 9.2 is calculated as $T_m = 526$ years based on the analysis of the 19 paleotsunami records spanning the past 10,000 years (Witter et al., 2013). The weight factors (w_M) for the three discretized moment magnitude events are derived from turbidities records (Goldfinger et al., 2012). For partial rupture events, the mean recurrence intervals for magnitudes M_w 8.1 and 8.5 are set to $T_m = 62.5$ years and 167 years, respectively, based on the GR relationship.

Tsunami propagation is generally considered a solved problem because the equations are well defined for long-wave propagation in the open ocean. However, the more precise propagation model requires accurate knowledge of the underlying bathymetry, which affects wave deformations through dispersion, diffraction, refraction, breaking, and shoaling. Tsunami inundation considers the flow of kinematics over dry land. This is not a trivial problem to solve because of the complex interaction of the flow with complicated bathymetry and the built/natural environments that are also changing due to the destructive nature of the overland flow. The various numerical flow models were developed and validated for tsunami runup and inundation problems (Synolakis et al., 2008). The shallow water equation is widely applied to tsunami inundation, including tsunami generation and propagation, while high-order fully nonlinear Boussinesq types models (Kim et al., 2009; Shi et al., 2012) and multilayered nonhydrostatic models (Ma et al., 2012; Yamazaki et al., 2011) are preferable at the complicated bathymetry.

Most inundation models assume the "bare earth" condition, which excludes any detailed macro roughness on the bathymetry of the numerical model by utilizing a digital elevation model in which the natural and built environments are removed. The modeling of tsunami inundation with nonbare earth conditions is still challenging because of the complicated flow fields due to fluid–structure interactions in the built environment (Park et al., 2013) and relatively long computation time. Considering the hundreds or thousands of scenarios in PTHA, the modeling of all scenarios accounting for the nonbare earth conditions is quite challenging and involves many uncertainties. The impacts of the existing vegetation and structures may be replaced by a suitable friction factor (Bricker et al., 2015), while the application of roughness is still limited.

To simulate tsunami generation, propagation, and inundation that result in tsunami hazard in Seaside, Oregon, two numerical models were applied: the ComMIT/MOST model (Titov et al., 2011) for tsunami generation and propagation from the source to the near coastal region, and the COULWAVE model (Lynett et al., 2002) for the inundation process in the near and onshore regions. The ComMIT/MOST model solves time-dependent nonlinear shallow water wave equations, and the COULWAVE model solves a Boussinesq-type equation with higher nonlinear terms (Kim et al., 2009).

To optimize the computation time, we developed three nested grids, named A, B, and C-grid, for our studies. The size and dimension of each grid were 1 arcminute (400×400), 3 arcseconds (800×800), and 24 m (416×390), respectively (Fig. 20.4). The constant bottom friction coefficient (Manning number, $n = 0.03$) over the entire computation domain was assigned to both the ComMIT/MOST and COULWAVE models. The tidal level was set to the mean high water as a default, which is 1.9 m above NAVD88, and maintained at a constant level throughout the simulation for conservative analysis.

As previously mentioned, the coseismic dislocation model proposed by Okada (1985) is used to determine the initial dislocation of surface elevation resulting from the fault rupture in the CSZ. In Fig. 20.4A, the 27 black rectangles represent the unit subfaults utilized in the ComMIT/MOST model for tsunami generation, and they correspond to identical geometrical conditions (e.g., subfault coordinates, area, dip angle, and strike direction) of the 27 subfaults. The geophysical information for each subfault is tabulated in Fig. 20.4. The distribution of fault slip at each subfault is determined using Eq. (20.5) following the logic tree model.

Fig. 20.5 provides the information on Seaside, Oregon, utilized in the tsunami and earthquake hazard model. Fig. 20.5A shows the aerial image, and Fig. 20.5B provides detailed bathymetry and topography of the study area. It reveals the presence of a coastal dune that serves as the foundation for the city, running parallel to the shoreline. Between the Necanicum River and the Neawanna Creek, there is a secondary rise in the topography (near Point 2). Moving eastward from the Neawanna Creek, there is a steep gradient leading to foothills, and a large headland is situated to the southwest. Fig. 20.5C illustrates the distribution of soil classes in the region. The steeper mountainous areas are classified as Class C, while most of the city area falls under Class D (ASCE, 2010). The distribution of soil classes also aligns roughly parallel to the shore.

20.2.2.2 Earthquake hazard model

Near-field ground motions in seismic hazard assessment are influenced by various factors, including the heterogeneity of earthquake rupture processes, such as slip distribution, multiple ruptures, rupture directivity, and the behavior of the rupture front. Accurate estimation of ground motion for such assessments necessitates considering and characterizing this heterogeneity. Hybrid broadband ground motion simulation procedures (Somerville et al., 2012) or 3D simulations of earthquake scenarios (Delorey et al., 2014; Olsen et al., 2008) are commonly employed to accomplish this, as they involve explicit source-to-site wave propagation and synthetic ground motion generation.

While these detailed simulations offer valuable insights and are particularly useful for generating synthetic waveforms, their execution requires significant computational resources, making them impractical for large-scale PSHA.

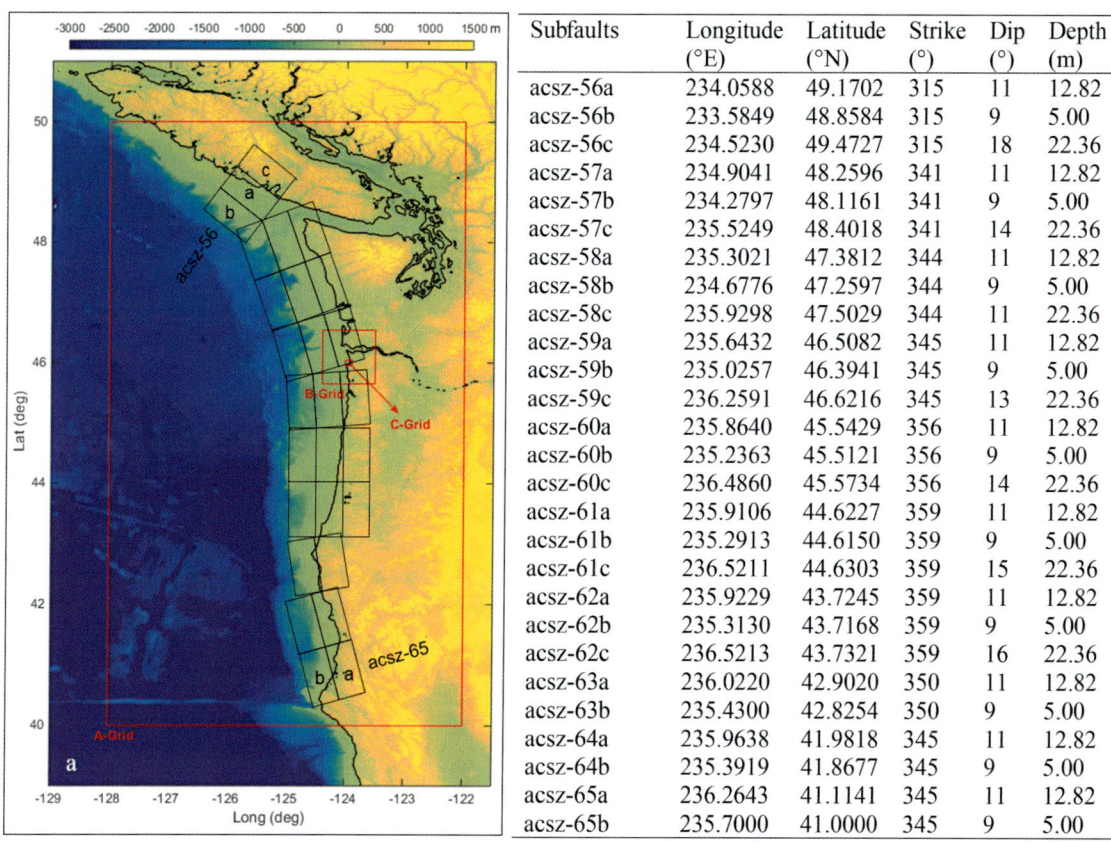

FIGURE 20.4 Bathymetry, grids, and subfault information. The left panel shows the bathymetry, numerical model grid setup, and location of the CSZ subfaults, and the right panel presents detailed subfault information. *CSZ*, Cascadia subduction zone. Map lines delineate study areas and do not necessarily depict accepted national boundaries.

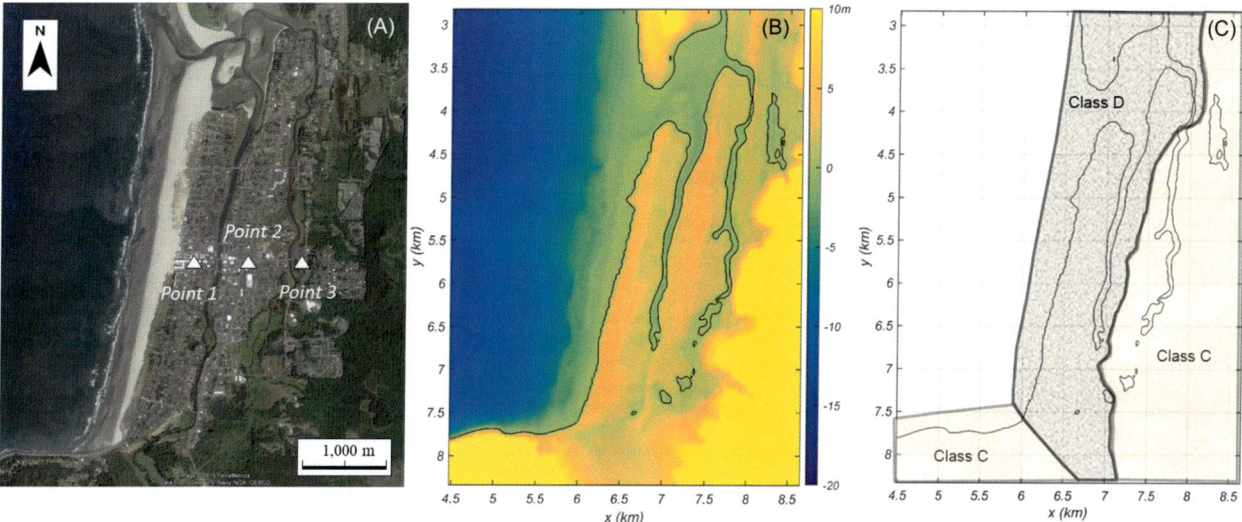

FIGURE 20.5 (A) Satellite image, (B) bathymetry/topography, and (C) soil class of Seaside, Oregon. Map lines delineate study areas and do not necessarily depict accepted national boundaries. *From Park, H., Cox, D. T., Alam, M. S., & Barbosa, A.R. (2017). Probabilistic seismic and tsunami hazard analysis conditioned on a megathrust rupture of the Cascadia subduction zone.* Frontiers in Built Environment, *3. https://www.frontiersin.org/articles/10.3389/fbuil.2017.00032/pdf. https://doi.org/10.3389/fbuil.2017.00032.*

Therefore, in practice, GMPEs remain widely used in such analyses. GMPEs provide a simplified and efficient approach for estimating ground motions by considering empirical relationships between ground motion parameters and earthquake characteristics.

GMPEs are dependent on local and regional site conditions. To calculate the source-to-site distance for GMPEs, the closest rupture distance between the site and the rupture surface is required. In this study, the BC Hydro GMPE (Abrahamson et al., 2016) is utilized, and the precomputed closest distances for the centroid of each building parcel in Seaside are obtained from the 27 subfault areas (Fig. 20.4). The rupture surface for different magnitude scenarios is then computed based on the magnitude scaling law by Murotani et al. (2013). The rupture surfaces are randomly positioned along the 27 subfaults corresponding to the approximately 1000 km long by 150 km wide for a full rupture. The precomputed closest distances are then interpolated for each scenario using the randomly positioned rupture surfaces. The calculated closest distance for all scenarios is 27.5 km, representing the fault area directly beneath the C-grid location. For partial rupture events, the rupture surface is randomly placed within the fault plane, resulting in variations in the computed closest rupture distance.

20.2.2.3 Calculating annual exceedance probability and hazard map

Once tsunami and earthquake hazards in Seaside are characterized, sharing the same logic tree model and characteristics of fault information, we can determine the *AEP* of the *IMs* resulting from earthquake ground motions or tsunami inundation flows. In this study, four *IMs* were selected, including the peak ground acceleration (*PGA*), spectral acceleration at a fundamental period of 0.3 seconds ($S_a(T_1 = 0.3$ seconds$)$), the maximum flow depth (d_{max}), and the maximum momentum flux (MF_{max}) because these values are often linked with structural damage and risk assessments.

Following the Poisson process (Cornell, 1968), the probability of exceedance of each *IM* for earthquakes or tsunamis at a particular site in time t is given by

$$P(IM > im|t) = 1 - \exp(-\lambda t) \quad (20.6)$$

where λ is the mean exceedance rate of occurrence at which the *IM* exceeds a specific *IM* value, *im*, at a given site, during the time, t. If $t = 1$ year, Eq. (20.6) provides the *AEP* of a specific *IM*, *im*. The mean exceedance rate of occurrence of each *IM* exceeding specific *im* is computed using the Total Probability Theorem by integrating the contributions of all possible sources of earthquakes for each of the sources and all values of moment magnitude considered, and is given by

$$\lambda_{IM>im}(im) = \sum_{i=1}^{N_{source}} \lambda_i(M_w \geq m_{min}) \int_{m_{min}}^{m_{max}} P(IM > im|\theta, M_w) S_{\theta_i}(\theta|M_w) f_{M_i}(m) dm \quad (20.7)$$

where $\lambda_i(M_w \geq m)$ corresponds to the TGR (Fig. 20.3) for the *i*th source and N_{source} is the total number of sources/scenarios considered in the logic tree model in Fig. 20.6. The variable m is used for earthquake magnitude and can be

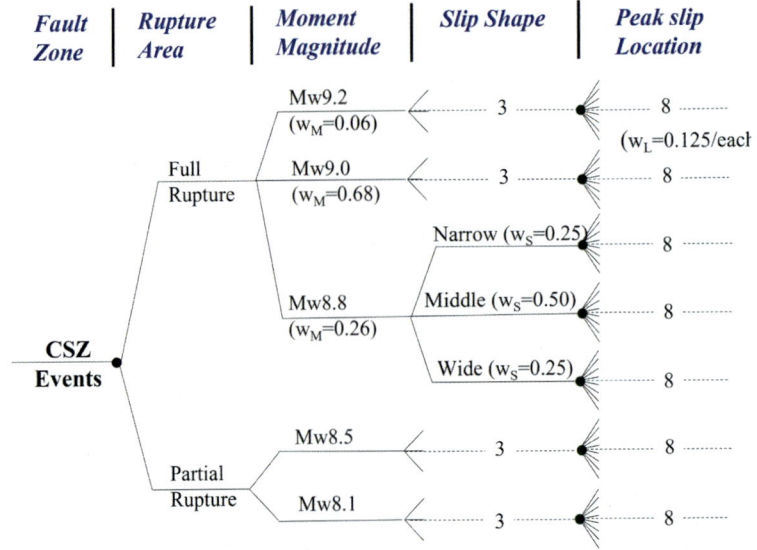

FIGURE 20.6 Logic tree model for the Cascadia subduction zone.

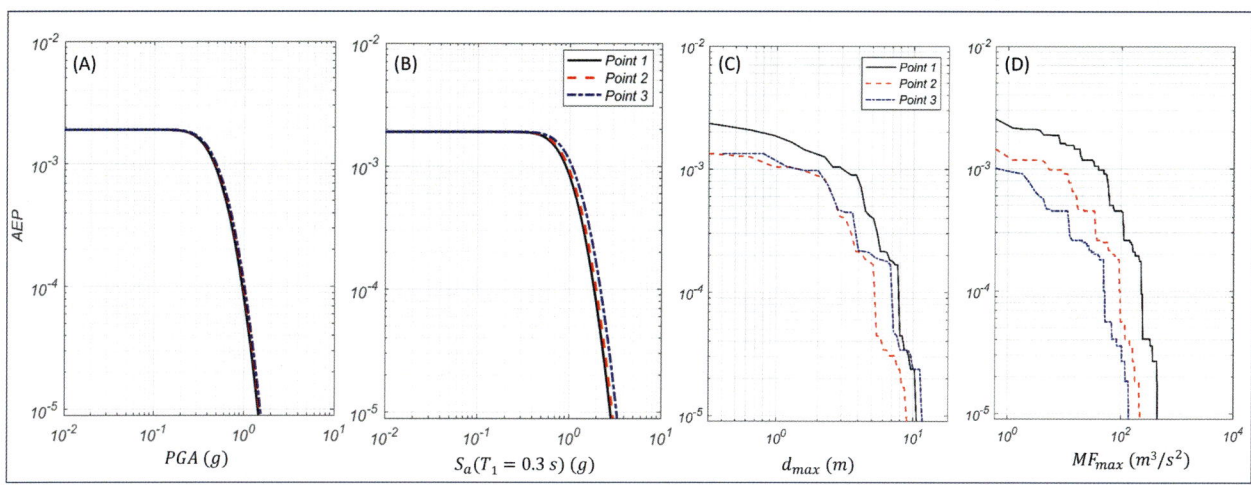

FIGURE 20.7 Example of hazard curves for (A) *PGA*, (B) $Sa(T_1 = 0.3$ seconds), (C) d_{max}, and (D) MF_{max} at three observation points. *PGA*, Peak ground acceleration. *From Park, H., Cox, D. T., Alam, M. S., & Barbosa, A.R. (2017). Probabilistic seismic and tsunami hazard analysis conditioned on a megathrust rupture of the Cascadia subduction zone.* Frontiers in Built Environment, *3. https://www.frontiersin.org/articles/10.3389/fbuil.2017.00032/pdf. https://doi.org/10.3389/fbuil.2017.00032.*

interchanged with the moment magnitude M_w. $P(IM > im|\theta, M_w)$ represents the CCDF of *IM* exceeding *im* conditional on seismic source parameters, source-to-site distance, and earthquake magnitude, and this term is evaluated using GMPEs for the ground motion shaking or using tsunami propagation and inundation modeling, depending on the *IM* considered. $S_{\theta_i}(\theta|M_w)$ is the distribution of the characteristic values of the fault and the source-to-site distance conditional on M_w. Lastly, $f_{Mi}(m)$ denotes the probability density function of the magnitude (*m*) given the *i*th source.

Figs. 20.7A and B show the *AEP* of *PGA* and $S_a(T_1 = 0.3$ seconds) at three representative observing points that were displayed in Fig. 20.5A. As per the soil site class map of Fig. 20.5C, Point 1 and Point 2 fall in site class D, whereas Point 3 is in site class C. A value $\lambda_m = 0.0019$ corresponding to $M_w \geq 8.8$ is used for the hazard curves and subsequent *AEP* computations. In *AEP* computations, the probability mass function considered for the magnitudes is consistent with values shown in the logic tree model, which was obtained from Goldfinger et al. (2012). Since each weight assigned to M_w 8.8, 9.0, and 9.2 is 5/19, 13/19, and 1/19, respectively, the M_w 9.0 is the dominant contributor to the hazard curves presented in Fig. 20.7 for both *PGA* and $S_a(T_1 = 0.3$ seconds). As shown in Fig. 20.7, the *AEP* of *PGA* at the three points is almost identical, whereas a slight variation in *AEP* for $S_a(T_1 = 0.3$ seconds) is observed for those three points. The slight variation in *AEP* for $S_a(T_1 = 0.3$ seconds) is due to the different shear wave velocities assigned to those points in GMPEs. Overall, the *AEP* of both *IMs* is mostly insensitive to their ground conditions.

Figs. 20.7C and D illustrate hazard curves of two tsunami *IMs*, d_{max} and MF_{max}, at the same observation points. Typically, as the distance from the shoreline increases, the tsunami *IMs* decrease due to significant energy dissipation caused by friction and gravity during the inundation process. For instance, both d_{max} and MF_{max} at Point 1 exhibit higher *AEP* compared to Points 2 and 3. Moreover, the clear sensitivity of the tsunami *IMs* to site-specific bathymetry conditions is observed for d_{max} and MF_{max}. Particularly, MF_{max}, as shown in Fig. 20.7D, exhibits three distinct curves with a general decreasing trend from the shoreline to the inundation limits, while d_{max} of Points 2 and 3 shows a similar curve when $d_{max} < 3$ m.

In summary, the variation of *AEP* for tsunami *IMs* among the three observation points differs significantly from the results obtained for earthquake *IMs*. Notably, there are strong spatial gradients for the tsunami *IMs* across the length scale of the city. This disparity is somewhat expected because the *AEP* of both *PGA* and S_a primarily depends on soil types and the rupture-to-site distance (R_{rup}), which exhibit relatively small variations over the study region in Seaside (about 1 km). Moreover, this distinction underscores the differences in the fundamental physics governing the propagation of seismic energy through the subsurface and the propagation of hydrodynamic tsunami energy. These distinct mechanisms contribute to the dissimilarity in the observed *AEP* variations for earthquake and tsunami *IMs*.

To generate hazard maps of earthquakes and tsunamis, we calculate *IMs* at specific *AEPs* across the study area. These maps show seismic and tsunami hazards at different recurrence periods (e.g., 500, 1000, and 2500 years). Fig. 20.8 shows hazard maps that exhibit the spatial distributions of the *IMs* from earthquakes and tsunamis for *AEP* = 0.001 (often referred to as the "1000-year event") for the CSZ. Each part of Figs. 20.8A−D presents the *PGA*, S_a

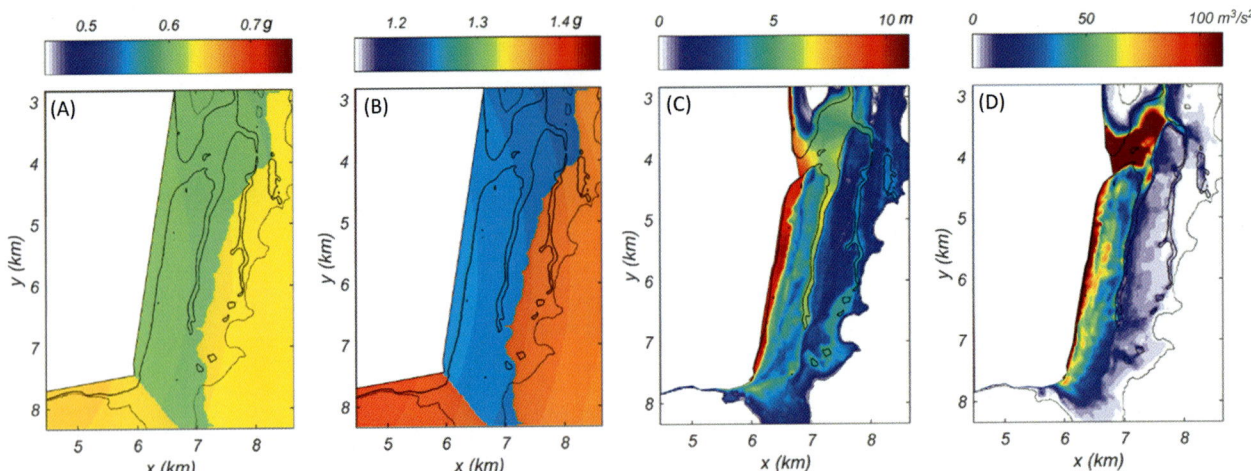

FIGURE 20.8 Hazard maps for Seabed, Oregon for the 1000-year event for (A) *PGA*, (B) $S_a(T_1 = 0.3$ seconds), (C) d_{max}, and (D) MF_{max}. *PGA*, Peak ground acceleration. Map lines delineate study areas and do not necessarily depict accepted national boundaries. *From Park, H., Cox, D. T., Alam, M. S., & Barbosa, A.R. (2017). Probabilistic seismic and tsunami hazard analysis conditioned on a megathrust rupture of the Cascadia subduction zone. Frontiers in Built Environment, 3. https://www.frontiersin.org/articles/10.3389/fbuil.2017.00032/pdf. https://doi.org/10.3389/fbuil.2017.00032.*

($T_1 = 0.3$ seconds), d_{max}, and MF_{max} that correspond to $AEP = 0.001$ at each grid point across the study area and are displayed over the dry land area only ($z > 0$). Here, the dotted contour lines in each panel show the maximum inundation limits, which are defined as $d_{max} > 0.3$ m. The spatial distribution of earthquake *IM*s (*PGA* and S_a) is generally uniform but they separate into two regions depending on the two different soil type conditions (Fig. 20.5C). In the case of tsunami *IM*s (d_{max} and MF_{max}), they are highly dominated by local bathymetry and elevation conditions. Therefore, they show relatively complicated distributions depending on the local bathymetry and elevation but d_{max} and MF_{max} mostly decrease from the shore toward land (i.e., in the positive *x*-direction) as the flow reaches higher grounds.

The results of PSTHA, including the *AEP* curves and hazard maps of both earthquake and tsunami *IM*s, are utilized in the subsequent subsection for probabilistic damage analysis. The fragility functions provide a quantitative representation of the relationship between the intensity of the hazard (*GS* or tsunami inundation) and the probability of different levels of damage occurring to structures or assets in Seaside. By combining the PSTHA results with the fragility functions, a comprehensive assessment of the potential damage and associated uncertainties can be performed for the target city, enabling a better understanding of the risk posed by seismic and tsunami events in the study area.

20.2.3 Probabilistic seismic and tsunami damage assessment

The built environment consists of an inventory of buildings and lifeline infrastructure systems, such as transportation, water, power, and communication networks. Each component of the built environment may show a different damage response, and the analysis may require different *IM*s and damage descriptions depending on the characteristics of each component. The main objective of PSTDA is to estimate the accumulated damage from the seismic and tsunami scenario events to develop probabilistic estimates of the damage accounting for both seismic and tsunami hazards that are considered in the PSTHA.

A general methodology of PSTDA is summarized, and the procedure involves:

Step 4: Collecting building inventory and fragility functions,
Step 5: Building damage assessments for earthquakes and tsunamis, and
Step 6: Combine damage probabilities for multihazards and risk assessments.

20.2.3.1 Building inventory in Seaside, Oregon

In general, it is difficult to collect all details of the building inventory at a community scale, and it is hard to analyze building damage considering all the building attributes collected. In this study, for realistic building inventory for Seaside, three sources of input (tax lot data, images from Google Street View [GSV], and a field survey based on an adapted version of the FEMA-154 Rapid Visual Screening [RVS]) were used to develop the building inventory in Seaside, Oregon. The tax lot data for Clatsop County, obtained from the 2012 dataset, offer valuable information about individual parcels, such as centroid coordinates, size, address, date of construction, owner, real market value, and land value. It also includes building information through a three-digit "stat class" code, indicating land use, occupancy, and additional details like the number of floors. While tax lot data allow for quick classification of building classes over a large area, it lacks specific information on building materials, complete floor levels, and building frame types. In addition, the data may be outdated due to remodeling or changes in building use. To complement the tax lot data and address its limitations, GSV was utilized to validate and modify the classified data by evaluating features, such as the number of floors, construction type, and general age of buildings (Park, Cox, & Barbosa, 2017). GSV has the potential to provide additional information, like the number of windows or openings that could impact tsunami forces. Combining GSV with satellite imagery could also aid in estimating building width, proximity to other structures, and debris hazards. Lastly, a short field survey using RVS was conducted to verify the tax lot data and GSV methods for 10 buildings, following FEMA P-154 guidelines.

Fig. 20.9 shows the overlapped parcel from the tax lot data. It provides information on building attributes, such as material or building types (left top panel), number of stories (left middle panel), and building seismic design code levels

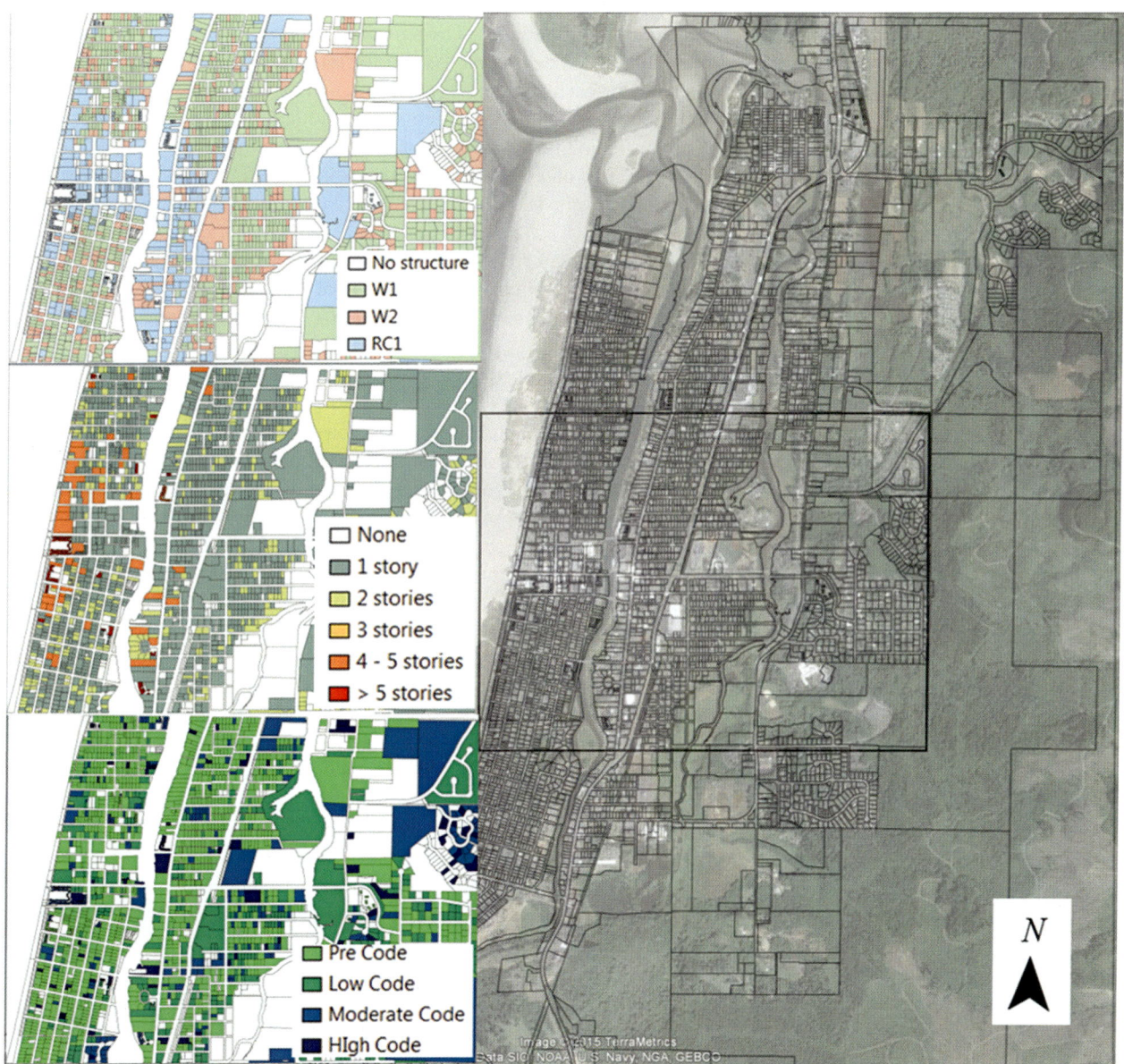

FIGURE 20.9 Maps of building inventories in Seaside, Oregon, consisting of a map of building parcels from tax lot data (right) and three inserted maps (left) for the building inventories, such as materials (top), number of stories (middle), and design codes (bottom). Map lines delineate study areas and do not necessarily depict accepted national boundaries.

based on the date of construction (left bottom panel). The data inventory for all types of buildings in Seaside, including infrastructure facilities, population, and PSTHA results, are available from the open storage from the DesignSafe-CI (Cox et al., 2022).

20.2.3.2 Fragility functions for Seaside, Oregon

The collected building inventory is matched with the fragility functions for each building's typology that were utilized to estimate the *DS* probability for earthquake shaking and tsunami inundation. The fragility function provides the probability of a damage measure exceeding a certain *DS* conditional on an *IM*. In the case of an earthquake, the development of fragility functions has been an active field of research in earthquake engineering over the past four decades (e.g., nuclear power plants, Kennedy et al., 1980; model building types in HAZUS, Kircher et al., 1997; wood structures, Rosowsky & Ellingwood, 2002; steel structures, Kinali & Ellingwood, 2007; and reinforced concrete (RC) structures, Alam & Barbosa, 2018; Goulet et al., 2007).

In the case of seismic damage at Seaside, Oregon, the lognormal fragility functions from FEMA (2011) were utilized to assess structural and nonstructural damage on each building. The structural damage assessment involves evaluating the 5% of critical damping linear response spectral acceleration ($Sa(T)$) for all buildings, obtained using GMPEs at specific characteristic periods (T) corresponding to each building type. These $Sa(T)$ values are subsequently converted to spectral displacement values ($Sd(T)$) for conducting structural damage due to *GS*. Similarly, structural damage due to permanent ground deformation (*GD*) was evaluated using the peak ground deformation (*PGD*) values that were converted from *PGA* utilizing GMPE and the guidelines of FEMA (2011). In addition, two sets of fragility functions are also provided in FEMA (2011), namely, drift-sensitive systems (*NDS*) and acceleration-sensitive systems (*NAS*) for the nonstructural damage assessment.

These fragility functions are defined by the median value of *IM*s and the standard deviation of the natural logarithm of the *IM*s for each building type and at four *DS*s (complete, extensive, moderate, and slight). In addition, these fragility functions corresponded to different seismic design levels, including high-, moderate-, low-, and precode, depending on the built year.

In the case of a tsunami, fragility functions had been developed through field data and numerical studies using various *IM*s for flows, such as the maximum flow depth (d_{max}), velocity (V_{max}), and MF_{max}. Among tsunami *IM*s, d_{max} has been widely accepted as a representative *IM* in developing tsunami fragility functions (Koshimura et al., 2009; Suppasri et al., 2013) because of its simplicity in predicting and availability from field observations. However, the application of those fragility functions developed from the field's studies (e.g., Japan or other Asian countries) is limited to apply in the United States because of the different building typologies (materials, stories, etc.) and potential uncertainties on d_{max} to represent damage in the different built environments and geometry conditions. For example, there are questions about the uncertainties and limits on d_{max} to solely predict structural damage on various types of buildings (De Risi et al., 2017). Furthermore, the physics-based tsunami fragility functions have been developed based on the response of finite element structural models subjected to tsunami loading (Alam et al., 2018; Attary et al., 2017; Macabuag et al., 2016; Park et al., 2012) and they highlighted that *IM*s that include the effects of flow speed and flow depth provide the best estimation efficiency in predicting structural damage.

In this study, the fragility function developed by FEMA (2013) is adopted, utilizing the maximum MF_{max} and maximum inundation depth (d_{max}) for the analysis of structural and nonstructural damage. Specifically, MF_{max} for the structural damage fragility function is an indicator of the lateral force induced by the inundation flow. The net maximum flood inundation depth (D_F) from the bottom of the 1st floor of each building is used to represent the impact of flooding on the nonstructural systems (NSS) and building content. Here, D_F is calculated as $D_F = d_{max} - d_F$, where d_F is the net flow depth to the base of the first floor of the building from the local ground elevation.

These fragility functions provide structural and nonstructural fragility functions for 36 model building typologies at varied seismic design code levels and *DS*s. These are the same model building typologies developed for seismic damage used in FEMA (2011) for evaluating the earthquake-induced damage for the study area. Therefore, the fragility functions of tsunamis from FEMA (2013) were adopted for this study and statistically combined with the building damage from earthquakes to represent accumulated damage from multihazards.

Fig. 20.10 shows the fragility functions for the structural damage from earthquakes (Fig. 20.10A) and tsunamis (Fig. 20.10B). At a particular building typology, four *DS*s (complete, extensive, moderate, and slight) and four seismic codes (pre, low, moderate, and high code) are considered. However, the slight *DS* of the tsunami fragility functions is defined as qualitative damage severity, so the tsunami fragility functions have three *DS*s defined as moderate, extensive, and complete damage rather than four *DS*s. However, each *DS* has a qualitative definition of damage associated with it,

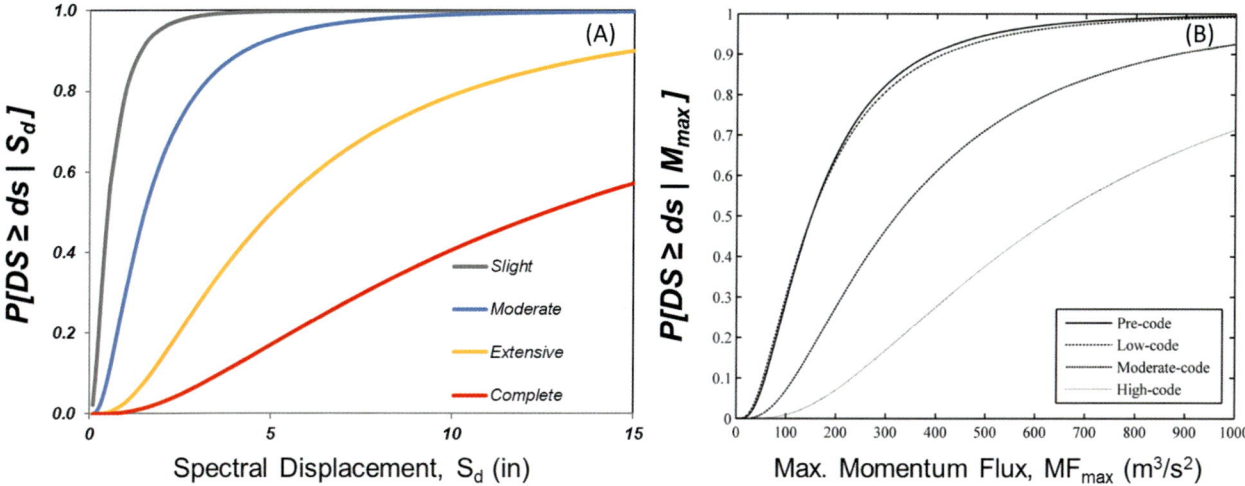

FIGURE 20.10 Example fragility functions for (A) earthquake (FEMA, 2011) and (B) tsunami (FEMA, 2013).

and default values for the *DS* parameters that define the fragility curves are also defined. For example, for a mid-rise building, there are three *DS*s defined for tsunami, including moderate, extensive, and complete.

20.2.3.3 Building damage assessment

Once building inventory and fragility functions of earthquakes and tsunamis are prepared, we can evaluate the probability of damage to buildings. The PSTDA formulation provides the mean annual rate at an infrastructure component that experiences seismic shaking-induced and tsunami inundation-induced damage exceeding a certain damage threshold given for a given location. Using the hazard maps in PSTHA, the mean annual rate of the damage exceeding and given damage threshold *ds* is given by

$$\lambda_{DS}(ds) = \int_{IM} P(DS \geq ds|IM)|d\lambda(im)| \tag{20.8}$$

where $P(DS \geq ds|IM)$ is a fragility function that provides the probability of exceeding a given *DS* for a given *IM* for earthquakes or tsunamis. As indicated in the previous section, the mean annual rate can be converted to the *AEP* using the Poisson process defined in Eq. (20.6).

Fig. 20.11 shows the spatial structural (left panels) and nonstructural (right panels) damage probability maps of achieving or exceeding the complete *DS* of buildings in Seaside, Oregon, for the 1000-year return period (*AEP* = 0.001). Each part of Fig. 20.11A, C, and E shows the structural damage from tsunami inundation ($im = MF_{max}$), *GD* ($im = PGD$) from the earthquake, and *GS* ($im = Sa(T)$) from the earthquake. Each part of Fig. 20.11B, D, and F shows the nonstructural damage from tsunami flooding ($im = D_F$), the nonstructural damage on *NAS* from earthquakes, and damage on *NDS* from earthquakes, respectively. Here, the nonstructural damage from the earthquake shaking is divided into acceleration-sensitive and drift-sensitive systems for seismic damage and loss estimations (Aslani & Miranda, 2005).

An interesting feature of the results is that the pattern of building damage in Seaside is quite different between earthquakes and tsunamis. Coastal areas (near the shoreline) experience relatively higher damage from tsunamis; however, there is relatively lower damage but uniform damage from *GD* and *GS*. This different aspect of the damage pattern is due to two reasons: (1) larger *IM*s (MF_{max} and d_{max}) at the shorelines, and these tsunami *IM*s are decreasing at the higher ground levels from the coast to inland as we observed in the hazard map (Fig. 20.8C and D) and (2) the relatively stronger resistance of wood structures against the *GS* compared to tsunami inundation flow and conversely larger resistance to tsunami loading of the taller nonductile to low-code RC structures near the shoreline, which are, however, more sensitive to larger shaking intensities.

In the case of nonstructural damage, less significant damage is observed from the tsunami. However, the most major damage is found at the shoreline and near the Necanicum River, as the hazard factor is relatively higher in those areas. For nonstructural damage from the tsunami, the intensity of the hazard depends on the net flow depth from the first floor of the building (D_F). Thus the damage could be sensitive to the specific building types and local ground level. In

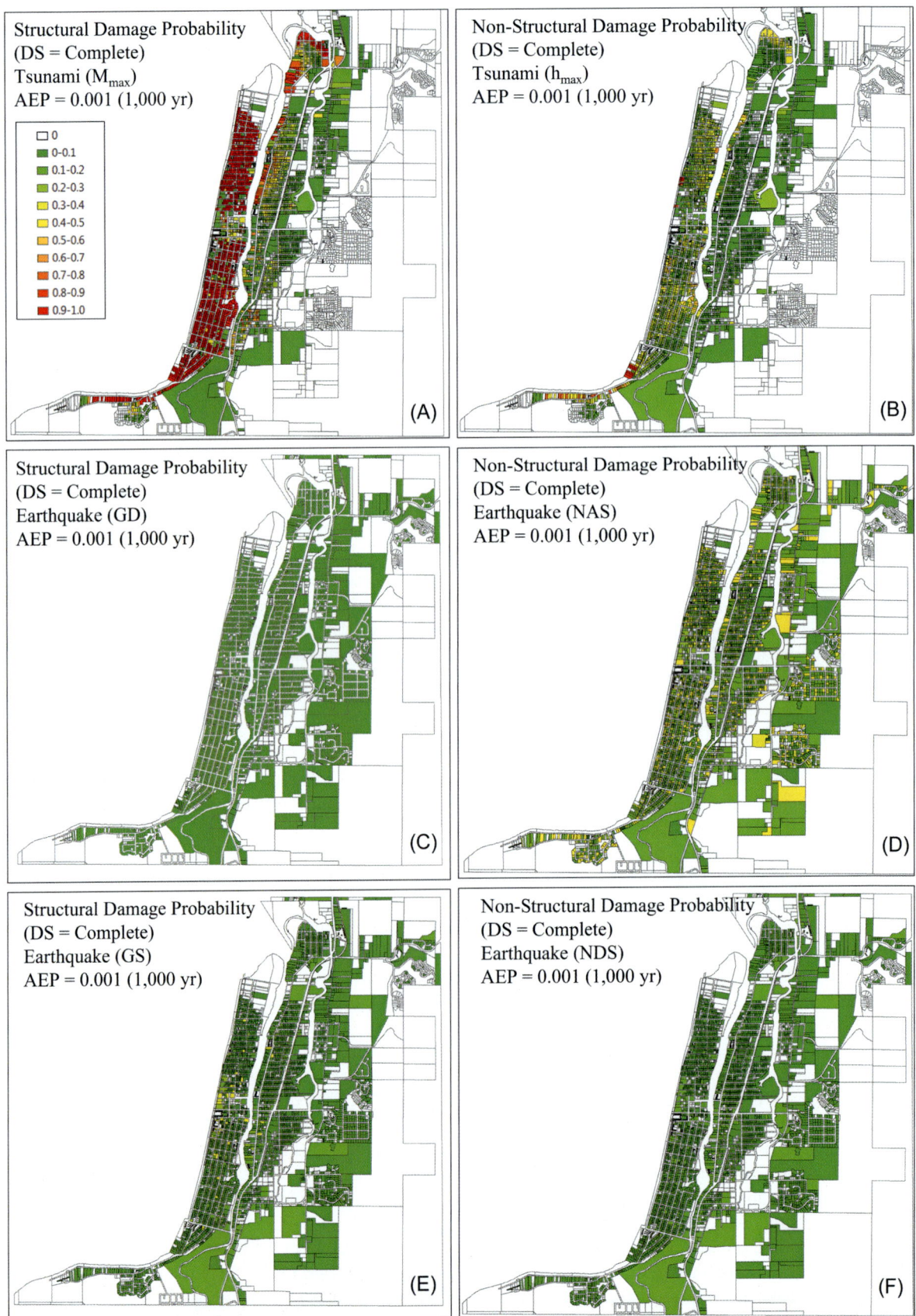

FIGURE 20.11 Damage probability maps from tsunami and earthquake at $AEP = 0.001$. (A, C, and E) Structural damage from inundation, ground deformation, and ground shaking and (B, D, and F) nonstructural damage from inundation, ground deformation, and ground shaking. *AEP*, Annual exceedance probability. Map lines delineate study areas and do not necessarily depict accepted national boundaries.

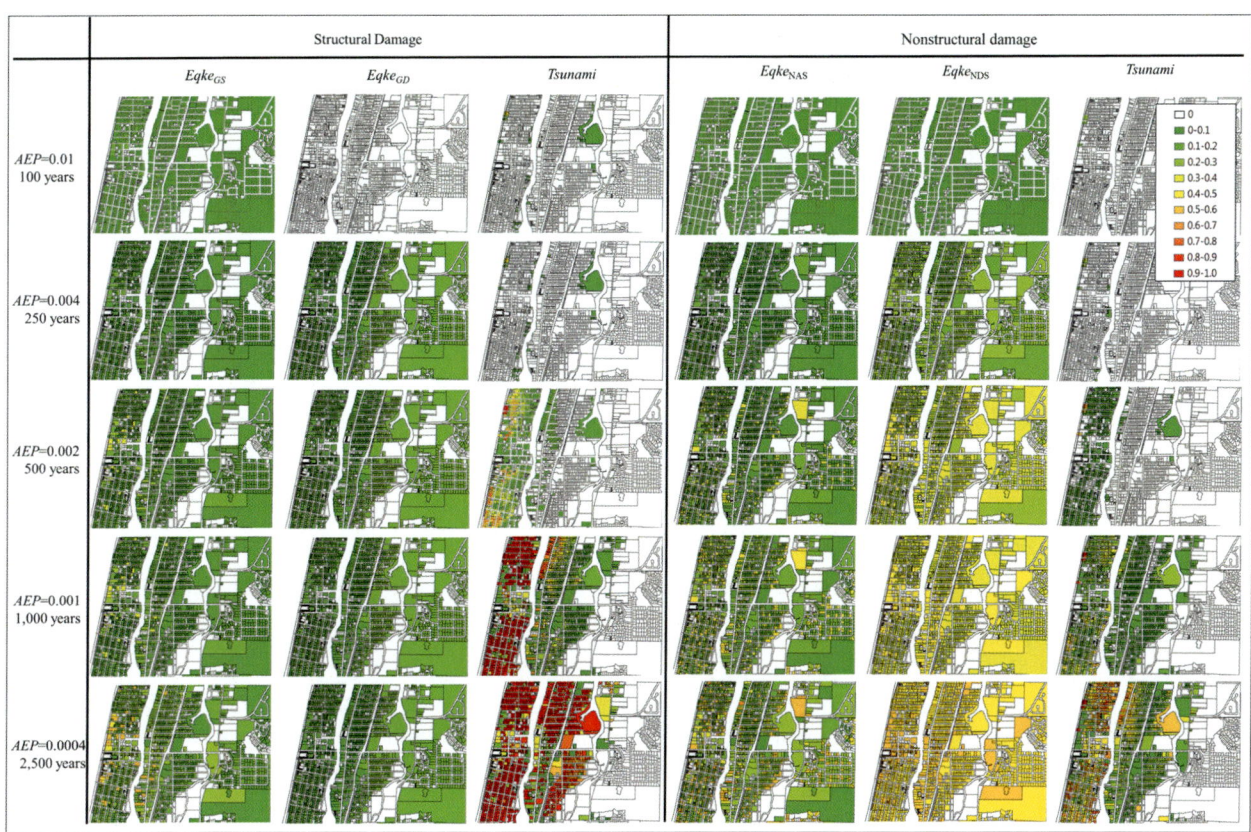

FIGURE 20.12 *AEP* structural and nonstructural damage probability maps for varied *AEP* due to a tsunami and earthquake. *AEP*, Annual exceedance probability.

contrast, quite uniform minor damage is observed from earthquakes for both *NAS* and *NDS* damage, similar to structural damage. The more significant nonstructural damage is observed for *NDS*.

Fig. 20.12 presents the probability of damage exceeding a complete *DS* for both structural and NSS under different *AEP* conditions. Lower *AEP* values, such as *AEP* = 0.0004 (2500-year return period), yield patterns similar to *AEP* = 0.001. However, overall damage probabilities increase for both tsunami and earthquake events.

For higher *AEP* conditions (*AEP* = 0.01 and 0.004), earthquakes show a similar pattern in Fig. 20.11, with damage occurring throughout Seaside, slightly lower than *AEP* = 0.001. In contrast, tsunami cases show significant differences, as *IM*s of tsunamis are highly sensitive to local bathymetry and elevation. At *AEP* = 0.01 and 0.004, almost no damage occurs due to lower inundation depth than the ground elevation. At *AEP* = 0.002 (500-year return period), the damage pattern is similar to *AEP* = 0.001, but the affected area is limited to the shoreline or near river streams (Zone A).

20.2.3.4 Combined damage probabilities for multihazards and risk assessment

Combining the damage incurred from the seismic motions and tsunami inundation is poorly understood due to the complex mechanisms of the two cascading damage processes and difficulties in post-disaster surveys by distinguishing between earthquake-induced damage and tsunami-induced damage to buildings. Goda and De Risi (2018) performed a combined seismic and tsunami damage analysis, treating each seismic and tsunami hazard as independent events, and selecting the higher probability of damage and risk between the seismic and tsunami damage analysis results to estimate the combined damage and risk. However, they excluded the possible accumulated damage from two subsequent events, thus neglecting the compounding damage effects.

In this study, the combined damage probability on a building due to two sequential events is calculated based on the assumption that each damage probability is statistically independent (FEMA, 2013). Both seismic and tsunami hazard maps are developed by accounting for site-specific characteristics. Therefore, the implemented methods do not carry over to the event-specific dependency of the seismic and tsunami hazards at different locations within an area. To overcome these limitations, an approach of combining the structural and nonstructural probabilities of exceeding a specific *DS* can be based on the basic axioms of probability theory, Boolean logic rules, and the assumption of statistical independence between damage incurred from earthquake and tsunami hazards. Table 20.1 provides the abbreviations and terminology used in the study for estimating the combined damage probability (P_{comb}).

There is an exception to reflect the potential accumulation of damage conditions. A specific level of *DS* could be generated from the joint lower-level *DS*. For example, the probability of combined complete damage *DS*(*Complete*) also includes the joint probability of extensive *DS*(*Extensive*) due to the tsunami and extensive damage due to the earthquake based on the assumption that simultaneous experience of extensive damage due to the tsunami and earthquake could result in overall complete damage to the structure. The combined damage probability of the structure for each *DS* is calculated as

$$P_{\text{Comb}}[DS = C_{Str}] = P[DS = C_{Str}|Eqke] + P[DS = C_{Str}|Tsu] - P[DS = C_{Str}|Eqke]P[DS = C_{Str}|Tsu] +$$
$$(P[DS \geq E_{Str}|Eqke] - P[DS = C_{Str}|Eqke])(P[DS \geq E_{Str}|Tsu] - P[DS = C_{Str}|Tsu]) \tag{20.9}$$

$$P_{\text{Comb}}[DS \geq E_{Str}] = P[DS \geq E_{Str}|Eqke] + P[DS \geq E_{Str}|Tsu] - P[DS \geq E_{Str}|Eqke] +$$
$$(P[DS \geq M_{Str}|Eqke] - P[DS \geq E_{Str}|Eqke])(P[DS \geq M_{Str}|Tsu] - P[DS \geq E_{Str}|Tsu]) \tag{20.10}$$

$$P_{\text{Comb}}[DS \geq M_{Str}] = P[DS \geq M_{Str}|Eqke] + P[DS \geq M_{Str}|Tsu] - P[DS \geq M_{Str}|Eqke]P[DS \geq M_{Str}|Tsu] \tag{20.11}$$

$$P_{\text{Comb}}[DS \geq S_{Str}] = P[DS \geq S_{Str}|Eqke] + P[DS \geq M_{Str}|Tsu] - P[DS \geq S_{Str}|Eqke]P[DS \geq M_{Str}|Tsu] \tag{20.12}$$

where there is no slight *DS*(*Slight*) for tsunami fragility functions. Instead, the moderate *DS* is substituted for Eq. (20.12).

There are two types of earthquake structural damage, such as *GD* and *GS* on buildings. In Eqs. (20.9–12), we used the combined damage of these two. We apply the same Boolean logic rules with the independent assumption between two damage situations.

$$P[DS = ds|Eqke] = P[DS = ds|Eqke(GD)] + P[DS = ds|Eqke(GS)] - P[DS = ds|Eqke(GD)]P[DS = ds|Eqke(GS)] \tag{20.13}$$

Fig. 20.13 shows the combined (accumulated) earthquake and tsunami damage probability map exceeding the complete *DS* for a 1000-year event (*AEP* = 0.001). Fig. 20.13A illustrates the merged structural damage from *GD* and *GS* using Eq. (20.13), while Fig. 20.13B shows the structural damage from tsunami inundation. The combined probability of structural damage (Fig. 20.13C) shows a pattern similar to the damage probability from the tsunami alone, as the overall intensity and corresponding damage probability of the tsunami dominate at *AEP* = 0.001. However, an overall increase in structural damage probabilities is observed over the entire city due to accumulated earthquake damage in Seaside. Notably, accumulated damage is concentrated in the city center (within the dashed box region), where RC buildings are prominent. Additionally, a wider range of damage is observed in Seaside compared to damage from the tsunami alone since we merge both the earthquake and tsunami damage. The combined damage for *NSS* is not presented in this study, but the combination rules can also be applied to *NSS* in the same manner.

The PSTDA can be expanded to encompass other systems within the built environment, such as transportation, power, water, and communication networks. These systems are crucial for evaluating the initial response and functionality of each component within the community and the overall functionality of the social and economic systems. The assessment of loss and damage based on PSTDA serves as fundamental input data for decision-making support on mitigation strategies and recovery studies.

TABLE 20.1 Abbreviations used in Eqs. (20.9–20.13) and corresponding descriptions.

Hazard	Structure	Damage state
Eqke Earthquake; *Tsu* Tsunami	*Str* Structure	C Complete E Extensive M Moderate S Slight

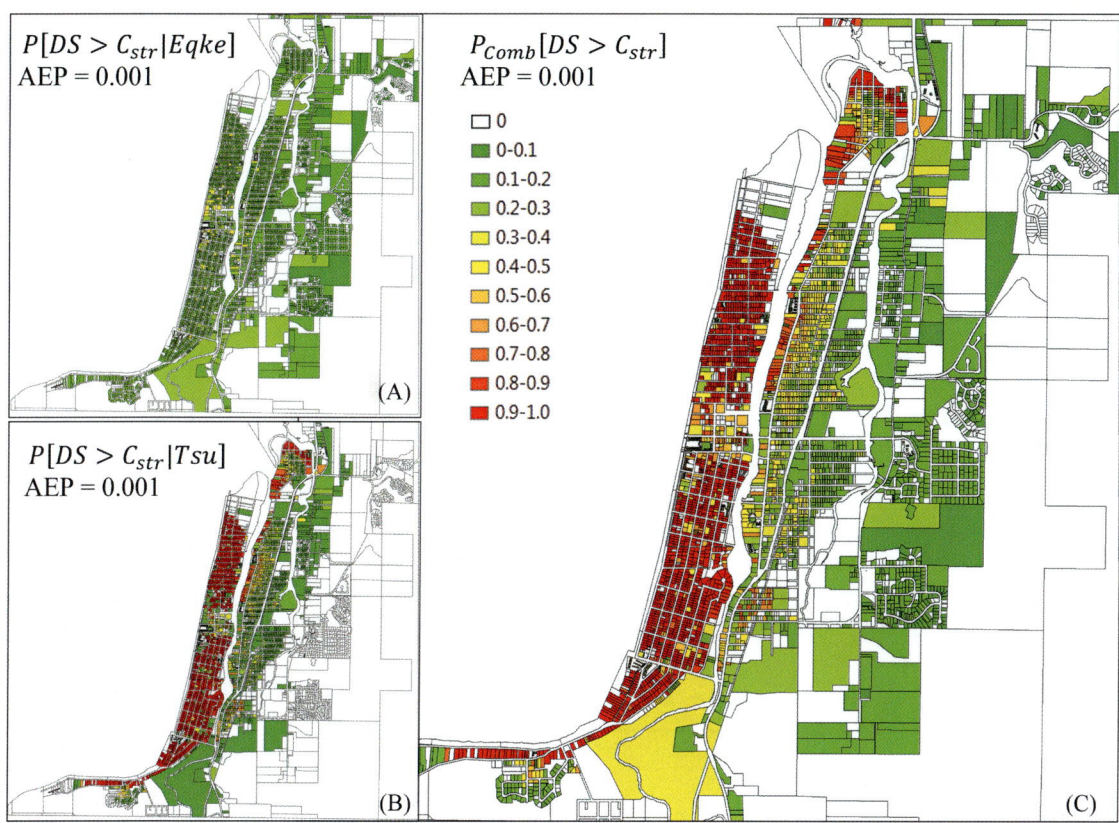

FIGURE 20.13 Combined structural damage probabilities for (A) earthquake only, (B) tsunami only, and (C) combined damage probability at *AEP* = 0.001 (1000-year event). *AEP*, Annual exceedance probability. Map lines delineate study areas and do not necessarily depict accepted national boundaries.

The current study assumes statistically independent events and utilizes independent fragility functions for each sequential damage analysis due to the earthquake and tsunami. The analysis does not currently consider ground failure due to potential liquefaction, scouring, or debris impacts during earthquake shaking, subsidence, and subsequent inundation flow by the tsunami.

20.2.4 Loss and risk assessment

The direct economic losses on structures and corresponding risks in coastal communities can be computed by using the probability of damage at each *DS* and the real market value that is required for the recovery. For example, Sanderson et al. (2021) demonstrated the economic losses and risks from multihazards (earthquakes and tsunamis) on buildings and three major infrastructure systems (water, electricity, and transportation) using the PSTHA and PSTDA results in Seaside.

The example results of disaggregated economic losses of the entire building portfolio from earthquakes, tsunamis, and combined earthquake and tsunami hazards are shown in Fig. 20.14 (Sanderson et al., 2021). The economic losses were computed using the HAZUS damage ratios (Hazus-MH 2.1, 2015), which provide a weighting factor for losses at each *DS*. The total economic losses are obtained by multiplying the HAZUS damage ratio by the economic loss value of buildings at each *DS* and summing all cases.

Fig. 20.14A shows the accumulated economic losses on the entire buildings in Seaside, Oregon, from earthquakes only (blue), tsunamis only (red), combined earthquakes and tsunamis (black solid), and the sum of earthquakes and tsunamis (black dashed line), which are obtained by simply summing earthquakes only and tsunamis only. The uncertainty associated with the economic losses is shown by the bands around each median curve, representing the 5th and 95th percentiles.

Each loss curve increases with the return period; however, the dominant hazards between earthquakes and tsunamis for building losses are changed with the recurrence interval (return period). Earthquakes are the dominant hazard at the short recurrence interval (less than 500 years), while tsunamis are the dominant hazard at the long recurrence interval. The results for the combined earthquake-tsunami loss and the summed loss of earthquakes and tsunamis highlight the potential overprediction of losses when considering multihazards independently.

Fig. 20.14B displays the economic risk of buildings from multihazards at different recurrence intervals. The concept of risk is defined as the multiplication of the loss and the probability of occurrence. Therefore, the expected value of losses is in the unit of dollars per year, as the probability provides the *AEP*. The concept of risk provides insight into events that result in significant economic losses and have a high probability of occurrence. The economic risk of buildings in Seaside, Oregon, shows the largest risk for a 250-year event, primarily dominated by earthquakes. Additionally, the second largest risk is observed for a 500-year event, with an equal contribution from earthquake and tsunami hazards.

20.3 Conclusions

This chapter presented a case study on multihazards, damage, and risk assessment in Seaside, Oregon, from earthquake and tsunami events. The common earthquake and tsunami scenarios from the CSZ were defined using a multilogic tree model that accounts for geophysical information and variations in the CSZ and recurrence information utilizing the TGR that was improved the classic GR from turbidite information for the last 10,000 years. A numerical tsunami model and GMPEs were utilized to evaluate the representative hazard from each tsunami and earthquake scenario from the logic tree model. The *AEP* and spatial *IM*s were calculated at a specific *AEP* condition.

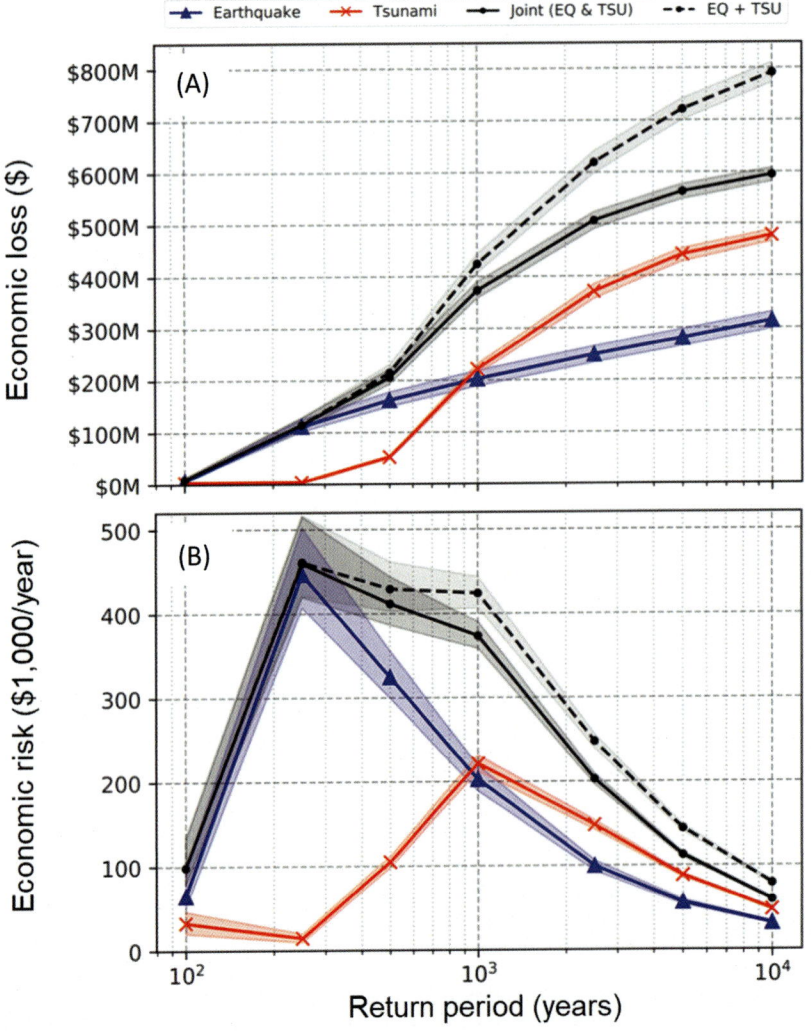

FIGURE 20.14 Loss and risk assessments of buildings in Seaside. (A) Median economic losses for buildings and (B) median economic risk from multihazards. The shaded region shows the 5th and 95th percentiles only. *From Sanderson, D., Kameshwar, S., Rosenheim, N., & Cox, D. (2021). Deaggregation of multi-hazard damages, losses, risks, and connectivity: An application to the joint seismic-tsunami hazard at Seaside, Oregon. Natural Hazards, 109(2), 1821–1847. http://www.wkap.nl/journalhome.htm/0921-030X. https://doi.org/10.1007/s11069-021-04900-9.*

The building inventory in Seaside was collected, classified, and matched with the available fragility functions for earthquakes and tsunamis that share the same building typologies. PGA and $Sa(T_1 = 0.3$ seconds) for earthquake and d_{max} and MF_{max} for tsunami were selected as IMs to estimate structural and nonstructural damage probabilities of buildings through a fragility analysis. Combined damage probabilities from earthquakes and tsunamis were evaluated using FEMA (2013), assuming statistical independence of damage. Economic losses on buildings and risk, considering the occurrence of hazard and intensity of the event at each DS, were evaluated for each hazard, including combined risk in a disaggregating format (Sanderson et al., 2021). The main conclusions are as follows:

1. The patterns of spatial annual exceedance probabilities on IMs from earthquakes and tsunamis are different in the two hazards. The IMs of tsunamis tend to decrease from shoreline to inland, while IMs of earthquakes show relatively uniform distributions over the whole study area.
2. At a lower return period, earthquake shaking is the dominant hazard for both wood and RC buildings over the entire study region. However, the proportion of tsunami-induced damage is increased with the increase in the recurrence interval.
3. Combined damage probabilities of earthquakes and tsunamis are significant throughout the study region. Especially for the predominant building typologies (RC or wood) for the study region.
4. The results of risk assessment are utilized to identify the recurrence intervals and multihazards that lead to substantial economic losses.

The current approach could be applied to other infrastructure systems and help understand the resilience of communities to future tsunamigenic earthquake events. However, the computation of combined damage probabilities from multihazards assumed statistical independence, and the adapted approach to combining the two hazard impacts may overpredict the results by not considering the intrinsic correlation between IMs. Moreover, the adapted fragility functions for tsunamis were developed from the seismic capacity of the building against the lateral force and do not consider complicated tsunami-driven loading and damage processes during real tsunamis, including tsunami-driven debris impact or damming loadings, scouring of the foundation, sheltering, and channeling effects. In addition, future work should explore sequential damage from earthquakes and tsunamis and develop the joint fragility functions for the multihazards (Attary et al., 2021; Xu et al., 2021) for various typologies of buildings and infrastructures.

References

Abrahamson, N., Gregor, N., & Addo, K. (2016). BC hydro ground motion prediction equations for subduction earthquakes. *Earthquake Spectra, 32*(1), 23–44. Available from https://doi.org/10.1193/051712EQS188MR, http://earthquakespectra.org/doi/pdf/10.1193/051712EQS188MR.

Alam, M. S., & Barbosa, A. R. (2018). Probabilistic seismic demand assessment accounting for finite element model class uncertainty: Application to a code-designed URM infilled reinforced concrete frame building. *Earthquake Engineering and Structural Dynamics, 47*(15), 2901–2920. Available from https://doi.org/10.1002/eqe.3113, http://onlinelibrary.wiley.com/journal/10.1002/(ISSN)1096-9845.

Alam, M. S., Barbosa, A. R., Scott, M. H., Cox, D. T., & van de Lindt, J. W. (2018). Development of physics-based tsunami fragility functions considering structural member failures. *Journal of Structural Engineering, 144*(3). Available from https://doi.org/10.1061/(asce)st.1943-541x.0001953.

Amini, M., Sanderson, D. R., Cox, D. T., Barbosa, A. R., & Rosenheim, N. (2023). Methodology to incorporate seismic damage and debris to evaluate strategies to reduce life safety risk for multi-hazard earthquake and tsunami. *Natural Hazards*. Available from https://doi.org/10.1007/s11069-023-05937-8, https://www.springer.com/journal/11069.

ASCE. (2010). *Minimum design loads for buildings and other structures*. American Society of Civil Engineering (pp. 7–10).

Aslani, H., & Miranda, E. (2005). Probability-based seismic response analysis. *Engineering Structures, 27*(8), 1151–1163. Available from https://doi.org/10.1016/j.engstruct.2005.02.015.

Attary, N., van de Lindt, J. W., Unnikrishnan, V. U., Barbosa, A. R., & Cox, D. T. (2017). Methodology for development of physics-based tsunami fragilities. *Journal of Structural Engineering, 143*(5). Available from https://doi.org/10.1061/(asce)st.1943-541x.0001715.

Attary, N., Lindt., Barbosa, A. R., Cox, D. T., & Unnikrishnan, V. (2021). Performance-based risk assessment of structures subjected to multi-hazard case of tsunamis following earthquakes. *Journal of Earthquake Engineering, 25*(10), 2065–2084.

Bazzurro, P. (1998). *Probabilistic seismic demand analysis*.

Bricker, J. D., Gibson, S., Takagi, H., & Imamura, F. (2015). On the need for larger Manning's roughness coefficients in depth-integrated tsunami inundation models. *Coastal Engineering Journal, 57*(2). Available from https://doi.org/10.1142/S0578563415500059, https://www.tandfonline.com/loi/tcej20.

Cornell, C. A. (1968). Engineering seismic risk analysis. *Bulletin of the Seismological Society of America, 58*(5), 1583–1606. Available from https://doi.org/10.1785/bssa0580051583.

Cox, D., Barbosa, A., Alam, M., Amini, M., Kameshwar, S., Park, H., & Sanderson, D. (2022). *Seaside testbed data inventory for infrastructure, population, and earthquake-tsunami hazard*. DesignSafe-CI.

De Risi, R., & Goda, K. (2016). Probabilistic earthquake−tsunami multi-hazard analysis: Application to the Tohoku Region, Japan. *Frontiers in Built Environment, 2*.

De Risi, R., Goda, K., Yasuda, T., & Mori, N. (2017). Is flow velocity important in tsunami empirical fragility modeling? *Earth-Science Reviews, 166*, 64−82. Available from https://doi.org/10.1016/j.earscirev.2016.12.015, http://www.sciencedirect.com/science/journal/00128252.

Delorey, A. A., Frankel, A. D., Liu, P., & Stephenson, W. J. (2014). Modeling the effects of source and path heterogeneity on ground motions of great earthquakes on the Cascadia subduction zone using 3D simulations. *Bulletin of the Seismological Society of America, 104*(3), 1430−1446. Available from https://doi.org/10.1785/0120130181, http://www.bssaonline.org/content/104/3/1430.full.pdf + html.

FEMA. (2011). *Multi-hazard loss estimation methodology: Earthquake model Hazus-MH MR5 technical manual*. Federal Emergency Management Agency.

FEMA. (2013). *Tsunami methodology technical manual*. National Institute of Building Sciences (NIBS) for the Federal Emergency Management Agency.

Goda, K., & De Risi, R. (2023). Future perspectives of earthquake-tsunami catastrophe modelling: From single-hazards to cascading and compounding multi-hazards. *Frontiers in Built Environment, 8*. Available from https://doi.org/10.3389/fbuil.2022.1022736, http://journal.frontiersin.org/journal/built-environment.

Goda, K., & De Risi, R. (2018). Multi-hazard loss estimation for shaking and tsunami using stochastic rupture sources. *International Journal of Disaster Risk Reduction, 28*, 539−554. Available from https://doi.org/10.1016/j.ijdrr.2018.01.002, http://www.journals.elsevier.com/international-journal-of-disaster-risk-reduction/.

Goda, K., De Risi, R., De Luca, F., Muhammad, A., Yasuda, T., & Mori, N. (2021). Multi-hazard earthquake-tsunami loss estimation of Kuroshio Town, Kochi Prefecture, Japan considering the Nankai-Tonankai megathrust rupture scenarios. *International Journal of Disaster Risk Reduction, 54*, 102050. Available from https://doi.org/10.1016/j.ijdrr.2021.102050.

Goldfinger, C., Nelson, C.H., Morey, A.E., Johnson, J.E., Patton, J.R., Karabanov, E., Patton, J., Gracia, E., Enkin, R., Dallimore, A., Dunhill, G., & Vallier T. (2012). *Turbidite event history: Methods and implications for Holocene paleoseismicity of the Cascadia subduction zone U.S.* Geological Survey.

González, F. I., Geist, E. L., Jaffe, B., Kânoğlu, U., Mofjeld, H., Synolakis, C. E., Titov, V. V., Arcas, D., Bellomo, D., Carlton, D., Horning, T., Johnson, J., Newman, J., Parsons, T., Peters, R., Peterson, C., Priest, G., Venturato, A., Weber, J., … Yalciner, A. (2009). Probabilistic tsunami hazard assessment at Seaside, Oregon, for near- and far-field seismic sources. *Journal of Geophysical Research, 114*(C11). Available from https://doi.org/10.1029/2008jc005132.

Goulet, C. A., Haselton, C. B., Mitrani-Reiser, J., Beck, J. L., Deierlein, G. G., Porter, K. A., & Stewart, J. P. (2007). Evaluation of the seismic performance of a code-conforming reinforced-concrete frame building—From seismic hazard to collapse safety and economic losses. *Earthquake Engineering and Structural Dynamics, 36*(13), 1973−1997. Available from https://doi.org/10.1002/eqe.694, http://onlinelibrary.wiley.com/journal/10.1002/(ISSN)1096-9845.

Hazus-MH 2.1. (2015). *Technical manual*. National Institute of Building Sciences and Federal Emergency Management Agency (NIBS and FEMA).

Heaton, T. H., & Hartzell, S. H. (1987). Earthquake hazards on the Cascadia subduction zone. *Science, 236*(4798), 162−168. Available from https://doi.org/10.1126/science.236.4798.162.

Kagan, Y. Y. (2002). Seismic moment distribution revisited: I. Statistical results. *Geophysical Journal International, 148*(3), 520−541. Available from https://doi.org/10.1046/j.1365-246x.2002.01594.x.

Kameshwar, S., Park, H., Cox, D. T., & Barbosa, A. R. (2021). Effect of disaster debris, floodwater pooling duration, and bridge damage on immediate post-tsunami connectivity. *International Journal of Disaster Risk Reduction, 56*. Available from https://doi.org/10.1016/j.ijdrr.2021.102119, http://www.journals.elsevier.com/international-journal-of-disaster-risk-reduction/.

Kennedy, R. P., Cornell, C. A., Campbell, R. D., Kaplan, S., & Perla, H. F. (1980). Probabilistic seismic safety study of an existing nuclear power plant. *Nuclear Engineering and Design, 59*(2), 315−338. Available from https://doi.org/10.1016/0029-5493(80)90203-4.

Kim, D. H., Lynett, P. J., & Socolofsky, S. A. (2009). A depth-integrated model for weakly dispersive, turbulent, and rotational fluid flows. *Ocean Modelling, 27*(3-4), 198−214. Available from https://doi.org/10.1016/j.ocemod.2009.01.005.

Kinali, K., & Ellingwood, B. R. (2007). Seismic fragility assessment of steel frames for consequence-based engineering: A case study for Memphis, TN. *Engineering Structures, 29*(6), 1115−1127. Available from https://doi.org/10.1016/j.engstruct.2006.08.017.

Kircher, C. A., Nassar, A. A., Kustu, O., & Holmes, W. T. (1997). Development of building damage functions for earthquake loss estimation. *Earthquake Spectra, 13*(4), 663−682. Available from https://doi.org/10.1193/1.1585974, https://journals.sagepub.com/home/eqs.

Koshimura, S., Oie, T., Yanagisawa, H., & Imamura, F. (2009). Developing fragility functions for tsunami damage estimation using numerical model and post-tsunami data from Banda Aceh, Indonesia. *Coastal Engineering Journal, 51*(3), 243−273. Available from https://doi.org/10.1142/S0578563409002004, https://www.tandfonline.com/loi/tcej20.

Lynett, P. J., Wu, T. R., & Liu, P. L. F. (2002). Modeling wave runup with depth-integrated equations. *Coastal Engineering, 46*(2), 89−107. Available from https://doi.org/10.1016/S0378-3839(02)00043-1.

Ma, G., Shi, F., & Kirby, J. T. (2012). Shock-capturing non-hydrostatic model for fully dispersive surface wave processes. *Ocean Modelling, 43-44*, 22−35. Available from https://doi.org/10.1016/j.ocemod.2011.12.002.

Macabuag, J., Rossetto, T., Ioannou, I., Suppasri, A., Sugawara, D., Adriano, B., Imamura, F., Eames, I., & Koshimura, S. (2016). A proposed methodology for deriving tsunami fragility functions for buildings using optimum intensity measures. *Natural Hazards, 84*(2), 1257−1285. Available from https://doi.org/10.1007/s11069-016-2485-8, http://www.wkap.nl/journalhome.htm/0921-030X.

Mori, N., Goda, K., & Cox, D. (2018). Recent process in probabilistic tsunami hazard analysis (PTHA) for mega thrust subduction earthquakes. *Advances in Natural and Technological Hazards Research, 47*. Available from https://doi.org/10.1007/978-3-319-58691-5_27, http://www.springer.com/series/6362?detailsPage = titles.

Murotani, S., Satake, K., & Fujii, Y. (2013). Scaling relations of seismic moment, rupture area, average slip, and asperity size for $M \sim 9$ subduction-zone earthquakes. *Geophysical Research Letters*, *40*(19), 5070–5074. Available from https://doi.org/10.1002/grl.50976.

Okada, Y. (1985). Surface deformation due to shear and tensile faults in a half-space. *Bulletin of the Seismological Society of America*, *75*(4), 1135–1154. Available from https://doi.org/10.1785/bssa0750041135.

Olsen, K. B., Stephenson, W. J., & Geisselmeyer, A. (2008). 3D crustal structure and long-period ground motions from a $M9.0$ megathrust earthquake in the Pacific Northwest region. *Journal of Seismology*, *12*(2), 145–159. Available from https://doi.org/10.1007/s10950-007-9082-y.

Park, H., Alam, M. S., Cox, D. T., Barbosa, A. R., & van de Lindt, J. W. (2019). Probabilistic seismic and tsunami damage analysis (PSTDA) of the Cascadia subduction zone applied to Seaside, Oregon. *International Journal of Disaster Risk Reduction*, *35*. Available from https://doi.org/10.1016/j.ijdrr.2019.101076, http://www.journals.elsevier.com/international-journal-of-disaster-risk-reduction/.

Park, H., Cox, D. T., Alam, M. S., & Barbosa, A. R. (2017). Probabilistic seismic and tsunami hazard analysis conditioned on a megathrust rupture of the Cascadia subduction zone. *Frontiers in Built Environment*, *3*. Available from https://doi.org/10.3389/fbuil.2017.00032, https://www.frontiersin.org/articles/10.3389/fbuil.2017.00032/pdf.

Park, H., Cox, D. T., & Barbosa, A. R. (2017). Comparison of inundation depth and momentum flux based fragilities for probabilistic tsunami damage assessment and uncertainty analysis. *Coastal Engineering*, *122*, 10–26. Available from https://doi.org/10.1016/j.coastaleng.2017.01.008, http://www.elsevier.com/inca/publications/store/5/0/3/3/2/5/.

Park, H., Cox, D. T., Lynett, P. J., Wiebe, D. M., & Shin, S. (2013). Tsunami inundation modeling in constructed environments: A physical and numerical comparison of free-surface elevation, velocity, and momentum flux. *Coastal Engineering*, *79*, 9–21. Available from https://doi.org/10.1016/j.coastaleng.2013.04.002.

Park, H., & Cox, D. T. (2016). Probabilistic assessment of near-field tsunami hazards: Inundation depth, velocity, momentum flux, arrival time, and duration applied to Seaside, Oregon. *Coastal Engineering*, *117*, 79–96. Available from https://doi.org/10.1016/j.coastaleng.2016.07.011, http://www.elsevier.com/inca/publications/store/5/0/3/3/2/5/.

Park, S., Van De Lindt, J. W., Cox, D., Gupta, R., & Aguiniga, F. (2012). Successive earthquake-tsunami analysis to develop collapse fragilities. *Journal of Earthquake Engineering*, *16*(6), 851–863. Available from https://doi.org/10.1080/13632469.2012.685209.

Power, W., Wang, X., Lane, E., & Gillibrand, P. (2013). A probabilistic tsunami hazard study of the Auckland Region, Part I: Propagation modelling and tsunami hazard assessment at the shoreline. *Pure and Applied Geophysics*, *170*(9-10), 1621–1634. Available from https://doi.org/10.1007/s00024-012-0543-z.

Priest, G. R., Goldfinger, C., Wang, K., Witter, R. C., Zhang, Y., & Baptista, A. M. (2010). Confidence levels for tsunami-inundation limits in northern Oregon inferred from a 10,000-year history of great earthquakes at the Cascadia subduction zone. *Natural Hazards*, *54*(1), 27–73. Available from https://doi.org/10.1007/s11069-009-9453-5.

Rong, Y., Jackson, D. D., Magistrale, H., & Goldfinger, C. (2014). Magnitude limits of subduction zone Earthquakes. *Bulletin of the Seismological Society of America*, *104*(5), 2359–2377. Available from https://doi.org/10.1785/0120130287, http://www.bssaonline.org/content/104/5/2359.full.pdf.

Rosowsky, D. V., & Ellingwood, B. R. (2002). Performance-based engineering of wood frame housing: Fragility analysis methodology. *Journal of Structural Engineering*, *128*(1), 32–38. Available from https://doi.org/10.1061/(ASCE)0733-9445(2002)128:1(32).

Sanderson, D., Kameshwar, S., Rosenheim, N., & Cox, D. (2021). Deaggregation of multi-hazard damages, losses, risks, and connectivity: An application to the joint seismic-tsunami hazard at Seaside, Oregon. *Natural Hazards*, *109*(2), 1821–1847. Available from https://doi.org/10.1007/s11069-021-04900-9, http://www.wkap.nl/journalhome.htm/0921-030X.

Satake, K., Wang, K., & Atwater, B. F. (2003). Fault slip and seismic moment of the 1700 Cascadia earthquake inferred from Japanese tsunami descriptions. *Journal of Geophysical Research: Solid Earth*, *108*(11). Available from http://onlinelibrary.wiley.com/journal/10.1002/(ISSN)2169-9356.

Shi, F., Kirby, J. T., Harris, J. C., Geiman, J. D., & Grilli, S. T. (2012). A high-order adaptive time-stepping TVD solver for Boussinesq modeling of breaking waves and coastal inundation. *Ocean Modelling*, *43-44*, 36–51. Available from https://doi.org/10.1016/j.ocemod.2011.12.004.

Skarlatoudis, A. A., Somerville, P. G., Thio, H. K., & Bayless, J. R. (2015). Broadband strong ground motion simulations of large subduction earthquakes. *Bulletin of the Seismological Society of America*, *105*(6), 3050–3067. Available from https://doi.org/10.1785/0120140322, http://www.bssaonline.org/content/105/6/3050.full.pdf.

Somerville, P., Skarlatoudis, A., & Li, W. (2012). *Ground motions and tsunamis from large Cascadia subduction earthquakes based on the 2011 Tokoku*.

Suppasri, A., Mas, E., Charvet, I., Gunasekera, R., Imai, K., Fukutani, Y., Abe, Y., & Imamura, F. (2013). Building damage characteristics based on surveyed data and fragility curves of the 2011 Great East Japan tsunami. *Natural Hazards*, *66*(2), 319–341. Available from https://doi.org/10.1007/s11069-012-0487-8, http://www.wkap.nl/journalhome.htm/0921-030X.

Synolakis, C. E., Bernard, E. N., Titov, V. V., Kânoğlu, U., & González, F. I. (2008). Validation and verification of tsunami numerical models. *Pure and Applied Geophysics*, *165*(11-12), 2197–2228. Available from https://doi.org/10.1007/s00024-004-0427-y.

Thio, H. K., & Somerville, P. (2009). A probabilistic tsunami hazard analysis of California. *TCLEE 2009: Lifeline Earthquake Engineering in a Multihazard Environment*, *357*, 57. Available from https://doi.org/10.1061/41050(357)57.

Titov, V. V., Moore, C. W., Greenslade, D. J. M., Pattiaratchi, C., Badal, R., Synolakis, C. E., & Kânoğlu, U. (2011). A new tool for inundation modeling: Community modeling interface for tsunamis (ComMIT). *Pure and Applied Geophysics*, *168*(11), 2121–2131. Available from https://doi.org/10.1007/s00024-011-0292-4.

Wang, H., Mostafizi, A., Cramer, L. A., Cox, D., & Park, H. (2016). An agent-based model of a multimodal near-field tsunami evacuation: Decision-making and life safety. *Transportation Research Part C: Emerging Technologies*, *64*, 86–100. Available from https://doi.org/10.1016/j.trc.2015.11.010, http://www.elsevier.com/inca/publications/store/1/3/0/.

Witter, R. C., Zhang, Y. J., Wang, K., Priest, G. R., Goldfinger, C., Stimely, L., English, J. T., & Ferro, P. A. (2013). Simulated tsunami inundation for a range of Cascadia megathrust earthquake scenarios at Bandon, Oregon, USA. *Geosphere*, *9*(6), 1783–1803. Available from https://doi.org/10.1130/GES00899.1.

Wood, N. J., Burton, C. G., & Cutter, S. L. (2010). Community variations in social vulnerability to Cascadia-related tsunamis in the U.S. Pacific Northwest. *Natural Hazards*, *52*(2), 369–389. Available from https://doi.org/10.1007/s11069-009-9376-1.

Xu, J. G., Wu, G., Feng, D. C., & Fan, J. J. (2021). Probabilistic multi-hazard fragility analysis of RC bridges under earthquake-tsunami sequential events. *Engineering Structures*, *238*. Available from https://doi.org/10.1016/j.engstruct.2021.112250, http://www.journals.elsevier.com/engineering-structures/.

Yamazaki, Y., Lay, T., Cheung, K. F., Yue, H., & Kanamori, H. (2011). Modeling near-field tsunami observations to improve finite-fault slip models for the 11 March 2011 Tohoku earthquake. *Geophysical Research Letters*, *38*(20). Available from https://doi.org/10.1029/2011GL049130, http://onlinelibrary.wiley.com/journal/10.1002/(ISSN)1944-8007/issues?year = 2012.

Chapter 21

Dynamic agent-based evacuation

Tomoyuki Takabatake[1] and Miguel Esteban[2]
[1]Department of Civil and Environmental Engineering, Kindai University, Higashiosaka, Japan, [2]Faculty of Science and Engineering, Waseda University, Tokyo, Japan

21.1 Introduction

Tsunamis are one of the most destructive types of natural hazards. Historically, these destructive waves have caused a considerable number of fatalities and devastating damage to coastal communities, as exemplified by the 2004 Indian Ocean Tsunami, the 2009 Samoa Tsunami, the 2010 Chile Tsunami, the 2011 Tohoku Tsunami, the 2018 Palu Tsunami, and the 2018 Sunda Strait Tsunami. While constructing coastal structures, such as dykes and seawalls, is an effective way to protect communities against tsunami waves, it is not realistic to expect that the residents of coastal areas can be fully protected using only hard measures. Therefore, evacuation to safe zones plays a crucial role in minimizing human losses from tsunamis.

While a number of studies have emphasized the importance of proper and swift evacuation of the population at risk during a tsunami event, it is not easy to implement this in practice. In fact, Cabinet Office of Japan (2012) reported that only approximately 50% of people at risk started evacuation within 30 minutes after the ground shaking for the case of the 2011 Tohoku Tsunami. During this event, there were even some evacuees who entered the inundation zone from outside (i.e., picking-up behavior, as described in Makinoshima et al. (2021)). Severe congestion of the roads, generated due to the use of vehicles by evacuees, was also observed in past tsunami events, such as during the 2011 Tohoku Tsunami (Suppasri et al., 2016). The congestion of pedestrian routes connected to refuge places was also reported during the 2018 Sunda Strait Tsunami (Takabatake, Shibayama, et al., 2019). Therefore, it is important to increase the awareness of the importance of prompt evacuation and clarify the potential difficulties that may arise during evacuation, such as the required evacuation time, the required capacity of tsunami shelters, and the locations of possible congested roads. While a practical approach to do these is to conduct evacuation drills, it is often challenging to make all stakeholders, including local residents and visitors, participate in evacuation drills. Therefore, tsunami evacuation simulations can serve as tools that can help evaluate the effectiveness of various countermeasures and behavior during the evacuation to increase their overall effectiveness and reduce the number of casualties.

Agent-based modeling is a methodology that can be used to simulate the behavior and interactions of a multitude of individual agents. In such models, each agent is represented as a decision-making entity that determines its own actions, such as its destination and speed, based on predefined rules (Dawson et al., 2011). As each agent moves independently, the modeler can observe a macro-level outcome, which is the result of the interactions between the agents (Epstein, 1999). While there are other approaches to model tsunami evacuation, including geographic information systems (GIS) (Sugimoto et al., 2003; Wood & Schmidtlein, 2013) and distinct element methods (Abustan et al., 2012), it is often difficult for them to express the complex evacuee's decision-making process and dynamic evacuation behavior (e.g., change in evacuation route and moving speed), as well as the interaction between the individual evacuees. As agent-based modeling can characterize the complexity observed during tsunami evacuation, it is considered ideal for simulating emergency events. This chapter first introduces a literature review of agent-based tsunami evacuation simulations and then explains how these can be developed and applied to a coastal area.

21.2 Literature review on agent-based tsunami evacuation simulations

Agent-based modeling has been widely used to simulate evacuation processes during various emergencies (i.e., not only for evacuation from a tsunami but also from other hazards). For instance, several studies employed this approach to simulate crowd evacuation during building fires or earthquakes (Liu et al., 2018; Owen et al., 1996; Pan et al., 2007; Pelechano & Badler, 2006). Some of them have also considered the complicated interactions among evacuees, such as information exchange (Pelechano & Badler, 2006) and competitive behavior (Pan et al., 2007). Liu et al. (2016) incorporated the effects of environmental factors on evacuation behavior, such as floor layout and damage caused by earthquakes. Agent-based modeling has also been actively used to simulate the evacuation from river flooding or storm surges. For instance, Dawson et al. (2011) developed an agent-based evacuation simulation model that integrated flooding inundation and evacuation behavior and investigated the effectiveness of flood management measures under different storm surge conditions. Using their agent-based evacuation model, Coates et al. (2019) and Coates et al. (2020) investigated how structural and social preparedness mitigation measures would improve the resilience of small and medium-sized companies against river flooding. Nakanishi et al. (2020) also performed an agent-based evacuation simulation of residents in Takamatsu, Japan, for storm surge flooding from the 2004 typhoon and assessed the effectiveness of the simulation by sharing the results with residents through workshops. Some recent articles (Anshuka et al., 2022; Simmonds et al., 2020) reported other recent applications of agent-based modeling for flooding.

Mas et al. (2015) indicated that one of the early instances of utilizing an agent-based model to simulate tsunami evacuation was introduced by Usuzawa et al. (1997). The study simulated evacuation from the Aonae District in Okushiri Island, Japan, during the 1993 Hokkaido Earthquake. Following the 2011 Tohoku Tsunami, the number of studies that have simulated tsunami evacuation using agent-based modeling has significantly increased. According to Mls et al. (2023), who conducted a thorough review of recent applications of agent-based modeling to tsunami evacuation, most of the relevant publications were published between 2012 and 2021.

Some researchers have verified actual evacuations by comparing agent-based tsunami evacuation simulation models with available observation data and survivor testimonies (Kumagai, 2014; Makinoshima et al., 2016; Mas et al., 2015; Nishihata et al., 2012; Takabatake, Fujisawa, et al., 2020). For instance, Kumagai (2014) validated their agent-based tsunami evacuation simulation model by comparing the simulation results with questionnaire responses from pedestrian evacuees in Kamaishi, Japan, during the 2011 Tohoku Tsunami. Although the evacuation route of each evacuee was determined by simply calculating the shortest route to the closest refugee location (Kumagai, 2014), the simulation results were shown to agree relatively well with the actual tsunami evacuation rate. In contrast, Makinoshima et al. (2016) considered the evacuation behavior of both car and pedestrian evacuees and showed that applying the assumption that those evacuating by vehicle would choose the shortest route did not successfully reproduce the locations of road congestion. Takabatake, Fujisawa, et al. (2020) incorporated the preference for those evacuating by vehicle to choose wider roads into their agent-based tsunami evacuation model, following the approach proposed by Jacob et al. (2014) and successfully replicated the observed road congestion during the 2011 Tohoku Tsunami.

Other studies have utilized agent-based tsunami evacuation models to estimate the potential number of casualties and identify issues associated with evacuation in a given coastal area (Takabatake et al., 2017; Takabatake et al., 2018; Takabatake, Chenxi, et al., 2022; Wang et al., 2016). Due to the inherent uncertainty in evacuation behavior, Wang et al. (2016) investigated the effects of behavioral changes on mortality rates to assess tsunami risks in their study area comprehensively. Takabatake et al. (2017) investigated the effects of the presence of visitors who were not familiar with a coastal area they visited and demonstrated that the estimated casualties were significantly different when considering the presence of visitors.

Agent-based models have also been actively used to assess the effectiveness of countermeasures for reducing mortality rates (Fathianpour et al., 2023; Johnstone & Lence, 2009, 2012; Koyanagi & Arikawa, 2016; León et al., 2023; Mostafizi et al., 2017; Mostafizi et al., 2019; Takabatake, Esteban, et al. 2022; Takabatake, Esteban, et al., 2020; Uno et al., 2015; Wang & Jia, 2022). For example, Johnstone and Lence (2009, 2012) developed an agent-based tsunami evacuation simulation model and examined the effects of early evacuation measures on reducing the tsunami mortality rates for Vancouver Island, Canada. The effectiveness of early evacuation has also been confirmed by many researchers. Kim et al. (2022) demonstrated that shortening tsunami alert times led to a significant increase in survival rates. Uno et al. (2015), Koyanagi and Arikawa (2016), and Takabatake, Esteban, et al. (2020) also explored the effectiveness of implementing hard measures (e.g., increasing seawall heights) and implementing soft measures (e.g., early evacuation). Mostafizi et al. (2017, 2019) used an agent-based tsunami evacuation simulation model to evaluate network vulnerability and optimize shelter locations. Takabatake, Esteban, et al. (2022) incorporated the effects of coastal forests into their agent-based model to study their effectiveness in reducing the loss of lives. Fathianpour et al. (2023) simulated two scenarios, evacuation on foot or by vehicle, and demonstrated that a smaller number of evacuees could reach designated refugee locations when they used vehicles for evacuation. Wang and Jia (2022) also showed that when more people used vehicles for evacuation, the overall risk of being affected by a tsunami increased. Recently, the effectiveness of tsunami vertical evacuation buildings has also been investigated by León et al. (2023), using the results of virtual reality experiments to understand the decision-making process.

Considering that the damage or collapse of buildings and structures (due to severe ground shaking) in a coastal area will lead to road blockage in the case of a near-field tsunami, the effects of road blockage have also been considered by some agent-based tsunami evacuation simulation models (Ito et al., 2020; Takabatake, Chenxi, et al., 2022; Wang & Jia, 2021). For instance, Takabatake, Chenxi, et al. (2022) applied the model to three coastal areas in Japan (Kamakura, Zushi, and Fujisawa) and demonstrated that when the effects of building collapse and road blockage were considered, mortality rates increased by 1.3–2.3 times (depending on the coastal areas).

21.3 Development of an agent-based tsunami evacuation simulation model

When performing an agent-based tsunami evacuation simulation, it is first necessary to determine the software tools to be used (i.e., an agent-based modeling platform or programming language). As shown by Mls et al. (2023), who summarized recent agent-based tsunami evacuation models, a number of different tools have been used to construct the models. However, as agent-based modeling platforms (e.g., NetLogo, AnyLogic, Artisoc, GAMA, and MATSim) have various functionalities and associated tools that allow users to model the complex behavior of agents easily, they are preferred to other general programming languages (e.g., Python and Fortran). Some of the studies have utilized traffic simulators (e.g., SUMO and NETSIM), which were developed initially to investigate traffic flows and estimate the required evacuation time. According to Mls et al. (2023), while approximately 30% of the articles they reviewed did not provide the name of the software they used, NetLogo was the most frequently used in the mentioned academic research articles.

In general, to save computational costs, the entire city that is focused on is not covered in the simulation domain. Instead, the domain of the tsunami evacuation simulation is often determined by considering the maximum extent of tsunami inundation, which can be obtained by utilizing tsunami propagation and inundation simulation models. However, this means that the simulations can neglect the influence of the presence and behavior of evacuees located outside the inundation zone. Problems associated with the limited extent of simulation domains may become more critical when considering those evacuating by vehicles.

Once the extent of the simulation area is determined, it is necessary to consider a number of scenarios, such as daytime, night-time, or high-season scenarios (in terms of the number of visitors), as the initial locations and behavior of evacuees during a tsunami event could differ significantly Takabatake, Esteban, et al. (2020). Then, the environment where the evacuation will take place, including the study area's road networks, buildings (including evacuation shelters), building collapse associated with road blockage, population distribution, and progress of tsunami inundation, needs to be prepared. Preferably, data from the government or local authorities of the study area should be used, but otherwise, it can be extracted from OpenStreetMap (León et al., 2023; Wang & Jia, 2021). Both horizontal and vertical evacuation shelters need to be inputted into the simulation model. Generally, the location of designated horizontal shelters can be determined from the website of local authorities or the simulated maximum inundation extent. Regarding the vertical evacuation shelters, both their locations and capacities need to be obtained. If the information about the capacity of shelters is not available, it may roughly be estimated by using GIS data based on the building size and considering how many people can be accommodated per square meters (Applied Technology Council, 2019).

There can be significant spatiotemporal variations in the number of people who should be considered as evacuees and their distribution (i.e., their initial location at the time of the earthquake or the moment when the tsunami alarm is issued; Mostafizi et al., 2017). While the number of local residents in the hazard zone can be roughly estimated based on census data, their distribution is difficult to determine. In addition, for coastal areas where many tourists may be present, it is necessary to include visitors as evacuees. Owing to the difficulty in determining the initial locations of evacuees (especially during the daytime), it would be ideal to consider different population scenarios in the simulations (Wang & Jia, 2021). To consider the uncertainty in population distribution, probabilistic models, such as a normal distribution, have also been adopted in some studies (Mostafizi et al., 2017; Wang & Jia, 2021, 2022). As the characteristics of evacuees (e.g., local residents or visitors, gender, and age) are essential for determining evacuation behavior (e.g., moving speed, stability against tsunami flow, and decision-making process), such information should also be obtained (generally, local authorities can provide some information about the sociodemographic of residents). If buildings in the study area are anticipated to be severely damaged or collapse due to an earthquake preceding the tsunami wave, it would also be better to incorporate their effects and associated road blockage information into the evacuation environment (Ito et al., 2020; Takabatake, Chenxi, et al., 2022).

The behavior of evacuees then needs to be determined based on the scenarios considered, particularly the departure time, path selection, and moving speed. The departure (or milling) time is the time when each evacuee starts to evacuate. Several studies indicated that this parameter has a crucial effect on estimating the number of casualties resulting from a tsunami (Mostafizi et al., 2017; Wang et al., 2016). To determine the departure time, a sigmoid curve or Rayleigh distribution, together with the mean departure time, have been frequently used in past studies (Kim et al., 2022; Mas et al., 2012; Takabatake et al., 2017; Wang & Jia, 2022; Wang et al., 2016). To model the variations in departure time more realistically among evacuees, Takabatake et al. (2018) considered different evacuation triggers that each evacuee could respond to based on questionnaire surveys on people's awareness and preparedness for tsunami evacuation.

To model the path selection and moving speed of evacuees, it is necessary to first establish the evacuation mode that each evacuee would use. While most studies have considered a single evacuation mode, generally either on foot or by vehicle, some studies (Makinoshima et al., 2016; Takabatake, Fujisawa, et al., 2020; Wang & Jia, 2021; Wang et al., 2016) considered multiple evacuation modes. If evacuees in the study area are expected to employ a variety of other evacuation modes, such as the use of motorbikes or bicycles, these should also be considered (Kim et al., 2022; Muhammad et al., 2021). In the case of pedestrian evacuees, they are usually assigned to follow the shortest path to the closest evacuation shelter, provided that they have enough knowledge of the study area. In fact, Kumagai (2014) confirmed that adopting such an assumption could reproduce relatively well the evacuation behavior observed during the 2011 Tohoku Tsunami. The shortest route can be calculated using well-known pathfinding algorithms, such as the Dijkstra or A* search algorithms (Dawson et al., 2011; Kim et al., 2022; Mas et al., 2012; Takabatake et al., 2017). However, the evacuation route that visitors who are not familiar with the study area would take may be different from the shortest path (Arce et al., 2017). In addition, changes to the evacuation route should also be incorporated if evacuees need to reroute when they encounter impassable roads (e.g., flooded roads due to the tsunami, or blocked roads due to debris generated from collapsed buildings; see Takabatake, Chenxi, et al., 2022). For those evacuating by vehicle, the assumption of the shortest path would not be appropriate, as in reality, they are more likely to choose a main road instead of the shortest path via a minor road (Goto et al., 2011). Such preferences of evacuees traveling by vehicle have been incorporated in some models (Makinoshima et al., 2016; Takabatake, Fujisawa, et al., 2020). While vehicle drivers would also change their evacuation routes according to road conditions (e.g., when encountering crowded, damaged, or flooded roads), modeling such behavior is still challenging.

The moving speed of pedestrian evacuees is often determined according to their age and road congestion (Kumagai, 2014; Takabatake, Esteban, et al., 2020) or considering the uncertainty in the characteristics (e.g., age and physical condition) of each evacuee (Wang et al., 2016). Wang et al. (2016) incorporated uncertainty by assuming a normal distribution and a mean and standard deviation of moving speed. While various studies have investigated the moving speed of a pedestrian, all of them indicate that the speed would decrease as the number of people on a given road increases. Therefore, the decrease in moving speed due to congestion should be considered when road congestion could be a factor. The moving speed of evacuees by vehicle is more difficult to determine, as it is more likely to be influenced by the surrounding environment. Several studies have attempted to determine the moving speed of a car in simulations (Chandler et al., 1958; Gipps, 1981; Krauss et al., 1997; Treiber et al., 2000). For instance, Treiber et al. (2000) proposed an Intelligent Driver Model in which the moving speed of a car is calculated from the information about the car itself, the vehicle in front of it, and the maximum velocity allowed on the road. Although many existing tsunami evacuation simulations have not considered the interactions between pedestrians and vehicles, this is an important factor. In this regard, Fathianpour et al. (2023) recently investigated this through the application of a well-known traffic simulator (SUMO).

As one of the aims of tsunami evacuation simulations is to estimate the number of casualties resulting from an earthquake and a tsunami, it is important to incorporate the progress of simulated tsunami inundation over the study area. Typically, to determine whether an agent becomes a casualty, either the inundation depth (Goto et al., 2011; Sugimoto et al., 2003; Wang & Jia, 2021; Wang et al., 2016) or depth-velocity (dv) product (Takabatake, Esteban, et al., 2020) at the mesh where the evacuee is located is considered. Ito et al. (2020) and Takabatake, Chenxi, et al., 2022 also considered the casualties resulting from building collapses based on statistical estimates, as per the guidelines by Central Disaster Management Council of Japan (2019).

As explained earlier, the agent-based tsunami evacuation simulation model intrinsically incorporates some uncertainty related to the evacuation behavior of each evacuee, which would lead to a variation in the simulated results if slightly different parameters or distributions of agents are considered. Thus each time a simulation is run, the simulated results are not exactly the same, even for the same scenario. Therefore, it is generally required to repeat each scenario multiple times until the results obtained converge onto a stable value.

Finally, it is essential to note that how precisely evacuation behavior is modeled should depend on the purpose of the simulations. If this purpose is to understand what would actually happen in the study area during an evacuation from a tsunami event, better results may be obtained by incorporating additional information into the simulations. For instance, while many existing studies assume that evacuees would go to the closest evacuation shelter, evacuees may prefer to go to other shelters. Collecting additional information about the actual expected behavior of evacuees would be crucial for developing a realistic agent-based tsunami evacuation simulation model. In contrast, if the purpose is to investigate the effectiveness of a countermeasure to reduce the number of casualties, some of the evacuation behavior may be simplified or idealized to independently evaluate the influence of the proposed intervention. However, in this case, it is important to clearly mention what assumptions were made to model evacuation behavior in the model.

21.4 Application of an agent-based tsunami evacuation simulation model

This section will provide an example of an agent-based tsunami evacuation simulation to investigate how the number of casualties would change due to buildings collapsing and causing road blockages. Mihama Town, located in Wakayama Prefecture, Japan, is expected to be affected by a future Nankai–Tonankai Earthquake Tsunami (Central Disaster Management Council of Japan, 2019) and was selected as the study area. The location of Mihama is shown in Fig. 21.1, together with the simulated water level 10 minutes after a future Nankai–Tonankai Earthquake Tsunami. In this simulation, Case 5 was selected among 11 tsunami simulation cases proposed by Central Disaster Management Council of Japan (2012), as it would result in the highest tsunami in the study area.

The model was constructed using Artisoc4. Artisoc4 is a modeling platform with various functionalities and tools that help perform agent-based simulations. This model has been widely used, particularly amongst the Japanese community (Jiang & Murao, 2017; Kosaka et al., 2017). The number of evacuees considered in the simulation was determined according to census data of residents living in Mihama and the maximum simulated extent of the tsunami inundation (the population density in each district of Mihama was multiplied by the simulated inundation area, which produced a total of 2875 evacuees). The age distribution of the population was determined based on the statistics for Mihama Town; 3.2% (age 0–4), 7.8% (age 5–14), 55.6% (age 15–64), 15.5% (age 65–74), and 17.9% (age 75–). The presence of visitors was considered to be negligible. Data on buildings and road networks in the study area were obtained from publicly available GIS data, and evacuees were assumed to start evacuation from one of the buildings. In addition to 14 designated evacuation shelters in Mihama, 40 intersections located in the road network and outside of the

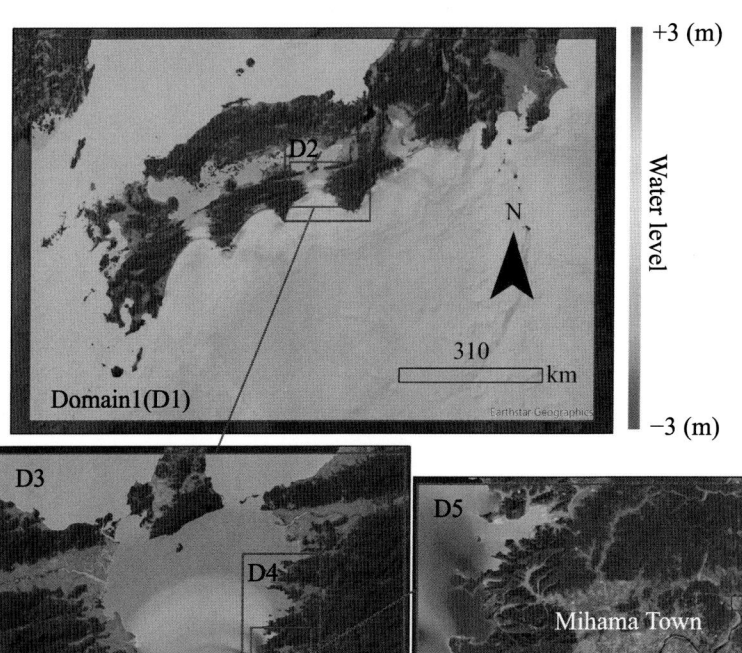

FIGURE 21.1 Location of Mihama Town and computational domains used in the tsunami simulations, together with the simulated water level 30 minutes after the earthquake. Map lines delineate study areas and do not necessarily depict accepted national boundaries.
Source of satellite imagery: Esri, Maxar, Earthstar Geographics, and the GIS User Community. *GIS*, Geographic information systems.

maximum extent of tsunami inundation were set as evacuation places where evacuees would go. The initial locations of evacuees and evacuation shelters are displayed in Fig. 21.2A.

Wakayama Prefecture (2014) estimated that 45% of the buildings in Mihama would collapse due to ground shaking generated by a future Nankai–Tonankai Earthquake. The simulation in this chapter adopted this estimation, and 45% of the buildings that were randomly selected among those located within the domain of the evacuation simulation were assumed to collapse (however, buildings that were categorized as "sturdy" in the GIS data were excluded). Then, the extent of debris generated from the collapsed buildings was calculated by multiplying the height of the collapsed building by a debris spreading coefficient of 0.6 (Takabatake, Chenxi, et al., 2022) and defined as a buffer zone utilizing the ArcGIS software (Fig. 21.2B). As shown in Fig. 21.2B, roads that were fully covered by the buffer zones were categorized as fully blocked roads and defined to be impassable. In contrast, partially covered roads were categorized as partially blocked roads, and evacuees were assumed to be able to pass through them but at a reduced moving speed. Aside from this, when the evacuees started the simulation inside collapsed buildings, it was assumed that 7% of them would become casualties immediately, 12% would not be able to start evacuation (trapped inside the collapsed building), 14% would be able to start the evacuation but be injured (and assigned a slower moving speed), and the remaining 67% would do so without being injured, following Central Disaster Management Council of Japan (2019).

The departure time of evacuees was determined by assuming that it follows the Rayleigh distribution with a mean departure time of 5 minutes (Fig. 21.2C). The shortest route from the initial location of each evacuee to the closest evacuation shelter was calculated by using the A* algorithm. However, when the evacuees encounter entirely blocked or flooded roads, they can reevaluate the shortest route to reach one of the evacuation shelters (without considering impassable roads). In addition, even when they reached a vertical evacuation shelter, if the capacity of the shelter was

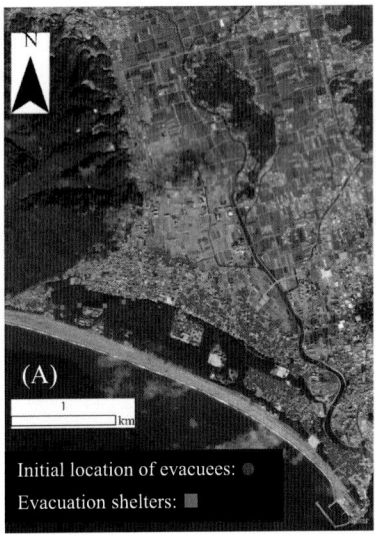
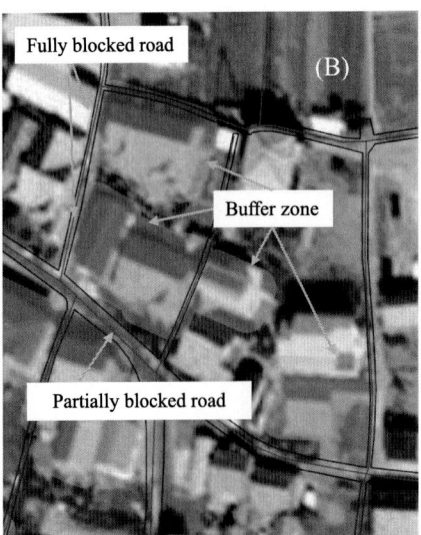
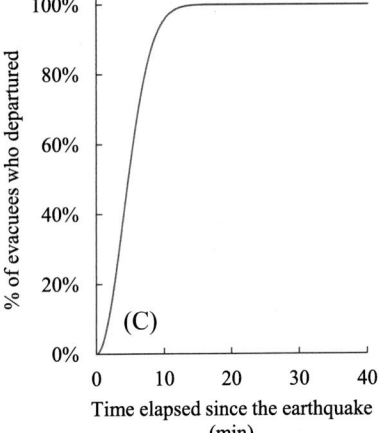
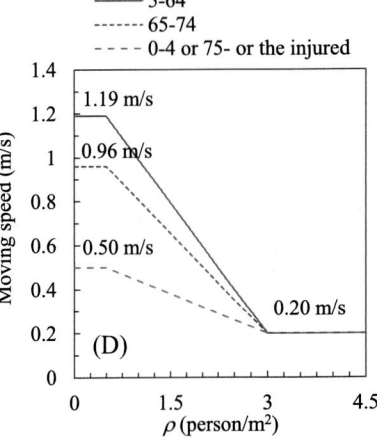

FIGURE 21.2 (A) Initial location of evacuees and evacuation shelters in Mihama, (B) an example of the fully blocked roads and partially blocked roads, (C) time history of the percentage of evacuees who initiated their evacuation, and (D) moving speed of evacuees.
Source of satellite imagery: Esri, Maxar, Earthstar Geographics, and the GIS User Community. GIS, Geographic information systems.

reached, the evacuee could reevaluate an alternative route to reach the next evacuation shelter. The information about fully blocked roads, flooded roads, and the shelters that had already reached their capacities were considered to be exchanged among evacuees encountering each other on the road. The moving speed of evacuees was determined according to their age and the agent density on the road, as shown in Fig. 21.2D. However, when agents evacuated through partially blocked roads, or they were injured, their moving speed was assumed to be halved.

The simulations assumed that 7% of the evacuees who were initially located inside collapsed buildings would immediately become casualties. In addition, those who were caught by the tsunami with a dv product that exceeded their stability limit were also assumed to become casualties. The stability limit is varied according to the age of the evacuees by setting $dv > 0$ m^2/second for ages 0–4 or over 65, $dv > 0.6$ m^2/second for ages 5–14, and $dv > 1.2$ m^2/second for ages 15–64. The dv values were simulated using a tsunami propagation and inundation simulation model, which is based on the shallow water equations (Takabatake, St-Germain, et al., 2019).

In the present simulations, two scenarios were prepared to investigate the impact of road blockage on the number of casualties, with scenario 1 not considering the effects of road blockage and scenario 2 considering it. Each scenario was repeated five times, and the average mortality rates were compared, as shown in Fig. 21.3. Both scenarios show the locations of evacuees who did not reach one of the evacuation shelters within 30 minutes after the earthquake, with a larger number of evacuees found still evacuating when considering road blockage. While the mortality rate was less than 6% when not considering road blockage, it increased to 45% when considering it. Thus it was found that the retrofitting of buildings with antiseismic countermeasures to prevent road blockages is extremely important for reducing casualties from a potential tsunami in the study area.

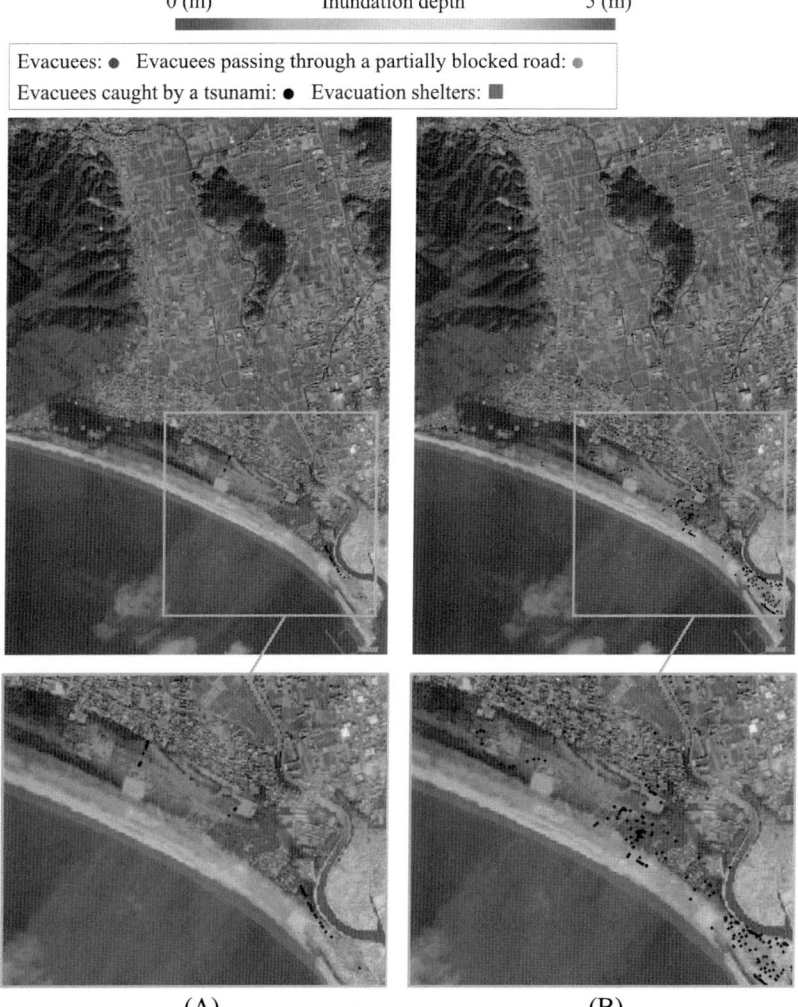

FIGURE 21.3 Snapshots of the tsunami evacuation simulations for (A) a scenario not considering road blockage and (B) that considering road blockage.
Source of satellite imagery: Esri, Maxar, Earthstar Geographics, and the GIS User Community. *GIS*, Geographic information systems.

21.5 Conclusions

Evacuation is an essential strategy for reducing the number of people who could be injured or killed by a tsunami. Therefore, it is important to formulate effective evacuation plans and increase the number of people following them. Agent-based tsunami evacuation simulation is a powerful tool to develop such strategies, as discussed in this chapter. However, it is also important to understand that various assumptions and considerable uncertainty are inherent in such simulations. One of the main challenges is related to the validation of evacuation simulation models, owing to the difficulties in obtaining the records of the evacuation behavior of evacuees at the time of a disaster. In addition, even if the model is validated for a certain coastal area for a given tsunami event, there is no guarantee that the model could universally simulate human evacuation behavior, as how people evacuate will significantly vary according to many sociocultural variables and also on the location and situation. Thus, agent-based tsunami evacuation models should be used as risk communication tools by showing stakeholders a possible scenario to raise their level of awareness and preparedness. Another effective way to utilize agent-based models relates to the analysis and prioritization of several mitigation counter-measures, as this can help evaluate their effectiveness in saving lives. The optimization of an evacuation plan for a given coastal area (e.g., to optimize the route and destination that each evacuee would take and go or the locations of evacuation shelters) would also be a promising use of an agent-based evacuation simulation model.

References

Abustan, M. S., Harada, E., & Gotoh, H. (2012). *Numerical simulation for evacuation process against tsunami disaster at Teluk Batik in Malaysia by multi-agent DEM model*. In Proceedings of Coastal Engineering, Japan Society of Civil Engineers (Vol. 3, pp. 56–60).

Anshuka, A., van Ogtrop, F. F., Sanderson, D., & Leao, S. Z. (2022). A systematic review of agent-based model for flood risk management and assessment using the ODD protocol. *Natural Hazards, 112*(3), 2739–2771. Available from https://doi.org/10.1007/s11069-022-05286-y, http://www.wkap.nl/journalhome.htm/0921-030X.

Applied Technology Council. (2019). *Guidelines for design of structures for vertical evacuation from tsunamis*. Federal Emergency Management Agency, FEMA (P–646). Washington. Available from: https://www.fema.gov/sites/default/files/2020-08/fema_earthquakes_guidelines_for_design-of-structures-for-vertical-evacuation-from-tsunamis-fema-p-646.pdf.

Arce, R. S. C., Onuki, M., Esteban, M., & Shibayama, T. (2017). Risk awareness and intended tsunami evacuation behaviour of international tourists in Kamakura City, Japan. *International Journal of Disaster Risk Reduction, 23*, 178–192. Available from https://doi.org/10.1016/j.ijdrr.2017.04.005, http://www.journals.elsevier.com/international-journal-of-disaster-risk-reduction/.

Cabinet Office of Japan. (2012). *Questionnaire survey of residents on earthquake and tsunami evacuation during the Great East Japan earthquake*. https://www.bousai.go.jp/jishin/tsunami/hinan/pdf/20121221_chousa1_1.pdf.

Central Disaster Management Council of Japan. (2012). *Tsunami fault modeling: Tsunami fault model, tsunami height, inundation area: Tsunami calculation results (tsunami height)*. https://www.bousai.go.jp/jishin/nankai/taisaku/pdf/20120829_2nd_report07.pdf.

Central Disaster Management Council of Japan. (2019). *Damage estimation for the Nankai Trough megathrust earthquake (damage to buildings and human casualties)*. https://www.bousai.go.jp/jishin/nankai/taisaku_wg/pdf/1_sanko2.pdf.

Central Disaster Management Council of Japan (CDMC). (2019). *Overview of damage estimation items and assumption methodology for Nankai Trough megathrust earthquake: Damage to buildings and human casualties*. https://www.bousai.go.jp/jishin/nankai/taisaku_wg/pdf/2_sanko2.pdf.

Chandler, R. E., Herman, R., & Montroll, E. W. (1958). Traffic dynamics: Studies in car following. *Operations Research, 6*(2), 165–184. Available from https://doi.org/10.1287/opre.6.2.165.

Coates, G., Alharbi, M., Li, C., Ahilan, S., & Wright, N. (2020). Evaluating the operational resilience of small and medium-sized enterprises to flooding using a computational modelling and simulation approach: A case study of the 2007 flood in Tewkesbury. *Philosophical Transactions of the Royal Society A: Mathematical, Physical and Engineering Sciences, 378*(2168). Available from https://doi.org/10.1098/rsta.2019.0210, http://rsta.royalsocietypublishing.org/.

Coates, G., Li, C., Ahilan, S., Wright, N., & Alharbi, M. (2019). Agent-based modeling and simulation to assess flood preparedness and recovery of manufacturing small and medium-sized enterprises. *Engineering Applications of Artificial Intelligence, 78*, 195–217. Available from https://doi.org/10.1016/j.engappai.2018.11.010.

Dawson, R. J., Peppe, R., & Wang, M. (2011). An agent-based model for risk-based flood incident management. *Natural Hazards, 59*(1), 167–189. Available from https://doi.org/10.1007/s11069-011-9745-4.

Epstein, J. M. (1999). Agent-based computational models and generative social science. *Complexity, 4*(5), 41–60. Available from https://doi.org/10.1002/(SICI)1099-0526(199905/06)4:5 < 41::AID-CPLX9 > 3.0.CO;2-F.

Fathianpour, A., Evans, B., Jelodar, M. B., & Wilkinson, S. (2023). Tsunami evacuation modelling via micro-simulation model. *Progress in Disaster Science, 17*. Available from https://doi.org/10.1016/j.pdisas.2023.100277, http://www.journals.elsevier.com/progress-in-disaster-science/.

Gipps, P. G. (1981). A behavioural car-following model for computer simulation. *Transportation Research Part B, 15*(2), 105–111. Available from https://doi.org/10.1016/0191-2615(81)90037-0.

Goto, Y., Mikami, T., & Nakabayashi, I. (2011). *Fact-finding about the evacuation from the unexpectedly large tsunami of March 11, 2011 in East Japan*. In Proceedings of the 15th World Conference on Earthquake Engineering.

Ito, E., Kawase, H., Matsushima, S., & Hatayama, M. (2020). Tsunami evacuation simulation considering road blockage by collapsed buildings evaluated from predicted strong ground motion. *Natural Hazards*, *101*(3), 959−980. Available from https://doi.org/10.1007/s11069-020-03903-2, http://www.wkap.nl/journalhome.htm/0921-030X.

Jacob, S., Aguilar, L., Wijerathne, L., Hori, M., Ichimura, T., & Tanaka, S. (2014). Agent based modeling and simulation of tsunami triggered mass evacuation considering changes of environment due to earthquake and inundation. *Journal of Japan Society of Civil Engineers, Ser. A2 (Applied Mechanics (AM))*, *70*(2), 671−680. Available from https://doi.org/10.2208/jscejam.70.i_671.

Jiang, D., & Murao, O. (2017). Consideration of tsunami risk reduction effect focusing on tourists' evacuation time in Katase-Nishihama and Kugenuma District, Fujisawa City. *Journal of Social Safety Science.*, *31*, 117−124. Available from https://doi.org/10.11314/jisss.31.117.

Johnstone, W. M., & Lence, B. J. (2009). Assessing the value of mitigation strategies in reducing the impacts of rapid-onset, catastrophic floods. *Journal of Flood Risk Management.*, *2*(3), 209−221. Available from https://doi.org/10.1111/j.1753-318X.2009.01035.x.

Johnstone, W. M., & Lence, B. J. (2012). Use of flood, loss, and evacuation models to assess exposure and improve a community Tsunami response plan: Vancouver Island. *Natural Hazards Review*, *13*(2), 162−171. Available from https://doi.org/10.1061/(ASCE)NH.1527-6996.0000056.

Kim, K., Kaviari, F., Pant, P., & Yamashita, E. (2022). An agent-based model of short-notice tsunami evacuation in Waikiki, Hawaii. *Transportation Research Part D: Transport and Environment*, *105*, 103239. Available from https://doi.org/10.1016/j.trd.2022.103239.

Kosaka, Y., Nomura, N., Oto, A., & Miyajima, M. (2017). Analysis on tsunami evacuation by using multi agent system: Case study of Wajima district in Wajima city. *Journal of Japan Society of Civil Engineers, Ser. A1 (Structural Engineering & Earthquake Engineering (SE/EE))*, *73*(4), 1010−1017. Available from https://doi.org/10.2208/jscejseee.73.i_1010.

Koyanagi, Y., & Arikawa, T. (2016). Study of the tsunami evacuation tower using the tsunami evacuation simulation. *Journal of Japan Society of Civil Engineers, Ser. B2 (Coastal Engineering)*, *2*, 1567−1572. Available from https://doi.org/10.2208/kaigan.72.I_1567.

Krauss, S., Wagner, P., & Gawron, C. (1997). Metastable states in a microscopic model of traffic flow. *Physical Review E—Statistical Physics, Plasmas, Fluids, and Related Interdisciplinary Topics*, *55*(5), 5597−5602. Available from https://doi.org/10.1103/PhysRevE.55.5597.

Kumagai, K. (2014). Validation of tsunami evacuation simulation to evacuation activity from the 2011 off the Pacific coast of Tohoku earthquake tsunami. *Journal of Japan Society of Civil Engineers, Ser. D3 (Infrastructure Planning and Management)*, *70*(5), 187−196. Available from https://doi.org/10.2208/jscejipm.70.I_187.

León, J., Ogueda, A., Gubler, A., Catalán, P., Correa, M., Castañeda, J., & Beninati, G. (2023). Increasing resilience to catastrophic near-field tsunamis: Systems for capturing, modelling, and assessing vertical evacuation practices. *Natural Hazards*. Available from https://doi.org/10.1007/s11069-022-05732-x, https://www.springer.com/journal/11069.

Liu, H., Liu, B., Zhang, H., Li, L., Qin, X., & Zhang, G. (2018). Crowd evacuation simulation approach based on navigation knowledge and two-layer control mechanism. *Information Sciences*, *436-437*, 247−267. Available from https://doi.org/10.1016/j.ins.2018.01.023, http://www.journals.elsevier.com/information-sciences/.

Liu, Z., Jacques, C. C., Szyniszewski, S., Guest, J. K., Schafer, B. W., Igusa, T., & Mitrani-Reiser, J. (2016). Agent-based simulation of building evacuation after an earthquake: Coupling human behavior with structural response. *Natural Hazards Review*, *17*(1). Available from https://doi.org/10.1061/(ASCE)NH.1527-6996.0000199, http://www.pubs.asce.org/journals/nh.html.

Makinoshima, F., Imamura, F., & Abe, Y. (2016). Behavior from tsunami recorded in the multimedia sources at Kesennuma City in the 2011 Tohoku Tsunami and its simulation by using the evacuation model with pedestrian-car interaction. *Coastal Engineering Journal*, *58*(4). Available from https://doi.org/10.1142/S0578563416400234, https://www.tandfonline.com/loi/tcej20.

Makinoshima, F., Oishi, Y., Nakagawa, M., Sato, S., & Imamura, F. (2021). Revealing complex tsunami evacuation process patterns induced by social interactions: A case study in Ishinomaki. *International Journal of Disaster Risk Reduction*, *58*, 102182. Available from https://doi.org/10.1016/j.ijdrr.2021.102182.

Mas, E., Koshimura, S., Imamura, F., Suppasri, A., Muhari, A., & Adriano, B. (2015). Recent advances in agent-based tsunami evacuation simulations: Case studies in Indonesia, Thailand, Japan and Peru. *Pure and Applied Geophysics*, *172*(12), 3409−3424. Available from https://doi.org/10.1007/s00024-015-1105-y, http://www.springer.com/birkhauser/geo + science/journal/24.

Mas, E., Suppasri, A., Imamura, F., & Koshimura, S. (2012). Agent-based simulation of the 2011 Great East Japan earthquake/tsunami evacuation: An integrated model of tsunami inundation and evacuation. *Journal of Natural Disaster Science*, *34*(1), 41−57. Available from https://doi.org/10.2328/jnds.34.41.

Mls, K., Kořínek, M., Štekerová, K., Tučník, P., Bureš, V., Čech, P., Husáková, M., Mikulecký, P., Nacházel, T., Ponce, D., Zanker, M., Babič, F., & Triantafyllou, I. (2023). Agent-based models of human response to natural hazards: Systematic review of tsunami evacuation. *Natural Hazards*, *115*(3), 1887−1908. Available from https://doi.org/10.1007/s11069-022-05643-x, https://www.springer.com/journal/11069.

Mostafizi, A., Wang, H., Cox, D., Cramer, L. A., & Dong, S. (2017). Agent-based tsunami evacuation modeling of unplanned network disruptions for evidence-driven resource allocation and retrofitting strategies. *Natural Hazards*, *88*(3), 1347−1372. Available from https://doi.org/10.1007/s11069-017-2927-y, http://www.wkap.nl/journalhome.htm/0921-030X.

Mostafizi, A., Wang, H., Cox, D., & Dong, S. (2019). An agent-based vertical evacuation model for a near-field tsunami: Choice behavior, logical shelter locations, and life safety. *International Journal of Disaster Risk Reduction*, *34*, 467−479. Available from https://doi.org/10.1016/j.ijdrr.2018.12.018, http://www.journals.elsevier.com/international-journal-of-disaster-risk-reduction/.

Muhammad, A., De Risi, R., De Luca, F., Mori, N., Yasuda, T., & Goda, K. (2021). Are current tsunami evacuation approaches safe enough? *Stochastic Environmental Research and Risk Assessment*, *35*(4), 759−779. Available from https://doi.org/10.1007/s00477-021-02000-5, http://link.springer-ny.com/link/service/journals/00477/index.htm.

Nakanishi, H., Wise, S., Suenaga, Y., & Manley, E. (2020). Simulating emergencies with transport outcomes Sim (SETOSim): Application of an agent-based decision support tool to community evacuation planning. *International Journal of Disaster Risk Reduction, 49*, 101657. Available from https://doi.org/10.1016/j.ijdrr.2020.101657.

Nishihata, T., Moriya, Y., Anno, K., & Imamura, F. (2012). Modeling of car evacuation from tsunami attaking and evaluation for accompanied traffic jam. *Journal of Japan Society of Civil Engineers, Ser. B2 (Coastal Engineering), 68*(2), 1316–1320. Available from https://doi.org/10.2208/kaigan.68.i_1316.

Owen, M., Galea, E. R., & Lawrence, P. J. (1996). The EXODUS evacuation model applied to building evacuation scenarios. *Journal of Fire Protection Engineering, 8*(2), 65–84. Available from https://doi.org/10.1177/104239159600800202.

Pan, X., Han, C. S., Dauber, K., & Law, K. H. (2007). A multi-agent based framework for the simulation of human and social behaviors during emergency evacuations. *AI and Society, 22*(2), 113–132. Available from https://doi.org/10.1007/s00146-007-0126-1.

Pelechano, N., & Badler, N. I. (2006). Modeling crowd and trained leader behavior during building evacuation. *IEEE Computer Graphics and Applications, 26*(6), 80–86. Available from https://doi.org/10.1109/MCG.2006.133.

Simmonds, J., Gómez, J. A., & Ledezma, A. (2020). The role of agent-based modeling and multi-agent systems in flood-based hydrological problems: A brief review. *Journal of Water and Climate Change, 11*(4), 1580–1602. Available from https://doi.org/10.2166/wcc.2019.108, https://watermark.silverchair.com/jwc0111603.pdf?.

Sugimoto, T., Murakami, H., Kozuki, Y., Nishikawa, K., & Shimada, T. (2003). A human damage prediction method for tsunami disasters incorporating evacuation activities. *Natural Hazards., 29*(3), 585–600.

Suppasri, A., Latcharote, P., Bricker, J. D., Leelawat, N., Hayashi, A., Yamashita, K., Makinoshima, F., Roeber, V., & Imamura, F. (2016). Improvement of tsunami countermeasures based on lessons from the 2011 Great East Japan Earthquake and Tsunami—Situation after five years. *Coastal Engineering Journal, 58*(4). Available from https://doi.org/10.1142/S0578563416400118, https://www.tandfonline.com/loi/tcej20.

Takabatake, T., Chenxi, D. H., Esteban, M., & Shibayama, T. (2022). Influence of road blockage on tsunami evacuation: A comparative study of three different coastal cities in Japan. *International Journal of Disaster Risk Reduction, 68*, 102684. Available from https://doi.org/10.1016/j.ijdrr.2021.102684.

Takabatake, T., Esteban, M., Nistor, I., Shibayama, T., & Nishizaki, S. (2020). Effectiveness of hard and soft tsunami countermeasures on loss of life under different population scenarios. *International Journal of Disaster Risk Reduction, 45*, 101491. Available from https://doi.org/10.1016/j.ijdrr.2020.101491.

Takabatake, T., Esteban, M., & Shibayama, T. (2022). Simulated effectiveness of coastal forests on reduction in loss of lives from a tsunami. *International Journal of Disaster Risk Reduction, 74*, 102954. Available from https://doi.org/10.1016/j.ijdrr.2022.102954.

Takabatake, T., Fujisawa, K., Esteban, M., & Shibayama, T. (2020). Simulated effectiveness of a car evacuation from a tsunami. *International Journal of Disaster Risk Reduction, 47*, 101532. Available from https://doi.org/10.1016/j.ijdrr.2020.101532.

Takabatake, T., Shibayama, T., Esteban, M., Achiari, H., Nurisman, N., Gelfi, M., Tarigan, T. A., Kencana, E. R., Fauzi, M. A. R., Panalaran, S., Harnantyari, A. S., & Kyaw, T. O. (2019). Field survey and evacuation behaviour during the 2018 Sunda Strait tsunami. *Coastal Engineering Journal, 61*(4), 423–443. Available from https://doi.org/10.1080/21664250.2019.1647963, https://www.tandfonline.com/loi/tcej20.

Takabatake, T., Shibayama, T., Esteban, M., & Ishii, H. (2018). Advanced casualty estimation based on tsunami evacuation intended behavior: Case study at Yuigahama Beach, Kamakura, Japan. *Natural Hazards, 92*(3), 1763–1788. Available from https://doi.org/10.1007/s11069-018-3277-0, http://www.wkap.nl/journalhome.htm/0921-030X.

Takabatake, T., Shibayama, T., Esteban, M., Ishii, H., & Hamano, G. (2017). Simulated tsunami evacuation behavior of local residents and visitors in Kamakura, Japan. *International Journal of Disaster Risk Reduction, 23*, 1–14. Available from https://doi.org/10.1016/j.ijdrr.2017.04.003, http://www.journals.elsevier.com/international-journal-of-disaster-risk-reduction/.

Takabatake, T., St-Germain, P., Nistor, I., Stolle, J., & Shibayama, T. (2019). Numerical modelling of coastal inundation from Cascadia subduction zone tsunamis and implications for coastal communities on western Vancouver Island, Canada. *Natural Hazards, 98*(1), 267–291. Available from https://doi.org/10.1007/s11069-019-03614-3, http://www.wkap.nl/journalhome.htm/0921-030X.

Treiber, M., Hennecke, A., & Helbing, D. (2000). Congested traffic states in empirical observations and microscopic simulations. *Physical Review E–Statistical Physics, Plasmas, Fluids, and Related Interdisciplinary Topics, 62*(2), 1805–1824. Available from https://doi.org/10.1103/PhysRevE.62.1805.

Uno, Y., Shigihara, Y., & Okayasu, A. (2015). Development of crowd evacuation simulation for risk evaluation of human damage by tsunami inundation. *Journal of Japan Society of Civil Engineers, Ser. B2 (Coastal Engineering), 71*(2), 1615–1620. Available from https://doi.org/10.2208/kaigan.71.I_1615.

Usuzawa, H., Imamura, F., & Shuto, N. (1997). *Development of the method for evacuation numerical simulation for tsunami events*. In Annual Meeting of the Tohoku Branch Technology Research Conference, Japan Society of Civil Engineers (pp. 430–431).

Wakayama Prefecture. (2014). *Wakayama Prefecture earthquake damage estimation survey report (summary version)*. https://www.pref.wakayama.lg.jp/prefg/011400/d00153668_d/fil/wakayama_higaisoutei.pdf.

Wang, H., Mostafizi, A., Cramer, L. A., Cox, D., & Park, H. (2016). An agent-based model of a multimodal near-field tsunami evacuation: Decision-making and life safety. *Transportation Research Part C: Emerging Technologies, 64*, 86–100. Available from https://doi.org/10.1016/j.trc.2015.11.010, http://www.elsevier.com/inca/publications/store/1/3/0/.

Wang, Z., & Jia, G. (2021). A novel agent-based model for tsunami evacuation simulation and risk assessment. *Natural Hazards, 105*(2), 2045–2071. Available from https://doi.org/10.1007/s11069-020-04389-8, http://www.wkap.nl/journalhome.htm/0921-030X.

Wang, Z., & Jia, G. (2022). Simulation-based and risk-informed assessment of the effectiveness of tsunami evacuation routes using agent-based modeling: A case study of Seaside, Oregon. *International Journal of Disaster Risk Science*, *13*(1), 66–86. Available from https://doi.org/10.1007/s13753-021-00387-x, http://www.springer.com/earth + sciences + and + geography/natural + hazards/journal/13753.

Wood, N. J., & Schmidtlein, M. C. (2013). Community variations in population exposure to near-field tsunami hazards as a function of pedestrian travel time to safety. *Natural Hazards*, *65*(3), 1603–1628. Available from https://doi.org/10.1007/s11069-012-0434-8.

Chapter 22

Sea-level rise and tsunami risk

Miguel Esteban[1], Tomoyuki Takabatake[2], Ryutaro Nagai[1], Kentaro Koyano[1] and Tomoya Shibayama[1]

[1]Faculty of Science and Engineering, Waseda University, Tokyo, Japan, [2]Department of Civil and Environmental Engineering, Kindai University, Higashiosaka, Japan

22.1 Introduction

Tsunamis can devastate the coastline, impacting not only the local economy but also the overall development of communities affected (Chapter 1). Several major tsunami events have occurred recently, including the 2004 Indian Ocean Earthquake and Tsunami, with an estimated moment magnitude M_w of 9.3. This event caused widespread destruction to countries around the Indian Ocean and resulted in over 283,000 casualties (Spence et al., 2009). On March 11, 2011, an M_w 9.0 earthquake off the Pacific coast of Japan overcame coastal defenses and destroyed large sections of the coastline. Along the Sendai Plain, the maximum inundation height reached 19.5 m, whereas the maximum recorded runup height was 40.4 m along the Sanriku Coast (Mikami et al., 2012; The 2011 Tohoku Earthquake Tsunami Joint Survey Group, 2011; Yamao et al., 2015), causing an estimated 25,000 casualties and over 41,000 people being displaced (National Police Agency of Japan, 2020). Other recent events include the 2018 Sulawesi Tsunami, generated by an earthquake that affected Palu Bay in Indonesia (Harnantyari et al., 2020; Mikami et al., 2019), and the 2018 Sunda Strait Tsunami, resulting from the flank collapse of the Anak Krakatau Volcano (Takabatake, Shibayama, et al., 2019).

The conventional method of assessing the risk of tsunamis typically focuses on the analysis of worst case scenarios of historical events (Teh et al., 2011). More recent approaches also involve the probabilistic estimation of the return periods of different inundation heights by using probabilistic tsunami hazard assessments (PTHA), as discussed by Becerra et al. (2020) and Park et al. (2017); see also Chapter 5. Through such methods, coastal risk managers can estimate, for a given return, the maximum inundation extent, possible casualties, and total financial damage, allowing for the formulation of appropriate countermeasures.

As a consequence of the 2011 Tohoku Earthquake and Tsunami, the Japanese coastal engineering community established two levels of tsunami design. Additionally, the idea that hard (physical) countermeasures can always protect coastal communities against tsunamis has been discarded (Shibayama et al., 2013). Level 1 tsunamis are considered to have a return period of several decades and up to around 100 years, with coastal structures being designed to protect people and properties against such events. Level 2 tsunamis are deemed to have a return period in the order of 1000 years, and associated evacuation measures should be designed such that all residents can safely escape in the case of such events. Following this reclassification of tsunami events, several PTHA have been conducted along the coastline of the Tohoku and Kanto regions (Annaka et al., 2007; Fukutani et al., 2018; Goda & Song, 2016; Goda et al., 2014; Kotani et al., 2020; Mori et al., 2017; Nobuoka & Onoue, 2017; Sugino et al., 2014).

However, when performing either probabilistic or historical worst case scenarios, little research has considered the influence that sea-level rise (SLR) will have on the overall tsunami risk. As global CO_2 emissions continue to rise, it appears likely that atmospheric temperatures will also continue to increase and that a significant SLR is inevitable due to the melting of glaciers and ice sheets/shelves and thermal expansion. Since 1993, sea levels have been rising at a global average rate of over 3 mm per year, as highlighted by the Intergovernmental Panel on Climate Change 6th Assessment Report (IPCC 6AR) (Church et al., 2013). The signing of the Paris Agreement in 2015, negotiated through the United Nations Framework Convention on Climate Change, has brought hopes that greenhouse emissions could eventually be curbed, yet, as of 2023, they continue to increase. Thus, SLR by the end of the 21st century could be higher than that forecasted in the IPCC 6AR, and work on probabilistic process-based models that take into account rapid losses in the Antarctic ice sheets indicates that global average SLR could be as high as 2.97–3.39 m under Representative Concentration Pathways (RCP) 8.5 scenarios (Kopp et al., 2017; Le Bars et al., 2017).

There is little literature that has considered the issue of future SLR when evaluating tsunami risks. Wang et al. (2016) and Li et al. (2018) simulated the effects of SLR concerning tsunami inundation in Macau and indicated that the frequency of flooding will increase even with a minor rise in sea levels. Arnaud et al. (2021) indicated how tsunamis originating from the La Palma Volcano could affect Guadeloupe when considering the effects of SLR, whereas Tursina et al. (2021) investigated the impact of SLR on future tsunamis in Banda Aceh.

Although there is significant uncertainty regarding the pace of SLR, it is clear that after a certain amount of it has taken place coastal cities will have to take actions to protect their citizens. Adaptation to SLR is no longer a hypothetical future problem, but rather it is something that coastal communities are already implementing as a result of the cumulative rise in sea levels experienced over the past decades, compounded sometimes by land subsidence, with examples including cities in Florida, USA (Grant et al., 2023), Taiwan (Chang et al., 2022), Ho Chi Minh City in Vietnam (Cao, Esteban, Onuki, et al., 2021), Manila in the Philippines (Cao, Esteban, Valenzuela, et al., 2021), wastewater treatment plans in Tohoku, Japan (Cao et al., 2020), small islands in the Philippines (Jamero et al., 2019), and Jakarta in Indonesia (Esteban et al., 2019).

One of the countries that are at high risk of tsunamis is Japan, and given the amount of resources that have been spent (and continue to be spent) to improve safety against these hazards and storm surges, the general culture of the Japanese people, and the importance of these areas to the economy of the country (and the world in general), it is almost certain that investment will be made in adaptation. Studying Japanese cases is thus pertinent, and this chapter will present and demonstrate the concept of "action thresholds", which can help coastal disaster risk managers identify the planning timing and horizon when defenses will have to be upgraded. Such "action thresholds" can be useful for disaster risk managers when formulating adaptation pathways (Haasnoot et al., 2019).

22.2 Methodology

Nagai et al. (2020) and Koyano et al. (2022) conducted numerical simulations to attempt to understand the potential increase in tsunami risk due to SLR in Japan, and the methodology that these authors followed will be briefly summarized below. These methodologies differ slightly, partly because Tokyo Bay, as considered by Nagai et al. (2020), has only a small opening to the south, while the eastern coastline of Japan modeled by Koyano et al. (2022) faces the open ocean.

22.2.1 Sea-level rise scenarios

Nagai et al. (2020) and Koyano et al. (2022) considered a series of SLR scenarios, as summarized in Table 22.1. These were based on the work of Kopp et al. (2017) for the RCP 8.5 scenario, which represents the most severe greenhouse gas concentration situations considered in the IPCC 6AR. Under this scenario, SLR would manifest sequentially between 2062 and 2107, see Table 22.1. While the pace of SLR may not accurately follow the predictions of such studies, having a hypothetical date for the worst SLR scenarios can help coastal planners envisage a timeframe of when actions might have to be taken (which helps with the formulation of adaptation pathway strategies). Mitigation counter measures being implemented by many countries may be insufficient and could result in this worst case scenario. Given the considerable risks presented by tsunamis in Japan, such an approach and assessment are warranted.

TABLE 22.1 Summary of sea-level rise scenarios considered (Kopp et al., 2017).

Scenario	Sea-level rise	Year
1	+0.5 m	2062
2	+1.0 m	2080
3	+1.5 m	2093
4	+2.0 m	2107

22.2.2 Simulation methodology

The simulations performed by Nagai et al. (2020) and Koyano et al. (2022) used the nonlinear shallow water equations, as shown in the following equations, and as described in Chapter 3.

$$\frac{\partial \eta}{\partial t} + \frac{\partial M}{\partial x} + \frac{\partial N}{\partial y} = 0 \tag{22.1}$$

$$\frac{\partial M}{\partial t} + \frac{\partial}{\partial x}\left(\frac{M^2}{D}\right) + \frac{\partial}{\partial y}\left(\frac{MN}{D}\right) + gD\frac{\partial \eta}{\partial x} + \frac{gn^2}{D^{7/3}}M\sqrt{(M^2+N^2)} = 0 \tag{22.2}$$

$$\frac{\partial N}{\partial t} + \frac{\partial}{\partial x}\left(\frac{MN}{D}\right) + \frac{\partial}{\partial y}\left(\frac{N^2}{D}\right) + gD\frac{\partial \eta}{\partial y} + \frac{gn^2}{D^{7/3}}M\sqrt{(M^2+N^2)} = 0 \tag{22.3}$$

where η is the water surface elevation, D refers to the total water depth, M and N represent the flux discharge in two horizontal directions, n is Manning's roughness coefficient, and g is the acceleration due to gravity. The computer code was developed by the authors of this chapter, following the study of Goto et al. (1997), and was verified extensively (Koyano et al., 2021; Kukita & Shibayama, 2012; Takabatake et al., 2020; Takabatake et al., 2019). SLR heights were set based on the mean sea level of Tokyo Bay (T.P.) and used as the initial condition of sea levels in this model (Hoshino et al., 2016).

22.2.2.1 Approach and simulations for the Tokyo Bay area

Several earthquake scenarios around the Kanto region have a high possibility of occurrence within the next 30 years, including the North part of Tokyo Bay (M_w 7.3), East Metropolitan (M_w 6.9), a coupled earthquake from these two, the Tonankai (M_w 8.0), and the Tokai earthquake (M_w 8.0); see Ohira and Shibayama (2013). Despite this, the possibility of a tsunami inundating the cities located around Tokyo Bay is relatively limited, given that the potential tsunami wave heights are low and the shoreline is protected by coastal barriers purposely designed against storm surges (Hoshino et al., 2016). These defenses are designed to withstand an event similar to the 1959 Isewan Typhoon (Vera), which resulted in a 3.5-m storm surge (Kawai et al., 2007). Most of the research on tsunami impacts on Tokyo Bay considers the possibility that an earthquake could destroy some of the storm surge gates and/or levees (which protect the entrance of rivers and canals). This would allow a tsunami to flood low-lying areas of Tokyo Bay and appear reasonable assumptions given that many of these structures are over 50 years old. As a consequence of the land subsidence that took place due to groundwater extraction during the 20th century, some parts of Tokyo are currently below the mean water level (Esteban et al., 2019). Furthermore, recent events, such as the Jebi Typhoon in 2018, have highlighted the need to reassess coastal defenses, given that even areas considered well protected could suffer from events higher than what they were designed for Takabatake et al. (2018).

Nagai et al. (2020) selected the Keicho earthquake scenario, which is based on a historical event that took place along the Nankai Trough in 1605, as the target seismic event among several earthquake scenarios considered by the Prefectural Government of Kanagawa (Kanagawa Prefecture, 2015). Tsunami hazard maps and evacuation plans for Yokohama City and Kawasaki City were created based on the maximum extent of inundation expected from this Keicho earthquake. The earthquake parameters (e.g., length, width, slip, and dip angle) proposed by Kanagawa Prefecture consider the worst case seismic scenario for this source location, which would be an M_w 8.5 event (Kanagawa Prefecture, 2015). According to their results, the tsunami generated by the Keicho seismic scenario would inundate some coastal areas of Kanagawa. This is currently regarded as one of the most hazardous scenarios for this prefecture.

The seafloor deformation was calculated using the set of formulas proposed by Mansinha and Smylie (1971) and the parameters provided by Kanagawa Prefecture (2015), see Fig. 22.1. Nagai et al. (2020) performed inundation simulations and processed them using ArcGIS to find the number of people at risk of being flooded by a tsunami as the 21st century advances.

To understand the level of threat posed by a tsunami, it is necessary to consider not only the depth of inundation but also the speed of the water flows as they run over land. Following research by Suga et al. (1995), Takagi et al. (2016), and Wright et al. (2010), a depth–velocity product parameter, where $dv = 1.2$ m²/second, was adopted as the safe limit for pedestrians, which could be seen as a conservative assumption (as some casualties could actually take place even for lower dv values). To calculate the extent of the areas affected by inundation and by the condition $dv > 1.2$ m²/second,

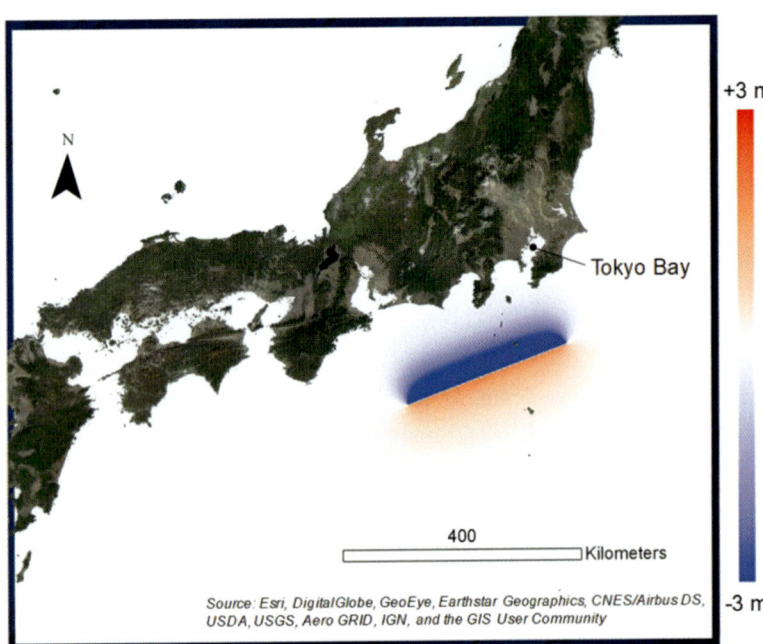

FIGURE 22.1 Map of Japan, showing the location of Tokyo Bay and the initial water surface conditions resulting from the Keicho earthquake scenario. Map lines delineate study areas and do not necessarily depict accepted national boundaries.

Nagai et al. (2020) analyzed the outcome of the computer simulations using ArcGIS software and census data (Statistics Bureau of Japan, 2015).

22.2.2.2 Approach and simulation for the eastern coast of the Kanto region

The 2011 Tohoku Earthquake and Tsunami caused significant damage along the coastline of the Tohoku and Kanto regions in Japan. In the case of the Kanto region, the first tsunami wave (8.7 m high) reached to Asahi City about 60 minutes after the earthquake, and larger waves were observed 150 minutes after the earthquake (The 2011 Tohoku Earthquake Tsunami Joint Survey Group, 2011), resulting in over 270 casualties.

Koyano et al. (2022) employed a PTHA method (see Chapters 2 and 5) to analyze the variation and probability of the maximum runup and inundation heights for a certain return period of the target earthquake zones by modeling the uncertainty in the source and rupture processes. The logic tree method (see Chapter 5) was used to analyze the uncertainty in tsunami height distribution along Sendai Port and Kujukuri Beach, following the methodology proposed by Annaka et al. (2007). In the adopted method, the aleatory uncertainty was determined from Aida's (1978) κ, obtained by calculating the ratios of observed to numerically calculated tsunami heights for eleven historical tsunami sources. The weights of the branches were determined based on a questionnaire survey of tsunami and earthquake experts and an error evaluation (Annaka et al., 2007; Japan Society of Civil Engineers, 2008). The earthquake source parameters used to model the 2011 Tohoku-level earthquake were based on Fujii et al. (2011), and the initial water level for each earthquake scenario was calculated using the formulas proposed by Okada (1985).

22.3 Simulation results

22.3.1 Increase in tsunami risk around Tokyo Bay due to sea-level rise

Yokohama and Kawasaki are located next to each other on the western side of Tokyo Bay (see Fig. 22.2). Figs. 22.3, 22.4, and 22.5 show the extent of the inundation for the present-day and future SLR scenarios, together with the spatial distribution of the dv product. It is worth noting that some inundation is expected even for the present-day conditions in the event of the Keicho tsunami, though the dv would be low enough not to be hazardous to adults. In the case of Yokohama, the flooded areas include some of the historical parts of the city, including the 19th century "Red Brick Warehouses," the areas between Chinatown and the sea, and the industrial warehouse area of Suzushigecho. Newer parts of the city, such as Minato Mirai 21, have been built on reclaimed land and would be unaffected (most of the coastline around Tokyo Bay has been reclaimed at different times over the past 300 years, although the majority of the

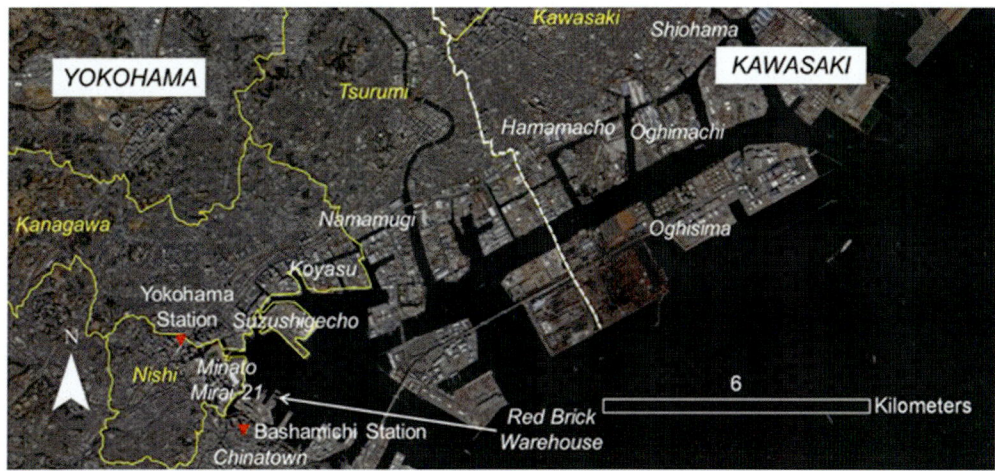

FIGURE 22.2 Coastal regions of Yokohama and Kawasaki. *Esri, DigitalGlobe, Earthstar Geographics, CNES/Airbus DS, USDA, USGA, Aero GRID, IGN, and the GIS User Community.*

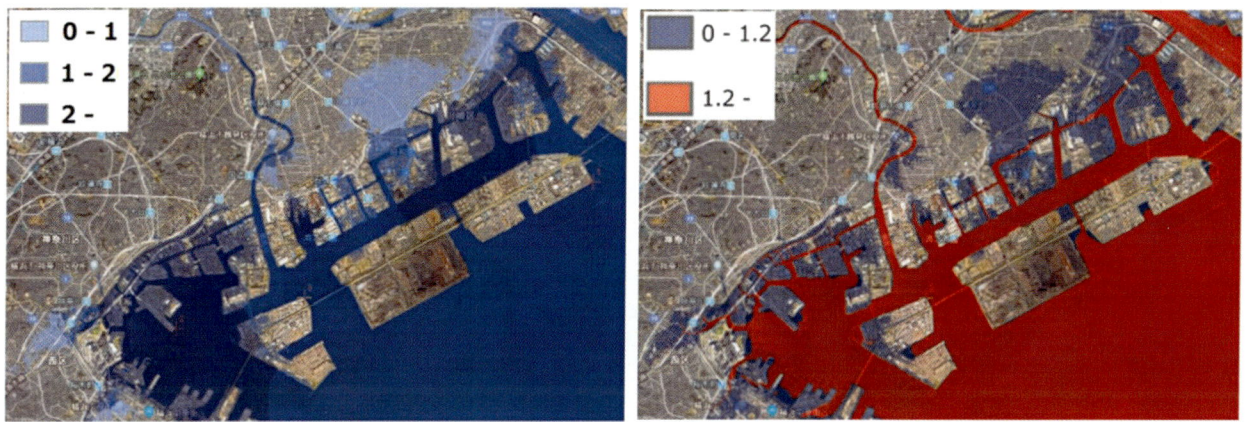

FIG. 22.3 Spatial distribution of inundation (left, in m) and *dv* (right, in m^2/second) values for present-day conditions in Kawasaki.

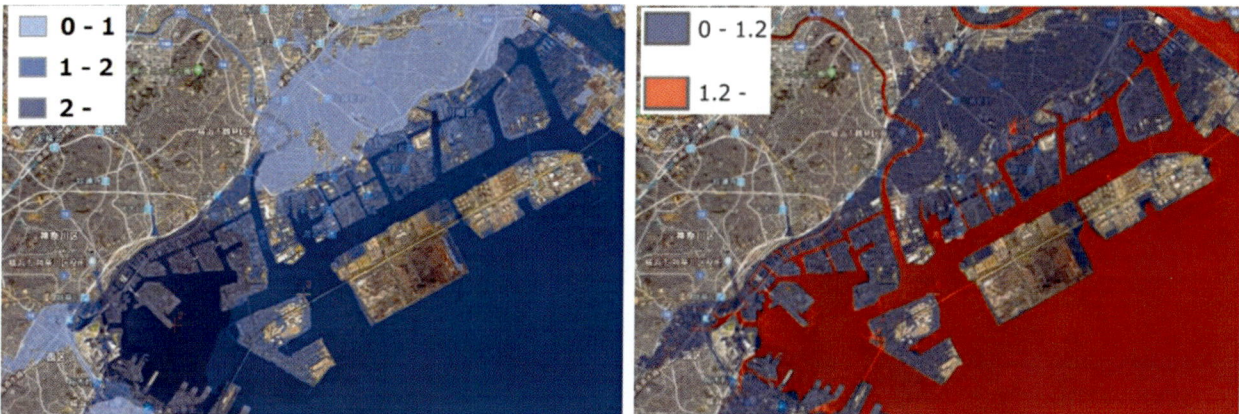

FIGURE 22.4 Spatial distribution of inundation (left, in m) and *dv* (right, in m^2/second) values results for +1.0 m SLR scenario in Kawasaki.

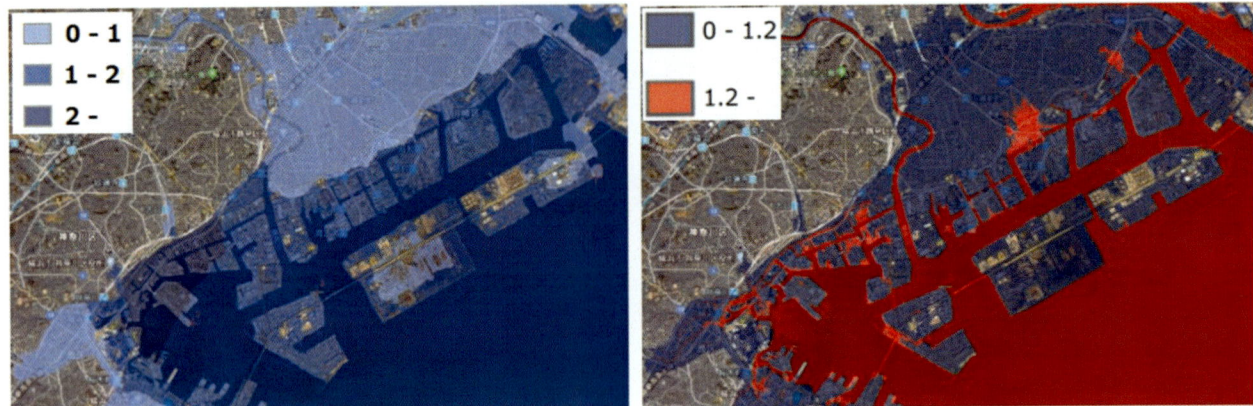

FIGURE 22.5 Spatial distribution of inundation (left, in m) and dv (right, in m^2/second) values for +2.0 m SLR scenario in Kawasaki.

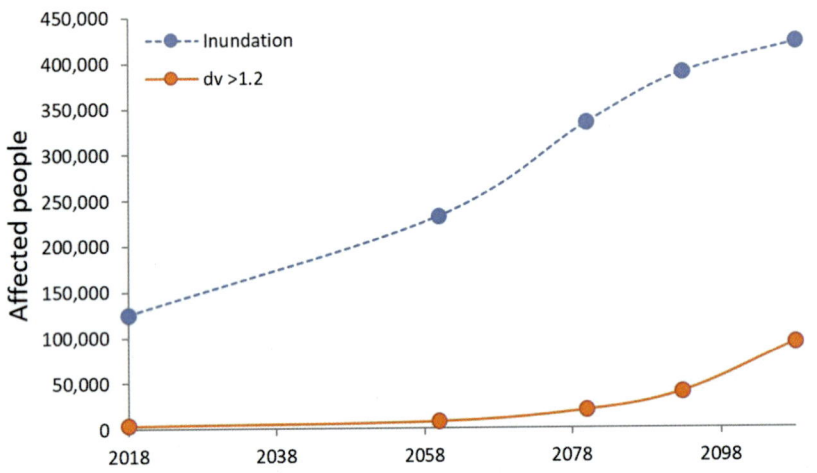

FIGURE 22.6 Evolution of the extent of inundation and $dv > 1.2$ m^2/second areas for the various SLR scenarios considered.

reclamation took place in the late 19th and 20th centuries). Similar reclamation patterns are present in the Kawasaki area, with newer parts being safer than older areas in the city.

Risk levels start to increase with SLR. For the +0.5 m SLR scenario, the extent of the flooded area around the main part of Kawasaki would increase significantly, although in most cases $dv > 1.2$ m^2/second values would not be reached. The extent of the inundation would continue to increase for the +1.0 m SLR scenario, and some limited areas could start reaching critical dv levels. While the extent of areas at risk in Yokohama (especially Kanagawa and Nishi Ward) does not change significantly for the +1.5 m SLR scenario (given that a large extent of the area located close to the shoreline is made up of small hills), the coastal areas of Kawasaki would be flooded (except for some of the newer reclaimed islands). Eventually, for the +2.0 m SLR scenario, large areas around Bashamichi station and Namamugi in Yokohama and Hamacho and Shiohama in Kawasaki Ward could be flooded in the event of a tsunami (simulated dv values exceeded 1.2 m^2/second in these areas).

The increase in the extent of the land under the threat of being inundated by tsunamis for various SLR scenarios is shown in Fig. 22.6. The figure highlights that hazardous dv values will only start to expand after 2080. Fig. 22.7 shows the increase in the number of people who will be affected by tsunami inundation and $dv > 1.2$ m^2/second with the progress of time. The threat to the population greatly increases after 2080, from 3% of the population of the wards studied in 2018 to 6% in 2080. It is worth noting that the simulations detailed in this chapter assumed that the topography, number and distribution of population, and land use would not change significantly from the present-day patterns. However, it is also worth noticing that, while the population in big cities, such as Tokyo and Yokohama, will probably not significantly decrease in the future, even if the overall Japanese population is declining (Esteban et al., 2010), the population of other coastal cities around the planet is generally increasing.

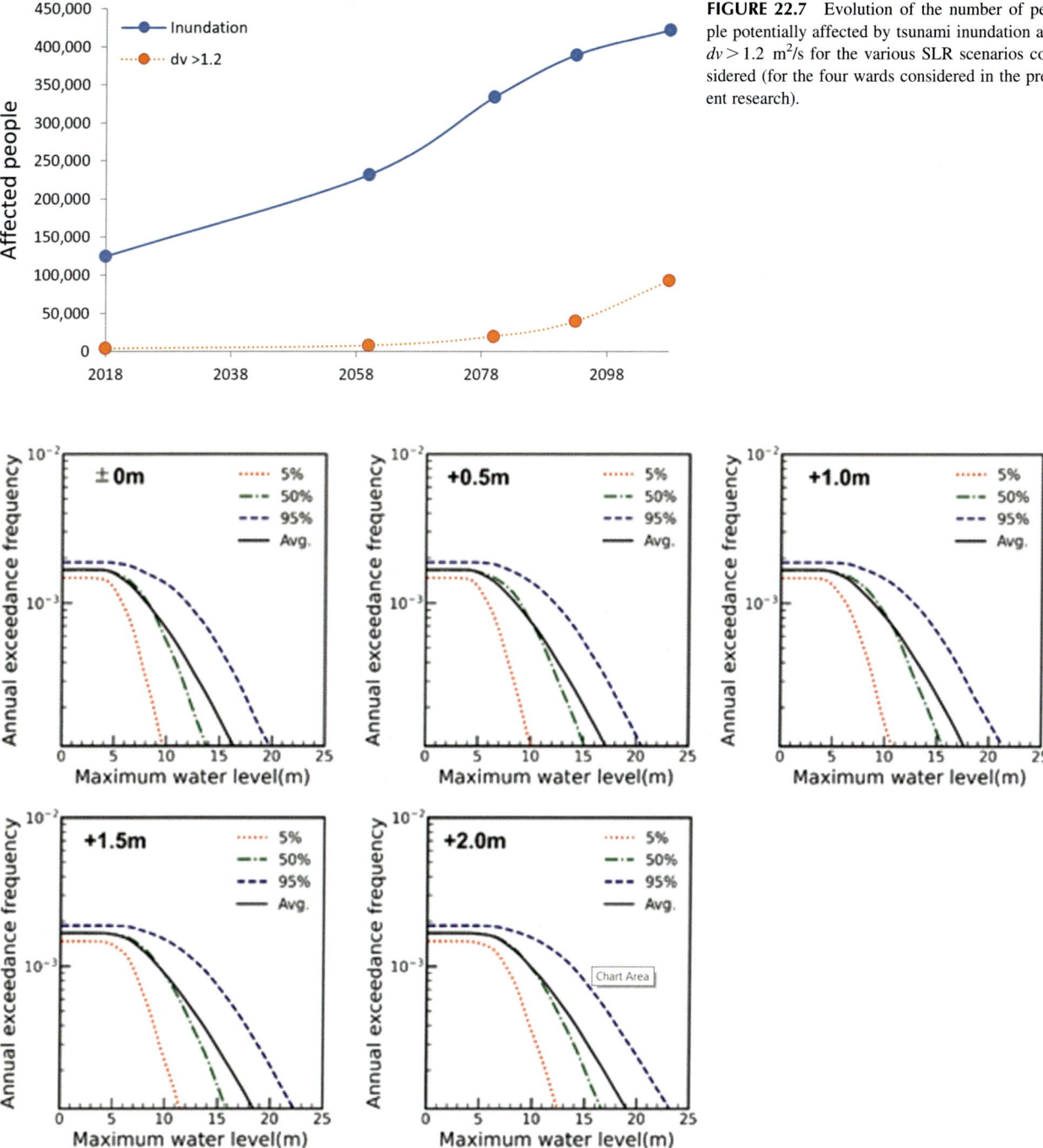

FIGURE 22.7 Evolution of the number of people potentially affected by tsunami inundation and $dv > 1.2$ m²/s for the various SLR scenarios considered (for the four wards considered in the present research).

FIGURE 22.8 Tsunami hazard curves (5%, 50%, and 95%, and their arithmetic average) of tsunami heights off Sendai Port based on SLR of between 0 and 2.0 m.

22.3.2 Increase in tsunami risk around the eastern coast of the Kanto region due to sea-level rise

The return period of the 2011 Tohoku Earthquake and Tsunami has been estimated to be 600 years (Fujiwara et al., 2013). Koyano et al. (2022) estimated that a Level 2 tsunami at Sendai Port at present is in a range of 5.6−11.7 m (from the 5th and 95th percentile curves), which would increase to 5.8−12.0 m (for the +0.5 m SLR scenario) and to 6.8−13.2 m (+1.5 m SLR), as shown in Fig. 22.8. Thus the results indicate that due to nonlinear effects the increase in tsunami height at the coastline would likely exceed the increase in the sea level. This highlights the importance of performing tsunami simulations to accurately estimate the SLR effects on the propagation and inundation of tsunamis.

Fig. 22.9 shows the distribution of tsunami heights corresponding to a return period of 1000 years (i.e., a Level 2 tsunami) at the 10 m deep digital water buoys (28 output points) along the eastern part of the Kanto region for each SLR scenario (for the 90% confidence intervals). The increase in the maximum water level for the +0.5 m SLR scenario exceeds 0.5 m at over 64% of the output points, indicating that there are nonlinear effects involved. Specifically, the increase at output points along the northern part of the studied buoy locations (point A in Fig. 22.9) and the southern part of Kujukuri Beach (points U and V) reached nearly 1.0 m. The 90% confidence intervals at most output points became narrower from the present day to the +1.5 m SLR scenario. For instance, the confidence intervals for locations around the northeastern coastline of the Kanto region (points A to D) and the middle part of Kujukuri Beach (point P) narrowed by over 1.0 m. The amount of the change in the 90% confidence interval differs depending on location, and around Asahi City (point J) was 0.8–1.9 m smaller than the average change at other points.

22.4 Discussion

SLR will likely increase the risks to tsunami-prone coastal cities, although the extent of the expected flooded areas and the number of affected people will vary depending on location. It is also important to emphasize that, for the case of sheltered areas, such as Tokyo Bay, while the extent of the potential inundation area might increase almost linearly, the surface areas covered by hazardous dv values might start to rapidly expand after an SLR of +1.0 m. The most onerous scenarios will likely take place after around 2080 (Kopp et al., 2017). This rapid change would also likely result in a significant increase in the number of casualties, compared to the current expectations. Disaster risk managers should keep an SLR +1.0 m in mind as the threshold for decisive actions to be taken when adaptation pathways are formulated (if human casualty risk levels do not increase substantially from what is acceptable at present). For the case of cities facing the open sea, such as locations along Kujukuri Beach, the effect that SLR has on potential tsunami heights might not be linear and would require detailed simulations to be carried out to identify optimal adaptation pathways.

It is important to note that as a consequence of a major earthquake land subsidence can occur, as evidenced along a wide stretch of the Japanese coastline after the 2011 Tohoku Earthquake and Tsunami. For instance, a land subsidence of up to −1.14 m was recorded at Ishinomaki City, causing widespread problems for coastal infrastructure (Cao et al., 2020) and necessitating extensive land elevation of coastal settlements (Esteban et al., 2015). Such effects should also be taken into account in simulations and compound the cumulative ongoing effects of SLR.

This chapter has mainly discussed the risk to coastal residents from tsunami events, although it is important to remember that these events also result in important economic losses. While countermeasures can be implemented to protect people, ports are more difficult to protect, and significant damage can be caused, for example, by floating objects (e.g., merchant vessels and shipping containers) (Naito et al., 2016; Stolle et al., 2019).

Despite the risk posed by SLR, cities around the planet will gradually adapt to it, likely in the same way that Tokyo adapted to land subsidence in the 20th century, and Jakarta is adapting at present (Esteban et al., 2019). Nevertheless, this will result in changes to the patterns of population density and distribution, and more people may live in areas at risk (as there is a long history of land reclamation around Tokyo Bay and elsewhere, which is still ongoing nowadays). Nevertheless, it is important that any adaptation to SLR carefully considers the risk of both storm surges and tsunamis. In this regard, the study of Tokyo Bay can provide many lessons to other parts of the world. At present, the northern side of Tokyo Bay is at the highest risk of being flooded by storm surges (Hoshino et al., 2016), and tsunami risks throughout the bay are generally thought to be lower. Hence, any newly reclaimed land areas around Tokyo Bay are built up to a ground level higher than the expected storm surge level, explaining why these areas are safer than older land. For example, in the case of Yokohama, the ground level was calculated according to simulations by the Kanagawa Prefectural Government, and thus for Minato Mirai 21 (an area that was reclaimed in the latter half of the 20th century), the ground level was set to be +3.1–5.0 m.

The implementation of adaptation measures to SLR should also carefully consider the problem of liquefaction, which could compound the problems of an earthquake and damage coastal defenses. In the case of Yokohama and Kawasaki, some dykes are present in the city, but many are old and recent research typically assumes that they would fail. This represents one of the major problems in disaster risk management not only in Tokyo Bay but elsewhere on the planet, as countermeasures are typically not designed to withstand multiple hazards in close succession. For example, dykes will probably withstand a storm surge, though if the defenses were damaged by an earthquake before it they could fail. It is thus important that efforts are made to consider the potential consequences of multiple hazards affecting a given city in close succession and the influence that SLR could have on them to formulate optimal adaptation pathway strategies that maximize resilience while minimizing lifecycle costs.

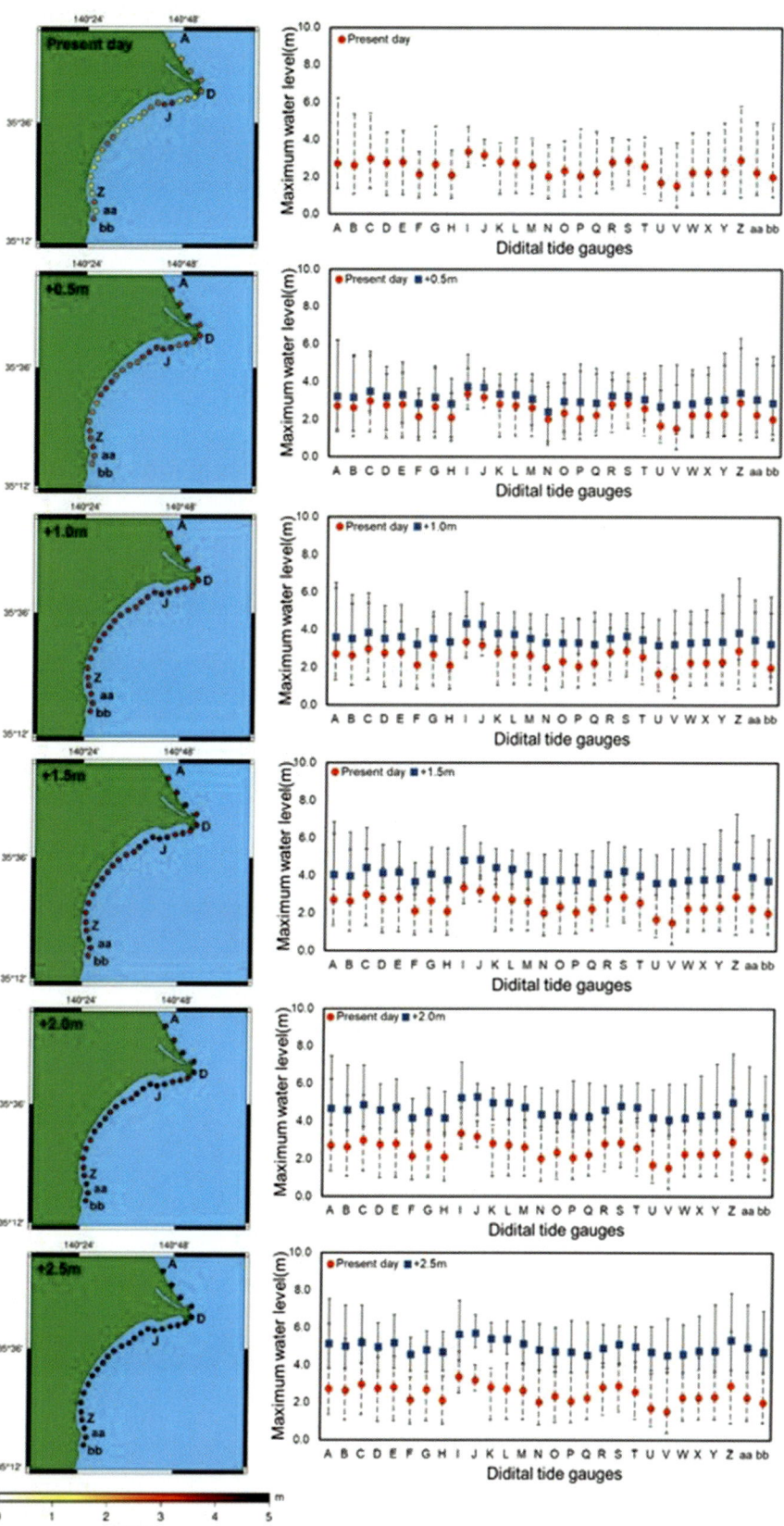

FIGURE 22.9 (Left) Distribution of the maximum water levels for Level 2 tsunamis at digital water buoys along the eastern part of the Kanto region. Right: Maximum water levels at these gauges, for the 90% confidence interval.

22.5 Conclusions

SLR will gradually increase the risk of tsunami flooding taking place around the planet and it is not a question of if, but rather of when the defenses of vulnerable coastlines will have to be strengthened. These effects, and the population that will be affected by them, will not be linear and will likely become more severe after an SLR of $+1.0$ m. As a result, adaptation measures will be necessary in the future to maintain (if not reduce) current tsunami risk levels. This can be achieved by improving coastal defenses and enhancing disaster evacuation strategies. Coastal risk managers should start to formulate adaptation pathways by taking such risks into account, implement precautionary strategies immediately, and incorporate them into the design with sufficient flexibility so that future protection infrastructure can be built as needed through the use of a long-term adaptation pathway strategy.

References

Aida, I. (1978). Reliability of a tsunami source model derived from fault parameters. *Journal of Physics of the Earth*, 26(1), 57–73. Available from https://doi.org/10.4294/jpe1952.26.57.

Annaka, T., Satake, K., Sakakiyama, T., Yanagisawa, K., & Shuto, N. (2007). Logic-tree approach for probabilistic tsunami hazard analysis and its applications to the Japanese coasts. *Pure and Applied Geophysics*, 164(2-3), 577–592. Available from https://doi.org/10.1007/s00024-006-0174-3.

Arnaud, G. E., Krien, Y., Abadie, S., Zahibo, N., & Dudon, B. (2021). How would the potential collapse of the Cumbre Vieja volcano in La Palma Canary Islands impact the Guadeloupe Islands?: Insights into the consequences of climate change. *Geosciences*, 11(2), 56. Available from https://doi.org/10.3390/geosciences11020056.

Becerra, I., Aránguiz, R., González, J., & Benavente, R. (2020). An improvement of tsunami hazard analysis in Central Chile based on stochastic rupture scenarios. *Coastal Engineering Journal*, 62(4), 473–488. Available from https://doi.org/10.1080/21664250.2020.1812943, https://www.tandfonline.com/loi/tcej20.

Cao, A., Esteban, M., & Mino, T. (2020). Adapting wastewater treatment plants to sea level rise: Learning from land subsidence in Tohoku, Japan. *Natural Hazards*, 103(1), 885–902. Available from https://doi.org/10.1007/s11069-020-04017-5, http://www.wkap.nl/journalhome.htm/0921-030X.

Cao, A., Esteban, M., Onuki, M., Nguyen, K., Nguyen, D. T., & Le, V. T. (2021). Decoupled formal and informal flooding adaptation and conflicts in coastal cities: A case study of Ho Chi Minh City. *Ocean and Coastal Management*, 209. Available from https://doi.org/10.1016/j.ocecoaman.2021.105654, http://www.elsevier.com/inca/publications/store/4/0/5/8/8/9.

Cao, A., Esteban, M., Valenzuela, V. P. B., Onuki, M., Takagi, H., Thao, N. D., & Tsuchiya, N. (2021). Future of Asian deltaic megacities under sea level rise and land subsidence: Current adaptation pathways for Tokyo, Jakarta, Manila, and Ho Chi Minh City. *Current Opinion in Environmental Sustainability*, 50, 87–97. Available from https://doi.org/10.1016/j.cosust.2021.02.010, http://www.elsevier.com/wps/find/journaldescription.cws_home/718675/description#description.

Chang, J., Mäll, M., Nakamura, R., Takabatake, T., Bricker, J., Esteban, M., & Shibayama, T. (2022). Estimating the influence of sea level rise and climate change on coastal defences in western Taiwan. *Journal of Coastal and Hydraulic Structures*, 2. Available from https://doi.org/10.48438/jchs.2022.0016, https://journals.open.tudelft.nl/jchs/article/view/6141.

Church, J.A., Clark, P.U., Cazenave, A., Gregory, J.M., Jevrejeva, S., Levermann, A., Merrifield, M.A., Milne, G.A., Nerem, R.S., Nunn, P.D., Payne, A.J., Pfeffer, W.T., Stammer, D., Unnikrishnan, A.S. (2013). *Climate change 2013: The physical science basis. Contribution of working group I to the fifth assessment report of the intergovernmental panel on climate change Sea level change*. Cambridge University Press.

Esteban, M., Jamero, M. L., Nurse, L., Yamamoto, L., Takagi, H., Thao, N. D., Mikami, T., Kench, P., Onuki, M., Nellas, A., Crichton, R., Valenzuela, V. P., Chadwick, C., Avelino, J. E., Tan, N., & Shibayama, T. (2019). Adaptation to sea level rise on low coral islands: Lessons from recent events. *Ocean and Coastal Management*, 168, 35–40. Available from https://doi.org/10.1016/j.ocecoaman.2018.10.031, http://www.elsevier.com/inca/publications/store/4/0/5/8/8/9.

Esteban, M., Onuki, M., Ikeda, I., & Akiyama, T. (2015). *Reconstruction following the 2011 Tohoku earthquake tsunami: Case study of Otsuchi Town in Iwate prefecture, Japan. Handbook of coastal disaster mitigation for engineers and planners* (pp. 615–631). Elsevier Inc. Available from http://www.sciencedirect.com/science/book/9780128010600, 10.1016/B978-0-12-801060-0.00029-0.

Esteban, M., Zhang, Q., Utama, A., Tezuka, T., & Ishihara, K. N. (2010). Methodology to estimate the output of a dual solar-wind renewable energy system in Japan. *Energy Policy*, 38(12), 7793–7802. Available from https://doi.org/10.1016/j.enpol.2010.08.039.

Fujii, Y., Satake, K., Sakai, S., Shinohara, M., & Kanazawa, T. (2011). Tsunami source of the 2011 off the pacific coast of Tohoku earthquake. *Earth, Planets and Space*, 63(7), 815–820. Available from https://doi.org/10.5047/eps.2011.06.010, http://rd.springer.com/journal/40623.

Fujiwara, H., Morikawa, N., & Okumura, T. (2013). Seismic hazard assessment for Japan: Reconsiderations after the 2011 Tohoku earthquake. *Journal of Disaster Research*, 8(5), 848–860. Available from https://doi.org/10.20965/jdr.2013.p0848, http://www.fujipress.jp/finder/access_check.php?pdf_filename = DSSTR000800050001.pdf&frompage = abst_page&errormode = Login&pid = &lang = English.

Fukutani, Y., Moriguchi, S., Kotani, T., & Terada, K. (2018). Probabilistic tsunami loss estimation using response surface method: Application to Sagami Trough earthquake. *Journal of Japan Society of Civil Engineers, Ser. B2 (Coastal Engineering).*, 74(2), 463–468. Available from https://doi.org/10.2208/kaigan.74.i_463.

Goda, K., Mai, P. M., Yasuda, T., & Mori, N. (2014). Sensitivity of tsunami wave profiles and inundation simulations to earthquake slip and fault geometry for the 2011 Tohoku earthquake. *Earth, Planets and Space*, 66(1). Available from https://doi.org/10.1186/1880-5981-66-105, http://rd.springer.com/journal/40623.

Goda, K., & Song, J. (2016). Uncertainty modeling and visualization for tsunami hazard and risk mapping: A case study for the 2011 Tohoku earthquake. *Stochastic Environmental Research and Risk Assessment*, *30*(8), 2271−2285. Available from https://doi.org/10.1007/s00477-015-1146-x, http://link.springer-ny.com/link/service/journals/00477/index.htm.

Goto, C., Ogawa, Y., Shuto, N., & Imamura, F. (1997). *Numerical method of tsunami simulation with the Leap-frog scheme*. IOC Manual.

Grant, E., Iliopoulos, N., Esteban, M., & Onuki, M. (2023). A tale of two (Florida) cities: Perceptions of flooding risk and adaptation in Tampa's Hyde Park and Saint Augustine. *Mitigation and adaptation strategies for global change*. Springer.

Haasnoot, M., Brown, S., Scussolini, P., Jimenez, J. A., Vafeidis, A. T., & Nicholls, R. J. (2019). Generic adaptation pathways for coastal archetypes under uncertain sea-level rise. *Environmental Research Communications*, *1*(7). Available from https://doi.org/10.1088/2515-7620/ab1871, https://iopscience.iop.org/article/10.1088/2515-7620/ab1871/pdf.

Harnantyari, A. S., Takabatake, T., Esteban, M., Valenzuela, P., Nishida, Y., Shibayama, T., Achiari, H., Rusli., Marzuki, A. G., Marzuki, M. F. H., Aránguiz, R., & Kyaw, T. O. (2020). Tsunami awareness and evacuation behaviour during the 2018 Sulawesi Earthquake tsunami. *International Journal of Disaster Risk Reduction*, *43*. Available from https://doi.org/10.1016/j.ijdrr.2019.101389, http://www.journals.elsevier.com/international-journal-of-disaster-risk-reduction/.

Hoshino, S., Esteban, M., Mikami, T., Takagi, H., & Shibayama, T. (2016). Estimation of increase in storm surge damage due to climate change and sea level rise in the Greater Tokyo area. *Natural Hazards*, *80*(1), 539−565. Available from https://doi.org/10.1007/s11069-015-1983-4, http://www.wkap.nl/journalhome.htm/0921-030X.

Jamero, M. L., Onuki, M., Esteban, M., Chadwick, C., Tan, N., Valenzuela, V. P., Crichton, R., & Avelino, J. E. (2019). In-situ adaptation against climate change can enable relocation of impoverished small islands. *Marine Policy*, *108*. Available from https://doi.org/10.1016/j.marpol.2019.103614, http://www.elsevier.com/inca/publications/store/3/0/4/5/3/.

Japan Society of Civil Engineers. (2008). *Questionnaire survey of the weights on the logic-tree*. http://committees.jsce.or.jp/ceofnp/system/files/Questionare_RT_PTHA_20141009_0.pdf.

Kanagawa Prefecture. (2015). *Explanatory document of potential tsunami inundation*. http://www.pref.kanagawa.jp/uploaded/attachment/774580.pdf.

Kawai, H., Hashimoto, N., & Matsuura, K. (2007). *Improvement of stochastic typhoon model for the purpose of simulating typhoons and storm surges Unger global warming*. In Proceedings of the Coastal Engineering Conference (pp. 1838−1850). 10.1142/9789812709554_0155 http://ascelibrary.org/

Kopp, R. E., DeConto, R. M., Bader, D. A., Hay, C. C., Horton, R. M., Kulp, S., Oppenheimer, M., Pollard, D., & Strauss, B. H. (2017). Evolving understanding of Antarctic ice-sheet physics and ambiguity in probabilistic sea-level projections. *Earth's Future*, *5*(12), 1217−1233. Available from https://doi.org/10.1002/2017EF000663, http://onlinelibrary.wiley.com/journal/10.1002/(ISSN)2328-4277.

Kotani, T., Tozato, K., Takase, S., Moriguchi, S., Terada, K., Fukutani, Y., Otake, Y., Nojima, K., Sakuraba, M., & Choe, Y. (2020). Probabilistic tsunami hazard assessment with simulation-based response surfaces. *Coastal Engineering*, *160*, 103719. Available from https://doi.org/10.1016/j.coastaleng.2020.103719.

Koyano, K., Takabatake, T., Esteban, M., & Shibayama, T. (2021). Influence of edge waves on tsunami characteristics along Kujukuri Beach, Japan. *Journal of Waterway, Port, Coastal and Ocean Engineering*, *147*(1). Available from https://doi.org/10.1061/(ASCE)WW.1943-5460.0000617, https://ascelibrary.org/journal/jwped5.

Koyano, K., Takabatake, T., Esteban, M., & Shibayama, T. (2022). Magnification of tsunami risks due to sea level rise along the eastern coastline of Japan. *Journal of Coastal and Hydraulic Structures*. Available from https://doi.org/10.48438/jchs.2022.00012.

Kukita, S., & Shibayama, T. (2012). Simulation and video analysis of the 2011 Tohoku Tsunami in Kesennuma. *Journal of Japan Society of Civil Engineers, Ser. B3 (Ocean Engineering).*, *68*(2), 49−54. Available from https://doi.org/10.2208/jscejoe.68.i_49.

Le Bars, D., Drijfhout, S., & de Vries, H. (2017). A high-end sea level rise probabilistic projection including rapid Antarctic ice sheet mass loss. *Environmental Research Letters*, *12*(4), 044013. Available from https://doi.org/10.1088/1748-9326/aa6512.

Li, L., Switzer, A. D., Wang, Y., Chan, C. H., Qiu, Q., & Weiss, R. (2018). A modest 0.5-m rise in sea level will double the tsunami hazard in Macau. *Science Advances*, *4*(8). Available from https://doi.org/10.1126/sciadv.aat1180, http://advances.sciencemag.org/content/4/8/eaat1180.

Mansinha, L., & Smylie, D. E. (1971). The displacement fields of inclined faults. *Bulletin of the Seismological Society of America*, *61*(5), 1433−1440. Available from https://doi.org/10.1785/bssa0610051433.

Mikami, T., Shibayama, T., Esteban, M., & Matsumaru, R. (2012). Field survey of the 2011 Tohoku earthquake and tsunami in Miyagi and Fukushima prefectures. *Coastal Engineering Journal*, *54*(1). Available from https://doi.org/10.1142/S0578563412500118, https://www.tandfonline.com/loi/tcej20.

Mikami, T., Shibayama, T., Esteban, M., Takabatake, T., Nakamura, R., Nishida, Y., Achiari, H., Rusli., Marzuki, A. G., Marzuki, M. F. H., Stolle, J., Krautwald, C., Robertson, I., Aránguiz, R., & Ohira, K. (2019). Field survey of the 2018 Sulawesi Tsunami: Inundation and run-up heights and damage to coastal communities. *Pure and Applied Geophysics*, *176*(8), 3291−3304. Available from https://doi.org/10.1007/s00024-019-02258-5, http://www.springer.com/birkhauser/geo + science/journal/24.

Mori, N., Mai, P. M., Goda, K., & Yasuda, T. (2017). Tsunami inundation variability from stochastic rupture scenarios: Application to multiple inversions of the 2011 Tohoku, Japan earthquake. *Coastal Engineering*, *127*, 88−105. Available from https://doi.org/10.1016/j.coastaleng.2017.06.013, http://www.elsevier.com/inca/publications/store/5/0/3/3/2/5/.

Nagai, R., Takabatake, T., Esteban, M., Ishii, H., & Shibayama, T. (2020). Tsunami risk hazard in Tokyo Bay: The challenge of future sea level rise. *International Journal of Disaster Risk Reduction*, *45*, 101321. Available from https://doi.org/10.1016/j.ijdrr.2019.101321.

Naito, C., Riggs, H. R., Wei, Y., & Cercone, C. (2016). Shipping-container impact assessment for tsunamis. *Journal of Waterway, Port, Coastal and Ocean Engineering*, *142*(5). Available from https://doi.org/10.1061/(ASCE)WW.1943-5460.0000348, http://ojps.aip.org/wwo/.

National Police Agency of Japan. (2020). *Damage report national police agency Japan* (In Japanese). Retrieved August 4, 2020. https://www.npa.go.jp/news/other/earthquake2011/pdf/higaijokyo.pdf.

Nobuoka, H., & Onoue, Y. (2017). Comparison on capacity of probabilistic tsunami hazard analysis from high frequency to low frequency. *Journal of Japan Society of Civil Engineers, Ser. B2 (Coastal Engineering).*, *73*(2), 1495–1500. Available from https://doi.org/10.2208/kaigan.73.i_1495.

Ohira, K., & Shibayama, T. (2013). *Wave behaviour in Tokyo Bay caused by a tsunami or long-period ground motions*. In Proceedings of the Coastal Structures 2011 Conference (pp. 1313–1324).

Okada, Y. (1985). Surface deformation due to shear and tensile faults in a half-space. *Bulletin of the Seismological Society of America*, *75*(4), 1135–1154. Available from https://doi.org/10.1785/bssa0750041135.

Park, H., Cox, D. T., Alam, M. S., & Barbosa, A. R. (2017). Probabilistic seismic and tsunami hazard analysis conditioned on a megathrust rupture of the Cascadia subduction zone. *Frontiers in Built Environment*, *3*. Available from https://doi.org/10.3389/fbuil.2017.00032, https://www.frontiersin.org/articles/10.3389/fbuil.2017.00032/pdf.

Shibayama, T., Esteban, M., Nistor, I., Takagi, H., Thao, N. D., Matsumaru, R., Mikami, T., Aranguiz, R., Jayaratne, R., & Ohira, K. (2013). Classification of tsunami and evacuation areas. *Natural Hazards*, *67*(2), 365–386. Available from https://doi.org/10.1007/s11069-013-0567-4, http://www.wkap.nl/journalhome.htm/0921-030X.

Spence, R., Palmer, J., & Potangaroa, R. (2009). Eyewitness reports of the 2004 Indian Ocean tsunami from Sri Lanka, Thailand and Indonesia. *Geotechnical, Geological and Earthquake Engineering.*, *7*, 473–495. Available from https://doi.org/10.1007/978-1-4020-8609-0_30, http://www.springerlink.com/content/1573-6059/.

Statistics Bureau of Japan. (2015). *2018 Census results*. https://www.stat.go.jp/data/kokusei/2015/kekka.html.

Stolle, J., Takabatake, T., Hamano, G., Ishii, H., Iimura, K., Shibayama, T., Nistor, I., Goseberg, N., & Petriu, E. (2019). Debris transport over a sloped surface in tsunami-like flow conditions. *Coastal Engineering Journal*, *61*(2), 241–255. Available from https://doi.org/10.1080/21664250.2019.1586288, https://www.tandfonline.com/loi/tcej20.

Suga, K., Uesaka, T., Yoshida, T., Hamaguchi, K., & Chen, Z. (1995). Preliminary study on feasible safe evacuation in flood disaster. *Proceedings of Hydraulic Engineering*, *39*, 879–882. Available from https://doi.org/10.2208/prohe.39.879.

Sugino, H., Iwabuchi, Y., Hashimoto, N., Matsusue, K., Ebisawa, K., Kameda, H., & Imamura, F. (2014). The characterizing model for tsunami source regarding the inter-plate earthquake tsunami. *Journal of Japan Association for Earthquake Engineering*, *14*(5), 1–18. Available from https://doi.org/10.5610/jaee.14.5_1.

Takabatake, T., Fujisawa, K., Esteban, M., & Shibayama, T. (2020). Simulated effectiveness of a car evacuation from a tsunami. *International Journal of Disaster Risk Reduction*, *47*101532. Available from https://doi.org/10.1016/j.ijdrr.2020.101532.

Takabatake, T., Mäll, M., Esteban, M., Nakamura, R., Kyaw, T. O., Ishii, H., Valdez, J. J., Nishida, Y., Noya, F., & Shibayama, T. (2018). Field survey of 2018 Typhoon Jebi in Japan: Lessons for disaster risk management. *Geosciences (Switzerland)*, *8*(11). Available from https://doi.org/10.3390/geosciences8110412, https://www.mdpi.com/2076-3263/8/11/412/pdf.

Takabatake, T., Shibayama, T., Esteban, M., Achiari, H., Nurisman, N., Gelfi, M., Tarigan, T. A., Kencana, E. R., Fauzi, M. A. R., Panalaran, S., Harnantyari, A. S., & Kyaw, T. O. (2019). Field survey and evacuation behaviour during the 2018 Sunda Strait tsunami. *Coastal Engineering Journal*, *61*(4), 423–443. Available from https://doi.org/10.1080/21664250.2019.1647963.

Takabatake, T., St-Germain, P., Nistor, I., Stolle, J., & Shibayama, T. (2019). Numerical modelling of coastal inundation from Cascadia subduction zone tsunamis and implications for coastal communities on western Vancouver Island, Canada. *Natural Hazards*, *98*(1), 267–291. Available from https://doi.org/10.1007/s11069-019-03614-3, http://www.wkap.nl/journalhome.htm/0921-030X.

Takagi, H., Li, S., de Leon, M., Esteban, M., Mikami, T., Matsumaru, R., Shibayama, T., & Nakamura, R. (2016). Storm surge and evacuation in urban areas during the peak of a storm. *Coastal Engineering*, *108*, 1–9. Available from https://doi.org/10.1016/j.coastaleng.2015.11.002, http://www.elsevier.com/inca/publications/store/5/0/3/3/2/5/.

Teh, S. Y., Koh, H. L., Moh, Y. T., De Angelis, D. L., & Jiang, J. (2011). Tsunami risk mapping simulation for Malaysia. *WIT Transactions on the Built Environment*, *119*, 3–14. Available from https://doi.org/10.2495/DMAN110011.

The 2011 Tohoku Earthquake Tsunami Joint Survey Group. (2011). Nationwide field survey of the 2011 off the Pacific coast of Tohoku earthquake tsunami. *Journal of Japan Society of Civil Engineers, Ser. B2 (Coastal Engineering)*, *67*(1), 63–66. Available from https://doi.org/10.2208/kaigan.67.63.

Tursina, S., Kato, S., & Afifuddin, M. (2021). Coupling sea-level rise with tsunamis: Projected adverse impact of future tsunamis on Banda Aceh city, Indonesia. *International Journal of Disaster Risk Reduction*, *55*, 102084. Available from https://doi.org/10.1016/j.ijdrr.2021.102084.

Wang, L., Huang, G., Zhou, W., & Chen, W. (2016). Historical change and future scenarios of sea level rise in Macau and adjacent waters. *Advances in Atmospheric Sciences*, *33*(4), 462–475. Available from https://doi.org/10.1007/s00376-015-5047-1, http://www.springerlink.com/content/0256-1530.

Wright, K., Doody, B.J., Becker, J., & McClure, J. (2010). *Pedestrian and motorist flood safety study: A review of behaviours in and around floodwater and strategies to enhance appropriate behaviour*. GNS Science Report.

Yamao, S., Esteban, M., Yun, N. Y., Mikami, T., & Shibayama, T. (2015). *Estimation of the current risk to human damage life posed by future tsunamis in Japan. Handbook of coastal disaster mitigation for engineers and planners* (pp. 257–275). Elsevier Inc. Available from http://www.sciencedirect.com/science/book/9780128010600, 10.1016/B978-0-12-801060-0.00013-7.

Chapter 23

Long-term tsunami risk considering time-dependent earthquake hazard and nonstationary sea-level rise

Katsuichiro Goda[1] and Raffaele De Risi[2]

[1]Department of Earth Sciences, Western University, London, ON, Canada, [2]School of Civil, Aerospace and Design Engineering, University of Bristol, Bristol, United Kingdom

23.1 Introduction

Buildings and infrastructure in active subduction zones are exposed to earthquake-triggered tsunami risks and influenced by rising sea levels due to climate change, which can lead to compounded consequences in the long run (Akiyama et al., 2020). Climate change exacerbates these circumstances gradually but steadily, amplifying the intensity of natural hazards and increasing the frequency of subsequent hazardous events. Fast-changing climates shorten the intervals between potentially damaging events, creating a chain of compounding events that result in severe disaster impacts. Because the population and infrastructure in coastal areas have increased due to economic reasons, geological and climate risks pose greater threats than ever (United Nations Office for Disaster Risk Reduction, 2022). Coastal communities urgently need planning guidelines and science-informed strategies/policies founded on accurate long-term risk assessments of the built environment under multihazard disturbances (De Risi et al., 2022; Tilloy et al., 2019).

Tsunami risks are nonstationary due to various factors. From tsunami source perspectives, the occurrence of tsunamigenic earthquakes is often regarded as time dependent (Headquarters for Earthquake Research Promotion, 2019; Kulkarni et al., 2013; Sykes & Menke, 2006), although the suitability of time-dependent earthquake occurrence models, compared to a time-independent Poisson model, is not always obvious due to the small sample size of historical events and their uncertainty/ambiguity (Griffin et al., 2020; Williams et al., 2019). By adopting renewal models (Abaimov et al., 2008), when the time since the last major earthquake is less than a mean recurrence period, the probability of earthquake occurrence given no event occurrence to date increases with the progress of time. Another contributing factor is the long-term rise of sea levels due to climate change (Church et al., 2013), which gradually alters the baseline sea level for tsunami inundation hazard assessments (Li et al., 2018). In this context, tidal levels during a major tsunami are variable. In tsunami hazard and risk assessments, the consideration of uncertain tidal levels has not been investigated rigorously, and some conservative assumptions are often made to account for the impacts of varying tides.

Probabilistic tsunami hazard and risk assessments explicitly incorporate uncertainties associated with tsunami sources and the physical vulnerability of the built environment (Behrens et al., 2021). The evaluation procedure is divided into hazard, exposure, vulnerability, and risk integration modules (Mitchell-Wallace et al., 2017). The hazard module characterizes the earthquake occurrence process, magnitude—frequency relationship, earthquake rupture process, tsunami generation, and tsunami propagation and inundation. The exposure-vulnerability modules involve the inventory of the population and built environment, damage assessment of the exposed entities, and estimation of the consequences of the tsunami (Goda & De Risi, 2017; Park et al., 2017). Using renewal occurrence models to generate stochastic event sets in the risk integration module facilitates the time-dependent tsunami impact assessment (Fukutani et al., 2021; Goda, 2019). Recent advances in probabilistic tsunami hazard and risk analysis methods include the consideration of sea-level rise effects caused by climate change (Alhamid et al., 2022; Sepúlveda et al., 2021), the use of heterogenous earthquake slips with variable fault-plane geometry (Melgar et al., 2019), the implementation of a logic tree to consider multiple alternatives of the models and parameters (Miyashita et al., 2020), and the development of tsunami fragility functions based on an extensive tsunami damage dataset (De Risi et al., 2017; Vescovo et al., 2023). Despite these refinements, studies that probabilistically evaluate long-term tsunami risks of a coastal community based on time-dependent earthquake hazards, nonstationary tidal and rising sea levels, numerous earthquake rupture scenarios, high-resolution topographic data, and high-quality building exposure data are lacking. All these elements are crucial for accurately assessing quantitative long-term tsunami risks.

This chapter presents a long-term probabilistic tsunami risk assessment for a Canadian coastal town, Tofino, that faces significant tsunami threats originating from the Cascadia subduction zone in the Pacific Northwest (Goldfinger et al., 2012; Walton et al., 2021). The earthquake-triggered tsunami hazards in the Cascadia subduction region are driven by the thrusting movements of the Juan de Fuca, Gorda, and Explorer Plates relative to the North American Plate. The Cascadia subduction zone hosted moment magnitude (M_w) 9.0 megathrust earthquakes in the past, and the most recent event occurred in 1700. Past studies of the tsunami hazard assessments for Canadian coastal locations considered a small number of full-rupture scenarios having M_w 8.7–9.3 (Gao et al., 2018; Takabatake et al., 2019), whereas the variable effects of tides and future sea-level rises were not incorporated in these studies. The limitations of the previous studies can be overcome by generating stochastic event sets based on time-dependent earthquake hazards and climate change-related nonstationary sea-level rises by evaluating conditional probability distributions of tsunami risk impact metrics based on different baseline sea levels and by integrating the stochastic event sets with the conditional probability distributions of the tsunami loss. The developed tsunami risk model opens a new avenue for conducting a long-term tsunami risk assessment for coastal communities and informing their long-term risk management strategies.

23.2 Tofino and physical environment

The probabilistic framework for long-term tsunami risk assessments focuses on Tofino. Tofino is one of the Canadian coastal towns most exposed to the Cascadia subduction earthquakes. First, Tofino's geography and physical/built environment are introduced, and then tidal levels and future relative sea-level rise scenarios for Tofino are summarized.

23.2.1 Cascadia subduction zone

The driving tectonic movement in the Cascadia region is the eastward subduction of the Juan de Fuca, Gorda, and Explorer Plates beneath the North American Plate (Fig. 23.1A). The Cascadia subduction zone extends 1100 km from British Columbia to Northern California and has a convergence rate of 30–45 mm/year (DeMets et al., 2010). Recurrence characteristics and rupture patterns of the past Cascadia subduction events have been investigated by collecting onshore subsidence records (Atwater et al., 2015) and offshore marine turbidites (Goldfinger et al., 2012). The mean recurrence period of megathrust earthquakes is about 530 years, but the recurrence interval of two successive events can vary widely between 120 and 1380 years (Goldfinger et al., 2012). Therefore, the possibility of experiencing a future Cascadia event in the 21st century is not negligible, noting that the current time elapsed since the 1700 rupture is 323 years.

23.2.2 District of Tofino

Tofino is located at Esowista Peninsula within Clayoquot Sound on Vancouver Island (Fig. 23.1B). Tofino's population is 2516 (2021 Census), while in the summer peak times, 5000 to 8.000 visitors and seasonal workers arrive and stay in Tofino. The main commercial area of Tofino is at the tip of the peninsula, and its elevation is 10 m above sea level (Fig. 23.1B); thus the tsunami risk in Tofino Town is relatively low. On the other hand, houses, resort hotels, and campsites are located in beach areas with elevations below 10 m (Fig. 23.1B); therefore, the tsunami risk is high in these areas. The beach areas are equipped with a siren system for tsunami warnings (Fig. 23.1C) and evacuation signages (Fig. 23.1D).

For earthquake-tsunami and coastal risk assessments, the District of Tofino conducted bathymetry surveys in the surrounding shallow water areas and LiDAR surveys to develop high-resolution bathymetry-elevation data. Tofino also developed a comprehensive inventory of buildings, cultural/historical sites, water sanitation facilities, potential environmental contamination sites, and infrastructures. As part of this exposure database development, a building-by-building inspection was conducted to determine the main structural features of the buildings (e.g., material, number of stories, and construction years) (Goda et al., 2023). The building inventory includes 1789 structures for residential, commercial, industrial, and civic occupancy (Fig. 23.1B), with a total asset value of 2.27 billion Canadian dollars (C$).

Long-term tsunami risk considering time-dependent earthquake hazard and nonstationary sea-level rise Chapter | 23

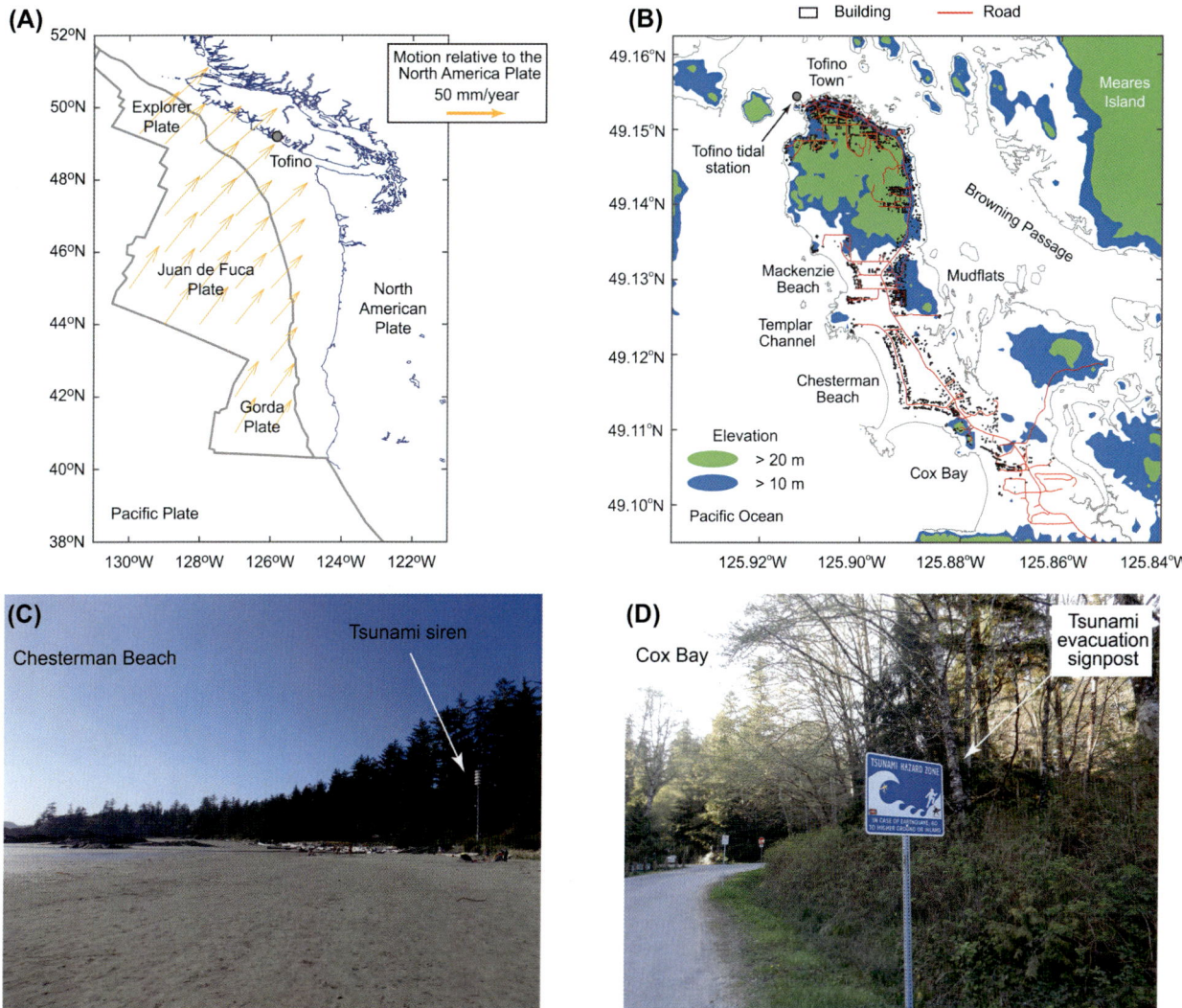

FIGURE 23.1 The District of Tofino and its surrounding environments: (A) seismotectonic environment of the Cascadia subduction region; (B) map of buildings and roads in Tofino; (C) photo at Chesterman Beach; and (D) photo at Cox Bay. Map lines delineate study areas and do not necessarily depict accepted national boundaries.

23.2.3 Tides and relative sea-level rise in Tofino

A tidal monitoring station has been operational in Tofino since 1909, and its location is shown in Fig. 23.1B. The hourly and daily average tide data since 1981 are obtained from the Government's data portal site (https://www.tides.gc.ca/en/stations/8615) and are shown in Fig. 23.2. The baseline (zero) tidal level is adjusted to the long-term average. The hourly tidal data vary between -2 and $+2$ m with respect to the long-term average tidal level. The fluctuating tide can be sampled from the hourly tide data in Tofino.

The baseline tidal level is subjected to global sea and land elevation changes (i.e., local/regional uplift and subsidence). Future projections of relative sea-level change are essential for coastal flooding predictions due to storm surges and tsunamis and infrastructure planning and maintenance. In southwestern British Columbia, considering the glacial isostatic adjustment is necessary for accurately characterizing the relative sea-level change. James et al. (2021) produced the latest national relative sea-level projections for Canada based on the Fifth Assessment Report of the Intergovernmental Panel on Climate Change (IPCC-AR5) (Church et al., 2013) by accounting for the vertical land motion for Canada based on GPS observations (Robin et al., 2020). The projections were generated at grids in latitude and longitude with an interval of 0.1 degrees across Canada by considering three Representative Concentration Pathways (RCP) scenarios (RCP2.6, RCP4.5, and RCP8.5) of the IPCC-AR5 and three percentiles (median (M), 5th percentile (L), and 95th percentile (U)) for each RCP scenario. Fig. 23.3 shows nine possible projections for the relative

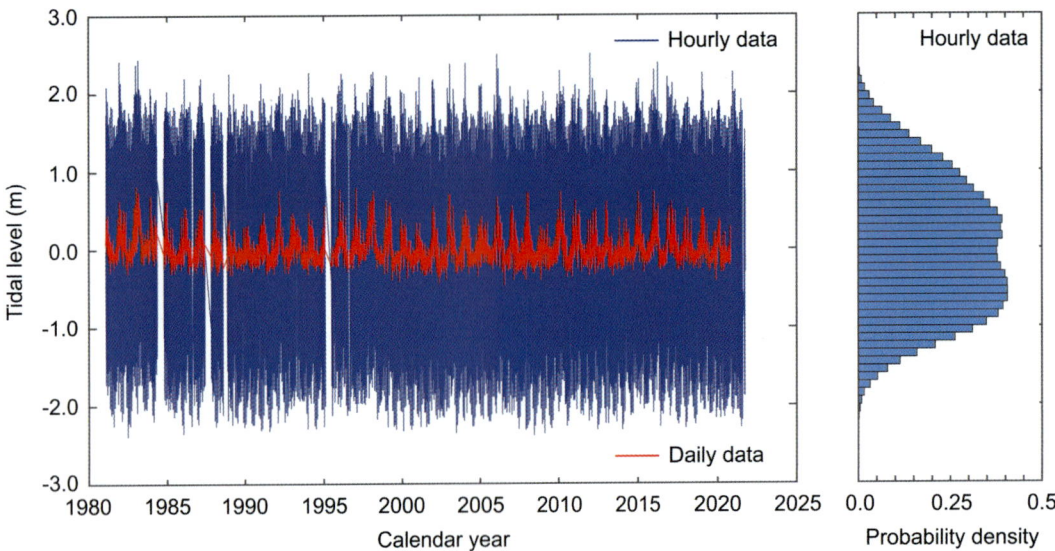

FIGURE 23.2 Tidal variation at the Tofino station since 1981.

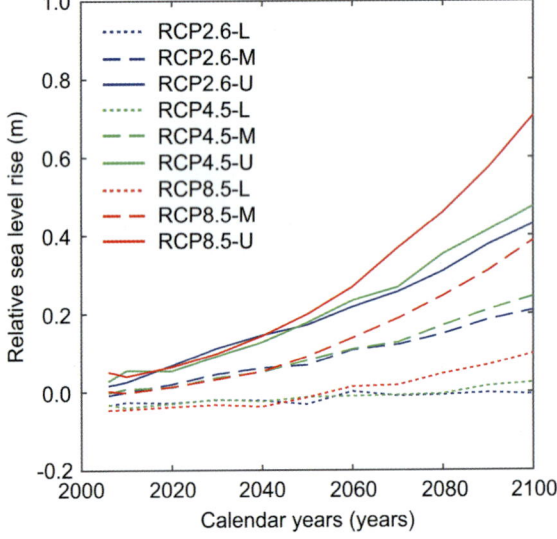

FIGURE 23.3 Relative sea-level rise scenarios (James et al., 2021). The data are extracted at (49.1°N and 125.9°W).

sea-level rise at the coordinate (49.1°N and 125.9°W) near Tofino. By 2100, the relative sea-level rise could reach +0.7 m with respect to the baseline sea level in 2006. The data shown in Fig. 23.3 can be used to determine the baseline sea level at a future time.

23.3 Long-term probabilistic tsunami risk model for Tofino

A long-term probabilistic tsunami risk model is developed for Tofino by focusing on the Cascadia subduction earthquakes as the primary tsunamigenic sources. The computational steps of the risk assessment can be divided into three parts. The time-dependent earthquake occurrence and magnitude models for the full-rupture Cascadia subduction events are explained first, which generates stochastic event sets for possible tsunamigenic events. Subsequently, evaluating the conditional probability distributions of tsunami risk metrics for Tofino is described by considering the effects of different baseline tidal levels. Finally, the stochastic event sets and the conditional probability distributions of the tsunami risk metrics are combined by accounting for nonstationary and variable sea levels based on tidal data and future projections of relative sea-level rise in Tofino.

23.3.1 Occurrence and magnitude models of megathrust Cascadia earthquakes

The offshore turbidite records collected by Goldfinger et al. (2012) exhibit time-dependent characteristics of the interarrival times between full-margin rupture events (Kulkarni et al., 2013). Out of the 40 events identified over the last 10,000 years, 19 events ruptured the entire length of the Cascadia subduction zone. The remaining events ruptured the middle and southern parts of the subduction zone only (i.e., Oregon and northern California). To account for inherent uncertainty associated with radiocarbon dating of the turbidite data, Goda (2023) performed Monte Carlo resampling of the Cascadia age data and obtained the interarrival times of the full-margin Cascadia subduction events. The resampled interarrival time data showed the short-term clustering and long-term gap in addition to the central mode. Goda (2023) used the three-component Gaussian mixture model to reflect these multimodal features of the interarrival time data, which outperformed conventional one-component renewal models with different interarrival time distributions. The central (dominant) component corresponds to earthquake recurrence with mean (μ) = 503 years and standard deviation (σ) = 139 years with the mixing proportion (π) = 0.646. The second component represents long gaps (μ = 905 years, σ = 224 years, and π = 0.240), whereas the third component represents short-term clustering (μ = 167 years, σ = 95 years, and π = 0.114). The preceding mixing proportions for the three components correspond to the elapsed time of 0 years (i.e., immediately after a major event), while when a longer elapsed time is considered, the mixing proportions can be modified for the situation that no event is yet observed till the date of the evaluation. Consequently, the probability of earthquake occurrence can be continuously updated for time-dependent tsunami hazard and risk assessments.

Such evolutionary earthquake occurrence of the Cascadia subduction earthquakes is illustrated in Fig. 23.4. The middle panel (Fig. 23.4B) shows the temporal variations of the mixing proportions of the three components, and the top-row panels show the conditional probability distributions of the interarrival time of the Cascadia subduction events at the elapsed times of 0, 200, and 400 years. With time, the mixing proportion for the short-term clustering decreases, while those for the central mode and long gap increase. This evolutionary trend can also be seen in the hazard rate function plot (i.e., the conditional probability density function upon survival up to the evaluation time) in Fig. 23.4C. The hazard rate function of the Gaussian mixture model is less than the hazard rate function of the Poisson process till the elapsed time of 390 years (i.e., Year 2090), indicating that the assumption of time-independent earthquake occurrence results in a conservative estimate of the earthquake and tsunami hazard until the end of the 21st century.

The earthquake magnitude distribution is critical for tsunami hazard assessments. The magnitudes of the Cascadia subduction events primarily depend on the rupture patterns and corresponding rupture areas. In the developed tsunami risk model, two end-member magnitude models are considered by assigning equal weights to the two models. The first one is the Gutenberg–Richter model with the b-value of 1. The second one is based on the characteristic magnitude model with a uniform distribution. Since the full-rupture scenarios are concerned, the minimum and maximum magnitudes are set to 8.7 and 9.1 for both magnitude models. The abovementioned earthquake occurrence model controls the occurrence probabilities of the full-margin megathrust Cascadia events.

23.3.2 Conditional tsunami risk curves

The other critical model component is the conditional probability distributions of tsunami risk metrics for different baseline tidal levels. Such distributions can be derived by implementing the stochastic rupture model, tsunami inundation model, building exposure model, tsunami fragility model, and tsunami damage-loss model. A graphical flowchart of the integration process of different models is shown in Fig. 23.5. The key aspects of these modules are explained below.

23.3.2.1 Stochastic rupture model

The fault plane for the Cascadia subduction zone is defined using the Slab2 model (Hayes et al., 2018) and is approximated by a set of 7452 subfaults that reach depths of 30 km, each having the size of 5.6 km along strike and 3.8 km along dip. To simulate the slip distributions stochastically, a scenario magnitude is specified with a 0.1 magnitude bin, and the magnitude value is simulated from the uniform distribution within the magnitude bin. The whole range of the earthquake magnitude spans from M_w 8.1 to 9.1. Subsequently, eight earthquake source parameters, that is, fault length, fault width, mean slip, maximum slip, Box-Cox parameter, along-strike correlation length, along-dip correlation length, and Hurst number, are generated from the scaling relationships (Goda, 2022). Once a suitable fault geometry is determined, the fault plane is placed randomly within the Cascadia fault plane.

Next, a heterogeneous earthquake slip distribution is generated from an anisotropic von Kármán wavenumber spectrum with its amplitude spectrum being parametrized by along-strike correlation length, along-dip correlation length,

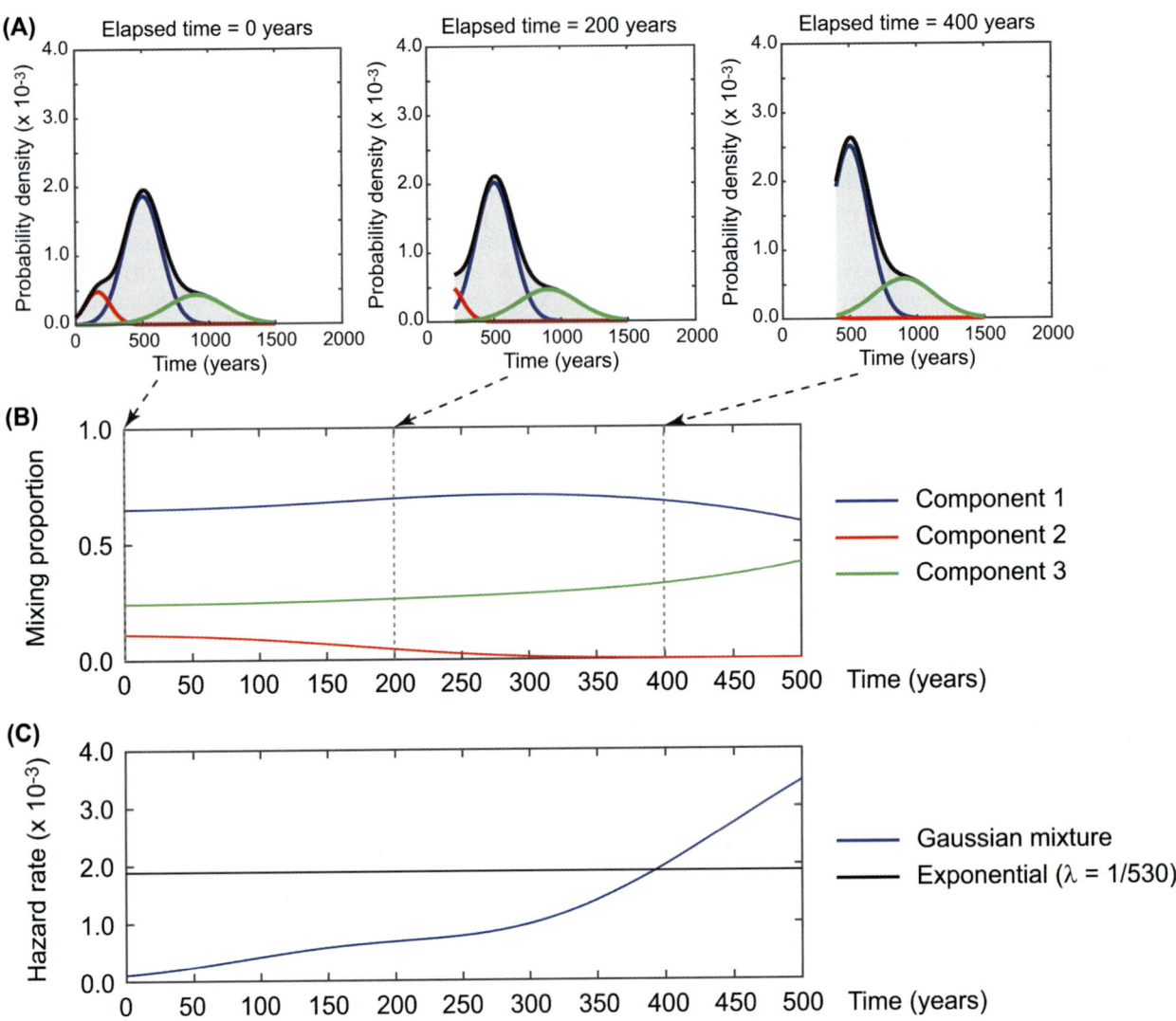

FIGURE 23.4 Time-dependent earthquake hazard model. (A) Probability distributions of interarrival time based on the Gaussian mixture model for the elapsed times of 0, 200, and 400 years. (B) Mixing proportions of three components of the Gaussian mixture model as a function of elapsed time. (C) Hazard rate functions of the Gaussian mixture model and exponential (time-independent Poisson) model as a function of elapsed time.

and Hurst number and its phase being randomly distributed between 0 and 2π (Mai & Beroza, 2002). The simulated slip distribution is modified via Box-Cox power transformation and scaled to the mean slip to achieve desirable slip characteristics over the fault plane (e.g., right skewness of the marginal distribution). To ensure that the simulated earthquake slip distribution has realistic characteristics for the target Cascadia events, major asperities are constrained to occur in the shallow part of the subduction interface to broadly coincide with the outer wedge of the accretionary prism. If the trial slip distribution does not meet the criteria, another trial model is generated, and this process is continued until an acceptable model is obtained. Repeating the above procedure 500 times for each of the 10 bins between M_w 8.1 and 9.1 generates a set of 5000 earthquake rupture models (Goda, 2022). The stochastic rupture models can represent different fault geometry, positions within the overall fault plane of the Cascadia subduction zone, and heterogenous earthquake slip distributions (Fig. 23.5A).

23.3.2.2 Tsunami inundation model

To evaluate tsunami inundations in Tofino by considering numerous stochastic earthquake rupture models, analytical formulae by Okada (1985) and Tanioka and Satake (1996) are implemented to calculate ground deformations due to

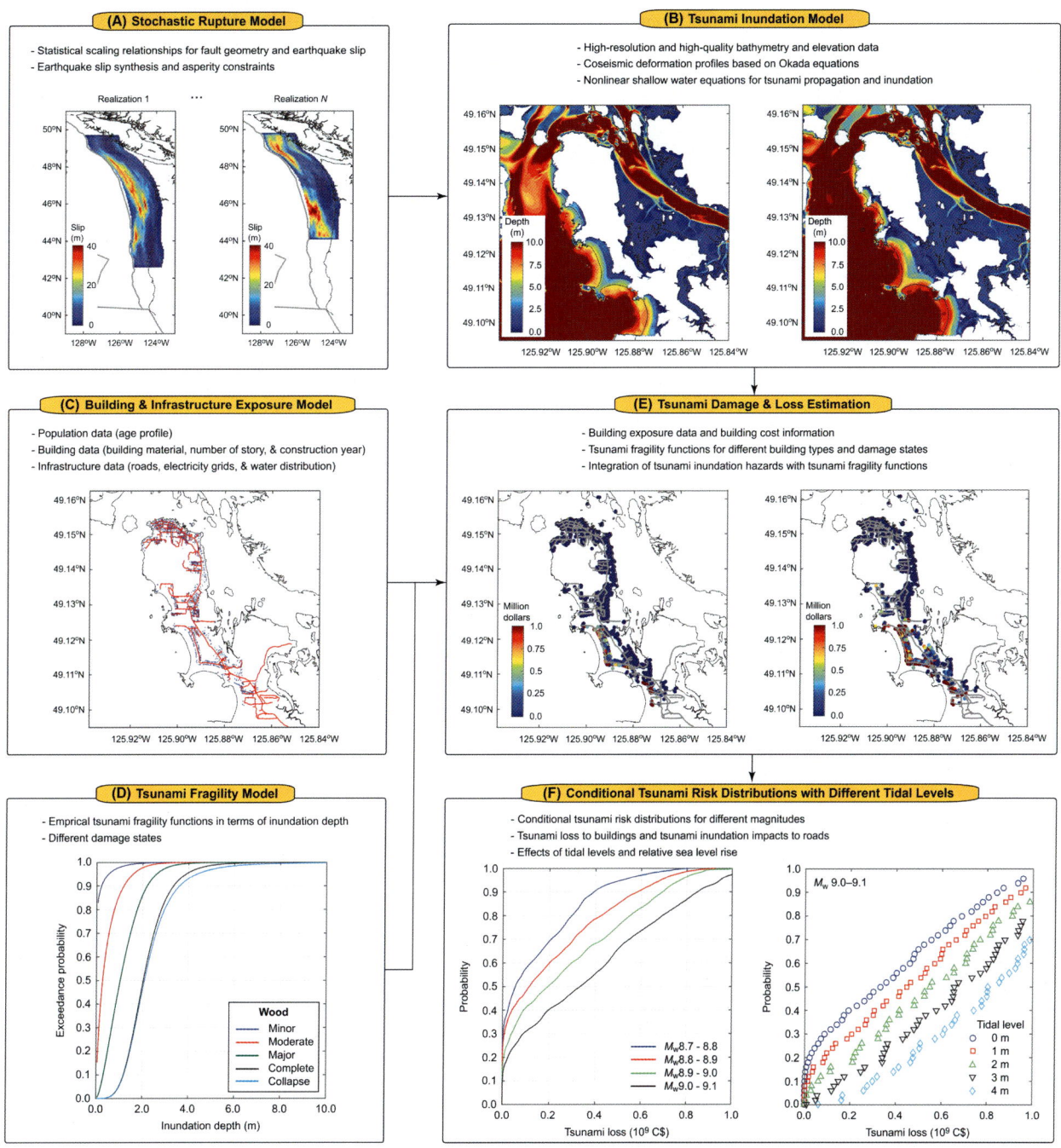

FIGURE 23.5 Development of the conditional tsunami risk distributions for different earthquake magnitude ranges and tidal levels: (A) stochastic rupture model; (B) tsunami inundation model; (C) building exposure model; (D) tsunami fragility model; (E) tsunami damage and loss estimation; and (F) conditional tsunami risk distributions with different baseline tidal levels. Map lines delineate study areas and do not necessarily depict accepted national boundaries.

earthquake ruptures, and then nonlinear shallow water equations are solved (Goto et al., 1997). For this purpose, nested grids of 810, 270, 90, 30, 10, and 5 m that cover the entire Cascadia subduction zone are set up by combining global bathymetry data (GEBCO-450-m), national bathymetry data (CHS-10-m), and LiDAR-derived local topographic contours (0.5-m). The 810-m grids cover the entire Cascadia subduction zone (Fig. 23.1A), while the 5-m grids cover Tofino (Fig. 23.1B). The vertical datum is set to the mean sea level for the base case and is increased to 1, 2, 3, and 4 m to evaluate the tidal effects on the tsunami inundation impacts. For all computational cells, the bottom friction and surface roughness are represented by a Manning's coefficient equal to $0.025 \text{ second/m}^{1/3}$, which is often used for agricultural land and ocean/water. In addition to the abovementioned baseline tidal levels, the effects of coseismic ground deformation are considered by adjusting the elevation data before each tsunami simulation run. Due to the strong directivity of radiated tsunami waves, the central and southern margin earthquake ruptures with M_w less than 8.7 do not cause large tsunami waves in Tofino (typically, less than 1 m in terms of maximum wave amplitude) (Goda, 2022). For this reason, the high-resolution tsunami inundation simulations using 5-m grids are performed for 2000 stochastic source scenarios with earthquake magnitudes between M_w 8.7 and 9.1. Under these conditions, tsunami simulations are run for a 2-hour duration, which is sufficient to model the most critical phase of tsunami waves for the full-margin Cascadia rupture scenarios (Fig. 23.5B).

23.3.2.3 Building exposure model

To carry out earthquake-tsunami risk assessments, the comprehensive inventory of buildings developed by the District of Tofino is utilized. Most buildings in Tofino Town are at high elevations and are protected from direct tsunami waves. In contrast, buildings along the McKenzie, Chesterman, and Cox Bay Beaches are at low elevations and open to the Pacific Ocean (Fig. 23.1B). The building data are related to the building classification scheme adopted by Natural Resources Canada (Hobbs et al., 2021). In the developed tsunami risk model, a portfolio of 1789 buildings is considered, excluding campsites, marina docks, and nonpermanent buildings (Fig. 23.5C). Most buildings are 1- to 2-story wooden houses constructed in the 1960s or afterward. The total replacement cost of a building, which consists of structural elements, nonstructural elements, and building contents, is typically less than C$ 2 million, with an average value of C$ 1.27 million.

23.3.2.4 Tsunami fragility model

To evaluate the tsunami damage extent to buildings in Tofino, an empirical tsunami fragility model by De Risi et al. (2017) is adopted, which was developed based on the tsunami damage data from the 2011 Tohoku Tsunami in Japan, containing more than 200,000 observations. This choice is due to the unavailability of local/regional tsunami fragility models for southwestern British Columbia and the strong preference for empirical tsunami damage data. The tsunami fragility model is characterized using multinomial logistic regression analysis by considering the structural typology (i.e., wood, concrete, steel, and masonry), number of stories, and topographical indicators (i.e., coastal plain and ria) as explanatory variables. Fig. 23.5D shows tsunami fragility functions of wooden buildings for five tsunami damage levels, that is, minor, moderate, extensive, complete, and collapse.

23.3.2.5 Tsunami damage and loss estimation

For tsunami loss estimation of buildings, these tsunami damage levels, identified using the tsunami fragility functions, are related to building damage ratio ranges of 0.03–0.1, 0.1–0.3, 0.3–0.5, 0.5–1.0, and 1.0, respectively (http://www.mlit.go.jp/toshi/toshi-hukkou-arkaibu.html). During the tsunami damage and loss analyses, the tsunami fragility functions are applied using flow depth values at individual buildings from tsunami inundation simulations. The damage states are assigned probabilistically by comparing a uniform random number between 0 and 1 with the corresponding tsunami damage probabilities. Suppose the random number falls within the range of tsunami damage probabilities for a specific damage level. In such a case, the tsunami damage state is selected, and the tsunami damage ratio is subsequently sampled within the suggested range. Finally, the tsunami loss value is determined by multiplying the total asset value of the property and the sampled damage ratio. The above procedure is repeated for all buildings for a given inundation scenario (Fig. 23.5E).

23.3.2.6 Conditional tsunami risk distributions with different baseline tidal levels

The development of conditional probability distributions for the aggregate building loss for a given magnitude range between M_w 8.7 and 9.1 with 0.1 intervals and different baseline tidal levels is conducted in two stages. In the first stage, all 2000 stochastic tsunami inundation results for the mean sea level are utilized to develop the conditional tsunami risk distributions. These distributions are shown on the left-hand side of Fig. 23.5F. In the second stage, because of highly demanding computation for carrying out high-resolution tsunami inundation simulations, for each magnitude range, the tsunami risk results for the mean sea level are sorted in ascending order, and 51 stochastic sources that correspond to the minimum, 49 intermediate (with 2% apart in terms of cumulative probability or every 10 stochastic sources out of the sorted 500 sources), and the maximum tsunami risk results are identified. Subsequently, the high-resolution tsunami inundation simulations are carried out for these 204 stochastic sources (i.e., 51 times 4) by considering the baseline tidal levels of 1, 2, 3, and 4 m. The minimum tidal level of 0 m is considered because the tidal condition below 0 m is rarely considered in tsunami hazard and risk assessments. On the other hand, the maximum tidal level of 4 m is sufficient to cover an extreme tidal condition (e.g., a high tide of 2 + m and a high relative sea-level rise of 1 + m in 2100). The M_w 9.0 to 9.1 results are shown on the right-hand side of Fig. 23.5F. For a given source model, the effects of the increased baseline tidal levels are monotonic in terms of the maximum inundation depths (i.e., a 2-m increase of the baseline leads to approximately a 2-m increase in the maximum inundation depth). Because of this reason, although the increased baseline tidal level significantly impacts the absolute values of the tsunami risk metrics, the order of the risk metric values for different source models is more or less maintained. This means that the reduced computational approaches to investigating the effects of the varied tidal levels can include a wide range of possible tsunami inundation cases (as captured by the original 2000 stochastic source models).

23.3.3 Long-term tsunami risk analysis with tidal variations and relative sea-level rises

To develop the exceedance probability curves of the tsunami risk metrics for Tofino, Monte Carlo simulations can be implemented by combining the stochastic event sets of time-dependent tsunami events with the conditional tsunami risk distributions for different baseline tidal levels of 0, 1, 2, 3, and 4 m. A graphical flowchart of this procedure is shown in Fig. 23.6. The earthquake occurrence and magnitude models will generate synthetic event catalogs containing information on the event occurrence time and magnitude (Fig. 23.6A). Each catalog is simulated by considering the specific starting and ending times; the starting time can be defined by the elapsed time since the last major event, while the difference between the starting and ending times establishes the duration of the tsunami risk assessment. Subsequently, the occurrence times of the events are used to determine the tidal levels at future times of these tsunami events. The tidal level can be sampled from nine possible relative sea-level rise scenarios at a given time by considering appropriate weights, and its variability can be sampled from the past tidal data in Tofino (Fig. 23.6B). For instance, the relative weights of the three RCP scenarios can be set to equal, while the relative weights of the low, median, and high branches can be determined by using the discrete approximations of the probability distribution (Miller & Rice, 1983). Using the earthquake magnitude information and the tidal level, samples of the aggregate tsunami loss can be extracted from the conditional tsunami risk distributions in terms of stochastic source models. Then, the final tsunami risk values are determined by interpolation (Fig. 23.6C). In this study, the negative tidal level is rounded up to zero (i.e., conservative estimate). Fig. 23.6D illustrates the stochastic event sets together with magnitudes, sea levels, and resulting tsunami risks. By generating numerous stochastic event sets and evaluating the nonstationary tsunami risks, the probability distribution of the tsunami risk metric can be derived and displayed in the form of an exceedance probability curve (Fig. 23.6E).

23.4 Long-term tsunami risk assessment for Tofino

This section presents results from long-term tsunami risk assessments for Tofino. The duration of each stochastic event set and the simulated number of stochastic event sets are fixed at 1 year and 10 million. The elapsed time since the last major event (i.e., the calendar year of the tsunami risk assessment) varies from 300 to 400 years to investigate the effects of time passed since the last event on the tsunami risk metrics and their distributions. More specifically, as of the Year 2023, the consideration of the elapsed times equal to 323, 349, 374, and 399 years corresponds to the present time, the Years 2049 − 50, 2074 − 75, and 2099 − 2100, respectively. For comparison, the time-independent Poisson model (with the exponential interarrival time distribution) is adopted as a benchmark as it is most often considered in probabilistic tsunami hazard and risk analysis. The tsunami simulation results for different baseline sea levels are first discussed. Subsequently, the effects of time-dependent earthquake hazards on the tsunami risk curves are investigated. Finally, the effects of the different relative sea-level rise scenarios are investigated by focusing on the exceedance probability curves of the risk metrics.

23.4.1 Effects of baseline sea levels

Before discussing the overall tsunami risk impacts due to the elapsed times and relative sea-level rise scenarios, it is important to appreciate the effects of the increased baseline sea levels on the tsunami inundation extent and their consequences on the tsunami risk metrics. Fig. 23.7 shows the tsunami inundation maps of Tofino for the two tsunami source models (50th percentile and 90th percentile) for the M_w 9.0−9.1 magnitude range by considering different baseline tidal levels of 0, 1, and 2 m. The two source models are identified based on the conditional probability distribution of the aggregate tsunami loss of the buildings in Tofino (see the black curve (left) or the blue circles (right) in Fig. 23.6C). The riskier tsunami scenario (i.e., 90th percentile source model) is associated with significantly more severe tsunami

534 SECTION | 2 Advanced topics and applications related to probabilistic tsunami hazard and risk analysis

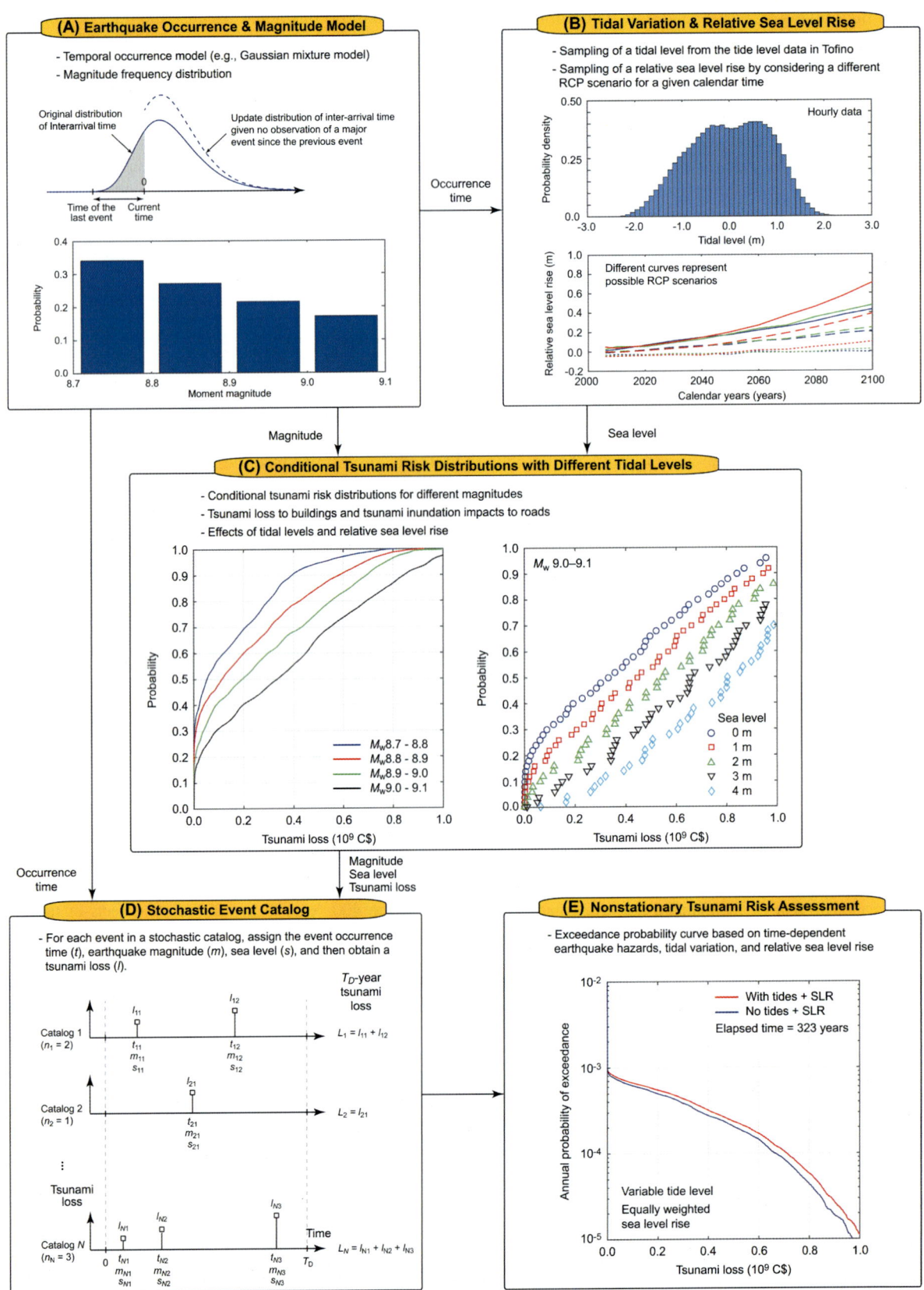

FIGURE 23.6 Long-term tsunami risk assessment procedure: (A) earthquake occurrence and magnitude model; (B) tidal variation and relative sea-level rise; (C) conditional tsunami risk distributions for different baseline tidal levels; (D) stochastic event catalog; and (E) nonstationary tsunami risk assessment.

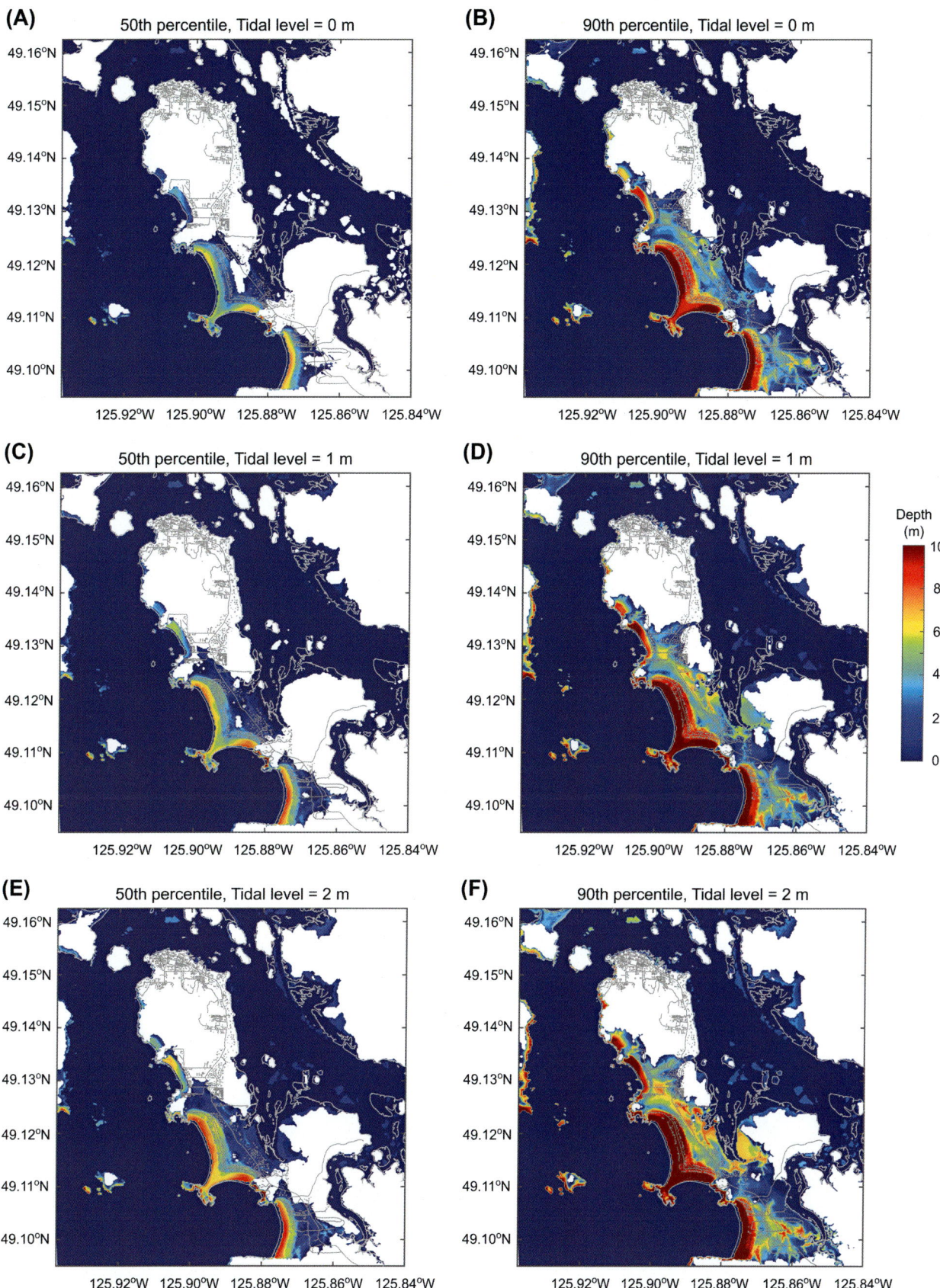

FIGURE 23.7 Tsunami inundation maps of Tofino for the two tsunami source models (50th percentile [A, C, E] and 90th percentile [B, D, F]) of the M_w 9.0–9.1 scenario by considering different baseline tidal levels of 0, 1, and 2 m. Map lines delineate study areas and do not necessarily depict accepted national boundaries.

inundation (i.e., flow depth and spatial extent) in Tofino than the median scenario. Under the median scenario, severe inundation occurs only in the low-lying part of Tofino, and the main road is mostly intact. In contrast, under the 90th percentile tsunami scenario, the low-lying part of Tofino will be completely inundated by the tsunami, and major damage to the main road may occur; the road may be impassable without major restoration work due to tsunami debris and foundation scouring. Moreover, the increased baseline sea level worsens the tsunami inundation risks to the buildings in the low-lying part of Tofino. For instance, the +2 m sea-level scenario of the 50th percentile tsunami source (Fig. 23.7E) will result in inundation situations similar to the 90th percentile tsunami scenario under the mean sea-level condition (Fig. 23.7B). Therefore, the considerations of both effects (i.e., variability of tsunami source scenarios and sea-level scenarios) are important.

To present the comprehensive results of the conditional tsunami risk assessments for Tofino, Fig. 23.8 shows the cumulative distribution functions of aggregate building tsunami loss by considering different magnitude ranges and

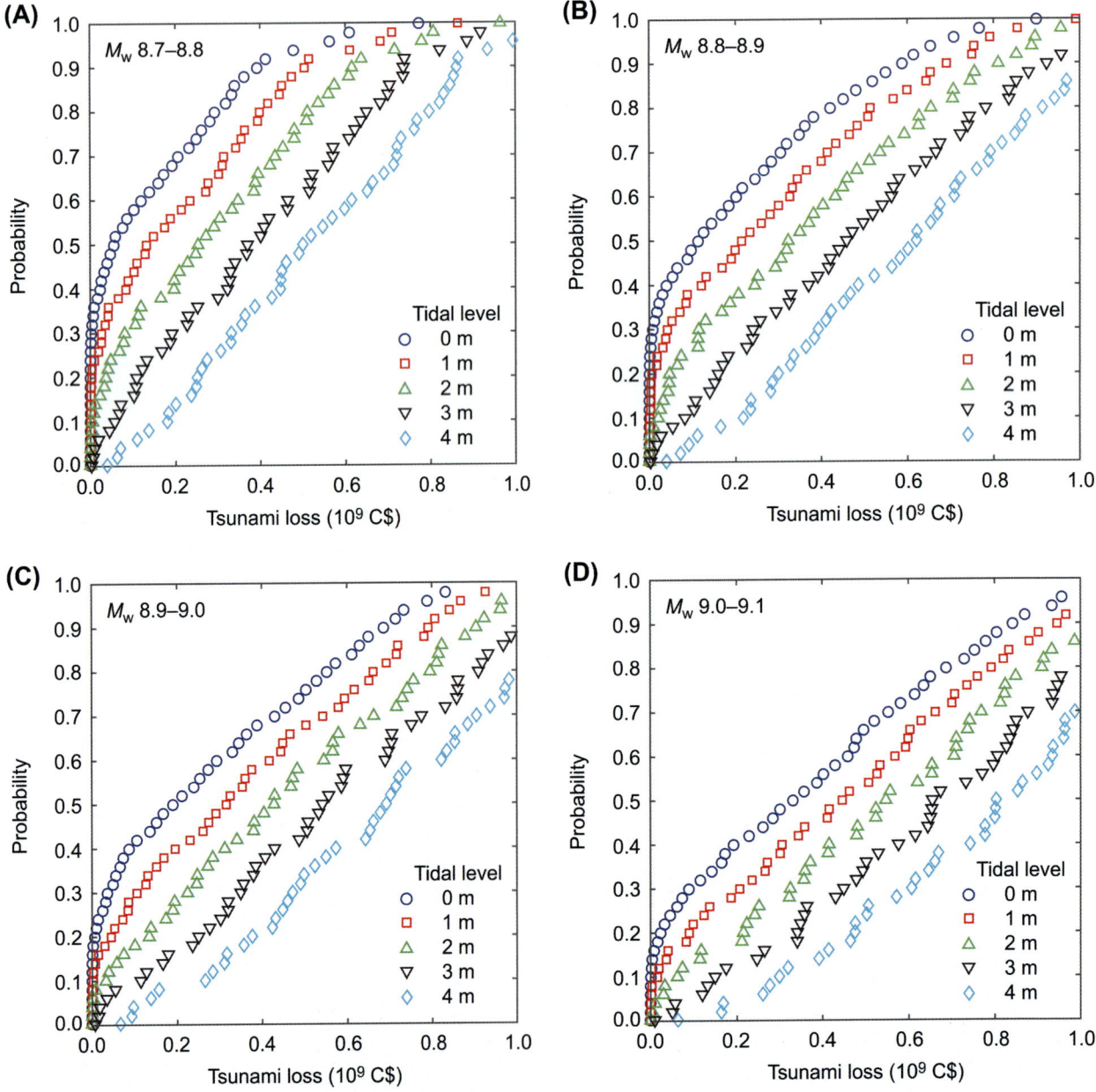

FIGURE 23.8 Cumulative distribution functions of aggregate building tsunami loss in Tofino by considering different baseline tidal levels of 0, 1, 2, 3, and 4 m: (A) M_w 8.7–8.8 scenario; (B) M_w 8.8–8.9 scenario; (C) M_w 8.9–9.0 scenario; and (D) M_w 9.0–9.1 scenario.

different baseline tidal levels of 0, 1, 2, 3, and 4 m. The general increasing trends of the conditional tsunami risk distributions with the increased baseline tidal levels are evident. At the higher baseline tidal levels, there are some abrupt changes in the tsunami loss values. These are the results of focusing on tsunami risks instead of conventional tsunami inundation depths or other parameters and are affected by several factors, including the spatial distribution of buildings, building asset values, and local topography.

23.4.2 Tsunami risk curves for different elapsed times and relative sea-level rise scenarios

The time-dependent tsunami risk analysis is iterated by considering a range of the elapsed times since the last major event. To make the analyzed cases relevant to actual tsunami risk management in Tofino, the elapsed time varies from 300 to 400 years with 5-year intervals corresponding to Years 2000 and 2100, respectively. The tidal level is set to 0 m. The obtained exceedance probability loss curves are shown in Fig. 23.9; the curves for different elapsed times or calendar years are displayed with different colors. The results displayed in Fig. 23.9 demonstrate the evolutionary aspects of tsunami risks over the 21st century.

Time-dependent earthquake hazards and nonstationary sea levels affect the tsunami risk curves. To investigate these combined effects quantitatively, exceedance probability curves of aggregate building tsunami loss in Tofino are compared in Fig. 23.10 by considering four elapsed times, that is, 323, 349, 374, and 399 years. For each elapsed time, two nonstationary tsunami risk cases and two time-independent tsunami risk cases are considered. The two nonstationary cases are based on time-dependent earthquake hazards (Fig. 23.4) by considering and ignoring tidal variations and relative sea-level rises (Figs. 23.2 and 23.3). When the relative sea-level rise is considered, three RCP scenarios are regarded as equally likely, while the low, median, and high scenarios are weighted with 1/6, 4/6, and 1/6 (Miller & Rice, 1983). The two time-independent tsunami risk cases are based on the Poisson earthquake occurrence model (see Fig. 23.4C) with 0 and 1 m baseline tide levels. Note that the time-independent tsunami risk curves shown in the figure panels of Fig. 23.10 are the same (i.e., independent of elapsed time) and thus serve as a benchmark for comparison.

First, by focusing on the present situation (Fig. 23.10A), considering the nonstationary and uncertain tidal effects results in greater aggregate tsunami risk in Tofino. The tsunami loss increase due to the tidal effects is typically between 5% and 10%. The impact is mainly due to the tide variations (Fig. 23.2) rather than the relative sea-level rise (Fig. 23.3). The nonstationary tsunami risk curves are lower than the time-independent tsunami risk curve under the mean sea-level condition. This comparison indicates that the conventional assumption of adopting the Poisson earthquake occurrence model results in the overestimation of the impact of the tsunami risk on Tofino. On the other hand, from a tsunami risk management perspective, this assumption is acceptable as it is sufficiently conservative with respect to more accurate and refined cases.

With the progressing time (Fig. 23.10B−D compared to Fig. 23.10A), the time-dependent tsunami risk curves increase because of the larger hazard rate for the Cascadia subduction events. At the elapsed time of 399 years, the

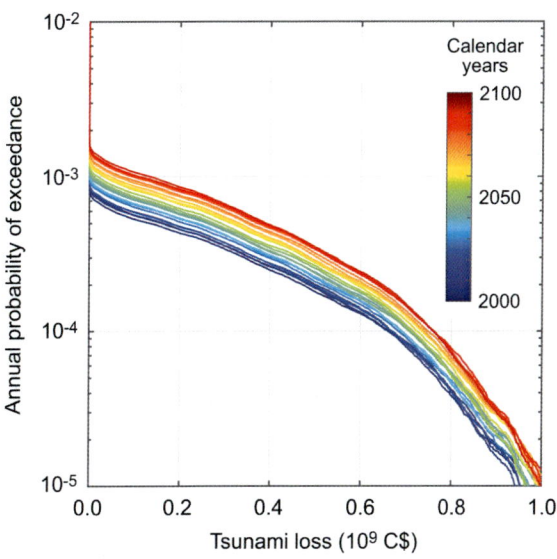

FIGURE 23.9 Exceedance probability tsunami loss curves for the time-dependent hazards for different elapsed times since the last major event in 1700. The different colors correspond to different calendar years or elapsed times. The tidal level is set to 0 m.

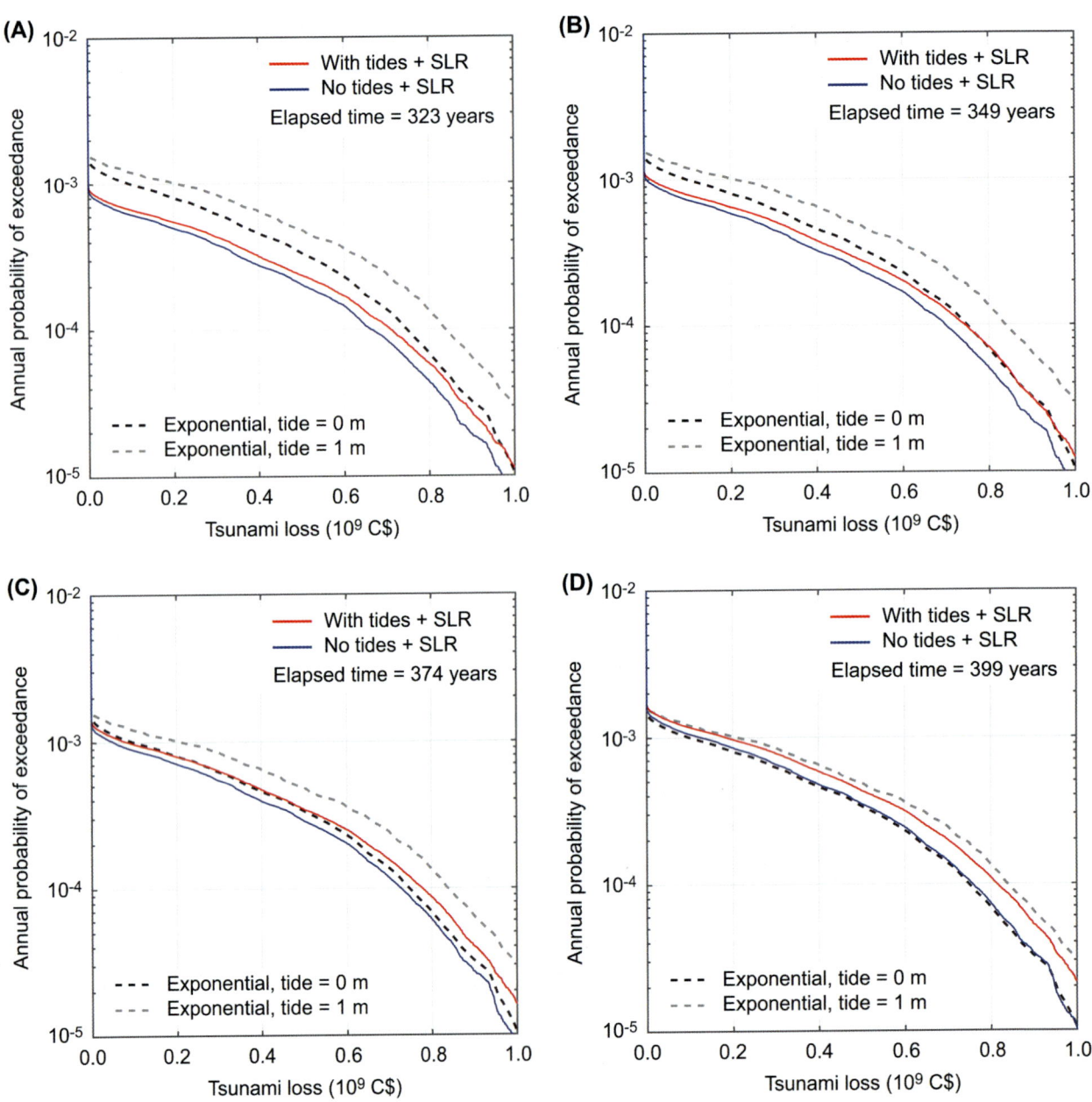

FIGURE 23.10 Comparison of exceedance probability curves of aggregate building tsunami loss in Tofino by considering and ignoring tidal variations and relative SLR: (A) elapsed time = 323 years; (B) elapsed time = 349 years; (C) elapsed time = 374 years; and (D) elapsed time = 399 years. As a benchmark, the exceedance probability curves for time-independent earthquake hazards (exponential interarrival time) with two baseline tidal levels of 0 and 1 m (*black* and *gray* broken curves) are included in the figures. *SLR*, Sea-level rise.

annual probability of experiencing the megathrust Cascadia subduction earthquake based on the Gaussian mixture model exceeds that based on the Poisson model (Fig. 23.4C). These effects can be observed more easily by examining the intersection points of the time-dependent tsunami risk curves with the vertical axis at zero tsunami loss. In addition to the evolutionary impacts of time-dependent tsunami hazards (Fig. 23.9), the effects of considering nonstationary relative sea-level rise become greater (i.e., differences between the blue and red curves). When the nonstationary tsunami risk curves are compared with the time-independent curves, at the end of the 21st century, the conventional assumption of the Poisson model and the mean sea-level condition are no longer conservative. More specifically, for the elapsed time of 399 years (Fig. 23.10D), the tsunami risk curves for the time-dependent earthquake hazard without tidal and sea-level rise effects (blue curve) and the time-independent earthquake hazard with the mean sea level (black broken

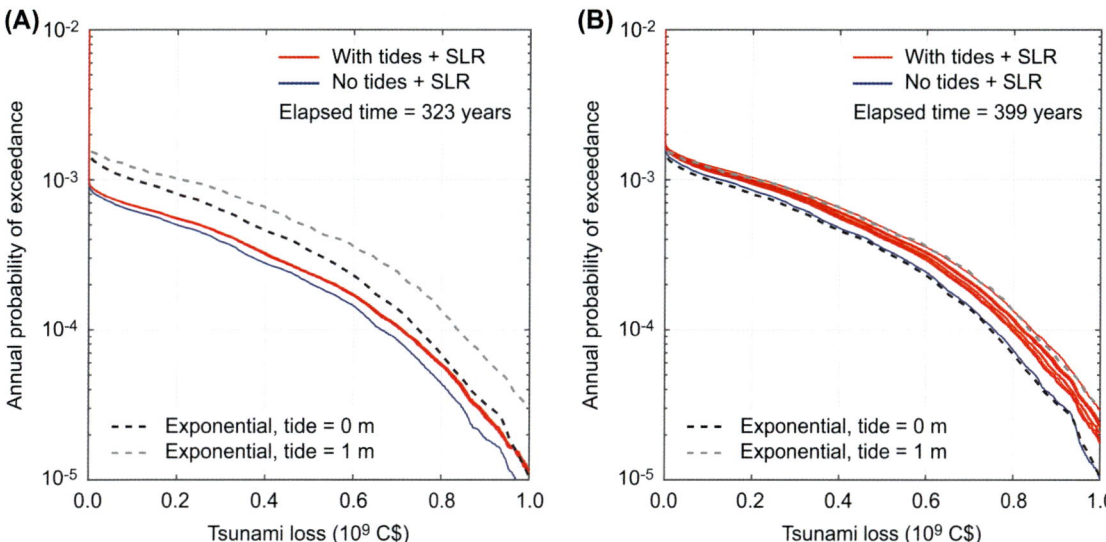

FIGURE 23.11 Comparison of exceedance probability curves of aggregate building tsunami loss in Tofino by considering tidal variations and nine relative SLR scenarios (*red* curves): (A) elapsed time = 323 years and (B) elapsed time = 399 years. As a benchmark, the exceedance probability curves for time-independent earthquake hazards (exponential interarrival time) with two baseline tidal levels of 0 and 1 m (*black* and *gray* broken curves), as well as the exceedance probability curve for time-dependent earthquake hazards (Gaussian mixture interarrival time) are included in the figures. *SLR*, Sea-level rise.

curve) are nearly identical (see Fig. 23.4C). On the other hand, the tsunami risk curve for the time-dependent earthquake hazard with tidal and sea-level rise effects (red curve) is slightly below the tsunami risk curve for the time-independent earthquake hazard with +1 m sea level. These results are relevant for emergency managers and city planners in coastal municipalities and local governments in British Columbia for taking actions related to long-term tsunami risk management.

Previously, the nonstationary tsunami risk curves for the tidal and sea-level rise situations are exemplified for the weighted case of the nine relative sea-level projections. It is important to evaluate the tsunami risk curves for different projections individually, especially for future situations. For this purpose, Fig. 23.11 shows the comparison of the tsunami risk curves by considering the elapsed times of 323 years (present time) and 399 years (Year 2100) in terms of aggregate building tsunami loss in Tofino. In each figure, nine tsunami risk curves with red color are included, each corresponding to one of the nine relative sea-level rise scenarios (Fig. 23.3). As expected, no noticeable variations are observed for the case of the present situation (Fig. 23.11A); the variability is originated from the Monte Carlo sampling. On the other hand, variations of the risk curves for Year 2100 are manifested in Fig. 23.11B. The risk curve with the pessimistic projection scenario RCP8.5-U, which predicts the relative sea-level rise of 0.7 m in the Year 2100, exceeds the time-independent Poisson case with +1 m baseline tide. The results shown in Fig. 23.11 are valuable from the long-term tsunami risk management.

23.5 Conclusions

This study developed a novel long-term tsunami risk model for Tofino, which is subjected to significant tsunami threats from the Cascadia subduction earthquakes. The evolutionary tsunami risk profiles of Tofino over the 21st century underscore the importance of considering nonstationary processes for future tsunami risk assessments. The conventional assumptions for tsunami hazard and risk assessments, typically based on deterministic tsunami source scenarios, time-independent hazard processes, and ad hoc tidal conditions, may not serve for all future situations that the existing and future built environment will experience. The long-term tsunami risk results for Tofino offer valuable insights related to tsunami risk management and emergency preparedness by local municipalities and governments.

Another critical evolutionary aspect of the tsunami risk for Tofino, which is not reflected in the current study, is the changing exposure characteristics. The population in Tofino has grown five times, from approximately 500 in the early 1970s to 2500 as of 2021 with annual growth rates between 2% and 3%. In addition, the town's main industry has transformed from a resource-oriented economy (e.g., logging, mining, and fishery) to a tourism-oriented economy (e.g., eco-tours, nature cruises, and surfing). The number of tourists and seasonal workers has increased significantly (the peak summer population exceeds 10,000). It is also essential to be aware that the increased population, when the available habitable land is limited, can stress the existing infrastructure, such as roads and water supply systems. The criticality of the infrastructures becomes more important in a situation of catastrophic events, such as megathrust tsunamis.

In the future, the quantitative tsunami risk assessments should be extended in several major ways. Firstly, shaking risks due to the Cascadia mainshock and aftershocks need to be included (Goda, 2024). The shaking damage precedes the arrival of tsunami waves; therefore, considering damage accumulation due to shaking-tsunami sequences is necessary. When megathrust earthquakes occur near Tofino, the ground level is likely to subside due to the coseismic deformation (by 1 to 2 m). This effect will cause compounding impacts on the risks posed by other coastal hazards, such as storm surges and coastal flooding. Future disaster risk impact assessments should address the issues related to cascading and compounding risks by considering hazard interaction and dynamic vulnerability in the multihazard context (De Risi et al., 2022).

References

Abaimov, S. G., Turcotte, D. L., Shcherbakov, R., Rundle, J. B., Yakovlev, G., Goltz, C., & Newman, W. I. (2008). Earthquakes: Recurrence and interoccurrence times. *Pure and Applied Geophysics*, 165(3-4), 777–795. Available from https://doi.org/10.1007/s00024-008-0331-y.

Akiyama, M., Frangopol, D. M., & Ishibashi, H. (2020). Toward life-cycle reliability-, risk- and resilience-based design and assessment of bridges and bridge networks under independent and interacting hazards: Emphasis on earthquake, tsunami and corrosion. *Structure and Infrastructure Engineering*, 16(1), 26–50. Available from https://doi.org/10.1080/15732479.2019.1604770, http://www.tandf.co.uk/journals/titles/15732479.asp.

Alhamid, A. K., Akiyama, M., Aoki, K., Koshimura, S., & Frangopol, D. M. (2022). Stochastic renewal process model of time-variant tsunami hazard assessment under nonstationary effects of sea-level rise due to climate change. *Structural Safety*, 99. Available from https://doi.org/10.1016/j.strusafe.2022.102263, http://www.elsevier.com/inca/publications/store/5/0/5/6/6/4/.

Atwater, B., Musumi-Rokkaku, S., Satake, K., Tsuji, Y., Ueda, K., & Yamaguchi, D. K. (2015). *The orphan tsunami of 1700—Japanese clues to a parent earthquake in North America* (144). University of Washington Press.

Behrens, J., Løvholt, F., Jalayer, F., Lorito, S., Salgado-Gálvez, M. A., Sørensen, M., Abadie, S., Aguirre-Ayerbe, I., Aniel-Quiroga, I., Babeyko, A., Baiguera, M., Basili, R., Belliazzi, S., Grezio, A., Johnson, K., Murphy, S., Paris, R., Rafliana, I., De Risi, R., ... Vyhmeister, E. (2021). Probabilistic tsunami hazard and risk analysis: A review of research gaps. *Frontiers in Earth Science*, 9. Available from https://doi.org/10.3389/feart.2021.628772, https://www.frontiersin.org/journals/earth-science.

Church, J. A., Clark, P. U., Cazenave, A., Gregory, J. M., Jevrejeva, S., Levermann, A., Merrifield, M. A., Milne, G. A., Nerem, R. S., Nunn, P. D., Payne, A. J., Pfeffer, W. T., Stammer, D., & Unnikrishnan, A. S. (2013). *Sea level change climate change. The physical science basis. Contribution of Working Group I to the fifth assessment report of the intergovernmental panel on climate change*. Cambridge University Press. Available from https://www.ipcc.ch/site/assets/uploads/2018/02/WG1AR5_Chapter13_FINAL.pdf.

DeMets, C., Gordon, R. G., & Argus, D. F. (2010). Geologically current plate motions. *Geophysical Journal International*, 181(1), 1–80. Available from https://doi.org/10.1111/j.1365-246X.2009.04491.x.

De Risi, R., Goda, K., Yasuda, T., & Mori, N. (2017). Is flow velocity important in tsunami empirical fragility modeling? *Earth-Science Reviews*, 166, 64–82. Available from https://doi.org/10.1016/j.earscirev.2016.12.015, http://www.sciencedirect.com/science/journal/00128252.

De Risi, R., Muhammad, A., De Luca, F., Goda, K., & Mori, N. (2022). Dynamic risk framework for cascading compounding climate-geological hazards: A perspective on coastal communities in subduction zones. *Frontiers in Earth Science*, 10. Available from https://doi.org/10.3389/feart.2022.1023018, https://www.frontiersin.org/journals/earth-science.

Fukutani, Y., Moriguchi, S., Terada, K., & Otake, Y. (2021). Time-dependent probabilistic tsunami inundation assessment using mode decomposition to assess uncertainty for an earthquake scenario. *Journal of Geophysical Research: Oceans*, 126(7). Available from https://doi.org/10.1029/2021JC017250, http://agupubs.onlinelibrary.wiley.com/agu/jgr/journal/10.1002/(ISSN)2169-9291/.

Gao, D., Wang, K., Insua, T. L., Sypus, M., Riedel, M., & Sun, T. (2018). Defining megathrust tsunami source scenarios for northernmost Cascadia. *Natural Hazards*, 94(1), 445–469. Available from https://doi.org/10.1007/s11069-018-3397-6, http://www.wkap.nl/journalhome.htm/0921-030X.

Goda, K. (2019). Time-dependent probabilistic tsunami hazard analysis using stochastic rupture sources. *Stochastic Environmental Research and Risk Assessment*, 33(2), 341–358. Available from https://doi.org/10.1007/s00477-018-1634-x, http://link.springer-ny.com/link/service/journals/00477/index.htm.

Goda, K. (2022). Stochastic source modeling and tsunami simulations of Cascadia subduction earthquakes for Canadian Pacific coast. *Coastal Engineering Journal*, 64(4), 575–596. Available from https://doi.org/10.1080/21664250.2022.2139918, https://www.tandfonline.com/loi/tcej20.

Goda, K. (2023). Probabilistic tsunami hazard analysis for Vancouver Island coast using stochastic rupture models for the Cascadia subduction earthquakes. *GeoHazards*, 4(3), 217–238. Available from https://doi.org/10.3390/geohazards4030013, https://www.mdpi.com/journal/geohazards.

Goda, K. (2024). Probabilistic earthquake-tsunami financial risk evaluation for the District of Tofino, British Columbia, Canada. *Georisk*. Available from https://doi.org/10.1080/17499518.2024.2346656, http://www.tandf.co.uk/journals/titles/17499518.asp.

Goda, K., Orchiston, K., Borozan, J., Novakovic, M., & Yenier, E. (2023). Evaluation of reduced computational approaches to assessment of tsunami hazard and loss using stochastic source models: Case study for Tofino, British Columbia, Canada, subjected to Cascadia megathrust earthquakes. *Earthquake Spectra*, 39(3), 1303–1327. Available from https://doi.org/10.1177/87552930231187407, https://journals.sagepub.com/home/eqs.

Goda, K., & De Risi, R. (2017). Probabilistic tsunami loss estimation methodology: Stochastic earthquake scenario approach. *Earthquake Spectra*, *33* (4), 1301−1323. Available from https://doi.org/10.1193/012617EQS019M, http://earthquakespectra.org/doi/pdf/10.1193/012617EQS019M.

Goldfinger, C., Nelson, C. H., Morey, A. E., Johnson, J. E., Patton, J., Karabanov, E., Gutierrez-Pastor, J., Eriksson, A. T., Gracia, E., Dunhill, G., Enkin, R. J., Dallimore, A., & Vallier, T. (2012). *Turbidite event history: Methods and implications for Holocene paleoseismicity of the Cascadia subduction zone. Professional Paper 1661-F*. Geological Survey. Available from https://pubs.usgs.gov/publication/pp1661F, http://doi.org/10.3133/pp1661F.

Goto, C., Ogawa, Y., Shuto, N., & Imamura, F. (1997). *Numerical method of tsunami simulation with the leap-frog scheme IOC Manual*. UNESCO. Available from https://www.jodc.go.jp/info/ioc_doc/Manual/122367eb.pdf.

Griffin, J. D., Stirling, M. W., & Wang, T. (2020). Periodicity and clustering in the long-term earthquake record. *Geophysical Research Letters*, *47* (22). Available from https://doi.org/10.1029/2020GL089272, http://agupubs.onlinelibrary.wiley.com/hub/journal/10.1002/(ISSN)1944-8007/.

Hayes, G. P., Moore, G. L., Portner, D. E., Hearne, M., Flamme, H., Furtney, M., & Smoczyk, G. M. (2018). Slab2, a comprehensive subduction zone geometry model. *Science (New York, N.Y.)*, *362*(6410), 58−61. Available from https://doi.org/10.1126/science.aat4723, http://science.sciencemag.org/content/362/6410/58/tab-pdf.

Headquarters for Earthquake Research Promotion. (2019). *Evaluation of long-term probability of earthquake occurrence along the Japan trench*.

Hobbs, T., Journeay, J. M., & LeSueur, P. (2021). *Developing a retrofit scheme for Canada's seismic risk model open file 8822*. Geological Survey of Canada. Available from https://ostrnrcan-dostrncan.canada.ca/entities/publication/549fa5f3-dce8-4f61-80be-510a5c5d7ebd, 10.4095/328860.

James, T. S., Robin, C., Henton, J. A., & Craymer, M. (2021). *Relative sea-level projections for Canada based on the IPCC fifth assessment report and the NAD83v70VG national crustal velocity model open file 8764*. Geological Survey of Canada. Available from https://geoscan.nrcan.gc.ca/text/geoscan/fulltext/of_8764.pdf, 10.4095/327878.

Kulkarni, R., Wong, I., Zachariasen, J., Goldfinger, C., & Lawrence, M. (2013). Statistical analyses of great earthquake recurrence along the Cascadia subduction zone. *Bulletin of the Seismological Society of America*, *103*(6), 3205−3221. Available from https://doi.org/10.1785/0120120105, http://www.bssaonline.org/content/103/6/3205.full.pdf + html, United States.

Li, L., Switzer, A. D., Wang, Y., Chan, C. H., Qiu, Q., & Weiss, R. (2018). A modest 0.5-m rise in sea level will double the tsunami hazard in Macau. *Science Advances*, *4*(8). Available from https://doi.org/10.1126/sciadv.aat1180, http://advances.sciencemag.org/content/4/8/eaat1180.

Mai, P. M., & Beroza, G. C. (2002). A spatial random field model to characterize complexity in earthquake slip. *Journal of Geophysical Research: Solid Earth*, *107*(B11). Available from 10.1029/2001jb000588.

Melgar, D., Williamson, A. L., & Salazar-Monroy, E. F. (2019). Differences between heterogenous and homogenous slip in regional tsunami hazards modelling. *Geophysical Journal International*, *219*(1), 553−562. Available from https://doi.org/10.1093/gji/ggz299, http://gji.oxfordjournals.org/.

Miller, A. C., & Rice, T. R. (1983). Discrete approximations of probability distributions. *Management Science*, *29*(3), 352−362. Available from https://doi.org/10.1287/mnsc.29.3.352.

Mitchell-Wallace, K., Jones, M., Hillier, J., & Foote, M. (2017). *Natural catastrophe risk management and modelling: A practitioner's guide*. Wiley-Blackwell.

Miyashita, T., Mori, N., & Goda, K. (2020). Uncertainty of probabilistic tsunami hazard assessment of Zihuatanejo (Mexico) due to the representation of tsunami variability. *Coastal Engineering Journal*, 413−428. Available from https://doi.org/10.1080/21664250.2020.1780676, https://www.tandfonline.com/loi/tcej20.

Okada, Y. (1985). Surface deformation due to shear and tensile faults in a half-space. *Bulletin of the Seismological Society of America*, *75*(4), 1135−1154. Available from https://doi.org/10.1785/bssa0750041135.

Park, H., Cox, D. T., & Barbosa, A. R. (2017). Comparison of inundation depth and momentum flux based fragilities for probabilistic tsunami damage assessment and uncertainty analysis. *Coastal Engineering*, *122*, 10−26. Available from https://doi.org/10.1016/j.coastaleng.2017.01.008, http://www.elsevier.com/inca/publications/store/5/0/3/3/2/5/.

Robin, C. M. I., Craymer, M., Ferland, R., James, T. S., Lapelle, E., Piraszewski, M., & Zhao, Y. (2020). NAD83v70VG: A new national crustal velocity model for Canada. *Geomatics Canada*. Available from https://doi.org/10.4095/327592.

Sepúlveda, I., Haase, J. S., Liu, P. L. F., Grigoriu, M., & Winckler, P. (2021). Non-stationary probabilistic tsunami hazard assessments incorporating climate-change-driven sea level rise. *Earth's Future*, *9*(6). Available from https://doi.org/10.1029/2021EF002007, http://onlinelibrary.wiley.com/journal/10.1002/(ISSN)2328-4277.

Sykes, L. R., & Menke, W. (2006). Repeat times of large earthquakes: Implications for earthquake mechanics and long-term prediction. *Bulletin of the Seismological Society of America*, *96*(5), 1569−1596. Available from https://doi.org/10.1785/0120050083.

Takabatake, T., St-Germain, P., Nistor, I., Stolle, J., & Shibayama, T. (2019). Numerical modelling of coastal inundation from Cascadia subduction zone tsunamis and implications for coastal communities on western Vancouver Island, Canada. *Natural Hazards*, *98*(1), 267−291. Available from https://doi.org/10.1007/s11069-019-03614-3, http://www.wkap.nl/journalhome.htm/0921-030X.

Tanioka, Y., & Satake, K. (1996). Tsunami generation by horizontal displacement of ocean bottom. *Geophysical Research Letters*, *23*(8), 861−864. Available from https://doi.org/10.1029/96GL00736, http://onlinelibrary.wiley.com/journal/10.1002/(ISSN)1944-8007/issues?year = 2012.

Tilloy, A., Malamud, B. D., Winter, H., & Joly-Laugel, A. (2019). A review of quantification methodologies for multi-hazard interrelationships. *Earth-Science Reviews*, *196*, 102881. Available from https://doi.org/10.1016/j.earscirev.2019.102881.

United Nations Office for Disaster Risk Reduction. (2022). *Technical guidance on comprehensive risk assessment and planning in the context of climate change*. https://www.undrr.org/publication/technical-guidance-comprehensive-risk-assessment-and-planning-context-climate-change.

Vescovo, R., Adriano, B., Mas, E., & Koshimura, S. (2023). Beyond tsunami fragility functions: Experimental assessment for building damage estimation. *Scientific Reports*, *13*(1). Available from https://doi.org/10.1038/s41598-023-41047-y, https://www.nature.com/srep/.

Walton, M. A. L., Staisch, L. M., Dura, T., Pearl, J. K., Sherrod, B., Gomberg, J., Engelhart, S., Tréhu, A., Watt, J., Perkins, J., Witter, R. C., Bartlow, N., Goldfinger, C., Kelsey, H., Morey, A. E., Sahakian, V. J., Tobin, H., Wang, K., Wells, R., ... Wirth, E. (2021). Toward an integrative geological and geophysical view of Cascadia subduction zone earthquakes. *Annual Review of Earth and Planetary Sciences*, *49*, 367–398. Available from https://doi.org/10.1146/annurev-earth-071620-065605, http://www.annualreviews.org/journal/earth.

Williams, R. T., Davis, J. R., & Goodwin, L. B. (2019). Do large earthquakes occur at regular intervals through time? A perspective from the geologic record. *Geophysical Research Letters*, *46*(14), 8074–8081. Available from https://doi.org/10.1029/2019GL083291, http://agupubs.onlinelibrary.wiley.com/hub/journal/10.1002/(ISSN)1944-8007/.

Chapter 24

Digital twin paradigm for coastal disaster risk reduction and resilience

Shunichi Koshimura[1,2,3], Nobuhito Mori[4], Naotaka Chikasada[5], Keiko Udo[3], Junichi Ninomiya[6], Yoshihiro Okumura[7] and Erick Mas[1,2,3]

[1]*International Research Institute of Disaster Science, Tohoku University, Sendai, Miyagi, Japan,* [2]*RTi-cast, Inc., Sendai, Miyagi, Japan,* [3]*Graduate School of Engineering, Tohoku University, Sendai, Miyagi, Japan,* [4]*Disaster Prevention Research Institute, Kyoto University, Uji, Kyoto, Japan,* [5]*National Research Institute for Earth Science and Disaster Resilience, Tsukuba, Ibaraki, Japan,* [6]*Institute of Science and Engineering, Kanazawa University, Kanazawa, Ishikawa, Japan,* [7]*Faculty of Societal Safety Sciences, Kansai University, Osaka, Osaka, Japan*

24.1 Introduction

The "digital twin" has been demonstrated in various fields of academia and industries, such as aerospace (Li et al., 2022), manufacturing (Lu et al., 2020; Tao et al., 2018), and earth sciences (European Commission, 2020) with the advances of sensing methods, large-scale and real-time monitoring data, high-performance computing and simulation methods, and data-driven sciences with machine learning. A digital twin is a technological system that copies (i.e., creates a virtual replica) the physical world's system from various sensors to the cyber world (in a computer) and executes simulations to forecast the sequences and consequences of target phenomena. Users have a complete view of the target through real-time feedback for their policy designs, responses, and adaptations. For example, in the aerospace industry, the digital twin is utilized as a virtual counterpart of physical aircraft of aerospace systems to perform design optimization, predictive maintenance, fuel efficiency, and aviation safety. In civil engineering, the actual application of digital twins is still mostly at the prototype stage, and there is a strong need for developing common procedures and standards tailored to the plan, design, construction procedures, and use cases (Pregnolato et al., 2022). In the field of urban planning, Toyota Motor Corporation envisions a new type of smart city by constructing a "Woven City" based on a digital twin paradigm (Toyoda, 2020). They are exploring new possibilities in the mobility of information, goods, and people through this grand project to create a new virtual city where automated driving, mobility as a service, personal mobility, robotics, smart home technology, and artificial intelligence (AI) to be verified in a realistic environment of people's lives. In Singapore, the entire national territory has been converted into a 3D virtual twin to visualize urban information in real time (Gobeawan et al., 2018; National Research Foundation, 2023). This country's authoritative platform can be used in simulations and virtual tests of new solutions to urban planning problems.

In short, digital twins use real-world operational data to run simulations and predict possible outcomes and consequences. By leveraging other technologies, such as AI, internet of things, and immersive technologies, such as augmented reality, virtual reality (VR), and mixed reality, digital twins bridge the gap between the digital and physical worlds.

Given the importance of the implications of constructing digital twin for various fields, we present a digital twin paradigm that facilitates the study, design, and search for optimal solutions to various coastal and marine disaster issues in the virtual world, which would be impossible in the physical world due to time and cost constraints, thereby creating a new horizon of coastal and marine disaster sciences.

24.2 Concept of coastal digital twin

Among the research community of coastal engineering, a project has been launched to establish a digital twin paradigm and to enhance coastal community's resilience by constructing "coastal digital twins (CDTs)." For example, CDTs can

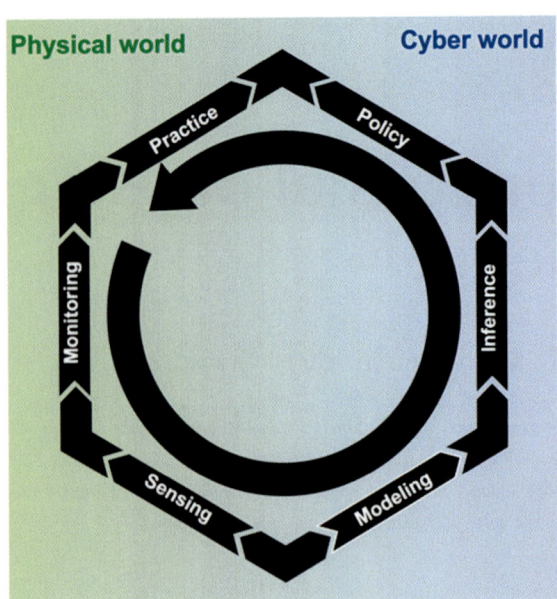

FIGURE 24.1 Framework of a cyber-physical system of a coastal digital twin and its feedback loop.

be defined as creating virtual replicas of coastal environments to analyze and evaluate various aspects of coastal issues, such as sea-level change, coastal erosion, and disaster risks, to develop efficient policies or adaptation measures.

Fig. 24.1 illustrates the critical elements of CDTs that are referred to as a "Hexagon" concept: "Sensing" and "Monitoring" are the deployment of sensors and the collection of real-time data to capture the physical world's environments, objects, or phenomena to improve our understanding and interpretation of the real environment. The outcomes of sensing can be inputs for "Modeling", like model parameter calibration, initial and boundary conditions, assimilations, and validations. "Modeling" capability provides forecasts of the target phenomena to foresee the effects, impacts, and consequences to the end-users. A data-driven approach including the application of machine learning combined with modeling would provide advanced "Inference" to be utilized for "Policy" design considering possible policies, such as response and adaptation measures. Social dynamics simulations in the cyber world can play an important role in policymaking and evaluation by capturing complex social, economic, and environmental interactions; hence, sensing and continuous monitoring must be combined with social dynamics simulations. Many examples of social dynamics simulations can be utilized in coastal policy contexts, including land-use planning and zoning (Lloyd et al., 2013), economic impact assessment, resilience and adaptation strategies (Xu et al., 2021), crisis and emergency management (Mas et al., 2022), and urban planning (Becu et al., 2017). Some of the possible policies addressed in the cyber world will be adopted in the physical world. Thus "Practice" is defined as efforts or actions gaining feedback from the cyber world to implement possible policies and actions. Practitioners and policymakers explore different scenarios, assess potential outcomes, and make informed decisions to enhance the sustainability and resilience of coastal communities. Those practices need to be monitored continuously and evaluated in evidence-based policymaking using rigorous, empirical evidence to inform and guide policy development, implementation, and evaluation. In short, the CDT is defined as a mixed initiative of a feedback loop made by the fusion of these components to gain knowledge and insights for optimal solutions in the physical world.

24.3 Tsunami disaster digital twin in Japan

A new digital twin project is underway to enhance the resilience of disaster response systems by creating a "tsunami disaster digital twin (TDDT)" to support the disaster response team in the anticipated tsunami disaster (Cabinet Office of the Government of Japan, 2023). The TDDT platform consists of a fusion of real-time hazard simulations, for example, tsunami inundation forecast, social sensing to identify the exposed population dynamically, and multiagent simulation of disaster response activities to find an optimal allocation or strategy of response efforts and to achieve the enhancement of disaster resilience. To attain the goal of innovating digital twin computing for enhancing disaster resilience, this chapter describes four objectives of the TDDT.

1. Developing a nationwide real-time tsunami inundation and damage forecast system. The priority target for forecasting is the Pacific coast of Japan, a region where a megathrust earthquake is likely to occur.
2. Establishing a real-time estimation of the number of exposed populations in the inundation zone and medical demand for the exposed population.
3. Developing a multiagent simulation of response activities in the affected areas with the use of damage information, relief demands, and resources to find optimal allocation and strategies of the response efforts.
4. Developing a digital twin computing platform to support disaster response activities through the integration of data, hazard simulation, and multiagent systems.

Fig. 24.2 illustrates the framework and functionality of the TDDT as a cyber-physical system. TDDT captures a disaster process in the physical world and creates a digital copy (twin) in the cyber world. Running simulations with the copied data to forecast consequences, acquire optimal solutions, and provide insights into the physical world's policy or decision. The technological elements are defined by the hexagon: sensing, monitoring, policymaking in the physical world, simulation, quantifying functionality loss and resilience, and policy evaluation in the cyber world.

Integrating sensing and simulation provides real-time estimates of disaster processes and consequences. These outputs will be utilized to quantify the functionality loss of the society, for example, human activities (Okajima et al., 2013), lifelines, medical services (Mas et al., 2022), traffic (Aoki et al., 2022; Ishibashi et al., 2020), and supply chains (Hachiya et al., 2022), using the resilience function F (Fig. 24.2). Continuous monitoring emphasizes the importance of quantifying resilience to capture the status of social systems and to decide how the resilience functions recover and how the targets and goals can be set for the recovery and reconstruction with various possible policies and countermeasures.

Given the consequences and possible response policies and measures, social response analysis and simulations in cyberspace are performed to foresee the consequences of implementing policies (Kosaka et al., 2024). This analysis will provide a large number of possible options for policies. The advantage of running simulations and analyses in cyberspace is to simulate the consequences of possible policy implementations through what-if analyses to evaluate their benefits and drawbacks. The evaluations are based on quantifying the resilience metric R by integrating the resilience function (Fig. 24.2) and obtaining the optimal solutions by maximizing R. Those outputs are fed back to the physical world in the form of policy implications and curations of the policy evaluations. The core technological factors and their preliminary achievements are described in the following subsections.

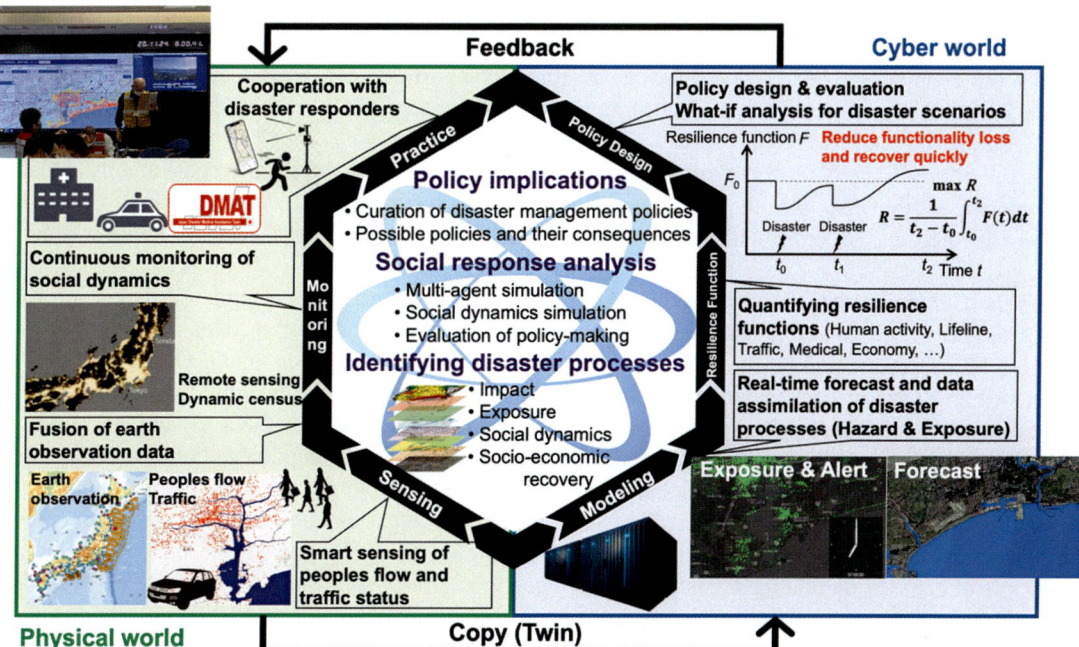

FIGURE 24.2 Framework of a cyber-physical system of the tsunami disaster digital twin.

24.3.1 Real-time tsunami inundation and damage forecast

The advances in real-time tsunami inundation forecast technologies are summarized in the literature (Harold et al., 2020; Koshimura et al., 2024). The 2011 Tohoku Earthquake and Tsunami revealed many problems in Japan's disaster management policies. These have been drastically reformed in the years after the disaster to promote initiatives for building national resilience, creating safe and secure regions, and improving society's strength and flexibility in the face of future disasters (Koshimura & Shuto, 2015; Koshimura et al., 2014). One of the critical challenges in the aftermath of any tsunami is identifying its impact and prioritizing disaster response and relief activities. Because of the widespread damage to infrastructure and communication networks, the impacted regions were hampered in addressing the overall damage, sometimes for months (Koshimura et al., 2014). This experience highlighted the need to develop technologies to forecast the regional impact of tsunamis. Recent advances in high-performance computing and large datasets, which comprise observations, tsunami generation, propagation, and effects, hold out the promise of dramatically improving our understanding of the whole picture of tsunami processes in real time (Koshimura et al., 2017; Mizutani et al., 2020).

Three approaches to real-time tsunami inundation forecasting methods have been proposed. The first is a so-called "data-driven" approach that searches tsunami forecast data from a precomputed database connected with off-shore tsunami observation facilities (Gusman et al., 2014; Igarashi et al., 2016; Makinoshima et al., 2021; Nomura et al., 2022; Yamamoto et al., 2016). The second is a "data assimilation" approach that integrates tsunami wave field data using dense off-shore tsunami observation networks (Gusman et al., 2016; Maeda et al., 2015; Tanioka, 2020; Wang et al., 2018). Both are highly dependent on the configuration of off-shore tsunami sensors. The third is the "real-time forward simulation" approach that runs simulations in real time with given tsunami source models based on seismic and geodetic observations (Koshimura et al., 2017; Musa et al., 2018; Musa et al., 2019; Ohta et al., 2018). The real-time forward approach has the most advantages in simulating tsunami inundation on land if reliable tsunami source model information is obtained. In this subsection, the development of the third model is focused upon.

Koshimura et al. (2020) established a method of real-time tsunami inundation forecasting and damage estimation to predict the impact of a tsunami. The technique has been verified through case studies of the 2011 Tohoku Earthquake and Tsunami regarding its forecasting reliability and capability. After the verification, in 2018, the system started operating as a critical component of the tsunami response system in the Cabinet Office of the Government of Japan. Additionally, a technology start-up, RTi-cast, Inc., is taking a role in offering and operating real-time tsunami inundation damage forecasting services across Japan (Koshimura et al., 2020). Currently, RTi-cast is a unique private tsunami warning provider licensed by the Japan Meteological Agency (JMA).

The vector-parallel supercomputer SX-Aurora is the core simulation architecture for the real-time forward simulation approach. These are installed at both Tohoku University and Osaka University and operated independently to enhance redundancy in an emergency. A simulation management system that achieves an optimal allocation of jobs between nodes has been established. This enables SX-Aurora to support urgent tasks, executing tsunami inundation simulation at the highest priority while suspending other active jobs, automatically resuming these as soon as the urgent simulation is completed.

Currently, the priority targets for forecasting at present cover most parts of the Pacific coast and the Japan Sea coast, where tsunamigenic earthquakes are likely to occur (Fig. 24.3). For instance, the Nankai Trough Earthquake is estimated to occur in the next 30 years with 80% probability, according to long-term evaluations of seismic activity in Japan.

Activating the tsunami propagation and inundation forecasting system requires receipt of seismic information from the JMA's Earthquake Early Warning service, which can happen just a few tens of seconds after initial observation of seismic waves. More precise information is then provided by real-time analysis of GEONET (Global Navigation Satellite System Earth Observation Network System) data that is supposed to be transmitted within 10 minutes after an earthquake occurs. We use the fault rupture estimation derived by the RAPiD (Ohta et al., 2012) and REGARD (Kawamoto et al., 2017) algorithms that include the estimates of the moment magnitude, fault geometry, focal mechanism, and slip distribution. Given the tsunami source information, the system moves on to model tsunami propagation/inundation simulations running on SX-Aurora together with off-shore/coastal tide gauges to determine tsunami travel and arrival times, the extent of the inundation zone, the maximum flow depth distribution, and potential loss. The implemented model is based on nonlinear, shallow-water equations discretized using a staggered, leap-frog finite difference method. Using SX-Aurora, the system completed tsunami source determination in 10 minutes and tsunami inundation modeling in 10 minutes at 10-m grid resolution (Musa et al., 2018; Musa et al., 2019). The simulations are now optimized to be more efficient using a new nested grid system and its message passing interface (MPI) parallelization

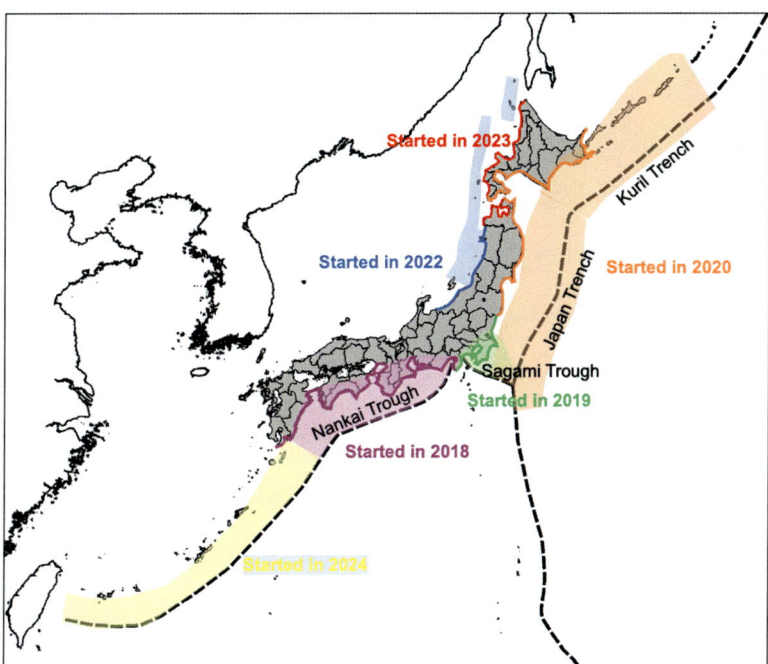

FIGURE 24.3 Tsunami inundation forecast areas in Japan (as of November 2023). Map lines delineate study areas and do not necessarily depict accepted national boundaries.

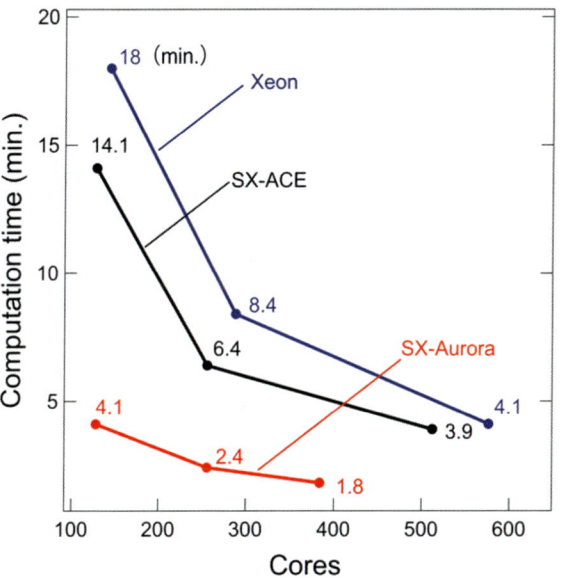

FIGURE 24.4 Performance of real-time tsunami inundation forecasts with multicomputational platforms. The computational time represents the time to complete the 6-h inundation forecast with 10-m grids in Kochi City, Japan.

(Fig. 24.4) and running on other high-performance computing processors to expand its capability to other areas and users.

Identifying a tsunami's impact and determining which areas are most likely to be devastated is critical for disaster response, recovery, and relief activities. Given the maximum flow depth distribution, the system performs Geographic Information System (GIS) analysis to determine the size of the exposed population and structures using census data, then estimates the numbers of potential deaths and damaged structures. The estimate is made by applying the tsunami fragility curves, representing structural damage probabilities as a function of tsunami flow depth (Koshimura, Namegaya, et al., 2009; Koshimura, Oie, et al., 2009). Recent advances in machine learning methods have changed the damage estimate approach. Vescovo et al. (2023) proposed a simple machine learning method trained on physical parameters inspired by, but expanded beyond, tsunami fragility function intensity measures.

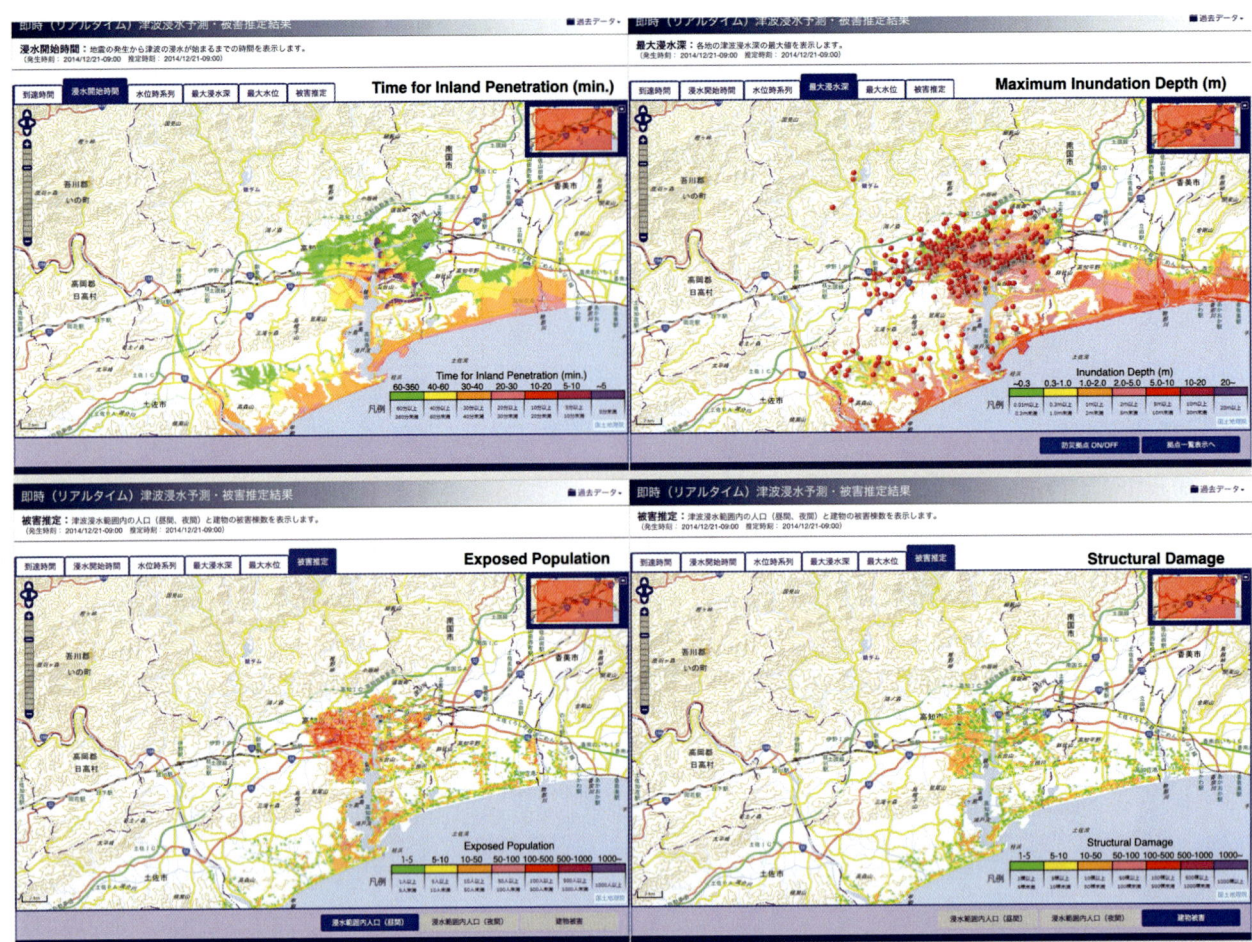

FIGURE 24.5 Tsunami inundation and damage forecast results for Kochi City. Top left: tsunami arrival time, top right: maximum flow depth, bottom left: structural damage, and bottom right: exposed populations (census-based). Map lines delineate study areas and do not necessarily depict accepted national boundaries.

The results are disseminated as mapping products (Fig. 24.5) to responders and stakeholders, for example, national, regional, and municipal authorities, and private sector actors, for use in their emergency response activities, for example, identifying the scale of the exposed population; assessing the potential damage to houses, road networks, and other critical infrastructure; and for search and rescue and recovery.

Utilizing the tsunami inundation and damage forecast information, Kosaka et al. (2023) introduced recovery levels, allowing areas that need immediate response to be more easily recognized for allocating human and physical resources. They evaluated their usefulness from the viewpoint of disaster responders by surveying users in a local government through a disaster response drill and an explanatory meeting (Fig. 24.6). Consequently, it was found that the recognition and understanding of the tsunami forecast improved, and many positive opinions were obtained about utilizing the forecast in the initial disaster response activity.

24.3.2 Tsunami exposure analysis using mobile spatial statistics

The population at risk of tsunami varies, depending on the season, day of the week, and morning or night. Thus the same tsunami hazard may pose different levels of risk for an area with high mobility of residents and tourists. In the context of the efforts to develop the TDDT in Japan, mobile spatial statistics (MSS) have been applied to real-time population exposure estimation (Okajima et al., 2013; Terada et al., 2013) and anomalous event detection (Mas & Koshimura, 2024). MSS is an estimation of aggregated population based on mobile phone tracks within the base station networks and considering the penetration rate of the mobile carrier in each region. This new type of data was developed

FIGURE 24.6 The use case of tsunami inundation forecast information in a local government's disaster response drill. Left: tsunami inundation forecast result displayed in the emergency operation room. Right: forecast products distributed at the explanatory meeting in the drill.

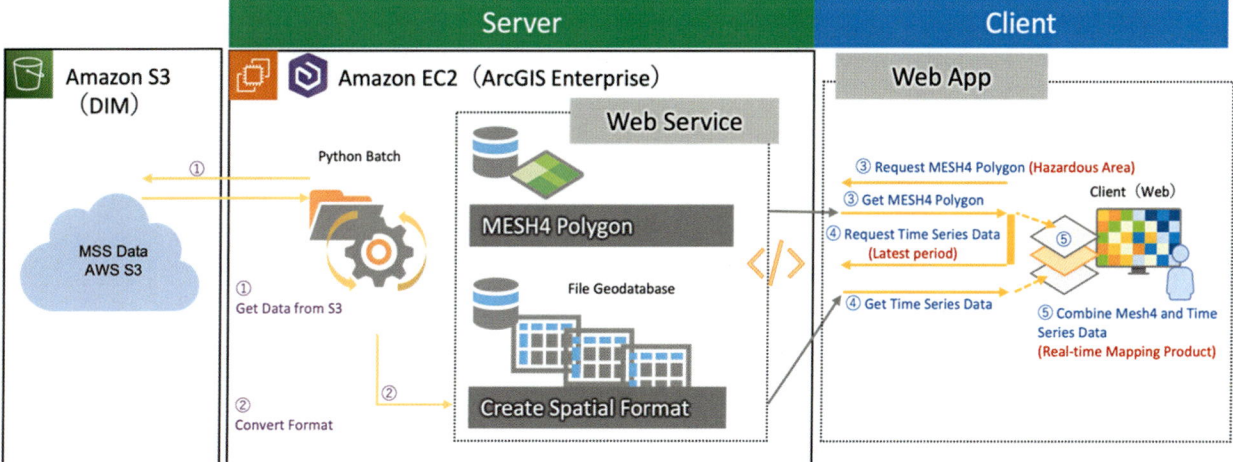

FIGURE 24.7 Structure of PoEMS. *PoEMS*, Population exposure mapping system.

by NTT DOCOMO. Population counts are available for the entire country of Japan on a 500-m by 500-m spatial and 1-hour temporal resolution for the whole year.

We combine MSS and real-time tsunami inundation forecast into a population exposure mapping system (PoEMS) to perform the rapid assessment of evacuation needs. By deploying estimation algorithms and visualization, we created a web-based system hosted on an ArcGIS Enterprise server that connects the MSS data stored in an Amazon S3 to a Web App for mapping and data query (Fig. 24.7). Python scripts are used to merge and create geospatial formats to store and display spatiotemporal data efficiently. A population exposure of two test areas from 2016 to 2022 is shown here. The system provides an interactive environment to query MSS data based on a customized area (i.e., hazard map polygon or other area of interest) with a user-defined period (including the latest hour data for real-time observations).

MSS data are available for the whole of Japan. The MSS-PoEMS approach is illustrated in Fig. 24.8 by focusing on two areas prone to the anticipated Nankai Trough Earthquake, that is, Katsurahama Beach and Tanezaki district in Kochi Prefecture. The areas near Katsurahama correspond to a low residency but highly touristic zone, while Tanezaki is a highly residential area with low tourist activity. The size of both sites is similar for comparison, and the expected tsunami runup is in the same order of magnitude. The tsunami inundation scenario is based on the ones anticipated by the Cabinet Office of Japan. We found significant variations in the expected number of people at risk of tsunami during a year. Population dynamics affect the necessary levels of response and assistance (Fig. 24.9).

In this example, Tanezaki (residential area) and Katsurahama (beach area) were compared across the entire data to show differences in population exposure. Clear peaks are observed on particular days (e.g., January 1st—first sunrise) or seasons (e.g., summer), with a clear difference in pre- and post-COVID conditions and the discrepancy with the census data. MSS data fusion with real-time inundation forecasting provides a way to grasp population exposure and address evacuation needs immediately.

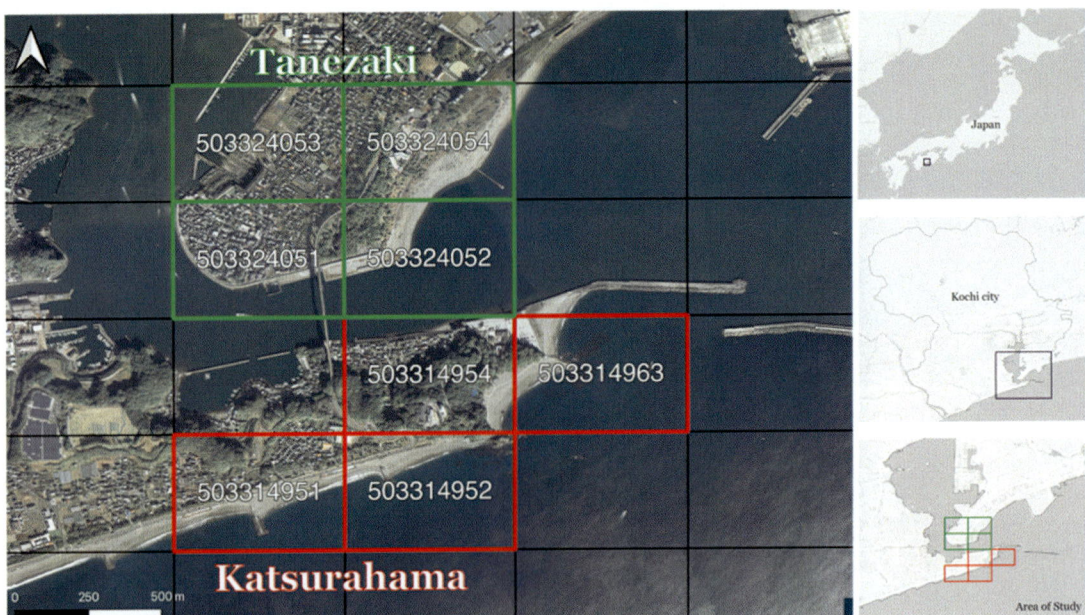

FIGURE 24.8 The target area (Kochi City) of the case study analyzes and estimates the exposed population using mobile spatial statistics. Map lines delineate study areas and do not necessarily depict accepted national boundaries.

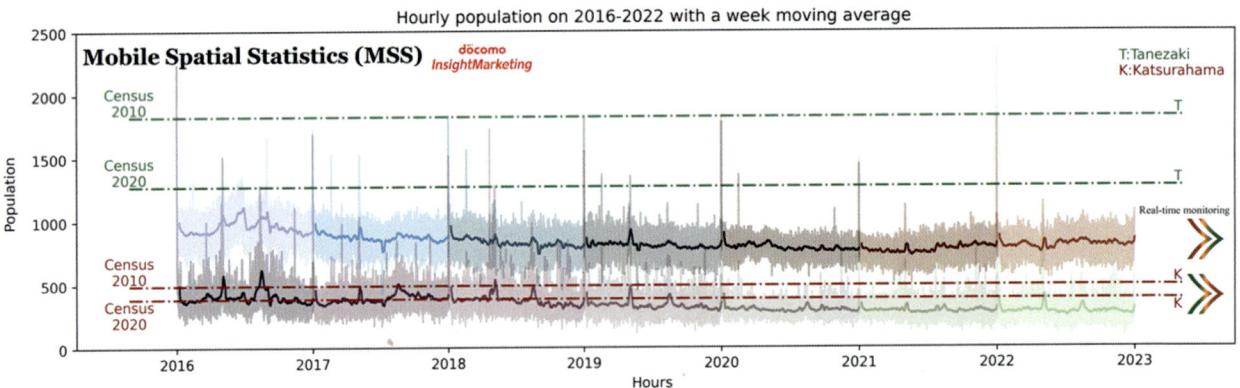

FIGURE 24.9 Time series of exposed population in the tsunami inundation zone.

24.3.3 Multiagent modeling of disaster response activities

In the aftermath of the 2011 Tohoku Earthquake and Tsunami, search and rescue operations were complex. Search and rescue activities started locally by firefighters and the police on land and were led by the Japan Coast Guard in sea. In particular circumstances, such as this one, the Japan Self-Defense Forces (JSDF) were also involved in the operations. Japan's Fire and Disaster Management Agency has conducted reviews of rescue activities in response to the 2011 disaster (Koresawa, 2013), where they summarized the lessons for search and rescue focusing on the activities at tsunami-hit areas during the critical 72 hours. The activities conducted by first responders in the fire services include shutting down water gates, delivering tsunami warnings to local residents, and guiding evacuation. Unfortunately, member volunteers of the fire corps died in closing tsunami gates in the area (Ranghieri & Ishiwatari, 2014).

Mas et al. (2022) developed an agent-based disaster response model incorporating search and rescue, medical assistance evaluation through the disaster medical assistance team, and transportation of patients within the first three days after the event. We summarize a case study on the central Ishinomaki area. Ishinomaki City in Miyagi Prefecture had the largest number of casualties and missing bodies in the 2011 event (3972). Moreover, medical centers were inundated and out of service due to the damage caused by the earthquake and tsunami. Within the Ishinomaki area, one of the main hospitals out of the inundation zone was the Ishinomaki Red Cross Hospital (IRCH), which remained active.

At the IRCH, the total number of patients in the first month was at least five times the regular number of patients before the earthquake (World Health Organization, 2012). The IRCH coordinated all medical teams from the Japanese Red Cross Society and other agencies at evacuation centers throughout Ishinomaki City (Ishii, 2011). Several predisaster preparedness activities conducted by the IRCH were critical to a better response during the 2011 disaster (Egawa et al., 2017). The IRCH reported the situation of the first days of transported patients and their conditions classified as minor, medium, severe, and serious injuries, as shown in Table 24.1.

We have simulated three days of disaster response activities, including search and rescue and transportation of patients to the IRCH. The reference event corresponds to the 2011 Tohoku Earthquake and Tsunami and the local time of the event was 2:46 pm (JST). We assumed 1 hour and 14 minutes for headquarters to be established and then search and rescue and transportation activities to get started. Thus our simulation starts from 4:00 p.m. until the end of the third day after the event; this is 56 hours of simulation. The target area for this analysis is bounded by the Old Kitakami River in the east and north, near the Fire Department Headquarters, the channel connecting Old Kitakami and Jo Rivers on the west side, and the inundated areas below the 2-m inundation depth. We focused our case study below the 2-m inundation depth since the areas over that level of inundation were heavily damaged and immediate access was reported to be limited due to water presence (Koresawa, 2013) (magenta dashed line in Fig. 24.10). We tested three cases of search and rescue patterns to evaluate the rapidity of finding and dispatching survivors from the affected area to the IRCH (Fig. 24.10). The rate of victims within the target area versus the whole inundated area (1448/2954 = 0.49) is used to estimate the number of patients transported to the IRCH from our study area resulting in values shown in Table 24.1. In our simulation, only "medium" and "severe" injury levels were transported by the medical agents in the model; thus we use these in our assessment.

Fig. 24.11 and Table 24.2 summarize the results of the simulations. We averaged the outcome values of 40 simulation runs. The outcomes are consistent with the total estimated in Table 24.1. Overall day-by-day discrepancies shown in Table 24.2 are balanced out in the total sum where the model shows good agreement. In particular, Fig. 24.11 shows the overall statistics (i.e., average, minimum, maximum, and standard deviation) of patients searched and rescued from the targeted tsunami-affected area for each case. There are slight differences among cases; however, Case A appears to be the pattern giving better results with a higher average value. Table 24.2 shows the disaggregated results of patients transported to hospitals based on search and rescue and medical resources and activities per day. At a finer level, the model presents underestimated results during the first 2 days of relief activities, while overestimation is found on the

TABLE 24.1 Estimated number of patients transported to the Ishinomaki Red Cross Hospital from the target area within the first 3 days after the event classified by injury level.

Day	Medium	Severe	Total
12 March	16	8	24
13 March	70	36	106
14 March	74	19	93
Total	160	63	223

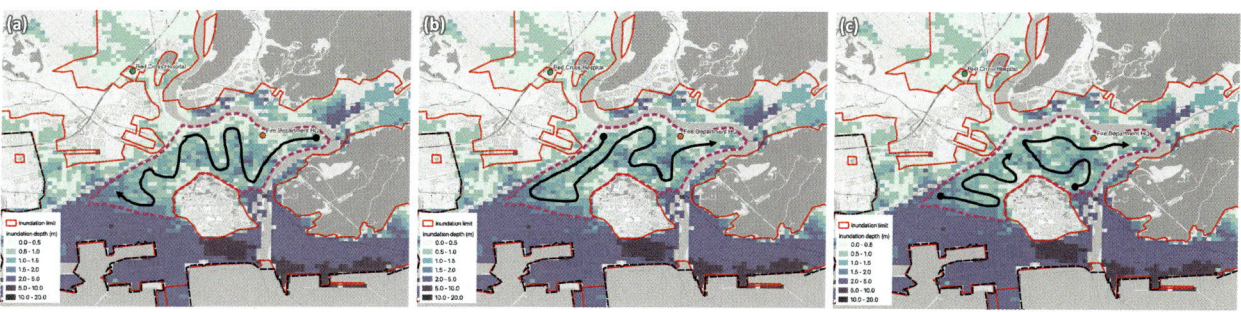

FIGURE 24.10 Three search patterns for search and rescue teams (see black arrows): (A) easy access prioritization, (B) closest transportation prioritization, and (C) deeper inundation prioritization. The magenta dashed line shows the target area for calculations. Map lines delineate study areas and do not necessarily depict accepted national boundaries.

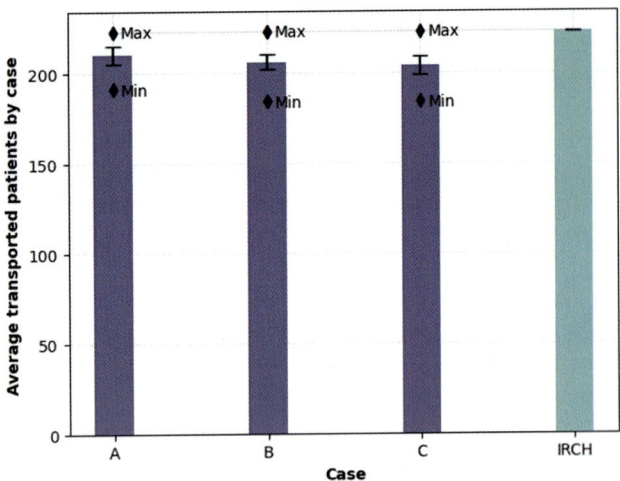

FIGURE 24.11 Average number of patients transported in simulation cases A, B, and C versus the estimated number transported to the Ishinomaki Red Cross Hospital.

TABLE 24.2 Number of patients transported to the Ishinomaki Red Cross Hospital according to triage level and the days after the event.

	Case A		Case B		Case C		IRCH records	
	Medium	Severe	Medium	Severe	Medium	Severe	Medium	Severe
Day1	2	1	2	1	2	1	16	8
Day2	54	21	51	19	49	20	70	36
Day3	94	37	95	38	96	36	74	19
Total	150	59	148	58	147	57	160	63

Three cases were compared to the record of patients and their medical level of need. The values listed in the table are the average of 40 repetitions.

third day. Limitations of the model, as well as a possible number of self-responses via private means of transportation, might explain such discrepancies. Further improvements of the model are necessary by incorporating other means of transportation; however, data for verification might be scarce in this case. Some of the reasons for such differences can be explained by the limitations of the model to represent other activities that contribute to the transportation of patients. For instance, the JSDF transported patients via helicopters, and several patients were transported by private vehicles without the assistance of emergency responders.

24.4 Towards establishing coastal digital twin paradigm

The status quo of establishing a digital twin in coastal engineering reflects the dynamic and diverse nature of the problem. A review of the scientific and technical literature that discusses CDT includes a conceptual model and its PoC (proof of concept) (Allen et al., 2022; Camastra et al., 2023; Cassottana et al., 2023; Chowdhury et al., 2023; Obonyo & Ouma, 2022; Riaz et al., 2023), integration of sensing methods and multimodal data fusion (Barbie et al., 2022; Bianucci et al., 2022; Duque & Brovelli, 2022; Grossmann et al., 2022; Zhang et al., 2020), advanced modeling (Jeong & Lee, 2023; Koshimura et al., 2020), and enhanced visualization/VR (Boletsis, 2022; Yavo-Ayalon et al., 2023). From the thematic perspectives, the literature covers issues and solutions in assessing storm surge risk (Cai et al., 2023; Takagi et al., 2022; Wang et al., 2022), environmental and ecosystem monitoring (De Leon et al., 2023; Ditria et al., 2022; Pillai et al., 2022; Prandi et al., 2022), urbanization (Arai et al., 2021; Hartmann et al., 2022; Qi et al., 2022; Zhao et al., 2022), climate change adaptation (Chowdhury et al., 2023; Rouja et al., 2022; Xiao et al., 2023; Karatvuo, et al., 2022), aquaculture (Ferreira et al., 2023), autonomous mobility (Dubey et al., 2022), and education (Chacón et al., 2018). In summary, the digital twin research in coastal engineering has become active since 2018, but it is still fragmented without holistic and comprehensive viewpoints.

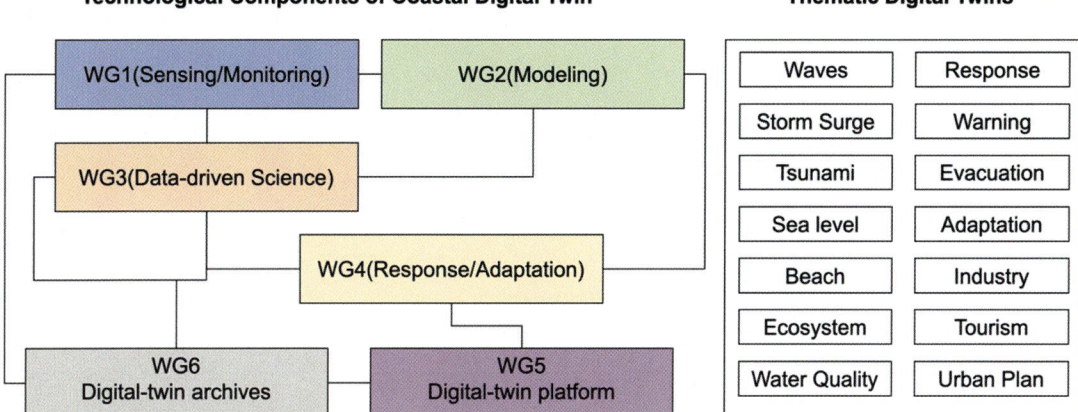

FIGURE 24.12 Structure of Coastal Disaster Digital Twin Research group.

In 2022, approximately 60 researchers and engineers from the Coastal Engineering Committee, Japan Society of Civil Engineers, formed the Coastal Disaster Digital Twin Research group (Fig. 24.12) and started its activities (Coastal Engineering Committee, 2022). The activity aims to introduce the national digital twin paradigm to various coastal and marine issues, examine the applicability of the paradigm to various issues, and create a new horizon of coastal and marine disaster prevention and mitigation science in the digital age through systematization of the technologies that constitute the components of the digital twin. In this section, preliminary activities and outcomes of the Coastal Disaster Digital Twin Research group toward the construction of a national CDT are described.

The research group is organized by the following seven working groups (WG).

- WG0 (Steering group) plans and decides the direction of the activities.
- WG1 (Sensing and monitoring WG) reviews and considers new sensing and monitoring technologies of the coastal environment to create a virtual replica in CDT.
- WG2 (Modeling WG) develops modeling techniques and methods as a basis for predictions and forecast capability in CDT.
- WG3 (Data-driven science WG) applies data-driven sciences to elucidate latent structures and processes behind high-dimensional sensing and monitoring data.
- WG4 (Social response WG) identifies users' needs for CDT and examines the methodologies to identify optimized response and adaptation measures in an exploratory manner.
- WG5 (Digital twin computing platform WG) develops the conceptual design of CDT and specifies the technological and facility requirements of CDT.
- WG6 (Digital twin archive WG) stores the review results, data, and findings created by all the WGs and explores the utilization of the digital archives in CDT.

The project was officially launched in January 2023, and the "Project Definition Workshop" took place to share the objectives of the project and define the research questions. In the workshop, participants collaborated to identify the important seven research questions to be tackled by the project.

1. How do we address the coastal and marine issues?
2. What are the potential users and stakeholders of CDT?
3. How do we establish the linkage between the physical and cyber systems?
4. How can the fusion of data and technological elements of CDT be realized?
5. How do we deal with uncertainties in forecasts and outputs of CDT?
6. How do we define disaster resilience indicators in coastal disasters?
7. How do we define CDT outputs?

The first outcome of the workshop mainly focused on the research questions 1 and 2. The coastal and marine issues are categorized by their time scale and summarized in Table 24.3. CDTs have a wide range of potential users and stakeholders due to their ability to provide comprehensive insights into coastal environments (the physical world). Potential users and stakeholders who can benefit from CDTs are listed below, along with Table 24.3.

TABLE 24.3 Preliminary targets of coastal digital twin research (thematic digital twins).

Category	Short-term themes	Long-term themes
Coastal disaster	Inundation, high surf, wave overtopping, salt damage, liquefaction, structural damage, warning, evacuation plan, and exposure estimation	Climate change and sea-level rise impact, extreme weather forecast, and mobility design
Coastal environment	Water quality and environment, ecosystem management, sediment transport and shoreline dynamics, suspended substances, microplastics, and vegetation	Coastal zone management, water quality management, ecosystem management, habitat degradation, and restoration
Industry, Tourism	Energy harvesting, beach loss estimation, and landscape	Marine energy, economy, supply chain, and environmental effect
Infrastructure	Functionality loss of coastal structures	Infrastructure decay and aging
Urban planning	Landscape design, land use management	Urbanization and land use change

- Emergency response and coastal disaster management organizations will benefit from CDTs to prepare for and respond to natural disasters, such as storm surges and tsunamis (Cai et al., 2023). Real-time data acquisition and forecasting enable timely decision-making during disasters (Koshimura et al., 2024).
- Insurance industries and risk assessment firms can leverage CDTs to visualize risks and impacts associated with coastal properties and infrastructures (Bakhtiari et al., 2024). Advanced modeling of coastal processes and their consequences aids in determining insurance premiums and planning risk management strategies.
- Environmental and conservation organizations use CDTs to monitor and assess the impact of human activities on coastal ecosystems (Ditria et al., 2022).
- In the maritime industry, port authorities optimize operations, enhance navigation safety, and plan for the impact of coastal hazards on port infrastructure using CDTs (Dubey et al., 2022).
- Energy industries identify possible inferences in the dynamic and interdependent marine environment around offshore wind farms (Schneider et al., 2023).
- The tourism industry utilizes CDTs to plan services and safety measures for tourists by understanding the impact on coastal areas for responsible tourism and environmental preservation (Macías et al., 2021).
- Infrastructure developers and investors employ CDTs to assess the viability and resilience of projects in coastal zones, evaluating the potential impact of coastal hazards on investments (Prandi et al., 2022).
- Urban planners are direct users of CDTs, leveraging the technology to simulate and analyze coastal processes, assess infrastructure resilience, and plan for sustainable urban development (Yavo-Ayalon et al., 2023).

24.5 Summary and future challenges

Research and practices of establishing CDTs present promising advantages for managing coastal environments, mitigating disaster risks, and enhancing community resilience. Through the activities among researchers in coastal science and engineering fields, we raised seven research questions to achieve the goal. Preliminary outcomes suggest that CDTs have a diverse range of potential users and applicable fields to solve issues (research questions 1 and 2). However, we still face challenges for the successful implementation of CDTs. The missing linkage or the gap between the physical and cyber worlds suggests the need to integrate diverse data sources into a coherent model (research question 3). Real-time data acquisition remains challenging due to the dynamic nature of coastal environments, necessitating robust sensing, monitoring, and communication systems. The fusion of data and technological elements of CDTs should be realized to accelerate data-driven sciences that support inference of future consequences (research question 4). Realization of uncertainties in the CDTs outputs leads to the need for validation and calibration efforts as the degree of integration of the models increases (research question 5).

To overcome these challenges associated with CDTs and their successful implementations for coastal community's resilience (research question 6), stronger collaborative efforts are necessary among researchers, policymakers, developers, and local communities. Other arguments are related to how the useful outputs from CDTs can be derived (research question 7). With more data added from the physical world, the digital twin of the cyber world becomes richer and more strongly linked to the physical world. In this sense, CDTs do not have to strongly rely only on social needs or practical missions but also lead to new scientific discoveries through the efforts of finding the hidden and latent structures in the data, interpreting the meaning and the reasoning of data, and enhancing visualizations.

Understanding the complex interface and cascading features among coastal processes requires strong partnerships between the research teams of various disciplines, sciences, engineering, data sciences, and social sciences, but with common goals to achieve. This effort will establish a collaborative research to produce scientific evidence and datasets applicable beyond the duration of this project. The efforts of establishing CDTs are in progress, but the collaborative processes and experiences summarized in this chapter are useful for solving diverse issues in coastal engineering.

References

Allen, T. R., McLeod, G., Richter, H., & Nielsen, A. (2022). Digitally twinning coastal resilience via multisensor imagery, in situ sensors, and geospatial analysis. *IGARSS 2022—2022 IEEE International Geoscience and Remote Sensing Symposium*, 4739–4742. Available from https://doi.org/10.1109/IGARSS46834.2022.9883133.

Aoki, K., Fuse, Y., Akiyama, M., Ishibashi, H., Koshimura, S., & Frangopol, D. M. (2022). *Probalilistic connectivity assessment of bridge networks under seismic hazard considering spatial correlation of ground motion-induced damage. Bridge Sagety, Maintenance, Management, Lice-Cycle, Resilience and Sustainability.* Taylor & Francis.

Arai, H., Inubushi, K., & Chiu, C. Y. (2021). Dynamics of methane in mangrove forest: Will it worsen with decreasing mangrove forests? *Forests, 12*(9), 1204. Available from https://doi.org/10.3390/f12091204. https://www.mdpi.com/1999-4907/12/9/1204/pdf.

Bakhtiari, V., Piadeh, F., Chen, A. S., & Behzadian, K. (2024). Stakeholder analysis in the application of cutting-edge digital visualisation technologies for urban flood risk management: A critical review. *Expert Systems with Applications, 236*. Available from https://doi.org/10.1016/j.eswa.2023.121426.

Barbie, A., Pech, N., Hasselbring, W., Flogel, S., Wenzhofer, F., Walter, M., Shchekinova, E., Busse, M., Turk, M., Hofbauer, M., & Sommer, S. (2022). Developing an underwater network of ocean observation systems with digital twin prototypes—A field report from the Baltic Sea. *IEEE Internet Computing, 26*(3), 33–42. Available from https://doi.org/10.1109/mic.2021.3065245.

Becu, N., Amalric, M., Anselme, B., Beck, E., Bertin, X., Delay, E., Long, N., Marilleau, N., Pignon-Mussaud, C., & Rousseaux, F. (2017). Participatory simulation to foster social learning on coastal flooding prevention. *Environmental Modelling & Software, 98*, 1–11. Available from https://doi.org/10.1016/j.envsoft.2017.09.003.

Bianucci, M., Merlino, S., Locritani, M., & Paterni, M. (2022). Monitoring sea current and marine litter transport using a low cost approaches. *2022 IEEE International Workshop on Metrology for the Sea*, 1–6. Available from https://doi.org/10.1109/MetroSea55331.2022.9950922.

Boletsis, C. (2022). *The Gaia system: A tabletop projection mapping system for raising environmental awareness in islands and coastal areas.* In Association for Computing Machinery, Norway—The 15th International Conference on PErvasive Technologies Related to Assistive Environments (pp. 50–54). Available from: http://portal.acm.org/, https://doi.org/10.1145/3529190.3535340.

Cabinet Office of the Government of Japan. (2023). *Cross-ministerial strategic innovation promotion program (SIP): Establishing smart disaster management network.* https://www8.cao.go.jp/cstp/panhu/sip_english/4-6.pdf.

Cai, Z., Newman, G., Lee, J., Ye, X., Retchless, D., Zou, L., & Ham, Y. (2023). Simulating the spatial impacts of a coastal barrier in Galveston Island, Texas: A three-dimensional urban modeling approach. *Geomatics, Natural Hazards and Risk, 14*(1). Available from https://doi.org/10.1080/19475705.2023.2192332.

Camastra, F., Ciaramella, A., Di Nardo, E., Ferone, A., Maratea, A., Montella, R., & Staiano, A. (2023). AI-based monitoring of coastal and marine environments. *CEUR-WS, Italy CEUR Workshop Proceedings, 3486*, 575–579. Available from http://ceur-ws.org/.

Cassottana, B., Roomi, M. M., Mashima, D., & Sansavini, G. (2023). Resilience analysis of cyber-physical systems: A review of models and methods. *Risk Analysis, 43*(11), 2359–2379. Available from https://doi.org/10.1111/risa.14089. http://onlinelibrary.wiley.com/journal/10.1111/(ISSN)1539-6924.

Chacón,R., Sánchez-Juny,M., Real,E., Gironella,F.X., Puigagut,J., & Ledesma,A. (2018). *Digital twins in civil and environmental engineering classrooms.* In EUCEET 2018—4th International Conference on Civil Engineering Education: Challenges for the Third Millennium (pp. 290–299).

Chowdhury, P., Lakku, N. K. G., Lincoln, S., Seelam, J. K., & Behera, M. R. (2023). Climate change and coastal morphodynamics: Interactions on regional scales. *Science of the Total Environment, 899*. Available from https://doi.org/10.1016/j.scitotenv.2023.166432, http://www.elsevier.com/locate/scitotenv.

Coastal Engineering Committee. (2022). *Research Sub-committee of coastal disaster digital twin.* https://www.jsce.or.jp/committee/cec/en/index_en.html.

De Leon, J. A. R., Concepcion, R. S., Billones, R. K. C., Baun, J. J. G., Custodio, J. M. F., Vicerra, R. R. P., Bandala, A. A., & Dadios, E. P. (2023). Digital twin concept utilizing electrical resistivity tomography for monitoring seawater intrusion. *Journal of Advanced Computational Intelligence and Intelligent Informatics, 27*(1), 12–18. Available from https://doi.org/10.20965/jaciii.2023.p0012, https://www.fujipress.jp/main/wp-content/themes/Fujipress/pdf_subscribed.php.

Ditria, E. M., Buelow, C. A., Gonzalez-Rivero, M., & Connolly, R. M. (2022). Artificial intelligence and automated monitoring for assisting conservation of marine ecosystems: A perspective. *Frontiers in Marine Science, 9.* Available from https://doi.org/10.3389/fmars.2022.918104, https://www.frontiersin.org/journals/marine-science#.

Dubey, A. C., Subramanian, A. V., & Jagadeesh Kumar, V. (2022). Steering model identification and control design of autonomous ship: A complete experimental study. *Ships and Offshore Structures, 17*(5), 992–1004. Available from https://doi.org/10.1080/17445302.2021.1889193, http://www.tandfonline.com/loi/tsos20.

Duque, J. P., & Brovelli, M. A. (2022). Building a digital twin of the Italian coasts. *International Archives of the Photogrammetry, Remote Sensing and Spatial Information Sciences—ISPRS Archives, 48*(4), 127–133. Available from https://doi.org/10.5194/isprs-archives-XLVIII-4-W1-2022-127-2022, http://www.isprs.org/proceedings/XXXVIII/4-W15/.

Egawa, S., Suda, T., Jones-Konneh, T. E. C., Murakami, A., & Sasaki, H. (2017). Nation-wide implementation of disaster medical coordinators in Japan. *Tohoku Journal of Experimental Medicine, 243*(1), 1–9. Available from https://doi.org/10.1620/tjem.243.1. https://www.jstage.jst.go.jp/article/tjem/243/1/243_1/_pdf.

European Commission. (2020). *Workshops reports on elements of digital twins on "weather-induced and geophysical extremes" and "climate change adaptation"*. https://digital-strategy.ec.europa.eu/en/library/workshops-reports-elements-digital-twins-weather-induced-and-geophysical-extremes-and-climate.

Ferreira, J. G., Bernard-Jannin, L., Cubillo, A., Lencart e Silva, J., Diedericks, G. P. J., Moore, H., Service, M., & Nunes, J. P. (2023). From soil to sea: An ecological modelling framework for sustainable aquaculture. *Aquaculture, 577.* Available from https://doi.org/10.1016/j.aquaculture.2023.739920. http://www.journals.elsevier.com/aquaculture/.

Gobeawan, L., Lin, E. S., Tandon, A., Yee, A. T. K., Khoo, V. H. S., Teo, S. N., Yi, S., Lim, C. W., Wong, S. T., Wise, D. J., Cheng, P., Liew, S. C., Huang, X., Li, Q. H., Teo, L. S., Fekete, G. S., & Poto, M. T. (2018). Modeling trees for virtual Singapore: From data acquisition to CityGML models. *The International Archives of the Photogrammetry, Remote Sensing and Spatial Information Sciences, XLII-4/W10*(4), 55–62. Available from https://doi.org/10.5194/isprs-archives-xlii-4-w10-55-2018.

Grossmann, V., Nakath, D., Urlaub, M., Oppelt, N., Koch, R., & Köser, K. (2022). Digital twinning in the ocean-challenges in multimodal sensing and multiscale fusion based on faithful visual models. *ISPRS Annals of the Photogrammetry, Remote Sensing and Spatial Information Sciences, 5*(4), 345–352. Available from https://doi.org/10.5194/isprs-Annals-V-4-2022-345-2022, http://www.isprs.org/publications/annals.aspx.

Gusman, A. R., Sheehan, A. F., Satake, K., Heidarzadeh, M., Mulia, I. E., & Maeda, T. (2016). Tsunami data assimilation of Cascadia seafloor pressure gauge records from the 2012 Haida Gwaii earthquake. *Geophysical Research Letters, 43*(9), 4189–4196. Available from https://doi.org/10.1002/2016GL068368, http://onlinelibrary.wiley.com/journal/10.1002/(ISSN)1944-8007/issues?year=2012.

Gusman, A. R., Tanioka, Y., Macinnes, B. T., & Tsushima, H. (2014). A methodology for near-field tsunami inundation forecasting: Application to the 2011 Tohoku tsunami. *Journal of Geophysical Research: Solid Earth, 119*(11), 8186–8206. Available from https://doi.org/10.1002/2014JB010958, http://onlinelibrary.wiley.com/journal/10.1002/(ISSN)2169-9356.

Harold, S. W., Brunelle, G., Chaturvedi, R., Hornung, J. W., Koshimura, S., Osoba, O. O., & Suga, C. (2020). *United States–Japan research exchange on artificial intelligence*. In Proceedings from a Pair of Conferences on the Impact of Artificial Intelligence on Work, Health, and Data Privacy and on Disaster Prediction, Resilience, and Recovery. Available from: https://www.rand.org/pubs/conf_proceedings/CFA521-1.html, https://doi.org/10.7249/CFA521-1.

Hachiya, D., Mas, E., & Koshimura, S. (2022). A reinforcement learning model of multiple UAVs for transporting emergency relief supplies. *Applied Sciences, 12*(20), 10427. Available from https://doi.org/10.3390/app122010427.

Hartmann, M. C., Koblet, O., Baer, M. F., & Purves, R. S. (2022). Automated motif identification: Analysing Flickr images to identify popular viewpoints in Europe's protected areas. *Journal of Outdoor Recreation and Tourism, 37.* Available from https://doi.org/10.1016/j.jort.2021.100479. http://www.journals.elsevier.com/journal-of-outdoor-recreation-and-tourism/.

Igarashi, Y., Hori, T., Murata, S., Sato, K., Baba, T., & Okada, M. (2016). Maximum tsunami height prediction using pressure gauge data by a Gaussian process at Owase in the Kii Peninsula, Japan. *Marine Geophysical Research, 37*(4), 361–370. Available from https://doi.org/10.1007/s11001-016-9286-z.

Ishibashi, H., Akiyama, M., Frangopol, D. M., Koshimura, S., Kojima, T., & Nanami, K. (2020). Framework for estimating the risk and resilience of road networks with bridges and embankments under both seismic and tsunami hazards. *Structure and Infrastructure Engineering, 17*(4), 494–514. Available from https://doi.org/10.1080/15732479.2020.1843503.

Ishii, T. (2011). Medical response to the Great East Japan Earthquake in Ishinomaki City. *Western Pacific Surveillance and Response Journal, 2*(4), 10–16. Available from https://doi.org/10.5365/wpsar.2011.2.4.005.

Jeong, J. S., & Lee, H. S. (2023). Unstructured grid-based river-coastal ocean circulation modeling towards a digital twin of the Seto Inland Sea. *Applied Sciences, 13*(14). Available from https://doi.org/10.3390/app13148143, http://www.mdpi.com/journal/applsci/.

Karatvuo, H., Linde, M., Dolatshah, A., & Mortensen, S. (2022). *Improved climate change adaptation in port of Brisbane using a digital twin cloud-based modelling approach*. In Proceedings of the International Conference on Offshore Mechanics and Arctic Engineering—OMAE (Vol. 1). Available from: http://www.asmedl.org/journals/doc/ASMEDL-home/proc/, https://doi.org/10.1115/OMAE2022-79613.

Kawamoto, S., Ohta, Y., Hiyama, Y., Todoriki, M., Nishimura, T., Furuya, T., Sato, Y., Yahagi, T., & Miyagawa, K. (2017). REGARD: A new GNSS-based real-time finite fault modeling system for GEONET. *Journal of Geophysical Research: Solid Earth, 122*(2), 1324–1349. Available from https://doi.org/10.1002/2016jb013485.

Koresawa, A. (2013). Evidence-based analysis of search and rescue operations following the Great East Japan earthquake. *Journal of Disaster Research, 8*, 746–755. Available from https://doi.org/10.20965/jdr.2013.p0746.

Kosaka, N., Koshimura, S., Terada, K., Murashima, Y., Kura, T., Koyama, A., & Matsubara, H. (2023). Decision-making support utilizing real-time tsunami inundation and damage forecast. *International Journal of Disaster Risk Reduction*, 94. Available from https://doi.org/10.1016/j.ijdrr.2023.103807.

Kosaka, N., Moriguchi, S., Shibayama, A., Kura, T., Shigematsu, N., Okumura, K., Mas, E., Okumura, M., Koshimura, S., Terada, K., Fujino, A., Matsubara, H., & Hisada, M. (2024). A study on digital model for decision-making in crisis response. *Journal of Disaster Research*, 19(3), 489–500. Available from https://doi.org/10.20965/jdr.2024.p0489.

Koshimura, S., Hayashi, S., & Gokon, H. (2014). The impact of the 2011 Tohoku earthquake tsunami disaster and implications to the reconstruction. *Soils and Foundations*, 54(4), 560–572. Available from https://doi.org/10.1016/j.sandf.2014.06.002.

Koshimura, S., Hino, R., Ohta, Y., Kobayashi, H., Murashima, Y., & Musa, A. (2017). *Advances of tsunami inundation forecasting and its future perspectives*. OCEANS 2017—Aberdeen (pp. 1–4). IEEE. Available from http://doi.org/10.1109/OCEANSE.2017.8084753.

Koshimura, S., Mas, E., & Musa, A. (2024). *Disaster digital twin for enhancing resilience of tsunami disaster response system*. In Proceedings of the 18th World Conference of Earthquake Engineering (pp. 1–12).

Koshimura, S., Namegaya, Y., & Yanagisawa, H. (2009). Tsunami fragility—A new measure to identify tsunami damage. *Journal of Disaster Research*, 4(6), 479–488. Available from https://doi.org/10.20965/jdr.2009.p0479.

Koshimura,S., Ohta,Y., Hino,R., Suzuki,T., Murashima,Y., Musa,A., Sato,Y., Kachi,M., Kobayashi,H. (2020). *Advances of real-time tsunami inundation and damage forecast—Present and Future*. Proceedings of the World Conference on Earthquake Engineering.

Koshimura, S., Oie, T., Yanagisawa, H., & Imamura, F. (2009). Developing fragility functions for tsunami damage estimation using numerical model and post-tsunami data from Banda Aceh, Indonesia. *Coastal Engineering Journal*, 51(3), 243–273. Available from https://doi.org/10.1142/S0578563409002004, https://www.tandfonline.com/loi/tcej20.

Koshimura, S., & Shuto, N. (2015). Response to the 2011 Great East Japan earthquake and tsunami disaster. *Philosophical Transactions of the Royal Society A: Mathematical, Physical and Engineering Sciences*, 373(2053). Available from https://doi.org/10.1098/rsta.2014.0373.

Li, L., Aslam, S., Wileman, A., & Perinpanayagam, S. (2022). *Digital twin in aerospace industry: A gentle introduction* (10, pp. 9543–9562). IEEE Access. Available from http://doi.org/10.1109/access.2021.3136458.

Lloyd, M. G., Peel, D., & Duck, R. W. (2013). Towards a social-ecological resilience framework for coastal planning. *Land Use Policy*, 30(1), 925–933. Available from https://doi.org/10.1016/j.landusepol.2012.06.012.

Lu, Y., Liu, C., Wang, K. I.-K., Huang, H., & Xu, X. (2020). Digital twin-driven smart manufacturing: Connotation, reference model, applications and research issues. *Robotics and Computer-Integrated Manufacturing*, 61. Available from https://doi.org/10.1016/j.rcim.2019.101837.

Macías, D., Prieto, L., & García-Gorriz, E. (2021). A model-based management tool to predict the spread of *Physalia physalis* in the Mediterranean Sea. Minimizing risks for coastal activities. *Ocean & Coastal Management*, 212(15), 105810. Available from https://doi.org/10.1016/j.ocecoaman.2021.105810.

Maeda, T., Obara, K., Shinohara, M., Kanazawa, T., & Uehira, K. (2015). Successive estimation of a tsunami wavefield without earthquake source data: A data assimilation approach toward real-time tsunami forecasting. *Geophysical Research Letters*, 42(19), 7923–7932. Available from https://doi.org/10.1002/2015gl065588.

Makinoshima, F., Oishi, Y., Yamazaki, T., Furumura, T., & Imamura, F. (2021). Early forecasting of tsunami inundation from tsunami and geodetic observation data with convolutional neural networks. *Nature Communications*, 12(1). Available from https://doi.org/10.1038/s41467-021-22348-0. http://www.nature.com/ncomms/index.html.

Mas, E., Egawa, S. M. D., Sasaki, H. M. D., & Koshimura, S. (2022). *Modeling search and rescue, medical disaster team response and transportation of patients in Ishinomaki city after tsunami disaster*, . The 13th of Aceh International Workshop and expo on sustainable tsunami disaster recovery (The 13th AIWEST-DR 2021) (340). . Available from http://www.e3s-conferences.org/, http://doi.org/10.1051/e3sconf/202234005001.

Mas, E., & Koshimura, S. (2024). Feasibility of anomalous detection based on Mobile Spatial Statistics: A study of six cases in Japan. *International Journal of Disaster Risk Reduction*, 110. Available from https://doi.org/10.1016/j.ijdrr.2024.104625.

Mizutani, A., Yomogida, K., & Tanioka, Y. (2020). Early tsunami detection with near-fault ocean-bottom pressure gauge records based on the comparison with seismic data. *Journal of Geophysical Research: Oceans*, 125(9). Available from https://doi.org/10.1029/2020JC016275, http://agupubs.onlinelibrary.wiley.com/agu/jgr/journal/10.1002/(ISSN)2169-9291/.

Musa, A., Abe, T., Kishitani, T., Inoue, T., Sato, M., Komatsu, K., Murashima, Y., Koshimura, S., & Kobayashi, H. (2019). Performance evaluation of tsunami inundation simulation on SX-Aurora TSUBASA. *Computational Science—ICCS 2019*, 11537, 363–376. Available from https://doi.org/10.1007/978-3-030-22741-8_26, https://www.springer.com/series/558.

Musa, A., Watanabe, O., Matsuoka, H., Hokari, H., Inoue, T., Murashima, Y., Ohta, Y., Hino, R., Koshimura, S., & Kobayashi, H. (2018). Real-time tsunami inundation forecast system for tsunami disaster prevention and mitigation. *The Journal of Supercomputing*, 74(7), 3093–3113. Available from https://doi.org/10.1007/s11227-018-2363-0.

National Research Foundation. (2023). *Singapore Land Authority Government Technology Agency 2023 Nov*. Virtual Singapore. https://www.sla.gov.sg/geospatial/gw/virtual-singapore.

Nomura, R., Fujita, S., Galbreath, J. M., Otake, Y., Moriguchi, S., Koshimura, S., LeVeque, R. J., & Terada, K. (2022). Sequential Bayesian update to detect the most likely tsunami scenario using observational wave sequences. *Journal of Geophysical Research: Oceans*, 127(10). Available from https://doi.org/10.1029/2021JC018324, http://agupubs.onlinelibrary.wiley.com/agu/jgr/journal/10.1002/(ISSN)2169-9291/.

Obonyo, E. A., & Ouma, L. A. (2022). Enhancing the resilience of low-income housing using emerging digital technologies. *IOP Conference Series: Earth and Environmental Science*, 1101(9). Available from https://doi.org/10.1088/1755-1315/1101/9/092013.

Ohta, Y., Inoue, T., Koshimura, S., Kawamoto, S., & Hino, R. (2018). Role of real-time GNSS in near-field tsunami forecasting. *Journal of Disaster Research*, *13*(3), 453–459. Available from https://doi.org/10.20965/jdr.2018.p0453.

Ohta, Y., Kobayashi, T., Tsushima, H., Miura, S., Hino, R., Takasu, T., Fujimoto, H., Iinuma, T., Tachibana, K., Demachi, T., Sato, T., Ohzono, M., & Umino, N. (2012). Quasi real-time fault model estimation for near-field tsunami forecasting based on RTK-GPS analysis: Application to the 2011 Tohoku-Oki earthquake (*M*w 9.0). *Journal of Geophysical Research: Solid Earth*, *117*(2). Available from https://doi.org/10.1029/2011JB008750, http://onlinelibrary.wiley.com/journal/10.1002/(ISSN)2169-9356.

Okajima, I., Tanaka, S., Terada, M., Ikeda, D., & Nagata, T. (2013). Supporting growth in society and industry using statistical data from mobile terminal networks—Overview of mobile spatial statistics. *NTT DOCOMO Technical Journal*, *14*, 4–9.

Pillai, U. P. A., Pinardi, N., Alessandri, J., Federico, I., Causio, S., Unguendoli, S., Valentini, A., & Staneva, J. (2022). A digital twin modelling framework for the assessment of seagrass nature based solutions against storm surges. *Science of the Total Environment*, *847*. Available from https://doi.org/10.1016/j.scitotenv.2022.157603, http://www.elsevier.com/locate/scitotenv.

Prandi, C., Cecilia, J. M., Manzoni, P., Peña-Haro, S., Pierson, D., Colom, W., Blanco, P., Garcìa, C. A., Navarro, I. J., & Senent, J. (2022). *On integrating intelligent infrastructure and participatory monitoring for environmental modelling: The SMARTLAGOON approach*. In GoodIT '22: Proceedings of the 2022 ACM Conference on Information Technology for Social Good (pp. 236–243). Available from: http://portal.acm.org/, https://doi.org/10.1145/3524458.3547228.

Pregnolato, M., Gunner, S., Voyagaki, E., De Risi, R., Carhart, N., Gavriel, G., Tully, P., Tryfonas, T., Macdonald, J., & Taylor, C. (2022). Towards civil engineering 4.0: Concept, workflow and application of digital twins for existing infrastructure. *Automation in Construction*, *141*. Available from https://doi.org/10.1016/j.autcon.2022.104421.

Qi, Y., Li, H., Pang, Z., Gao, W., & Liu, C. (2022). A case study of the relationship between vegetation coverage and urban heat island in a coastal city by applying digital twins. *Frontiers in Plant Science*, *13*. Available from https://doi.org/10.3389/fpls.2022.861768.

Ranghieri, F., & Ishiwatari, M. (2014). *Learning from megadisasters: lessons learnt from the Great East Japan earthquake and tsunami*. Available from https://doi.org/10.1596/978-1-4648-0153-2, https://elibrary.worldbank.org/doi/abs/10.1596/978-1-4648-0153-2.

Riaz, K., McAfee, M., & Gharbia, S. S. (2023). Management of climate resilience: Exploring the potential of digital twin technology, 3D city modelling, and early warning systems. *Sensors (Basel, Switzerland)*, *23*(5). Available from https://doi.org/10.3390/s23052659, http://www.mdpi.com/journal/sensors.

Rouja, P. M., Schneider, C. W., Rissolo, D., Blasco, S. M., Petrovic, V., Lo, E., Lightbourne, M. A., Tucker, W. S., & Kuester, F. (2022). The Royal Naval Dockyard in Bermuda: Use of a novel biological indicator and historical photographs for measuring local sea-level rise. *International Journal of Maritime History*, *34*(4), 634–657. Available from https://doi.org/10.1177/08438714221143297. https://journals.sagepub.com/home/IJH.

Schneider, J., Klüner, A., & Zielinski, O. (2023). Towards digital twins of the oceans: The potential of machine learning for monitoring the impacts of offshore wind farms on marine environments. *Sensors (Basel, Switzerland)*, *23*(10). Available from https://doi.org/10.3390/s23104581.

Takagi, M., Ninomiya, J., Mori, N., Shimura, T., & Miyashita, T. (2022). Impacts of wave-induced ocean surface turbulent kinetic energy flux on typhoon characteristics. *Coastal Engineering Journal*, *64*(1), 151–168. Available from https://doi.org/10.1080/21664250.2021.2017191.

Tanioka, Y. (2020). Improvement of near-field tsunami forecasting method using ocean-bottom pressure sensor network (S-net). *Earth, Planets and Space*, *72*(1). Available from https://doi.org/10.1186/s40623-020-01268-1, http://rd.springer.com/journal/40623.

Tao, F., Cheng, J., Qi, Q., Zhang, M., Zhang, H., & Sui, F. (2018). Digital twin-driven product design, manufacturing and service with big data. *The International Journal of Advanced Manufacturing Technology*, *94*(9-12), 3563–3576. Available from https://doi.org/10.1007/s00170-017-0233-1.

Terada, M., Nagata, T., & Kobayashi, M. (2013). Population estimation technology for mobile spatial statistics. *NTT DOCOMO Technical Journal*, *14*, 10–15. Available from https://www.docomo.ne.jp/english/corporate/technology/rd/technical_journal/bn/vol14_3/.

Toyoda A. (2020). *Woven City, a prototype city where people, buildings, and vehicles are connected through data and sensors. We welcome you all to join us in our quest to create an ever-better way of mobility for all*. CES2020. https://global.toyota/en/newsroom/corporate/31221914.html.

Vescovo, R., Adriano, B., Mas, E., & Koshimura, S. (2023). Beyond tsunami fragility functions: Experimental assessment for building damage estimation. *Scientific Reports*, *13*(1). Available from https://doi.org/10.1038/s41598-023-41047-y.

Wang, Y., Chen, X., & Wang, L. (2022). Differential semi-quantitative urban risk assessment of storm surge inundation. *ISPRS Annals of the Photogrammetry, Remote Sensing and Spatial Information Sciences*, *10*(3), 177–185. Available from https://doi.org/10.5194/isprs-annals-X-3-W1-2022-177-2022, http://www.isprs.org/publications/annals.aspx.

Wang, Y., Satake, K., Maeda, T., & Gusman, A. R. (2018). Data assimilation with dispersive tsunami model: A test for the Nankai Trough. *Earth, Planets and Space*, *70*(1). Available from https://doi.org/10.1186/s40623-018-0905-6, http://rd.springer.com/journal/40623.

World Health Organization. (2012). *The Great East Japan earthquake. A story of a devastating natural disaster, a tale of human compassion*. Available from: https://www.who.int/publications/i/item/the-great-east-japan-earthquake-a-story-of-a-devastating-natural-disaster-a-tale-of-human-compassion.

Xiao, S., Udo, K., & Zhang, Y. (2023). Projection of future beach loss along the Chinese coastline due to sea level rise. *Coastal Engineering Journal*, *65*(4), 620–637. Available from https://doi.org/10.1080/21664250.2023.2265683.

Xu, L., Cui, S., Wang, X., Tang, J., Nitivattananon, V., Ding, S., & Nguyen Nguyen, M. (2021). Dynamic risk of coastal flood and driving factors: Integrating local sea level rise and spatially explicit urban growth. *Journal of Cleaner Production*, *321*. Available from https://doi.org/10.1016/j.jclepro.2021.129039.

Yamamoto, N., Aoi, S., Hirata, K., Suzuki, W., Kunugi, T., & Nakamura, H. (2016). Multi-index method using offshore ocean-bottom pressure data for real-time tsunami forecast. *Earth, Planets and Space*, *68*(1). Available from https://doi.org/10.1186/s40623-016-0500-7.

Yavo-Ayalon, S., Joshi, S., Zhang, Y., Han, R., Mahyar, N., & Ju, W. (2023). Building community resiliency through immersive communal extended reality (CXR). *Multimodal Technologies and Interaction*, *7*(5). Available from https://doi.org/10.3390/mti7050043.

Zhang, X., Liu, Z., & Han, B. (2020). *Toward digital twins based marine SCADA system*. In Proceedings of 2020 IEEE International Conference on Artificial Intelligence and Computer Applications, ICAICA 2020 (pp. 1049–1053). Available from: http://ieeexplore.ieee.org/xpl/mostRecentIssue.jsp?punumber = 9169742, https://doi.org/10.1109/ICAICA50127.2020.9182549.

Zhao, D., Li, X., Wang, X., Shen, X., & Gao, W. (2022). Applying digital twins to research the relationship between urban expansion and vegetation coverage: A case study of natural preserve. *Frontiers in Plant Science*, *13*. Available from https://doi.org/10.3389/fpls.2022.840471.

Appendix

This appendix summarizes numerical examples that interested readers can perform to understand the content of the book more deeply. There are 5 examples in Chapter 2, 2 examples in Chapter 3, and 14 examples in Chapter 5. These examples are fundamental building blocks to conducting probabilistic tsunami hazard analysis and probabilistic tsunami risk analysis. Each example is accompanied by a MATLAB® Live Script (file extension .mlx), datasets, or related programs. Using this functionality of MATLAB, readers can view the source codes and outputs (numerical and graphical) in an integrated way. The following shows the names of files and variables in *italics*.

A.1 Chapter 2

A.1.1 Example 2.1 (Section 2.3.1)

Objective:

- Develop a Gutenberg-Richter magnitude recurrence relationship (Gutenberg & Richter, 1956) for the Tohoku region of Japan using an earthquake catalog from the USGS NEIC (https://earthquake.usgs.gov/earthquakes/search/) and generate stochastic event catalogs based on the fitted Gutenberg-Richter magnitude recurrence relationship.

Input and datasets:

- Extracted coastal line data from the GSHHG database (*Japan_coastline_GSHHG_Elsevier.mat*)
- Extracted earthquake data from the USGS NEIC catalog (*NEICcatalog_Elsevier.mat*)

Method:

- In Step 1, screening (selection) of the earthquake catalog data is performed based on the spatial and temporal ranges.
- In Step 2, the fitting of the Gutenberg-Richter relationship is conducted based on the maximum likelihood estimation method.
- In Step 3, the fitted model is compared with the existing models developed by Headquarters for Earthquake Research Promotion (2019).
- In Step 4, the discretized version of the fitted Gutenberg-Richter relationship is obtained.
- In Step 5, for a specified duration T, multiple stochastic earthquake catalogs (*NumSimu*) are generated by sampling the number of events during the specified duration and by sampling the magnitude values from the Gutenberg-Richter relationship. For the temporal process of earthquake occurrence, the Poisson process is considered.

Output:

- The fitted Gutenberg-Richter relationship is obtained (*GRpara*).
- The variable *EQcatalogSimu* contains (1) stochastic catalog ID number (1 to *NumSimu*), (2) earthquake occurrence time (0 to T), and (3) earthquake magnitude.

A.1.2 Example 2.2 (Section 2.3.2)

Objective:

- Demonstrate random sampling of the earthquake occurrence time from four interarrival time distributions, that is, exponential, lognormal, Brownian Passage Time, and Weibull.

Input and datasets:

- All parameters are specified within the live script.

Method:

- In Step 1, specify the interarrival time distribution and its parameters (mean, coefficient of variation, and elapsed time since the last event).
- In Step 2, earthquake occurrence times are simulated from the four interarrival time distributions. The simulated data are compared with the theoretical models in terms of the probability density function and the cumulative distribution function.
- In Step 3, hazard rate functions of the four interarrival time distributions are compared.

Output:

- The simulated samples of the interarrival time are stored in *t_exp*, *t_logn*, *t_bpt*, and *t_weibull*.

A.1.3 Example 2.3 (Section 2.3.3)

Objective:

- Implement a characteristic earthquake magnitude model based on Youngs and Coppersmith (1985) and generate a stochastic event catalog.

Input and datasets:

- All parameters are specified within the live script.

Method:

- In Step 1, parameters for the fault model and magnitude recurrence characteristics are specified. The parameters include rigidity, fault plane area, slip rate, Gutenberg-Richter slope parameter, maximum magnitude, minimum magnitude, magnitude range for the exponential distribution part, and magnitude range for the characteristic distribution part.
- In Step 2, a stochastic catalog of events is generated based on the characteristic earthquake magnitude model. The length of the earthquake catalog is specified by *num_simu*.

Output:

- The stochastic event catalog based on the characteristic model is stored in *SEC_char*.
- The stochastic event catalog based on the Gutenberg-Richter model is stored in *SEC_exp*.

A.1.4 Example 2.4 (Section 2.4.2)

Objective:

- Demonstrate the estimation of stochastic source parameters of finite-fault slip models by considering the von Karman power spectrum.

Input and datasets:

- The finite-fault model developed by Shao and Ji (2005) for the 2005 Sumatra Earthquake (*s2005SUMATR01SHAO_Elsevier.fsp*) is obtained from the SRCMOD database (http://equake-rc.info/srcmod/).

Method:

- The estimation of the von Karman power spectrum of the earthquake slip distribution is based on the *RUPGEN* package developed by Dr. Martin Mai (http://equake-rc.info/cers-software/).

- In Step 1, parameters for spectral analysis are set up.
- In Step 2, read the finite-fault model and store the information in *FFinfo*, *slipinfo*, and *slip*. The variable *slip* is a cell-based representation of the 2D earthquake slip distribution.
- In Step 3, an effective dimension analysis of the earthquake slip distribution is performed (*effdim_Elsevier.m*). Calculate the effective dimensions and update the finite-fault model information based on the effective dimensions (fault length, fault width, fault area, mean slip, and maximum slip).
- In Step 4, the Box-Cox transformation of the earthquake slip values is performed, and the Box-Cox power parameter is determined.
- In Step 5, the cell-based slip distribution is transformed into the corner-based slip distribution (*SRCMOD_slip_corner_Elsevier.m*).
- In Step 6, perform the 2D Fast Fourier Transform (FFT) and compute the normalized power spectrum of the slip distribution.
- In Steps 7 and 8, the circular average power spectrum is computed and the von Karman spectral parameters, that is, Hurst number and correlation lengths along strike and dip, are estimated (*vonKarman2d_Elsevier.m*). Two approaches are considered in fitting the von Karman spectral model to the (modified) earthquake slip distribution. The first approach is a two-step approach where the Hurst number is obtained based on the circular average power spectrum. Subsequently, the correlation lengths for the strike and dip directions are evaluated for the common Hurst number. The second approach fits the von Karman spectral model to the along-strike and along-dip power spectral values using the 2D grid search. In this case, the Hurst numbers for the strike and dip directions can be different.

Output:

- The earthquake source parameters are obtained for the analyzed finite-fault model.
- The effective dimensions are stored in the 29th and 30th columns of *FFinfo*.
- The mean slip based on the effective dimensions is stored in the 37th column of *FFinfo*.
- The maximum slip is stored in the 25th column of *FFinfo*.
- The spectral parameters based on the von Karman model are stored in *ParaSlip*. The 5th, 7th, and 8th columns of *ParaSlip* contain the Hurst number, correlation length in the dip direction, and correlation length in the strike direction, respectively. The 9th, 10th, 11th, and 12th columns of *ParaSlip* contain the Hurst number in the dip direction, Hurst number in the strike direction, correlation length in the dip direction, and correlation length in the strike direction, respectively.
- The Box-Cox parameter is stored in *best_power_para*. The same information is also stored in the 12th to 14th columns of *ParaSlip*.

A.1.5 Example 2.5 (Section 2.5.2)

Objective:

- Demonstrate the simulation of stochastic source models and the calculation of coseismic deformations based on the Okada equations.

Input and datasets:

- Extracted coastal line data for Indonesia (*coastline_Indonesia_Elsevier.mat*)
- Global plate boundary data by Bird (2003) (*Plate_Bird2003_Elsevier.mat*)
- Spectral analysis results of the SRCMOD database by Goda et al. (2016) (*SRCMOD_FinalResult_Combined_Elsevier.mat*).
- The finite-fault model developed by Shao and Ji (2005) for the 2005 Sumatra Earthquake (*s2005SUMATR01SHAO_Elsevier.fsp*) is obtained from the SRCMOD database (http://equake-rc.info/srcmod/).

Method:

- In Step 1, the relevant map data are loaded for visualization (Indonesian coastal line and global plate boundaries).
- In Step 2, extract the spectral analysis results for three finite-fault models that were developed for the 2005 Sumatra Earthquake based on Goda et al. (2016) and compare them with the predicted parameters based on the tsunamigenic event scaling relationships by Goda et al. (2016).
- In Step 3, read the finite-fault model by Shao and Ji (2005) to be used as a base in defining the geometry of the stochastic source models.

564 Appendix

- In Step 4, define a fault plane model for stochastic source modeling based on the Shao and Ji finite-fault model. The dimensions of the subfaults are set to 20 km by 20 km.
- In Step 5, set up the parameters for stochastic source modeling. The number of stochastic source models to be simulated is specified by *Num*, whereas the earthquake magnitude range of the stochastic source models is specified by *Para_Mw*.
- In Step 6, stochastic source models are generated. There are five substeps: (1) generation of source model parameters from the scaling relationships by Goda et al. (2016), (2) spectral synthesis (random field generation) based on Mai and Beroza (2002), (3) evaluation of candidate stochastic source models based on constraints, (4) kinematic rupture modeling based on Mai et al. (2005) and Melgar and Hayes (2017), and (5) generation of hypocenter locations and determination of rupture trigger times.
- In Step 7, compare the simulated earthquake source parameters with the scaling relationships by Goda et al. (2016), and display the average slip model based on the simulated stochastic sources.
- In Step 8, calculate the elastic deformations (Okada, 1985); *okada85_Elsevier.m*; the original MATLAB code was written by Dr. François Beauducel for the baseline finite-fault model by Shao and Ji (2005), generated stochastic source models, and the average source model.

Output:

- The fault plane information is stored in *FAULTPLANE*.
- The information on the accepted stochastic source models is stored in *SIMU_PARA*. The simulated slip distributions are stored in *ParaSlipSimu1*, *ParaSlipSimu2*, and *ParaSlipSimu3*, in different formats.
- The calculated coseismic deformation profiles (vertical components) are stored in *DEFORM*.

A.2 Chapter 3

In Example 3.1 and Example 3.2, linear and nonlinear tsunami simulations are performed using COMCOT. Files for Example 3.1 are stored in *Ex3_1_linear_tsunami_simulation_Elsevier*, whereas files for Example 3.2 are stored in *Ex3_2_nonlinear_tsunami_simulation_Elsevier*. MATLAB scripts that can be used for visualizing the tsunami simulation results are stored in *COMCOT_MATLAB_Codes_Elsevier*.

A.2.1 Example 3.1 (Section 3.5.2.1)

Objective:

- Run a linear tsunami simulation for the 2021 Raoul Island Earthquake (M_w 8.1) in the Kermadec subduction zone using COMCOT.

Input and datasets:

- The main MATLAB script is *Ex3_1_linear_tsunami_simulation_Elsevier.mlx*.
- A sea surface displacement source model previously estimated by Romano et al. (2021) is implemented (*Initial_disp.asc*).
- A global bathymetry digital elevation model with a grid size of 5 arc-min is obtained from ETOPO2 (https://www.ncei.noaa.gov/products/etopo-global-relief-model) (*global_5min.asc*).
- A list of stations at which tsunami time series are simulated is defined (*ts_location.dat*).
- For model validation, observed tsunami waveforms are used (*NZB.txt*, *NZC.txt*, *NZE.txt*, *NZG.txt*, and *NZI.txt*).

Method:

- In Step 1, model parameters, such as simulation time, source model, modeling domains, nested grids, boundary conditions, and the selection of linear shallow water equations, are configured in *comcot.ctl*.
- In Step 2, the tsunami is simulated with COMCOT by solving the linear shallow water equations. This step takes some time and will generate numerous COMCOT output files in the current working directory.
- In Step 3, COMCOT outputs are postprocessed to produce tsunami animation and plots using MATLAB scripts stored in *COMCOT_MATLAB_Codes_Elsevier*.

Output:

- Simulated maximum tsunami amplitude distribution.

- Simulated travel time map for tsunami propagation.
- Comparison of observed and simulated tsunami waveforms at DART stations.
- Animation of tsunami propagation (animated gif).

A.2.2 Example 3.2 (Section 3.5.3.4)

Objective:

- Run a nonlinear tsunami simulation from earthquake and landslide source models during the 2018 Palu Earthquake (M_w 7.5) in Indonesia using COMCOT.

Input and datasets:

- The main MATLAB script is *Ex3_2_nonlinear_tsunami_simulation_Elsevier.mlx*.
- The sea surface displacement source model and landslide source model parameters, which were previously estimated by Gusman et al. (2019), are implemented (*Vertical_disp.asc*).
- The bathymetry data and topography digital elevation model with a grid size of ~20 m for Palu Bay are used (Gusman et al., 2019) (*demPaluBayV5.asc*).
- A list of stations at which tsunami time series are simulated is defined (*ts_location.dat*).
- For model validation, observed tsunami waveforms and limit of inundations are used (*Pantoloan.txt* and *LimitInundation.csv*).

Method:

- In Step 1, model parameters, such as simulation time, source models, modeling domains, nested grids, boundary conditions, and the selection of nonlinear shallow water equations with inundation, are configured in *comcot.ctl*.
- In Step 2, tsunami propagation and inundation with COMCOT are simulated by solving the nonlinear shallow water equations with a moving boundary scheme. This step takes some time and will generate numerous COMCOT output files in the current working directory.
- In Step 3, COMCOT outputs are postprocessed to produce plots using MATLAB scripts stored in *COMCOT_MATLAB_Codes_Elsevier*.

Output:

- Simulated maximum tsunami amplitude distribution.
- Simulated tsunami inundation.
- Comparison of observed and simulated tsunami waveforms at the Pantoloan station.

A.3 Chapter 5

A.3.1 Example 5.1 (Section 5.2.4)

Objective:

- Understand how the hazard curve can be represented in the linear and log-log spaces and in terms of mean annual rate and annual exceedance probability.

Input and datasets:

- The values of the intensity measure in terms of tsunami wave height and the corresponding mean annual rates are provided at the beginning of the live script.

Method:

- In Step 1, the hazard curve in terms of tsunami wave height is provided in terms of the mean annual rate of exceedance.
- In Step 2, the hazard curve in terms of the mean annual rate of exceedance is converted to the annual probability of exceedance.
- In Step 3, the hazard curve in terms of tsunami wave height is provided in terms of the annual probability of exceedance.

Output:

- Plots of the tsunami hazard curves are represented in the linear and log-log spaces in terms of mean annual rate and annual exceedance probability.

A.3.2 Example 5.2 (Section 5.2.5)

Objective:

- Understand how to input a logic tree and how to compute all the possible combinations of the branches and their associated probabilities.

Input and datasets:

- A graphical representation of a logic tree by Fukutani et al. (2015) is provided directly in the live script.

Method:

- In Step 1, the variables of the logic tree are reported in a vector format to be processed in MATLAB.
- In Step 2, all possible combinations of variables are defined by combining the items of each vector defined in the previous step.
- In Step 3, the probability associated with each combination is computed by multiplying the probability associated with each option defined in the previous step.

Output:

- Tables provide the combination of options and the associated probabilities. Also, a numerical check that the final probabilities sum to 1 is performed.

A.3.3 Example 5.3 (Section 5.2.5)

Objective:

- Understand how to generate samples from a logic tree with a simulation-based approach.

Input and datasets:

- A graphical representation of a logic tree from Fukutani et al. (2015) is provided directly in the live script.

Method:

- In Step 1, the variables of the logic tree are reported in a vector format to be processed in MATLAB.
- In Step 2, the weights (i.e., PMF) are transformed into probabilities of not being exceeded (i.e., CDF).
- In Step 3, five consecutive random numbers are simulated a *Number_of_simulations* number of times.
- In Step 4, the simulations are represented with histograms.

Output:

- A table provides the simulated values, and the histograms show the simulations graphically.

A.3.4 Example 5.4 (Section 5.2.5)

Objective:

- Understand how to use the logic tree to compute a weighted average and to compute percentile hazard curves based on the ensemble approach.

Input and datasets:

- A hypothetical logic tree with 10 possible outcomes having specific weights is directly provided in the live script.
- The medians and the logarithmic standard deviations for the 10 hazard curves associated with the possible outcomes of the logic tree are provided in the live script.

Method:

- In Step 1, the hypothetical hazard curves are plotted.

- In Step 2, the weighted average is computed and represented.
- In Step 3, the ensemble approach is employed and the percentile hazard curves are computed.

    Output:

- Graphical representations of the hypothetical hazard curves, the weighted average, and the ensemble approach.

A.3.5 Example 5.5 (Section 5.3.4)

Objective:

- Understand how to represent an empirical tsunami hazard curve and the curves corresponding to the 16th and 84th percentiles.

    Input and datasets:

- A set of simulated tsunami wave heights (in meters) computed for a specific location on the Sendai coast with a magnitude of 9 is provided at the beginning of the live script.

    Method:

- In Step 1, the hazard curves are computed with the function *Empirical_CCDF* provided at the end of the script.

    Output:

- A graphical representation of the hazard curves.

A.3.6 Example 5.6 (Section 5.3.4)

Objective:

- Understand how to convolute tsunami hazard curves corresponding to different magnitude scenarios to evaluate the PTHA integral.

    Input and datasets:

- A set of seven vectors of simulated tsunami wave heights (in meters) computed for a specific location on the Sendai coast corresponding to seven values of magnitude (i.e., 7.5, 7.75, 8, 8.25, 8.5, 8.75, and 9) is provided at the beginning of the live script.
- The Gutenberg-Richter recurrence information for the seven values of magnitude is specified in the live script.

    Method:

- In Step 1, the conditional hazard curves for each value of magnitude are computed and represented similarly to Example 5.5.
- In Step 2, the conditional hazard curves are multiplied by the probability values provided by the Gutenberg-Richter relationship for the seven values of magnitude.
- In Step 3, the hazard curves obtained in the previous step are summed up and premultiplied by the mean annual rate of observing a magnitude larger than the minimum value.

    Output:

- Graphical representations of the hazard curves for the different steps provided above.

A.3.7 Example 5.7 (Section 5.3.5)

Objective:

- Understand how to derive the tsunami hazard intensity for specific values of return period.

    Input and datasets:

- A set of simulated tsunami wave heights (in meters) computed for a specific location on the Sendai coast with a magnitude of 9 is provided directly at the beginning of the live script.
- A vector of four possible return periods is specified in the live script.

Method:

- In Step 1, an interpolation of the hazard curves is carried out for the predefined return period.

Output:

- Graphical representations of the operation above.

A.3.8　Example 5.8 (Section 5.3.6)

Objective:

- Understand how to perform a bootstrap analysis to check how many simulations are enough for a stable probabilistic result.

Input and datasets:

- A set of 500 simulated tsunami wave heights (in meters) computed for a specific location on the Sendai coast with a magnitude of 8.5 is provided at the beginning of the live script.
- The desired number of samples is specified in the live script.

Method:

- In Step 1, the bootstrap is performed and multiple hazard percentiles are computed.
- In Step 2, the results are plotted.

Output:

- Graphical representations of the operation above.

A.3.9　Example 5.9 (Section 5.4.1)

Objective:

- Understand how to define the exposure data for low-rise timber buildings in Onagawa, Miyagi Prefecture, Japan.

Input and datasets:

- Probabilistic information on the construction costs is given in the live script.
- Damage ratios for different damage states are defined in the live script.

Method:

- In Step 1, the distribution of the value at stake is computed.
- In Step 2, the distribution is represented, and the damage ratios are shown.

Output:

- Graphical representations of the operation above.

A.3.10　Example 5.10 (Section 5.4.1)

Objective:

- Understand how to simulate the exposed value and the possible damage ratio for a specific damage state based on a simulation-based procedure using the exposure data for low-rise timber buildings in Onagawa, Miyagi Prefecture, Japan.

Input and datasets:

- Probabilistic information on the construction costs is given in the live script.
- Damage ratios for different damage states are defined in the live script.
- The desired number of simulations is specified in the live script.
- The assumed damage state is specified in the live script.

Method:

- In Step 1, the simulation is performed.
- In Step 2, the distribution and the simulated values are graphically represented.
- In Step 3, a damage ratio is simulated for the assumed damage ratio.

Output:

- Graphical representations of the operation above.

A.3.11 Example 5.11 (Section 5.4.2)

Objective:

- Understand how to plot tsunami fragility curves.

Input and datasets:

- The parameters for the lognormal model are obtained from Suppasri et al. (2013).
- The parameters and the function for the multinomial model are obtained from De Risi et al. (2017).

Method:

- In Step 1, the lognormal model is computed and plotted.
- In Step 2, the multinomial model is computed and plotted.

Output:

- Graphical representations of the operation above.

A.3.12 Example 5.12 (Section 5.4.2)

Objective:

- Understand how to simulate the possible damage state for a specific level of intensity measure.

Input and datasets:

- The parameters for the lognormal model provided are obtained from Suppasri et al. (2013).
- The intensity measure of interest is specified in the live script.

Method:

- In Step 1, the lognormal model is computed, and the damage state is simulated for the assumed value of intensity measure.

Output:

- Graphical representations of the operation above.

A.3.13 Example 5.13 (Section 5.4.3)

Objective:

- Understand how to convolute the risk integral using a single tsunami scenario corresponding to a single value of earthquake magnitude.

Input and datasets:

- Information on the exposed asset, that is, floor area and number of stories, is given in the live script.
- The parameters for the lognormal model for the vulnerability are obtained from Suppasri et al. (2013).
- Probabilistic information on the construction costs is given in the live script.
- Damage ratios for different damage states are defined in the live script.

- A set of 500 simulated tsunami wave heights (in meters) computed for a specific location on the Sendai coast with a magnitude of 8.25 is provided at the beginning of the live script.
- The function *Empirical_CCDF* is provided at the end of the script.

Method:

- In Step 1, the fragility functions are defined according to Example 5.10.
- In Step 2, the simulation-based approach is implemented.
- In Step 3, the loss curve and the corresponding confidence interval are computed.

Output:

- Graphical representations of the operation above.

A.3.14 Example 5.14 (Section 5.4.3)

Objective:

- Understand how to convolute the risk integral using multiple tsunami scenarios corresponding to seven values of earthquake magnitude (i.e., 7.5, 7.75, 8, 8.25, 8.5, 8.75, and 9).

Input and datasets:

- Information on the exposed asset, that is, floor area and number of stories, is given in the live script.
- The parameters for the lognormal model for the vulnerability are obtained from Suppasri et al. (2013).
- Probabilistic information on the construction costs is given in the live script.
- Damage ratios for different damage states are defined in the live script.
- A set of seven vectors of simulated tsunami wave heights (in meters) computed for a specific location on the Sendai coast corresponding to seven values of earthquake magnitude (i.e., 7.5, 7.75, 8, 8.25, 8.5, 8.75, and 9) is provided at the beginning of the live script.
- The function *Empirical_CCDF* is provided at the end of the script.

Method:

- In Step 1, the fragility functions are defined according to Example 5.11.
- In Step 2, the consequence functions are fully defined.
- In Step 3, the simulation-based approach is implemented.
- In Step 4, the loss curves and the corresponding confidence interval are computed for the seven values of earthquake magnitude.
- In Step 5, the conditional loss curves are multiplied by the probability values provided by the Gutenberg-Richter relationship for the seven values of earthquake magnitude.
- In Step 6, the loss curves obtained in the previous step are summed up and premultiplied by the mean annual rate of observing a magnitude larger than the minimum value.

Output:

- Graphical representations of the operation above.

Exercises with demo computer codes and datasets that are linked to the Appendix are available in the below companion site. https://www.elsevier.com/books-and-journals/book-companion/9780443189876.

References

Bird, P. (2003). An updated digital model of plate boundaries. *Geochemistry, Geophysics, Geosystems, 4*(3). Available from https://doi.org/10.1029/2001GC000252.

Fukutani, Y., Suppasri, A., & Imamura, F. (2015). Stochastic analysis and uncertainty assessment of tsunami wave height using a random source parameter model that targets a Tohoku-type earthquake fault. *Stochastic Environmental Research and Risk Assessment, 29*(7), 1763–1779. Available from https://doi.org/10.1007/s00477-014-0966-4, http://link.springer-ny.com/link/service/journals/00477/index.htm.

Goda, K., Yasuda, T., Mori, N., & Maruyama, T. (2016). New scaling relationships of earthquake source parameters for stochastic tsunami simulation. *Coastal Engineering Journal, 58*(3). Available from https://doi.org/10.1142/S0578563416500108, https://www.tandfonline.com/loi/tcej20.

Gusman, A. R., Supendi, P., Nugraha, A. D., Power, W., Latief, H., Sunendar, H., Widiyantoro, S., Daryono, S. H., Wiyono, A., Hakim, A., Muhari, X., Wang, D., Burbidge, K., Palgunadi, I., Hamling, M. R., & Daryono. (2019). Source model for the tsunami inside Palu Bay following the 2018 Palu earthquake, Indonesia. *Geophysical Research Letters*, *46*(15), 8721−8730. Available from https://doi.org/10.1029/2019GL082717, http://agupubs.onlinelibrary.wiley.com/hub/journal/10.1002/(ISSN)1944-8007/.

Gutenberg, B., & Richter, C. F. (1956). Magnitude and energy of earthquakes. *Annals of Geophysics*, *9*(1), 15.

Headquarters for Earthquake Research Promotion (2019), Evaluation of long-term probability of earthquake occurrence along the Japan Trench [Unpublished content].

Mai, P. M., & Beroza, G. C. (2002). A spatial random field model to characterize complexity in earthquake slip. *Journal of Geophysical Research: Solid Earth*, *107*(11). Available from https://doi.org/10.1029/2001JB000588, http://onlinelibrary.wiley.com/journal/10.1002/(ISSN)2169-9356.

Mai, P. M., Spudich, P., & Boatwright, J. (2005). Hypocenter locations in finite-source rupture models. *Bulletin of the Seismological Society of America*, *95*(3), 965−980. Available from https://doi.org/10.1785/0120040111.

Melgar, D., & Hayes, G. P. (2017). Systematic observations of the slip pulse properties of large earthquake ruptures. *Geophysical Research Letters*, *44*(19), 9691−9698. Available from https://doi.org/10.1002/2017GL074916, http://onlinelibrary.wiley.com/journal/10.1002/(ISSN)1944-8007/issues?year = 2012.

Okada, Y. (1985). Surface deformation due to shear and tensile faults in a half-space. *Bulletin of the Seismological Society of America*, *75*(4), 1135−1154. Available from https://doi.org/10.1785/bssa0750041135.

De Risi, R., Goda, K., Yasuda, T., & Mori, N. (2017). Is flow velocity important in tsunami empirical fragility modeling? *Earth-Science Reviews*, *166*, 64−82. Available from https://doi.org/10.1016/j.earscirev.2016.12.015, http://www.sciencedirect.com/science/journal/00128252.

Romano, F., Gusman, A. R., Power, W., Piatanesi, A., Volpe, M., Scala, A., & Lorito, S. (2021). Tsunami source of the 2021 Mw 8.1 Raoul Island earthquake from DART and tide-gauge data inversion. *Geophysical Research Letters*, *48*(17). Available from https://doi.org/10.1029/2021GL094449, http://agupubs.onlinelibrary.wiley.com/hub/journal/10.1002/(ISSN)1944-8007/.

Shao G., Ji C. 2005 Preliminary result of the Mar 28, 2005 Mw 8.68 Nias earthquake, http://www.geol.ucsb.edu/faculty/ji/big_earthquakes/2005/03/smooth/nias.html

Suppasri, A., Mas, E., Charvet, I., Gunasekera, R., Imai, K., Fukutani, Y., Abe, Y., & Imamura, F. (2013). Building damage characteristics based on surveyed data and fragility curves of the 2011 Great East Japan tsunami. *Natural Hazards*, *66*(2), 319−341. Available from https://doi.org/10.1007/s11069-012-0487-8, http://www.wkap.nl/journalhome.htm/0921-030X.

Youngs, R. R., & Coppersmith, K. J. (1985). Implications of fault slip rates and earthquake recurrence models to probabilistic seismic hazard estimates. *Bulletin of the Seismological Society of America*, *75*(4), 939−964. Available from https://doi.org/10.1785/BSSA0750040939.

Index

Note: Page numbers followed by "*f*" and "*t*" refer to figures and tables, respectively.

A

Absorbing coefficients, 101–102
Absorption zone, 101–102
Accretionary prisms, 277
Accretionary wedges, 284
Agent-based disaster response model, 550–551
Agent-based evacuation models, 17
Agent-based modeling, 501–502
Agent-based tsunami evacuation simulations, 501–503
    application of, 505–507
    development of, 503–505
Aggregated intensity data, 438
AIC. *See* Akaike information criterion (AIC)
Aida's *k*., 162–163
Air pressure wave, 70
Air pressure wave source, 107–108
Akaike information criterion (AIC), 146, 150, 209–210, 444
Alaskan earthquakes, 277
Aleatory uncertainty, 375, 462
Amazon, 333
American Society of Civil Engineers (ASCE), 121, 125, 127, 361
    framework of tsunami design zone, 360–367
Anak Krakatau volcano, 11–12
Analytical fragility functions, 150–151
Analytical tsunami fragility curves, 449
    development of analytical fragility curves, 462
    structural modeling for tsunami analysis, 458–459
    tsunami actions on buildings, 450–454
        characterization of tsunami inundation flows, 450
        sustained lateral loads, 451–453
        sustained vertical loads, 453–454
        tsunami intensity measures for fragility analysis, 454
    tsunami damage assessment, 459–462
    tsunami structural assessment approaches, 454–457
        tsunami nonlinear static analyses, 456–457
        tsunami nonlinear time-history dynamic analysis, 455
Andean region, 422–423
Angular frequency, 83
Ansei–Nankai Earthquake, 245
ArcGIS software, 506, 515–516
Artificial intelligence (AI), 326, 444, 543

Artificial neural networks, 326
ASCE7-22, 450, 452–453
ASCE-VDPO2, 457
As Low As Reasonably Practicable (ALARP), 203
Asset-to-asset variability, 439
ASTARTE Paleotsunami Deposits Database, 399
Atmospheric pressure, 403
Augmented reality, 543
Autocorrelation function model, 42
Average annual loss (AAL), 3, 157, 178–179, 217–218

B

Back-projection method, 305
Balearic islands, 403
Bare-earth topography, 354
Baseline tidal level, 527–528
Bathymetry distribution, 468
Bathymetry mapping, 259
Bayesian inference, 444–445
Bayesian information criterion (BIC), 146, 150
Bayesian model class selection, 444
Bayesian trans-dimensional approach, 423
Benefit–cost analysis, 192
Bernoulli's principle for fluid dynamics, 83
B/h ratio, 128
Binned regression, 148*f*
Binomial logistic method, 147–148
Boolean logic rules, 493–494
Bootstrap method, 445
Bottom pressure recorder (BPR), 340
Boussinesq-type approach, 93
Box-Cox analysis, 42
Box-Cox transformation, 44–45, 53–54
Brownian passage time (BPT) distribution, 33, 471
Building exposure model, 532
Bulking factor, 292–293
Buoyancy, 126, 453–454
b-value, 30

C

Cabled bottom pressure gauges, 236–237
Caldera collapse, 67, 70–71
Campi Flegrei caldera, 397
Caribbean and adjacent regions (CARIBE-EWS), 350

Caribbean and Adjacent Regions Tsunami Sources and Models (CATSAM), 358, 359*f*
Caribbean Sea, 4, 358
Cartesian coordinates, 88–89, 93, 101–102
Cascadia megathrust events, 196
Cascadia subduction earthquakes, 529
Cascadia subduction events, 197–198, 537–539
Cascadia subduction zone (CSZ), 30, 359, 480–481, 483, 496
Central Disaster Management Council's model, 51–52, 283–284
Central processing unit (CPU), 344–345, 347*t*
Centroid moment tensor (CMT), 28–29, 28*f*, 344–345
Chart datum (CD), 113–114
Chatham Islands, 373
Chile, 417
    tsunami hazard assessments in, 420–428
        boundary conditions and local hydrodynamics, 420–421
        initial conditions, 421–423
        other susceptibility estimation studies, 425–426
        scenario-based tsunami hazard assessment, 423
        seismic probabilistic tsunami hazard assessment, 423–425
        short-term assessment, 426–428
    tsunami occurrence in, 419–420
Chilean coasts, 421
Chilean margin, 419
Chilean tectonic context, 417–418
Chilean Tsunami Warning System, 426–427
Climate change, 404, 525
Cloglog, 441
Closure coefficient, 128, 451
Clusters, 305
Coastal communities, 123–124, 525
Coastal digital twins (CDTs), 543–544
    fusion of data and technological elements of, 554
    implementation of, 554
    towards establishing coastal digital twin paradigm, 552–554
    tsunami disaster digital twin in Japan, 544–552
        multiagent modeling of disaster response activities, 550–552

574 Index

Coastal digital twins (CDTs) (*Continued*)
    real-time tsunami inundation and damage forecast, 546–548
    tsunami exposure analysis using mobile spatial statistics, 548–549
Coastal engineering, 79
Coastal sea-level measurements, 235–236
Coastal topography, 450
Coastal variability, 397
COMCOT, 389
    tsunami simulation program, 65
ComMIT/MOST model, 484
Commonwealth of the Northern Mariana Islands (CNMI), 359
Community model interface for tsunamis (ComMIT), 343
Complementary cumulative distribution function (CCDF), 160–161, 482
Compound and cascading hazard modeling, 473–474
Compounding damage effects, 493
Computational efficiency, 330–331
Computational fluid dynamics, 450
Conditional tsunami risk curves, 529–533
    building exposure model, 532
    conditional tsunami risk distributions with different baseline tidal levels, 532–533
    stochastic rupture model, 529–530
    tsunami damage and loss estimation, 532
    tsunami fragility model, 532
    tsunami inundation model, 530–532
Conditional VaR (CVaR), 218–219
Consequence assessment, 193
Consequence model, 161
Constant depth pushover (CDPO), 151, 456, 456f
Constant Height Pushover, 456
Constant Roughness Model (CRM), 112–113
Continental margin, 420
Continuity equation, 90
Contouring algorithm, 292
Coriolis effects, 93
Coriolis force coefficient, 93
Correlation coefficient matrix, 180
Correlation functions, 307–308
Coulomb friction coefficient, 64–65, 107
COULWAVE model, 484
Courant–Friedrichs–Lewy (CFL) condition, 97
Courant number, 97
Covariance matrix, 307
Creep, 259
Crossing fragility curves, 441–442
Crustal fault earthquake models, 381
CSZ. *See* Cascadia subduction zone (CSZ)
Cumulative distribution function (CDF), 439, 482
Curvature effects, 298

D

Damage ratios (Drs), 175–176
Damage scale for earthquakes and tsunami (DET), 459, 460t
Damage states (DSs), 121, 161

Darcy friction factor, 83
Data assimilation approach, 306–312, 325, 546
    Green's function–based tsunami data assimilation, 308–310
    near-field tsunami data assimilation, 310–312
    tsunami data assimilation, 306–308
Data-driven approach, 544, 546
Data quality, 176
Debris impact loads, 134–142
    alternative detailed approach, 140–141
    impact by shipping containers, 138–140
    impact by vehicles, 137–138
    impact force by general objects, 136–137
    simplified approach, 141
    site hazard assessment, 135–136
Decision-making process, 331–332, 502–503
Decision variables (DV), 160
Deep learning approaches, 206–207, 211
Deep-ocean assessment and reporting of tsunamis (DART), 89–90, 238–239, 239f, 339–340, 341f
Deep ocean measurements, 238–239
DEMs. *See* Digital elevation models (DEMs)
Dense and light pyroclastic flow models, 69–70
Dense Ocean floor Network system for earthquakes and tsunamis (DONET), 237, 303–304
Dense tsunami monitoring system
    accurate earthquake source estimation, 304–305
    data assimilation, 306–312
        Green's function–based tsunami data assimilation, 308–310
        near-field tsunami data assimilation, 310–312
        tsunami data assimilation, 306–308
    observed tsunamis from other sources, 316–321
        analysis of 2022 Tonga Tsunami, 317–321
        meteotsunamis, 316–317
    tsunami early warning, 312–315
        tsunami forecasting method based on inversion for initial sea-surface height, 312–315
        tsunami forecast using multiindex method for a scenario database, 315
Deterministic assessments, 356
Deterministic tsunami hazard assessment, 356–359
    deterministic hazards assessment based on sensitivity tests using synthetic tsunamis, 358–359
    worst-case scenarios of crustal faults in the Pacific Northwest, 356
    worst-credible scenarios for Hawaii and the Caribbean Sea, 356–358
Diagnostic analysis, 150
Diffuse permanent deformation, 270
Digital elevation models (DEMs), 82–83, 157, 223, 342, 354, 355f, 383
Digital terrain model, 157
Digital twin, 543

Disaggregation, 172–173
Disaster cycle framework, 4f
Disaster resilience, 191
Disaster risk management (DRM), 3, 18, 191
Disaster risk reduction (DRR), 3, 18, 191
Discrete probability, 167–168
Dispersive tsunami model, 93–94
Donggala Port, 10
Drag coefficients, 128, 451
    for rectilinear structures, 128t
    for structural components, 130t
Dry cells, 102
Dynamic agent-based evacuation
    on agent-based tsunami evacuation simulations, 501–503
    application of, 505–507
    development of, 503–505
Dynamic analysis, 140
Dynamic phase, 400
Dynamic rupture models, 263

E

Earthquake hazard model, 484–486
Earthquake rupture modeling, 13–14, 39–50, 525–526
Earthquakes, 1, 163, 289
Earthquake source modeling, 378–381, 471–472
    crustal fault earthquake models, 381
    plate interface earthquake models, 378–381
Earth's rising temperature, 404
Elastic buried-rupture model, 270
Electromagnetic techniques, 260
Element drag forces, 452
Empirical fragility assessment, 439
Empirical fragility curves, 437
    correlations and dependences, 446
    definitions and limitations of, 438–441
        class of asset at risk, 439
        for damage scale, 440–441, 440t
        tsunami damage scale, 439–440
        tsunami intensity measure, 439
    lack of standard damage scales, 446
    methods for empirical tsunami fragility assessment, 441–445
        sources of uncertainties in tsunami intensity characterization, 442–443
        sources of uncertainties related to fragility model parameters, 444–445
        sources of uncertainties related to the choice of fragility models, 444
        sources of uncertainties related to tsunami damage evaluation, 443–444
    scale issues, 446
    spatial and temporal gaps, 446
    surveyors' bias, 446
    uncertainty in IM, 446
Empirical fragility functions, 145–150
    binomial logistic method, 147–148
    functional forms of link function, 150
    lognormal method, 146–147
    model selection, 150
    multinomial logistic method, 148–149

Energy-based stress drop, 304–305
Energy Grade-line Analysis method, 450
Energy trapping, 421
Engineering demand parameters (EDPs), 142, 455, 459
Episodic tremor and slip (ETS), 261
Epistemic uncertainty, 162, 375
Equation of mass conservation, 88–89
Equation of motion, 89
Equivalent Roughness Model (ERM), 112–113
Euler's equation, 89
Evacuation, 508
Event loss table (ELT), 218
Exceedance probability (EP) curves, 3, 193
Explanatory variables, 145–146
Exponential model, 35–36

F

Fast Fourier Transform (FFT), 45
Fast marching method (FMM), 305
Fault slip model, 483
Fault zone rheology, 259–260
Federal Emergency Management Agency, 459–461
FEM. *See* Finite element modeling (FEM)
Fiber-based numerical techniques, 458
Fiber-based structural modeling, 458
Financial risk diversification, 221–222
Finite area uniform slip (FAUS), 423
Finite difference equations, 307
Finite element modeling (FEM), 79, 96–99, 151, 263–265, 286, 458
Finite-fault models, 39–42, 41*f*, 55*f*
First-order approximation, 400
Flank collapse model, 104
Flat-surface dislocation model, 263–265
Flow velocity sign function, 452
Fluid dynamics principles, 128
F–N curve, 16
Forward modeling, 233
Forward problem, 39
Fourier coefficients, 53–54
Fourier integral method, 168
Fourier spectral analysis, 42
Fourth generation (4G), 340
Fourth-order Runge–Kutta method, 84
Fragility curves, 244*f*, 437
 for different damage levels, 441*f*
Fragility functions, 449, 473, 490–491
Fragility modeling, 15–16, 176–177, 442, 472–473
Frequency dispersion, 93
Frequency–magnitude modeling, 13–14
Friction law, 259
Frontal thrust, 270
Froude number, 400, 452–454
Froude number constant, 151
Fukushima Earthquake, 304–305

G

Gaussian function, 307–308, 313
Gaussian mixture model, 529

Gaussian random values, 427
Gauss–Seidel method, 98–99
GED4GEM database, 175
General Bathymetric Chart of the Oceans (GEBCO), 109, 354
Generalized linear model (GLM), 146
Generated rupture models, 14–15
GeoClaw model, 425
Geographic information systems (GIS), 501
Geological hazard monitoring system (GeoNET), 351
Geophysical imaging, 260
Geospatial Information System, 196
Giant tsunamis, 465
Global Earthquake Model, 377
Global navigation satellite system (GNSS), 328, 344–345, 348–349
Global Navigation Satellite System Earth Observation Network System (GEONET), 546–547
Global Positioning System (GPS), 343
Global sea level (GSL), 404
Global Sea Level Observation System (GLOSS), 235
Global warming, 404
Google Cloud, 333
Governing equations, 93, 104–105, 468–470
Graphic processor unit (GPU) technology, 344–345, 347*t*
Gravitational acceleration, 311
Green's function–based tsunami data assimilation, 308–310
Green's functions, 233–234, 263–265, 333, 362
Green's law, 82
Greenwood's formula, 170
Grid cells, 102
Grounding limit, 136
Ground motion models (GMMs), 163
Ground motion prediction equations (GMPEs), 480, 486
Ground shaking (GS), 479
Gutenberg–Richter (G–R) relationship, 27, 39*f*, 167, 167*f*, 471, 482

H

Haida Gwaii Earthquake, 307–308
Harmonic wave function, 83
Hazard analysis, 289
Hazard assessment, 193, 272
Hazard curves, 160, 474
Hazard maps, 170–172
HAZUS earthquake model, 459
Hellenic Arc, 397
Heterogeneous earthquake slip distribution, 529–530
High-density pyroclastic flow, 104–106
Highest astronomical tide (HAT), 113–114
High water level (HWL), 472
Hunga Tonga–Hunga Ha'apai volcano explosion, 108
Hurst number, 45–47
Huygens's principle, 87, 327–328
Hydrodynamic forces, 142, 452–453

Hydrodynamic loads, 127–134
 drag force on elevated horizontal slabs, 131–134
 drag force on individual components, 129
 drag force on perforated walls, 130–131
 drag force on vertical structural components, 130
 overall drag force, 128
 simplified approach, 127–128
Hydrostatic forces, 141–142
Hydrostatic loads, 126–127
 buoyancy, 126
 residual water surcharge load, 127
 unbalanced lateral hydrostatic force, 126–127

I

Indian Ocean Tsunami Warning and Mitigation System (IOTWMS), 350, 352
Inference, 544
Informing megathrust tsunami source models
 slip distribution in predictive source scenarios, 262–271
  along-strike variability of rupture behavior, 271
  off-megathrust permanent deformation, 269–270
  slip distribution in dip direction, 265–269
  static kinematic models informed by fault mechanics, 263–265
 subduction megathrust and tsunamigenic earthquakes, 258–262
  fault zone rheology and fault friction, 259–260
  fault zone structure affected by seafloor morphology and sediment subduction, 258–259
  intriguing strike dimension, 262
  updip and downdip limits of megathrust rupture, 260–261
Input variables, 145–146
In situ surveys, 443
Insurance-linked securities (ILS), 217–218
Insurance-reinsurance system for tsunami risk, 220–223, 220*f*
 financial risk diversification, 221–222
 policyholders, insurers, and governments, 220–221
 risk-based insurance rate making for tsunamis, 223
 risk perception and behavioral issues, 222–223
 risk transfer for tsunami, 224–226
Intensity measures (IMs), 159–160, 437–438, 465
 observed IM, 442
 simulated IM, 442
Interarrival time (IAT) distribution, 31
Interclass correlations, 442
Interfacial shear stress, 105–106
Intergovernmental Oceanographic Commission (IOC), 235
Intergovernmental Panel on Climate Change (IPCC-AR5), 527–528

Intergovernmental Panel on Climate Change 6th Assessment Report (IPCC 6AR), 513
International Hydrographic Organization (IHO), 354
Internet of Things (IoT), 326
Intriguing strike dimension, 262
Inundation flow velocity, 450
Inundation forecast models, 342–343
Inverse Box-Cox transformation, 44–45
Inverse problem, 39
Inverse transformation method, 33, 36
Inversion algorithm, 341–342
Inversion analysis, 325
Ishinomaki Red Cross Hospital (IRCH), 550–551

J
JAGURS, 99
Japanese MLIT damage scale, 439
Japanese Red Cross Society, 550–551
Japan Meteorological Agency (JMA), 205, 235, 303
Japan Reconstruction Agency, 192
Japan Sea coast, 546
Japan Self-Defense Forces (JSDF), 550
Japan's tsunami, 351
JMA. See Japan Meteorological Agency (JMA)
Jogan Earthquake, 247–248
Joint Australian TWC (JATWC), 351–352

K
Kaikoura earthquake, 243, 373
Kajiura filter, 60–62
Kaplan–Meier estimator, 170
Karhunen–Loéve expansion (KL), 425
Kermadec subduction zones, 87–88, 109–110, 384
Kernel functional forms, 443
Kinematic finite-fault models, 42
Kinematic stochastic tsunami source modeling, 54–57
Knee-point method, 210
Kumano mega-splay fault, 279–280
Kumano splay fault, 283–284

L
Lamb waves, 108, 320–321
Lame's constants, 59–60
Lampedusa island, 403
Landslide model, 298
Landslide PTHA (L-PTHA), 289, 291, 399–400
Landslide-triggered tsunamis, 397
Landslide tsunamis, 289
Latin Hypercube algorithm, 462
Leap-frog scheme, 96
Lehigh University, 137
Limit Equilibrium Method, 400
Linear correlation coefficients, 49t
Linear generation model, 293–294
Linear long-wave approximation, 305
Linear regression, 441
Linear shallow-water equations, 108–109
Linear superposition, 405
Linear tsunami models, 90–92, 108–109, 298
Linear tsunami simulation, 110
Link functions, 441
Lisbon earthquake, 244
Local crustal faults, 384
Local subduction zones, 384
Logic tree, 474
  approach, 162–163
  framework, 362
  model, 481–482
Logistic regression, 147
Logit function, 147–148
Log–log space, 161–162
Lognormal method, 146–147
Long-term hazard assessment, 352–367
  challenges for, 352–353
  deterministic tsunami hazard assessment, 356–359
  global and regional digital elevation models and model grids, 354
  literature review and data collection, 353–354
  model validation using historical data, 355–356
  probabilistic tsunami hazard assessment, 360–367
  tsunami model benchmarking, 354–355
Long-term probabilistic tsunami risk model, 526
  long-term tsunami risk assessment for Tofino, 533–539
    effects of baseline sea levels, 533–537
    tsunami risk curves for different elapsed times and relative sea-level rise scenarios, 537–539
  for Tofino, 528–533
    conditional tsunami risk curves, 529–533
    long-term tsunami risk analysis with tidal variations and relative sea-level rises, 533
    occurrence and magnitude models of megathrust Cascadia earthquakes, 529
  Tofino and physical environment, 526–528
    Cascadia subduction zone, 526
    District of Tofino, 526, 527f
    tides and relative sea-level rise in Tofino, 527–528
Long-wave theory, 84–86
Loss curve, 178–179
Loss metrics, 160
Low-density pyroclastic flow (LDF) model, 106–107
Lowest astronomical tide (LAT), 113–114

M
Machine-learning algorithms, 17–18, 146
Machine learning-based tsunami forecasting, 325
  advantages, limitations, and future direction, 330–333
  existing studies, 327–330
  theoretical framework, 326–327
  tsunami inundation waveform forecasting using, 328
Magnitude–frequency relationship, 167, 525–526
Magnitude scaling law, 486
Makran plate boundary, 278
Makran region, 278, 282
Makran subduction zone, 278–279, 281
Manning's roughness coefficient, 82–83, 92–93, 105–107, 112, 389–390, 450, 470, 515
Mantle wedge corner (MWC), 261
Marrobbio., 402–403
Mass-wave interactions, 397
Maximum likelihood estimation (MLE), 147–148
Mazaro River, 403
Mean high water neaps (MHWN), 113–114, 116f
Mean high water springs (MHWS), 113, 116f, 390
Mean low water springs (MLWS), 113–114, 116f
Mean sea level (MSL), 113, 116f
Mean squared errors (MSE), 211, 211f
Mean water level (MWL), 472
Mediterranean basin, 406
Mediterranean coasts, 404–405
Mediterranean Sea, 4, 333, 397, 407–410
  multisource probabilistic tsunami hazard analysis in Gulf of Naples, 406
  potential tsunamigenic sources in, 398f
  regional tsunami hazard, 398–406
    effects of climate change and vertical land movements on tsunamis hazard, 404–406
    meteotsunami hazard, 402–404
    seismic tsunami hazard, 399
    submarine and subaerial landslides tsunami hazard, 399–401
    volcanic tsunami hazard, 401–402
  regional tsunami risk index, 406–407
Mega-splay, 269
Mega-splay thrust system, 279–280
Megathrust fault gouge, 260
Meiji Sanriku Tsunami, 122, 245
Memory-less Poisson process, 33
Meteotsunami hazard, 402–404
Meteotsunami hot spots, 402–403
Meteotsunamis, 70, 90, 316–317, 321, 402, 404
Method of Splitting Tsunami (MOST) model, 341–342, 354
Microsoft Azure, 333
Ministry of Land, Infrastructure, Transport, and Tourism (MLIT), 123, 143f, 438
Missing damage levels, 442
Mixed reality, 543
Modifiers, 174
Monte Carlo analysis, 375, 399
Monte Carlo sampling, 178, 203, 375, 539
Monte Carlo simulations, 32–33, 172, 185, 462
Multihazard risk assessments
  at Seaside, Oregon, 480–496

building damage assessment, 491–493
building inventory in, 488–490
calculating annual exceedance probability and hazard map, 486–488
combined damage probabilities for, 493–495
earthquake hazard model, 484–486
fragility functions for, 490–491
loss and risk assessment, 495–496
study site, 480–481
tsunami hazard model, 483–484
Multinomial logit regression, 148–149
Multinomial probit regression, 149
Multiple faulting, 270
Multiple linear regression model, 209–210
Multivariate normal distribution, 168

N

Nankai Earthquake, 277
Nankai–Tonankai Earthquake Tsunami, 505
Nankai–Tonankai Trough region, 52
Nankai–Tonankai Trough tsunami, 284
Nankai Trough Earthquake, 546, 549
Nankai Trough zone, 279–280
National Centers for Environmental Information (NCEI), 244, 342
National Crash Analysis Center, 137–138
National Data Buoy Center (NDBC), 340–341
National Earthquake Information Center, 422–423
National Oceanic and Atmospheric Administration (NOAA), 87, 380
National Tsunami Hazard Mitigation Program (NTHMP), 354
Natural geographical barriers, 135–136
Navier–Stokes equations, 401
Navier–Stokes type models, 299
Near-field tsunami data assimilation, 310–312
Near-field tsunami hazard analysis, 263
Nearshore and offshore tsunami modeling, 108–110
 linear tsunami simulation, 110
Necanicum River, 481
Neural network, 211
Newton's second law, 89
New Zealand's National Tsunami Hazard Model (NTHM), 373
Nonlinear dynamic analysis method, 450
Nonlinear shallow water equations, 515
Nonlinear tsunami model, 92–93
Nonstructural elements, 126
Nonstructural systems (NSS), 490
North-Eastern Atlantic, Mediterranean, and connected seas (NEAMTWS), 350, 352
Numerical modeling, 79, 291
Numerical simulation, 233
Numerical tsunami models, 118

O

Observed tsunami waveforms, 109–110
Ocean bottom pressure (OBP), 236
Ocean bottom seismometers (OBSs), 237
Oceanographic radar measurements, 240
Ocean waves, 79–80, 80f
Office of Emergency of the Ministry of Interior (ONEMI), 417–418
Off-megathrust permanent deformation, 269–270
Offshore tsunami measurements, 236
Okada equations, 61
One-dimensional slip functions, 42–44
OpenStreetMap, 215, 503
Operational machine learning models, 333
Ordinary least squares method (OLSM), 146
Oregon State University (OSU), 365–367
Orthogonal regressions, 422–423
Outermost grid boundary condition, 100–102
 absorbing boundary condition, 101–102
 forced boundary condition, 102
 radiation boundary condition, 100–101

P

Pacific coast, 546
Pacific Earthquake Engineering Research (PEER), 157–158
Pacific Marine Environmental Laboratory (PMEL), 339–340, 342f
Palu–Koro fault, 9–11
Pantoloan Port, 10–11
Pathfinding algorithms, 504
Path selection, 504
Peak Coastal Tsunami Amplitude (PCTA), 423
Peak ground acceleration (PGA), 181–182, 465, 486
Peak ground velocity (PGV), 465
Perceived risk, 193–194
Performance-based engineering methodology, 219–220
Personal mobility, 543
Phase spectrum, 427
Phase velocity, 86
Physics-based tsunami model, 328
Plastic-hinge-based structural modeling, 458
Plate interface earthquake models, 378–381
Plate motion velocities, 30
Poisson distribution, 471
Poisson earthquake occurrence model, 537
Poisson occurrence model, 27
Poisson process, 14, 32–33, 439, 486–487
Polynomial function, 235
Population exposure mapping system (PoEMS), 549, 549f
Popup structure, 270
Posttsunami field surveys, 242
Pressure sensors, 307–308
Primary hazards, 479
Probabilistic approach, 34t, 159–160, 360–361
Probabilistic seismic hazard analysis (PSHA), 30, 157–158, 479
Probabilistic treatment, 291–292
Probabilistic tsunami forecast (PTF), 426–427
Probabilistic tsunami hazard analysis (PTHA), 3, 27, 36, 51, 191, 196, 398–399, 479
 flowchart of, 14f
 2004 Indian Ocean Tsunami, 6–7
 2018 Indonesia tsunamis in Sulawesi and Sunda Strait, 9–12
 2010 Maule Chile tsunami, 7–8
 nonexponential magnitude distribution in, 14
 2024 Noto Peninsula Tsunami, 12, 13f
 2011 Tohoku Japan Tsunami, 8–9
 tsunami disaster risk reduction and management, 16–18
  mitigation, 16–17
  preparedness, 17
  recovery, 18
  response, 17–18
 tsunami hazard assessment, 12–15
 tsunami risk assessment, 15–16
Probabilistic tsunami hazard assessment (PTHA), 157–158, 161f, 162f, 340, 360–367, 417, 425f, 428, 465, 466f, 468
 derivation of offshore wave amplitudes, 362–364
 derivation of probabilistic 2475-year offshore tsunami amplitudes, 361–362
 examples of developing tsunami design zone maps for building design, 365–367
 extension to multihazard risks, 179–186
  compound earthquake–tsunami hazard and risk assessment, 179
  development of joint earthquake–tsunami hazard curves, 180–182
  earthquake simulation, 179–180
  earthquake–tsunami disaggregation, 182–183
  earthquake–tsunami risk, 183–186
  earthquake–tsunami uniform hazard maps, 182
 overview of tsunami hazard and risk assessment, 158–163
  general probabilistic formulation, 160–162
  logic tree approach, 162–163
  probabilistic approach, 159–160
  sensitivity analysis, 158
  worst case scenario approach, 158–159
 probabilistic seismic tsunami hazard analysis, 163–173
  disaggregation, 172–173
  empirical tsunami hazard curve based on tsunami simulation results, 170
  hazard maps, 170–172
  optimal number of simulations, 172
  scaling relationships of earthquake source parameters and stochastic source models, 168
  source characterization and magnitude-frequency distribution, 164–168
  tsunami modeling, 169–170
 probabilistic seismic tsunami risk analysis, 173–179
  exposure characterization, 174–176
  numerical evaluation of tsunami risk equation, 177–178
  risk metrics, 178–179
  vulnerability characterization, 176–177
 updates of tsunami design zone maps, 364–365

Probabilistic tsunami hazard assessment for New Zealand
  earthquake source models, 378–381
    crustal fault earthquake models, 381
    plate interface earthquake models, 378–381
  overview of statistical modeling approach, 374–377
  tsunami hazard model results, 384–391
    tsunami hazard curves and disaggregation results, 385
    tsunami inundation hazard modeling, 385–391
  tsunami height estimation, 383–384
    local crustal faults, 384
    local subduction zones, 384
    regional and distant sources, 383–384
  tsunami numerical simulation, 381–383
Probabilistic tsunami hazard assessments (PTHA), 513
Probabilistic tsunami risk assessment (PTRA), 3, 27, 51, 157–158, 191, 196, 465, 468
Probability density function, 31, 35f, 472
Probable maximum loss (PML), 218–219
PSHA. See Probabilistic seismic hazard analysis (PSHA)
PTHA. See Probabilistic tsunami hazard analysis (PTHA); Probabilistic tsunami hazard assessment (PTHA)
PTRA. See Probabilistic tsunami risk assessment (PTRA)
Pyroclastic density currents (PDCs), 397
Pyroclastic flows, 69–70, 401–402

Q
Qualitative risk matrix, 194t
Quantitative risk assessment, 15, 157f
Quantitative tsunami risk analysis, 121
Quantitative tsunami risk assessments, 219–220

R
Random forest, 206–207, 210
Random phase model, 471
Rapid tsunami impact assessment methods, 17–18
Rate weakening, 259–260
Real-time automatic detection method for permanent displacement (RAPiD), 314–315
Realtime forward simulation approach, 546
Real-time kinematic GPS technique, 236
Reciprocity principle, 234–235
Recurrence models, 482
Reference set of assets, 442
Regional and distant sources, 383–384
Regional tsunami hazard, 398–406
  effects of climate change and vertical land movements on tsunamis hazard, 404–406
  meteotsunami hazard, 402–404
  seismic tsunami hazard, 399
  submarine and subaerial landslides tsunami hazard, 399–401
  volcanic tsunami hazard, 401–402
Regional tsunami risk index, 406–407
Reinforced concrete, 175
Reinforced concrete elements, 458
Reinforcement learning, 326
Rejection method, 36
Relative entropy, 444
Relative risk ranking method, 192
Relative risk scoring, 194–196
Relative SLR (RSLR), 405
Remote sensing, 443–444
Renewal process, 31
Representative concentration pathways (RCP), 513, 527–528
Research gaps, 474–475
Resilience, 3–4
Response analysis, 158
Response variables, 145–146
Risk assessment, 193, 194f
Risk estimation, 193
Risk evaluation, 193
Risk metrics, 178–179
Risk transfer for tsunami, 224–226
Robotics, 543

S
Satellite observations, 240
Scenario-based tsunami hazard assessments (SBTHA), 397–399, 417, 428
Seafloor deformation, 515
Sea-level rise (SLR), 397, 513
  increase in tsunami risk around the eastern coast of Kanto region due to, 519–520
  increase in tsunami risk around Tokyo Bay due to, 516–518
  scenarios, 514, 514t
  simulation methodology, 515–516
    eastern coast of Kanto region, 516
    Tokyo Bay area, 515–516
  socioeconomic impacts of, 404–405
Secondary hazards, 479
Sediment transport modeling, 250–251
Seismic data, 305
Seismic gaps, 428
Seismic probabilistic tsunami hazard assessment, 423–425
Seismic reflection lines, 278
Seismic tsunami hazard, 399
Seismic waves, 473
Sendai coast in Japan, 172
Sensitivity analysis, 158, 452
Sequential feature selection algorithm, 209–210
Shallow megathrust, 260
Shallow-water gravity wave models, 282
Shallow-water wave equations, 79, 89–94
  dispersive tsunami model, 93–94
  linear tsunami model, 90–92
  nonlinear tsunami model, 92–93
Shallow-water wave theory, 88
Shear modulus, 28–29, 38

Shock spectrum, 137
Short-term assessment, 426–428
Short-term clustering, 529
Short-term hazard assessment, 340–352
  deep-ocean assessment and reporting of Tsunamis buoy network, 340–341
  enhancement in developing a global tsunami propagation database, 349–350
  enhancement in model computational speed, 344–345
  enhancement in modeling a broader range of earthquake sources, 345–347
  enhancements toward more rapid tsunami detection, 348–349
  inundation forecast models, 342–343
  March 11, 2011 Japan Tsunami, 343
  in other countries and regions, 350–352
  tsunami source inversion using precomputed propagation database, 341–342
Simple matrix calculation, 308
Single-degree-of-freedom system, 140
Site hazard assessment, 135–136
Sliding solid block model, 64
Slip distributions, 265–269, 304–305, 304f, 468
  kinematic function of, 265–267
  parameters, 483
  rigidity variations in upper plate, 268–269
  trench-breaching slip, 267–268
Smart home technology, 543
S-net project, 206
South America Risk Assessment, 422–423
Southwestern Sicily coast, 403
Spatial averaging, 443
Spatial correlations, 471
Spatial grid refinements, 99–100
Spectral synthesis method, 168
Spherical coordinate systems, 96
Splay faults, 277, 281–282
  numerical example, 282–284
  seismic expression of offshore splay faults, 278–280
    Makran subduction zone, 278–279
    Nankai Trough zone, 279–280
  steeper dip angles for, 282
  tectonic background, 277–278
  and tsunami generation, 281–282
Sponge, 101–102
SRCMOD database, 39–40, 47
Static stress-driven models, 263
Steady-state drag expression, 129
Stochastic reduced order model (SROM), 425
Stochastic rupture model, 529–530
Stochastic tsunami source modeling, 53–54
Stress–strain relationship, 458
Structural analysis, 151
Subaerial landslides
  implications for future methods, 298–299
  method for subaerial landslide probabilistic hazard analysis, 291–294
    numerical modeling, 291
    probabilistic treatment, 291–292
    shortcomings, 292–293

suggested improved methodology using hindcasting to calibrate parameter probabilities, 293–294
model calibration for constraining uncertainty analysis, 294–298
Subduction region, 29
Submarine and subaerial landslides tsunami hazard, 399–401
Submarine earthquakes, 397
Sumatra earthquakes, 39, 262
Supervised learning, 326
Sustainability, 3–4

T

Tafjord simulations, 298
TanDEM-X, 354
Tapered GR (TGR) distribution, 482
Target magnitude, 168
Tectonic regimes, 29
Thermodynamics, 88
Three-dimensional fluid dynamic behavior, 468
Three-dimensional (3D) model, 289–291
Tide, 468
Time-dependent earthquake occurrence models, 525
Time-dependent occurrence model using renewal process, 33–37
Time–history (TH) curves, 125
Time-history dynamic analyses (TDA), 454–455
Time reversal imaging, 234–235
Tofino, 528–533
  conditional tsunami risk curves, 529–533
  long-term tsunami risk analysis with tidal variations and relative sea-level rises, 533
  long-term tsunami risk assessment for, 533–539
    effects of baseline sea levels, 533–537
    tsunami risk curves for different elapsed times and relative sea-level rise scenarios, 537–539
  monthly variation of average population in, 204f
  occurrence and magnitude models of megathrust Cascadia earthquakes, 529
  and physical environment, 526–528
    Cascadia subduction zone, 526
    District of Tofino, 526, 527f
    tides and relative sea-level rise in Tofino, 527–528
  tsunami fatality risk curve in, 203–204
  tsunami loss maps for, 201f
  tsunami planning map for, 200f
Tofino and physical environment, 526–528
Tohoku earthquake, 233–234, 267–268, 304
Tohoku region, 208–209
Tohoku seismicity data, 32
Tohoku-type earthquakes, 162–163
Tonankai region, 284
Top-edge depth effects, 61–62
Topographic Model (TM), 112–113

Toyota Motor Corporation, 543
Trans-Pacific tsunami propagation, 241
Trapezoidal lateral load distributions, 452–453
Trench-breaching slip, 267–268
  mechanism, 268
  propensity, 267–268
  tsunamigenic effects, 267
Trigger phase, 400
Truncated exponential model, 31
Tsunami
  amplitudes, 81
  characteristics, 79–88
    harmonic wave, 83
    tsunami ray tracing, 84–86
    tsunami travel time, 87–88
  damage and loss estimation, 532
  data assimilation, 306–308
  debris, 472
  earthquake, 246
  numerical method, 95–102
    finite difference scheme, 96–99
    moving boundary scheme, 102
    outermost grid boundary condition, 100–102
    spatial grid refinements, 99–100
  propagation and inundation modeling, 88–94
    equation of mass conservation, 88–89
    equation of motion, 89
    shallow-water wave equations, 89–94
  simulations, 102–117
    nearshore and offshore tsunami modeling, 108–110
    tsunami inundation modeling, 110–117
    tsunami source, 102–108
Tsunami design zone (TDZ), 361
Tsunami disaster digital twin (TDDT), 544–552, 545f
  in Japan, 544–552
    multiagent modeling of disaster response activities, 550–552
    real-time tsunami inundation and damage forecast, 546–548
    tsunami exposure analysis using mobile spatial statistics, 548–549
Tsunami disaster risk reduction and management
  emergency response, 205–217
    tsunami early warning system, 205–212
    tsunami evacuation, 212–217
    tsunami risk financing, 217–226
  tsunami hazard–risk mapping for disaster preparedness planning, 192–205
    risk assessment procedure and relative risk scoring, 192–196
    risk-based critical tsunami scenarios, 200–202
    tsunami fatality risk and societal risk tolerability, 202–205
    tsunami hazard and exposure, 197
    tsunami hazard and risk mapping, 197–200
  tsunami risk financing, 217–226

financial risk metrics for natural catastrophe risk management, 218–220
insurance-reinsurance system for tsunami risk, 220–223
Tsunami early warning system, 205–212
  network design and optimization for developing tsunami early warning models, 212
  risk-based tsunami early warning and rapid tsunami loss estimation, 212
  statistical approaches for developing tsunami early warning models, 209–211
  use of synthetic tsunami datasets for developing tsunami early warning models, 207–209
Tsunami effects on built environment
  differences among guidelines, standards, and codes, 122
  differences in tsunami-induced loads between American and Japanese design codes, 141–142
    debris impact loads, 142
    hydrodynamic forces, 142
    hydrostatic forces, 141–142
    loading conditions, 141
  early development of tsunami guidelines, standards, and codes, 122–123
    Japan, 123
    other countries, 124
  tsunami-induced loads and effects, 124–141
    debris impact loads, 134–141
    hydrodynamic loads, 127–134
    hydrostatic loads, 126–127
  tsunami structural vulnerability assessment, 142–151
    analytical fragility functions, 150–151
    empirical fragility functions, 145–150
    forms and typologies of vulnerability functions, 144–145
    tsunami damage states, 143–144
  United States of America, 123–124
Tsunami evacuation, 212–217
  model setup, 213–215
  simulation results, 216–217
Tsunami forecasting method based on inversion for initial sea-surface height (tFISH), 312–313
Tsunami forecasting methods, 326
Tsunami fragility, 437
  curve, 244
  functions, 145
  model, 532
Tsunami generation phase, 400
Tsunami generation processes, 27
  earthquake occurrence, 30–38
    characteristic earthquake model for large events, 37–38
    statistical analysis of earthquake catalog and time-independent occurrence model, 31–33
    time-dependent occurrence model using renewal process, 33–37
  earthquake rupture model, 39–50

Tsunami generation processes (*Continued*)
   analysis of finite-fault models, 42–47
   finite-fault models, 39–42, 41*f*
   prediction of earthquake source parameters, 47–50
  nonseismic sources of tsunamis, 63–72
   air pressure wave, 70
   caldera collapse, 70–71
   pyroclastic flow, 69–70
   sources from landslides, 64–67
   underwater explosion, 68–69
   volcanic earthquake, 71–72
  seismotectonic characteristics of active seismic regions, 28–30
  source modeling for tsunamigenic earthquakes, 51–63
   characterization of seismic tsunami sources, 51–52
   kinematic stochastic tsunami source modeling, 54–57
   seafloor displacement due to earthquake rupture, 57–63
   stochastic tsunami source modeling, 53–54
Tsunami genesis, 260, 289
Tsunamigenic earthquakes, 169*f*
Tsunamigenic models, 49–50
Tsunami hazard analysis, 2–3
Tsunami hazard and risk assessments, 525
Tsunami hazard assessments, 12–15, 79, 367, 418, 420–428
  boundary conditions and local hydrodynamics, 420–421
  initial conditions, 421–423
  other susceptibility estimation studies, 425–426
  scenario-based tsunami hazard assessment, 423
  seismic probabilistic tsunami hazard assessment, 423–425
  short-term assessment, 426–428
Tsunami hazard curves, 519*f*
Tsunami hazard maps, 515
Tsunami hazard model, 483–484
Tsunami hazard, 340
  long-term hazard assessment, 352–367
   challenges for, 352–353
   deterministic tsunami hazard assessment, 356–359
   global and regional digital elevation models and model grids, 354
   literature review and data collection, 353–354
   model validation using historical data, 355–356
   probabilistic tsunami hazard assessment, 360–367
   tsunami model benchmarking, 354–355
  short-term hazard assessment, 340–352
   deep-ocean assessment and reporting of Tsunamis buoy network, 340–341
   enhancement in developing a global tsunami propagation database, 349–350
   enhancement in model computational speed, 344–345
   enhancement in modeling a broader range of earthquake sources, 345–347
   enhancements toward more rapid tsunami detection, 348–349
   inundation forecast models, 342–343
   March 11, 2011 Japan Tsunami, 343
   in other countries and regions, 350–352
   tsunami source inversion using precomputed propagation database, 341–342
Tsunami height estimation, 383–384
  local crustal faults, 384
  local subduction zones, 384
  regional and distant sources, 383–384
Tsunami intensity, 437
  measure, 439
Tsunami inundation depths, 468–470
Tsunami inundation flows, 450
Tsunami inundation hazard modeling, 385–391
Tsunami inundation modeling, 110–117, 530–532
  bottom roughness and built area, 112–113
  coseismic deformation in coastal areas, 111–112
  tidal state and sea-level rise, 113–114
Tsunami-like ocean waves, 316
Tsunami loss curve, 161–162
Tsunami magnitude, 71
Tsunami model benchmarking, 354–355
Tsunami modeling, 108–109, 169–170, 472
Tsunami nonlinear static analyses, 456–457
Tsunami nonlinear time-history dynamic analysis, 455
Tsunami numerical simulation, 79, 381–383
Tsunami propagation, 284, 484
  and impact on coast, 401
Tsunami ray tracing method, 84–86
Tsunami risk assessment, 15–16
Tsunami risk exposures, 203
Tsunami risk financing, 217–226
Tsunami risks, 525
Tsunami runup, 82, 82*f*
Tsunamis, 233, 501, 513
  field surveys to measure tsunami heights, 242–244
  forward and inverse problems, 233*f*
  geological data, 248–251
   sediment transport modeling, 250–251
   tsunami deposits in Sendai Plain, 249–250
  historical data, 244–248
   869 Jogan Tsunami, 247–248
   1611 Keicho Tohoku Tsunami, 247, 248*f*
   1854 Ansei Earthquakes and Tsunamis along Nankai Trough, 245
   1896 Sanriku Tsunami, 245–246
  source and heights, 242*f*
  source models of the 2011 Tohoku Earthquake based on inversion of tsunami data, 241–242
  waveform inversions, 234*f*
  waveforms recorded on instruments, 235–240
   cabled bottom pressure gauges, 236–237
   coastal sea-level measurements, 235–236
   deep ocean measurements, 238–239
   oceanographic radar measurements, 240
   offshore tsunami measurements, 236
   satellite observations, 240
Tsunami simulation, 472
Tsunami source, 102–108
  air pressure wave source, 107–108
  two-layered landslide-tsunami model, 103–104
  two-layered pyroclastic flow-tsunami model, 104–107
Tsunami vulnerability, 176–177, 407
Tsunami warning and mitigation systems, 350
Tsunami warning system (TWS), 339–340
Tsunami wavelengths, 89–90
2D depth-averaged method, 103
2D model, 458
Two-dimensional slip distribution, 471
2-D static deformation models, 264*f*
Two-layered landslide-tsunami model, 103–104
Two-layered pyroclastic flow-tsunami model, 104–107
  high-density pyroclastic flow, 104–106
  low-density pyroclastic flow, 106–107
Two-phase nonlinear static analysis, 457

U

Unbinned regression, 148*f*
Uncertainty
  classification, 467–470
   related to exposure and damage, 470
   related to tsunami inundation, 470
   related to tsunami propagation, 468–470
   related to tsunami source, 467–468
  propagation, 470–474
   compound and cascading hazard modeling, 473–474
   earthquake source modeling, 471–472
   fragility modeling, 472–473
   incorporating different uncertainty models, 474
   tsunami modeling, 472
Underwater explosions (UWEs), 397
UNESCO Intergovernmental Oceanographic Commission (IOC) facilitates, 350
Uniform slip models, 381
Universal gas constant, 108
Universal nonlinear function approximator, 328
Unsupervised learning, 326
US National Oceanographic and Atmospheric Administration (NOAA), 339–340

V

Value at risk (VaR), 16, 217–218
Variable depth pushover (VDPO) approach, 151, 456, 456*f*

VDPO-BI, 457
Vector-parallel supercomputer SX-Aurora, 546
Vertical compensated-linear-vector-dipole (vertical-CLVD) earthquakes, 71
Vertical land movements (VLMs), 397, 404–405
Virtual reality (VR), 543
VoellmyClaw simulations, 294–295
Voellmy rheological model, 298
Voellmy-type landslide model, 293–294
Volcanic activity, 397
Volcanic earthquake, 71–72
Volcanic eruptions, 67
Volcanic tsunami hazard, 401–402

VolcFlow code, 103
von Kármán model, 168
von Karman-type spectra, 471
von Kármán wavenumber spectrum, 45–46
Vulnerability analysis, 144
Vulnerability component, 15
Vulnerability curve, 439

W

Wastewater treatment plans, 514
Water depth coefficient, 142, 468
Wave generation, 298
Wavelength, 80
Wavenumber, 45–46
Weak creep, 259
Weibull distribution, 33
Weighted interpolation equation, 311–312
Welch's t-test, 425
World Health Organization, 550–551

Y

Year loss table (YLT), 218

Z

Zero tsunami loss, 537–539

Printed in the United States
by Baker & Taylor Publisher Services